Electrical Engineering
Principles and Applications

SIXTH EDITION

Allan R. Hambley

Department of Electrical and Computer Engineering
Michigan Technological University

PEARSON

Delhi • Chennai

ISBN 978-93-3256-330-8

First Impression

This edition is manufactured in India and is authorized for sale only in India, Bangladesh, Bhutan, Pakistan, Nepal, Sri Lanka and the Maldives. Circulation of this edition outside of these territories is UNAUTHORIZED.

Published by Pearson India Education Services Pvt. Ltd, CIN: U72200TN2005PTC057128.
Formerly known as TutorVista Global Pvt. Ltd, licensee of Pearson Education in South Asia.

Head Office: A-8 (A), 7th Floor, Knowledge Boulevard, Sector 62, Noida 201 309, Uttar Pradesh, India.
Registered Office: Fourth Floor, Software Block, Elnet Software City, TS-140, Block 2 & 9,
Rajiv Gandhi Salai, Taramani, Chennai 600 113, Tamil Nadu, India.
Fax: 080-30461003, Phone: 080-30461060
www.pearson.co.in, Email id: companysecretary.india@pearson.com

Printed in India by HT Media Ltd., Noida

To Judy, Tony, Pam, and Mason

Practical Applications of Electrical Engineering Principles

Contents

List of Examples

Chapter 10

Chapter 11

Chapter 12

Chapter 13

Chapter 14

Chapter 15

Chapter 16

Chapter 17

Appendix A

List of Tables

Preface

As in the previous editions, my guiding philosophy in writing this book has three elements. The first element is my belief that in the long run students are best served by learning basic concepts in a general setting. Second, I believe that students need to be motivated by seeing how the principles apply to specific and interesting problems in their own fields. The third element of my philosophy is to take every opportunity to make learning free of frustration for the student.

This book covers circuit analysis, digital systems, electronics, and electromechanics at a level appropriate for either electrical-engineering students in an introductory course or nonmajors in a survey course. The only essential prerequisites are basic physics and single-variable calculus. Teaching a course using this book offers opportunities to develop theoretical and experimental skills and experiences in the following areas:

- Basic circuit analysis and measurement
- First- and second-order transients
- Steady-state ac circuits
- Resonance and frequency response
- Digital logic circuits
- Microcontrollers
- Computer-based instrumentation, including LabVIEW
- Diode circuits
- Electronic amplifiers
- Field-effect and bipolar junction transistors
- Operational amplifiers
- Transformers
- Ac and dc machines
- Computer-aided circuit analysis using MATLAB

While the emphasis of this book is on basic concepts, a key feature is the inclusion of short articles scattered throughout showing how electrical-engineering concepts are applied in other fields. The subjects of these articles include anti-knock signal processing for internal combustion engines, a cardiac pacemaker, active noise control, and the use of RFID tags in fisheries research, among others.

I welcome comments from users of this book. Information on how the book could be improved is especially valuable and will be taken to heart in future revisions. My e-mail address is `arhamble@mtu.edu`

INSTRUCTOR RESOURCES

Resources for instructors include:

- A complete Instructor's Solutions Manual
- PowerPoint slides with all the figures from the book

Instructor Resources are available for download by users of this book at the Pearson website: www.pearsoned.co.in/hambley. If you are in need of a log in and password, please contact your local Pearson representative.

WHAT'S NEW IN THIS EDITION

- We have continued the popular Practice Tests that students can use in preparing for course exams at the end of each chapter. Answers for the Practice Tests appear in Appendix D and complete solutions are included in the on-line Student Solutions Manual files.
- We have updated the coverage of MATLAB and the Symbolic Toolbox for network analysis in Chapters 2 through 6.
- Approximately 200 problems are new to this edition, replacing some of the problems from the previous edition, and many other problems have been modified.
- In Chapter 2, we have added an explanation of how the Wheatstone bridge is used in strain measurements.
- Sections 3.8 and 4.6 have been updated, deleting the coverage of piecewise linear functions which are problematic with recent versions of the Symbolic Toolbox.
- Chapter 8 has been extensively updated and now uses the Freescale Semiconductor HCS12/9S12 family as an example of microcontrollers.
- Section 9.4 has been updated to the most recent version of LabVIEW.
- Relatively minor corrections and improvements appear throughout the book.

PREREQUISITES

The essential prerequisites for a course from this book are basic physics and single-variable calculus. A prior differential equations course would be helpful but is not essential. Differential equations are encountered in Chapter 4 on transient analysis, but the skills needed are developed from basic calculus.

PEDAGOGICAL FEATURES

The book includes various pedagogical features designed with the goal of stimulating student interest, eliminating frustration, and engendering an awareness of the relevance of the material to their chosen profession. These features are:

- Statements of learning objectives open each chapter.
- Comments in the margins emphasize and summarize important points or indicate common pitfalls that students need to avoid.
- Short boxed articles demonstrate how electrical-engineering principles are applied in other fields of engineering. For example, see the articles on active noise cancellation (page 305) and electronic pacemakers (starting on page 403).
- Step-by-step problem solving procedures. For example, see the step-by-step summary of node-voltage analysis (on pages 94–95) or the summary of Thévenin equivalents (on page 113).
- A Practice Test at the end of each chapter gives students a chance to test their knowledge. Answers appear in Appendix D.
- Complete solutions to the in-chapter exercises and Practice Tests, included as PDF files on-line.
- Summaries of important points at the end of each chapter provide references for students.
- Key equations are highlighted in the book to draw attention to important results.

MEETING ABET-DIRECTED OUTCOMES

Courses based on this book provide excellent opportunities to meet many of the directed outcomes for accreditation. The Criteria for Accrediting Engineering Programs require that graduates of accredited programs have "an ability to apply knowledge of mathematics, science, and engineering" and "an ability to identify, formulate, and solve engineering problems." This book, in its entirety, is aimed at developing these abilities.

Also, graduates must have "an ability to design and conduct experiments, as well as analyze and interpret data." Chapter 9, Computer-Based Instrumentation Systems, helps to develop this ability. If the course includes a laboratory, this ability can be developed even further.

Furthermore, the criteria require "an ability to function on multi-disciplinary teams" and "an ability to communicate effectively." Courses based on this book contribute to these abilities by giving nonmajors the knowledge and vocabulary to communicate effectively with electrical engineers. The book also helps to inform electrical engineers about applications in other fields of engineering. To aid in communication skills, end-of-chapter problems that ask students to explain electrical-engineering concepts in their own words are included.

CONTENT AND ORGANIZATION

Basic Circuit Analysis

Chapter 1 defines current, voltage, power, and energy. Kirchhoff's laws are introduced. Voltage sources, current sources, and resistance are defined.

Chapter 2 treats resistive circuits. Analysis by network reduction, node voltages, and mesh currents is covered. Thévenin equivalents, superposition, and the Wheatstone bridge are treated.

Capacitance, inductance, and mutual inductance are treated in Chapter 3.

Transients in electrical circuits are discussed in Chapter 4. First-order *RL* and *RC* circuits and time constants are covered, followed by a discussion of second-order circuits.

Chapter 5 considers sinusoidal steady-state circuit behavior. (A review of complex arithmetic is included in Appendix A.) Power calculations, ac Thévenin and Norton equivalents, and balanced three-phase circuits are treated.

Chapter 6 covers frequency response, Bode plots, resonance, filters, and digital signal processing. The basic concept of Fourier theory (that signals are composed of sinusoidal components having various amplitudes, phases, and frequencies) is qualitatively discussed.

Digital Systems

Chapter 7 introduces logic gates and the representation of numerical data in binary form. It then proceeds to discuss combinatorial and sequential logic. Boolean algebra, De Morgan's laws, truth tables, Karnaugh maps, coders, decoders, flip-flops, and registers are discussed.

Chapter 8 treats microcomputers with emphasis on embedded systems using the Freescale Semiconductor HCS12/9S12 as the primary example. Computer organization and memory types are discussed. Digital process control using microcontrollers is described in general terms. Finally, selected instructions and addressing modes for the CPU12 are described. Assembly language programming is treated very briefly.

Chapter 9 discusses computer-based instrumentation systems including measurement concepts, sensors, signal conditioning, and analog-to-digital conversion. The chapter ends with a discussion of LabVIEW, including an example virtual instrument that students can duplicate using an evaluation version on their own computers.

Electronic Devices and Circuits

Chapter 10 presents the diode, its various models, load-line analysis, and diode circuits, such as rectifiers, Zener-diode regulators, and wave shapers.

In Chapter 11, the specifications and imperfections of amplifiers that need to be considered in applications are discussed from a users perspective. These include gain, input impedance, output impedance, loading effects, frequency response, pulse response, nonlinear distortion, common-mode rejection, and dc offsets.

Chapter 12 covers the MOS field-effect transistor, its characteristic curves, load-line analysis, large-signal and small-signal models, bias circuits, the common-source amplifier, and the source follower.

Chapter 13 gives a similar treatment for bipolar transistors. If desired, the order of Chapters 12 and 13 can be reversed. Another possibility is to skip most of both chapters so more time can be devoted to other topics.

Chapter 14 treats the operational amplifier and many of its applications. Nonmajors can learn enough from this chapter to design and use op-amp circuits for instrumentation applications in their own fields.

Electromechanics

Chapter 15 reviews basic magnetic field theory, analyzes magnetic circuits, and presents transformers.

DC machines and ac machines are treated in Chapters 16 and 17, respectively. The emphasis is on motors rather than generators because the nonelectrical engineer applies motors much more often than generators. In Chapter 16, an overall view of motors in general is presented before considering DC machines, their equivalent circuits, and performance calculations. The universal motor and its applications are discussed.

Chapter 17 deals with AC motors, starting with the three-phase induction motor. Synchronous motors and their advantages with respect to power-factor correction are analyzed. Small motors including single-phase induction motors are also discussed. A section on stepper motors and brushless dc motors ends the chapter.

ACKNOWLEDGMENTS

I wish to thank my colleagues, past and present, in the Electrical and Computer Engineering Department at Michigan Technological University, all of whom have given me help and encouragement at one time or another in writing this book and in my other projects.

I have received much excellent advice from professors at other institutions who reviewed the manuscript in various stages over the years. This advice has improved the final result a great deal, and I am grateful for their help.

Current and past reviewers include:

Ibrahim Abdel-Motaled, Northwestern University
William Best, Lehigh University
Steven Bibyk, Ohio State University
D. B. Brumm, Michigan Technological University
Karen Butler-Purry, Texas A&M University
Robert Collin, Case Western University
Joseph A. Coppola, Syracuse University
Norman R. Cox, University of Missouri at Rolla
W.T. Easter, North Carolina State University
Zoran Gajic, Rutgers University
Edwin L. Gerber, Drexel University
Victor Gerez, Montana State University
Walter Green, University of Tennessee
Elmer Grubbs, New Mexico Highlands University
Jasmine Henry, University of Western Australia
Ian Hutchinson, MIT
David Klemer, University of Wisconsin, Milwaukee
Richard S. Marleau, University of Wisconsin
Sunanda Mitra, Texas Tech University
Phil Noe, Texas A&M University
Edgar A. O'Hair, Texas Tech University
John Pavlat, Iowa State University
Clifford Pollock, Cornell University
Michael Reed, Carnegie Mellon University
Gerald F. Reid, Virginia Polytechnic Institute
Selahattin Sayil, Lamar University

William Sayle II, Georgia Institute of Technology
Len Trombetta, University of Houston
John Tyler, Texas A&M University
Belinda B. Wang, University of Toronto
Carl Wells, Washington State University
Al Wicks, Virginia Tech
Edward Yang, Columbia University
Subbaraya Yuvarajan, North Dakota State University
Rodger E. Ziemer, University of Colorado, Colorado Springs

Over the years, many students and faculty using my books at Michigan Technological University and elsewhere have made many excellent suggestions for improving the books and correcting errors. I thank them very much.

I am indebted to Andrew Gilfillan and Tom Robbins, my present and past editors at Pearson, for keeping me pointed in the right direction and for many excellent suggestions that have improved my books a great deal. A very special thank you, also, to Scott Disanno for a great job of managing the production of this and past editions of this book.

Thanks are extended to National Instruments which provided many excellent suggestions. Thanks are also extended to Pavithra Jayapaul of Jouve India for her excellent work on this edition.

Also, I want to thank Tony and Pam for their continuing encouragement and valuable insights. I thank Judy for many good things much too extensive to list.

ALLAN R. HAMBLEY

The publishers would like to thank Mr. S. K. Sahdev, Associate Dean, Faculty of Science & Technology, Lovely Proffessional University, Jalandhar, for his valuable suggestions and inputs for enchancing the content of this book to suit the requirement of Indian Universities.

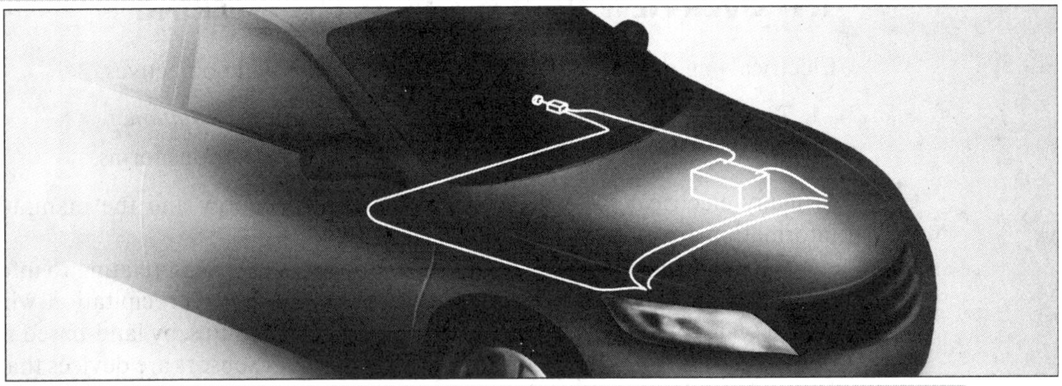

Chapter 1

Introduction

Study of this chapter will enable you to:

- Recognize interrelationships between electrical engineering and other fields of science and engineering.
- List the major subfields of electrical engineering.
- List several important reasons for studying electrical engineering.
- Define current, voltage, and power, including their units.
- Calculate power and energy and determine whether energy is supplied or absorbed by a circuit element.
- State and apply Kirchhoff's current and voltage laws.
- Recognize series and parallel connections.
- Identify and describe the characteristics of voltage and current sources.
- State and apply Ohm's law.
- Solve for currents, voltages, and powers in simple circuits.

Introduction to this chapter:

In this chapter, we introduce electrical engineering, define circuit variables (current, voltage, power, and energy), study the laws that these circuit variables obey, and meet several circuit elements (current sources, voltage sources, and resistors).

1.1 OVERVIEW OF ELECTRICAL ENGINEERING

Electrical engineers design systems that have two main objectives:

1. To gather, store, process, transport, and present *information*.
2. To distribute, store, and convert *energy* between various forms.

In many electrical systems, the manipulation of energy and the manipulation of information are interdependent.

For example, numerous aspects of electrical engineering relating to information are applied in weather prediction. Data about cloud cover, precipitation, wind speed, and so on are gathered electronically by weather satellites, by land-based radar stations, and by sensors at numerous weather stations. (Sensors are devices that convert physical measurements to electrical signals.) This information is transported by electronic communication systems and processed by computers to yield forecasts that are disseminated and displayed electronically.

In electrical power plants, energy is converted from various sources to electrical form. Electrical distribution systems transport the energy to virtually every factory, home, and business in the world, where it is converted to a multitude of useful forms, such as mechanical energy, heat, and light.

No doubt you can list scores of electrical engineering applications in your daily life. Increasingly, electrical and electronic features are integrated into new products. Automobiles and trucks provide just one example of this trend. The electronic content of the average automobile is growing rapidly in value. Auto designers realize that electronic technology is a good way to provide increased functionality at lower cost. Table 1.1 shows some of the applications of electrical engineering in automobiles.

You may find it interesting to search the web for sites related to "mechatronics."

As another example, we note that many common household appliances contain keypads for operator control, sensors, electronic displays, and computer chips, as well as more conventional switches, heating elements, and motors. Electronics have become so intimately integrated with mechanical systems that the name **mechatronics** is used for the combination.

Subdivisions of Electrical Engineering

Next, we give you an overall picture of electrical engineering by listing and briefly discussing eight of its major areas.

1. Communication systems transport information in electrical form. Cellular phone, radio, satellite television, and the Internet are examples of communication systems. It is possible for virtually any two people (or computers) on the globe to communicate almost instantaneously. A climber on a mountaintop in Nepal can call or send e-mail to friends whether they are hiking in Alaska or sitting in a New York City office. This kind of connectivity affects the way we live, the way we conduct business, and the design of everything we use. For example, communication systems will change the design of highways because traffic and road-condition information collected by roadside sensors can be transmitted to central locations and used to route traffic. When an accident occurs, an electrical signal can be emitted automatically when the airbags deploy, giving the exact location of the vehicle, summoning help, and notifying traffic-control computers.

Computers that are part of products such as appliances and automobiles are called *embedded computers*.

2. Computer systems process and store information in digital form. No doubt you have already encountered computer applications in your own field. Besides the computers of which you are aware, there are many in unobvious places, such as household appliances and automobiles. A typical modern automobile contains several

Table 1.1. Current and Emerging Electronic/Electrical Applications in Automobiles and Trucks

Safety
Antiskid brakes
Inflatable restraints
Collision warning and avoidance
Blind-zone vehicle detection (especially for large trucks)
Infrared night vision systems
Heads-up displays
Automatic accident notification
Rear-view cameras
Communications and entertainment
AM/FM radio
Digital audio broadcasting
CD/DVD player
Cellular phone
Computer/e-mail
Satellite radio
Convenience
Electronic GPS navigation
Personalized seat/mirror/radio settings
Electronic door locks
Emissions, performance, and fuel economy
Vehicle instrumentation
Electronic ignition
Tire inflation sensors
Computerized performance evaluation and maintenance scheduling
Adaptable suspension systems
Alternative propulsion systems
Electric vehicles
Advanced batteries
Hybrid vehicles

dozen special-purpose computers. Chemical processes and railroad switching yards are routinely controlled through computers.

3. Control systems gather information with sensors and use electrical energy to control a physical process. A relatively simple control system is the heating/cooling system in a residence. A sensor (thermostat) compares the temperature with the desired value. Control circuits operate the furnace or air conditioner to achieve the desired temperature. In rolling sheet steel, an electrical control system is used to obtain the desired sheet thickness. If the sheet is too thick (or thin), more (or less) force is applied to the rollers. The temperatures and flow rates in chemical processes are controlled in a similar manner. Control systems have even been installed in tall buildings to reduce their movement due to wind.

4. Electromagnetics is the study and application of electric and magnetic fields. The device (known as a magnetron) used to produce microwave energy in an oven is one application. Similar devices, but with much higher power levels, are employed in manufacturing sheets of plywood. Electromagnetic fields heat the glue between layers of wood so that it will set quickly. Cellular phone and television antennas are also examples of electromagnetic devices.

5. Electronics is the study and application of materials, devices, and circuits used in amplifying and switching electrical signals. The most important electronic devices are transistors of various kinds. They are used in nearly all places where electrical information or energy is employed. For example, the cardiac pacemaker is an electronic circuit that senses heart beats, and if a beat does not occur when it should, applies a minute electrical stimulus to the heart, forcing a beat. Electronic instrumentation and electrical sensors are found in every field of science and engineering. Many of the aspects of electronic amplifiers studied later in this book have direct application to the instrumentation used in your field of engineering.

Electronic devices are based on controlling electrons. Photonic devices perform similar functions by controlling photons.

6. Photonics is an exciting new field of science and engineering that promises to replace conventional computing, signal-processing, sensing, and communication devices based on manipulating electrons with greatly improved products based on manipulating photons. Photonics includes light generation by lasers and light-emitting diodes, transmission of light through optical components, as well as switching, modulation, amplification, detection, and steering light by electrical, acoustical, and photon-based devices. Current applications include readers for DVD disks, holograms, optical signal processors, and fiber-optic communication systems. Future applications include optical computers, holographic memories, and medical devices. Photonics offers tremendous opportunities for nearly all scientists and engineers.

7. Power systems convert energy to and from electrical form and transmit energy over long distances. These systems are composed of generators, transformers, distribution lines, motors, and other elements. Mechanical engineers often utilize electrical motors to empower their designs. The selection of a motor having the proper torque–speed characteristic for a given mechanical application is another example of how you can apply the information in this book.

8. Signal processing is concerned with information-bearing electrical signals. Often, the objective is to extract useful information from electrical signals derived from sensors. An application is machine vision for robots in manufacturing. Another application of signal processing is in controlling ignition systems of internal combustion engines. The timing of the ignition spark is critical in achieving good performance and low levels of pollutants. The optimum ignition point relative to crankshaft rotation depends on fuel quality, air temperature, throttle setting, engine speed, and other factors.

If the ignition point is advanced slightly beyond the point of best performance, *engine knock* occurs. Knock can be heard as a sharp metallic noise that is caused by rapid pressure fluctuations during the spontaneous release of chemical energy in the combustion chamber. A combustion-chamber pressure pulse displaying knock is shown in Figure 1.1. At high levels, knock will destroy an engine in a very short time. Prior to the advent of practical signal-processing electronics for this application, engine timing needed to be adjusted for distinctly suboptimum performance to avoid knock under varying combinations of operating conditions.

By connecting a sensor through a tube to the combustion chamber, an electrical signal proportional to pressure is obtained. Electronic circuits process this signal to determine whether the rapid pressure fluctuations characteristic of knock are present. Then electronic circuits continuously adjust ignition timing for optimum performance while avoiding knock.

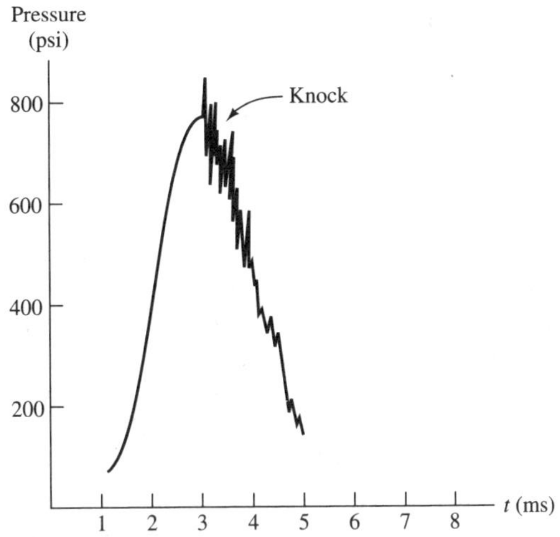

Figure 1.1 Pressure versus time for an internal combustion engine experiencing knock. Sensors convert pressure to an electrical signal that is processed to adjust ignition timing for minimum pollution and good performance.

Why You Need to Study Electrical Engineering

As a reader of this book, you may be majoring in another field of engineering or science and taking a required course in electrical engineering. Your immediate objective is probably to meet the course requirements for a degree in your chosen field. However, there are several other good reasons to learn and retain some basic knowledge of electrical engineering:

1. *To pass the Fundamentals of Engineering (FE) Examination as a first step in becoming a Registered Professional Engineer.* In the United States, before performing engineering services for the public, you will need to become registered as a Professional Engineer (PE). This book gives you the knowledge to answer questions relating to electrical engineering on the registration examinations. Save this book and course notes to review for the FE examination. (See Appendix C for more on the FE exam.)

Save this book and course notes to review for the FE exam.

2. *To have a broad enough knowledge base so that you can lead design projects in your own field.* Increasingly, electrical engineering is interwoven with nearly all scientific experiments and design projects in other fields of engineering. Industry has repeatedly called for engineers who can see the big picture and work effectively in teams. Engineers or scientists who narrow their focus strictly to their own field are destined to be directed by others. (Electrical engineers are somewhat fortunate in this respect because the basics of structures, mechanisms, and chemical processes are familiar from everyday life. On the other hand, electrical engineering concepts are somewhat more abstract and hidden from the casual observer.)

3. *To be able to operate and maintain electrical systems, such as those found in control systems for manufacturing processes.* The vast majority of electrical-circuit malfunctions can be readily solved by the application of basic electrical-engineering principles. You will be a much more versatile and valuable engineer or scientist if you can apply electrical-engineering principles in practical situations.

4. *To be able to communicate with electrical-engineering consultants.* Very likely, you will often need to work closely with electrical engineers in your career. This book will give you the basic knowledge needed to communicate effectively.

Content of This Book

Circuit theory is the electrical engineer's fundamental tool.

Electrical engineering is too vast to cover in one or two courses. Our objective is to introduce the underlying concepts that you are most likely to need. Circuit theory is the electrical engineer's fundamental tool. That is why the first six chapters of this book are devoted to circuits.

Embedded computers, sensors, and electronic circuits will be an increasingly important part of the products you design and the instrumentation you use as an engineer or scientist. Chapters 7, 8, and 9 treat digital systems with emphasis on embedded computers and instrumentation. Chapters 10 through 14 deal with electronic devices and circuits.

As a mechanical, chemical, civil, industrial, or other engineer, you will very likely need to employ energy-conversion devices. The last three chapters relate to electrical energy systems treating transformers, generators, and motors.

Because this book covers many basic concepts, it is also sometimes used in introductory courses for electrical engineers. Just as it is important for other engineers and scientists to see how electrical engineering can be applied to their fields, it is equally important for electrical engineers to be familiar with these applications.

1.2 CIRCUITS, CURRENTS, AND VOLTAGES

Overview of an Electrical Circuit

Before we carefully define the terminology of electrical circuits, let us gain some basic understanding by considering a simple example: the headlight circuit of an automobile. This circuit consists of a battery, a switch, the headlamps, and wires connecting them in a closed path, as illustrated in Figure 1.2.

The battery voltage is a measure of the energy gained by a unit of charge as it moves through the battery.

Chemical forces in the battery cause electrical charge (electrons) to flow through the circuit. The charge gains energy from the chemicals in the battery and delivers energy to the headlamps. The battery voltage (nominally, 12 volts) is a measure of the energy gained by a unit of charge as it moves through the battery.

Electrons readily move through copper but not through plastic insulation.

The wires are made of an excellent electrical conductor (copper) and are insulated from one another (and from the metal auto body) by electrical insulation (plastic) coating the wires. Electrons readily move through copper but not through the plastic insulation. Thus, the charge flow (electrical current) is confined to the wires until it reaches the headlamps. Air is also an insulator.

The switch is used to control the flow of current. When the conducting metallic parts of the switch make contact, we say that the switch is **closed** and current flows through the circuit. On the other hand, when the conducting parts of the switch do not make contact, we say that the switch is **open** and current does not flow.

Electrons experience collisions with the atoms of the tungsten wires, resulting in heating of the tungsten.

The headlamps contain special tungsten wires that can withstand high temperatures. Tungsten is not as good an electrical conductor as copper, and the electrons experience collisions with the atoms of the tungsten wires, resulting in heating of the tungsten. We say that the tungsten wires have electrical resistance. Thus, energy is transferred by the chemical action in the battery to the electrons and then to the tungsten, where it appears as heat. The tungsten becomes hot enough so that copious light is emitted. We will see that the power transferred is equal to the product of current (rate of flow of charge) and the voltage (also called electrical potential) applied by the battery.

Energy is transferred by the chemical action in the battery to the electrons and then to the tungsten.

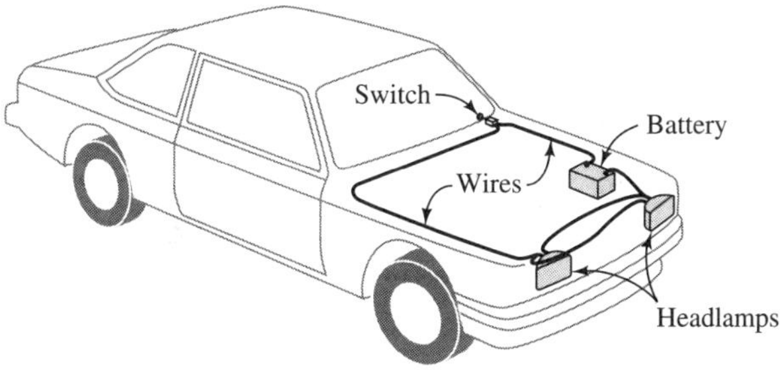

(a) Physical configuration

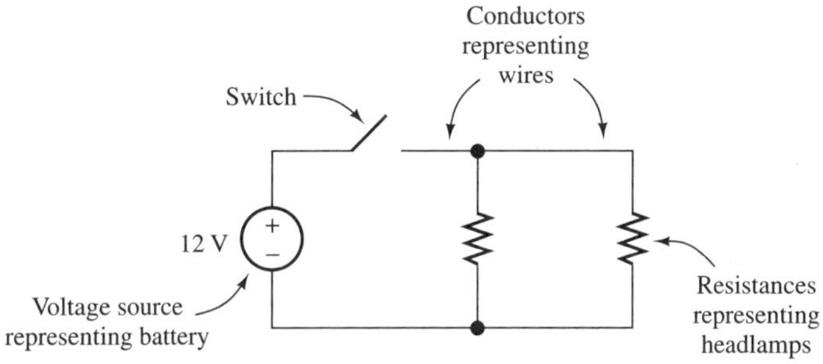

(b) Circuit diagram

Figure 1.2 The headlight circuit. (a) The actual physical layout of the circuit. (b) The circuit diagram.

(Actually, the simple description of the headlight circuit we have given is most appropriate for older cars. In more modern automobiles, sensors provide information to an embedded computer about the ambient light level, whether or not the ignition is energized, and whether the transmission is in park or drive. The dashboard switch merely inputs a logic level to the computer, indicating the intention of the operator with regard to the headlights. Depending on these inputs, the computer controls the state of an electronic switch in the headlight circuit. When the ignition is turned off and if it is dark, the computer keeps the lights on for a few minutes so the passengers can see to exit and then turns them off to conserve energy in the battery. This is typical of the trend to use highly sophisticated electronic and computer technology to enhance the capabilities of new designs in all fields of engineering.)

Fluid-Flow Analogy

Electrical circuits are analogous to fluid-flow systems. The battery is analogous to a pump, and charge is analogous to the fluid. Conductors (usually copper wires) correspond to frictionless pipes through which the fluid flows. Electrical current is the counterpart of the flow rate of the fluid. Voltage corresponds to the pressure differential between points in the fluid circuit. Switches are analogous to valves. Finally, the electrical resistance of a tungsten headlamp is analogous to a constriction in a fluid system that results in turbulence and conversion of energy to heat. Notice that current is a measure of the flow of charge *through* the cross section of a circuit element, whereas voltage is measured *across* the ends of a circuit element or *between* any other two points in a circuit.

The fluid-flow analogy can be very helpful initially in understanding electrical circuits.

Now that we have gained a basic understanding of a simple electrical circuit, we will define the concepts and terminology more carefully.

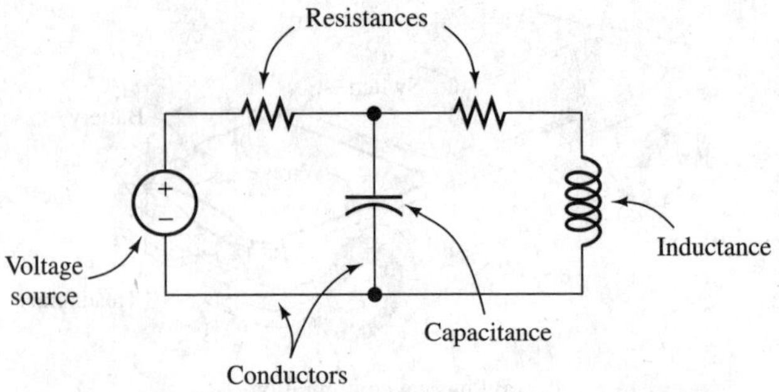

Figure 1.3 An electrical circuit consists of circuit elements, such as voltage sources, resistances, inductances, and capacitances, connected in closed paths by conductors.

Electrical Circuits

An electrical circuit consists of various types of circuit elements connected in closed paths by conductors.

An **electrical circuit** consists of various types of circuit elements connected in closed paths by conductors. An example is illustrated in Figure 1.3. The circuit elements can be resistances, inductances, capacitances, and voltage sources, among others. The symbols for some of these elements are illustrated in the figure. Eventually, we will carefully discuss the characteristics of each type of element.

Charge flows easily through conductors.

Charge flows easily through conductors, which are represented by lines connecting circuit elements. Conductors correspond to connecting wires in physical circuits. Voltage sources create forces that cause charge to flow through the conductors and other circuit elements. As a result, energy is transferred between the circuit elements, resulting in a useful function.

Electrical Current

Current is the time rate of flow of electrical charge. Its units are amperes (A), which are equivalent to coulombs per second (C/s).

Electrical current is the time rate of flow of electrical charge through a conductor or circuit element. The units are amperes (A), which are equivalent to coulombs per second (C/s). (The charge on an electron is -1.602×10^{-19} C.)

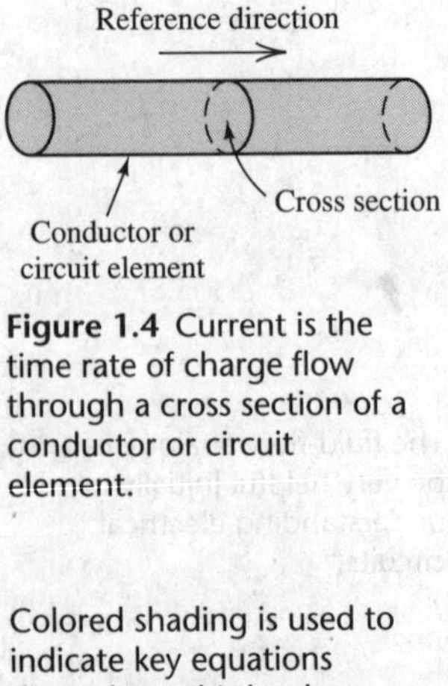

Figure 1.4 Current is the time rate of charge flow through a cross section of a conductor or circuit element.

Conceptually, to find the current for a given circuit element, we first select a cross section of the circuit element roughly perpendicular to the flow of current. Then, we select a **reference direction** along the direction of flow. Thus, the reference direction points from one side of the cross section to the other. This is illustrated in Figure 1.4.

Next, suppose that we keep a record of the net charge flow through the cross section. Positive charge crossing in the reference direction is counted as a positive contribution to net charge. Positive charge crossing opposite to the reference is counted as a negative contribution. Furthermore, negative charge crossing in the reference direction is counted as a negative contribution, and negative charge against the reference direction is a positive contribution to charge.

Thus, in concept, we obtain a record of the net charge in coulombs as a function of time in seconds denoted as $q(t)$. The electrical current flowing through the element in the reference direction is given by

Colored shading is used to indicate key equations throughout this book.

$$i(t) = \frac{dq(t)}{dt} \quad (1.1)$$

A constant current of one ampere means that one coulomb of charge passes through the cross section each second.

To find charge given current, we must integrate. Thus, we have

$$q(t) = \int_{t_0}^{t} i(t)\,dt + q(t_0) \qquad (1.2)$$

in which t_0 is some initial time at which the charge is known. (Throughout this book, we assume that time t is in seconds unless stated otherwise.)

Current flow is the same for all cross sections of a circuit element. (We reexamine this statement when we introduce the capacitor in Chapter 3.) The current that enters one end flows through the element and exits through the other end.

Example 1.1 Determining Current Given Charge

Suppose that charge versus time for a given circuit element is given by

$$q(t) = 0 \qquad \text{for } t < 0$$

and

$$q(t) = 2 - 2e^{-100t}\,\text{C} \qquad \text{for } t > 0$$

Sketch $q(t)$ and $i(t)$ to scale versus time.

Solution First we use Equation 1.1 to find an expression for the current:

$$\begin{aligned} i(t) &= \frac{dq(t)}{dt} \\ &= 0 \qquad \text{for } t < 0 \\ &= 200e^{-100t}\,\text{A} \qquad \text{for } t > 0 \end{aligned}$$

Plots of $q(t)$ and $i(t)$ are shown in Figure 1.5. ■

Reference Directions

In analyzing electrical circuits, we may not initially know the *actual direction* of current flow in a particular circuit element. Therefore, we start by assigning current

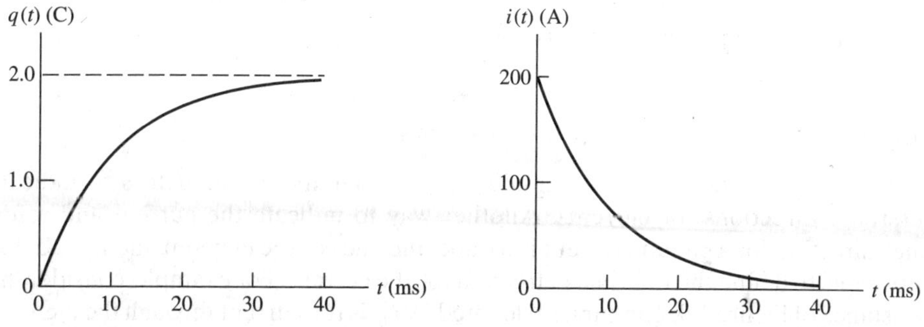

Figure 1.5 Plots of charge and current versus time for Example 1.1. *Note*: The time scale is in milliseconds (ms). One millisecond is equivalent to 10^{-3} seconds.

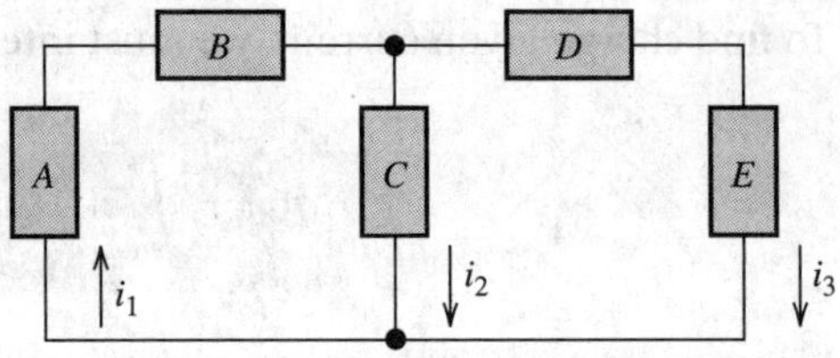

Figure 1.6 In analyzing circuits, we frequently start by assigning current variables i_1, i_2, i_3, and so forth.

variables and arbitrarily selecting a *reference direction* for each current of interest. It is customary to use the letter i for currents and subscripts to distinguish different currents. This is illustrated by the example in Figure 1.6, in which the boxes labeled A, B, and so on represent circuit elements. After we solve for the current values, we may find that some currents have negative values. For example, suppose that $i_1 = -2$ A in the circuit of Figure 1.6. Because i_1 has a negative value, we know that current actually flows in the direction opposite to the reference initially selected for i_1. Thus, the actual current is 2 A flowing downward through element A.

Direct Current and Alternating Current

Dc currents are constant with respect to time, whereas ac currents vary with time.

When a current is constant with time, we say that we have **direct current**, abbreviated as dc. On the other hand, a current that varies with time, reversing direction periodically, is called **alternating current**, abbreviated as ac. Figure 1.7 shows the values of a dc current and a sinusoidal ac current versus time. When $i_b(t)$ takes a negative value, the actual current direction is opposite to the reference direction for $i_b(t)$. The designation ac is used for other types of time-varying currents, such as the triangular and square waveforms shown in Figure 1.8.

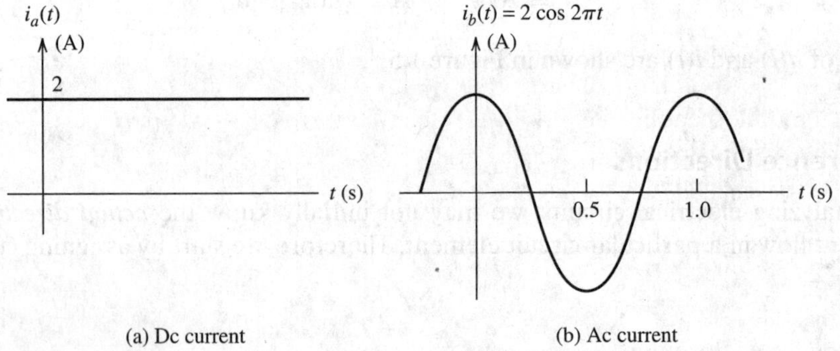

Figure 1.7 Examples of dc and ac currents versus time.

Double-Subscript Notation for Currents

So far we have used arrows alongside circuit elements or conductors to indicate reference directions for currents. Another way to indicate the current and reference direction for a circuit element is to label the ends of the element and use double subscripts to define the reference direction for the current. For example, consider the resistance of Figure 1.9. The current denoted by i_{ab} is the current through the element with its reference direction pointing from a to b. Similarly, i_{ba} is the current with its reference directed from b to a. Of course, i_{ab} and i_{ba} are the same in magnitude and

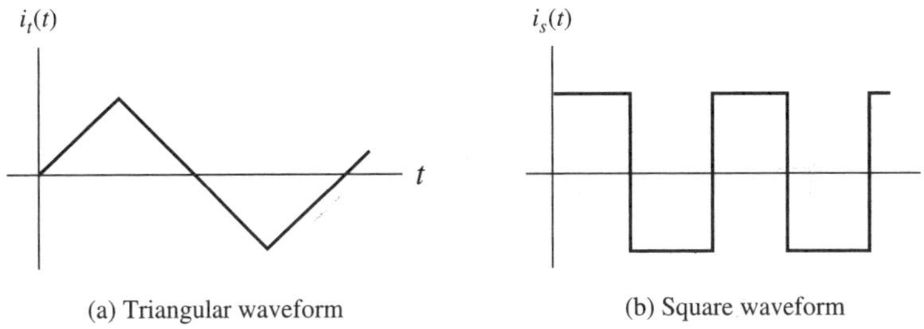

Figure 1.8 Ac currents can have various waveforms.

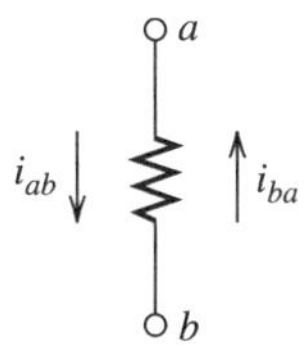

Figure 1.9 Reference directions can be indicated by labeling the ends of circuit elements and using double subscripts on current variables. The reference direction for i_{ab} points from a to b. On the other hand, the reference direction for i_{ba} points from b to a.

opposite in sign, because they denote the same current but with opposite reference directions. Thus, we have

$$i_{ab} = -i_{ba}$$

Exercise 1.1 A constant current of 2 A flows through a circuit element. In 10 seconds (s), how much net charge passes through the element?

Answer 20 C. □

Exercise 1.2 The charge that passes through a circuit element is given by $q(t) = 0.01 \sin(200t)$ C, in which the angle is in radians. Find the current as a function of time.

Answer $i(t) = 2\cos(200t)$ A. □

Exercise 1.3 In Figure 1.6, suppose that $i_2 = 1$ A and $i_3 = -3$ A. Assuming that the current consists of positive charge, in which direction (upward or downward) is charge moving in element C? In element E?

Answer Downward in element C and upward in element E. □

Voltages

When charge moves through circuit elements, energy can be transferred. In the case of automobile headlights, stored chemical energy is supplied by the battery and absorbed by the headlights where it appears as heat and light. The **voltage** associated with a circuit element is the energy transferred per unit of charge that flows through the element. The units of voltage are volts (V), which are equivalent to joules per coulomb (J/C).

Voltage is a measure of the energy transferred per unit of charge when charge moves from one point in an electrical circuit to a second point.

For example, consider the storage battery in an automobile. The voltage across its terminals is (nominally) 12 V. This means that 12 J are transferred to or from the battery for each coulomb that flows through it. When charge flows in one direction, energy is supplied by the battery, appearing elsewhere in the circuit as heat or light or perhaps as mechanical energy at the starter motor. If charge moves through the battery in the opposite direction, energy is absorbed by the battery, where it appears as stored chemical energy.

Notice that voltage is measured across the ends of a circuit element, whereas current is a measure of charge flow through the element.

Voltages are assigned polarities that indicate the direction of energy flow. If positive charge moves from the positive polarity through the element toward the negative polarity, the element absorbs energy that appears as heat, mechanical energy, stored chemical energy, or as some other form. On the other hand, if positive charge moves from the negative polarity toward the positive polarity, the element supplies energy. This is illustrated in Figure 1.10. For negative charge, the direction of energy transfer is reversed.

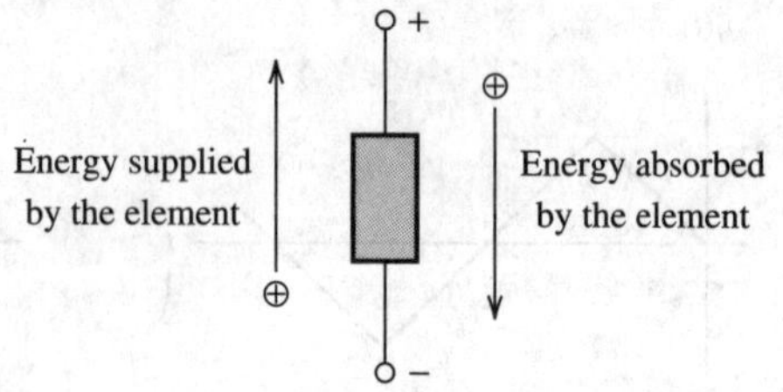

Figure 1.10 Energy is transferred when charge flows through an element having a voltage across it.

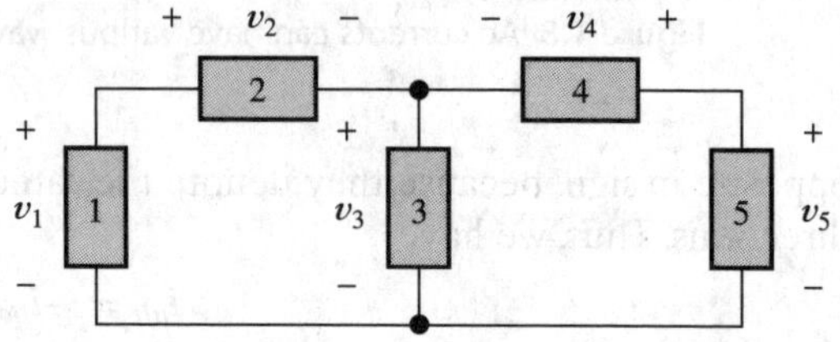

Figure 1.11 If we do not know the voltage values and polarities in a circuit, we can start by assigning voltage variables choosing the reference polarities arbitrarily. (The boxes represent unspecified circuit elements.)

Reference Polarities

When we begin to analyze a circuit, we often do not know the actual polarities of some of the voltages of interest in the circuit. Then, we simply assign voltage variables choosing *reference* polarities arbitrarily. (Of course, the *actual* polarities are not arbitrary.) This is illustrated in Figure 1.11. Next, we apply circuit principles (discussed later), obtaining equations that are solved for the voltages. If a given voltage has an actual polarity opposite to our arbitrary choice for the reference polarity, we obtain a negative value for the voltage. For example, if we find that $v_3 = -5$ V in Figure 1.11, we know that the voltage across element 3 is 5 V in magnitude and its actual polarity is opposite to that shown in the figure (i.e., the actual polarity is positive at the bottom end of element 3 and negative at the top).

In circuit analysis, we frequently assign reference polarities for voltages arbitrarily. If we find at the end of the analysis that the value of a voltage is negative, then we know that the true polarity is opposite of the polarity selected initially.

We usually do not put much effort into trying to assign "correct" references for current directions or voltage polarities. If we have doubt about them, we make arbitrary choices and use circuit analysis to determine true directions and polarities (as well as the magnitudes of the currents and voltages).

Voltages can be constant with time or they can vary. Constant voltages are called **dc voltages**. On the other hand, voltages that change in magnitude and alternate in polarity with time are said to be **ac voltages**. For example,

$$v_1(t) = 10 \text{ V}$$

is a dc voltage. It has the same magnitude and polarity for all time. On the other hand,

$$v_2(t) = 10 \cos(200\pi t) \text{V}$$

is an ac voltage that varies in magnitude and polarity. When $v_2(t)$ assumes a negative value, the actual polarity is opposite the reference polarity. (We study sinusoidal ac currents and voltages in Chapter 5.)

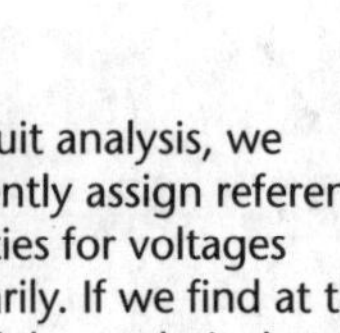

Figure 1.12 The voltage v_{ab} has a reference polarity that is positive at point a and negative at point b.

Double-Subscript Notation for Voltages

Another way to indicate the reference polarity of a voltage is to use double subscripts on the voltage variable. We use letters or numbers to label the terminals between which the voltage appears, as illustrated in Figure 1.12. For the resistance shown in the figure, v_{ab} represents the voltage between points a and b with the positive reference at point a. The two subscripts identify the points between which the voltage appears, and the first subscript is the positive reference. Similarly, v_{ba} is the voltage between a and b with the positive reference at point b. Thus, we can write

$$v_{ab} = -v_{ba} \tag{1.3}$$

because v_{ba} has the same magnitude as v_{ab} but has opposite polarity.

Still another way to indicate a voltage and its reference polarity is to use an arrow, as shown in Figure 1.13. The positive reference corresponds to the head of the arrow.

Figure 1.13 The positive reference for v is at the head of the arrow.

Switches

Switches control the currents in circuits. When an ideal switch is open, the current through it is zero and the voltage across it is determined by the remainder of the circuit. When an ideal switch is closed, the voltage across it is zero and the current through it is determined by the remainder of the circuit.

Exercise 1.4 The voltage across a given circuit element is $v_{ab} = 20$ V. A positive charge of 2 C moves through the circuit element from terminal b to terminal a. How much energy is transferred? Is the energy supplied by the circuit element or absorbed by it?

Answer 40 J are supplied by the circuit element. □

1.3 POWER AND ENERGY

Consider the circuit element shown in Figure 1.14. Because the current i is the rate of flow of charge and the voltage v is a measure of the energy transferred per unit of charge, the product of the current and the voltage is the rate of energy transfer. In other words, the product of current and voltage is power:

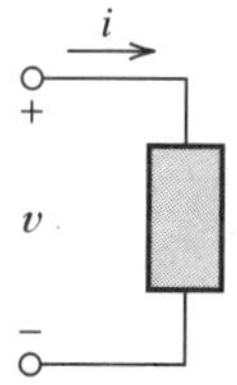

Figure 1.14 When current flows through an element and voltage appears across the element, energy is transferred. The rate of energy transfer is $p = vi$.

$$p = vi \tag{1.4}$$

The physical units of the quantities on the right-hand side of this equation are

$$\begin{aligned} \text{volts} \times \text{amperes} &= \\ (\text{joules/coulomb}) \times (\text{coulombs/second}) &= \\ \text{joules/second} &= \\ \text{watts} & \end{aligned}$$

Passive Reference Configuration

Now we may ask whether the power calculated by Equation 1.4 represents energy supplied by or absorbed by the element. Refer to Figure 1.14 and notice that the current reference enters the positive polarity of the voltage. We call this arrangement the **passive reference configuration**. Provided that the references are picked in this manner, a positive result for the power calculation implies that energy is being absorbed by the element. On the other hand, a negative result means that the element is supplying energy to other parts of the circuit.

If the current reference enters the negative end of the reference polarity, we compute the power as

$$p = -vi \tag{1.5}$$

Then, as before, a positive value for p indicates that energy is absorbed by the element, and a negative value shows that energy is supplied by the element.

If the circuit element happens to be an electrochemical battery, positive power means that the battery is being charged. In other words, the energy absorbed by the battery is being stored as chemical energy. On the other hand, negative power indicates that the battery is being discharged. Then the energy supplied by the battery is delivered to some other element in the circuit.

Sometimes currents, voltages, and powers are functions of time. To emphasize this fact, we can write Equation 1.4 as

$$p(t) = v(t)i(t) \tag{1.6}$$

Example 1.2 Power Calculations

Consider the circuit elements shown in Figure 1.15. Calculate the power for each element. If each element is a battery, is it being charged or discharged?

Solution In element A, the current reference enters the positive reference polarity. This is the passive reference configuration. Thus, power is computed as

$$p_a = v_a i_a = 12\text{ V} \times 2\text{ A} = 24\text{ W}$$

Because the power is positive, energy is absorbed by the device. If it is a battery, it is being charged.

In element B, the current reference enters the negative reference polarity. (Recall that the current that enters one end of a circuit element must exit from the other end, and vice versa.) This is opposite to the passive reference configuration. Hence, power is computed as

$$p_b = -v_b i_b = -(12\text{ V}) \times 1\text{ A} = -12\text{ W}$$

Since the power is negative, energy is supplied by the device. If it is a battery, it is being discharged.

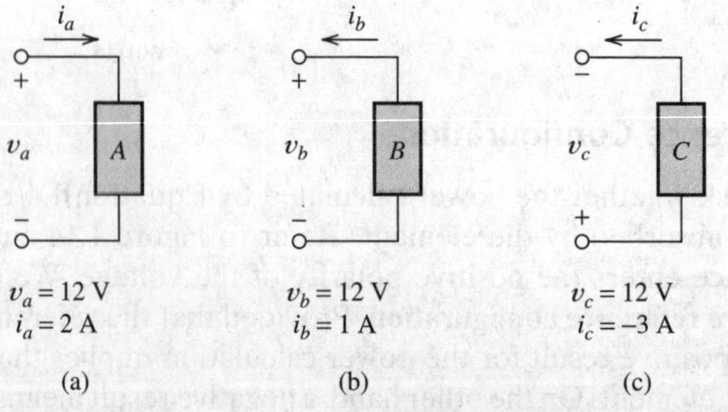

Figure 1.15 Circuit elements for Example 1.2.

In element C, the current reference enters the positive reference polarity. This is the passive reference configuration. Thus, we compute power as

$$p_c = v_c i_c = 12 \text{ V} \times (-3 \text{ A}) = -36 \text{ W}$$

Since the result is negative, energy is supplied by the element. If it is a battery, it is being discharged. (Notice that since i_c takes a negative value, current actually flows downward through element C.) ■

Energy Calculations

To calculate the energy w delivered to a circuit element between time instants t_1 and t_2, we integrate power:

$$w = \int_{t_1}^{t_2} p(t)\, dt \tag{1.7}$$

Here we have explicitly indicated that power can be a function of time by using the notation $p(t)$.

Example 1.3 Energy Calculation

Find an expression for the power for the voltage source shown in Figure 1.16. Compute the energy for the interval from $t_1 = 0$ to $t_2 = \infty$.

Solution The current reference enters the positive reference polarity. Thus, we compute power as

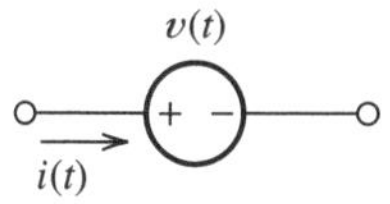

$v(t) = 12$ V
$i(t) = 2e^{-t}$ A

Figure 1.16 Circuit element for Example 1.3.

$$\begin{aligned} p(t) &= v(t)i(t) \\ &= 12 \times 2e^{-t} \\ &= 24e^{-t} \text{ W} \end{aligned}$$

Subsequently, the energy transferred is given by

$$\begin{aligned} w &= \int_0^{\infty} p(t)\, dt \\ &= \int_0^{\infty} 24e^{-t}\, dt \\ &= \left[-24e^{-t}\right]_0^{\infty} = -24e^{-\infty} - (-24e^0) = 24 \text{ J} \end{aligned}$$

Because the energy is positive, it is absorbed by the source. ■

Prefixes

In electrical engineering, we encounter a tremendous range of values for currents, voltages, powers, and other quantities. We use the prefixes shown in Table 1.2 when working with very large or small quantities. For example, 1 milliampere (1 mA) is equivalent to 10^{-3} A, 1 kilovolt (1 kV) is equivalent to 1000 V, and so on.

Table 1.2. Prefixes Used for Large or Small Physical Quantities

Prefix	Abbreviation	Scale Factor
giga-	G	10^9
meg- or mega-	M	10^6
kilo-	k	10^3
milli-	m	10^{-3}
micro-	μ	10^{-6}
nano-	n	10^{-9}
pico-	p	10^{-12}
femto-	f	10^{-15}

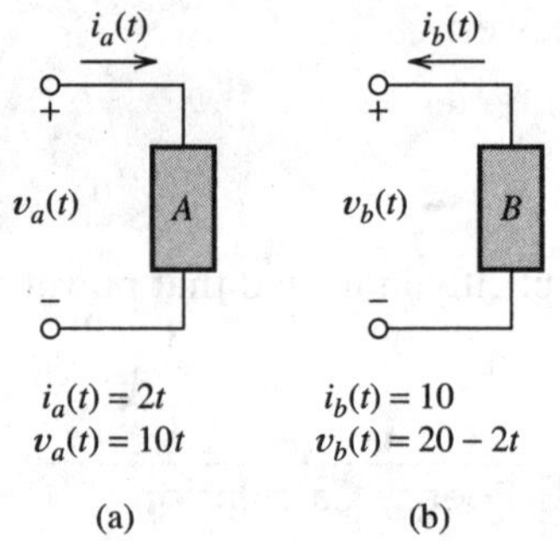

Figure 1.17 See Exercise 1.6.

Exercise 1.5 The ends of a circuit element are labeled a and b, respectively. Are the references for i_{ab} and v_{ab} related by the passive reference configuration? Explain.
Answer The reference direction for i_{ab} enters terminal a, which is also the positive reference for v_{ab}. Therefore, the current reference direction enters the positive reference polarity, so we have the passive reference configuration. □

Exercise 1.6 Compute the power as a function of time for each of the elements shown in Figure 1.17. Find the energy transferred between $t_1 = 0$ and $t_2 = 10$ s. In each case is energy supplied or absorbed by the element?
Answer **a.** $p_a(t) = 20t^2$ W, $w_a = 6667$ J; since w_a is positive, energy is absorbed by element A. **b.** $p_b(t) = 20t - 200$ W, $w_b = -1000$ J; since w_b is negative, energy is supplied by element B. □

1.4 KIRCHHOFF'S CURRENT LAW

Kirchhoff's current law states that the net current entering a node is zero.

A **node** in an electrical circuit is a point at which two or more circuit elements are joined together. Examples of nodes are shown in Figure 1.18.

An important principle of electrical circuits is **Kirchhoff's current law**: *The net current entering a node is zero.* To compute the *net* current entering a node, we add the currents entering and subtract the currents leaving. For illustration, consider the nodes of Figure 1.18. Then, we can write:

$$\text{Node } a: \quad i_1 + i_2 - i_3 = 0$$

$$\text{Node } b: \quad i_3 - i_4 = 0$$

$$\text{Node } c: \quad i_5 + i_6 + i_7 = 0$$

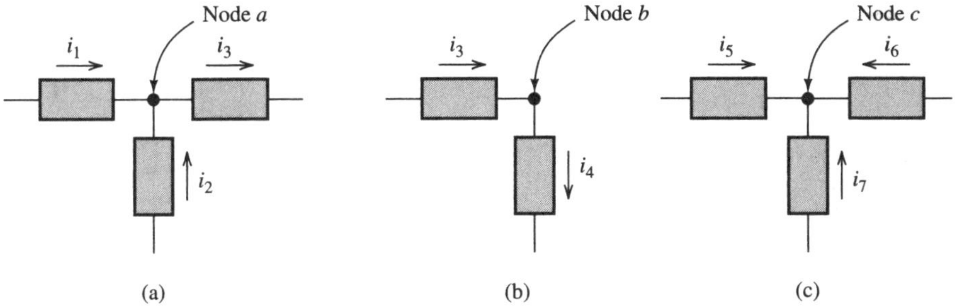

Figure 1.18 Partial circuits showing one node each to illustrate Kirchhoff's current law.

Notice that for node b, Kirchhoff's current law requires that $i_3 = i_4$. In general, if only two circuit elements are connected at a node, their currents must be equal. The current flows into the node through one element and out through the other. Usually, we will recognize this fact and assign a single current variable for both circuit elements.

For node c, either all of the currents are zero or some are positive while others are negative.

We abbreviate Kirchhoff's current law as KCL. There are two other equivalent ways to state KCL. One way is: *The net current leaving a node is zero.* To compute the net current leaving a node, we add the currents leaving and subtract the currents entering. For the nodes of Figure 1.18, this yields the following:

$$\text{Node } a: \quad -i_1 - i_2 + i_3 = 0$$

$$\text{Node } b: \quad -i_3 + i_4 = 0$$

$$\text{Node } c: \quad -i_5 - i_6 - i_7 = 0$$

Of course, these equations are equivalent to those obtained earlier.

Another way to state KCL is: *The sum of the currents entering a node equals the sum of the currents leaving a node.* Applying this statement to Figure 1.18, we obtain the following set of equations:

An alternative way to state Kirchhoff's current law is that the sum of the currents entering a node is equal to the sum of the currents leaving a node.

$$\text{Node } a: \quad i_1 + i_2 = i_3$$

$$\text{Node } b: \quad i_3 = i_4$$

$$\text{Node } c: \quad i_5 + i_6 + i_7 = 0$$

Again, these equations are equivalent to those obtained earlier.

Physical Basis for Kirchhoff's Current Law

An appreciation of why KCL is true can be obtained by considering what would happen if it were violated. Suppose that we could have the situation shown in Figure 1.18(a), with $i_1 = 3$ A, $i_2 = 2$ A, and $i_3 = 4$ A. Then, the net current entering the node would be

$$i_1 + i_2 - i_3 = 1 \text{ A} = 1 \text{ C/s}$$

In this case, 1 C of charge would accumulate at the node during each second. After 1 s, we would have +1 C of charge at the node, and −1 C of charge somewhere else in the circuit.

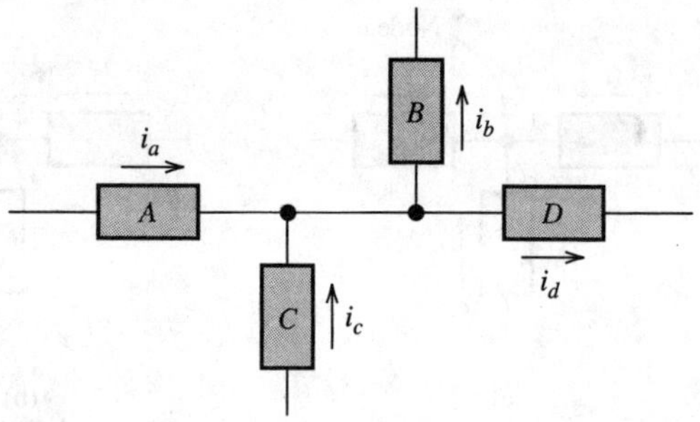

Figure 1.19 Elements A, B, C, and D can be considered to be connected to a common node, because all points in a circuit that are connected directly by conductors are electrically equivalent to a single point.

Suppose that these charges are separated by a distance of one meter (m). Recall that unlike charges experience a force of attraction. The resulting force turns out to be approximately 8.99×10^9 newtons (N) (equivalent to 2.02×10^9 pounds). Very large forces are generated when charges of this magnitude are separated by moderate distances. In effect, KCL states that such forces prevent charge from accumulating at the nodes of a circuit.

All points in a circuit that are connected directly by conductors can be considered to be a single node.

All points in a circuit that are connected directly by conductors can be considered to be a single node. For example, in Figure 1.19, elements $A, B, C,$ and D are connected to a common node. Applying KCL, we can write

$$i_a + i_c = i_b + i_d$$

Series Circuits

We make frequent use of KCL in analyzing circuits. For example, consider the elements A, B, and C shown in Figure 1.20. When elements are connected end to end, we say that they are connected in **series**. *In order for elements A and B to be in series, no other path for current can be connected to the node joining A and B. Thus, all elements in a series circuit have identical currents.* For example, writing Kirchhoff's current law at node 1 for the circuit of Figure 1.20, we have

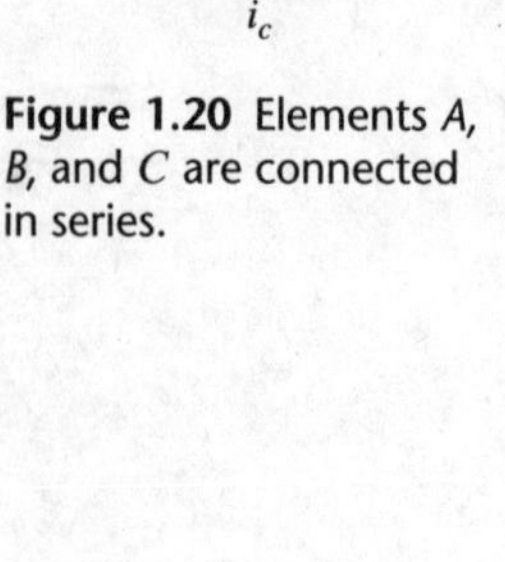

Figure 1.20 Elements A, B, and C are connected in series.

$$i_a = i_b$$

At node 2, we have

$$i_b = i_c$$

Thus, we have

$$i_a = i_b = i_c$$

The current that enters a series circuit must flow through each element in the circuit.

Exercise 1.7 Use KCL to determine the values of the unknown currents shown in Figure 1.21.

Answer $i_a = 4\text{ A}, i_b = -2\text{ A}, i_c = -8\text{ A}.$ □

Exercise 1.8 Consider the circuit of Figure 1.22. Identify the groups of circuit elements that are connected in series.

Answer Elements A and B are in series; elements $E, F,$ and G form another series combination. □

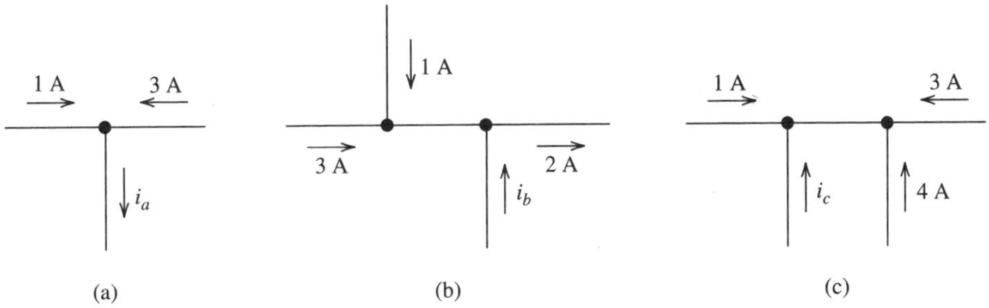

Figure 1.21 See Exercise 1.7.

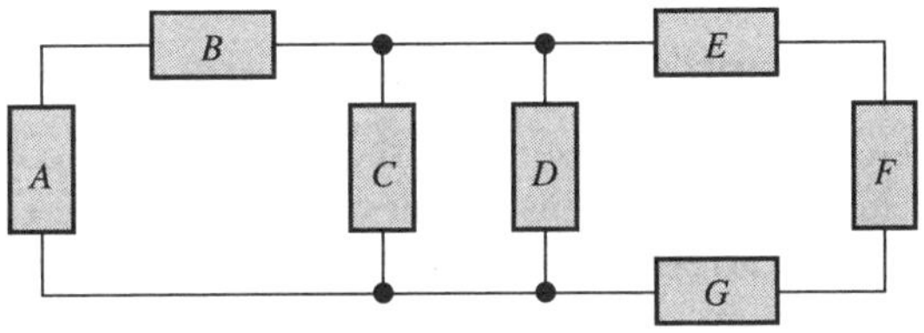

Figure 1.22 Circuit for Exercise 1.8.

1.5 KIRCHHOFF'S VOLTAGE LAW

A **loop** in an electrical circuit is a closed path starting at a node and proceeding through circuit elements, eventually returning to the starting node. Frequently, several loops can be identified for a given circuit. For example, in Figure 1.22, one loop consists of the path starting at the top end of element A and proceeding clockwise through elements B and C, returning through A to the starting point. Another loop starts at the top of element D and proceeds clockwise through E, F, and G, returning to the start through D. Still another loop exists through elements A, B, E, F, and G around the periphery of the circuit.

Kirchhoff's voltage law (KVL) states that the algebraic sum of the voltages equals zero for any closed path (loop) in an electrical circuit.

Kirchhoff's voltage law (KVL) states: *The algebraic sum of the voltages equals zero for any closed path (loop) in an electrical circuit.* In traveling around a loop, we encounter various voltages, some of which carry a positive sign while others carry a negative sign in the algebraic sum. A convenient convention is to use the first polarity mark encountered for each voltage to decide if it should be added or subtracted in the algebraic sum. If we go through the voltage from the positive polarity reference to the negative reference, it carries a plus sign. If the polarity marks are encountered in the opposite direction (minus to plus), the voltage carries a negative sign. This is illustrated in Figure 1.23.

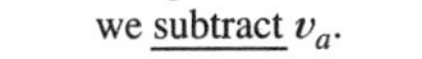

Figure 1.23 In applying KVL to a loop, voltages are added or subtracted depending on their reference polarities relative to the direction of travel around the loop.

For the circuit of Figure 1.24, we obtain the following equations:

$$\text{Loop 1:} \quad -v_a + v_b + v_c = 0$$

$$\text{Loop 2:} \quad -v_c - v_d + v_e = 0$$

$$\text{Loop 3:} \quad v_a - v_b + v_d - v_e = 0$$

Notice that v_a is subtracted for loop 1, but it is added for loop 3, because the direction of travel is different for the two loops. Similarly, v_c is added for loop 1 and subtracted for loop 2.

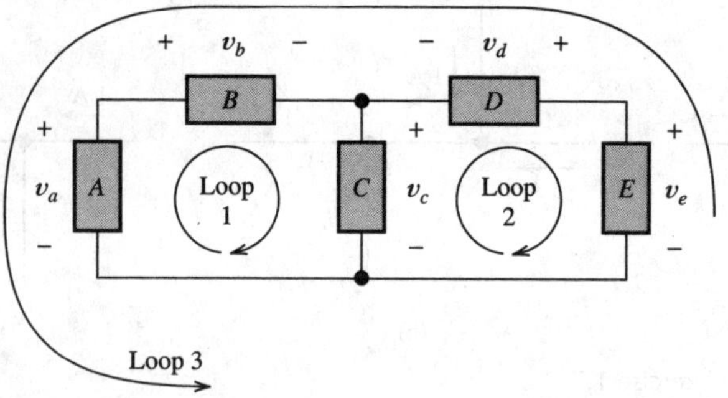

Figure 1.24 Circuit used for illustration of Kirchhoff's voltage law.

Kirchhoff's Voltage Law Related to Conservation of Energy

KVL is a consequence of the law of energy conservation. Consider the circuit shown in Figure 1.25. This circuit consists of three elements connected in series. Thus, the same current i flows through all three elements. The power for each of the elements is given by

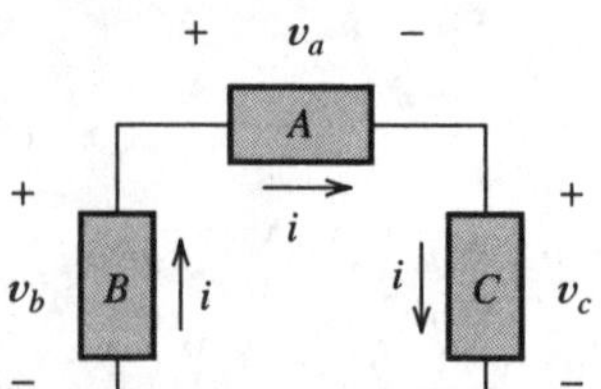

Figure 1.25 In this circuit, conservation of energy requires that $v_b = v_a + v_c$.

$$\text{Element } A\text{:}\quad p_a = v_a i$$

$$\text{Element } B\text{:}\quad p_b = -v_b i$$

$$\text{Element } C\text{:}\quad p_c = v_c i$$

Notice that the current and voltage references have the passive configuration (the current reference enters the plus polarity mark) for elements A and C. For element B, the relationship is opposite to the passive reference configuration. That is why we have a negative sign in the calculation of p_b.

At a given instant, the sum of the powers for all of the elements in a circuit must be zero. Otherwise, for an increment of time taken at that instant, more energy would be absorbed than is supplied by the circuit elements (or vice versa):

$$p_a + p_b + p_c = 0$$

Substituting for the powers, we have

$$v_a i - v_b i + v_c i = 0$$

Canceling the current i, we obtain

$$v_a - v_b + v_c = 0$$

This is exactly the same equation that is obtained by adding the voltages around the loop and setting the sum to zero for a clockwise loop in the circuit of Figure 1.25.

One way to check our results after solving for the currents and voltages in a circuit is the check to see that the power adds to zero for all of the elements.

Two circuit elements are connected in parallel if both ends of one element are connected directly (i.e., by conductors) to corresponding ends of the other.

Parallel Circuits

We say that two circuit elements are connected in **parallel** if both ends of one element are connected directly (i.e., by conductors) to corresponding ends of the other. For

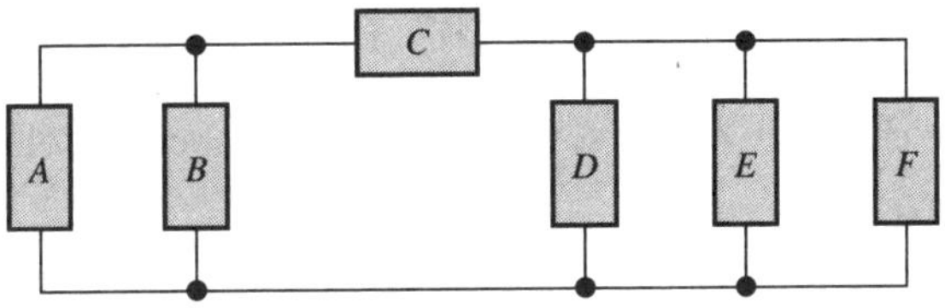

Figure 1.26 In this circuit, elements A and B are in parallel. Elements D, E, and F form another parallel combination.

example, in Figure 1.26, elements A and B are in parallel. Similarly, we say that the three circuit elements D, E, and F are in parallel. Element B is *not* in parallel with D because the top end of B is not *directly* connected to the top end of D.

The voltages across parallel elements are equal in magnitude and have the same polarity. For illustration, consider the partial circuit shown in Figure 1.27. Here elements A, B, and C are connected in parallel. Consider a loop from the bottom end of A upward and then down through element B back to the bottom of A. For this clockwise loop, we have $-v_a + v_b = 0$. Thus, KVL requires that

$$v_a = v_b$$

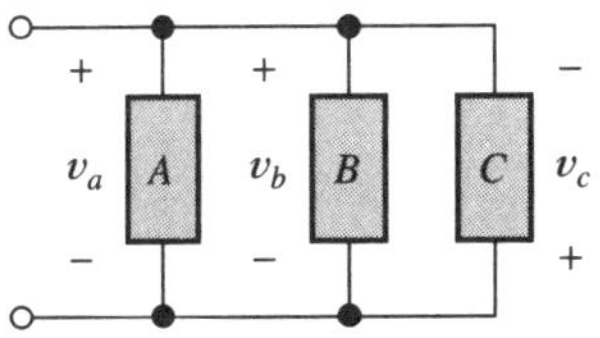

Figure 1.27 For this circuit, we can show that $v_a = v_b = -v_c$. Thus, the magnitudes and *actual* polarities of all three voltages are the same.

Next, consider a clockwise loop through elements A and C. For this loop, KVL requires that

$$-v_a - v_c = 0$$

This implies that $v_a = -v_c$. In other words, v_a and v_c have opposite algebraic signs. Furthermore, one or the other of the two voltages must be negative (unless both are zero). Therefore, one of the voltages has an actual polarity opposite to the reference polarity shown in the figure. Thus, the actual polarities of the voltages are the same (either both are positive at the top of the circuit or both are positive at the bottom).

Usually, when we have a parallel circuit, we simply use the same voltage variable for all of the elements as illustrated in Figure 1.28.

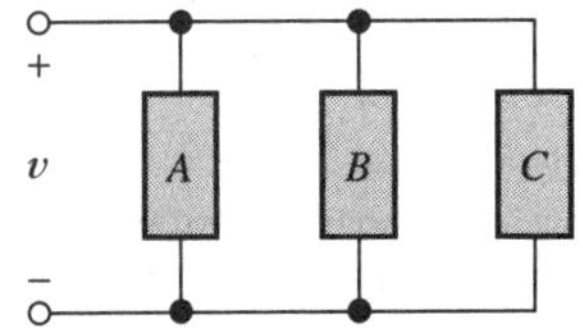

Figure 1.28 Analysis is simplified by using the same voltage variable and reference polarity for elements that are in parallel.

Exercise 1.9 Use repeated application of KVL to find the values of v_c and v_e for the circuit of Figure 1.29.

Answer $v_c = 8$ V, $v_e = -2$ V. □

Exercise 1.10 Identify elements that are in parallel in Figure 1.29. Identify elements in series.

Answer Elements E and F are in parallel; elements A and B are in series. □

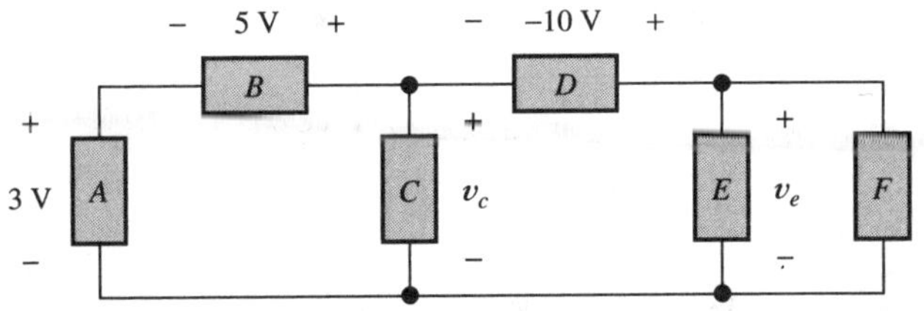

Figure 1.29 Circuit for Exercises 1.9 and 1.10.

1.6 INTRODUCTION TO CIRCUIT ELEMENTS

In this section, we carefully define several types of ideal circuit elements:

Conductors
Voltage sources
Current sources
Resistors

Later in the book, we will encounter additional elements, including inductors and capacitors. Eventually, we will be able to use these idealized circuit elements to describe (model) complex real-world electrical devices.

Conductors

We have already encountered conductors. Ideal conductors are represented in circuit diagrams by unbroken lines between the ends of other circuit elements. We define ideal circuit elements in terms of the relationship between the voltage across the element and the current through it.

The voltage between the ends of an ideal conductor is zero regardless of the current flowing through the conductor.

All points in a circuit that are connected by ideal conductors can be considered as a single node.

The voltage between the ends of an ideal conductor is zero regardless of the current flowing through the conductor. When two points in a circuit are connected together by an ideal conductor, we say that the points are **shorted** together. Another term for an ideal conductor is **short circuit**. All points in a circuit that are connected by ideal conductors can be considered as a single node.

If no conductors or other circuit elements are connected between two parts of a circuit, we say that an **open circuit** exists between the two parts of the circuit. No current can flow through an ideal open circuit.

Independent Voltage Sources

An ideal independent voltage source maintains a specified voltage across its terminals.

An **ideal independent voltage source** maintains a specified voltage across its terminals. The voltage across the source is independent of other elements that are connected to it and of the current flowing through it. We use a circle enclosing the reference polarity marks to represent independent voltage sources. The value of the voltage is indicated alongside the symbol. The voltage can be constant or it can be a function of time. Several voltage sources are shown in Figure 1.30.

In Figure 1.30(a), the voltage across the source is constant. Thus, we have a dc voltage source. On the other hand, the source shown in Figure 1.30(b) is an ac voltage source having a sinusoidal variation with time. We say that these are *independent* sources because the voltages across their terminals are independent of all other voltages and currents in the circuit.

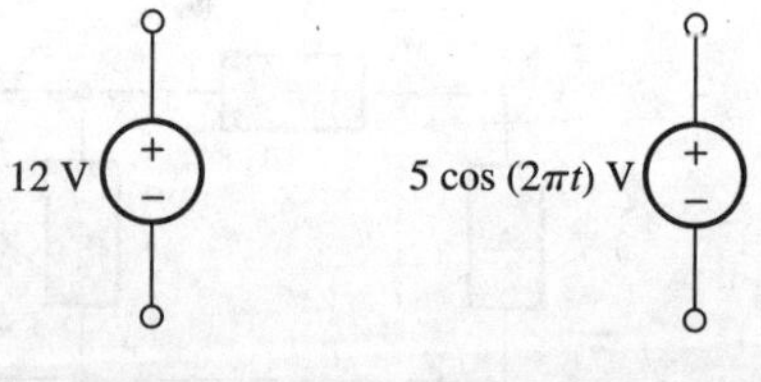

Figure 1.30 Independent voltage sources.

(a) Constant or dc voltage source

(b) Ac voltage source

Ideal Circuit Elements versus Reality

Here we are giving definitions of *ideal* circuit elements. It is possible to draw ideal circuits in which the definitions of various circuit elements conflict. For example, Figure 1.31 shows a 12-V voltage source with a conductor connected across its terminals. In this case, the definition of the voltage source requires that $v_x = 12$ V. On the other hand, the definition of an ideal conductor requires that $v_x = 0$. In our study of ideal circuits, we avoid such conflicts.

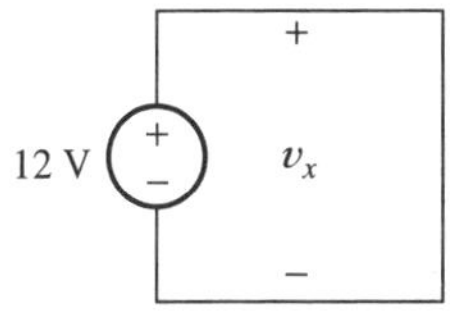

Figure 1.31 We avoid self-contradictory circuit diagrams such as this one.

In the real world, an automobile battery is nearly an ideal 12-V voltage source, and a short piece of heavy-gauge copper wire is nearly an ideal conductor. If we place the wire across the terminals of the battery, a very large current flows through the wire, stored chemical energy is converted to heat in the wire at a very high rate, and the wire will probably melt or the battery be destroyed.

When we encounter a contradictory idealized circuit model, we often have an undesirable situation (such as a fire or destroyed components) in the real-world counterpart to the model. In any case, a contradictory circuit model implies that we have not been sufficiently careful in choosing circuit models for the real circuit elements. For example, an automobile battery is not exactly modeled as an ideal voltage source. We will see that a better model (particularly if the currents are very large) is an ideal voltage source in series with a resistance. (We will discuss resistance very soon.) A short piece of copper wire is not modeled well as an ideal conductor, in this case. Instead, we will see that it is modeled better as a small resistance. If we have done a good job at picking circuit models for real-world circuits, we will not encounter contradictory circuits, and the results we calculate using the model will match reality very well.

Dependent Voltage Sources

A **dependent** or **controlled voltage source** is similar to an independent source except that the voltage across the source terminals is a function of other voltages or currents in the circuit. Instead of a circle, it is customary to use a diamond to represent controlled sources in circuit diagrams. Two examples of dependent sources are shown in Figure 1.32.

A **voltage-controlled voltage source** is a voltage source having a voltage equal to a constant times the voltage across a pair of terminals elsewhere in the network.

A voltage-controlled voltage source maintains a voltage across its terminals equal to a constant times a voltage elsewhere in the circuit.

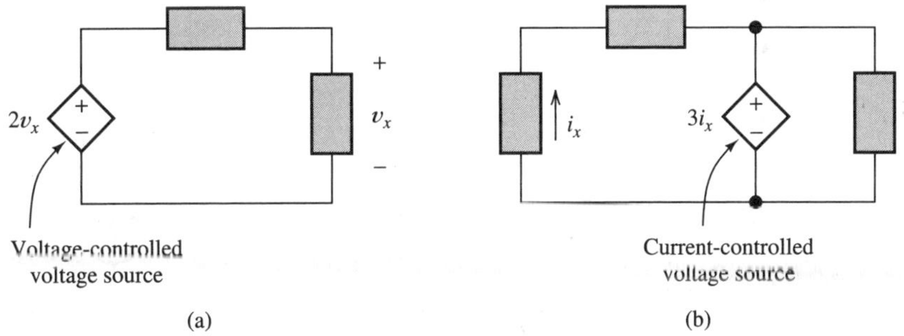

Figure 1.32 Dependent voltage sources (also known as controlled voltage sources) are represented by diamond-shaped symbols. The voltage across a controlled voltage source depends on a current or voltage that appears elsewhere in the circuit.

An example is shown in Figure 1.32(a). The dependent voltage source is the diamond symbol. The reference polarity of the source is indicated by the marks inside the diamond. The voltage v_x determines the value of the voltage produced by the source. For example, if it should turn out that $v_x = 3$ V, the source voltage is $2v_x = 6$ V. If v_x should equal -7 V, the source produces $2v_x = -14$ V (in which case, the actual positive polarity of the source is at the bottom end).

A current-controlled voltage source maintains a voltage across its terminals equal to a constant times a current flowing through some other element in the circuit.

A **current-controlled voltage source** is a voltage source having a voltage equal to a constant times the current through some other element in the circuit. An example is shown in Figure 1.32(b). In this case, the source voltage is three times the value of the current i_x. The factor multiplying the current is called the **gain parameter**. We assume that the voltage has units of volts and the current is in amperes. Thus, the gain parameter [which is 3 in Figure 1.32(b)] has units of volts per ampere (V/A). (Shortly, we will see that the units V/A are the units of resistance and are called ohms.)

Returning our attention to the voltage-controlled voltage source in Figure 1.32(a), we note that the gain parameter is 2 and is unitless (or we could say that the units are V/V).

Later in the book, we will see that controlled sources are very useful in modeling transistors, amplifiers, and electrical generators, among other things.

Independent Current Sources

An ideal independent current source forces a specified current to flow through itself.

An ideal **independent current source** forces a specified current to flow through itself. The symbol for an independent current source is a circle enclosing an arrow that gives the reference direction for the current. The current through an independent current source is independent of the elements connected to it and of the voltage across it. Figure 1.33 shows the symbols for a dc current source and for an ac current source.

If an open circuit exists across the terminals of a current source, we have a contradictory circuit. For example, consider the 2-A dc current source shown in Figure 1.33(a). This current source is shown with an open circuit across its terminals. By definition, the current flowing into the top node of the source is 2 A. Also by definition, no current can flow through the open circuit. Thus, KCL is not satisfied at this node. In good models for actual circuits, this situation does not occur. Thus, we will avoid current sources with open-circuited terminals in our discussion of ideal networks.

A battery is a good example of a voltage source, but an equally familiar example does not exist for a current source. However, current sources are useful in constructing theoretical models. Later, we will see that a good approximation to an ideal current source can be achieved with electronic amplifiers.

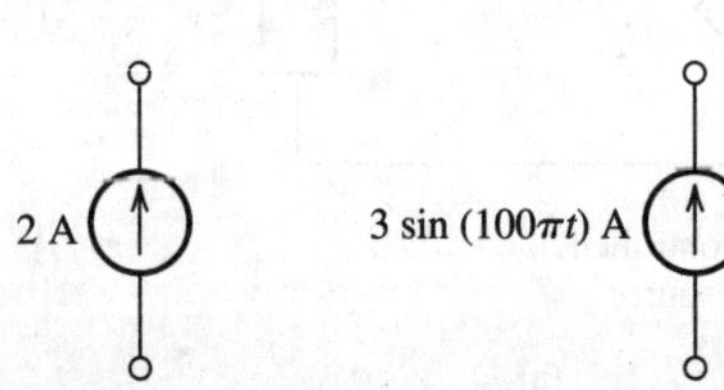

Figure 1.33 Independent current sources. (a) Dc current source (b) Ac current source

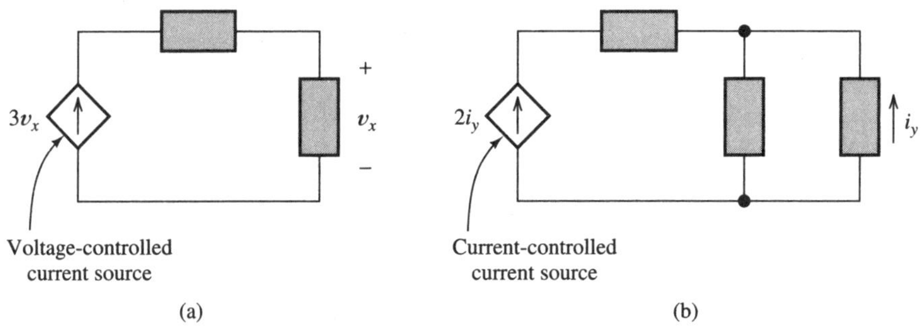

Figure 1.34 Dependent current sources. The current through a dependent current source depends on a current or voltage that appears elsewhere in the circuit.

Dependent Current Sources

The current flowing through a dependent current source is determined by a current or voltage elsewhere in the circuit.

The current flowing through a **dependent current source** is determined by a current or voltage elsewhere in the circuit. The symbol is a diamond enclosing an arrow that indicates the reference direction. Two types of controlled current sources are shown in Figure 1.34.

In Figure 1.34(a), we have a **voltage-controlled current source**. The current through the source is three times the voltage v_x. The gain parameter of the source (3 in this case) has units of A/V (which we will soon see are equivalent to siemens or inverse ohms). If it turns out that v_x has a value of 5 V, the current through the controlled current source is $3v_x = 15$ A.

Figure 1.34(b) illustrates a **current-controlled current source**. In this case, the current through the source is twice the value of i_y. The gain parameter, which has a value of 2 in this case, has units of A/A (i.e., it is unitless).

Like controlled voltage sources, controlled current sources are useful in constructing circuit models for many types of real-world devices, such as electronic amplifiers, transistors, transformers, and electrical machines. If a controlled source is needed for some application, it can be implemented by using electronic amplifiers. In sum, these are the four kinds of controlled sources:

1. Voltage-controlled voltage sources
2. Current-controlled voltage sources
3. Voltage-controlled current sources
4. Current-controlled current sources

Resistors and Ohm's Law

The voltage v across an ideal **resistor** is proportional to the current i through the resistor. The constant of proportionality is the resistance R. The symbol used for a resistor is shown in Figure 1.35(a). Notice that the current reference and voltage polarity reference conform to the passive reference configuration. In other words, the reference direction for the current is into the positive polarity mark and out of the negative polarity mark. In equation form, the voltage and current are related by **Ohm's law**:

$$v = iR$$

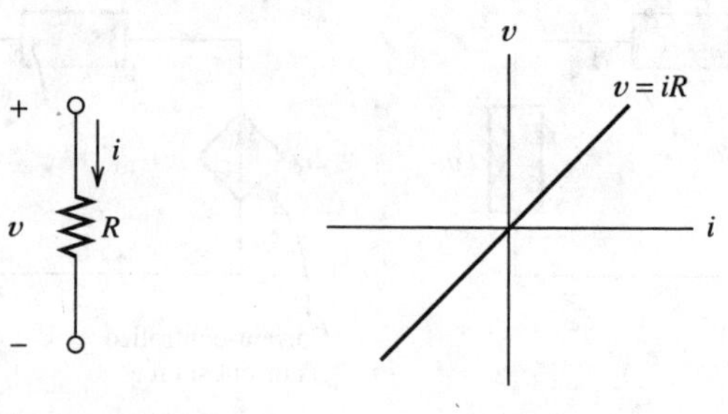

(a) Resistance symbol (b) Ohm's law

Figure 1.35 Voltage is proportional to current in an ideal resistor. Notice that the references for v and i conform to the passive reference configuration.

The units of resistance are V/A, which are called ohms. The uppercase Greek letter omega (Ω) represents ohms. In practical circuits, we encounter resistances ranging from milliohms (mΩ) to megohms (MΩ).

Except for rather unusual situations, the resistance R assumes positive values. (In certain types of electronic circuits, we can encounter negative resistance, but for now we assume that R is positive.) In situations for which the current reference direction enters the *negative* reference of the voltage, Ohm's law becomes

$$v = -iR$$

This is illustrated in Figure 1.36.

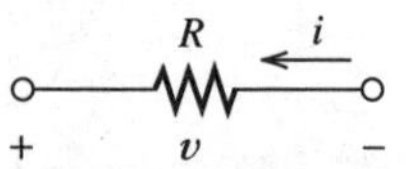

Figure 1.36 If the references for v and i are opposite to the passive configuration, we have $v = -Ri$.

The relationship between current direction and voltage polarity can be neatly included in the equation for Ohm's law if double-subscript notation is used. (Recall that to use double subscripts, we label the ends of the element under consideration, which is a resistance in this case.) If the order of the subscripts is the same for the current as for the voltage (i_{ab} and v_{ab}, for example), the current reference direction enters the first terminal and the positive voltage reference is at the first terminal. Thus, we can write

$$v_{ab} = i_{ab}R$$

On the other hand, if the order of the subscripts is not the same, we have

$$v_{ab} = -i_{ba}R$$

Conductance

Solving Ohm's law for current, we have

$$i = \frac{1}{R}v$$

We call the quantity $1/R$ a **conductance**. It is customary to denote conductances with the letter G:

$$G = \frac{1}{R} \qquad (1.8)$$

Conductances have the units of inverse ohms (Ω^{-1}), which are called siemens (abbreviated S). Thus, we can write Ohm's law as

$$i = Gv \tag{1.9}$$

Resistors

It turns out that we can construct nearly ideal resistors by attaching terminals to many types of conductive materials. This is illustrated in Figure 1.37. Conductive materials that can be used to construct resistors include most metals, their alloys, and carbon.

On a microscopic level, current in metals consists of electrons moving through the material. (On the other hand, in solutions of ionic compounds, current is carried partly by positive ions.) The applied voltage creates an electric field that accelerates the electrons. The electrons repeatedly collide with the atoms of the material and lose their forward momentum. Then they are accelerated again. The net effect is a constant average velocity for the electrons. At the macroscopic level, we observe a current that is proportional to the applied voltage.

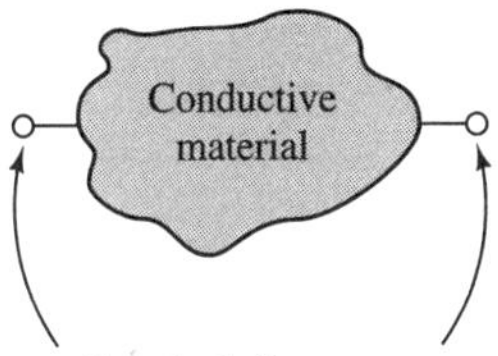

Figure 1.37 We construct resistors by attaching terminals to a piece of conductive material.

Resistance Related to Physical Parameters

The dimensions and geometry of the resistor as well as the particular material used to construct a resistor influence its resistance. We consider only resistors that take the form of a long cylinder or bar with terminals attached at the ends, as illustrated in Figure 1.38. The cross-sectional area A is constant along the length of the cylinder or bar. If the length L of the resistor is much greater than the dimensions of its cross section, the resistance is approximately given by

$$R = \frac{\rho L}{A} \tag{1.10}$$

in which ρ is the *resistivity* of the material used to construct the resistor. The units of resistivity are ohm meters (Ωm).

Materials can be classified as conductors, semiconductors, or insulators, depending on their resistivity. **Conductors** have the lowest resistivity and easily conduct electrical current. **Insulators** have very high resistivity and conduct very little current (at least for moderate voltages). **Semiconductors** fall between conductors and insulators. We will see in Chapters 10, 12, and 13 that certain semiconductors are very useful in constructing electronic devices. Table 1.3 gives approximate values of resistivity for several materials.

Figure 1.38 Resistors often take the form of a long cylinder (or bar) in which current enters one end and flows along the length.

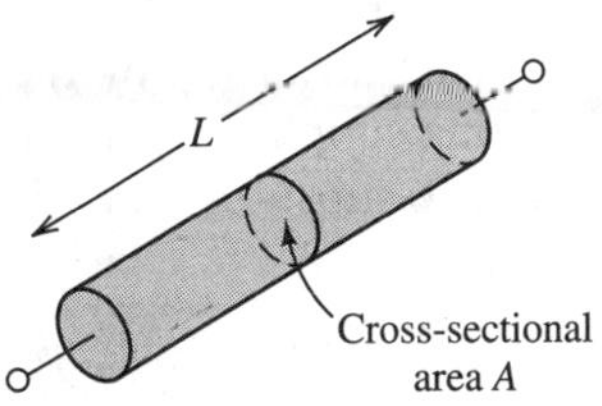

Table 1.3. Resistivity Values (Ωm) for Selected Materials at 300 K

Conductors	
Aluminum	2.73×10^{-8}
Carbon (amorphous)	3.5×10^{-5}
Copper	1.72×10^{-8}
Gold	2.27×10^{-8}
Nichrome	1.12×10^{-6}
Silver	1.63×10^{-8}
Tungsten	5.44×10^{-8}
Semiconductors	
Silicon (device grade) depends on impurity concentration	10^{-5} to 1
Insulators	
Fused quartz	$> 10^{21}$
Glass (typical)	1×10^{12}
Teflon	1×10^{19}

Example 1.4 Resistance Calculation

Compute the resistance of a copper wire having a diameter of 2.05 mm and a length of 10 m.

Solution First, we compute the cross-sectional area of the wire:

$$A = \frac{\pi d^2}{4} = \frac{\pi(2.05 \times 10^{-3})^2}{4} = 3.3 \times 10^{-6}\ \text{m}^2$$

Then, the resistance is given by

$$R = \frac{\rho L}{A} = \frac{1.72 \times 10^{-8} \times 10}{3.3 \times 10^{-6}} = 0.052\ \Omega$$

These are the approximate dimensions of a piece of 12-gauge copper wire that we might find connecting an electrical outlet to the distribution box in a residence. Of course, two wires are needed for a complete circuit. ■

Power Calculations for Resistances

Recall that we compute power for a circuit element as the product of the current and voltage:

$$p = vi \tag{1.11}$$

If v and i have the passive reference configuration, a positive sign for power means that energy is being absorbed by the device. Furthermore, a negative sign means that energy is being supplied by the device.

If we use Ohm's law to substitute for v in Equation 1.11, we obtain

$$p = Ri^2 \tag{1.12}$$

On the other hand, if we solve Ohm's law for i and substitute into Equation 1.11, we obtain

$$p = \frac{v^2}{R} \tag{1.13}$$

Notice that power for a resistance is positive regardless of the sign of v or i (assuming that R is positive, which is ordinarily the case). Thus, power is absorbed by resistances. If the resistance results from collisions of electrons with the atoms of the material composing a resistor, this power shows up as heat.

Some applications for conversion of electrical power into heat are heating elements for ovens, water heaters, cooktops, and space heaters. In a typical space heater, the heating element consists of a nichrome wire that becomes red hot in operation. (Nichrome is an alloy of nickel, chromium, and iron.) To fit the required length of wire in a small space, it is coiled rather like a spring.

PRACTICAL APPLICATION 1.1

Using Resistance to Measure Strain

Civil and mechanical engineers routinely employ the dependence of resistance on physical dimensions of a conductor to measure strain. These measurements are important in experimental stress–strain analysis of mechanisms and structures. (Strain is defined as fractional change in length, given by $\epsilon = \Delta L/L$.)

A typical resistive strain gauge consists of nickel–copper alloy foil that is photoetched to obtain multiple conductors aligned with the direction of the strain to be measured. This is illustrated in Figure PA1.1. Typically, the conductors are bonded to a thin polyimide (a tough flexible plastic) backing, which in turn is attached to the structure under test by a suitable adhesive, such as cyanoacrylate cement.

The resistance of a conductor is given by

$$R = \frac{\rho L}{A}$$

As strain is applied, the length and area change, resulting in changes in resistance. The strain and the change in resistance are related by the gauge factor:

$$G = \frac{\Delta R/R_0}{\epsilon}$$

in which R_0 is the resistance of the gauge before strain. A typical gauge has $R_0 = 350\ \Omega$ and $G = 2.0$. Thus, for a strain of 1%, the change in resistance is $\Delta R = 7\ \Omega$. Usually, a Wheatstone bridge (discussed in Chapter 2) is used to measure the small changes in resistance associated with accurate strain determination.

Sensors for force, torque, and pressure are constructed by using resistive strain gauges.

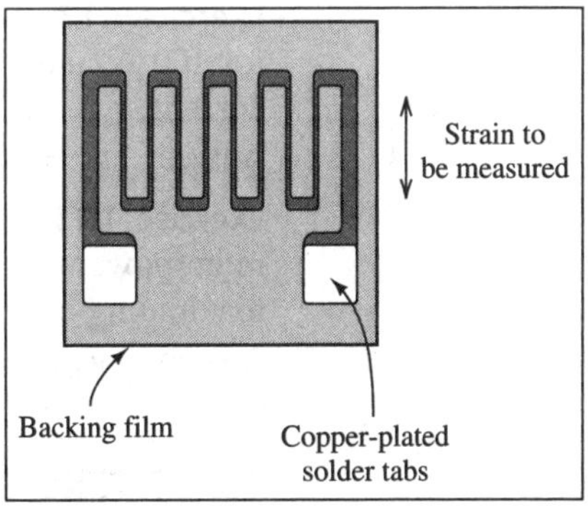

Figure PA1.1

Resistors versus Resistances

As an aside, we mention that resistance is often useful in modeling devices in which electrical power is converted into forms other than heat. For example, a loudspeaker appears to have a resistance of 8 Ω. Part of the power delivered to the loudspeaker is converted to acoustic power. Another example is a transmitting antenna having a resistance of 50 Ω. The power delivered to an antenna is radiated, traveling away as an electromagnetic wave.

There is a slight distinction between the terms *resistor* and *resistance*. A resistor is a two-terminal device composed of a conductive material. Resistance is a circuit property for which voltage is proportional to current. Thus, resistors have the property of resistance. However, resistance is also useful in modeling antennas and loudspeakers, which are quite different from resistors. Often, we are not careful about this distinction in using these terms.

Example 1.5 Determining Resistance for Given Power and Voltage Ratings

A certain electrical heater is rated for 1500 W when operated from 120 V. Find the resistance of the heater element and the operating current. (Resistance depends on temperature, and we will find the resistance at the operating temperature of the heater.)

Solution Solving Equation 1.13 for resistance, we obtain

$$R = \frac{v^2}{p} = \frac{120^2}{1500} = 9.6\ \Omega$$

Then, we use Ohm's law to find the current:

$$i = \frac{v}{R} = \frac{120}{9.6} = 12.5\ \text{A}$$

■

Exercise 1.11 The 9.6-Ω resistance of Example 1.5 is in the form of a nichrome wire having a diameter of 1.6 mm. Find the length of the wire. (*Hint:* The resistivity of nichrome is given in Table 1.3.)

Answer $L = 17.2$ m. □

Exercise 1.12 Suppose we have a typical incandescent electric light bulb that is rated for 100 W and 120 V. Find its resistance (at operating temperature) and operating current.

Answer $R = 144\ \Omega, i = 0.833$ A. □

Exercise 1.13 A 1-kΩ resistor used in a television receiver is rated for a maximum power of 1/4 W. Find the current and voltage when the resistor is operated at maximum power.

Answer $v_{max} = 15.8$ V, $i_{max} = 15.8$ mA. □

1.7 INTRODUCTION TO CIRCUITS

In this chapter, we have defined electrical current and voltage, discussed Kirchhoff's laws, and introduced several ideal circuit elements: voltage sources, current sources, and resistances. Now we illustrate these concepts by considering a few relatively simple circuits. In the next chapter, we consider more complex circuits and analysis techniques.

Consider the circuit shown in Figure 1.39(a). Suppose that we want to know the current, voltage, and power for each element. To obtain these results, we apply

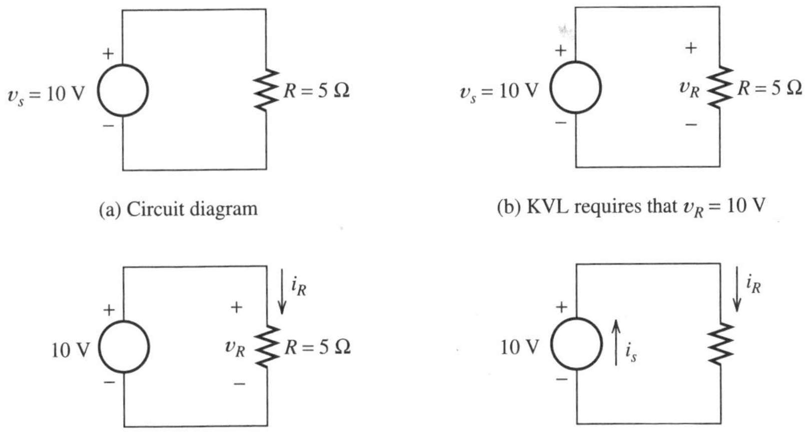

(a) Circuit diagram

(b) KVL requires that $v_R = 10$ V

(c) Ohm's law yields $i_R = v_R/R = 2$ A

(d) KCL requires that $i_s = i_R$

Figure 1.39 A circuit consisting of a voltage source and a resistance.

the basic principles introduced in this chapter. At first, we proceed in small, methodical steps. Furthermore, for ease of understanding, we initially select reference polarities and directions that agree with the actual polarities and current directions.

KVL requires that the sum of the voltages around the circuit shown in Figure 1.39 must equal zero. Thus, traveling around the circuit clockwise, we have $v_R - v_s = 0$. Consequently, $v_R = v_s$, and the voltage across the resistor v_R must have an actual polarity that is positive at the top end and a magnitude of 10 V.

An alternative way of looking at the voltages in this circuit is to notice that the voltage source and the resistance are in parallel. (The top ends of the voltage source and the resistance are connected, and the bottom ends are also connected.) Recall that when elements are in parallel, the voltage magnitude and polarity are the same for all elements.

Now consider Ohm's law. Because 10 V appears across the 5-Ω resistance, the current is $i_R = 10/5 = 2$ A. This current flows through the resistance from the positive polarity to the negative polarity. Thus, $i_R = 2$ A flows downward through the resistance, as shown in Figure 1.39(c).

According to KCL, the sum of the currents entering a given node must equal the sum of the currents leaving. There are two nodes for the circuit of Figure 1.39: one at the top and one at the bottom. The current i_R leaves the top node through the resistance. Thus, an equal current must enter the top node through the voltage source. The actual direction of current flow is upward through the voltage source, as shown in Figure 1.39(d).

Another way to see that the currents i_s and i_R are equal is to notice that the voltage source and the resistance are in series. In a series circuit, the current that flows in one element must continue through the other element. (Notice that for this circuit the voltage source and the resistance are in parallel and they are also in series. A two-element circuit is the only case for which this occurs. If more than two elements are interconnected, a pair of elements that are in parallel cannot also be in series, and vice versa.)

Notice that in Figure 1.39, the current in the voltage source flows from the negative polarity toward the positive polarity. It is only for resistances that the current is required to flow from plus to minus. For a voltage source, the current can flow in either direction, depending on the circuit to which the source is connected.

It is only for resistances that the current is required to flow from plus to minus. Current may flow in either direction for a voltage source depending on the other elements in the circuit.

Now let us calculate the power for each element. For the resistance, we have several ways to compute power:

$$p_R = v_R i_R = 10 \times 2 = 20 \text{ W}$$

$$p_R = i_R^2 R = 2^2 \times 5 = 20 \text{ W}$$

$$p_R = \frac{v_R^2}{R} = \frac{10^2}{5} = 20 \text{ W}$$

Of course, all the calculations yield the same result. Energy is delivered to the resistance at the rate of 20 J/s.

To find the power for the voltage source, we have

$$p_s = -v_s i_s$$

where the minus sign is used because the reference direction for the current enters the negative voltage reference (opposite to the passive reference configuration). Substituting values, we obtain

$$p_s = -v_s i_s = -10 \times 2 = -20 \text{ W}$$

Because p_s is negative, we understand that energy is being delivered by the voltage source.

As a check, if we add the powers for all the elements in the circuit, the result should be zero, because energy is neither created nor destroyed in an electrical circuit. Instead, it is transported and changed in form. Thus, we can write

$$p_s + p_R = -20 + 20 = 0$$

Using Arbitrary References

In the previous discussion, we selected references that agree with actual polarities and current directions. This is not always possible at the start of the analysis of more complex circuits. Fortunately, it is not necessary. We can pick the references in an arbitrary manner. Application of circuit laws will tell us not only the magnitudes of the currents and voltages but the true polarities and current directions as well.

Example 1.6 Circuit Analysis Using Arbitrary References

Analyze the circuit of Figure 1.39 using the current and voltage references shown in Figure 1.40. Verify that the results are in agreement with those found earlier.

Solution Traveling clockwise and applying KVL, we have

$$-v_s - v_x = 0$$

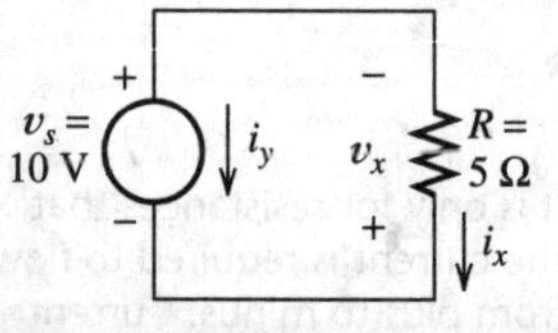

Figure 1.40 Circuit for Example 1.6.

This yields $v_x = -v_s = -10$ V. Since v_x assumes a negative value, the actual polarity is opposite to the reference. Thus, as before, we conclude that the voltage across the resistance is actually positive at the top end.

According to Ohm's law,

$$i_x = -\frac{v_x}{R}$$

where the minus sign appears because v_x and i_x have references opposite to the passive reference configuration. Substituting values, we get

$$i_x = -\frac{-10}{5} = 2\text{ A}$$

Since i_x assumes a positive value, the actual current direction is downward through the resistance.

Next, applying KCL at the bottom node of the circuit, we have

$$\text{total current entering} = \text{total current leaving}$$

$$i_y + i_x = 0$$

Thus, $i_y = -i_x = -2$ A, and we conclude that a current of 2 A actually flows upward through the voltage source.

The power for the voltage source is

$$p_s = v_s i_y = 10 \times (-2) = -20\text{ W}$$

Finally, the power for the resistance is given by

$$p_R = -v_x i_x$$

where the minus sign appears because the references for v_x and i_x are opposite to the passive reference configuration. Substituting, we find that $p_R = -(-10) \times (2) = 20$ W. Because p_R has a positive value, we conclude that energy is delivered to the resistance. ■

Sometimes circuits can be solved by repeated application of Kirchhoff's laws and Ohm's law. We illustrate with an example.

Example 1.7 Using KVL, KCL, and Ohm's Law to Solve a Circuit

Solve for the source voltage in the circuit of Figure 1.41 in which we have a current-controlled current source and we are given that the voltage across the 5-Ω resistance is 15 V.

Solution First, we use Ohm's Law to determine the value of i_y:

$$i_y = \frac{15\text{ V}}{5\ \Omega} = 3\text{ A}$$

Next, we apply KCL at the top end of the controlled source:

$$i_x + 0.5 i_x = i_y$$

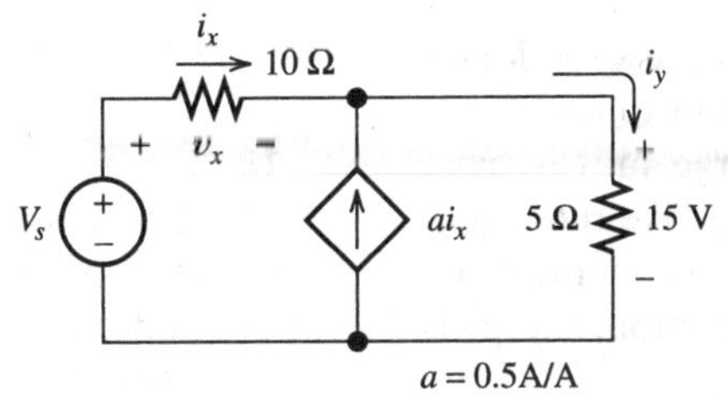

Figure 1.41 Circuit for Example 1.7.

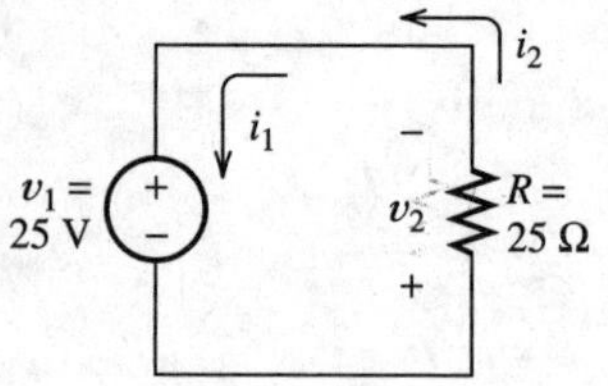

Figure 1.42 Circuit for Exercise 1.14.

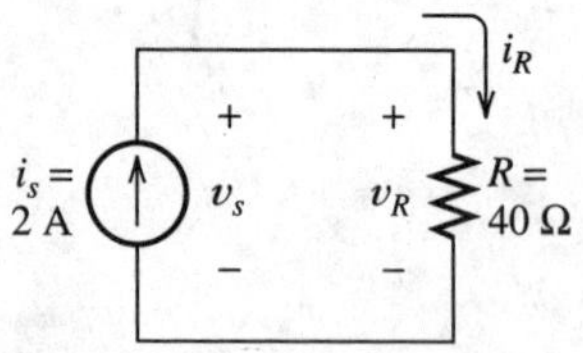

Figure 1.43 Circuit for Exercise 1.15.

Substituting the value found for i_y and solving, we determine that $i_x = 2$ A. Then Ohm's law yields $v_x = 10i_x = 20$ V. Applying KCL around the periphery of the circuit gives

$$V_s = v_x + 15$$

Finally, substituting the value found for v_x yields $V_s = 35$ V. ■

Exercise 1.14 Analyze the circuit shown in Figure 1.42 to find the values of i_1, i_2, and v_2. Use the values found to compute the power for each element.
Answer $i_1 = i_2 = -1$ A, $v_2 = -25$ V, $p_R = 25$ W, $p_s = -25$ W. □

Exercise 1.15 Figure 1.43 shows an independent current source connected across a resistance. Analyze to find the values of i_R, v_R, v_s, and the power for each element.
Answer $i_R = 2$ A, $v_s = v_R = 80$ V, $p_s = -160$ W, $p_R = 160$ W. □

Summary

1. Electrical and electronic features are increasingly integrated into the products and systems designed by engineers in other fields. Furthermore, instrumentation in all fields of engineering and science is based on the use of electrical sensors, electronics, and computers.
2. Some of the main areas of electrical engineering are communication systems, computer systems, control systems, electromagnetics, photonics, electronics, power systems, and signal processing.
3. Some important reasons to learn basic electrical engineering principles are to pass the Fundamentals of Engineering Examination, to have a broad enough knowledge base to lead design projects in your own field, to be able to identify and correct simple malfunctions in electrical systems, and to be able to communicate efficiently with electrical engineering consultants.
4. Current is the time rate of flow of electrical charge. Its units are amperes (A), which are equivalent to coulombs per second (C/s).
5. The voltage associated with a circuit element is the energy transferred per unit of charge that flows through the element. The units of voltages are volts (V), which are equivalent to joules per coulomb (J/C). If positive charge moves from the positive reference to the negative reference, energy is absorbed by the circuit element. If the charge moves in the opposite direction, energy is delivered by the element.
6. In the passive reference configuration, the current reference direction enters the positive reference polarity.
7. If the references have the passive configuration, power for a circuit element is computed as the product of the current through the element and the voltage across it:

$$p = vi$$

If the references are opposite to the passive configuration, we have

$$p = -vi$$

In either case, if p is positive, energy is being absorbed by the element.
8. A node in an electrical circuit is a point at which two or more circuit elements are joined together. All points joined by ideal conductors are electrically equivalent and constitute a single node.
9. Kirchhoff's current law (KCL) states that the sum of the currents entering a node equals the sum of the currents leaving.
10. Elements connected end to end are said to be in series. For two elements to be in series, no other current path can be connected to their common node. The current is identical for all elements in a series connection.

11. A loop in an electrical circuit is a closed path starting at a node and proceeding through circuit elements eventually returning to the starting point.
12. Kirchhoff's voltage law (KVL) states that the algebraic sum of the voltages in a loop must equal zero. If the positive polarity of a voltage is encountered first in going around the loop, the voltage carries a plus sign in the sum. On the other hand, if the negative polarity is encountered first, the voltage carries a minus sign.
13. Two elements are in parallel if both ends of one element are directly connected to corresponding ends of the other element. The voltages of parallel elements are identical.
14. The voltage between the ends of an ideal conductor is zero regardless of the current flowing through the conductor. All points in a circuit that are connected by ideal conductors can be considered as a single point.
15. An ideal independent voltage source maintains a specified voltage across its terminals independent of other elements that are connected to it and of the current flowing through it.
16. For a controlled voltage source, the voltage across the source terminals depends on other voltages or currents in the circuit. A voltage-controlled voltage source is a voltage source having a voltage equal to a constant times the voltage across a pair of terminals elsewhere in the network. A current-controlled voltage source is a voltage source having a voltage equal to a constant times the current through some other element in the circuit.
17. An ideal independent current source forces a specified current to flow through itself, independent of other elements that are connected to it and of the voltage across it.
18. For a controlled current source, the current depends on other voltages or currents in the circuit. A voltage-controlled current source produces a current equal to a constant times the voltage across a pair of terminals elsewhere in the network. A current-controlled current source produces a current equal to a constant times the current through some other element in the circuit.
19. For constant resistances, voltage is proportional to current. If the current and voltage references have the passive configuration, Ohm's law states that $v = Ri$. For references opposite to the passive configuration, $v = -Ri$.

Problems

Section 1.1: Overview of Electrical Engineering

P1.1. Broadly speaking, what are the two main objectives of electrical systems?

P1.2. Name eight subdivisions of electrical engineering.

P1.3. Briefly describe four important reasons that other engineering students need to learn the fundamentals of electrical engineering.

P1.4. Write a few paragraphs describing an interesting application of electrical engineering in your field. Consult engineering journals and trade magazines such as the *IEEE Spectrum*, *Automotive Engineering*, *Chemical Engineering*, or *Civil Engineering* for ideas.

Section 1.2: Circuits, Currents, and Voltages

P1.5. Carefully define or explain each of the following terms in your own words giving units where appropriate: **a.** Electrical current; **b.** Voltage; **c.** An open switch; **d.** A closed switch; **e.** Direct current; **f.** Alternating current.

P1.6. In the fluid-flow and electric circuit analogy, suggest the analogous quantities for: **a.** a conductor; **b.** a constriction; **c.** a pump; and **d.** current flow.

***P1.7.** The ends of a length of wire are labeled *a* and *b*. If the current in the wire is $i_{ab} = -3$ A, are electrons moving toward *a* or *b*? How much charge passes through a cross section of the wire in 3 seconds?

* Denotes that answers are contained in the Student Solutions files. See Appendix E for more information about accessing the Student Solutions.

***P1.8.** A net charge of $q(t) = 4 + 2t + 5t^2 C$ passes through a cross section of a circuit element in t seconds. Find the current through the element, and the time required to cross a 5 cm length of the element.

***P1.9.** The current through a given circuit element is given by $i(t) = 2e^{-t}$ A. As usual, time t is in seconds. Find the net charge that passes through the element in the interval from $t =$ 0 to $t = \infty$. (*Hint:* Current is the rate of flow of charge. Thus, to find charge, we must integrate current with respect to time.)

***P1.10.** A certain lead-acid storage battery has a mass of 30 kg. Starting from a fully charged state, it can supply 5 A for 24 hours with a terminal voltage of 12 V before it is totally discharged. **a.** If the energy stored in the fully charged battery is used to lift the battery with 100-percent efficiency, what height is attained? Assume that the acceleration due to gravity is 9.8 m/s^2 and is constant with height. **b.** If the stored energy is used to accelerate the battery with 100 percent efficiency, what velocity is attained? **c.** Gasoline contains about 4.5×10^7 J/kg. Compare this with the energy content per unit mass for the fully charged battery.

***P1.11.** A typical "deep-cycle" battery (used for electric trolling motors for fishing boats) is capable of delivering 12.6 V and 10 A for a period of 10 hours. How much charge flows through the battery in this interval? How much energy does the battery deliver?

P1.12. An ac current given by $i(t) = 5\sin(200\pi t)$ A, in which t is in seconds and the angle is in radians, flows through an element of an electrical circuit. **a.** Sketch $i(t)$ to scale versus time for t ranging from 0 to 15 ms. **b.** Determine the net charge that passes through the element between $t = 0$ and $t = 10$ ms. **c.** Repeat for the interval from $t = 0$ to $t = 15$ ms.

P1.13. Consider the headlight circuit of Figure 1.2 on page 25. For current to flow through the headlight, should the switch be open or closed? In the fluid-flow analogy for the circuit, would the valve corresponding to the switch be open or closed? What state for a valve, open or closed, is analogous to an open switch?

P1.14. What is the net number of electrons per second that pass through the cross-section of a wire carrying 5 A of dc current? The current flow is due to electrons, and the magnitude of the charge of each electron is 1.60×10^{-19} C.

P1.15. The circuit element shown in Figure P1.15 carries a current of –5 A, and consumes a power of 60 W. What is the value of voltage v? What should be its correct polarity: v_{ba} or v_{ab}?

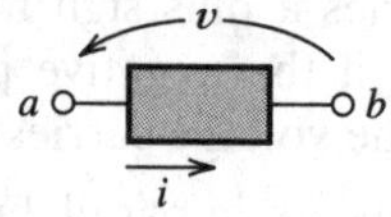

Figure P1.15

P1.16. Suppose that the net charge passing through the cross section of a certain circuit element is given by $q(t) = 2t + 3t^2$ C, with time t in seconds. Determine the current through the element as a function of time.

P1.17. In typical residential wiring, the copper wire has a diameter of 2.05 mm and carries a maximum current of $15\sqrt{2}$ A due solely to electrons, each of which has a charge of 1.60×10^{-19} C. Given that the free electron (these are the electrons capable of moving through the copper) concentration in copper is 10^{29} electrons/m^3, find the average velocity of the electrons in the wire when the maximum current is flowing.

P1.18. The charge carried by an electron is -1.60×10^{-19} C. Suppose that an electron moves through a voltage of 120 V from the negative polarity to the positive polarity. How much energy is transferred? Does the electron gain or lose energy?

P1.19. Consider a circuit element, with terminals a and b, that has $v_{ab} = -12$ V and $i_{ab} =$ 3 A. Over a period of 2 seconds, how much charge moves through the element? If electrons carry the charge, which terminal do they enter? How much energy is transferred? Is it delivered to the element or taken from it?

Section 1.3: Power and Energy

P1.20. What does the term *passive reference configuration* imply? When do we have this configuration if we are using double subscript notation for an element having terminals a and b?

***P1.21.** Compute the power for each element shown in Figure P1.21. For each element, state whether energy is being absorbed by the element or supplied by it.

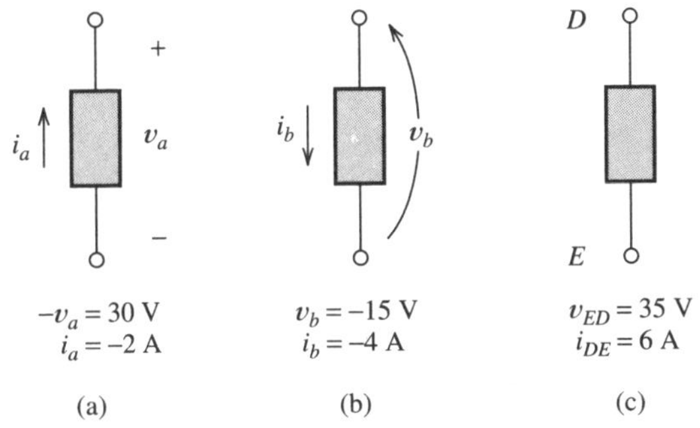

Figure P1.21

***P1.22.** A certain battery has terminals labeled a and b. The battery voltage is $v_{ab} = 12$ V. To increase the chemical energy stored in the battery by 600 J, how much charge must move through the battery? Should electrons move from a to b or from b to a?

P1.23. The terminals of an electrical device are labeled a and b. If $v_{ab} = 25$ V, how much energy is exchanged when a positive charge of 4 C moves through the device from a to b? Is the energy delivered to the device or taken from it?

P1.24. Consider the element shown in Figure P1.24 which has $v(t) = -15$ V and $i(t) = 3e^{-2t}$ A. Compute the power for the circuit element. Find the energy transferred between $t = 0$ and $t = \infty$. Is this energy absorbed by the element or supplied by it?

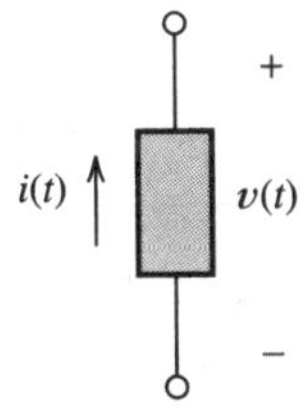

Figure P1.24

P1.25. Suppose that the cost of electrical energy is \$0.12 per kilowatt hour and that your electrical bill for 30 days is \$60. Assume that the power delivered is constant over the entire 30 days. What is the power in watts? If a voltage of 120 V supplies this power, what current flows? Part of your electrical load is a 60-W light that is on continuously. By what percentage can your energy consumption be reduced by turning this light off?

P1.26. An electrical device has $i_{ab}(t) = 2$ A and $v_{ab}(t) = 10\sin(100\pi t)$ V, in which the angle is in radians. **a.** Find the power delivered to the device and sketch it to scale versus time for t ranging from 0 to 30 ms. **b.** Determine the energy delivered to the device for the interval from $t = 0$ to $t = 10$ ms. **c.** Repeat for the interval from $t = 0$ to $t = 20$ ms.

P1.27. A fully charged deep-cycle lead–acid storage battery is rated for 12.6 V and 100 ampere hours. (The ampere-hour rating of the battery is the operating time to discharge the battery multiplied by the current.) This battery is used aboard a sailboat to power the electronics which consume 30 W. Assume that the battery voltage is constant during the discharge. For how many hours can the electronics be operated from the battery without recharging? How much energy in kilowatt hours is initially stored in the battery? If the battery costs \$95 and has a life of 250 charge–discharge cycles, what is the cost of the energy in dollars per kilowatt hour? Neglect the cost of recharging the battery.

***P1.28.** Figure P1.28 shows an ammeter (AM) and voltmeter (VM) connected to measure the current and voltage, respectively, for circuit element A. When current actually enters the + terminal of the ammeter the reading is positive, and when current leaves the + terminal the reading is negative. If the actual voltage polarity is positive at the + terminal of the VM, the reading is positive; otherwise, it is negative. (Actually, for the connection shown, the ammeter reads the sum of the current in element A and the very small current taken by the voltmeter. For purposes of this problem, assume that the current taken by the voltmeter is negligible.) Find the power for element A and state whether energy is being delivered to element A or taken from it if: **a.** the ammeter reading is +2 A and the voltmeter reading is −25 V; **b.** the ammeter reading is −2 A and the voltmeter reading is +25 V; **c.** the ammeter reading is −2 A and the voltmeter reading is −25 V.

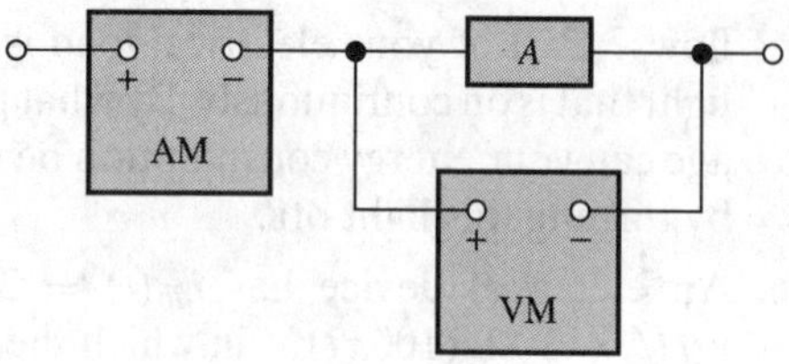

Figure P1.28

P1.29. Repeat Problem P1.28 with the meters connected as shown in Figure P1.29.

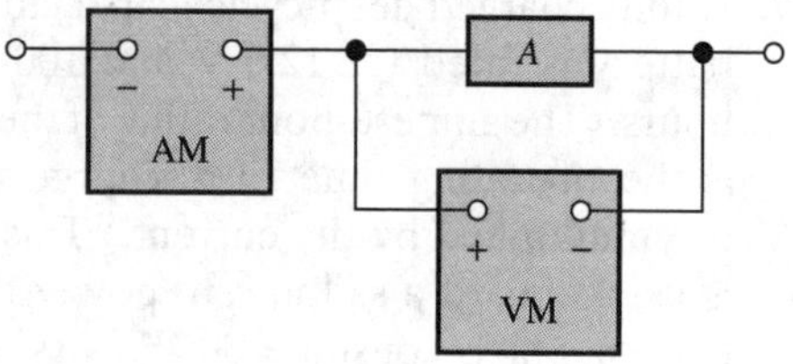

Figure P1.29

P1.30. Determine the cost per kilowatt hour for the energy delivered by a typical alkaline 9-V "transistor" battery that costs \$1.95 and is capable of delivering a current of 50 mA for a period of 10 hours. (For comparison, the approximate cost of energy purchased from electric utilities in the United States is \$0.12 per kilowatt hour.)

Section 1.4: Kirchhoff's Current Law

P1.31. Define the term *node* as it applies in electrical circuits. Identify the nodes in the circuit of Figure P1.31. Keep in mind that all points

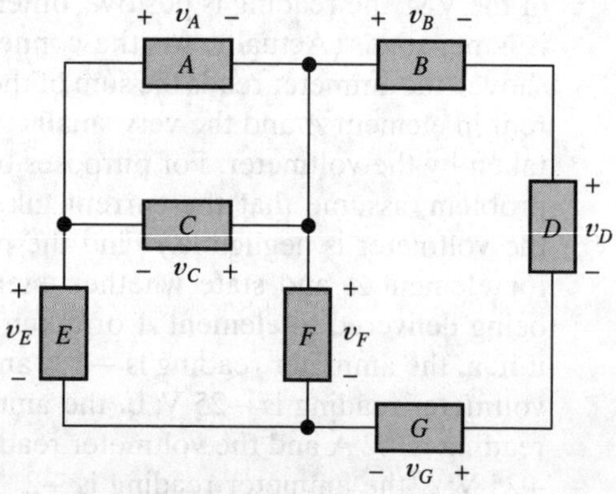

Figure P1.31

connected by ideal conductors are considered to be a single node in electrical circuits.

P1.32. Find the values of the branch currents in Figure P1.32. Use KCL. For the constant node current 3A, what will happen if 'C' is removed?

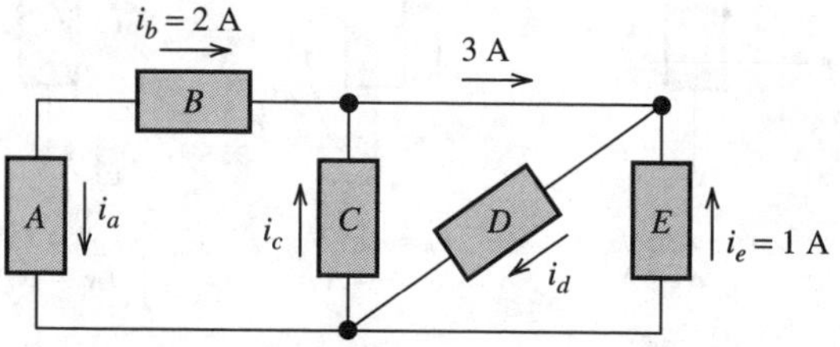

Figure P1.32

P1.33. What can you say about the currents through series-connected elements in an electrical circuit?

P1.34. State Kirchhoff's current law in your own words. Why is it true?

***P1.35.** Identify the series-parallel elements in Figure P1.31.

P1.36. In the fluid-flow analogy for an electrical circuit, the analog of electrical current is volumetric flow rate with units of cm^3/s. For a proper analogy to electrical circuits, must the fluid be compressible or incompressible? Must the walls of the pipes be elastic or inelastic? Explain your answers.

***P1.37.** Given that $i_b = 3$ A; $i_c = 5$ A; $i_g = 6$ A, and $i_e = 5$ A, determine the values of other currents in Figure P1.37. Comment on the role of 'H'.

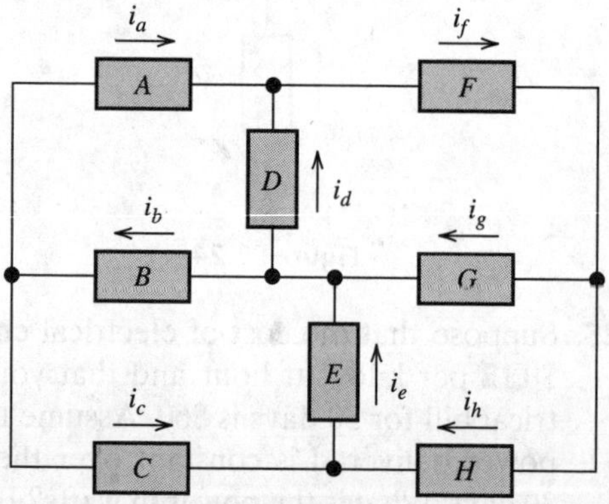

Figure P1.37

P1.38. Determine the values of the other currents in Figure P1.37, given that $i_a = 2$ A, $i_c = -3$ A, $i_g = 6$ A, and $i_h = 1$ A.

P1.39. **a.** Which elements are in series and which are in parallel in Figure P1.39? **b.** What is the relationship between i_a and i_e? **c.** Given that $i_b = 8$ A, and $i_c = -2$ A, determine the values of i_a and i_d.

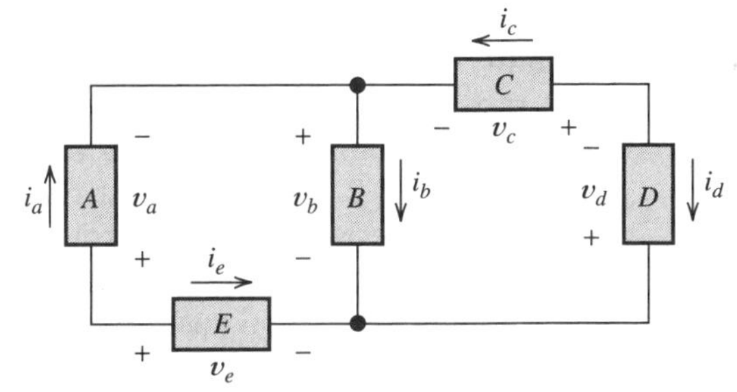

Figure P1.39

Section 1.5: Kirchhoff's Voltage Law

P1.40. Explain Kirchhoff's voltage law in your own words. Why is it true?

***P1.41.** Use KVL and KCL to solve for the labeled currents and voltages in Figure P1.41. Compute the power for each element and show that power is conserved (i.e., the algebraic sum of the powers is zero).

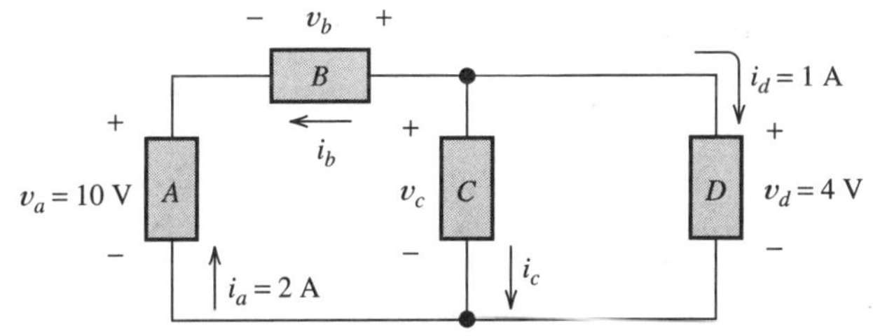

Figure P1.41

***P1.42.** Use KVL to solve for the voltages v_a, v_b, and v_c in Figure P1.42.

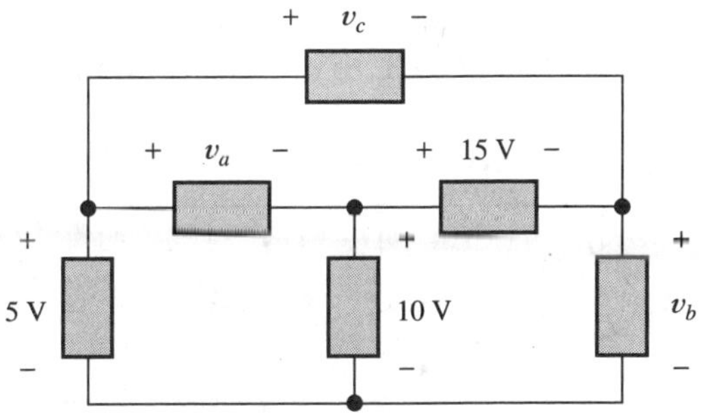

Figure P1.42

P1.43. We know that $v_a = 10$ V, $v_b = -3$ V, $v_f = 12$ V, and $v_h = 5$ V, solve for the other voltages shown in Figure P1.43.

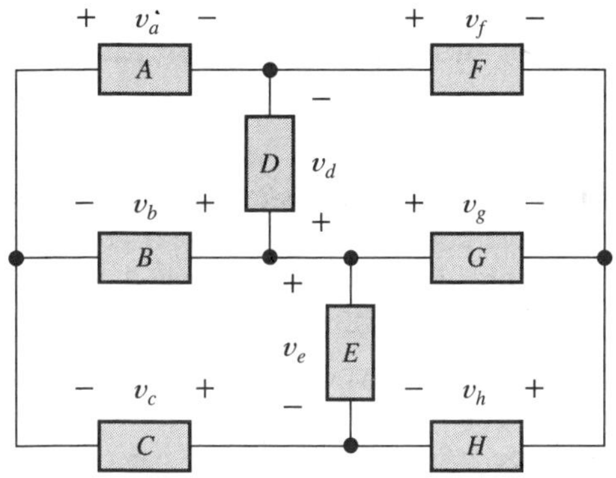

Figure P1.43

P1.44. Consider the circuit shown in Figure P1.31. **a.** Which elements are in parallel? **b.** What is the relationship between v_A and v_C? **c.** Given that $v_A = 8$ V and $v_E = 4$ V, determine the values of v_C and v_F.

P1.45. Identify the elements that are in parallel: **a.** in Figure P1.32, **b.** in Figure P1.43.

P1.46. Identify the nodes in Figure P1.46. Which elements are in series? Which are in parallel?

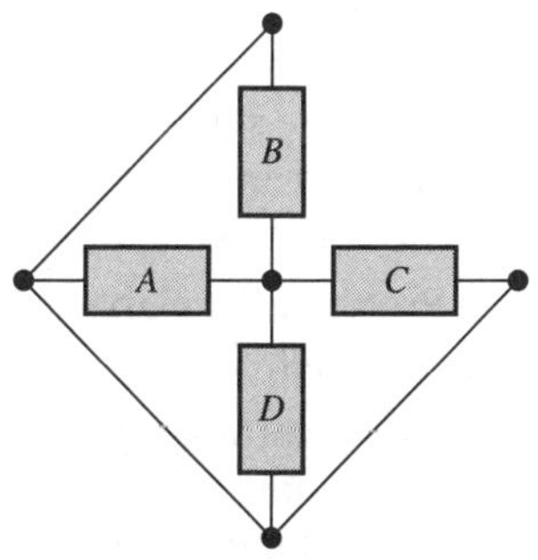

Figure P1.46

P1.47. Consider a circuit containing four nodes labeled a, b, c, and d. Also, we know that $v_{ab} = 12$ V, $v_{cb} = -4$ V, and $v_{da} = 8$ V. Determine the values of v_{ac} and v_{cd}. (*Hint:* Draw a picture showing the nodes and the known voltages.)

P1.48. A typical golf cart uses a number of 6-V batteries (which, for the purposes of this problem, can be modeled as ideal 6-V voltage sources). If the motor requires 36 V, what is the minimum number of batteries needed? How should they be connected? Sketch a diagram for the battery connections showing the polarity of each battery.

Section 1.6: Introduction to Circuit Elements

P1.49. Explain Ohm's law in your own words, including references.

P1.50. Define these terms in your own words: **a.** an ideal conductor; **b.** an ideal voltage source; **c.** an ideal current source; **d.** short circuit; **e.** open circuit.

P1.51. Name four types of dependent sources and give the units for the gain parameter of each type.

P1.52. **a.** Show that for wires of identical dimensions but made of different materials, such as copper and aluminum, the resistance is proportional to the resistivity of the material. (See Table 1.3 on page 46 for resistivity values of some materials used in electrical wiring.) **b.** We know that the resistance of a certain copper wire is 1.5 Ω. Determine the resistance of an aluminum wire having the same dimensions as the copper wire.

***P1.53.** Draw a circuit that contains a 5-Ω resistance, a 10-V independent voltage source, and a 2-A independent current source. Connect all three elements in series. (Because the polarity of the voltage source and reference direction for the current source are not specified, several correct answers are possible.)

P1.54. Repeat Problem P1.53, placing all three elements in parallel.

P1.55. Suppose that a certain wire has a resistance of 10 Ω. Find the new resistance: **a.** if the length of the wire is doubled; **b.** if the diameter of the wire is doubled.

P1.56. If a copper wire and an aluminum wire of the same length and cross-sectional area are placed in parallel with a current source of 10 A, calculate the ratio in which the current will be divided in the two paths.

***P1.57.** A power of 100 W is delivered to a certain resistor when the applied voltage is 100 V. Find the resistance. Suppose that the voltage is reduced by 10 percent (to 90 V). By what percentage is the power reduced? Assume that the resistance remains constant.

P1.58. Sketch the diagram of a circuit that contains a 3-Ω resistor, a 10-V voltage source, and a current-controlled current source having a gain constant of 0.4 A/A. Assume that the current through the resistor is the control current for the controlled source. Place all three elements in parallel. Several answers are possible, depending on the polarities and reference directions chosen.

P1.59. Sketch the diagram of a circuit that contains a 3-Ω resistor, a 2-A current source, and a voltage-controlled current source having a gain constant of 2 S. Assume that the voltage across the resistor is the control voltage for the controlled source. Place all three elements in parallel. Several answers are possible, depending on the polarities and reference directions chosen.

P1.60. The voltage across a 40-Ω resistor is given by $v(t) = 80\exp(-2t)\ V$. Determine the energy delivered to the resistor between time $t = 10$ and $t = \infty$.

P1.61. The voltage across a 20-Ω resistor is given by $v(t) = 120\sin(2\pi t)$ V. Calculate the energy delivered to the resistor between $t = 0$ and $t = 5$ s. The argument of the sine function, $2\pi t$, is in radians.

Section 1.7: Introduction to Circuits

***P1.62.** Which of the following are self-contradictory combinations of circuit elements? **a.** A 12-V voltage source in parallel with a 2-A current source. **b.** A 2-A current source in series with a 3-A current source. **c.** A 2-A current source in parallel with a short circuit. **d.** A 2-A current source in series with an open circuit. **e.** A 5-V voltage source in parallel with a short circuit.

***P1.63.** Consider the circuit shown in Figure P1.63. Use repeated applications of Ohm's law, KVL, and KCL to eventually find V_x.

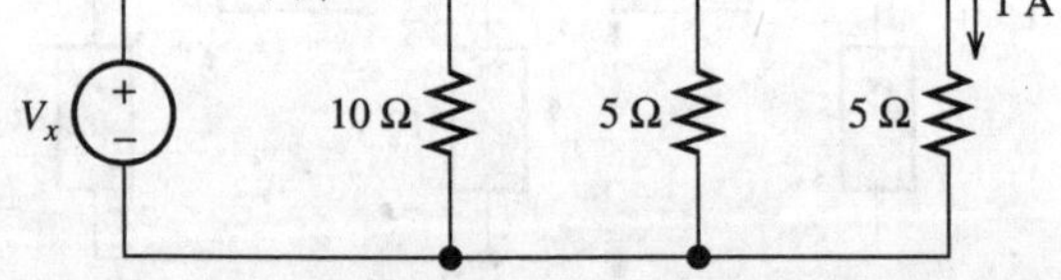

Figure P1.63

***P1.64.** Consider the circuit shown in Figure P1.64. Find the current i_R flowing through the resistor. Find the power for each element in the circuit. Which elements are absorbing power?

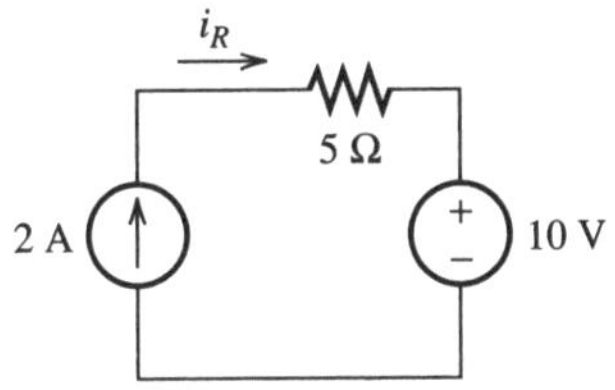

Figure P1.64

P1.65. Plot, to scale, i_{ab} versus v_{ab} for each of the parts of Figure P1.65.

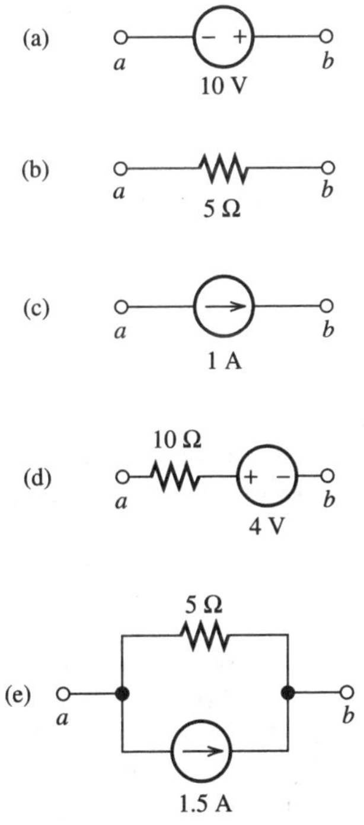

Figure P1.65

P1.66. Consider the circuit shown in Figure P1.66. **a.** Which elements are in series? **b.** Which elements are in parallel? **c.** Apply Ohm's and Kirchhoff's laws to solve for v_x.

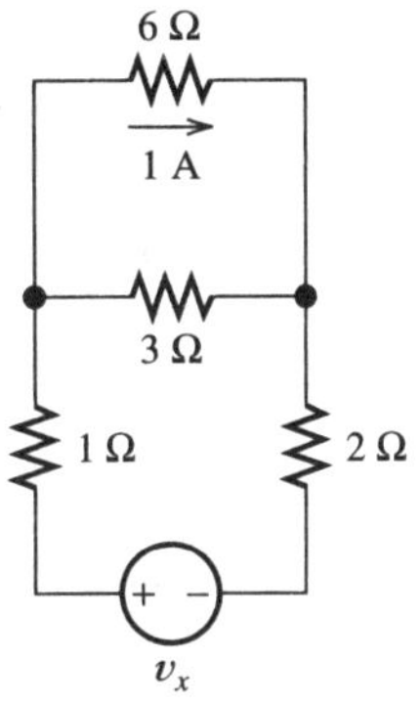

Figure P1.66

P1.67. Given the circuit shown in Figure P1.67, find the power for each source. Which source is absorbing power? Which is delivering power?

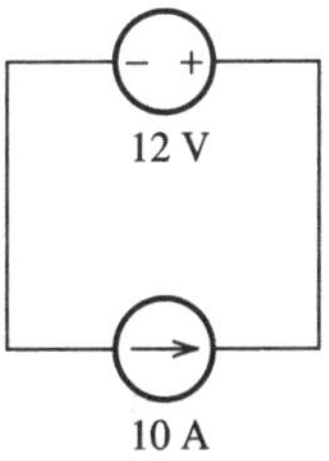

Figure P1.67

P1.68. For the circuit shown in Figure P1.68, find the current i_R. If the current $i_R = -4$ A, what should be the new value of the voltage source to replace the 12 V source?

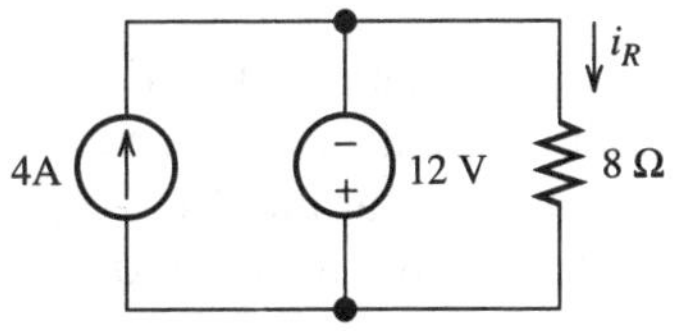

Figure P1.68

P1.69. Use repeated applications of Ohm's law, KVL, and KCL to eventually find the value of I_x in the circuit of Figure P1.69.

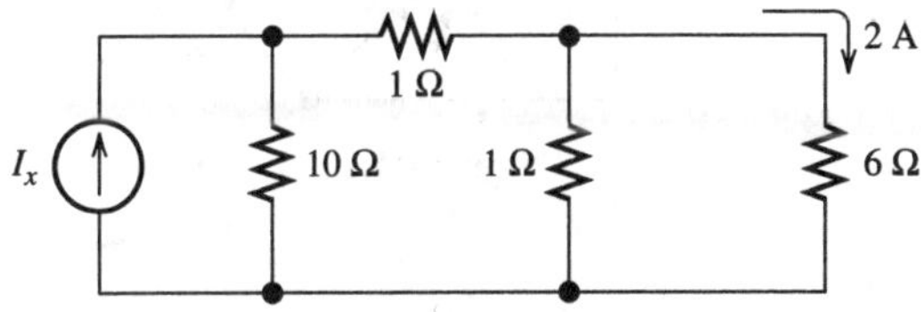

Figure P1.69

***P1.70.** The circuit shown in Figure P1.70 contains a voltage-controlled voltage source. **a.** Use KVL to write an equation relating the voltages and solve for v_x. **b.** Use Ohm's law to find the current i_x. **c.** Find the power for each element in the circuit and verify that power is conserved.

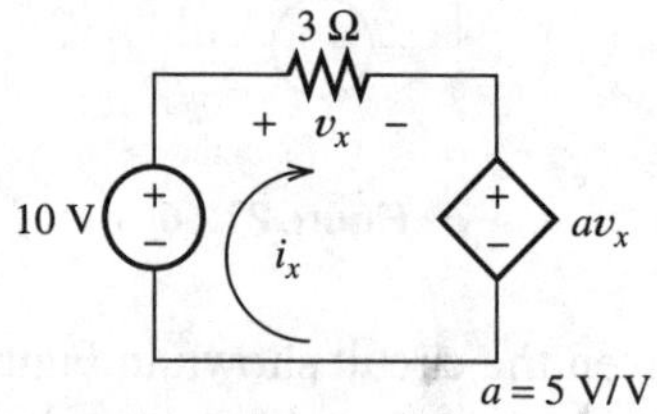

Figure P1.70

***P1.71.** What type of controlled source is shown in the circuit of Figure P1.71? Solve for v_s.

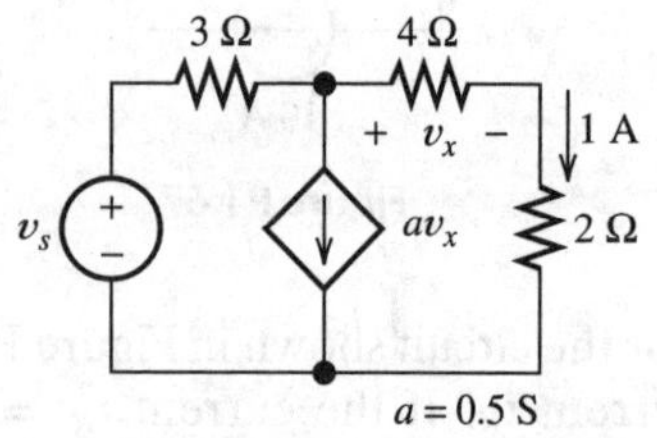

Figure P1.71

P1.72. Consider the circuit shown in Figure P1.72. **a.** Which elements are in series? **b.** Which elements are in parallel? **c.** Apply Ohm's and Kirchhoff's laws to solve for R_x.

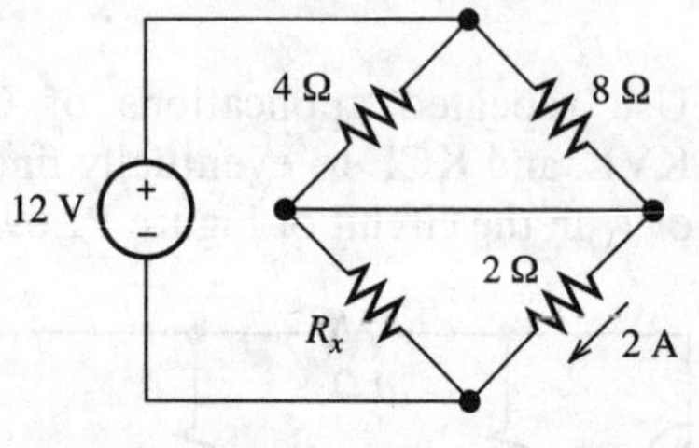

Figure P1.72

P1.73. Solve for the currents shown in Figure P1.73.

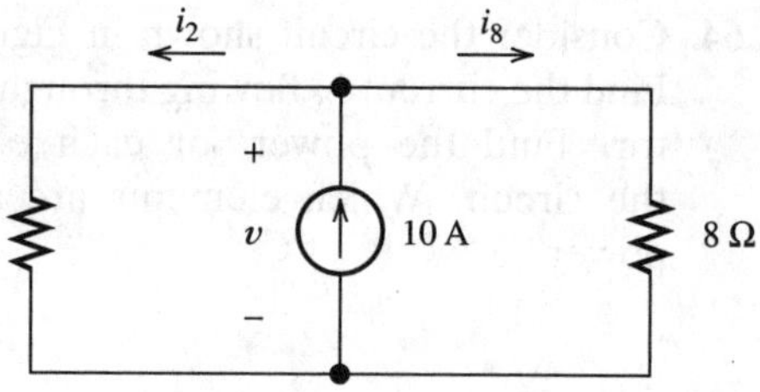

Figure P1.73

P1.74. Figure P1.74 is the electrical model for an electronic megaphone, in which the 16-Ω resistance models a loudspeaker, the source I_x and the 5-Ω resistance represent a microphone, and the remaining elements model an amplifier. What is the name of the type of controlled source shown? Given that the power delivered to the 16-Ω resistance is 16 W, determine the current flowing in the controlled source. Also, determine the value of the microphone current I_x.

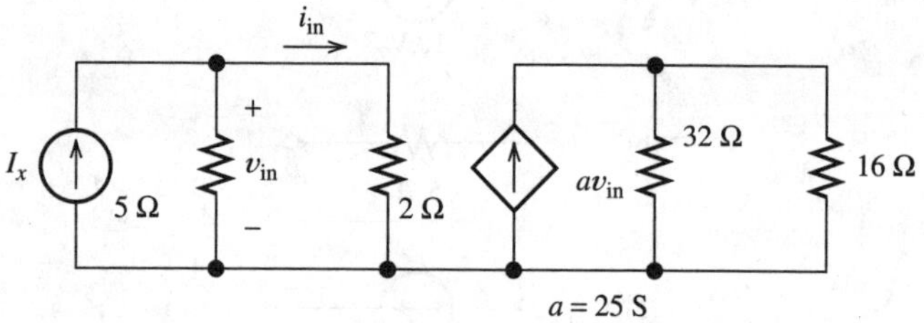

Figure P1.74

P1.75. What type of controlled source appears in the circuit of Figure P1.75? Determine the values of v_x and i_y.

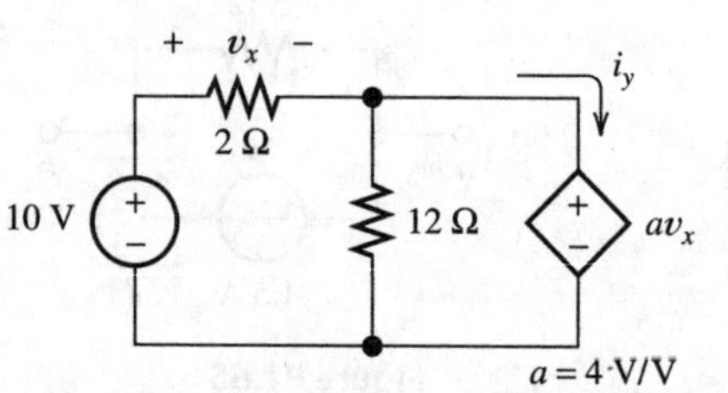

Figure P1.75

P1.76. A 10-A independent current source is connected in series with a 2-Ω resistance between terminals a and b. What can you say about

i_{ab} and v_{ab}? What single circuit element is equivalent to this series combination?

P1.77. A 10-V independent voltage source is connected in parallel with a 2-Ω resistance between terminals a and b. What can you say about i_{ab} and v_{ab}? What single circuit element is equivalent to this parallel combination?

P1.78. Consider the circuit shown in Figure P1.78, given $R_1 = 15\ \Omega$, $R_2 = 10\ \Omega$, and $I_s = 4$ A. **a.** Use KCL to write an equation relating the currents. **b.** Use Ohm's law to write equations relating i_1 and i_2 to the voltage v. **c.** Substitute the equations from part (b) into the equation from part (a) and solve for v. **d.** Find the power for each element in the circuit and verify that power is conserved.

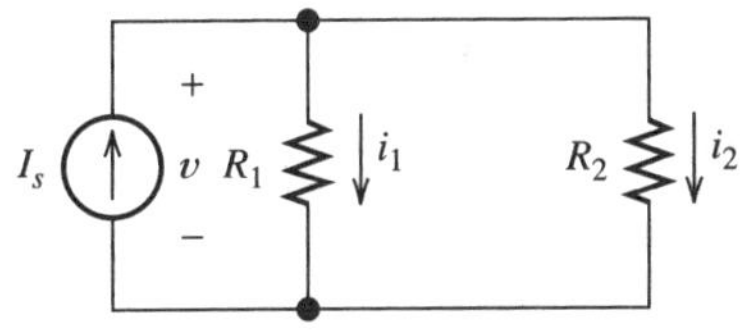

Figure P1.78

P1.79. Consider the circuit shown in Figure P1.79, given $R_1 = 3\ \Omega$, $R_2 = 4\ \Omega$, and $I_s = 2$ A. **a.** Use KVL to write an equation relating the voltages. **b.** Use Ohm's law to write equations relating v_1 and v_2 to the current I_s. **c.** Substitute the equations from part (b) into the equation from part (a) and solve for v. **d.** Find the power for each element in the circuit and verify that power is conserved.

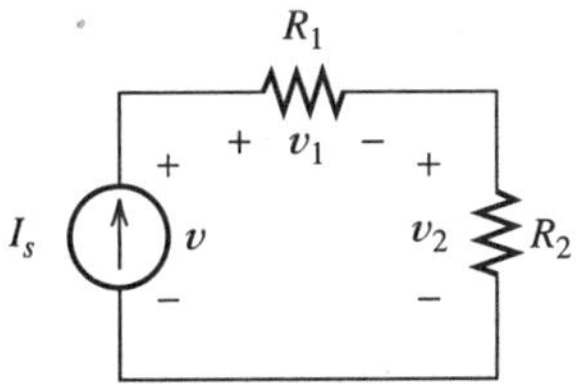

Figure P1.79

P1.80. What types of sources are present in the circuit shown in Figure P1.80? Solve for I_s and the voltage v.

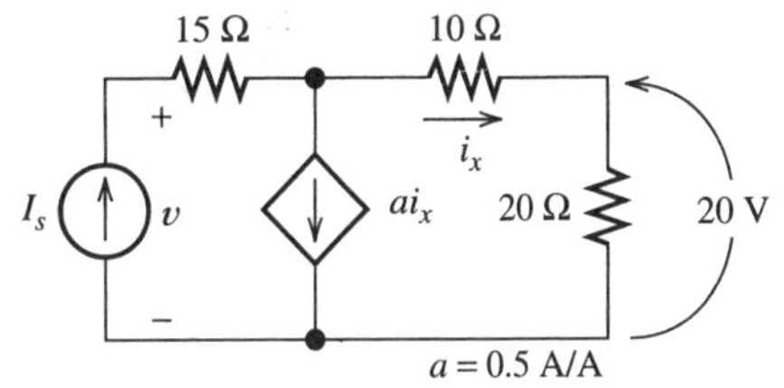

Figure P1.80

P1.81. KVL is a consequence of KCL and conservation of power in electrical circuits. This problem illustrates this fact for the circuit of Figure P1.81. **a.** Write the KCL equation for the circuit. **b.** Apply conservation of power to the circuit, summing the powers for the elements (observing the proper algebraic signs) and setting the sum to zero. **c.** Use the KCL equation to substitute for i_3 in the equation of part (b) and put the result into the form $Ai_1 + Bi_2 = 0$, in which A and B are algebraic sums of the voltages. **d.** Now, suppose that we have control of the current sources. For any values we select for i_1 and i_2, the equation of part (c) must be satisfied. This can only happen if we have $A = 0$ and $B = 0$. Show that this gives the KVL equations for the circuit.

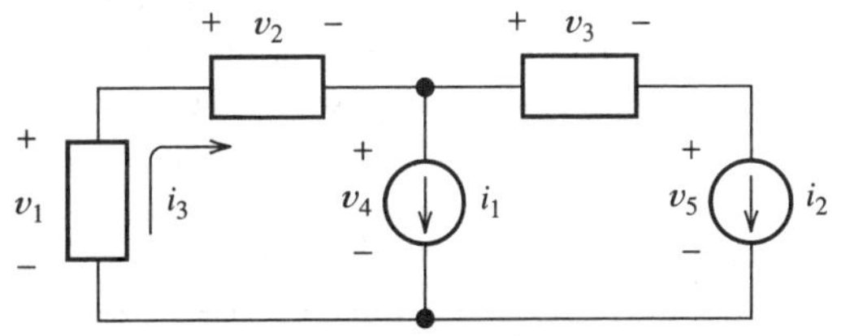

Figure P1.81

P1.82. What types of sources are present in the circuit of Figure P1.82? Solve for the current i_x.

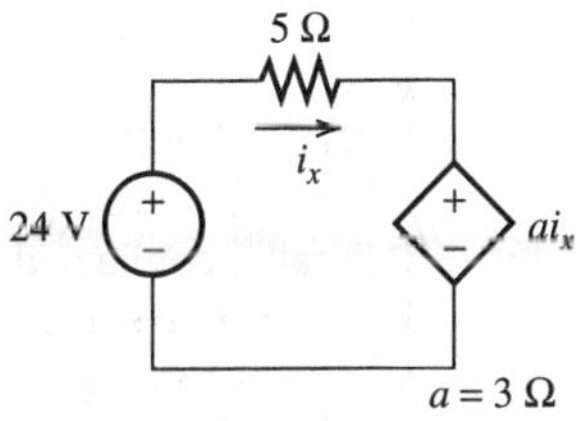

Figure P1.82

Frequently Asked Questions

1. What are the major areas of Electrical Engineering?
2. Why do you need to study electrical engineering?
3. What do you understand by an electric circuit?
4. Define and explain electric current.
5. How will you distinguish between direct current and alternating current?
6. What do you mean by voltage (electric potential), how polarities are assigned to voltages?
7. What do you mean by reference polarity?
8. What do you understand by double subscript notation, how it is useful to represent voltages?
9. Define electric power and energy.
10. Define and explain kirchhoff's current law (KCL).
11. Define and explain kirchhoff's voltage law (KVL).
12. What do you mean by series circuits? What are their characteristics?
13. What do you mean by parallel circuits? What are their characteristics?
14. How Kirchhoff's voltage law is related to law of conservation of energy?
15. Distinguish between independent and dependent voltage sources.
16. What do you mean by independent and dependent currents? How will you distinguish between them?
17. What do you mean by electric resistance? What are the physical parameters on which it depends?
18. Define and explain Ohm's Law. What are its limitations?
19. How is a resistance used to measure strain? Explain.
20. How will you distinguish between a resistor and resistance?

Practice Test

Here is a practice test you can use to check your comprehension of the most important concepts in this chapter. Answers can be found in Appendix D and complete solutions are included in the Student Solutions files. See Appendix E for more information about the Student Solutions.

T1.1. Match each entry in Table T1.1(a) with the best choice from the list given in

Table T1.1

Item	Best Match
(a)	
a. Node	
b. Loop	
c. KVL	
d. KCL	
e. Ohm's law	
f. Passive reference configuration	
g. Ideal conductor	
h. Opcn circuit	
i. Current source	
j. Parallel connected elements	
k. Controlled source	
l. Units for voltage	
m. Units for current	
n. Units for resistance	
o. Series connected elements	

Table T1.1. (*Cont.*)

(b)
1. $v_{ab} = Ri_{ab}$
2. The current reference for an element enters the positive voltage reference
3. A path through which no current can flow
4. Points connected by ideal conductors
5. An element that carries a specified current
6. An element whose current or voltage depends on a current or voltage elsewhere in the circuit
7. A path starting at a node and proceeding from node to node back to the starting node
8. An element for which the voltage is zero
9. A/V
10. V/A
11. J/C
12. C/V
13. C/s
14. Elements connected so their currents must be equal
15. Elements connected so their voltages must be equal
16. The algebraic sum of voltages for a closed loop is zero
17. The algebraic sum of the voltages for elements connected to a node is zero
18. The sum of the currents entering a node equals the sum of those leaving

Table T1.1(b). [Items in Table T1.1(b) may be used more than once or not at all.]

T1.2. Consider the circuit of Figure T1.2 with $I_s = 3\,\text{A}$, $R = 2\ \Omega$, and $V_s = 10\,\text{V}$. **a.** Determine the value of v_R. **b.** Determine the magnitude of the power for the voltage source and state whether the voltage source is absorbing energy or delivering it. **c.** How many nodes does this circuit have? **d.** Determine the magnitude of the power for the current source and state whether the current source is absorbing energy or delivering it.

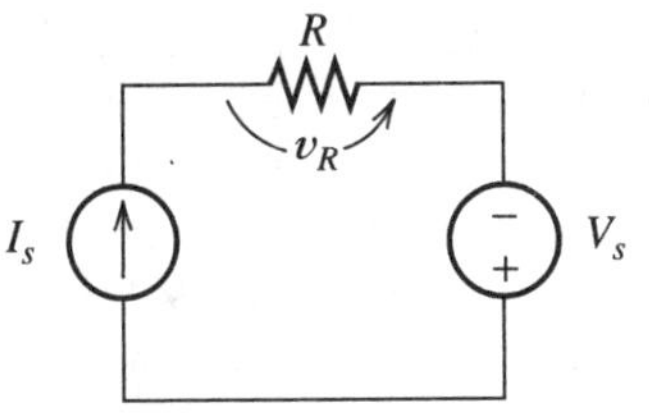

Figure T1.2

T1.3. The circuit of Figure T1.3 has $I_1 = 3\,\text{A}$, $I_2 = 1\,\text{A}$, $R_1 = 12\,\Omega$, and $R_2 = 6\,\Omega$. **a.** Determine the value of v_{ab}. **b.** Determine the power for each current source and state whether it is absorbing energy or delivering it. **c.** Compute the power absorbed by R_1 and by R_2.

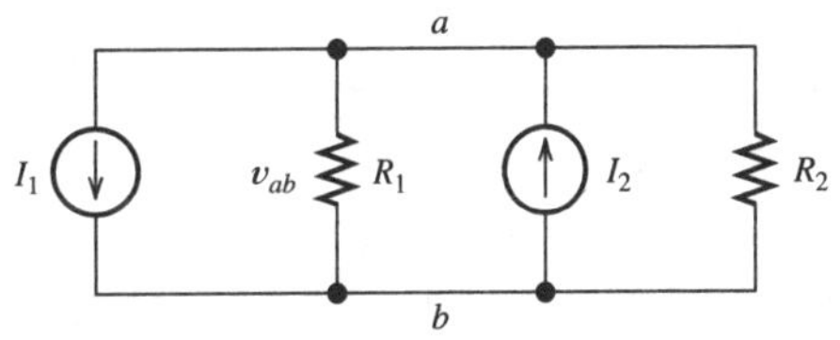

Figure T1.3

T1.4. The circuit shown in Figure T1.4 has $V_s = 12\,\text{V}$, $v_2 = 4\,\text{V}$, and $R_1 = 4\ \Omega$. **a.** Find the values of: **a.** v_1; **b.** i; **c.** R_2.

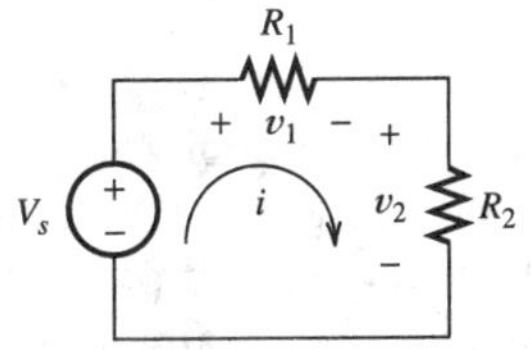

Figure T1.4

T1.5. We are given $V_s = 15\text{V}$, $R = 10\ \Omega$, and $a = 0.3\,\text{S}$ for the circuit of Figure T1.5. Find the value of the current i_{sc} flowing through the short circuit.

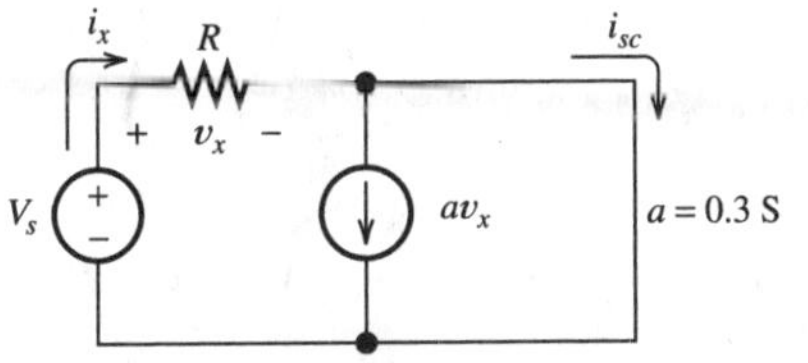

Figure T1.5

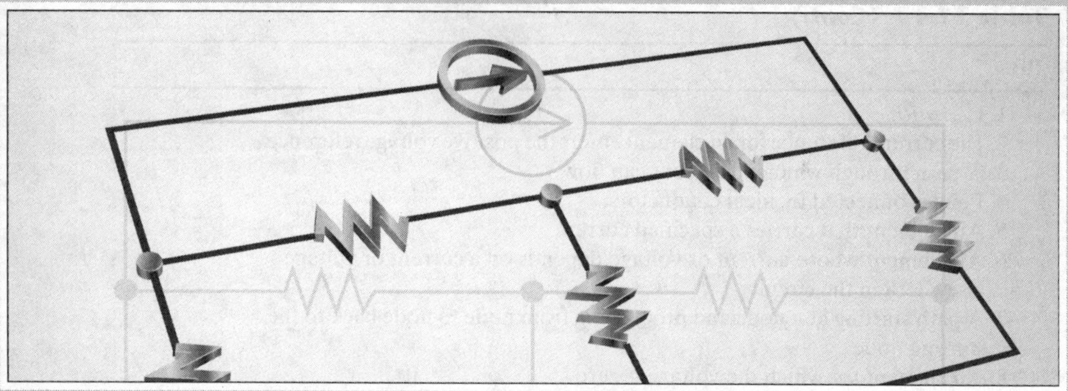

Chapter 2

Resistive Circuits

Study of this chapter will enable you to:

- Solve circuits (i.e., find currents and voltages of interest) by combining resistances in series and parallel.
- Apply the voltage-division and current-division principles.
- Solve circuits by the node-voltage technique.
- Solve circuits by the mesh-current technique.
- Find Thévenin and Norton equivalents and apply source transformations.
- Use MATLAB® to solve circuit equations numerically and symbolically.
- Understand and apply the superposition principle.
- Draw the circuit diagram and state the principles of operation for the Wheatstone bridge.

Introduction to this chapter:

In applications of electrical engineering, we often face circuit-analysis problems for which the structure of a circuit, including element values, is known and the currents, voltages, and powers need to be found. In this chapter, we examine techniques for analyzing circuits composed of resistances, voltage sources, and current sources. Later, we extend many of these concepts to circuits containing inductance and capacitance.

Over the years, you will meet many applications of electrical engineering in your field of engineering or science. This chapter will give you the skills needed to work effectively with the electronic instrumentation and other circuits that you will encounter. The material in this book will help you to answer questions on the Fundamentals of Engineering Examination and become a Registered Professional Engineer.

2.1 RESISTANCES IN SERIES AND PARALLEL

In this section, we show how to replace series or parallel combinations of resistances by equivalent resistances. Then, we demonstrate how to use this knowledge in solving circuits.

Series Resistances

Consider the series combination of three resistances shown in Figure 2.1(a). Recall that in a series circuit the elements are connected end to end and that the same current flows through all of the elements. By Ohm's law, we can write

$$v_1 = R_1 i \tag{2.1}$$

$$v_2 = R_2 i \tag{2.2}$$

and

$$v_3 = R_3 i \tag{2.3}$$

Using KVL, we can write

$$v = v_1 + v_2 + v_3 \tag{2.4}$$

Substituting Equations 2.1, 2.2, and 2.3 into Equation 2.4, we obtain

$$v = R_1 i + R_2 i + R_3 i \tag{2.5}$$

Factoring out the current i, we have

$$v = (R_1 + R_2 + R_3)i \tag{2.6}$$

Now, we define the equivalent resistance R_{eq} to be the sum of the resistances in series:

$$R_{eq} = R_1 + R_2 + R_3 \tag{2.7}$$

Using this to substitute into Equation 2.6, we have

$$v = R_{eq} i \tag{2.8}$$

Thus, we conclude that the three resistances in series can be replaced by the equivalent resistance R_{eq} shown in Figure 2.1(b) with no change in the relationship between

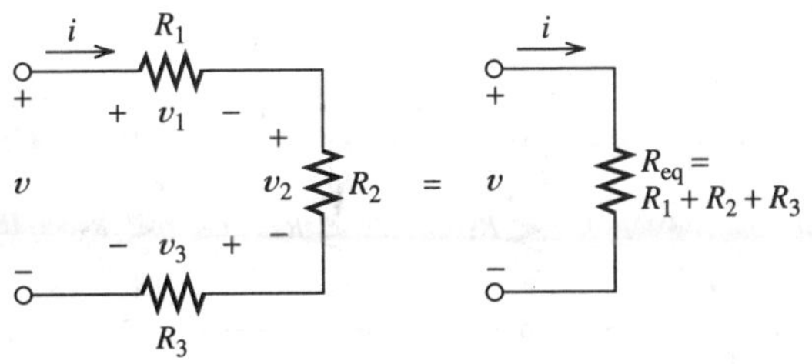

(a) Three resistances in series

(b) Equivalent resistance

Figure 2.1 Series resistances can be combined into an equivalent resistance.

the voltage v and current i. If the three resistances are part of a larger circuit, replacing them by a single equivalent resistance would make no changes in the currents or voltages in other parts of the circuit.

A series combination of resistances has an equivalent resistance equal to the sum of the original resistances.

This analysis can be applied to any number of resistances. For example, two resistances in series can be replaced by a single resistance equal to the sum of the original two. To summarize, *a series combination of resistances has an equivalent resistance equal to the sum of the original resistances.*

Parallel Resistances

Figure 2.2(a) shows three resistances in parallel. In a parallel circuit, the voltage across each element is the same. Applying Ohm's law in Figure 2.2(a), we can write

$$i_1 = \frac{v}{R_1} \tag{2.9}$$

$$i_2 = \frac{v}{R_2} \tag{2.10}$$

$$i_3 = \frac{v}{R_3} \tag{2.11}$$

The top ends of the resistors in Figure 2.2(a) are connected to a single node. (Recall that all points in a circuit that are connected by conductors constitute a node.) Thus, we can apply KCL to the top node of the circuit and obtain

$$i = i_1 + i_2 + i_3 \tag{2.12}$$

Now using Equations 2.9, 2.10, and 2.11 to substitute into Equation 2.12, we have

$$i = \frac{v}{R_1} + \frac{v}{R_2} + \frac{v}{R_3} \tag{2.13}$$

Factoring out the voltage, we obtain

$$i = \left(\frac{1}{R_1} + \frac{1}{R_2} + \frac{1}{R_3}\right) v \tag{2.14}$$

Now, we define the equivalent resistance as

$$R_{eq} = \frac{1}{1/R_1 + 1/R_2 + 1/R_3} \tag{2.15}$$

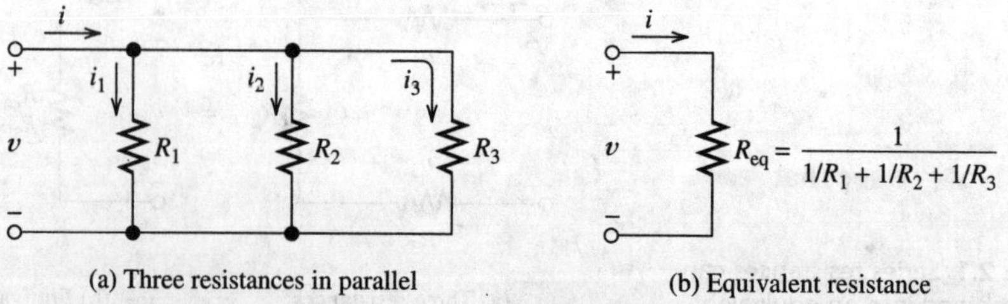

Figure 2.2 Parallel resistances can be combined into an equivalent resistance.

In terms of the equivalent resistance, Equation 2.14 becomes

$$i = \frac{1}{R_{eq}} v \tag{2.16}$$

Comparing Equations 2.14 and 2.16, we see that i and v are related in the same way by both equations provided that R_{eq} is given by Equation 2.15. Therefore, a parallel combination of resistances can be replaced by its equivalent resistance without changing the currents and voltages in other parts of the circuit. The equivalence is illustrated in Figure 2.2(b).

A parallel combination of resistances can be replaced by its equivalent resistance without changing the currents and voltages in other parts of the circuit.

This analysis can be applied to any number of resistances in parallel. For example, if four resistances are in parallel, the equivalent resistance is

$$R_{eq} = \frac{1}{1/R_1 + 1/R_2 + 1/R_3 + 1/R_4} \tag{2.17}$$

Similarly, for two resistances, we have

$$R_{eq} = \frac{1}{1/R_1 + 1/R_2} \tag{2.18}$$

This can be put into the form

$$R_{eq} = \frac{R_1 R_2}{R_1 + R_2} \tag{2.19}$$

(Notice that Equation 2.19 applies only for two resistances. The product over the sum does not apply for more than two resistances.)

The product over the sum does not apply for more than two resistances.

Sometimes, resistive circuits can be reduced to a single equivalent resistance by repeatedly combining resistances that are in series or parallel.

Example 2.1 Combining Resistances in Series and Parallel

Find a single equivalent resistance for the network shown in Figure 2.3(a).

Solution First, we look for a combination of resistances that is in series or in parallel. In Figure 2.3(a), R_3 and R_4 are in series. (In fact, as it stands, no other two resistances in this network are either in series or in parallel.) Thus, our first step is to combine R_3 and R_4, replacing them by their equivalent resistance. Recall that for a series combination, the equivalent resistance is the sum of the resistances in series:

1. Find a series or parallel combination of resistances.
2. Combine them.
3. Repeat until the network is reduced to a single resistance (if possible).

$$R_{eq1} = R_3 + R_4 = 5 + 15 = 20\ \Omega$$

Figure 2.3(b) shows the network after replacing R_3 and R_4 by their equivalent resistance. Now we see that R_2 and R_{eq1} are in parallel. The equivalent resistance for this combination is

$$R_{eq2} = \frac{1}{1/R_{eq1} + 1/R_2} = \frac{1}{1/20 + 1/20} = 10\ \Omega$$

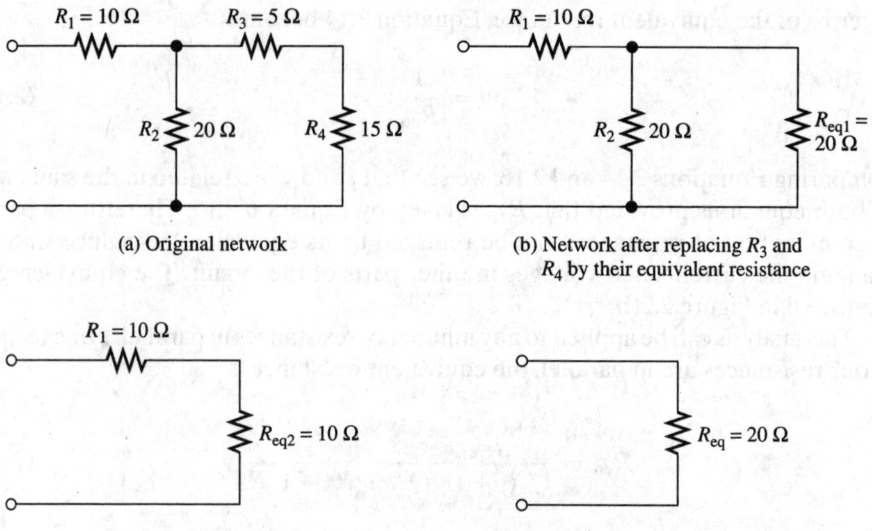

(a) Original network

(b) Network after replacing R_3 and R_4 by their equivalent resistance

(c) Network after replacing R_2 and R_{eq1} by their equivalent

(d) Combining R_1 and R_{eq2} in series yields the equivalent resistance of the entire network

Figure 2.3 Resistive network for Example 2.1.

Making this replacement gives the equivalent network shown in Figure 2.3(c).

Finally, we see that R_1 and R_{eq2} are in series. Thus, the equivalent resistance for the entire network is

$$R_{eq} = R_1 + R_{eq2} = 10 + 10 = 20\ \Omega$$ ■

Exercise 2.1 Find the equivalent resistance for each of the networks shown in Figure 2.4. [*Hint for part (b):* R_3 and R_4 are in parallel.]
Answer **a.** 3 Ω; **b.** 5 Ω; **c.** 52.1 Ω; **d.** 1.5 kΩ. □

Conductances in Series and Parallel

Recall that conductance is the reciprocal of resistance. Using this fact to change resistances to conductances for a series combination of n elements, we readily obtain:

Combine conductances in series as you would resistances in parallel. Combine conductances in parallel as you would resistances in series.

$$G_{eq} = \frac{1}{1/G_1 + 1/G_2 + \cdots + 1/G_n} \tag{2.20}$$

Thus, we see that conductances in series combine as do resistances in parallel. For two conductances in series, we have:

$$G_{eq} = \frac{G_1 G_2}{G_1 + G_2}$$

For n conductances in parallel, we can show that

$$G_{eq} = G_1 + G_2 + \cdots + G_n \tag{2.21}$$

Conductances in parallel combine as do resistances in series.

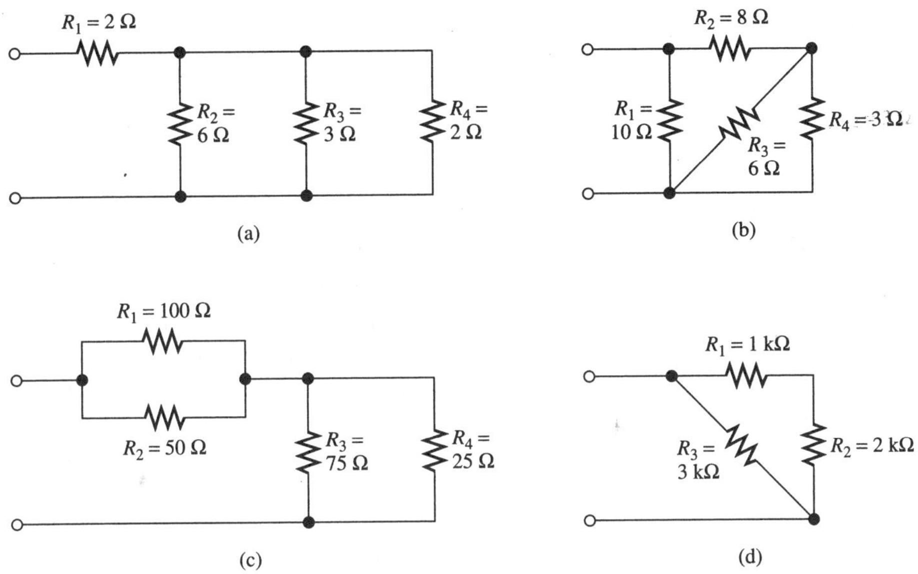

Figure 2.4 Resistive networks for Exercise 2.1.

Series versus Parallel Circuits

An element such as a toaster or light bulb that absorbs power is called a **load**. When we want to distribute power from a single voltage source to various loads, we usually place the loads in parallel. A switch in series with each load can break the flow of current to that load without affecting the voltage supplied to the other loads.

When we want to distribute power from a single voltage source to various loads, we usually place the loads in parallel.

Sometimes, to save wire, strings of Christmas lights consist of bulbs connected in series. The bulbs tend to fail or "burn out" by becoming open circuits. Then the entire string is dark and the defective bulb can be found only by trying each in turn. If several bulbs are burned out, it can be very tedious to locate the failed units. In a parallel connection, only the failed bulbs are dark.

2.2 NETWORK ANALYSIS BY USING SERIES AND PARALLEL EQUIVALENTS

An electrical **network** (or electrical circuit) consists of circuit elements, such as resistances, voltage sources, and current sources, connected together to form closed paths. **Network analysis** is the process of determining the current, voltage, and power for each element, given the circuit diagram and the element values. In this and the sections that follow, we study several useful techniques for network analysis.

An electrical network consists of circuit elements such as resistances, voltage sources, and current sources, connected together to form closed paths.

Sometimes, we can determine the currents and voltages for each element in a resistive circuit by repeatedly replacing series and parallel combinations of resistances by their equivalent resistances. Eventually, this may reduce the circuit sufficiently that the equivalent circuit can be solved easily. The information gained from the simplified circuit is transferred to the previous steps in the chain of equivalent circuits. In the end, we gain enough information about the original circuit to determine all the currents and voltages.

Circuit Analysis Using Series/Parallel Equivalents

Here are the steps in solving circuits using series/parallel equivalents:

1. Begin by locating a combination of resistances that are in series or parallel. Often the place to start is farthest from the source.
2. Redraw the circuit with the equivalent resistance for the combination found in step 1.
3. Repeat steps 1 and 2 until the circuit is reduced as far as possible. Often (but not always) we end up with a single source and a single resistance.
4. Solve for the currents and voltages in the final equivalent circuit. Then, transfer results back one step and solve for additional unknown currents and voltages. Again transfer the results back one step and solve. Repeat until all of the currents and voltages are known in the original circuit.
5. Check your results to make sure that KCL is satisfied at each node, KVL is satisfied for each loop, and the powers add to zero.

Some good advice for beginners: Don't try to combine steps. Be very methodical and do one step at a time. Take the time to redraw each equivalent carefully and label unknown currents and voltages consistently in the various circuits. The slow methodical approach will be faster and more accurate when you are learning. Walk now—later you will be able to run.

Example 2.2 Circuit Analysis Using Series/Parallel Equivalents

Find the current, voltage, and power for each element of the circuit shown in Figure 2.5(a).

Steps 1, 2, and 3.

Solution First, we combine resistances in series and parallel. For example, in the original circuit, R_2 and R_3 are in parallel. Replacing R_2 and R_3 by their parallel equivalent, we obtain the circuit shown in Figure 2.5(b). Next, we see that R_1 and R_{eq1} are in series. Replacing these resistances by their sum, we obtain the circuit shown in Figure 2.5(c).

After we have reduced a network to an equivalent resistance connected across the source, we solve the simplified network. Then, we transfer results back through the chain of equivalent circuits. We illustrate this process in Figure 2.6. (Figure 2.6 is identical to Figure 2.5, except for the currents and voltages shown in Figure 2.6.

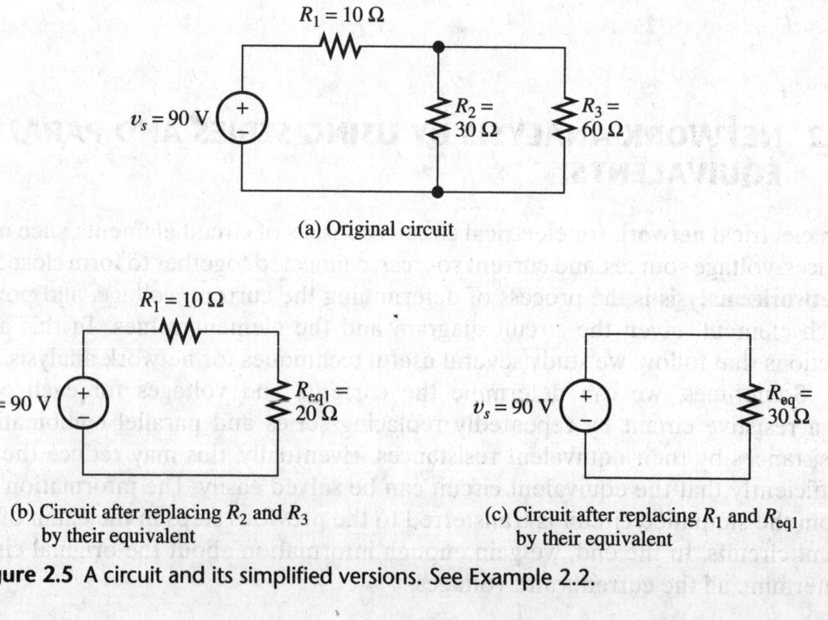

Figure 2.5 A circuit and its simplified versions. See Example 2.2.

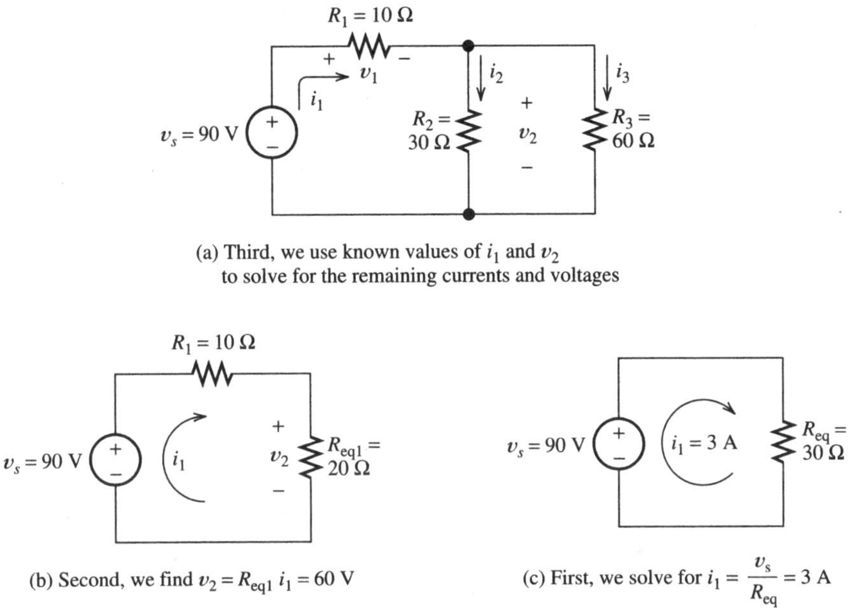

(a) Third, we use known values of i_1 and v_2 to solve for the remaining currents and voltages

(b) Second, we find $v_2 = R_{eq1}\, i_1 = 60$ V

(c) First, we solve for $i_1 = \dfrac{v_s}{R_{eq}} = 3$ A

Figure 2.6 After reducing the circuit to a source and an equivalent resistance, we solve the simplified circuit. Then, we transfer results back to the original circuit. Notice that the logical flow in solving for currents and voltages starts from the simplified circuit in (c).

Usually, in solving a network by this technique, we first draw the chain of equivalent networks and then write results on the same drawings. However, this might be confusing in our first example.)

Step 4.

First, we solve the simplified network shown in Figure 2.6(c). Because R_{eq} is in parallel with the 90-V voltage source, the voltage across R_{eq} must be 90 V, with its positive polarity at the top end. Thus, the current flowing through R_{eq} is given by

$$i_1 = \frac{v_s}{R_{eq}} = \frac{90\ \text{V}}{30\ \Omega} = 3\ \text{A}$$

We know that this current flows downward (from plus to minus) through R_{eq}. Since v_s and R_{eq} are in series in Figure 2.6(c), the current must also flow upward through v_s. Thus, $i_1 = 3$ A flows clockwise around the circuit, as shown in Figure 2.6(c).

Because R_{eq} is the equivalent resistance seen by the source in all three parts of Figure 2.6, the current through v_s must be $i_1 = 3$ A, flowing upward in all three equivalent circuits. In Figure 2.6(b), we see that i_1 flows clockwise through v_s, R_1, and R_{eq1}. The voltage across R_{eq1} is given by

$$v_2 = R_{eq1} i_1 = 20\ \Omega \times 3\ \text{A} = 60\ \text{V}$$

Because R_{eq1} is the equivalent resistance for the parallel combination of R_2 and R_3, the voltage v_2 also appears across R_2 and R_3 in the original network.

At this point, we have found that the current through v_s and R_1 is $i_1 = 3$ A. Furthermore, the voltage across R_2 and R_3 is 60 V. This information is shown in

Figure 2.6(a). Now, we can compute the remaining values desired:

$$i_2 = \frac{v_2}{R_2} = \frac{60\text{ V}}{30\ \Omega} = 2\text{ A}$$

$$i_3 = \frac{v_2}{R_3} = \frac{60\text{ V}}{60\ \Omega} = 1\text{ A}$$

(As a check, we can use KCL to verify that $i_1 = i_2 + i_3$.)

Next, we can use Ohm's law to compute the value of v_1:

$$v_1 = R_1 i_1 = 10\ \Omega \times 3\text{ A} = 30\text{ V}$$

(As a check, we use KVL to verify that $v_s = v_1 + v_2$.)

Step 5. Now, we compute the power for each element. For the voltage source, we have

$$p_s = -v_s i_1$$

We have included the minus sign because the references for v_s and i_1 are opposite to the passive configuration. Substituting values, we have

$$p_s = -(90\text{ V}) \times 3\text{ A} = -270\text{ W}$$

Because the power for the source is negative, we know that the source is supplying energy to the other elements in the circuit.

The powers for the resistances are

$$p_1 = R_1 i_1^2 = 10\ \Omega \times (3\text{ A})^2 = 90\text{ W}$$

$$p_2 = \frac{v_2^2}{R_2} = \frac{(60\text{ V})^2}{30\ \Omega} = 120\text{ W}$$

$$p_3 = \frac{v_2^2}{R_3} = \frac{(60\text{ V})^2}{60\ \Omega} = 60\text{ W}$$

(As a check, we verify that $p_s + p_1 + p_2 + p_3 = 0$, showing that power is conserved.) ■

Power Control by Using Heating Elements in Series or Parallel

Resistances are commonly used as heating elements for the reaction chamber of chemical processes. For example, the catalytic converter of an automobile is not effective until its operating temperature is achieved. Thus, during engine warm-up, large amounts of pollutants are emitted. Automotive engineers have proposed and studied the use of electrical heating elements to heat the converter more quickly, thereby reducing pollution. By using several heating elements that can be operated individually, in series, or in parallel, several power levels can be achieved. This is useful in controlling the temperature of a chemical process.

Exercise 2.2 Find the currents labeled in Figure 2.7 by combining resistances in series and parallel.

Answer **a.** $i_1 = 1.04$ A, $i_2 = 0.480$ A, $i_3 = 0.320$ A, $i_4 = 0.240$ A; **b.** $i_1 = 1$ A, $i_2 = 1$ A; **c.** $i_1 = 1$ A, $i_2 = 0.5$ A, $i_3 = 0.5$ A. □

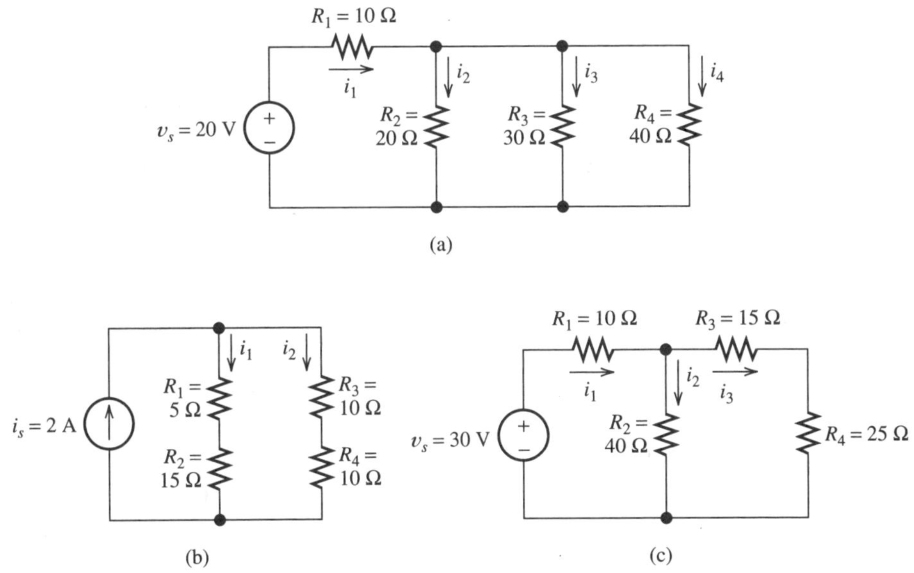

Figure 2.7 Circuits for Exercise 2.2.

2.3 VOLTAGE-DIVIDER AND CURRENT-DIVIDER CIRCUITS

Voltage Division

When a voltage is applied to a series combination of resistances, a fraction of the voltage appears across each of the resistances. Consider the circuit shown in Figure 2.8. The equivalent resistance seen by the voltage source is

$$R_{eq} = R_1 + R_2 + R_3 \tag{2.22}$$

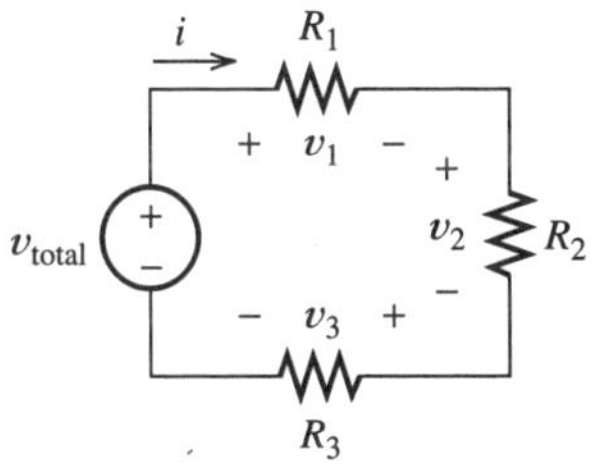

Figure 2.8 Circuit used to derive the voltage-division principle.

The current is the total voltage divided by the equivalent resistance:

$$i = \frac{v_{total}}{R_{eq}} = \frac{v_{total}}{R_1 + R_2 + R_3} \tag{2.23}$$

Furthermore, the voltage across R_1 is

$$v_1 = R_1 i = \frac{R_1}{R_1 + R_2 + R_3} v_{total} \tag{2.24}$$

Similarly, we have

$$v_2 = R_2 i = \frac{R_2}{R_1 + R_2 + R_3} v_{total} \tag{2.25}$$

and

$$v_3 = R_3 i = \frac{R_3}{R_1 + R_2 + R_3} v_{total} \tag{2.26}$$

Of the total voltage, the fraction that appears across a given resistance in a series circuit is the ratio of the given resistance to the total series resistance.

We can summarize these results by the statement: *Of the total voltage, the fraction that appears across a given resistance in a series circuit is the ratio of the given resistance to the total series resistance.* This is known as the **voltage-division principle**.

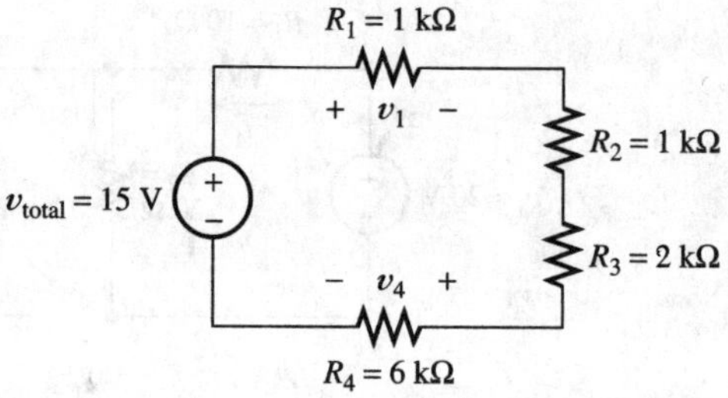

Figure 2.9 Circuit for Example 2.3.

We have derived the voltage-division principle for three resistances in series, but it applies for any number of resistances as long as they are connected in series.

Example 2.3 Application of the Voltage-Division Principle

Find the voltages v_1 and v_4 in Figure 2.9.

Solution Using the voltage-division principle, we find that v_1 is the total voltage times the ratio of R_1 to the total resistance:

$$v_1 = \frac{R_1}{R_1 + R_2 + R_3 + R_4} v_{\text{total}}$$

$$= \frac{1000}{1000 + 1000 + 2000 + 6000} \times 15 = 1.5 \text{ V}$$

Similarly,

$$v_4 = \frac{R_4}{R_1 + R_2 + R_3 + R_4} v_{\text{total}}$$

$$= \frac{6000}{1000 + 1000 + 2000 + 6000} \times 15 = 9 \text{ V}$$

Notice that the largest voltage appears across the largest resistance in a series circuit. ■

Current Division

The total current flowing into a parallel combination of resistances divides, and a fraction of the total current flows through each resistance. Consider the circuit shown in Figure 2.10. The equivalent resistance is given by

$$R_{\text{eq}} = \frac{R_1 R_2}{R_1 + R_2} \tag{2.27}$$

The voltage across the resistances is given by

$$v = R_{\text{eq}} i_{\text{total}} = \frac{R_1 R_2}{R_1 + R_2} i_{\text{total}} \tag{2.28}$$

Now, we can find the current in each resistance:

$$i_1 = \frac{v}{R_1} = \frac{R_2}{R_1 + R_2} i_{\text{total}} \tag{2.29}$$

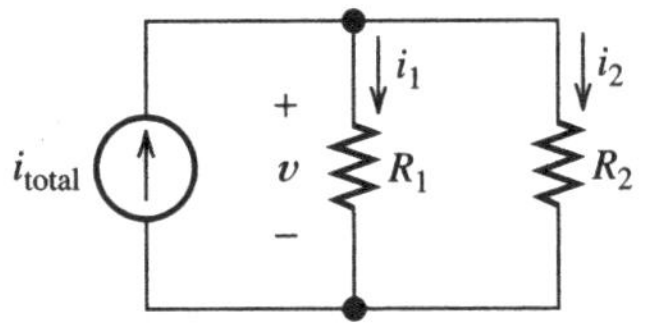

Figure 2.10 Circuit used to derive the current-division principle.

and

$$i_2 = \frac{v}{R_2} = \frac{R_1}{R_1 + R_2} i_{total} \tag{2.30}$$

We can summarize these results by stating the **current-division principle**: *For two resistances in parallel, the fraction of the total current flowing in a resistance is the ratio of the other resistance to the sum of the two resistances.* Notice that this principle applies only for two resistances. If we have more than two resistances in parallel, we should combine resistances so we only have two before applying the current-division principle.

For two resistances in parallel, the fraction of the total current flowing in a resistance is the ratio of the other resistance to the sum of the two resistances.

An alternative approach is to work with conductances. For n conductances in parallel, it can be shown that

$$i_1 = \frac{G_1}{G_1 + G_2 + \cdots + G_n} i_{total}$$

$$i_2 = \frac{G_2}{G_1 + G_2 + \cdots + G_n} i_{total}$$

Current division using conductances uses a formula with the same form as the formula for voltage division using resistances.

and so forth. In other words, current division using conductances uses a formula with the same form as the formula for voltage division using resistances.

Example 2.4 Applying the Current- and Voltage-Division Principles

Use the voltage-division principle to find the voltage v_x in Figure 2.11(a). Then find the source current i_s and use the current-division principle to compute the current i_3.

Solution The voltage-division principle applies only for resistances in series. Therefore, we first must combine R_2 and R_3. The equivalent resistance for the parallel

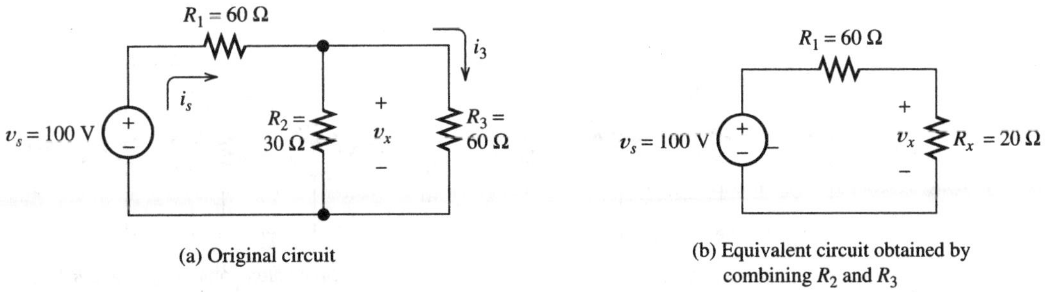

Figure 2.11 Circuit for Example 2.4.

combination of R_2 and R_3 is

$$R_x = \frac{R_2R_3}{R_2 + R_3} = \frac{30 \times 60}{30 + 60} = 20\ \Omega$$

The equivalent network is shown in Figure 2.11(b).

Now, we can apply the voltage-division principle to find v_x. The voltage v_x is equal to the total voltage times R_x divided by the total series resistance:

$$v_x = \frac{R_x}{R_1 + R_x} v_s = \frac{20}{60 + 20} \times 100 = 25\ \text{V}$$

The source current i_s is given by

$$i_s = \frac{v_s}{R_1 + R_x} = \frac{100}{60 + 20} = 1.25\ \text{A}$$

Now, we can use the current-division principle to find i_3. The fraction of the source current i_s that flows through R_3 is $R_2/(R_2 + R_3)$. Thus, we have

$$i_3 = \frac{R_2}{R_2 + R_3} i_s = \frac{30}{30 + 60} \times 1.25 = 0.417\ \text{A}$$

As a check, we can also compute i_3 another way:

$$i_3 = \frac{v_x}{R_3} = \frac{25}{60} = 0.417\ \text{A}$$ ■

Example 2.5 Application of the Current-Division Principle

Use the current-division principle to find the current i_1 in Figure 2.12(a).

The current-division principle applies for *two* resistances in parallel. Therefore, our first step is to combine R_2 and R_3.

Solution The current-division principle applies for two resistances in parallel. Therefore, our first step is to combine R_2 and R_3:

$$R_{\text{eq}} = \frac{R_2R_3}{R_2 + R_3} = \frac{30 \times 60}{30 + 60} = 20\ \Omega$$

The resulting equivalent circuit is shown in Figure 2.12(b). Applying the current-division principle, we have

$$i_1 = \frac{R_{\text{eq}}}{R_1 + R_{\text{eq}}} i_s = \frac{20}{10 + 20} 15 = 10\ \text{A}$$

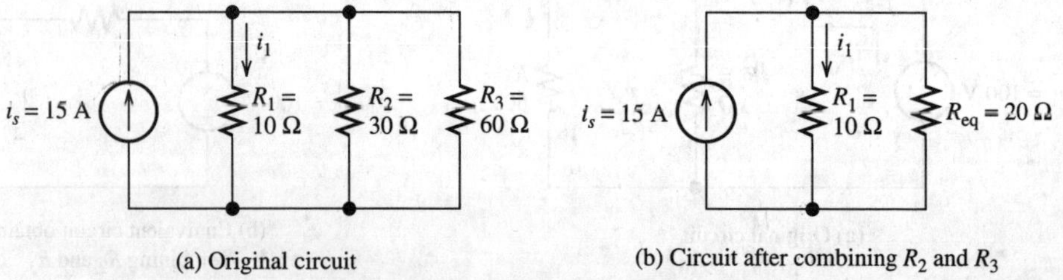

Figure 2.12 Circuit for Example 2.5.

Reworking the calculations using conductances, we have

$$G_1 = \frac{1}{R_1} = 100\,\text{mS}, \quad G_2 = \frac{1}{R_2} = 33.33\,\text{mS}, \quad \text{and} \quad G_3 = \frac{1}{R_3} = 16.67\,\text{mS}$$

Then, we compute the current

$$i_1 = \frac{G_1}{G_1 + G_2 + G_3} i_s = \frac{100}{100 + 33.33 + 16.67} 15 = 10\,\text{A}$$

which is the same value that we obtained working with resistances. ■

Position Transducers Based on the Voltage-Division Principle

Transducers are used to produce a voltage (or sometimes a current) that is proportional to a physical quantity of interest, such as distance, pressure, or temperature. For example, Figure 2.13 shows how a voltage that is proportional to the rudder angle of a boat or aircraft can be obtained. As the rudder turns, a sliding contact moves along a resistance such that R_2 is proportional to the rudder angle θ. The total resistance $R_1 + R_2$ is fixed. Thus, the output voltage is

$$v_o = v_s \frac{R_2}{R_1 + R_2} = K\theta$$

where K is a constant of proportionality that depends on the source voltage v_s and the construction details of the transducer. Many examples of transducers such as this are employed in all areas of science and engineering.

Exercise 2.3 Use the voltage-division principle to find the voltages labeled in Figure 2.14.

Answer **a.** $v_1 = 10$ V, $v_2 = 20$ V, $v_3 = 30$ V, $v_4 = 60$ V; **b.** $v_1 = 6.05$ V, $v_2 = 5.88$ V, $v_4 = 8.07$ V. □

Exercise 2.4 Use the current-division principle to find the currents labeled in Figure 2.15.

Answer **a.** $i_1 = 1$ A, $i_3 = 2$ A; **b.** $i_1 = i_2 = i_3 = 1$ A. □

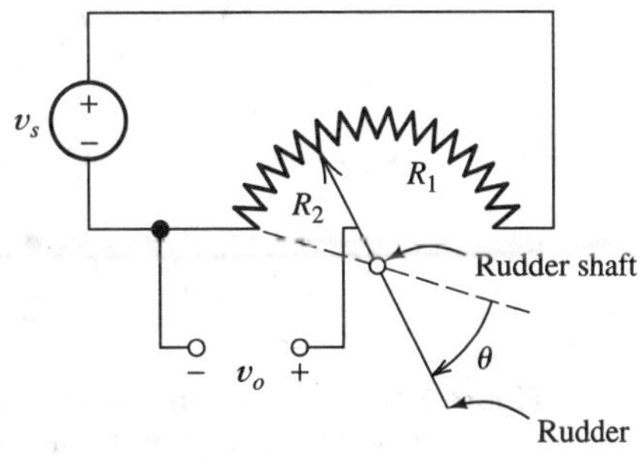

Figure 2.13 The voltage-division principle forms the basis for some position sensors. This figure shows a transducer that produces an output voltage v_o proportional to the rudder angle θ.

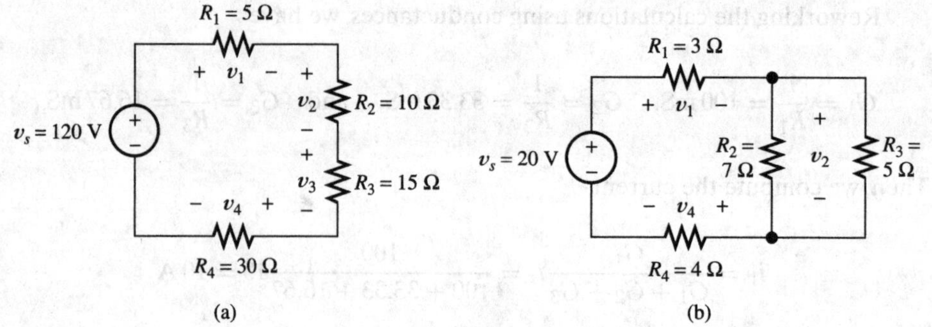

Figure 2.14 Circuits for Exercise 2.3.

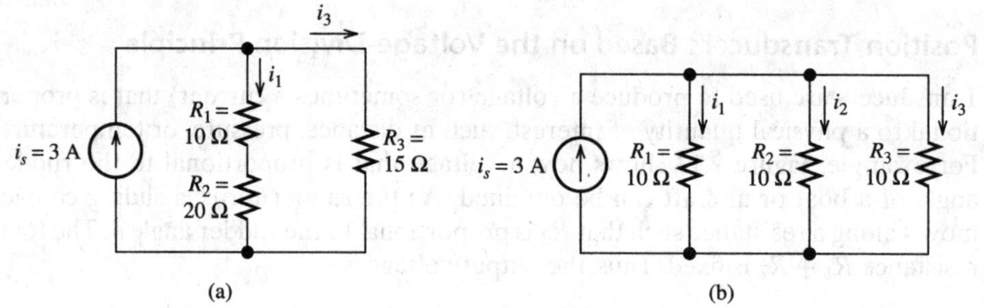

Figure 2.15 Circuits for Exercise 2.4.

2.4 NODE-VOLTAGE ANALYSIS

Although they are very important concepts, series/parallel equivalents and the current/voltage division principles are not sufficient to solve all circuits.

The network analysis methods that we have studied so far are useful, but they do not apply to all networks. For example, consider the circuit shown in Figure 2.16. We cannot solve this circuit by combining resistances in series and parallel because no series or parallel combination of resistances exists in the circuit. Furthermore, the voltage-division and current-division principles cannot be applied to this circuit. In this section, we learn **node-voltage analysis**, which is a general technique that can be applied to any circuit.

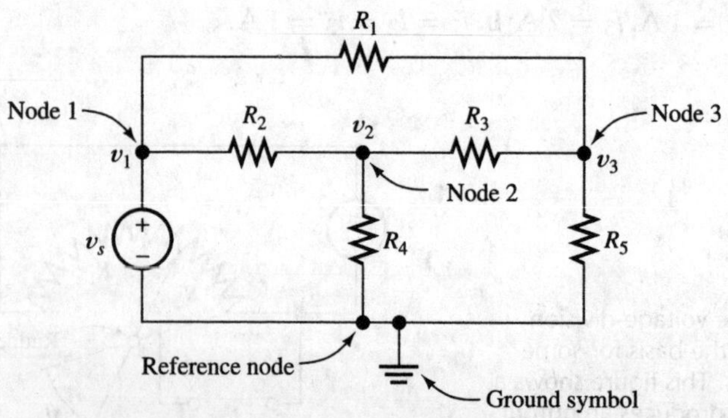

Figure 2.16 The first step in node analysis is to select a reference node and label the voltages at each of the other nodes.

Selecting the Reference Node

A **node** is a point at which two or more circuit elements are joined together. In node-voltage analysis, we first select one of the nodes as the **reference node**. In principle, any node can be picked to be the reference node. However, the solution is usually facilitated by selecting one end of a voltage source as the reference node. We will see why this is true as we proceed.

For example, the circuit shown in Figure 2.16 has four nodes. Let us select the bottom node as the reference node. We mark the reference node by the **ground symbol**, as shown in the figure.

Assigning Node Voltages

Next, we label the voltages at each of the other nodes. For example, the voltages at the three nodes are labeled v_1, v_2, and v_3 in Figure 2.16. The voltage v_1 is the voltage between node 1 and the reference node. The reference polarity for v_1 is positive at node 1 and negative at the reference node. Similarly, v_2 is the voltage between node 2 and the reference node. The reference polarity for v_2 is positive at node 2 and negative at the reference node. *In fact, the negative reference polarity for each of the node voltages is at the reference node.* We say that v_1 is the voltage at node 1 with respect to the reference node.

The negative reference polarity for each of the node voltages is at the reference node.

Finding Element Voltages in Terms of the Node Voltages

In node-voltage analysis, we write equations and eventually solve for the node voltages. Once the node voltages have been found, it is relatively easy to find the current, voltage, and power for each element in the circuit.

For example, suppose that we know the values of the node voltages and we want to find the voltage across R_3 with its positive reference on the left-hand side. To avoid additional labels in Figure 2.16, we have made a second drawing of the circuit, which is shown in Figure 2.17. The node voltages and the voltage v_x across R_3 are shown in Figure 2.17, where we have used arrows to indicate reference polarities. (Recall that the positive reference is at the head of the arrow.) Notice that v_2, v_x, and v_3 are the voltages encountered in traveling around the closed path through R_4, R_3, and R_5. Thus, these voltages must obey Kirchhoff's voltage law. Traveling around the loop

Once the node voltages have been determined, it is relatively easy to determine other voltages and currents in the circuit.

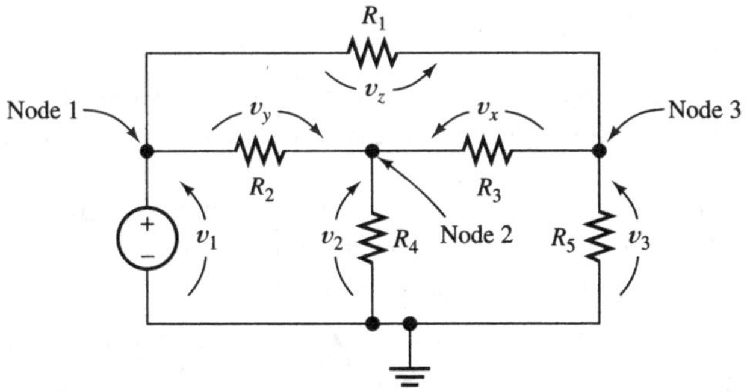

Figure 2.17 Assuming that we can determine the node voltages v_1, v_2, and v_3, we can use KVL to determine v_x, v_y, and v_z. Then using Ohm's law, we can find the current in each of the resistances. Thus, the key problem is in determining the node voltages.

This is the same circuit shown in Figure 2.16. We have redrawn it simply to avoid cluttering the original diagram with the voltages v_x, v_y, and v_z that are not involved in the final node equations.

clockwise and summing voltages, we have

$$-v_2 + v_x + v_3 = 0$$

Solving for v_x, we obtain

$$v_x = v_2 - v_3$$

Thus, we can find the voltage across any element in the network as the difference between node voltages. (If one end of an element is connected to the reference node, the voltage across the element is a node voltage.)

After the voltages are found, Ohm's law and KCL can be used to find the current in each element. Then, power can be computed by taking the product of the voltage and current for each element.

Exercise 2.5 In the circuit of Figure 2.17, find expressions for v_y and v_z in terms of the node voltages v_1, v_2, and v_3.

Answer $v_y = v_2 - v_1, v_z = v_3 - v_1$. □

Writing KCL Equations in Terms of the Node Voltages

After choosing the reference node and assigning the voltage variables, we write equations that can be solved for the node voltages.

After choosing the reference node and assigning the voltage variables, we write equations that can be solved for the node voltages. We demonstrate by continuing with the circuit of Figure 2.16.

In Figure 2.16, the voltage v_1 is the same as the source voltage v_s:

$$v_1 = v_s$$

(In this case, one of the node voltages is known without any effort. This is the advantage in selecting the reference node at one end of an independent voltage source.)

Therefore, we need to determine the values of v_2 and v_3, and we must write two independent equations. We usually start by trying to write current equations at each of the nodes corresponding to an unknown node voltage. For example, at node 2 in Figure 2.16, the current leaving through R_4 is given by

$$\frac{v_2}{R_4}$$

This is true because v_2 is the voltage across R_4 with its positive reference at node 2. Thus, the current v_2/R_4 flows from node 2 toward the reference node, which is away from node 2.

Next, referring to Figure 2.17, we see that the current flowing out of node 2 through R_3 is given by v_x/R_3. However, we found earlier that $v_x = v_2 - v_3$. Thus, the current flowing out of node 2 through R_3 is given by

$$\frac{v_2 - v_3}{R_3}$$

To find the current flowing out of node n through a resistance toward node k, we subtract the voltage at node k from the voltage at node n and divide the difference by the resistance.

At this point, we pause in our analysis to make a useful observation. *To find the current flowing out of node n through a resistance toward node k, we subtract the voltage at node k from the voltage at node n and divide the difference by the resistance.*

Thus, if v_n and v_k are the node voltages and R is the resistance connected between the nodes, the current flowing from node n toward node k is given by

$$\frac{v_n - v_k}{R}$$

Applying this observation in Figure 2.16 to find the current flowing out of node 2 through R_2, we have

$$\frac{v_2 - v_1}{R_2}$$

[In Exercise 2.5, we found that $v_y = v_2 - v_1$ (see Figure 2.17). The current flowing to the left through R_2 is v_y/R_2. Substitution yields the aforementioned expression.]

Of course, if the resistance is connected between node n and the reference node, the current away from node n toward the reference node is simply the node voltage v_n divided by the resistance. For example, as we noted previously, the current leaving node 2 through R_4 is given by v_2/R_4.

Now we apply KCL, adding all of the expressions for the currents leaving node 2 and setting the sum to zero. Thus, we obtain

$$\frac{v_2 - v_1}{R_2} + \frac{v_2}{R_4} + \frac{v_2 - v_3}{R_3} = 0$$

Writing the current equation at node 3 is similar. We try to follow the same pattern in writing each equation. Then, the equations take a familiar form, and mistakes are less frequent. We usually write expressions for the currents leaving the node under consideration and set the sum to zero. Applying this approach at node 3 of Figure 2.16, we have

$$\frac{v_3 - v_1}{R_1} + \frac{v_3}{R_5} + \frac{v_3 - v_2}{R_3} = 0$$

In many networks, we can obtain all of the equations needed to solve for the node voltages by applying KCL to the nodes at which the unknown voltages appear.

Example 2.6 Node-Voltage Analysis

Write equations that can be solved for the node voltages v_1, v_2, and v_3 shown in Figure 2.18.

Solution We use KCL to write an equation at node 1:

$$\frac{v_1}{R_1} + \frac{v_1 - v_2}{R_2} + i_s = 0$$

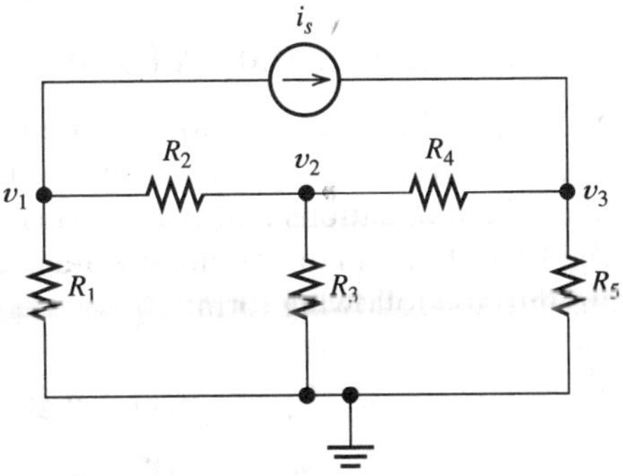

Figure 2.18 Circuit for Example 2.6.

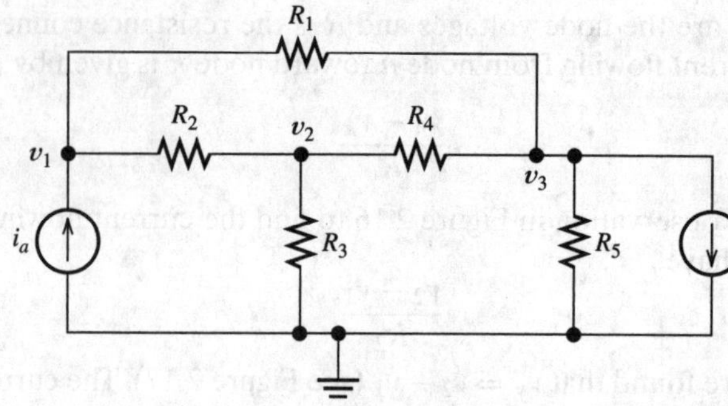

Figure 2.19 Circuit for Exercise 2.6.

Each term on the left-hand side of this equation represents a current leaving node 1. Summing the currents leaving node 2, we have

$$\frac{v_2 - v_1}{R_2} + \frac{v_2}{R_3} + \frac{v_2 - v_3}{R_4} = 0$$

Similarly, at node 3, we get

$$\frac{v_3}{R_5} + \frac{v_3 - v_2}{R_4} = i_s$$

Here, the currents leaving node 3 are on the left-hand side and the current entering is on the right-hand side. ■

Exercise 2.6 Use KCL to write equations at each node (except the reference node) for the circuit shown in Figure 2.19.

Answer

$$\text{Node 1:} \quad \frac{v_1 - v_3}{R_1} + \frac{v_1 - v_2}{R_2} = i_a$$

$$\text{Node 2:} \quad \frac{v_2 - v_1}{R_2} + \frac{v_2}{R_3} + \frac{v_2 - v_3}{R_4} = 0$$

$$\text{Node 3:} \quad \frac{v_3}{R_5} + \frac{v_3 - v_2}{R_4} + \frac{v_3 - v_1}{R_1} + i_b = 0$$ □

Circuit Equations in Standard Form

Once we have written the equations needed to solve for the node voltages, we put the equations into standard form. We group the node-voltage variables on the left-hand sides of the equations and place terms that do not involve the node voltages on the right-hand sides. For two node voltages, this eventually puts the node-voltage equations into the following form:

$$g_{11}v_1 + g_{12}v_2 = i_1 \tag{2.31}$$

$$g_{21}v_1 + g_{22}v_2 = i_2 \tag{2.32}$$

If we have three unknown node voltages, the equations can be put into the form

$$g_{11}v_1 + g_{12}v_2 + g_{13}v_3 = i_1 \tag{2.33}$$

$$g_{21}v_1 + g_{22}v_2 + g_{23}v_3 = i_2 \tag{2.34}$$

$$g_{31}v_1 + g_{32}v_2 + g_{33}v_3 = i_3 \tag{2.35}$$

We have chosen the letter g for the node-voltage coefficients because they are often (but not always) conductances with units of siemens. Similarly, we have used i for the terms on the right-hand sides of the equations because they are often currents.

In matrix form, the equations can be written as

$$\mathbf{GV} = \mathbf{I}$$

in which we have

$$\mathbf{G} = \begin{bmatrix} g_{11} & g_{12} \\ g_{21} & g_{22} \end{bmatrix} \quad \text{or} \quad \mathbf{G} = \begin{bmatrix} g_{11} & g_{12} & g_{13} \\ g_{21} & g_{22} & g_{23} \\ g_{31} & g_{32} & g_{33} \end{bmatrix}$$

depending on whether we have two or three unknown node voltages. Also, **V** and **I** are column vectors:

$$\mathbf{V} = \begin{bmatrix} v_1 \\ v_2 \end{bmatrix} \quad \text{or} \quad \mathbf{V} = \begin{bmatrix} v_1 \\ v_2 \\ v_3 \end{bmatrix} \quad \text{and} \quad \mathbf{I} = \begin{bmatrix} i_1 \\ i_2 \end{bmatrix} \quad \text{or} \quad \mathbf{I} = \begin{bmatrix} i_1 \\ i_2 \\ i_3 \end{bmatrix}$$

As the number of nodes and node voltages increases, the dimensions of the matrices increase.

One way to solve for the node voltages is to find the inverse of **G** and then compute the solution vector as:

$$\mathbf{V} = \mathbf{G}^{-1}\mathbf{I}$$

A Shortcut to Writing the Matrix Equations

If we put the node equations for the circuit of Exercise 2.6 (Figure 2.19) into matrix form, we obtain

$$\begin{bmatrix} \frac{1}{R_1} + \frac{1}{R_2} & -\frac{1}{R_2} & -\frac{1}{R_1} \\ -\frac{1}{R_2} & \frac{1}{R_2} + \frac{1}{R_3} + \frac{1}{R_4} & -\frac{1}{R_4} \\ -\frac{1}{R_1} & -\frac{1}{R_4} & \frac{1}{R_1} + \frac{1}{R_4} + \frac{1}{R_5} \end{bmatrix} \begin{bmatrix} v_1 \\ v_2 \\ v_3 \end{bmatrix} = \begin{bmatrix} i_a \\ 0 \\ -i_b \end{bmatrix}$$

Let us take a moment to compare the circuit in Figure 2.19 with the elements in this equation. First, look at the elements on the diagonal of the **G** matrix, which are

$$g_{11} = \frac{1}{R_1} + \frac{1}{R_2} \quad g_{22} = \frac{1}{R_2} + \frac{1}{R_3} + \frac{1}{R_4} \quad \text{and} \quad g_{33} = \frac{1}{R_1} + \frac{1}{R_4} + \frac{1}{R_5}$$

We see that the diagonal elements of **G** are equal to the sums of the conductances connected to the corresponding nodes. Next, notice the off diagonal terms:

$$g_{12} = -\frac{1}{R_2} \quad g_{13} = -\frac{1}{R_1} \quad g_{21} = -\frac{1}{R_2} \quad g_{23} = -\frac{1}{R_4} \quad g_{31} = -\frac{1}{R_1} \quad g_{32} = -\frac{1}{R_4}$$

In each case, g_{jk} is equal to the negative of the conductance connected between node j and k. The terms in the **I** matrix are the currents pushed into the corresponding nodes by the current sources. These observations hold whenever the network consists of resistances and independent current sources, assuming that we follow our usual pattern in writing the equations.

Thus, if a circuit consists of resistances and independent current sources, we can use the following steps to rapidly write the node equations directly in matrix form.

This is a shortcut way to write the node equations in matrix form, *provided that the circuit contains only resistances and independent current sources.*

1. Make sure that the circuit contains only resistances and independent current sources.
2. The diagonal terms of **G** are the sums of the conductances connected to the corresponding nodes.
3. The off diagonal terms of **G** are the negatives of the conductances connected between the corresponding nodes.
4. The elements of **I** are the currents pushed into the corresponding nodes by the current sources.

Keep in mind that if the network contains voltage sources or controlled sources this pattern does not hold.

Exercise 2.7 Working directly from Figure 2.18 on page 81, write its node-voltage equations in matrix form.

Answer

$$\begin{bmatrix} \frac{1}{R_1}+\frac{1}{R_2} & -\frac{1}{R_2} & 0 \\ -\frac{1}{R_2} & \frac{1}{R_2}+\frac{1}{R_3}+\frac{1}{R_4} & -\frac{1}{R_4} \\ 0 & -\frac{1}{R_4} & \frac{1}{R_4}+\frac{1}{R_5} \end{bmatrix} \begin{bmatrix} v_1 \\ v_2 \\ v_3 \end{bmatrix} = \begin{bmatrix} -i_s \\ 0 \\ i_s \end{bmatrix}$$ □

Example 2.7 Node-Voltage Analysis

Write the node-voltage equations in matrix form for the circuit of Figure 2.20.

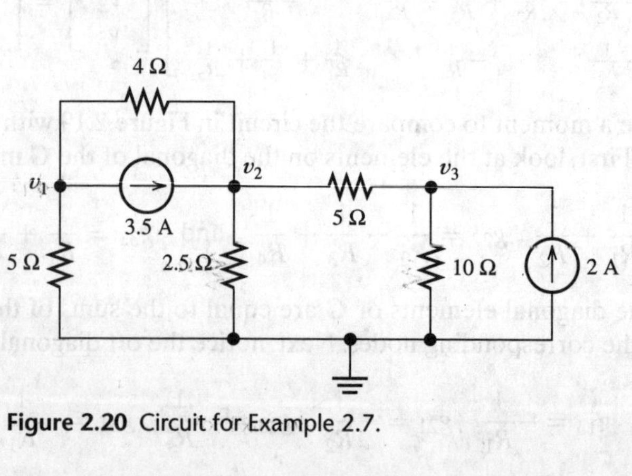

Figure 2.20 Circuit for Example 2.7.

Solution Writing KCL at each node, we have

$$\frac{v_1}{5} + \frac{v_1 - v_2}{4} + 3.5 = 0$$

$$\frac{v_2 - v_1}{4} + \frac{v_2}{2.5} + \frac{v_2 - v_3}{5} = 3.5$$

$$\frac{v_3 - v_2}{5} + \frac{v_3}{10} = 2$$

Manipulating the equations into standard form, we have

$$0.45v_1 - 0.25v_2 = -3.5$$

$$-0.25v_1 + 0.85v_2 - 0.2v_3 = 3.5$$

$$-0.2v_2 + 0.35v_3 = 2$$

Then, in matrix form, we obtain

$$\begin{bmatrix} 0.45 & -0.25 & 0 \\ -0.25 & 0.85 & -0.20 \\ 0 & -0.20 & 0.30 \end{bmatrix} \begin{bmatrix} v_1 \\ v_2 \\ v_3 \end{bmatrix} = \begin{bmatrix} -3.5 \\ 3.5 \\ 2 \end{bmatrix} \tag{2.36}$$

Because the circuit contains no voltage sources or controlled sources, we could have used the shortcut method to write the matrix form directly. For example, $g_{11} = 0.45$ is the sum of the conductances connected to node 1, $g_{12} = -0.25$ is the negative of the conductance connected between nodes 1 and 2, $i_3 = 2$ is the current pushed into node 3 by the 2-A current source, and so forth. ■

Solving the Network Equations

After we have obtained the equations in standard form, we can solve them by a variety of methods, including substitution, Gaussian elimination, and determinants. As an engineering student, you may own a powerful calculator such as the TI-84 or TI-89 that has the ability to solve systems of linear equations. You should learn to do this by practicing on the exercises and the problems at the end of this chapter.

In some situations, you may not be allowed to use one of the more advanced calculators or a notebook computer. For example, only fairly simple scientific calculators are allowed on the Fundamentals of Engineering (FE) Examination, which is the first step in becoming a registered professional engineer in the United States. The calculator policy for the professional engineering examinations can be found at `http://ncees.org/`. Thus, even if you own an advanced calculator, you may wish to practice with one of those allowed in the FE Examination.

Exercise 2.8 Use your calculator to solve Equation 2.36.
Answer $v_1 = -5$ V, $v_2 = 5$ V, $v_3 = 10$ V. □

Using MATLAB to Solve Network Equations

When you have access to a computer and MATLAB software, you have a very powerful system for engineering and scientific calculations. This software is available to students at many engineering schools and is very likely to be encountered in some of your other courses.

In this and the next several chapters, we illustrate the application of MATLAB to various aspects of circuit analysis, but we cannot possibly cover all of its many useful features in this book. If you are new to MATLAB, and it is available to you, use the video and/or getting started documentation that appear at the top of the command screen and/or in the Help menu to gain some initial familiarity with the software. Furthermore, you can gain access to a variety of online interactive tutorials at `http://www.mathworks.com/academia/student_center/tutorials/register.html`. If you have already used the program, the MATLAB commands we present may be familiar to you. In either case, you should be able to easily modify the examples we present to work out similar circuit problems.

Next, we illustrate the solution for Equation 2.36 using MATLAB. Instead of using $\mathbf{V} = \mathbf{G}^{-1}\mathbf{I}$ to compute node voltages, MATLAB documentation recommends using the command V = G\I which invokes a more accurate algorithm for computing solutions to systems of linear equations.

The comments following the % sign are ignored by MATLAB. For improved clarity, we use a **bold** font for the input commands, a regular font for comments, and a color font for the responses from MATLAB, otherwise the following has the appearance of the MATLAB command screen for this problem. (>> is the MATLAB command prompt.)

```
>> clear  % First we clear the work space.
>> % Then, we enter the coefficient matrix of Equation 2.36 with
>> % spaces between elements in each row and semicolons between rows.
>> G = [0.45 -0.25 0; -0.25 0.85 -0.2; 0 -0.2 0.30]
G =
    0.4500   -0.2500         0
   -0.2500    0.8500   -0.2000
         0   -0.2000    0.3000
>> % Next, we enter the column vector for the right-hand side.
>> I = [-3.5; 3.5; 2]
I =
   -3.5000
    3.5000
    2.0000
>>  % The MATLAB documentation recommends computing the node
>>  % voltages using V = G\I instead of using V = inv(G)*I.
>> V = G\I
V =
   -5.0000
    5.0000
   10.0000
```

Thus, we have $v_1 = -5$ V, $v_2 = 5$ V, and $v_3 = 10$ V, as you found when working Exercise 2.8 with your calculator.

LabVIEW MathScript

If you have access to LabVIEW software including the RT (Real Time) Module, you can use its MathScript option to solve equations. To do this, first start a Blank VI in LabVIEW. This will bring up two windows: a Block Diagram window and a Front Panel window. Left click on **Tools** in the menu bar of either window and then left click on **MathScript Window...**, which will open a window in which you can use many commands identical to those of MATLAB. At this point, you can close the Block Diagram and Front Panel windows. Enter the commands in the Command

Window, and the results will appear in the Output Window. If you wish to try this software alternative, start by trying to duplicate the solution to the example of the previous several pages. Simply enter the commands

```
clear
G = [0.45 -0.25 0; -0.25 0.85 -0.2; 0 -0.2 0.30]
I = [-3.5; 3.5; 2]
V = G\I
```

in the Command Window and the results will appear in the Output Window.

Example 2.8 Node-Voltage Analysis

Solve for the node voltages shown in Figure 2.21 and determine the value of the current i_x.

Solution Our first step in solving a circuit is to select the reference node and assign the node voltages. This has already been done, as shown in Figure 2.21.

Next, we write equations. In this case, we can write a current equation at each node. This yields

$$\text{Node 1:}\quad \frac{v_1}{10}+\frac{v_1-v_2}{5}+\frac{v_1-v_3}{20}=0$$

$$\text{Node 2:}\quad \frac{v_2-v_1}{5}+\frac{v_2-v_3}{10}=10$$

$$\text{Node 3:}\quad \frac{v_3}{5}+\frac{v_3-v_2}{10}+\frac{v_3-v_1}{20}=0$$

Next, we place these equations into standard form:

$$\begin{aligned} 0.35v_1-0.2v_2-0.05v_3&=0\\ -0.2v_1+0.3v_2-0.10v_3&=10\\ -0.05v_1-0.10v_2+0.35v_3&=0 \end{aligned}$$

In matrix form, the equations are

$$\begin{bmatrix} 0.35 & -0.2 & -0.05 \\ -0.2 & 0.3 & -0.1 \\ -0.05 & -0.1 & 0.35 \end{bmatrix}\begin{bmatrix} v_1 \\ v_2 \\ v_3 \end{bmatrix}=\begin{bmatrix} 0 \\ 10 \\ 0 \end{bmatrix}$$

or $\mathbf{GV}=\mathbf{I}$

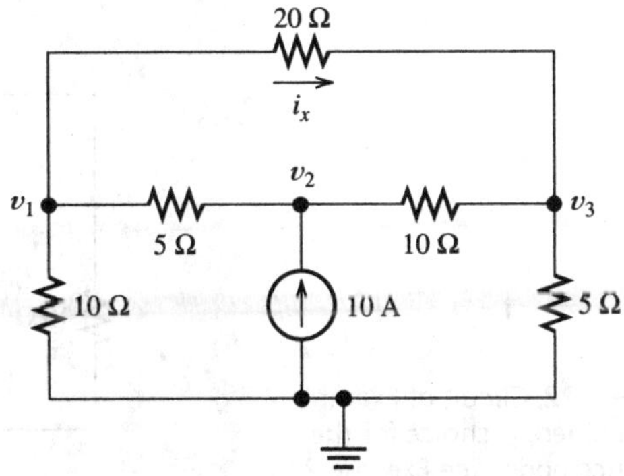

Figure 2.21 Circuit for Example 2.8.

in which **G** represents the coefficient matrix of conductances, **V** is the column vector of node voltages, and **I** is the column vector of currents on the right-hand side.

Here again, we could write the equations directly in standard or matrix form using the short cut method because the circuit contains only resistances and independent current sources.

The MATLAB solution is:

```
>> clear
>> G = [0.35 -0.2 -0.05; -0.2 0.3 -0.1; -0.05 -0.1 0.35];
>> % A semicolon at the end of a command suppresses the
>> % MATLAB response.
>> I = [0; 10; 0];
>> V = G\I
V =
    45.4545
    72.7273
    27.2727
>> % Finally, we calculate the current.
>> Ix = (V(1) - V(3))/20
Ix =
     0.9091
```

Alternatively, you can use the same commands with LabVIEW MathScript to obtain the answers. ■

Exercise 2.9 Repeat the analysis of the circuit of Example 2.8, using the reference node and node voltages shown in Figure 2.22. **a.** First write the network equations. **b.** Put the network equations into standard form. **c.** Solve for v_1, v_2, and v_3. (The values will be different than those we found in Example 2.8 because v_1, v_2, and v_3 are not the same voltages in the two figures.) **d.** Find i_x. (Of course, i_x is the same in both figures, so it should have the same value.)

Answer

a.

$$\frac{v_1 - v_3}{20} + \frac{v_1}{5} + \frac{v_1 - v_2}{10} = 0$$

$$\frac{v_2 - v_1}{10} + 10 + \frac{v_2 - v_3}{5} = 0$$

$$\frac{v_3 - v_1}{20} + \frac{v_3}{10} + \frac{v_3 - v_2}{5} = 0$$

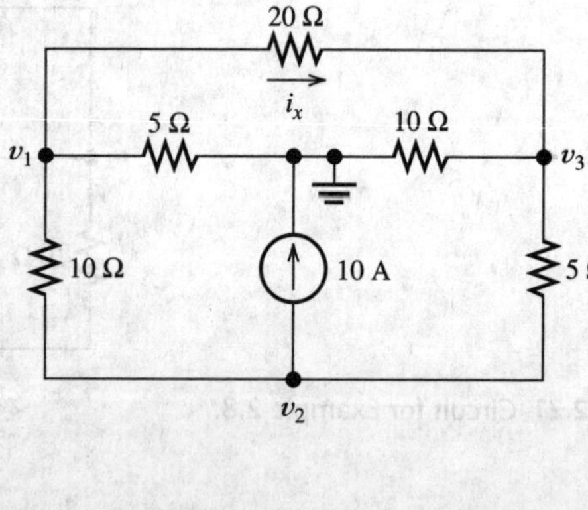

Figure 2.22 Circuit of Example 2.8 with a different choice for the reference node. See Exercise 2.9.

b.

$$\begin{aligned} 0.35v_1 - 0.10v_2 - 0.05v_3 &= 0 \\ -0.10v_1 + 0.30v_2 - 0.20v_3 &= -10 \\ -0.05v_1 - 0.20v_2 + 0.35v_3 &= 0 \end{aligned}$$

c. $v_1 = -27.27, v_2 = -72.73, v_3 = -45.45$

d. $i_x = 0.909$ A

□

Circuits with Voltage Sources

When a circuit contains a single voltage source, we can often pick the reference node at one end of the source, and then we have one less unknown node voltage for which to solve.

Example 2.9 Node-Voltage Analysis

Write the equations for the network shown in Figure 2.23 and put them into standard form.

Solution Notice that we have selected the reference node at the bottom end of the voltage source. Thus, the voltage at node 3 is known to be 10 V, and we do not need to assign a variable for that node.

Writing current equations at nodes 1 and 2, we obtain

$$\frac{v_1 - v_2}{5} + \frac{v_1 - 10}{2} = 1$$

$$\frac{v_2}{5} + \frac{v_2 - 10}{10} + \frac{v_2 - v_1}{5} = 0$$

Now if we group terms and place the constants on the right-hand sides of the equations, we have

$$0.7v_1 - 0.2v_2 = 6$$

$$-0.2v_1 + 0.5v_2 = 1$$

Thus, we have obtained the equations needed to solve for v_1 and v_2 in standard form. ■

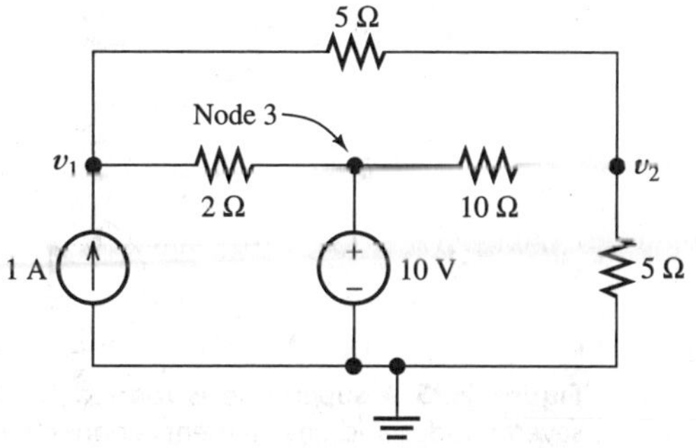

Figure 2.23 Circuit for Example 2.9.

Exercise 2.10 Solve the equations of Example 2.9 for v_1 and v_2.
Answer $v_1 = 10.32$ V, $v_2 = 6.129$ V. □

Exercise 2.11 Solve for the node voltages v_1 and v_2 in the circuit of Figure 2.24.
Answer $v_1 = 6.77$ V, $v_2 = 4.19$ V. □

Sometimes, the pattern for writing node-voltage equations that we have illustrated so far must be modified. For example, consider the network and node voltages shown in Figure 2.25. Notice that $v_3 = -15$ V because of the 15-V source connected between node 3 and the reference node. Therefore, we need two equations relating the unknowns v_1 and v_2.

If we try to write a current equation at node 1, we must include a term for the current through the 10-V source. We could assign an unknown for this current, but then we would have a higher-order system of equations to solve. Especially if we are solving the equations manually, we want to minimize the number of unknowns. For this circuit, it is not possible to write a current equation in terms of the node voltages for any single node (even the reference node) because a voltage source is connected to each node.

Another way to obtain a current equation is to form a **supernode**. This is done by drawing a dashed line around several nodes, including the elements connected between them. This is shown in Figure 2.25. Two supernodes are indicated, one enclosing each of the voltage sources.

Another way to state Kirchhoff's current law is that the net current flowing through any closed surface must equal zero.

We can state Kirchhoff's current law in a slightly more general form than we have previously: *The net current flowing through any closed surface must equal zero.* Thus, we can apply KCL to a supernode. For example, for the supernode enclosing

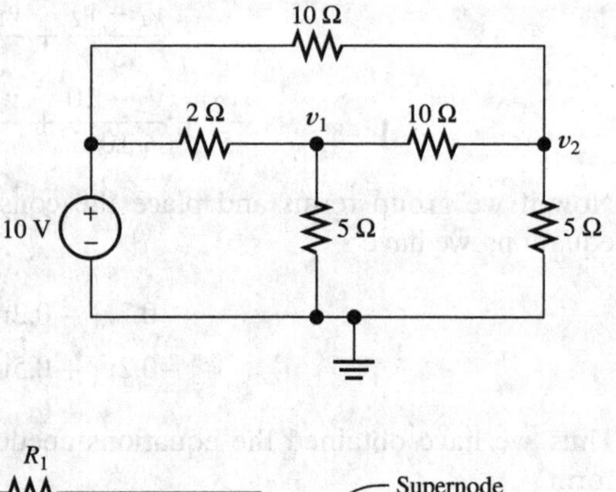

Figure 2.24 Circuit for Exercise 2.11.

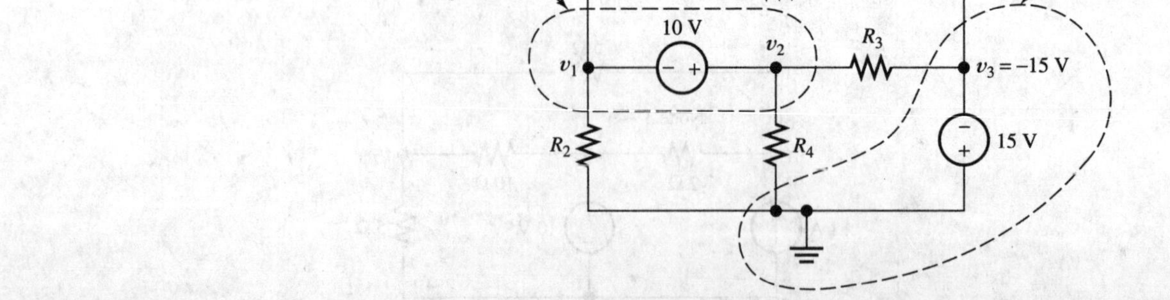

Figure 2.25 A supernode is formed by drawing a dashed line enclosing several nodes and any elements connected between them.

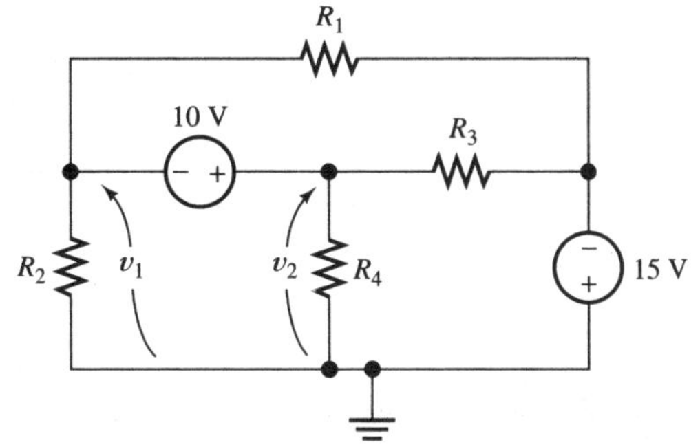

Figure 2.26 Node voltages v_1 and v_2 and the 10-V source form a closed loop to which KVL can be applied. (This is the same circuit as that of Figure 2.25.)

the 10-V source, we sum currents leaving and obtain

$$\frac{v_1}{R_2} + \frac{v_1 - (-15)}{R_1} + \frac{v_2}{R_4} + \frac{v_2 - (-15)}{R_3} = 0 \quad (2.37)$$

Each term on the left-hand side of this equation represents a current leaving the supernode through one of the resistors. Thus, by enclosing the 10-V source within the supernode, we have obtained a current equation without introducing a new variable for the current in the source.

Next, we might be tempted to write another current equation for the other supernode. However, we would find that the equation is equivalent to the one already written. *In general, we obtain dependent equations if we use all of the nodes in writing current equations.* Nodes 1 and 2 were part of the first supernode, while node 3 and the reference node are part of the second supernode. Thus, in writing equations for both supernodes, we would have used all four nodes in the network.

We obtain dependent equations if we use all of the nodes in a network to write KCL equations.

If we tried to solve for the node voltages by using substitution, at some point all of the terms would drop out of the equations and we would not be able to solve for those voltages. In MATLAB, you will receive a warning that the G matrix is singular, in other words, its determinant is zero. If this happens, we know that we should return to writing equations and find another equation to use in the solution. This will not happen if we avoid using all of the nodes in writing current equations.

There is a way to obtain an independent equation for the network under consideration. We can use KVL because v_1, the 10-V source, and v_2 form a closed loop. This is illustrated in Figure 2.26, where we have used arrows to indicate the polarities of v_1 and v_2. Traveling clockwise and summing the voltages around the loop, we obtain

When a voltage source is connected between nodes so that current equations cannot be written at the individual nodes, first write a KVL equation, including the voltage source, and then enclose the voltage source in a supernode and write a KCL equation for the supernode.

$$-v_1 - 10 + v_2 = 0 \quad (2.38)$$

Equations 2.37 and 2.38 form an independent set that can be used to solve for v_1 and v_2 (assuming that the resistance values are known).

Exercise 2.12 Write the current equation for the supernode that encloses the 15-V source in Figure 2.25. Show that your equation is equivalent to Equation 2.37. □

Exercise 2.13 Write a set of independent equations for the node voltages shown in Figure 2.27.

Answer

KVL:

$$-v_1 + 10 + v_2 = 0$$

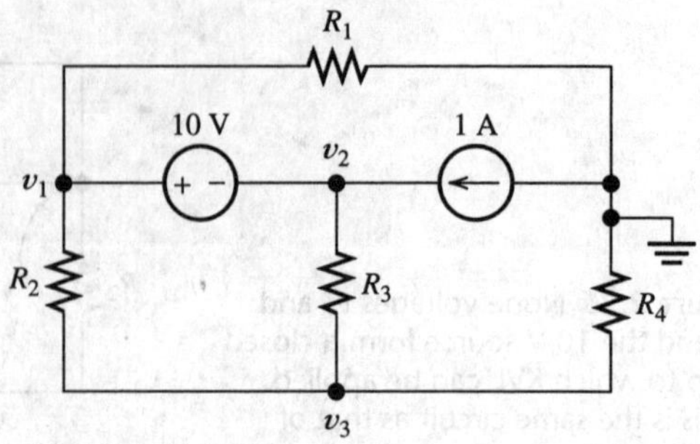

Figure 2.27 Circuit for Exercise 2.13.

KCL for the supernode enclosing the 10-V source:

$$\frac{v_1}{R_1} + \frac{v_1 - v_3}{R_2} + \frac{v_2 - v_3}{R_3} = 1$$

KCL for node 3:

$$\frac{v_3 - v_1}{R_2} + \frac{v_3 - v_2}{R_3} + \frac{v_3}{R_4} = 0$$

KCL at the reference node:

$$\frac{v_1}{R_1} + \frac{v_3}{R_4} = 1$$

For independence, the set must include the KVL equation. Any two of the three KCL equations can be used to complete the three-equation set. (The three KCL equations use all of the network nodes and, therefore, do not form an independent set.) □

Circuits with Controlled Sources

Controlled sources present a slight additional complication of the node-voltage technique. (Recall that the value of a controlled source depends on a current or voltage elsewhere in the network.) In applying node-voltage analysis, first we write equations exactly as we have done for networks with independent sources. Then, we express the controlling variable in terms of the node-voltage variables and substitute into the network equations. We illustrate with two examples.

Example 2.10 Node-Voltage Analysis with a Dependent Source

Write an independent set of equations for the node voltages shown in Figure 2.28.

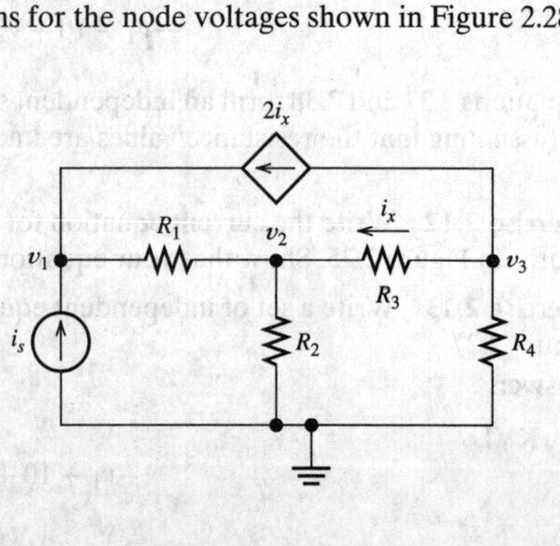

Figure 2.28 Circuit containing a current-controlled current source. See Example 2.10.

Solution First, we write KCL equations at each node, including the current of the controlled source just as if it were an ordinary current source:

$$\frac{v_1 - v_2}{R_1} = i_s + 2i_x \tag{2.39}$$

$$\frac{v_2 - v_1}{R_1} + \frac{v_2}{R_2} + \frac{v_2 - v_3}{R_3} = 0 \tag{2.40}$$

$$\frac{v_3 - v_2}{R_3} + \frac{v_3}{R_4} + 2i_x = 0 \tag{2.41}$$

Next, we find an expression for the controlling variable i_x in terms of the node voltages. Notice that i_x is the current flowing away from node 3 through R_3. Thus, we can write

$$i_x = \frac{v_3 - v_2}{R_3} \tag{2.42}$$

Finally, we use Equation 2.42 to substitute into Equations 2.39, 2.40, and 2.41. Thus, we obtain the required equation set:

$$\frac{v_1 - v_2}{R_1} = i_s + 2\frac{v_3 - v_2}{R_3} \tag{2.43}$$

$$\frac{v_2 - v_1}{R_1} + \frac{v_2}{R_2} + \frac{v_2 - v_3}{R_3} = 0 \tag{2.44}$$

$$\frac{v_3 - v_2}{R_3} + \frac{v_3}{R_4} + 2\frac{v_3 - v_2}{R_3} = 0 \tag{2.45}$$

Assuming that the value of i_s and the resistances are known, we could put this set of equations into standard form and solve for v_1, v_2, and v_3. ■

Example 2.11 Node-Voltage Analysis with a Dependent Source

Write an independent set of equations for the node voltages shown in Figure 2.29.

Solution First, we ignore the fact that the voltage source is a dependent source and write equations just as we would for a circuit with independent sources. We cannot write a current equation at either node 1 or node 2, because of the voltage source connected between them. However, we can write a KVL equation:

$$-v_1 + 0.5v_x + v_2 = 0 \tag{2.46}$$

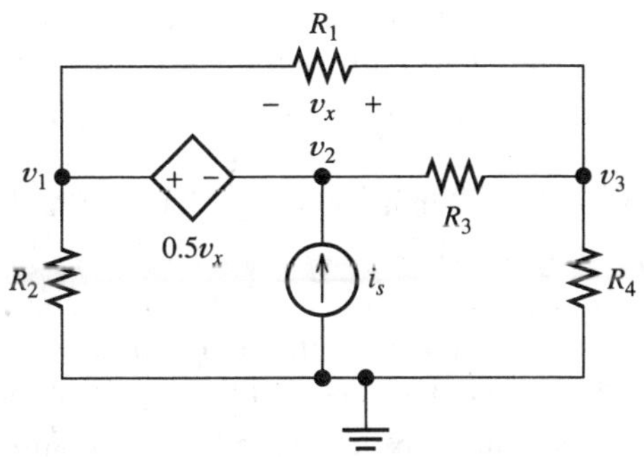

Figure 2.29 Circuit containing a voltage-controlled voltage source. See Example 2.11.

Then, we use KCL to write current equations. For a supernode enclosing the controlled voltage source,

$$\frac{v_1}{R_2} + \frac{v_1 - v_3}{R_1} + \frac{v_2 - v_3}{R_3} = i_s$$

For node 3,

$$\frac{v_3}{R_4} + \frac{v_3 - v_2}{R_3} + \frac{v_3 - v_1}{R_1} = 0 \tag{2.47}$$

For the reference node,

$$\frac{v_1}{R_2} + \frac{v_3}{R_4} = i_s \tag{2.48}$$

Of course, these current equations are dependent because we have used all four nodes in writing them. We must use Equation 2.46 and two of the KCL equations to form an independent set. However, Equation 2.46 contains the controlling variable v_x, which must be eliminated before we have equations in terms of the node voltages.

Thus, our next step is to write an expression for the controlling variable v_x in terms of the node voltages. Notice that v_1, v_x, and v_3 form a closed loop. Traveling clockwise and summing voltages, we have

$$-v_1 - v_x + v_3 = 0$$

Solving for v_x, we obtain

$$v_x = v_3 - v_1$$

Now if we substitute into Equation 2.46, we get

$$v_1 = 0.5(v_3 - v_1) + v_2 \tag{2.49}$$

Equation 2.49 along with any two of the KCL equations forms an independent set that can be solved for the node voltages. ■

Using the principles we have discussed in this section, we can write node-voltage equations for any network consisting of sources and resistances. Thus, given a computer or calculator to help in solving the equations, we can compute the currents and voltages for any network.

Next, we summarize the steps in analyzing circuits by the node-voltage technique:

Here is a convenient step-by-step guide to node-voltage analysis.

1. Select a reference node and assign variables for the unknown node voltages. If the reference node is chosen at one end of an independent voltage source, one node voltage is known at the start, and fewer need to be computed.
2. Write network equations. First, use KCL to write current equations for nodes and supernodes. Write as many current equations as you can without using all of the nodes. Then if you do not have enough equations because of voltage sources connected between nodes, use KVL to write additional equations.
3. If the circuit contains dependent sources, find expressions for the controlling variables in terms of the node voltages. Substitute into the network equations, and obtain equations having only the node voltages as unknowns.
4. Put the equations into standard form and solve for the node voltages.

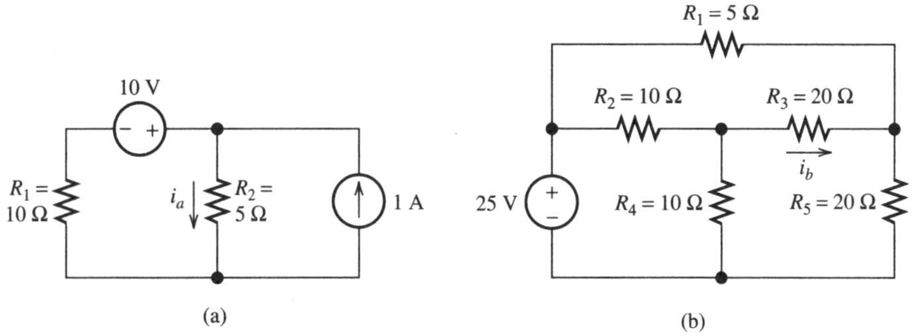

Figure 2.30 Circuits for Exercise 2.14.

5. Use the values found for the node voltages to calculate any other currents or voltages of interest.

Exercise 2.14 Use the node-voltage technique to solve for the currents labeled in the circuits shown in Figure 2.30.

Answer **a.** $i_a = 1.33$ A; **b.** $i_b = -0.259$ A. □

Exercise 2.15 Use the node-voltage technique to solve for the values of i_x and i_y in Figure 2.31.

Answer $i_x = 0.5$ A, $i_y = 2.31$ A. □

Using the MATLAB Symbolic Toolbox to Obtain Symbolic Solutions

If the Symbolic Toolbox is included with your version of MATLAB, you can use it to solve node voltage and other equations symbolically. (LabVIEW MathScript does not have symbolic mathematics capabilities.) We illustrate by solving Equations 2.43, 2.44, and 2.45 from Example 2.10 on page 92.

For help with a command such as "solve" simply type "help solve" at the command prompt.

```
>> clear
>> % First we clear the workspace, then we enter the equations into
>> % the solve command followed by the variables for which we want
>> % to solve.
>> [V1, V2, V3] = solve('(V1 - V2)/R1 = Is + 2*(V3 - V2)/R3', ...
                  '(V2 - V1)/R1 + V2/R2 + (V2 - V3)/R3 = 0', ...
                  '(V3 - V2)/R3 + V3/R4 + 2*(V3 - V2)/R3 = 0',...
                  'V1','V2','V3')
```

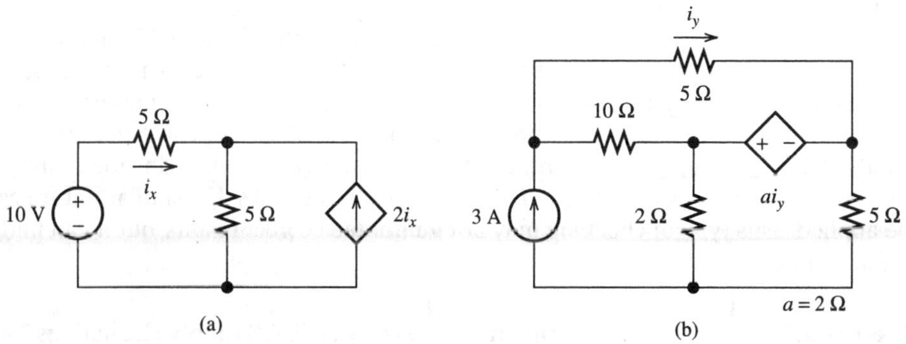

Figure 2.31 Circuits for Exercise 2.15.

```
V1 =
(Is*R1*R2 + Is*R1*R3 + 3*Is*R1*R4 + Is*R2*R3 + 3*Is*R2*R4)/(3*R2 + R3 +
3*R4)
V2 =
(Is*R2*R3 + 3*Is*R2*R4)/(3*R2 + R3 + 3*R4)
V3 =
(3*Is*R2*R4)/(3*R2 + R3 + 3*R4)
>> % The solve command gives the answers, but in a form that is
>> % somewhat difficult to read.
>> % A more readable version of the answers is obtained using the
>> % pretty command. We combine the three commands on one line
>> % by placing commas between them.
>> pretty(V1), pretty(V2), pretty(V3)
  Is R1 R2 + Is R1 R3 + 3 Is R1 R4 + Is R2 R3 + 3 Is R2 R4
  --------------------------------------------------------
                     3 R2 + R3 + 3 R4

  Is R2 R3 + 3 Is R2 R4
  ---------------------
     3 R2 + R3 + 3 R4

    3 Is R2 R4
  ----------------
  3 R2 + R3 + 3 R4
```

(Here we have shown the results obtained using MATLAB version R2008b; other versions may give results different in appearance but equivalent mathematically.) In more standard mathematical format, the results are:

$$v_1 = \frac{i_s R_1 R_2 + i_s R_1 R_3 + 3 i_s R_1 R_4 + i_s R_2 R_3 + 3 i_s R_2 R_4}{3R_2 + R_3 + 3R_4}$$

$$v_2 = \frac{i_s R_2 R_3 + 3 i_s R_2 R_4}{3R_2 + R_3 + 3R_4}$$

$$\text{and } v_3 = \frac{3 i_s R_2 R_4}{3R_2 + R_3 + 3R_4}$$

Checking Answers

As usual, it is a good idea to apply some checks to the answers. First of all, make sure that the answers have proper units, which are volts in this case. If the units don't check, look to see if any of the numerical values entered in the equations have units. Referring to the circuit (Figure 2.28 on page 92), we see that the only numerical parameter entered into the equations was the gain of the current-controlled current source, which has no units.

Again referring to the circuit diagram, we can see that we should have $v_2 = v_3$ for $R_3 = 0$, and we check the results to see that this is the case. Another check is obtained by observing that we should have $v_3 = 0$ for $R_4 = 0$. Still another check of the results comes from observing that, in the limit as R_3 approaches infinity, we should have $i_x = 0$, (so the controlled current source becomes an open circuit), $v_3 = 0$, $v_1 = i_s(R_1 + R_2)$, and $v_2 = i_s R_2$. Various other checks of a similar nature can be applied. This type of checking may not guarantee correct results, but it can find a lot of errors.

Exercise 2.16 Use the symbolic math features of MATLAB to solve Equations 2.47, 2.48, and 2.49 for the node voltages in symbolic form.

Answer

$$v_1 = \frac{2i_sR_1R_2R_3 + 3i_sR_1R_2R_4 + 2i_sR_2R_3R_4}{3R_1R_2 + 2R_1R_3 + 3R_1R_4 + 2R_2R_3 + 2R_3R_4}$$

$$v_2 = \frac{3i_sR_1R_2R_3 + 3i_sR_1R_2R_4 + 2i_sR_2R_3R_4}{3R_1R_2 + 2R_1R_3 + 3R_1R_4 + 2R_2R_3 + 2R_3R_4}$$

$$v_3 = \frac{3i_sR_1R_2R_4 + 2i_sR_2R_3R_4}{3R_1R_2 + 2R_1R_3 + 3R_1R_4 + 2R_2R_3 + 2R_3R_4}$$

Depending on the version of MATLAB and the Symbolic Toolbox that you use, your answers may have a different appearance but should be algebraically equivalent to these. □

2.5 MESH-CURRENT ANALYSIS

In this section, we show how to analyze networks by using another general technique, known as mesh-current analysis. Networks that can be drawn on a plane without having one element (or conductor) crossing over another are called **planar networks**. On the other hand, circuits that must be drawn with one or more elements crossing others are said to be **nonplanar**. We consider only planar networks.

Let us start by considering the planar network shown in Figure 2.32(a). Suppose that the source voltages and resistances are known and that we wish to solve for the currents. We first write equations for the currents shown in Figure 2.32(a), which are called branch currents because a separate current is defined in each branch of the network. However, we will eventually see that using the mesh currents illustrated in Figure 2.32(b) makes the solution easier.

Three independent equations are needed to solve for the three branch currents shown in Figure 2.32(a). In general, the number of independent KVL equations that can be written for a planar network is equal to the number of open areas defined by the network layout. For example, the circuit of Figure 2.32(a) has two open areas: one defined by v_A, R_1, and R_3, while the other is defined by R_3, R_2, and v_B. Thus, for this network, we can write only two independent KVL equations. We must employ KCL to obtain the third equation.

Application of KVL to the loop consisting of v_A, R_1, and R_3 yields

$$R_1i_1 + R_3i_3 = v_A \tag{2.50}$$

Similarly, for the loop consisting of R_3, R_2, and v_B, we get

$$-R_3i_3 + R_2i_2 = -v_B \tag{2.51}$$

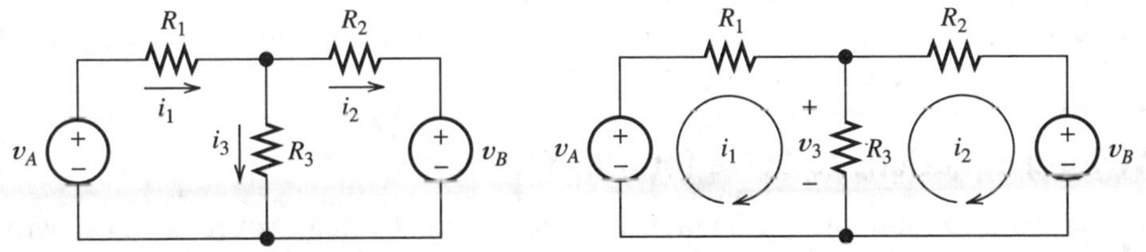

(a) Circuit with branch currents (b) Circuit with mesh currents

Figure 2.32 Circuit for illustrating the mesh-current method of circuit analysis.

Applying KCL to the node at the top end of R_3, we have

$$i_1 = i_2 + i_3 \tag{2.52}$$

Next, we solve Equation 2.52 for i_3 and substitute into Equations 2.50 and 2.51. This yields the following two equations:

$$R_1 i_1 + R_3(i_1 - i_2) = v_A \tag{2.53}$$

$$-R_3(i_1 - i_2) + R_2 i_2 = -v_B \tag{2.54}$$

Thus, we have used the KCL equation to reduce the KVL equations to two equations in two unknowns.

When several mesh currents flow through one element, we consider the current in that element to be the algebraic sum of the mesh currents.

Now, consider the mesh currents i_1 and i_2 shown in Figure 2.32(b). As indicated in the figure, mesh currents are considered to flow around closed paths. Hence, mesh currents automatically satisfy KCL. *When several mesh currents flow through one element, we consider the current in that element to be the algebraic sum of the mesh currents.* Thus, assuming a reference direction pointing downward, the current in R_3 is $(i_1 - i_2)$. Thus, $v_3 = R_3(i_1 - i_2)$. Now if we follow i_1 around its loop and apply KVL, we get Equation 2.53 directly. Similarly, following i_2, we obtain Equation 2.54 directly.

Because mesh currents automatically satisfy KCL, some work is saved in writing and solving the network equations. The circuit of Figure 2.32 is fairly simple, and the advantage of mesh currents is not great. However, for more complex networks, the advantage can be quite significant.

Choosing the Mesh Currents

For a planar circuit, we can choose the current variables to flow through the elements around the periphery of each of the open areas of the circuit diagram. For consistency, we usually define the mesh currents to flow clockwise.

We usually choose the current variables to flow clockwise around the periphery of each of the open areas of the circuit diagram.

Two networks and suitable choices for the mesh currents are shown in Figure 2.33. When a network is drawn with no crossing elements, it resembles a window, with each open area corresponding to a pane of glass. Sometimes it is said that the mesh currents are defined by "soaping the window panes."

Keep in mind that, if two mesh currents flow through a circuit element, we consider the current in that element to be the algebraic sum of the mesh currents. For example, in Figure 2.33(a), the current in R_2 referenced to the left is $i_3 - i_1$. Furthermore, the current referenced upward in R_3 is $i_2 - i_1$.

Exercise 2.17 Consider the circuit shown in Figure 2.33(b). In terms of the mesh currents, find the current in **a.** R_2 referenced upward; **b.** R_4 referenced to the right; **c.** R_8 referenced downward; **d.** R_8 referenced upward.

Answer **a.** $i_4 - i_1$; **b.** $i_2 - i_1$; **c.** $i_3 - i_4$; **d.** $i_4 - i_3$. [Notice that the answer for part (d) is the negative of the answer for part (c).] □

Writing Equations to Solve for Mesh Currents

If a network contains only resistances and independent voltage sources, we can write the required equations by following each current around its mesh and applying KVL. (We do not need to apply KCL because the mesh currents flow out of each node that they flow into.)

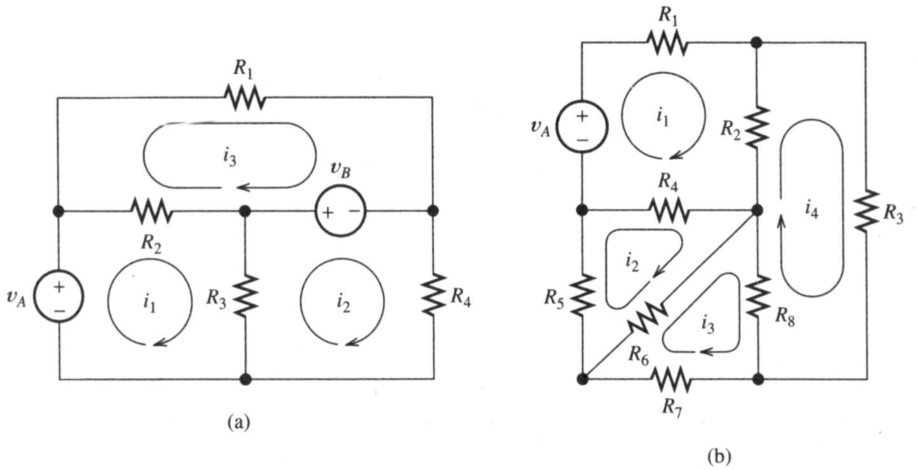

Figure 2.33 Two circuits and their mesh-current variables.

Example 2.12 Mesh-Current Analysis

Write the equations needed to solve for the mesh currents in Figure 2.33(a).

Solution Using a pattern in solving networks by the mesh-current method helps to avoid errors. Part of the pattern that we use is to select the mesh currents to flow clockwise. Then, we write a KVL equation for each mesh, going around the meshes clockwise. As usual, we add a voltage if its positive reference is encountered first in traveling around the mesh, and we subtract the voltage if the negative reference is encountered first. Our pattern is always to take the first end of each resistor encountered as the positive reference for its voltage. Thus, we are always adding the resistor voltages.

If a network contains only resistances and independent voltage sources, we can write the required equations by following each current around its mesh and applying KVL.

For example, in mesh 1 of Figure 2.33(a), we first encounter the left-hand end of R_2. The voltage across R_2 referenced positive on its left-hand end is $R_2(i_1 - i_3)$. Similarly, we encounter the top end of R_3 first, and the voltage across R_3 referenced positive at the top end is $R_3(i_1 - i_2)$. By using this pattern, we add a term for each resistor in the KVL equation, consisting of the resistance times the current in the mesh under consideration minus the current in the adjacent mesh (if any). Using this pattern for mesh 1 of Figure 2.33(a), we have

$$R_2(i_1 - i_3) + R_3(i_1 - i_2) - v_A = 0$$

Similarly, for mesh 2, we obtain

$$R_3(i_2 - i_1) + R_4 i_2 + v_B = 0$$

Finally, for mesh 3, we have

$$R_2(i_3 - i_1) + R_1 i_3 - v_B = 0$$

Notice that we have taken the positive reference for the voltage across R_3 at the top in writing the equation for mesh 1 and at the bottom for mesh 3. This is not an error because the terms for R_3 in the two equations are opposite in sign.

In standard form, the equations become:

$$(R_2 + R_3)i_1 - R_3 i_2 - R_2 i_3 = v_A$$
$$-R_3 i_1 + (R_3 + R_4)i_2 = -v_B$$
$$-R_2 i_1 + (R_1 + R_2)i_3 = v_B$$

In matrix form, we have

$$\begin{bmatrix} (R_2+R_3) & -R_3 & -R_2 \\ -R_3 & (R_3+R_4) & 0 \\ -R_2 & 0 & (R_1+R_2) \end{bmatrix} \begin{bmatrix} i_1 \\ i_2 \\ i_3 \end{bmatrix} = \begin{bmatrix} v_A \\ -v_B \\ v_B \end{bmatrix}$$

Often, we use **R** to represent the coefficient matrix, **I** to represent the column vector of mesh currents, and **V** to represent the column vector of the terms on the right-hand sides of the equations in standard form. Then, the mesh-current equations are represented as:

$$\mathbf{RI} = \mathbf{V}$$

We refer to the element of the *i*th row and *j*th column of **R** as r_{ij}. ■

Exercise 2.18 Write the equations for the mesh currents in Figure 2.32(b) and put them into matrix form.

Answer Following each mesh current in turn, we obtain

$$R_1 i_1 + R_2(i_1 - i_4) + R_4(i_1 - i_2) - v_A = 0$$
$$R_5 i_2 + R_4(i_2 - i_1) + R_6(i_2 - i_3) = 0$$
$$R_7 i_3 + R_6(i_3 - i_2) + R_8(i_3 - i_4) = 0$$
$$R_3 i_4 + R_2(i_4 - i_1) + R_8(i_4 - i_3) = 0$$

$$\begin{bmatrix} (R_1+R_2+R_4) & -R_4 & 0 & -R_2 \\ -R_4 & (R_4+R_5+R_6) & -R_6 & 0 \\ 0 & -R_6 & (R_6+R_7+R_8) & -R_8 \\ -R_2 & 0 & -R_8 & (R_2+R_3+R_8) \end{bmatrix} \begin{bmatrix} i_1 \\ i_2 \\ i_3 \\ i_4 \end{bmatrix} = \begin{bmatrix} v_A \\ 0 \\ 0 \\ 0 \end{bmatrix} \quad (2.55)$$

□

Solving Mesh Equations

After we write the mesh-current equations, we can solve them by using the methods that we discussed in Section 2.4 for the node-voltage approach. We illustrate with a simple example.

Example 2.13 Mesh-Current Analysis

Solve for the current in each element of the circuit shown in Figure 2.34.

Solution First, we select the mesh currents. Following our standard pattern, we define the mesh currents to flow clockwise around each mesh of the circuit. Then, we write a KVL equation around mesh 1:

$$20(i_1 - i_3) + 10(i_1 - i_2) - 70 = 0 \quad (2.56)$$

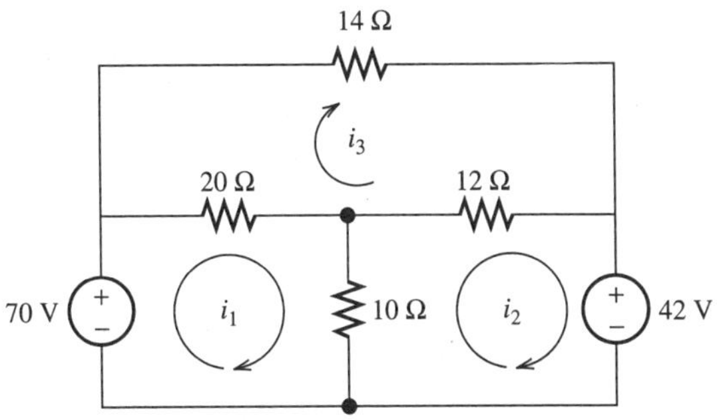

Figure 2.34 Circuit of Example 2.13.

For meshes 2 and 3, we have:

$$10(i_2 - i_1) + 12(i_2 - i_3) + 42 = 0 \tag{2.57}$$

$$20(i_3 - i_1) + 14i_3 + 12(i_3 - i_2) = 0 \tag{2.58}$$

Putting the equations into standard form, we have:

$$30i_1 - 10i_2 - 20i_3 = 70 \tag{2.59}$$

$$-10i_1 + 22i_2 - 12i_3 = -42 \tag{2.60}$$

$$-20i_1 - 12i_2 + 46i_3 = 0 \tag{2.61}$$

In matrix form, the equations become:

$$\begin{bmatrix} 30 & -10 & -20 \\ -10 & 22 & -12 \\ -20 & -12 & 46 \end{bmatrix} \begin{bmatrix} i_1 \\ i_2 \\ i_3 \end{bmatrix} = \begin{bmatrix} 70 \\ -42 \\ 0 \end{bmatrix}$$

These equations can be solved in a variety of ways. We will demonstrate using MATLAB. (The same results can be obtained by using these same commands in LabVIEW MathScript.) We use **R** for the coefficient matrix, because the coefficients often are resistances. Similarly, we use **V** for the column vector for the right-hand side of the equations and **I** for the column vector of the mesh currents. The commands and results are:

```
>> R = [30 -10 -20; -10 22 -12; -20 -12 46];
>> V = [70; -42; 0];
>> I = R\V % Try to avoid using i, which represents the square root of
>> % -1 in MATLAB.
I =
    4.0000
    1.0000
    2.0000
```

Thus, the values of the mesh currents are $i_1 = 4\,\text{A}$, $i_2 = 1\,\text{A}$, and $i_3 = 2\,\text{A}$. Next, we can find the current in any element. For example, the current flowing downward in the 10-Ω resistance is $i_1 - i_2 = 3\,\text{A}$. ■

Exercise 2.19 Use mesh currents to solve for the current flowing through the 10-Ω resistance in Figure 2.35. Check your answer by combining resistances in series and parallel to solve the circuit. Check a second time by using node voltages.

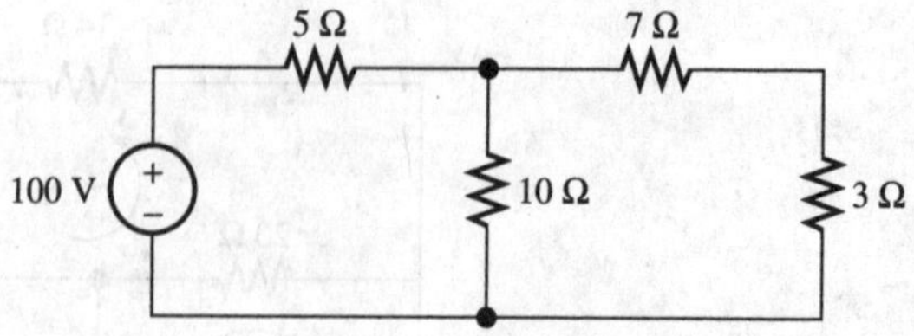

Figure 2.35 Circuit of Exercise 2.19.

Answer The current through the 10-Ω resistance is 5 A. □

Exercise 2.20 Use mesh currents to solve for the current flowing through the 2-Ω resistance in Figure 2.24 on page 90.

Answer The current is 1.613 A directed toward the right. □

Writing Mesh Equations Directly in Matrix Form

If a circuit contains only resistances and independent voltage sources, and if we select the mesh currents flowing clockwise, the mesh equations can be obtained directly in matrix form using these steps:

This is a shortcut way to write the mesh equations in matrix form, *provided that the circuit contains only resistances and independent voltage sources.*

1. Make sure that the circuit contains only resistances and independent voltage sources. Select all of the mesh currents to flow in the clockwise direction.

2. Write the sum of the resistances contained in each mesh as the corresponding element on the main diagonal of **R**. In other words, r_{jj} equals the sum of the resistances encountered in going around mesh j.

3. Insert the negatives of the resistances common to the corresponding meshes as the off diagonal terms of **R**. Thus, for $i \neq j$, the elements r_{ij} and r_{ji} are the same and are equal to negative of the sum of the resistances common to meshes i and j.

4. For each element of the **V** matrix, go around the corresponding mesh clockwise, *subtracting* the values of voltage sources for which we encounter the positive reference first and *adding* the values of voltage sources for which we encounter the negative reference first. (We have reversed the rules for adding or subtracting the voltage source values from what we used when writing KVL equations because the elements of **V** correspond to terms on the opposite side of the KVL equations.)

Keep in mind that this procedure does not apply to circuits having current sources or controlled sources.

Example 2.14 Writing Mesh Equations Directly in Matrix Form

Write the mesh equations directly in matrix form for the circuit of Figure 2.36.

Solution The matrix equation is:

$$\begin{bmatrix} (R_2+R_4+R_5) & -R_2 & -R_5 \\ -R_2 & (R_1+R_2+R_3) & -R_3 \\ -R_5 & -R_3 & (R_3+R_5+R_6) \end{bmatrix} \begin{bmatrix} i_1 \\ i_2 \\ i_3 \end{bmatrix} = \begin{bmatrix} -v_A+v_B \\ v_A \\ -v_B \end{bmatrix}$$

Notice that mesh 1 includes R_2, R_4, and R_5, so the r_{11} element of **R** is the sum of these resistances. Similarly, mesh 2 contains R_1, R_2, and R_3, so r_{22} is the sum of these resistances. Because R_2 is common to meshes 1 and 2, we have $r_{12} = r_{21} = -R_2$. Similar observations can be made for the other elements of **R**.

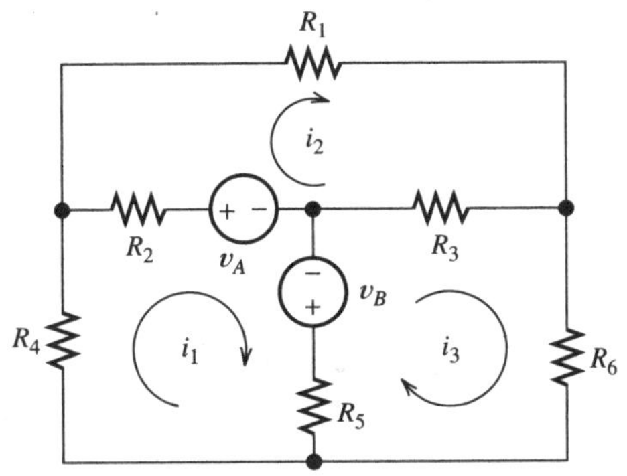

Figure 2.36 Circuit of Example 2.14.

As we go around mesh 1 clockwise, we encounter the positive reference for v_A first and the negative reference for v_B first, so we have $v_1 = -v_A + v_B$, and so forth. ■

Exercise 2.21 Examine the circuit of Figure 2.33(a) on page 99, and write its mesh equations directly in matrix form.

Answer

$$\begin{bmatrix} (R_2+R_3) & -R_3 & -R_2 \\ -R_3 & (R_3+R_4) & 0 \\ -R_2 & 0 & (R_1+R_2) \end{bmatrix} \begin{bmatrix} i_1 \\ i_2 \\ i_3 \end{bmatrix} = \begin{bmatrix} v_A \\ -v_B \\ v_B \end{bmatrix}$$ □

Mesh Currents in Circuits Containing Current Sources

Recall that a current source forces a specified current to flow through its terminals, but the voltage across its terminals is not predetermined. Instead, the voltage across a current source depends on the circuit to which the source is connected. Often, it is not easy to write an expression for the voltage across a current source. *A common mistake made by beginning students is to assume that the voltages across current sources are zero.*

A common mistake made by beginning students is to assume that the voltages across current sources are zero.

Consequently, when a circuit contains a current source, we must depart from the pattern that we use for circuits consisting of voltage sources and resistances. First, consider the circuit of Figure 2.37. As usual, we have defined the mesh currents flowing clockwise. If we were to try to write a KVL equation for mesh 1, we would need to include an unknown for the voltage across the current source. Because we do not wish to increase the number of unknowns in our equations, we avoid writing KVL equations for loops that include current sources. In the circuit in Figure 2.37, we have defined the current in the current source as i_1. However, we know that this

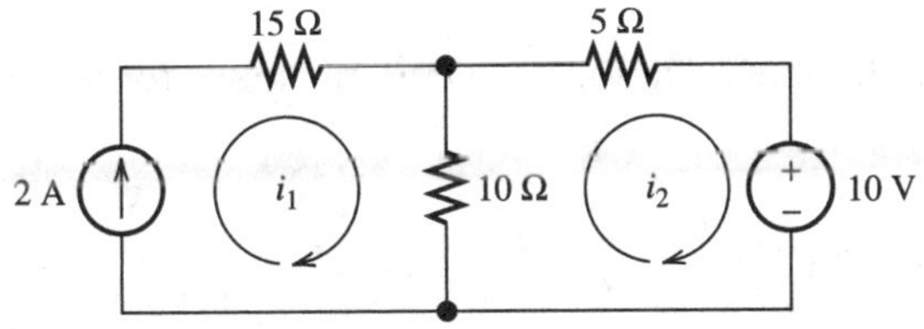

Figure 2.37 In this circuit, we have $i_1 = 2$ A.

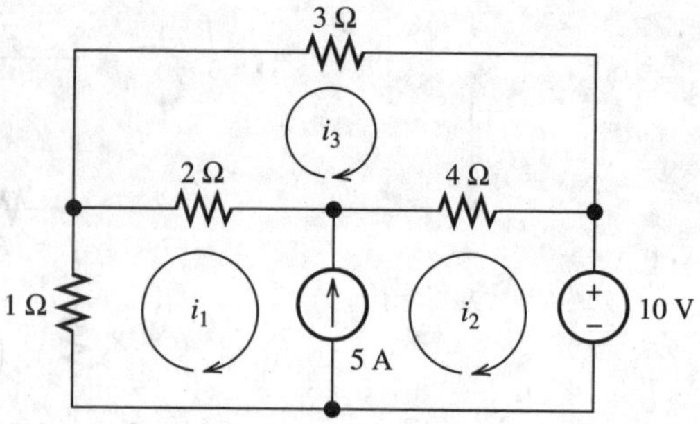

Figure 2.38 A circuit with a current source common to two meshes.

current is 2 A. Thus, we can write

$$i_1 = 2 \text{ A} \tag{2.62}$$

The second equation needed can be obtained by applying KVL to mesh 2, which yields

$$10(i_2 - i_1) + 5i_2 + 10 = 0 \tag{2.63}$$

Equations 2.62 and 2.63 can readily be solved for i_2. Notice that in this case the presence of a current source facilitates the solution.

Now let us consider the somewhat more complex situation shown in Figure 2.38. As usual, we have defined the mesh currents flowing clockwise. We cannot write a KVL equation around mesh 1 because the voltage across the 5-A current source is unknown (and we do not want to increase the number of unknowns in our equations). A solution is to combine meshes 1 and 2 into a **supermesh**. In other words, we write a KVL equation around the periphery of meshes 1 and 2 combined. This yields

$$i_1 + 2(i_1 - i_3) + 4(i_2 - i_3) + 10 = 0 \tag{2.64}$$

Next, we can write a KVL equation for mesh 3:

$$3i_3 + 4(i_3 - i_2) + 2(i_3 - i_1) = 0 \tag{2.65}$$

Finally, we recognize that we have defined the current in the current source referenced upward as $i_2 - i_1$. However, we know that the current flowing upward through the current source is 5 A. Thus, we have

$$i_2 - i_1 = 5 \tag{2.66}$$

It is important to realize that Equation 2.66 is not a KCL equation. Instead, it simply states that we have defined the current referenced upward through the current source in terms of the mesh currents as $i_2 - i_1$, but this current is known to be 5 A. Equations 2.64, 2.65, and 2.66 can be solved for the mesh currents.

It is important to realize that Equation 2.66 is not a KCL equation.

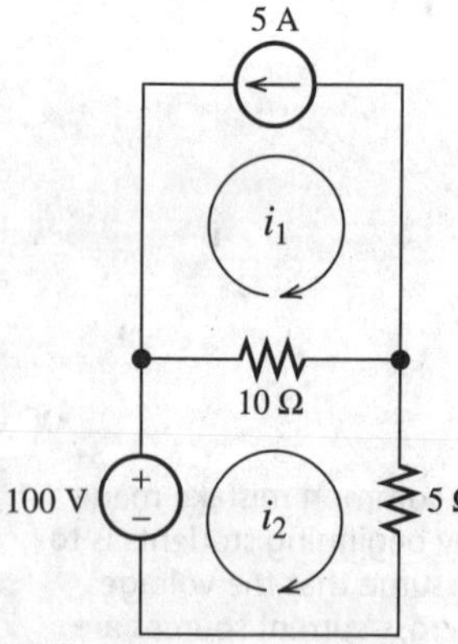

Figure 2.39 The circuit for Exercise 2.22.

Exercise 2.22 Write the equations needed to solve for the mesh currents in Figure 2.39.

Answer

$$i_1 = -5 \text{ A}$$

$$10(i_2 - i_1) + 5i_2 - 100 = 0$$

□

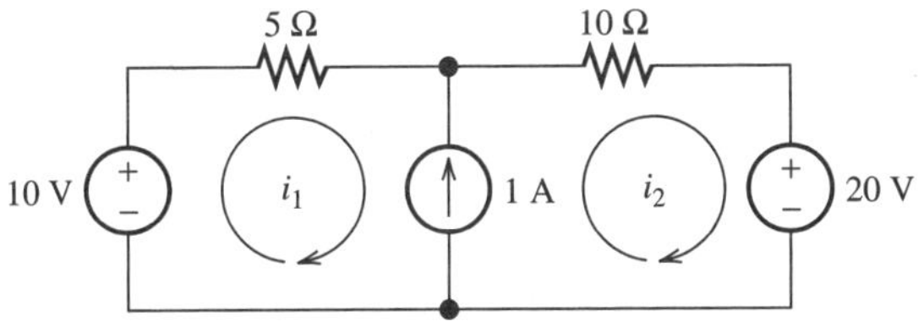

Figure 2.40 The circuit for Exercise 2.23.

Exercise 2.23 Write the equations needed to solve for the mesh currents in Figure 2.40. Then solve for the currents.

Answer The equations are $i_2 - i_1 = 1$ and $5i_1 + 10i_2 + 20 - 10 = 0$. Solving, we have $i_1 = -4/3$ A and $i_2 = -1/3$ A. □

Circuits with Controlled Sources

Controlled sources present a slight additional complication to the mesh-current technique. First, we write equations exactly as we have done for networks with independent sources. Then, we express the controlling variables in terms of the mesh-current variables and substitute into the network equations. We illustrate with an example.

Example 2.15 Mesh-Current Analysis with Controlled Sources

Solve for the currents in the circuit of Figure 2.41(a), which contains a voltage-controlled current source common to the two meshes.

Solution First, we write equations for the mesh currents as we have done for independent sources. Since there is a current source common to mesh 1 and mesh 2, we start by combining the meshes to form a supermesh and write a voltage equation:

$$-20 + 4i_1 + 6i_2 + 2i_2 = 0 \tag{2.67}$$

Then, we write an expression for the source current in terms of the mesh currents:

$$av_x = 0.25v_x = i_2 - i_1 \tag{2.68}$$

Next, we see that the controlling voltage is

$$v_x = 2i_2 \tag{2.69}$$

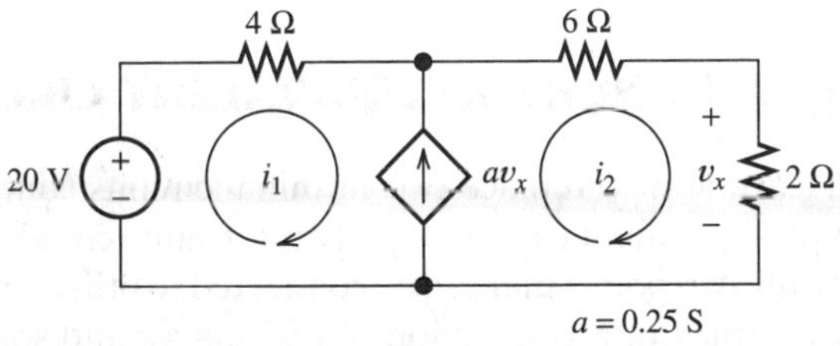

Figure 2.41 A circuit with a voltage-controlled current source. See Example 2.15.

Using Equation 2.58 to substitute for v_x in Equation 2.57, we have

$$\frac{i_2}{2} = i_2 - i_1 \tag{2.70}$$

Finally, we put Equations 2.67 and 2.70 into standard form, resulting in

$$4i_1 + 8i_2 = 20 \tag{2.71}$$

$$i_1 - \frac{i_2}{2} = 0 \tag{2.72}$$

Solving these equations yields $i_1 = 1$ A and $i_2 = 2$ A. ■

Using the principles we have discussed in this section, we can write mesh-current equations for any planar network consisting of sources and resistances.

Next, we summarize the steps in analyzing planar circuits by the mesh-current technique:

Here is a convenient step-by-step guide to mesh-current analysis.

1. If necessary, redraw the network without crossing conductors or elements. Then, define the mesh currents flowing around each of the open areas defined by the network. For consistency, we usually select a clockwise direction for each of the mesh currents, but this is not a requirement.
2. Write network equations, stopping after the number of equations is equal to the number of mesh currents. First, use KVL to write voltage equations for meshes that do not contain current sources. Next, if any current sources are present, write expressions for their currents in terms of the mesh currents. Finally, if a current source is common to two meshes, write a KVL equation for the supermesh.
3. If the circuit contains dependent sources, find expressions for the controlling variables in terms of the mesh currents. Substitute into the network equations, and obtain equations having only the mesh currents as unknowns.
4. Put the equations into standard form. Solve for the mesh currents by use of determinants or other means.
5. Use the values found for the mesh currents to calculate any other currents or voltages of interest.

Exercise 2.24 Use the mesh-current technique to solve for the currents labeled in the circuits shown in Figure 2.30 on page 95.

Answer **a.** $i_a = 1.33$ A; **b.** $i_b = -0.259$ A. □

Exercise 2.25 Use the mesh-current technique to solve for the values of i_x and i_y in Figure 2.31 on page 95.

Answer $i_x = 0.5$ A, $i_y = 2.31$ A. □

2.6 THÉVENIN AND NORTON EQUIVALENT CIRCUITS

In this section, we learn how to replace two-terminal circuits containing resistances and sources by simple equivalent circuits. By a two-terminal circuit, we mean that the original circuit has only two points that can be connected to other circuits. The original circuit can be any complex interconnection of resistances and sources. However, a restriction is that the controlling variables for any controlled sources must appear inside the original circuit.

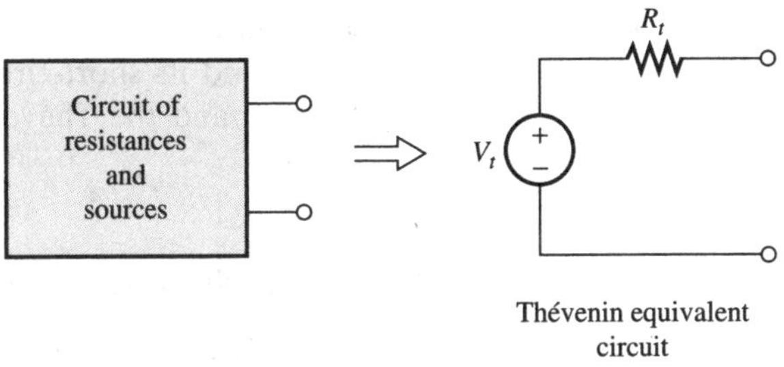

Figure 2.42 A two-terminal circuit consisting of resistances and sources can be replaced by a Thévenin equivalent circuit.

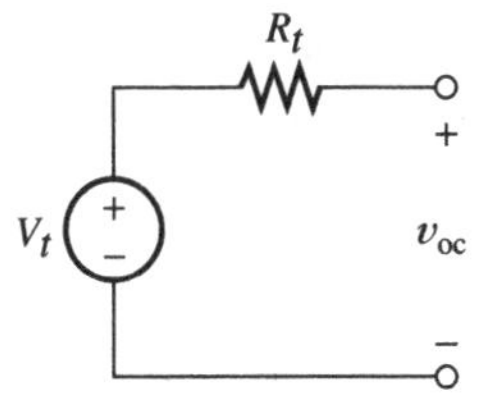

Figure 2.43 Thévenin equivalent circuit with open-circuited terminals. The open-circuit voltage v_{oc} is equal to the Thévenin voltage V_t.

Thévenin Equivalent Circuits

One type of equivalent circuit is the **Thévenin equivalent**, which consists of an independent voltage source in series with a resistance. This is illustrated in Figure 2.42.

The Thévenin equivalent circuit consists of an independent voltage source in series with a resistance.

Consider the Thévenin equivalent with open-circuited terminals as shown in Figure 2.43. By definition, no current can flow through an open circuit. Therefore, no current flows through the Thévenin resistance, and the voltage across the resistance is zero. Applying KVL, we conclude that

$$V_t = v_{oc}$$

Both the original circuit and the equivalent circuit are required to have the same open-circuit voltage. *Thus, the Thévenin source voltage V_t is equal to the open-circuit voltage of the original network.*

Now, consider the Thévenin equivalent with a short circuit connected across its terminals as shown in Figure 2.44. The current flowing in this circuit is

$$i_{sc} = \frac{V_t}{R_t}$$

The Thévenin voltage v_t is equal to the open-circuit voltage of the original network.

The short-circuit current i_{sc} is the same for the original circuit as for the Thévenin equivalent. Solving for the Thévenin resistance, we have

$$R_t = \frac{V_t}{i_{sc}} \tag{2.73}$$

Using the fact that the Thévenin voltage is equal to the open-circuit voltage of the network, we have

$$R_t = \frac{v_{oc}}{i_{sc}} \tag{2.74}$$

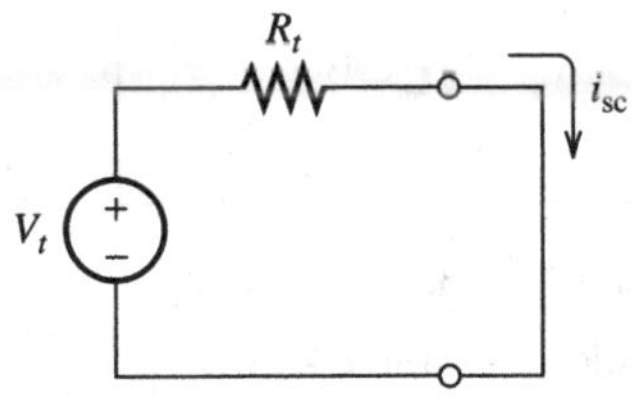

Figure 2.44 Thévenin equivalent circuit with short-circuited terminals. The short-circuit current is $i_{sc} = V_t/R_t$.

The Thévenin resistance is equal to the open-circuit voltage divided by the short-circuit current.

Thus, to determine the Thévenin equivalent circuit, we can start by analyzing the original network for its open-circuit voltage and its short-circuit current. The Thévenin voltage equals the open-circuit voltage, and the Thévenin resistance is given by Equation 2.74.

Example 2.16 Determining the Thévenin Equivalent Circuit

Find the Thévenin equivalent for the circuit shown in Figure 2.45(a).

Solution First, we analyze the circuit with open-circuited terminals. This is shown in Figure 2.45(b). The resistances R_1 and R_2 are in series and have an equivalent resistance of $R_1 + R_2$. Therefore, the current circulating is

$$i_1 = \frac{v_s}{R_1 + R_2} = \frac{15}{100 + 50} = 0.10\text{ A}$$

The open-circuit voltage is the voltage across R_2:

$$v_{oc} = R_2 i_1 = 50 \times 0.10 = 5\text{ V}$$

Thus, the Thévenin voltage is $V_t = 5$ V.

Now, we consider the circuit with a short circuit connected across its terminals as shown in Figure 2.45(c). By definition, the voltage across a short circuit is zero. Hence, the voltage across R_2 is zero, and the current through it is zero, as shown in the figure. Therefore, the short-circuit current i_{sc} flows through R_1. The source voltage v_s appears across R_1, so we can write

$$i_{sc} = \frac{v_s}{R_1} = \frac{15}{100} = 0.15\text{ A}$$

Now, we can use Equation 2.74 to determine the Thévenin resistance:

$$R_t = \frac{v_{oc}}{i_{sc}} = \frac{5\text{ V}}{0.15\text{ A}} = 33.3\ \Omega$$

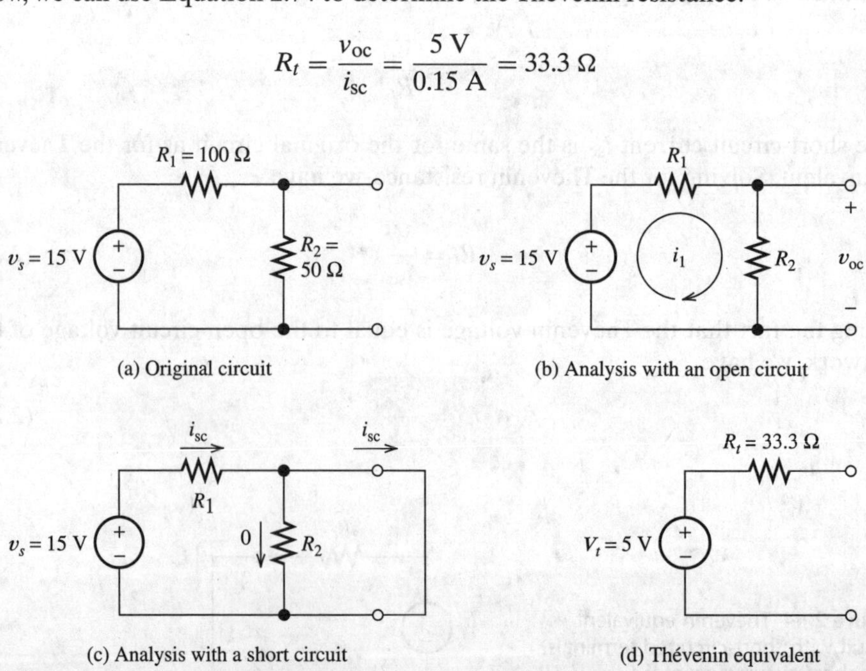

Figure 2.45 Circuit for Example 2.16.

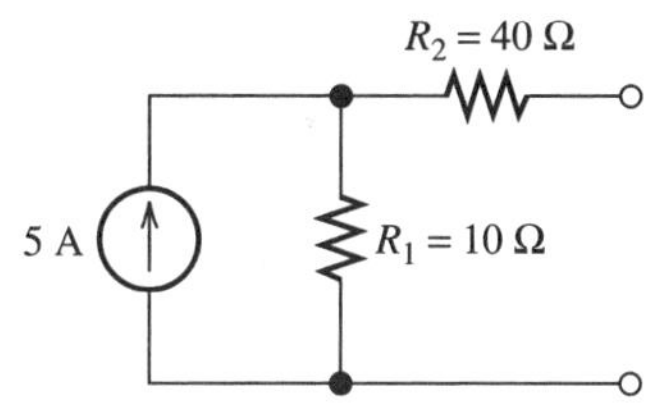

Figure 2.46 Circuit for Exercise 2.26.

The Thévenin equivalent circuit is shown in Figure 2.45(d). ■

Exercise 2.26 Find the Thévenin equivalent circuit for the circuit shown in Figure 2.46.
Answer $V_t = 50$ V, $R_t = 50\ \Omega$. □

Finding the Thévenin Resistance Directly. If a network contains no dependent sources, there is an alternative way to find the Thévenin resistance. First, we *zero* the sources in the network. In zeroing a voltage source, we reduce its voltage to zero. A voltage source with zero voltage is equivalent to a short circuit.

In zeroing a current source, we reduce its current to zero. By definition, an element that always carries zero current is an open circuit. *Thus, to zero the independent sources, we replace voltage sources with short circuits and replace current sources with open circuits.*

When zeroing a current source, it becomes an open circuit. When zeroing a voltage source, it becomes a short circuit.

Figure 2.47 shows a Thévenin equivalent before and after zeroing its voltage source. Looking back into the terminals after the source is zeroed, we see the Thévenin resistance. *Thus, we can find the Thévenin resistance by zeroing the sources in the original network and then computing the resistance between the terminals.*

We can find the Thévenin resistance by zeroing the sources in the original network and then computing the resistance between the terminals.

Example 2.17 Zeroing Sources to Find Thévenin Resistance

Find the Thévenin resistance for the circuit shown in Figure 2.48(a) by zeroing the sources. Then, find the short-circuit current and the Thévenin equivalent circuit.

Solution To zero the sources, we replace the voltage source by a short circuit and replace the current source by an open circuit. The resulting circuit is shown in Figure 2.48(b).

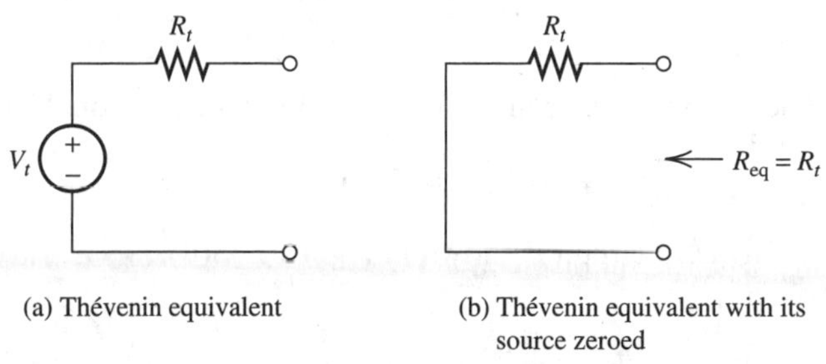

Figure 2.47 When the source is zeroed, the resistance seen from the circuit terminals is equal to the Thévenin resistance.

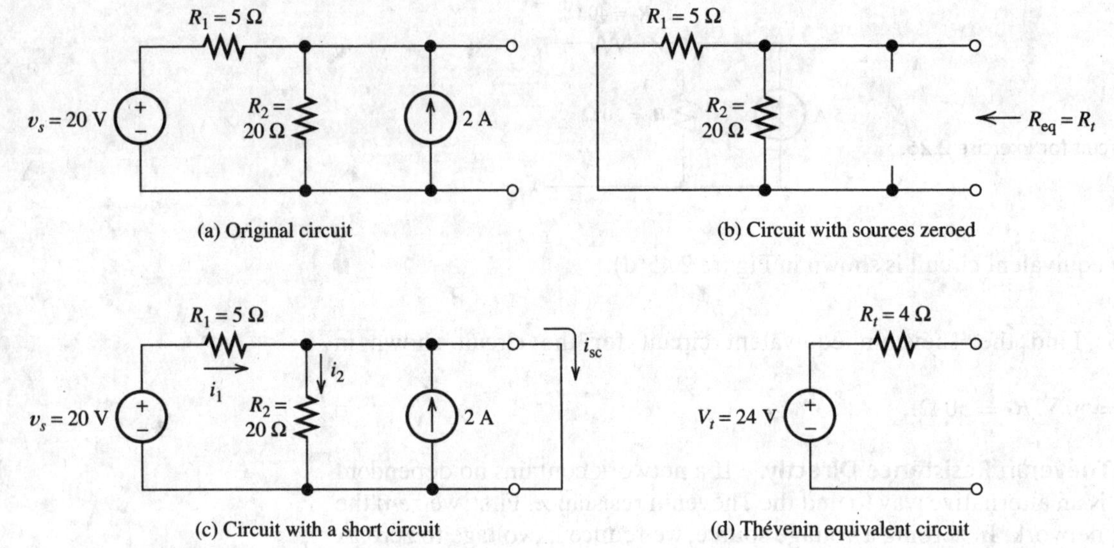

Figure 2.48 Circuit for Example 2.17.

The Thévenin resistance is the equivalent resistance between the terminals. This is the parallel combination of R_1 and R_2, which is given by

$$R_t = R_{eq} = \frac{1}{1/R_1 + 1/R_2} = \frac{1}{1/5 + 1/20} = 4\,\Omega$$

Next, we find the short-circuit current for the circuit. The circuit is shown in Figure 2.48(c). In this circuit, the voltage across R_2 is zero because of the short circuit. Thus, the current through R_2 is zero:

$$i_2 = 0$$

Furthermore, the voltage across R_1 is equal to 20 V. Thus, the current is

$$i_1 = \frac{v_s}{R_1} = \frac{20}{5} = 4\text{ A}$$

Finally, we write a current equation for the node joining the top ends of R_2 and the 2-A source. Setting the sum of the currents entering equal to the sum of the currents leaving, we have

$$i_1 + 2 = i_2 + i_{sc}$$

This yields $i_{sc} = 6$ A.

Now, the Thévenin voltage can be found. Applying Equation 2.74, we get

$$V_t = R_t i_{sc} = 4 \times 6 = 24\text{ V}$$

The Thévenin equivalent circuit is shown in Figure 2.48(d). ■

Exercise 2.27 Use node-voltage analysis of the circuit shown in Figure 2.48(a) to show that the open-circuit voltage is equal to the Thévenin voltage found in Example 2.17. □

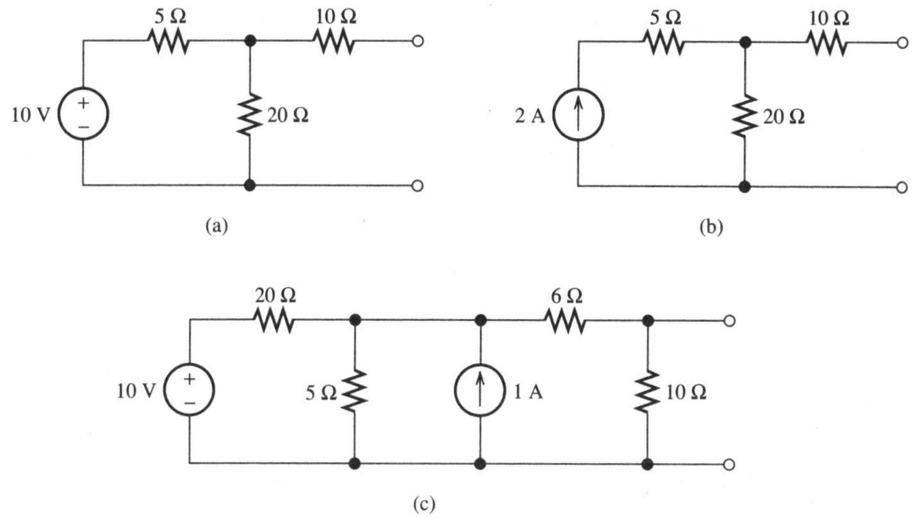

Figure 2.49 Circuits for Exercise 2.28.

Exercise 2.28 Find the Thévenin resistance for each of the circuits shown in Figure 2.49 by zeroing the sources.

Answer **a.** $R_t = 14\ \Omega$; **b.** $R_t = 30\ \Omega$; **c.** $R_t = 5\ \Omega$. □

We complete our discussion of Thévenin equivalent circuits with one more example.

Example 2.18 Thévenin Equivalent of a Circuit with a Dependent Source

Find the Thévenin equivalent for the circuit shown in Figure 2.50(a).

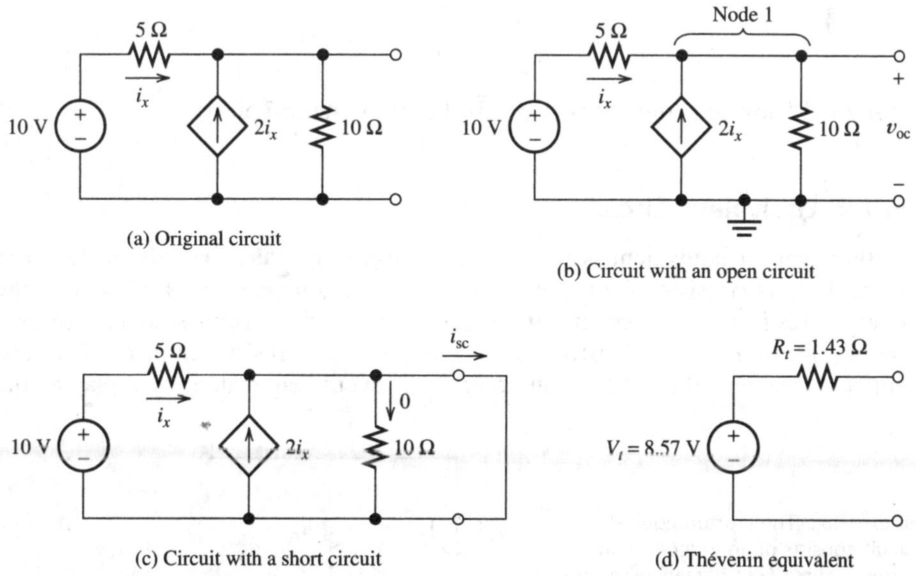

Figure 2.50 Circuit for Example 2.18.

Because this circuit contains a dependent source, we cannot find the Thévenin resistance by zeroing the sources and combining resistances in series and parallel.

Solution Because this circuit contains a dependent source, we cannot find the Thévenin resistance by zeroing the sources and combining resistances in series and parallel. Thus, we must analyze the circuit to find the open-circuit voltage and the short-circuit current.

We start with the open-circuit voltage. Consider Figure 2.50(b). We use node-voltage analysis, picking the reference node at the bottom of the circuit. Then, v_{oc} is the unknown node-voltage variable. First, we write a current equation at node 1.

$$i_x + 2i_x = \frac{v_{oc}}{10} \tag{2.75}$$

Next, we write an expression for the controlling variable i_x in terms of the node voltage v_{oc}:

$$i_x = \frac{10 - v_{oc}}{5}$$

Substituting this into Equation 2.75, we have

$$3\frac{10 - v_{oc}}{5} = \frac{v_{oc}}{10}$$

Solving, we find that $v_{oc} = 8.57$ V.

Now, we consider short-circuit conditions as shown in Figure 2.50(c). In this case, the current through the 10-Ω resistance is zero. Furthermore, we get

$$i_x = \frac{10 \text{ V}}{5 \text{ }\Omega} = 2 \text{ A}$$

and

$$i_{sc} = 3i_x = 6 \text{ A}$$

Next, we use Equation 2.74 to compute the Thévenin resistance:

$$R_t = \frac{v_{oc}}{i_{sc}} = \frac{8.57 \text{ V}}{6 \text{ A}} = 1.43 \text{ }\Omega$$

Finally, the Thévenin equivalent circuit is shown in Figure 2.50(d). ■

Norton Equivalent Circuit

Another type of equivalent, known as the **Norton equivalent circuit**, is shown in Figure 2.51. It consists of an independent current source I_n in parallel with the Thévenin resistance. Notice that if we zero the Norton current source, replacing it by an open circuit, the Norton equivalent becomes a resistance of R_t. This also happens if we zero the voltage source in the Thévenin equivalent by replacing the

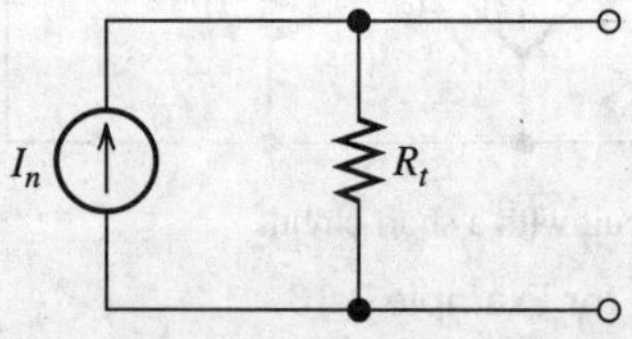

Figure 2.51 The Norton equivalent circuit consists of an independent current source I_n in parallel with the Thévenin resistance R_t.

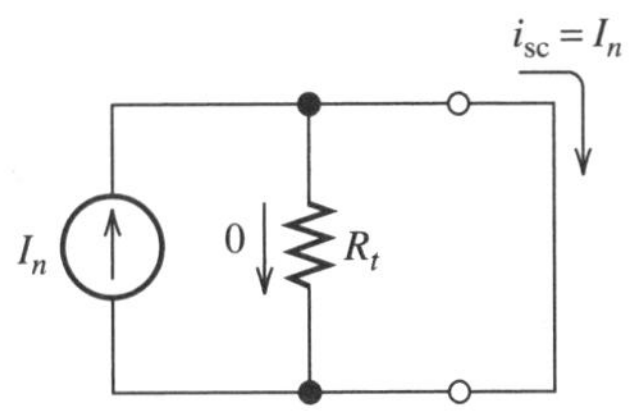

Figure 2.52 The Norton equivalent circuit with a short circuit across its terminals.

voltage source by a short circuit. Thus, the resistance in the Norton equivalent is the same as the Thévenin resistance.

Consider placing a short circuit across the Norton equivalent as shown in Figure 2.52. In this case, the current through R_t is zero. *Therefore, the Norton current is equal to the short-circuit current*:

$$I_n = i_{sc}$$

We can find the Norton equivalent by using the same techniques as we used for the Thévenin equivalent.

Step-by-Step Thévenin/Norton-Equivalent-Circuit Analysis

1. Perform two of these:
  a. Determine the open-circuit voltage $V_t = v_{oc}$.
  b. Determine the short-circuit current $I_n = i_{sc}$.
  c. Zero the independent sources and find the Thévenin resistance R_t looking back into the terminals. Do not zero dependent sources.

2. Use the equation $V_t = R_t I_n$ to compute the remaining value.

3. The Thévenin equivalent consists of a voltage source V_t in series with R_t.

4. The Norton equivalent consists of a current source I_n in parallel with R_t.

Example 2.19 Norton Equivalent Circuit

Find the Norton equivalent for the circuit shown in Figure 2.53(a).

Solution Because the circuit contains a controlled source, we cannot zero the sources and combine resistances to find the Thévenin resistance. First, we consider the circuit with an open circuit as shown in Figure 2.53(a). We treat v_{oc} as a node-voltage variable. Writing a current equation at the top of the circuit, we have

$$\frac{v_x}{4} + \frac{v_{oc} - 15}{R_1} + \frac{v_{oc}}{R_2 + R_3} = 0 \qquad (2.76)$$

Next, we use the voltage-divider principle to write an expression for v_x in terms of resistances and v_{oc}:

$$v_x = \frac{R_3}{R_2 + R_3} v_{oc} = 0.25 v_{oc}$$

Substituting into Equation 2.76, we find that

$$\frac{0.25 v_{oc}}{4} + \frac{v_{oc} - 15}{R_1} + \frac{v_{oc}}{R_2 + R_3} = 0$$

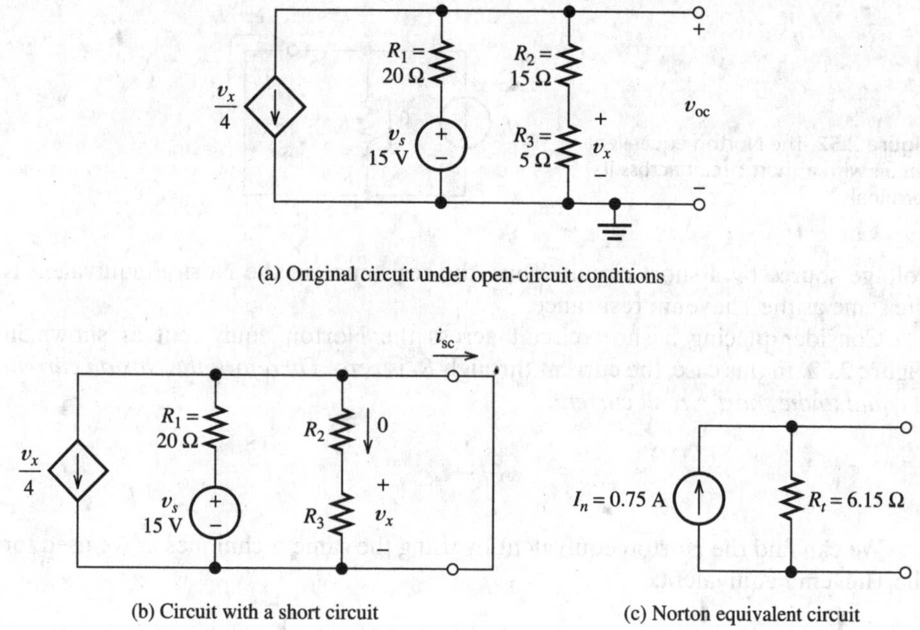

Figure 2.53 Circuit of Example 2.19.

Substituting resistance values and solving, we observe that $v_{oc} = 4.62$ V.

Next, we consider short-circuit conditions as shown in Figure 2.53(b). In this case, the current through R_2 and R_3 is zero. Thus, $v_x = 0$, and the controlled current source appears as an open circuit. The short-circuit current is given by

$$i_{sc} = \frac{v_s}{R_1} = \frac{15\text{ V}}{20\ \Omega} = 0.75\text{ A}$$

Now, we can find the Thévenin resistance:

$$R_t = \frac{v_{oc}}{i_{sc}} = \frac{4.62}{0.75} = 6.15\ \Omega$$

The Norton equivalent circuit is shown in Figure 2.53(c). ■

Exercise 2.29 Find the Norton equivalent for each of the circuits shown in Figure 2.54.

Answer **a.** $I_n = 1.67$ A, $R_t = 9.375$ Ω; **b.** $I_n = 2$ A, $R_t = 15$ Ω. □

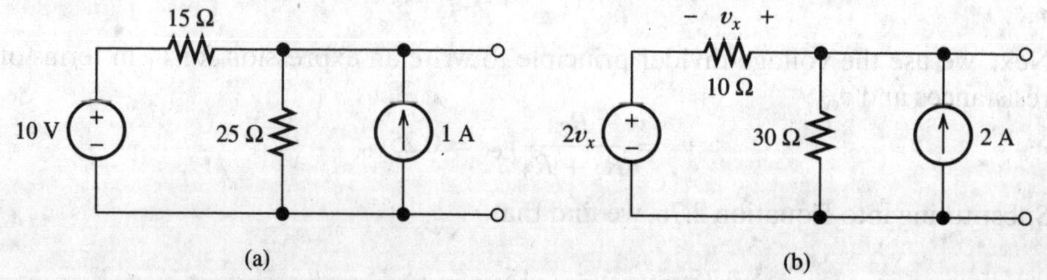

Figure 2.54 Circuits for Exercise 2.29.

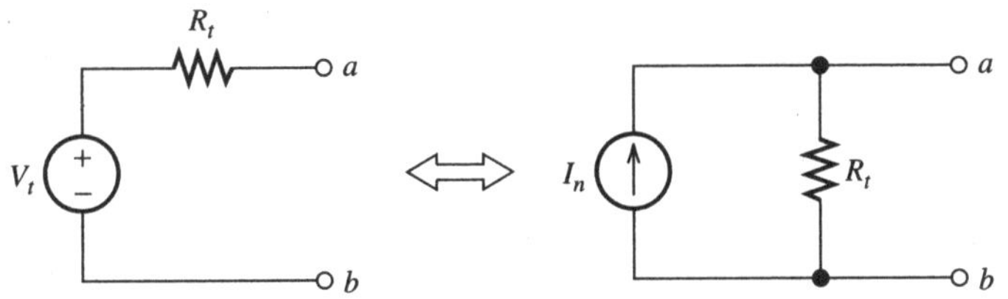

Figure 2.55 A voltage source in series with a resistance is externally equivalent to a current source in parallel with the resistance, provided that $I_n = V_t/R_t$.

Source Transformations

We can replace a voltage source in series with a resistance by a Norton equivalent circuit, which consists of a current source in parallel with the resistance. This is called a **source transformation** and is illustrated in Figure 2.55. The two circuits are identical in terms of their external behavior. In other words, the voltages and currents at terminals a and b remain the same after the transformation is made. However, in general, the current flowing through R_t is different for the two circuits. For example, suppose that the two circuits shown in Figure 2.55 are open circuited. Then no current flows through the resistor in series with the voltage source, but the current I_n flows through the resistance in parallel with the current source.

In making source transformations it is very important to maintain the proper relationship between the reference direction for the current source and the polarity of the voltage source. If the positive polarity is closest to terminal a, the current reference must point toward terminal a, as shown in Figure 2.55.

Sometimes, we can simplify the solution of a circuit by source transformations. This is similar to solving circuits by combining resistances in series or parallel. We illustrate with an example.

Here is a "trick" question that you might have some fun with: Suppose that the circuits of Figure 2.55 are placed in identical black boxes with the terminals accessible from outside the box. How could you determine which box contains the Norton equivalent? An answer can be found at the end of the chapter summary on the bottom of page 126.

Example 2.20 Using Source Transformations

Use source transformations to aid in solving for the currents i_1 and i_2 shown in Figure 2.56(a).

Solution Several approaches are possible. One is to transform the 1-A current source and R_2 into a voltage source in series with R_2. This is shown in Figure 2.56(b). Notice that the positive polarity of the 10-V source is at the top, because the 1-A source reference points upward. The single-loop circuit of Figure 2.56(b) can be solved by writing a KVL equation. Traveling clockwise and summing voltages, we have

$$R_1 i_1 + R_2 i_1 + 10 - 20 = 0$$

Solving and substituting values, we get

$$i_1 = \frac{10}{R_1 + R_2} = 0.667 \text{ A}$$

Then in the original circuit, we can write a current equation at the top node and solve for i_2:

$$i_2 = i_1 + 1 = 1.667 \text{ A}$$

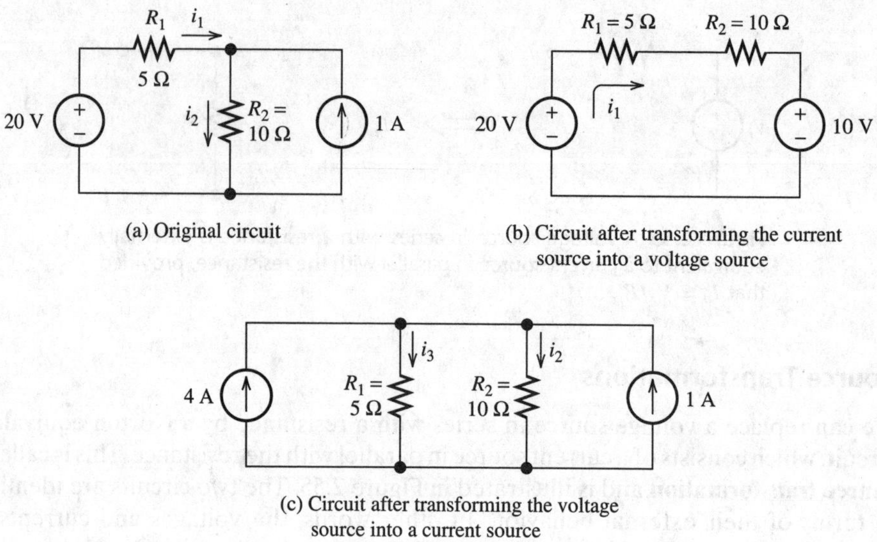

Figure 2.56 Circuit for Example 2.20.

Another approach is to transform the voltage source and R_1 into a current source in parallel with R_1. Making this change to the original circuit yields the circuit shown in Figure 2.56(c). Notice that we have labeled the current through R_1 as i_3 rather than i_1. This is because the current in the resistance of the transformed source is not the same as in the original circuit. Now, in Figure 2.56(c), we see that a total current of 5 A flows into the parallel combination of R_1 and R_2. Using the current-division principle, we find the current through R_2:

$$i_2 = \frac{R_1}{R_1 + R_2} i_{\text{total}} = \frac{5}{5 + 10}(5) = 1.667 \text{ A}$$

This agrees with our previous result. ■

Exercise 2.30 Use source transformations to solve for the values of i_1 and i_2 in Figure 2.57. First, transform the current source and R_1 into a voltage source in series with R_1. (Make sure in making the transformation that the polarity of the voltage source bears the correct relationship to the current reference direction.) Solve the circuit by a second approach. Starting with the original circuit, transform the 10-V source and R_2 into a current source in parallel with R_2. Of course, the answers should be the same for both approaches.

Answer $i_1 = -0.667$ A, $i_2 = 1.333$ A. □

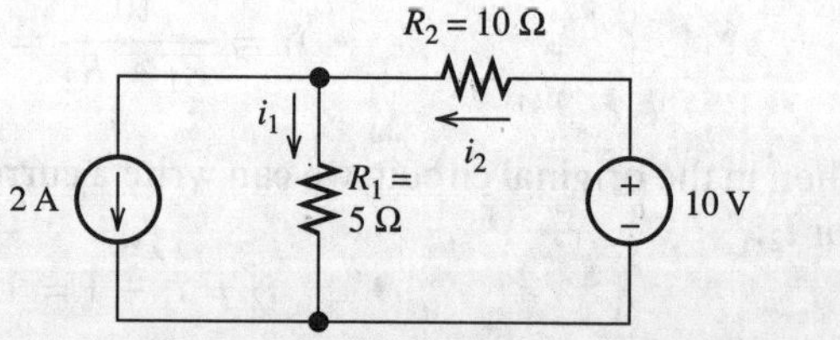

Figure 2.57 Circuit for Exercise 2.30.

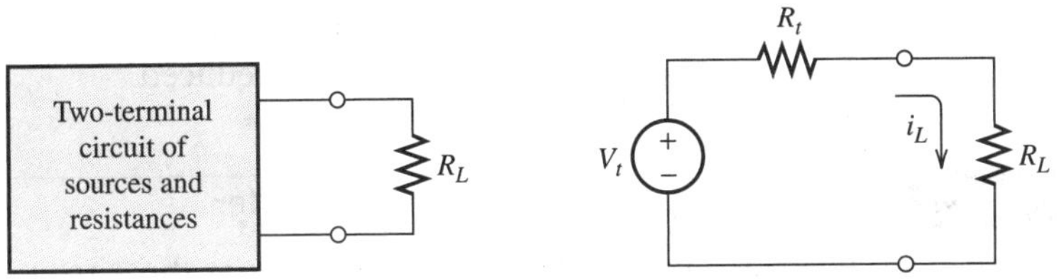

(a) Original circuit with load

(b) Thévenin equivalent circuit with load

Figure 2.58 Circuits for analysis of maximum power transfer.

Maximum Power Transfer

Suppose that we have a two-terminal circuit and we want to connect a load resistance R_L such that the maximum possible power is delivered to the load. This is illustrated in Figure 2.58(a). To analyze this problem, we replace the original circuit by its Thévenin equivalent as shown in Figure 2.58(b). The current flowing through the load resistance is given by

$$i_L = \frac{V_t}{R_t + R_L}$$

The power delivered to the load is

$$p_L = i_L^2 R_L$$

Substituting for the current, we have

$$p_L = \frac{V_t^2 R_L}{(R_t + R_L)^2} \qquad (2.77)$$

To find the value of the load resistance that maximizes the power delivered to the load, we set the derivative of p_L with respect to R_L equal to zero:

$$\frac{dp_L}{dR_L} = \frac{V_t^2(R_t + R_L)^2 - 2V_t^2 R_L (R_t + R_L)}{(R_t + R_L)^4} = 0$$

Solving for the load resistance, we have

$$R_L = R_t$$

Thus, the load resistance that absorbs the maximum power from a two-terminal circuit is equal to the Thévenin resistance. The maximum power is found by substituting $R_L = R_t$ into Equation 2.77. The result is

The load resistance that absorbs the maximum power from a two-terminal circuit is equal to the Thévenin resistance.

$$P_{L\,\max} = \frac{V_t^2}{4R_t} \qquad (2.78)$$

An All-Too-Common Example. You may have had difficulty in starting your car on a frigid morning. The battery in your car can be represented by a Thévenin equivalent circuit. It turns out that the Thévenin voltage of the battery does not change greatly with temperature. However, when the battery is very cold, the chemical reactions occur much more slowly and its Thévenin resistance is much higher. Thus, the power that the battery can deliver to the starter motor is greatly reduced.

Example 2.21 Determining Maximum Power Transfer

Find the load resistance for maximum power transfer from the circuit shown in Figure 2.59. Also, find the maximum power.

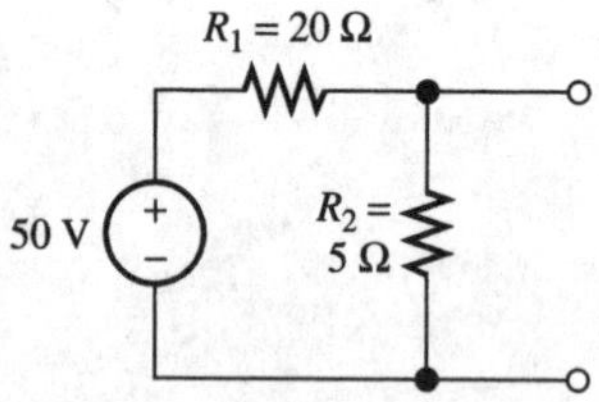

Figure 2.59 Circuit for Example 2.21.

Solution First, we must find the Thévenin equivalent circuit. Zeroing the voltage source, we find that the resistances R_1 and R_2 are in parallel. Thus, the Thévenin resistance is

$$R_t = \frac{1}{1/R_1 + 1/R_2} = \frac{1}{1/20 + 1/5} = 4\ \Omega$$

The Thévenin voltage is equal to the open-circuit voltage. Using the voltage-division principle, we find that

$$V_t = v_{oc} = \frac{R_2}{R_1 + R_2}(50) = \frac{5}{5 + 20}(50) = 10\ \text{V}$$

Hence, the load resistance that receives maximum power is

$$R_L = R_t = 4\ \Omega$$

and the maximum power is given by Equation 2.78:

$$P_{L\max} = \frac{V_t^2}{4R_t} = \frac{10^2}{4 \times 4} = 6.25\ \text{W}$$

■

PRACTICAL APPLICATION 2.1

An Important Engineering Problem: Energy-Storage Systems for Electric Vehicles

Imagine pollution-free electric vehicles with exciting performance and 500-mile range. They do not exist, but they are the target of an ongoing large-scale engineering effort to which you may contribute. Such electric vehicles (EVs) are a worthwhile goal because they can be very efficient in their use of energy, particularly in stop-and-go traffic. Kinetic energy can be recovered during braking and saved for later use during acceleration. Furthermore, EVs emit little pollution into crowded urban environments.

So far, EV range and performance remains less than ideal. The availability of suitable energy-storage devices is the key stumbling block in achieving better EVs (and a multitude of other highly desirable devices, such as laptop computers that do not need recharging for a week).

In Chapter 3, we will see that capacitors and inductors are capable of storing electrical energy. However, it turns out that their energy content per unit volume is too small to make them a practical solution for EVs. The energy content of modern rechargeable batteries is better but still not on a par with the energy content of gasoline, which is approximately 10,000 watt-hours/liter (Wh/L). In contrast, the energy content of nickel-metal hydride batteries used in current EVs is about 175 Wh/L. Lithium-ion batteries under current development are expected to increase this to about 300 Wh/L. Thus, even allowing for the relative inefficiency of the internal combustion engine in converting chemical energy to mechanical energy, much more usable energy can be obtained from gasoline than from current batteries of comparable volume.

Although EVs do not emit pollutants at the point of use, the mining, refining, and disposal of metals pose grave environmental dangers. We must always consider the entire environmental (as well as economic) impact of the systems we design. As an engineer, you can do a great service to humanity by accepting the challenge to develop safe, clean systems for storing energy in forms that are readily converted to and from electrical form.

Naturally, one possibility currently under intense development is improved electrochemical batteries based on nontoxic chemicals. Another option is a mechanical flywheel system that would be coupled through an electrical generator to electric drive motors. Still another solution is a hybrid vehicle that uses a small internal combustion engine, an electrical generator, an energy-storage system, and electrical drive motors. The engine achieves low pollution levels by being optimized to run at a constant load while charging a relatively small energy-storage system. When the storage capacity becomes full, the engine shuts down automatically and the vehicle runs on stored energy. The engine is just large enough to keep up with energy demands under high-speed highway conditions.

Whatever form the ultimate solution to vehicle pollution may take, we can anticipate that it will include elements from mechanical, chemical, manufacturing, and civil engineering in close combination with electrical-engineering principles.

Application of Maximum Power Transfer. When a load resistance equals the internal Thévenin resistance of the source, half of the power is dissipated in the source resistance and half is delivered to the load. In higher power applications for which efficiency is important, we do not usually design for maximum power transfer. For example, in designing an electric vehicle, we would want to deliver the energy stored in the batteries mainly to the drive motors and minimize the power loss in the resistance of the battery and wiring. This system would approach maximum power transfer rarely when maximum acceleration is needed.

On the other hand, when small amounts of power are involved, we would design for maximum power transfer. For example, we would design a radio receiver to extract the maximum signal power from the receiving antenna. In this application, the power is very small, typically much less than one microwatt, and efficiency is not a consideration.

2.7 SUPERPOSITION PRINCIPLE

Suppose that we have a circuit composed of resistances, linear dependent sources, and n independent sources. (We will explain the term *linear* dependent source shortly.) The current flowing through a given element (or the voltage across it) is called a **response**, because the currents and voltages appear in response to the independent sources.

Recall that we zeroed the independent sources as a method for finding the Thévenin resistance of a two-terminal circuit. To zero a source, we reduce its value to zero. Then, current sources become open circuits, and voltage sources become short circuits.

The superposition principle states that any response in a linear circuit is the sum of the responses for each independent source acting alone with the other independent sources zeroed. When zeroed, current sources become open circuits and voltage sources become short circuits.

Now, consider zeroing all of the independent sources except the first, observe a particular response (a current or voltage), and denote the value of that response as r_1. (We use the symbol r rather than i or v because the response could be either a current or a voltage.) Similarly, with only source 2 activated, the response is denoted as r_2, and so on. The response with all the sources activated is called the total response, denoted as r_T. The **superposition principle** states that the total response is the sum of the responses to each of the independent sources acting individually. In equation

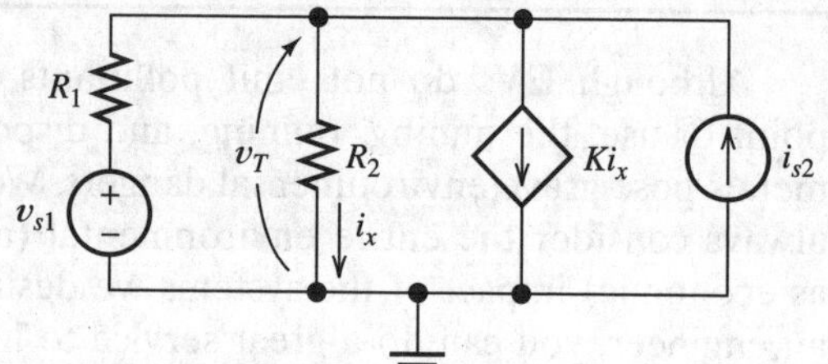

Figure 2.60 Circuit used to illustrate the superposition principle.

form, this is

$$r_T = r_1 + r_2 + \cdots + r_n \tag{2.79}$$

Next, we illustrate the validity of superposition for the example circuit shown in Figure 2.60. In this circuit, there are two independent sources: the first, the voltage source v_{s1}, and the second, the current source i_{s2}. Suppose that the response of interest is the voltage across the resistance R_2.

First, we solve for the total response v_T by solving the circuit with both sources in place. Writing a current equation at the top node, we obtain

$$\frac{v_T - v_{s1}}{R_1} + \frac{v_T}{R_2} + Ki_x = i_{s2} \tag{2.80}$$

The control variable i_x is given by

$$i_x = \frac{v_T}{R_2} \tag{2.81}$$

Substituting Equation 2.81 into Equation 2.80 and solving for the total response, we get

$$v_T = \frac{R_2}{R_1 + R_2 + KR_1} v_{s1} + \frac{R_1 R_2}{R_1 + R_2 + KR_1} i_{s2} \tag{2.82}$$

If we set i_{s2} to zero, we obtain the response to v_{s1} acting alone:

$$v_1 = \frac{R_2}{R_1 + R_2 + KR_1} v_{s1} \tag{2.83}$$

Similarly, if we set v_{s1} equal to zero in Equation 2.82, the response due to i_{s2} is givenby

$$v_2 = \frac{R_1 R_2}{R_1 + R_2 + KR_1} i_{s2} \tag{2.84}$$

Comparing Equations 2.82, 2.83, and 2.84, we see that

$$v_T = v_1 + v_2$$

Thus, as expected from the superposition principle, the total response is equal to the sum of the responses for each of the independent sources acting individually.

Dependent sources do not contribute a separate term to the total response, and we must not zero dependent sources in applying superposition.

Notice that if we zero both of the independent sources ($v_{s1} = 0$ and $i_{s2} = 0$), the response becomes zero. Hence, the dependent source does not contribute to the total response. However, the dependent source affects the contributions of the two independent sources. This is evident because the gain parameter K of the dependent source appears in the expressions for both v_1 and v_2. *In general, dependent sources do not contribute a separate term to the total response, and we must not zero dependent sources in applying superposition.*

Linearity

If we plot voltage versus current for a resistance, we have a straight line. This is illustrated in Figure 2.61. Thus, we say that Ohm's law is a **linear equation**. Similarly, the current in the controlled source shown in Figure 2.60 is given by $i_{cs} = Ki_x$, which is also a linear equation. In this book, the term **linear controlled source** means a source whose value is a constant times a control variable that is a current or a voltage appearing in the network.

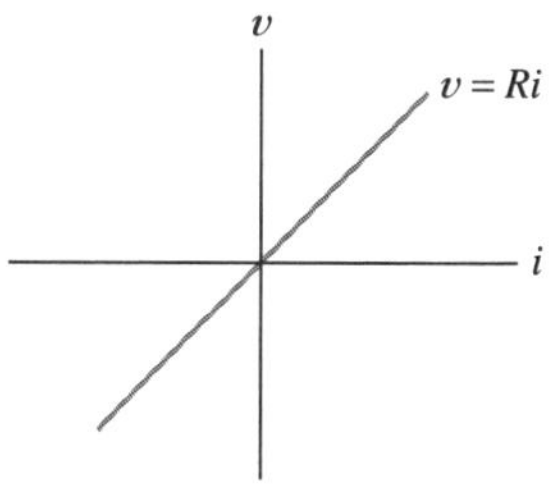

Figure 2.61 A resistance that obeys Ohm's law is linear.

Some examples of nonlinear equations are

$$v = 10i^2$$

$$i_{cs} = K\cos(i_x)$$

and

$$i = e^v$$

The superposition principle does not apply to any circuit that has element(s) described by nonlinear equation(s). We will encounter nonlinear elements later in our study of electronic circuits.

The superposition principle does not apply to any circuit that has element(s) described by nonlinear equation(s).

Furthermore, superposition does not apply for power in resistances, because $P = v^2/R$ and $P = i^2R$ are nonlinear equations.

Using Superposition to Solve Circuits

We can apply superposition in circuit analysis by analyzing the circuit for each source separately. Then, we add the individual responses to find the total response. Sometimes, the analysis of a circuit is simplified by considering each independent source separately. We illustrate with an example.

Example 2.22 Circuit Analysis Using Superposition

Use superposition in solving the circuit shown in Figure 2.62(a) for the voltage v_T.

Solution We analyze the circuit with only one source activated at a time and add the responses. Figure 2.62(b) shows the circuit with only the voltage source active. The response can be found by applying the voltage-division principle:

$$v_1 = \frac{R_2}{R_1 + R_2} v_s = \frac{5}{5 + 10}(15) = 5 \text{ V}$$

Next, we analyze the circuit with only the current source active. The circuit is shown in Figure 2.62(c). In this case, the resistances R_1 and R_2 are in parallel, and the equivalent resistance is

$$R_{eq} = \frac{1}{1/R_1 + 1/R_2} = \frac{1}{1/10 + 1/5} = 3.33 \ \Omega$$

The voltage due to the current source is given by

$$v_2 = i_s R_{eq} = 2 \times 3.33 = 6.66 \text{ V}$$

Finally, we obtain the total response by adding the individual responses:

$$v_T = v_1 + v_2 = 5 + 6.66 = 11.66$$

■

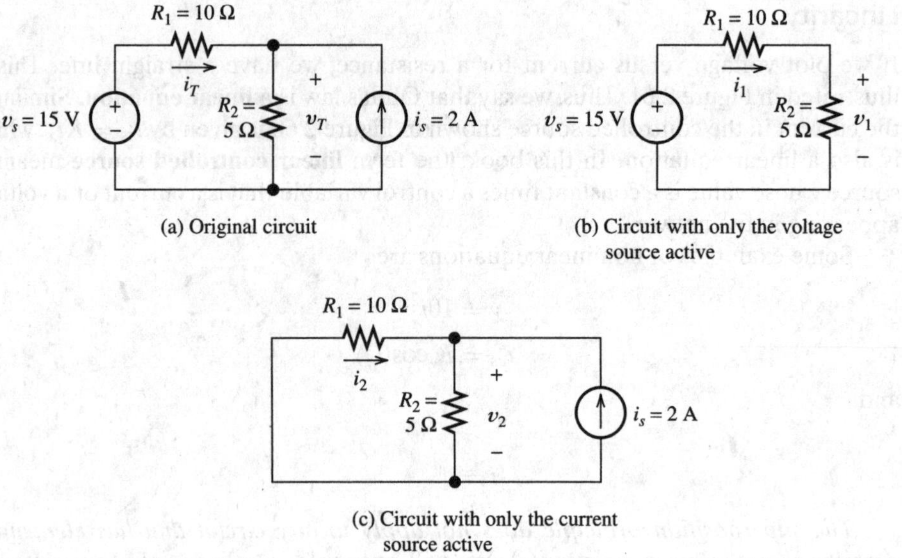

Figure 2.62 Circuit for Example 2.22 and Exercise 2.31.

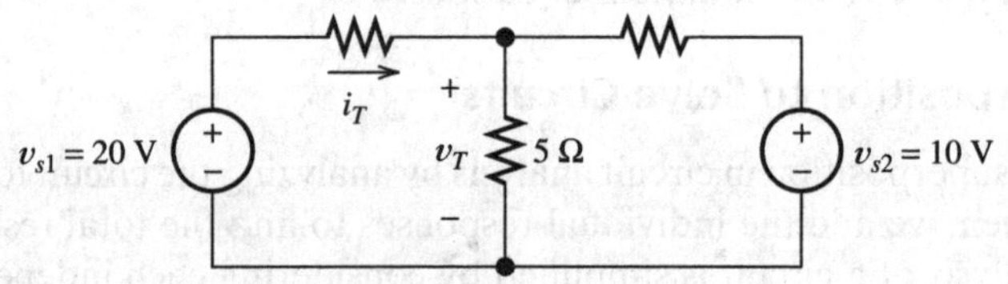

Figure 2.63 Circuit for Exercise 2.32.

Exercise 2.31 Find the responses i_1, i_2, and i_T for the circuit of Figure 2.62.
Answer $i_1 = 1$ A, $i_2 = -0.667$ A, $i_T = 0.333$ A. □

Exercise 2.32 Use superposition to find the responses v_T and i_T for the circuit shown in Figure 2.63.
Answer $v_1 = 5.45$ V, $v_2 = 1.82$ V, $v_T = 7.27$ V, $i_1 = 1.45$ A, $i_2 = -0.181$ A, $i_T = 1.27$ A. □

2.8 WHEATSTONE BRIDGE

The Wheatstone bridge is used by mechanical and civil engineers to measure the resistances of strain gauges in experimental stress studies of machines and buildings.

The **Wheatstone bridge** is a circuit used to measure unknown resistances. For example, it is used by mechanical and civil engineers to measure the resistances of strain gauges in experimental stress studies of machines and buildings. The circuit is shown in Figure 2.64. The circuit consists of a dc voltage source v_s, a detector, the unknown resistance to be measured R_x, and three precision resistors, R_1, R_2, and R_3. Usually, R_2 and R_3 are adjustable resistances, which is indicated in the figure by the arrow drawn through the resistance symbols.

The detector is capable of responding to very small currents (less than one microampere). However, it is not necessary for the detector to be calibrated. It is only necessary for the detector to indicate whether or not current is flowing through

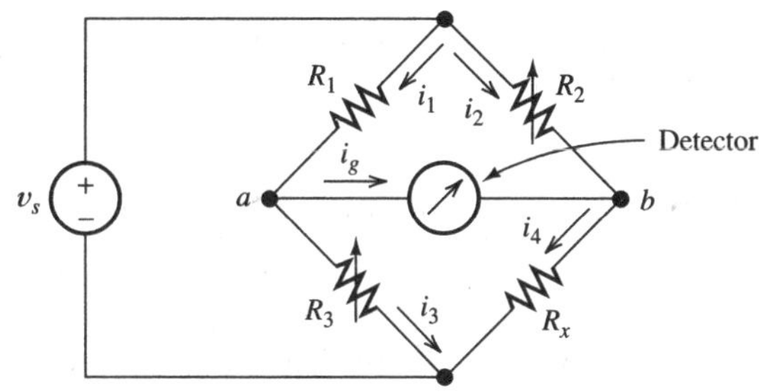

Figure 2.64 The Wheatstone bridge. When the Wheatstone bridge is balanced, $i_g = 0$ and $v_{ab} = 0$.

it. Often, the detector is a d'Arsonval galvanometer, which has a pointer that deflects one way or the other, depending on the direction of the current through it.

In operation, the resistors R_2 and R_3 are adjusted in value until the detector indicates zero current. In this condition, we say that the bridge is **balanced**. Then, the current i_g and the voltage across the detector v_{ab} are zero.

Applying KCL at node a (Figure 2.64) and using the fact that $i_g = 0$, we have

$$i_1 = i_3 \tag{2.85}$$

Similarly, at node b, we get

$$i_2 = i_4 \tag{2.86}$$

Writing a KVL equation around the loop formed by R_1, R_2, and the detector, we obtain

$$R_1 i_1 + v_{ab} = R_2 i_2 \tag{2.87}$$

However, when the bridge is balanced, $v_{ab} = 0$, so that

$$R_1 i_1 = R_2 i_2 \tag{2.88}$$

Similarly, for the loop consisting of R_3, R_4, and the detector, we have

$$R_3 i_3 = R_x i_4 \tag{2.89}$$

Using Equations 2.85 and 2.86 to substitute into Equation 2.89, we obtain

$$R_3 i_1 = R_x i_2 \tag{2.90}$$

Dividing each side of Equation 2.90 by the respective sides of Equation 2.88, we find that

$$\frac{R_3}{R_1} = \frac{R_x}{R_2}$$

Finally, solving for the unknown resistance, we have

$$R_x = \frac{R_2}{R_1} R_3 \tag{2.91}$$

Often, in commercial bridges, a multiposition switch selects an order-of-magnitude scale factor R_2/R_1 by changing the value of R_2. Then, R_3 is adjusted by means of calibrated switches until balance is achieved. Finally, the unknown resistance R_x is the scale factor times the value of R_3.

Example 2.23 Using a Wheatstone Bridge to Measure Resistance

In a certain commercial Wheatstone bridge, R_1 is a fixed 1-kΩ resistor, R_3 can be adjusted in 1-Ω steps from 0 to 1100 Ω, and R_2 can be selected to be 1 kΩ, 10 kΩ, 100 kΩ, or 1 MΩ. **a.** Suppose that the bridge is balanced with $R_3 = 732\ \Omega$ and $R_2 = 10\ \text{k}\Omega$. What is the value of R_x? **b.** What is the largest value of R_x for which the bridge can be balanced? **c.** Suppose that $R_2 = 1\ \text{M}\Omega$. What is the increment between values of R_x for which the bridge can be precisely balanced?

Solution

1. From Equation 2.91, we have

$$R_x = \frac{R_2}{R_1}R_3 = \frac{10\ \text{k}\Omega}{1\ \text{k}\Omega} \times 732\ \Omega = 7320\ \Omega$$

Notice that R_2/R_1 is a scale factor that can be set at 1, 10, 100, or 1000, depending on the value selected for R_2. The unknown resistance is the scale factor times the value of R_3 needed to balance the bridge.

2. The maximum resistance for which the bridge can be balanced is determined by the largest values available for R_2 and R_3. Thus,

$$R_{x\,\max} = \frac{R_{2\,\max}}{R_1}R_{3\,\max} = \frac{1\ \text{M}\Omega}{1\ \text{k}\Omega} \times 1100\ \Omega = 1.1\ \text{M}\Omega$$

3. The increment between values of R_x for which the bridge can be precisely balanced is the scale factor times the increment in R_3:

$$R_{x\text{inc}} = \frac{R_2}{R_1}R_{3\text{inc}} = \frac{1\ \text{M}\Omega}{1\ \text{k}\Omega} \times 1\ \Omega = 1\ \text{k}\Omega$$ ■

Strain Measurements

The Wheatstone bridge circuit configuration is often employed with strain gauges in measuring strains of beams and other mechanical structures. (See the Practical Application on page 47 for more information about strain gauges.)

For example, consider the cantilevered beam subject to a downward load force at its outer end as shown in Figure 2.65(a). Two strain gauges are attached to the top of the beam where they are stretched, increasing their resistance by ΔR when the load is applied. The change in resistance is given by

$$\Delta R = R_0 G \frac{\Delta L}{L} \tag{2.92}$$

in which $\Delta L/L$ is the strain for the surface of the beam to which the gauge is attached, R_0 is the gauge resistance before strain is applied, and G is the **gauge factor** which is typically about 2. Similarly, two gauges on the bottom of the beam are compressed, reducing their resistance by ΔR with load. (For simplicity, we have assumed that the strain magnitude is the same for all four gauges.)

The four gauges are connected in a Wheatstone bridge as shown in Figure 2.65(b).

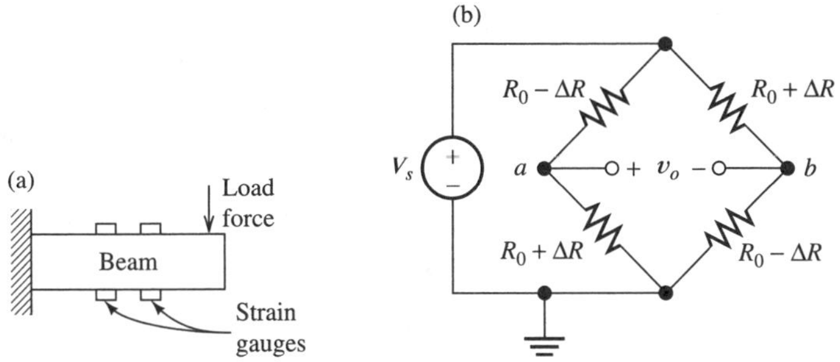

Figure 2.65 Strain measurements using the Wheatstone bridge.

The resistances labeled $R_0 + \Delta R$ are the gauges on the top of the beam and are being stretched, and those labeled $R_0 - \Delta R$ are those on the bottom and are being compressed. Before the load is applied, all four resistances have a value of R_0, the Wheatstone bridge is balanced, and the output voltage v_o is zero.

It can be shown that the output voltage v_o from the bridge is given by

$$v_o = V_s \frac{\Delta R}{R_0} = V_s G \frac{\Delta L}{L} \tag{2.93}$$

Thus, the output voltage is proportional to the strain of the beam.

In principle, the resistance of one of the gauges could be measured and the strain determined from the resistance measurements. However, the changes in resistance are very small, and the measurements would need to be very precise. Furthermore, gauge resistance changes slightly with temperature. In the bridge arrangement with the gauges attached to the beam, the temperature changes tend to track very closely and have very little effect on v_o.

Usually, v_o is amplified by an instrumentation-quality differential amplifier such as that discussed in Section 14.8 which starts on page 703. The amplified voltage can be converted to digital form and input to a computer or relayed wirelessly to a remote location for monitoring.

Summary

1. Series resistances have an equivalent resistance equal to their sum. For n resistances in series, we have

$$R_{eq} = R_1 + R_2 + \cdots + R_n$$

2. Parallel resistances have an equivalent resistance equal to the reciprocal of the sum of their reciprocals. For n resistances in parallel, we get

$$R_{eq} = \frac{1}{1/R_1 + 1/R_2 + \cdots + 1/R_n}$$

3. Some resistive networks can be solved by repeatedly combining resistances in series or parallel. The simplified network is solved, and results are transferred back through the chain of equivalent circuits. Eventually, the currents and voltages of interest in the original circuit are found.

4. The voltage-division principle applies when a voltage is applied to several resistances in series. A fraction of the total voltage appears across each resistance. The fraction that appears across a given resistance is the ratio of the given resistance to the total series resistance.

5. The current-division principle applies when current flows through two resistances in parallel. A fraction of the total current flows through each resistance. The fraction of the total current flowing through R_1 is equal to $R_2/(R_1 + R_2)$.
6. The node-voltage method can be used to solve for the voltages in any resistive network. A step-by-step summary of the method is given starting on page 94.
7. A step-by-step procedure to write the node-voltage equations directly in matrix form for circuits consisting of resistances and independent current sources appears on page 84.
8. The mesh-current method can be used to solve for the currents in any planar resistive network. A step-by-step summary of the method is given on page 106.
9. A step-by-step procedure to write the mesh-current equations directly in matrix form for circuits consisting of resistances and independent voltage sources appears on page 102. For this method to apply, all of the mesh currents must flow in the clockwise direction.
10. A two-terminal network of resistances and sources has a Thévenin equivalent that consists of a voltage source in series with a resistance. The Thévenin voltage is equal to the open-circuit voltage of the original network. The Thévenin resistance is the open-circuit voltage divided by the short-circuit current of the original network. Sometimes, the Thévenin resistance can be found by zeroing the independent sources in the original network and combining resistances in series and parallel. When independent voltage sources are zeroed, they are replaced by short circuits. Independent current sources are replaced by open circuits. Dependent sources must not be zeroed.
11. A two-terminal network of resistances and sources has a Norton equivalent that consists of a current source in parallel with a resistance. The Norton current is equal to the short-circuit current of the original network. The Norton resistance is the same as the Thévenin resistance. A step-by-step procedure for determining Thévenin and Norton equivalent circuits is given on page 113.
12. Sometimes source transformations (i.e., replacing a Thévenin equivalent with a Norton equivalent or vice versa) are useful in solving networks.
13. For maximum power from a two-terminal network, the load resistance should equal the Thévenin resistance.
14. The superposition principle states that the total response in a resistive circuit is the sum of the responses to each of the independent sources acting individually. The superposition principle does not apply to any circuit that has element(s) described by nonlinear equation(s).
15. The Wheatstone bridge is a circuit used to measure unknown resistances. The circuit consists of a voltage source, a detector, three precision calibrated resistors, of which two are adjustable, and the unknown resistance. The resistors are adjusted until the bridge is balanced, and then the unknown resistance is given in terms of the three known resistances.

Here's the answer to the trick question on page 115: Suppose that we open circuit the terminals. Then, no current flows through the Thévenin equivalent, but a current I_n circulates in the Norton equivalent. Thus, the box containing the Norton equivalent will become warm because of power dissipation in the resistance. The point of this question is that the circuits are equivalent in terms of their terminal voltage and current, not in terms of their internal behavior.

Note: **You can check the answers to many of the problems in this chapter by using a computer-aided circuit-analysis program such as Multisim from National Instruments or OrCAD Capture from Cadence Inc.**

Problems

Section 2.1: Resistances in Series and Parallel

***P2.1.** Reduce each of the networks shown in Figure P2.1 to a single equivalent resistance by combining resistances in series and parallel.

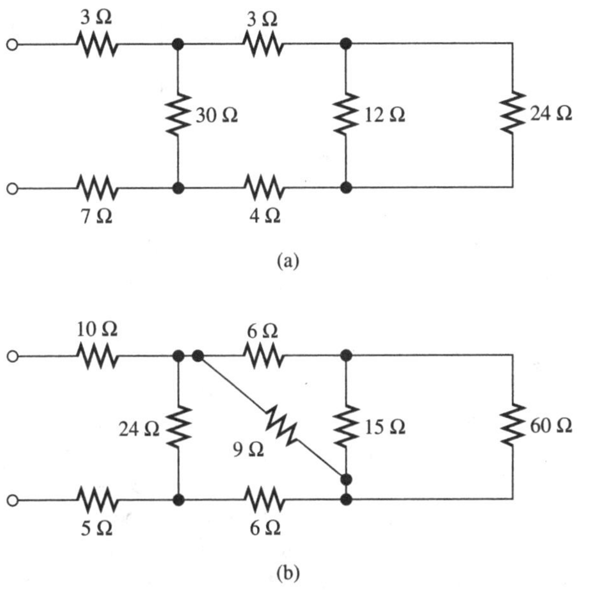

Figure P2.1

***P2.2.** A 20-Ω resistance is in parallel with a series combination of a 12-Ω resistance, and an unknown resistance R_x. The equivalent resistance for the network is 10 Ω. Determine the value of R_x.

***P2.3.** Find the equivalent resistance between terminals *a* and *b* in Figure P2.3.

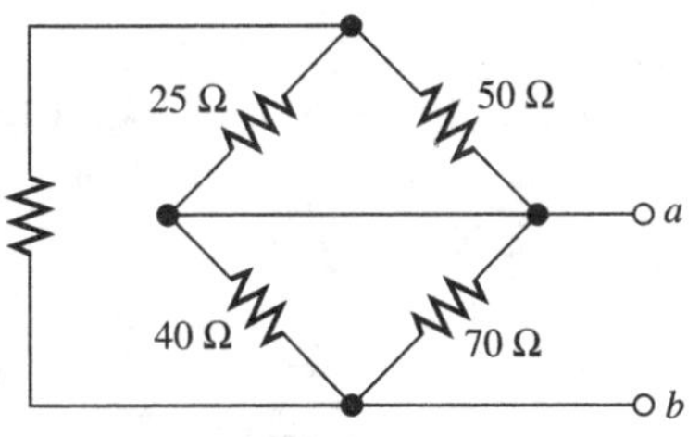

Figure P2.3

***P2.4.** Find the equivalent resistance looking into terminals *a* and *b* in Figure P2.4.

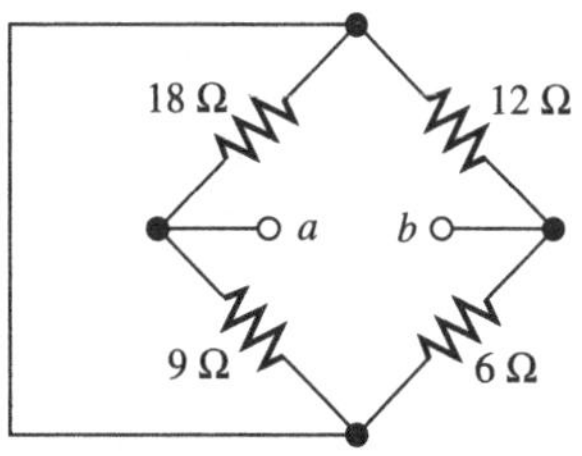

Figure P2.4

***P2.5.** Suppose that we need a resistance of 4.5 kΩ and you have a box of 2-kΩ resistors. Devise a network of 2-kΩ resistors so the equivalent resistance is 4.5 kΩ. Repeat for an equivalent resistance of 8.8 kΩ.

P2.6. **a.** Determine the resistance between terminals *c* and *d* for the network shown in Figure P2.6. **b.** Repeat after removing the short circuit between terminals *a* and *b*.

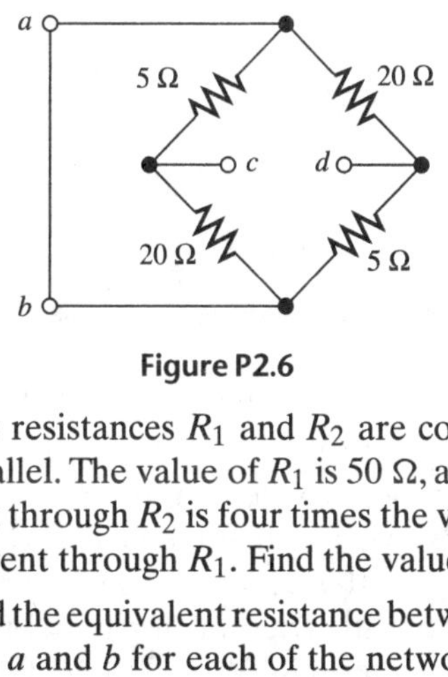

Figure P2.6

P2.7. Two resistances R_1 and R_2 are connected in parallel. The value of R_1 is 50 Ω, and the current through R_2 is four times the value of the current through R_1. Find the value of R_2.

P2.8. Find the equivalent resistance between terminals *a* and *b* for each of the networks shown in Figure P2.8.

P2.9. What conductance in series with a 70 ℧ conductance results in an equivalent conductance of 20 ℧?

P2.10. Two resistances having values of $2R$ and $3R$ are in parallel. R and the equivalent resistance are both integers. What are the possible values for R?

P2.11. A network connected between terminals *a* and *b* consists of two parallel combinations in series. The first parallel combination has three resistances having values 10 Ω, 20 Ω,

* Denotes that answers are contained in the Student Solutions files. See Appendix E for more information about accessing the Student Solutions.

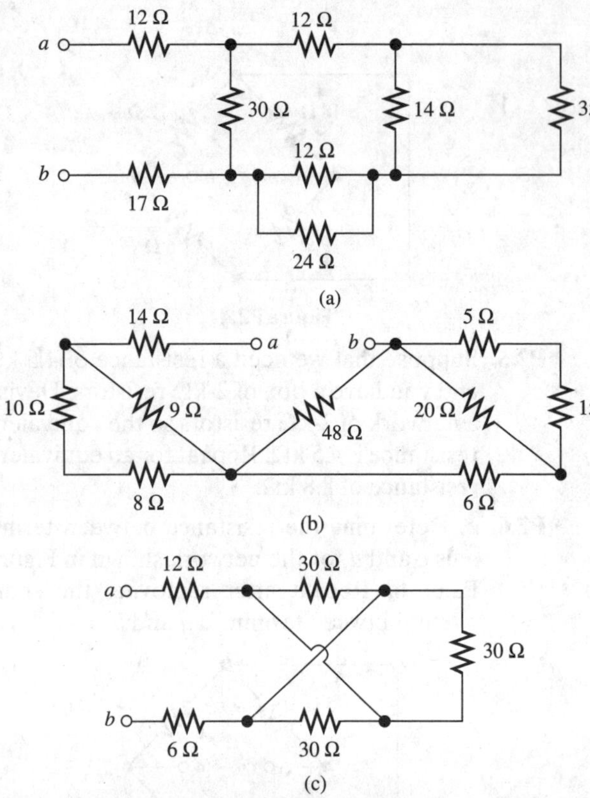

Figure P2.8

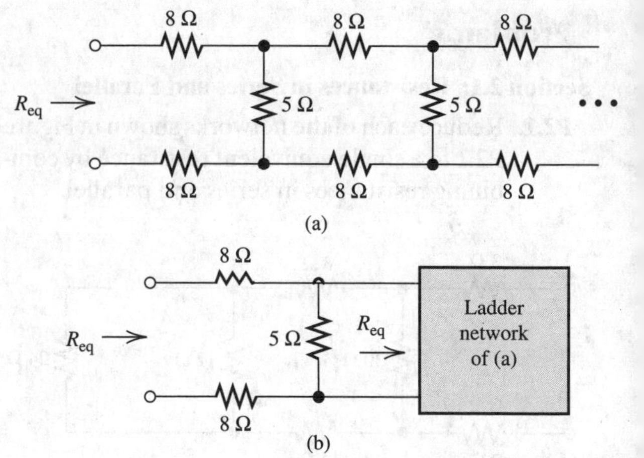

Figure P2.13

and 30 Ω respectively. The second combination has 2 resistances of 20 Ω and 40 Ω respectively. Determine the equivalent resistance, and draw the network.

P2.12. The heating element of an electric cook top has two resistive elements, $R_1 = 40\ \Omega$ and $R_2 = 100\ \Omega$, which can be operated separately, in series, or in parallel from voltages of either 120 V or 240 V. For the lowest power, R_1 is in series with R_2, and the combination is operated from 120 V. What is the lowest power? For the highest power, how should the elements be operated? What power results? List three more modes of operation and the resulting power for each.

P2.13. Find the equivalent resistance for the infinite network shown in Figure P2.13(a). Because of its form, this network is called a semi-infinite ladder. (*Hint:* If another section is added to the ladder as shown in Figure P2.13(b), the equivalent resistance is the same. Thus, working from Figure P2.13(b), we can write an expression for R_{eq} in terms of R_{eq}. Then, we can solve for R_{eq}.)

P2.14. If we connect n 1000-Ω resistances in parallel, what value is the equivalent resistance?

P2.15. We are designing an electric space heater to operate from 120 V. Two heating elements with resistances R_1 and R_2 are to be used that can be operated in parallel, separately, or in series. The highest power is to be 960 watts, and the lowest power is to be 180 watts. What values are needed for R_1 and R_2? What intermediate power settings are available?

P2.16. The equivalent resistance between terminals a and b in Figure P2.16 is $R_{ab} = 60\ \Omega$. Determine the value of R.

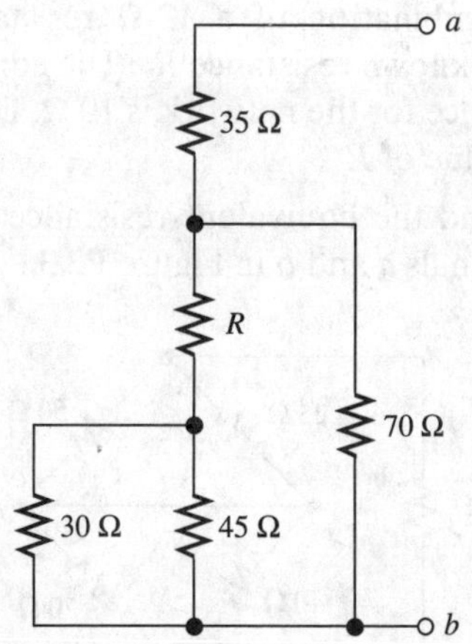

Figure P2.16

P2.17. Sometimes, we can use symmetry considerations to find the resistance of a circuit that cannot be reduced by series or parallel combinations. A classic problem of this type is illustrated in Figure P2.17. Twelve 1-Ω resistors are arranged on the edges of a cube, and terminals a and b are connected to diagonally opposite corners of the cube. The problem is to find the resistance between the terminals. Approach the problem this way: Assume that 1 A of current enters terminal a and exits through terminal b. Then, the voltage between terminals a and b is equal to the unknown resistance. By symmetry considerations, we can find the current in each resistor. Then, using KVL, we can find the voltage between a and b.

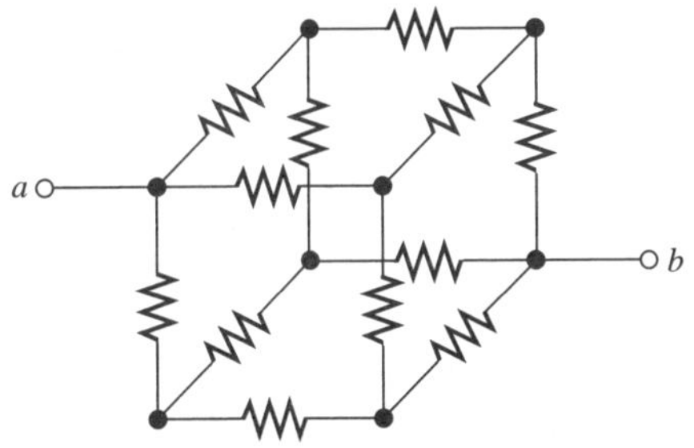

Figure P2.17

P2.18. **a.** Three conductances G_1, G_2, and G_3 are in series. Write an expression for the equivalent conductance $G_{eq} = 1/R_{eq}$ in terms of G_1, G_2, and G_3. **b.** Repeat part (a) with the conductances in parallel.

P2.19. Most sources of electrical power behave as (approximately) ideal voltage sources. In this case, if we have several loads that we want to operate independently, we place the loads in parallel with a switch in series with each load. Thereupon, we can switch each load on or off without affecting the power delivered to the other loads.

How would we connect the loads and switches if the source were an ideal independent current source? Draw the diagram of the current source and three loads with on–off switches such that each load can be switched on or off without affecting the power supplied to the other loads. To turn a load off, should the corresponding switch be opened or closed? Explain.

P2.20. Often, we encounter delta-connected loads, such as that illustrated in Figure P2.20, in three-phase power distribution systems (which are treated in Section 5.7). If we only have access to the three terminals, a method for determining the resistances is to repeatedly short two terminals together and measure the resistance between the shorted terminals and the third terminal. Then, the resistances can be calculated from the three measurements. Suppose that the measurements are $R_{as} = 24\ \Omega$, $R_{bs} = 30\ \Omega$, and $R_{cs} = 40\ \Omega$, where R_{as} is the resistance between terminal a and the short between b and c, etc. Determine the values of R_a, R_b, and R_c. (*Hint*: You may find the equations easier to deal with if you work in terms of conductances rather than resistances. Once the conductances are known, you can easily invert their values to find the resistances.)

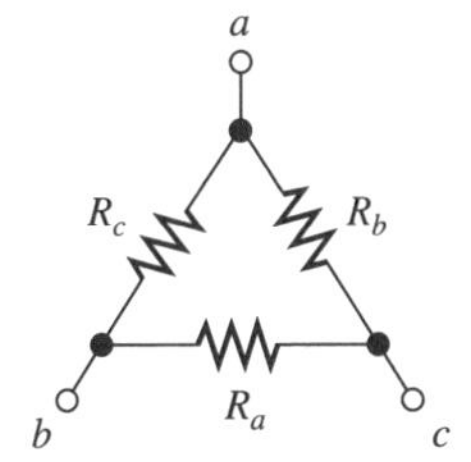

Figure P2.20

P2.21. The resistance between terminals a and b with c open circuited for the network shown in Figure P2.21 is $R_{ab} = 30\ \Omega$. Similarly, the resistance between terminals b and c with a open is $R_{bc} = 50\ \Omega$, and between c and a with b open, the resistance is $R_{ca} = 40\ \Omega$. Now, suppose that a short circuit is connected from terminal b to terminal c, and determine the resistance between terminal a and the shorted terminals b–c.

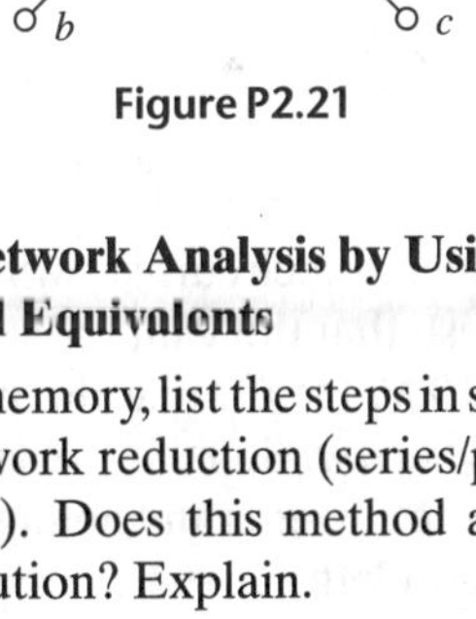

Figure P2.21

Section 2.2: Network Analysis by Using Series and Parallel Equivalents

P2.22. From memory, list the steps in solving a circuit by network reduction (series/parallel combinations). Does this method always provide the solution? Explain.

***P2.23.** Find the values of i_1 and i_2 in Figure P2.23.

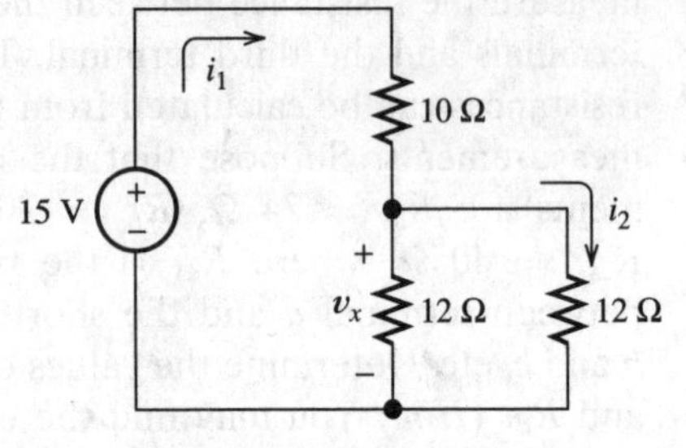

Figure P2.23

***P2.24.** Find the values of i_1 and i_2 in Figure P2.24. Find the power for each element in the circuit, and state whether each is absorbing or delivering energy. Verify that the total power absorbed equals the total power delivered.

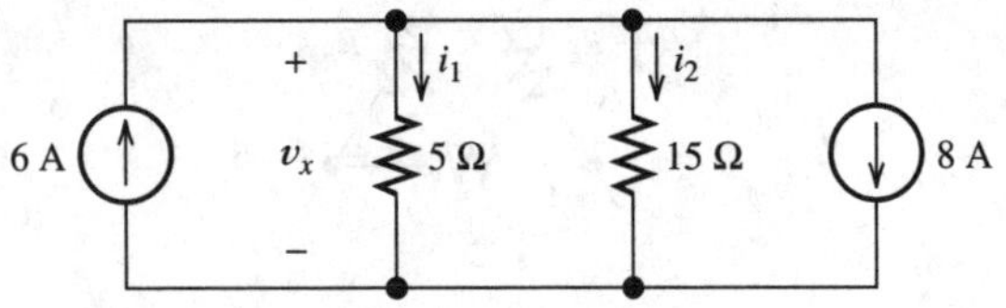

Figure P2.24

***P2.25.** Find the values of v and i in Figure P2.25.

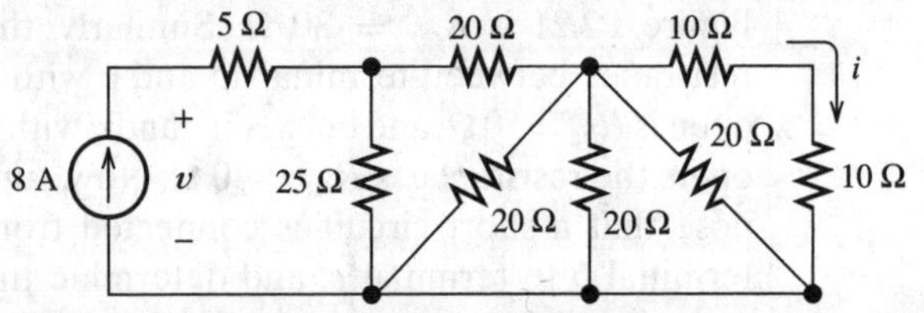

Figure P2.25

***P2.26.** Find the voltages v_1 and v_2 for the circuit shown in Figure P2.26 by combining resistances in series and parallel.

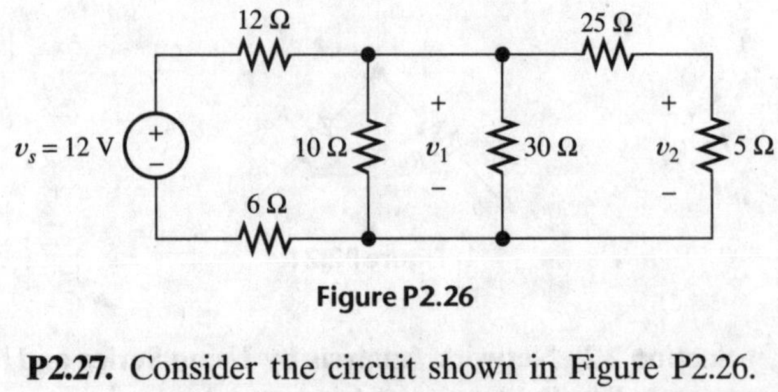

Figure P2.26

P2.27. Consider the circuit shown in Figure P2.26. Suppose that the value of v_s is adjusted until $v_1 = 10$ V. Determine the new values for v_2 and v_s. (*Hint*: Start at the location of v_2 and compute currents and voltages, moving to the right and left.)

P2.28. Find the values of v_s, v_1, and i_2 in Figure P2.28.

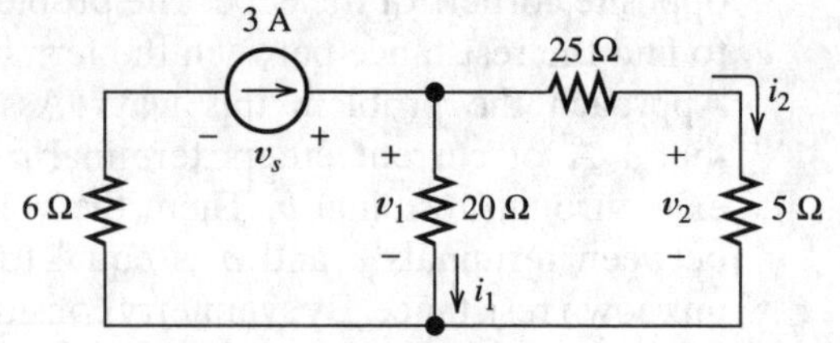

Figure P2.28

P2.29. Determine the values of i_1 and i_2 in Figure P2.29.

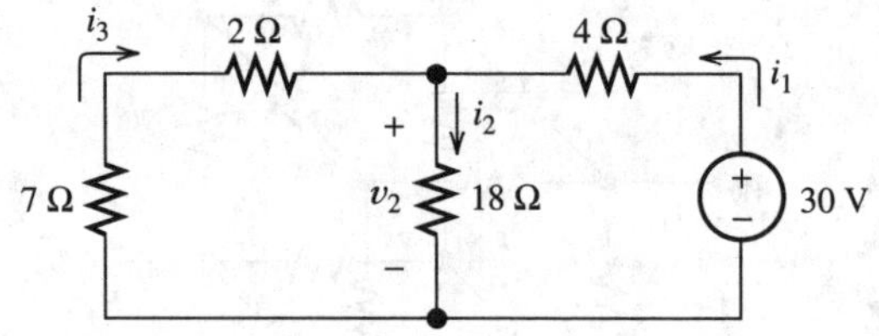

Figure P2.29

P2.30. Find the voltage v and the currents i_1 and i_2 for the circuit shown in Figure P2.30.

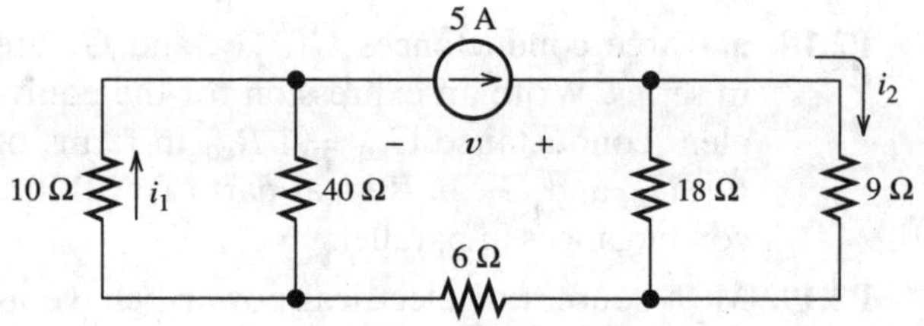

Figure P2.30

P2.31. Solve for the values of i_1, i_2, and the powers for the sources in Figure P2.31. Is the current source absorbing energy or delivering it? Is the voltage source absorbing energy or delivering it? Check to see that power is conserved in the circuit.

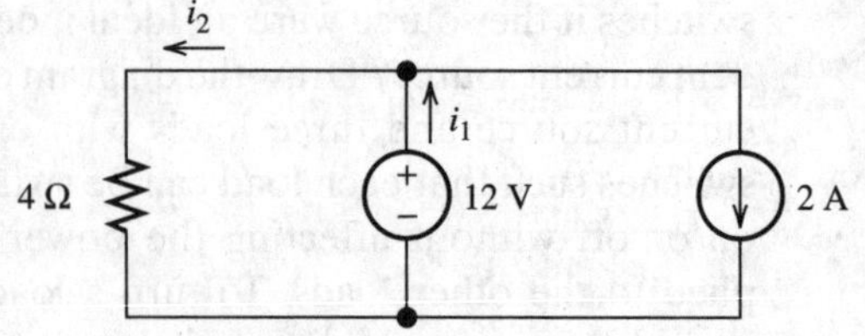

Figure P2.31

P2.32. Consider the circuit shown in Figure P2.32. With the switch open, we have $v_2 = 10$ V. On the other hand, with the switch closed, we have $v_2 = 8$ V. Determine the values of R_2 and R_L.

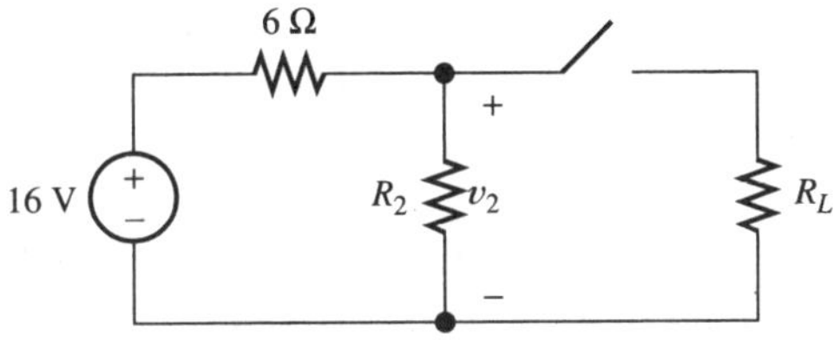

Figure P2.32

P2.33. Consider the circuit shown in Figure P2.33. Find the values of v_1, v_2, v_{ab}, v_{bc}, and v_{ca}.

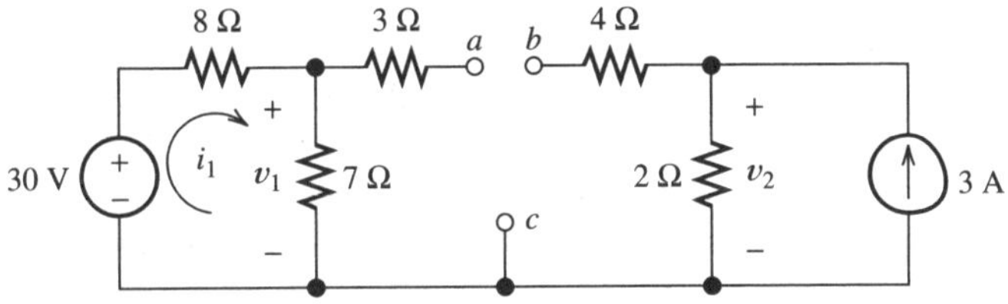

Figure P2.33

P2.34. The 15-V source in Figure P2.34 delivers 4.5 W of power. All four resistors have same value R. Find the value of R.

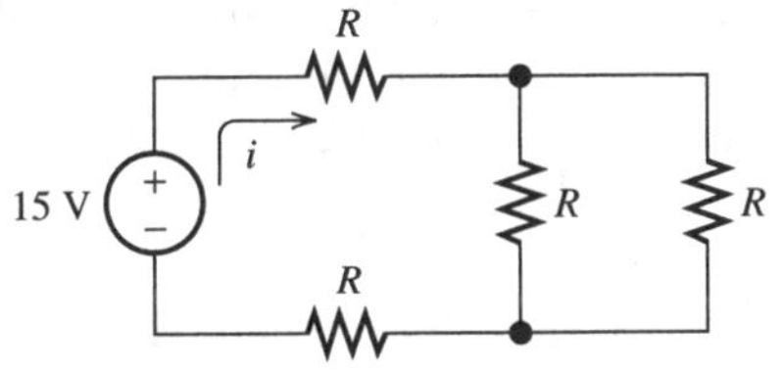

Figure P2.34

***P2.35.** Find the values of i_1 and i_2 in Figure P2.35.

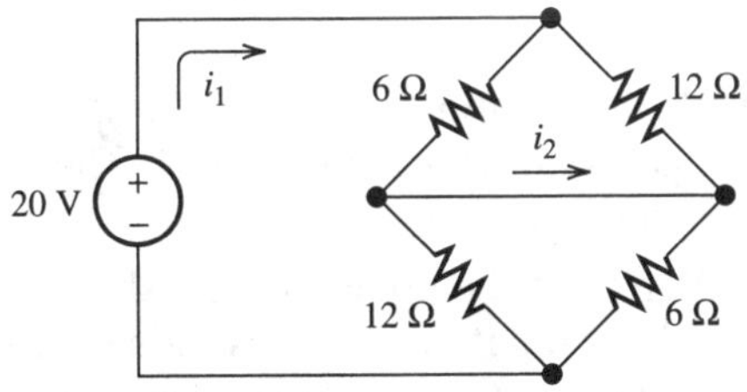

Figure P2.35

Section 2.3: Voltage-Divider and Current-Divider Circuits

***P2.36.** Use the voltage-division principle to calculate v_1, v_2, and v_3 in Figure P2.36.

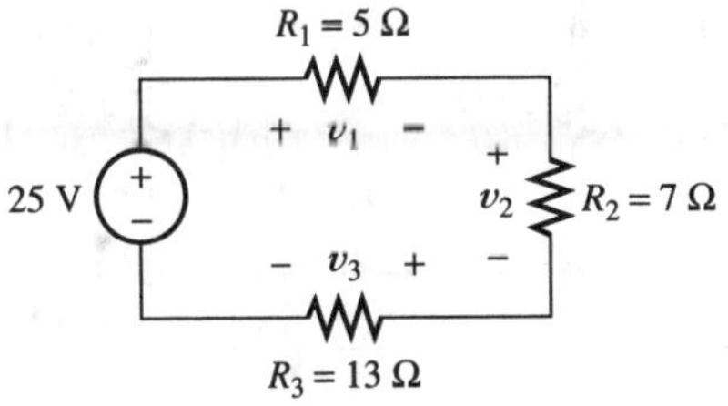

Figure P2.36

***P2.37.** Use the current-division principle to calculate i_1 and i_2 in Figure P2.37.

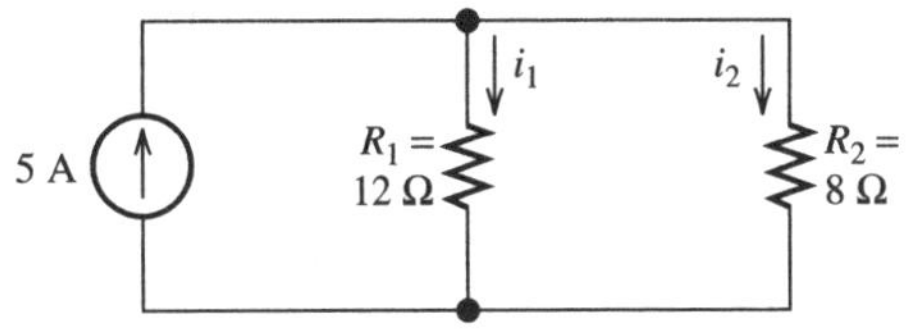

Figure P2.37

***P2.38.** Use the voltage-division principle to calculate v in Figure P2.38.

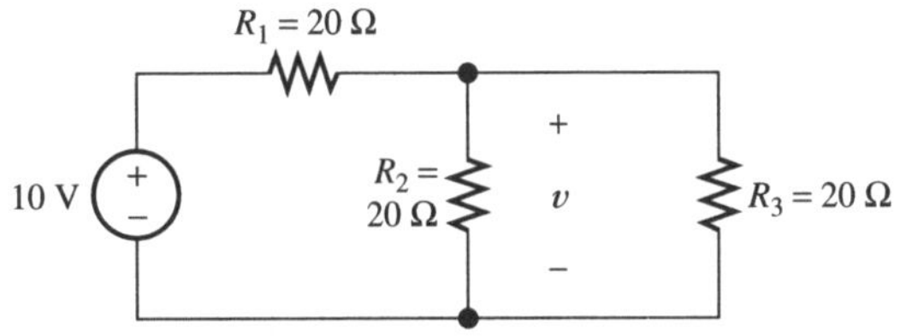

Figure P2.38

P2.39. Use the current-division principle to calculate the value of i_3 in Figure P2.39.

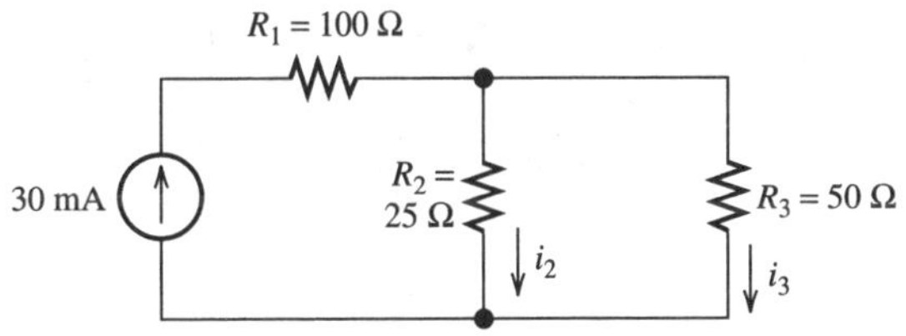

Figure P2.39

P2.40. We want to design a voltage-divider circuit to provide an output voltage $v_o = 5$ V from a 9-V battery as shown in Figure P2.40. The current taken from the 9-V source with no load connected is to be 10 mA. **a.** Find the values of R_1 and R_2. **b.** Now suppose that a load resistance of 1 kΩ is connected across the output terminals (i.e., in parallel with R_2). Find the loaded value of v_o. **c.** How could we change the design so the voltage remains closer to 5 V when the load is connected? How would this affect the life of the battery?

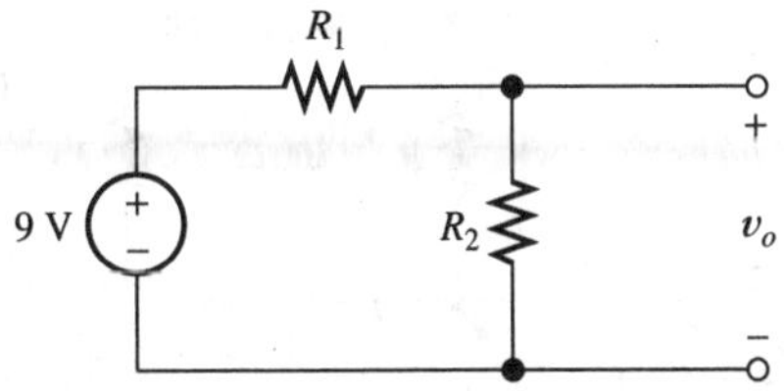

Figure P2.40

P2.41. A series-connected circuit has a 120-V voltage source, a 15-Ω resistance, a 10-Ω resistance, and an unknown resistance R_x. The voltage across the 10-Ω resistance is 45 V. Determine the value of the unknown resistance.

P2.42. A parallel circuit (i.e., all elements are in parallel with one another) has a 60-Ω resistance, a 20-Ω resistance, an unknown resistance R_x, and 30 mA current source. The current through the unknown resistance is 10 mA. Determine the value of R_x.

P2.43. The circuit of Figure P2.43 is similar to networks used in some digital-to-analog converters. For this problem, assume that the circuit continues indefinitely to the right. Find the values of i_1, i_2, i_3, and i_4. How is i_{n+2} related to i_n? What is the value of i_{18}? (*Hint:* See Problem P2.13 for hints on how to handle semi-infinite networks.)

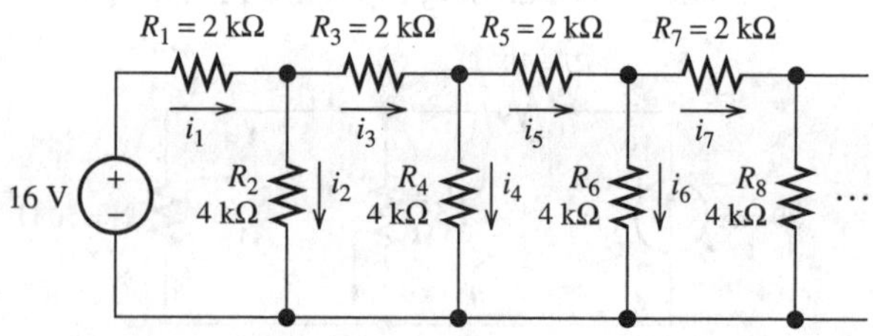

Figure P2.43

***P2.44.** A worker is standing on a wet concrete floor, holding an electric drill having a metallic case. The metallic case is connected through the ground wire of a three-terminal power outlet to power-system ground. The resistance of the ground wire is R_g. The resistance of the worker's body is $R_w = 500\ \Omega$. Due to faulty insulation in the drill, a current of 2 A flows into its metallic case. The circuit diagram for this situation is shown in Figure P2.44. Find the maximum value of R_g so that the current through the worker does not exceed 0.1 mA.

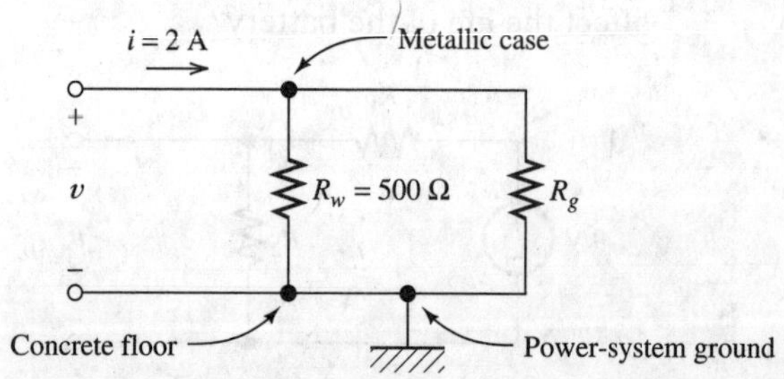

Figure P2.44

P2.45. We have a 15-V source and a load that absorbs power and requires a current varying between 0 and 100 mA. The voltage across the load must remain between 4.7 and 5.0 V for all values of load current. Design a voltage-divider network to supply the load. You may assume that resistors of any value desired are available. Also, give the minimum power rating for each resistor.

P2.46. A load resistance of 250 Ω needs to be supplied with 9 V. A 30-V voltage source and resistors of any value needed are available. Draw a suitable circuit consisting of the voltage source, the load, and one additional resistor. Specify the value of the resistor.

P2.47. Suppose that we wish to supply 500 mW to a 200-Ω load resistance R_L. A 100-mA current source and resistors of any value needed are available. Draw a suitable circuit consisting of the current source, the load, and one additional resistor. Specify the value of the resistor.

Section 2.4: Node-Voltage Analysis

P2.48. On your own, using analytical thinking and memory, list the steps to follow in analyzing a general circuit with the node-voltage technique.

***P2.49.** Write equations and solve for the node voltages shown in Figure P2.49. Then, find the value of i_1.

P2.50. Solve for the node voltages shown in Figure P2.50. What are the new values of the node voltages after the direction of the current source is reversed? How are the values related?

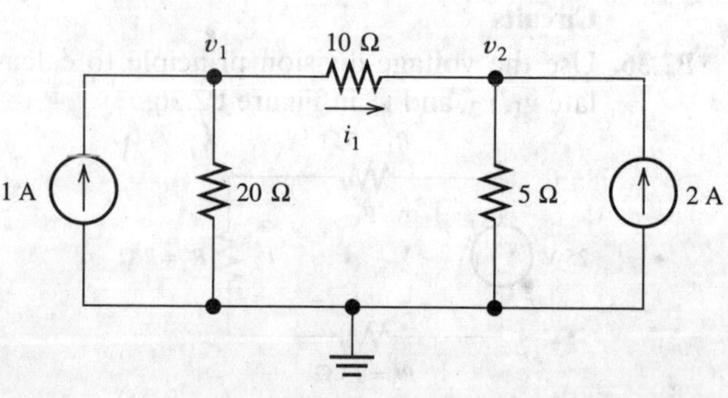

FigureP2.49

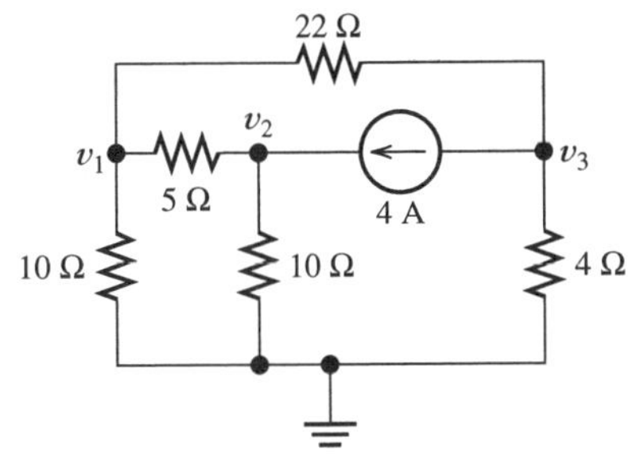

Figure P2.50

P2.51. Given $R_1 = 4\ \Omega$, $R_2 = 5\ \Omega$, $R_3 = 8\ \Omega$, $R_4 = 6\ \Omega$, $R_5 = 8\ \Omega$, and $I_s = 4$ A, solve for the node voltages shown in Figure P2.51.

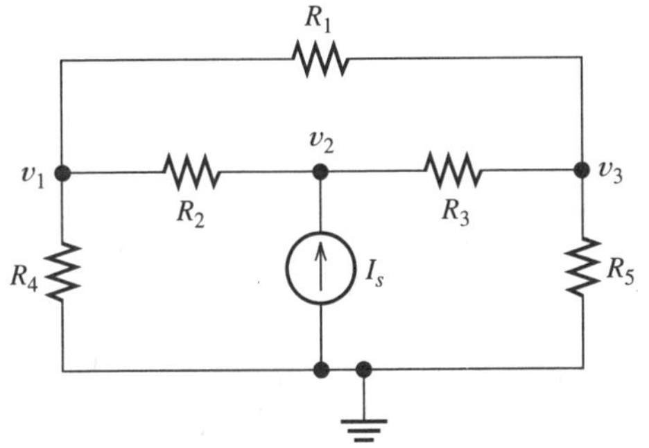

Figure P2.51

P2.52. Given $R_1 = 20\ \Omega$, $R_2 = 8\ \Omega$, $R_3 = 20\ \Omega$, $R_4 = 12\ \Omega$, $R_5 = 10\ \Omega$, $R_6 = 5\ \Omega$, and $I_s = 4$ A. Solve for the node voltages shown in Figure P2.52.

***P2.53.** Solve for the node voltages shown in Figure P2.53. Then, find the value of i_s.

P2.54. Determine the value of i_1 in Figure P2.54 using node voltages to solve the circuit. Select the location of the reference node to minimize the number of unknown node voltages. What effect does the 17-Ω resistance have on the answer? Explain.

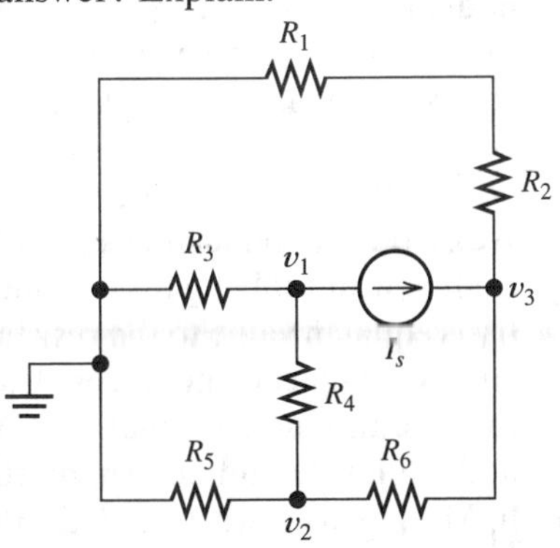

Figure P2.52

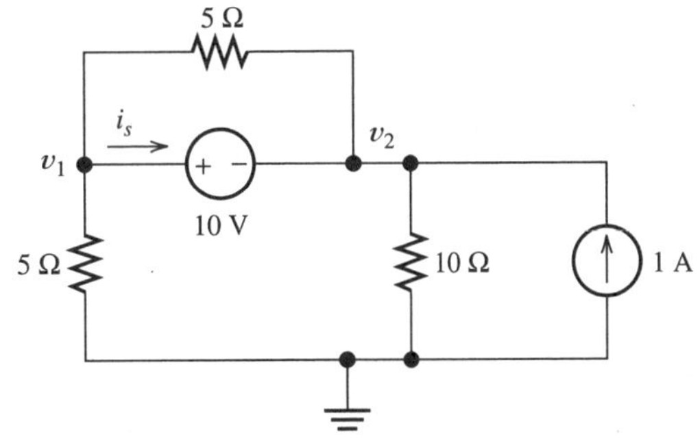

Figure P2.53

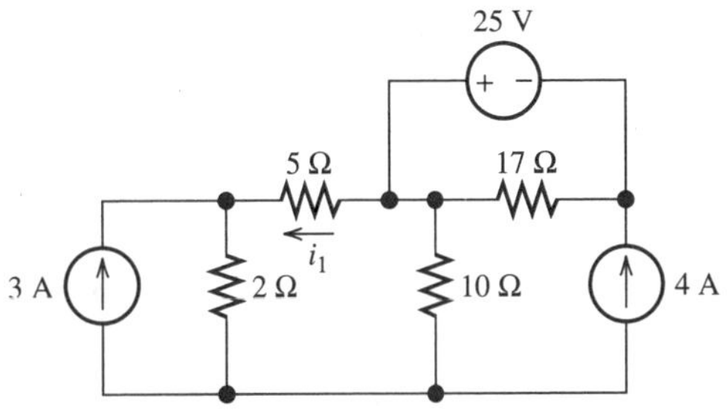

Figure P2.54

P2.55. In solving a network, what rule must you observe when writing KCL equations? Why?

P2.56. Use the symbolic features of MATLAB to find an expression for the equivalent resistance for the network shown in Figure P2.56. (*Hint:* First, connect a 1-A current source across terminals a and b. Then, solve the network by the node-voltage technique. The voltage across the current source is equal in value to the equivalent resistance.) Finally, use the subs command to evaluate for $R_1 = 15\ \Omega$, $R_2 = 15\ \Omega$, $R_3 = 15\ \Omega$, $R_4 = 10\ \Omega$, and $R_5 = 10\ \Omega$.

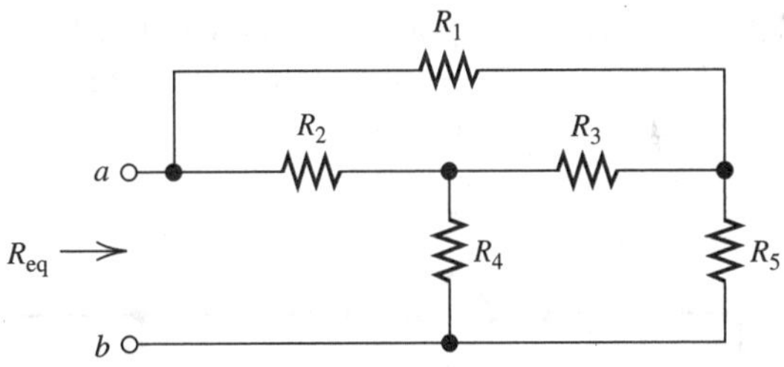

Figure P2.56

***P2.57.** Solve for the values of the node voltages shown in Figure P2.57. Then, find the value of i_x.

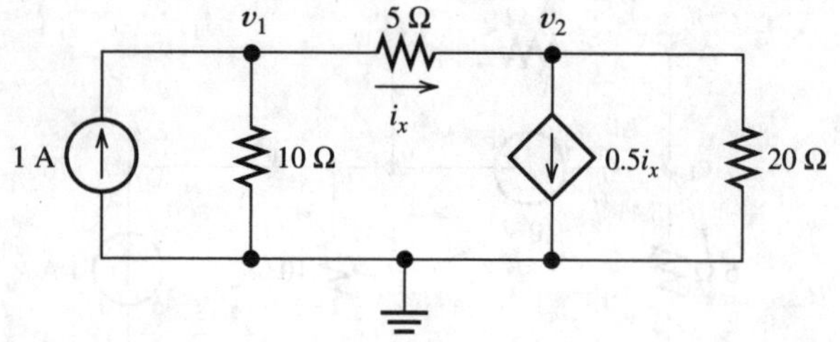

Figure P2.57

***P2.58.** Solve for the node voltages shown in Figure P2.58.

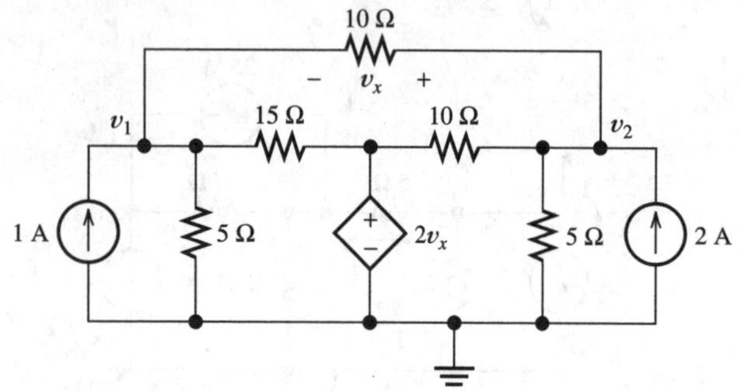

Figure P2.58

P2.59. Solve for the node voltages shown in Figure P2.59.

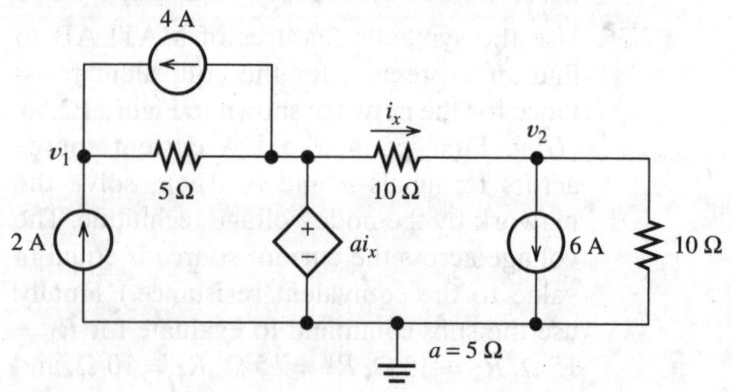

Figure P2.59

P2.60. Solve for the power delivered to the 8-Ω resistance and solve for the node voltages shown in Figure P2.60.

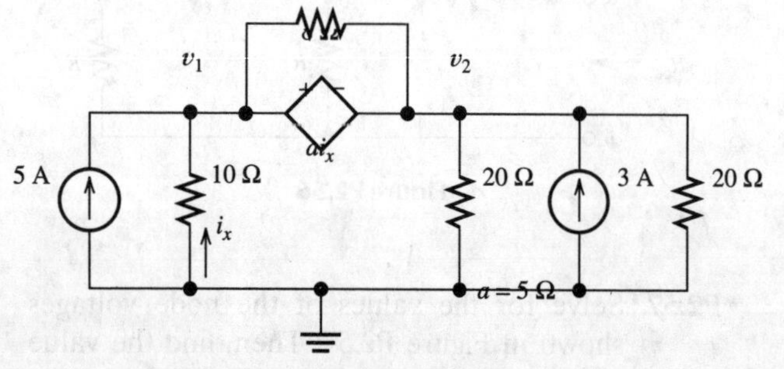

Figure P2.60

P2.61. Find the equivalent resistance looking into terminals a–b for the network shown in Figure P2.61. (*Hint:* First, connect a 1-A current source across terminals a and b. Then, solve the network by the node-voltage technique. The voltage across the current source is equal in value to the equivalent resistance.)

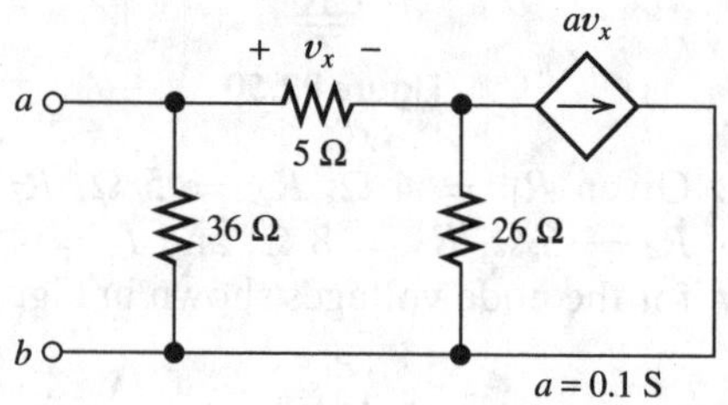

Figure P2.61

P2.62. Find the equivalent resistance looking into terminals a–b for the network shown in Figure P2.62. (*Hint:* First, connect a 1-A current source across terminals a and b. Then, solve the network by the node-voltage technique. The voltage across the current source is equal in value to the equivalent resistance.)

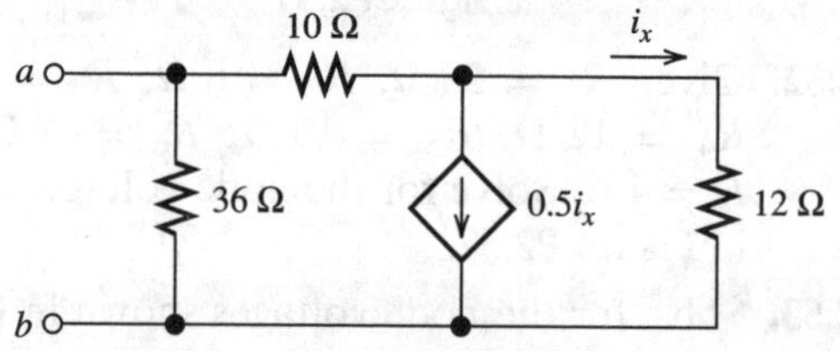

Figure P2.62

P2.63. We have a cube with 1-Ω resistances along each edge as illustrated in Figure P2.63, in which we are looking into the front face which has corners at nodes 1, 2, 7, and the reference node. Nodes 3, 4, 5, and 6 are the corners on the rear face of the cube. (Alternatively, you can consider it to be a planar network.) We want to find the resistance between adjacent nodes, such as node 1 and the reference node. We do this by connecting a 1-A current source as shown and solving for v_1, which by symmetry is equal in value to the resistance between any two adjacent nodes. **a.** Use MATLAB to solve the matrix equation **GV** = **I** for the node voltages and determine the resistance. **b.** Modify your work to determine the resistance between nodes at the ends of a diagonal across a face, such as node 2 and the reference node. **c.** Finally, find the resistance between opposite corners of the cube. (*Comment*: Part (c) is the same as Problem 2.17 in which we suggested using symmetry to solve for the resistance. Parts (a) and (b) can also be solved by use of symmetry and the fact that nodes having the same value of voltage can be connected by short circuits without changing the currents and voltages. With the shorts in place, the resistances can be combined in series and parallel to obtain the answers. Of course, if the resistors have arbitrary values, the MATLAB approach will still work, but considerations of symmetry will not.)

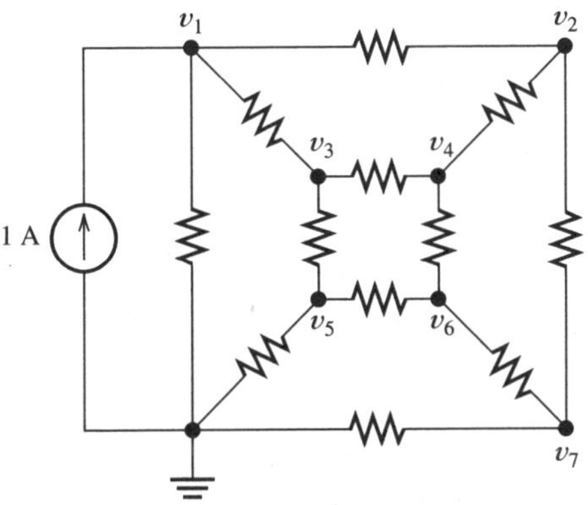

Figure P2.63 Each resistance is 1 Ω.

P2.64. Figure P2.64 shows an unusual voltage-divider circuit. Use node-voltage analysis and the symbolic math commands in MATLAB to solve for the voltage-division ratio V_{out}/V_{in} in terms of the resistances. Notice that the node voltage variables are V_1, V_2, and V_{out}.

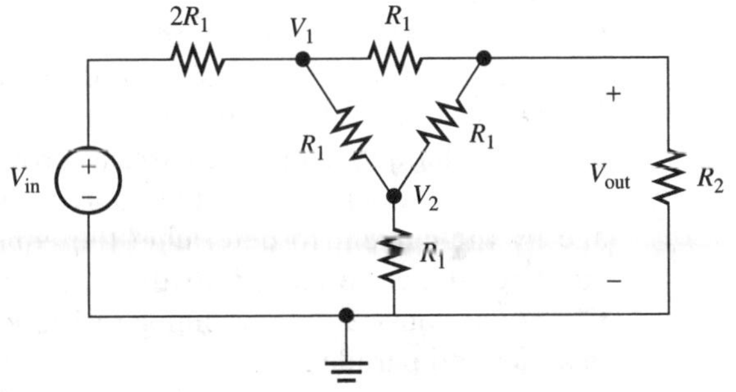

Figure P2.64

P2.65. Solve for the node voltages in the circuit of Figure P2.65. (Disregard the mesh currents, i_1, i_2, i_3, and i_4 when working with the node voltages.)

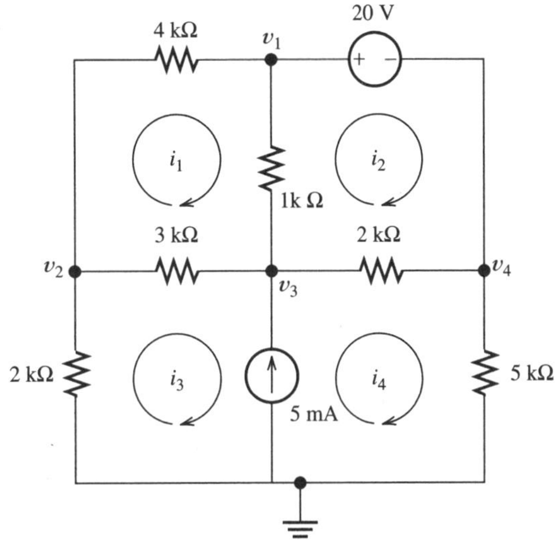

Figure P2.65

Section 2.5: Mesh-Current Analysis

P2.66. List the steps in analyzing a planar network using the mesh-current method. Attempt this as a "closed-book" exam problem.

***P2.67.** Solve for the power delivered to the 24-Ω resistor and for the mesh currents shown in Figure P2.67.

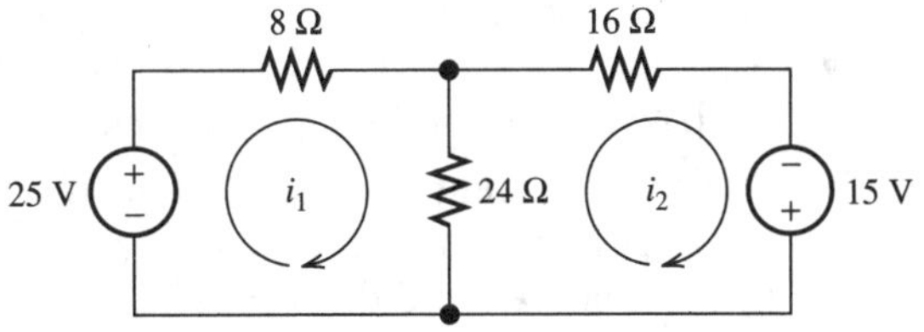

Figure P2.67

***P2.68.** Determine the value of v_2 and the power delivered by the source in the circuit of Figure P2.26 by using mesh-current analysis.

***P2.69.** Use mesh-current analysis to find the value of i_1 in the circuit of Figure P2.49.

P2.70. Solve for the power delivered by the voltage source in Figure P2.70, using the mesh-current method.

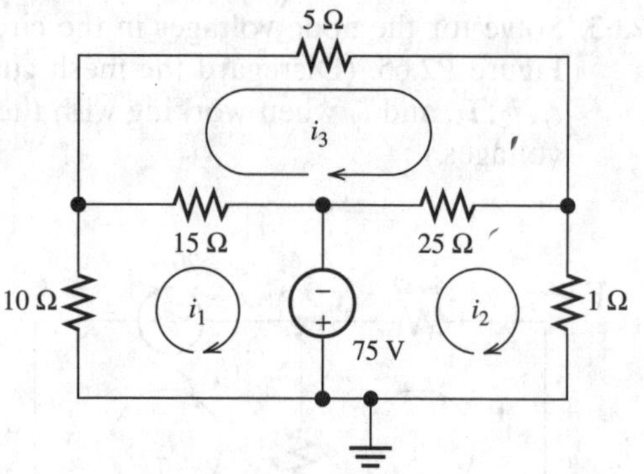

Figure P2.70

P2.71. Use mesh-current analysis to find the value of v in the circuit of Figure P2.38.

P2.72. Use mesh-current analysis to find the value of i_3 in the circuit of Figure P2.39. (Choose your mesh-current variables as i_A and i_B to avoid confusion with the current labels on the circuit diagram.)

P2.73. Use mesh-current analysis to find the values of i_1 and i_2 in Figure P2.30. Select i_1 clockwise around the left-hand mesh, i_2 clockwise around the right-hand mesh, and i_3 clockwise around the center mesh.

P2.74. Find the power delivered by the source and the values of i_1 and i_2 in the circuit of Figure P2.23, using mesh-current analysis.

P2.75. Use mesh-current analysis to find the values of i_1 and i_2 in Figure P2.29. First, select i_A clockwise around the left-hand mesh and i_B clockwise around the right-hand mesh. After solving for the mesh currents, i_A and i_B, determine the values of i_1 and i_2.

P2.76. Use mesh-current analysis to find the values of i_1 and i_2 in Figure P2.28. First, select i_A clockwise around the left-hand mesh and i_B clockwise around the right-hand mesh. After solving for the mesh currents, i_A and i_B, determine the values of i_1 and i_2.

P2.77. The circuit shown in Figure P2.77 is the dc equivalent of a simple residential power distribution system. Each of the resistances labeled R_1 and R_2 represents various parallel-connected loads, such as lights or devices plugged into outlets that nominally operate at 120 V, while R_3 represents a load, such as the heating element in an oven that nominally operates at 240 V. The resistances labeled R_w represent the resistances of wires. R_n represents the "neutral" wire. **a.** Use mesh-current analysis to determine the voltage magnitude for each load and the current in the neutral wire. **b.** Now, suppose that due to a fault in the wiring at the distribution panel, the neutral wire becomes an open circuit. Again, compute the voltages across the loads and comment on the probable outcome for a sensitive device such as a computer or plasma television that is part of the 20-Ω load.

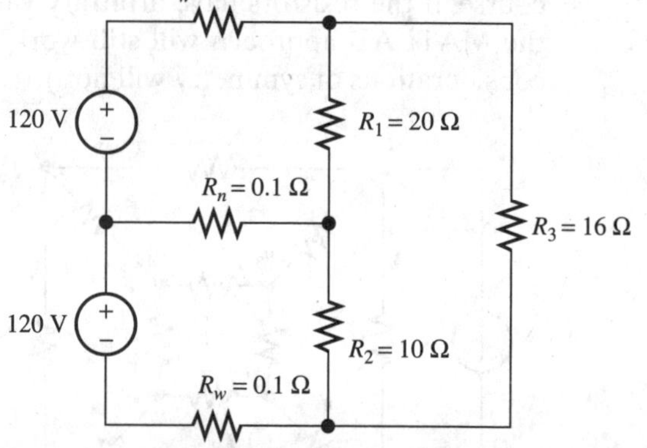

Figure P2.77

P2.78. Use MATLAB and mesh-current analysis to determine the value of v_2 in the circuit of Figure P2.51. The component values are $R_1 = 4\ \Omega, R_2 = 5\ \Omega, R_3 = 8\ \Omega, R_4 = 6\ \Omega, R_5 = 8\ \Omega$, and $I_s = 4$ A.

P2.79. Connect a 1-V voltage source across terminals *a* and *b* of the network shown in Figure P2.56. Then, solve the network by the mesh-current technique to find the current through the source. Finally, divide the source voltage by the current to determine the equivalent resistance looking into terminals *a* and *b*. The resistance values are $R_1 = 15\ \Omega, R_2 = 15\ \Omega, R_3 = 15\ \Omega, R_4 = 10\ \Omega$, and $R_5 = 10\ \Omega$.

P2.80. Connect a 1-V voltage source across the terminals of the network shown in Figure P2.1(a). Then, solve the network by the mesh-current technique to find the current through the source. Finally, divide the source voltage by the current to determine the equivalent resistance looking into the terminals. Check your answer by combining resistances in series and parallel.

P2.81. Use MATLAB to solve for the mesh currents in Figure P2.65.

Section 2.6: Thévenin and Norton Equivalent Circuits

P2.82. List the steps in determining the Thévenin and Norton equivalent circuits for a general two-terminal circuit. Try this as if it were a "closed-book" exam question.

***P2.83.** Find the Thévenin and Norton equivalent circuits for the two-terminal circuit shown in Figure P2.83.

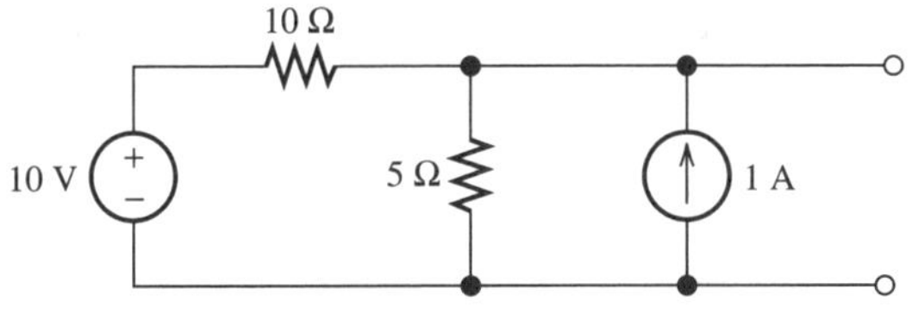

Figure P2.83

***P2.84.** We can model a certain battery as a voltage source in series with a resistance. The open-circuit voltage of the battery is 9 V. When a 100-Ω resistor is placed across the terminals of the battery, the voltage drops to 6 V. Determine the internal resistance (Thévenin resistance) of the battery.

P2.85. Find the Thévenin and Norton equivalent circuits for the circuit shown in Figure P2.85. Take care that you orient the polarity of the voltage source and the direction of the current source correctly relative to terminals *a* and *b*. What effect does the 9-Ω resistor have on the equivalent circuits? Explain your answer.

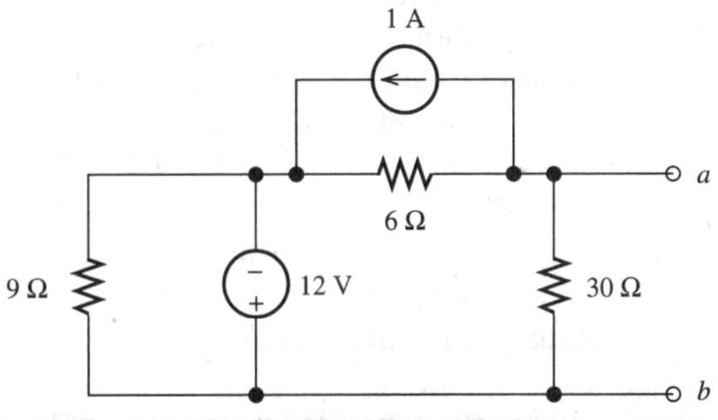

Figure P2.85

P2.86. Find the Thévenin and Norton equivalent circuits for the circuit shown in Figure P2.86.

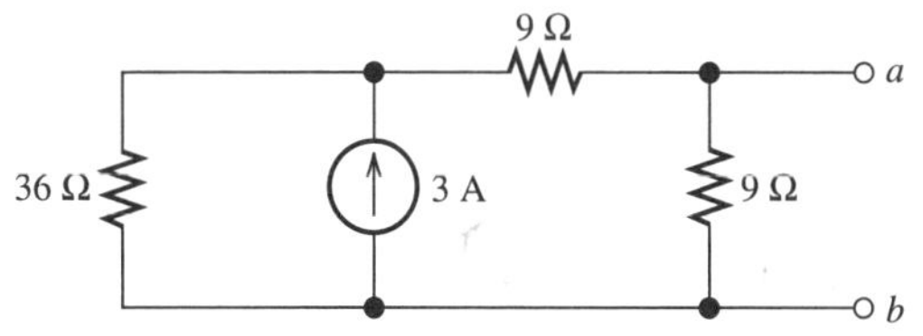

Figure P2.86

P2.87. Find the Thévenin and Norton equivalent circuits for the two-terminal circuit shown in Figure P2.87.

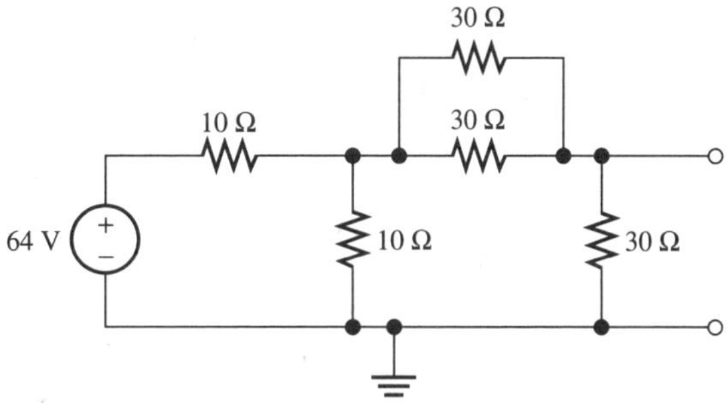

Figure P2.87

P2.88. A somewhat discharged automotive battery has an open-circuit voltage of 12.5 V and supplies 50 A when a 0.1-Ω resistance is connected across the battery terminals. Draw the Thévenin and Norton equivalent circuits, including values for the circuit parameters. What current can this battery deliver to a short circuit? Considering that the energy stored in the battery remains constant under open-circuit conditions, which of these equivalent circuits seems more realistic? Explain.

P2.89. The open circuit voltage of a certain two-terminal circuit is 15 V. When a 150-Ω load is connected, the voltage across the load is 12 V. Determine the Thévenin resistance for the circuit.

P2.90. If we measure the voltage at the terminals of a two-terminal network with two known (and different) resistive loads attached, we can determine the Thévenin and Norton equivalent circuits. When a 1-kΩ load is attached to a two-terminal circuit, the load voltage is 8 V. When the load is increased to 2 kΩ, the load voltage becomes 10 V. Find the Thévenin voltage and resistance for the circuit.

P2.91. Find the Thévenin and Norton equivalent circuits for the circuit shown in Figure P2.91.

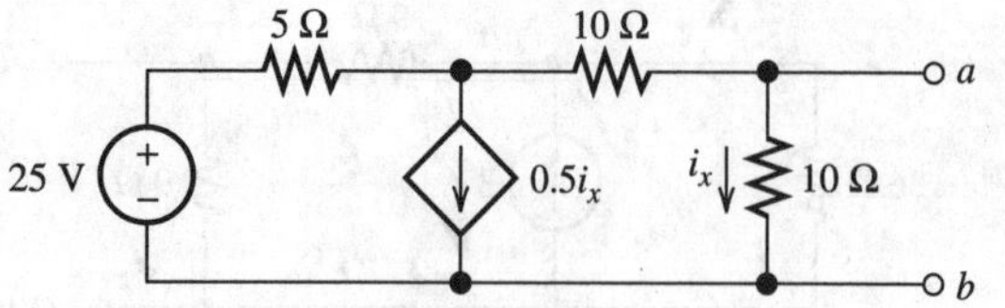

Figure P2.91

P2.92. Find the maximum power that can be delivered to a resistive load by the circuit shown in Figure P2.83. For what value of load resistance is the power maximum?

P2.93. Find the maximum power that can be delivered to a resistive load by the circuit shown in Figure P2.86. For what value of load resistance is the power maximum?

P2.94. A battery can be modeled by a voltage source V_t in series with a resistance R_t. Assuming that the load resistance is selected to maximize the power delivered, what percentage of the power taken from the voltage source V_t is actually delivered to the load? Suppose that $R_L = 9R_t$; what percentage of the power taken from V_t is delivered to the load? Usually, we want to design battery-operated systems so that nearly all of the energy stored in the battery is delivered to the load. Should we design for maximum power transfer?

***P2.95.** Figure P2.95 shows a resistive load R_L connected to a Thévenin equivalent circuit. For what value of Thévenin resistance is the power delivered to the load maximized? Find the maximum power delivered to the load. (*Hint:* Be careful; this is a tricky question if you don't stop to think about it.)

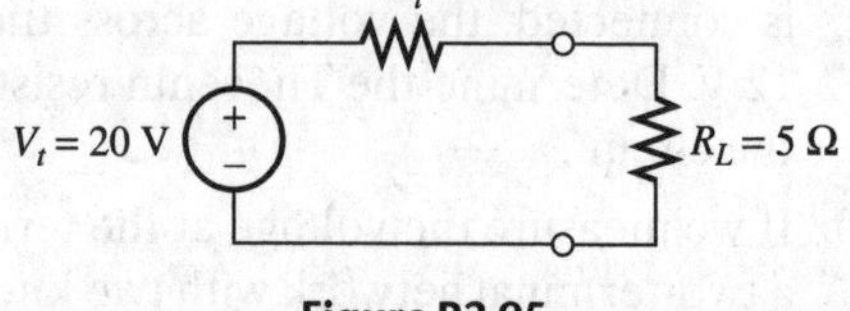

Figure P2.95

P2.96. Starting from the Norton equivalent circuit with a resistive load R_L attached, find an expression for the power delivered to the load in terms of I_n, R_t, and R_L. Assuming that I_n and R_t are fixed values and that R_L is a variable, show that maximum power is delivered for $R_L = R_t$. Find an expression for maximum power delivered to the load in terms of I_n and R_t.

Section 2.7: Superposition Principle

***P2.97.** Find the current i in Figure P2.97 using the principle of superposition. First, find the current i_v caused by the voltage source alone using voltage division. Then, find the value i_c caused by the current source alone using current division. Finally add the result algebraically.

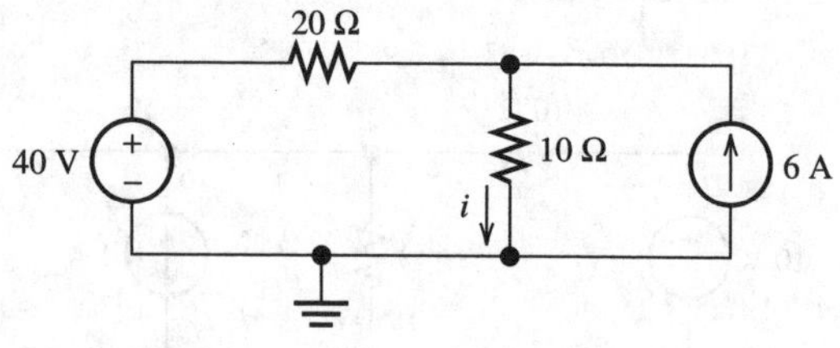

Figure P2.97

***P2.98.** Solve for i_s in Figure P2.53 by using superposition.

P2.99. Solve the circuit shown in Figure P2.49 by using superposition. First, zero the 1-A source and find the value of i_1 with only the 2-A source activated. Then, zero the 2-A source and find the value of i_1 with only the 1-A source activated. Finally, find the total value of i_1, with both sources activated, by algebraically adding the previous results.

P2.100. Solve for i_1 in Figure P2.24 by using superposition.

P2.101. Another method of solving the circuit of Figure P2.26 is to start by assuming that $v_2 = 1$ V. Accordingly, we work backward toward the source, using Ohm's law, KCL, and KVL to find the value of v_s. Since we know that v_2 is proportional to the value of v_s, and since we have found the value of v_s that produces $v_2 = 1$ V, we can calculate the value of v_2 that results when $v_s = 12$ V. Solve for v_2 by using this method.

P2.102. Use the method of Problem P2.101 for the circuit of Figure P2.23, starting with the assumption that $i_2 = 1$ A.

P2.103. Solve for the actual value of i_6 for the circuit of Figure P2.103 with $V_s = 10$ V, starting with the assumption that $i_6 = 1$ A. Work back through the circuit to find the value of V_s that results in $i_6 = 1$ A. Then, use proportionality to determine the value of i_6 that results for $V_s = 10$ V.

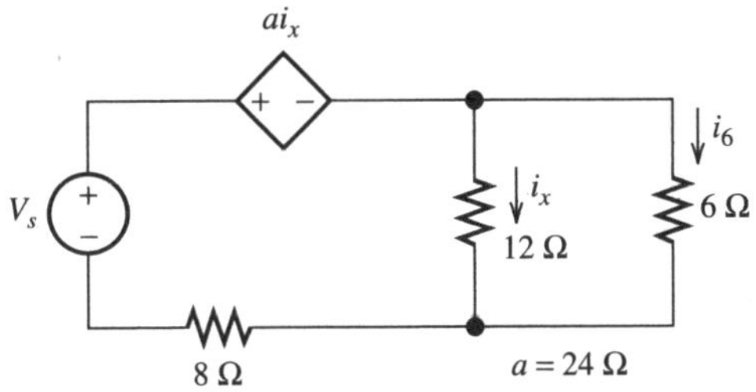

Figure P2.103

P2.104. Device A shown in Figure P2.104 has $v = 2i^3$. **a.** Solve for v with the 2-A source active and the 1-A source zeroed. **b.** Solve for v with the 1-A source active and the 2-A source zeroed. **c.** Solve for v with both sources active. Why doesn't superposition apply?

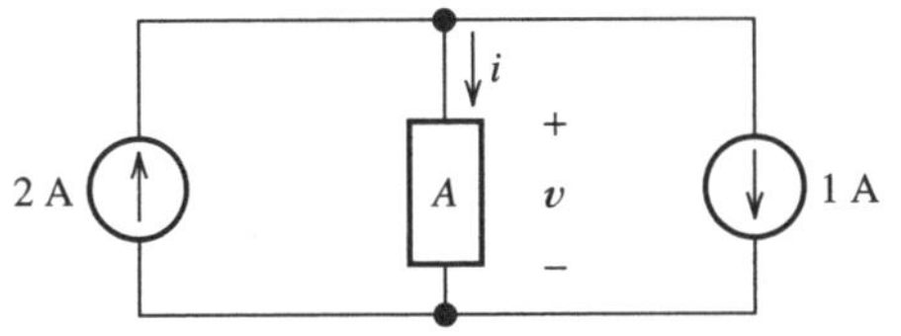

Figure P2.104

Section 2.8: Wheatstone Bridge

P2.105. **a.** The Wheatstone bridge shown in Figure 2.64 on page 123 is balanced with $R_1 = 1$ kΩ, $R_3 = 3419$ Ω, and $R_2 = 1$ kΩ. Find R_x. **b.** Repeat if R_2 is 100 kΩ and the other values are unchanged.

***P2.106.** The Wheatstone bridge shown in Figure 2.64 on page 123 has $v_s = 10$ V, $R_1 = 10$ kΩ, $R_2 = 10$ kΩ, and $R_x = 5932$ Ω. The detector can be modeled as a 5-kΩ resistance. **a.** What value of R_3 is required to balance the bridge? **b.** Suppose that R_3 is 1 Ω higher than the value found in part (a). Find the current through the detector. (*Hint:* Find the Thévenin equivalent for the circuit with the detector removed. Then, place the detector across the Thévenin equivalent and solve for the current.) Comment.

P2.107. In theory, any values can be used for R_1 and R_3 in the Wheatstone bridge of Figure 2.64 on page 123. For the bridge to balance, it is only the *ratio* R_3/R_1 that is important. What practical problems might occur if the values are very small? What practical problems might occur if the values are very large?

P2.108. Derive expressions for the Thévenin voltage and resistance "seen" by the detector in the Wheatstone bridge in Figure 2.64 on page 123. (In other words, remove the detector from the circuit and determine the Thévenin resistance for the remaining two-terminal circuit.) What is the value of the Thévenin voltage when the bridge is balanced?

P2.109. Derive Equation 2.93 for the bridge circuit of Figure 2.65 on page 125.

P2.110. Consider a strain gauge in the form of a long thin wire having a length L and a cross-sectional area A before strain is applied. After the strain is applied, the length increases slightly to $L + \Delta L$ and the area is reduced so the volume occupied by the wire is constant. Assume that $\Delta L/L << 1$ and that the resistivity ρ of the wire material is constant. Determine the gauge factor

$$G = \frac{\Delta R/R_0}{\Delta L/L}$$

(*Hint:* Make use of Equation 1.10.)

P2.111. Explain what would happen if, in wiring the bridge circuit of Figure 2.65 on page 125, the gauges in tension (i.e., those labeled $R + \Delta R$) were both placed on the top of the bridge circuit diagram, shown in part (b) of the figure, and those in compression were both placed at the bottom of the bridge circuit diagram.

Frequently Asked Questions

1. In a dc series circuit show that $R_{eq} = R_1 + R_2 + R_3 + \cdots + R_m$ where $R_1, R_2, R_3, \cdots R_m$ are the resistors connected in series

2. In a dc parallel circuit show that

$$R_{eq} = \frac{1}{1/R_1 + 1/R_2 + 1/R_3 + \cdots + 1/R_m}$$

where $R_1, R_2, R_3 \cdots R_m$ are the resistors connected in parallel.

3. In a dc series circuit show that

$$G_{eq} = \frac{1}{1/G_1 + 1/G_2 + 1/G_3 + \cdots + 1/G_m}$$

where $G_1, G_2, G_3, \cdots G_m$ are the conductances connected in series

4. In a dc parallel circuit show that $G_{eq} = G_1 + G_2 + G_3 + \cdots G_m$ where $G_1, G_2, G_3, \cdots G_m$ are the conductances connected in parallel.
5. What are the practical applications of series and parallel circuits?
6. Mention the important steps to be taken while analyzing a circuit using series parallel equivalent's?
7. How power (heating) in controlled by using heating elements in different series and parallel combinations?
8. Draw and explain a
   (i) Voltage divider circuit
   (ii) Current divider circuit
9. When two resistors R_1 and R_2 are connected in parallel, how current in divided between them?
10. When three resistors R_1, R_2 and R_3 are connected in parallel, how current is divided between them?
11. How a resistor can be used as potential-divider, explain with neat diagram?
12. How Node voltage analysis method is used to determine current in various branches of an electric network?
13. How MATLAB is used to solve network equations?
14. How mesh current analysis technique is used to determine current in various branches of an electric network?
15. State and explain Thevenin theorem
16. State and explain Norton theorem
17. How a Thevenin equivalent circuit is converted to Norton equivalent circuit and vice versa?
18. How a voltage source is converted into its current equivalent source and vice versa?
19. State and explain Maximum power transfer theorem, Give its applications.
20. Explain superposition theorem.
21. What is wheatstone bridge? How it is used to measure the value of an unknown resistance.

Practice Test

Here is a practice test you can use to check your comprehension of the most important concepts in this chapter. Answers can be found in Appendix D and complete solutions are included in the Student Solutions files. See Appendix E for more information about the Student Solutions.

T2.1. Match each entry in Table T2.1(a) with the best choice from the list given in Table T2.1(b)

Table T2.1

Item	Best Match
(a)	
a. The equivalent resistance of parallel-connected resistances...	
b. Resistances in parallel combine as do...	
c. Loads in power distribution systems are most often connected...	
d. Solving a circuit by series/parallel combinations applies to...	
e. The voltage-division principle applies to...	
f. The current-division principle applies to...	
g. The superposition principle applies to...	
h. Node-voltage analysis can be applied to...	
i. In this book, mesh-current analysis is applied to...	
j. The Thévenin resistance of a two-terminal circuit equals...	
k. The Norton current source value of a two-terminal circuit equals...	
l. A voltage source in parallel with a resistance is equivalent to...	
(b)	
1. conductances in parallel	
2. in parallel	
3. all circuits	

Table T2.1 (*Cont.*)

4. resistances or conductances in parallel
5. is obtained by summing the resistances
6. is the reciprocal of the sum of the reciprocals of the resistances
7. some circuits
8. planar circuits
9. a current source in series with a resistance
10. conductances in series
11. circuits composed of linear elements
12. in series
13. resistances or conductances in series
14. a voltage source
15. the open-circuit voltage divided by the short-circuit current
16. a current source
17. the short-circuit current

for circuits composed of sources and resistances. [Items in Table T2.1(b) may be used more than once or not at all.]

T2.2. Consider the circuit of Figure T2.2 with $v_s = 96\,\text{V}$, $R_1 = 6\,\Omega$, $R_2 = 48\,\Omega$, $R_3 = 16\,\Omega$, and $R_4 = 60\,\Omega$. Determine the values of i_s and i_4.

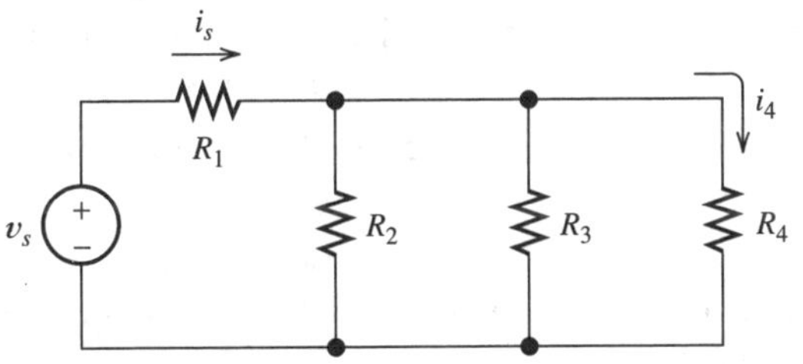

Figure T2.2

T2.3. Write MATLAB code to solve for the node voltages for the circuit of Figure T2.3.

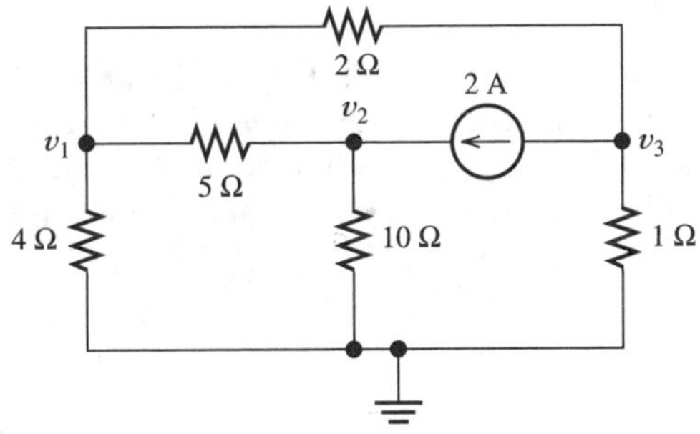

Figure T2.3

T2.4. Write a set of equations that can be used to solve for the mesh currents of Figure T2.4. Be sure to indicate which of the equations you write form the set.

T2.5. Determine the Thévenin and Norton equivalent circuits for the circuit of Figure T2.5. Draw the equivalent circuits labeling the terminals to correspond with the original circuit.

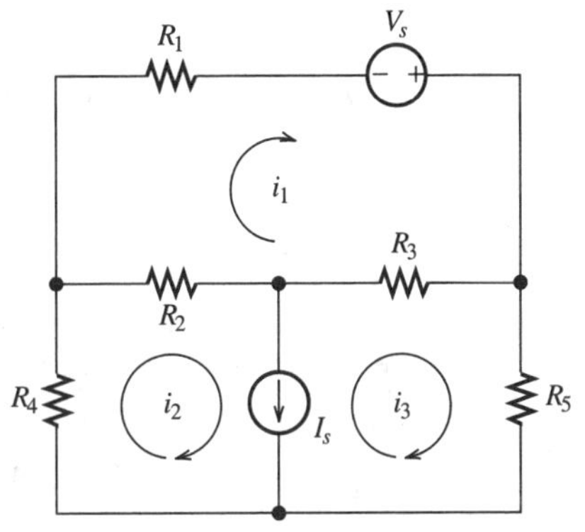

FigueT2.4

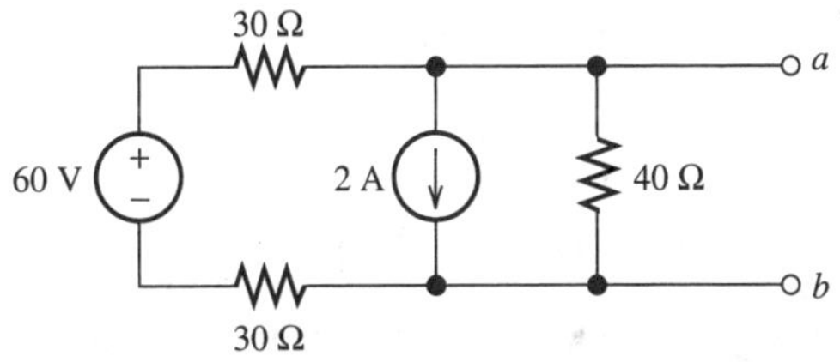

Figure T2.5

T2.6. According to the superposition principle, what percentage of the total current flowing through the 5-Ω resistance in the circuit of Figure T2.6 results from the 5-V source? What percentage of the power supplied to the 5-Ω resistance is supplied by the 5-V source? Assume that both sources are active when answering both questions.

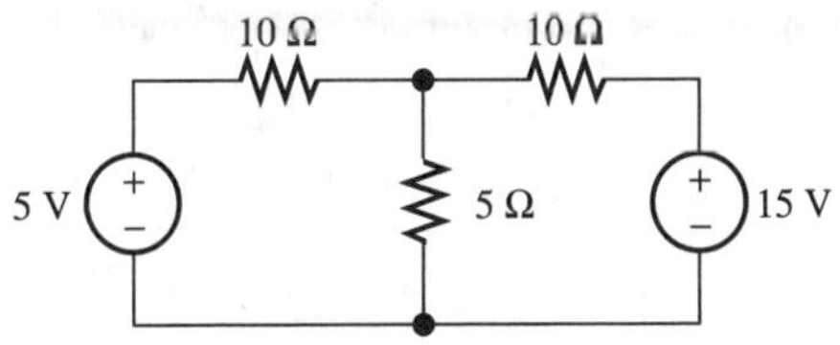

Figure T2.6

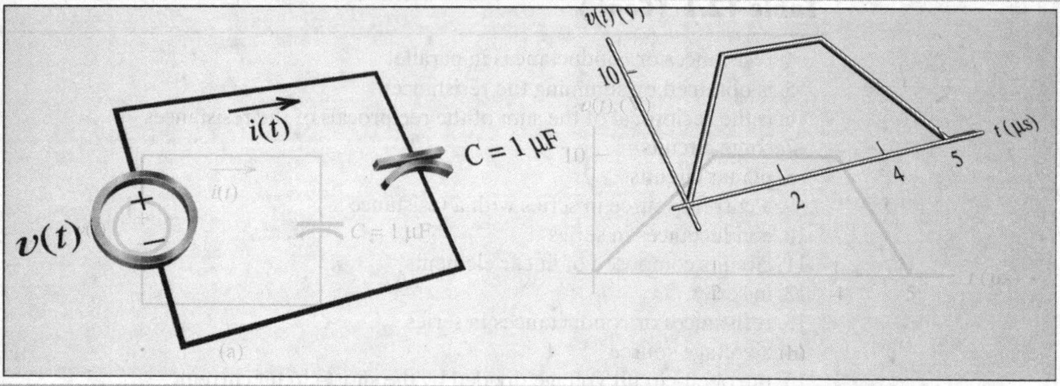

Chapter 3

Inductance and Capacitance

Study of this chapter will enable you to:

- Find the current (voltage) for a capacitance or inductance given the voltage (current) as a function of time.
- Compute the capacitances of parallel-plate capacitors.
- Compute the energies stored in capacitances or inductances.
- Describe typical physical construction of capacitors and inductors and identify parasitic effects.
- Find the voltages across mutually coupled inductances in terms of the currents.
- Apply the MATLAB Symbolic Toolbox to the current–voltage relationships for capacitances and inductances.

Introduction to this chapter:

Previously, we studied circuits composed of resistances and sources. In this chapter, we discuss two additional circuit elements: inductors and capacitors. Whereas resistors convert electrical energy into heat, inductors and capacitors are **energy-storage elements**. They can store energy and later return it to the circuit. Capacitors and inductors do not generate energy—only the energy that has been put into these elements can be extracted. Thus, like resistors, they are said to be **passive** elements.

Electromagnetic field theory is the basic approach to the study of the effects of electrical charge. However, circuit theory is a simplification of field theory that is much easier to apply. Capacitance is the circuit property that accounts for energy stored in electric fields. Inductance accounts for energy stored in magnetic fields.

We will learn that the voltage across an ideal inductor is proportional to the time derivative of the current. On the other hand, the voltage across an ideal capacitor is proportional to the time integral of the current.

We will also study mutual inductance, a circuit property that accounts for magnetic fields that are mutual to several inductors. In Chapter 15, we will see that mutual inductance forms the basis for transformers, which are critical to the transmission of electrical power over long distances.

Several types of transducers are based on inductance and capacitance. For example, one type of microphone is basically a capacitor in which the capacitance changes with sound pressure. An application of mutual inductance is the linear variable differential transformer in which position of a moving iron core is converted into a voltage.

Sometimes an electrical signal that represents a physical variable such as displacement is noisy. For example, in an active (electronically controlled) suspension for an automobile, the position sensors are affected by road roughness as well as by the loading of the vehicle. To obtain an electrical signal representing the displacement of each wheel, the rapid fluctuations due to road roughness must be eliminated. Later, we will see that this can be accomplished using inductance and capacitance in circuits known as filters.

After studying this chapter, we will be ready to extend the basic circuit-analysis techniques learned in Chapter 2 to circuits having inductance and capacitance.

3.1 CAPACITANCE

Capacitors are constructed by separating two sheets of conductor, which is usually metallic, by a thin layer of insulating material. In a parallel-plate capacitor, the sheets are flat and parallel as shown in Figure 3.1. The insulating material between the plates, called a **dielectric**, can be air, Mylar®, polyester, polypropylene, mica, or a variety of other materials.

Capacitors are constructed by separating two conducting plates, which are usually metallic, by a thin layer of insulating material.

Let us consider what happens as current flows through a capacitor. Suppose that current flows downward, as shown in Figure 3.2(a). In most metals, current consists of electrons moving, and conventional current flowing downward represents electrons actually moving upward. As electrons move upward, they collect on the lower plate of the capacitor. Thus, the lower plate accumulates a net negative charge that produces an electric field in the dielectric. This electric field forces electrons to leave the upper plate at the same rate that they accumulate on the lower plate. Therefore, current appears to flow through the capacitor. As the charge builds up, voltage appears across the capacitor.

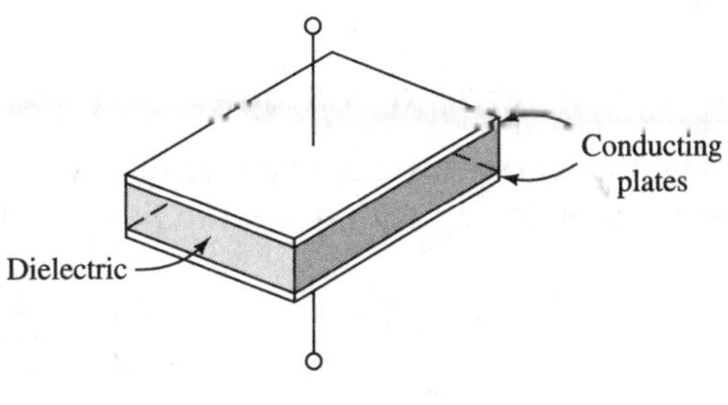

Figure 3.1 A parallel-plate capacitor consists of two conductive plates separated by a dielectric layer.

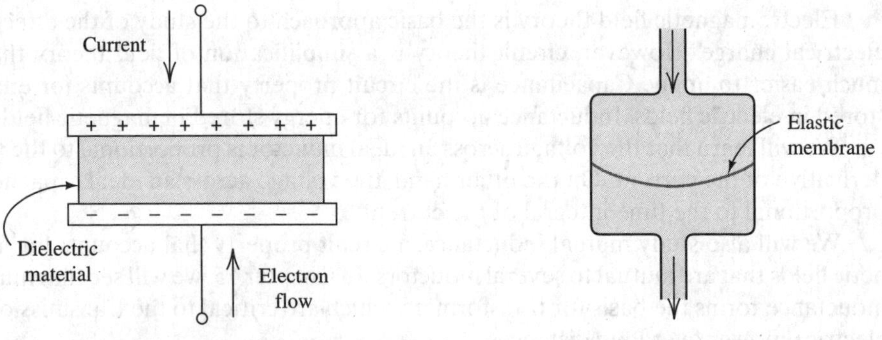

(a) As current flows through a capacitor, charges of opposite signs collect on the respective plates

(b) Fluid-flow analogy for capacitance

Figure 3.2 A capacitor and its fluid-flow analogy.

Positive charge on one plate is balanced by negative charge of equal magnitude on the other plate.

We say that the charge accumulated on one plate is stored in the capacitor. However, the total charge on both plates is always zero, because positive charge on one plate is balanced by negative charge of equal magnitude on the other plate.

Fluid-Flow Analogy

In terms of the fluid-flow analogy, a capacitor represents a reservoir with an elastic membrane separating the inlet and outlet.

In terms of the fluid-flow analogy, a capacitor represents a reservoir with an elastic membrane separating the inlet and outlet as shown in Figure 3.2(b). As the fluid flows into the inlet, the membrane is stretched, creating a force (analogous to capacitor voltage) that opposes further flow. The displaced fluid volume starting from the unstretched membrane position is analogous to the charge stored on one plate of the capacitor.

Stored Charge in Terms of Voltage

In an ideal capacitor, the stored charge q is proportional to the voltage between the plates.

In an ideal capacitor, the stored charge q is proportional to the voltage between the plates:

$$q = Cv \tag{3.1}$$

The constant of proportionality is the capacitance C, which has units of farads (F). Farads are equivalent to coulombs per volt.

To be more precise, the charge q is the net charge on the plate corresponding to the positive reference for v. Thus, if v is positive, there is positive charge on the plate corresponding to the positive reference for v. On the other hand, if v is negative, there is negative charge on the plate corresponding to the positive reference.

In most applications, we deal with capacitances in the range from a few picofarads up to perhaps 0.01 F.

A farad is a very large amount of capacitance. In most applications, we deal with capacitances in the range from a few picofarads (1 pF $= 10^{-12}$ F) up to perhaps 0.01 F. Capacitances in the femtofarad (1 fF $= 10^{-15}$ F) range are responsible for limiting the performance of computer chips.

Current in Terms of Voltage

Recall that current is the time rate of flow of charge. Taking the derivative of each side of Equation 3.1 with respect to time, we have

$$i = \frac{dq}{dt} = \frac{d}{dt}(Cv) \tag{3.2}$$

Ordinarily, capacitance is not a function of time. (An exception is the capacitor microphone mentioned earlier.) Thus, the relationship between current and voltage becomes

$$i = C\frac{dv}{dt} \tag{3.3}$$

Equations 3.1 and 3.3 show that as voltage increases, current flows through the capacitance and charge accumulates on each plate. If the voltage remains constant, the charge is constant and the current is zero. Thus, a capacitor appears to be an open circuit for a steady dc voltage.

Capacitors act as open circuits for steady dc voltages.

The circuit symbol for capacitance and the references for v and i are shown in Figure 3.3. Notice that the references for the voltage and current have the passive configuration. In other words, the current reference direction points into the positive reference polarity. If the references were opposite to the passive configuration, Equation 3.3 would have a minus sign:

$$i = -C\frac{dv}{dt} \tag{3.4}$$

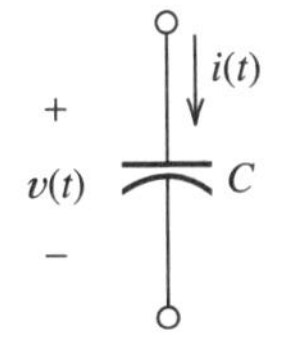

Figure 3.3 The circuit symbol for capacitance, including references for the current $i(t)$ and voltage $v(t)$.

Sometimes, we emphasize the fact that in general the voltage and current are functions of time by denoting them as $v(t)$ and $i(t)$.

Example 3.1 Determining Current for a Capacitance Given Voltage

Suppose that the voltage $v(t)$ shown in Figure 3.4(b) is applied to a 1-μF capacitance. Plot the stored charge and the current through the capacitance versus time.

Solution The charge stored on the top plate of the capacitor is given by Equation 3.1. [We know that $q(t)$ represents the charge on the top plate because that is the plate corresponding to the positive reference for $v(t)$.] Thus,

$$q(t) = Cv(t) = 10^{-6}v(t)$$

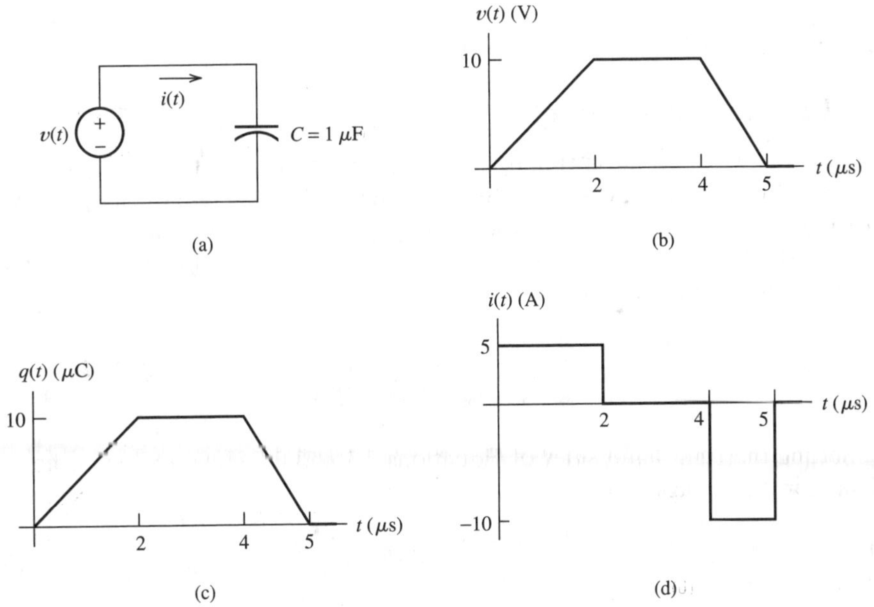

Figure 3.4 Circuit and waveforms for Example 3.1.

This is shown in Figure 3.4(c).

The current flowing through the capacitor is given by Equation 3.3:

$$i(t) = C\frac{dv(t)}{dt} = 10^{-6}\frac{dv(t)}{dt}$$

Of course, the derivative of the voltage is the slope of the voltage versus time plot. Hence, for t between 0 and 2 μs, we have

$$\frac{dv(t)}{dt} = \frac{10 \text{ V}}{2 \times 10^{-6} \text{ s}} = 5 \times 10^{6} \text{ V/s}$$

and

$$i(t) = C\frac{dv(t)}{dt} = 10^{-6} \times 5 \times 10^{6} = 5 \text{ A}$$

Between $t = 2$ and 4 μs, the voltage is constant ($dv/dt = 0$) and the current is zero. Finally, between $t = 4$ and 5 μs, we get

$$\frac{dv(t)}{dt} = \frac{-10 \text{ V}}{10^{-6} \text{ s}} = -10^{7} \text{ V/s}$$

and

$$i(t) = C\frac{dv(t)}{dt} = 10^{-6} \times (-10^{7}) = -10 \text{ A}$$

A plot of $i(t)$ is shown in Figure 3.4(d).

Notice that as the voltage increases, current flows through the capacitor and charges accumulate on the plates. For constant voltage, the current is zero and the charge is constant. When the voltage decreases, the direction of the current reverses, and the stored charge is removed from the capacitor. ■

Exercise 3.1 The charge on a 2-μF capacitor is given by

$$q(t) = 10^{-6}\sin(10^{5}t) \text{ C}$$

Find expressions for the voltage and for the current. (The angle is in radians.)

Answer $v(t) = 0.5\sin(10^{5}t)$ V, $i(t) = 0.1\cos(10^{5}t)$ A. □

Voltage in Terms of Current

Suppose that we know the current $i(t)$ flowing through a capacitance C and we want to compute the charge and voltage. Since current is the time rate of charge flow, we must integrate the current to compute charge. Often in circuit analysis problems, action starts at some initial time t_0, and the initial charge $q(t_0)$ is known. Then, charge as a function of time is given by

$$q(t) = \int_{t_0}^{t} i(t)\,dt + q(t_0) \tag{3.5}$$

Setting the right-hand sides of Equations 3.1 and 3.5 equal to each other and solving for the voltage $v(t)$, we have

$$v(t) = \frac{1}{C}\int_{t_0}^{t} i(t)\,dt + \frac{q(t_0)}{C} \tag{3.6}$$

However, the initial voltage across the capacitance is given by

$$v(t_0) = \frac{q(t_0)}{C} \tag{3.7}$$

Substituting this into Equation 3.6, we have

$$v(t) = \frac{1}{C}\int_{t_0}^{t} i(t)\,dt + v(t_0) \tag{3.8}$$

Usually, we take the initial time to be $t_0 = 0$.

Example 3.2 Determining Voltage for a Capacitance Given Current

After $t_0 = 0$, the current in a 0.1-μF capacitor is given by

$$i(t) = 0.5\sin(10^4 t)$$

(The argument of the sin function is in radians.) The initial charge on the capacitor is $q(0) = 0$. Plot $i(t)$, $q(t)$, and $v(t)$ to scale versus time.

Solution First, we use Equation 3.5 to find an expression for the charge:

$$\begin{aligned} q(t) &= \int_0^t i(t)\,dt + q(0) \\ &= \int_0^t 0.5\sin(10^4 t)\,dt \\ &= -0.5 \times 10^{-4}\cos(10^4 t)\Big|_0^t \\ &= 0.5 \times 10^{-4}[1 - \cos(10^4 t)] \end{aligned}$$

Solving Equation 3.1 for voltage, we have

$$\begin{aligned} v(t) &= \frac{q(t)}{C} = \frac{q(t)}{10^{-7}} \\ &= 500[1 - \cos(10^4 t)] \end{aligned}$$

Plots of $i(t)$, $q(t)$, and $v(t)$ are shown in Figure 3.5. Immediately after $t = 0$, the current is positive and $q(t)$ increases. After the first half-cycle, $i(t)$ becomes negative and $q(t)$ decreases. At the completion of one cycle, the charge and voltage have returned to zero. ■

Stored Energy

The power delivered to a circuit element is the product of the current and the voltage (provided that the references have the passive configuration):

$$p(t) = v(t)i(t) \tag{3.9}$$

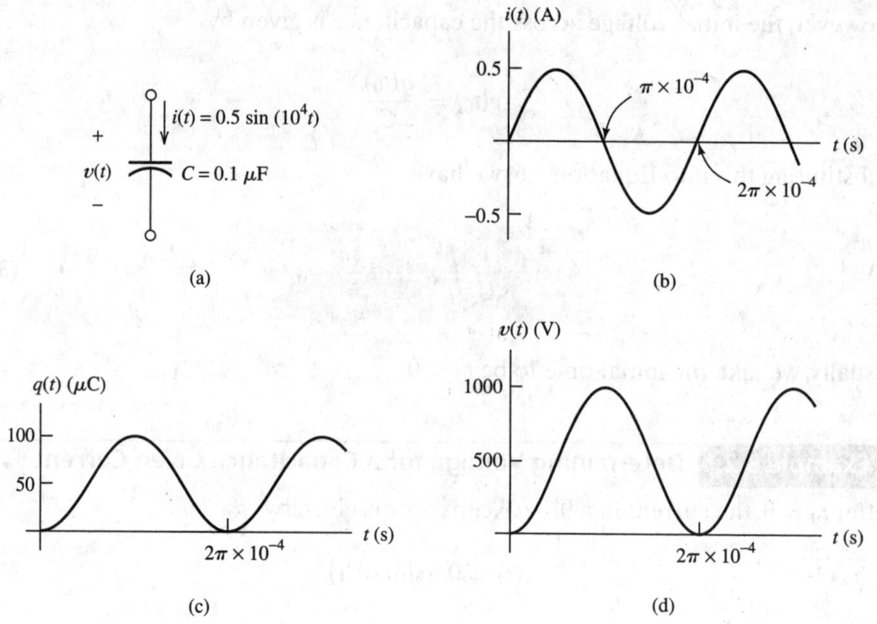

Figure 3.5 Waveforms for Example 3.2.

Using Equation 3.3 to substitute for the current, we have

$$p(t) = Cv\frac{dv}{dt} \tag{3.10}$$

Suppose we have a capacitor that initially has $v(t_0) = 0$. Then the initial stored electrical energy is zero, and we say that the capacitor is uncharged. Furthermore, suppose that between time t_0 and some later time t the voltage changes from 0 to $v(t)$ volts. As the voltage magnitude increases, energy is delivered to the capacitor, where it is stored in the electric field between the plates.

If we integrate the power delivered from t_0 to t, we find the energy delivered:

$$w(t) = \int_{t_0}^{t} p(t)\,dt \tag{3.11}$$

Using Equation 3.10 to substitute for power, we find that

$$w(t) = \int_{t_0}^{t} Cv\frac{dv}{dt}\,dt \tag{3.12}$$

Canceling differential time and changing the limits to the corresponding voltages, we have

$$w(t) = \int_{0}^{v(t)} Cv\,dv \tag{3.13}$$

Integrating and evaluating, we get

$$w(t) = \frac{1}{2}Cv^2(t) \tag{3.14}$$

This represents energy stored in the capacitance that can be returned to the circuit.

Solving Equation 3.1 for $v(t)$ and substituting into Equation 3.14, we can obtain two alternative expressions for the stored energy:

$$w(t) = \frac{1}{2}v(t)q(t) \tag{3.15}$$

$$w(t) = \frac{q^2(t)}{2C} \tag{3.16}$$

Example 3.3 Current, Power, and Energy for a Capacitance

Suppose that the voltage waveform shown in Figure 3.6(a) is applied to a 10-μF capacitance. Find and plot the current, the power delivered, and the energy stored for time between 0 and 5 s.

Solution First, we write expressions for the voltage as a function of time:

$$v(t) = \begin{cases} 1000t \text{ V} & \text{for } 0 < t < 1 \\ 1000 \text{ V} & \text{for } 1 < t < 3 \\ 500(5 - t) \text{ V} & \text{for } 3 < t < 5 \end{cases}$$

Using Equation 3.3, we obtain expressions for the current:

$$i(t) = C\frac{dv(t)}{dt}$$

$$i(t) = \begin{cases} 10 \times 10^{-3} \text{ A} & \text{for } 0 < t < 1 \\ 0 \text{ A} & \text{for } 1 < t < 3 \\ -5 \times 10^{-3} \text{ A} & \text{for } 3 < t < 5 \end{cases}$$

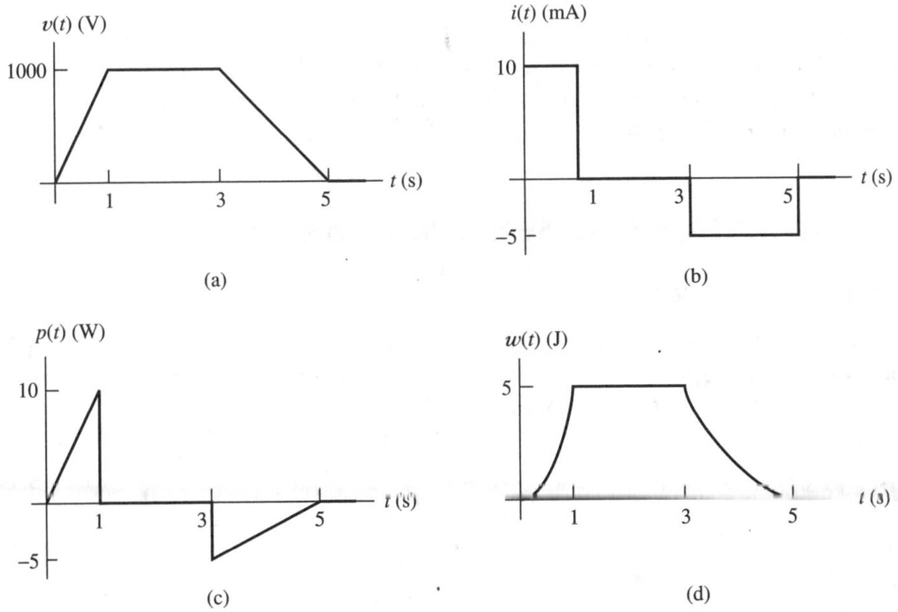

Figure 3.6 Waveforms for Example 3.3.

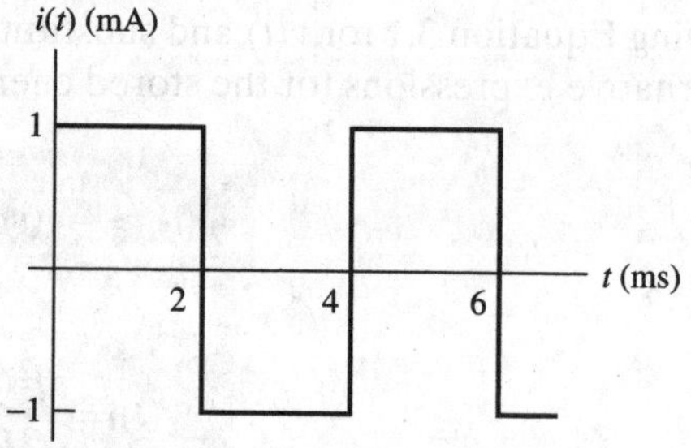

Figure 3.7 Square-wave current for Exercise 3.2.

The plot of $i(t)$ is shown in Figure 3.6(b).

Next, we find expressions for power by multiplying the voltage by the current:

$$p(t) = v(t)i(t)$$

$$p(t) = \begin{cases} 10t \text{ W} & \text{for } 0 < t < 1 \\ 0 \text{ W} & \text{for } 1 < t < 3 \\ 2.5(t - 5) \text{ W} & \text{for } 3 < t < 5 \end{cases}$$

The plot of $p(t)$ is shown in Figure 3.6(c). Notice that between $t = 0$ and $t = 1$ power is positive, showing that energy is being delivered to the capacitance. Between $t = 3$ and $t = 5$, energy flows out of the capacitance back into the rest of the circuit.

Next, we use Equation 3.14 to find expressions for the stored energy:

$$w(t) = \frac{1}{2}Cv^2(t)$$

$$w(t) = \begin{cases} 5t^2 \text{ J} & \text{for } 0 < t < 1 \\ 5 \text{ J} & \text{for } 1 < t < 3 \\ 1.25(5 - t)^2 \text{ J} & \text{for } 3 < t < 5 \end{cases}$$

The plot of $w(t)$ is shown in Figure 3.6(d). ■

Exercise 3.2 The current through a 0.1-μF capacitor is shown in Figure 3.7. At $t_0 = 0$, the voltage across the capacitor is zero. Find the charge, voltage, power, and stored energy as functions of time and plot them to scale versus time.

Answer The plots are shown in Figure 3.8. □

3.2 CAPACITANCES IN SERIES AND PARALLEL

Capacitances in Parallel

Suppose that we have three capacitances in parallel as shown in Figure 3.9. Of course, the same voltage appears across each of the elements in a parallel circuit. The currents are related to the voltage by Equation 3.3. Thus, we can write

$$i_1 = C_1\frac{dv}{dt} \tag{3.17}$$

$$i_2 = C_2\frac{dv}{dt} \tag{3.18}$$

$$i_3 = C_3\frac{dv}{dt} \tag{3.19}$$

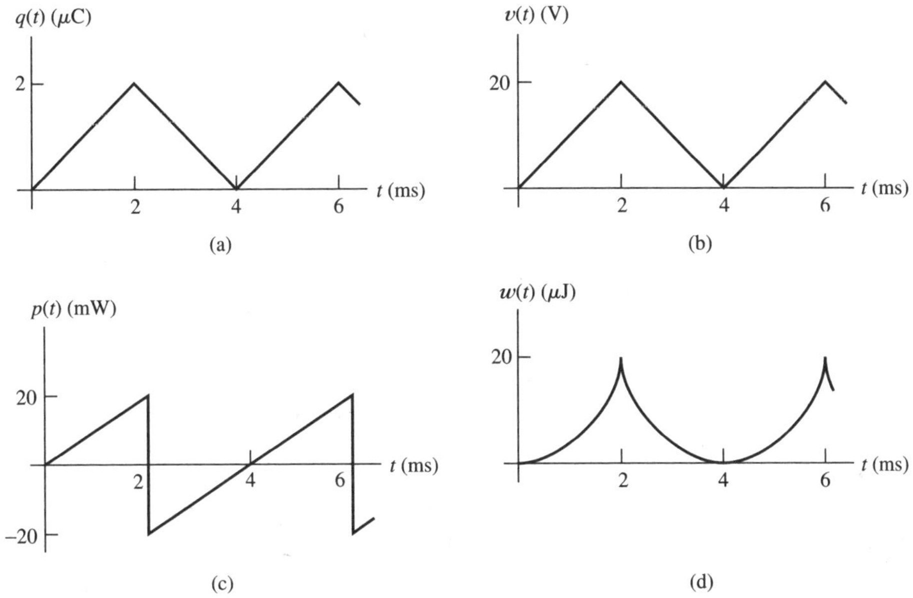

Figure 3.8 Answers for Exercise 3.2.

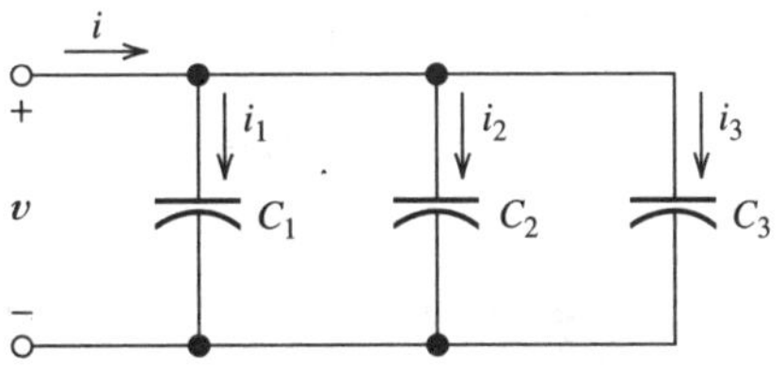

$C_{eq} = C_1 + C_2 + C_3$

Figure 3.9 Three capacitances in parallel.

Applying KCL at the top node of the circuit, we have

$$i = i_1 + i_2 + i_3 \tag{3.20}$$

Using Equations 3.17, 3.18, and 3.19 to substitute into Equation 3.20, we obtain

$$i = C_1 \frac{dv}{dt} + C_2 \frac{dv}{dt} + C_3 \frac{dv}{dt} \tag{3.21}$$

This can be written as

$$i = (C_1 + C_2 + C_3) \frac{dv}{dt} \tag{3.22}$$

Now, we define the equivalent capacitance as the sum of the capacitances in parallel:

$$C_{eq} = C_1 + C_2 + C_3 \tag{3.23}$$

We add parallel capacitances to find the equivalent capacitance.

Using this definition in Equation 3.22, we find that

$$i = C_{eq} \frac{dv}{dt} \tag{3.24}$$

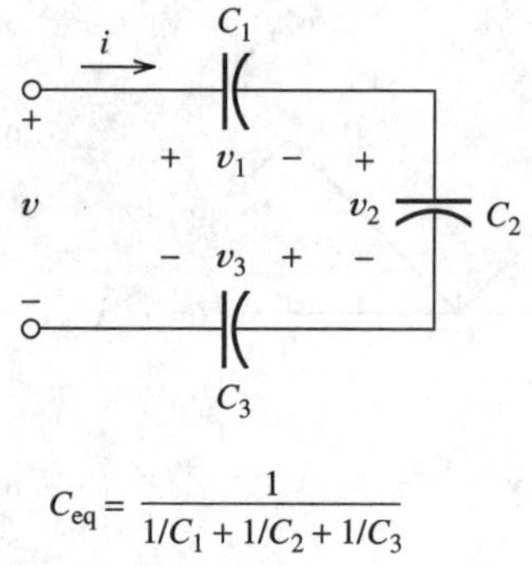

Figure 3.10 Three capacitances in series.

Thus, the current in the equivalent capacitance is the same as the total current flowing through the parallel circuit.

In sum, we add parallel capacitances to find the equivalent capacitance. Recall that for resistances, the resistances are added if they are in *series* rather than parallel. Thus, we say that capacitances in parallel are combined like resistances in series.

Capacitances in parallel are combined like resistances in series.

Capacitances in Series

By a similar development, it can be shown that the equivalent capacitance for three series capacitances is

Capacitances in series are combined like resistances in parallel.

$$C_{eq} = \frac{1}{1/C_1 + 1/C_2 + 1/C_3} \tag{3.25}$$

We conclude that capacitances in series are combined like resistances in parallel.

A technique for obtaining high voltages from low-voltage sources is to charge n capacitors in parallel with the source, and then to switch them to a series combination. The resulting voltage across the series combination is n times the source voltage. For example, in some cardiac pacemakers, a 2.5-V battery is used, but 5 V need to be applied to the heart muscle to initiate a beat. This is accomplished by charging two capacitors from the 2.5-V battery. The capacitors are then connected in series to deliver a brief 5-V pulse to the heart.

Exercise 3.3 Derive Equation 3.25 for the three capacitances shown in Figure 3.10. □

Exercise 3.4 **a.** Two capacitances of 2 μF and 1 μF are in series. Find the equivalent capacitance. **b.** Repeat if the capacitances are in parallel.

Answer **a.** 2/3 μF; **b.** 3 μF. □

3.3 PHYSICAL CHARACTERISTICS OF CAPACITORS

Capacitance of the Parallel-Plate Capacitor

A parallel-plate capacitor is shown in Figure 3.11, including dimensions. The area of each plate is denoted as A. (Actually, A is the area of one side of the plate.) The rectangular plate shown has a width W, length L, and area $A = W \times L$. The plates are parallel, and the distance between them is denoted as d.

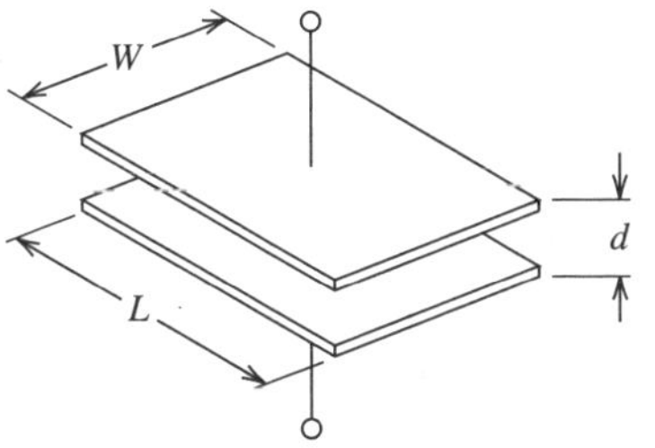

Figure 3.11 A parallel-plate capacitor, including dimensions.

Table 3.1. Relative Dielectric Constants for Selected Materials

Air	1.0
Diamond	5.5
Mica	7.0
Polyester	3.4
Quartz	4.3
Silicon dioxide	3.9
Water	78.5

If the distance d between the plates is much smaller than both the width and the length of the plates, the capacitance is approximately given by

$$C = \frac{\epsilon A}{d} \tag{3.26}$$

in which ϵ is the **dielectric constant** of the material between the plates. For vacuum, the dielectric constant is

$$\epsilon = \epsilon_0 \cong 8.85 \times 10^{-12} \text{ F/m}$$

Dielectric constant of vacuum.

For other materials, the dielectric constant is

$$\epsilon = \epsilon_r \epsilon_0 \tag{3.27}$$

where ϵ_r is the **relative dielectric constant**. Values of the relative dielectric constant for selected materials are given in Table 3.1.

Example 3.4 Calculating Capacitance Given Physical Parameters

Compute the capacitance of a parallel-plate capacitor having rectangular plates 10 cm by 20 cm separated by a distance of 0.1 mm. The dielectric is air. Repeat if the dielectric is mica.

Solution First, we compute the area of a plate:

$$A = L \times W = (10 \times 10^{-2}) \times (20 \times 10^{-2}) = 0.02 \text{ m}^2$$

From Table 3.1, we see that the relative dielectric constant of air is 1.00. Thus, the dielectric constant is

$$\epsilon = \epsilon_r \epsilon_0 = 1.00 \times 8.85 \times 10^{-12} \text{ F/m}$$

Then, the capacitance is

$$C = \frac{\epsilon A}{d} = \frac{8.85 \times 10^{-12} \times 0.02}{10^{-4}} = 1770 \times 10^{-12}\ \text{F}$$

For a mica dielectric, the relative dielectric constant is 7.0. Thus, the capacitance is seven times larger than for air or vacuum:

$$C = 12{,}390 \times 10^{-12}\ \text{F}$$ ■

Exercise 3.5 We want to design a 1-μF capacitor. Compute the length required for rectangular plates of 2-cm width if the dielectric is polyester of 15-μm thickness.
Answer $L = 24.93$ m. □

Practical Capacitors

To achieve capacitances on the order of a microfarad, the dimensions of parallel-plate capacitors are too large for compact electronic circuits such as portable computers or cellular telephones. Frequently, capacitors are constructed by alternating the plates with two layers of dielectric, which are then rolled to fit in a smaller area. By staggering the plates before rolling, electrical contact can be made with the plates from the ends of the roll. This type of construction is illustrated in Figure 3.12.

Real capacitors have maximum voltage ratings.

An engineering trade-off exists between compact size and high voltage rating.

To achieve small-volume capacitors, a very thin dielectric having a high dielectric constant is desirable. However, dielectric materials break down and become conductors when the electric field intensity (volts per meter) is too high. Thus, real capacitors have maximum voltage ratings. For a given voltage, the electric field intensity becomes higher as the dielectric layer becomes thinner. Clearly, an engineering trade-off exists between compact size and voltage rating.

Electrolytic Capacitors

In **electrolytic capacitors**, one of the plates is metallic aluminum or tantalum, the dielectric is an oxide layer on the surface of the metal, and the other "plate" is an electrolytic solution. The oxide-coated metallic plate is immersed in the electrolytic solution.

Only voltages of the proper polarity should be applied to electrolytic capacitors.

This type of construction results in high capacitance per unit volume. However, only one polarity of voltage should be applied to electrolytic capacitors. For the opposite polarity, the dielectric layer is chemically attacked, and a conductive path appears between the plates. (Usually, the allowed polarity is marked on the outer case.) On the other hand, capacitors constructed with polyethylene, Mylar®, and so on can be used in applications where the voltage polarity reverses. When the application results in voltages of only one polarity and a large-value capacitance is required, designers frequently use electrolytic capacitors.

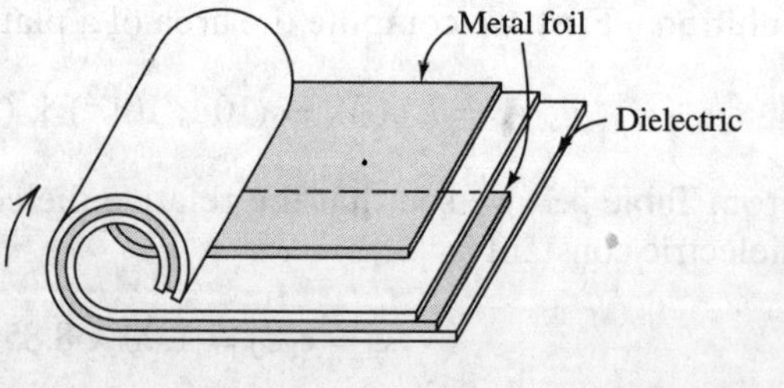

Figure 3.12 Practical capacitors can be constructed by interleaving the plates with two dielectric layers and rolling them up. By staggering the plates, connection can be made to one plate at each end of the roll.

Parasitic Effects

Real capacitors are not always well modeled simply as a capacitance. A more complete circuit model for a capacitor is shown in Figure 3.13. In addition to the capacitance C, series resistance R_s appears because of the resistivity of the material composing the plates. A series inductance L_s (we discuss inductance later in this chapter) occurs because the current flowing through the capacitor creates a magnetic field. Finally, no practical material is a perfect insulator, and the resistance R_p represents conduction through the dielectric.

We call R_s, L_s, and R_p **parasitic elements**. We design capacitors to minimize the effects of parasitic circuit elements consistent with other requirements such as physical size and voltage rating. However, parasitics are always present to some degree. In designing circuits, care must be used to select components for which the parasitic effects do not prevent proper operation of the circuit.

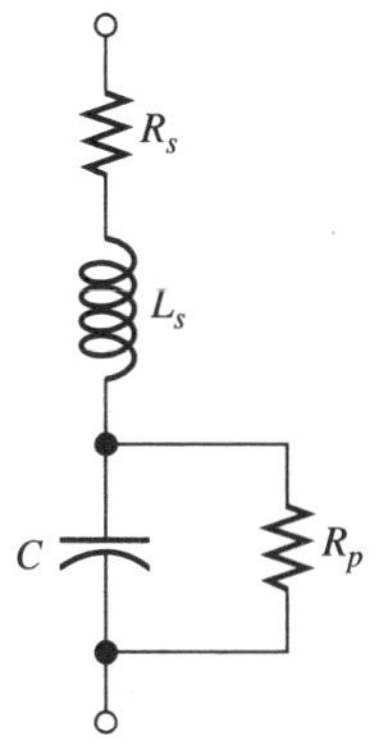

Figure 3.13 The circuit model for a capacitor, including the parasitic elements R_s, L_s, and R_p.

Example 3.5 What Happened to the Missing Energy?

Consider the situation shown in Figure 3.14. Prior to $t = 0$, the capacitor C_1 is charged to a voltage of $v_1 = 100$ V and the other capacitor has no charge (i.e., $v_2 = 0$). At $t = 0$, the switch closes. Compute the total energy stored by both capacitors before and after the switch closes.

Solution The initial stored energy for each capacitor is

$$w_1 = \frac{1}{2}C_1v_1^2 = \frac{1}{2}(10^{-6})(100)^2 = 5 \text{ mJ}$$

$$w_2 = 0$$

and the total energy is

$$w_{\text{total}} = w_1 + w_2 = 5 \text{ mJ}$$

To find the voltage and stored energy after the switch closes, we make use of the fact that the total charge on the top plates cannot change when the switch closes. This is true because there is no path for electrons to enter or leave the upper part of the circuit.

The charge stored on the top plate of C_1 prior to $t = 0$ is given by

$$q_1 = C_1v_1 = 1 \times 10^{-6} \times 100 = 100\ \mu\text{C}$$

Furthermore, the initial charge on C_2 is zero:

$$q_2 = 0$$

Thus, after the switch closes, the charge on the equivalent capacitance is

$$q_{\text{eq}} = q_1 + q_2 = 100\ \mu\text{C}$$

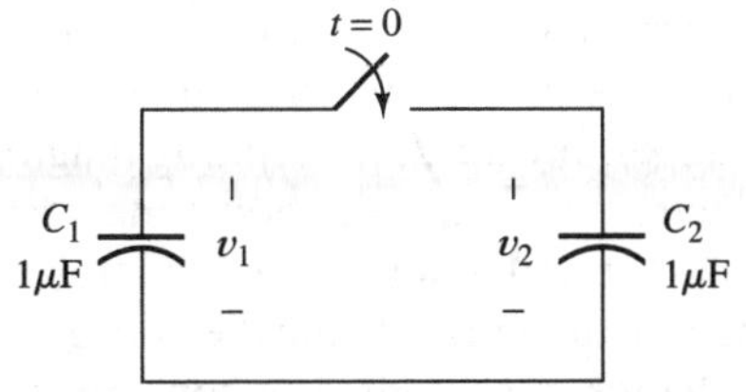

Figure 3.14 See Example 3.5.

Also, notice that after the switch is closed, the capacitors are in parallel and have an equivalent capacitance of

$$C_{eq} = C_1 + C_2 = 2\ \mu F$$

The voltage across the equivalent capacitance is

$$v_{eq} = \frac{q_{eq}}{C_{eq}} = \frac{100\ \mu C}{2\ \mu F} = 50\ V$$

Of course, after the switch is closed, $v_1 = v_2 = v_{eq}$.

Now, we compute the stored energy with the switch closed:

$$w_1 = \frac{1}{2}C_1 v_{eq}^2 = \frac{1}{2}(10^{-6})(50)^2 = 1.25\ mJ$$

$$w_2 = \frac{1}{2}C_2 v_{eq}^2 = \frac{1}{2}(10^{-6})(50)^2 = 1.25\ mJ$$

The total stored energy with the switch closed is

$$w_{total} = w_1 + w_2 = 2.5\ mJ$$

Thus, we see that the stored energy after the switch is closed is half of the value before the switch is closed. What happened to the missing energy?

Usually, the missing energy is absorbed in the parasitic resistances.

Usually, the answer to this question is that it is absorbed in the parasitic resistances. It is impossible to construct capacitors that do not have some parasitic effects. Even if we use superconductors for the wires and capacitor plates, there would be parasitic inductance. If we included the parasitic inductance in the circuit model, we would not have missing energy. (We study LC circuits with time-varying voltages and currents in Chapter 4.)

A physical circuit that is modeled exactly by Figure 3.14 does not exist.

To put it another way, a physical circuit that is modeled exactly by Figure 3.14 does not exist. Invariably, if we use a realistic model for an actual circuit, we can account for all of the energy. ■

3.4 INDUCTANCE

Inductors are usually constructed by coiling wire around a form.

An inductor is usually constructed by coiling a wire around some type of form. Several examples of practical construction are illustrated in Figure 3.15. Current flowing through the coil creates a magnetic field or flux that links the coil. Frequently, the coil form is composed of a magnetic material such as iron or iron oxides that increases the magnetic flux for a given current. (Iron cores are often composed of thin sheets called **laminations**. We discuss the reason for this construction technique in Chapter 15.)

When the current changes in value, the resulting magnetic flux changes. According to Faraday's law of electromagnetic induction, time-varying magnetic flux linking a coil induces voltage across the coil. For an ideal inductor, the voltage is proportional to the time rate of change of the current. Furthermore, the polarity of the voltage is such as to oppose the change in current. The constant of proportionality is called inductance, usually denoted by the letter L.

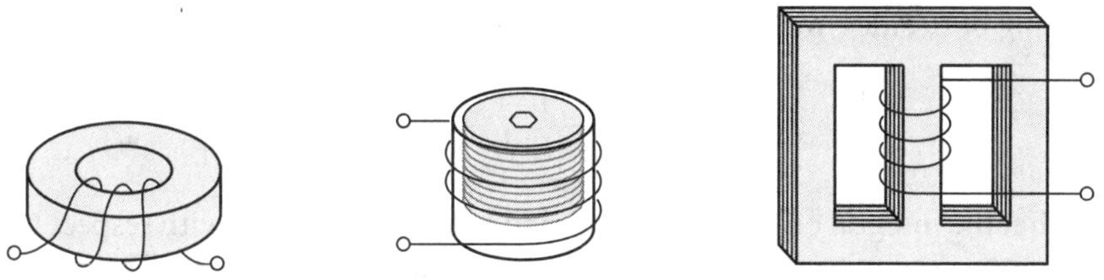

(a) Toroidal inductor

(b) Coil with an iron-oxide slug that can be screwed in or out to adjust the inductance

(c) Inductor with a laminated iron core

Figure 3.15 An inductor is constructed by coiling a wire around some type of form.

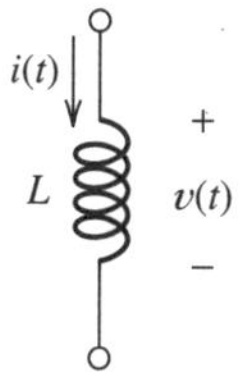

$$v(t) = L\frac{di}{dt}$$

Figure 3.16 Circuit symbol and the $v - i$ relationship for inductance.

The circuit symbol for inductance is shown in Figure 3.16. In equation form, the voltage and current are related by

$$v(t) = L\frac{di}{dt} \tag{3.28}$$

As usual, we have assumed the passive reference configuration. In case the references are opposite to the passive configuration, Equation 3.28 becomes

$$v(t) = -L\frac{di}{dt} \tag{3.29}$$

Inductance has units of henries (H), which are equivalent to volt seconds per ampere. Typically, we deal with inductances ranging from a fraction of a microhenry (μH) to several tens of henries.

Inductance has units of henries (H), which are equivalent to volt seconds per ampere.

Fluid-Flow Analogy

The fluid-flow analogy for inductance is the inertia of the fluid flowing through a *frictionless* pipe of constant diameter. The pressure differential between the ends of the pipe is analogous to voltage, and the flow rate or velocity is analogous to current. Thus, the acceleration of the fluid is analogous to rate of change of current. A pressure differential exists between the ends of the pipe only when the flow rate is increasing or decreasing.

The fluid-flow analogy for inductance is the inertia of the fluid flowing through a frictionless pipe of constant diameter.

One place where the inertia of flowing fluid is encountered is when a valve (typically operated by an electrical solenoid) closes suddenly, cutting off the flow. For example, in a washing machine, the sudden change in velocity of the water flow can cause high pressure, resulting in a bang and vibration of the plumbing. This is similar to electrical effects that occur when current in an inductor is suddenly interrupted. An application for the high voltage that appears when current is suddenly interrupted is in the ignition system for a gasoline-powered internal combustion engine.

Current in Terms of Voltage

Suppose that we know the initial current $i(t_0)$ and the voltage $v(t)$ across an inductance. Furthermore, suppose that we need to compute the current for $t > t_0$. Rearranging Equation 3.28, we have

$$di = \frac{1}{L}v(t)\,dt \tag{3.30}$$

Integrating both sides, we find that

$$\int_{i(t_0)}^{i(t)} di = \frac{1}{L}\int_{t_0}^{t} v(t)\,dt \tag{3.31}$$

Notice that the integral on the right-hand side of Equation 3.31 is with respect to time. Furthermore, the limits are the initial time t_0 and the time variable t. The integral on the left-hand side is with respect to current with limits that correspond to the time limits on the right-hand side. Integrating, evaluating, and rearranging, we have

$$i(t) = \frac{1}{L}\int_{t_0}^{t} v(t)\,dt + i(t_0) \tag{3.32}$$

Notice that as long as $v(t)$ is finite, $i(t)$ can change only by an incremental amount in a time increment. Thus, $i(t)$ must be continuous with no instantaneous jumps in value (i.e., discontinuities). (Later, we encounter idealized circuits in which infinite voltages appear briefly, and then the current in an inductance can change instantaneously.)

Stored Energy

Assuming that the references have the passive configuration, we compute the power delivered to a circuit element by taking the product of the current and the voltage:

$$p(t) = v(t)i(t) \tag{3.33}$$

Using Equation 3.28 to substitute for the voltage, we obtain

$$p(t) = Li(t)\frac{di}{dt} \tag{3.34}$$

Consider an inductor having an initial current $i(t_0) = 0$. Then, the initial electrical energy stored is zero. Furthermore, assume that between time t_0 and some later time t, the current changes from 0 to $i(t)$. As the current magnitude increases, energy is delivered to the inductor, where it is stored in the magnetic field.

Integrating the power from t_0 to t, we find the energy delivered:

$$w(t) = \int_{t_0}^{t} p(t)\,dt \tag{3.35}$$

Using Equation 3.34 to substitute for power, we have

$$w(t) = \int_{t_0}^{t} Li\frac{di}{dt}\,dt \tag{3.36}$$

Canceling differential time and changing the limits to the corresponding currents, we get

$$w(t) = \int_{0}^{i(t)} Li\,di \tag{3.37}$$

Integrating and evaluating, we obtain

$$w(t) = \frac{1}{2}Li^2(t) \tag{3.38}$$

This represents energy stored in the inductance that is returned to the circuit if the current changes back to zero.

Example 3.6 Voltage, Power, and Energy for an Inductance

The current through a 5-H inductance is shown in Figure 3.17(a). Plot the voltage, power, and stored energy to scale versus time for t between 0 and 5 s.

Solution We use Equation 3.28 to compute voltages:

$$v(t) = L\frac{di}{dt}$$

The time derivative of the current is the slope (rise over run) of the current versus time plot. For t between 0 and 2 s, we have $di/dt = 1.5$ A/s and thus, $v = 7.5$ V. For t between 2 and 4 s, $di/dt = 0$, and therefore, $v = 0$. Finally, between 4 and 5 s, $di/dt = -3$ A/s and $v = -15$ V. A plot of the voltage versus time is shown in Figure 3.17(b).

Next, we obtain power by taking the product of current and voltage at each point in time. The resulting plot is shown in Figure 3.17(c).

Finally, we use Equation 3.38 to compute the stored energy as a function of time:

$$w(t) = \frac{1}{2}Li^2(t)$$

The resulting plot is shown in Figure 3.17(d).

Notice in Figure 3.17 that as current magnitude increases, power is positive and stored energy accumulates. When the current is constant, the voltage is zero, the

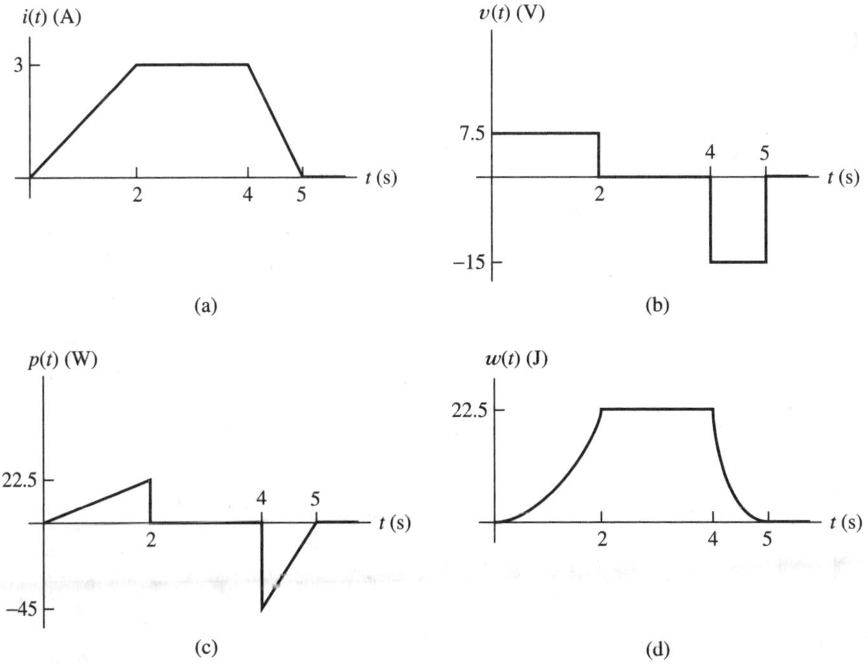

Figure 3.17 Waveforms for Example 3.6.

power is zero, and the stored energy is constant. When the current magnitude falls toward zero, the power is negative, showing that energy is being returned to the other parts of the circuit. ■

Example 3.7 Inductor Current with Constant Applied Voltage

Consider the circuit shown in Figure 3.18(a). In this circuit, we have a switch that closes at $t = 0$, connecting a 10-V source to a 2-H inductance. Find the current as a function of time.

Solution Notice that because the voltage applied to the inductance is finite, the current must be continuous. Prior to $t = 0$, the current must be zero. (Current cannot flow through an open switch.) Thus, the current must also be zero immediately after $t = 0$.

The voltage across the inductance is shown in Figure 3.18(b). To find the current, we employ Equation 3.32:

$$i(t) = \frac{1}{L}\int_{t_0}^{t} v(t)\,dt + i(t_0)$$

In this case, we take $t_0 = 0$, and we have $i(t_0) = i(0) = 0$. Substituting values, we get

$$i(t) = \frac{1}{2}\int_{0}^{t} 10\,dt$$

where we have assumed that t is greater than zero. Integrating and evaluating, we obtain

$$i(t) = 5t \text{ A} \qquad \text{for } t > 0$$

A plot of the current is shown in Figure 3.18(c).

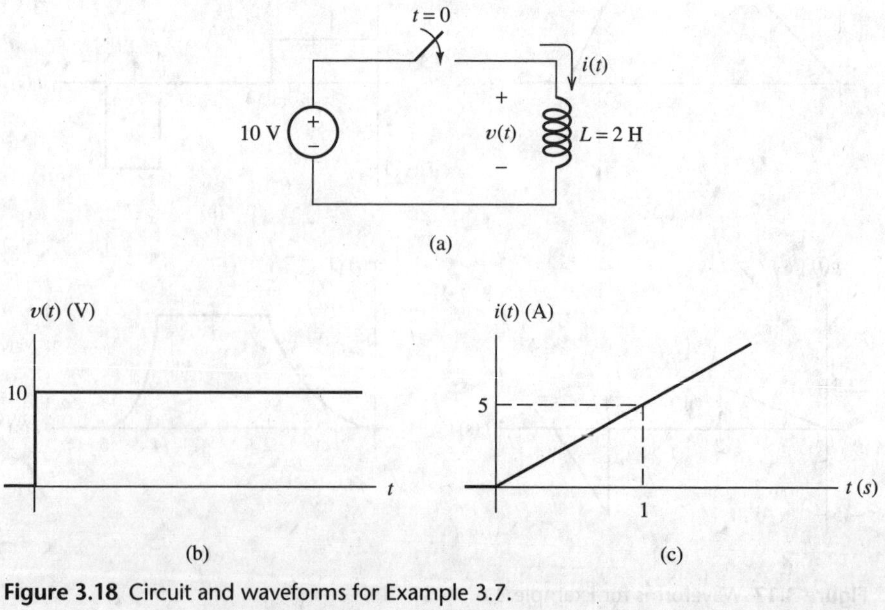

Figure 3.18 Circuit and waveforms for Example 3.7.

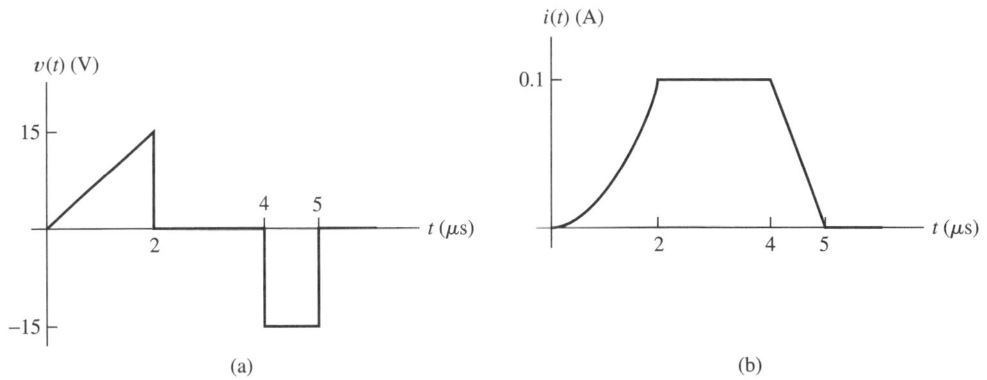

Figure 3.19 See Exercise 3.7.

Notice that the current in the inductor gradually increases after the switch is closed. Because a constant voltage is applied after $t = 0$, the current increases at a steady rate as predicted by Equation 3.28, which is repeated here for convenience:

$$v(t) = L\frac{di}{dt}$$

If $v(t)$ is constant, the rate of change of the current di/dt is constant. ■

Suppose that at $t = 1$ s, we open the switch in the circuit of Figure 3.18. Ideally, current cannot flow through an open switch. Hence, we expect the current to fall abruptly to zero at $t = 1$ s. However, the voltage across the inductor is proportional to the time rate of change of the current. For an abrupt change in current, this principle predicts infinite voltage across the inductor. This infinite voltage would last for only the instant at which the current falls. Later, we introduce the concept of an impulse function to describe this situation (and similar ones). For now, we simply point out that very large voltages can appear when we switch circuits that contain inductances.

If we set up a real circuit corresponding to Figure 3.18(a) and open the switch at $t = 1$ s, we will probably find that the high voltage causes an arc across the switch contacts. The arc persists until the energy in the inductor is used up. If this is repeated, the switch will soon be destroyed.

Exercise 3.6 The current through a 10-mH inductance is $i(t) = 0.1\cos(10^4 t)$ A. Find the voltage and stored energy as functions of time. Assume that the references for $v(t)$ and $i(t)$ have the passive configuration. (The angle is in radians.)
Answer $v(t) = -10\sin(10^4 t)$ V, $w(t) = 50\cos^2(10^4 t)$ μJ. □

Exercise 3.7 The voltage across a 150-μH inductance is shown in Figure 3.19(a). The initial current is $i(0) = 0$. Find and plot the current $i(t)$ to scale versus time. Assume that the references for $v(t)$ and $i(t)$ have the passive configuration.
Answer The current is shown in Figure 3.19(b). □

3.5 INDUCTANCES IN SERIES AND PARALLEL

It can be shown that the equivalent inductance for a series circuit is equal to the sum of the inductances connected in series. On the other hand, for inductances in parallel, we find the equivalent inductance by taking the reciprocal of the sum of the

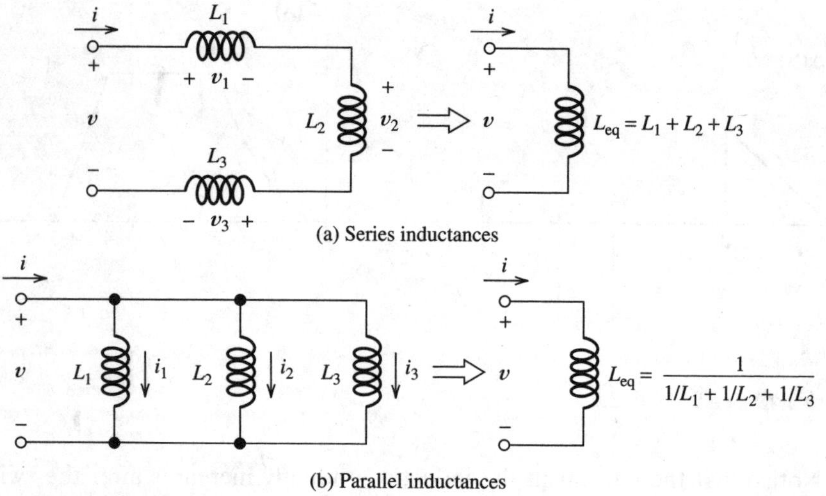

Figure 3.20 Inductances in series and parallel are combined in the same manner as resistances.

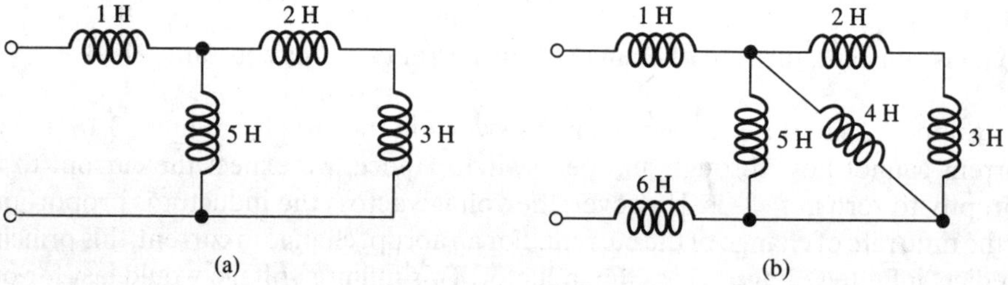

Figure 3.21 See Exercise 3.10.

Inductances in series and parallel are combined by using the same rules as for resistances: series inductances are added; parallel inductances are combined by taking the reciprocal of the sum of the reciprocals of the individual inductances.

reciprocals of the parallel inductances. Series and parallel equivalents for inductances are illustrated in Figure 3.20. Notice that inductances are combined in exactly the same way as are resistances. These facts can be proven by following the pattern used earlier in this chapter to derive the equivalents for series capacitances.

Exercise 3.8 Prove that inductances in series are added to find the equivalent inductance. □

Exercise 3.9 Prove that inductances in parallel are combined according to the formula given in Figure 3.20(b). □

Exercise 3.10 Find the equivalent inductance for each of the circuits shown in Figure 3.21.

Answer **a.** 3.5 H; **b.** 8.54 H. □

3.6 PRACTICAL INDUCTORS

Real inductors take a variety of appearances, depending on their inductance and the application. (Some examples were shown earlier in Figure 3.15.) For example, a 1-μH inductor could consist of 25 turns of fine (say, number 28) wire wound on an iron oxide toroidal (doughnut-shaped) core having an outside diameter of 1/2 cm.

On the other hand, a typical 5-H inductor consists of several hundred turns of number 18 wire on an iron form having a mass of 1 kg.

Usually, metallic iron forms, also called *cores*, are made of thin sheets called *laminations*. [See Figure 3.15(c) for an example.] This is necessary because voltages are induced in the core by the changing magnetic field. These voltages cause **eddy currents** to flow in the core, dissipating energy. Usually, this **core loss** is undesirable. Using laminations that are insulated from one another helps to reduce eddy-current loss. The laminations are arranged perpendicular to the expected current direction.

Another way to defeat eddy currents is to use a core composed of **ferrites**, which are oxides of iron that are electrical insulators. Still another approach is to combine powdered iron with an insulating binder.

PRACTICAL APPLICATION 3.1

Electronic Photo Flash

Figure PA3.1 shows the electrical circuit of an electronic photo flash such as you may have seen on a camera. The objective of the unit is to produce a bright flash of light by passing a high current through the flash tube while the camera shutter is open. As much as 1000 W is supplied to the flash tube during the flash, which lasts for less than a millisecond. Although the power level is quite high, the total energy delivered is not great because of the short duration of the flash. (The energy is on the order of a joule.)

It is not possible to deliver the power directly from the battery to the flash tube for several reasons. First, practical batteries supply a few tens of volts at most, while several hundred volts are needed to operate the flash tube. Second, applying the principle of maximum power transfer, the maximum power available from the battery is limited to 1 W by its internal Thévenin resistance. (See Equation 2.67 and the related discussion.) This does not nearly meet the needs of the flash tube. Instead, energy is delivered by the battery over a period of several seconds and stored in the capacitor. The stored energy can be quickly extracted from the capacitor because the parasitic resistance in series with the capacitor is very low.

The electronic switch alternates between open and closed approximately 10,000 times per second. (In some units, you can hear a high-pitched whistle resulting from incidental conversion of some of the energy to acoustic form.) While the electronic switch is closed, the battery causes the current in the inductor to build up. Then when the switch opens, the inductor forces current to flow through the diode, charging the capacitor. (Recall that the current in an inductor cannot change instantaneously.)

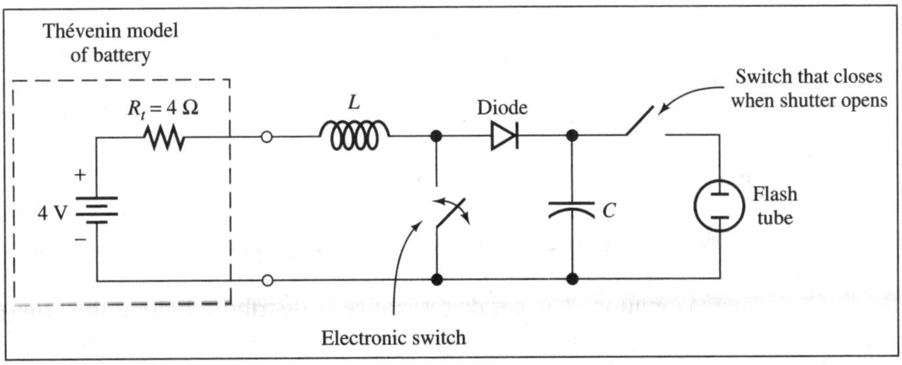

Figure PA3.1

Current can flow through the diode only in the direction of the arrow. Thus, the diode allows charge to flow into the capacitor when the electronic switch is open and prevents charge from flowing off the capacitor when the electronic switch is closed. Thus, the charge stored on the capacitor increases each time the electronic switch opens. Eventually, the voltage on the capacitor reaches several hundred volts. When the camera shutter is opened, another switch is closed, allowing the capacitor to discharge through the flash tube.

A friend of the author has a remote cabin on the north shore of Lake Superior that has an unusual water system (illustrated in Figure PA3.2) analogous to the electronic flash circuit. Water flows through a large pipe immersed in the river. Periodically, a valve on the bottom end of the pipe suddenly closes, stopping the flow. The inertia of the flowing water creates a pulse of high pressure when the valve closes. This high pressure forces water through a one-way ball valve into a storage tank. Air trapped in the storage tank is compressed and forces water to flow to the cabin as needed.

Can you identify the features in Figure PA3.2 that are analogous to each of the circuit elements in Figure PA3.1?

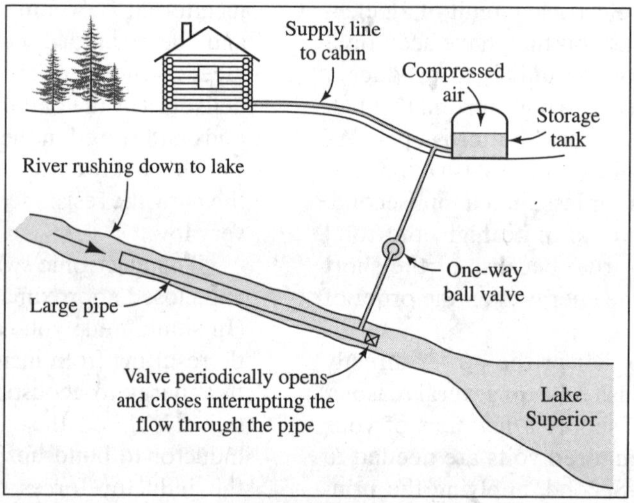

Figure PA3.2

Parasitic Effects for Real Inductors

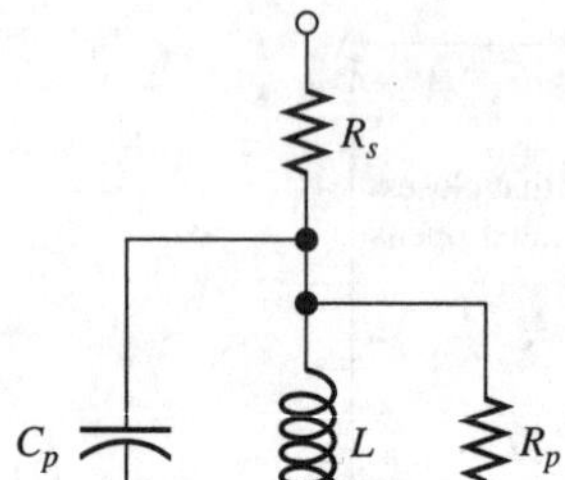

Figure 3.22 Circuit model for real inductors including several parasitic elements.

Real inductors have parasitic effects in addition to the desired inductance. A circuit model for a real inductor is shown in Figure 3.22. The series resistance R_s is caused by the resistivity of the material composing the wire. (This parasitic effect can be avoided by using wire composed of a superconducting material, which has zero resistivity.) The parallel capacitance is associated with the electric field in the dielectric (insulation) between the coils of wire. It is called **interwinding capacitance**. The parallel resistance R_p represents core loss due, in part, to eddy currents in the core.

Actually, the circuit model for a real inductor shown in Figure 3.22 is an approximation. The series resistance is distributed along the length of the wire, as is the interwinding capacitance. A more accurate model for a real inductor would break each of the parasitic effects into many segments (possibly, an infinite number). Ultimately, we could abandon circuit models altogether and use electromagnetic field theory directly.

Rarely is this degree of detail necessary. Usually, modeling a real inductor as an inductance, including at most a few parasitic effects, is sufficiently accurate. Of course, computer-aided circuit analysis allows us to use more complex models and achieve more accurate results than traditional mathematical analysis.

3.7 MUTUAL INDUCTANCE

Sometimes, several coils are wound on the same form so that magnetic flux produced by one coil links the others. Then a time-varying current flowing through one coil induces voltages in the other coils. The circuit symbols for two mutually coupled inductances are shown in Figure 3.23. The **self inductances** of the two coils are denoted as L_1 and L_2, respectively. The **mutual inductance** is denoted as M, which also has units of henries. Notice that we have selected the passive reference configuration for each coil in Figure 3.23.

The equations relating the voltages to the currents are also shown in Figure 3.23. The mutual terms, $M\,di_1/dt$ and $M\,di_2/dt$, appear because of the mutual coupling of the coils. The self terms, $L_1\,di_1/dt$ and $L_2\,di_2/dt$, are the voltages induced in each coil due to its own current.

The magnetic flux produced by one coil can either aid or oppose the flux produced by the other coil.

The magnetic flux produced by one coil can either aid or oppose the flux produced by the other coil. The dots on the ends of the coils indicate whether the fields are aiding or opposing. If one current enters a dotted terminal and the other leaves, the fields oppose one another. For example, if both i_1 and i_2 have positive values in Figure 3.23(b), the fields are opposing. If both currents enter the respective dots (or if both leave), the fields aid. Thus, if both i_1 and i_2 have positive values in Figure 3.23(a), the fields are aiding.

The signs of the mutual terms in the equations for the voltages depend on how the currents are referenced with respect to the dots. If both currents are referenced into (or if both are referenced out of) the dotted terminals, as in Figure 3.23(a), the mutual term is positive. If one current is referenced into a dot and the other out, as in Figure 3.23(b), the mutual term carries a negative sign.

Linear Variable Differential Transformer

An application of mutual inductance can be found in a position transducer known as the linear variable differential transformer (LVDT), illustrated in Figure 3.24. An ac source connected to the center coil sets up a magnetic field that links both halves of

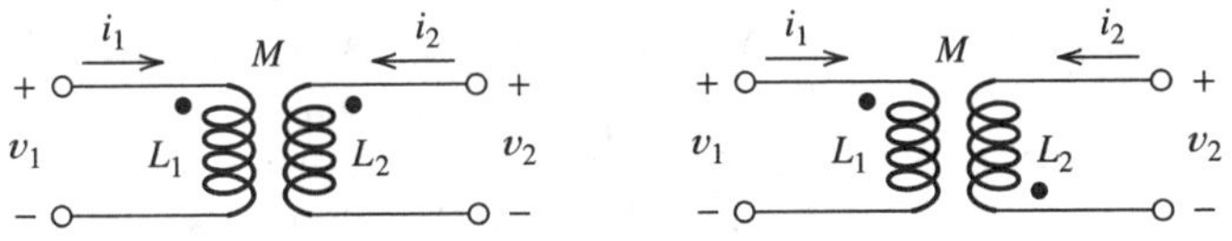

(a)

$$v_1 = L_1 \frac{di_1}{dt} + M \frac{di_2}{dt}$$

$$v_2 = M \frac{di_1}{dt} + L_2 \frac{di_2}{dt}$$

(b)

$$v_1 = L_1 \frac{di_1}{dt} - M \frac{di_2}{dt}$$

$$v_2 = -M \frac{di_1}{dt} + L_2 \frac{di_2}{dt}$$

Figure 3.23 Circuit symbols and $v - i$ relationships for mutually coupled inductances.

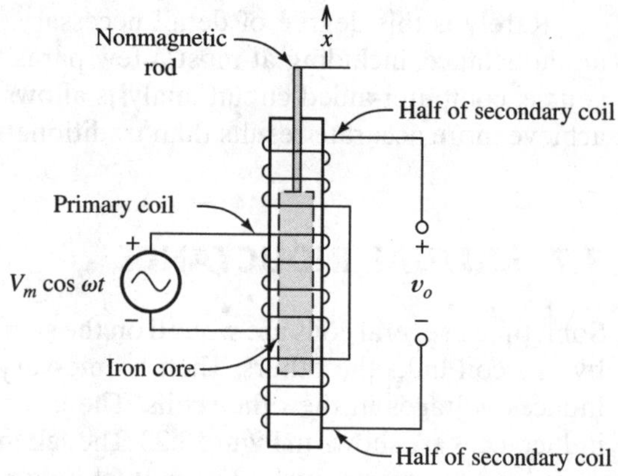

Figure 3.24 A linear variable differential transformer used as a position transducer.

the secondary coil. When the iron core is centered in the coils, the voltages induced in the two halves of the secondary cancel so that $v_o(t) = 0$. (Notice that the two halves of the secondary winding are wound in opposite directions.) As the core moves up or down, the couplings between the primary and the halves of the secondary change. The voltage across one half of the coil becomes smaller, and the voltage across the other half becomes greater. Ideally, the output voltage is given by

$$v_o(t) = Kx \cos(\omega t)$$

where x is the displacement of the core. LVDTs are used in applications such as automated manufacturing operations to measure displacements.

3.8 SYMBOLIC INTEGRATION AND DIFFERENTIATION USING MATLAB

As we have seen, finding the current given the voltage (or vice versa) for an energy storage element involves integration or differentiation. Thus, we may sometimes need to find symbolic answers for integrals or derivatives of complex functions, which can be very difficult by traditional methods. Then, we can resort to using symbolic mathematics software. Several programs are available including Maple™ from Maplesoft Corporation, Mathematica™ from Wolfram Research, and the Symbolic Toolbox which is an optional part of MATLAB from Mathsoft. Each of these programs has its strengths and weaknesses, and when a difficult problem warrants the effort, all of them should be tried. Because MATLAB is widely used in Electrical Engineering, we confine our brief discussion to the Symbolic Toolbox.

One note of caution: We have developed the examples, exercises, and problems using MATLAB version R2011b. Keep in mind that if you use versions other than R2011b, you may not be able to reproduce our results. Try running our example m-files before sinking a lot of time into solving the problems. Hopefully, your instructor can give you some guidance on what to expect with the MATLAB versions available to you.

In the following, we assume that you have some familiarity with MATLAB. A variety of online interactive tutorials are available at `https://www.mathworks.com/`. However, you should find it easy to write MATLAB instructions for the exercises and problems in this chapter by modeling your solutions after the code in our examples.

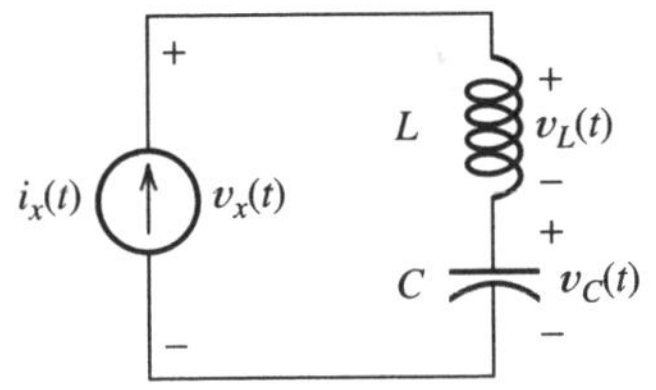

Figure 3.25 Circuit of Example 3.8.

Example 3.8 Integration and Differentiation Using the MATLAB Symbolic Toolbox

Use MATLAB to find expressions for the three voltages shown in Figure 3.25 given $v_C(0) = 0$ and

$$\begin{aligned} i_x(t) &= kt^2 \exp(-at)\sin(\omega t) \text{ for } t \geq 0 \\ &= 0 \text{ for } t < 0 \end{aligned} \tag{3.39}$$

Also, plot the current and the voltages for $k = 3, a = 2, \omega = 1, L = 0.5$ H, $C = 1$ F, and $t \geq 0$. (These values have been chosen mainly to facilitate the demonstration of MATLAB capabilities.) The currents are in amperes, voltages are in volts, ωt is in radians, and time t is in seconds.

Solution At first, we use symbols to represent the various parameters (k, a, ω, L, and C), denoting the current and the voltages as ix, vx, vL, and vC. Then, we substitute the numerical values for the symbols and denote the results as ixn, vxn, vLn, and vCn. (The letter "n" is selected to suggest that the "numerical" values of the parameters have been substituted into the expressions.)

We show the commands in **boldface**, comments in regular font, and MATLAB responses in color. Comments (starting with the % sign) are ignored by MATLAB. We present the work as if we were entering the commands and comments one at a time in the MATLAB command window, however, it is usually more convenient to place all of the commands in an m-file and execute them as a group.

To start, we define the various symbols as symbolic objects in MATLAB, define the current ix, and substitute the numerical values of the parameters to obtain ixn.

```
>> clear all % Clear work area of previous work.
>> syms vx ix vC vL vxn ixn vCn vLn k a w t L C
>> % Names for symbolic objects must start with a letter and
>> % contain only alpha-numeric characters.
>> % Next, we define ix.
>> ix=k*t^2*exp(-a*t)*sin(w*t)
     ix =
     (k*t^2*sin(w*t))/exp(a*t)
>> % Next, we substitute k=3, a=2, and w=1
>> % into ix and denote the result as ixn.
>> ixn = subs(ix,[k a w],[3 2 1])
     ixn =
     (3*t^2*sin(t))/exp(a*t)
```

Next, we want to plot the current versus time. We need to consider what range of t should be used for the plot. In standard mathematical typesetting, the expression we need to plot is

$$\begin{aligned} i_x(t) &= 3t^2 \exp(-2t)\sin(t) \text{ for } t \geq 0 \\ &= 0 \text{ for } t < 0 \end{aligned}$$

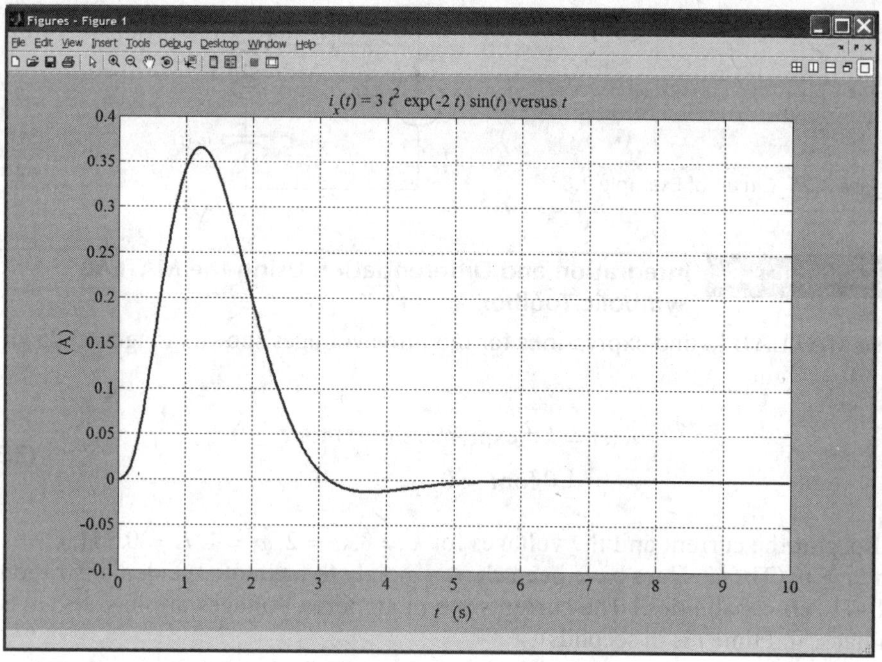

Figure 3.26 Plot of $i_x(t)$ produced by MATLAB.

Thoughtful examination of this expression (perhaps supplemented with a little work with a calculator) reveals that the current is zero at $t = 0$, builds up quickly after $t = 0$ because of the t^2 term, and decays to relatively small values after about $t = 10$ s because of the exponential term. Thus, we select the range from $t = 0$ to $t = 10$ s for the plot. Continuing in MATLAB, we have

```
>> % Next, we plot ixn for t ranging from 0 to 10 s.
>> ezplot(ixn,[0,10])
```

This opens a window with a plot of the current versus time as shown in Figure 3.26. As expected, the current increases rapidly after $t = 0$ and decays to insignificant values by $t = 10$ s. (We have used various Edit menu commands to improve the appearance of the plot for inclusion in this book.)

Next, we determine the inductance voltage, which is given by

$$v_L(t) = L\frac{di_x(t)}{dt}$$

in which the parameters, a, k, and ω are treated as constants. The corresponding MATLAB command and the result are:

```
>> vL=L*diff(ix,t) % L times the derivative of ix with respect to t.
   vL =
   L*((2*k*t*sin(t*w))/exp(a*t) - (a*k*t^2*sin(t*w))/exp(a*t) +
   (k*t^2*w*cos(t*w))/exp(a*t))
>> % A nicer display for vL is produced with the commands:
>> vL = simple(vL);
>> pretty(vL)
   L k t (2 sin(t w) - a t sin(t w) + t w cos(t w))
   ------------------------------------------------
                      exp(a t)
```

In more standard mathematical typesetting, this becomes

$$v_L(t) = Lkt\exp(-at)[2\sin(\omega t) - at\sin(\omega t) + \omega t\cos(\omega t)]$$

which we can verify by manually differentiating the right-hand side of Equation 3.39 and multiplying by L. Next, we determine the voltage across the capacitance.

$$v_C(t) = \frac{1}{C}\int_0^t i_x(t)dt + v_C(0) \text{ for } t \geq 0$$

Substituting the expressions for the current and initial voltage we obtain,

$$v_C(t) = \frac{1}{C}\int_0^t kt^2\exp(-at)\sin(\omega t)dt \text{ for } t \geq 0$$

This is not a simple integration to perform by hand, but we can accomplish it easily with MATLAB:

```
>> % Integrate ix with respect to t with limits from 0 to t.
>> vC=(1/C)*int(ix,t,0,t);
>> % We included the semicolon to suppress the output, which is
>> % much too complex for easy interpretation.
>> % Next, we find the total voltage vx.
>> vx = vC + vL;
>> % Now we substitute numerical values for the parameters.
>> vLn=subs(vL,[k a w L C],[3 2 1 0.5 1]);
>> vCn=subs(vC,[k a w L C],[3 2 1 0.5 1]);
>> vxn=subs(vx,[k a w L C],[3 2 1 0.5 1]);
>> % Finally, we plot all three voltages in the same window.
>> figure % Open a new figure for this plot.
>> ezplot(vLn,[0,10])
>> hold on % Hold so the following two plots are on the same axes.
>> ezplot(vCn,[0,10])
>> ezplot(vxn,[0,10])
```

The resulting plot is shown in Figure 3.27. (Here again, we have used various items on the Edit menu to change the scale of the vertical axis and dress up the plot for inclusion in this book.)

The commands for this example are included as an m-file named Example_3_8 in the MATLAB files. (See Appendix E for information about accessing these MATLAB files.) If you copy the file and place it in a folder in the MATLAB path for your computer, you can run the file and experiment with it. For example, after running the m-file, if you enter the command

```
>> simple(vC)
```

you will see the rather complicated symbolic mathematical expression for the voltage across the capacitance. ■

Exercise 3.11 Use MATLAB to work Example 3.2 on page 147 resulting in plots like those in Figure 3.5.

Answer The MATLAB commands including some explanatory comments are:

```
clear % Clear the work area.
% We avoid using i alone as a symbol for current because
% we reserve i for the square root of -1 in MATLAB. Thus, we
```

```
% will use iC for the capacitor current.
syms t iC qC vC % Define t, iC, qC and vC as symbolic objects.
iC = 0.5*sin((1e4)*t);
ezplot(iC, [0 3*pi*1e-4])
qC=int(iC,t,0,t); % qC equals the integral of iC.
figure % Plot the charge in a new window.
ezplot(qC, [0 3*pi*1e-4])
vC = 1e7*qC;
figure % Plot the voltage in a new window.
ezplot(vC, [0 3*pi*1e-4])
```

The plots are very similar to those of Figure 3.5 on page 148. An m-file (named Exercise_3_11) can be found in the MATLAB folder. □

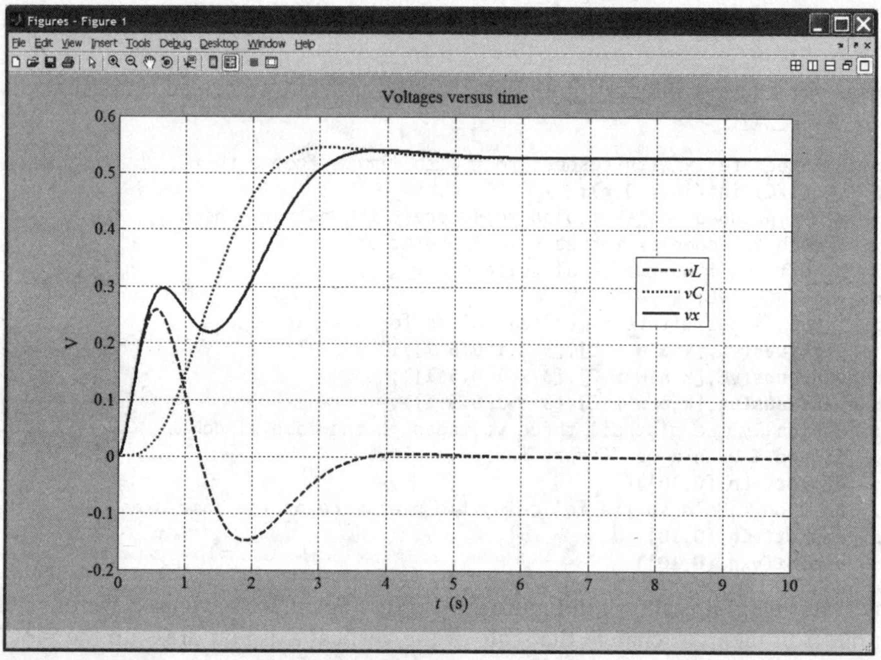

Figure 3.27 Plots of the voltages for Example 3.8.

Summary

1. Capacitance is the circuit property that accounts for electric-field effects. The units of capacitance are farads (F), which are equivalent to coulombs per volt.
2. The charge stored by a capacitance is given by $q = Cv$.
3. The relationships between current and voltage for a capacitance are

$$i = C\frac{dv}{dt}$$

and

$$v(t) = \frac{1}{C}\int_{t_0}^{t} i(t)\,dt + v(t_0)$$

4. The energy stored by a capacitance is given by

$$w(t) = \frac{1}{2}Cv^2(t)$$

5. Capacitances in series are combined in the same manner as resistances in parallel.

6. Capacitances in parallel are combined in the same manner as resistances in series.
7. The capacitance of a parallel-plate capacitor is given by

$$C = \frac{\epsilon A}{d}$$

For vacuum, the dielectric constant is $\epsilon = \epsilon_0 \cong 8.85 \times 10^{-12}$ F/m. For other materials, the dielectric constant is $\epsilon = \epsilon_r \epsilon_0$, where ϵ_r is the relative dielectric constant.
8. Real capacitors have several parasitic effects.
9. Inductance accounts for magnetic-field effects. The units of inductance are henries (H).
10. The relationships between current and voltage for an inductance are

$$v(t) = L\frac{di}{dt}$$

and

$$i(t) = \frac{1}{L}\int_{t_0}^{t} v(t)\,dt + i(t_0)$$

11. The energy stored in an inductance is given by

$$w(t) = \frac{1}{2}Li^2(t)$$

12. Inductances in series or parallel are combined in the same manner as resistances.
13. Real inductors have several parasitic effects.
14. Mutual inductance accounts for mutual coupling of magnetic fields between coils.
15. MATLAB is a powerful tool for symbolic integration, differentiation, and plotting of functions.

Problems

Section 3.1: Capacitance

P3.1. Describe the internal construction of capacitors.

P3.2. What current flows through an ideal capacitor if the voltage across the capacitor is constant with time? To what circuit element is an ideal capacitor equivalent in circuits for which the currents and voltages are constant with time?

P3.3. Explain the properties of dielectric materials giving two examples. What is permittivity?

P3.4. How can electrical current flow "through" a capacitor even though a nonconducting layer separates the metallic parts?

***P3.5.** The voltage across a 10-μF capacitor is given by $v(t) = 100\sin(1000t)$. Assume that the argument of the sin function is in radians. Find expressions for the current, power, and stored energy. Sketch the waveforms to scale versus time for time ranging from 0 to 2π ms.

***P3.6.** A 10000-μF capacitor is initially in a discharged condition. How long will it take to charge it to a value of 10 V if a steady current of 100 μA is used?

***P3.7.** A constant (dc) current $i(t) = 3$ mA flows into a 50-μF capacitor. The voltage at $t = 0$ is $v(0) = -20$ V. The references for $v(t)$ and $i(t)$ have the passive configuration. Find the power at $t = 0$ and state whether the power flow is into or out of the capacitor. Repeat for $t = 1$ s.

***P3.8.** We want to store sufficient energy in a 0.01-F capacitor to supply 5 horsepower (hp) for 1 hour. To what voltage must the capacitor be charged? (*Note:* One horsepower is equivalent to 745.7 watts.) Does this seem to be a practical method for storing this amount of energy? Do you think that an electric automobile design based on capacitive energy storage is feasible?

P3.9. Suppose that we have a 5-μF capacitor with 100 V between its terminals. Determine the magnitude of the net charge stored on each plate and the total net charge on both the plates.

P3.10. Suppose that a current given by $i(t) = I_m\cos(\omega t)$ flows through a capacitance C and the voltage is zero at $t = 0$. Assume that ωt has units of radians. Furthermore, ω is very large, ideally approaching infinity. For this current, does the capacitance approximate either an open or a short circuit? Explain.

* Denotes that answers are contained in the Student Solutions files. See Appendix E for more information about accessing the Student Solutions.

P3.11. The voltage across a 5-μF capacitor is given by $v(t) = 50e^{-100t}$. Find the expressions for the current, power, and the stored energy. Sketch the waveforms to scale versus time.

P3.12. We are given a 5-μF capacitor that is charged to 200 V. Determine the initial stored charge and energy. If this capacitor is discharged to 0 V in a time interval of 1 μs, find the average power delivered by the capacitor during the discharge interval.

P3.13. Prior to $t = 0$, a 100-μF capacitance is uncharged. Starting at $t = 0$, the voltage across the capacitor is increased linearly with time to 100 V in 2 s. Then, the voltage remains constant at 100 V. Sketch the voltage, current, power, and stored energy to scale versus time.

P3.14. The current through a 0.5-μF capacitor is shown in Figure P3.14. At $t = 0$, the voltage is zero. Sketch the voltage, power, and stored energy to scale versus time.

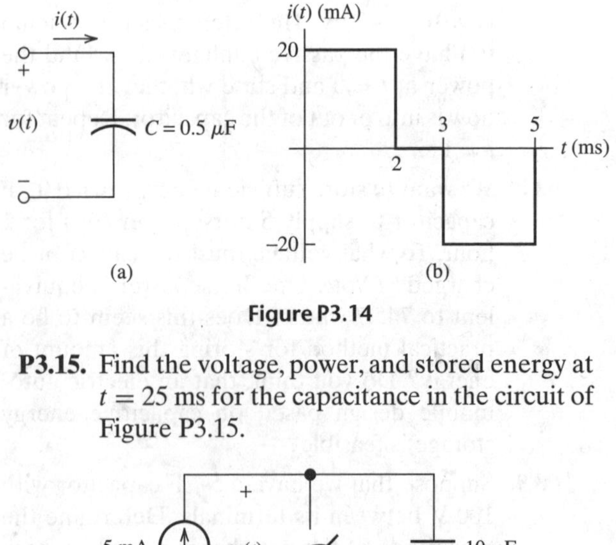

Figure P3.14

P3.15. Find the voltage, power, and stored energy at $t = 25$ ms for the capacitance in the circuit of Figure P3.15.

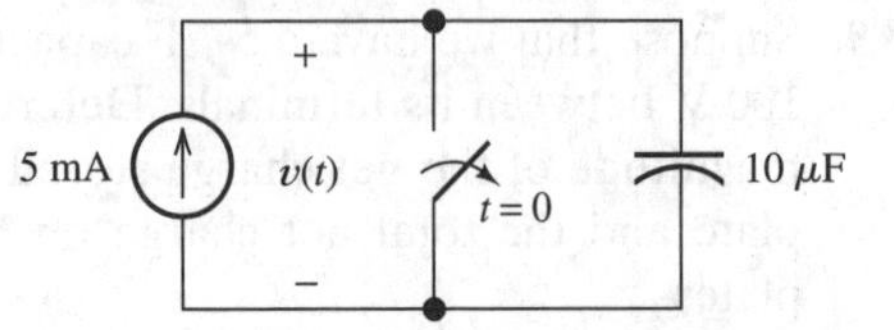

Figure P3.15

P3.16. A capacitance and the current through it are shown in Figure P3.16. At $t = 0$, the voltage is $v_C(0) = 10$ V. Sketch the voltage, power, and stored energy to scale versus time.

P3.17. At $t = 5$ s, the energy stored in a 10-μF capacitor is 200 J and is decreasing at 500 J/s. Determine the voltage magnitude and current magnitude at $t = 5$ s. Does the current enter or leave the positive terminal of the capacitor?

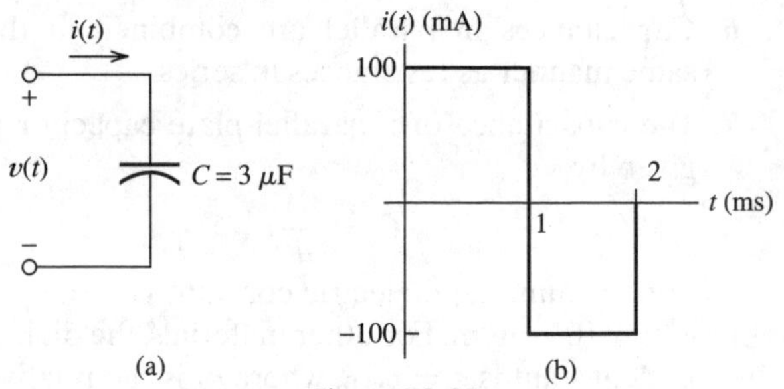

Figure P3.16

P3.18. Consider a very large capacitance (ideally, infinite) charged to 10 V. What other circuit element has the same current–voltage relationship? Explain your answer.

P3.19. A certain parallel-plate capacitor, which has one plate rotating so the overlap of the plates is a function of time, has a capacitance given by

$$C = 2 + \sin(200t)\ \mu\text{F}$$

in which the argument of the sine function is in radians. A constant voltage of 100 V is applied to this capacitor. Determine the current as a function of time.

P3.20. At $t = t_0$ the voltage across a certain capacitance C is zero. A pulse of current flows through the capacitance between t_0 and $t_0 + \Delta t$, and the voltage across the capacitance increases to V_f. What can you say about the peak amplitude I_m and area under the pulse waveform (i.e., current versus time)? What are the units and physical significance of the area under the pulse? What must happen to the peak amplitude and area under the pulse as Δt approaches zero, assuming that V_f remains the same?

P3.21. First, consider a resistor. What resistance corresponds to a short circuit? Next, consider an uncharged capacitor. What value of capacitance corresponds to a short circuit? Explain your answers. Repeat for an open circuit.

P3.22. A 20-μF capacitor has a voltage given by

$$v(t) = 3\cos(10^5 t) + 2\sin(10^5 t)$$

Assume that the arguments of the sine and cosine functions are in radians. Find the power at $t = 0$ and state whether the power flow is into or out of the capacitor. Repeat for $t_2 = (\pi/2) \times 10^{-5}$ s.

Section 3.2: Capacitances in Series and Parallel

P3.23. Describe how capacitances are combined in series and in parallel. Compare with how resistances are combined.

***P3.24.** Find the equivalent capacitance for each of the circuits shown in Figure P3.24.

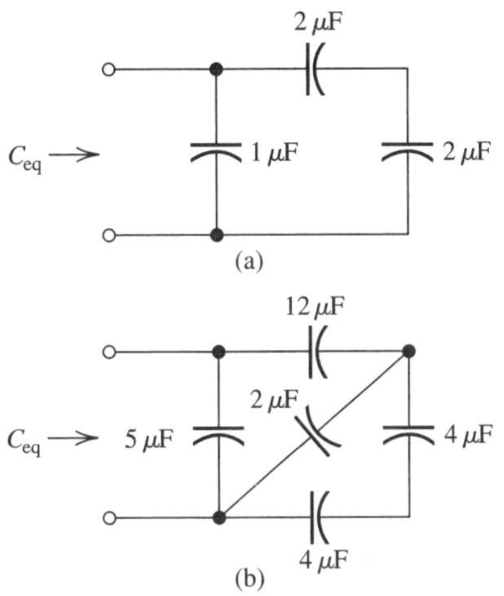

Figure P3.24

***P3.25.** Suppose that we are designing a cardiac pacemaker circuit. The circuit is required to deliver pulses of 1-ms duration to the heart, which can be modeled as a 500-Ω resistance. The peak amplitude of the pulses is required to be 5 V. However, the battery delivers only 2.5 V. Therefore, we decide to charge two equal-value capacitors in parallel from the 2.5-V battery and then switch the capacitors in series with the heart during the 1-ms pulse. What is the minimum value of the capacitances required so the output pulse amplitude remains between 4.9 V and 5.0 V throughout its 1-ms duration? If the pulses occur once every second, what is the average current drain from the battery? Use approximate calculations, assuming constant current during the output pulse. Find the ampere-hour rating of the battery so it lasts for five years.

P3.26. Find the equivalent capacitance between terminals x and y for each of the circuits shown in Figure P3.26.

P3.27. When two capacitors are connected in parallel, the total capacitance is 6 μF. When they are connected in series, the capacitance is 1.33 μF. Find the value of the two capacitances.

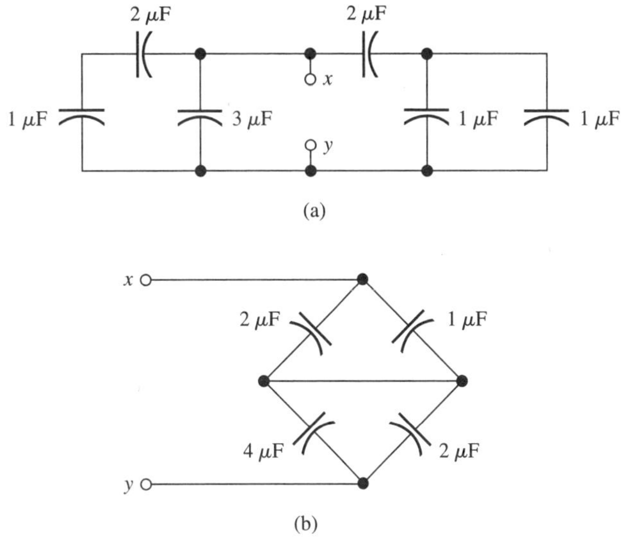

Figure P3.26

P3.28. Consider two initially uncharged capacitors $C_1 = 15$ μF and $C_2 = 10$ μF connected in series. Then, a 50-V source is connected to the series combination, as shown in Figure P3.28. Find the voltages v_1 and v_2 after the source is applied. (*Hint:* The charges stored on the two capacitors must be equal, because the current is the same for both the capacitors.)

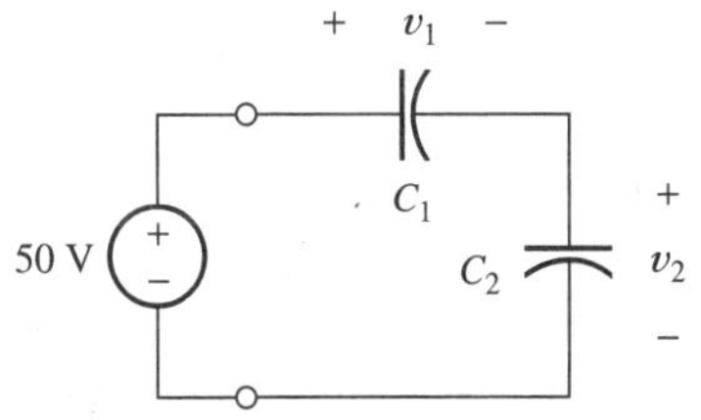

Figure P3.28

P3.29. We have capacitor $C_1 = 200$ μF, which is charged to an initial voltage of 50 V, and capacitor $C_2 = 200$ μF, which is charged to 100 V. If they are placed in series with the positive terminal of the first connected to the negative terminal of the second, determine the equivalent capacitance and its initial voltage. Now, compute the total energy stored in the two capacitors. Compute the energy stored in the equivalent capacitance. Why is it less than the total energy stored in the original capacitors?

P3.30. Suppose the capacitors C_1 and C_2 of Figure P3.28 are placed in parallel instead of series. Find the charge deposited on each if the voltage source of 50 V is connected across the parallel combination.

Section 3.3: Physical Characteristics of Capacitors

***P3.31.** Determine the capacitance of a parallel-plate capacitor having plates 100 mm by 0.3 m separated by 0.04 mm. The dielectric constant $\epsilon = 100$ pF/m.

P3.32. We have a 100-pF parallel-plate capacitor, with each plate having a width W and a length L. The plates are separated by air with a distance d. Assume that L and W are both much larger than d. What is the new capacitance if: **a.** both L and W are doubled and the other parameters are unchanged? **b.** the separation d is halved and the other parameters are unchanged from their initial values? **c.** the air dielectric is replaced with oil having a relative dielectric constant of 35 and the other parameters are unchanged from their initial values?

***P3.33.** Suppose that we have a 1000-pF parallel-plate capacitor with air dielectric charged to 1000 V. The capacitor terminals are open circuited. Find the stored energy. If the plates are moved farther apart so that d is doubled, determine the new voltage on the capacitor and the new stored energy. Where did the extra energy come from?

P3.34. We have a parallel-plate capacitor, with each plate having a width W and a length L. The plates are separated by air with a distance d. Assume that L and W are both much larger than d. The maximum voltage that can be applied is limited to $V_{\max} = Kd$, in which K is called the breakdown strength of the dielectric. Derive an expression for the maximum energy that can be stored in the capacitor in terms of K and the volume of the dielectric. If we want to store the maximum energy per unit volume, does it matter what values are chosen for L, W, and d? What parameters are important?

P3.35. One type of microphone is formed from a parallel-plate capacitor arranged so the acoustic pressure of the sound wave affects the distance between the plates. Suppose we have such a microphone in which the plates have an area of 10 cm^2, the dielectric is air and the distance between the plates is a function of time given by

$$d(t) = 100 + 0.3\cos(1000t)\ \mu\text{m}$$

A constant voltage of 200 V is applied to the plates. Determine the current through the capacitance as a function of time by using the approximation $1/(1+x) \cong 1-x$ for $x << 1$. (The argument of the sinusoid is in radians.)

P3.36. Consider a liquid-level sensor that consists of two parallel plates of conductor immersed in an insulating liquid, as illustrated in Figure P3.36. When the tank is empty (i.e., $x = 0$), the capacitance of the plates is 200 pF. The relative dielectric constant of the liquid is 15. Determine an expression for the capacitance C as a function of the height x of the liquid.

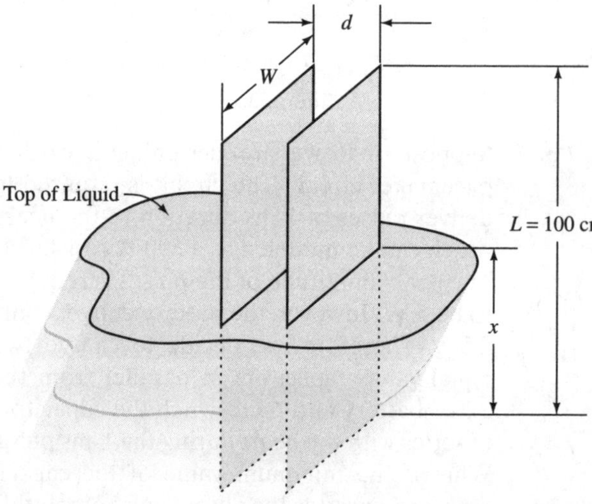

Figure P3.36

P3.37. A parallel-plate capacitor like that shown in Figure P3.36 has a capacitance of 2500 pF when the tank is full so the plates are totally immersed in the insulating liquid. (The dielectric constant of the fluid and the plate dimensions are different for this problem than for Problem P3.36.) The capacitance is 100 pF when the tank is empty and the space between the plates is filled with air. Suppose that the tank is full and the capacitance is charged to 1000 V. Then, the capacitance is open circuited so the charge on the plates cannot change, and the tank is drained. Compute the voltage after the tank is drained and the electrical energy stored in the capacitor before and after the tank is drained. With the plates open circuited, there is no electrical source for the extra energy. Where could it have come from?

P3.38. A 0.1-μF capacitor has a parasitic series resistance of 10 Ω as shown in Figure P3.38. Suppose that the voltage across the capacitance is $v_c(t) = 10\cos(100t)$; find the voltage across the resistance. In this situation, to find the total voltage $v(t) = v_r(t) + v_c(t)$ to within 1% accuracy, is it necessary to include the parasitic resistance? Repeat if $v_c(t) = 0.1\cos(10^7 t)$. The argument of the cosine function is in radians.

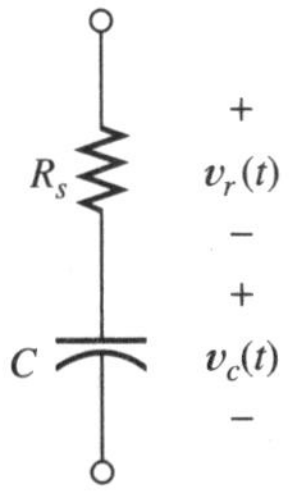

Figure P3.38

P3.39. Suppose that a parallel-plate capacitor has a dielectric that breaks down if the electric field exceeds K V/m. Thus, the maximum voltage rating of the capacitor is $V_{\max} = Kd$, where d is the thickness of the dielectric. In working Problem P3.34, we find that the maximum energy that can be stored before breakdown is $w_{\max} = 1/2\epsilon_r\epsilon_0 K^2(\text{Vol})$, in which Vol is the volume of the dielectric. Air has approximately $K = 32 \times 10^5$ V/m and $\epsilon_r = 1$. Find the minimum volume of air (as a dielectric in a parallel-plate capacitor) needed to store the energy content of one U.S. gallon of gasoline, which is approximately 132 MJ. What thickness should the air dielectric have if we want the voltage for maximum energy storage to be 1000 V?

P3.40. As shown in Figure P3.40, two 10-μF capacitors have an initial voltage of 100 V before the switch is closed. Find the total stored energy before the switch is closed. Find the voltage across each capacitor and the total stored energy after the switch is closed. What could have happened to the energy?

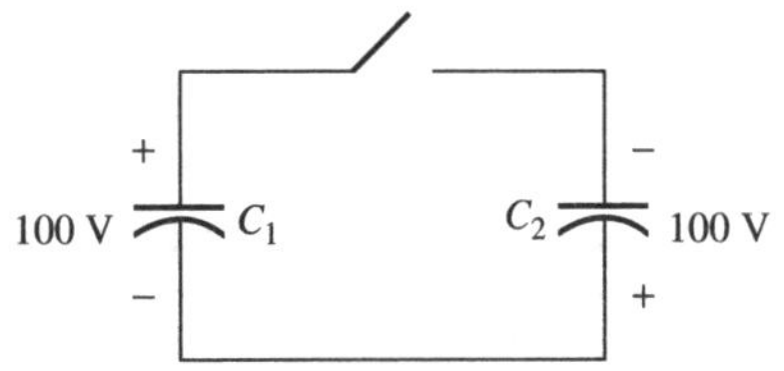

Figure P3.40

Section 3.4: Inductance

P3.41. Explain, in a few sentences, the fluid-flow analogy for an inductor.

P3.42. How are inductors constructed?

***P3.43.** The voltage applied across a 2-H inductor is shown in Figure P3.43. Sketch the current, power, and stored energy versus time.

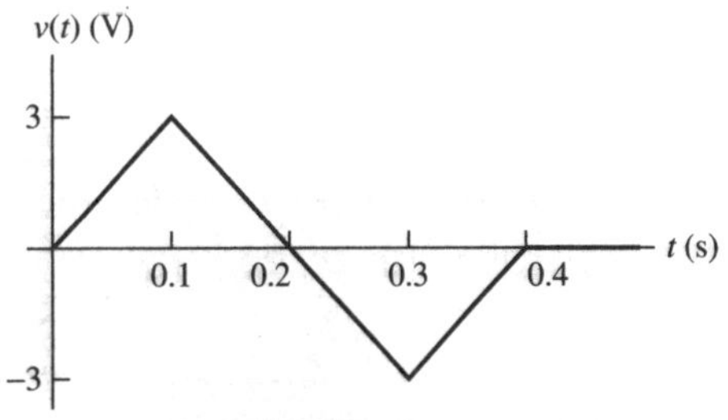

Figure P3.43

***P3.44.** At $t = 0$, the current flowing in a 0.5-H inductance is 2 A. What constant voltage must be applied to increase the current to 4 A at $t = 0.4$ s?

***P3.45.** A constant voltage of 10 V is applied to a 50-μH inductance, as shown in Figure P3.45. The current in the inductance at $t = 0$ is −100 mA. At what time t_x does the current reach +100 mA?

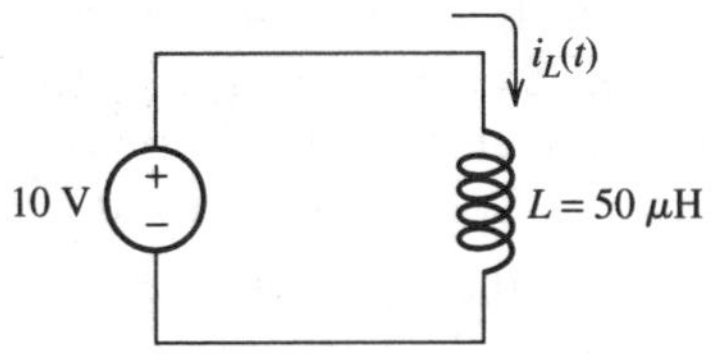

Figure P3.45

***P3.46.** A dc voltage of 20 V is applied to a 5-H inductor at $t = 0$, starting with an initial current of −0.2 A. Determine the current, power, and stored energy at $t_1 = 1$ s and $t_2 = 6$ s.

P3.47. The current flowing through an inductor is decreasing in magnitude. Is energy flowing into or out of the inductor? What if the current is constant with time?

P3.48. What is the value of the voltage across an ideal inductor if the current through it is constant with time? Comment. To what circuit element is an ideal inductor equivalent for circuits with constant currents and voltages?

P3.49. The current in a 100-mH inductance is given by 0.5 sin(1000t) A. Find expressions and sketch the waveforms to scale for the voltage, power, and stored energy, allowing t to range from 0 to 3π ms. The argument of the sine function is in radians.

P3.50. The current flowing through a 300-mH inductance is given by 5exp($-200t$) A. Find expressions for the voltage, power, and stored energy. Sketch the waveforms to scale for $0 < t < 20$ ms.

P3.51. The voltage across a 2-H inductance is shown in Figure P3.51. The initial current in the inductance is $i(0) = 0$. Sketch the current, power, and stored energy to scale versus time.

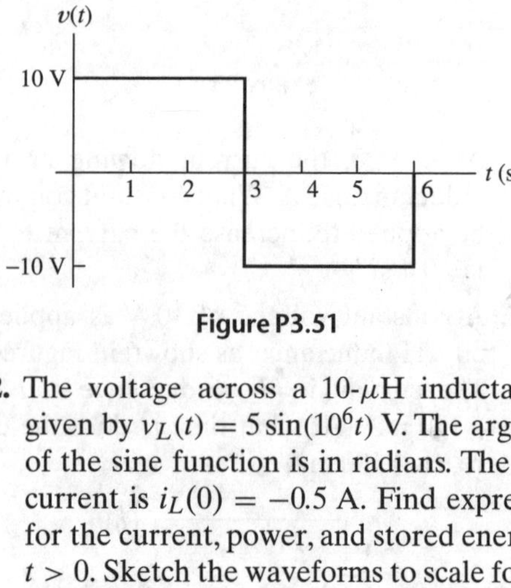

Figure P3.51

P3.52. The voltage across a 10-μH inductance is given by $v_L(t) = 5\sin(10^6 t)$ V. The argument of the sine function is in radians. The initial current is $i_L(0) = -0.5$ A. Find expressions for the current, power, and stored energy for $t > 0$. Sketch the waveforms to scale for time ranging from zero to 2π μs.

P3.53. What is the equivalent circuit element for a very large (ideally, infinite) inductance having an initial current of 10 A. Explain your answer.

P3.54. The energy stored in a 2-H inductor is 100 J and is increasing at 200 J/s at $t = 4$ s. Determine the voltage magnitude and current magnitude at $t = 4$ s. Does the current enter or leave the positive terminal of the inductor?

P3.55. What value of inductance corresponds to an open circuit, assuming zero initial current? Explain your answer. Repeat for a short circuit.

P3.56. Before $t = 0$, the current in a 2-H inductance is zero. Starting at $t = 0$, the current is increased linearly with time to 5 A in 1 s. Then, the current remains constant at 5 A. Sketch the voltage, current, power, and stored energy to scale versus time.

P3.57. The voltage across an inductance is given by $v(t) = V_m\cos(\omega t)$ and the initial current in the inductor is zero. Suppose that ω is very large—ideally, approaching infinity. For this voltage, does the inductance approximate either an open or a short circuit? Explain.

P3.58. Suppose that at $t = t_0$ the current through a certain inductance is zero. A voltage pulse is applied to the inductance between t_0 and $t_0 + \Delta t$, and the current through the inductance increases to I_f. What can you say about the peak amplitude V_m and the area under the pulse waveform (i.e., voltage versus time)? What are the units of the area under the pulse? What must happen to the peak amplitude and the area under the pulse as Δt approaches zero, assuming that I_f remains the same?

Section 3.5: Inductances in Series and Parallel

P3.59. Describe how inductances are combined in series and in parallel. Compare with how resistances are combined.

***P3.60.** Determine the equivalent inductance for each of the series and parallel combinations shown in Figure P3.60.

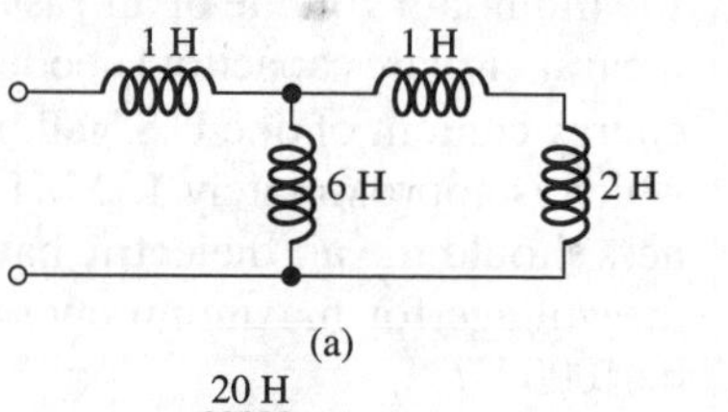

FigureP3.60

P3.61. Find the equivalent inductance for each of the series and parallel combinations shown in Figure P3.61.

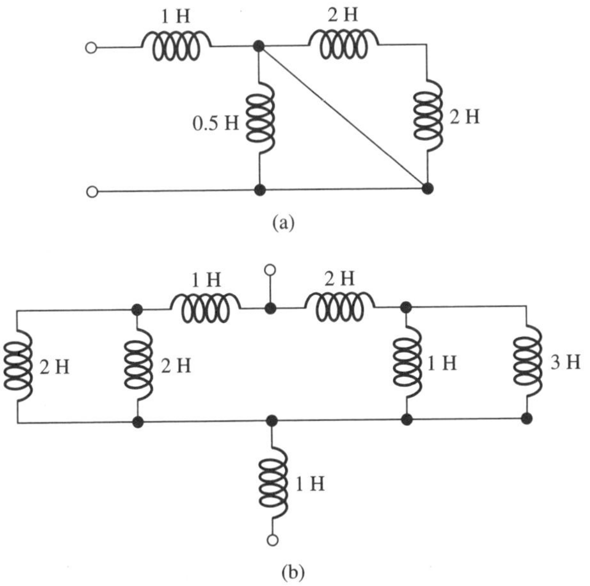

Figure P3.61

P3.62. The maximum value obtained by connecting 4 identical inductances in series is 16 H. What is the value of each inductor? What is the minimum value achieved when they are connected in parallel?

***P3.63.** Two inductances $L_1 = 2$ H and $L_2 = 3$ H are connected in parallel as shown in Figure P3.63. The initial currents in both the inductors are zero at $t = 0$. Find an expression for $i_2(t)$ in terms of $i(t)$, L_1 and L_2. Repeat for $i_1(t)$. Comment.

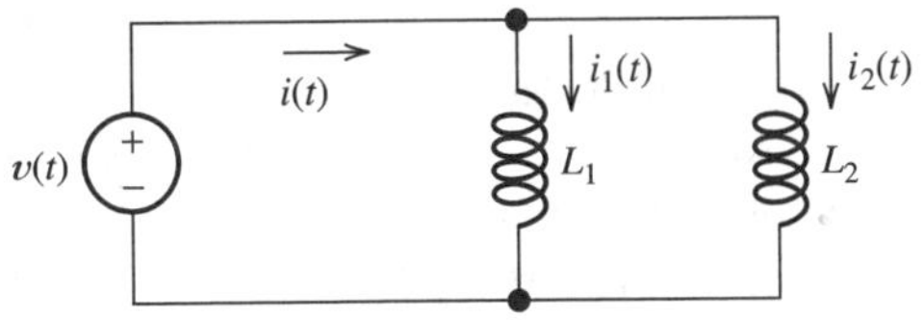

Figure P3.63

P3.64. We need to combine (in series or in parallel) an unknown inductance L with a second inductance of 4 H to attain an equivalent inductance of 3 H. Should L be placed in series or in parallel with the original inductance? What value is required for L?

P3.65. An unknown inductance L in series with a 4-H inductor is connected in parallel with a 10-H inductor. The effective inductance is 5-H. Find the value of L.

Section 3.6: Practical Inductors

P3.66. Draw the equivalent circuit for a real inductor, including three parasitic effects.

P3.67. A 10-mH inductor has a parasitic series resistance of $R_s = 1\ \Omega$ as shown in Figure P3.67. **a.** The current is given by $i(t) = 0.1\cos(10^5 t)$. The argument of the cosine function is in radians. Find $v_R(t)$, $v_L(t)$, and $v(t)$. In this case, for 1% accuracy in computing $v(t)$, could the resistance be neglected? **b.** Repeat if $i(t) = 0.1\cos(10t)$.

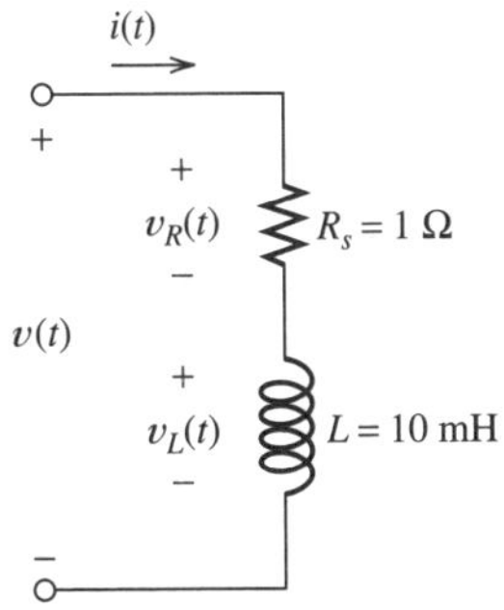

Figure P3.67

P3.68. A constant (dc) current of 100 mA flows through a real inductor and the voltage across its external terminals is 400 mV. Which of the circuit parameters of Figure 3.22 on page 164 can be deduced from this information and what is its value? Explain how you arrived at your answer.

P3.69. Find $v(t)$, $i_C(t)$, $i(t)$, the energy stored in the capacitance, the energy stored in the inductance, and the total stored energy for the circuit of Figure P3.69, given that $i_L(t) = \sin(10^4 t)$ A. (The argument of the sine function is in radians.) Show that the total stored energy is constant with time. Comment on the results.

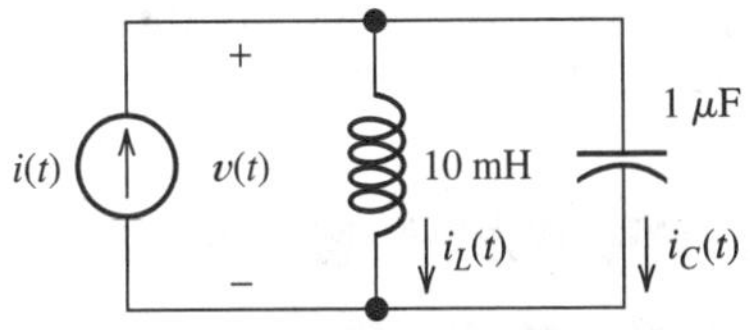

Figure P3.69

P3.70. For the circuit of Figure P3.70, determine $i(t)$, $v_L(t)$, $v(t)$, the energy stored in the capacitance, the energy stored in the inductance, and the total stored energy, given that

$v_C(t) = 40\cos(1000t)$ V. (The argument of the cosine function is in radians.) Show that the total stored energy is constant with time. Comment on the results.

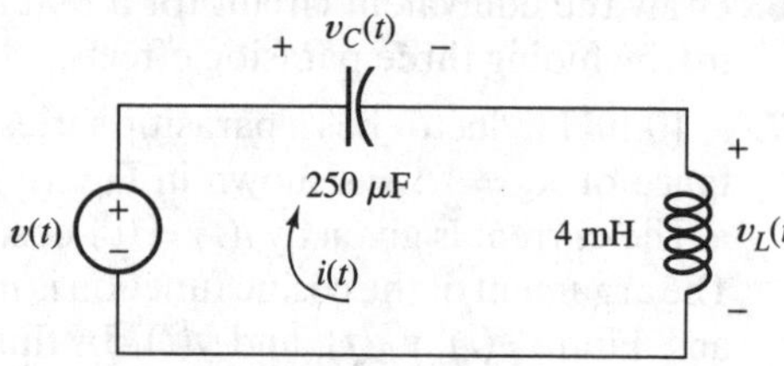

Figure P3.70

Section 3.7: Mutual Inductance

P3.71. Describe briefly the physical basis for mutual inductance.

***P3.72.** **a.** Derive an expression for the equivalent inductance for the circuit shown in Figure P3.72. **b.** Repeat if the dot for L_2 is moved to the bottom end.

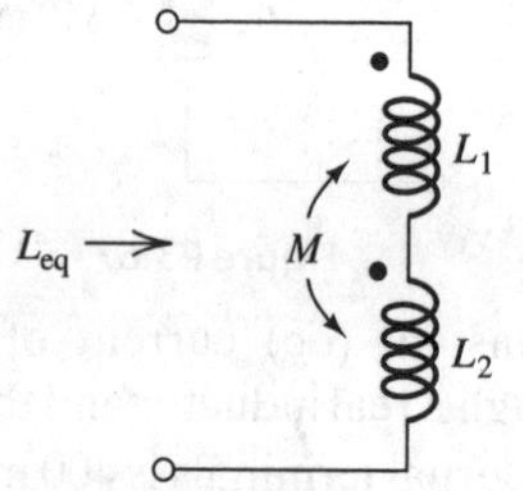

Figure P3.72

P3.73. The mutually coupled inductances in Figure P3.73 have $L_1 = 2$ H, $L_2 = 3$ H, and $M = 1$ H. Furthermore, $i_1(t) = \sin(20t)$ A and $i_2(t) = 0.5\cos(30t)$ A. (The arguments of the sine and cosine functions are in radians.) **a.** Find expressions for $v_1(t)$ and $v_2(t)$. **b.** Repeat with the dot placed at the bottom of L_2.

P3.74. A pair of mutually coupled inductances has $L_1 = 2$ H, $L_2 = 1$ H, $i_1 = 2\cos(1000t)$ A, $i_2 = 0$, and $v_2 = 2000\sin(1000t)$ V. (The arguments of the sine and cosine functions are in radians.) Find $v_1(t)$ and the magnitude of the mutual inductance.

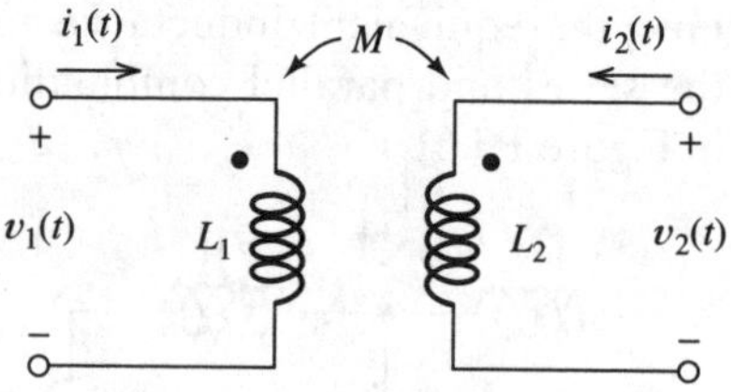

Figure P3.73

P3.75. Suppose we place a short circuit across the terminals of L_2 in the mutually coupled inductors shown in Figure 3.23(a) on page 165. Derive an expression for the equivalent inductance seen looking into the terminals of L_1.

P3.76. Consider the parallel inductors shown in Figure P3.63, with mutual coupling and the dots at the top end of L_1 and the bottom end of L_2. Derive an expression for the equivalent inductance seen by the source in terms of L_1, L_2, and M. [*Hint*: Write the circuit equations and manipulate them to obtain an expression of the form $v(t) = L_{eq} di(t)/dt$ in which L_{eq} is a function of L_1, L_2, and M.]

Section 3.8: Symbolic Integration and Differentiation using MATLAB

P3.77. The current through a 200-mH inductance is given by $i_L(t) = \exp(-2t)\sin(4\pi t)$ A in which the angle is in radians. Using your knowledge of calculus, find an expression for the voltage across the inductance. Then, use MATLAB to verify your answer for the voltage and to plot both the current and the voltage for $0 \le t \le 2$ s.

P3.78. A 1-H inductance has $i_L(0) = 0$ and $v_L(t) = t\exp(-t)$ for $0 \le t$. Using your calculus skills, find an expression for $i_L(t)$. Then, use MATLAB to verify your answer for $i_L(t)$ and to plot $v_L(t)$ and $i_L(t)$ for $0 \le t \le 10$ s.

Frequently Asked Questions

1. What is a capacitor? Draw the sketch of a parallel plate capacitors, also draw its symbol.
2. How will you represent a capacitor on the basis of fluid-flow analogy?
3. What are the factors on which the charge stored in a capacitor depends upon?
4. Represent a current in terms of voltage at any instant during charging of a capacitor.
5. Derive an equation for voltage built-up across a capacitor in terms of current.
6. Derive an expression for the energy stored in a capacitor.

7. What will be the total capacitance, if three capacitors each of capacitance C_1, C_2 and C_3 are connected in series? Derive the relation.
8. Derive a relation for the total capacitance of the grouping if three capacitor each of capacitance C_1, C_2 and C_3 are connected in parallel.
9. Derive a relation for the total capacitance of the grouping if two capacitors each of capacitance C_1 and C_2 are connected in parallel and this combination is connected in series with two more capacitors each of capacitance C_3 and C_4 connected in series.
10. In a parallel plate capacitor having circular plates, if the distance between the plates is reduced to half and the diameter of the plates is also reduced to half but the dielectric medium between the plates is kept the same, what will be the effect on its capacitance ?
11. Explain the construction of an electrolytic capacitor.
12. What do you mean by parasitic effects in a capacitor?
13. What is an inductor? Draw its symbol. How will you differentiate between inductor and inductance?
14. Derive a relation for the self induced emf in a coil when alternating current flows through it
15. Derive a relation for the energy stored in a coil of inductance L henry.
16. What will be the effect if number of inductors are connected in (i) series (ii) parallel?
17. What do understand by parasitic effects in inductors?
18. What do you mean by mutual inductance?
19. Give the construction, working and application of Linear variable differential transformer (LVDT).
20. How will you differentiate between self and mutual inductance?

Practice Test

Here is a practice test you can use to check your comprehension of the most important concepts in this chapter. Answers can be found in Appendix D and complete solutions are included in the Student Solutions files. See Appendix E for more information about the Student Solutions.

T3.1. The current flowing through a 10-μF capacitor having terminals labeled a and b is $i_{ab} = 0.3\exp(-2000t)$ A for $t \geq 0$. Given that $v_{ab}(0) = 0$, find an expression for $v_{ab}(t)$ for $t \geq 0$. Then, find the energy stored in the capacitor for $t = \infty$.

T3.2. Determine the equivalent capacitance C_{eq} for Figure T3.2.

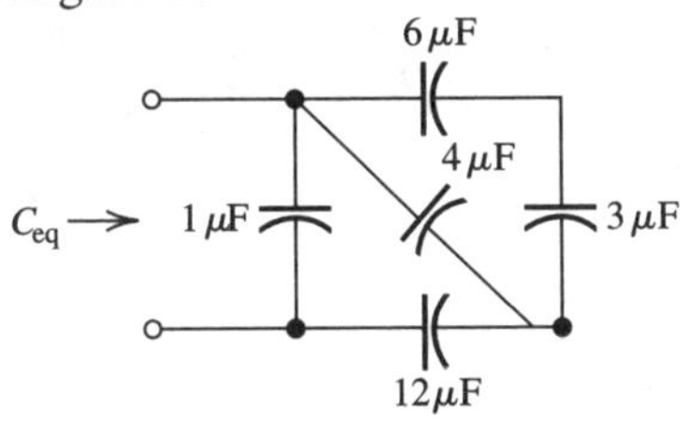

Figure T3.2

T3.3. A certain parallel-plate capacitor has plate length of 2 cm and width of 3 cm. The dielectric has a thickness of 0.1 mm and a relative dielectric constant of 80. Determine the capacitance.

T3.4. A 2-mH inductance has $i_{ab} = 0.3\sin(2000t)$ A. Find an expression for $v_{ab}(t)$. Then, find the peak energy stored in the inductance.

T3.5. Determine the equivalent inductance L_{eq} between terminals a and b in Figure T3.5.

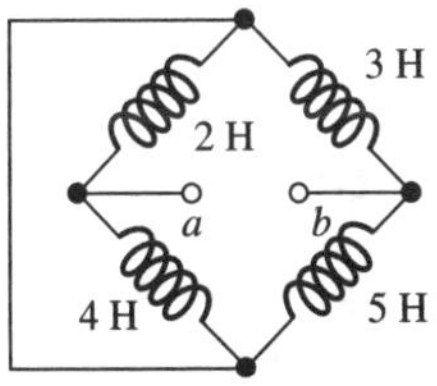

Figure T3.5

T3.6. Figure T3.6 has $L_1 = 40$ mH, $M = 20$ mH, and $L_2 = 30$ mH. Find expressions for $v_1(t)$ and $v_2(t)$.

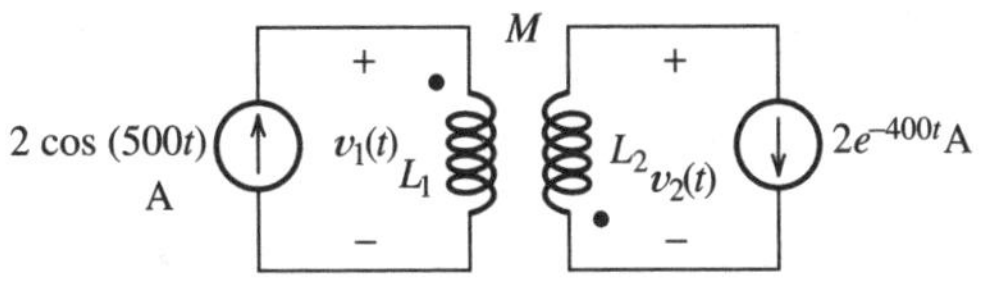

Figure T3.6

T3.7. The current flowing through a 20-μF capacitor having terminals labeled a and b is $i_{ab} = 3 \times 10^5 t^2 \exp(-2000t)$ A for $t \geq 0$. Given that $v_{ab}(0) = 5$ V, write a sequence of MATLAB commands to find the expression for $v_{ab}(t)$ for $t \geq 0$ and to produce plots of the current and voltage for $0 \leq t \leq 5$ ms.

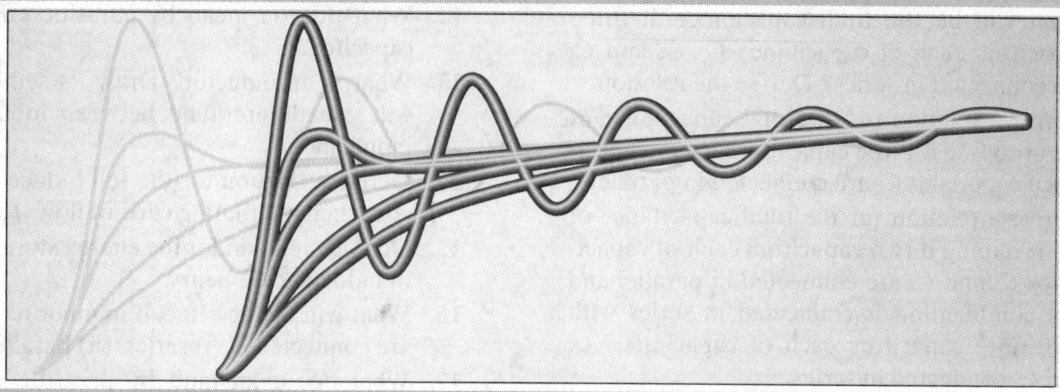

Chapter 4

Transients

Study of this chapter will enable you to:

- Solve first-order *RC* or *RL* circuits.
- Understand the concepts of transient response and steady-state response.
- Relate the transient response of first-order circuits to the time constant.
- Solve *RLC* circuits in dc steady-state conditions.
- Solve second-order circuits.
- Relate the step response of a second-order system to its natural frequency and damping ratio.
- Use the MATLAB Symbolic Toolbox to solve differential equations.

Introduction to this chapter:

In this chapter, we study circuits that contain sources, switches, resistances, inductances, and capacitances. The time-varying currents and voltages resulting from the sudden application of sources, usually due to switching, are called **transients**.

In transient analysis, we start by writing circuit equations using concepts developed in Chapter 2, such as KCL, KVL, node-voltage analysis, and mesh-current analysis. Because the current–voltage relationships for inductances and capacitances involve integrals and derivatives, we obtain integrodifferential equations. These equations can be converted to pure differential equations by differentiating with respect to time. Thus, the study of transients requires us to solve differential equations.

4.1 FIRST-ORDER *RC* CIRCUITS

In this section, we consider transients in circuits that contain independent dc sources, resistances, and a single capacitance.

Discharge of a Capacitance through a Resistance

As a first example, consider the circuit shown in Figure 4.1(a). Prior to $t = 0$, the capacitor is charged to an initial voltage V_i. Then, at $t = 0$, the switch closes and current flows through the resistor, discharging the capacitor.

Writing a current equation at the top node of the circuit after the switch is closed yields

$$C\frac{dv_C(t)}{dt} + \frac{v_C(t)}{R} = 0$$

Multiplying by the resistance gives

$$RC\frac{dv_C(t)}{dt} + v_C(t) = 0 \tag{4.1}$$

As expected, we have obtained a differential equation.

Equation 4.1 indicates that the solution for $v_C(t)$ must be a function that has the same form as its first derivative. Of course, a function with this property is an exponential. Thus, we anticipate that the solution is of the form

Equation 4.1 indicates that the solution for $v_C(t)$ must be a function that has the same form as its first derivative. The function with this property is an exponential.

$$v_C(t) = Ke^{st} \tag{4.2}$$

in which K and s are constants to be determined.

Using Equation 4.2 to substitute for $v_C(t)$ in Equation 4.1, we have

$$RCKse^{st} + Ke^{st} = 0 \tag{4.3}$$

Solving for s, we obtain

$$s = \frac{-1}{RC} \tag{4.4}$$

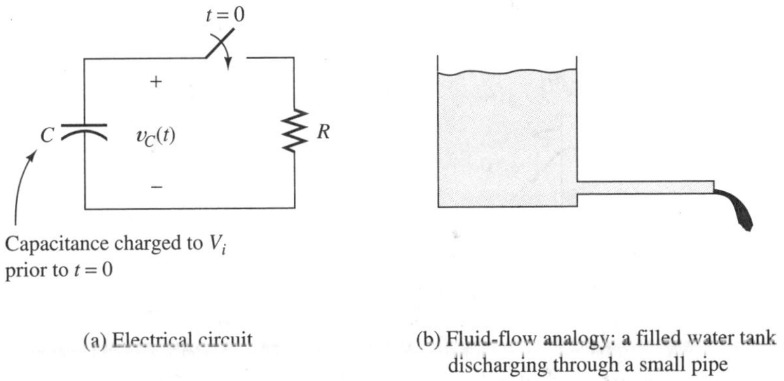

Figure 4.1 A capacitance discharging through a resistance and its fluid-flow analogy. The capacitor is charged to V_i prior to $t = 0$ (by circuitry that is not shown). At $t = 0$, the switch closes and the capacitor discharges through the resistor.

Substituting this into Equation 4.2, we see that the solution is

$$v_C(t) = Ke^{-t/RC} \tag{4.5}$$

Because the current is finite, the voltage across the capacitor cannot change instantaneously when the switch closes.

Referring to Figure 4.1(a), we reason that the voltage across the capacitor cannot change instantaneously when the switch closes. This is because the current through the capacitance is $i_C(t) = C\,dv_C/dt$. In order for the voltage to change instantaneously, the current would have to be infinite. Since the voltage is finite, the current in the resistance must be finite, and we conclude that the voltage across the capacitor must be continuous. Thus, we write

$$v_C(0+) = V_i \tag{4.6}$$

in which $v_C(0+)$ represents the voltage immediately after the switch closes. Substituting into Equation 4.5, we have

$$v_C(0+) = V_i = Ke^0 = K \tag{4.7}$$

Hence, we conclude that the constant K equals the initial voltage across the capacitor. Finally, the solution for the voltage is

$$v_C(t) = V_i e^{-t/RC} \tag{4.8}$$

A plot of the voltage is shown in Figure 4.2. Notice that the capacitor voltage decays exponentially to zero.

The time interval $\tau = RC$ is called the time constant of the circuit.

The time interval

$$\tau = RC \tag{4.9}$$

is called the **time constant** of the circuit. In one time constant, the voltage decays by the factor $e^{-1} \cong 0.368$. After about five time constants, the voltage remaining on the capacitor is negligible compared with the initial value.

At one time constant, the voltage across a capacitance discharging through a resistance is $e^{-1} \cong 0.368$ times its initial value. After about three to five time constants, the capacitance is almost totally discharged.

An analogous fluid-flow system is shown in Figure 4.1(b). The tank initially filled with water is analogous to the charged capacitor. Furthermore, the small pipe is analogous to the resistor. At first, when the tank is full, the flow is large and the water level drops fast. As the tank empties, the flow decreases.

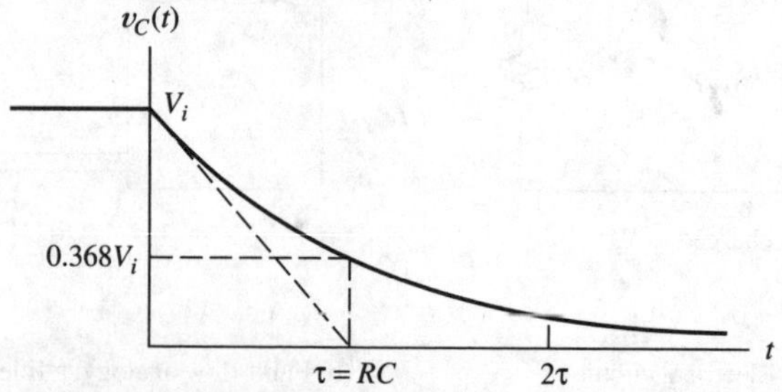

Figure 4.2 Voltage versus time for the circuit of Figure 4.1(a). When the switch is closed, the voltage across the capacitor decays exponentially to zero. At one time constant, the voltage is equal to 36.8 percent of its initial value.

In the past, engineers have frequently applied *RC* circuits in timing applications. For example, suppose that when a garage door opens or closes, a light is to be turned on and is to remain on for 30 s. To achieve this objective, we could design a circuit consisting of (1) a capacitor that is charged to an initial voltage V_i while the door opener is energized, (2) a resistor through which the capacitor discharges, and (3) a sensing circuit that keeps the light on as long as the capacitor voltage is larger than $0.368V_i$. If we choose the time constant $\tau = RC$ to be 30 s, the desired operation is achieved.

Charging a Capacitance from a DC Source through a Resistance

Next, consider the circuit shown in Figure 4.3. The source voltage V_s is constant—in other words, we have a dc source. The source is connected to the *RC* circuit by a switch that closes at $t = 0$. We assume that the initial voltage across the capacitor just before the switch closes is $v_C(0-) = 0$. Let us solve for the voltage across the capacitor as a function of time.

We start by writing a current equation at the node that joins the resistor and the capacitor. This yields

$$C\frac{dv_C(t)}{dt} + \frac{v_C(t) - V_s}{R} = 0 \tag{4.10}$$

The first term on the left-hand side is the current referenced downward through the capacitor. The second term is the current referenced toward the left through the resistor. KCL requires that the currents leaving the node sum to zero.

Rearranging Equation 4.10, we obtain

$$RC\frac{dv_C(t)}{dt} + v_C(t) = V_s \tag{4.11}$$

$v_C(0-)$ is the voltage across the capacitor the instant before the switch closes (at $t = 0$). Similarly, $v_C(0+)$ is the voltage across the capacitor the instant after the switch closes.

As expected, we have obtained a linear first-order differential equation with constant coefficients. As in the previous circuit, the voltage across the capacitance cannot change instantaneously because the voltages are finite, and thus, the current through the resistance (and therefore through the capacitance) is finite. Infinite current is required to change the voltage across a capacitance in an instant. Thus, we have

$$v_C(0+) = v_C(0-) = 0 \tag{4.12}$$

Now, we need to find a solution for $v_C(t)$ that (1) satisfies Equation 4.11 and (2) matches the initial conditions of the circuit stated in Equation 4.12. Notice that Equation 4.11 is the same as Equation 4.1, except for the constant on the right-hand side. Thus, we expect the solution to be the same as for Equation 4.1, except for an added constant term. Thus, we are led to try the solution

$$v_C(t) = K_1 + K_2e^{st} \tag{4.13}$$

in which K_1, K_2, and s are constants to be determined.

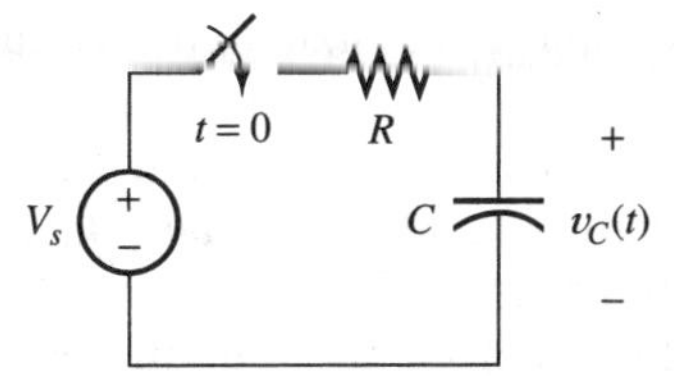

Figure 4.3 Capacitance charging through a resistance. The switch closes at $t = 0$, connecting the dc source V_s to the circuit.

If we use Equation 4.13 to substitute for $v_C(t)$ in Equation 4.11, we obtain

$$(1 + RCs)K_2e^{st} + K_1 = V_s \tag{4.14}$$

For equality, the coefficient of e^{st} must be zero. This leads to

$$s = \frac{-1}{RC} \tag{4.15}$$

From Equation 4.14, we also have

$$K_1 = V_s \tag{4.16}$$

Using Equations 4.15 and 4.16 to substitute into Equation 4.13, we obtain

$$v_C(t) = V_s + K_2e^{-t/RC} \tag{4.17}$$

in which K_2 remains to be determined.

Now, we use the initial condition (Equation 4.12) to find K_2. We have

$$v_C(0+) = 0 = V_s + K_2e^0 = V_s + K_2 \tag{4.18}$$

from which we find $K_2 = -V_s$. Finally, substituting into Equation 4.17, we obtain the solution

$$v_C(t) = V_s - V_se^{-t/RC} \tag{4.19}$$

When a dc source is contained in the circuit, the total response contains two parts: forced (or steady-state) and transient.

The second term on the right-hand side is called the **transient response**, which eventually decays to negligible values. The first term on the right-hand side is the **steady-state response**, also called the **forced response**, which persists after the transient has decayed.

Here again, the product of the resistance and capacitance has units of seconds and is called the time constant $\tau = RC$. Thus, the solution can be written as

$$v_C(t) = V_s - V_se^{-t/\tau} \tag{4.20}$$

In the case of a capacitance charging from a dc source through a resistance, a straight line tangent to the start of the transient reaches the final value at one time constant.

A plot of $v_C(t)$ is shown in Figure 4.4. Notice that $v_C(t)$ starts at 0 and approaches the final value V_s asymptotically as t becomes large. After one time constant, $v_C(t)$ has reached 63.2 percent of its final value. For practical purposes, $v_C(t)$ is equal to its final value V_s after about five time constants. Then, we say that the circuit has reached steady state.

It can be shown that if the initial slope of v_C is extended, it intersects the final value at one time constant as shown in Figure 4.4.

RC transients are the main limitation on the speed at which computer chips can operate.

We have seen in this section that several time constants are needed to charge or discharge a capacitance. This is the main limitation on the speed at which digital computers can process data. In a typical computer, information is represented by voltages that nominally assume values of either +1.8 or 0 V, depending on the data represented. When the data change, the voltages must change. It is impossible to build circuits that do not have some capacitance that is charged or discharged when voltages change in value. Furthermore, the circuits always have nonzero resistances that limit the currents available for charging or discharging the capacitances. Therefore, a nonzero time constant is associated with each circuit in the computer, limiting its speed. We will learn more about digital computer circuits in later chapters.

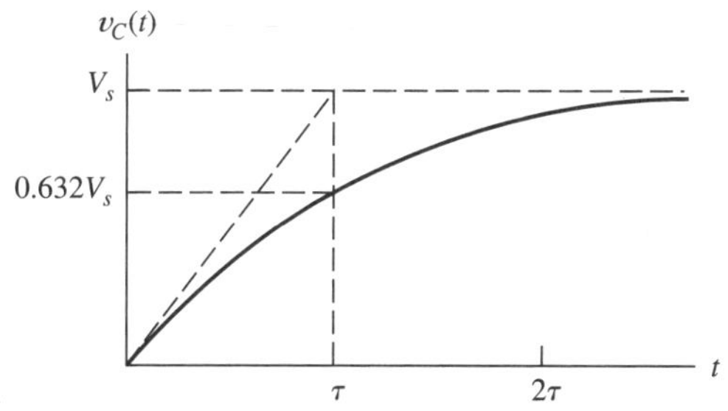

Figure 4.4 The charging transient for the *RC* circuit of Figure 4.3.

Exercise 4.1 Suppose that $R = 5000\ \Omega$ and $C = 1\ \mu\text{F}$ in the circuit of Figure 4.1(a). Find the time at which the voltage across the capacitor reaches 1 percent of its initial value.

Answer $t = -5\ln(0.01)$ ms $\cong 23$ ms. □

Exercise 4.2 Show that if the initial slope of $v_C(t)$ is extended, it intersects the final value at one time constant, as shown in Figure 4.4. [The expression for $v_C(t)$ is given in Equation 4.20.] □

4.2 DC STEADY STATE

The transient terms in the expressions for currents and voltages in *RLC* circuits decay to zero with time. (An exception is *LC* circuits having no resistance.) For dc sources, the steady-state currents and voltages are also constant.

The transient terms in the expressions for currents and voltages in *RLC* circuits decay to zero with time.

Consider the equation for current through a capacitance:

$$i_C(t) = C\frac{dv_C(t)}{dt}$$

If the voltage $v_C(t)$ is constant, the current is zero. In other words, the capacitance behaves as an open circuit. Thus, we conclude that *for steady-state conditions with dc sources, capacitances behave as open circuits.*

Similarly, for an inductance, we have

$$v_L(t) = L\frac{di_L(t)}{dt}$$

When the current is constant, the voltage is zero. Thus, we conclude that *for steady-state conditions with dc sources, inductances behave as short circuits.*

These observations give us another approach to finding the steady-state solutions to circuit equations for *RLC* circuits with constant sources. First, we replace the capacitors by open circuits and the inductors by short circuits. The circuit then consists of dc sources and resistances. Finally, we solve the equivalent circuit for the steady-state currents and voltages.

The steps in determining the forced response for *RLC* circuits with dc sources are

1. Replace capacitances with open circuits.
2. Replace inductances with short circuits.
3. Solve the remaining circuit.

Example 4.1 Steady-State DC Analysis

Find v_x and i_x for the circuit shown in Figure 4.5(a) for $t >> 0$.

Solution After the switch has been closed a long time, we expect the transient response to have decayed to zero. Then the circuit is operating in dc steady-state

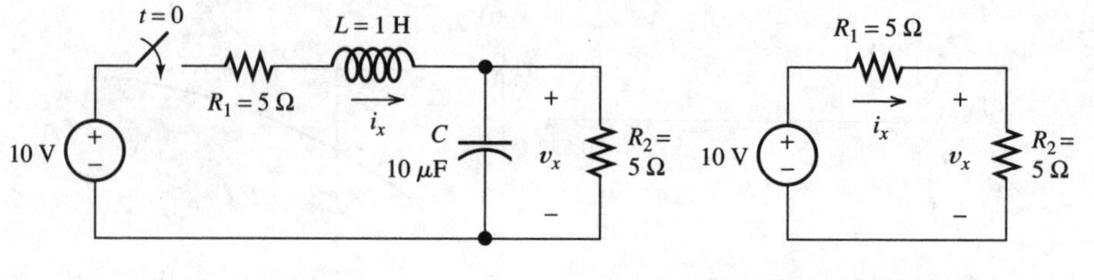

(a) Original circuit (b) Equivalent circuit for steady state

Figure 4.5 The circuit and its dc steady-state equivalent for Example 4.1.

Steps 1 and 2.

conditions. We start our analysis by replacing the inductor by a short circuit and the capacitor by an open circuit. The equivalent circuit is shown in Figure 4.5(b).

Step 3.

This resistive circuit is readily solved. The resistances R_1 and R_2 are in series. Thus, we have

$$i_x = \frac{10}{R_1 + R_2} = 1\text{ A}$$

and

$$v_x = R_2 i_x = 5\text{ V}$$ ■

Sometimes, we are only interested in the steady-state operation of circuits with dc sources. For example, in analyzing the headlight circuits in an automobile, we are concerned primarily with steady state. On the other hand, we must consider transients in analyzing the operation of the ignition system.

In other applications, we are interested in steady-state conditions with sinusoidal ac sources. For sinusoidal sources, the steady-state currents and voltages are also sinusoidal. In Chapter 5, we study a method for solving sinusoidal steady-state circuits that is similar to the method we have presented here for dc steady state. Instead of short and open circuits, we will replace inductances and capacitances by impedances, which are like resistances, except that impedances can have imaginary values.

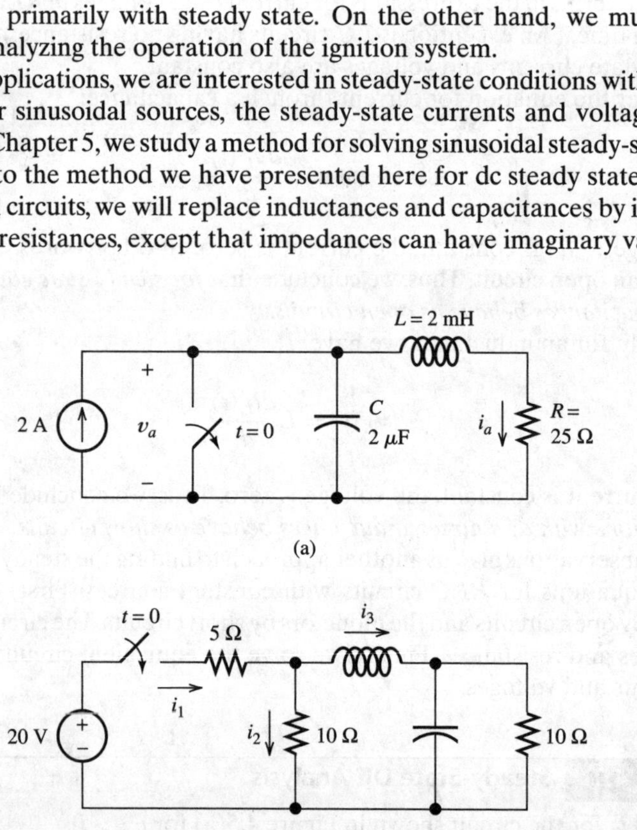

(b)

Figure 4.6 Circuits for Exercise 4.3.

Exercise 4.3 Solve for the steady-state values of the labeled currents and voltages for the circuits shown in Figure 4.6.

Answer **a.** $v_a = 50$ V, $i_a = 2$ A; **b.** $i_1 = 2$ A, $i_2 = 1$ A, $i_3 = 1$ A. □

4.3 *RL* CIRCUITS

In this section, we consider circuits consisting of dc sources, resistances, and a single inductance. The methods and solutions are very similar to those we studied for *RC* circuits in Section 4.1.

The steps involved in solving simple circuits containing dc sources, resistances, and one energy-storage element (inductance or capacitance) are as follows:

1. Apply Kirchhoff's current and voltage laws to write the circuit equation.
2. If the equation contains integrals, differentiate each term in the equation to produce a pure differential equation.
3. Assume a solution of the form $K_1 + K_2 e^{st}$.
4. Substitute the solution into the differential equation to determine the values of K_1 and s. (Alternatively, we can determine K_1 by solving the circuit in steady state as discussed in Section 4.2.)
5. Use the initial conditions to determine the value of K_2.
6. Write the final solution.

Example 4.2 *RL* Transient Analysis

Consider the circuit shown in Figure 4.7. Find the current $i(t)$ and the voltage $v(t)$.

Solution First, we find the current $i(t)$. Of course, prior to $t = 0$, the switch is open and the current is zero:

$$i(t) = 0 \qquad \text{for } t < 0 \tag{4.21}$$

After the switch is closed, the current increases in value eventually reaching a steady-state value.

Writing a KVL equation around the loop, we have

Step 1.

$$Ri(t) + L\frac{di}{dt} = V_s \tag{4.22}$$

Step 2 is not needed in this case.

This is very similar to Equation 4.11, and we are, therefore, led to try a solution of the same form as that given by Equation 4.13. Thus, our trial solution is

Step 3.

$$i(t) = K_1 + K_2 e^{st} \tag{4.23}$$

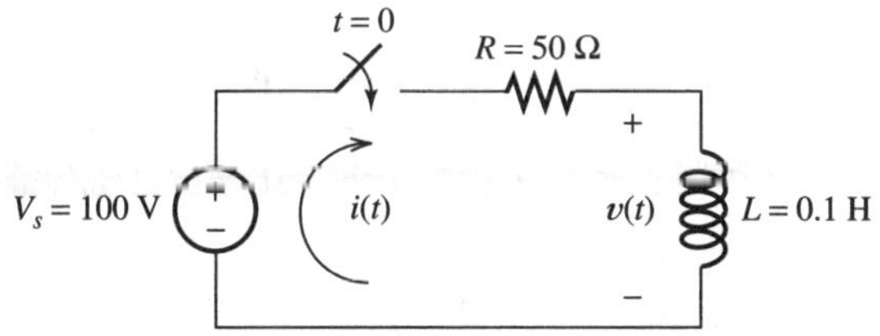

Figure 4.7 The circuit analyzed in Example 4.2.

in which K_1, K_2, and s are constants that need to be determined. Following the

Step 4. procedure used in Section 4.1, we substitute the trial solution into the differential equation, resulting in

$$RK_1 + (RK_2 + sLK_2)e^{st} = V_s \tag{4.24}$$

from which we obtain

$$K_1 = \frac{V_s}{R} = 2 \tag{4.25}$$

and

$$s = \frac{-R}{L} \tag{4.26}$$

Substituting these values into Equation 4.23 results in

$$i(t) = 2 + K_2 e^{-tR/L} \tag{4.27}$$

Step 5. Next, we use the initial conditions to determine the value of K_2. The current in the inductor is zero prior to $t = 0$ because the switch is open. The applied voltage is finite, and the inductor current must be continuous (because $v_L = L\,di/dt$). Thus, immediately after the switch is closed, the current must be zero. Hence, we have

$$i(0+) = 0 = 2 + K_2 e^0 = 2 + K_2 \tag{4.28}$$

Solving, we find that $K_2 = -2$.

Substituting into Equation 4.27, we find that the solution for the current is

Step 6.

$$i(t) = 2 - 2e^{-t/\tau} \quad \text{for } t > 0 \tag{4.29}$$

in which the time constant is given by

$$\tau = \frac{L}{R} \tag{4.30}$$

A plot of the current versus time is shown in Figure 4.8(a). Notice that the current increases from zero to the steady-state value of 2 A. After five time constants, the current is within 99 percent of the final value. As a check, we verify that the steady-state current is 2 A. (As we saw in Section 4.2, this value can be obtained directly by treating the inductor as a short circuit.)

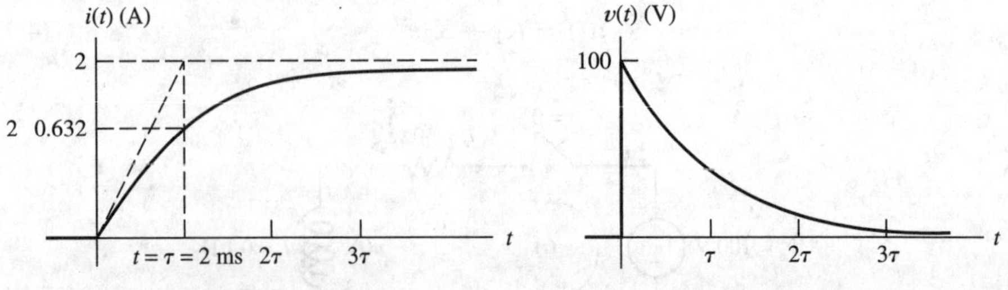

(a) (b)

Figure 4.8 Current and voltage versus time for the circuit of Figure 4.7.

Now, we consider the voltage $v(t)$. Prior to $t = 0$, with the switch open, the voltage is zero.

$$v(t) = 0 \qquad \text{for } t < 0 \tag{4.31}$$

After $t = 0$, $v(t)$ is equal to the source voltage minus the drop across R. Thus, we have

$$v(t) = 100 - 50i(t) \qquad \text{for } t > 0 \tag{4.32}$$

Substituting the expression found earlier for $i(t)$, we obtain

$$v(t) = 100e^{-t/\tau} \tag{4.33}$$

A plot of $v(t)$ is shown in Figure 4.8(b).

At $t = 0$, the voltage across the inductor jumps from 0 to 100 V. As the current gradually increases, the drop across the resistor increases, and the voltage across the inductor falls. In steady state, we have $v(t) = 0$ because the inductor behaves as a short circuit. ■

After solving several circuits with a single energy-storage element, we can use our experience to skip some of the steps listed earlier in the section. We illustrate this in the next example.

Example 4.3 *RL* Transient Analysis

Consider the circuit shown in Figure 4.9 in which V_s is a dc source. Assume that the circuit is in steady state with the switch closed prior to $t = 0$. Find expressions for the current $i(t)$ and the voltage $v(t)$.

Solution Prior to $t = 0$, the inductor behaves as a short circuit. Thus, we have

First, we use dc steady-state analysis to determine the current before the switch opens.

$$v(t) = 0 \qquad \text{for } t < 0$$

and

$$i(t) = \frac{V_s}{R_1} \qquad \text{for } t < 0$$

Before the switch opens, current circulates clockwise through V_s, R_1, and the inductance. When the switch opens, current continues to flow through the inductance, but the return path is through R_2. Then, a voltage appears across R_2 and the inductance, causing the current to decay.

Since there are no sources driving the circuit after the switch opens, the steady-state solution is zero for $t > 0$. Hence, the solution for $i(t)$ is given by

After the switch opens, the source is disconnected from the circuit, so the steady-state solution for $t > 0$ is zero.

$$i(t) = Ke^{-t/\tau} \qquad \text{for } t > 0 \tag{4.34}$$

in which the time constant is

$$\tau = \frac{L}{R_2} \tag{4.35}$$

Unless an infinite voltage appears across the inductance, the current must be continuous. Recall that prior to $t = 0$, $i(t) = V_s/R_1$. Consequently, just after the switch opens, we have

$$i(0+) = \frac{V_s}{R_1} = Ke^{-0} = K$$

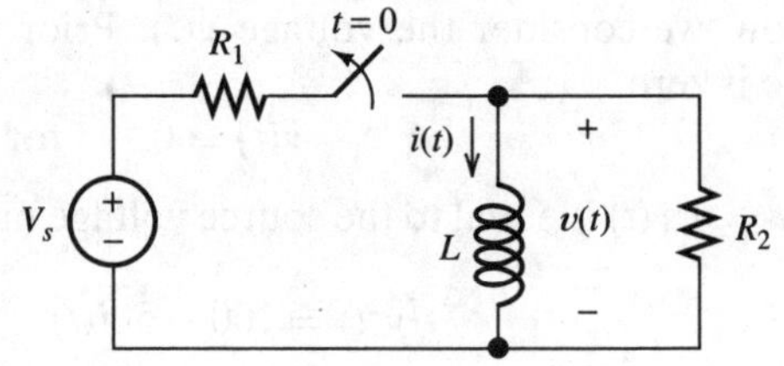

Figure 4.9 The circuit analyzed in Example 4.3.

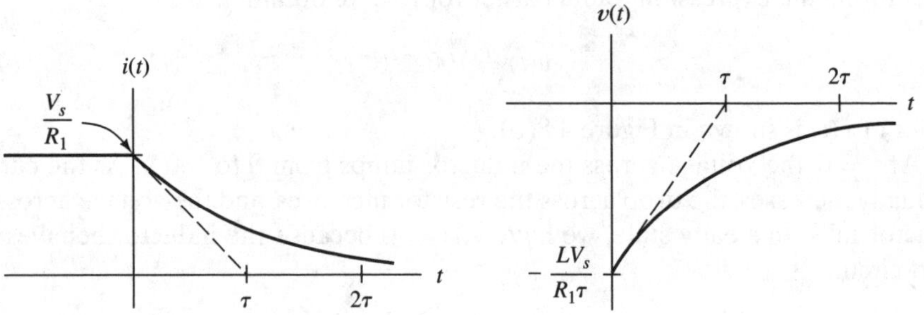

Figure 4.10 The current and voltage for the circuit of Figure 4.9.

Substituting the value of K into Equation 4.34, we find that the current is

$$i(t) = \frac{V_s}{R_1} e^{-t/\tau} \quad \text{for } t > 0 \tag{4.36}$$

The voltage is given by

$$\begin{aligned} v(t) &= L\frac{di(t)}{dt} \\ &= 0 \quad \text{for } t < 0 \\ &= -\frac{LV_s}{R_1\tau} e^{-t/\tau} \quad \text{for } t > 0 \end{aligned}$$

Plots of the voltage and current are shown in Figure 4.10. ■

Exercise 4.4 For the circuit of Example 4.3 (Figure 4.9), assume that $V_s = 15$ V, $R_1 = 10\ \Omega$, $R_2 = 100\ \Omega$, and $L = 0.1$ H. **a.** What is the value of the time constant (after the switch opens)? **b.** What is the maximum magnitude of $v(t)$? **c.** How does the maximum magnitude of $v(t)$ compare to the source voltage? **d.** Find the time t at which $v(t)$ is one-half of its value immediately after the switch opens.

Answer **a.** $\tau = 1$ ms; **b.** $|v(t)|_{max} = 150$ V; **c.** the maximum magnitude of $v(t)$ is 10 times the value of V_s; **d.** $t = \tau \ln(2) = 0.693$ ms. □

Exercise 4.5 Consider the circuit shown in Figure 4.11, in which the switch opens at $t = 0$. Find expressions for $v(t)$, $i_R(t)$, and $i_L(t)$ for $t > 0$. Assume that $i_L(t)$ is zero before the switch opens.

Answer $v(t) = 20e^{-t/0.2}$, $i_R(t) = 2e^{-t/0.2}$, $i_L(t) = 2 - 2e^{-t/0.2}$. □

Exercise 4.6 Consider the circuit shown in Figure 4.12. Assume that the switch has been closed for a very long time prior to $t = 0$. Find expressions for $i(t)$ and $v(t)$.

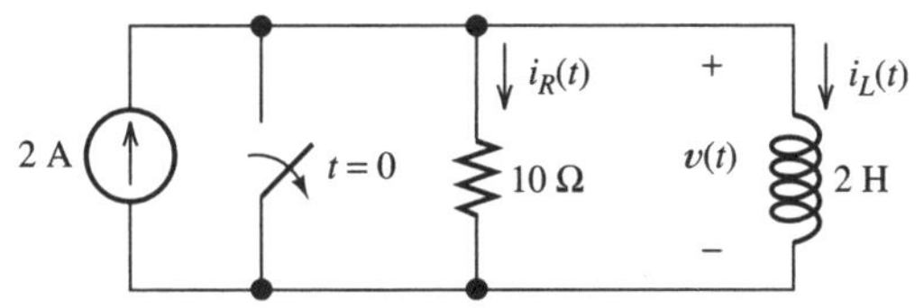

Figure 4.11 The circuit for Exercise 4.5.

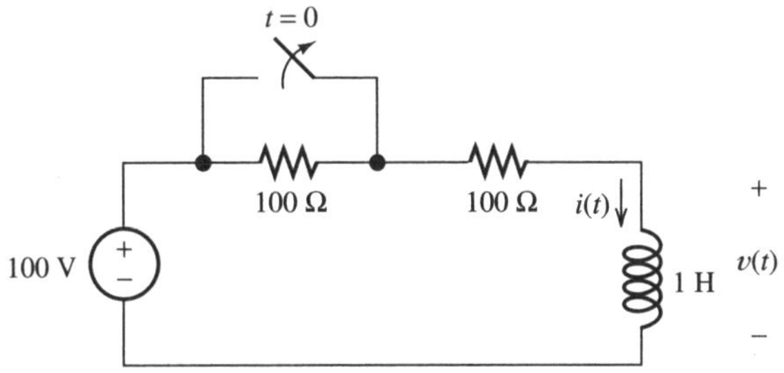

Figure 4.12 The circuit for Exercise 4.6.

Answer

$$\begin{aligned} i(t) &= 1.0 \qquad \text{for } t < 0 \\ &= 0.5 + 0.5e^{-t/\tau} \qquad \text{for } t > 0 \\ v(t) &= 0 \qquad \text{for } t < 0 \\ &= -100e^{-t/\tau} \qquad \text{for } t > 0 \end{aligned}$$

where the time constant is $\tau = 5$ ms. □

4.4 *RC* AND *RL* CIRCUITS WITH GENERAL SOURCES

Now that we have gained some familiarity with *RL* and *RC* circuits, we discuss their solution in general. In this section, we treat circuits that contain one energy-storage element, either an inductance or a capacitance.

Consider the circuit shown in Figure 4.13(a). The circuit inside the box can be any combination of resistances and sources. The single inductance L is shown explicitly. Recall that we can find a Thévenin equivalent for circuits consisting of sources and resistances. The Thévenin equivalent is an independent voltage source $v_t(t)$ in series with the Thévenin resistance R. Thus, any circuit composed of sources, resistances, and one inductance has the equivalent circuit shown in Figure 4.13(b). (Of course, we could reduce any circuit containing sources, resistances, and a single capacitance in a similar fashion.)

Writing a KVL equation for Figure 4.13(b), we obtain

$$L\frac{di(t)}{dt} + Ri(t) = v_t(t) \tag{4.37}$$

If we divide through by the resistance R, we have

$$\frac{L}{R}\frac{di(t)}{dt} + i(t) = \frac{v_t(t)}{R} \tag{4.38}$$

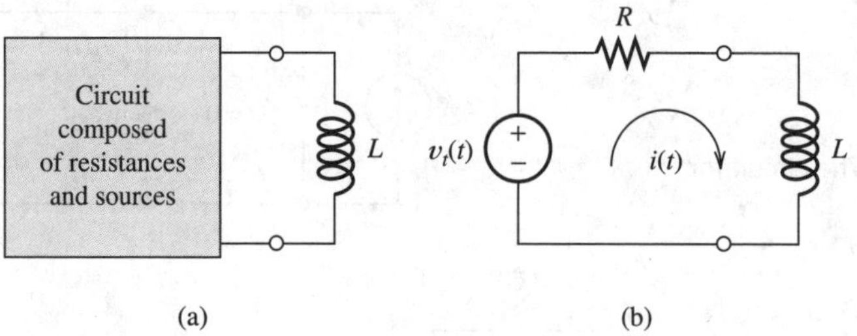

Figure 4.13 A circuit consisting of sources, resistances, and one inductance has an equivalent circuit consisting of a voltage source and a resistance in series with the inductance.

In general, the equation for any circuit containing one inductance or one capacitance can be put into the form

$$\tau \frac{dx(t)}{dt} + x(t) = f(t), \qquad (4.39)$$

in which $x(t)$ represents the current or voltage for which we are solving. Then, we need to find solutions to Equation 4.39 that are consistent with the initial conditions (such as the initial current in the inductance).

The constant τ (which turns out to be the time constant) is a function of only the resistances and the inductance (or capacitance). The sources result in the term $f(t)$, which is called the **forcing function**. If we have a circuit without sources (such as Figure 4.1), the forcing function is zero. For dc sources, the forcing function is constant.

Equation 4.39 is called a first-order differential equation because the highest-order derivative is first order. It is a linear equation because it does not involve powers or other nonlinear functions of $x(t)$ or its derivatives. Thus, to solve an *RL* (or *RC*) circuit, we must find the general solution of a linear first-order differential equation with constant coefficients.

Solution of the Differential Equation

The general solution to Equation 4.39 consists of two parts.

An important result in differential equations states that the general solution to Equation 4.39 consists of two parts. The first part is called the **particular solution** $x_p(t)$ and is any expression that satisfies Equation 4.39. Thus,

$$\tau \frac{dx_p(t)}{dt} + x_p(t) = f(t) \qquad (4.40)$$

The particular solution (also called the forced response) is any expression that satisfies the equation.

The particular solution is also called the **forced response** because it depends on the forcing function (which in turn is due to the independent sources).

In order to have a solution that satisfies the initial conditions, we must add the complementary solution to the particular solution.

(Even though the particular solution satisfies the differential equation, it may not be consistent with the initial conditions, such as the initial voltage on a capacitance or current through an inductance. By adding another term, known as the complementary solution, we obtain a general solution that satisfies both the differential equation and meets the initial conditions.)

For the forcing functions that we will encounter, we can often select the form of the particular solution by inspection. Usually, the particular solution includes terms with the same functional forms as the terms found in the forcing function and its derivatives.

Sinusoidal functions of time are one of the most important types of forcing functions in electrical engineering. For example, consider the forcing function

$$f(t) = 10\cos(200t)$$

Because the derivatives of sine and cosine functions are also sine and cosine functions, we would try a particular solution of the form

$$x_p(t) = A\cos(200t) + B\sin(200t)$$

where A and B are constants that must be determined. We find these constants by substituting the proposed solution into the differential equation and requiring the two sides of the equation to be identical. This leads to equations that can be solved for A and B. (In Chapter 5, we study shortcut methods for solving for the forced response of circuits with sinusoidal sources.)

The second part of the general solution is called the **complementary solution** $x_c(t)$ and is the solution of the **homogeneous equation**

$$\tau\frac{dx_c(t)}{dt} + x_c(t) = 0 \tag{4.41}$$

The homogeneous equation is obtained by setting the forcing function to zero.

We obtain the homogeneous equation by setting the forcing function to zero. Thus, the form of the complementary solution does not depend on the sources. It is also called the **natural response** because it depends on the passive circuit elements. The complementary solution must be added to the particular solution in order to obtain a general solution that matches the initial values of the currents and voltages.

The complementary solution (also called the natural response) is obtained by solving the homogeneous equation.

We can rearrange the homogeneous equation into this form:

$$\frac{dx_c(t)/dt}{x_c(t)} = \frac{-1}{\tau} \tag{4.42}$$

Integrating both sides of Equation 4.42, we have

$$\ln[x_c(t)] = \frac{-t}{\tau} + c \tag{4.43}$$

in which c is the constant of integration. Equation 4.43 is equivalent to

$$x_c(t) = e^{(-t/\tau + c)} = e^c e^{-t/\tau}$$

Then, if we define $K = e^c$, we have the complementary solution

$$x_c(t) = Ke^{-t/\tau} \tag{4.44}$$

Step-by-Step Solution

Next, we summarize an approach to solving circuits containing a resistance, a source, and an inductance (or a capacitance):

1. Write the circuit equation and reduce it to a first-order differential equation.
2. Find a particular solution. The details of this step depend on the form of the forcing function. We illustrate several types of forcing functions in examples, exercises, and problems.

3. Obtain the complete solution by adding the particular solution to the complementary solution given by Equation 4.44, which contains the arbitrary constant K.
4. Use initial conditions to find the value of K.

We illustrate this procedure with an example.

Example 4.4 Transient Analysis of an *RC* Circuit with a Sinusoidal Source

Solve for the current in the circuit shown in Figure 4.14. The capacitor is initially charged so that $v_C(0+) = 1$ V.

Step 1: Write the circuit equation and reduce it to a first-order differential equation.

Solution First, we write a voltage equation for $t > 0$. Traveling clockwise and summing voltages, we obtain

$$Ri(t) + \frac{1}{C}\int_0^t i(t)\,dt + v_C(0) - 2\sin(200t) = 0$$

We convert this to a differential equation by taking the derivative of each term. Of course, the derivative of the integral is simply the integrand. Because $v_C(0)$ is a constant, its derivative is zero. Thus, we have

$$R\frac{di(t)}{dt} + \frac{1}{C}i(t) = 400\cos(200t) \tag{4.45}$$

Multiplying by C, we get

$$RC\frac{di(t)}{dt} + i(t) = 400\,C\cos(200t) \tag{4.46}$$

Substituting values for R and C, we obtain

$$5 \times 10^{-3}\frac{di(t)}{dt} + i(t) = 400 \times 10^{-6}\cos(200t) \tag{4.47}$$

Step 2: Find a particular solution.

The second step is to find a particular solution $i_p(t)$. Often, we start by guessing at the form of $i_p(t)$, possibly including some unknown constants. Then, we substitute our guess into the differential equation and solve for the constants. In the present case, since the derivatives of $\sin(200t)$ and $\cos(200t)$ are $200\cos(200t)$ and $-200\sin(200t)$, respectively, we try a particular solution of the form

The particular solution for a sinusoidal forcing function always has the form given by Equation 4.48.

$$i_p(t) = A\cos(200t) + B\sin(200t) \tag{4.48}$$

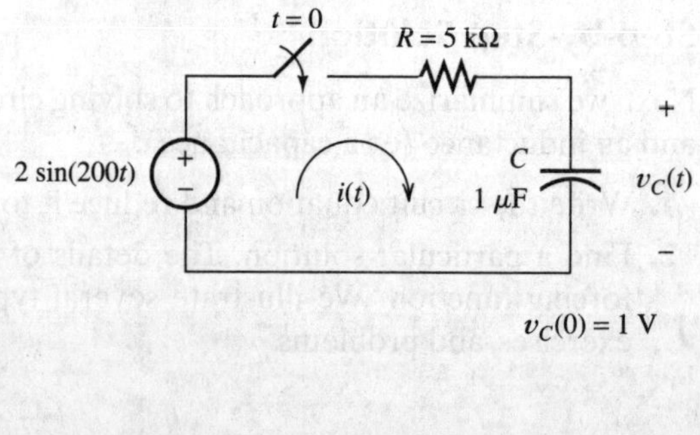

Figure 4.14 A first-order *RC* circuit with a sinusoidal source. See Example 4.4.

where A and B are constants to be determined so that i_p is indeed a solution to Equation 4.47.

Substituting the proposed solution into Equation 4.47, we obtain

We substitute Equation 4.48 into the differential equation, and solve for A and B.

$$-A\sin(200t) + B\cos(200t) + A\cos(200t) + B\sin(200t)$$
$$= 400 \times 10^{-6}\cos(200t)$$

However, the left-hand side of this equation is required to be identical to the right-hand side. Equating the coefficients of the sine functions, we have

$$-A + B = 0 \tag{4.49}$$

Equating the coefficients of the cosine functions, we get

$$B + A = 400 \times 10^{-6} \tag{4.50}$$

These equations can be readily solved, yielding

$$A = 200 \times 10^{-6} = 200\ \mu\text{A}$$

and

$$B = 200 \times 10^{-6} = 200\ \mu\text{A}$$

Substituting these values into Equation 4.48, we obtain the particular solution

$$i_p(t) = 200\cos(200t) + 200\sin(200t)\ \mu\text{A} \tag{4.51}$$

which can also be written as

$$i_p(t) = 200\sqrt{2}\cos(200t - 45°)$$

(In Chapter 5, we will learn shortcut methods for combining sine and cosine functions.)

We obtain the homogeneous equation by substituting 0 for the forcing function in Equation 4.46. Thus, we have

$$RC\frac{di(t)}{dt} + i(t) = 0 \tag{4.52}$$

The complementary solution is

$$i_c(t) = Ke^{-t/RC} = Ke^{-t/\tau} \tag{4.53}$$

Adding the particular solution and the complementary solution, we obtain the general solution

Step 3: Obtain the complete solution by adding the particular solution to the complementary solution.

$$i(t) = 200\cos(200t) + 200\sin(200t) + Ke^{-t/RC}\ \mu\text{A} \tag{4.54}$$

Finally, we determine the value of the constant K by using the initial conditions. The voltages and currents immediately after the switch closes are shown in Figure 4.15. The source voltage is 0 V and the voltage across the capacitor is

Step 4: Use initial conditions to find the value of K.

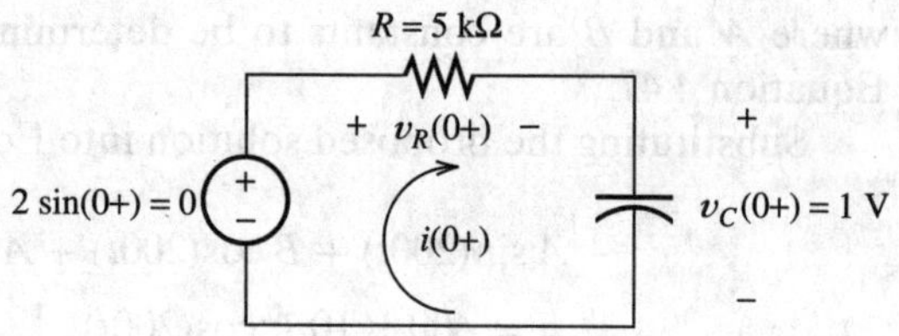

Figure 4.15 The voltages and currents for the circuit of Figure 4.14 immediately after the switch closes.

$v_C(0+) = 1$. Consequently, the voltage across the resistor must be $v_R(0+) = -1$ V. Thus, we get

$$i(0+) = \frac{v_R(0+)}{R} = \frac{-1}{5000} = -200\ \mu\text{A}$$

Substituting $t = 0$ into Equation 4.54, we obtain

$$i(0+) = -200 = 200 + K\ \mu\text{A} \tag{4.55}$$

Solving, we find that $K = -400\ \mu$A. Substituting this into Equation 4.54, we have the solution

$$i(t) = 200\cos(200t) + 200\sin(200t) - 400e^{-t/RC}\ \mu\text{A} \tag{4.56}$$

Plots of the particular solution and of the complementary solution are shown in Figure 4.16. The time constant for this circuit is $\tau = RC = 5$ ms. Notice that the natural response decays to negligible values in about 25 ms. As expected, the natural response has decayed in about five time constants. Furthermore, notice that for a sinusoidal forcing function, the forced response is also sinusoidal and persists after the natural response has decayed.

Notice that the forced response is sinusoidal for a sinusoidal forcing function.

A plot of the complete solution is shown in Figure 4.17. ■

Exercise 4.7 Repeat Example 4.4 if the source voltage is changed to $2\cos(200t)$ and the initial voltage on the capacitor is $v_C(0) = 0$. The circuit with these changes is shown in Figure 4.18.

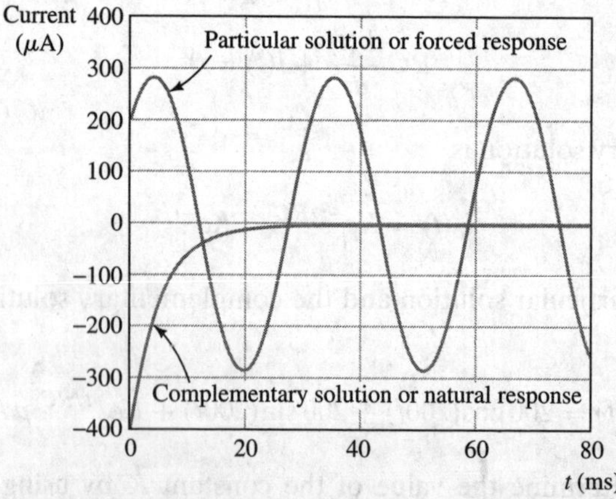

Figure 4.16 The complementary solution and the particular solution for Example 4.4.

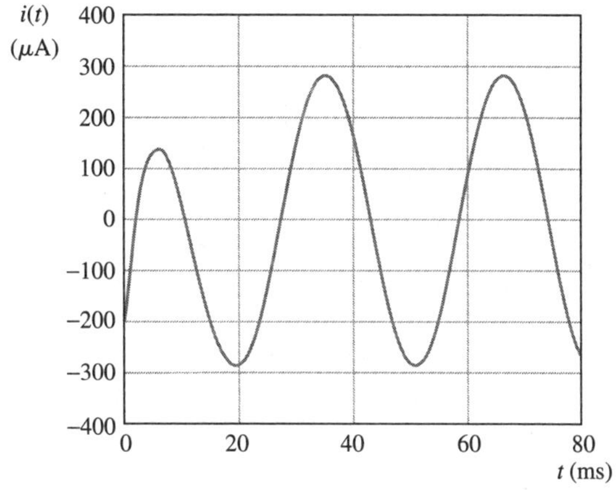

Figure 4.17 The complete solution for Example 4.4.

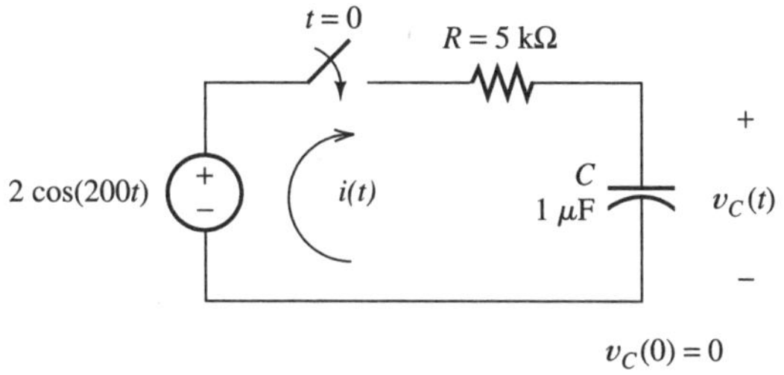

Figure 4.18 The circuit for Exercise 4.7.

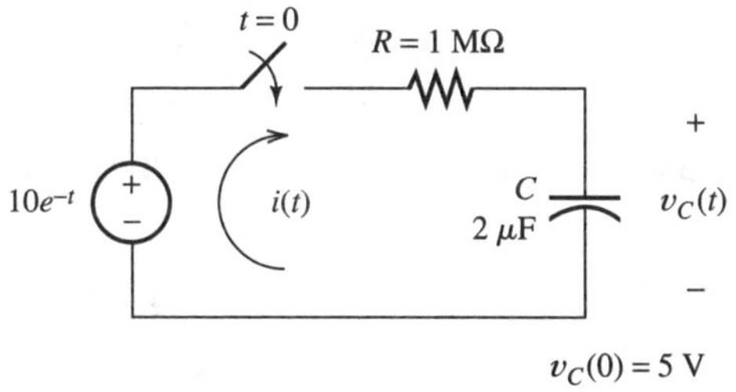

Figure 4.19 The circuit for Exercise 4.8.

Answer $i(t) = -200\sin(200t) + 200\cos(200t) + 200e^{-t/RC}$ μA, in which $\tau = RC =$ 5 ms. □

Exercise 4.8 Solve for the current in the circuit shown in Figure 4.19 after the switch closes. [*Hint:* Try a particular solution of the form $i_p(t) = Ae^{-t}$.]

Answer $i(t) = 20e^{-t} - 15e^{-t/2}$ μA. □

4.5 SECOND-ORDER CIRCUITS

In this section, we consider circuits that contain two energy-storage elements. In particular, we look at circuits that have one inductance and one capacitance, either in series or in parallel.

Differential Equation

To derive the general form of the equations that we encounter in circuits with two energy-storage elements, consider the series circuit shown in Figure 4.20(a). Writing

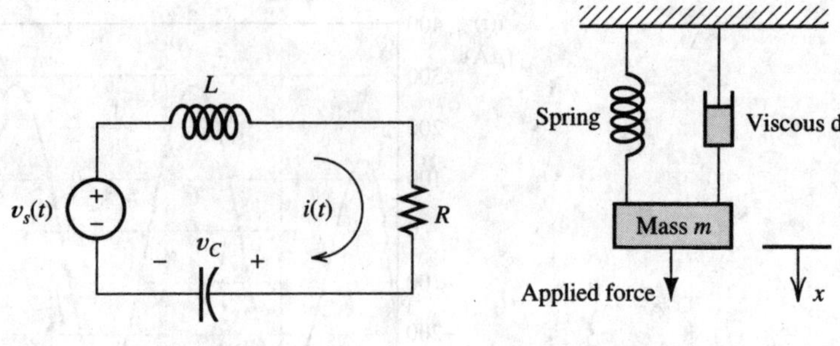

(a) Electrical circuit (b) Mechanical analog

Figure 4.20 The series *RLC* circuit and its mechanical analog.

a KVL equation, we have

$$L\frac{di(t)}{dt} + Ri(t) + \frac{1}{C}\int_0^t i(t)dt + v_C(0) = v_s(t) \tag{4.57}$$

We convert the integrodifferential equation to a pure differential equation by differentiating with respect to time.

Taking the derivative with respect to time, we get

$$L\frac{d^2i(t)}{dt^2} + R\frac{di(t)}{dt} + \frac{1}{C}i(t) = \frac{dv_s(t)}{dt} \tag{4.58}$$

Dividing through by L, we obtain

$$\frac{d^2i(t)}{dt^2} + \frac{R}{L}\frac{di(t)}{dt} + \frac{1}{LC}i(t) = \frac{1}{L}\frac{dv_s(t)}{dt} \tag{4.59}$$

Now, we define the **damping coefficient** as

$$\alpha = \frac{R}{2L} \tag{4.60}$$

and the **undamped resonant frequency** as

$$\omega_0 = \frac{1}{\sqrt{LC}} \tag{4.61}$$

The **forcing function** is

$$f(t) = \frac{1}{L}\frac{dv_s(t)}{dt} \tag{4.62}$$

Using these definitions, we find that Equation 4.59 can be written as

$$\frac{d^2i(t)}{dt^2} + 2\alpha\frac{di(t)}{dt} + \omega_0^2 i(t) = f(t) \tag{4.63}$$

If a circuit contains two energy-storage elements (after substituting all possible series or parallel equivalents), the circuit equations can always be reduced to the form given by Equation 4.63.

This is a linear second-order differential equation with constant coefficients. Thus, we refer to circuits having two energy-storage elements as second-order circuits. (An exception occurs if we can combine the energy-storage elements in series or parallel. For example, if we have two capacitors in parallel, we can combine them into a single equivalent capacitance, and then we would have a first-order circuit.)

Mechanical Analog

The mechanical analog of the series RLC circuit is shown in Figure 4.20(b). The displacement x of the mass is analogous to electrical charge, the velocity dx/dt is analogous to current, and force is analogous to voltage. The mass plays the role of the inductance, the spring plays the role of the capacitance, and the damper plays the role of the resistance. The equation of motion for the mechanical system can be put into the form of Equation 4.63.

Based on an intuitive consideration of Figure 4.20, we can anticipate that the sudden application of a constant force (dc voltage) can result in a displacement (current) that either approaches steady-state conditions asymptotically or oscillates before settling to the steady-state value. The type of behavior depends on the relative values of the mass, spring constant, and damping coefficient.

Solution of the Second-Order Equation

We will see that the circuit equations for currents and voltages in circuits having two energy-storage elements can always be put into the form of Equation 4.63. Thus, let us consider the solution of

$$\frac{d^2x(t)}{dt^2} + 2\alpha\frac{dx(t)}{dt} + \omega_0^2 x(t) = f(t) \tag{4.64}$$

where we have used $x(t)$ for the variable, which could represent either a current or a voltage.

Here again, the general solution $x(t)$ to this equation consists of two parts: a particular solution $x_p(t)$ plus the complementary solution $x_c(t)$ and is expressed as

$$x(t) = x_p(t) + x_c(t) \tag{4.65}$$

Particular Solution. The particular solution is any expression $x_p(t)$ that satisfies the differential equation

$$\frac{d^2x_p(t)}{dt^2} + 2\alpha\frac{dx_p(t)}{dt} + \omega_0^2 x_p(t) = f(t) \tag{4.66}$$

The particular solution is also called the **forced response**. (Usually, we eliminate any terms from $x_p(t)$ that produce a zero net result when substituted into the left-hand side of Equation 4.66. In other words, we eliminate any terms that have the same form as the homogeneous solution.)

We will be concerned primarily with either constant (dc) or sinusoidal (ac) forcing functions. For dc sources, we can find the particular solution directly from the circuit by replacing the inductances by short circuits, replacing the capacitances by open circuits, and solving. This technique was discussed in Section 4.2. In Chapter 5, we will learn efficient methods for finding the forced response due to sinusoidal sources.

For dc sources, we can find the particular solution by performing a dc steady-state analysis as discussed in Section 4.2.

Complementary Solution. The complementary solution $x_c(t)$ is found by solving the homogeneous equation, which is obtained by substituting 0 for the forcing function $f(t)$. Thus, the homogeneous equation is

$$\frac{d^2x_c(t)}{dt^2} + 2\alpha\frac{dx_c(t)}{dt} + \omega_0^2 x_c(t) = 0 \tag{4.67}$$

In finding the solution to the homogeneous equation, we start by substituting the trial solution $x_c(t) = Ke^{st}$. This yields

$$s^2Ke^{st} + 2\alpha sKe^{st} + \omega_0^2Ke^{st} = 0 \tag{4.68}$$

Factoring, we obtain

$$(s^2 + 2\alpha s + \omega_0^2)Ke^{st} = 0 \tag{4.69}$$

Since we want to find a solution Ke^{st} that is nonzero, we must have

$$s^2 + 2\alpha s + \omega_0^2 = 0 \tag{4.70}$$

This is called the **characteristic equation.**

The **damping ratio** is defined as

$$\zeta = \frac{\alpha}{\omega_0} \tag{4.71}$$

The form of the complementary solution depends on the value of the damping ratio.

The form of the complementary solution depends on the value of the damping ratio. The roots of the characteristic equation are given by

$$s_1 = -\alpha + \sqrt{\alpha^2 - \omega_0^2} \tag{4.72}$$

and

$$s_2 = -\alpha - \sqrt{\alpha^2 - \omega_0^2} \tag{4.73}$$

We have three cases depending on the value of the damping ratio ζ compared with unity.

If the damping ratio is greater than unity, we say that the circuit is overdamped, the roots of the characteristic equation are real, and the complementary solution has the form given in Equation 4.74.

1. *Overdamped case* ($\zeta > 1$). If $\zeta > 1$ (or equivalently, if $\alpha > \omega_0$), the roots of the characteristic equation are real and distinct. Then the complementary solution is

$$x_c(t) = K_1e^{s_1t} + K_2e^{s_2t} \tag{4.74}$$

In this case, we say that the circuit is **overdamped.**

If the damping ratio equals unity, the circuit is critically damped, the roots of the characteristic equation are real and equal, and the complementary solution has the form given in Equation 4.75.

2. *Critically damped case* ($\zeta = 1$). If $\zeta = 1$ (or equivalently, if $\alpha = \omega_0$), the roots are real and equal. Then, the complementary solution is

$$x_c(t) = K_1e^{s_1t} + K_2te^{s_1t} \tag{4.75}$$

In this case, we say that the circuit is **critically damped.**

If the damping ratio is less than unity, the roots of the characteristic equation are complex conjugates, and the complementary solution has the form given in Equation 4.77.

3. *Underdamped case* ($\zeta < 1$). Finally, if $\zeta < 1$ (or equivalently, if $\alpha < \omega_0$), the roots are complex. (By the term *complex*, we mean that the roots involve the imaginary number $\sqrt{-1}$.) In other words, the roots are of the form

$$s_1 = -\alpha + j\omega_n \quad \text{and} \quad s_2 = -\alpha - j\omega_n$$

in which $j = \sqrt{-1}$ and the **natural frequency** is given by

$$\omega_n = \sqrt{\omega_0^2 - \alpha^2} \tag{4.76}$$

(In electrical engineering, we use j rather than i to stand for the imaginary number $\sqrt{-1}$ because we use i for current.)

For complex roots, the complementary solution is of the form

$$x_c(t) = K_1 e^{-\alpha t} \cos(\omega_n t) + K_2 e^{-\alpha t} \sin(\omega_n t) \tag{4.77}$$

In this case, we say that the circuit is **underdamped**.

Example 4.5 Analysis of a Second-Order Circuit with a DC Source

A dc source is connected to a series RLC circuit by a switch that closes at $t = 0$ as shown in Figure 4.21. The initial conditions are $i(0) = 0$ and $v_C(0) = 0$. Write the differential equation for $v_C(t)$. Solve for $v_C(t)$ if $R = 300, 200,$ and $100\ \Omega$.

Solution First, we can write an expression for the current in terms of the voltage across the capacitance:

First, we write the circuit equations and reduce them to the form given in Equation 4.63.

$$i(t) = C\frac{dv_C(t)}{dt} \tag{4.78}$$

Then, we write a KVL equation for the circuit:

$$L\frac{di(t)}{dt} + Ri(t) + v_C(t) = V_s \tag{4.79}$$

Using Equation 4.78 to substitute for $i(t)$, we get

$$LC\frac{d^2 v_C(t)}{dt^2} + RC\frac{dv_C(t)}{dt} + v_C(t) = V_s \tag{4.80}$$

Dividing through by LC, we have

$$\frac{d^2 v_C(t)}{dt^2} + \frac{R}{L}\frac{dv_C(t)}{dt} + \frac{1}{LC}v_C(t) = \frac{V_s}{LC} \tag{4.81}$$

As expected, the differential equation for $v_C(t)$ has the same form as Equation 4.63.

Next, we find the particular solution. Since we have a dc source, we can find this part of the solution by replacing the inductance by a short circuit and the capacitance by an open circuit. This is shown in Figure 4.22. Then the current is zero, the drop across the resistance is zero, and the voltage across the capacitance (open circuit) is equal to the dc source voltage. Therefore, the particular solution is

Next, we find the particular solution by solving the circuit for dc steady-state conditions.

$$v_{Cp}(t) = V_s = 10\text{ V} \tag{4.82}$$

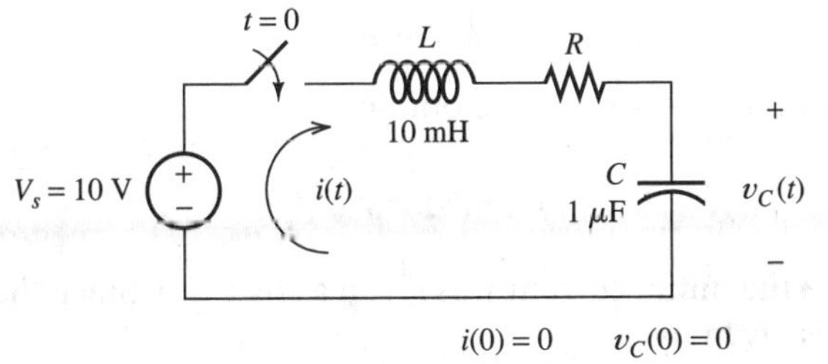

Figure 4.21 The circuit for Example 4.5.

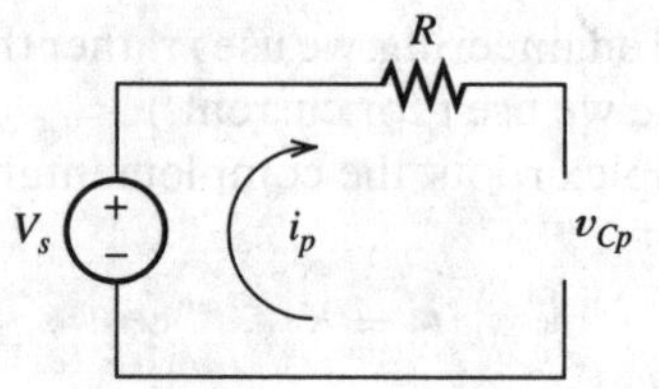

Figure 4.22 The equivalent circuit for Figure 4.21 under steady-state conditions. The inductor has been replaced by a short circuit and the capacitor by an open circuit.

(It can be verified that this is a particular solution by substituting it into Equation 4.81.) Notice that in this circuit the particular solution for $v_C(t)$ is the same for all three values of resistance.

Next, we find the complementary solution for each value of R. For each resistance value, we

1. Determine the damping ratio and roots of the characteristic equation.
2. Select the appropriate form for the homogeneous solution, depending on the value of the damping ratio.
3. Add the homogeneous solution to the particular solution and determine the values of the coefficients (K_1 and K_2), based on the initial conditions.

Next, we find the homogeneous solution and general solution for each value of R. For all three cases, we have

$$\omega_0 = \frac{1}{\sqrt{LC}} = 10^4 \tag{4.83}$$

Case I ($R = 300\ \Omega$)

In this case, we get

$$\alpha = \frac{R}{2L} = 1.5 \times 10^4 \tag{4.84}$$

The damping ratio is $\zeta = \alpha/\omega_0 = 1.5$. Because we have $\zeta > 1$, this is the overdamped case. The roots of the characteristic equation are given by Equations 4.72 and 4.73. Substituting values, we find that

$$\begin{aligned} s_1 &= -\alpha + \sqrt{\alpha^2 - \omega_0^2} \\ &= -1.5 \times 10^4 + \sqrt{(1.5 \times 10^4)^2 - (10^4)^2} \\ &= -0.3820 \times 10^4 \end{aligned}$$

and

$$\begin{aligned} s_2 &= -\alpha - \sqrt{\alpha^2 - \omega_0^2} \\ &= -2.618 \times 10^4 \end{aligned}$$

The homogeneous solution has the form of Equation 4.74. Adding the particular solution given by Equation 4.82 to the homogeneous solution, we obtain the general solution

$$v_C(t) = 10 + K_1 e^{s_1 t} + K_2 e^{s_2 t} \tag{4.85}$$

Now, we must find values of K_1 and K_2 so the solution matches the known initial conditions in the circuit. It was given that the initial voltage on the capacitance is zero. Hence,

$$v_C(0) = 0$$

Evaluating Equation 4.85 at $t = 0$, we obtain

$$10 + K_1 + K_2 = 0 \tag{4.86}$$

Furthermore, the initial current was given as $i(0) = 0$. Since the current through the capacitance is given by

$$i(t) = C\frac{dv_C(t)}{dt}$$

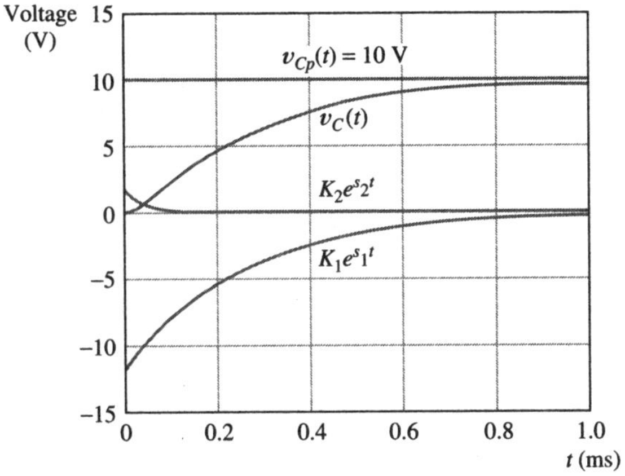

Figure 4.23 Solution for $R = 300\ \Omega$.

we conclude that

$$\frac{dv_C(0)}{dt} = 0$$

Taking the derivative of Equation 4.85 and evaluating at $t = 0$, we have

$$s_1K_1 + s_2K_2 = 0 \tag{4.87}$$

Now, we can solve Equations 4.86 and 4.87 for the values of K_1 and K_2. The results are $K_1 = -11.708$ and $K_2 = 1.708$. Substituting these values into Equation 4.85, we have the solution

$$v_C(t) = 10 - 11.708e^{s_1t} + 1.708e^{s_2t}$$

Plots of each of the terms of this equation and the complete solution are shown in Figure 4.23.

Case II ($R = 200\ \Omega$)

Now, we repeat the steps for $R = 200\ \Omega$.

In this case, we get

$$\alpha = \frac{R}{2L} = 10^4 \tag{4.88}$$

Because $\zeta = \alpha/\omega_0 = 1$, this is the critically damped case. The roots of the characteristic equation are given by Equations 4.72 and 4.73. Substituting values, we have

$$s_1 = s_2 = -\alpha + \sqrt{\alpha^2 - \omega_0^2} = -\alpha = -10^4$$

The homogeneous solution has the form of Equation 4.75. Adding the particular solution (Equation 4.82) to the homogeneous solution, we find that

$$v_C(t) = 10 + K_1e^{s_1t} + K_2te^{s_1t} \tag{4.89}$$

As in case I, the initial conditions require $v_C(0) = 0$ and $dv_C(0)/dt = 0$. Thus, substituting $t = 0$ into Equation 4.89, we get

$$10 + K_1 = 0 \tag{4.90}$$

Differentiating Equation 4.89 and substituting $t = 0$ yields

$$s_1K_1 + K_2 = 0 \tag{4.91}$$

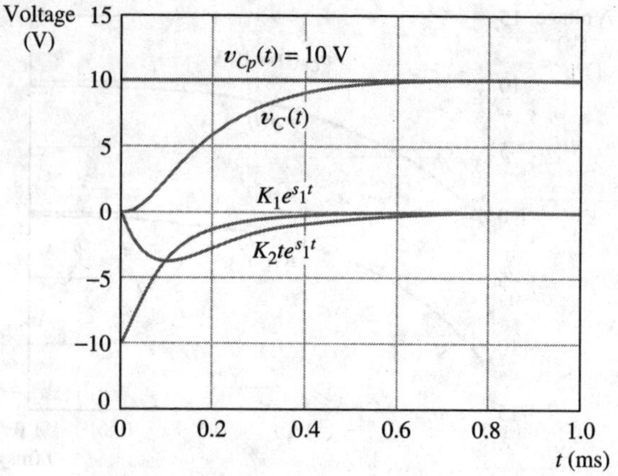

Figure 4.24 Solution for $R = 200\ \Omega$.

Solving Equations 4.90 and 4.91 yields $K_1 = -10$ and $K_2 = -10^5$. Thus, the solution is

$$v_C(t) = 10 - 10e^{s_1 t} - 10^5 t e^{s_1 t} \tag{4.92}$$

Plots of each of the terms of this equation and the complete solution are shown in Figure 4.24.

Case III ($R = 100\ \Omega$)

Finally, we repeat the solution for $R = 100\ \Omega$.

For this value of resistance, we have

$$\alpha = \frac{R}{2L} = 5000 \tag{4.93}$$

Because $\zeta = \alpha/\omega_0 = 0.5$, this is the underdamped case. Using Equation 4.76, we compute the natural frequency:

$$\omega_n = \sqrt{\omega_0^2 - \alpha^2} = 8660 \tag{4.94}$$

The homogeneous solution has the form of Equation 4.77. Adding the particular solution found earlier to the homogeneous solution, we obtain the general solution:

$$v_C(t) = 10 + K_1 e^{-\alpha t}\cos(\omega_n t) + K_2 e^{-\alpha t}\sin(\omega_n t) \tag{4.95}$$

As in the previous cases, the initial conditions are $v_C(0) = 0$ and $dv_C(0)/dt = 0$. Evaluating Equation 4.95 at $t = 0$, we obtain

$$10 + K_1 = 0 \tag{4.96}$$

Differentiating Equation 4.95 and evaluating at $t = 0$, we have

$$-\alpha K_1 + \omega_n K_2 = 0 \tag{4.97}$$

Solving Equations 4.96 and 4.97, we obtain $K_1 = -10$ and $K_2 = -5.774$. Thus, the complete solution is

$$v_C(t) = 10 - 10e^{-\alpha t}\cos(\omega_n t) - 5.774e^{-\alpha t}\sin(\omega_n t) \tag{4.98}$$

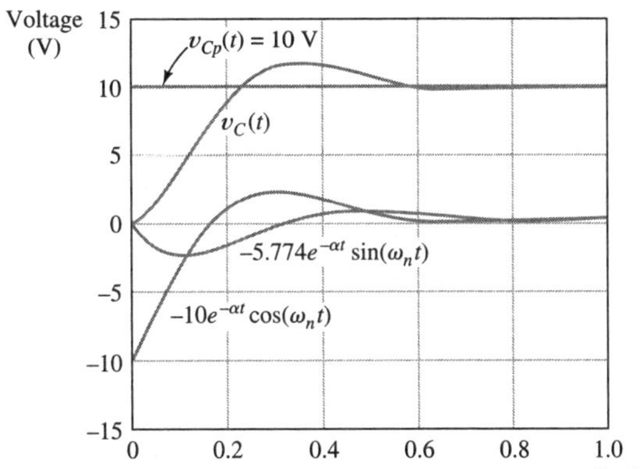

Figure 4.25 Solution for $R = 100\ \Omega$.

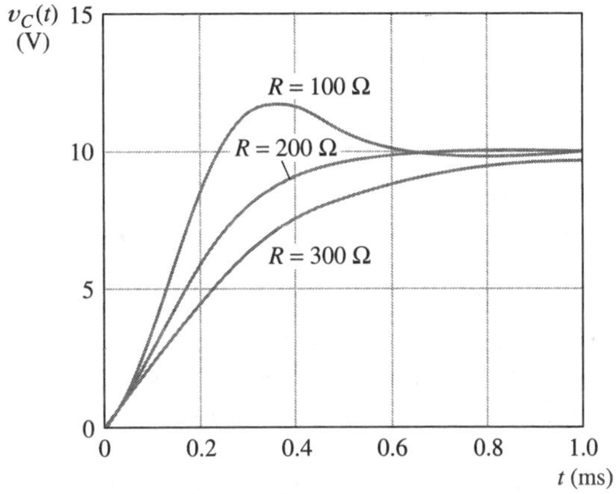

Figure 4.26 Solutions for all three resistances.

Plots of each of the terms of this equation and the complete solution are shown in Figure 4.25.

Figure 4.26 shows the complete response for all three values of resistance. ■

Normalized Step Response of Second-Order Systems

When we suddenly apply a constant source to a circuit, we say that the forcing function is a **step function**. A unit step function, denoted by $u(t)$, is shown in Figure 4.27. By definition, we have

$$\begin{aligned} u(t) &= 0 \qquad t < 0 \\ &= 1 \qquad t \geq 0 \end{aligned}$$

For example, if we apply a dc voltage of A volts to a circuit by closing a switch, the applied voltage is a step function, given by

$$v(t) = Au(t)$$

This is illustrated in Figure 4.28.

We often encounter situations, such as Example 4.5, in which step forcing functions are applied to second-order systems described by a differential equation

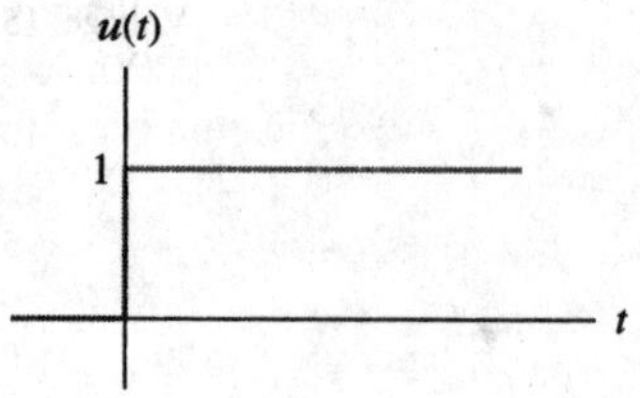

Figure 4.27 A unit step function $u(t)$. For $t < 0$, $u(t) = 0$. For $t \geq 0$, $u(t) = 1$.

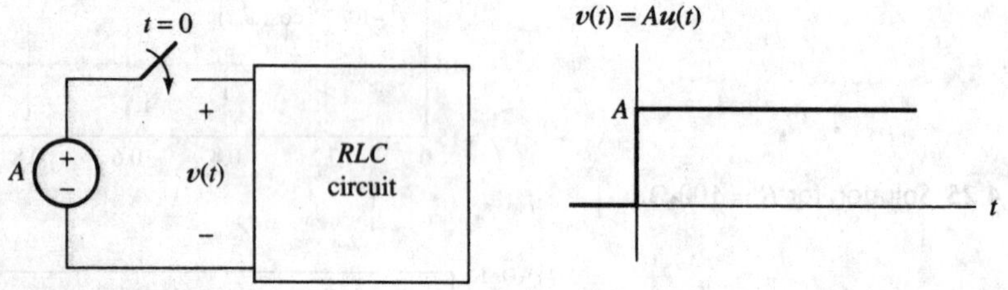

Figure 4.28 Applying a dc voltage by closing a switch results in a forcing function that is a step function.

of the form

$$\frac{d^2x(t)}{dt^2} + 2\alpha\frac{dx(t)}{dt} + \omega_0^2 x(t) = Au(t) \tag{4.99}$$

The differential equation is characterized by its undamped resonant frequency ω_0 and damping ratio $\zeta = \alpha/\omega_0$. [Of course, the solution for $x(t)$ also depends on the initial conditions.] Normalized solutions are shown in Figure 4.29 for the initial conditions $x(0) = 0$ and $x'(0) = 0$.

The system response for small values of the damping ratio ζ displays **overshoot** and **ringing** before settling to the steady-state value. On the other hand, if the damping ratio is large (compared to unity), the response takes a relatively long time to closely approach the final value.

Frequently, electrical control systems and mechanical systems are best designed with a damping ratio close to unity. For example, when the suspension system on your automobile becomes severely underdamped, it is time for new shock absorbers.

Sometimes, we want to design a second-order system that quickly settles to steady state. Then we try to design for a damping ratio close to unity. For example, the control system for a robot arm could be a second-order system. When a step signal calls for the arm to move, we probably want it to achieve the final position in the minimum time without excessive overshoot and ringing.

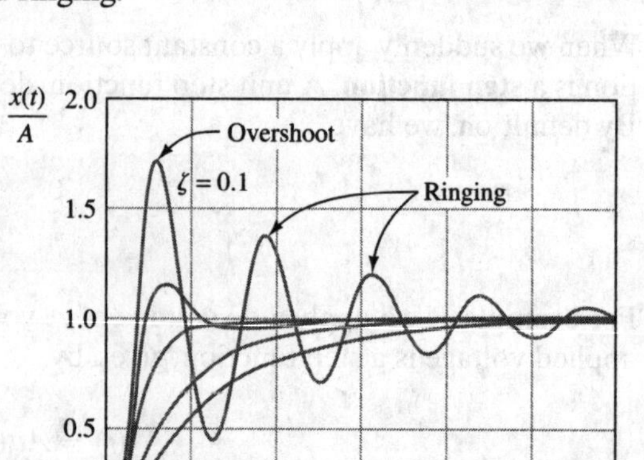

Figure 4.29 Normalized step responses for second-order systems described by Equation 4.99 with damping ratios of $\zeta = 0.1, 0.5, 1, 2,$ and 3. The initial conditions are assumed to be $x(0) = 0$ and $x'(0) = 0$.

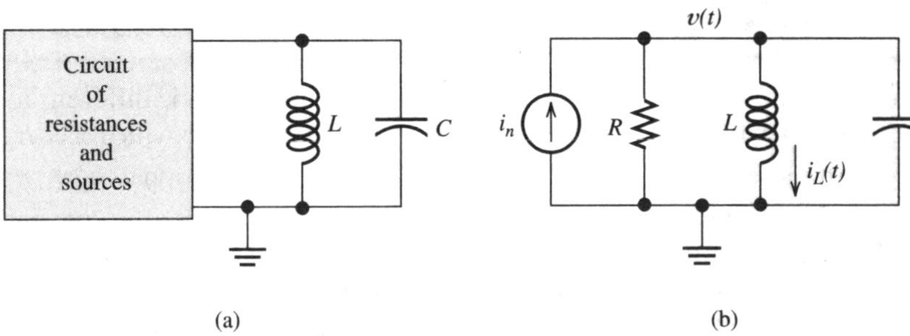

Figure 4.30 Any circuit consisting of sources, resistances, and a parallel LC combination can be reduced to the equivalent circuit shown in (b).

Circuits with Parallel *L* and *C*

The solution of circuits having an inductance and capacitance in parallel is very similar to the series case. Consider the circuit shown in Figure 4.30(a). The circuit inside the box is assumed to consist of sources and resistances. As we saw in Section 2.6, we can find a Norton equivalent circuit for any two-terminal circuit composed of resistances and sources. The equivalent circuit is shown in Figure 4.30(b).

We can analyze this circuit by writing a KCL equation at the top node of Figure 4.30(b) which results in

$$C\frac{dv(t)}{dt} + \frac{1}{R}v(t) + \frac{1}{L}\int_0^t v(t)\,dt + i_L(0) = i_n(t) \tag{4.100}$$

This can be converted into a pure differential equation by taking the derivative with respect to time:

$$C\frac{d^2v(t)}{dt^2} + \frac{1}{R}\frac{dv(t)}{dt} + \frac{1}{L}v(t) = \frac{di_n(t)}{dt} \tag{4.101}$$

Dividing through by the capacitance, we have

$$\frac{d^2v(t)}{dt^2} + \frac{1}{RC}\frac{dv(t)}{dt} + \frac{1}{LC}v(t) = \frac{1}{C}\frac{di_n(t)}{dt} \tag{4.102}$$

Now, if we define the damping coefficient

$$\alpha = \frac{1}{2RC} \tag{4.103}$$

the undamped resonant frequency

$$\omega_0 = \frac{1}{\sqrt{LC}} \tag{4.104}$$

and the forcing function

$$f(t) = \frac{1}{C}\frac{di_n(t)}{dt} \tag{4.105}$$

the differential equation can be written as

$$\frac{d^2v(t)}{dt^2} + 2\alpha\frac{dv(t)}{dt} + \omega_0^2 v(t) = f(t) \tag{4.106}$$

Notice that the equation for the damping coefficient of the parallel RLC circuit is different from that for the series circuit.

This equation has exactly the same form as Equation 4.64. Therefore, transient analysis of circuits with parallel LC elements is very similar to that of series LC circuits. However, notice that the equation for the damping coefficient α is different for the parallel circuit (in which $\alpha = 1/2RC$) than for the series circuit (in which $\alpha = R/2L$).

$v(0-)$ and $i_L(0-)$ are the voltage and current values immediately before the switch opens.

Exercise 4.9 Consider the circuit shown in Figure 4.31 on page 209 with $R = 25\ \Omega$. **a.** Compute the undamped resonant frequency, the damping coefficient, and the damping ratio. **b.** The initial conditions are $v(0-) = 0$ and $i_L(0-) = 0$. Show that this requires that $v'(0+) = 10^6$ V/s. **c.** Find the particular solution for $v(t)$. **d.** Find the general solution for $v(t)$, including the numerical values of all parameters.

Answer **a.** $\omega_0 = 10^5, \alpha = 2 \times 10^5$, and $\zeta = 2$; **b.** KCL requires that $i_C(0) = 0.1\text{ A} = Cv'(0)$, thus $v'(0) = 10^6$; **c.** $v_p(t) = 0$; **d.** $v(t) = 2.89(e^{-0.268 \times 10^5 t} - e^{-3.73 \times 10^5 t})$. □

PRACTICAL APPLICATION 4.1

Electronics and the Art of Automotive Maintenance

Throughout much of the history of the automobile, ignition systems have been designed as a straightforward application of electrical transients. The basic ignition system used for many years is shown in Figure PA4.1. The coil is a pair of mutually coupled inductors known as the primary and the secondary. The points form a switch that opens and closes as the engine rotates, opening at the instant that an ignition spark is needed by one of the cylinders. While the points are closed, current builds up relatively slowly in the primary winding of the coil. Then, when the points open, the current is rapidly interrupted. The resulting high rate of change of current induces a large voltage across the secondary winding, which is connected to the appropriate spark plug by the distributor. The resistance is needed to limit the current in case the engine stops with the points closed.

The capacitor prevents the voltage across the points from rising too rapidly when they open. (Recall that the voltage across a capacitance cannot change instantaneously.) Otherwise, arcing would occur across the points, causing them to become burned and pitted. By slowing the rise of voltage, the capacitor gives the gap between the points time to become wide enough to withstand the voltage across them. (Even so, the peak voltage across the points is many times the battery voltage.)

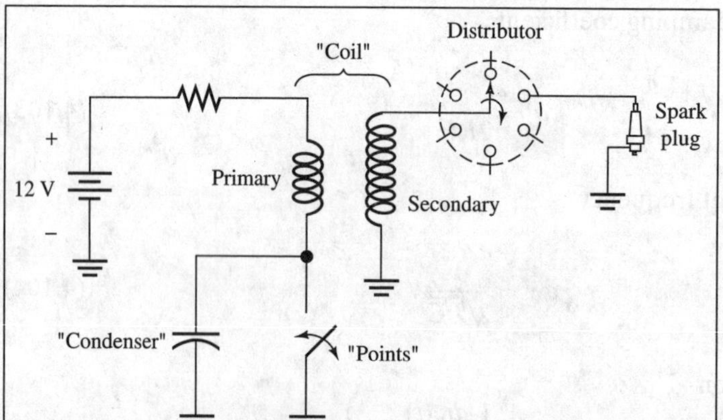

Figure PA4.1 Classic ignition for an internal-combustion engine.

The primary inductance, current-limiting resistance, and capacitance form an underdamped series *RLC* circuit. Thus, an oscillatory current flows through the primary when the points open, inducing the requisite voltage in the secondary.

In its early forms, the ignition system had mechanical or vacuum systems to make adjustments to the timing, depending on engine speed and throttle setting. In more recent years, the availability of complex electronics at reasonable costs plus the desire to adjust the ignition to obtain good performance and low pollution levels with varying air temperature, fuel quality, air pressure, engine temperature, and other factors have greatly affected the design of ignition systems. The basic principles remain the same as in the days of the classic automobile, but a complex network of electrical sensors, a digital computer, and an electronic switch have replaced the points and simple vacuum advance.

The complexity of modern engineering designs has become somewhat intimidating, even to practicing engineers. In the 1960s, as a new engineering graduate, one could study the design of an ignition system, a radio, or a home appliance, readily spotting and repairing malfunctions with the aid of a few tools and standard parts. Nowadays, if my car should fail to start due to ignition malfunction, at the end of a fishing trip into the backwoods of northern Michigan, I might very well have to walk back to civilization. Nevertheless, the improvements in performance provided by modern electronics make up for its difficulty of repair.

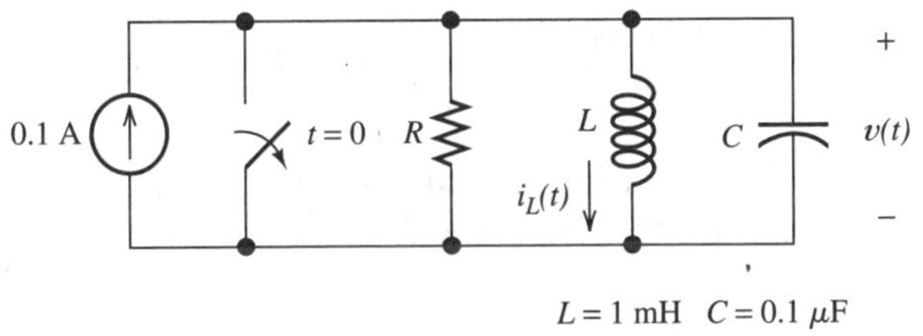

Figure 4.31 Circuit for Exercises 4.9, 4.10, and 4.11.

Exercise 4.10 Repeat Exercise 4.9 for $R = 50\ \Omega$.

Answer **a.** $\omega_0 = 10^5$, $\alpha = 10^5$, and $\zeta = 1$; **b.** KCL requires that $i_C(0) = 0.1\text{ A} = Cv'(0)$, thus $v'(0) = 10^6$; **c.** $v_p(t) = 0$; **d.** $v(t) = 10^6 te^{-10^5 t}$. □

Exercise 4.11 Repeat Exercise 4.9 for $R = 250\ \Omega$.

Answer **a.** $\omega_0 = 10^5$, $\alpha = 0.2 \times 10^5$, and $\zeta = 0.2$; **b.** KCL requires that $i_C(0) = 0.1\text{ A} = Cv'(0)$, thus $v'(0) = 10^6$; **c.** $v_p(t) = 0$; **d.** $v(t) = 10.21e^{-2\times 10^4 t}\sin(97.98 \times 10^3 t)$. □

4.6 TRANSIENT ANALYSIS USING THE MATLAB SYMBOLIC TOOLBOX

The MATLAB Symbolic Toolbox greatly facilitates the solution of transients in electrical circuits. It makes the solution of systems of differential equations almost as easy as arithmetic using a calculator. A step-by-step process for solving a circuit in this manner is

1. Write the differential-integral equations for the mesh currents, node voltages, or other circuit variables of interest.
2. If necessary, differentiate the equations to eliminate integrals.
3. Analyze the circuit at $t = 0+$ (i.e., immediately after switches operate) to determine initial conditions for the circuit variables and their derivatives. For a first-order equation, we need the initial value of the circuit variable. For a second-order equation we need the initial values of the circuit variable and its first derivative.
4. Enter the equations and initial values into the `dsolve` command in MATLAB.

We illustrate with a few examples.

Example 4.6 Computer-Aided Solution of a First-Order Circuit

Solve for $v_L(t)$ in the circuit of Figure 4.32(a).

Solution First, we write a KCL equation at the node joining the resistance and inductance.

$$\frac{v_L(t) - 20\cos(100t)}{R} + \frac{1}{L}\int_0^t v_L(t)dt + i_L(0) = 0$$

Taking the derivative of the equation to eliminate the integral, multiplying each term by R, and substituting values, we eventually obtain

$$\frac{dv_L(t)}{dt} + 100v_L(t) = -2000\sin(100t)$$

Next, we need to determine the initial value of v_L. Because the switch is open prior to $t = 0$, the initial current in the inductance is zero prior to $t = 0$. Furthermore, the current cannot change instantaneously in this circuit. Thus, we have $i_L(0+) = 0$. Immediately after the switch closes, the voltage source has a value of 20 V, and the current flowing in the circuit is zero, resulting in zero volts across the resistor. Then KVL yields $v_L(0+) = 20$ V. This is illustrated in Figure 4.32(b).

Now, we can write the MATLAB commands. As usual, we show the commands in **boldface**, comments in regular font, and MATLAB responses in color.

```
>> clear all
>> syms VL t
>> % Enter the equation and initial value in the dsolve command.
>> % DVL represents the derivative of VL with respect to time.
>> VL = dsolve('DVL + 100*VL = -2000*sin(100*t)', 'VL(0) = 20');
>> % Print answer with 4 decimal place accuracy for the constants:
>> vpa(VL,4)
     ans =
     10.0*cos(100.0*t)-10.0*sin(100.0*t)+10.0*exp(-100.0*t)
```

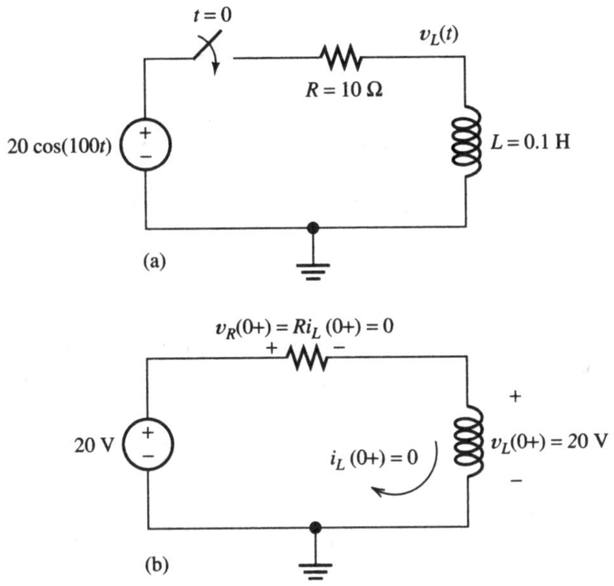

Figure 4.32 (a) Circuit of Example 4.6. (b) Circuit conditions at $t = 0+$.

In standard mathematical notation, the result becomes

$$v_L(t) = 10\cos(100t) - 10\sin(100t) + 10\exp(-100t)$$

An m-file named Example_4_6 containing the commands for this example can be found in the MATLAB folder. (See Appendix E for information about access to this folder.) ■

Example 4.7 Computer-Aided Solution of a Second-Order Circuit

The switch in the circuit of Figure 4.33(a) is closed for a long time prior to $t = 0$. Assume that $i_L(0+) = 0$. Use MATLAB to solve for $i_L(t)$ and plot the result for $0 \le t \le 2$ ms. (*Note:* The code for this example takes several minutes to run on the author's laptop using MATLAB version R2011b. It may not run at all with other versions.)

Solution Because this circuit contains two nodes and three meshes, node-voltage analysis is simpler than mesh analysis. We will solve for $v(t)$ and then take $1/L$ times the integral of the voltage to obtain the current through the inductance.

We start the node-voltage analysis by writing the KCL equation at the top node of the circuit (with the switch open).

$$C\frac{dv(t)}{dt} + \frac{v(t)}{R} + \frac{1}{L}\int_0^t v(t)dt + i_L(0+) = 0.2\exp(-1000t)$$

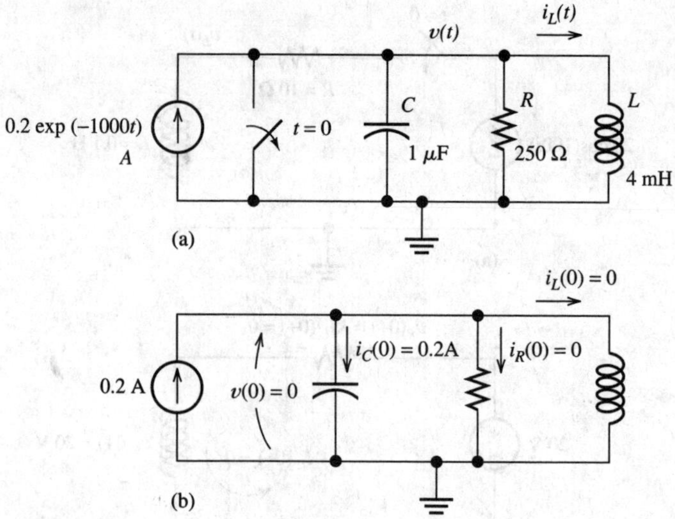

Figure 4.33 (a) Circuit of Example 4.7. (b) Circuit conditions at $t = 0+$.

Taking the derivative of the equation to eliminate the integral and substituting values, we eventually obtain

$$10^{-6}\frac{d^2v(t)}{dt^2} + 4 \times 10^{-3}\frac{dv(t)}{dt} + 250v(t) = -200\exp(-1000t)$$

Because this is a second-order equation, we need the initial value for both $v(t)$ and its first derivative. The circuit conditions at $t = 0+$ are shown in Figure 4.33(b). The problem states that the initial current in the inductance is zero. The initial voltage $v(0+)$ is zero, because, with the switch closed, the capacitor is shorted. When the switch opens, the voltage remains zero, because an infinite current would be required to change the capacitor voltage instantaneously. Furthermore, the current flowing through the resistor is zero because the voltage across it is zero. Thus, the 0.2 A from the source must flow through the capacitor, and we have

$$C\frac{dv(0+)}{dt} = 0.2$$

We have established that $v(0+) = 0$ and $v'(0+) = dv(0+)/dt = 0.2 \times 10^6$ V/s.

After the voltage is found, the current is given by

$$i_L(t) = \frac{1}{L}\int_0^t v(t)dt = 250\int_0^t v(t)dt$$

We use the following MATLAB commands to obtain the solution.

```
>> clear all
>> syms ILV t
>> % Enter the equation and initial values in the dsolve command.
>> % D2V represents the second derivative ofV.
>> V = dsolve('(1e-6)*D2V + (4e-3)*DV + 250*V = -200*exp(-1000*t)',...
        'DV(0)=0.2e6', 'V(0)=0');
>> % Calculate the inductor current by integratingV with respect to t
>> % from 0 to t and multiplying by 1/L:
>> IL = (250)*int(V,t,0,t);
```

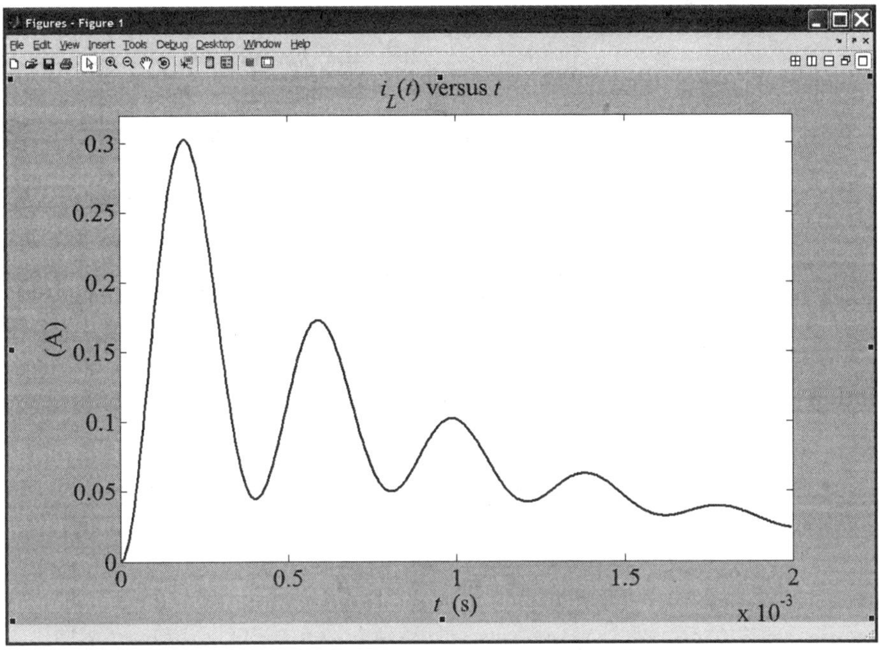

Figure 4.34 Plot of $i_L(t)$ versus t.

```
>> % Display the expression for current to 4 decimal place accuracy:
>> vpa(IL,4)
      ans =
      -(0.0008229*(246.0*cos(15688.0*t) - 246.0*exp(1000.0*t) +
         15.68*sin(15688.0*t)))/exp(2000.0*t)
>> ezplot(IL,[0 2e-3])
```

In standard mathematical notation, the result is

$$i_L(t) = -0.2024\exp(-2000t)\cos(15680t) - \\ 0.01290\exp(-2000t)\sin(15680t) + 0.2024\exp(-1000t)$$

The plot (after some editing to dress it up) is shown in Figure 4.34. An m-file named Example_4_7 containing the commands for this example can be found in the MATLAB folder. (See Appendix E for information about accessing this folder.) ■

Solving Systems of Linear Differential Equations

So far in this chapter, each of our examples has involved a single differential equation. Circuits that require two or more circuit variables (such as node voltages or mesh currents) result in systems of differential equations. While these systems can be rather formidable to solve by traditional methods, the MATLAB Symbolic Toolbox can solve them with relative ease.

Example 4.8 Computer-Aided Solution of a System of Differential Equations

Use MATLAB to solve for the node voltages in the circuit of Figure 4.35. The circuit has been connected for a long time prior to $t = 0$ with the switch open, so the initial values of the node voltages are zero.

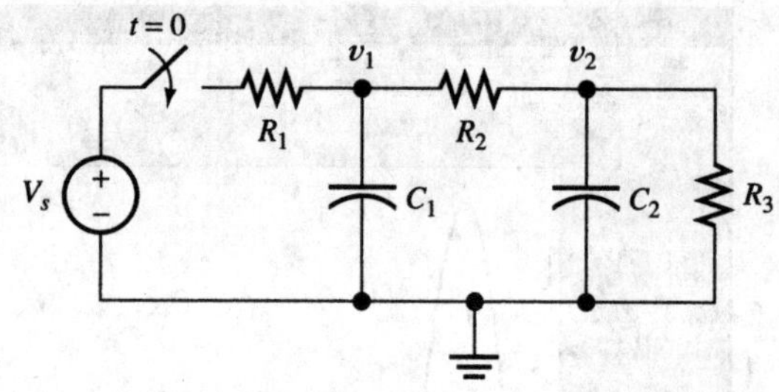

Figure 4.35 Circuit of Example 4.8. $V_s = 10\text{ V}$ $R_1 = R_2 = R_3 = 1\text{ M}\Omega$ $C_1 = C_2 = 1\ \mu\text{F}$

Solution First, we write the KCL equations at nodes 1 and 2.

$$C_1\frac{dv_1(t)}{dt} + \frac{v_1(t) - V_s}{R_1} + \frac{v_1(t) - v_2(t)}{R_2} = 0$$

$$C_2\frac{dv_2(t)}{dt} + \frac{v_2(t) - v_1(t)}{R_2} + \frac{v_2(t)}{R_3} = 0$$

Now substituting values, multiplying each term by 10^6, and rearranging terms, we have

$$\frac{dv_1(t)}{dt} + 2v_1(t) - v_2(t) = 10$$

$$\frac{dv_2(t)}{dt} + 2v_2(t) - v_1(t) = 0$$

The MATLAB commands and results are:

```
>> clear all
>> syms v1 v2 t
>> [v1 v2] = dsolve('Dv1 + 2*v1 - v2 = 10','Dv2 + 2*v2 -v1 = 0',...
           'v1(0) = 0','v2(0)= 0');
>> v1
   v1 =
   ((5*exp(3*t))/3 + exp(2*t)*(5*exp(t) - 5) - 5/3)/exp(3*t)
>> v2
   v2 =
   (exp(2*t)*(5*exp(t) - 5) - (5*exp(3*t))/3 + 5/3)/exp(3*t)
```

Thus, the node voltages are given by

$$v_1(t) = 20/3 - 5\exp(-t) - (5/3)\exp(-3t)$$

$$v_2(t) = 10/3 - 5\exp(-t) + (5/3)\exp(-3t)$$

It is always a good idea to perform a few checks on our answers. First, we can verify that the MATLAB results are both zero at $t = 0$ as required by the initial conditions. Furthermore, at $t = \infty$, the capacitors act as open circuits, and the voltage division principle yields $v_1(\infty) = 20/3\text{ V}$ and $v_2(\infty) = 10/3$. The expressions delivered by MATLAB also yield these values. ■

Exercise 4.12 Use the MATLAB Symbolic Toolbox to solve Example 4.4, obtaining the result given in Equation 4.56 and a plot similar to Figure 4.17 on page 197.

Answer A sequence of commands that produces the solution and the plot is:

```
clear all
syms ix t R C vCinitial w
ix = dsolve('(R*C)*Dix + ix = (w*C)*2*cos(w*t)', 'ix(0)=-vCinitial/R');
ians =subs(ix,[R C vCinitial w],[5000 1e-6 1 200]);
vpa(ians, 4)
ezplot(ians,[0 80e-3])
```

An m-file named Exercise_4_12 containing these commands can be found in the MATLAB folder. (See Appendix E for information about accessing this folder.) □

Exercise 4.13 Use the MATLAB Symbolic Toolbox to solve Example 4.5 obtaining the results given in the example for $v_C(t)$ and a plot similar to Figure 4.26 on page 205.

Answer A list of commands that produces the solution and the plot is:

```
clear all
syms vc t
% Case I, R = 300:
vc = dsolve('(1e-8)*D2vc + (1e-6)*300*Dvc+ vc =10', 'vc(0) = 0','Dvc(0)=0');
vpa(vc,4)
ezplot(vc, [0 1e-3])
hold on % Turn hold on so all plots are on the same axes
% Case II, R = 200:
vc = dsolve('(1e-8)*D2vc + (1e-6)*200*Dvc+ vc =10', 'vc(0) = 0','Dvc(0)=0');
vpa(vc,4)
ezplot(vc, [0 1e-3])
% Case III, R = 100:
vc = dsolve('(1e-8)*D2vc + (1e-6)*100*Dvc+ vc =10', 'vc(0) = 0','Dvc(0)=0');
vpa(vc,4)
ezplot(vc, [0 1e-3])
```

An m-file named Exercise_4_13 containing these commands resides in the MATLAB folder. (See Appendix E for information about accessing this folder.) □

Summary

1. The transient part of the response for a circuit containing sources, resistances, and a single energy-storage element (L or C) is of the form $Ke^{-t/\tau}$. The time constant is given by $\tau = RC$ or by $\tau = L/R$, where R is the Thévenin resistance seen looking back into the circuit from the terminals of the energy-storage element.
2. In dc steady-state conditions, inductors behave as short circuits and capacitors behave as open circuits. We can find the steady-state (forced) response for dc sources by analyzing the dc equivalent circuit.
3. To find the transient currents and voltages, we must solve linear differential equations with constant coefficients. The solutions are the sum of two parts. The particular solution, also called the forced response, depends on the sources, as well as the other circuit elements. The homogeneous solution, also called the natural response, depends on the passive elements (R, L, and C), but not on the sources. In circuits that contain resistances, the natural response eventually decays to zero.
4. The natural response of a second-order circuit containing a series or parallel combination of inductance and capacitance depends on the damping ratio and undamped resonant frequency.

   If the damping ratio is greater than unity, the circuit is overdamped, and the natural response is of the form

$$x_C(t) = K_1e^{s_1t} + K_2e^{s_2t}$$

If the damping ratio equals unity, the circuit is critically damped, and the natural response is of the form

$$x_C(t) = K_1e^{s_1t} + K_2te^{s_1t}$$

If the damping ratio is less than unity, the circuit is underdamped, and the natural response is of the form

$$x_C(t) = K_1e^{-\alpha t}\cos(\omega_n t) + K_2e^{-\alpha t}\sin(\omega_n t)$$

The normalized step response for second-order systems is shown in Figure 4.29 on page 206 for several values of the damping ratio.

5. The MATLAB Symbolic Toolbox is a powerful tool for solving the equations for transient circuits. A step-by-step procedure is given on page 210.

***Note:* You can check the answers to many of the problems in this chapter by using a computer-aided circuit-analysis program such as Multisim from National Instruments or OrCAD Capture from Cadence Inc.**

Problems

Section 4.1: First-Order *RC* Circuits

P4.1. A capacitance C discharges through a resistance R. Define and give an expression for the time constant. What is the effect of the time constant on the discharging of the capacitor?

P4.2. An initially charged capacitance discharges through a resistance. At a time constant of five, what percentage of the initial voltage remains? What percentage of the initial stored energy remains?

P4.3. The initial voltage across the capacitor shown in Figure P4.3 is $v_C(0+) = 0$. Find an expression for the voltage across the capacitor as a function of time, and sketch to scale versus time.

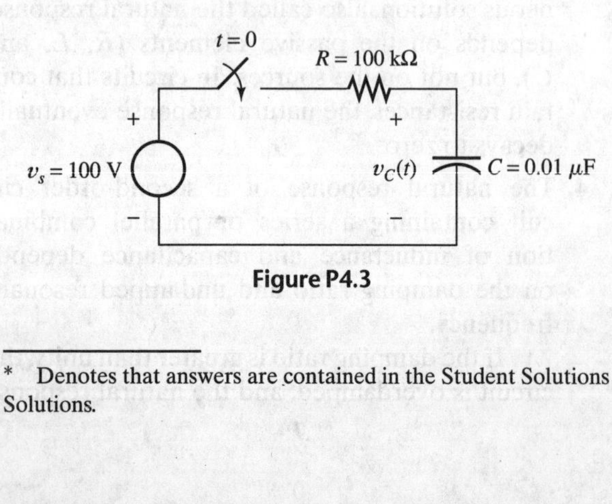

Figure P4.3

***P4.4.** Repeat Problem P4.3 for an initial voltage $v_C(0+) = -50$ V.

***P4.5.** The dielectric materials used in real capacitors are not perfect insulators. A resistance called a leakage resistance in parallel with the capacitance can model this imperfection. A 100-μF capacitor is initially charged to 100 V. We want 90 percent of the initial energy to remain after one minute. What is the limit on the leakage resistance for this capacitor?

P4.6. At $t = 0$, a 0.2-μF capacitance is charged to an unknown voltage V_i. The capacitance is in parallel with a 3-kΩ resistance. At $t = 1$ ms, the voltage across the capacitance is 10 V. Determine the value of V_i.

P4.7. The capacitor shown in Figure P4.7 is charged to a voltage of 50 V prior to $t = 0$. **a.** Find expressions for the voltage across the capacitor $v_C(t)$ and the voltage across the resistor $v_R(t)$. **b.** Find an expression for the power delivered to the resistor. **c.** Integrate the power from $t = 0$ to $t = \infty$ to find the energy delivered. **d.** Show that the energy delivered to the resistor is equal to the energy stored in the capacitor prior to $t = 0$.

* Denotes that answers are contained in the Student Solutions files. See Appendix E for more information about accessing the Student Solutions.

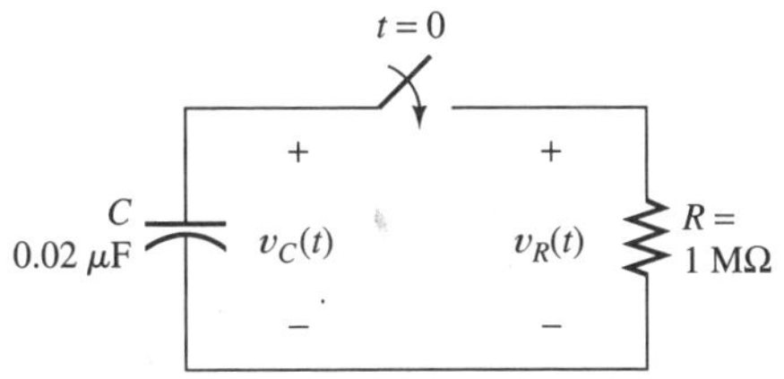

Figure P4.7

P4.8. We use the time constant to characterize transients in electric circuits. In physics, the half-life is often used to characterize the exponential decay of physical quantities such as radioactive substances. The half-life is the time required for the quantity to decay to half of its initial value. The time constant for the voltage on a capacitance discharging through a resistance is $\tau = RC$. Find an expression for the half-life of the voltage in terms of R and C.

P4.9. Find an expression for $v(t)$ for the circuit shown in Figure P4.9 and sketch $v(t)$ to scale versus time.

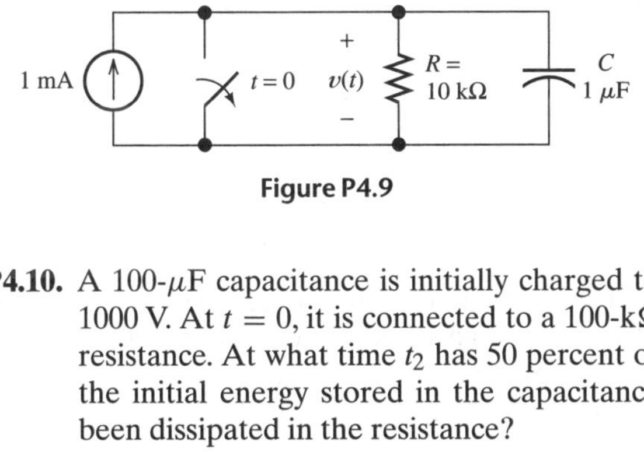

Figure P4.9

***P4.10.** A 100-μF capacitance is initially charged to 1000 V. At $t = 0$, it is connected to a 100-kΩ resistance. At what time t_2 has 50 percent of the initial energy stored in the capacitance been dissipated in the resistance?

***P4.11.** At $t = 0$, a charged 10-μF capacitance is connected to a voltmeter, as shown in Figure P4.11. The meter can be modeled as a resistance. At $t = 0$, the meter reads 50 V. At $t =$ 30 s, the reading is 25 V. Find the resistance of the voltmeter.

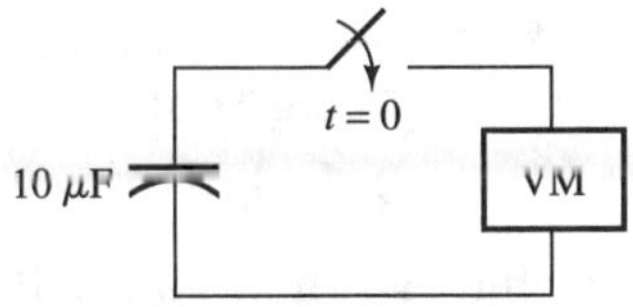

Figure P4.11

P4.12. The purchasing power P of a certain unit of currency declines by 3 percent per year. Determine the time constant associated with the purchasing power of this currency.

P4.13. At $t = 0$, an initially uncharged 10-μF capacitance is connected to a charging circuit consisting of a 1000-V voltage source in series with a 1-MΩ resistance. At $t = 25$ s, the capacitor is disconnected from the charging circuit and connected in parallel with a 2-MΩ resistor. Determine the voltage across the capacitor at $t = 25$ s and at $t = 50$ s. (*Hint*: You may find it convenient to redefine the time variable to be $t' = t - 25$ for the discharge interval so that the discharge starts at $t' = 0$.)

P4.14. A person shuffling across a dry carpet can be approximately modeled as a charged 100-pF capacitance with one end grounded. If the person touches a grounded metallic object such as a water faucet, the capacitance is discharged and the person experiences a brief shock. Typically, the capacitance may be charged to 20,000 V and the resistance (mainly of one's finger) is 100 Ω. Determine the peak current during discharge and the time constant of the current.

P4.15. A capacitance C is charged to an initial voltage V_i. At $t = 0$, a resistance R is connected across the capacitance. Write an expression for the current. Then, integrate the current from $t = 0$ to $t = \infty$, and show that the result is equal to the initial charge stored on the capacitance.

***P4.16.** At time t_1, a capacitance C is charged to a voltage of V_1. Then, the capacitance discharges through a resistance R. Write an expression for the voltage across the capacitance as a function of time for $t > t_1$ in terms of R, C, V_1, and t_1.

P4.17. In the circuit of Figure P4.17, the switch instantaneously moves back and forth between contacts A and B, spending 1 s in each position. Thus, the capacitor repeatedly charges for 1 s and then discharges for 1 s. Assume that $v_C(0) = 0$ and that the switch moves to position A at $t = 0$. Determine $v_C(1), v_C(2), v_C(3)$, and $v_C(4)$.

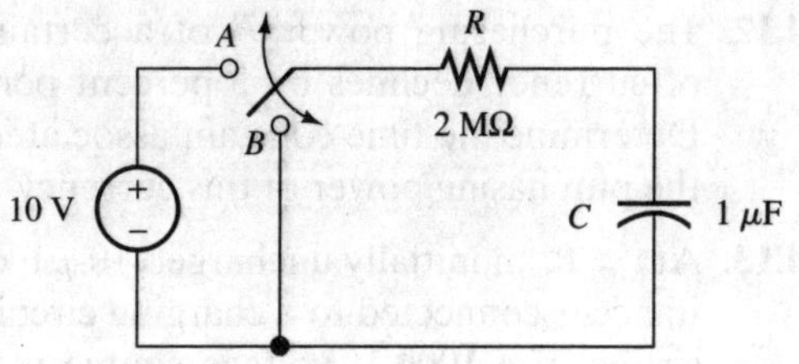

Figure P4.17

P4.18. Consider the circuit shown in Figure P4.18. Prior to $t = 0$, $v_1 = 100$ V and $v_2 = 0$. **a.** Immediately after the switch is closed, what is the value of the current [i.e., what is the value of $i(0+)$]? **b.** Write the KVL equation for the circuit in terms of the current and initial voltages. Take the derivative to obtain a differential equation. **c.** What is the value of the time constant in this circuit? **d.** Find an expression for the current as a function of time. **e.** Find the value that v_2 approaches as t becomes very large.

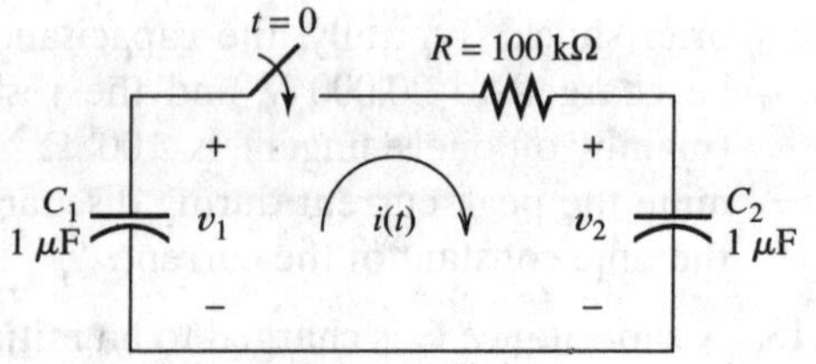

Figure P4.18

Section 4.2: DC Steady State

P4.19. List the steps for a dc steady-state analysis of *RLC* circuits.

P4.20. Explain dc steady-state analysis to replace capacitances and inductances with other suitable circuits.

***P4.21.** Solve for the steady-state values of i_1, i_2, and i_3 for the circuit shown in Figure P4.21.

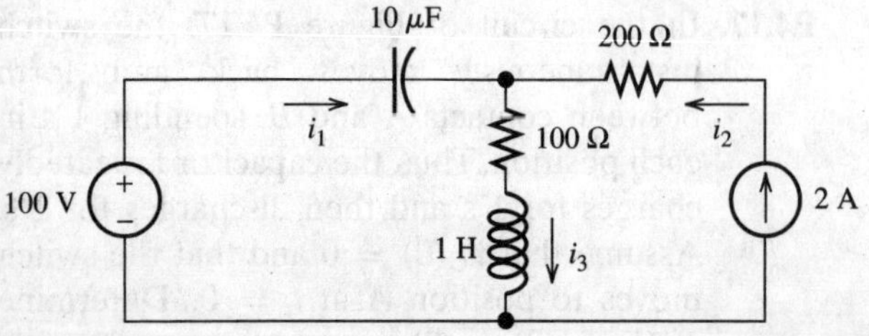

Figure P4.21

***P4.22.** Consider the circuit shown in Figure P4.22. What is the steady-state value of v_C after the switch opens? Determine how long it takes after the switch opens before v_C is within 1 percent of its steady-state value.

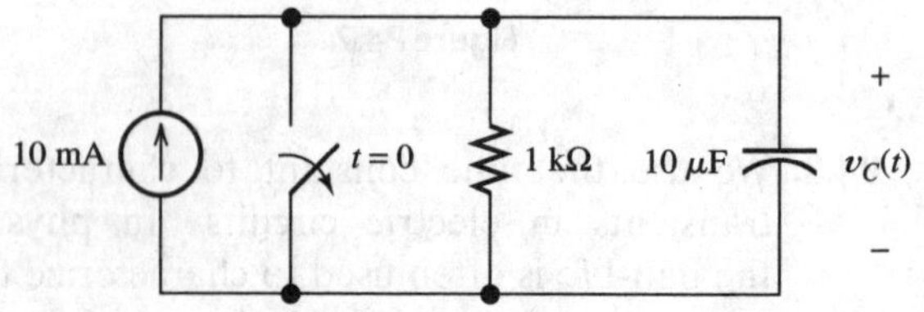

Figure P4.22

P4.23. Solve for the steady-state values of i_1, i_2, i_3, i_4, and v_C for the circuit shown in Figure P4.23, after the switch has been closed for a long time.

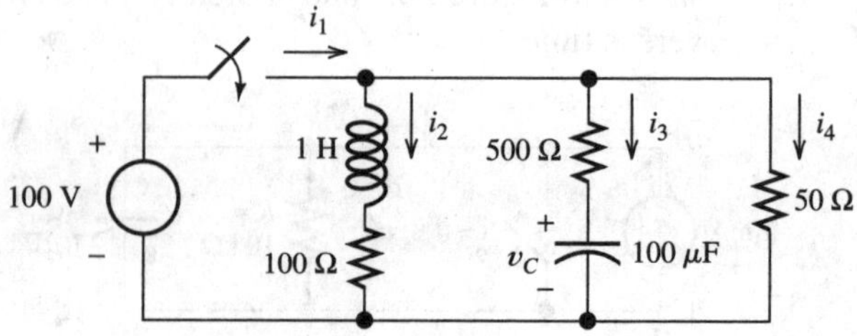

Figure P4.23

P4.24. The circuit shown in Figure P4.24 is operating in steady state. Determine the values of i_L, v_x, and v_C.

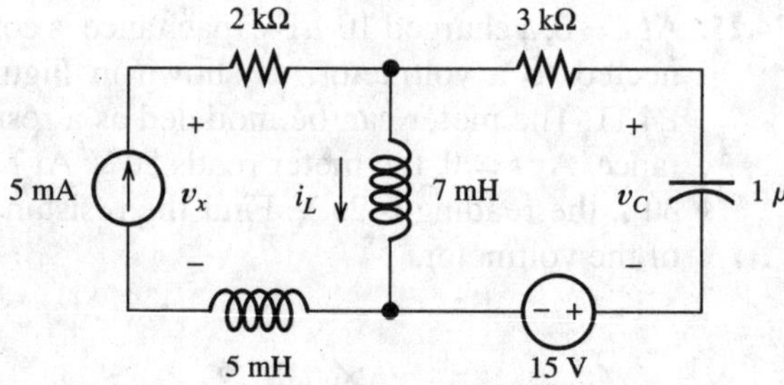

Figure P4.24

P4.25. The circuit shown in Figure P4.25 has been set up for a long time prior to $t = 0$ with the switch closed. Find the value of v_C prior to

$t = 0$. Find the steady-state value of v_C after the switch has been open for a long time.

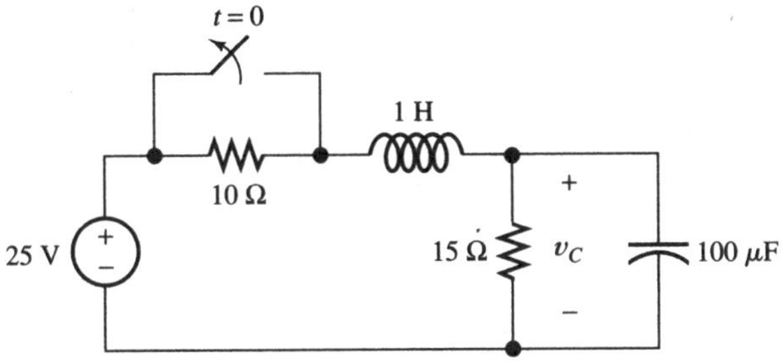

Figure P4.25

P4.26. Consider the circuit of Figure P4.26 in which the switch has been closed for a long time prior to $t = 0$. Determine the values of $v_C(t)$ just before $t = 0$ and a long time after $t = 0$. Also, determine the time constant after the switch opens and expressions for $v_C(t)$. Sketch $v_C(t)$ to scale versus time for $-0.2 \le t \le 1.0$ s.

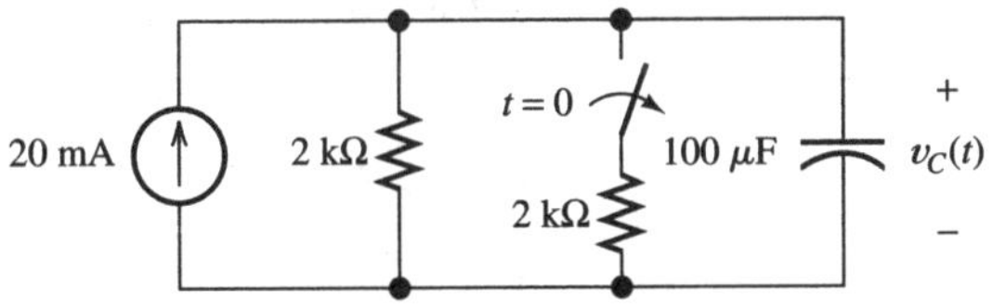

Figure P4.26

P4.27. Consider the circuit of Figure P4.27 in which the switch has been open for a long time prior to $t = 0$. Determine the values of $v_C(t)$ before $t = 0$ and a long time after $t = 0$. Also, determine the time constant after the switch closes and expressions for $v_C(t)$. Sketch $v_C(t)$ to scale versus time for $-2 \le t \le 5$ s.

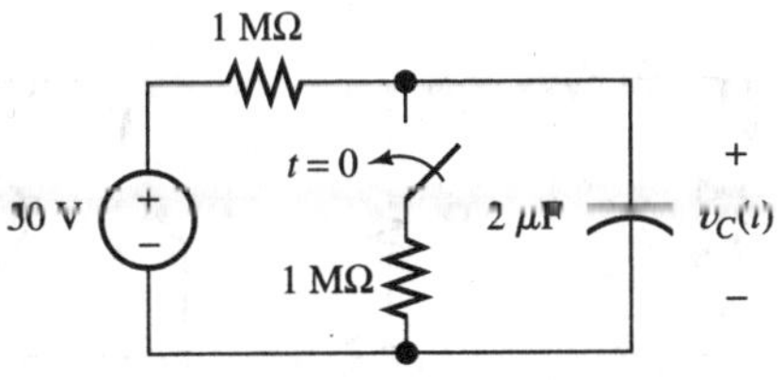

Figure P4.27

P4.28. The circuit of Figure P4.28 has been connected for a very long time. Determine the values of v_C and i_R.

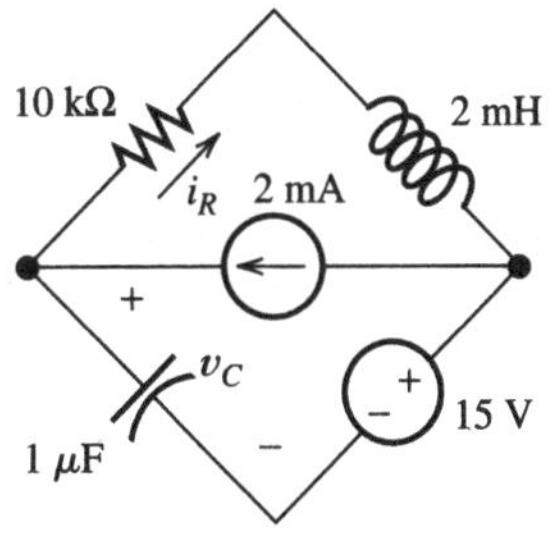

Figure P4.28

***P4.29.** In the circuit of Figure P4.29, the switch is in position A for a long time prior to $t = 0$. Find expressions for $v_R(t)$ and sketch it to scale for $-2 \le t \le 10$ s.

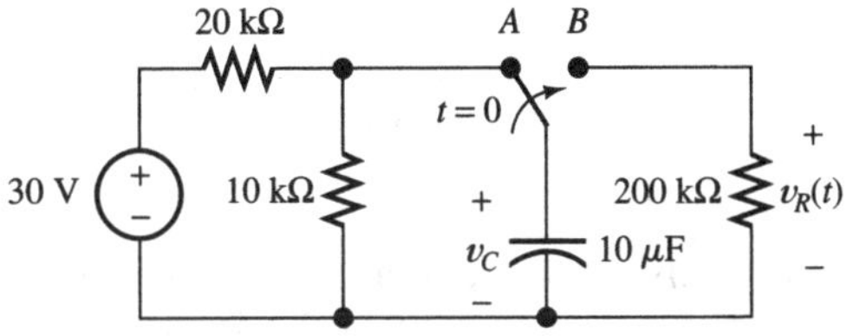

Figure P4.29

P4.30. For the circuit shown in Figure P4.30, the switch is closed for a long time prior to $t = 0$. Find expressions for $v_C(t)$ and sketch it to scale for $-6 \le t \le 30$ ms.

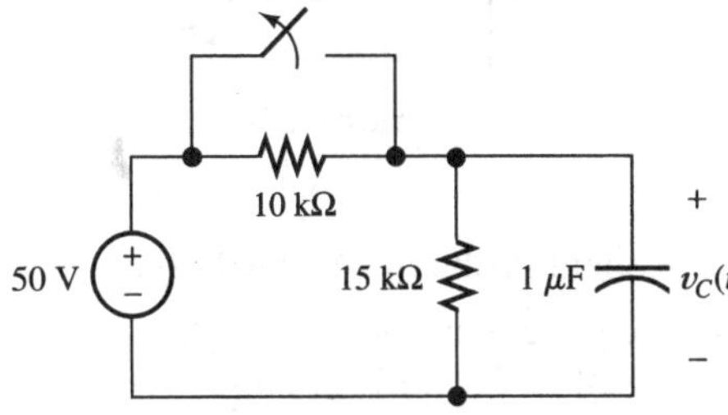

Figure P4.30

Section 4.3: *RL* Circuits

P4.31. Give the expression for the time constant of a circuit consisting of an inductance with an initial current in series with a resistance R. To attain a long time constant, do we need large or small values for R? For L?

P4.32. A circuit consists of switches that open or close at $t = 0$, resistances, dc sources, and a single-energy storage element—either an inductance or a capacitance. We wish to solve for a current or a voltage $x(t)$ as a function of time for $t \geq 0$. Write the general form for the solution. How is each unknown in the solution determined?

***P4.33.** The circuit shown in Figure P4.33 is operating in steady state with the switch closed prior to $t = 0$. Find $i(t)$ for $t < 0$ and for $t \geq 0$.

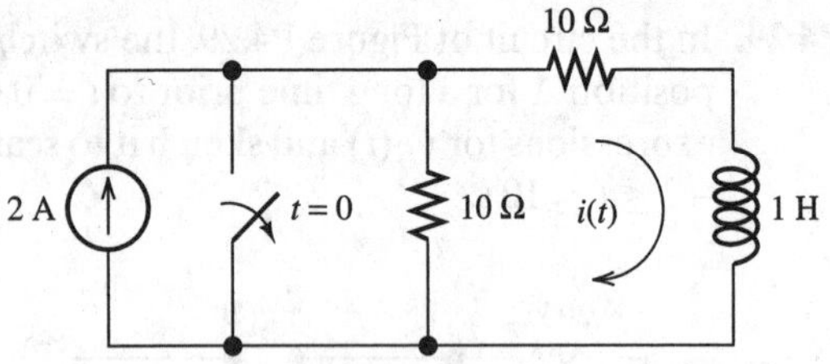

Figure P4.33

P4.34. Consider the circuit shown in Figure P4.34. The initial current in the inductor is $i_L(0-) = 0$. Find expressions for $i_L(t)$ and $v(t)$ for $t \geq 0$ and sketch to scale versus time.

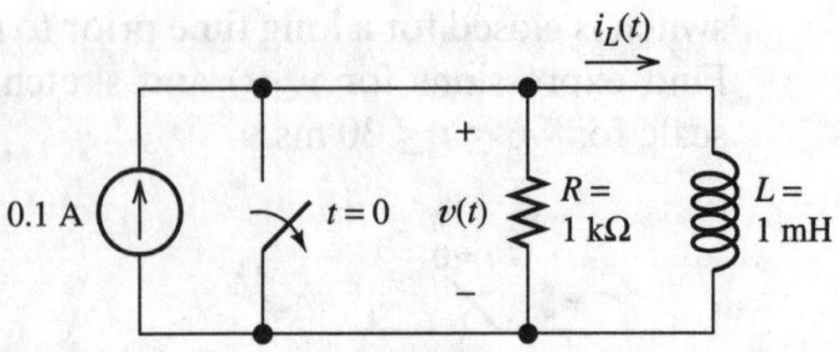

Figure P4.34

***P4.35.** Repeat Problem P4.34 with $i_L(0-) = -0.2$ A.

P4.36. For the circuit shown in Figure P4.36, the values of R, L, and V are unknown. The expression for the current $i_L(t)$ and voltage $v_L(t)$ is given by $i_L(t) = 0.5 - 0.5\,e^{-200t}$ *and* $v_L(t) = 100e^{-200t}$. Find values of R, L and V.

P4.37. The circuit shown in Figure P4.37 is operating in steady state with the switch open prior to $t = 0$. Find expressions for $i(t)$ for $t < 0$ and for $t \geq 0$. Sketch $i(t)$ to scale versus time.

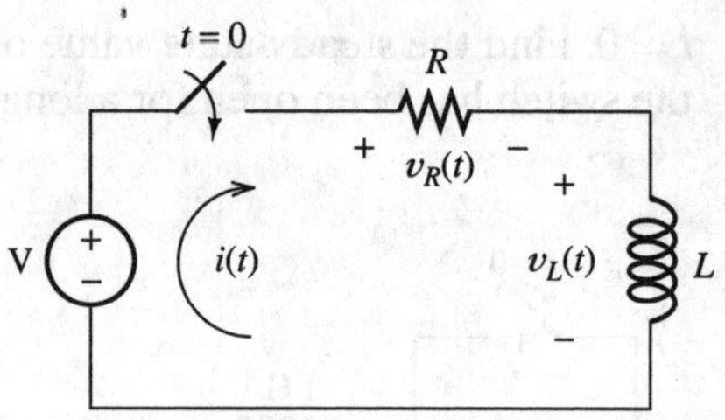

Figure P4.36

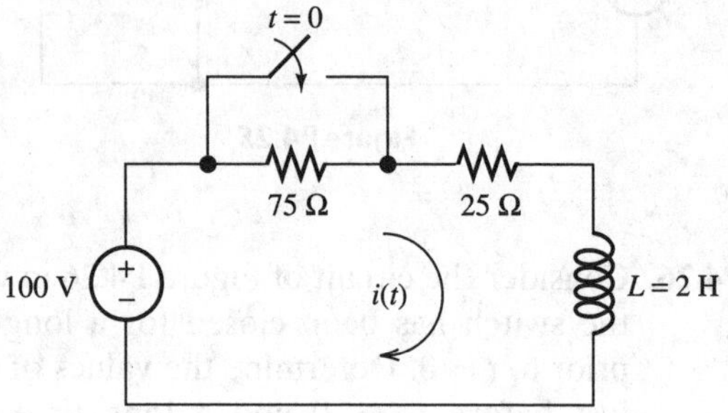

Figure P4.37

P4.38. Refer to the circuit of Figure P4.38. The current through the inductor is zero before $t = 0$. Determine expressions for and sketch $i(t)$ to scale versus time for $-0.2 \leq t \leq 1.0$ s.

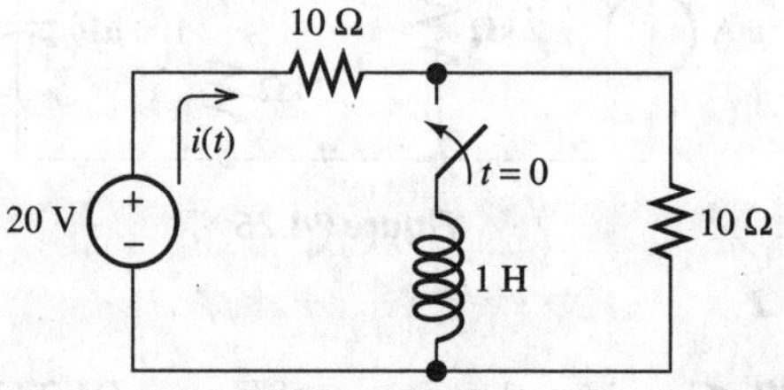

Figure P4.38

P4.39. The switch shown in Figure P4.39 has been closed for a long time prior to $t = 0$, then it opens at $t = 0$ and closes again at $t = 1$ s. Find $i_L(t)$ for all t.

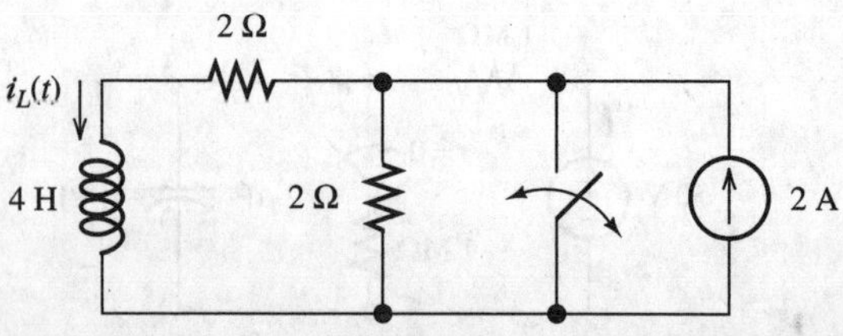

Figure P4.39

P4.40. We know that the circuit shown in Figure P4.40 has an initial current $i(0) = I_i$. **a.** Write an expression for $i(t)$ for $t \geq 0$. **b.** Find an expression for the power delivered to the resistance as a function of time. **c.** Integrate the power delivered to the resistance from $t = 0$ to $t = \infty$, and show that the result is equal to the initial energy stored in the inductance.

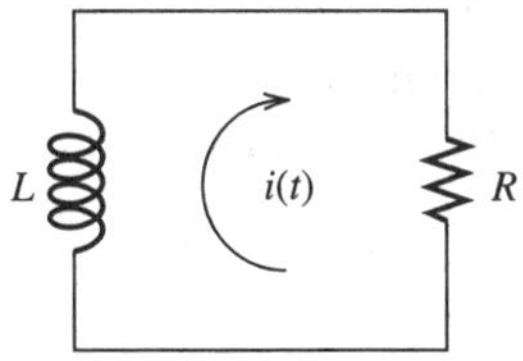

Figure P4.40

P4.41. Determine expressions for and sketch $v_R(t)$ to scale versus time for the circuit of Figure P4.41. The circuit is operating in steady state with the switch closed prior to $t = 0$. Consider the time interval $-2 \leq t \leq 10$ ms.

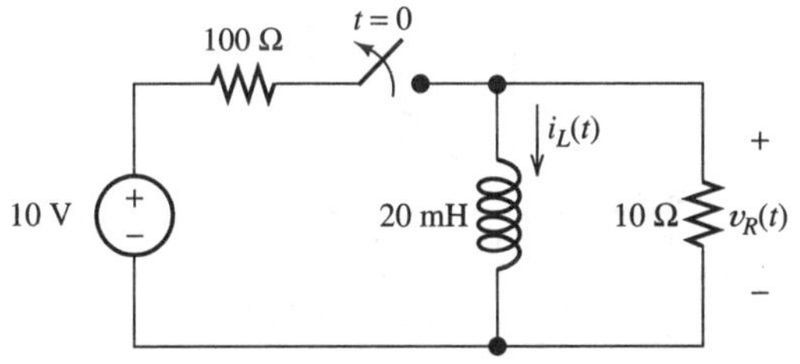

Figure P4.41

P4.42. Consider the circuit shown in Figure P4.42. A voltmeter VM is connected across the inductance. The switch has been closed for a long time. When the switch is opened, an arc appears across the switch contacts. Explain why. Assuming an ideal switch and inductor, what voltage appears across the inductor when the switch is opened? What could happen to the voltmeter when the switch opens?

***P4.43.** Real inductors have series resistance associated with the wire used to wind the coil. Suppose that we want to store energy in a 10-H inductor. Determine the limit on the series

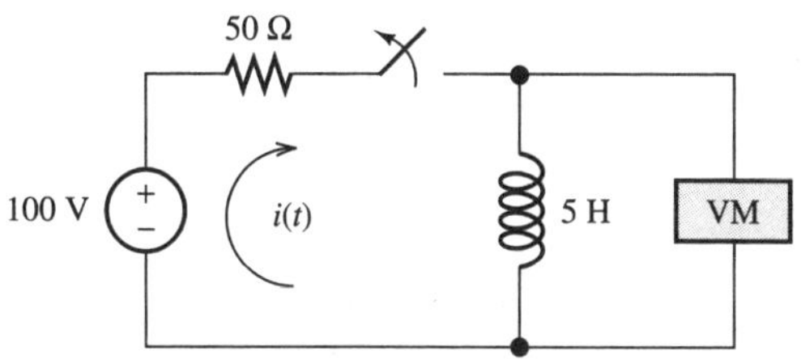

Figure P4.42

resistance so the energy remaining after one hour is at least 75 percent of the initial energy.

Section 4.4: *RC* and *RL* Circuits with General Sources

P4.44. What are the steps in solving a circuit having a resistance, a source, and an inductance (or capacitance)?

***P4.45.** Write the differential equation for $i(t)$ and find the complete solution for the circuit of Figure P4.45. [*Hint:* Try a particular solution of the form $i_p(t) = Ae^{-t}$.]

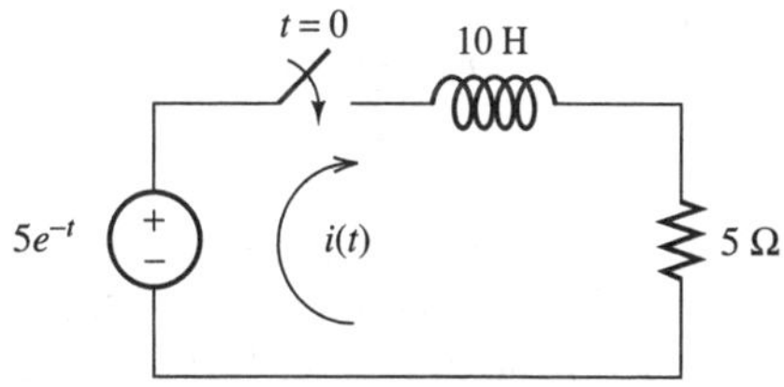

Figure P4.45

P4.46. Consider the circuit shown in Figure P4.46. The voltage source is known as a **ramp function,** which is defined by

$$v(t) = \begin{cases} 0 & \text{for } t < 0 \\ t & \text{for } t \geq 0 \end{cases}$$

Assume that $v_C(0) = 0$. Derive an expression for $v_C(t)$ for $t \geq 0$. Sketch $v_C(t)$ to scale versus time. [*Hint:* Write the differential equation for $v_C(t)$ and assume a particular solution of the form $v_{Cp}(t) = A + Bt$.]

***P4.47.** Solve for $v_C(t)$ for $t > 0$ in the circuit of Figure P4.47. [*Hint*: Try a particular solution of the form $v_{Cp}(t) = Ae^{-3t}$.]

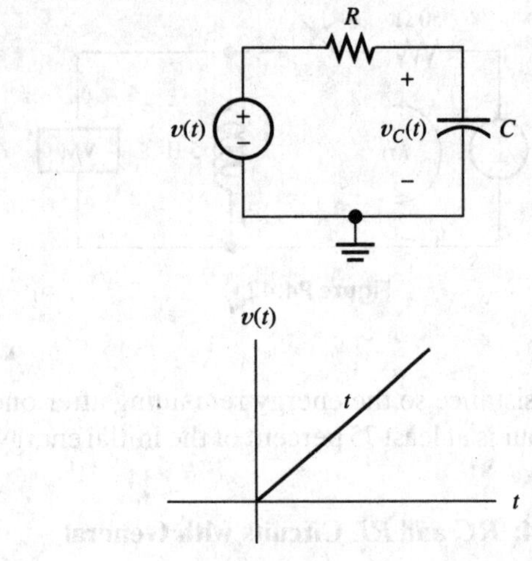

Figure P4.46

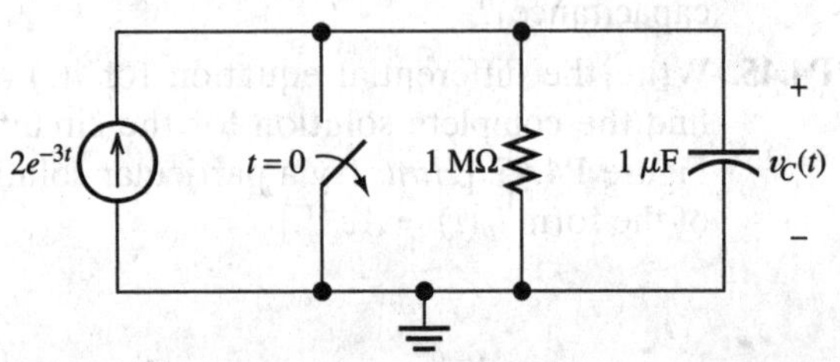

Figure P4.47

***P4.48.** Solve for $v(t)$ for $t > 0$ in the circuit of Figure P4.48, given that the inductor current is zero prior to $t = 0$. [*Hint*: Try a particular solution of the form $v_p = A\cos(10t) + B\sin(10t)$.]

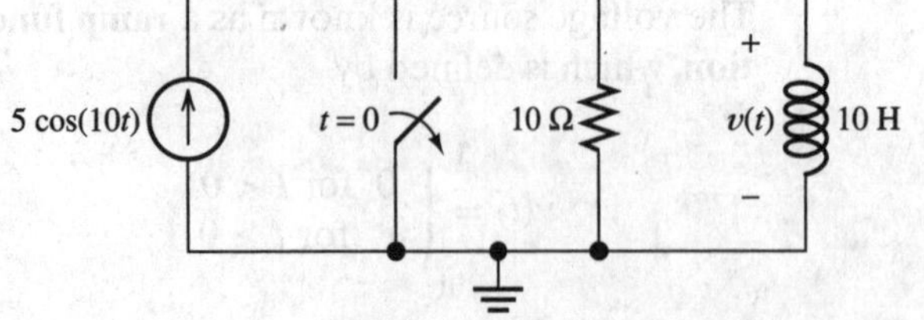

Figure P4.48

P4.49. Consider the circuit shown in Figure P4.49. The initial current in the inductor is $i(0+) = 0$. Write the differential equation for $i(t)$ and solve. [*Hint:* Try a particular solution of the form $i_p(t) = A\cos(300t) + B\sin(300t)$.]

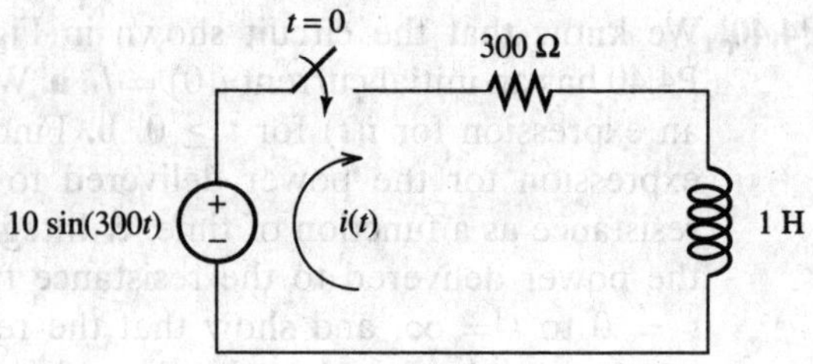

Figure P4.49

P4.50. The voltage source shown in Figure P4.50 is called a **ramp function**. Assume that $i(0) = 0$. Write the differential equation for $i(t)$, and find the complete solution. [*Hint:* Try a particular solution of the form $i_p(t) = A + Bt$.]

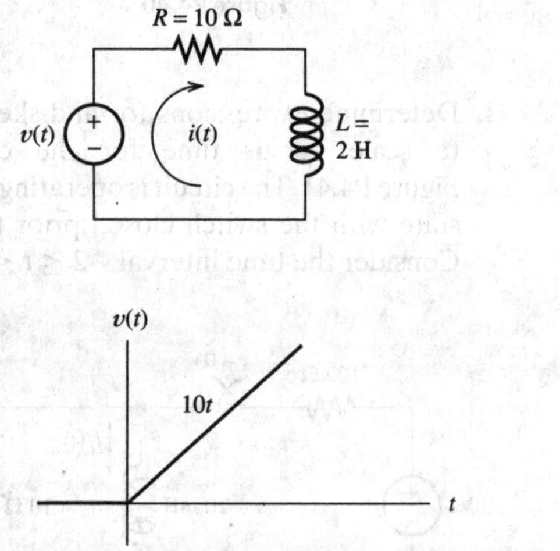

Figure P4.50

P4.51. Solve for $i_L(t)$ for $t > 0$ in the circuit of Figure P4.51. You will need to make an educated guess as to the form of the particular solution. (*Hint:* The particular solution includes terms with the same functional forms as the terms found in the forcing function and its derivatives.)

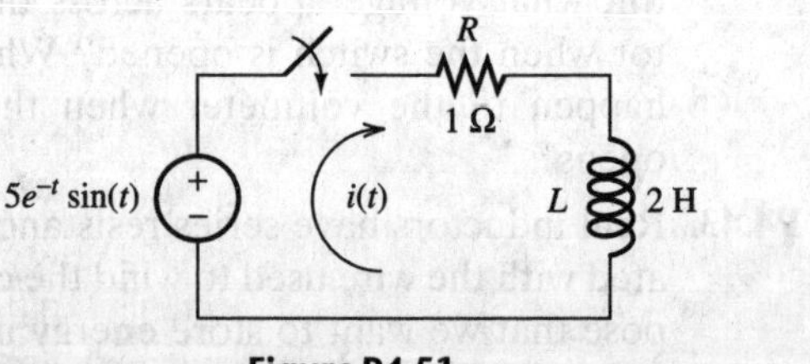

Figure P4.51

P4.52. Determine what form you would try for the particular solution for the differential equation

$$2\frac{dv(t)}{dt} + v(t) = 5t\sin(t)$$

Find the particular solution. (*Hint:* The particular solution includes terms with the same functional forms as the terms found in the forcing function and its derivatives.)

P4.53. Determine the form of the particular solution for the differential equation

$$\frac{dv(t)}{dt} + 3v(t) = t^2\exp(-t)$$

Then, find the particular solution. (*Hint:* The particular solution includes terms with the same functional forms as the terms found in the forcing function and its derivatives.)

P4.54. Consider the circuit shown in Figure P4.54. **a.** Write the differential equation for $i(t)$. **b.** Find the time constant and the form of the complementary solution. **c.** Usually, for an exponential-forcing function like this, we would try a particular solution of the form $i_p(t) = K\exp(-2t)$. Why doesn't that work in this case? **d.** Find the particular solution. [*Hint*: Try a particular solution of the form $i_p(t) = Kt\exp(-2t)$.] **e.** Find the complete solution for $i(t)$.

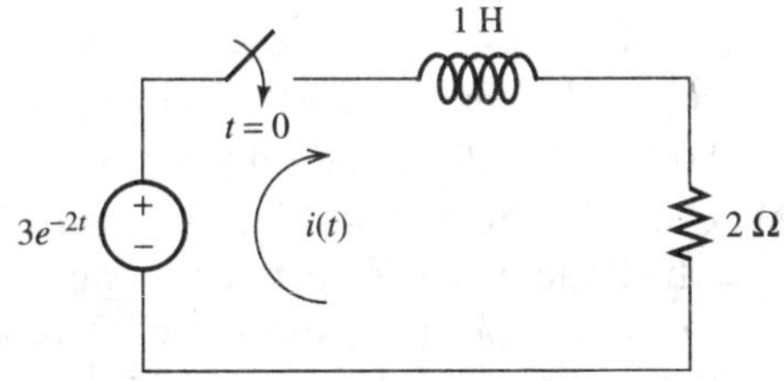

Figure P4.54

P4.55. Consider the circuit shown in Figure P4.55. **a.** Write the differential equation for $v(t)$. **b.** Find the time constant and the form of the complementary solution. **c.** Usually, for an exponential-forcing function like this, we would try a particular solution of the form $v_p(t) = K\exp(-10t)$. Why doesn't that work in this case? **d.** Find the particular solution. [*Hint*: Try a particular solution of the form $v_p(t) = Kt\exp(-10t)$.] **e.** Find the complete solution for $v(t)$.

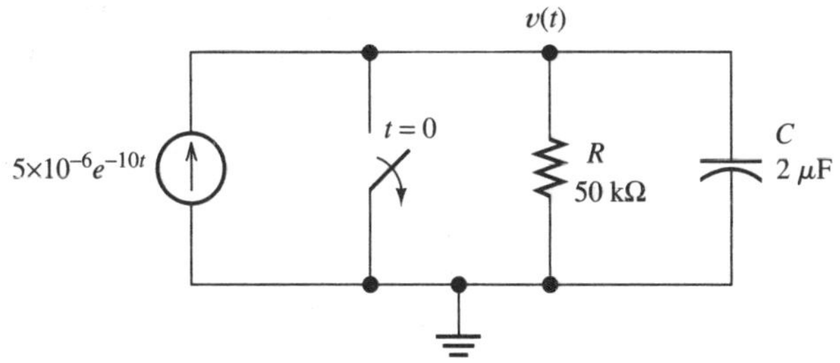

Figure P4.55

Section 4.5: Second-Order Circuits

P4.56. How can an underdamped second-order system be identified? What form does its complementary solution take? Repeat for a critically damped system and for an overdamped system.

P4.57. In the mechanical analog of a series RLC circuit, what quantity is analogous to voltage? To current? To charge?

P4.58. How can first- or second-order circuits be identified?

P4.59. Explain how a DC unity voltage is a unit step function.

P4.60. Sketch a step response for a second-order system that displays considerable overshoot and ringing. In what types of circuits do we find pronounced overshoot and ringing?

***P4.61.** A dc source is connected to a series *RLC* circuit by a switch that closes at $t = 0$, as shown in Figure P4.61. The initial conditions are $i(0+) = 0$ and $v_C(0+) = 0$. Write the differential equation for $v_C(t)$. Solve for $v_C(t)$ given that $R = 80\ \Omega$.

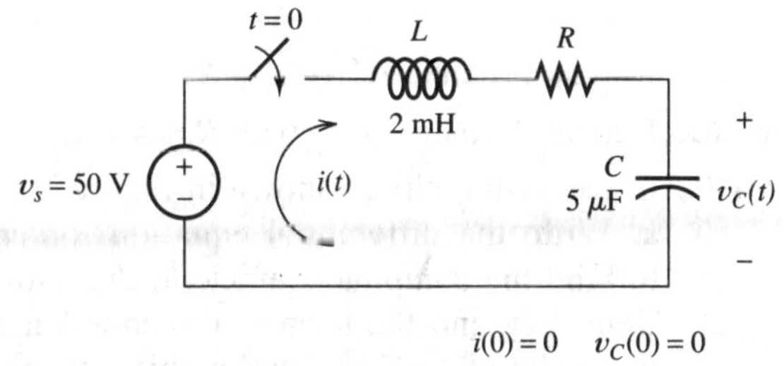

Figure P4.61

***P4.62.** Repeat Problem P4.61 for $R = 40\ \Omega$.

***P4.63.** Repeat Problem P4.61 for $R = 20\ \Omega$.

P4.64. Consider the circuit shown in Figure P4.64, with $R = 25\ \Omega$. **a.** Compute the undamped resonant frequency, the damping coefficient, and the damping ratio. **b.** The initial conditions are $v(0+) = 0$ and $i_L(0+) = 0$. Show that this requires that $v'(0+) = 10^9$ V/s. **c.** Find the particular solution for $v(t)$. **d.** Find the general solution for $v(t)$, including the numerical values of all parameters.

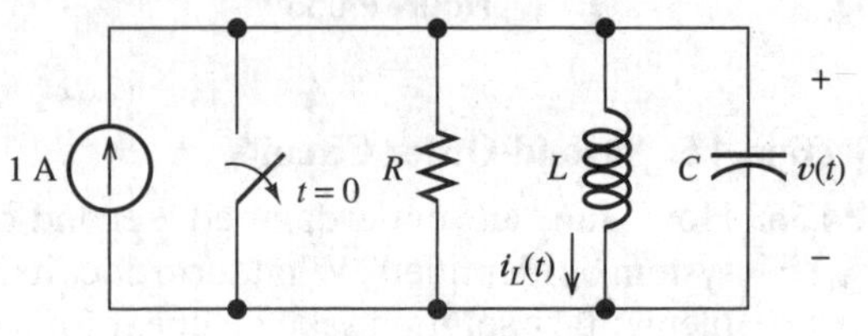

$L = 10\ \mu$H $C = 1000$ pF

Figure P4.64

P4.65. Repeat Problem P4.64 for $R = 50\ \Omega$.

P4.66. Repeat Problem P4.64 for $R = 500\ \Omega$.

P4.67. Solve for $i(t)$ for $t > 0$ in the circuit of Figure P4.67, with $R = 50\ \Omega$, given that the inductor current and capacitor voltage are both zero prior to $t = 0$. [*Hint:* Try a particular solution of the form $i_p(t) = A\cos(100t) + B\sin(100t)$.]

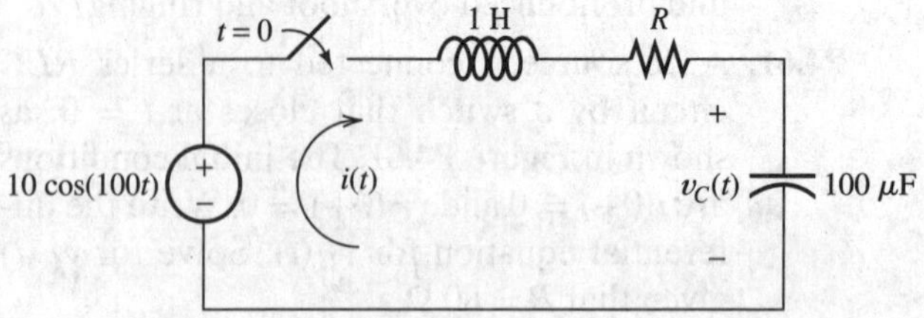

Figure P4.67

P4.68. Repeat Problem P4.67 with $R = 200\ \Omega$.

P4.69. Repeat Problem P4.67 with $R = 400\ \Omega$.

P4.70. Consider the circuit shown in Figure P4.70. **a.** Write the differential equation for $v(t)$. **b.** Find the damping coefficient, the natural frequency, and the form of the complementary solution. **c.** Usually, for this sinusoidal forcing function, we try a particular solution of the form $v_p(t) = A\cos(10^4 t) + B\sin(10^4 t)$. Why doesn't that work in this case? **d.** Find the particular solution. [*Hint:* Try a particular solution of the form $v_p(t) = At\cos(10^4 t) + Bt\sin(10^4 t)$.] **e.** Find the complete solution for $v(t)$.

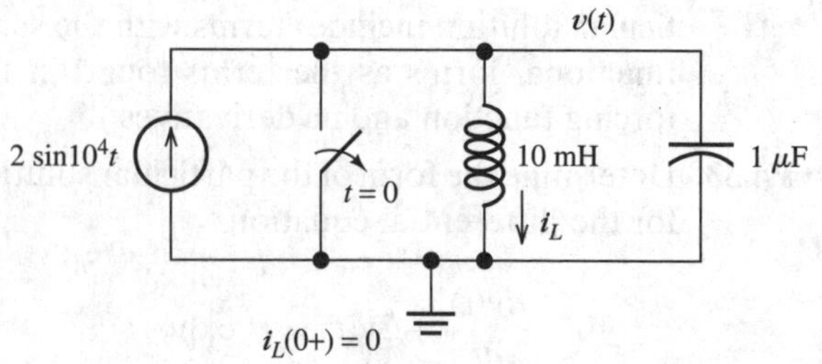

Figure P4.70

Section 4.6: Transient Analysis using the MATLAB Symbolic Toolbox

P4.71. Use MATLAB to derive an expression for $v(t)$ in the circuit of Figure P4.9 and plot $v(t)$ versus time for $0 < t < 50$ ms.

P4.72. Consider the circuit shown in Figure P4.46. The voltage source is known as a **ramp function,** which is defined by

$$v(t) = \begin{cases} 0 & \text{for } t < 0 \\ t & \text{for } t \geq 0 \end{cases}$$

Use MATLAB to derive an expression for $v_C(t)$ in terms of R, C, and t. Next, substitute $R = 1$ MΩ and $C = 1\ \mu$F. Then, plot $v_C(t)$ and $v(t)$ on the same axes for $0 < t < 5$ s.

P4.73. Consider the circuit shown in Figure P4.49 in which the switch is open for a long time prior to $t = 0$. (Because lowercase "i" represents the square root of −1 in MATLAB, we need to avoid using it as the name of a variable, so denote the current as $i_s(t)$ instead of $i(t)$ for this problem and use "Is" for the current in MATLAB.) The initial current is $i_s(0+) = 0$. Write the differential equation for $i_s(t)$ and use MATLAB to solve. Then, plot $i_s(t)$ for t ranging from 0 to 80 ms.

P4.74. Consider the circuit shown in Figure P4.74 in which the switch has been open for a long time prior to $t = 0$ and we are given $R = 25\ \Omega$.

a. Write the differential equation for $v(t)$.
b. Assume that the capacitor is initially charged by a 50-V dc source not shown in the figure, so we have $v(0+) = 50$ V. Determine the values of $i_L(0+)$ and $v'(0+)$.
c. Use MATLAB to find the general solution for $v(t)$.

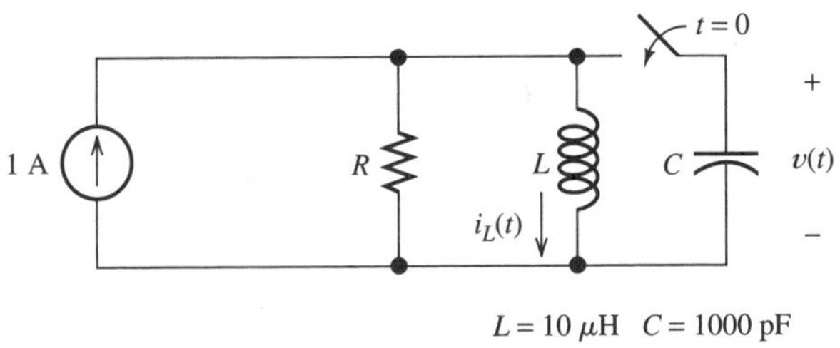

Figure P4.74

P4.75. Consider the circuit shown in Figure P4.70.
a. Write the differential equation for $v(t)$.
b. Determine the values for $v(0+)$ and $v'(0+)$.

c. Use MATLAB to find the complete solution for $v(t)$.

P4.76. Use MATLAB to solve for the mesh currents in the circuit of Figure P4.76. The circuit has been connected for a long time prior to $t = 0$ with the switch open, so the initial values of the inductor currents are zero.

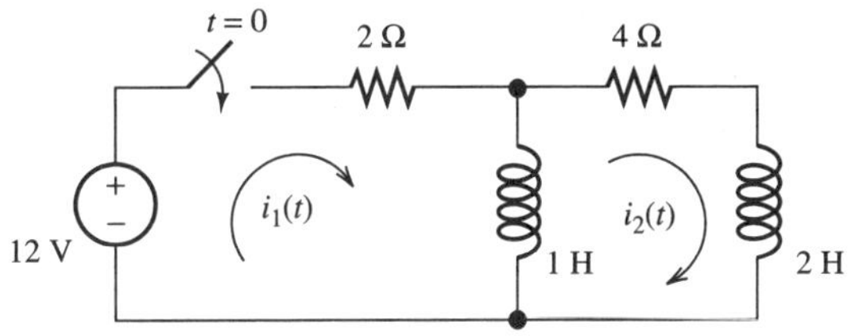

Figure P4.76

Frequently Asked Questions

1. What are the factors on which discharging of a capacitor depends?
2. If a capacitor in charged at V volt, what will be the voltage across the capacitor at any instant if it is discharged through a resistance R?
3. What do you mean by time constant for a capacitor?
4. If a capacitor is to be charged by a source of voltage V, what will be the value of charging current at any instant, considering the circuit resistance to be R ohm.
5. What do you understand by transient response?
6. Transient conditions exist only for a small duration, then what is its need to study?
7. What do you understand by independent and dependent initial valucs after switching?
8. What do you mean by zero and non-zero initial values?
9. How an inductor, connected in a circuit, behaves at initial and final conditions?
10. How a capacitor, connected in a circuit behaves at initial and final conditions?
11. What are transients? How are these produced?
12. Explain the behaviour of resistor and inductor when (i) step voltage, (ii) step current is applied to them.
13. Derive an expression for $V_R(t)$, $i(t)$ and $V_L(t)$ when an RL series circuit is switched-on to a step voltage V.
14. Discuss step voltage response of RL series circuit.
15. Derive an expression for $i(t)$, $V_R(t)$ and $V_C(t)$ when an RC series circuit is switched-on to a step voltage V.
16. Derive the differential equation

$$\frac{d^2v(t)}{dt^2} + 2\propto\frac{dv(t)}{dt} + \omega_0^2 v(t) = f(t)$$

for circuit with L and C parallel to each other.

Practice Test

Here is a practice test you can use to check your comprehension of the most important concepts in this chapter. Answers can be found in Appendix D and complete solutions are included in the Student Solutions files. See Appendix E for more information about the Student Solutions.

T4.1. Consider the circuit shown in Figure T4.1. The circuit has been operating for a long time with the switch closed prior to $t = 0$. **a.** Determine the values of i_L, i_1, i_2, i_3, and v_C just before the switch opens. **b.** Determine the values of i_L, i_1, i_2, i_3, and v_C immediately after the switch opens. **c.** Find $i_L(t)$ for $t > 0$. **d.** Find $v_C(t)$ for $t > 0$.

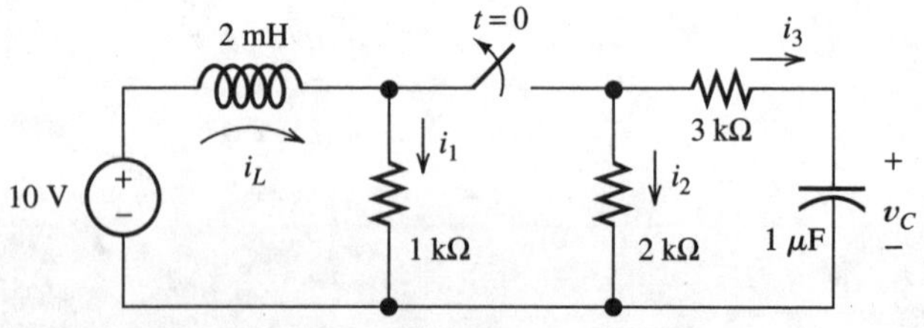

Figure T4.1

T4.2. Consider the circuit shown in Figure T4.2.

a. Write the differential equation for $i(t)$.

b. Find the time constant and the form of the complementary solution.

c. Find the particular solution.

d. Find the complete solution for $i(t)$.

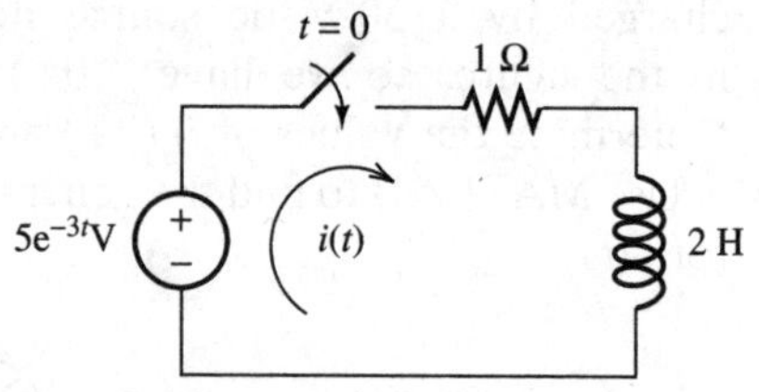

Figure T4.2

T4.3. Consider the circuit shown in Figure T4.3 in which the initial inductor current and capacitor voltage are both zero.

a. Write the differential equation for $v_C(t)$.

b. Find the particular solution.

c. Is this circuit overdamped, critically damped, or underdamped? Find the form of the complementary solution.

d. Find the complete solution for $v_C(t)$.

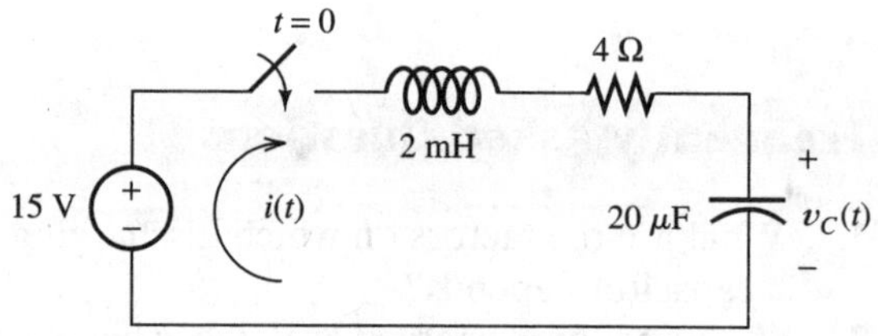

Figure T4.3

T4.4. Write the MATLAB commands to obtain the solution for the differential equation of question T4.3 with four decimal place accuracy for the constants.

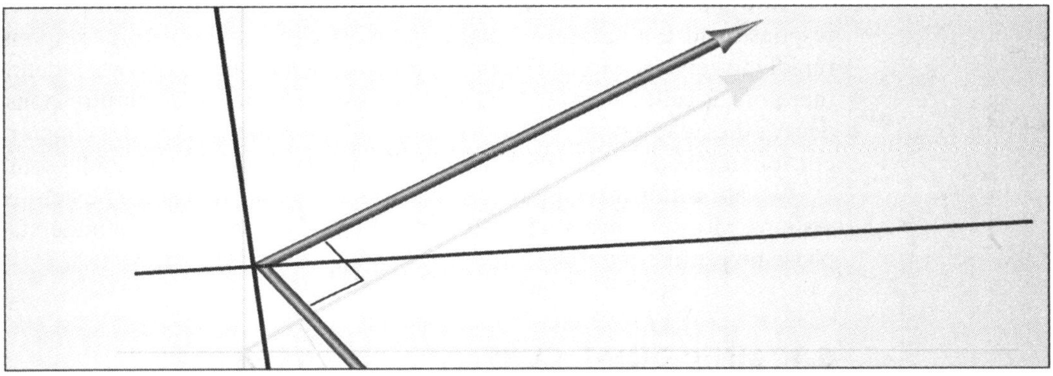

Chapter 5

Steady-State Sinusoidal Analysis

Study of this chapter will enable you to:

- Identify the frequency, angular frequency, peak value, rms value, and phase of a sinusoidal signal.
- Determine the root-mean-square (rms) value of any periodic current or voltage.
- Solve steady-state ac circuits, using phasors and complex impedances.
- Compute power for steady-state ac circuits.
- Find Thévenin and Norton equivalent circuits.
- Determine load impedances for maximum power transfer.
- Discuss the advantages of three-phase power distribution.
- Solve balanced three-phase circuits.
- Use MATLAB to facilitate ac circuit calculations.

Introduction to this chapter:

Circuits with sinusoidal sources have many important applications. For example, electric power is distributed to residences and businesses by sinusoidal currents and voltages. Furthermore, sinusoidal signals have many uses in radio communication. Finally, a branch of mathematics known as Fourier analysis shows that all signals of practical interest are composed of sinusoidal components. Thus, the study of circuits with sinusoidal sources is a central theme in electrical engineering.

In Chapter 4, we saw that the response of a network has two parts: the forced response and the natural response. In most circuits, the natural response decays rapidly to zero. The forced response for sinusoidal sources persists indefinitely and, therefore, is called the steady-state response. Because the natural response quickly decays, the steady-state response is often of highest interest. In this chapter, we learn efficient methods for finding the steady-state responses for sinusoidal sources.

We also study three-phase circuits, which are used in electric power-distribution systems. Most engineers who work in industrial settings need to understand three-phase power distribution.

5.1 SINUSOIDAL CURRENTS AND VOLTAGES

A sinusoidal voltage is shown in Figure 5.1 and is given by

$$v(t) = V_m \cos(\omega t + \theta) \tag{5.1}$$

where V_m is the **peak value** of the voltage, ω is the **angular frequency** in radians per second, and θ is the **phase angle**.

Sinusoidal signals are periodic, repeating the same pattern of values in each **period** T. Because the cosine (or sine) function completes one cycle when the angle increases by 2π radians, we get

$$\omega T = 2\pi \tag{5.2}$$

The **frequency** of a periodic signal is the number of cycles completed in one second. Thus, we obtain

$$f = \frac{1}{T} \tag{5.3}$$

We refer to ω as angular frequency with units of radians per second and f simply as frequency with units of hertz (Hz).

The units of frequency are hertz (Hz). (Actually, the physical units of hertz are equivalent to inverse seconds.) Solving Equation 5.2 for the angular frequency, we have

$$\omega = \frac{2\pi}{T} \tag{5.4}$$

Using Equation 5.3 to substitute for T, we find that

$$\omega = 2\pi f \tag{5.5}$$

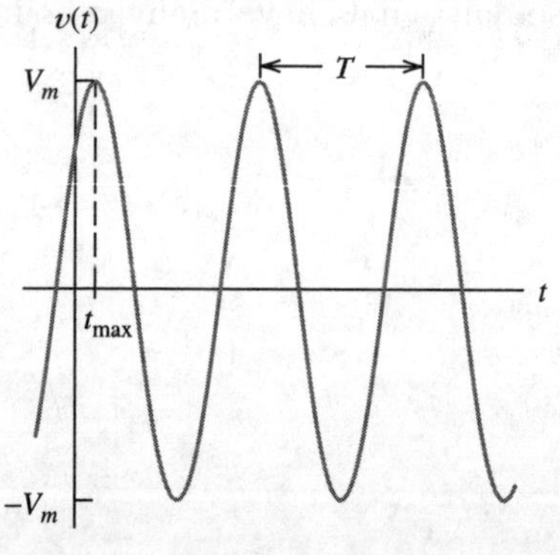

Figure 5.1 A sinusoidal voltage waveform given by $v(t) = V_m \cos(\omega t + \theta)$. *Note:* Assuming that θ is in degrees, we have $t_{max} = \frac{-\theta}{360} \times T$. For the waveform shown, θ is $-45°$.

Throughout our discussion, the argument of the cosine (or sine) function is of the form

$$\omega t + \theta$$

We assume that the angular frequency ω has units of radians per second (rad/s). However, we sometimes give the phase angle θ in degrees. Then, the argument has mixed units. If we wanted to evaluate $\cos(\omega t + \theta)$ for a particular value of time, we would have to convert θ to radians before adding the terms in the argument. Usually, we find it easier to visualize an angle expressed in degrees, and mixed units are not a problem.

Electrical engineers often write the argument of a sinusoid in mixed units: ωt is in radians and the phase angle θ is in degrees.

For uniformity, we express sinusoidal functions by using the cosine function rather than the sine function. The functions are related by the identity

$$\sin(z) = \cos(z - 90°) \tag{5.6}$$

For example, when we want to find the phase angle of

$$v_x(t) = 10\sin(200t + 30°)$$

we first write it as

$$\begin{aligned} v_x(t) &= 10\cos(200t + 30° - 90°) \\ &= 10\cos(200t - 60°) \end{aligned}$$

Thus, we state that the phase angle of $v_x(t)$ is $-60°$.

Root-Mean-Square Values

Consider applying a periodic voltage $v(t)$ with period T to a resistance R. The power delivered to the resistance is given by

$$p(t) = \frac{v^2(t)}{R} \tag{5.7}$$

Furthermore, the energy delivered in one period is given by

$$E_T = \int_0^T p(t)\, dt \tag{5.8}$$

The average power P_{avg} delivered to the resistance is the energy delivered in one cycle divided by the period. Thus,

$$P_{avg} = \frac{E_T}{T} = \frac{1}{T}\int_0^T p(t)\, dt \tag{5.9}$$

Using Equation 5.7 to substitute into Equation 5.9, we obtain

$$P_{avg} = \frac{1}{T}\int_0^T \frac{v^2(t)}{R}\, dt \tag{5.10}$$

This can be rearranged as

$$P_{avg} = \frac{\left[\sqrt{\frac{1}{T}\int_0^T v^2(t)\,dt}\right]^2}{R} \tag{5.11}$$

Now, we define the **root-mean-square** (rms) value of the periodic voltage $v(t)$ as

$$V_{rms} = \sqrt{\frac{1}{T}\int_0^T v^2(t)\,dt} \tag{5.12}$$

Using this equation to substitute into Equation 5.11, we get

$$P_{avg} = \frac{V_{rms}^2}{R} \tag{5.13}$$

Power calculations are facilitated by using rms values for voltage or current.

Thus, if the rms value of a periodic voltage is known, it is relatively easy to compute the average power that the voltage can deliver to a resistance. The rms value is also called the **effective value**.

Similarly for a periodic current $i(t)$, we define the rms value as

$$I_{rms} = \sqrt{\frac{1}{T}\int_0^T i^2(t)\,dt} \tag{5.14}$$

and the average power delivered if $i(t)$ flows through a resistance is given by

$$P_{avg} = I_{rms}^2 R \tag{5.15}$$

RMS Value of a Sinusoid

Consider a sinusoidal voltage given by

$$v(t) = V_m \cos(\omega t + \theta) \tag{5.16}$$

To find the rms value, we substitute into Equation 5.12, which yields

$$V_{rms} = \sqrt{\frac{1}{T}\int_0^T V_m^2 \cos^2(\omega t + \theta)\,dt} \tag{5.17}$$

Next, we use the trigonometric identity

$$\cos^2(z) = \frac{1}{2} + \frac{1}{2}\cos(2z) \tag{5.18}$$

to write Equation 5.17 as

$$V_{rms} = \sqrt{\frac{V_m^2}{2T}\int_0^T [1 + \cos(2\omega t + 2\theta)]\,dt} \tag{5.19}$$

Integrating, we get

$$V_{\text{rms}} = \sqrt{\frac{V_m^2}{2T}\left[t + \frac{1}{2\omega}\sin(2\omega t + 2\theta)\right]_0^T} \quad (5.20)$$

Evaluating, we have

$$V_{\text{rms}} = \sqrt{\frac{V_m^2}{2T}\left[T + \frac{1}{2\omega}\sin(2\omega T + 2\theta) - \frac{1}{2\omega}\sin(2\theta)\right]} \quad (5.21)$$

Referring to Equation 5.2, we see that $\omega T = 2\pi$. Thus, we obtain

$$\begin{aligned}\frac{1}{2\omega}\sin(2\omega T + 2\theta) - \frac{1}{2\omega}\sin(2\theta) &= \frac{1}{2\omega}\sin(4\pi + 2\theta) - \frac{1}{2\omega}\sin(2\theta)\\ &= \frac{1}{2\omega}\sin(2\theta) - \frac{1}{2\omega}\sin(2\theta)\\ &= 0\end{aligned}$$

Therefore, Equation 5.21 reduces to

$$V_{\text{rms}} = \frac{V_m}{\sqrt{2}} \quad (5.22)$$

This is a useful result that we will use many times in dealing with sinusoids.

Usually in discussing sinusoids, the rms or effective value is given rather than the peak value. For example, ac power in residential wiring is distributed as a 60-Hz 115-V rms sinusoid (in the United States). Most people are aware of this, but probably few know that 115 V is the rms value and that the peak value is $V_m = V_{\text{rms}} \times \sqrt{2} = 115 \times \sqrt{2} \cong 163$ V. (Actually, 115 V is the nominal residential distribution voltage. It can vary from approximately 105 to 130 V.)

The rms value for a sinusoid is the peak value divided by the square root of two. This is not true for other periodic waveforms such as square waves or triangular waves.

Keep in mind that $V_{\text{rms}} = V_m/\sqrt{2}$ applies to sinusoids. To find the rms value of other periodic waveforms, we would need to employ the definition given by Equation 5.12.

Example 5.1 Power Delivered to a Resistance by a Sinusoidal Source

Suppose that a voltage given by $v(t) = 100\cos(100\pi t)$ V is applied to a 50-Ω resistance. Sketch $v(t)$ to scale versus time. Find the rms value of the voltage and the average power delivered to the resistance. Find the power as a function of time and sketch to scale.

Solution By comparison of the expression given for $v(t)$ with Equation 5.1, we see that $\omega = 100\pi$. Using Equation 5.5, we find that the frequency is $f = \omega/2\pi = 50$ Hz. Then, the period is $T = 1/f = 20$ ms. A plot of $v(t)$ versus time is shown in Figure 5.2(a).

The peak value of the voltage is $V_m = 100$ V. Thus, the rms value is $V_{\text{rms}} = V_m/\sqrt{2} = 70.71$ V. Then, the average power is

$$P_{\text{avg}} = \frac{V_{\text{rms}}^2}{R} = \frac{(70.71)^2}{50} = 100 \text{ W}$$

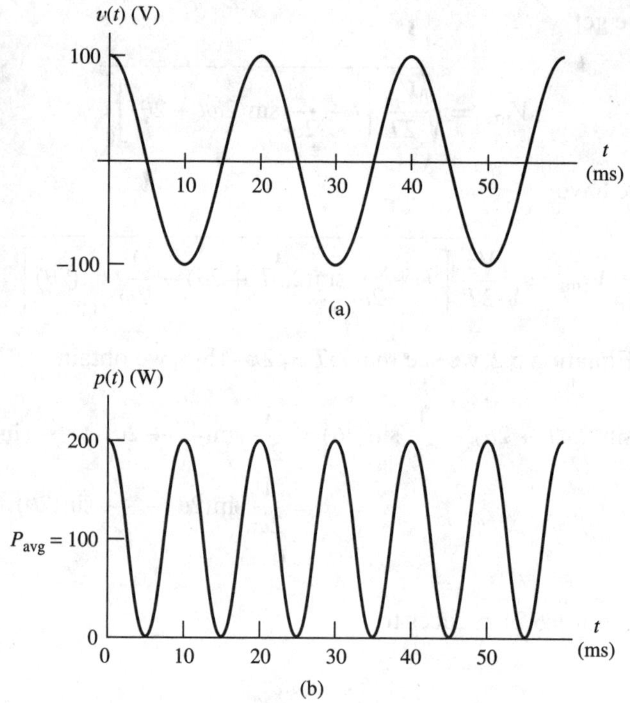

Figure 5.2 Voltage and power versus time for Example 5.1.

The power as a function of time is given by

$$p(t) = \frac{v^2(t)}{R} = \frac{100^2 \cos^2(100\pi t)}{50} = 200 \cos^2(100\pi t) \text{ W}$$

A plot of $p(t)$ versus time is shown in Figure 5.2(b). Notice that the power fluctuates from 0 to 200 W. However, the average power is 100 W, as we found by using the rms value. ■

For a sinusoidal current flowing in a resistance, power fluctuates periodically from zero to twice the average value.

RMS Values of Nonsinusoidal Voltages or Currents

Sometimes we need to determine the rms values of periodic currents or voltages that are not sinusoidal. We can accomplish this by applying the definition given by Equation 5.12 or 5.14 directly.

Example 5.2 RMS Value of a Triangular Voltage

The voltage shown in Figure 5.3(a) is known as a triangular waveform. Determine its rms value.

Solution First, we need to determine the equations describing the waveform between $t = 0$ and $t = T = 2$ s. As illustrated in Figure 5.3(b), the equations for the first period of the triangular wave are

$$v(t) = \begin{cases} 3t & \textit{for} \quad 0 \le t \le 1 \\ 6 - 3t & \textit{for} \quad 1 \le t \le 2 \end{cases}$$

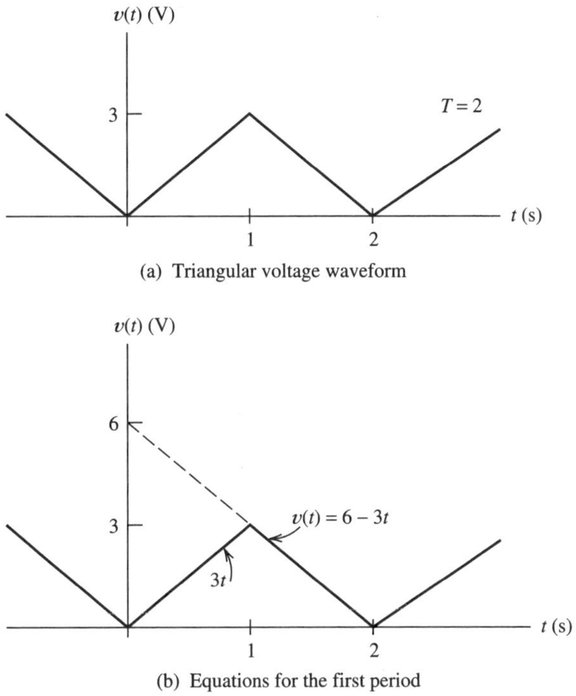

Figure 5.3 Triangular voltage waveform of Example 5.2.

Equation 5.12 gives the rms value of the voltage.

$$V_{\text{rms}} = \sqrt{\frac{1}{T}\int_0^T v^2(t)dt}$$

Dividing the interval into two parts and substituting for $v(t)$, we have

$$V_{\text{rms}} = \sqrt{\frac{1}{2}\left[\int_0^1 9t^2 dt + \int_1^2 (6-3t)^2 dt\right]}$$

$$V_{\text{rms}} = \sqrt{\frac{1}{2}\left[3t^3|_{t=0}^{t=1} + (36t - 18t^2 + 3t^3)|_{t=1}^{t=2}\right]}$$

Evaluating, we find

$$V_{\text{rms}} = \sqrt{\frac{1}{2}[3 + (72 - 36 - 72 + 18 + 24 - 3)]} = \sqrt{3}\text{ V}$$

■

The integrals in this example are easy to carry out manually. However, when the integrals are more difficult, we can sometimes obtain answers using the MATLAB Symbolic Toolbox. Here are the MATLAB commands needed to perform the integrals in this example:

```
>> syms Vrms t
>> Vrms = sqrt((1/2)*(int(9*t^2,t,0,1) + int((6-3*t)^2,t,1,2)))
     Vrms =
     3^(1/2)
```

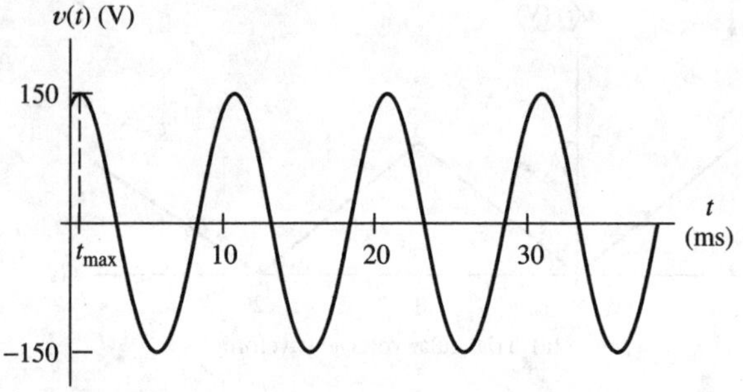

Figure 5.4 Answer for Exercise 5.1(c).

Exercise 5.1 Suppose that a sinusoidal voltage is given by

$$v(t) = 150\cos(200\pi t - 30°)\text{ V}$$

a. Find the angular frequency, the frequency in hertz, the period, the peak value, and the rms value. Also, find the first value of time t_{max} after $t = 0$ such that $v(t)$ attains its positive peak. **b.** If this voltage is applied to a 50-Ω resistance, compute the average power delivered. **c.** Sketch $v(t)$ to scale versus time.

Answer **a.** $\omega = 200\pi$, $f = 100$ Hz, $T = 10$ ms, $V_m = 150$ V, $V_{rms} = 106.1$ V, $t_{max} = \frac{30°}{360°} \times T = 0.833$ ms; **b.** $P_{avg} = 225$ W; **c.** a plot of $v(t)$ versus time is shown in Figure 5.4. □

Exercise 5.2 Express $v(t) = 100\sin(300\pi t + 60°)$ V as a cosine function.

Answer $v(t) = 100\cos(300\pi t - 30°)$ V. □

Exercise 5.3 Suppose that the ac line voltage powering a computer has an rms value of 110 V and a frequency of 60 Hz, and the peak voltage is attained at $t = 5$ ms. Write an expression for this ac voltage as a function of time.

Answer $v(t) = 155.6\cos(377t - 108°)$ V. □

5.2 PHASORS

In the next several sections, we will see that sinusoidal steady-state analysis is greatly facilitated if the currents and voltages are represented as vectors (called **phasors**) in the complex-number plane. In preparation for this material, you may wish to study the review of complex-number arithmetic in Appendix A.

We start with a study of convenient methods for adding (or subtracting) sinusoidal waveforms. We often need to do this in applying Kirchhoff's voltage law (KVL) or Kirchhoff's current law (KCL) to ac circuits. For example, in applying KVL to a network with sinusoidal voltages, we might obtain the expression

$$v(t) = 10\cos(\omega t) + 5\sin(\omega t + 60°) + 5\cos(\omega t + 90°) \quad (5.23)$$

To obtain the peak value of $v(t)$ and its phase angle, we need to put Equation 5.23 into the form

$$v(t) = V_m\cos(\omega t + \theta) \quad (5.24)$$

This could be accomplished by repeated substitution, using standard trigonometric identities. However, that method is too tedious for routine work. Instead, we will see that we can represent each term on the right-hand side of Equation 5.23 by a vector in the complex-number plane known as a **phasor**. Then, we can add the phasors with relative ease and convert the sum into the desired form.

Phasor Definition

For a sinusoidal voltage of the form

$$v_1(t) = V_1 \cos(\omega t + \theta_1)$$

we define the phasor as

$$\mathbf{V}_1 = V_1 \angle \theta_1$$

Thus, the phasor for a sinusoid is a complex number having a magnitude equal to the peak value and having the same phase angle as the sinusoid. We use boldface letters for phasors. (Actually, engineers are not consistent in choosing the magnitudes of phasors. In this chapter and in Chapter 6, we take the peak values for the magnitudes of phasors, which is the prevailing custom in circuit-analysis courses for electrical engineers. However, later in Chapters 15, 16, and 17, we will take the rms values for the phasor magnitudes as power-system engineers customarily do. We will take care to label rms phasors as such when we encounter them. In this book, if phasors are not labeled as rms, you can assume that they are peak values.)

Phasors are complex numbers that represent sinusoidal voltages or currents. The magnitude of a phasor equals the peak value and the angle equals the phase of the sinusoid (written as a cosine).

If the sinusoid is of the form

$$v_2(t) = V_2 \sin(\omega t + \theta_2)$$

we first convert to a cosine function by using the trigonometric identity

$$\sin(z) = \cos(z - 90°) \tag{5.25}$$

Thus, we have

$$v_2(t) = V_2 \cos(\omega t + \theta_2 - 90°)$$

and the phasor is

$$\mathbf{V}_2 = V_2 \angle \theta_2 - 90°$$

Phasors are obtained for sinusoidal currents in a similar fashion. Thus, for the currents

$$i_1(t) = I_1 \cos(\omega t + \theta_1)$$

and

$$i_2(t) = I_2 \sin(\omega t + \theta_2)$$

the phasors are

$$\mathbf{I}_1 = I_1 \angle \theta_1$$

and

$$\mathbf{I}_2 = I_2 \angle \theta_2 - 90°$$

respectively.

Adding Sinusoids Using Phasors

Now, we illustrate how we can use phasors to combine the terms of the right-hand side of Equation 5.23. In this discussion, we proceed in small logical steps to illustrate clearly why sinusoids can be added by adding their phasors. Later, we streamline the procedure for routine work.

Our first step in combining the terms in Equation 5.23 is to write all the sinusoids as cosine functions by using Equation 5.25. Thus, Equation 5.23 can be written as

$$v(t) = 10\cos(\omega t) + 5\cos(\omega t + 60^\circ - 90^\circ) + 5\cos(\omega t + 90^\circ) \tag{5.26}$$

$$v(t) = 10\cos(\omega t) + 5\cos(\omega t - 30^\circ) + 5\cos(\omega t + 90^\circ) \tag{5.27}$$

Referring to Euler's formula (Equation A.8) in Appendix A, we see that we can write

$$\cos(\theta) = \text{Re}\left(e^{j\theta}\right) = \text{Re}[\cos(\theta) + j\sin(\theta)] \tag{5.28}$$

where the notation Re( ) means that we retain only the real part of the quantity inside the parentheses. Thus, we can rewrite Equation 5.27 as

$$v(t) = 10\,\text{Re}\left[e^{j\omega t}\right] + 5\,\text{Re}\left[e^{j(\omega t - 30^\circ)}\right] + 5\,\text{Re}\left[e^{j(\omega t + 90^\circ)}\right] \tag{5.29}$$

When we multiply a complex number Z by a real number A, both the real and imaginary parts of Z are multiplied by A. Thus, Equation 5.29 becomes

$$v(t) = \text{Re}\left[10e^{j\omega t}\right] + \text{Re}\left[5e^{j(\omega t - 30^\circ)}\right] + \text{Re}\left[5e^{j(\omega t + 90^\circ)}\right] \tag{5.30}$$

Next, we can write

$$v(t) = \text{Re}\left[10e^{j\omega t} + 5e^{j(\omega t - 30^\circ)} + 5e^{j(\omega t + 90^\circ)}\right] \tag{5.31}$$

because the real part of the sum of several complex quantities is equal to the sum of the real parts. If we factor out the common term $e^{j\omega t}$, Equation 5.31 becomes

$$v(t) = \text{Re}\left[\left(10 + 5e^{-j30^\circ} + 5^{j90^\circ}\right)e^{j\omega t}\right] \tag{5.32}$$

Putting the complex numbers into polar form, we have

$$v(t) = \text{Re}\left[(10\,\angle 0^\circ + 5\,\angle -30^\circ + 5\,\angle 90^\circ)e^{j\omega t}\right] \tag{5.33}$$

Now, we can combine the complex numbers as

$$\begin{aligned} 10\,\angle 0^\circ + 5\,\angle -30^\circ + 5\,\angle 90^\circ &= 10 + 4.33 - j2.50 + j5 \\ &= 14.33 + j2.5 \\ &= 14.54\,\angle 9.90^\circ \\ &= 14.54e^{j9.90^\circ} \end{aligned} \tag{5.34}$$

Using this result in Equation 5.33, we have

$$v(t) = \text{Re}\left[\left(14.54e^{j9.90^\circ}\right)e^{j\omega t}\right]$$

which can be written as

$$v(t) = \text{Re}\left[14.54e^{j(\omega t+9.90°)}\right] \quad (5.35)$$

Now, using Equation 5.28, we can write this as

$$v(t) = 14.54\cos(\omega t + 9.90°) \quad (5.36)$$

Thus, we have put the original expression for $v(t)$ into the desired form. The terms on the left-hand side of Equation 5.34 are the phasors for the terms on the right-hand side of the original expression for $v(t)$. Notice that the essential part of the work needed to combine the sinusoids is to add the phasors.

Streamlined Procedure for Adding Sinusoids

From now on, to add sinusoids, we will first write the phasor for each term in the sum, add the phasors by using complex-number arithmetic, and then write the simplified expression for the sum.

To add sinusoids, we find the phasor for each term, add the phasors by using complex-number arithmetic, express the sum in polar form, and then write the corresponding sinusoidal time function.

Example 5.3 Using Phasors to Add Sinusoids

Suppose that

$$v_1(t) = 20\cos(\omega t - 45°)$$
$$v_2(t) = 10\sin(\omega t + 60°)$$

In using phasors to add sinusoids, all of the terms must have the same frequency.

Reduce the sum $v_s(t) = v_1(t) + v_2(t)$ to a single term.

Solution The phasors are

Step 1: Determine the phasor for each term.

$$\mathbf{V}_1 = 20\ \underline{/-45°}$$
$$\mathbf{V}_2 = 10\ \underline{/-30°}$$

Notice that we have subtracted 90° to find the phase angle for $\mathbf{V}_2$ because $v_2(t)$ is a sine function rather than a cosine function.

Next, we use complex-number arithmetic to add the phasors and convert the sum to polar form:

Step 2: Use complex arithmetic to add the phasors.

$$\begin{aligned}\mathbf{V}_s &= \mathbf{V}_1 + \mathbf{V}_2\\ &= 20\ \underline{/-45°} + 10\ \underline{/-30°}\\ &= 14.14 - j14.14 + 8.660 - j5\\ &= 22.80 - j19.14\\ &= 29.77\ \underline{/-40.01°}\end{aligned}$$

Step 3: Convert the sum to polar form.

Now, we write the time function corresponding to the phasor $\mathbf{V}_s$.

Step 4: Write the result as a time function.

$$v_s(t) = 29.77\cos(\omega t - 40.01°)$$

■

Exercise 5.4 Reduce the following expressions by using phasors:

$$v_1(t) = 10\cos(\omega t) + 10\sin(\omega t)$$
$$i_1(t) = 10\cos(\omega t + 30°) + 5\sin(\omega t + 30°)$$
$$i_2(t) = 20\sin(\omega t + 90°) + 15\cos(\omega t - 60°)$$

Answer

$$v_1(t) = 14.14\cos(\omega t - 45°)$$
$$i_1(t) = 11.18\cos(\omega t + 3.44°)$$
$$i_2(t) = 30.4\cos(\omega t - 25.3°)$$

□

Phasors as Rotating Vectors

Consider a sinusoidal voltage given by

$$v(t) = V_m \cos(\omega t + \theta)$$

In developing the phasor concept, we write

$$v(t) = \text{Re}\left[V_m e^{j(\omega t + \theta)}\right]$$

The complex quantity inside the brackets is

$$V_m e^{j(\omega t + \theta)} = V_m \angle \omega t + \theta$$

Sinusoids can be visualized as the real-axis projection of vectors rotating in the complex plane. The phasor for a sinusoid is a snapshot of the corresponding rotating vector at $t = 0$.

This can be visualized as a vector of length V_m that rotates counterclockwise in the complex plane with an angular velocity of ω rad/s. Furthermore, the voltage $v(t)$ is the real part of the vector, which is illustrated in Figure 5.5. As the vector rotates, its projection on the real axis traces out the voltage as a function of time. The phasor is simply a "snapshot" of this rotating vector at $t = 0$.

Phase Relationships

We will see that the phase relationships between currents and voltages are often important. Consider the voltages

$$v_1(t) = 3\cos(\omega t + 40°)$$

To determine phase relationships from a phasor diagram, consider that the phasors rotate counterclockwise. Then, when standing at a fixed point, if $\mathbf{V}_1$ arrives first followed by $\mathbf{V}_2$ after a rotation of θ, we say that $\mathbf{V}_1$ leads $\mathbf{V}_2$ by θ. Alternatively, we could say that $\mathbf{V}_2$ lags $\mathbf{V}_1$ by θ. (Usually, we take θ as the smaller angle between the two phasors.)

and

$$v_2(t) = 4\cos(\omega t - 20°)$$

The corresponding phasors are

$$\mathbf{V}_1 = 3 \angle 40°$$

and

$$\mathbf{V}_2 = 4 \angle -20°$$

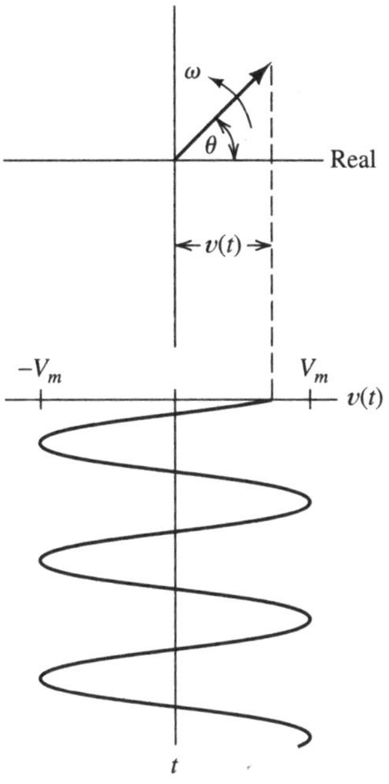

Figure 5.5 A sinusoid can be represented as the real part of a vector rotating counterclockwise in the complex plane.

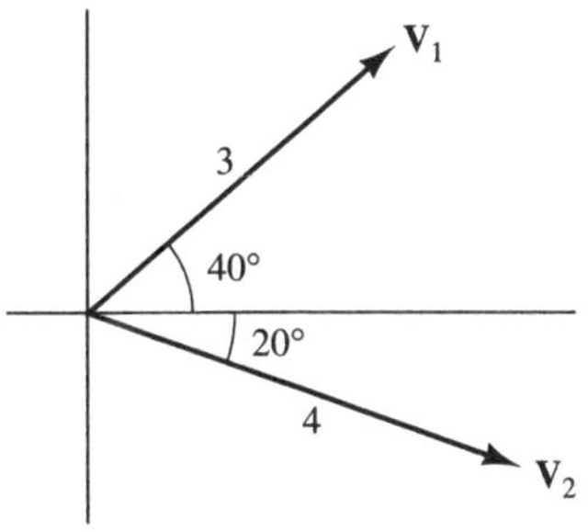

Figure 5.6 Because the vectors rotate counterclockwise, $\mathbf{V}_1$ leads $\mathbf{V}_2$ by 60° (or, equivalently, $\mathbf{V}_2$ lags $\mathbf{V}_1$ by 60°).

The phasor diagram is shown in Figure 5.6. Notice that the angle between $\mathbf{V}_1$ and $\mathbf{V}_2$ is 60°. Because the complex vectors rotate counterclockwise, we say that $\mathbf{V}_1$ *leads* $\mathbf{V}_2$ by 60°. (An alternative way to state the phase relationship is to state that $\mathbf{V}_2$ *lags* $\mathbf{V}_1$ by 60°.)

We have seen that the voltages versus time can be obtained by tracing the real part of the rotating vectors. The plots of $v_1(t)$ and $v_2(t)$ versus ωt are shown in Figure 5.7. Notice that $v_1(t)$ reaches its peak 60° earlier than $v_2(t)$. This is the meaning of the statement that $v_1(t)$ leads $v_2(t)$ by 60°.

To determine phase relationships between sinusoids from their plots versus time, find the shortest time interval t_p between positive peaks of the two waveforms. Then, the phase angle is $\theta = (t_p/T) \times 360°$. If the peak of $v_1(t)$ occurs first, we say that $v_1(t)$ leads $v_2(t)$ or that $v_2(t)$ lags $v_1(t)$.

Exercise 5.5 Consider the voltages given by

$$v_1(t) = 10\cos(\omega t - 30°)$$
$$v_2(t) = 10\cos(\omega t + 30°)$$
$$v_3(t) = 10\sin(\omega t + 45°)$$

State the phase relationship between each pair of the voltages. (*Hint:* Find the phasor for each voltage and draw the phasor diagram.)

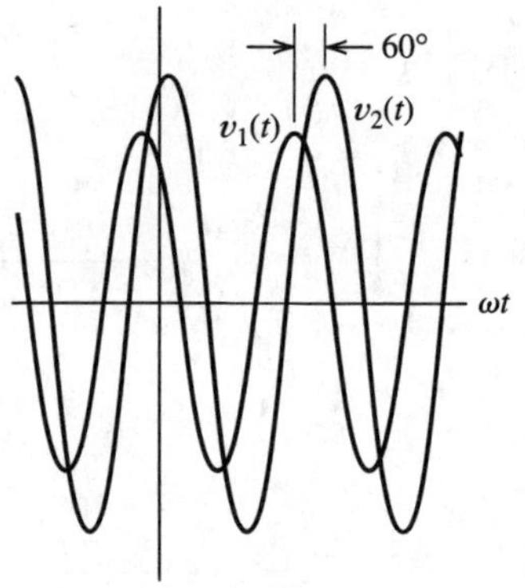

Figure 5.7 The peaks of $v_1(t)$ occur 60° before the peaks of $v_2(t)$. In other words, $v_1(t)$ leads $v_2(t)$ by 60°.

Answer

v_1 lags v_2 by 60° (or v_2 leads v_1 by 60°)
v_1 leads v_3 by 15° (or v_3 lags v_1 by 15°)
v_2 leads v_3 by 75° (or v_3 lags v_2 by 75°)

□

5.3 COMPLEX IMPEDANCES

In this section, we learn that by using phasors to represent sinusoidal voltages and currents, we can solve sinusoidal steady-state circuit problems with relative ease compared with the methods of Chapter 4. Except for the fact that we use complex arithmetic, sinusoidal steady-state analysis is virtually the same as the analysis of resistive circuits, which we studied in Chapter 2.

Inductance

Consider an inductance in which the current is a sinusoid given by

$$i_L(t) = I_m \sin(\omega t + \theta) \tag{5.37}$$

Recall that the voltage across an inductance is

$$v_L(t) = L\frac{di_L(t)}{dt} \tag{5.38}$$

Substituting Equation 5.37 into Equation 5.38 and reducing, we obtain

$$v_L(t) = \omega L I_m \cos(\omega t + \theta) \tag{5.39}$$

Now, the phasors for the current and voltage are

$$\mathbf{I}_L = I_m \angle \theta - 90° \tag{5.40}$$

and

$$\mathbf{V}_L = \omega L I_m \angle \theta = V_m \angle \theta \tag{5.41}$$

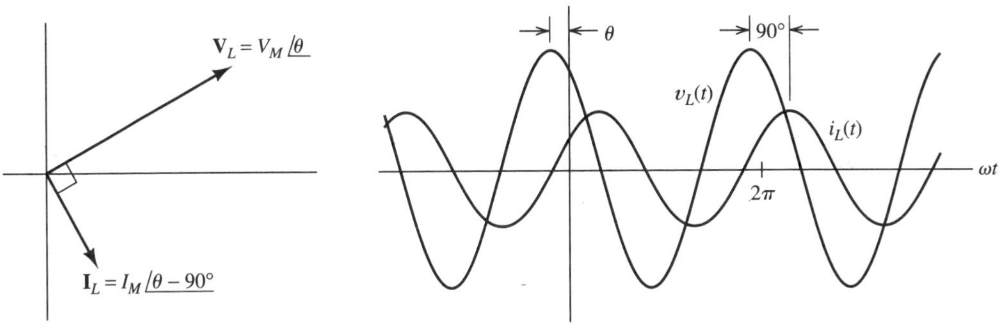

(a) Phasor diagram (b) Current and voltage versus time

Figure 5.8 Current lags voltage by 90° in a pure inductance.

The phasor diagram of the current and voltage is shown in Figure 5.8(a). The corresponding waveforms of current and voltage are shown in Figure 5.8(b). *Notice that the current lags the voltage by* 90° *for a pure inductance.*

Current lags voltage by 90° for a pure inductance.

Equation 5.41 can be written in the form

$$\mathbf{V}_L = (\omega L \angle 90°) \times I_m \angle \theta - 90° \tag{5.42}$$

Using Equation 5.40 to substitute into Equation 5.42, we find that

$$\mathbf{V}_L = (\omega L \angle 90°) \times \mathbf{I}_L \tag{5.43}$$

which can also be written as

$$\mathbf{V}_L = j\omega L \times \mathbf{I}_L \tag{5.44}$$

We refer to the term $j\omega L = \omega L \angle 90°$ as the **impedance** of the inductance and denote it as Z_L. Thus, we have

$$Z_L = j\omega L = \omega L \angle 90° \tag{5.45}$$

and

$$\mathbf{V}_L = Z_L \mathbf{I}_L \tag{5.46}$$

Thus, the phasor voltage is equal to the impedance times the phasor current. This is Ohm's law in phasor form. However, for an inductance, the impedance is an imaginary number, whereas resistance is a real number. (Impedances that are pure imaginary are also called **reactances**.)

Equation 5.46 shows that phasor voltage and phasor current for an inductance are related in a manner analogous to Ohm's law.

Capacitance

In a similar fashion for a capacitance, we can show that if the current and voltage are sinusoidal, the phasors are related by

$$\mathbf{V}_C = Z_C \mathbf{I}_C \tag{5.47}$$

in which the impedance of the capacitance is

$$Z_C = -j\frac{1}{\omega C} = \frac{1}{j\omega C} = \frac{1}{\omega C} \angle -90° \tag{5.48}$$

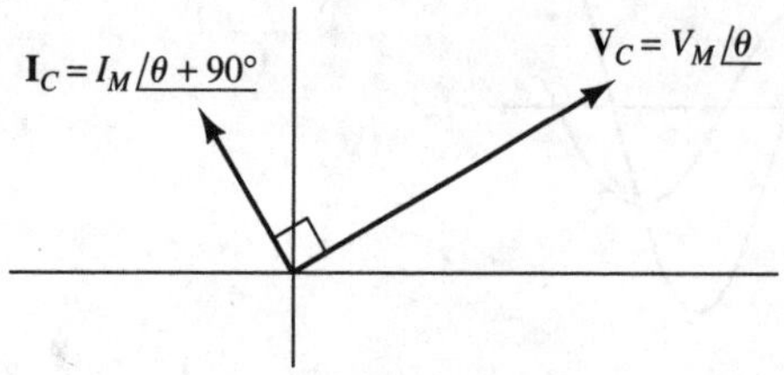

(a) Phasor diagram

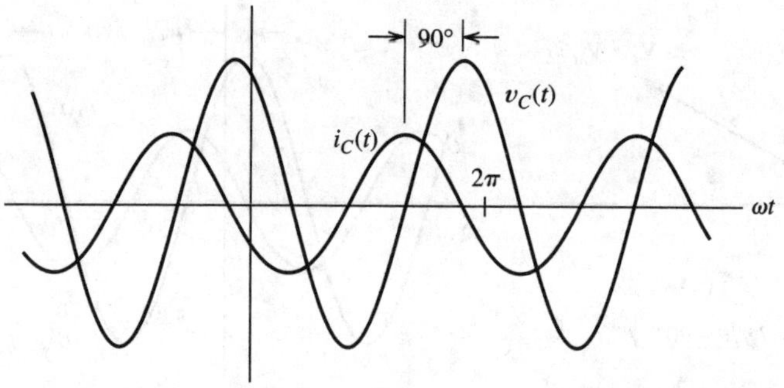

(b) Current and voltage versus time

Figure 5.9 Current leads voltage by 90° in a pure capacitance.

Notice that the impedance of a capacitance is also a pure imaginary number.

Suppose that the phasor voltage is

$$\mathbf{V}_C = V_m \angle \theta$$

Then, the phasor current is

$$\mathbf{I}_C = \frac{\mathbf{V}_C}{Z_C} = \frac{V_m \angle \theta}{(1/\omega C) \angle -90°} = \omega C V_m \angle \theta + 90°$$

$$\mathbf{I}_C = I_m \angle \theta + 90°$$

Current leads voltage by 90° for a pure capacitance.

where $I_m = \omega C V_m$. The phasor diagram for current and voltage in a pure capacitance is shown in Figure 5.9(a). The corresponding plots of current and voltage versus time are shown in Figure 5.9(b). Notice that the current leads the voltage by 90°. (On the other hand, current lags voltage for an inductance. This is easy to remember if you know *ELI* the *ICE* man. The letter *E* is sometimes used to stand for *electromotive force*, which is another term for voltage, *L* and *C* are used for inductance and capacitance, respectively, and *I* is used for current.)

Resistance

For a resistance, the phasors are related by

$$\mathbf{V}_R = R\mathbf{I}_R \tag{5.49}$$

Current and voltage are in phase for a resistance.

Because resistance is a real number, the current and voltage are in phase, as illustrated in Figure 5.10.

Exercise 5.6 A voltage $v_L(t) = 100\cos(200t)$ is applied to a 0.25-H inductance. (Notice that $\omega = 200$.) **a.** Find the impedance of the inductance, the phasor current, and the phasor voltage. **b.** Draw the phasor diagram.

Answer **a.** $Z_L = j50 = 50 \angle 90°$, $\mathbf{I}_L = 2 \angle -90°$, $\mathbf{V}_L = 100 \angle 0°$; **b.** the phasor diagram is shown in Figure 5.11(a). □

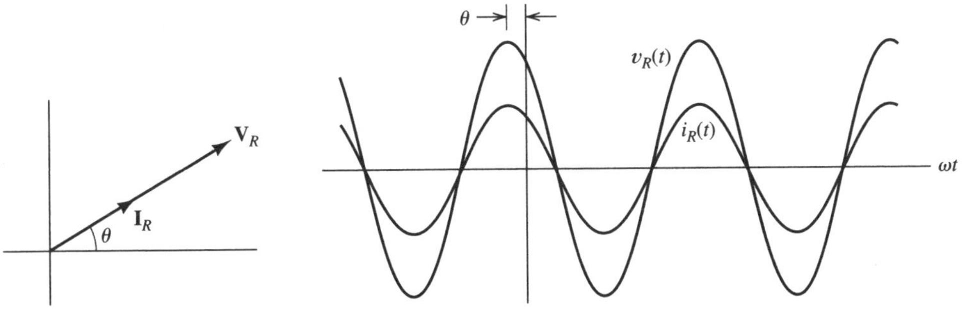

(a) Phasor diagram (b) Current and voltage versus time

Figure 5.10 For a pure resistance, current and voltage are in phase.

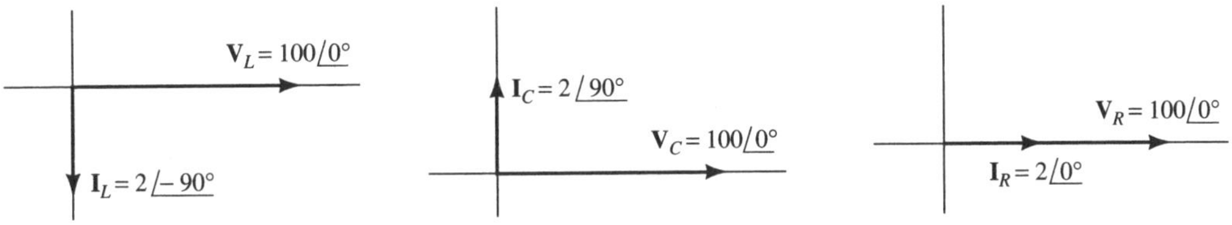

(a) Exercise 5.6 (0.25 H inductance) (b) Exercise 5.7 (100 μF capacitance) (c) Exercise 5.8 (50 Ω resistance)

Figure 5.11 Answers for Exercises 5.6, 5.7, and 5.8. The scale has been expanded for the currents compared with the voltages so the current phasors can be easily seen.

Exercise 5.7 A voltage $v_C(t) = 100\cos(200t)$ is applied to a 100-μF capacitance. **a.** Find the impedance of the capacitance, the phasor current, and the phasor voltage. **b.** Draw the phasor diagram.

Answer **a.** $Z_C = -j50 = 50\angle -90°$, $\mathbf{I}_C = 2\angle 90°$, $\mathbf{V}_C = 100\angle 0°$; **b.** the phasor diagram is shown in Figure 5.11(b). □

Exercise 5.8 A voltage $v_R(t) = 100\cos(200t)$ is applied to a 50-Ω resistance. **a.** Find the phasor for the current and the phasor voltage. **b.** Draw the phasor diagram.

Answer **a.** $\mathbf{I}_R = 2\angle 0°$, $\mathbf{V}_R = 100\angle 0°$; **b.** the phasor diagram is shown in Figure 5.11(c). □

5.4 CIRCUIT ANALYSIS WITH PHASORS AND COMPLEX IMPEDANCES

Kirchhoff's Laws in Phasor Form

Recall that KVL requires that the voltages sum to zero for any closed path in an electrical network. A typical KVL equation is

$$v_1(t) + v_2(t) - v_3(t) = 0 \tag{5.50}$$

If the voltages are sinusoidal, they can be represented by phasors. Then, Equation 5.50 becomes

$$\mathbf{V}_1 + \mathbf{V}_2 - \mathbf{V}_3 = 0 \tag{5.51}$$

Thus, we can apply KVL directly to the phasors. The sum of the phasor voltages equals zero for any closed path.

Similarly, KCL can be applied to currents in phasor form. The sum of the phasor currents entering a node must equal the sum of the phasor currents leaving.

Circuit Analysis Using Phasors and Impedances

We have seen that phasor currents and voltages are related by complex impedances, and Kirchhoff's laws apply in phasor form. Except for the fact that the voltages, currents, and impedances can be complex, the equations are exactly like those of resistive circuits.

A step-by-step procedure for steady-state analysis of circuits with sinusoidal sources is

1. Replace the time descriptions of the voltage and current sources with the corresponding phasors. (All of the sources must have the same frequency.)
2. Replace inductances by their complex impedances $Z_L = j\omega L = \omega L \angle 90°$. Replace capacitances by their complex impedances $Z_C = 1/(j\omega C) = (1/\omega C) \angle -90°$. Resistances have impedances equal to their resistances.
3. Analyze the circuit by using any of the techniques studied in Chapter 2, and perform the calculations with complex arithmetic.

Example 5.4 Steady-State AC Analysis of a Series Circuit

Find the steady-state current for the circuit shown in Figure 5.12(a). Also, find the phasor voltage across each element and construct a phasor diagram.

Step 1: Replace the time description of the voltage source with the corresponding phasor.

Solution From the expression given for the source voltage $v_s(t)$, we see that the peak voltage is 100 V, the angular frequency is $\omega = 500$, and the phase angle is 30°. The phasor for the voltage source is

$$\mathbf{V}_s = 100 \angle 30°$$

Step 2: Replace inductances and capacitances with their complex impedances.

The complex impedances of the inductance and capacitance are

$$Z_L = j\omega L = j500 \times 0.3 = j150\ \Omega$$

and

$$Z_C = -j\frac{1}{\omega C} = -j\frac{1}{500 \times 40 \times 10^{-6}} = -j50\ \Omega$$

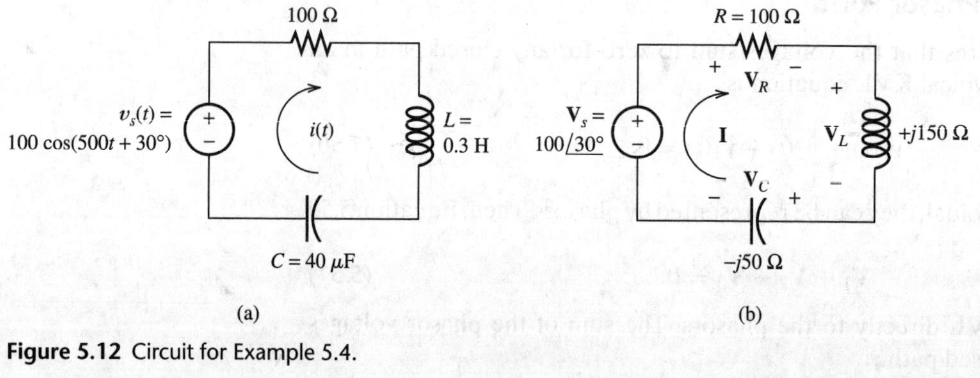

Figure 5.12 Circuit for Example 5.4.

The transformed circuit is shown in Figure 5.12(b). All three elements are in series. Thus, we find the equivalent impedance of the circuit by adding the impedances of all three elements:

Step 3: Use complex arithmetic to analyze the circuit.

$$Z_{eq} = R + Z_L + Z_C$$

Substituting values, we have

$$Z_{eq} = 100 + j150 - j50 = 100 + j100$$

Converting to polar form, we obtain

$$Z_{eq} = 141.4 \angle 45^\circ$$

Now, we can find the phasor current by dividing the phasor voltage by the equivalent impedance, resulting in

$$\mathbf{I} = \frac{\mathbf{V}_s}{Z} = \frac{100 \angle 30^\circ}{141.4 \angle 45^\circ} = 0.707 \angle -15^\circ$$

As a function of time, the current is

$$i(t) = 0.707 \cos(500t - 15^\circ)$$

Next, we can find the phasor voltage across each element by multiplying the phasor current by the respective impedance:

$$\mathbf{V}_R = R \times \mathbf{I} = 100 \times 0.707 \angle -15^\circ = 70.7 \angle -15^\circ$$

$$\mathbf{V}_L = j\omega L \times \mathbf{I} = \omega L \angle 90^\circ \times \mathbf{I} = 150 \angle 90^\circ \times 0.707 \angle -15^\circ$$

$$= 106.1 \angle 75^\circ$$

$$\mathbf{V}_C = -j\frac{1}{\omega C} \times \mathbf{I} = \frac{1}{\omega C} \angle -90^\circ \times \mathbf{I} = 50 \angle -90^\circ \times 0.707 \angle -15^\circ$$

$$= 35.4 \angle -105^\circ$$

The phasor diagram for the current and voltages is shown in Figure 5.13. Notice that the current **I** lags the source voltage $\mathbf{V}_s$ by 45°. As expected, the voltage $\mathbf{V}_R$ and current **I** are in phase for the resistance. For the inductance, the voltage $\mathbf{V}_L$ leads the current **I** by 90°. For the capacitance, the voltage $\mathbf{V}_C$ lags the current by 90°. ■

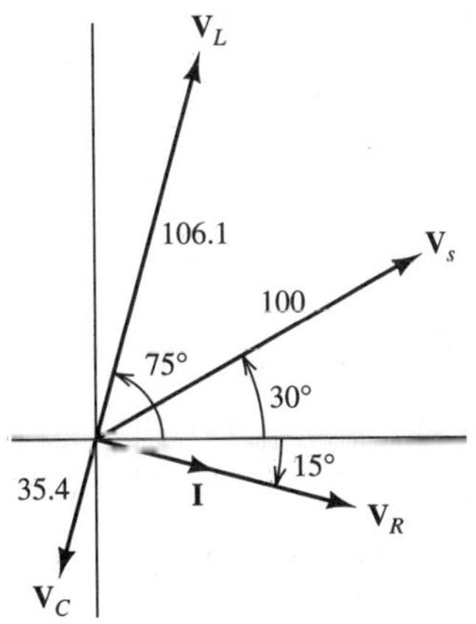

Figure 5.13 Phasor diagram for Example 5.4.

Example 5.5 Series and Parallel Combinations of Complex Impedances

Consider the circuit shown in Figure 5.14(a). Find the voltage $v_C(t)$ in steady state. Find the phasor current through each element, and construct a phasor diagram showing the currents and the source voltage.

Step 1: Replace the time description of the voltage source with the corresponding phasor.

Solution The phasor for the voltage source is $\mathbf{V}_s = 10\ \angle -90°$. [Notice that $v_s(t)$ is a sine function rather than a cosine function, and it is necessary to subtract 90° from the phase.] The angular frequency of the source is $\omega = 1000$. The impedances of the inductance and capacitance are

Step 2: Replace inductances and capacitances with their complex impedances.

$$Z_L = j\omega L = j1000 \times 0.1 = j100\ \Omega$$

and

$$Z_C = -j\frac{1}{\omega C} = -j\frac{1}{1000 \times 10 \times 10^{-6}} = -j100\ \Omega$$

Step 3: Use complex arithmetic to analyze the circuit.

The transformed network is shown in Figure 5.14(b).

To find $\mathbf{V}_C$, we will first combine the resistance and the impedance of the capacitor in parallel. Then, we will use the voltage-division principle to compute the voltage across the RC combination. The impedance of the parallel RC circuit is

$$Z_{RC} = \frac{1}{1/R + 1/Z_C} = \frac{1}{1/100 + 1/(-j100)}$$

$$= \frac{1}{0.01 + j0.01} = \frac{1\ \angle 0°}{0.01414\ \angle 45°} = 70.71\ \angle -45°$$

Converting to rectangular form, we have

$$Z_{RC} = 50 - j50$$

The equivalent network is shown in Figure 5.14(c).

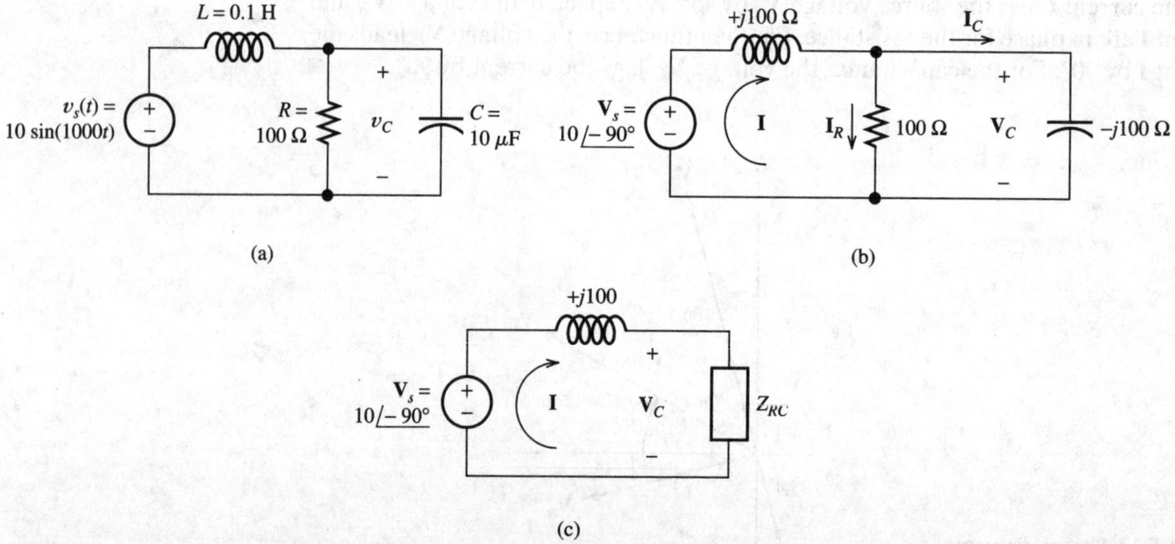

Figure 5.14 Circuit for Example 5.5.

Now, we use the voltage-division principle to obtain

$$\mathbf{V}_C = \mathbf{V}_s \frac{Z_{RC}}{Z_L + Z_{RC}} = 10\ \angle{-90^\circ}\ \frac{70.71\ \angle{-45^\circ}}{j100 + 50 - j50}$$

$$= 10\ \angle{-90^\circ}\ \frac{70.71\ \angle{-45^\circ}}{50 + j50} = 10\ \angle{-90^\circ}\ \frac{70.71\ \angle{-45^\circ}}{70.71\ \angle{45^\circ}}$$

$$= 10\ \angle{-180^\circ}$$

Converting the phasor to a time function, we have

$$v_C(t) = 10\cos(1000t - 180^\circ) = -10\cos(1000t)$$

Next, we compute the current in each element yielding

$$\mathbf{I} = \frac{\mathbf{V}_s}{Z_L + Z_{RC}} = \frac{10\ \angle{-90^\circ}}{j100 + 50 - j50} = \frac{10\ \angle{-90^\circ}}{50 + j50}$$

$$= \frac{10\ \angle{-90^\circ}}{70.71\ \angle{45^\circ}} = 0.1414\ \angle{-135^\circ}$$

$$\mathbf{I}_R = \frac{\mathbf{V}_C}{R} = \frac{10\ \angle{-180^\circ}}{100} = 0.1\ \angle{-180^\circ}$$

$$\mathbf{I}_C = \frac{\mathbf{V}_C}{Z_C} = \frac{10\ \angle{-180^\circ}}{-j100} = \frac{10\ \angle{-180^\circ}}{100\ \angle{-90^\circ}} = 0.1\ \angle{-90^\circ}$$

The phasor diagram is shown in Figure 5.15. ■

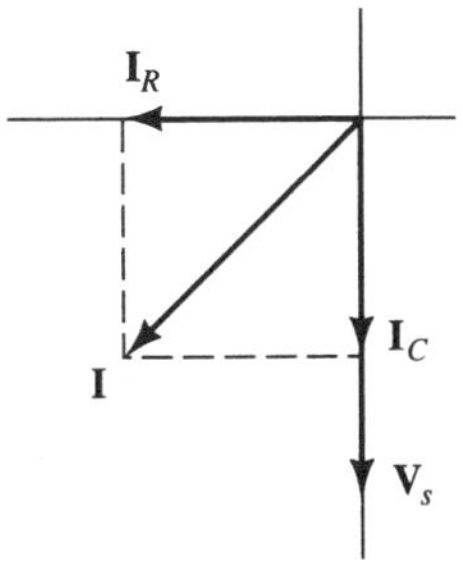

Figure 5.15 Phasor diagram for Example 5.5.

Node-Voltage Analysis

We can perform node-voltage analysis by using phasors in a manner parallel to that of Chapter 2. We illustrate with an example.

Example 5.6 Steady-State AC Node-Voltage Analysis

Use the node-voltage technique to find $v_1(t)$ in steady state for the circuit shown in Figure 5.16(a).

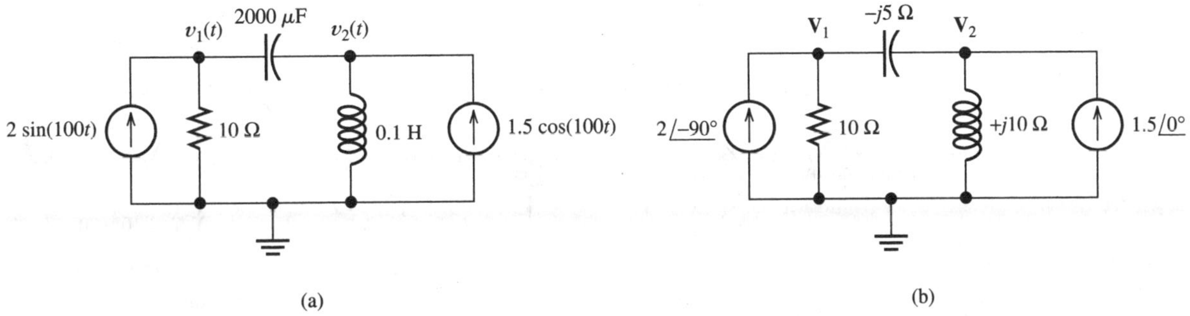

Figure 5.16 Circuit for Example 5.6.

Solution The transformed network is shown in Figure 5.16(b). We obtain two equations by applying KCL at node 1 and at node 2. This yields

$$\frac{\mathbf{V}_1}{10} + \frac{\mathbf{V}_1 - \mathbf{V}_2}{-j5} = 2\ \angle -90^\circ$$

$$\frac{\mathbf{V}_2}{j10} + \frac{\mathbf{V}_2 - \mathbf{V}_1}{-j5} = 1.5\ \angle 0^\circ$$

These equations can be put into the standard form

$$(0.1 + j0.2)\mathbf{V}_1 - j0.2\mathbf{V}_2 = -j2$$

$$-j0.2\mathbf{V}_1 + j0.1\mathbf{V}_2 = 1.5$$

Now, we solve for $\mathbf{V}_1$ yielding

$$\mathbf{V}_1 = 16.1\ \angle 29.7^\circ$$

Then, we convert the phasor to a time function and obtain

$$v_1(t) = 16.1\cos(100t + 29.7^\circ)$$

■

Mesh-Current Analysis

In a similar fashion, you can use phasors to carry out mesh-current analysis in ac circuits. Exercise 5.11 gives you a chance to try this approach.

Exercise 5.9 Consider the circuit shown in Figure 5.17(a). **a.** Find $i(t)$. **b.** Construct a phasor diagram showing all three voltages and the current. **c.** What is the phase relationship between $v_s(t)$ and $i(t)$?

Answer **a.** $i(t) = 0.0283\cos(500t - 135^\circ)$; **b.** the phasor diagram is shown in Figure 5.17(b); **c.** $i(t)$ lags $v_s(t)$ by 45°. □

Exercise 5.10 Find the phasor voltage and the phasor current through each element in the circuit of Figure 5.18.

Answer $\mathbf{V} = 277\ \angle -56.3^\circ$, $\mathbf{I}_C = 5.55\ \angle 33.7^\circ$, $\mathbf{I}_L = 1.39\ \angle -146.3^\circ$, $\mathbf{I}_R = 2.77\ \angle -56.3^\circ$. □

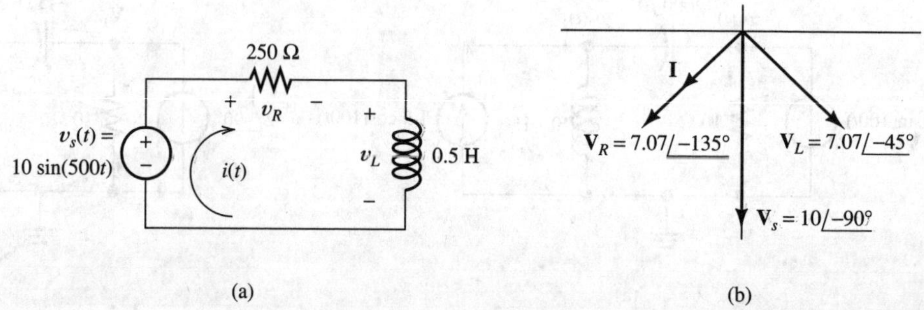

Figure 5.17 Circuit and phasor diagram for Exercise 5.9.

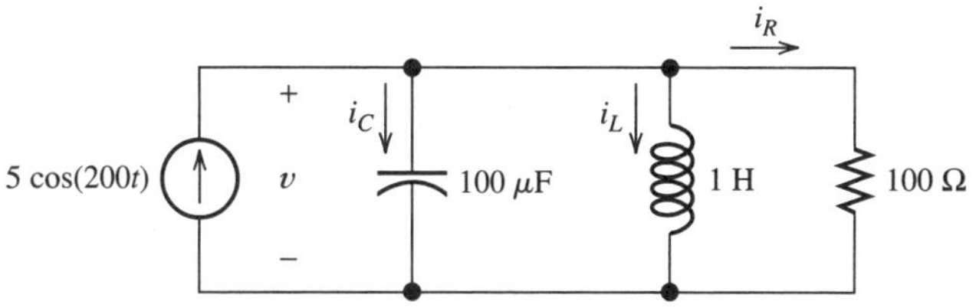

Figure 5.18 Circuit for Exercise 5.10.

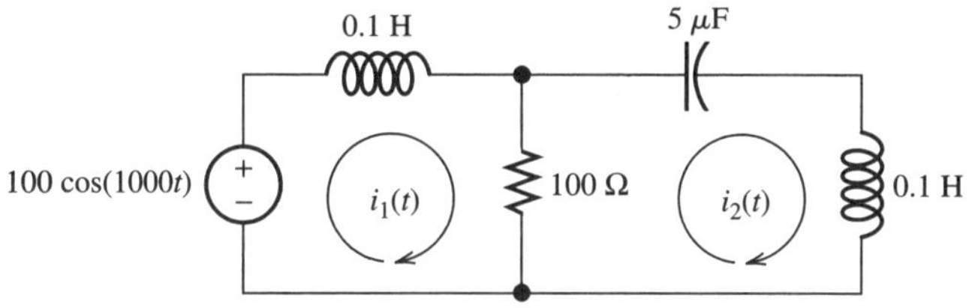

Figure 5.19 Circuit for Exercise 5.11.

Exercise 5.11 Solve for the mesh currents shown in Figure 5.19.

Answer $i_1(t) = 1.414\cos(1000t - 45°), i_2(t) = \cos(1000t)$. □

5.5 POWER IN AC CIRCUITS

Consider the situation shown in Figure 5.20. A voltage $v(t) = V_m \cos(\omega t)$ is applied to a network composed of resistances, inductances, and capacitances (i.e., an *RLC* network). The phasor for the voltage source is $\mathbf{V} = V_m \angle 0°$, and the equivalent impedance of the network is $Z = |Z| \angle \theta = R + jX$. The phasor current is

$$\mathbf{I} = \frac{\mathbf{V}}{Z} = \frac{V_m \angle 0°}{|Z| \angle \theta} = I_m \angle -\theta \tag{5.52}$$

where we have defined

$$I_m = \frac{V_m}{|Z|} \tag{5.53}$$

Before we consider the power delivered by the source to a general load, it is instructive to consider a pure resistive load, a purc inductive load, and a pure capacitive load.

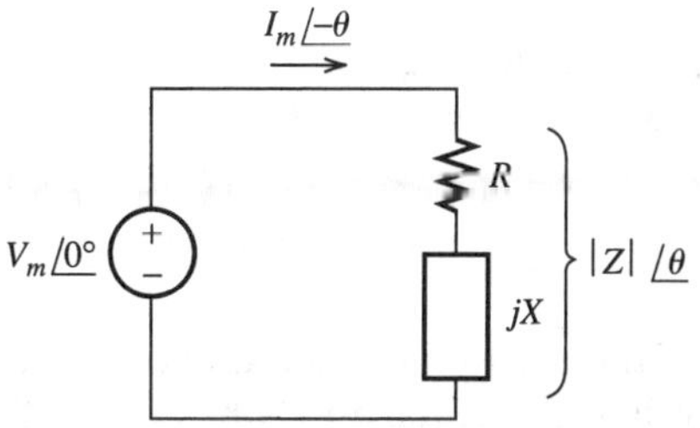

Figure 5.20 A voltage source delivering power to a load impedance $Z = R + jX$.

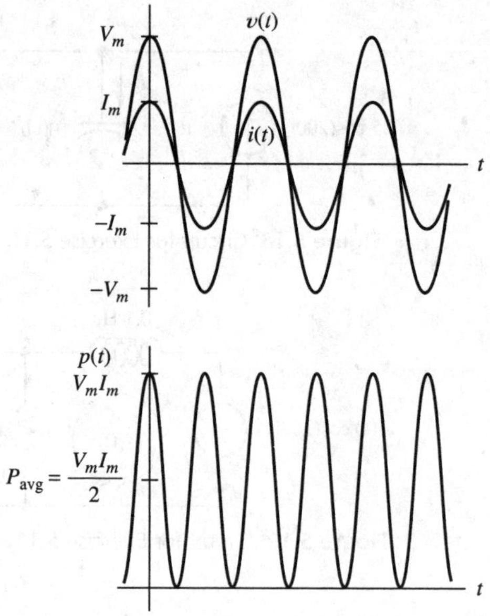

Figure 5.21 Current, voltage, and power versus time for a purely resistive load.

Current, Voltage, and Power for a Resistive Load

First, consider the case in which the network is a pure resistance. Then, $\theta = 0$, and we have

$$v(t) = V_m \cos(\omega t)$$
$$i(t) = I_m \cos(\omega t)$$
$$p(t) = v(t)i(t) = V_m I_m \cos^2(\omega t)$$

Average power is absorbed by resistances in ac circuits.

Plots of these quantities are shown in Figure 5.21. Notice that the current is in phase with the voltage (i.e., they both reach their peak values at the same time). Because $p(t)$ is positive at all times, we conclude that energy flows continually in the direction from the source to the load (where it is converted to heat). Of course, the value of the power rises and falls with the voltage (and current) magnitude.

Current, Voltage, and Power for an Inductive Load

Next, consider the case in which the load is a pure inductance for which $Z = \omega L \angle 90°$. Thus, $\theta = 90°$, and we get

$$v(t) = V_m \cos(\omega t)$$
$$i(t) = I_m \cos(\omega t - 90°) = I_m \sin(\omega t)$$
$$p(t) = v(t)i(t) = V_m I_m \cos(\omega t)\sin(\omega t)$$

Using the trigonometric identity $\cos(x)\sin(x) = (1/2)\sin(2x)$, we find that the expression for the power becomes

$$p(t) = \frac{V_m I_m}{2}\sin(2\omega t)$$

Power surges into and out of inductances in ac circuits. The average power absorbed by inductances is zero.

Plots of the current, voltage, and power are shown in Figure 5.22(a). Notice that the current lags the voltage by 90°. Half of the time the power is positive, showing

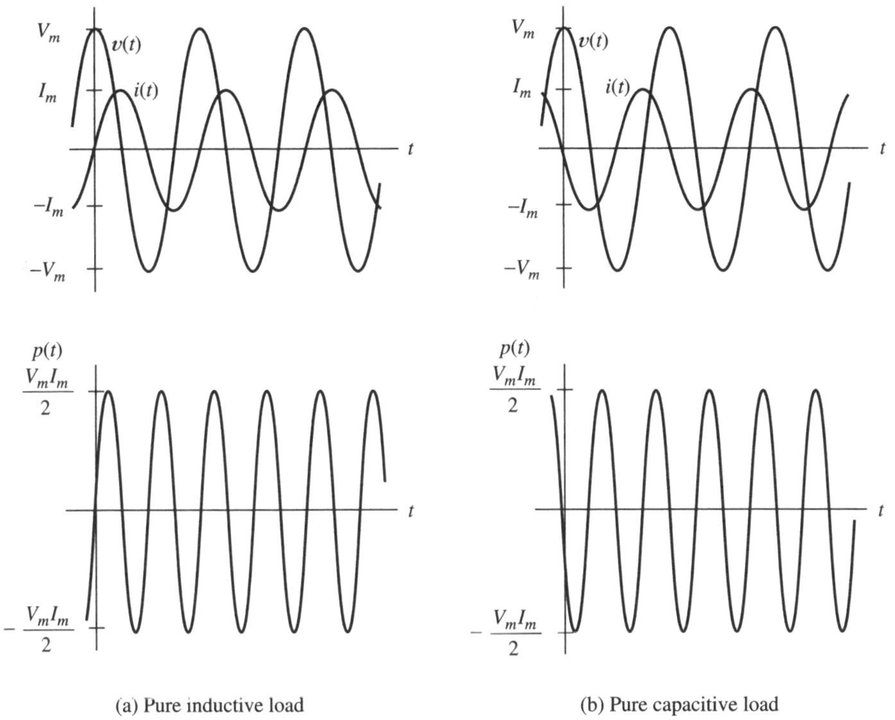

Figure 5.22 Current, voltage, and power versus time for pure energy-storage elements.

that energy is delivered to the inductance, where it is stored in the magnetic field. For the other half of the time, power is negative, showing that the inductance returns energy to the source. Notice that the average power is zero. In this case, we say that **reactive power** flows from the source to the load.

Current, Voltage, and Power for a Capacitive Load

Next, consider the case in which the load is a pure capacitance for which $Z = (1/\omega C)\ \underline{/-90^\circ}$. Then, $\theta = -90^\circ$, and we have

$$\begin{aligned} v(t) &= V_m \cos(\omega t) \\ i(t) &= I_m \cos(\omega t + 90^\circ) = -I_m \sin(\omega t) \\ p(t) &= v(t)i(t) = -V_m I_m \cos(\omega t)\sin(\omega t) \\ &= -\frac{V_m I_m}{2} \sin(2\omega t) \end{aligned}$$

Power surges into and out of capacitances in ac circuits. The average power absorbed by capacitances is zero.

Plots of the current, voltage, and power are shown in Figure 5.22(b). Here again, the average power is zero, and we say that reactive power flows. Notice, however, that the power for the capacitance carries the opposite sign as that for the inductance. Thus, we say that reactive power is positive for an inductance and is negative for a capacitance. If a load contains both inductance and capacitance with reactive powers of equal magnitude, the reactive powers cancel.

Importance of Reactive Power

The power flow back and forth to inductances and capacitances is called reactive power. Reactive power flow is important because it causes power dissipation in the lines and transformers of a power distribution system.

Even though no average power is consumed by a pure energy-storage element (inductance or capacitance), reactive power is still of concern to power-system engineers because transmission lines, transformers, fuses, and other elements must be capable of withstanding the current associated with reactive power. It is possible to have loads composed of energy-storage elements that draw large currents requiring heavy-duty wiring, even though little average power is consumed. Therefore, electric-power companies charge their industrial customers for reactive power (but at a lower rate) as well as for total energy delivered.

Power Calculations for a General Load

Now, let us consider the voltage, current, and power for a general *RLC* load for which the phase θ can be any value from $-90°$ to $+90°$. We have

$$v(t) = V_m \cos(\omega t) \tag{5.54}$$

$$i(t) = I_m \cos(\omega t - \theta) \tag{5.55}$$

$$p(t) = V_m I_m \cos(\omega t)\cos(\omega t - \theta) \tag{5.56}$$

Using the trigonometric identity

$$\cos(\omega t - \theta) = \cos(\theta)\cos(\omega t) + \sin(\theta)\sin(\omega t)$$

we can put Equation 5.56 into the form

$$p(t) = V_m I_m \cos(\theta)\cos^2(\omega t) + V_m I_m \sin(\theta)\cos(\omega t)\sin(\omega t) \tag{5.57}$$

Using the identities

$$\cos^2(\omega t) = \frac{1}{2} + \frac{1}{2}\cos(2\omega t)$$

and

$$\cos(\omega t)\sin(\omega t) = \frac{1}{2}\sin(2\omega t)$$

we find that Equation 5.57 can be written as

$$p(t) = \frac{V_m I_m}{2}\cos(\theta)[1 + \cos(2\omega t)] + \frac{V_m I_m}{2}\sin(\theta)\sin(2\omega t) \tag{5.58}$$

Notice that the terms involving $\cos(2\omega t)$ and $\sin(2\omega t)$ have average values of zero. Thus, the average power P is given by

$$P = \frac{V_m I_m}{2}\cos(\theta) \tag{5.59}$$

Using the fact that $V_{\text{rms}} = V_m/\sqrt{2}$ and $I_{\text{rms}} = I_m/\sqrt{2}$, we can write the expression for average power as

$$P = V_{\text{rms}} I_{\text{rms}} \cos(\theta) \tag{5.60}$$

As usual, the units of power are watts (W).

Power Factor

The term $\cos(\theta)$ is called the **power factor**:

$$\mathrm{PF} = \cos(\theta) \tag{5.61}$$

To simplify our discussion, we assumed a voltage having zero phase. In general, the phase of the voltage may have a value other than zero. Then, θ should be taken as the phase of the voltage θ_v minus the phase of the current θ_i, or

Power factor is the cosine of the angle θ by which the current lags the voltage. (If the current leads the voltage, the angle is negative.)

$$\theta = \theta_v - \theta_i \tag{5.62}$$

Sometimes, θ is called the **power angle**.

Often, power factor is stated as a percentage. Also, it is common to state whether the current leads (capacitive load) or lags (inductive load) the voltage. A typical power factor would be stated to be 90 percent lagging, which means that $\cos(\theta) = 0.9$ and that the current lags the voltage.

Often, power factor is expressed as a percentage.

If the current lags the voltage, the power factor is said to be inductive or lagging. If the current leads the voltage, the power factor is said to be capacitive or leading.

Reactive Power

In ac circuits, energy flows into and out of energy storage elements (inductances and capacitances). For example, when the voltage magnitude across a capacitance is increasing, energy flows into it, and when the voltage magnitude decreases, energy flows out. Similarly, energy flows into an inductance when the current flowing through it increases in magnitude. Although instantaneous power can be very large, the net energy transferred per cycle is zero for either an ideal capacitance or inductance.

When a capacitance and an inductance are in parallel (or series) energy flows into one, while it flows out of the other. Thus, the power flow of a capacitance tends to cancel that of an inductance at each instant in time.

The peak instantaneous power associated with the energy storage elements contained in a general load is called **reactive power** and is given by

$$Q = V_{\mathrm{rms}} I_{\mathrm{rms}} \sin(\theta) \tag{5.63}$$

where θ is the power angle given by Equation 5.62, V_{rms} is the effective (or rms) voltage across the load, and I_{rms} is the effective current through the load. (Notice that if we had a purely resistive load, we would have $\theta = 0$ and $Q = 0$.)

The physical units of reactive power are watts. However, to emphasize the fact that Q does not represent the flow of net energy, its units are usually given as *V*olt *A*mperes *R*eactive (VARs).

The units of reactive power Q are VARs.

Apparent Power

Another quantity of interest is the **apparent power**, which is defined as the product of the effective voltage and the effective current, or

Apparent power equals the product of rms current and rms voltage. The units for apparent power are stated as volt-amperes (VA).

$$\text{apparent power} = V_{\mathrm{rms}} I_{\mathrm{rms}}$$

Its units are volt-amperes (VA).

Using Equations 5.60 and 5.63, we can write

$$P^2 + Q^2 = (V_{\mathrm{rms}} I_{\mathrm{rms}})^2 \cos^2(\theta) + (V_{\mathrm{rms}} I_{\mathrm{rms}})^2 \sin^2(\theta)$$

However, $\cos^2(\theta) + \sin^2(\theta) = 1$, so we have

$$P^2 + Q^2 = (V_{\rm rms}I_{\rm rms})^2 \tag{5.64}$$

Units

Often, the units given for a quantity indicate whether the quantity is power (W), reactive power (VAR), or apparent power (VA). For example, if we say that we have a 5-kW load, this means that $P = 5$ kW. On the other hand, if we have a 5-kVA load, $V_{\rm rms}I_{\rm rms} = 5$ kVA. If we say that a load absorbs 5 kVAR, then $Q = 5$ kVAR.

Power Triangle

The power triangle is a compact way to represent ac power relationships.

The relationships between real power P, reactive power Q, apparent power $V_{\rm rms}I_{\rm rms}$, and the power angle θ can be represented by the **power triangle**. The power triangle is shown in Figure 5.23(a) for an inductive load, in which case θ and Q are positive. The power triangle for a capacitive load is shown in Figure 5.23(b), in which case θ and Q are negative.

Additional Power Relationships

The impedance Z is

$$Z = |Z|\ \angle\theta = R + jX$$

in which R is the resistance of the load and X is the reactance. This is illustrated in Figure 5.24. We can write

$$\cos(\theta) = \frac{R}{|Z|} \tag{5.65}$$

and

$$\sin(\theta) = \frac{X}{|Z|} \tag{5.66}$$

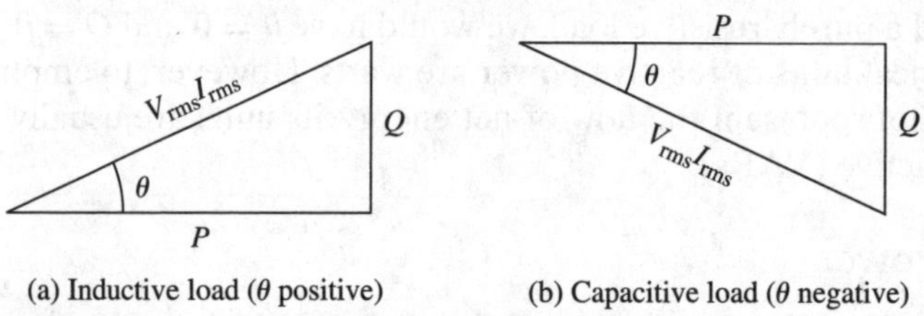

Figure 5.23 Power triangles for inductive and capacitive loads.

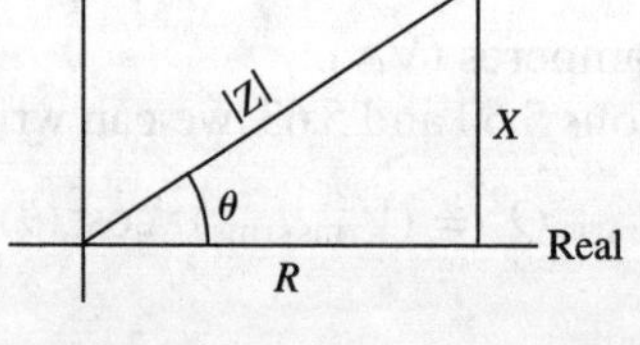

Figure 5.24 The load impedance in the complex plane.

Substituting Equation 5.65 into Equation 5.59, we find that

$$P = \frac{V_m I_m}{2} \times \frac{R}{|Z|} \tag{5.67}$$

However, Equation 5.53 states that $I_m = V_m/|Z|$, so we have

$$P = \frac{I_m^2}{2} R \tag{5.68}$$

Using the fact that $I_{\text{rms}} = I_m/\sqrt{2}$, we get

$$P = I_{\text{rms}}^2 R \tag{5.69}$$

In Equation 5.69, R is the real part of the impedance through which the current flows.

In a similar fashion, we can show that

$$Q = I_{\text{rms}}^2 X \tag{5.70}$$

In Equation 5.70, X is the imaginary part (including the algebraic sign) of the impedance through which the current flows.

In applying Equation 5.70, we retain the algebraic sign of X. For an inductive load, X is positive, whereas for a capacitive load, X is negative. This is not hard to remember if we keep in mind that Q is positive for inductive loads and negative for capacitive loads.

Reactive power Q is positive for inductive loads and negative for capacitive loads.

Furthermore, in Section 5.1, we showed that the average power delivered to a resistance is

$$P = \frac{V_{R\text{rms}}^2}{R} \tag{5.71}$$

In Equation 5.71, $V_{R\text{rms}}$ is the rms voltage across the resistance.

where $V_{R\text{rms}}$ is the rms value of the voltage *across the resistance.* (Notice in Figure 5.20 that the source voltage does not appear across the resistance, because the reactance is in series with the resistance.)

Similarly, we have

$$Q = \frac{V_{X\text{rms}}^2}{X} \tag{5.72}$$

In Equation 5.72, $V_{X\text{rms}}$ is the rms voltage across the reactance.

where $V_{X\text{rms}}$ is the rms value of the voltage *across the reactance*. Here again, X is positive for an inductance and negative for a capacitance.

Complex Power

Consider the portion of a circuit shown in Figure 5.25. The **complex power**, denoted as **S**, delivered to this circuit is defined as one half the product of the phasor voltage **V** and the complex conjugate of the phasor current **I***.

$$\mathbf{S} = \frac{1}{2}\mathbf{V}\mathbf{I}^* \tag{5.73}$$

The phasor voltage is $\mathbf{V} = V_m \angle \theta_v$ in which V_m is the peak value of the voltage and θ_v is the phase angle of the voltage. Furthermore, the phasor current is $\mathbf{I} = I_m \angle \theta_i$

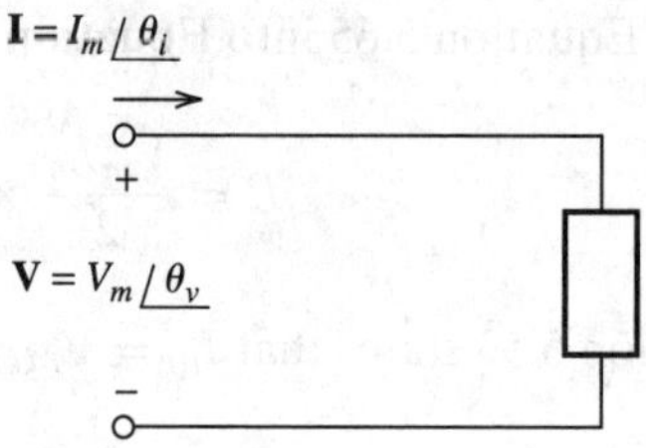

Figure 5.25 The complex power delivered to this circuit element is $\mathbf{S} = \frac{1}{2}\mathbf{VI}^*$.

where I_m is the peak value and θ_i is the phase angle of the current. Substituting into Equation 5.73, we have

$$\mathbf{S} = \frac{1}{2}\mathbf{VI}^* = \frac{1}{2}(V_m \angle \theta_v) \times (I_m \angle -\theta_i) = \frac{V_m I_m}{2} \angle \theta_v - \theta_i = \frac{V_m I_m}{2} \angle \theta \quad (5.74)$$

where, as before, $\theta = \theta_v - \theta_i$ is the power angle. Expanding the right-hand term of Equation 5.74 into real and imaginary parts, we have

$$\mathbf{S} = \frac{V_m I_m}{2}\cos(\theta) + j\frac{V_m I_m}{2}\sin(\theta)$$

However, the first term on the right-hand side is the average power P delivered to the circuit and the second term is j times the reactive power. Thus, we can write:

$$\mathbf{S} = \frac{1}{2}\mathbf{VI}^* = P + jQ \quad (5.75)$$

If we know the complex power $\mathbf{S}$, then we can find the power, reactive power, and apparent power:

$$P = \text{Re}(\mathbf{S}) = \text{Re}\left(\frac{1}{2}\mathbf{VI}^*\right) \quad (5.76)$$

$$Q = \text{Im}(\mathbf{S}) = \text{Im}\left(\frac{1}{2}\mathbf{VI}^*\right) \quad (5.77)$$

$$\text{apparent power} = |\mathbf{S}| = \left|\frac{1}{2}\mathbf{VI}^*\right| \quad (5.78)$$

where Re($\mathbf{S}$) denotes the real part of $\mathbf{S}$ and Im($\mathbf{S}$) denotes the imaginary part of $\mathbf{S}$.

Example 5.7 AC Power Calculations

Compute the power and reactive power taken from the source for the circuit of Example 5.5. Also, compute the power and reactive power delivered to each element in the circuit. For convenience, the circuit and the currents that were computed in Example 5.5 are shown in Figure 5.26.

Solution To find the power and reactive power for the source, we must first find the power angle which is given by Equation 5.62:

$$\theta = \theta_v - \theta_i$$

The angle of the source voltage is $\theta_v = -90°$, and the angle of the current delivered by the source is $\theta_i = -135°$. Therefore, we have

$$\theta = -90° - (-135°) = 45°$$

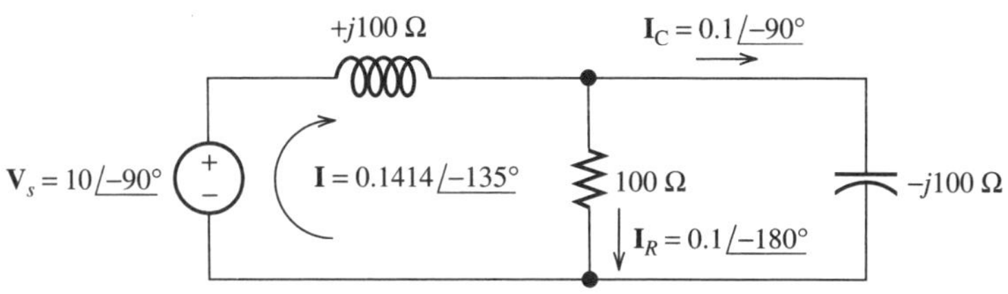

Figure 5.26 Circuit and currents for Example 5.7.

The rms source voltage and current are

$$V_{srms} = \frac{|\mathbf{V}_s|}{\sqrt{2}} = \frac{10}{\sqrt{2}} = 7.071 \text{ V}$$

$$I_{rms} = \frac{|\mathbf{I}|}{\sqrt{2}} = \frac{0.1414}{\sqrt{2}} = 0.1 \text{ A}$$

Now, we use Equations 5.60 and 5.63 to compute the power and reactive power delivered by the source:

$$\begin{aligned} P &= V_{srms} I_{rms} \cos(\theta) \\ &= 7.071 \times 0.1 \cos(45°) = 0.5 \text{ W} \\ Q &= V_{srms} I_{rms} \sin(\theta) \\ &= 7.071 \times 0.1 \sin(45°) = 0.5 \text{ VAR} \end{aligned}$$

An alternative and more compact method for computing P and Q is to first find the complex power and then take the real and imaginary parts:

$$\begin{aligned} \mathbf{S} &= \frac{1}{2}\mathbf{V}_s\mathbf{I}^* = \frac{1}{2}(10 \angle -90°)(0.1414 \angle 135°) = 0.707 \angle 45° = 0.5 + j0.5 \\ P &= \text{Re}(\mathbf{S}) = 0.5 \text{ W} \\ Q &= \text{Im}(\mathbf{S}) = 0.5 \text{ VAR} \end{aligned}$$

We can use Equation 5.70 to compute the reactive power delivered to the inductor, yielding

$$Q_L = I_{rms}^2 X_L = (0.1)^2(100) = 1.0 \text{ VAR}$$

For the capacitor, we have

$$Q_C = I_{Crms}^2 X_C = \left(\frac{0.1}{\sqrt{2}}\right)^2 (-100) = -0.5 \text{ VAR}$$

Notice that we have used the rms value of the current through the capacitor in this calculation. Furthermore, notice that the reactance X_C of the capacitance is negative. As expected, the reactive power is negative for a capacitance. The reactive power for the resistance is zero. As a check, we can verify that the reactive power delivered by the source is equal to the sum of the reactive powers absorbed by the inductance and capacitance. This is demonstrated by

$$Q = Q_L + Q_C$$

The power delivered to the resistance is

$$P_R = I_{R\text{rms}}^2 R = \left(\frac{|\mathbf{I}_R|}{\sqrt{2}}\right)^2 R = \left(\frac{0.1}{\sqrt{2}}\right)^2 100$$
$$= 0.5 \text{ W}$$

The power absorbed by the capacitance and inductance is given by

$$P_L = 0$$
$$P_C = 0$$

Thus, all of the power delivered by the source is absorbed by the resistance. ■

In power distribution systems, we typically encounter much larger values of power, reactive power, and apparent power than the small values of the preceding example. For example, a large power plant may generate 1000 MW. A 100-hp motor used in an industrial application absorbs approximately 85 kW of electrical power under full load.

A typical residence absorbs a *peak* power in the range of 10 to 40 kW. The *average* power for my home (which is of average size, has two residents, and does not use electrical heating) is approximately 600 W. It is interesting to keep your average power consumption and the power used by various appliances in mind because it gives you a clear picture of the economic and environmental impact of turning off lights, computers, and so on, that are not being used.

Example 5.8 Using Power Triangles

Consider the situation shown in Figure 5.27. Here, a voltage source delivers power to two loads connected in parallel. Find the power, reactive power, and power factor for the source. Also, find the phasor current **I**.

Solution By the units given in the figure, we see that load A has an *apparent power* of 10 kVA. On the other hand, the *power* for load B is specified as 5 kW.

Furthermore, load A has a power factor of 0.5 leading, which means that the current leads the voltage in load A. Another way to say this is that load A is capacitive. Similarly, load B has a power factor of 0.7 lagging (or inductive).

Our approach is to find the power and reactive power for each load. Then, we add these values to find the power and reactive power for the source. Finally, we compute the power factor for the source and then find the current.

Calculations for load A

Because load A has a leading (capacitive) power factor, we know that the reactive power Q_A and power angle θ_A are negative. The power triangle for load A is shown

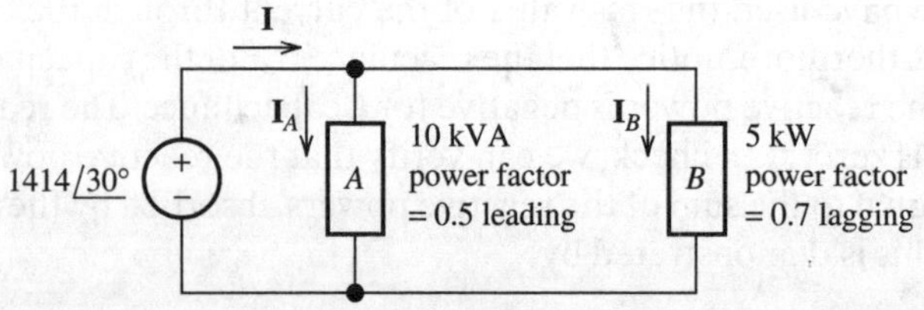

Figure 5.27 Circuit for Example 5.8.

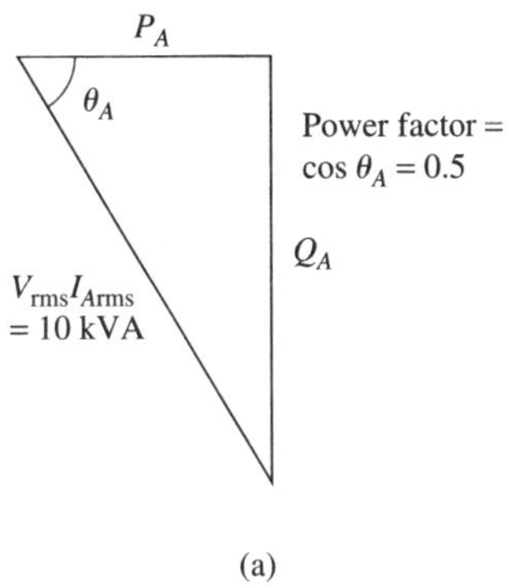

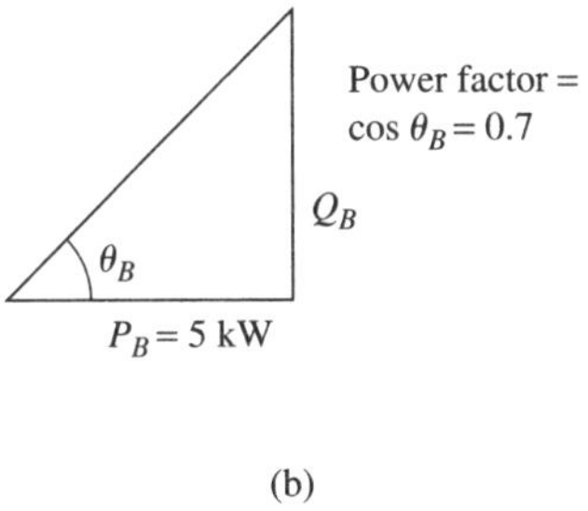

Figure 5.28 Power triangles for loads *A* and *B* of Example 5.8.

in Figure 5.28(a). The power factor is

$$\cos(\theta_A) = 0.5$$

The power is

$$P_A = V_{rms}I_{Arms}\cos(\theta_A) = 10^4(0.5) = 5\text{ kW}$$

Solving Equation 5.64 for reactive power, we have

$$\begin{aligned} Q_A &= \sqrt{(V_{rms}I_{Arms})^2 - P_A^2} \\ &= \sqrt{(10^4)^2 - (5000)^2} \\ &= -8.660\text{ kVAR} \end{aligned}$$

Notice that we have selected the negative value for Q_A, because we know that reactive power is negative for a capacitive (leading) load.

Calculations for load *B*

The power triangle for load *B* is shown in Figure 5.28(b). Since load *B* has a lagging (inductive) power factor, we know that the reactive power Q_B and power angle θ_B are positive. Thus,

$$\theta_B = \arccos(0.7) = 45.57°$$

Applying trigonometry, we can write

$$Q_B = P_B\tan(\theta_B) = 5000\tan(45.57°)$$

$$Q_B = 5.101\text{ kVAR}$$

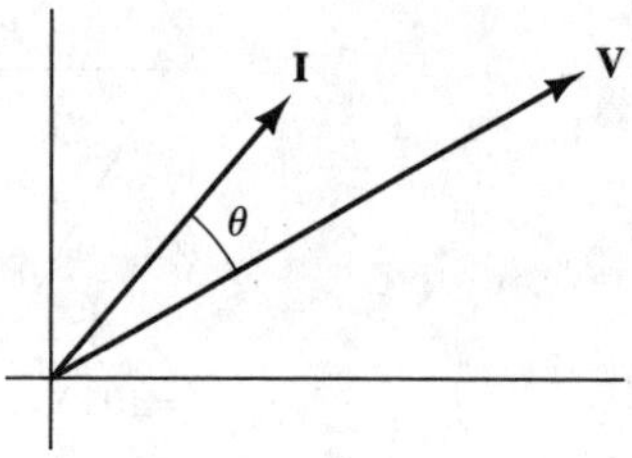

Figure 5.29 Phasor diagram for Example 5.8.

Total power is obtained by adding the powers for the various loads. Similarly, the reactive powers are added.

At this point, as shown here we can find the power and reactive power delivered by the source:

$$P = P_A + P_B = 5 + 5 = 10 \text{ kW}$$

$$Q = Q_A + Q_B = -8.660 + 5.101 = -3.559 \text{ kVAR}$$

Power calculations for the source.

Because Q is negative, we know that the power angle is negative. Thus, we have

$$\theta = \arctan\left(\frac{Q}{P}\right) = \arctan\left(\frac{-3.559}{10}\right) = -19.59°$$

The power factor is

$$\cos(\theta) = 0.9421$$

Power-system engineers frequently express power factors as percentages and would state this power factor as 94.21 percent leading.

The complex power delivered by the source is

$$\mathbf{S} = P + jQ = 10 - j3.559 = 10.61 \angle -19.59° \text{ kVA}$$

Thus, we have

$$\mathbf{S} = \frac{1}{2}\mathbf{V}_s\mathbf{I}^* = \frac{1}{2}(1414 \angle 30°)\mathbf{I}^* = 10.61 \times 10^3 \angle -19.59° \text{ kVA}$$

Solving for the phasor current, we obtain:

$$\mathbf{I} = 15.0 \angle 49.59° \text{ A}$$

The phasor diagram for the current and voltage is shown in Figure 5.29. Notice that the current is leading the voltage. ■

Power-Factor Correction

Power-factor correction can provide a significant economic advantage for consumers of large amounts of electrical energy.

We have seen that large currents can flow in energy-storage devices (inductance and capacitance) without average power being delivered. In heavy industry, many loads are partly inductive, and large amounts of reactive power flow. This reactive power causes higher currents in the power distribution system. Consequently, the lines and transformers must have higher ratings than would be necessary to deliver the same average power to a resistive (100 percent power factor) load.

Energy rates charged to industry depend on the power factor, with higher charges for energy delivered at lower power factors. (Power factor is not taken into account for residential customers.) Therefore, it is advantageous to choose loads that operate at near unity power factor. A common approach is to place capacitors in parallel with an inductive load to increase the power factor.

Example 5.9 **Power-Factor Correction**

A 50-kW load operates from a 60-Hz 10-kV-rms line with a power factor of 60 percent lagging. Compute the capacitance that must be placed in parallel with the load to achieve a 90 percent lagging power factor.

Solution First, we find the load power angle:

$$\theta_L = \arccos(0.6) = 53.13°$$

Then, we use the power-triangle concept to find the reactive power of the load. Hence,

$$Q_L = P_L \tan(\theta_L) = 66.67 \text{ kVAR}$$

After adding the capacitor, the power will still be 50 kW and the power angle will become

$$\theta_{\text{new}} = \arccos(0.9) = 25.84°$$

The new value of the reactive power will be

$$Q_{\text{new}} = P_L \tan(\theta_{\text{new}}) = 24.22 \text{ kVAR}$$

Thus, the reactive power of the capacitance must be

$$Q_C = Q_{\text{new}} - Q_L = -42.45 \text{ kVAR}$$

Now, we find that the reactance of the capacitor is

$$X_C = -\frac{V_{\text{rms}}^2}{Q_C} = \frac{(10^4)^2}{42{,}450} = -2356\ \Omega$$

Finally, the angular frequency is

$$\omega = 2\pi 60 = 377.0$$

and the required capacitance is

$$C = \frac{1}{\omega |X_C|} = \frac{1}{377 \times 2356} = 1.126\ \mu\text{F}$$

■

Exercise 5.12 **a.** A voltage source $\mathbf{V} = 707.1\angle 40°$ delivers 5 kW to a load with a power factor of 100 percent. Find the reactive power and the phasor current. **b.** Repeat if the power factor is 20 percent lagging. **c.** For which power factor would the current ratings of the conductors connecting the source to the load be higher? In which case could the wiring be a lower cost?

Answer **a.** $Q = 0, \mathbf{I} = 14.14\ \angle 40°$; **b.** $Q = 24.49$ kVAR, $\mathbf{I} = 70.7\ \angle -38.46°$; **c.** The current ratings for the conductors would need to be five times higher for part (b) than for part (a). Clearly, the wiring could be a lower cost for 100 percent power factor. □

Exercise 5.13 A 1-kV-rms 60-Hz voltage source delivers power to two loads in parallel. The first load is a 10-μF capacitor, and the second load absorbs an apparent power of 10 kVA with an 80 percent lagging power factor. Find the total power, the total reactive power, the power factor for the source, and the rms source current.

Answer $P = 8$ kW, $Q = 2.23$ kVAR, PF $= 96.33$ percent lagging, $I_{\text{rms}} = 8.305$ A. □

5.6 THÉVENIN AND NORTON EQUIVALENT CIRCUITS

Thévenin Equivalent Circuits

In Chapter 2, we saw that a two-terminal network composed of sources and resistances has a Thévenin equivalent circuit consisting of a voltage source in series with a resistance. We can apply this concept to circuits composed of sinusoidal sources (all having a common frequency), resistances, inductances, and capacitances. Here, the Thévenin equivalent consists of a phasor voltage source in series with a complex impedance as shown in Figure 5.30. Recall that phasors and complex impedances apply only for steady-state operation; therefore, these Thévenin equivalents are valid for only steady-state operation of the circuit.

The Thévenin voltage is equal to the open-circuit phasor voltage of the original circuit.

As in resistive circuits, the Thévenin voltage is equal to the open-circuit voltage of the two-terminal circuit. In ac circuits, we use phasors, so we can write

$$\mathbf{V}_t = \mathbf{V}_{\text{oc}} \tag{5.79}$$

We can find the Thévenin impedance by zeroing the independent sources and determining the impedance looking into the circuit terminals.

The Thévenin impedance Z_t can be found by zeroing the *independent* sources and looking back into the terminals to find the equivalent impedance. (Recall that in zeroing a voltage source, we reduce its voltage to zero, and it becomes a short circuit. On the other hand, in zeroing a current source, we reduce its current to zero, and it becomes an open circuit.) Also, keep in mind that we must not zero the *dependent* sources.

The Thévenin impedance equals the open-circuit voltage divided by the short-circuit current.

Another approach to determining the Thévenin impedance is first to find the short-circuit phasor current $\mathbf{I}_{\text{sc}}$ and the open-circuit voltage $\mathbf{V}_{\text{oc}}$. Then, the Thévenin impedance is given by

$$Z_t = \frac{\mathbf{V}_{\text{oc}}}{\mathbf{I}_{\text{sc}}} = \frac{\mathbf{V}_t}{\mathbf{I}_{\text{sc}}} \tag{5.80}$$

Thus, except for the use of phasors and complex impedances, the concepts and procedures for Thévenin equivalents of steady-state ac circuits are the same as for resistive circuits.

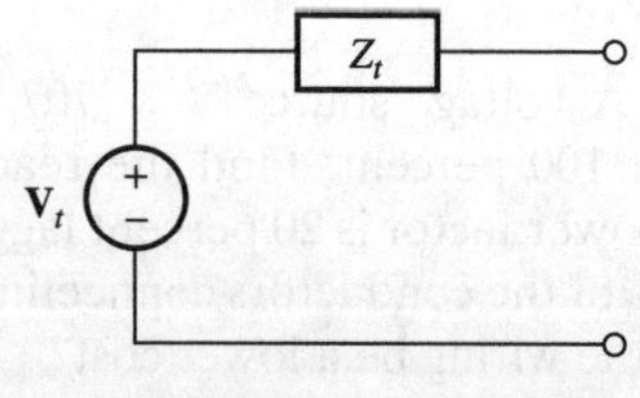

Figure 5.30 The Thévenin equivalent for an ac circuit consists of a phasor voltage source $\mathbf{V}_t$ in series with a complex impedance Z_t.

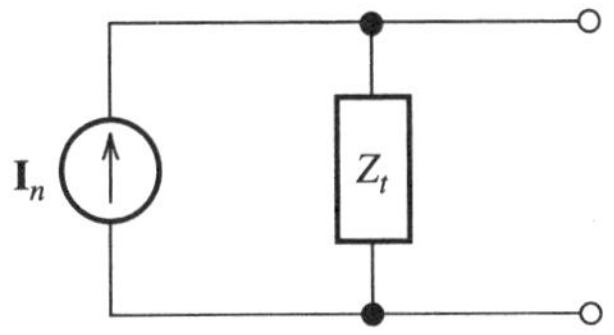

Figure 5.31 The Norton equivalent circuit consists of a phasor current source $\mathbf{I}_n$ in parallel with the complex impedance Z_t.

Norton Equivalent Circuits

Another equivalent for a two-terminal steady-state ac circuit is the Norton equivalent, which consists of a phasor current source $\mathbf{I}_n$ in parallel with the Thévenin impedance. This is shown in Figure 5.31. The Norton current is equal to the short-circuit current of the original circuit:

$$\mathbf{I}_n = \mathbf{I}_{sc} \tag{5.81}$$

Example 5.10 Thévenin and Norton Equivalents

Find the Thévenin and Norton equivalent circuits for the circuit shown in Figure 5.32(a).

Solution We must find two of the three quantities: $\mathbf{V}_{oc}$, $\mathbf{I}_{sc}$, or Z_t. Often, it pays to look for the two that can be found with the least amount of work. In this case, we elect to start by zeroing the sources to find Z_t. After that part of the problem is finished, we will find the short-circuit current.

First, look to see which two of the three quantities $\mathbf{V}_{oc}$, $\mathbf{I}_{sc}$, or Z_t are easiest to determine.

If we zero the sources, we obtain the circuit shown in Figure 5.32(b). The Thévenin impedance is the impedance seen looking back into terminals *a*–*b*. This is the parallel

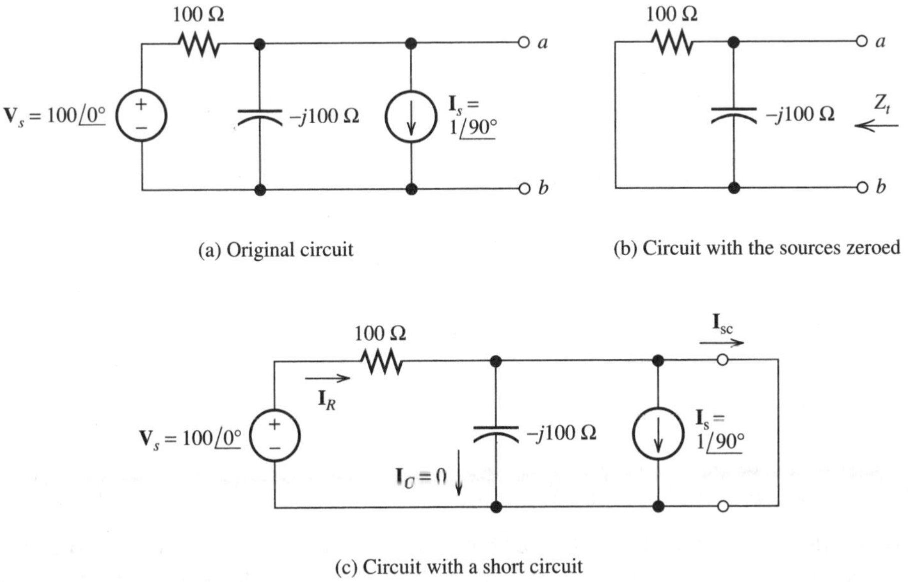

Figure 5.32 Circuit of Example 5.10.

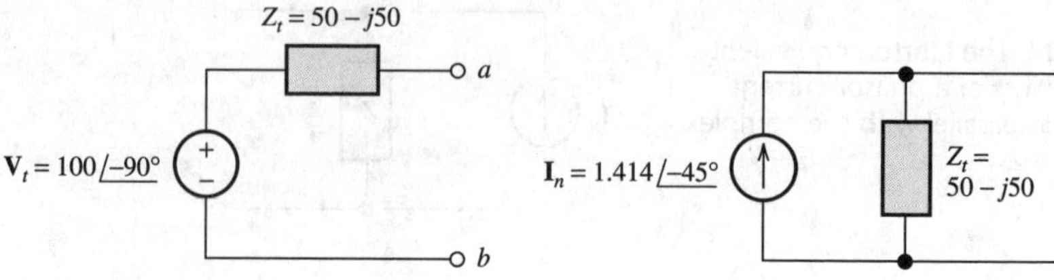

(a) Thévenin equivalent (b) Norton equivalent

Figure 5.33 Thévenin and Norton equivalents for the circuit of Figure 5.32(a).

combination of the resistance and the impedance of the capacitance. Thus, we have

$$\begin{aligned} Z_t &= \frac{1}{1/100 + 1/(-j100)} \\ &= \frac{1}{0.01 + j0.01} \\ &= \frac{1}{0.01414 \angle 45^\circ} \\ &= 70.71 \angle -45^\circ \\ &= 50 - j50\,\Omega \end{aligned}$$

Now, we apply a short circuit to terminals a–b and find the current, which is shown in Figure 5.32(c). With a short circuit, the voltage across the capacitance is zero. Therefore, $\mathbf{I}_C = 0$. Furthermore, the source voltage $\mathbf{V}_s$ appears across the resistance, so we have

$$\mathbf{I}_R = \frac{\mathbf{V}_s}{100} = \frac{100}{100} = 1 \angle 0^\circ \text{ A}$$

Then applying KCL, we can write

$$\mathbf{I}_{sc} = \mathbf{I}_R - \mathbf{I}_s = 1 - 1 \angle 90^\circ = 1 - j = 1.414 \angle -45^\circ \text{ A}$$

Next, we can solve Equation 5.80 for the Thévenin voltage:

$$\mathbf{V}_t = \mathbf{I}_{sc} Z_t = 1.414 \angle -45^\circ \times 70.71 \angle -45^\circ = 100 \angle -90^\circ \text{ V}$$

Finally, we can draw the Thévenin and Norton equivalent circuits, which are shown in Figure 5.33. ■

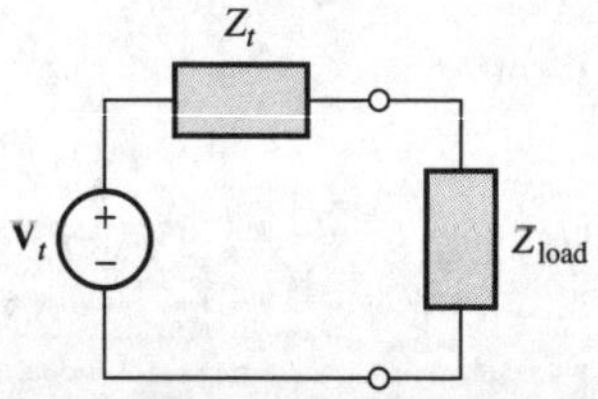

Figure 5.34 The Thévenin equivalent of a two-terminal circuit delivering power to a load impedance.

Maximum Average Power Transfer

Sometimes we are faced with the problem of adjusting a load impedance to extract the maximum average power from a two-terminal circuit. This situation is shown in Figure 5.34, in which we have represented the two-terminal circuit by its Thévenin equivalent. Of course, the power delivered to the load depends on the load impedance. A short-circuit load receives no power because the voltage across it is zero. Similarly, an open-circuit load receives no power because the current through it is zero. Furthermore, a pure reactive load (inductance or capacitance) receives no power because the load power factor is zero.

Two situations are of interest. First, suppose that the load impedance can take any complex value. Then, it turns out that the load impedance for maximum-power transfer is the complex conjugate of the Thévenin impedance:

If the load can take on any complex value, maximum-power transfer is attained for a load impedance equal to the complex conjugate of the Thévenin impedance.

$$Z_{\text{load}} = Z_t^*$$

Let us consider why this is true. Suppose that the Thévenin impedance is

$$Z_t = R_t + jX_t$$

Then, the load impedance for maximum-power transfer is

$$Z_{\text{load}} = Z_t^* = R_t - jX_t$$

Of course, the total impedance seen by the Thévenin source is the sum of the Thévenin impedance and the load impedance:

$$\begin{aligned} Z_{\text{total}} &= Z_t + Z_{\text{load}} \\ &= R_t + jX_t + R_t - jX_t \\ &= 2R_t \end{aligned}$$

Thus, the reactance of the load cancels the internal reactance of the two-terminal circuit. Maximum power is transferred to a given load resistance by maximizing the current. For given resistances, maximum current is achieved by choosing the reactance to minimize the total impedance magnitude. Of course, for fixed resistances, the minimum impedance magnitude occurs for zero total reactance.

Having established the fact that the total reactance should be zero, we have a resistive circuit. We considered this resistive circuit in Chapter 2, where we showed that maximum power is transferred for $R_{\text{load}} = R_t$.

If the load is required to be a pure resistance, maximum-power transfer is attained for a load resistance equal to the magnitude of the Thévenin impedance.

The second case of interest is a load that is constrained to be a pure resistance. In this case, it can be shown that the load resistance for maximum-power transfer is equal to the magnitude of the Thévenin impedance:

$$Z_{\text{load}} = R_{\text{load}} = |Z_t|$$

Example 5.11 Maximum Power Transfer

Determine the maximum power that can be delivered to a load by the two-terminal circuit of Figure 5.32(a) if **a.** the load can have any complex value and **b.** the load must be a pure resistance.

Solution In Example 5.10, we found that the circuit has the Thévenin equivalent shown in Figure 5.33(a). The Thévenin impedance is

$$Z_t = 50 - j50\ \Omega$$

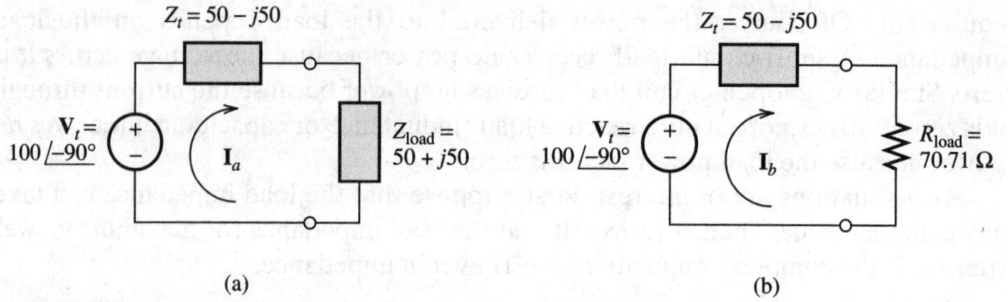

Figure 5.35 Thévenin equivalent circuit and loads of Example 5.11.

a. The complex load impedance that maximizes power transfer is

$$Z_{\text{load}} = Z_t^* = 50 + j50$$

The Thévenin equivalent with this load attached is shown in Figure 5.35(a). The current is

$$\mathbf{I}_a = \frac{\mathbf{V}_t}{Z_t + Z_{\text{load}}}$$

$$= \frac{100 \angle -90^\circ}{50 - j50 + 50 + j50}$$

$$= 1 \angle -90^\circ \text{ A}$$

The rms load current is $I_{arms} = 1/\sqrt{2}$. Finally, the power delivered to the load is

$$P = I_{arms}^2 R_{\text{load}} = \left(\frac{1}{\sqrt{2}}\right)^2 (50) = 25 \text{ W}$$

b. The purely resistive load for maximum power transfer is

$$R_{\text{load}} = |Z_t|$$

$$= |50 - j50|$$

$$= \sqrt{50^2 + (-50)^2}$$

$$= 70.71\ \Omega$$

The Thévenin equivalent with this load attached is shown in Figure 5.35(b). The current is

$$\mathbf{I}_b = \frac{\mathbf{V}_t}{Z_t + Z_{\text{load}}}$$

$$= \frac{100 \angle -90^\circ}{50 - j50 + 70.71}$$

$$= \frac{100 \angle -90^\circ}{130.66 \angle -22.50^\circ}$$

$$= 0.7654 \angle -67.50^\circ \text{ A}$$

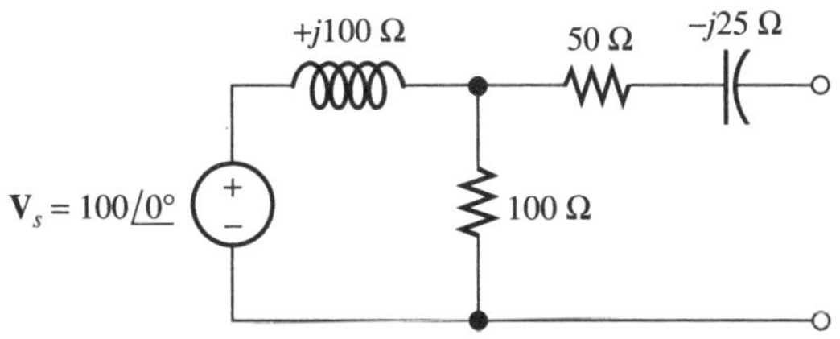

Figure 5.36 Circuit of Exercises 5.14 and 5.15.

The power delivered to this load is

$$\begin{aligned} P &= I_{brms}^2 R_{load} \\ &= \left(\frac{0.7653}{\sqrt{2}}\right)^2 70.71 \\ &= 20.71 \text{ W} \end{aligned}$$

Notice that the power available to a purely resistive load is less than that for a complex load. ■

Exercise 5.14 Find the Thévenin impedance, the Thévenin voltage, and the Norton current for the circuit shown in Figure 5.36.
Answer $Z_t = 100 + j25\ \Omega$, $\mathbf{V}_t = 70.71\ \angle -45°$, $\mathbf{I}_n = 0.686\ \angle -59.0°$. □

Exercise 5.15 Determine the maximum power that can be delivered to a load by the two-terminal circuit of Figure 5.36 if **a.** the load can have any complex value and **b.** the load must be a pure resistance.
Answer **a.** 6.25 W; **b.** 6.16 W. □

5.7 BALANCED THREE-PHASE CIRCUITS

Much of the power used by business and industry is supplied by three-phase distribution systems. Plant engineers need to be familiar with three-phase power.

We will see that there are important advantages in generating and distributing power with multiple ac voltages having different phases. We consider the most common case: three equal-amplitude ac voltages having phases that are 120° apart. This is known as a **balanced three-phase source**, an example of which is illustrated in Figure 5.37. [Recall that in double-subscript notation for voltages the first subscript is the positive reference. Thus, $v_{an}(t)$ is the voltage between nodes a and n with the positive reference at node a.] In Chapter 17, we will learn how three-phase voltages are generated.

The source shown in Figure 5.37(a) is said to be **wye connected (Y connected)**. Later in this chapter, we consider another configuration, known as the delta (Δ) connection.

The three voltages shown in Figure 5.37(b) are given by

$$v_{an}(t) = V_Y \cos(\omega t) \tag{5.82}$$

$$v_{bn}(t) = V_Y \cos(\omega t - 120°) \tag{5.83}$$

$$v_{cn}(t) = V_Y \cos(\omega t + 120°) \tag{5.84}$$

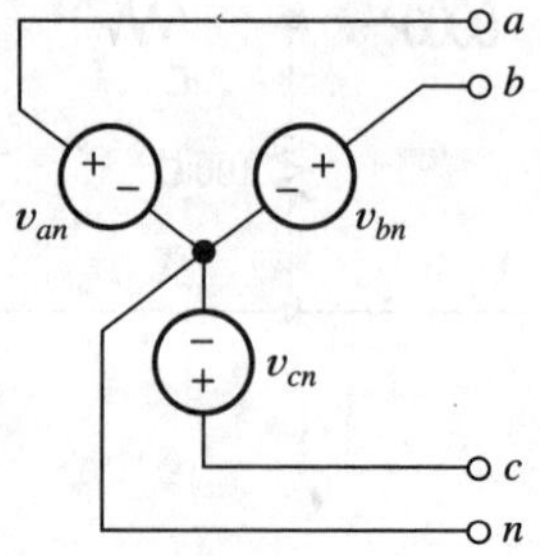

(a) Three-phase source

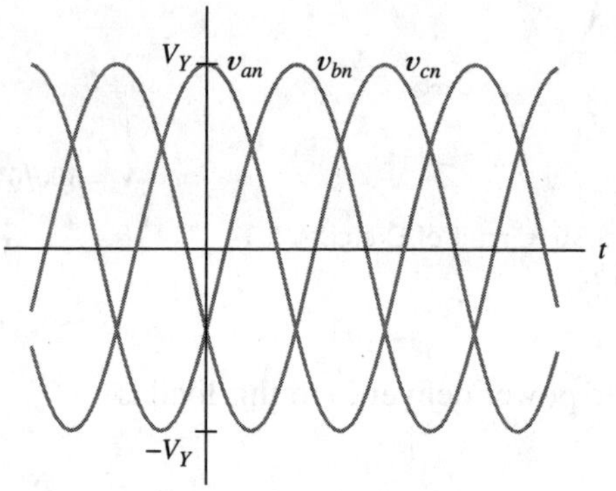

(b) Voltages versus time

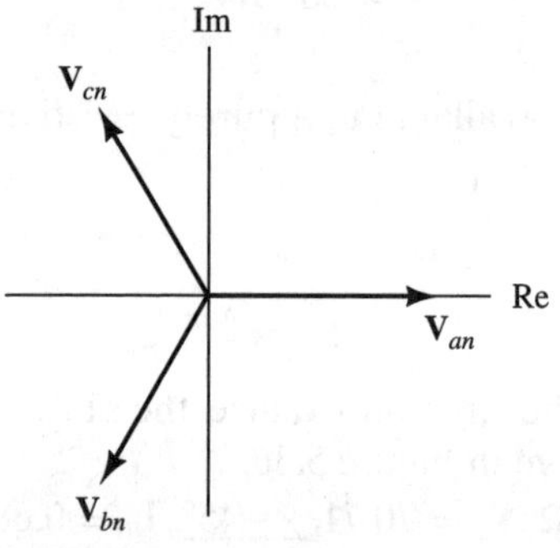

(c) Phasor diagram

Figure 5.37 A balanced three-phase voltage source.

where V_Y is the magnitude of each source in the wye-connected configuration. The corresponding phasors are

$$\mathbf{V}_{an} = V_Y \angle 0^\circ \tag{5.85}$$

$$\mathbf{V}_{bn} = V_Y \angle -120^\circ \tag{5.86}$$

$$\mathbf{V}_{cn} = V_Y \angle 120^\circ \tag{5.87}$$

The phasor diagram is shown in Figure 5.37(c).

Phase Sequence

Three-phase sources can have either a positive or negative phase sequence.

This set of voltages is said to have a **positive phase sequence** because the voltages reach their peak values in the order *abc*. Refer to Figure 5.37(c) and notice that v_{an} leads v_{bn}, which in turn leads v_{cn}. (Recall that we think of the phasors as rotating counterclockwise in determining phase relationships.) If we interchanged *b* and *c*, we would have a **negative phase sequence**, in which the order is *acb*.

We will see later in the book that the direction of rotation of certain three-phase motors can be reversed by changing the phase sequence.

Phase sequence can be important. For example, if we have a three-phase induction motor, the direction of rotation is opposite for the two phase sequences. To reverse the direction of rotation of such a motor, we would interchange the *b* and *c* connections. (You may find this piece of information useful if you ever work with three-phase motors, which are very common in industry.) Because circuit analysis is very similar for both phase sequences, we consider only the positive phase sequence in most of the discussion that follows.

Wye–Wye Connection

Consider the three-phase source connected to a balanced three-phase load shown in Figure 5.38. The wires *a–A*, *b–B*, and *c–C* are called **lines**, and the wire *n–N* is called the **neutral**. This configuration is called a wye–wye (Y–Y) connection with neutral. By the term *balanced load*, we mean that the three load impedances are equal. (In this book, we consider only balanced loads.)

Three-phase sources and loads can be connected either in a wye configuration or in a delta configuration.

Later, we will see that other configurations are useful. For example, the neutral wire *n–N* can be omitted. Furthermore, the source and load can be connected in the form of a delta. We will see that currents, voltages, and power can be computed for these other configurations by finding an equivalent wye–wye circuit. Thus, the key to understanding three-phase circuits is a careful examination of the wye–wye circuit.

The key to understanding the various three-phase configurations is a careful examination of the wye–wye circuit.

Often, we use the term *phase* to refer to part of the source or the load. Thus, phase A of the source is $v_{an}(t)$, and phase A of the load is the impedance connected between A and N. We refer to V_Y as the **phase voltage** or as the **line-to-neutral voltage** of the wye-connected source. (Power-systems engineers usually specify rms values rather than peak magnitudes. Unless stated otherwise, we use phasors having magnitudes equal to the peak values rather than the rms values.) Furthermore, $\mathbf{I}_{aA}$, $\mathbf{I}_{bB}$, and $\mathbf{I}_{cC}$ are called **line currents**. (Recall that in the double-subscript notation for currents, the reference direction is from the first subscript to the second. Thus, $\mathbf{I}_{aA}$ is the current referenced from node a to node A, as illustrated in Figure 5.38.)

In Chapters 5 and 6, we take the magnitude of a phasor to be the peak value. Power-systems engineers often use the rms value as the magnitude for phasors, which we do in Chapters 15, 16, and 17. We will label rms phasors as rms.

The current in phase A of the load is given by

$$\mathbf{I}_{aA} = \frac{\mathbf{V}_{an}}{Z\angle\theta} = \frac{V_Y\angle 0^\circ}{Z\angle\theta} = I_L\angle -\theta$$

where $I_L = V_Y/Z$ is the magnitude of the line current. Because the load impedances are equal, all of the line currents are the same, except for phase. Thus, the currents are given by

$$i_{aA}(t) = I_L\cos(\omega t - \theta) \tag{5.88}$$

$$i_{bB}(t) = I_L\cos(\omega t - 120^\circ - \theta) \tag{5.89}$$

$$i_{cC}(t) = I_L\cos(\omega t + 120^\circ - \theta) \tag{5.90}$$

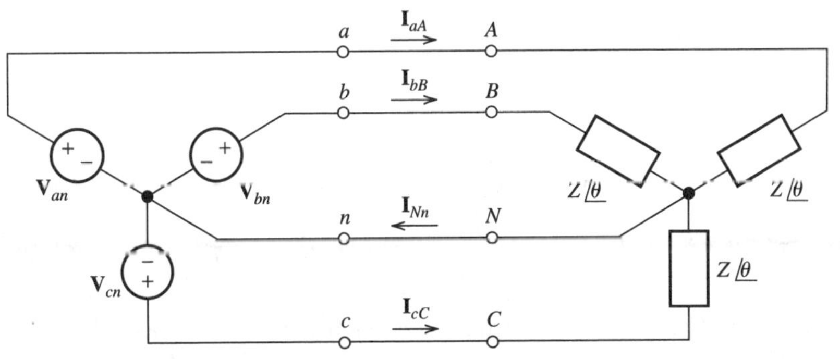

Figure 5.38 A three-phase wye–wye connection with neutral.

The neutral current in Figure 5.38 is given by

$$i_{Nn}(t) = i_{aA}(t) + i_{bB}(t) + i_{cC}(t)$$

In terms of phasors, this is

$$\mathbf{I}_{Nn} = \mathbf{I}_{aA} + \mathbf{I}_{bB} + \mathbf{I}_{cC}$$
$$= I_L \angle -\theta + I_L \angle -120^\circ - \theta + I_L \angle 120^\circ - \theta$$
$$= I_L \angle -\theta \times (1 + 1 \angle -120^\circ + 1 \angle 120^\circ)$$
$$= I_L \angle -\theta \times (1 - 0.5 - j0.866 - 0.5 + j0.866)$$
$$= 0$$

The sum of three equal magnitude phasors 120° apart in phase is zero.

The neutral current is zero in a balanced wye–wye system. Thus in theory, the neutral wire can be inserted or removed without affecting load currents or voltages. This is *not* true if the load is unbalanced, which is often the case in real power distribution systems.

Thus, the sum of three phasors with equal magnitudes and 120° apart in phase is zero. (We make use of this fact again later in this section.)

We have shown that the neutral current is zero in a balanced three-phase system. Consequently, the neutral wire can be eliminated without changing any of the voltages or currents. Then, the three source voltages are delivered to the three load impedances with three wires.

An important advantage of three-phase systems compared with single phase is that the wiring for connecting the sources to the loads is less expensive. As shown in Figure 5.39, it would take six wires to connect three single-phase sources to three loads separately, whereas only three wires (four if the neutral wire is used) are needed for the three-phase connection to achieve the same power transfer.

Power

Another advantage of balanced three-phase systems, compared with single-phase systems, is that the total power is constant (as a function of time) rather than pulsating. (Refer to Figure 5.2 on page 232 to see that power pulsates in the single-phase case.) To show that the power is constant for the balanced wye–wye connection shown in Figure 5.38, we write an expression for the total power. The power delivered to phase A of the load is $v_{an}(t)i_{aA}(t)$. Similarly, the power for each of the other phases of the load is the product of the voltage and the current. Thus, the total power is

$$p(t) = v_{an}(t)i_{aA}(t) + v_{bn}(t)i_{bB}(t) + v_{cn}(t)i_{cC}(t) \tag{5.91}$$

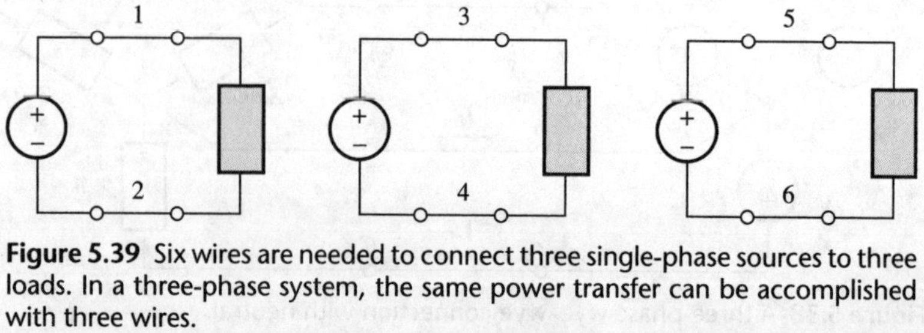

Figure 5.39 Six wires are needed to connect three single-phase sources to three loads. In a three-phase system, the same power transfer can be accomplished with three wires.

Using Equations 5.82, 5.83, and 5.84 to substitute for the voltages and Equations 5.88, 5.89, and 5.90 to substitute for the currents, we obtain

$$\begin{aligned} p(t) &= V_Y \cos(\omega t) I_L \cos(\omega t - \theta) \\ &+ V_Y \cos(\omega t - 120°) I_L \cos(\omega t - \theta - 120°) \\ &+ V_Y \cos(\omega t + 120°) I_L \cos(\omega t - \theta + 120°) \end{aligned} \quad (5.92)$$

Using the trigonometric identity

$$\cos(x)\cos(y) = \frac{1}{2}\cos(x - y) + \frac{1}{2}\cos(x + y)$$

we find that Equation 5.92 can be written as

$$\begin{aligned} p(t) = 3\frac{V_Y I_L}{2}\cos(\theta) + \frac{V_Y I_L}{2}[&\cos(2\omega t - \theta) \\ &+ \cos(2\omega t - \theta - 240°) + \cos(2\omega t - \theta + 480°)] \end{aligned} \quad (5.93)$$

However, the term in brackets is

$$\begin{aligned} &\cos(2\omega t - \theta) + \cos(2\omega t - \theta - 240°) + \cos(2\omega t - \theta + 480°) \\ &= \cos(2\omega t - \theta) + \cos(2\omega t - \theta + 120°) + \cos(2\omega t - \theta - 120°) \\ &= 0 \end{aligned}$$

(Here, we have used the fact, established earlier, that the sum is zero for three sine waves of equal amplitude and 120° apart in phase.) Thus, the expression for power becomes

$$p(t) = 3\frac{V_Y I_L}{2}\cos(\theta) \quad (5.94)$$

Notice that the total power is constant with respect to time. A consequence of this fact is that the torque required to drive a three-phase generator connected to a balanced load is constant, and vibration is lessened. Similarly, the torque produced by a three-phase motor is constant rather than pulsating as it is for a single-phase motor.

In balanced three-phase systems, total power flow is constant with respect to time.

The rms voltage from each line to neutral is

$$V_{Y\text{rms}} = \frac{V_Y}{\sqrt{2}} \quad (5.95)$$

Similarly, the rms value of the line current is

$$I_{L\text{rms}} = \frac{I_L}{\sqrt{2}} \quad (5.96)$$

Using Equations 5.95 and 5.96 to substitute into Equation 5.94, we find that

$$P_{\text{avg}} = p(t) = 3V_{Y\text{rms}}\, I_{L\text{rms}} \cos(\theta) \quad (5.97)$$

In Equations 5.97 and 5.98, $V_{Y\text{rms}}$ is the rms line-to-neutral voltage, $I_{L\text{rms}}$ is the rms line current, and θ is the angle of the load impedances.

Reactive Power

As in single-phase circuits, power flows back and forth between the sources and energy-storage elements contained in a three-phase load. This power is called *reactive power*. The higher currents that result because of the presence of reactive power require wiring and other power-distribution components having higher ratings. The reactive power delivered to a balanced three-phase load is given by

$$Q = 3\frac{V_Y I_L}{2}\sin(\theta) = 3V_{Y\text{rms}}I_{L\text{rms}}\sin(\theta) \tag{5.98}$$

Line-to-Line Voltages

As we have mentioned earlier, the voltages between terminals $a, b,$ or c and the neutral point n are called **line-to-neutral voltages**. On the other hand, voltages between a and b, b and c, or a and c are called **line-to-line voltages** or, more simply, **line voltages**. Thus $\mathbf{V}_{an}$, $\mathbf{V}_{bn}$, and $\mathbf{V}_{cn}$ are line-to-neutral voltages, whereas $\mathbf{V}_{ab}$, $\mathbf{V}_{bc}$, and $\mathbf{V}_{ca}$ are line-to-line voltages. (For consistency, we choose the subscripts cyclically in the order *abcabc*.) Let us consider the relationships between line-to-line voltages and line-to-neutral voltages.

We can obtain the following relationship by applying KVL to Figure 5.38:

$$\mathbf{V}_{ab} = \mathbf{V}_{an} - \mathbf{V}_{bn}$$

Using Equations 5.85 and 5.86 to substitute for $\mathbf{V}_{an}$ and $\mathbf{V}_{bn}$, we obtain

$$\mathbf{V}_{ab} = V_Y \angle 0^\circ - V_Y \angle -120^\circ \tag{5.99}$$

which is equivalent to

$$\mathbf{V}_{ab} = V_Y \angle 0^\circ + V_Y \angle 60^\circ \tag{5.100}$$

This relationship is illustrated in Figure 5.40. It can be shown that Equation 5.100 reduces to

$$\mathbf{V}_{ab} = \sqrt{3}V_Y \angle 30^\circ \tag{5.101}$$

We denote the magnitude of the line-to-line voltage as V_L. The magnitude of the line-to-line voltage is $\sqrt{3}$ times the magnitude of the line-to-neutral voltage:

$$V_L = \sqrt{3}V_Y \tag{5.102}$$

Figure 5.40 Phasor diagram showing the relationship between the line-to-line voltage $\mathbf{V}_{ab}$ and the line-to-neutral voltages $\mathbf{V}_{an}$ and $\mathbf{V}_{bn}$.

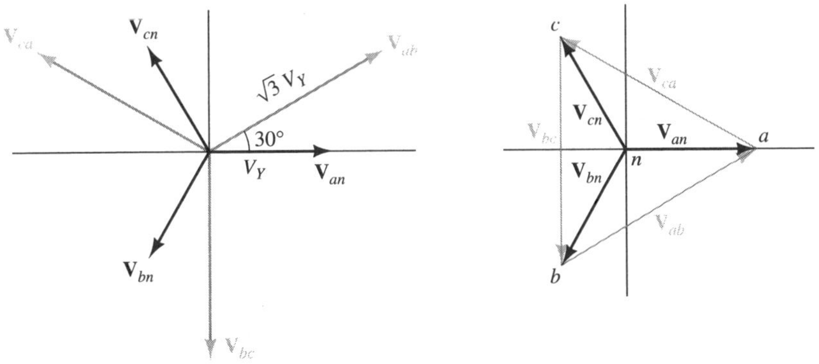

(a) All phasors starting from the origin

(b) A more intuitive way to draw the phasor diagram

Figure 5.41 Phasor diagram showing line-to-line voltages and line-to-neutral voltages.

Figure 5.41(b) provides a convenient way to remember the phase relationships between line-to-line and line-to-neutral voltages.

Thus, the relationship between the line-to-line voltage $\mathbf{V}_{ab}$ and the line-to-neutral voltage $\mathbf{V}_{an}$ is

$$\mathbf{V}_{ab} = \mathbf{V}_{an} \times \sqrt{3}\ \underline{/30^\circ} \tag{5.103}$$

Similarly, it can be shown that

$$\mathbf{V}_{bc} = \mathbf{V}_{bn} \times \sqrt{3}\ \underline{/30^\circ} \tag{5.104}$$

and

$$\mathbf{V}_{ca} = \mathbf{V}_{cn} \times \sqrt{3}\ \underline{/30^\circ} \tag{5.105}$$

These voltages are shown in Figure 5.41.

Example 5.12 Analysis of a Wye–Wye System

A balanced positive-sequence wye-connected 60-Hz three-phase source has line-to-neutral voltages of $V_Y = 1000$ V. This source is connected to a balanced wye-connected load. Each phase of the load consists of a 0.1-H inductance in series with a 50-Ω resistance. Find the line currents, the line-to-line voltages, the power, and the reactive power delivered to the load. Draw a phasor diagram showing the line-to-neutral voltages, the line-to-line voltages, and the line currents. Assume that the phase angle of $\mathbf{V}_{an}$ is zero.

Solution First, by computing the complex impedance of each phase of the load, we find that

$$\begin{aligned} Z &= R + j\omega L = 50 + j2\pi(60)(0.1) = 50 + j37.70 \\ &= 62.62\ \underline{/37.02^\circ} \end{aligned}$$

Next, we draw the circuit as shown in Figure 5.42(a). In balanced wye–wye calculations, we can assume that n and N are connected. (The currents and voltages are the same whether or not the neutral connection actually exists.) Thus, $\mathbf{V}_{an}$ appears across phase A of the load, and we can write

$$\mathbf{I}_{aA} = \frac{\mathbf{V}_{an}}{Z} = \frac{1000\ \underline{/0^\circ}}{62.62\ \underline{/37.02^\circ}} = 15.97\ \underline{/-37.02^\circ}$$

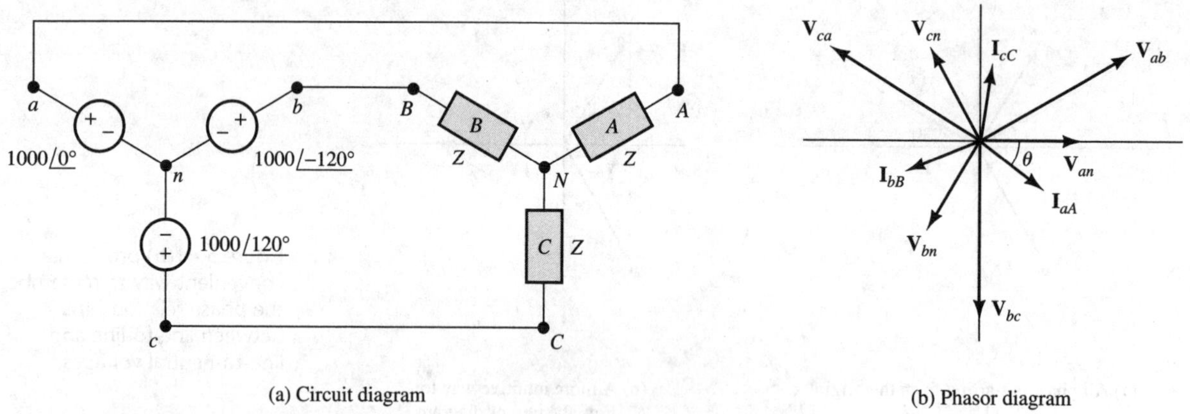

Figure 5.42 Circuit and phasor diagram for Example 5.12.

Similarly,

$$\mathbf{I}_{bB} = \frac{\mathbf{V}_{bn}}{Z} = \frac{1000\ \angle -120°}{62.62\ \angle 37.02°} = 15.97\ \angle -157.02°$$

$$\mathbf{I}_{cC} = \frac{\mathbf{V}_{cn}}{Z} = \frac{1000\ \angle 120°}{62.62\ \angle 37.02°} = 15.97\ \angle 82.98°$$

We use Equations 5.103, 5.104, and 5.105 to find the line-to-line phasors:

$$\mathbf{V}_{ab} = \mathbf{V}_{an} \times \sqrt{3}\ \angle 30° = 1732\ \angle 30°$$

$$\mathbf{V}_{bc} = \mathbf{V}_{bn} \times \sqrt{3}\ \angle 30° = 1732\ \angle -90°$$

$$\mathbf{V}_{ca} = \mathbf{V}_{cn} \times \sqrt{3}\ \angle 30° = 1732\ \angle 150°$$

The power delivered to the load is given by Equation 5.94:

$$P = 3\frac{V_Y I_L}{2}\cos(\theta) = 3\left(\frac{1000 \times 15.97}{2}\right)\cos(37.02°) = 19.13\ \text{kW}$$

The reactive power is given by Equation 5.98:

$$Q = 3\frac{V_Y I_L}{2}\sin(\theta) = 3\left(\frac{1000 \times 15.97}{2}\right)\sin(37.02°) = 14.42\ \text{kVAR}$$

The phasor diagram is shown in Figure 5.42(b). As usual, we have chosen a different scale for the currents than for the voltages. ■

Exercise 5.16 A balanced positive-sequence wye-connected 60-Hz three-phase source has line-to-line voltages of $V_L = 1000$ V. This source is connected to a balanced wye-connected load. Each phase of the load consists of a 0.2-H inductance in series with a 100-Ω resistance. Find the line-to-neutral voltages, the line currents, the power, and the reactive power delivered to the load. Assume that the phase of $\mathbf{V}_{an}$ is zero.

Answer $\mathbf{V}_{an} = 577.4\ \underline{/0^\circ}, \mathbf{V}_{bn} = 577.4\ \underline{/-120^\circ}, \mathbf{V}_{cn} = 577.4\ \underline{/120^\circ};$ $\mathbf{I}_{aA} = 4.61\ \underline{/-37^\circ}, \mathbf{I}_{bB} = 4.61\ \underline{/-157^\circ}, \mathbf{I}_{cC} = 4.61\ \underline{/83^\circ}; P = 3.19\ \text{kW};$ $Q = 2.40\ \text{kVAR}.$ □

Delta-Connected Sources

A set of balanced three-phase voltage sources can be connected in the form of a delta, as shown in Figure 5.43. Ordinarily, we avoid connecting voltage sources in closed loops. However, in this case, it turns out that the sum of the voltages is zero:

$$\mathbf{V}_{ab} + \mathbf{V}_{bc} + \mathbf{V}_{ca} = 0$$

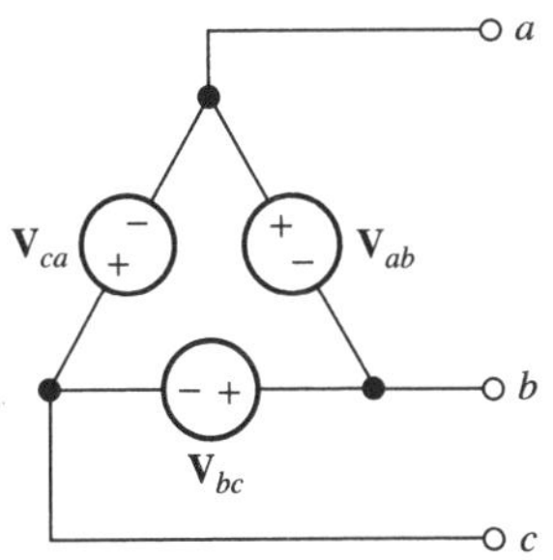

Figure 5.43 Delta-connected three-phase source.

Thus, the current circulating in the delta is zero. (Actually, this is a first approximation. There are many subtleties of power distribution systems that are beyond the scope of our discussion. For example, the voltages in actual power distribution systems are not exactly sinusoidal; instead, they are the sum of several harmonic components. The behavior of harmonic components is an important factor in making a choice between wye- and delta-connected sources or loads.)

For a given delta-connected source, we can find an equivalent wye-connected source (or vice versa) by using Equations 5.103 through 5.105. Clearly, a delta-connected source has no neutral point, so a four-wire connection is possible for only a wye-connected source.

Wye- and Delta-Connected Loads

Load impedances can be either wye connected or delta connected, as shown in Figure 5.44. It can be shown that the two loads are equivalent if

$$Z_\Delta = 3Z_Y \tag{5.106}$$

Thus, we can convert a delta-connected load to an equivalent wye-connected load, or vice versa.

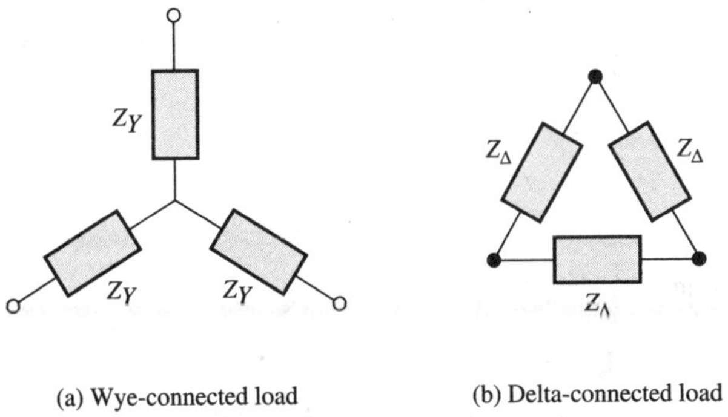

Figure 5.44 Loads can be either wye connected or delta connected.

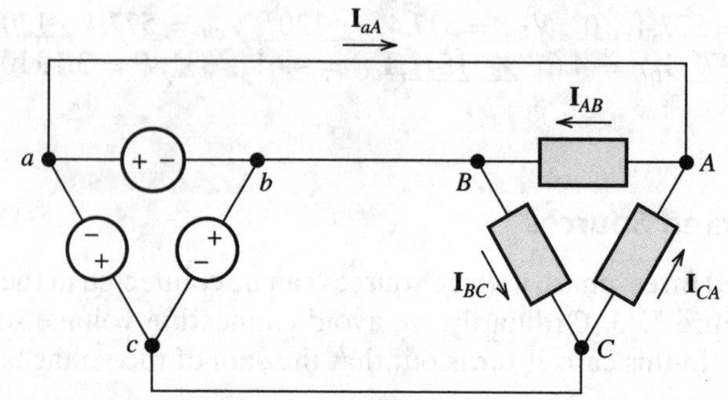

Figure 5.45 A delta-connected source delivering power to a delta-connected load.

Delta–Delta Connection

Figure 5.45 shows a delta-connected source delivering power to a delta-connected load. We assume that the source voltages are given by

$$\mathbf{V}_{ab} = V_L \angle 30° \tag{5.107}$$

$$\mathbf{V}_{bc} = V_L \angle -90° \tag{5.108}$$

$$\mathbf{V}_{ca} = V_L \angle 150° \tag{5.109}$$

These phasors are shown in Figure 5.41. (We have chosen the phase angles of the delta-connected source to be consistent with our earlier discussion.)

If the impedances of the connecting wires are zero, the line-to-line voltages at the load are equal to those at the source. Thus $\mathbf{V}_{AB} = \mathbf{V}_{ab}$, $\mathbf{V}_{BC} = \mathbf{V}_{bc}$, and $\mathbf{V}_{CA} = \mathbf{V}_{ca}$.

We assume that the impedance of each phase of the load is $Z_\Delta \angle \theta$. Then, the load current for phase AB is

$$\mathbf{I}_{AB} = \frac{\mathbf{V}_{AB}}{Z_\Delta \angle \theta} = \frac{\mathbf{V}_{ab}}{Z_\Delta \angle \theta} = \frac{V_L \angle 30°}{Z_\Delta \angle \theta} = \frac{V_L}{Z_\Delta} \angle 30° - \theta$$

We define the magnitude of the current as

$$I_\Delta = \frac{V_L}{Z_\Delta} \tag{5.110}$$

Hence,

$$\mathbf{I}_{AB} = I_\Delta \angle 30° - \theta \tag{5.111}$$

Similarly,

$$\mathbf{I}_{BC} = I_\Delta \angle -90° - \theta \tag{5.112}$$

$$\mathbf{I}_{CA} = I_\Delta \angle 150° - \theta \tag{5.113}$$

The current in line a–A is

$$\begin{aligned}\mathbf{I}_{aA} &= \mathbf{I}_{AB} - \mathbf{I}_{CA} \\ &= I_\Delta \angle 30° - \theta - I_\Delta \angle 150° - \theta \\ &= (I_\Delta \angle 30° - \theta) \times (1 - 1 \angle 120°)\end{aligned}$$

$$= (I_\Delta \angle 30° - \theta) \times (1.5 - j0.8660)$$
$$= (I_\Delta \angle 30° - \theta) \times (\sqrt{3} \angle -30°)$$
$$= \mathbf{I}_{AB} \times \sqrt{3} \angle -30°$$

The magnitude of the line current is

$$I_L = \sqrt{3} I_\Delta \tag{5.114}$$

For a balanced delta-connected load, the line-current magnitude is equal to the square root of three times the current magnitude in any arm of the delta.

Example 5.13 Analysis of a Balanced Delta–Delta System

Consider the circuit shown in Figure 5.46(a). A delta-connected source supplies power to a delta-connected load through wires having impedances of $Z_{\text{line}} = 0.3 + j0.4\ \Omega$. The load impedances are $Z_\Delta = 30 + j6$. The source voltages are

$$\mathbf{V}_{ab} = 1000 \angle 30°$$
$$\mathbf{V}_{bc} = 1000 \angle -90°$$
$$\mathbf{V}_{ca} = 1000 \angle 150°$$

Find the line current, the line-to-line voltage at the load, the current in each phase of the load, the power delivered to the load, and the power dissipated in the line.

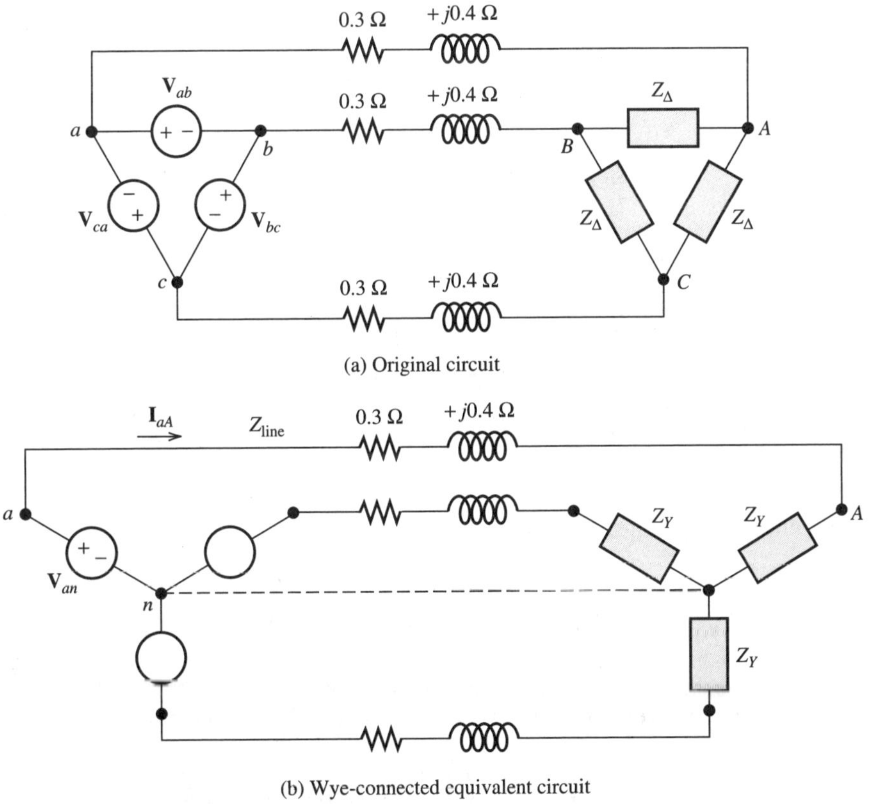

Figure 5.46 Circuit of Example 5.13.

Often, it is convenient to start an analysis by finding the wye–wye equivalent of a system.

Solution First, we find the wye-connected equivalents for the source and the load. (Actually, we only need to work with one third of the circuit because the other two thirds are the same except for phase angles.) We choose to work with the A phase of the wye-equivalent circuit. Solving Equation 5.103 for $\mathbf{V}_{an}$, we find that

$$\mathbf{V}_{an} = \frac{\mathbf{V}_{ab}}{\sqrt{3}\ \angle 30^\circ} = \frac{1000\ \angle 30^\circ}{\sqrt{3}\ \angle 30^\circ} = 577.4\ \angle 0^\circ$$

Using Equation 5.106, we have

$$Z_Y = \frac{Z_\Delta}{3} = \frac{30 + j6}{3} = 10 + j2$$

Now, we can draw the wye-equivalent circuit, which is shown in Figure 5.46(b).

In a balanced wye–wye system, we can consider the neutral points to be connected together as shown by the dashed line in Figure 5.46(b). This reduces the three-phase circuit to three single-phase circuits. For phase A of Figure 5.46(b), we can write

$$\mathbf{V}_{an} = (Z_{\text{line}} + Z_Y)\mathbf{I}_{aA}$$

Therefore,

$$\begin{aligned}\mathbf{I}_{aA} &= \frac{\mathbf{V}_{an}}{Z_{\text{line}} + Z_Y} = \frac{577.4\ \angle 0^\circ}{0.3 + j0.4 + 10 + j2}\\ &= \frac{577.4\ \angle 0^\circ}{10.3 + j2.4} = \frac{577.4\ \angle 0^\circ}{10.58\ \angle 13.12^\circ}\\ &= 54.60\ \angle -13.12^\circ\end{aligned}$$

To find the line-to-neutral voltage at the load, we write

$$\begin{aligned}\mathbf{V}_{An} &= \mathbf{I}_{Aa}Z_Y = 54.60\ \angle -13.12^\circ \times (10 + j2)\\ &= 54.60\ \angle -13.12^\circ \times 10.20\ \angle 11.31^\circ\\ &= 556.9\ \angle -1.81^\circ\end{aligned}$$

Now, we compute the line-to-line voltage at the load:

$$\begin{aligned}\mathbf{V}_{AB} &= \mathbf{V}_{An} \times \sqrt{3}\ \angle 30^\circ = 556.9\ \angle -1.81^\circ \times \sqrt{3}\ \angle 30^\circ\\ &= 964.6\ \angle 28.19^\circ\end{aligned}$$

The current through phase AB of the load is

$$\begin{aligned}\mathbf{I}_{AB} &= \frac{\mathbf{V}_{AB}}{Z_\Delta} = \frac{964.6\ \angle 28.19^\circ}{30 + j6} = \frac{964.6\ \angle 28.19^\circ}{30.59\ \angle 11.31^\circ}\\ &= 31.53\ \angle 16.88^\circ\end{aligned}$$

The power delivered to phase AB of the load is the rms current squared times the resistance:

$$P_{AB} = I_{AB\text{rms}}^2 R = \left(\frac{31.53}{\sqrt{2}}\right)^2 (30) = 14.91\ \text{kW}$$

The powers delivered to the other two phases of the load are the same, so the total power is

$$P = 3P_{AB} = 44.73 \text{ kW}$$

The power lost in line A is

$$P_{\text{line}A} = I_{a A\text{rms}}^2 R_{\text{line}} = \left(\frac{54.60}{\sqrt{2}}\right)^2 (0.3) = 0.447 \text{ kW}$$

The power lost in the other two lines is the same, so the total line loss is

$$P_{\text{line}} = 3 \times P_{\text{line}A} = 1.341 \text{ kW}$$ ■

Exercise 5.17 A delta-connected source has voltages given by

$$\mathbf{V}_{ab} = 1000 \angle 30^\circ$$

$$\mathbf{V}_{bc} = 1000 \angle -90^\circ$$

$$\mathbf{V}_{ca} = 1000 \angle 150^\circ$$

This source is connected to a delta-connected load consisting of 50-Ω resistances. Find the line currents and the power delivered to the load.

Answer $\mathbf{I}_{aA} = 34.6 \angle 0^\circ, \mathbf{I}_{bB} = 34.6 \angle -120^\circ, \mathbf{I}_{cC} = 34.6 \angle 120^\circ; P = 30$ kW. □

5.8 AC ANALYSIS USING MATLAB

In this section, we will illustrate how MATLAB can greatly facilitate the analysis of complicated ac circuits. In fact, a practicing engineer working at a computer might have little use for a calculator, as it is easy to keep a MATLAB window open for all sorts of engineering calculations. Of course, you will probably need to use calculators for course exams and when you take the Professional Engineer (PE) exams. The PE exams allow only fairly simple scientific calculators, and you should practice with one of those allowed before attempting the exams.

Complex Data in MATLAB

By default, MATLAB assumes that $i = j = \sqrt{-1}$. However, I have encountered at least one bug in the software attributable to using j instead of i, and therefore I recommend using i in MATLAB and the Symbolic Toolbox. We need to be careful to avoid using i for other purposes when using MATLAB to analyze ac circuits. For example, if we were to use i as the name of a current or other variable, we would later experience errors if we also used i for the imaginary unit without reassigning its value.

Complex numbers are represented in rectangular form (such as 3 + 4i or alternatively 3 + i*4) in MATLAB.

We can use the fact that $M \angle \theta = M \exp(j\theta)$ to enter polar data. In MATLAB, angles are assumed to be in radians, so we need to multiply angles that are expressed in degrees by $\pi/180$ to convert to radians before entering them. For example, we use the following command to enter the voltage $V_s = 5\sqrt{2} \angle 45^\circ$:

```
>> Vs = 5*sqrt(2)*exp(i*45*pi/180)
Vs =
   5.0000 + 5.0000i
```

We can readily verify that MATLAB has correctly computed the rectangular form of $5\sqrt{2}\ \underline{/45^\circ}$.

Alternatively, we could use Euler's formula

$$M\ \underline{/\theta} = M\exp(j\theta) = M\cos(\theta) + jM\sin(\theta)$$

to enter polar data, again with angles in radians. For example, $V_s = 5\sqrt{2}\ \underline{/45^\circ}$ can be entered as:

```
>> Vs = 5*sqrt(2)*cos(45*pi/180) + i*5*sqrt(2)*sin(45*pi/180)
Vs =
   5.0000 + 5.0000i
```

Values that are already in rectangular form can be entered directly. For example, to enter $Z = 3 + j4$, we use the command:

```
>> Z = 3 + i*4
Z =
   3.0000 + 4.0000i
```

Then, if we enter

```
>> Ix = Vs/Z
Ix =
   1.4000 - 0.2000i
```

MATLAB performs the complex arithmetic and gives the answer in rectangular form.

Finding the Polar Form of MATLAB Results

Frequently, we need the polar form of a complex value calculated by MATLAB. We can find the magnitude using the abs command and the angle in radians using the angle command. To obtain the angle in degrees, we must convert the angle from radians by multiplying by $180/\pi$. Thus, to obtain the magnitude and angle in degrees for Vs, we would enter the following commands:

```
>> abs(Vs) % Find the magnitude of Vs.
ans =
   7.0711
>> (180/pi)*angle(Vs) % Find the angle of Vs in degrees.
ans =
   45.0000
```

Adding New Functions to MATLAB

Because we often want to enter values or see results in polar form with the angles in degrees, it is convenient to add two new functions to MATLAB. Thus, we write an m-file, named pin.m, containing the commands to convert from polar to rectangular form, and store it in our working MATLAB folder. The commands in the m-file are:

```
function z = pin(magnitude, angleindegrees)
z = magnitude*exp(i*angleindegrees*pi/180)
```

Then, we can enter $\text{Vs} = 5\sqrt{2}\,\angle 45^\circ$ simply by typing the command:

```
>> Vs = pin(5*sqrt(2),45)
Vs =
   5.0000 + 5.0000i
```

We have chosen pin as the name of this new function to suggest "polar input." This file is included in the MATLAB folder. (See Appendix E for information about accessing this folder.)

Similarly, to obtain the polar form of an answer, we create a new function, named pout (to suggest "polar out"), with the commands:

```
function [y] = pout(x);
magnitude = abs(x);
angleindegrees = (180/pi)*angle(x);
y = [magnitude angleindegrees];
```

which are stored in the m-file named pout.m. Then, to find the polar form of a result, we can use the new function. For example,

```
>> pout(Vs)
ans =
    7.0711   45.0000
```

Here is another simple example:

```
>> pout(i*200)
ans =
    200    90
```

Solving Network Equations with MATLAB

We can readily solve node voltage or mesh equations and perform other calculations for ac circuits in MATLAB. The steps are:

1. Write the mesh current or node voltage equations.
2. Put the equations into matrix form, which is $\mathbf{ZI} = \mathbf{V}$ for mesh currents, in which $\mathbf{Z}$ is the coefficient matrix, $\mathbf{I}$ is the column vector of mesh current variables to be found, and $\mathbf{V}$ is the column vector of constant terms. For node voltages, the matrix equations take the form $\mathbf{YV} = \mathbf{I}$ in which $\mathbf{Y}$ is the coefficient matrix, $\mathbf{V}$ is the column vector of node voltage variables to be determined, and $\mathbf{I}$ is the column vector of constants.
3. Enter the matrices into MATLAB and compute the mesh currents or node voltages using the inverse matrix approach. $\mathbf{I} = \text{inv}(\mathbf{Z}) \times \mathbf{V}$ for mesh currents or $\mathbf{V} = \text{inv}(\mathbf{Y}) \times \mathbf{I}$ for node voltages, where inv denotes the matrix inverse.
4. Use the results to compute any other quantities of interest.

Example 5.14 Phasor Mesh-Current Analysis with MATLAB

Determine the values for the mesh currents, the real power supplied by $\mathbf{V}_1$, and the reactive power supplied by $\mathbf{V}_1$ in the circuit of Figure 5.47.

Solution First, we apply KVL to each loop obtaining the mesh-current equations:

$$(5 + j3)\mathbf{I}_1 + (50\,\angle -10^\circ)(\mathbf{I}_1 - \mathbf{I}_2) = 2200\sqrt{2}$$

$$(50\,\angle -10^\circ)(\mathbf{I}_2 - \mathbf{I}_1) + (4 + j)\mathbf{I}_2 + 2000\sqrt{2}\,\angle 30 = 0$$

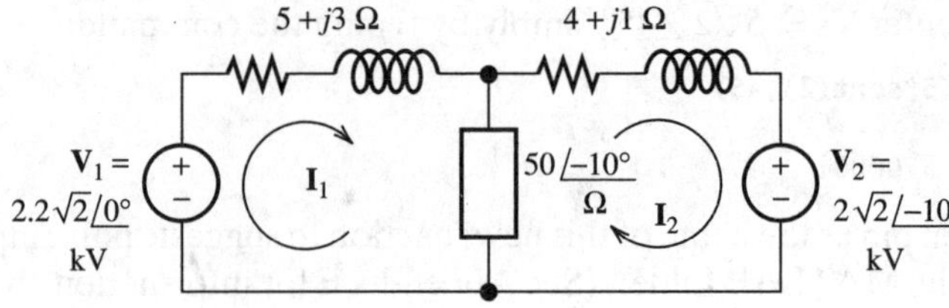

Figure 5.47 Cirucit for Example 5.14.

In matrix form, these equations become

$$\begin{bmatrix} (5 + j3 + 50\ \angle -10^\circ) & -50\ \angle -10^\circ \\ -50\ \angle -10^\circ & (4 + j + 50\ \angle -10^\circ) \end{bmatrix} \begin{bmatrix} \mathbf{I}_1 \\ \mathbf{I}_2 \end{bmatrix} = \begin{bmatrix} 2200\sqrt{2} \\ -2000\sqrt{2}\ \angle -10^\circ \end{bmatrix}$$

We will solve these equations for $\mathbf{I}_1$ and $\mathbf{I}_2$. Then, we will compute the complex power delivered by $\mathbf{V}_1$

$$\mathbf{S}_1 = \frac{1}{2}\mathbf{V}_1\mathbf{I}_1^*$$

Finally, the power is the real part of $\mathbf{S}_1$ and the reactive power is the imaginary part.

We enter the coefficient matrix **Z** and the voltage matrix **V** into MATLAB, making use of our new pin function to enter polar values. Then, we calculate the current matrix.

```
>> Z = [(5 + i*3 + pin(50,-10)) (-pin(50,-10));...
    (-pin(50,-10)) (4 + i + pin(50,-10))];
>> V = [2200*sqrt(2); -pin(2000*sqrt(2),-10)];
>> I = inv(Z)*V
I =
   74.1634 + 29.0852i
   17.1906 + 26.5112i
```

This has given us the values of the mesh currents in rectangular form. Next, we obtain the polar form for the mesh currents, making use of our new pout function:

```
>> pout(I(1))
ans =
   79.6628  21.4140
>> pout(I(2))
ans =
   31.5968  57.0394
```

Thus, the currents are $\mathbf{I}_1 = 79.66\ \angle 21.41^\circ$ A and $\mathbf{I}_2 = 31.60\ \angle 57.04^\circ$ A, rounded to two decimal places. Next, we compute the complex power, real power, and reactive power for the first source.

$$\mathbf{S}_1 = \frac{1}{2}\mathbf{V}_1\mathbf{I}_1^*$$

```
>> S1 = (1/2)*(2200*sqrt(2))*conj(I(1));
>> P1 = real(S1)
P1 =
  1.1537e + 005
>> Q1 = imag(S1)
Q1 =
  -4.5246e + 004
```

Thus, the power supplied by $\mathbf{V}_1$ is 115.37 kW and the reactive power is −45.25 kVAR. The commands for this example appear in the m-file named Example_5_14. ■

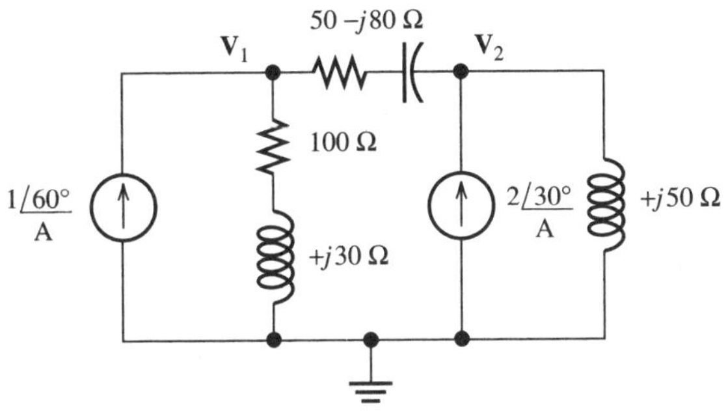

Figure 5.48 Circuit for Exercise 5.18.

Exercise 5.18 Use MATLAB to solve for the phasor node voltages in polar form for the circuit of Figure 5.48.

Answer The MATLAB commands are:

```
clear
Y = [(1/(100+i*30)+1/(50-i*80)) (-1/(50-i*80));...
      (-1/(50-i*80)) (1/(i*50)+1/(50-i*80))];
I = [pin(1,60); pin(2,30)];
V = inv(Y)*I;
pout(V(1))
pout(V(2))
```

and the results are $\mathbf{V}_1 = 79.98\,\angle 106.21^\circ$ and $\mathbf{V}_2 = 124.13\,\angle 116.30^\circ$. □

Summary

1. A sinusoidal voltage is given by $v(t) = V_m \cos(\omega t + \theta)$, where V_m is the peak value of the voltage, ω is the angular frequency in radians per second, and θ is the phase angle. The frequency in hertz is $f = 1/T$, where T is the period. Furthermore, $\omega = 2\pi f$.
2. For uniformity, we express sinusoidal voltages in terms of the cosine function. A sine function can be converted to a cosine function by use of the identity $\sin(z) = \cos(z - 90^\circ)$.
3. The root-mean-square (rms) value (or effective value) of a periodic voltage $v(t)$ is

$$V_{\mathrm{rms}} = \sqrt{\frac{1}{T}\int_0^T v^2(t)\,dt}$$

The average power delivered to a resistance by $v(t)$ is

$$P_{\mathrm{avg}} = \frac{V_{\mathrm{rms}}^2}{R}$$

Similarly, for a current $i(t)$, we have

$$I_{\mathrm{rms}} = \sqrt{\frac{1}{T}\int_0^T i^2(t)\,dt}$$

and the average power delivered if $i(t)$ flows through a resistance is

$$P_{\mathrm{avg}} = I_{\mathrm{rms}}^2 R$$

For a sinusoid, the rms value is the peak value divided by $\sqrt{2}$.

4. We can represent sinusoids with phasors. The magnitude of the phasor is the peak value of the sinusoid. The phase angle of the phasor is the phase angle of the sinusoid (assuming that we have written the sinusoid in terms of a cosine function).
5. We can add (or subtract) sinusoids by adding (or subtracting) their phasors.
6. The phasor voltage for a passive circuit is the phasor current times the complex impedance of the circuit. For a resistance, $\mathbf{V}_R = R\mathbf{I}_R$, and the voltage is in phase with the current. For an

inductance, $\mathbf{V}_L = j\omega L\mathbf{I}_L$, and the voltage leads the current by 90°. For a capacitance, $\mathbf{V}_C = -j(1/\omega C)\mathbf{I}_C$, and the voltage lags the current by 90°.

7. Many techniques learned in Chapter 2 for resistive circuits can be applied directly to sinusoidal circuits if the currents and voltages are replaced by phasors and the passive circuit elements are replaced by their complex impedances. For example, complex impedances can be combined in series or parallel in the same way as resistances (except that complex arithmetic must be used). Node voltages, the current-division principle, and the voltage-division principle also apply to ac circuits.

8. When a sinusoidal current flows through a sinusoidal voltage, the average power delivered is $P = V_{\text{rms}} I_{\text{rms}} \cos(\theta)$, where θ is the power angle, which is found by subtracting the phase angle of the current from the phase angle of the voltage (i.e., $\theta = \theta_v - \theta_i$). The power factor is $\cos(\theta)$.

9. Reactive power is the flow of energy back and forth between the source and energy-storage elements (L and C). We define reactive power to be positive for an inductance and negative for a capacitance. The net energy transferred per cycle by reactive power flow is zero. Reactive power is important because a power distribution system must have higher current ratings if reactive power flows than would be required for zero reactive power.

10. Apparent power is the product of rms voltage and rms current. Many useful relationships between power, reactive power, apparent power, and the power angle can be obtained from the power triangles shown in Figure 5.23 on page 254.

11. In steady state, a network composed of resistances, inductances, capacitances, and sinusoidal sources (all of the same frequency) has a Thévenin equivalent consisting of a phasor voltage source in series with a complex impedance. The Norton equivalent consists of a phasor current source in parallel with the Thévenin impedance.

12. For maximum-power transfer from a two-terminal ac circuit to a load, the load impedance is selected to be the complex conjugate of the Thévenin impedance. If the load is constrained to be a pure resistance, the value for maximum power transfer is equal to the magnitude of the Thévenin impedance.

13. Because of savings in wiring, three-phase power distribution is more economical than single phase. The power flow in balanced three-phase systems is smooth, whereas power pulsates in single-phase systems. Thus, three-phase motors generally have the advantage of producing less vibration than single-phase motors.

Problems

Section 5.1: Sinusoidal Currents and Voltages

P5.1. Define angular frequency, ω, frequency, f, and time period T. What is the relationship between f and T? Between f and ω?

P5.2. In terms of physical units, such as m, kg, C, and s, what are the units of radians? What are the *physical* units for angular frequency?

P5.3. Consider the plot of the sinusoidal voltage $v(t) = V_m \cos(\omega t + \theta)$ shown in Figure 5.1 on page 228. Which of the numbered statements below best describes **a.** Increasing the peak amplitude V_m? **b.** Increasing the frequency f? **c.** Increasing θ? **d.** Decreasing the angular frequency ω? **e.** Decreasing the period?

1. Stretches the sinusoidal curve horizontally.
2. Compresses the sinusoidal curve horizontally.
3. Translates the sinusoidal curve to the right.
4. Translates the sinusoidal curve to the left.
5. Stretches the sinusoidal curve vertically.
6. Compresses the sinusoidal curve vertically.

* Denotes that answers are contained in the Student Solutions files. See Appendix E for more information about accessing the Student Solutions.

***P5.4.** A voltage is given by $v(t) = 10\sin(1000\pi t + 30°)$ V. First, use a cosine function to express $v(t)$. Then, find the angular frequency, the frequency in hertz, the phase angle, the period, and the rms value. Find the power that this voltage delivers to a 50-Ω resistance. Find the first value of time after $t = 0$ that $v(t)$ reaches its peak value. Sketch $v(t)$ to scale versus time.

P5.5. Repeat Problem P5.4 for $v(t) = 5\sin(500\pi t + 150°)$.

***P5.6.** A sinusoidal voltage $v(t)$ has an rms value of 20 V, a period of 100 μs, and reaches a positive peak at $t = 20$ μs. Write an expression for $v(t)$.

P5.7. We have a sinusoidal current $i(t)$ that has an rms value of 20 A, a period of 1 ms, and reaches a positive peak at $t = 0.3$ ms. Write an expression for $i(t)$.

P5.8. The voltage $v(t) = 10\sin(250\pi t)$ V appears across a 20-Ω resistance. Sketch $v(t)$ and $p(t)$ to scale versus time. Find the average power delivered to the resistance.

P5.9. A sinusoidal voltage has a peak value of 15 V, a frequency of 500 Hz, and crosses zero with positive slope at $t = 0.1$ ms. Write an expression for the voltage.

P5.10. A current $i(t) = 4\cos(1000\pi t + 30°)$ A flows through a 10-Ω resistance. Sketch $i(t)$ and $p(t)$ to scale versus time. Find the average power delivered to the resistance.

P5.11. Is the rms value of a periodic waveform always equal to the peak value divided by the square root of two? When is it?

***P5.12.** Find the rms value of the current waveform shown in Figure P5.12.

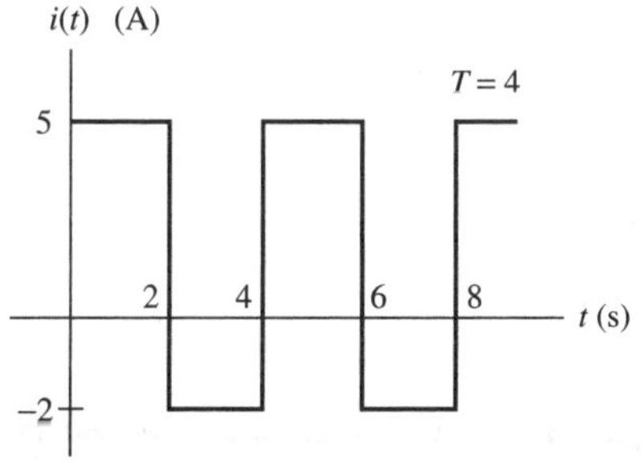

Figure P5.12

P5.13. Find the rms value of the voltage waveform shown in Figure P5.13.

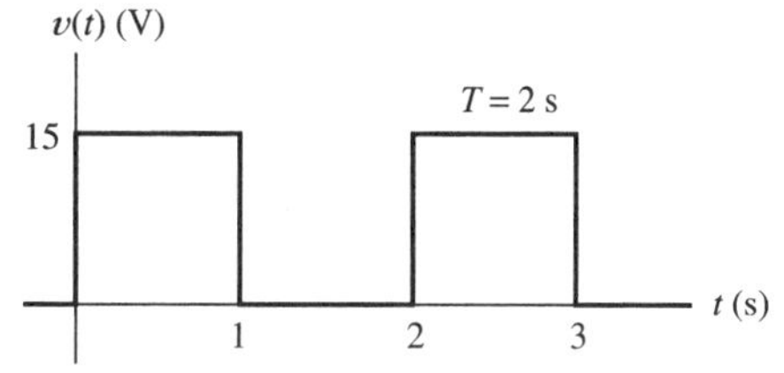

Figure P5.13

P5.14. Compute the rms value of the periodic waveform shown in Figure P5.14.

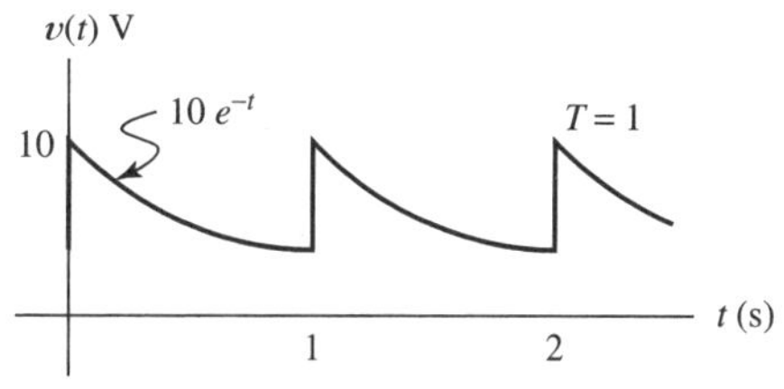

Figure P5.14

P5.15. Determine the rms value of $v(t) = A + B\cos(2\pi t) + 2C\sin(2\pi t)$.

P5.16. Calculate the rms value of the full-wave rectified sinusoidal wave shown in Figure P5.16, which is given by $v(t) = 4|\cos(20\pi t)|$.

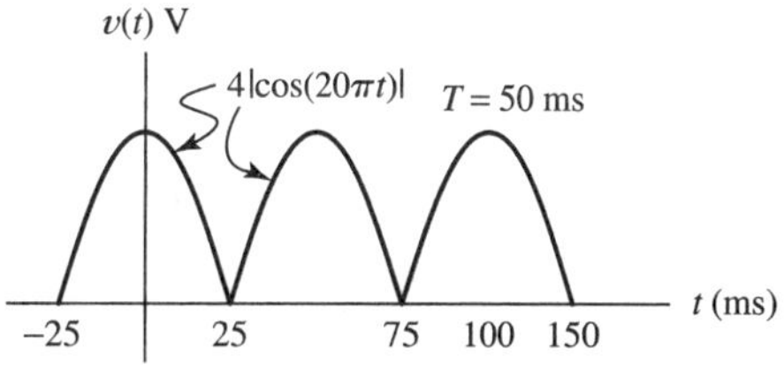

Figure P5.16

P5.17. Find the rms value of the voltage waveform shown in Figure P5.17.

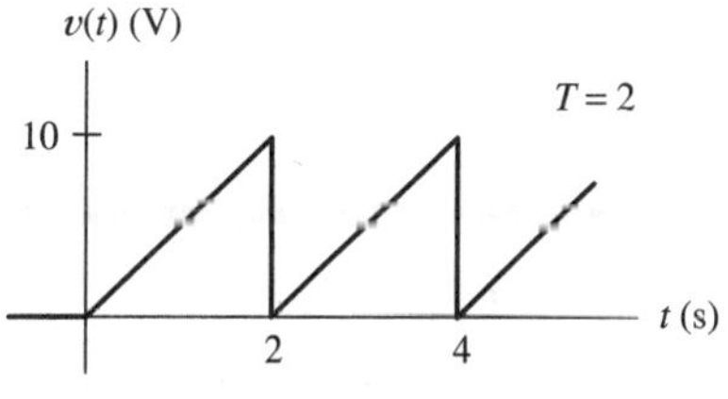

Figure P5.17

P5.18. Determine the rms value of $v(t) = 5 + 4\sqrt{2}\sin(20\pi t + 60°)$ V.

Section 5.2: Phasors

P5.19. What are the steps we follow in adding sinusoidal currents or voltages? What must be true of the sinusoids?

P5.20. **a.** Explain how the phase relationship between two sinusoids of the same frequency can be determined working from the phasor diagram. **b.** Explain how to determine the phase relationship between two sinusoids of the same frequency working from plots of the sinusoids versus time.

***P5.21.** Reduce $5\cos(\omega t + 75°) - 3\cos(\omega t - 75°) + 4\sin(\omega t)$ to the form $V_m \cos(\omega t + \theta)$.

***P5.22.** Suppose that $v_1(t) = 100\cos(\omega t)$ and $v_2(t) = 100\sin(\omega t)$. Use phasors to reduce the sum $v_s(t) = v_1(t) + v_2(t)$ to a single term of the form $V_m \cos(\omega t + \theta)$. Draw a phasor diagram, showing $\mathbf{V}_1, \mathbf{V}_2$, and $\mathbf{V}_s$. State the phase relationships between each pair of these phasors.

***P5.23.** Consider the phasors shown in Figure P5.23. The frequency of each signal is $f = 200$ Hz. Write a time-domain expression for each voltage in the form $V_m \cos(\omega t + \theta)$. State the phase relationships between pairs of these phasors.

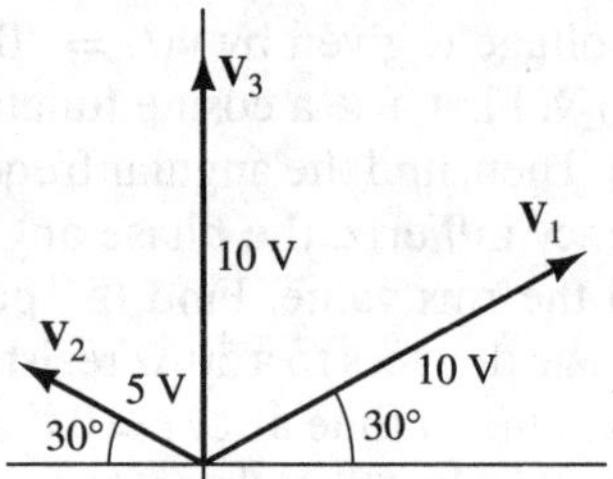

Figure P5.23

P5.24. Suppose we have two sinusoidal voltages of the same frequency with rms values of 12 V and 7 V, respectively. The phase angles are unknown. What is the smallest rms value that the sum of these voltages could have? The largest? Justify your answers.

P5.25. Find an expression for the sinusoid shown in Figure P5.25 of the form $v(t) = V_m \cos(\omega t + \theta)$, giving the numerical values of V_m, ω, and θ. Also, determine the phasor and the rms value of $v(t)$.

P5.26. A sinusoidal current $i_1(t)$ has a phase angle of $-30°$. Furthermore, $i_1(t)$ attains its positive peak 0.25 ms later than current $i_2(t)$ does. Both the currents have a frequency of 1000 Hz. Determine the phase angle of $i_2(t)$.

P5.27. Suppose that $v_1(t) = 90\cos(\omega t - 15°)$ and $v_2(t) = 50\sin(\omega t - 60°)$. Use phasors to

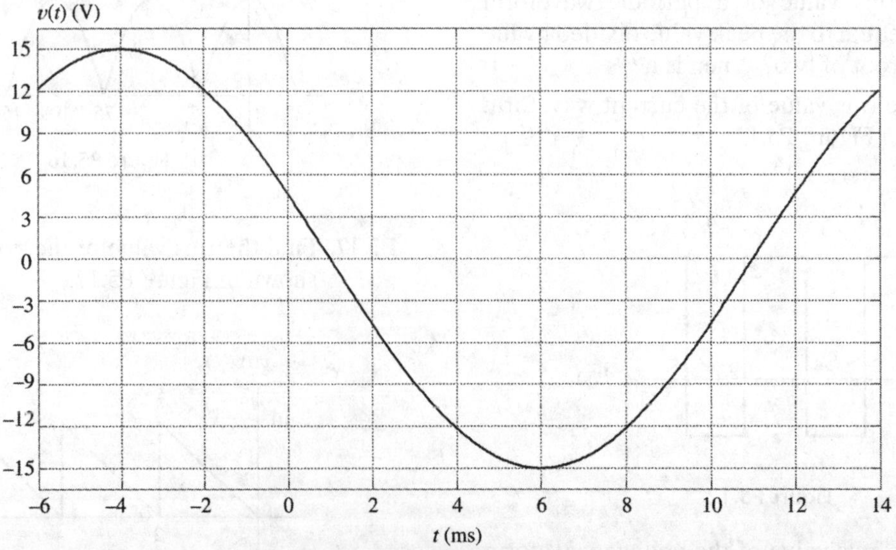

Figure P5.25

reduce the difference $v_s(t) = v_1(t) - v_2(t)$ to a single term of the form $V_m \cos(\omega t + \theta)$. State the phase relationships between each pair of these phasors.

P5.28. Reduce the expression

$$5 \sin(\omega t) + 15 \cos(\omega t - 30°) + 20 \cos(\omega t - 120°)$$

to the form $V_m \cos(\omega t + \theta)$.

P5.29. Suppose we have a circuit in which the voltage is $v_1(t) = 15 \cos(\omega t + 30°)$ V. Furthermore, the current $i_1(t)$ has an rms value of 5 A and lags $v_1(t)$ by 40°. (The current and the voltage have the same frequency.) Draw a phasor diagram and write an expression for $i_1(t)$ of the form $I_m \cos(\omega t + \theta)$.

Section 5.3: Complex Impedances

P5.30. What is the phase relationship between current and voltage for a pure resistance? For an inductance? For a capacitance?

P5.31. Write the relationship between the phasor voltage and phasor current for an inductance. Repeat for capacitance. Repeat for a resistance.

***P5.32.** A voltage $v_L(t) = 10 \cos(2000\pi t)$ is applied to a 100-mH inductance. Find the complex impedance of the inductance. Find the phasor voltage and current, and construct a phasor diagram. Write the current as a function of time. Sketch the voltage and current to scale versus time. State the phase relationship between the current and voltage.

***P5.33.** A voltage $v_C(t) = 10 \cos(2000\pi t)$ is applied to a 10-μF capacitance. Find the complex impedance of the capacitance. Find the phasor voltage and current, and construct a phasor diagram. Write the current as a function of time. Sketch the voltage and current to scale versus time. State the phase relationship between the current and voltage.

P5.34. A certain circuit element is known to be a pure resistance, a pure inductance, or a pure capacitance. Determine the type and value (in ohms, henrys, or farads) of the element if the voltage and current for the element are given by: **a.** $v(t) = 100 \cos(200t + 30°)$ V, $i(t) = 2.5 \sin(200t + 30°)$ A; **b.** $v(t) = 100 \sin(200t + 30°)$ V, $i(t) = 4 \cos(200t + 30°)$ A; **c.** $v(t) = 100 \cos(100t + 30°)$ V, $i(t) = 5 \cos(100t + 30°)$ A.

P5.35. **a.** The current and voltage for a certain circuit element are shown in Figure P5.35(a). Determine the nature and value of the element. **b.** Repeat for Figure P5.35(b).

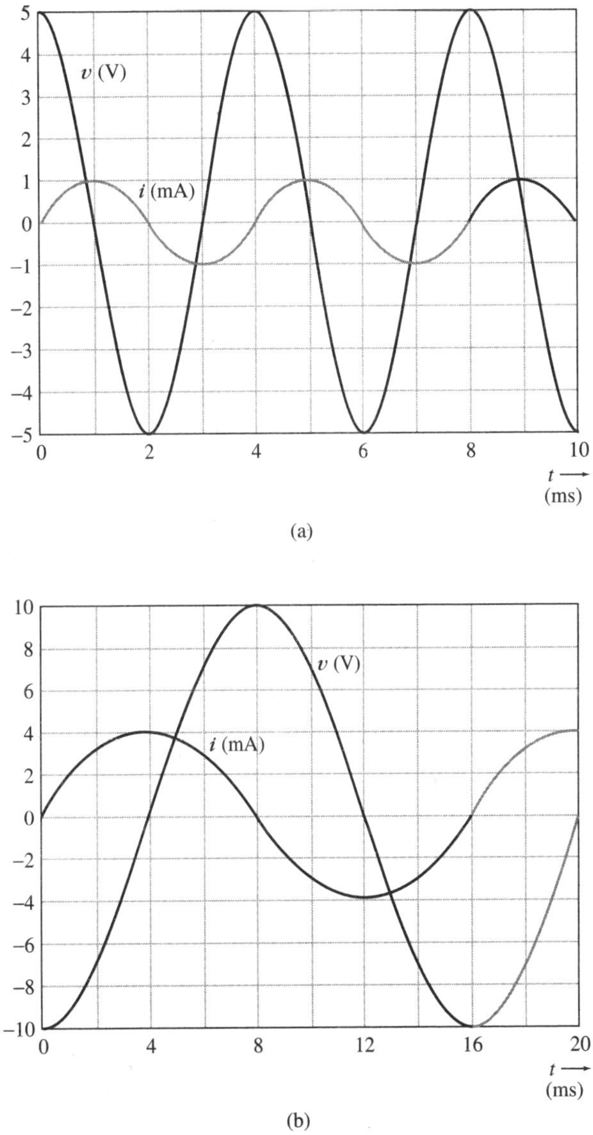

Figure P5.35

P5.36. **a.** A certain element has a phasor voltage of of $\mathbf{V} = 100 \angle 30°$ V and current of $\mathbf{I} = 5 \angle 120°$ A. The angular frequency is 1000 rad/s. Determine the nature and value of the element. **b.** Repeat for $\mathbf{V} = 20 \angle -45°$ V and current of $\mathbf{I} = 2 \angle -135°$ A. **c.** Repeat for $\mathbf{V} = 100 \angle 150°$ V and current of $\mathbf{I} = 5 \angle 150°$ A.

Section 5.4: Circuit Analysis with Phasors and Complex Impedances

P5.37. Explain the step-by-step procedure for steady-state analysis of circuits with sinusoidal sources. What condition must be true of the sources?

***P5.38.** Find the phasors for the current and for the voltages of the circuit shown in Figure P5.38. Construct a phasor diagram showing $\mathbf{V}_s$, $\mathbf{I}$, $\mathbf{V}_R$, and $\mathbf{V}_L$. What is the phase relationship between $\mathbf{V}_s$ and $\mathbf{I}$?

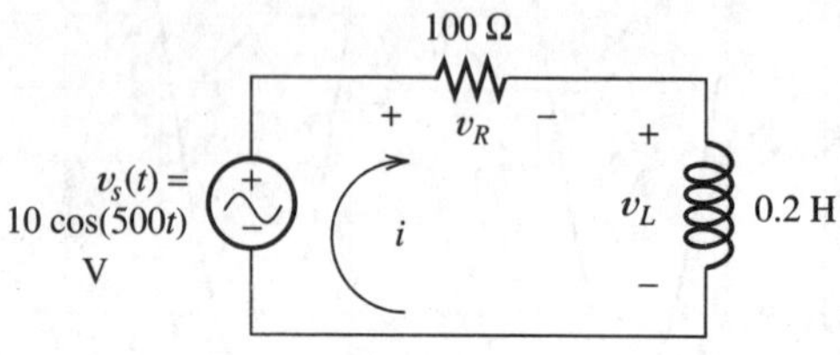

Figure P5.38

P5.39. Change the inductance to 0.4 H, and repeat Problem P5.38.

***P5.40.** Find the phasors for the current and the voltages for the circuit shown in Figure P5.40. Construct a phasor diagram showing $\mathbf{V}_s$, $\mathbf{I}$, $\mathbf{V}_R$, and $\mathbf{V}_C$. What is the phase relationship between $\mathbf{V}_s$ and $\mathbf{I}$?

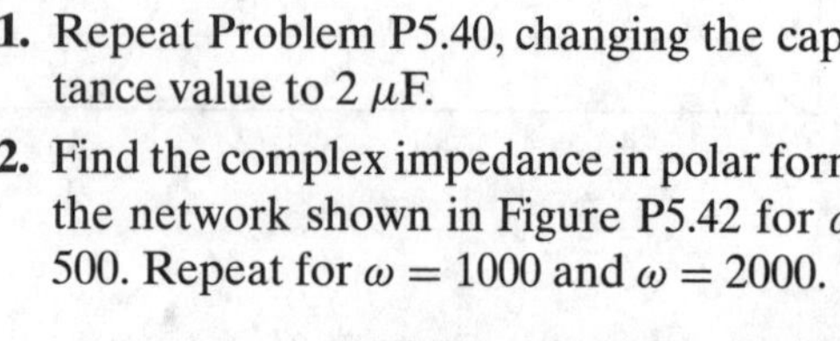

Figure P5.40

P5.41. Repeat Problem P5.40, changing the capacitance value to 2 μF.

***P5.42.** Find the complex impedance in polar form of the network shown in Figure P5.42 for $\omega = 500$. Repeat for $\omega = 1000$ and $\omega = 2000$.

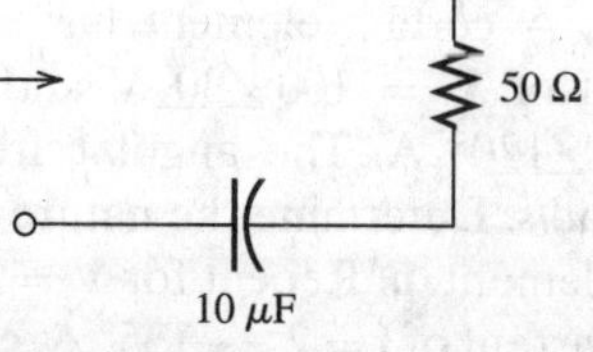

Figure P5.42

P5.43. Compute the complex impedance of the network shown in Figure P5.43 for $\omega = 500$. Repeat for $\omega = 1000$ and $\omega = 2000$. Give the answers in both polar and rectangular forms.

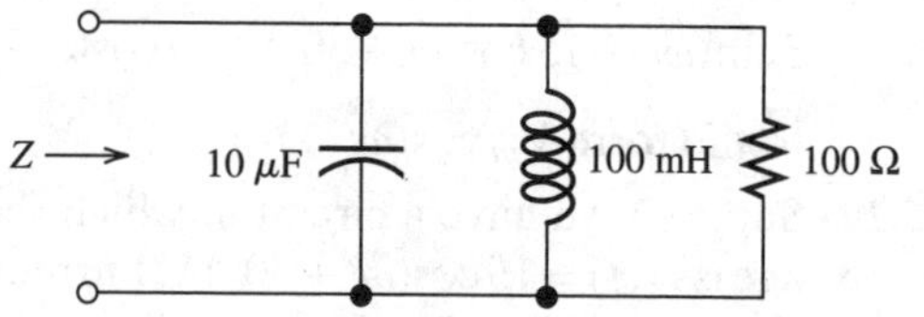

Figure P5.43

P5.44. A 100-μF capacitance is connected in parallel with the series combination of a 10-mH inductance and a 5-Ω resistance. Calculate the impedance of the combination in rectangular form and in polar form for angular frequencies of 500, 1000, and 2000 radians per second.

***P5.45.** Consider the circuit shown in Figure P5.45. Find the phasors $\mathbf{I}_s$, $\mathbf{V}$, $\mathbf{I}_R$, $\mathbf{I}_L$, and $\mathbf{I}_C$. Compare the peak value of $i_L(t)$ with the peak value of $i_s(t)$. Do you find the answer surprising? Explain.

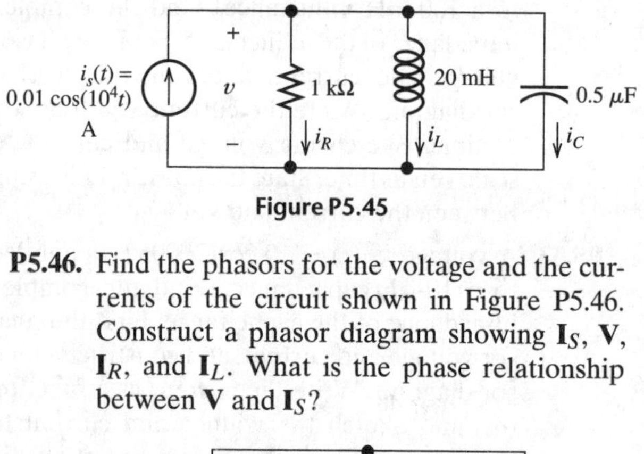

Figure P5.45

P5.46. Find the phasors for the voltage and the currents of the circuit shown in Figure P5.46. Construct a phasor diagram showing $\mathbf{I}_S$, $\mathbf{V}$, $\mathbf{I}_R$, and $\mathbf{I}_L$. What is the phase relationship between $\mathbf{V}$ and $\mathbf{I}_S$?

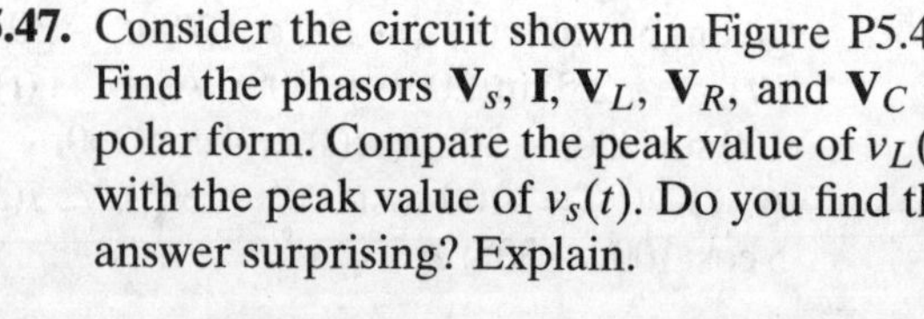

Figure P5.46

P5.47. Consider the circuit shown in Figure P5.47. Find the phasors $\mathbf{V}_s$, $\mathbf{I}$, $\mathbf{V}_L$, $\mathbf{V}_R$, and $\mathbf{V}_C$ in polar form. Compare the peak value of $v_L(t)$ with the peak value of $v_s(t)$. Do you find the answer surprising? Explain.

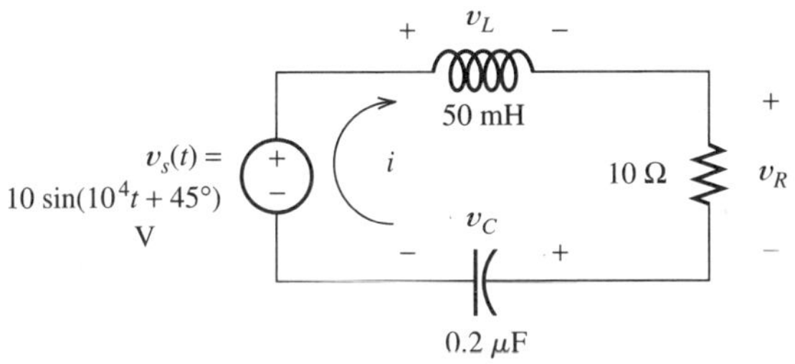

Figure P5.47

***P5.48.** Find the phasors for the voltage and the currents for the circuit shown in Figure P5.48. Construct a phasor diagram showing $\mathbf{I}_s$, $\mathbf{V}$, $\mathbf{I}_R$, and $\mathbf{I}_C$. What is the phase relationship between $\mathbf{V}$ and $\mathbf{I}_s$?

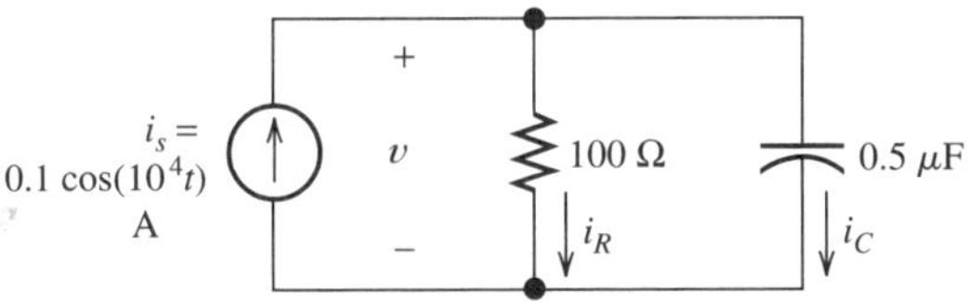

Figure P5.48

P5.49. Consider the circuit shown in Figure P5.49. Find the phasors $\mathbf{V}_1, \mathbf{V}_2, \mathbf{V}_R, \mathbf{V}_L$, and $\mathbf{I}$. Draw the phasor diagram to scale. What is the phase relationship between $\mathbf{I}$ and $\mathbf{V}_1$? Between $\mathbf{I}$ and $\mathbf{V}_L$?

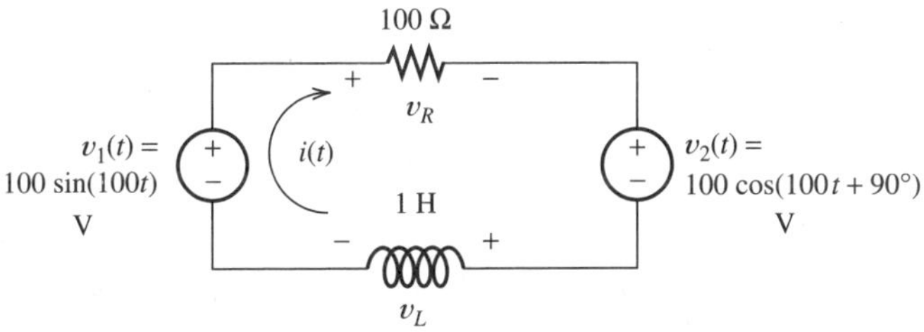

Figure P5.49

P5.50. Solve for the node voltage shown in Figure P5.50.

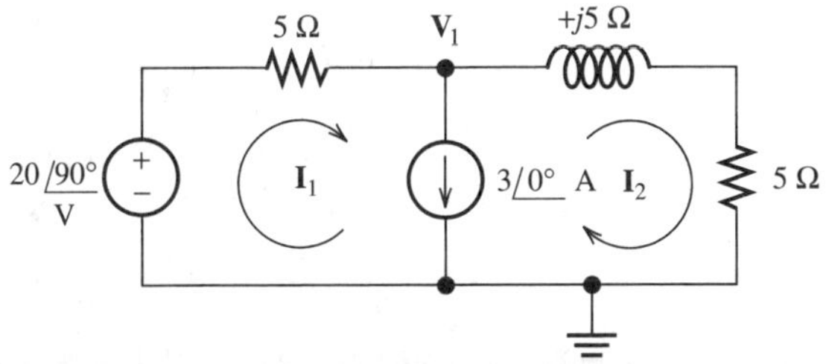

Figure P5.50

P5.51. Find the phasors $\mathbf{I}$, $\mathbf{I}_R$, and $\mathbf{I}_C$ for the circuit shown in Figure P5.51. Also, draw the phasor diagram for the currents.

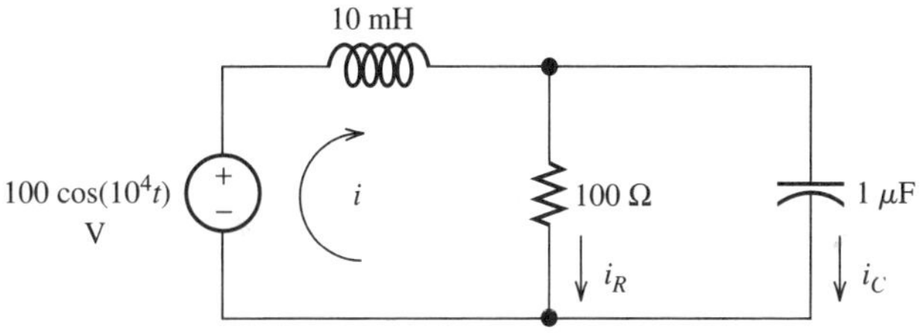

Figure P5.51

P5.52. a. At what frequency or frequencies is the series combination of elements shown in Figure P5.52 equivalent to an open circuit? A short circuit? **b.** Repeat with the elements in parallel.

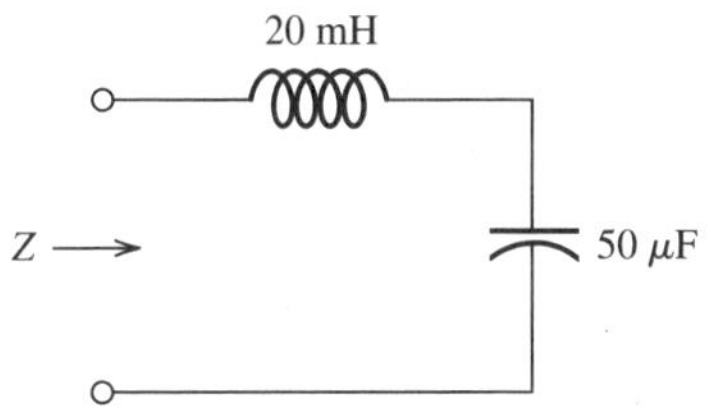

Figure P5.52

P5.53. Solve for the node voltage shown in Figure P5.53.

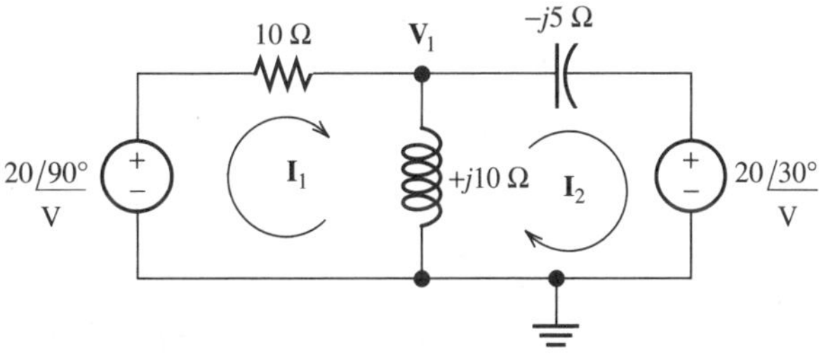

Figure P5.53

P5.54. a. At what frequency or frequencies is the series combination of elements shown in Figure P5.54 equivalent to an open circuit? A short circuit? **b.** Repeat with the elements in parallel.

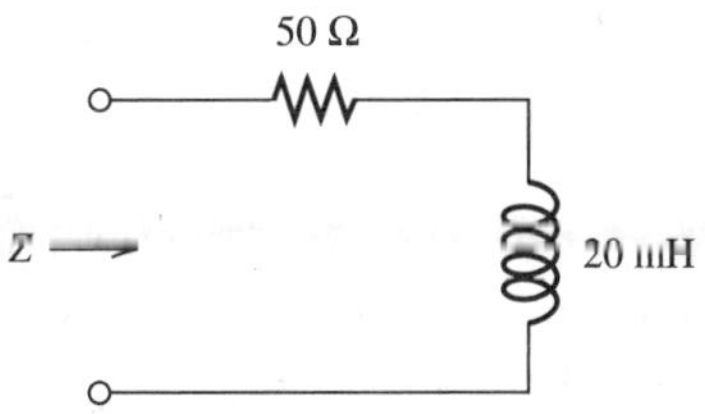

Figure P5.54

Section 5.5: Power in AC Circuits

P5.55. Define real power, reactive power, and apparent power. What are customary units for them?

P5.56. How do we compute the complex power delivered to a circuit component? How are average power and reactive power related to complex power?

P5.57. Explain how power factor and power angle are related. How is the power angle determined if we know the phasors for the current through and voltage across a load?

P5.58. A load is said to have a leading power factor. Is it capacitive or inductive? Is the reactive power positive or negative? Repeat for a load with lagging power factor.

P5.59. Assuming that a nonzero ac source is applied, state whether the power and reactive power are positive, negative, or zero for: **a.** a pure resistance; **b.** a pure inductance; **c.** a pure capacitance.

P5.60. Define what we mean by "power-factor correction." For power-factor correction of an inductive load, what type of element should we place in parallel with the load?

P5.61. **a.** Sketch a power triangle for an inductive load, label the sides, and show the power angle. **b.** Repeat for a capacitive load.

P5.62. Discuss why power plant and distribution system engineers are concerned with: **a.** the real power absorbed by a load; **b.** with the reactive power.

***P5.63.** Consider the circuit shown in Figure P5.63. Find the phasor current **I**. Find the power, reactive power, and apparent power delivered by the source. Find the power factor and state whether it is lagging or leading.

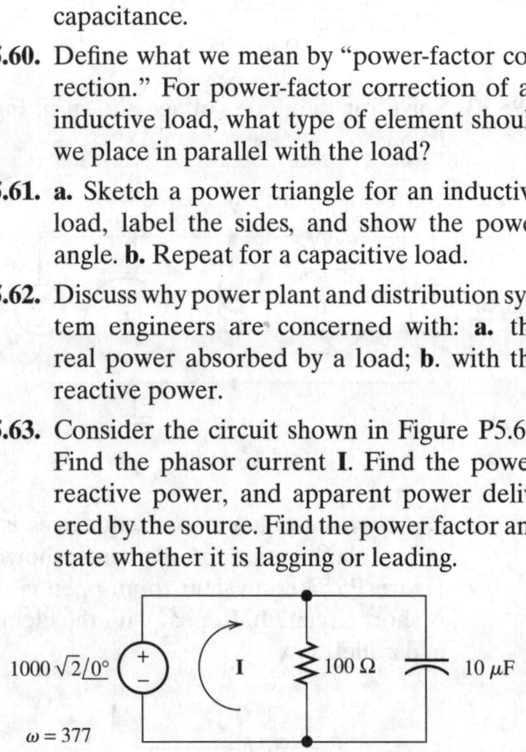

Figure P5.63

P5.64. Repeat Problem P5.63, replacing the capacitance by a 1-H inductance and replacing the voltage angle by 90°.

***P5.65.** Consider a load that has an impedance given by $Z = 100 - j50\ \Omega$. The current flowing through this load is $\mathbf{I} = 15\sqrt{2}\,\angle 30°$ A. Is the load inductive or capacitive? Determine the power factor, power, reactive power, and apparent power delivered to the load.

P5.66. Determine the complex power, power factor, power, reactive power, and apparent power delivered to the load shown in Figure P5.66, given that $\mathbf{I} = 15\sqrt{2}\,\angle 75°$ A. Also, determine the load impedance. Is the power factor leading or lagging?

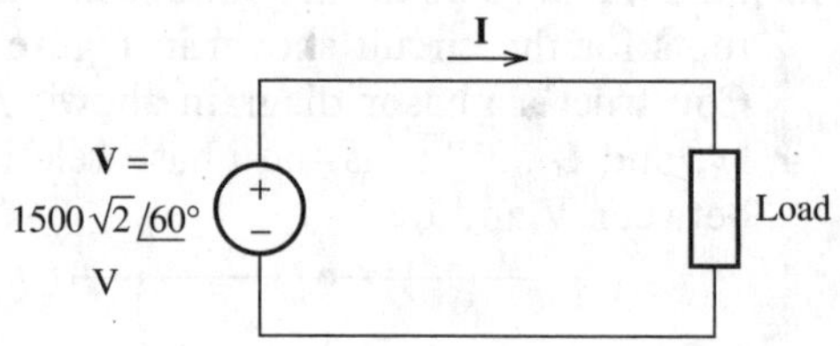

Figure P5.66

P5.67. Determine the power for each element, including the sources, shown in Figure P5.67. Also, state whether each element is delivering or absorbing energy.

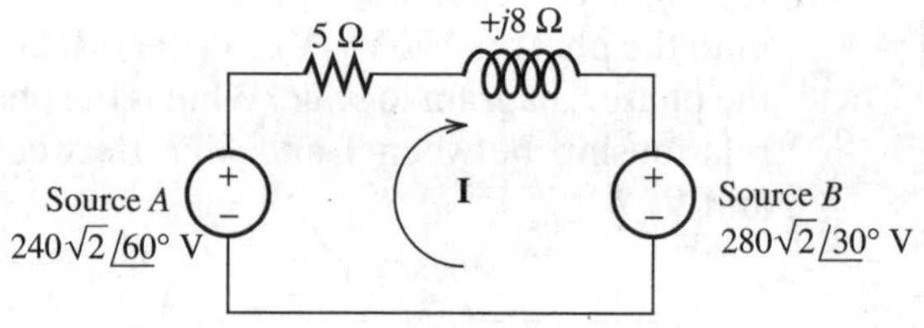

Figure P5.67

P5.68. Suppose that the load impedance shown in Figure P5.66 is given by $Z = 40 - j30\Omega$. (This is a different value from that of Problem P5.66.) Is the load inductive or capacitive? Determine the power factor, complex power, real power, reactive power, and apparent power delivered to the load.

P5.69. The voltage across a load is $v(t) = 10^4\sqrt{2}\cos(\omega t + 75°)$ V, and the current through the load is $i(t) = 25\sqrt{2}\cos(\omega t + 30°)$ A. The reference direction for the current points into the positive reference for the voltage. Determine the complex power, the power factor, the real power, the reactive power, and the apparent power of the load. Is this load inductive or capacitive?

P5.70. Given that a nonzero ac voltage source is applied, state whether the power and reactive power are positive, negative, or zero for: **a.** a pure capacitance; **b.** a resistance in series with an inductance; **c.** a resistance in series with a capacitance; **d.** a pure resistance. (Assume that the resistances, inductance, and capacitance are nonzero and finite in value.)

P5.71. Given that a nonzero ac voltage source is applied, what can you say about whether the power and reactive power are positive, negative, or zero for a pure capacitance in series with a pure inductance? Consider cases in which the impedance magnitude of the capacitance is greater than, equal to, or less than the impedance magnitude of the inductance.

P5.72. Repeat Problem P5.71 for the inductance and capacitance in parallel.

P5.73. A 60-Hz 240-V-rms source supplies power to a load consisting of a resistance in series with an inductance. The real power is 1500 W, and the apparent power is 2500 VA. Determine the value of the resistance, the value of the inductance, and the power factor.

P5.74. Determine the power for each element, including the sources, shown in Figure P5.74. Also, state whether each element is delivering or absorbing energy.

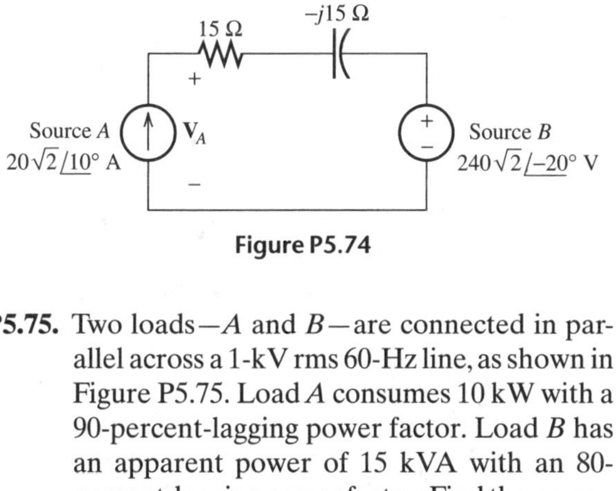

Figure P5.74

***P5.75.** Two loads—A and B—are connected in parallel across a 1-kV rms 60-Hz line, as shown in Figure P5.75. Load A consumes 10 kW with a 90-percent-lagging power factor. Load B has an apparent power of 15 kVA with an 80-percent-lagging power factor. Find the power, reactive power, and apparent power delivered by the source. What is the power factor seen by the source?

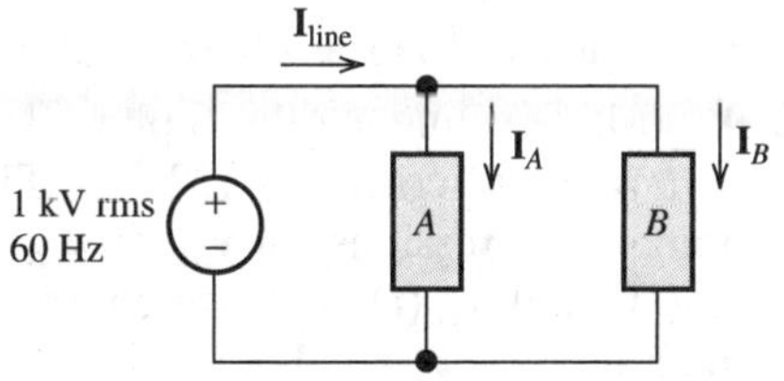

Figure P5.75

P5.76. Repeat Problem P5.75 given that load A consumes 50 kW with a 60-percent-lagging power factor and load B consumes 75 kW with an 80-percent-leading power factor.

P5.77. Find the power, reactive power, and apparent power delivered by the source in Figure P5.77. Find the power factor seen by the source and state whether it is leading or lagging.

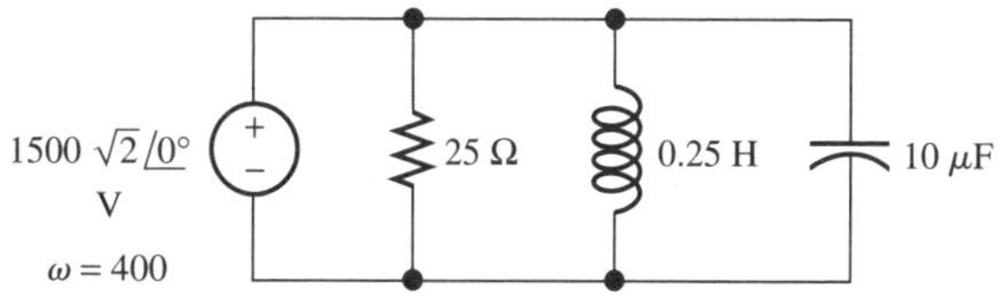

Figure P5.77

P5.78. Repeat Problem P5.77 with the resistance, inductance, and capacitance connected in series rather than in parallel.

***P5.79.** Consider the situation shown in Figure P5.79. A 1000-V rms source delivers power to a load. The load consumes 100 kW with a power factor of 25 percent lagging. **a.** Find the phasor **I**, assuming that the capacitor is not connected to the circuit. **b.** Find the value of the capacitance that must be connected in parallel with the load to achieve a power factor of 100 percent. Usually, power-systems engineers rate capacitances used for power-factor correction in terms of their reactive power rating. What is the rating of this capacitance in kVAR? Assuming that this capacitance is connected, find the new value for the phasor **I**. **c.** Suppose that the source is connected to the load by a long distance. What are the potential advantages and disadvantages of connecting the capacitance across the load?

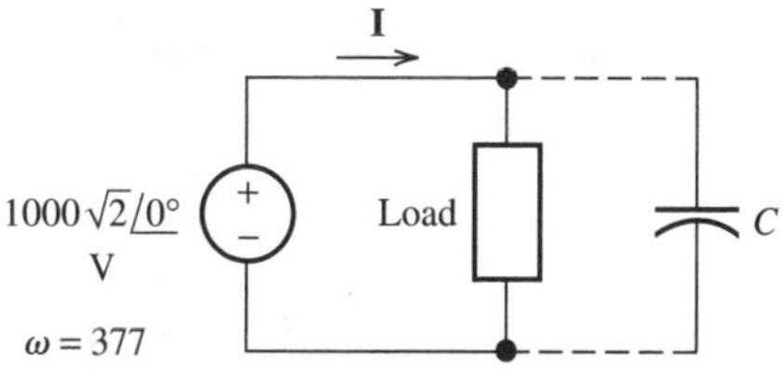

Figure P5.79

Section 5.6: Thévenin and Norton Equivalent Circuits

P5.80. Of what does an ac steady-state Thévenin equivalent circuit consist? A Norton equivalent circuit? How are the values of the parameters of these circuits determined?

P5.81. For an ac circuit consisting of a load connected to a Thévenin circuit, is it possible for the load voltage to exceed the Thévenin voltage in magnitude? If not, why not? If so, under what conditions is it possible? Explain.

P5.82. To attain maximum power delivered to a load, what value of load impedance is required if: **a.** the load can have any complex value; **b.** the load must be pure resistance?

***P5.83.** **a.** Find the Thévenin and Norton equivalent circuits for the circuit shown in Figure P5.83. **b.** Find the maximum power that this circuit can deliver to a load if the load can have any complex impedance; **c.** if the load is purely resistive.

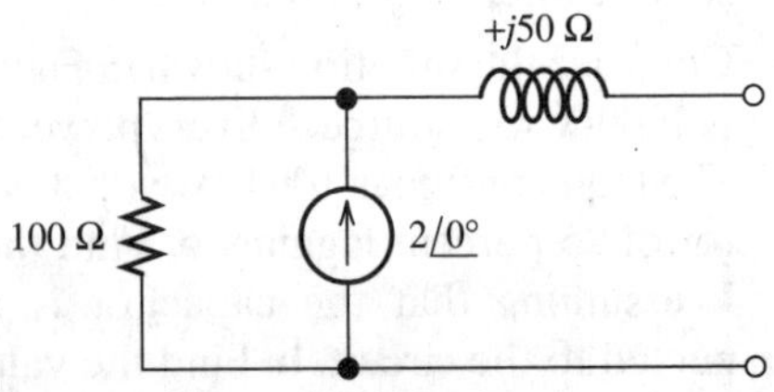

Figure P5.83

P5.84. Draw the Thévenin and Norton equivalent circuits for Figure P5.84, labeling the elements and terminals.

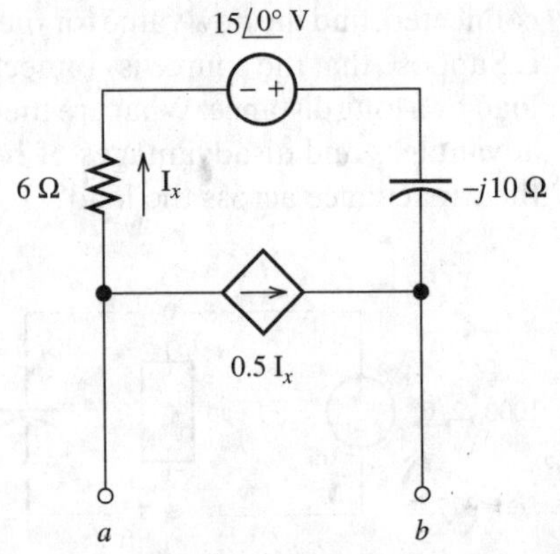

Figure P5.84

P5.85. Find the Thévenin voltage, Thévenin impedance, and Norton current for the two-terminal circuit shown in Figure P5.85.

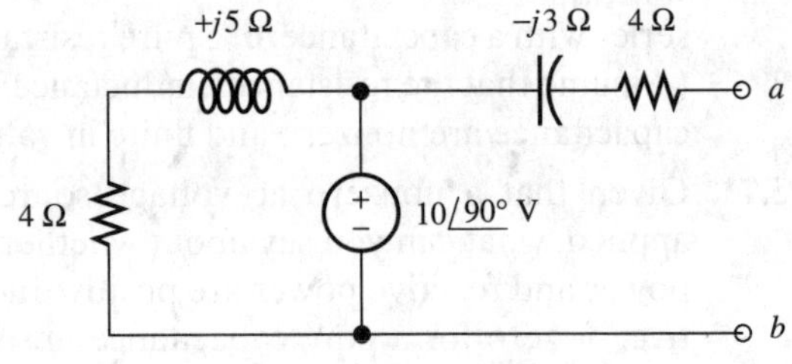

Figure P5.85

P5.86. Find the Thévenin and Norton equivalent circuits for the circuit shown in Figure P5.86. Find the maximum power that this circuit can deliver to a load if the load can have any complex impedance. Repeat if the load must be purely resistive.

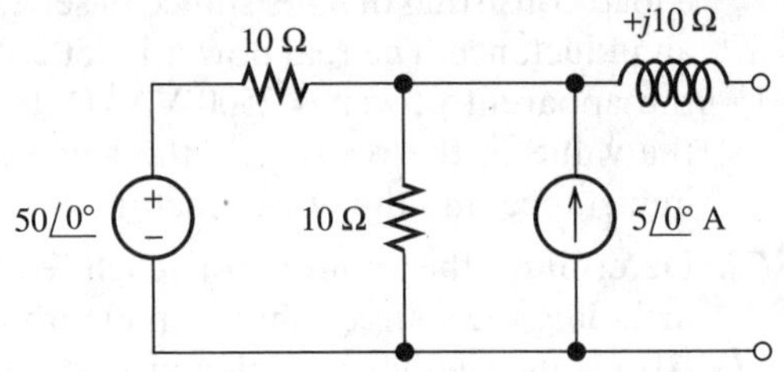

Figure P5.86

***P5.87.** The Thévenin equivalent of a two-terminal network is shown in Figure P5.87. The frequency is $f = 60$ Hz. We wish to connect a load across terminals *a*–*b* that consists of a resistance and a capacitance in parallel such that the power delivered to the resistance is maximized. Find the value of the resistance and the value of the capacitance.

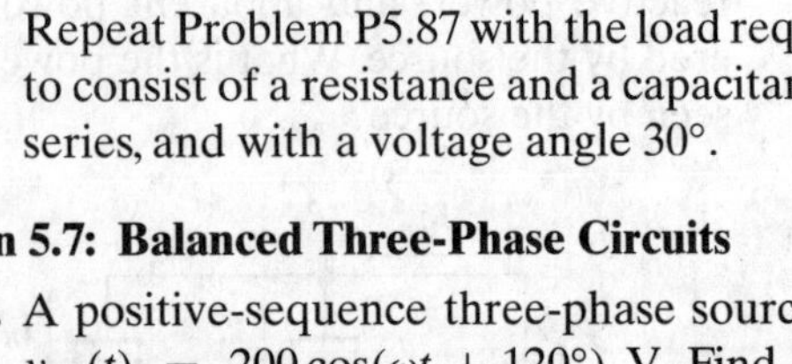

Figure P5.87

P5.88. Repeat Problem P5.87 with the load required to consist of a resistance and a capacitance in series, and with a voltage angle 30°.

Section 5.7: Balanced Three-Phase Circuits

P5.89. A positive-sequence three-phase source has $v_{an}(t) = 200\cos(\omega t + 120°)$ V. Find time-domain expressions for $v_{bn}(t)$, $v_{cn}(t)$, $v_{ab}(t)$, $v_{bc}(t)$, and $v_{ca}(t)$ and sketch their phasor diagram.

P5.90. We have a balanced positive-sequence three-phase source for which

$$v_{an}(t) = 150\cos(400\pi t)\text{ V}.$$

a. Find the frequency of this source in Hz.
b. Give expressions for $v_{bn}(t)$ and $v_{cn}(t)$.
c. Repeat part (b) for a negative-sequence source.

***P5.91.** Each phase of a wye-connected load consists of a 50-Ω resistance in parallel with a 100-μF capacitance. Find the impedance of each phase of an equivalent delta-connected load. The frequency of operation is 60 Hz.

***P5.92.** A balanced wye-connected three-phase source has line-to-neutral voltages of 440 V rms. Find the rms line-to-line voltage magnitude. If this source is applied to a wye-connected load composed of three 30-Ω resistances, find the rms line-current magnitude and the total power delivered.

P5.93. What can you say about the flow of power as a function of time between a balanced three-phase source and a balanced load? Is this true of a single-phase source and a load? How is this a potential advantage for the three-phase system? What is another advantage of three-phase power distribution compared with single phase?

P5.94. A delta-connected source delivers power to a delta-connected load, as shown in Figure P5.94. The rms line-to-line voltage at the source is $V_{ab\text{rms}} = 440$ V. The load impedance is $Z_\Delta = 12 + j3$. Find $\mathbf{I}_{aA}, \mathbf{V}_{AB}, \mathbf{I}_{AB}$, the total power delivered to the load, and the power lost in the line.

***P5.95.** Repeat Problem P5.94, with $Z_\Delta = 15 - j6$.

P5.96. A balanced wye-connected three-phase source has line-to-neutral voltages of 277 V rms. Find the rms line-to-line voltage. This source is applied to a delta-connected load, each arm of which consists of a 15-Ω resistance in parallel with a $+j30$-Ω reactance. Determine the rms line current magnitude, the power factor, and the total power delivered.

P5.97. A negative-sequence wye-connected source has line-to-neutral voltages $V_{an} = V_Y\ \underline{/180^\circ}$, $V_{bn} = V_Y\ \underline{/-60^\circ}$, and $V_{cn} = V_Y\ \underline{/60^\circ}$. Find the line-to-line voltages $\mathbf{V}_{ab}$, $\mathbf{V}_{bc}$, and $\mathbf{V}_{ca}$. Construct a phasor diagram showing both sets of voltages and compare with Figure 5.41 on page 273.

P5.98. A balanced positive-sequence wye-connected 60-Hz three-phase source has line-to-line voltages of $V_L = 208$ V rms. This source is connected to a balanced wye-connected load. Each phase of the load consists of an impedance of $30 + j40$ Ω. Find the line-to-neutral voltage phasors, the line-to-line voltage phasors, the line-current phasors, the power, and the reactive power delivered to the load. Assume that the phase of $\mathbf{V}_{an}$ is zero.

P5.99. In this chapter, we have considered balanced loads only. However, it is possible to determine an equivalent wye for an unbalanced delta, and vice versa. Consider the equivalent circuits shown in Figure P5.99. Derive

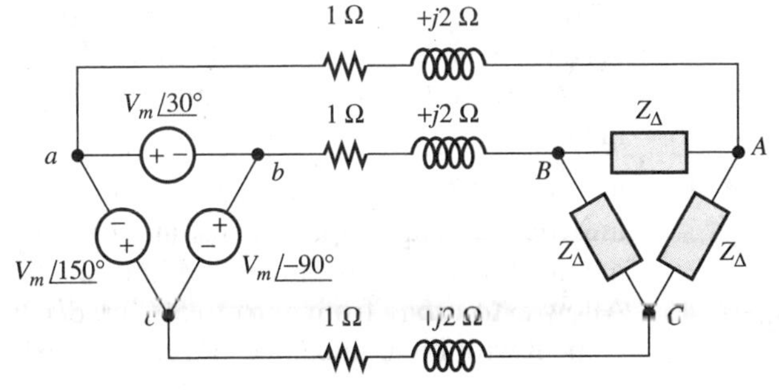

Figure P5.94

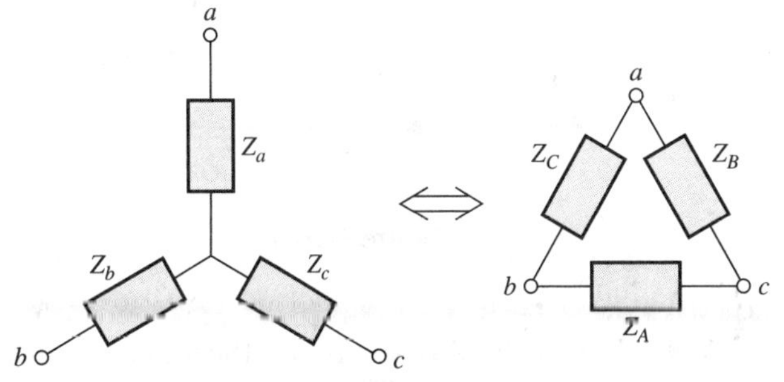

Figure P5.99

formulas for the impedances of the wye in terms of the impedances of the delta. (*Hint*: Equate the impedances between corresponding pairs of terminals of the two circuits with the third terminal open. Then, solve the equations for Z_a, Z_b, and Z_c in terms of Z_A, Z_B, and Z_C. Take care in distinguishing between upper- and lowercase subscripts.)

P5.100. Repeat Problem P5.99, but solve for the impedances of the delta in terms of those of the wye. [*Hint*: Start by working in terms of the admittances of the delta ($Y_A, Y_B,$ and Y_C) and the impedances of the wye (Z_a, Z_b, and Z_c). Short terminals b and c for each circuit. Then, equate the admittances between terminal a and the shorted terminals for the two circuits. Repeat this twice more with shorts between the remaining two pairs of terminals. Solve the equations to determine Y_A, Y_B, and Y_C in terms of Z_a, Z_b, and Z_c. Finally, invert the equations for Y_A, Y_B, and Y_C to obtain equations relating the impedances. Take care in distinguishing between upper- and lowercase subscripts.]

Section 5.8: AC Analysis using MATLAB

***P5.101.** Use MATLAB to solve for the node voltages shown in Figure P5.101.

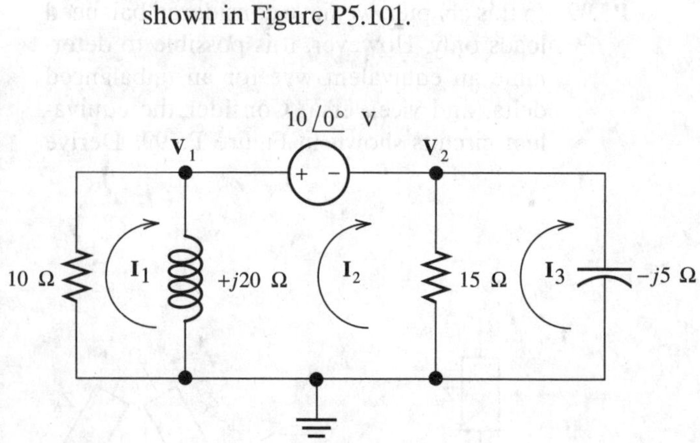

Figure P5.101

P5.102. Use MATLAB to solve for the mesh currents shown in Figure P5.101.

***P5.103.** Use MATLAB to solve for the mesh currents shown in Figure P5.53.

P5.104. Use MATLAB to solve for the mesh currents shown in Figure P5.50.

P5.105. Use MATLAB to solve for the node voltages shown in Figure P5.105.

P5.106. A **Lissajous figure** results if one sinusoid is plotted versus another. Consider $x(t) = \cos(\omega_x t)$ and $y(t) = \cos(\omega_y t + \theta)$. Use MATLAB to generate values of x and y for 20 seconds at 100 points per second and obtain a plot of y versus x for: **a.** $\omega_x = \omega_y = 2\pi$ and $\theta = 0°$; **b.** $\omega_x = \omega_y = 2\pi$ and $\theta = 45°$; **c.** $\omega_x = \omega_y = 2\pi$ and $\theta = 90°$; **d.** $\omega_x = 6\pi$, $\omega_y = 2\pi$, and $\theta = 90°$.

P5.107. Use the MATLAB Symbolic Toolbox to determine the rms value of $v(t)$ which has a period of 1 s and is given by $v(t) = 10\exp(-5t)\sin(20\pi t)$ V for $0 \le t \le 1$ s.

P5.108. Use MATLAB to obtain a plot of $v(t) = \cos(19\pi t) + \cos(21\pi t)$ for t ranging from 0 to 2 seconds. Explain why the terms in this expression cannot be combined by using phasors. Then, considering that the two terms can be represented as the real projection of the sum of two vectors rotating at different speeds in the complex plane, comment on the plot.

P5.109. Use MATLAB or manually produce plots of the magnitudes of the impedances of a 20-mH inductance, a 10-μF capacitance, and a 40-Ω resistance to scale versus frequency for the range from zero to 1000 Hz.

P5.110. **a.** Use MATLAB to produce a plot of the impedance magnitude versus angular frequency for the circuit of Figure P5.52. Allow ω to range from zero to 2000 rad/s and the vertical axis to range from 0 to 100 Ω. **b.** Repeat with the inductance and capacitance in parallel.

P5.111. **a.** Use MATLAB to produce a plot of the impedance magnitude versus angular frequency for the circuit shown in Figure P5.54. Allow ω to range from zero to 5000 rad/s. **b.** Repeat with the inductance and resistance in parallel.

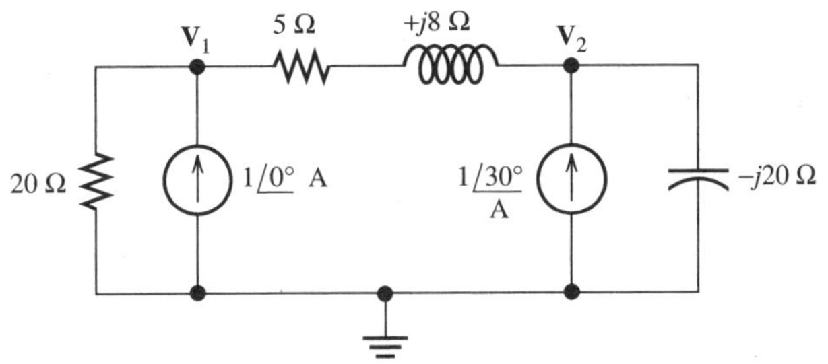

Figure P5.105

Frequently Asked Questions

1. Explain the terms amplitude, cycle, time period, frequency and phase difference as applied to alternating wave forms. **(UPTU)**
2. Define r.m.s. value and average value of an alternating quantity.
3. Show that the r.m.s. value of a sinusoidal a.c. voltage of amplitude V_m is $V_m/\sqrt{2}$.
4. Derive the expression for average and r.m.s. value of a sinusoidally varying quantity.
5. Show that the form factor of a sinusoidal wave form is 1.11.
6. Derive the values of form factor and peak factor of a sinusoidally varying quantity.
7. Explain how sinusoidal quantities can be represented by vectors.
8. Explain the component method for addition of alternating quantities. **(UPTU)**
9. Distinguish between
   (i) Time period and frequency
   (ii) Cycle and alternation
   (iii) Peak factor and form factor
   (iv) Average and rms value of an alternating current.
10. Explain why the phasor of voltage across an inductor leads its current by 90° and the phasor of voltage across a capacitor lags its current by 90°.
11. Determine phase angle relationship between alternating voltage and current in a purely inductive and a purely capacitance circuit under steady condition.
12. Explain with mathematical expression that power consumed in a pure inductance in zero. **(PTU)**
13. Explain with mathematical expression that power consumed in a pure capacitance is zero.
14. Show that power consumed in a pure resistive circuit is not constant but it is fluctuating.
15. Develop the expression for the mean power consumed over a cycle of a single-phase sinusoidal supply delivering power to load comprising of a resistance R in series with an inductance *L*.
16. Define the term 'watt-less current'.
17. What do you understand by real power, reactive power and apparent power? **(UPTU)**
18. What is power factor? Discuss the practical importance of p.f. **(UPTU)**
19. Explain the use of capacitance for p.f. improvement.
20. What do you understand by the operator *j*?
21. How will you determine the Thevenin equivalent of an ac network?
22. How will you determine the Norton equivalent of an ac network?
23. Determine the conditions for maximum power transfer to a load by an ac network.
24. What do you mean by phase sequence in a 3-phase ac system?
25. If the voltage of *R*-phase is represented by $v_R = V_m \sin \omega t$, write down similar expressions for the voltage of *Y*-phase and *B*-phase. The phase sequence being *RBY*.
26. What are the advantages of 3-phase system over 1-phase system? **(UPTU)**
27. What do you understand by 3 phase power supply? Describe star and delta-connections by showing line voltages and phase voltages, line currents and phase currents. **(UPTU)**

28. Derive the relation between phase and line voltages and currents for a balanced 3-phase star-connected system.

Or **(UPTU, 2004-05)**

For star connected system in a 3-phase circuit prove that $V_L = \sqrt{3}\, V_{ph}$ and $I_L = I_{plt}$.

Or **(UPTU, 2006-07)**

Derive the relationship between the line and phase voltages of an alternator.

(UPTU, 2006-07)

29. Write down the relationship between line voltage and line current with phase voltage and phase current in star-connected and delta connected circuits.

30. Derive expression for power in 3-phase star and delta connections. **(UTU, 2005-06)**

31. Obtain the line and phase relationships for voltages and currents for a delta-connected 3-phase power system. Also obtain the expression of total power and per phase power.

Practice Test

Here is a practice test you can use to check your comprehension of the most important concepts in this chapter. Answers can be found in Appendix D and complete solutions are included in the Student Solutions files. See Appendix E for more information about the Student Solutions.

T5.1. Determine the rms value of the current shown in Figure T5.1 and the average power delivered to the 50-Ω resistance.

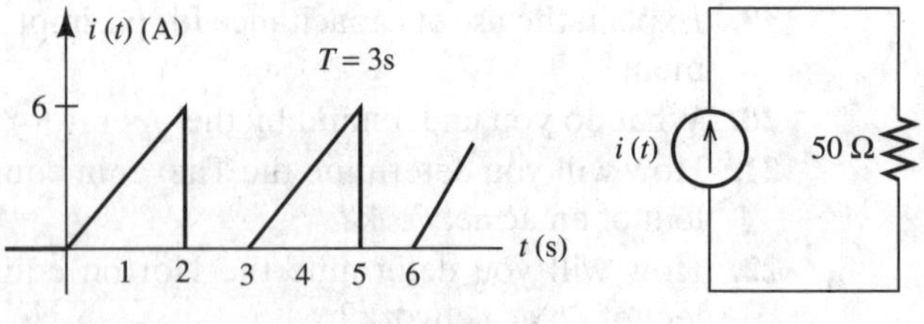

Figure T5.1

T5.2. Reduce the expression

$$v(t) = 5\sin(\omega t + 45°) + 5\cos(\omega t - 30°)$$

to the form $V_m \cos(\omega t + \theta)$.

T5.3. We have two voltages $v_1(t) = 15\sin(400\pi t + 45°)$ V and $v_2(t) = 5\cos(400\pi t - 30°)$ V. Determine (including units): **a.** the rms value of $v_1(t)$; **b.** the frequency of the voltages; **c.** the angular frequency of the voltages; **d.** the period of the voltages; **e.** the phase relationship between $v_1(t)$ and $v_2(t)$.

T5.4. Find the phasor values of $\mathbf{V}_R$, $\mathbf{V}_L$, and $\mathbf{V}_C$ in polar form for the circuit of Figure T5.4.

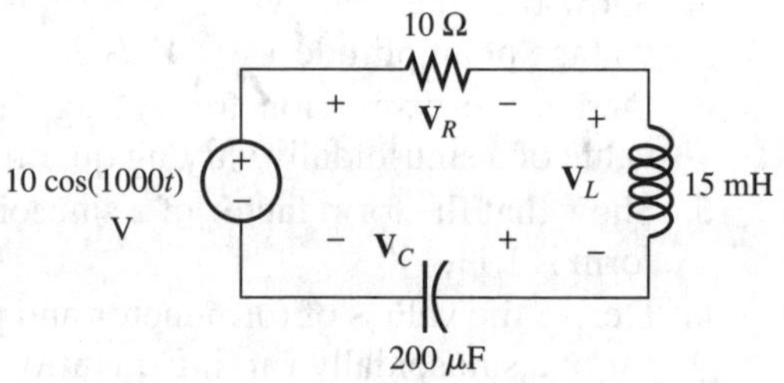

Figure T5.4

T5.5. Determine the complex power, power, reactive power, and apparent power absorbed by the load in Figure T5.5. Also, determine the power factor for the load.

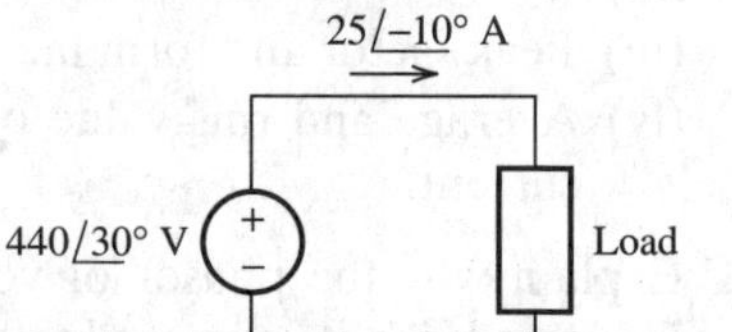

Figure T5.5

T5.6. Determine the line current $\mathbf{I}_{aA}$ in polar form for the circuit of Figure T5.6. This is a positive-sequence, balanced, three-phase system with $\mathbf{V}_{an} = 208\angle 30°$ V and $Z_\Delta = 6 + j8\ \Omega$.

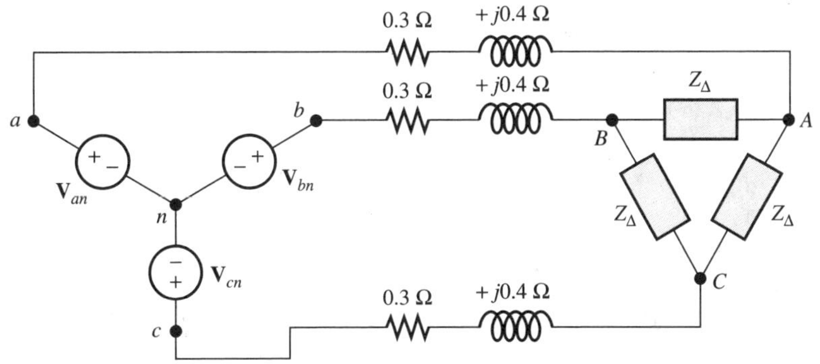

Figure T5.6

T5.7. Write the MATLAB commands to obtain the values of the mesh currents of Figure T5.7 in polar form. You may use the pin and pout functions defined in this chapter if you wish.

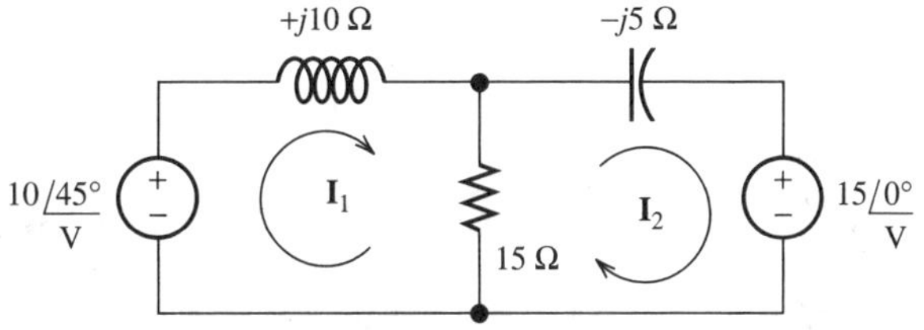

Figure T5.7

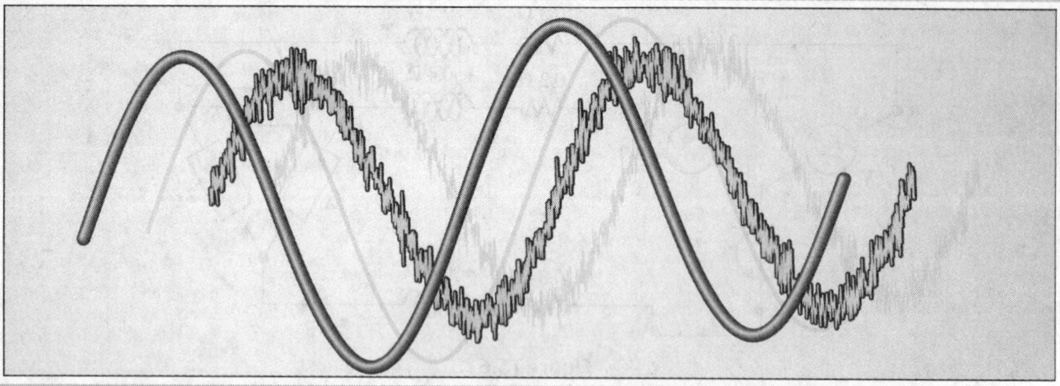

Chapter 6

Frequency Response, Bode Plots, and Resonance

Study of this chapter will enable you to:

- State the fundamental concepts of Fourier analysis.
- Use a filter's transfer function to determine its output for a given input consisting of sinusoidal components.
- Use circuit analysis to determine the transfer functions of simple circuits.
- Draw first-order lowpass or highpass filter circuits and sketch their transfer functions.
- Understand decibels, logarithmic frequency scales, and Bode plots.
- Draw the Bode plots for transfer functions of first-order filters.
- Calculate parameters for series- and parallel-resonant circuits.
- Select and design simple filter circuits.
- Use MATLAB to derive and plot network functions.
- Design simple digital signal-processing systems.

Introduction to this chapter:

Much of electrical engineering is concerned with information-bearing currents and voltages that we call **signals**. For example, transducers on an internal combustion engine provide electrical signals that represent temperature, speed, throttle position, and the rotational position of the crankshaft. These signals are **processed** (by electrical circuits) to determine the optimum firing instant for each cylinder. Finally, electrical pulses are generated for each spark plug.

Surveyors can measure distances by using an instrument that emits a pulse of light that is reflected by a mirror at the point of interest. The return light pulse is converted to an electrical signal that is processed by circuits to determine the round-trip time delay between the instrument and the mirror. Finally, the delay is converted to distance and displayed.

Another example of signal processing is the electrocardiogram, which is a plot of the electrical signal generated by the human heart. In a cardiac-care unit, circuits and computers are employed to extract information concerning the behavior of a patient's heart. A physician or nurse is alerted when the patient needs attention.

In general, **signal processing** is concerned with manipulating signals to extract information and using that information to generate other useful electrical signals. It is an important and far-reaching subject. In this chapter, we consider several simple but, nevertheless, useful circuits from a signal-processing point of view.

Recall that in Chapter 5 we learned how to analyze circuits containing sinusoidal sources, all of which have a common frequency. An important application is electrical power systems. However, most real-world information-bearing electrical signals are not sinusoidal. Nevertheless, we will see that phasor concepts can be very useful in understanding how circuits respond to nonsinusoidal signals. This is true because nonsinusoidal signals can be considered to be the sum of sinusoidal components having various frequencies, amplitudes, and phases.

6.1 FOURIER ANALYSIS, FILTERS, AND TRANSFER FUNCTIONS

Fourier Analysis

As mentioned in the introduction to this chapter, most information-bearing signals are not sinusoidal. For example, the waveform produced by a microphone for speech or music is a complex nonsinusoidal waveform that is not predictable in advance. Figure 6.1(a) shows a (very) short segment of a music signal.

Even though many interesting signals are not sinusoidal, it turns out that we can construct any waveform by adding sinusoids that have the proper amplitudes, frequencies, and phases. For illustration, the waveform shown in Figure 6.1(a) is the sum of the sinusoids shown in Figure 6.1(b). The waveform shown in Figure 6.1 is relatively simple because it is composed of only three components. Most natural

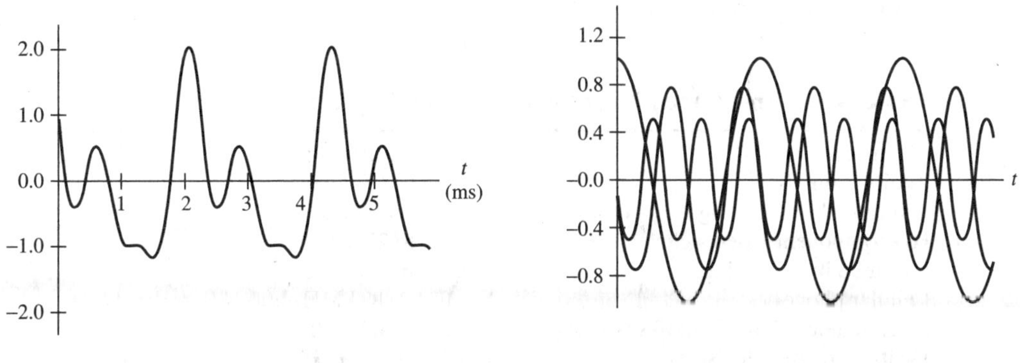

(a) Music waveform (b) Sinusoidal components

Figure 6.1 The short segment of a music waveform shown in (a) is the sum of the sinusoidal components shown in (b).

signals contain thousands of components. (In theory, the number is infinite in many cases.)

When we listen to music, our ears respond differently to the various frequency components. Some combinations of amplitudes and frequencies are pleasing, whereas other combinations are not. Thus, in the design of signal-processing circuits (such as amplifiers) for audio signals, we must consider how the circuits respond to components having different frequencies.

Fourier analysis is a mathematical technique for finding the amplitudes, frequencies, and phases of the components of a given waveform. Aside from mentioning some of the results of Fourier analysis, we will not develop the theory in detail. *The important point is that all real-world signals are sums of sinusoidal components.*

All real-world signals are sums of sinusoidal components having various frequencies, amplitudes, and phases.

The range of the frequencies of the components depends on the type of signal under consideration. The frequency ranges for several types of signals are given in Table 6.1. Thus, electrocardiograms are composed of numerous sinusoidal components with frequencies ranging from 0.05 Hz to 100 Hz.

Fourier Series of a Square Wave. As another example, consider the signal shown in Figure 6.2(a), which is called a **square wave**. Fourier analysis shows that the square wave can be written as an infinite series of sinusoidal components,

$$v_{sq}(t) = \frac{4A}{\pi}\sin(\omega_0 t) + \frac{4A}{3\pi}\sin(3\omega_0 t) + \frac{4A}{5\pi}\sin(5\omega_0 t) + \cdots \tag{6.1}$$

in which $\omega_0 = 2\pi/T$ is the called the **fundamental angular frequency** of the square wave.

Figure 6.2(b) shows several of the terms in this series and the result of summing the first five terms. Clearly, even the sum of the first five terms is a fairly good approximation to the square wave, and the approximation becomes better as more components are added. Thus, the square wave is composed of an infinite number of sinusoidal components. The frequencies of the components are odd integer multiples of the fundamental frequency, the amplitudes decline with increasing frequency, and the phases of all components are $-90°$. Unlike the square wave, the components of real-world signals are confined to finite ranges of frequency, and their amplitudes are not given by simple mathematical expressions.

The components of real-world signals are confined to finite ranges of frequency.

Sometimes a signal contains a component that has a frequency of zero. For zero frequency, a general sinusoid of the form $A\cos(\omega t + \theta)$ becomes simply $A\cos(\theta)$, which is constant for all time. Recall that we refer to constant voltages as dc, so zero frequency corresponds to dc.

Zero frequency corresponds to dc.

Table 6.1. Frequency Ranges of Selected Signals

Electrocardiogram	0.05 to 100 Hz
Audible sounds	20 Hz to 15 kHz
AM radio broadcasting	540 to 1600 kHz
HD component video signals	Dc to 25 MHz
FM radio broadcasting	88 to 108 MHz
Cellular phone	824 to 894 MHz and 1850 to 1990 MHz
Satellite television downlinks (C-band)	3.7 to 4.2 GHz
Digital satellite television	12.2 to 12.7 GHz

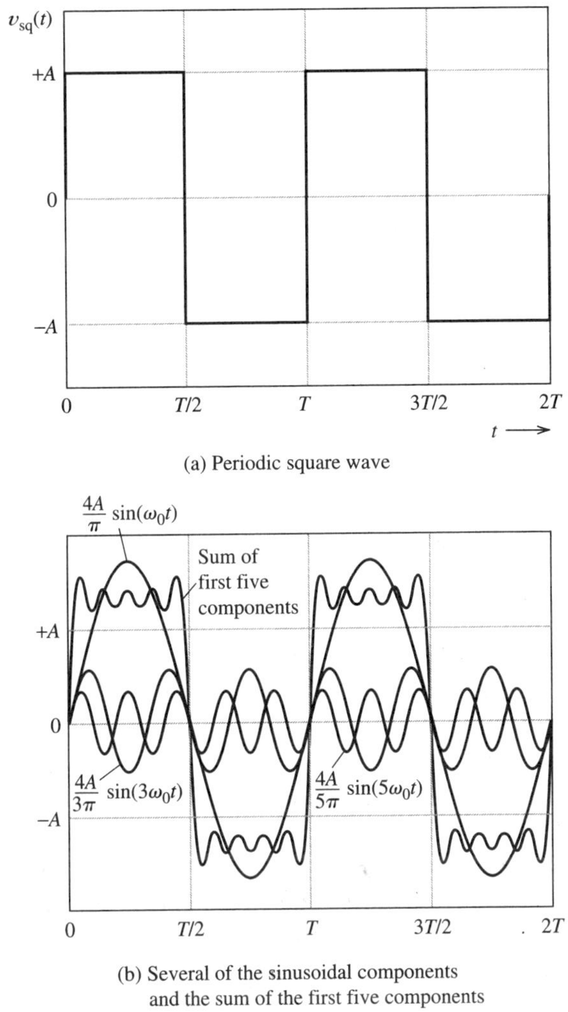

Figure 6.2 A square wave and some of its components.

In sum, the fact that all signals are composed of sinusoidal components is a fundamental idea in electrical engineering. The frequencies of the components, as well as their amplitudes and phases, for a given signal can be determined by theoretical analysis or by laboratory measurements (using an instrument called a *spectrum analyzer*). Very often, the design of a system for processing information-bearing signals is based on considerations of how the system should respond to components of various frequencies.

... the fact that all signals are composed of sinusoidal components is a fundamental idea in electrical engineering.

Filters

There are many applications in which we want to retain components in a given range of frequencies and discard the components in another range. This can be accomplished by the use of electrical circuits called **filters**. (Actually, filters can take many forms, but we limit our discussion to a few relatively simple *RLC* circuits.)

Usually, filter circuits are **two-port networks**, an example of which is illustrated in Figure 6.3. The signal to be filtered is applied to the input port and (ideally) only

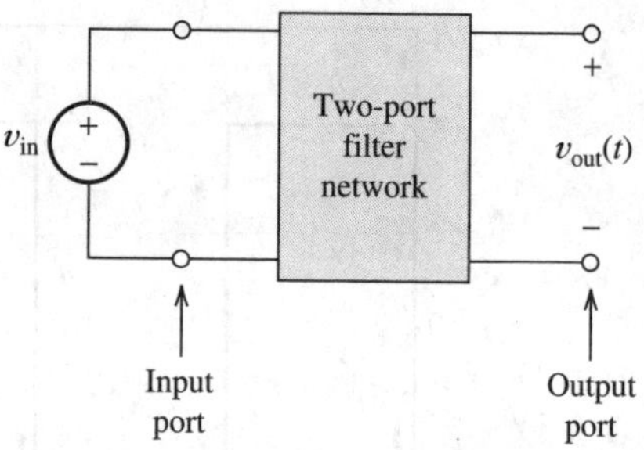

Figure 6.3 When an input signal $v_{in}(t)$ is applied to the input port of a filter, some components are passed to the output port while others are not, depending on their frequencies. Thus, $v_{out}(t)$ contains some of the components of $v_{in}(t)$, but not others. Usually, the amplitudes and phases of the components are altered in passing through the filter.

Filters process the sinusoid components of an input signal differently depending on the frequency of each component. Often, the goal of the filter is to retain the components in certain frequency ranges and to reject components in other ranges.

RLC circuits provide one way to realize filters.

the components in the frequency range of interest appear at the output port. For example, an FM radio antenna produces a voltage composed of signals from many transmitters. By using a filter that retains the components in the frequency range from 88 to 108 MHz and discards everything else, we can select the FM radio signals and reject other signals that could interfere with the process of extracting audio information.

As we learned in Chapter 5, the impedances of inductances and capacitances change with frequency. For example, the impedance of an inductance is $Z_L = \omega L \angle 90^\circ = 2\pi f L \angle 90^\circ$. Thus, the high-frequency components of a voltage signal applied to an inductance experience a higher impedance magnitude than do the low-frequency components. Consequently, electrical circuits can respond selectively to signal components, depending on their frequencies. Thus, *RLC* circuits provide one way to realize electrical filters. We consider several specific examples later in this chapter.

Transfer Functions

Consider the two-port network shown in Figure 6.3. Suppose that we apply a sinusoidal input signal having a frequency denoted as f and having a phasor $\mathbf{V}_{in}$. In steady state, the output signal is sinusoidal and has the same frequency as the input. The output phasor is denoted as $\mathbf{V}_{out}$.

The **transfer function** $H(f)$ of the two-port filter is defined to be the ratio of the phasor output voltage to the phasor input voltage as a function of frequency:

$$H(f) = \frac{\mathbf{V}_{out}}{\mathbf{V}_{in}} \tag{6.2}$$

The transfer function $H(f)$ of the two-port filter is defined to be the ratio of the phasor output voltage to the phasor input voltage as a function of frequency.

The magnitude of the transfer function shows how the amplitude of each frequency component is affected by the filter. Similarly, the phase of the transfer function shows how the phase of each frequency component is affected by the filter.

Because phasors are complex, the transfer function is a complex quantity having both magnitude and phase. Furthermore, both the magnitude and the phase can be functions of frequency.

The transfer-function magnitude is the ratio of the output amplitude to the input amplitude. The phase of the transfer function is the output phase minus the input phase. Thus, the magnitude of the transfer function shows how the amplitude of each frequency component is affected by the filter. Similarly, the phase of the transfer function shows how the phase of each frequency component is affected by the filter.

Example 6.1 Using the Transfer Function to Determine the Output

The transfer function $H(f)$ of a filter is shown in Figure 6.4. [Notice that the magnitude $|H(f)|$ and phase $\angle H(f)$ are shown separately in the figure.] If the input signal is given by

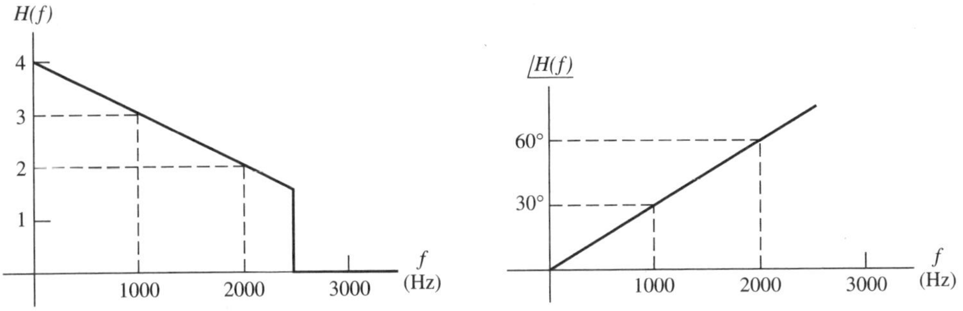

Figure 6.4 The transfer function of a filter. See Examples 6.1 and 6.2.

$$v_{\text{in}}(t) = 2\cos(2000\pi t + 40°)$$

find an expression (as a function of time) for the output of the filter.

Solution By inspection, the frequency of the input signal is $f = 1000$ Hz. Referring to Figure 6.4, we see that the magnitude and phase of the transfer function are $|H(1000)| = 3$ and $\angle H(1000) = 30°$, respectively. Thus, we have

$$H(1000) = 3\angle 30° = \frac{\mathbf{V}_{\text{out}}}{\mathbf{V}_{\text{in}}}$$

The phasor for the input signal is $\mathbf{V}_{\text{in}} = 2\angle 40°$, and we get

$$\mathbf{V}_{\text{out}} = H(1000) \times \mathbf{V}_{\text{in}} = 3\angle 30° \times 2\angle 40° = 6\angle 70°$$

Thus, the output signal is

$$v_{\text{out}}(t) = 6\cos(2000\pi t + 70°)$$

In this case, the amplitude of the input is tripled by the filter. Furthermore, the signal is phase shifted by 30°. Of course, this is evident from the values shown in the plots of the transfer function at $f = 1000$. ■

Exercise 6.1 Repeat Example 6.1 if the input signal is given by: **a.** $v_{\text{in}}(t) = 2\cos(4000\pi t)$; and **b.** $v_{\text{in}}(t) = 1\cos(6000\pi t - 20°)$.

Answer **a.** $v_{\text{out}}(t) = 4\cos(4000\pi t + 60°)$; **b.** $v_{\text{out}}(t) = 0$. □

Notice that the effect of the filter on the magnitude and phase of the signal depends on signal frequency.

Example: Graphic Equalizer. You may own a stereo audio system that has a *graphic equalizer*, which is a filter that has an adjustable transfer function. Usually, the controls of the equalizer are arranged so their positions give an approximate representation of the transfer-function magnitude versus frequency. (Actually, the equalizer in a stereo system contains two filters—one for the left channel and one for the right channel—and the controls are ganged together.) Users can adjust the transfer function to achieve the mix of amplitudes versus frequency that is most pleasing to them.

Input Signals with Multiple Components. If the input signal to a filter contains several frequency components, we can find the output for each input component separately and then add the output components. This is an application of the superposition principle first introduced in Section 2.7.

A step-by-step procedure for determining the output of a filter for an input with multiple components is as follows:

1. Determine the frequency and phasor representation for each input component.
2. Determine the (complex) value of the transfer function for each component.
3. Obtain the phasor for each output component by multiplying the phasor for each input component by the corresponding transfer-function value.
4. Convert the phasors for the output components into time functions of various frequencies. Add these time functions to produce the output.

Example 6.2 Using the Transfer Function with Several Input Components

Suppose that the input signal for the filter of Figure 6.4 is given by

$$v_{\text{in}}(t) = 3 + 2\cos(2000\pi t) + \cos(4000\pi t - 70^\circ)$$

Find an expression for the output signal.

Step 1.

Solution We start by breaking the input signal into its components. The first component is

$$v_{\text{in1}}(t) = 3$$

the second component is

$$v_{\text{in2}}(t) = 2\cos(2000\pi t)$$

and the third component is

$$v_{\text{in3}}(t) = \cos(4000\pi t - 70^\circ)$$

Step 2.

By inspection, we see that the frequencies of the components are 0, 1000, and 2000 Hz, respectively. Referring to the transfer function shown in Figure 6.4, we find that

$$H(0) = 4$$

$$H(1000) = 3\angle 30^\circ$$

and

$$H(2000) = 2\angle 60^\circ$$

The constant (dc) output term is simply $H(0)$ times the dc input:

$$v_{\text{out1}} = H(0)v_{\text{in1}} = 4 \times 3 = 12$$

Step 3.

The phasor outputs for the two input sinusoids are

$$\mathbf{V}_{\text{out2}} = H(1000) \times \mathbf{V}_{\text{in2}} = 3\angle 30^\circ \times 2\angle 0^\circ = 6\angle 30^\circ$$

$$\mathbf{V}_{\text{out3}} = H(2000) \times \mathbf{V}_{\text{in3}} = 2\angle 60^\circ \times 1\angle -70^\circ = 2\angle -10^\circ$$

Next, we can write the output components as functions of time:

Step 4.

$$v_{out1}(t) = 12$$
$$v_{out2}(t) = 6\ \cos(2000\pi t + 30°)$$

and

$$v_{out3}(t) = 2\ \cos(4000\pi t - 10°)$$

Finally, we add the output components to find the output voltage:

$$v_{out}(t) = v_{out1}(t) + v_{out2}(t) + v_{out3}(t)$$

and

$$v_{out}(t) = 12 + 6\ \cos(2000\pi t + 30°) + 2\ \cos(4000\pi t - 10°)$$

■

Notice that we did not add the phasors $\mathbf{V}_{out2}$ and $\mathbf{V}_{out3}$ in Example 6.2. The phasor concept was developed for sinusoids, all of which have the same frequency. *Hence, convert the phasors back into time-dependent signals before adding the components.*

We must convert the phasors back into time-dependent signals before adding the components.

Real-world information-bearing signals contain thousands of components. In principle, the output of a given filter for any input signal could be found by using the procedure of Example 6.2. However, it would usually be much too tedious to carry out. Fortunately, we will not need to do this. *It is the principle that is most important.* In summary, we can say that linear circuits (or any other systems for which the relationship between input and output can be described by linear time-invariant differential equations) behave as if they

1. Separate the input signal into components having various frequencies.
2. Alter the amplitude and phase of each component depending on its frequency.
3. Add the altered components to produce the output signal.

This process is illustrated in Figure 6.5.

The transfer function of a filter is important because it shows how the components are altered in amplitude and phase.

Experimental Determination of the Transfer Function. To determine the transfer function of a filter experimentally, we connect a sinusoidal source to the input port, measure the amplitudes and phases of both the input signal and the resulting output signal, and divide the output phasor by the input phasor. This is repeated for each frequency of interest. The experimental setup is illustrated in Figure 6.6. Various instruments, such as voltmeters and oscilloscopes, can be employed to measure the amplitudes and phases.

In the next few sections of this chapter, we use mathematical analysis to investigate the transfer functions of several relatively simple electrical circuits.

Exercise 6.2 Consider the transfer function shown in Figure 6.4. The input signal is given by

$$v_{in}(t) = 2\ \cos(1000\pi t + 20°) + 3\ \cos(3000\pi t)$$

Find an expression for the output signal.

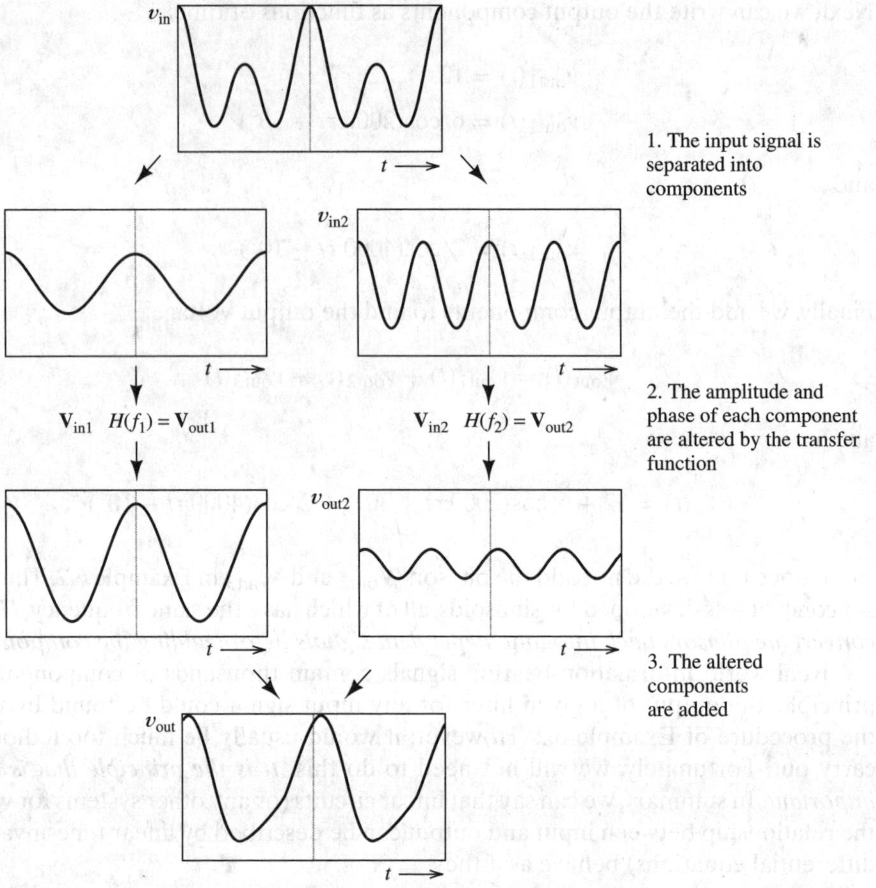

Figure 6.5 Filters behave as if they separate the input into components, modify the amplitudes and phases of the components, and add the altered components to produce the output.

Figure 6.6 To measure the transfer function we apply a sinusoidal input signal, measure the amplitudes and phases of input and output in steady state, and then divide the phasor output by the phasor input. The procedure is repeated for each frequency of interest.

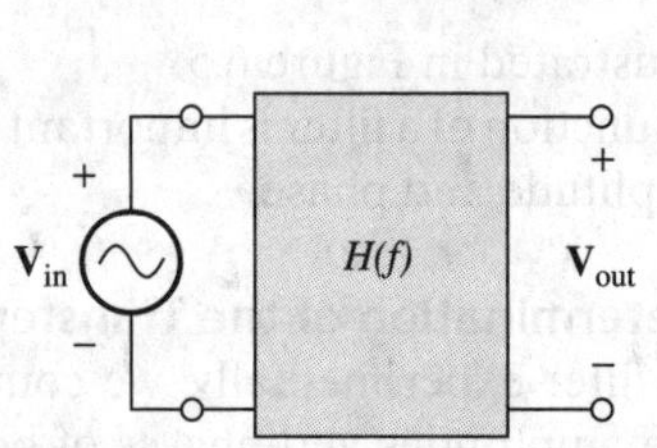

Answer $v_{out}(t) = 7\cos(1000\pi t + 35°) + 7.5\cos(3000\pi t + 45°)$. □

Exercise 6.3 Consider the transfer function shown in Figure 6.4. The input signal is given by

$$v_{in}(t) = 1 + 2\cos(2000\pi t) + 3\cos(6000\pi t)$$

Find an expression for the output signal.

Answer $v_{out}(t) = 4 + 6\cos(2000\pi t + 30°)$. Notice that the 3-kHz component is totally eliminated (rejected) by the filter. □

PRACTICAL APPLICATION 6.1

Active Noise Cancellation

Noise and vibration are annoying to passengers in helicopters and other aircraft. Traditional sound-absorbing materials can be very effective in reducing noise levels, but are too bulky and massive for application in aircraft. An alternative approach is an electronic system that cancels noise. The diagram of such a system is shown in Figure PA6.1. A microphone near the sources of the noise, such as the engines, samples the noise before it enters the passenger area. The resulting electrical signal passes through a filter whose transfer function is continuously adjusted by a special-purpose computer to match the transfer function of the sound path. Finally, an inverted version of the signal is applied to loudspeakers. The sound waves from the speaker are out of phase with those from the noise source, resulting in partial cancellation. Another set of microphones on the headrest monitors the sound experienced by the passenger so that the computer can determine the filter adjustments needed to best cancel the sound.

Recently, noise-canceling systems based on these principles have appeared that contain all of the system elements in a lightweight headset. Many passengers on commercial aircraft wear these headsets to provide themselves with a quieter, more restful trip.

Active noise cancellation systems can effectively replace sound absorbing materials weighing a great deal more. As a result, active noise cancellation is very attractive for use in aircraft and automobiles. You can find many research reports and popular articles on this topic with an internet search.

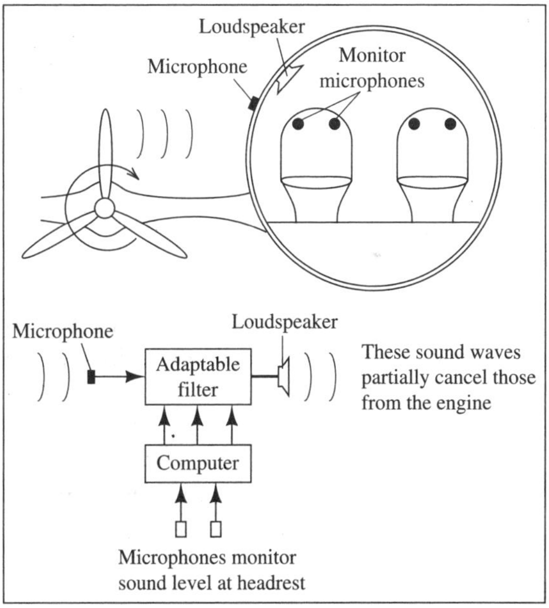

Figure PA6.1

6.2 FIRST-ORDER LOWPASS FILTERS

Consider the circuit shown in Figure 6.7. We will see that this circuit tends to pass low-frequency components and reject high-frequency components. (In other words, for low frequencies, the output amplitude is nearly the same as the input. For high frequencies, the output amplitude is much less than the input.) In Chapter 4, we saw that a first-order differential equation describes this circuit. Because of these facts, the circuit is called a **first-order lowpass filter**.

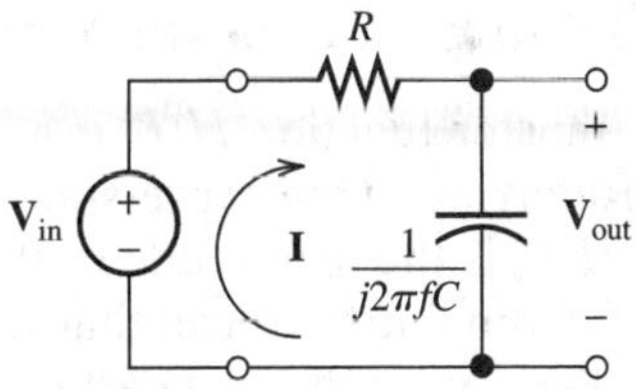

Figure 6.7 A first-order lowpass filter.

We can determine the transfer functions of RLC circuits by using steady-state analysis with complex impedances as a function of frequency.

To determine the transfer function, we apply a sinusoidal input signal having a phasor $\mathbf{V}_{\text{in}}$, and then we analyze the behavior of the circuit as a function of the source frequency f.

The phasor current is the input voltage divided by the complex impedance of the circuit. This is given by

$$\mathbf{I} = \frac{\mathbf{V}_{\text{in}}}{R + 1/j2\pi fC} \tag{6.3}$$

The phasor for the output voltage is the product of the phasor current and the impedance of the capacitance, illustrated by

$$\mathbf{V}_{\text{out}} = \frac{1}{j2\pi fC}\mathbf{I} \tag{6.4}$$

Using Equation 6.3 to substitute for **I**, we have

$$\mathbf{V}_{\text{out}} = \frac{1}{j2\pi fC} \times \frac{\mathbf{V}_{\text{in}}}{R + 1/j2\pi fC} \tag{6.5}$$

Recall that the transfer function $H(f)$ is defined to be the ratio of the output phasor to the input phasor:

$$H(f) = \frac{\mathbf{V}_{\text{out}}}{\mathbf{V}_{\text{in}}} \tag{6.6}$$

Rearranging Equation 6.5, we have

$$H(f) = \frac{\mathbf{V}_{\text{out}}}{\mathbf{V}_{\text{in}}} = \frac{1}{1 + j2\pi fRC} \tag{6.7}$$

Next, we define the parameter:

$$f_B = \frac{1}{2\pi RC} \tag{6.8}$$

Then, the transfer function can be written as

$$H(f) = \frac{1}{1 + j(f/f_B)} \tag{6.9}$$

Magnitude and Phase Plots of the Transfer Function

As expected, the transfer function $H(f)$ is a complex quantity having a magnitude and phase angle. Referring to the expression on the right-hand side of Equation 6.9, the magnitude of $H(f)$ is the magnitude of the numerator (which is unity) over the magnitude of the denominator. Recall that the magnitude of a complex quantity is the square root of the sum of the real part squared and the imaginary part squared.

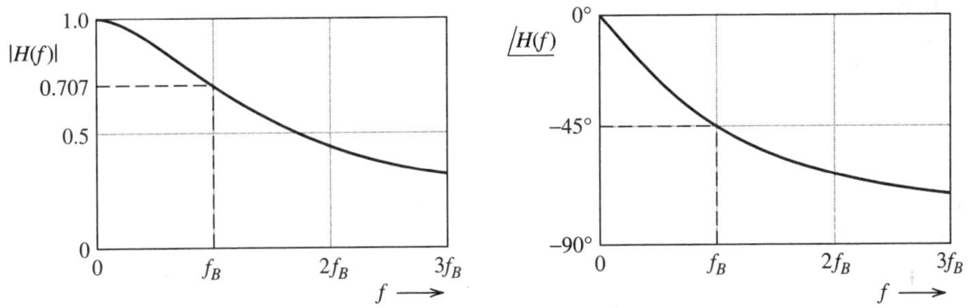

Figure 6.8 Magnitude and phase of the first-order lowpass transfer function versus frequency.

Thus, the magnitude is given by

$$|H(f)| = \frac{1}{\sqrt{1 + (f/f_B)^2}} \tag{6.10}$$

Referring to the expression on the right-hand side of Equation 6.9, the phase angle of the transfer function is the phase of the numerator (which is zero) minus the phase of the denominator. This is given by

$$\angle H(f) = -\arctan\left(\frac{f}{f_B}\right) \tag{6.11}$$

Plots of the magnitude and phase of the transfer function are shown in Figure 6.8. For low frequencies (f approaching zero), the magnitude is approximately unity and the phase is nearly zero, which means that the amplitudes and phases of low-frequency components are affected very little by this filter. The low-frequency components are passed to the output almost unchanged in amplitude or phase.

On the other hand, for high frequencies ($f >> f_B$), the magnitude of the transfer function approaches zero. Thus, the amplitude of the output is much smaller than the amplitude of the input for the high-frequency components. We say that the high-frequency components are rejected by the filter. Furthermore, at high frequencies, the phase of the transfer function approaches −90°. Thus, as well as being reduced in amplitude, the high-frequency components are phase shifted.

Notice that for $f = f_B$, the magnitude of the output is $1/\sqrt{2} \cong 0.707$ times the magnitude of the input signal. When the amplitude of a voltage is multiplied by a factor of $1/\sqrt{2}$, the power that the voltage can deliver to a given resistance is multiplied by a factor of one-half (because power is proportional to voltage squared). Thus, f_B is called the **half-power frequency**.

At the half-power frequency, the transfer-function magnitude is $1/\sqrt{2} \cong 0.707$ times its maximum value.

Applying the Transfer Function

As we saw in Section 6.1, if an input signal to a filter consists of several components of different frequencies, we can use the transfer function to compute the output for each component separately. Then, we can find the complete output by adding the separate components.

Example 6.3 Calculation of *RC* Lowpass Output

Suppose that an input signal given by

$$v_{in}(t) = 5\cos(20\pi t) + 5\cos(200\pi t) + 5\cos(2000\pi t)$$

is applied to the lowpass *RC* filter shown in Figure 6.9. Find an expression for the output signal.

Solution The filter has the form of the lowpass filter analyzed in this section. The half-power frequency is given by

$$f_B = \frac{1}{2\pi RC} = \frac{1}{2\pi \times (1000/2\pi) \times 10 \times 10^{-6}} = 100 \text{ Hz}$$

The first component of the input signal is

$$v_{in1}(t) = 5\cos(20\pi t)$$

For this component, the phasor is $\mathbf{V}_{in1} = 5\angle 0°$, and the angular frequency is $\omega = 20\pi$. Therefore, $f = \omega/2\pi = 10$. The transfer function of the circuit is given by Equation 6.9, which is repeated here for convenience:

$$H(f) = \frac{1}{1 + j(f/f_B)}$$

Evaluating the transfer function for the frequency of the first component ($f = 10$), we have

$$H(10) = \frac{1}{1 + j(10/100)} = 0.9950\angle -5.71°$$

The output phasor for the $f = 10$ component is simply the input phasor times the transfer function. Thus, we obtain

$$\mathbf{V}_{out1} = H(10) \times \mathbf{V}_{in1}$$
$$= (0.9950\angle -5.71°) \times (5\angle 0°) = 4.975\angle -5.71°$$

Hence, the output for the first component of the input signal is

$$v_{out1}(t) = 4.975\cos(20\pi t - 5.71°)$$

Similarly, the second component of the input signal is

$$v_{in2}(t) = 5\cos(200\pi t)$$

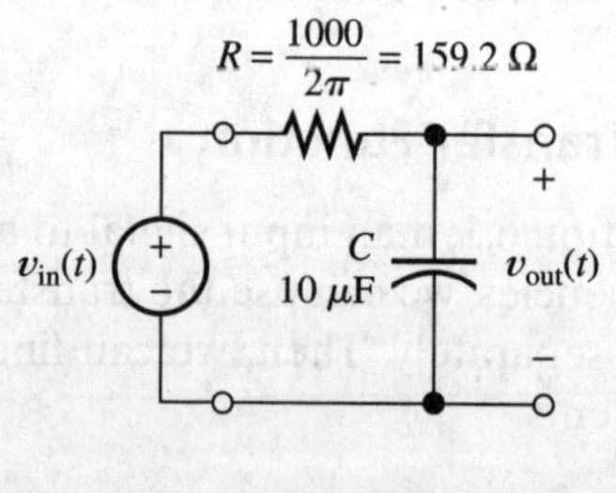

Figure 6.9 Circuit of Example 6.3. The resistance has been picked so the break frequency turns out to be a convenient value.

and we have

$$\mathbf{V}_{\text{in2}} = 5\angle 0^\circ$$

The frequency of the second component is $f = 100$:

$$H(100) = \frac{1}{1 + j(100/100)} = 0.7071\angle -45^\circ$$

$$\mathbf{V}_{\text{out2}} = H(100) \times \mathbf{V}_{\text{in2}}$$

$$= (0.7071\angle -45^\circ) \times (5\angle 0^\circ) = 3.535\angle -45^\circ$$

Therefore, the output for the second component of the input signal is

$$v_{\text{out2}}(t) = 3.535 \cos(200\pi t - 45^\circ)$$

Finally, for the third and last component, we have

$$v_{\text{in3}}(t) = 5 \cos(2000\pi t)$$

$$\mathbf{V}_{\text{in3}} = 5\angle 0^\circ$$

$$H(1000) = \frac{1}{1 + j(1000/100)} = 0.0995\angle -84.29^\circ$$

$$\mathbf{V}_{\text{out3}} = H(1000) \times \mathbf{V}_{\text{in3}}$$

$$= (0.0995\angle -84.29^\circ) \times (5\angle 0^\circ) = 0.4975\angle -84.29^\circ$$

Consequently, the output for the third component of the input signal is

$$v_{\text{out3}}(t) = 0.4975 \cos(2000\pi t - 84.29^\circ)$$

Now, we can write an expression for the output signal by adding the output components:

$$v_{\text{out}}(t) = 4.975 \cos(20\pi t - 5.71^\circ) + 3.535 \cos(200\pi t - 45^\circ) + 0.4975 \cos(2000\pi t - 84.29^\circ)$$

Notice that each component of the input signal $v_{\text{in}}(t)$ is treated differently by this filter. The $f = 10$ component is nearly unaffected in amplitude and phase. The $f = 100$ component is reduced in amplitude by a factor of 0.7071 and phase shifted by -45°. The amplitude of the $f = 1000$ component is reduced by approximately an order of magnitude. Thus, the filter discriminates against the high-frequency components. ■

Application of the First-Order Lowpass Filter

A simple application of the first-order lowpass filter is the tone control on an old-fashion AM radio. The tone control adjusts the resistance and, therefore, the break frequency of the filter. Suppose that we are listening to an interesting news item from a distant radio station with an AM radio and lightning storms are causing electrical noise. It turns out that the components of voice signals are concentrated in the low end of the audible frequency range. On the other hand, the noise caused by lightning has roughly equal-amplitude components at all frequencies. In this situation, we could adjust the tone control to lower the break frequency. Then, the high-frequency noise components would be rejected, while most of the voice components would be passed. In this way, we can improve the ratio of desired signal power to noise power produced by the loudspeaker and make the news more intelligible.

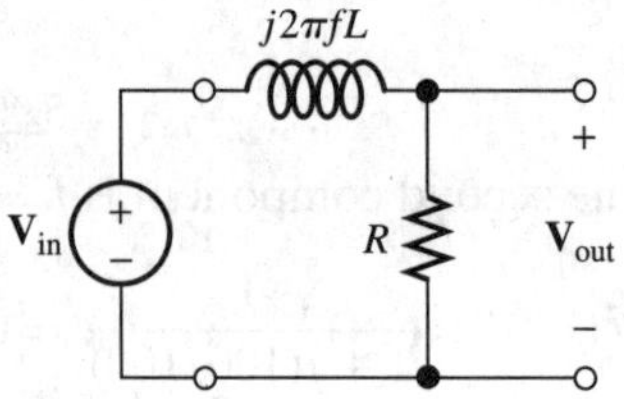

Figure 6.10 Another first-order lowpass filter; see Exercise 6.4.

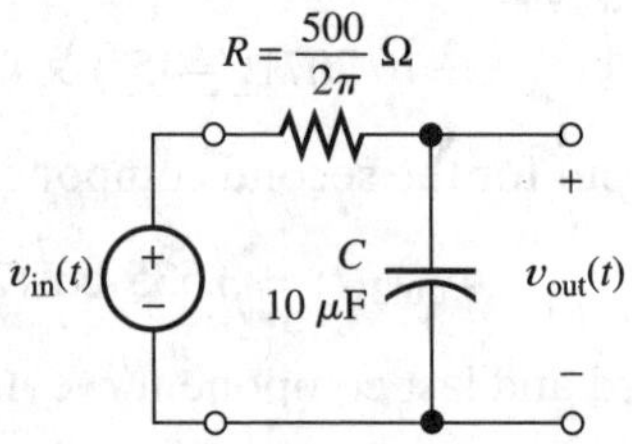

Figure 6.11 Circuit for Exercise 6.5.

Using Phasors with Components of Different Frequencies

We should not add the phasors for components with different frequencies.

Recall that phasors can be combined only for sinusoids with the same frequency. It is important to understand that *we should not add the phasors for components with different frequencies*. Thus, in the preceding example, we used phasors to find the output components as functions of time, which we then added.

Exercise 6.4 Derive an expression for the transfer function $H(f) = \mathbf{V}_{out}/\mathbf{V}_{in}$ of the filter shown in Figure 6.10. Show that $H(f)$ takes the same form as Equation 6.9 if we define $f_B = R/2\pi L$. □

Exercise 6.5 Suppose that the input signal for the circuit shown in Figure 6.11 is given by

$$v_{in}(t) = 10\cos(40\pi t) + 5\cos(1000\pi t) + 5\cos(2\pi 10^4 t)$$

Find an expression for the output signal $v_{out}(t)$.

Answer

$$v_{out}(t) = 9.95\cos(40\pi t - 5.71°) + 1.86\cos(1000\pi t - 68.2°) + 0.100\cos(2\pi 10^4 t - 88.9°)$$

□

6.3 DECIBELS, THE CASCADE CONNECTION, AND LOGARITHMIC FREQUENCY SCALES

In comparing the performance of various filters, it is helpful to express the magnitudes of the transfer functions in **decibels**. To convert a transfer-function magnitude to decibels, we multiply the common logarithm (base 10) of the transfer-function magnitude by 20:

$$|H(f)|_{dB} = 20\log|H(f)| \tag{6.12}$$

(A transfer function is a ratio of voltages and is converted to decibels as 20 times the logarithm of the ratio. On the other hand, ratios of powers are converted to decibels by taking 10 times the logarithm of the ratio.)

Table 6.2. Transfer-Function Magnitudes and Their Decibel Equivalents

$\|H(f)\|$	$\|H(f)\|_{dB}$
100	40
10	20
2	6
$\sqrt{2}$	3
1	0
$1/\sqrt{2}$	−3
1/2	−6
0.1	−20
0.01	−40

Table 6.2 shows the decibel equivalents for selected values of transfer-function magnitude. Notice that the decibel equivalents are positive for magnitudes greater than unity, whereas the decibel equivalents are negative for magnitudes less than unity.

In many applications, the ability of a filter to strongly reject signals in a given frequency band is of primary importance. For example, a common problem associated with audio signals is that a small amount of the ac power line voltage can inadvertently be added to the signal. When applied to a loudspeaker, this 60-Hz component produces a disagreeable hum. (Actually, this problem is rapidly becoming a thing of the past as digital technologies replace analog.)

Usually, we approach this problem by trying to eliminate the electrical path by which the power line voltage is added to the desired audio signal. However, this is sometimes not possible. Then, we could try to design a filter that rejects the 60-Hz component and passes components at other frequencies. The magnitude of a filter transfer function to accomplish this is shown in Figure 6.12(a). A filter such as this, designed to eliminate components in a narrow range of frequencies, is called a **notch filter**.

It turns out that to reduce a loud hum (as loud as a heated conversation) to be barely audible, the transfer function must be −80 dB or less for the 60-Hz component, which corresponds to $|H(f)| = 10^{-4}$ or smaller. On the other hand, the transfer-function magnitude should be close to unity for the components to be passed by the filter. We refer to the range of frequencies to be passed as the **passband**.

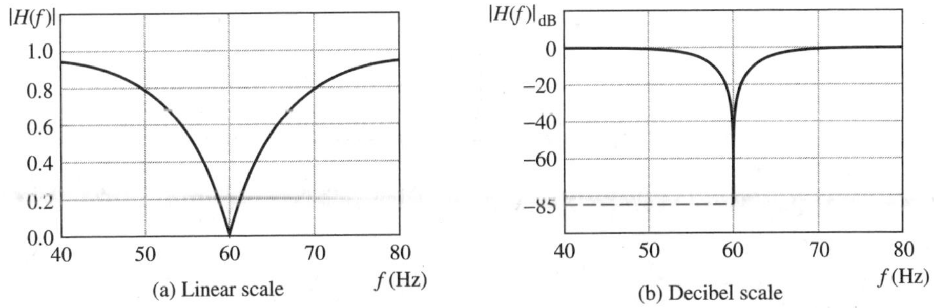

Figure 6.12 Transfer-function magnitude of a notch filter used to reduce hum in audio signals.

When we plot $|H(f)|$ without converting to decibels, it is difficult to show both values clearly on the same plot. If we choose a scale that shows the passband magnitude, we cannot see whether the magnitude is sufficiently small at 60 Hz. This is the case for the plot shown in Figure 6.12(a). On the other hand, if we choose a linear scale that clearly shows the magnitude at 60 Hz, the magnitude would be way off scale at other frequencies of interest.

However, when the magnitude is converted to decibels, both parts of the magnitude are readily seen. For example, Figure 6.12(b) shows the decibel equivalent for the magnitude plot shown in Figure 6.12(a). On this plot, we can see that the passband magnitude is approximately unity (0 dB) and that at 60 Hz, the magnitude is sufficiently small (less than −80 dB).

One of the advantages of converting transfer-function magnitudes to decibels before plotting is that very small and very large magnitudes can be displayed clearly on a single plot.

Thus, one of the advantages of converting transfer-function magnitudes to decibels before plotting is that very small and very large magnitudes can be displayed clearly on a single plot. We will see that another advantage is that decibel plots for many filter circuits can be approximated by straight lines (provided that a logarithmic scale is used for frequency). Furthermore, to understand some of the jargon used by electrical engineers, we must be familiar with decibels.

Cascaded Two-Port Networks

In the cascade connection, the output of one filter is connected to the input of a second filter.

When we connect the output terminals of one two-port circuit to the input terminals of another two-port circuit, we say that we have a **cascade** connection. This is illustrated in Figure 6.13. Notice that the output voltage of the first two-port network is the input voltage of the second two-port. The overall transfer function is

$$H(f) = \frac{\mathbf{V}_{\text{out}}}{\mathbf{V}_{\text{in}}}$$

However, the output voltage of the cascade is the output of the second two-port (i.e., $\mathbf{V}_{\text{out}} = \mathbf{V}_{\text{out2}}$). Furthermore, the input to the cascade is the input to the first two-port (i.e., $\mathbf{V}_{\text{in}} = \mathbf{V}_{\text{in1}}$). Thus,

$$H(f) = \frac{\mathbf{V}_{\text{out2}}}{\mathbf{V}_{\text{in1}}}$$

Multiplying and dividing by $\mathbf{V}_{\text{out1}}$, we have

$$H(f) = \frac{\mathbf{V}_{\text{out1}}}{\mathbf{V}_{\text{in1}}} \times \frac{\mathbf{V}_{\text{out2}}}{\mathbf{V}_{\text{out1}}}$$

Now, the output voltage of the first two-port is the input to the second two-port (i.e., $\mathbf{V}_{\text{out1}} = \mathbf{V}_{\text{in2}}$). Hence,

$$H(f) = \frac{\mathbf{V}_{\text{out1}}}{\mathbf{V}_{\text{in1}}} \times \frac{\mathbf{V}_{\text{out2}}}{\mathbf{V}_{\text{in2}}}$$

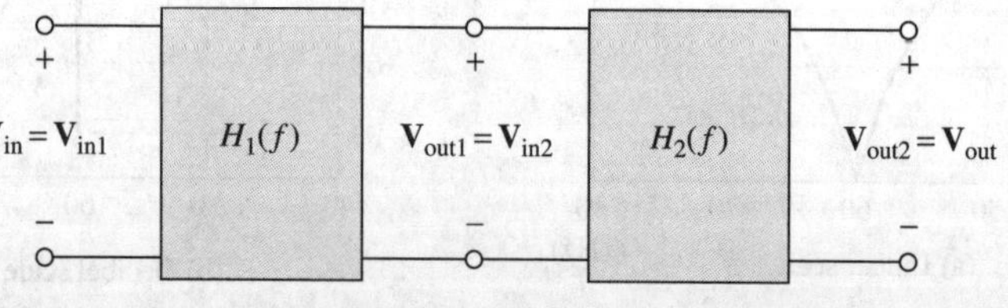

Figure 6.13 Cascade connection of two two-port circuits.

Finally, we can write

$$H(f) = H_1(f) \times H_2(f) \tag{6.13}$$

Thus, the transfer function of the cascade connection is the product of the transfer functions of the individual two-port networks. This fact can be extended to three or more two-ports connected in cascade.

A potential source of difficulty in applying Equation 6.13 is that the transfer function of a two-port usually depends on what is attached to its output terminals. *Thus, in applying Equation 6.13, we must find $H_1(f)$ with the second two-port attached.*

In applying Equation 6.13, we must find $H_1(f)$ with the second two-port attached.

Taking the magnitudes of the terms on both sides of Equation 6.13 and expressing in decibels, we have

$$20 \log |H(f)| = 20 \log [|H_1(f)| \times |H_2(f)|] \tag{6.14}$$

Using the fact that the logarithm of a product is equal to the sum of the logarithms of the terms in the product, we have

$$20 \log |H(f)| = 20 \log |H_1(f)| + 20 \log |H_2(f)| \tag{6.15}$$

which can be written as

$$|H(f)|_{dB} = |H_1(f)|_{dB} + |H_2(f)|_{dB} \tag{6.16}$$

Thus, in decibels, the individual transfer-function magnitudes are added to find the overall transfer-function magnitude for a cascade connection.

In decibels, the individual transfer-function magnitudes are added to find the overall transfer-function magnitude for a cascade connection.

Logarithmic Frequency Scales

We often use a **logarithmic scale** for frequency when plotting transfer functions. On a logarithmic scale, the variable is *multiplied* by a given factor for equal increments of length along the axis. (On a linear scale, equal lengths on the scale correspond to *adding* a given amount to the variable.) For example, a logarithmic frequency scale is shown in Figure 6.14.

On a logarithmic scale, the variable is multiplied by a given factor for equal increments of length along the axis.

A **decade** is a range of frequencies for which the ratio of the highest frequency to the lowest is 10. The frequency range from 2 to 20 Hz is one decade. Similarly, the range from 50 to 5000 Hz is two decades. (50 to 500 Hz is one decade, and 500 to 5000 Hz is another decade.)

An **octave** is a two-to-one change in frequency. For example, the range 10 to 20 Hz is one octave. The range 2 to 16 kHz is three octaves.

Suppose that we have two frequencies f_1 and f_2 for which $f_2 > f_1$. The number of decades between f_1 and f_2 is given by

$$\text{number of decades} = \log\left(\frac{f_2}{f_1}\right) \tag{6.17}$$

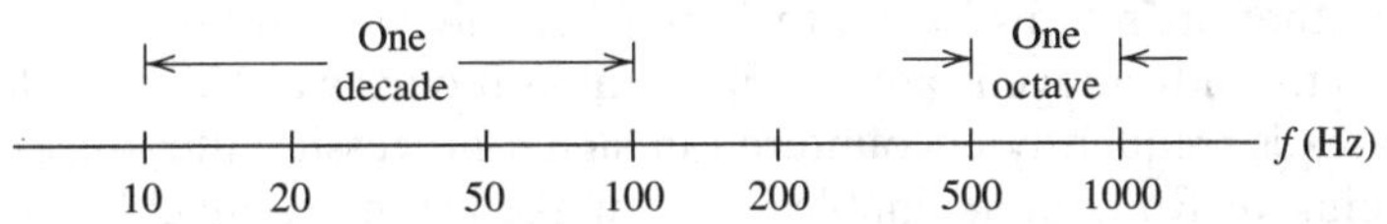

Figure 6.14 Logarithmic frequency scale.

in which we assume that the logarithm is base 10. The number of octaves between the two frequencies is

$$\text{number of octaves} = \log_2\left(\frac{f_2}{f_1}\right) = \frac{\log(f_2/f_1)}{\log(2)} \tag{6.18}$$

The advantage of a logarithmic frequency scale compared with a linear scale is that the variations in the magnitude or phase of a transfer function for a low range of frequency such as 10 to 20 Hz, as well as the variations in a high range such as 10 to 20 MHz, can be clearly shown on a single plot. With a linear scale, either the low range would be severely compressed or the high range would be off scale.

Exercise 6.6 Suppose that $|H(f)| = 50$. Find the decibel equivalent.

Answer $|H(f)|_{\text{dB}} = 34$ dB. □

Exercise 6.7 **a.** Suppose that $|H(f)|_{\text{dB}} = 15$ dB. Find $|H(f)|$. **b.** Repeat for $|H(f)|_{\text{dB}} = 30$ dB.

Answer **a.** $|H(f)| = 5.62$; **b.** $|H(f)| = 31.6$. □

Exercise 6.8 **a.** What frequency is two octaves higher than 1000 Hz? **b.** Three octaves lower? **c.** Two decades higher? **d.** One decade lower?

Answer **a.** 4000 Hz is two octaves higher than 1000 Hz; **b.** 125 Hz is three octaves lower than 1000 Hz; **c.** 100 kHz is two decades higher than 1000 Hz; **d.** 100 Hz is one decade lower than 1000 Hz. □

Exercise 6.9 **a.** What frequency is halfway between 100 and 1000 Hz on a logarithmic frequency scale? **b.** On a linear frequency scale?

Answer **a.** 316.2 Hz is halfway between 100 and 1000 Hz on a logarithmic scale; **b.** 550 Hz is halfway between 100 and 1000 Hz on a linear frequency scale. □

Exercise 6.10 **a.** How many decades are between $f_1 = 20$ Hz and $f_2 = 15$ kHz? (This is the approximate range of audible frequencies.) **b.** How many octaves?

Answer

a. $\text{Number of decades} = \log\left(\frac{15\text{ kHz}}{20\text{ Hz}}\right) = 2.87$

b. $\text{Number of octaves} = \frac{\log(15000/20)}{\log(2)} = 9.55$ □

6.4 BODE PLOTS

A Bode plot is a plot of the decibel magnitude of a network function versus frequency using a logarithmic scale for frequency.

A **Bode plot** is a plot of the decibel magnitude of a network function versus frequency using a logarithmic scale for frequency. Because it can clearly illustrate very large and very small magnitudes for a wide range of frequencies on one plot, the Bode plot is particularly useful for displaying transfer functions. Furthermore, it turns out that Bode plots of network functions can often be closely approximated by straight-line segments, so they are relatively easy to draw. (Actually, we now use computers to plot functions, so this advantage is not as important as it once was.) Terminology related to these plots is frequently encountered in signal-processing literature. Finally, an understanding of Bode plots enables us to make estimates quickly when dealing with transfer functions.

To illustrate Bode plot concepts, we consider the first-order lowpass transfer function of Equation 6.9, repeated here for convenience:

$$H(f) = \frac{1}{1 + j(f/f_B)}$$

The magnitude of this transfer function is given by Equation 6.10, which is

$$|H(f)| = \frac{1}{\sqrt{1 + (f/f_B)^2}}$$

To convert the magnitude to decibels, we take 20 times the logarithm of the magnitude:

$$|H(f)|_{\text{dB}} = 20 \log |H(f)|$$

Substituting the expression for the transfer-function magnitude, we get

$$|H(f)|_{\text{dB}} = 20 \log \frac{1}{\sqrt{1 + (f/f_B)^2}}$$

Using the properties of the logarithm, we obtain

$$|H(f)|_{\text{dB}} = 20 \log(1) - 20 \log \sqrt{1 + \left(\frac{f}{f_B}\right)^2}$$

Of course, the logarithm of unity is zero. Therefore,

$$|H(f)|_{\text{dB}} = -20 \log \sqrt{1 + \left(\frac{f}{f_B}\right)^2}$$

Finally, since $\log(\sqrt{x}) = \frac{1}{2}\log(x)$, we have

$$|H(f)|_{\text{dB}} = -10 \log \left[1 + \left(\frac{f}{f_B}\right)^2\right] \tag{6.19}$$

The low-frequency asymptote is constant at 0 dB.

Notice that the value given by Equation 6.19 is approximately 0 dB for $f << f_B$. Thus, for low frequencies, the transfer-function magnitude is approximated by the horizontal straight line shown in Figure 6.15, labeled as the **low-frequency asymptote**.

On the other hand, for $f >> f_B$, Equation 6.19 is approximately

$$|H(f)|_{\text{dB}} \cong -20 \log \left(\frac{f}{f_B}\right) \tag{6.20}$$

The high-frequency asymptote slopes downward at 20 dB/decade, starting from 0 dB at f_B.

Notice that the two straight-line asymptotes intersect at the half-power frequency f_B.

Evaluating for various values of f, we obtain the results shown in Table 6.3. Plotting these values results in the straight line shown sloping downward on the right-hand side of Figure 6.15, labeled as the **high-frequency asymptote**. Notice that the two straight-line asymptotes intersect at the half-power frequency f_B. For this reason, f_B is also known as the **corner frequency** or as the **break frequency**.

Also, notice that the slope of the high-frequency asymptote is −20 dB per decade of frequency. (This slope can also be stated as −6 dB per octave.)

If we evaluate Equation 6.19 at $f = f_B$, we find that

$$|H(f_B)|_{\text{dB}} = -3 \text{ dB}$$

The asymptotes are in error by only 3 dB at the corner frequency f_B.

Thus, the asymptotes are in error by only 3 dB at the corner frequency. The actual curve for $|H(f)|_{\text{dB}}$ is also shown in Figure 6.15.

Table 6.3. Values of the Approximate Expression (Equation 6.20) for Selected Frequencies

f	$\|H(f)\|_{dB}$
f_B	0
$2f_B$	−6
$10f_B$	−20
$100f_B$	−40
$1000f_B$	−60

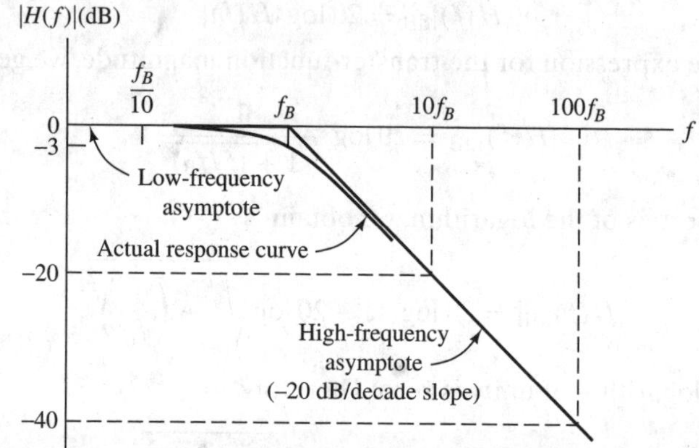

Figure 6.15 Magnitude Bode plot for the first-order lowpass filter.

Phase Plot

The phase of the first-order lowpass transfer function is given by Equation 6.11, which is repeated here for convenience:

$$\angle H(f) = -\arctan\left(\frac{f}{f_B}\right)$$

Evaluating, we find that the phase approaches zero at very low frequencies, equals −45° at the break frequency, and approaches −90° at high frequencies.

Figure 6.16 shows a plot of phase versus frequency. Notice that the curve can be approximated by the following straight-line segments:

1. A horizontal line at zero for $f < f_B/10$.
2. A sloping line from zero phase at $f_B/10$ to −90° at $10f_B$.
3. A horizontal line at −90° for $f > 10f_B$.

The actual phase curve departs from these straight-line approximations by less than 6°. Hence, working by hand, we could easily construct an approximate plot of phase.

Many circuit functions can be plotted by the methods we have demonstrated for the simple lowpass *RC* circuit; however, we will not try to develop your skill at this to a high degree. Bode plots of amplitude and phase for *RLC* circuits are easily produced by computer programs. We have shown the manual approach to analyzing and drawing the Bode plot for the *RC* lowpass filter mainly to present the concepts and terminology.

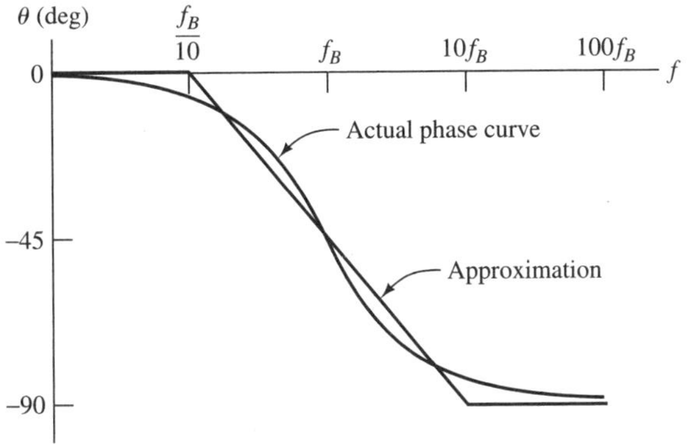

Figure 6.16 Phase Bode plot for the first-order lowpass filter.

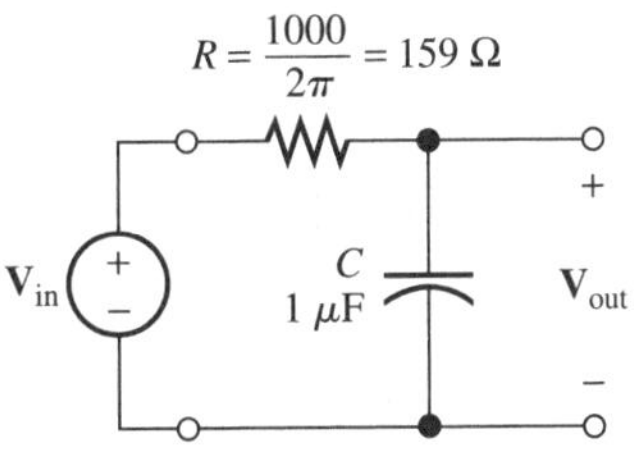

Figure 6.17 Circuit for Exercise 6.11.

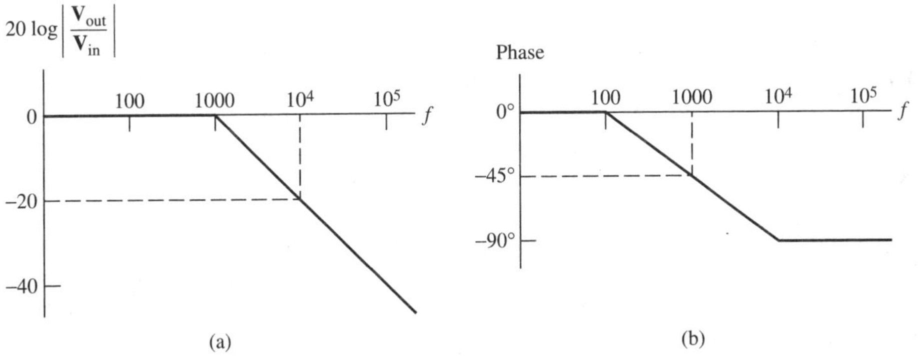

Figure 6.18 Answers for Exercise 6.11.

Exercise 6.11 Sketch the approximate straight-line Bode magnitude and phase plots to scale for the circuit shown in Figure 6.17.

Answer See Figure 6.18. □

6.5 FIRST-ORDER HIGHPASS FILTERS

The circuit shown in Figure 6.19 is called a **first-order highpass filter**. It can be analyzed in much the same manner as the lowpass circuit considered earlier in this chapter. The resulting transfer function is given by

$$H(f) = \frac{\mathbf{V}_{out}}{\mathbf{V}_{in}} = \frac{j(f/f_B)}{1 + j(f/f_B)} \tag{6.21}$$

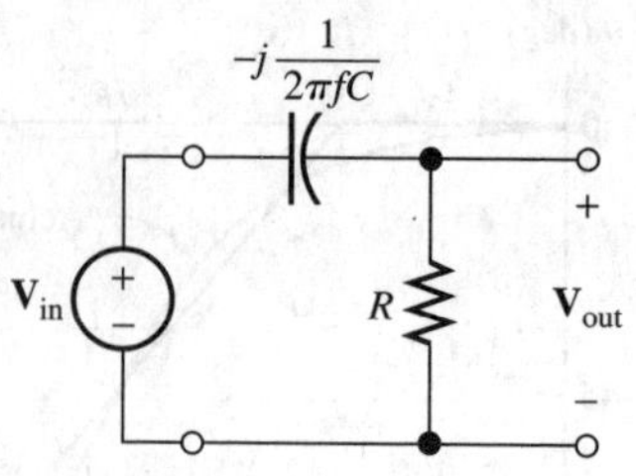

Figure 6.19 First-order highpass filter.

in which

$$f_B = \frac{1}{2\pi RC} \tag{6.22}$$

Exercise 6.12 Use circuit analysis to derive the transfer function for the circuit of Figure 6.19, and show that it can be put into the form of Equations 6.21 and 6.22. □

Magnitude and Phase of the Transfer Function

The magnitude of the transfer function is given by

$$|H(f)| = \frac{f/f_B}{\sqrt{1+(f/f_B)^2}} \tag{6.23}$$

This is plotted in Figure 6.20(a). Notice that the transfer-function magnitude goes to zero for dc ($f = 0$). For high frequencies ($f >> f_B$), the transfer-function magnitude approaches unity. Thus, this filter passes high-frequency components and tends to reject low-frequency components. That is why the circuit is called a highpass filter.

Highpass filters are useful whenever we want to retain high-frequency components and reject low-frequency components.

Highpass filters are useful whenever we want to retain high-frequency components and reject low-frequency components. For example, suppose that we want to record warbler songs in a noisy environment. It turns out that bird calls fall in the high-frequency portion of the audible range. The audible range of frequencies is from 20 Hz to 15 kHz (approximately), and the calls of warblers fall (mainly) in the range above 2 kHz. On the other hand, the noise may be concentrated at lower frequencies. For example, heavy trucks rumbling down a bumpy road would produce strong noise components lower in frequency than 2 kHz. To record singing warblers in the vicinity of such a noise source, a highpass filter would be helpful. We would select R and C

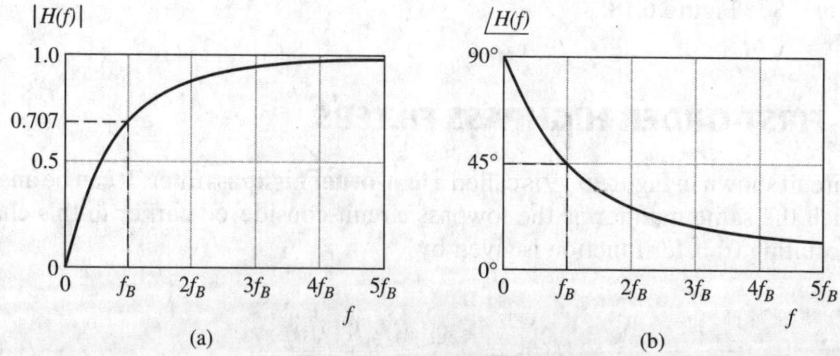

Figure 6.20 Magnitude and phase for the first-order highpass transfer function.

to achieve a half-power frequency f_B of approximately 2 kHz. Then, the filter would pass the songs and reject some of the noise.

Recall that if the amplitude of a component is multiplied by a factor of $1/\sqrt{2}$, the power that the component can deliver to a resistance is multiplied by a factor of 1/2. For $f = f_B$, $|H(f)| = 1/\sqrt{2} \cong 0.707$, so that, as in the case of the lowpass filter, f_B is called the *half-power frequency*. (Here again, several alternative names are *corner frequency*, *3-dB frequency*, and *break frequency*.)

The phase of the highpass transfer function (Equation 6.21) is given by

$$\angle H(f) = 90° - \arctan\left(\frac{f}{f_B}\right) \tag{6.24}$$

A plot of the phase shift of the highpass filter is shown in Figure 6.20(b).

Bode Plots for the First-Order Highpass Filter

As we have seen, a convenient way to plot transfer functions is to use the Bode plot, in which the magnitude is converted to decibels and a logarithmic frequency scale is used. In decibels, the magnitude of the highpass transfer function is

$$|H(f)|_{\text{dB}} = 20 \log \frac{f/f_B}{\sqrt{1 + (f/f_B)^2}}$$

This can be written as

$$|H(f)|_{\text{dB}} = 20 \log \left(\frac{f}{f_B}\right) - 10 \log \left[1 + \left(\frac{f}{f_B}\right)^2\right] \tag{6.25}$$

For $f << f_B$, the second term on the right-hand side of Equation 6.25 is approximately zero. Thus, for $f << f_B$, we have

$$|H(f)|_{\text{dB}} \cong 20 \log \left(\frac{f}{f_B}\right) \qquad \text{for } f << f_B \tag{6.26}$$

Evaluating this for selected values of f, we find the values given in Table 6.4. Plotting these values, we obtain the low-frequency asymptote shown on the left-hand side of Figure 6.21(a). Notice that the low-frequency asymptote slopes downward to the left at a rate of 20 dB per decade.

For $f >> f_B$, the magnitude given by Equation 6.25 is approximately 0 dB. Hence,

$$|H(f)|_{\text{dB}} \cong 0 \qquad \text{for } f >> f_B \tag{6.27}$$

Table 6.4. Values of the Approximate Expression Given in Equation 6.26 for Selected Frequencies

f	$\|H(f)\|_{\text{dB}}$
f_B	0
$f_B/2$	−6
$f_B/10$	−20
$f_B/100$	−40

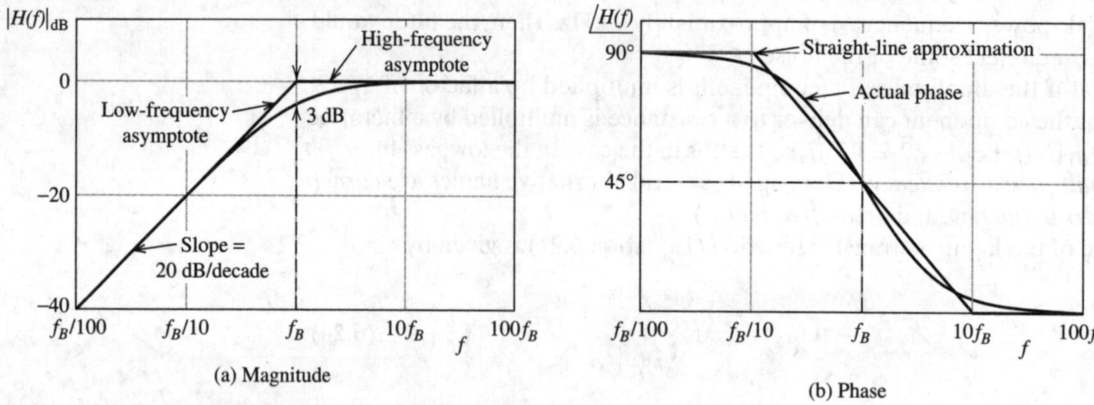

Figure 6.21 Bode plots for the first-order highpass filter.

This is plotted as the high-frequency asymptote in Figure 6.21(a). Notice that the high-frequency asymptote and the low-frequency asymptote meet at $f = f_B$. (That is why f_B is sometimes called the *break frequency*.)

The actual values of $|H(f)|_{\text{dB}}$ are also plotted in Figure 6.21(a). Notice that the actual value at $f = f_B$ is $|H(f_B)|_{\text{dB}} = -3$ dB. Thus, the actual curve is only 3 dB from the asymptotes at $f = f_B$. For other frequencies, the actual curve is closer to the asymptotes. The Bode phase plot is shown in Figure 6.21(b) along with straight-line approximations.

Example 6.4 Determination of the Break Frequency for a Highpass Filter

Suppose that we want a first-order highpass filter that has a transfer-function magnitude of −30 dB at $f = 60$ Hz. Find the break frequency for this filter.

Solution Recall that the low-frequency asymptote slopes at a rate of 20 dB/decade. Thus, we must select f_B to be

$$\frac{30 \text{ dB}}{20 \text{ dB/decade}} = 1.5 \text{ decades}$$

higher than 60 Hz. Employing Equation 6.17, we have

$$\log\left(\frac{f_B}{60}\right) = 1.5$$

This is equivalent to

$$\frac{f_B}{60} = 10^{1.5} = 31.6$$

which yields

$$f_B \cong 1900 \text{ Hz}$$

■

We often need a filter that greatly reduces the amplitude of a component at a given frequency, but has a negligible effect on components at nearby frequencies.

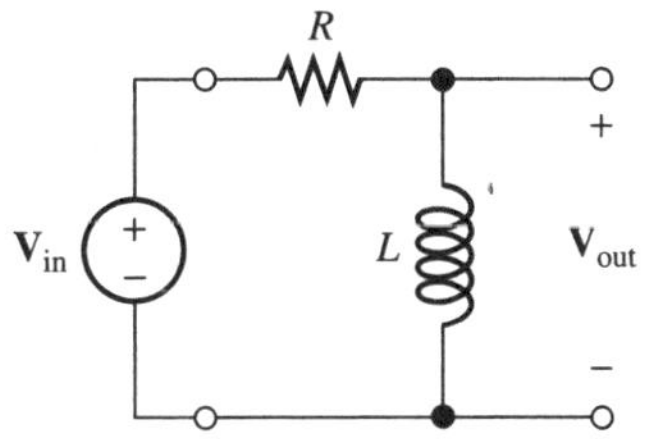

Figure 6.22 Circuit for Exercise 6.13.

The preceding example shows that to reduce the amplitude of a given component by a large factor by using a first-order filter, we must place the break frequency far from the component to be rejected. Then, components at other frequencies are also affected. This is a problem that can only be solved by using more complex (higher order) filter circuits. We consider second-order filters later in the chapter.

Exercise 6.13 Consider the circuit shown in Figure 6.22. Show that the transfer function of this filter is given by Equation 6.21 if the half-power frequency is defined to be $f_B = R/2\pi L$. □

Exercise 6.14 Suppose that we need a first-order *RC* highpass filter that reduces the amplitude of a component at a frequency of 1 kHz by 50 dB. The resistance is to be 1 kΩ. Find the half-power frequency and the capacitance.
Answer $f_B = 316$ kHz, $C = 503$ pF. □

6.6 SERIES RESONANCE

In this section and the next, we consider resonant circuits. These circuits form the basis for filters that have better performance (in passing desired signals and rejecting undesired signals that are relatively close in frequency) than first-order filters. Such filters are useful in radio receivers, for example. Another application is a notch filter to remove 60-Hz interference from audio signals. Resonance is a phenomenon that can be observed in mechanical systems as well as in electrical circuits. For example, a guitar string is a resonant mechanical system.

Resonance is a phenomenon that can be observed in mechanical systems and electrical circuits.

We will see that when a sinusoidal source of the proper frequency is applied to a resonant circuit, voltages much larger than the source voltage can appear in the circuit. The familiar story of opera singers using their voices to break wine goblets is an example of a mechanically resonant structure (the goblet) driven by an approximately sinusoidal source (the sound), resulting in vibrations in the glass of sufficient magnitude to cause fracture. Another example is the Tacoma Narrows Bridge collapse in 1940. Driven by wind forces, a resonance of the bridge structure resulted in oscillations that tore the bridge apart. Some other examples of mechanical resonant systems are the strings of musical instruments, bells, the air column in an organ pipe, and a mass suspended by a spring.

You can find a short video clip of the bridge in motion on the internet.

Consider the series circuit shown in Figure 6.23. The impedance seen by the source in this circuit is given by

$$Z_s(f) = j2\pi fL + R - j\frac{1}{2\pi fC} \qquad (6.28)$$

The resonant frequency f_0 is defined to be the frequency at which the impedance is purely resistive (i.e., the total reactance is zero).

The **resonant frequency** f_0 is defined to be the frequency at which the impedance is purely resistive (i.e., the total reactance is zero). For the reactance to equal zero,

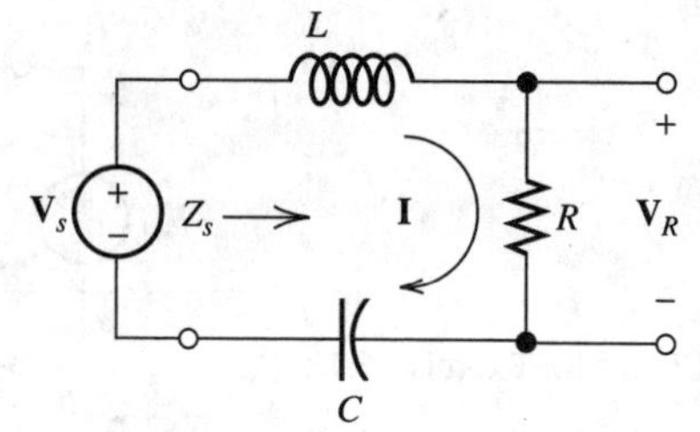

Figure 6.23 The series resonant circuit.

the impedance of the inductance must equal the impedance of the capacitance in magnitude. Thus, we have

$$2\pi f_0 L = \frac{1}{2\pi f_0 C} \tag{6.29}$$

Solving for the resonant frequency, we get

$$f_0 = \frac{1}{2\pi\sqrt{LC}} \tag{6.30}$$

The quality factor Q_s of a series circuit is defined to be the ratio of the reactance of the inductance at the resonant frequency to the resistance.

The **quality factor** Q_s is defined to be the ratio of the reactance of the inductance at the resonant frequency to the resistance:

$$Q_s = \frac{2\pi f_0 L}{R} \tag{6.31}$$

Solving Equation 6.29 for L and substituting into Equation 6.31, we obtain

$$Q_s = \frac{1}{2\pi f_0 CR} \tag{6.32}$$

Using Equations 6.30 and 6.31 to substitute into Equation 6.28, we can eventually reduce the equation for the impedance to

$$Z_s(f) = R\left[1 + jQ_s\left(\frac{f}{f_0} - \frac{f_0}{f}\right)\right] \tag{6.33}$$

Thus, the series resonant circuit is characterized by its quality factor Q_s and resonant frequency f_0.

Plots of the normalized magnitude and the phase of the impedance versus normalized frequency f/f_0 are shown in Figure 6.24. Notice that the impedance magnitude is minimum at the resonant frequency. As the quality factor becomes larger, the minimum becomes sharper.

Series Resonant Circuit as a Bandpass Filter

Referring to Figure 6.23, the current is given by

$$\mathbf{I} = \frac{\mathbf{V}_s}{Z_s(f)}$$

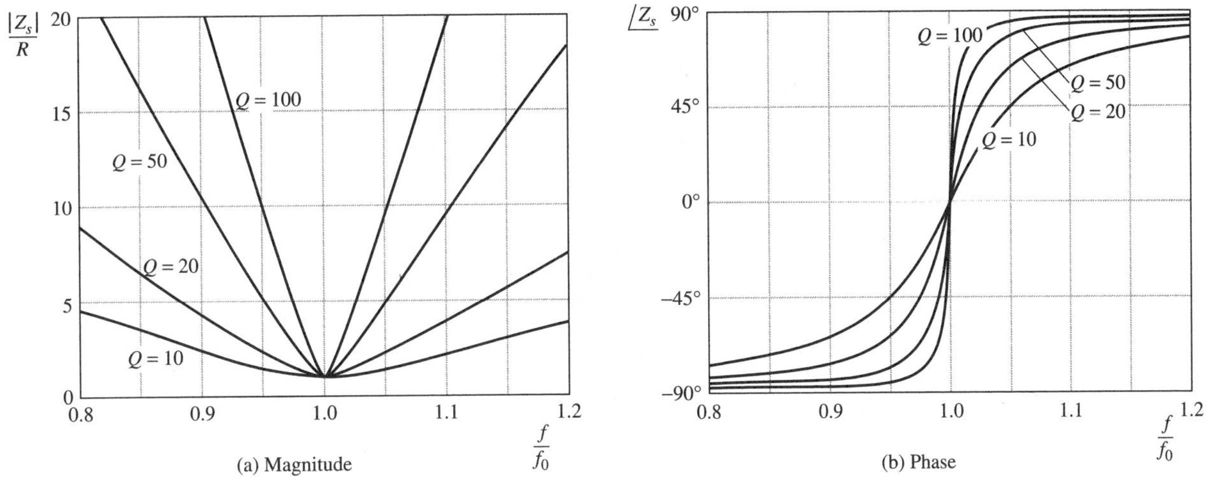

Figure 6.24 Plots of normalized magnitude and phase for the impedance of the series resonant circuit versus frequency.

Using Equation 6.33 to substitute for the impedance, we have

$$\mathbf{I} = \frac{\mathbf{V}_s/R}{1 + jQ_s(f/f_0 - f_0/f)}$$

The voltage across the resistance is

$$\mathbf{V}_R = R\mathbf{I} = \frac{\mathbf{V}_s}{1 + jQ_s(f/f_0 - f_0/f)}$$

Dividing by $\mathbf{V}_s$, we obtain the transfer function

$$\frac{\mathbf{V}_R}{\mathbf{V}_s} = \frac{1}{1 + jQ_s(f/f_0 - f_0/f)}$$

Plots of the magnitude of $\mathbf{V}_R/\mathbf{V}_s$ versus f are shown in Figure 6.25 for various values of Q_s.

Consider a (sinusoidal) source of constant amplitude and variable frequency. At low frequencies, the impedance magnitude of the capacitance is large, the current $\mathbf{I}$ is small in magnitude, and $\mathbf{V}_R$ is small in magnitude (compared with $\mathbf{V}_s$). At resonance, the total impedance magnitude reaches a minimum (because the reactances of the inductance and the capacitance cancel), the current magnitude is maximum, and $\mathbf{V}_R = \mathbf{V}_s$. At high frequencies, the impedance of the inductance is large, the current magnitude is small, and $\mathbf{V}_R$ is small in magnitude.

Now, suppose that we apply a source signal having components ranging in frequency about the resonant frequency. The components of the source that are close to the resonant frequency appear across the resistance with little change in amplitude. However, components that are higher or lower in frequency are significantly reduced in amplitude. Thus, a band of components centered at the resonant frequency is passed while components farther from the resonant frequency are (partly) rejected. We say that the resonant circuit behaves as a **bandpass filter**.

The resonant circuit behaves as a bandpass filter.

Recall that the half-power frequencies of a filter are the frequencies for which the transfer-function magnitude has fallen from its maximum by a factor of $1/\sqrt{2} \cong 0.707$.

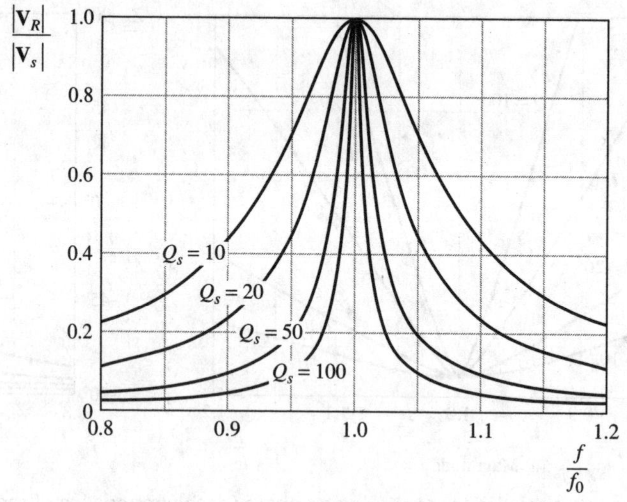

Figure 6.25 Plots of the transfer-function magnitude $|\mathbf{V}_R/\mathbf{V}_s|$ for the series resonant bandpass-filter circuit.

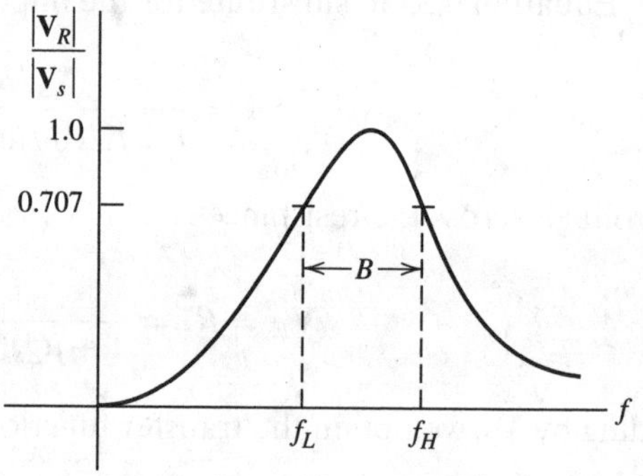

Figure 6.26 The bandwidth B is equal to the difference between the half-power frequencies.

For the series resonant circuit, there are two half-power frequencies f_L and f_H. This is illustrated in Figure 6.26.

The **bandwidth** B of this filter is the difference between the half-power frequencies:

$$B = f_H - f_L \tag{6.34}$$

For the series resonant circuit, it can be shown that

$$B = \frac{f_0}{Q_s} \tag{6.35}$$

Furthermore, for $Q_s >> 1$, the half-power frequencies are given by the approximate expressions

$$f_H \cong f_0 + \frac{B}{2} \tag{6.36}$$

and

$$f_L \cong f_0 - \frac{B}{2} \tag{6.37}$$

Example 6.5 Series Resonant Circuit

Consider the series resonant circuit shown in Figure 6.27. Compute the resonant frequency, the bandwidth, and the half-power frequencies. Assuming that the frequency of the source is the same as the resonant frequency, find the phasor voltages across the elements and draw a phasor diagram.

Solution First, we use Equation 6.30 to compute the resonant frequency:

$$f_0 = \frac{1}{2\pi\sqrt{LC}} = \frac{1}{2\pi\sqrt{0.1592 \times 0.1592 \times 10^{-6}}} = 1000 \text{ Hz}$$

The quality factor is given by Equation 6.31:

$$Q_s = \frac{2\pi f_0 L}{R} = \frac{2\pi \times 1000 \times 0.1592}{100} = 10$$

The bandwidth is given by Equation 6.35:

$$B = \frac{f_0}{Q_s} = \frac{1000}{10} = 100 \text{ Hz}$$

Next, we use Equations 6.36 and 6.37 to find the approximate half-power frequencies:

$$f_H \cong f_0 + \frac{B}{2} = 1000 + \frac{100}{2} = 1050 \text{ Hz}$$

$$f_L \cong f_0 - \frac{B}{2} = 1000 - \frac{100}{2} = 950 \text{ Hz}$$

At resonance, the impedance of the inductance and capacitance are

$$Z_L = j2\pi f_0 L = j2\pi \times 1000 \times 0.1592 = j1000\ \Omega$$

$$Z_C = -j\frac{1}{2\pi f_0 C} = -j\frac{1}{2\pi \times 1000 \times 0.1592 \times 10^{-6}} = -j1000\ \Omega$$

As expected, the reactances are equal in magnitude at the resonant frequency. The total impedance of the circuit is

$$Z_s = R + Z_L + Z_C = 100 + j1000 - j1000 = 100\ \Omega$$

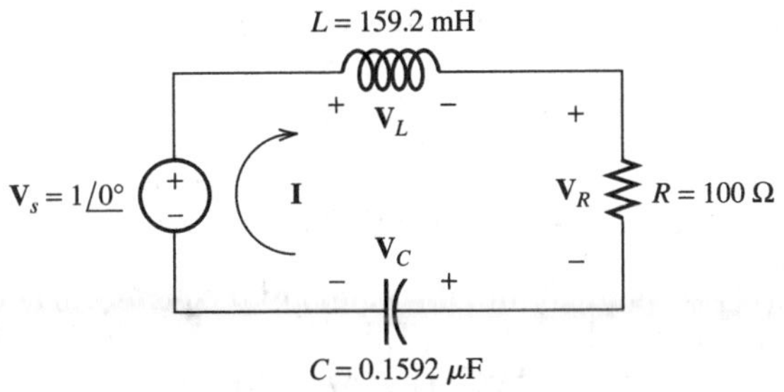

Figure 6.27 Series resonant circuit of Example 6.5. (The component values have been selected so the resonant frequency and Q_s turn out to be round numbers.)

The phasor current is given by

$$\mathbf{I} = \frac{\mathbf{V}_s}{Z_s} = \frac{1\angle 0^\circ}{100} = 0.01\angle 0^\circ$$

The voltages across the elements are

$$\mathbf{V}_R = R\mathbf{I} = 100 \times 0.01\angle 0^\circ = 1\angle 0^\circ$$

$$\mathbf{V}_L = Z_L\mathbf{I} = j1000 \times 0.01\angle 0^\circ = 10\angle 90^\circ$$

$$\mathbf{V}_C = Z_C\mathbf{I} = -j1000 \times 0.01\angle 0^\circ = 10\angle -90^\circ$$

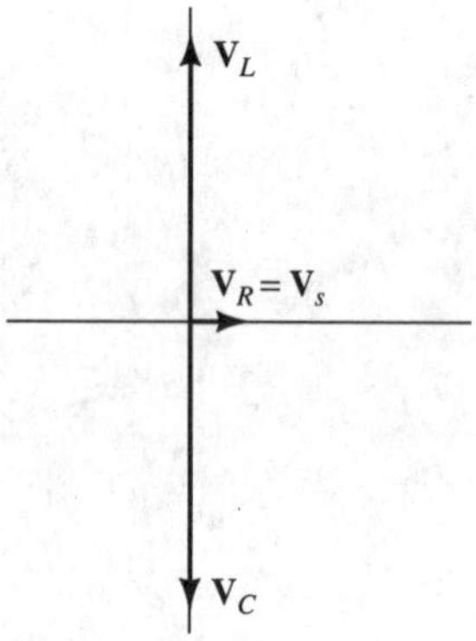

Figure 6.28 Phasor diagram for Example 6.5.

The phasor diagram is shown in Figure 6.28. Notice that the voltages across the inductance and capacitance are much larger than the source voltage in magnitude. Nevertheless, Kirchhoff's voltage law is satisfied because $\mathbf{V}_L$ and $\mathbf{V}_C$ are out of phase and cancel. ■

In Example 6.5, we found that the voltage magnitudes across the inductance and capacitance are Q_s times higher than the source voltage. Thus, a higher quality factor leads to higher voltage magnification. This is similar to the large vibrations that can be caused in a wine goblet by an opera singer's voice.

Exercise 6.15 Determine the R and C values for a series resonant circuit that has $L = 10\ \mu\text{H}$, $f_0 = 1$ MHz, and $Q_s = 50$. Find the bandwidth and approximate half-power frequencies of the circuit.

Answer $C = 2533$ pF, $R = 1.257\ \Omega$, $B = 20$ kHz, $f_L \cong 990$ kHz, $f_H \cong 1010$ kHz. □

Exercise 6.16 Suppose that a voltage $\mathbf{V}_s = 1\angle 0^\circ$ at a frequency of 1 MHz is applied to the circuit of Exercise 6.15. Find the phasor voltages across the resistance, capacitance, and inductance.

Answer $\mathbf{V}_R = 1\angle 0^\circ$, $\mathbf{V}_C = 50\angle -90^\circ$, $\mathbf{V}_L = 50\angle 90^\circ$. □

Exercise 6.17 Find the R and L values for a series resonant circuit that has $C = 470$ pF, a resonant frequency of 5 MHz, and a bandwidth of 200 kHz.

Answer $R = 2.709\ \Omega$, $L = 2.156\ \mu\text{H}$. □

6.7 PARALLEL RESONANCE

Another type of resonant circuit known as a **parallel resonant circuit** is shown in Figure 6.29. The impedance of this circuit is given by

$$Z_p = \frac{1}{1/R + j2\pi fC - j(1/2\pi fL)} \tag{6.38}$$

As in the series resonant circuit, the **resonant frequency** f_0 is the frequency for which the impedance is purely resistive. This occurs when the imaginary parts of the denominator of Equation 6.38 cancel. Thus, we have

$$2\pi f_0 C = \frac{1}{2\pi f_0 L} \tag{6.39}$$

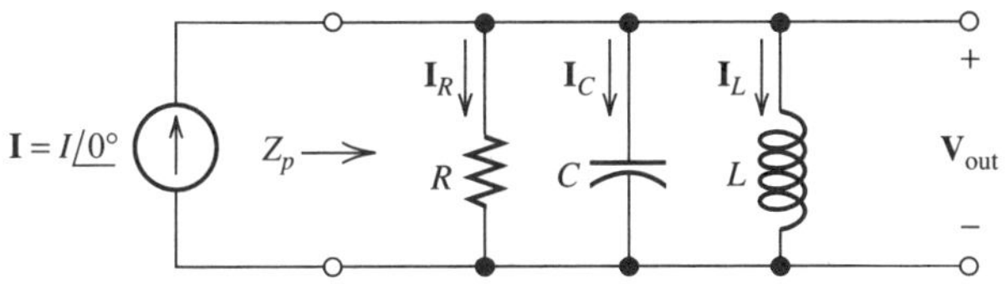

Figure 6.29 The parallel resonant circuit.

Solving for the resonant frequency, we get

$$f_0 = \frac{1}{2\pi\sqrt{LC}} \tag{6.40}$$

which is exactly the same as the expression for the resonant frequency of the series circuit discussed in Section 6.6.

For the parallel circuit, we define the quality factor Q_p as the ratio of the resistance to the reactance of the inductance at resonance, given by

$$Q_p = \frac{R}{2\pi f_0 L} \tag{6.41}$$

Notice that this is the reciprocal of the expression for the quality factor Q_s of the series resonant circuit. Solving Equation 6.40 for L and substituting into Equation 6.41, we obtain another expression for the quality factor:

Notice that the formula for Q_p of a parallel circuit in terms of the circuit elements is the reciprocal of the formula for Q_s of a series circuit.

$$Q_p = 2\pi f_0 CR \tag{6.42}$$

If we solve Equations 6.41 and 6.42 for L and C, respectively, and then substitute into Equation 6.38, we eventually obtain

$$Z_p = \frac{R}{1 + jQ_p(f/f_0 - f_0/f)} \tag{6.43}$$

The voltage across the parallel resonant circuit is the product of the phasor current and the impedance:

$$\mathbf{V}_{out} = \frac{\mathbf{I}R}{1 + jQ_p(f/f_0 - f_0/f)} \tag{6.44}$$

Suppose that we hold the current constant in magnitude and change the frequency. Then, the magnitude of the voltage is a function of frequency. A plot of voltage magnitude for the parallel resonant circuit is shown in Figure 6.30. Notice that the voltage magnitude reaches its maximum $V_{o\,max} = RI$ at the resonant frequency. These curves have the same shape as the curves shown in Figures 6.25 and 6.26 for the voltage transfer function of the series resonant circuit.

The half-power frequencies f_L and f_H are defined to be the frequencies at which the voltage magnitude reaches the maximum value times $1/\sqrt{2}$. The bandwidth of the circuit is given by

$$B = f_H - f_L \tag{6.45}$$

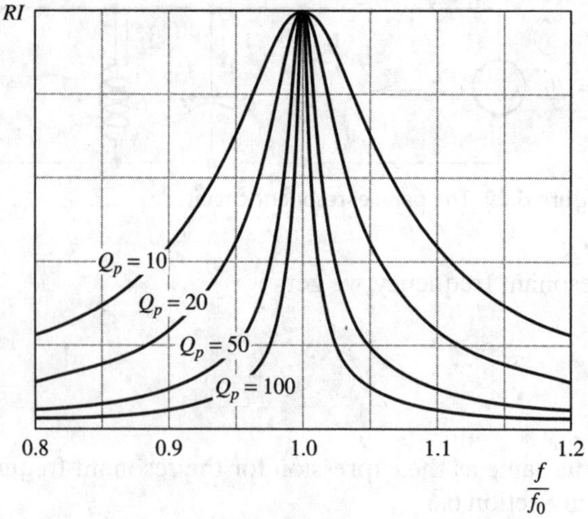

Figure 6.30 Voltage across the parallel resonant circuit for a constant-amplitude variable-frequency current source.

It can be shown that the bandwidth is related to the resonant frequency and quality factor by the expression

$$B = \frac{f_0}{Q_p} \tag{6.46}$$

Example 6.6 Parallel Resonant Circuit

Find the L and C values for a parallel resonant circuit that has $R = 10\,\text{k}\Omega$, $f_0 = 1\,\text{MHz}$, and $B = 100\,\text{kHz}$. If $\mathbf{I} = 10^{-3}\angle 0°$ A, draw the phasor diagram showing the currents through each of the elements in the circuit at resonance.

Solution First, we compute the quality factor of the circuit. Rearranging Equation 6.46 and substituting values, we have

$$Q_p = \frac{f_0}{B} = \frac{10^6}{10^5} = 10$$

Solving Equation 6.41 for the inductance and substituting values, we get

$$L = \frac{R}{2\pi f_0 Q_p} = \frac{10^4}{2\pi \times 10^6 \times 10} = 159.2\ \mu\text{H}$$

Similarly, using Equation 6.42, we find that

$$C = \frac{Q_p}{2\pi f_0 R} = \frac{10}{2\pi \times 10^6 \times 10^4} = 159.2\ \text{pF}$$

At resonance, the voltage is given by

$$\mathbf{V}_{\text{out}} = \mathbf{I}R = (10^{-3}\angle 0°) \times 10^4 = 10\angle 0°\ \text{V}$$

and the currents are given by

$$\mathbf{I}_R = \frac{\mathbf{V}_{\text{out}}}{R} = \frac{10\angle 0^\circ}{10^4} = 10^{-3}\angle 0^\circ \text{ A}$$

$$\mathbf{I}_L = \frac{\mathbf{V}_{\text{out}}}{j2\pi f_0 L} = \frac{10\angle 0^\circ}{j10^3} = 10^{-2}\angle -90^\circ \text{ A}$$

$$\mathbf{I}_C = \frac{\mathbf{V}_{\text{out}}}{-j/2\pi f_0 C} = \frac{10\angle 0^\circ}{-j10^3} = 10^{-2}\angle 90^\circ \text{ A}$$

The phasor diagram is shown in Figure 6.31. Notice that the currents through the inductance and capacitance are larger in magnitude than the applied source current. However, since $\mathbf{I}_C$ and $\mathbf{I}_L$ are out of phase, they cancel. ■

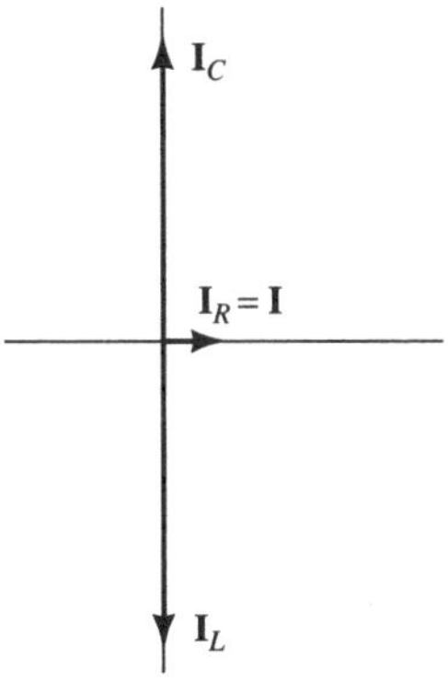

Figure 6.31 Phasor diagram for Example 6.6.

Exercise 6.18 A parallel resonant circuit has $R = 10$ kΩ, $L = 100$ μH, and $C = 500$ pF. Find the resonant frequency, quality factor, and bandwidth.
Answer $f_0 = 711.8$ kHz, $Q_p = 22.36$, $B = 31.83$ kHz. □

Exercise 6.19 A parallel resonant circuit has $f_0 = 10$ MHz, $B = 200$ kHz, and $R = 1$ kΩ. Find L and C.
Answer $L = 0.3183$ μH, $C = 795.8$ pF. □

6.8 IDEAL AND SECOND-ORDER FILTERS

Ideal Filters

In discussing filter performance, it is helpful to consider ideal filters. An ideal filter passes components in the desired frequency range with no change in amplitude or phase and totally rejects the components in the undesired frequency range. Depending on the locations of the frequencies to be passed and rejected, we have different types of filters: lowpass, highpass, bandpass, and band reject. The transfer functions $H(f) = \mathbf{V}_{\text{out}}/\mathbf{V}_{\text{in}}$ of the four types of ideal filters are shown in Figure 6.32.

- An **ideal lowpass filter** [Figure 6.32(a)] passes components below its cutoff frequency f_H and rejects components higher in frequency than f_H.
- An **ideal highpass filter** [Figure 6.32(b)] passes components above its cutoff frequency f_L and rejects components lower in frequency than f_L.
- An **ideal bandpass filter** [Figure 6.32(c)] passes components that lie between its cutoff frequencies (f_L and f_H) and rejects components outside that range.

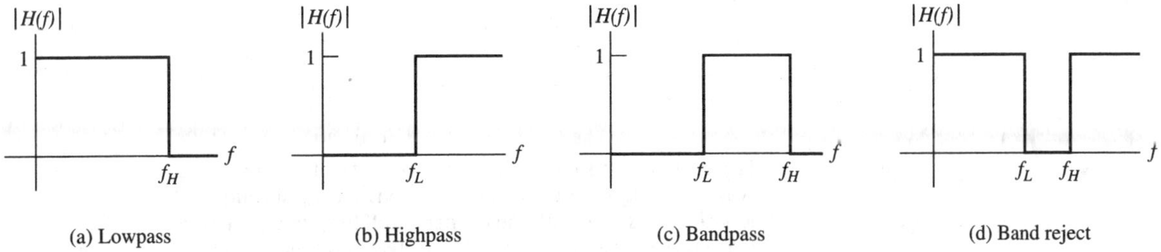

Figure 6.32 Transfer functions of ideal filters.

- An **ideal band-reject filter** [Figure 6.32(d)], which is also called a **notch filter**, rejects components that lie between its cutoff frequencies (f_L and f_H) and passes components outside that range.

As we have seen earlier in this chapter, filters are useful whenever a signal contains desired components in one range of frequency and undesired components in another range of frequency. For example, Figure 6.33(a) shows a 1-kHz sine wave that has been corrupted by high-frequency noise. By passing this noisy signal through a lowpass filter, the noise is eliminated.

Unfortunately, it is not possible to construct ideal filters—they can only be approximated by real circuits. As the circuits are allowed to increase in complexity, it is possible to design filters that do a better job of rejecting unwanted components and retaining the desired components. Thus, we will see that second-order circuits

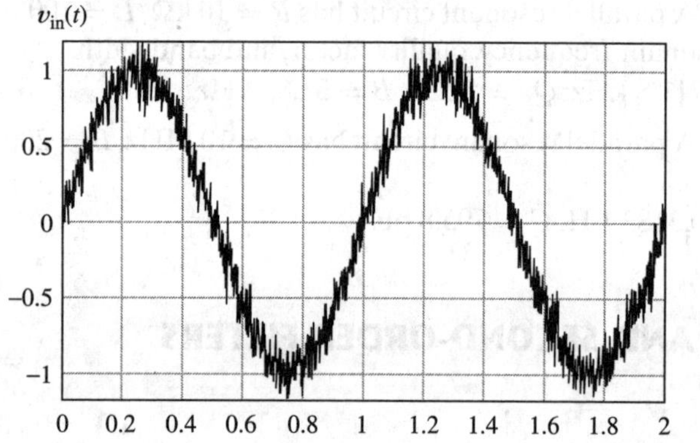

Figure 6.33 The input signal v_{in} consists of a 1-kHz sine wave plus high-frequency noise. By passing v_{in} through an ideal lowpass filter with the proper cutoff frequency, the sine wave is passed and the noise is rejected, resulting in a clean output signal.

perform better (i.e., closer to ideal) than the first-order circuits considered earlier in this chapter.

Second-Order Lowpass Filter

Figure 6.34(a) shows a second-order lowpass filter based on the series resonant circuit of Section 6.6. The filter is characterized by its resonant frequency f_0 and quality factor Q_s, which are given by Equations 6.30 and 6.31. It can be shown that the transfer function for this circuit is given by

$$H(f) = \frac{\mathbf{V}_{\text{out}}}{\mathbf{V}_{\text{in}}} = \frac{-jQ_s(f_0/f)}{1 + jQ_s(f/f_0 - f_0/f)} \tag{6.47}$$

Bode plots of the transfer-function magnitude are shown in Figure 6.34(c). Notice that for $Q_s >> 1$, the transfer-function magnitude reaches a high peak in the vicinity of the resonant frequency. Usually, in designing a filter, we want the gain to be approximately constant in the passband, and we select $Q_s \cong 1$. (Actually, $Q_s = 0.707$ is the highest value for which the transfer-function magnitude does not display an increase before rolling off. The transfer function for this value of Q_s is said to be *maximally flat*, is also known as a *Butterworth function*, and is often used for lowpass filters.)

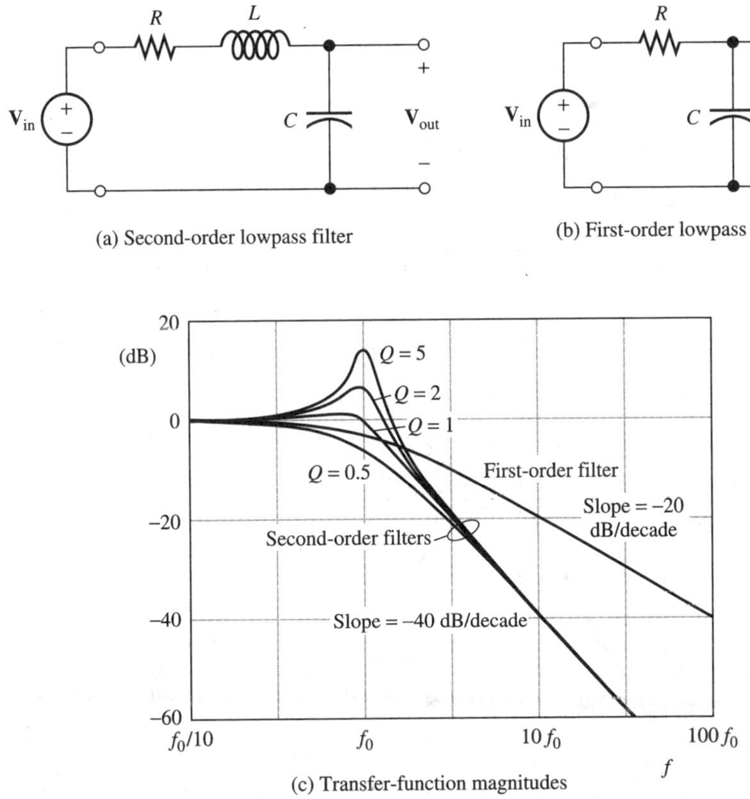

Figure 6.34 Lowpass filter circuits and their transfer-function magnitudes versus frequency.

The transfer-function magnitude of a second-order lowpass filter declines 40 dB per decade well above the break frequency, whereas the transfer-function magnitude for the first-order filter declines at only 20 dB per decade. Thus, the second-order filter is a better approximation to an ideal lowpass filter.

Comparison of First- and Second-Order Filters

For comparison, a first-order lowpass filter is shown in Figure 6.34(b), and the Bode plot of its transfer function is shown in Figure 6.34(c). The first-order circuit is characterized by its half-power frequency $f_B = 1/(2\pi RC)$. (We have selected $f_B = f_0$ in making the comparison.) Notice that above f_0 the magnitude of the transfer function falls more rapidly for the second-order filter than for the first-order filter (−40 dB/decade versus −20 dB/decade).

Second-Order Highpass Filter

A second-order highpass filter is shown in Figure 6.35(a), and its magnitude Bode plot is shown in Figure 6.35(b). Here again, we usually want the magnitude to be as nearly constant as possible in the passband, so we select $Q_s \cong 1$. (In other words, we usually want to design the filter to approximate an ideal filter as closely as possible.)

Second-Order Bandpass Filter

A second-order bandpass filter is shown in Figure 6.36(a), and its magnitude Bode plot is shown in Figure 6.36(b). The half-power bandwidth B is given by Equations 6.34 and 6.35, which state that

$$B = f_H - f_L$$

and

$$B = \frac{f_0}{Q_s}$$

Second-Order Band-Reject (Notch) Filter

A second-order band-reject filter is shown in Figure 6.37(a) and its magnitude Bode plot is shown in Figure 6.37(b). In theory, the magnitude of the transfer function is zero for $f = f_0$. [In decibels, this corresponds to $|H(f_0)| = -\infty$ dB.] However, real

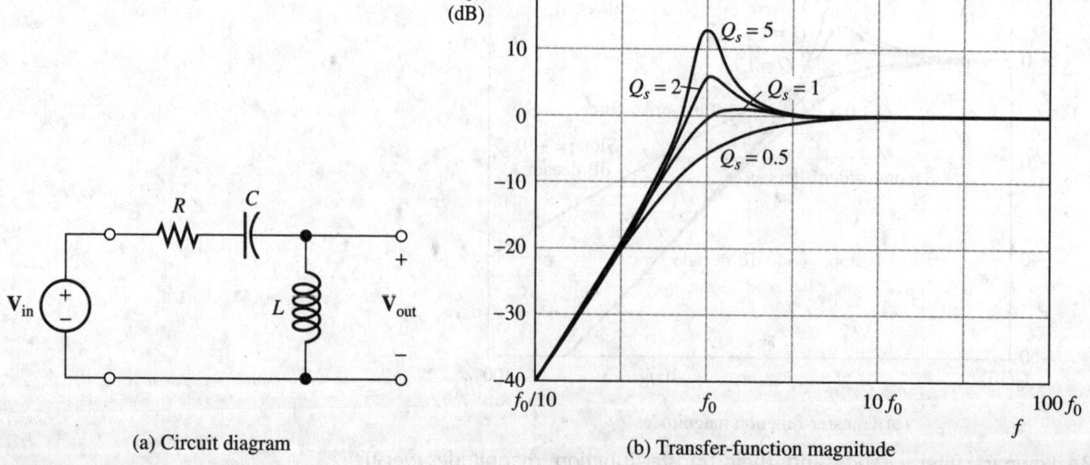

Figure 6.35 Second-order highpass filter and its transfer-function magnitude versus frequency for several values of Q_s.

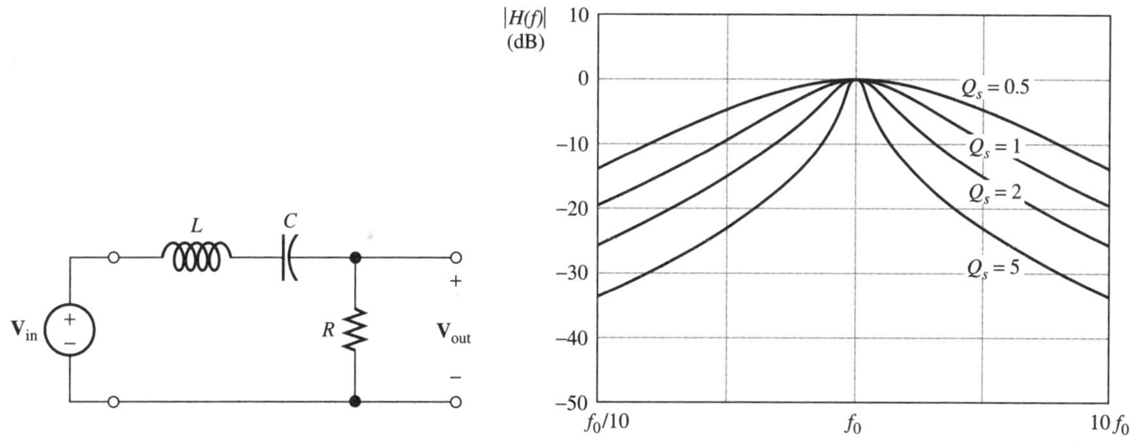

(a) Circuit diagram (b) Transfer-function magnitude

Figure 6.36 Second-order bandpass filter and its transfer-function magnitude versus frequency for several values of Q_s.

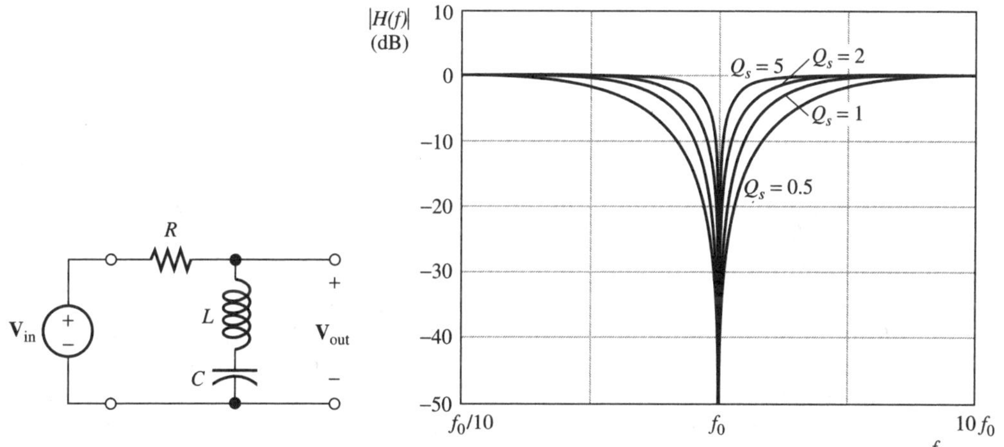

(a) Circuit diagram (b) Transfer-function magnitude

Figure 6.37 Second-order band-reject filter and its transfer-function magnitude versus frequency for several values of Q_s.

inductors contain series resistance, so rejection of the f_0 component is not perfect for actual circuits.

Example 6.7 Filter Design

Suppose that we need a filter that passes components higher in frequency than 1 kHz and rejects components lower than 1 kHz. Select a suitable second-order circuit configuration, choose $L = 50$ mH, and specify the values required for the other components.

Solution We need to pass high-frequency components and reject low-frequency components. Therefore, we need a highpass filter. The circuit diagram for a second-order highpass filter is shown in Figure 6.35(a), and the corresponding transfer-function magnitude plots are shown in Figure 6.35(b). Usually, we want the transfer function to be approximately constant in the passband. Thus, we choose $Q_s \cong 1$. We select $f_0 \cong 1$ kHz, so the components above 1 kHz are passed, while

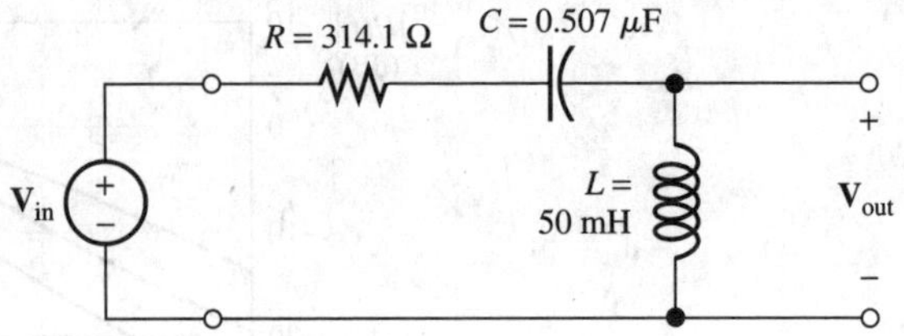

Figure 6.38 Filter designed in Example 6.7.

lower-frequency components are (at least partly) rejected. Solving Equation 6.30 for the capacitance and substituting values, we have

$$C = \frac{1}{(2\pi)^2 f_0^2 L} = \frac{1}{(2\pi)^2 \times 10^6 \times 50 \times 10^{-3}}$$
$$= 0.507\ \mu\text{F}$$

Solving Equation 6.31 for the resistance and substituting values, we get

$$R = \frac{2\pi f_0 L}{Q_s} = \frac{2\pi \times 1000 \times 50 \times 10^{-3}}{1} = 314.1\ \Omega$$

The circuit and values are shown in Figure 6.38. ■

There are several reasons why we might not use the exact values that we calculated for the components in the last example. First, fixed-value capacitors and resistors are readily available only in certain standard values. Furthermore, the design called for a filter to reject components lower than 1 kHz and pass components higher than 1 kHz. We arbitrarily selected $f_0 = 1$ kHz. Depending on whether it is more important to reject the low frequencies or to pass the high frequencies without change in amplitude, a slightly higher or lower value for f_0 could be better. Finally, our choice of Q_s was somewhat arbitrary. In practice, we could choose variable components by using the calculations as a starting point. Then, we would adjust the filter experimentally for the most satisfactory performance.

Exercise 6.20 Suppose that we need a filter that passes components lower in frequency than 5 kHz and rejects components higher than 5 kHz. Select a suitable second-order circuit configuration, choose $L = 5$ mH, and specify the values required for the other components.
Answer See Figure 6.39. □

Exercise 6.21 Suppose that we want a filter that passes components between $f_L = 45$ kHz and $f_H = 55$ kHz. Higher and lower frequencies are to be rejected. Design a circuit using a 1-mH inductance.

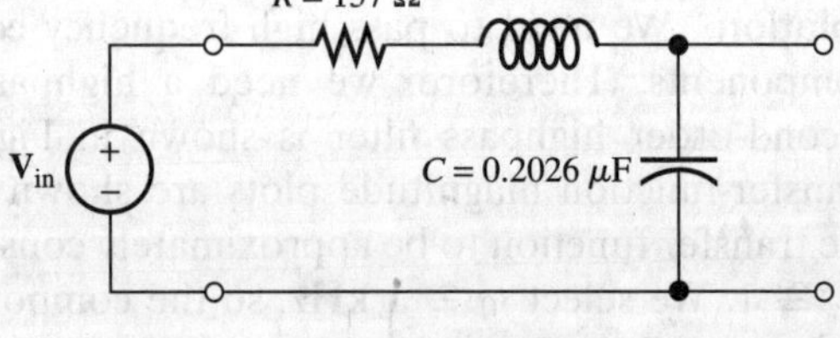

Figure 6.39 Answer for Exercise 6.20.

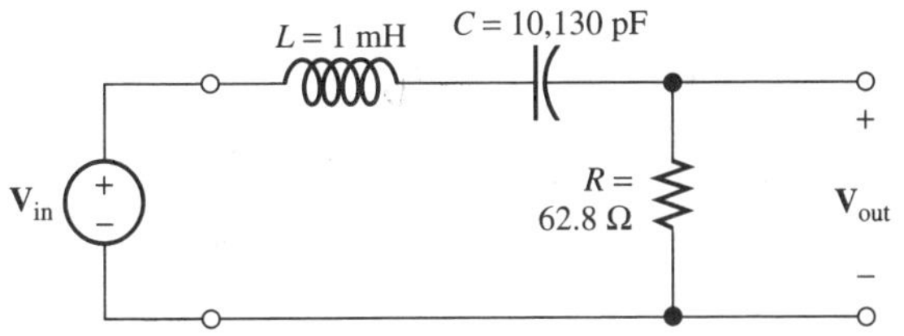

Figure 6.40 Answer for Exercise 6.21.

Answer We need a bandpass filter with $f_0 \cong 50$ kHz and $Q_s = 5$. The resulting circuit is shown in Figure 6.40. □

6.9 TRANSFER FUNCTIONS AND BODE PLOTS WITH MATLAB

So far in this chapter, we have used manual methods to illustrate Bode-plot concepts for simple filters. While manual methods can be extended to more complex circuits, it is often quicker and more accurate to use computer software to derive transfer functions and produce Bode plots.

Because subtle programming errors can result in grossly erroneous results, it is good practice to employ independent checks on computer-generated Bode plots. For example, a complex circuit can often be readily analyzed manually at very high and at very low frequencies. At very low frequencies, the inductances behave as short circuits and the capacitances behave as open circuits, as we discussed in Section 4.2. Thus, we can replace the inductances by shorts and the capacitances by opens and analyze the simplified circuit to determine the value of the transfer function at low frequencies, providing an independent check on the plots produced by a computer.

Manual analysis at dc and very high frequencies often provides some easy checks on computer-aided Bode plots.

Similarly, at very high frequencies, the inductances become open circuits, and the capacitances become shorts. Next, we illustrate this approach with an example.

Example 6.8 Computer-Generated Bode Plot

The circuit of Figure 6.41 is a notch filter. Use MATLAB to generate a magnitude Bode plot of the transfer function $H(f) = \mathbf{V}_{\text{out}}/\mathbf{V}_{\text{in}}$ with frequency ranging from 10 Hz to 100 kHz. Then, analyze the circuit manually at very high and very low frequencies to provide checks on the plot. Use the plot to determine the frequency of maximum attenuation and the value of the transfer function at that frequency.

Solution Using the voltage-divider principle, we can write the transfer function for the filter as

$$H(f) = \frac{\mathbf{V}_{\text{out}}}{\mathbf{V}_{\text{in}}} = \frac{R_3}{R_1 + R_3 + 1/[j\omega C + 1/(R_2 + j\omega L)]}$$

A MATLAB m-file that produces the Bode plot is:

```
clear
% Enter the component values:
R1 = 90; R2 = 10; R3 = 100;
L = 0.1; C = 1e-7;
% The following command generates 1000 frequency values
% per decade, evenly spaced from 10^1 to 10^5 Hz
% on a logarithmic scale:
```

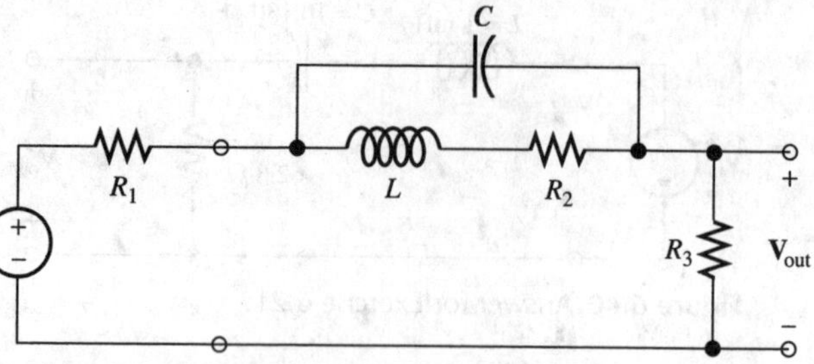

Figure 6.41 Filter of Example 6.8.

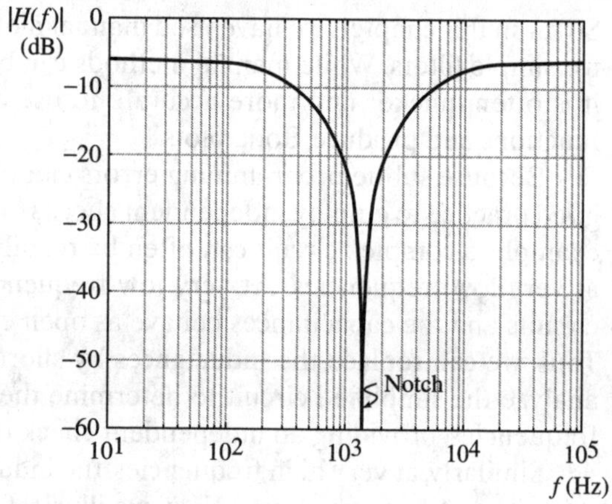

Figure 6.42 Bode plot for Example 6.8 produced using MATLAB.

```
f = logspace(1,5,4000);
w = 2*pi*f;
% Evaluate the transfer function for each frequency.
% As usual, we are using i in place of j:
H = R3./(R1+R3+1./(i*w*C + 1./(R2 + i*w*L)));
% Convert the magnitude values to decibels and plot:
semilogx(f,20*log10(abs(H)))
```

The resulting plot is shown in Figure 6.42. This circuit is called a notch filter because it strongly rejects components in the vicinity of 1591 Hz while passing higher and lower frequencies. The maximum attenuation is 60 dB.

The m-file is named Example_6_8 and appears in the MATLAB folder, and if you have access to MATLAB, you can run it to see the result. (See Appendix E for information on how to access the MATLAB folder.) Then, you can use the toolbar on the figure screen to magnify a portion of the plot and obtain the notch frequency and maximum attenuation with excellent accuracy.

The command

```
f = logspace(1,5,4000)
```

generates an array of 4000 frequency values, starting at 10^1 Hz and ending at 10^5 Hz, evenly spaced on a logarithmic scale with 1000 points per decade. (Typically, we might start with 100 points per decade, but this transfer function changes very rapidly in the vicinity of 1590 Hz, so we increased the number of points to more accurately determine the location and depth of the notch.)

As a partial check on our analysis and program, we analyze the circuit at $f = 0$ (dc) to determine the transfer function at very low frequencies. To do so, we replace the inductance by a short and the capacitance by an open circuit. Then, the circuit becomes a simple resistive voltage divider consisting of R_1, R_2, and R_3. Therefore, we have

$$H(0) = \frac{\mathbf{V}_{\text{out}}}{\mathbf{V}_{\text{in}}} = \frac{R_3}{R_1 + R_2 + R_3} = 0.5$$

In decibels, this becomes

$$H_{\text{dB}}(0) = 20\log(0.5) = -6 \text{ dB}$$

which agrees very well with the plotted value at 10 Hz.

For a second check, we replace the capacitance by a short circuit and the inductance by an open circuit to determine the value of the transfer function at very high frequencies. Then, the circuit again becomes a simple resistive voltage divider consisting of R_1 and R_3. Thus, we have

$$H(\infty) = \frac{R_3}{R_1 + R_3} = 0.5263$$

In decibels, this becomes

$$H_{\text{dB}}(\infty) = 20\log(0.5263) = -5.575 \text{ dB}$$

which agrees very closely with the value plotted at 100 kHz. ■

Using the Symbolic Toolbox to Derive Transfer Functions

In the previous example, we could readily write the transfer function by applying the voltage-divider principle. For more complex circuits, manual analysis can be very time consuming. An alternative is to use the MATLAB Symbolic Toolbox to help derive the desired transfer function. Assuming that the transfer function of interest is the output voltage divided by the input voltage, the step-by-step procedure is:

1. Set the input voltage $\mathbf{V}_{\text{in}}$ to 1 V.
2. Choose node-voltage variables for the circuit, including the output voltage $\mathbf{V}_{\text{out}}$ as one of the variables.
3. Write the node equations.
4. Use the `solve` command to solve for the output voltage $\mathbf{V}_{\text{out}}$ (which is actually equal to the transfer function, because we assumed that the input voltage is 1 V).
5. Define a row vector that contains the list of frequencies for which we want to evaluate the transfer function and use the `subs` command to substitute the frequency values into the transfer function.
6. Plot the results.

   (This procedure can be readily adapted to other transfer functions that may be of interest, such as output current divided by input current.) We illustrate the procedure with an example.

Example 6.9 Bode Plot Using the MATLAB Symbolic Toolbox

Obtain a magnitude bode plot of the transfer function $H(f) = \mathbf{V}_{\text{out}}/\mathbf{V}_{\text{in}}$ for the circuit of Figure 6.43 with frequency ranging from 100 kHz to 10 MHz. Manually check the plotted values for high and low frequencies.

Solution We assume that the input voltage $\mathbf{V}_{\text{in}}$ is 1 V. Then, we choose node voltage variables $\mathbf{V}_1$, $\mathbf{V}_2$, and $\mathbf{V}_3$ as shown in the figure. Notice that $\mathbf{V}_3$ and $\mathbf{V}_{\text{out}}$ are the same voltage and are equal to the desired transfer function. The node equations are obtained by applying KCL at each node.

$$\frac{\mathbf{V}_1 - 1}{R_s} + j\omega C_1 \mathbf{V}_1 + \frac{\mathbf{V}_1 - \mathbf{V}_2}{j\omega L_1} = 0$$

$$\frac{\mathbf{V}_2 - \mathbf{V}_1}{j\omega L_1} + j\omega C_2 \mathbf{V}_2 + \frac{\mathbf{V}_2 - \mathbf{V}_3}{j\omega L_2} = 0$$

$$\frac{\mathbf{V}_3 - \mathbf{V}_2}{j\omega L_2} + j\omega C_3 \mathbf{V}_3 + \frac{\mathbf{V}_3}{R_L} = 0$$

An m-file that produces the desired Bode plot is:

```
clear
% Construct the symbolic objects that appear in the circuit:
syms V1 V2 V3
syms w Rs RL C1 C2 C3 L1 L2 real
% Notice that V1, V2 and V3 are complex quantities
% while w, Rs, etc. are real.
% Solve the node voltage equations for V1, V2, and V3:
% Use i rather than j.
[V1 V2 V3] = solve('(V1-1)/Rs + i*w*C1*V1 + (V1-V2)/(i*w*L1) = 0',...
    '(V2-V1)/(i*w*L1) + i*w*C2*V2 + (V2-V3)/(i*w*L2) = 0',...
    '(V3-V2)/(i*w*L2) + i*w*C3*V3 + V3/RL = 0',...
    'V1','V2','V3');
% Enter the component values:
C1 = 1.967e-9; C2 = 6.366e-9; C3 = 1.967e-9;
L1 = 12.88e-6; L2 = 12.88e-6; Rs = 50; RL = 50;
% Substitute the component values into the solution for V3
% and define result as the transfer function H:
H = subs(V3); % Recall that the transfer function is the same as the
% output voltage V3.
% Next, set up a row matrix of logarithmically equally spaced
% frequencies at 100 points per decade from 10^5 to 10^7 Hz:
f = logspace(5,7,200);
```

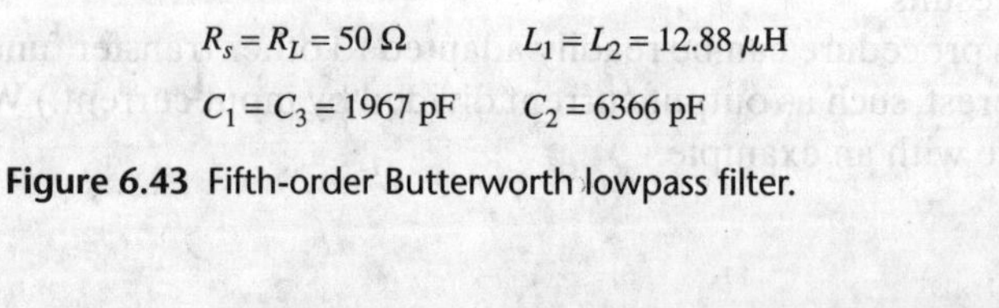

Figure 6.43 Fifth-order Butterworth lowpass filter.

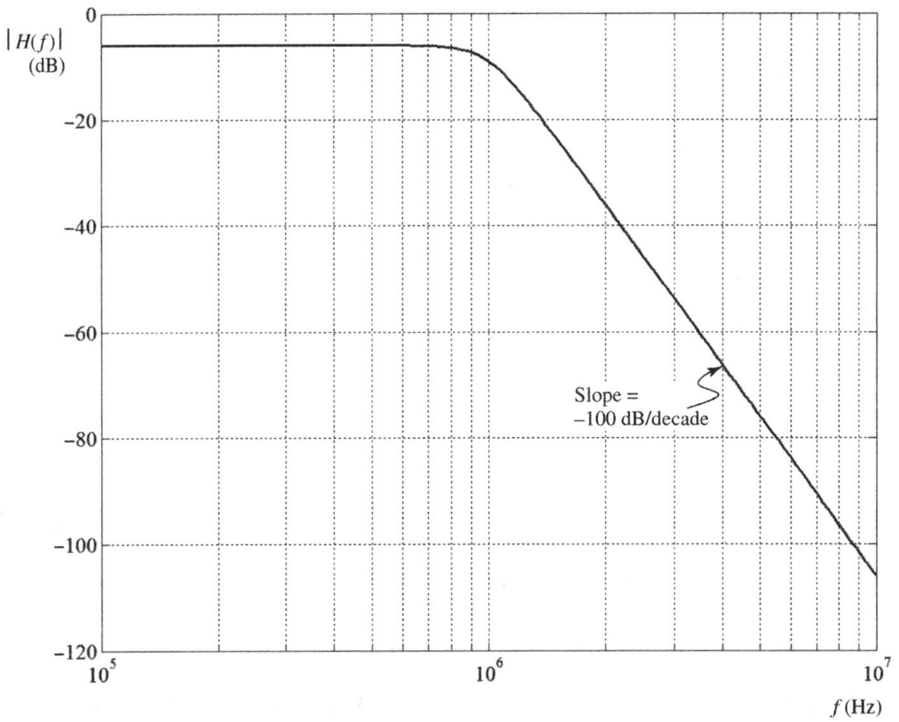

Figure 6.44 Bode plot for Example 6.9.

```
wn = 2*pi*f;
% Substitute the frequency values into the transfer function
% and convert to numeric form by using the double command:
H = double(subs(H,w,wn));
% Convert the transfer function magnitude to dB and plot:
HmagdB = 20*log10(abs(H));
semilogx(f,HmagdB)
```

The resulting plot is shown in Figure 6.44. The m-file is named Example_6_9 and can be found in the MATLAB folder.

To check the plotted transfer function at very low frequencies, we replace the inductances by shorts and the capacitors with opens. Then, the circuit becomes a simple voltage divider and the transfer function is

$$H(0) = \frac{R_L}{R_s + R_L} = 0.5$$

which is equivalent to −6 dB, agreeing very well with the value plotted at 100 kHz.

At very high frequencies, the capacitors become shorts and the inductors become opens. Then the output voltage tends toward zero, and the transfer function tends toward $-\infty$ dB. This agrees with the trend of the plot at high frequencies. ■

The filter of the previous example is known as a fifth-order Butterworth lowpass filter. Books on radio-frequency design typically have component values for normalized versions of this and many other types of filters listed in design tables. The values from the tables can be scaled to obtain filters with the desired half-power frequencies and impedance levels. So in this case, the designer would know exactly what the transfer function should be before doing the analysis. However, the parasitic effects of the components can significantly change the performance in unpredictable ways. A designer could include these parasitics in the analysis and use a MATLAB analysis to see their effects. However, such detailed analysis is beyond the scope of this book. Our aim is to illustrate the power of computer-aided analysis.

Exercise 6.22 If you have access to MATLAB and the Symbolic Toolbox, run the m-files Example_6_8 and Example_6_9 that are contained in the MATLAB folder. (**Note:** We have run these files successfully with MATLAB versions R2008a and R2008b. However, they may not run with other versions.)

Answer The resulting plots should be very similar to Figures 6.42 and 6.44, respectively. □

6.10 DIGITAL SIGNAL PROCESSING

So far, we have introduced the concepts related to filters in the context of *RLC* circuits. However, many modern systems make use of a more sophisticated technology called **digital signal processing** (DSP). In using DSP to filter a signal, the analog input signal $x(t)$ is converted to digital form (a sequence of numbers) by an **analog-to-digital converter** (ADC). A digital computer then uses the digitized input signal to compute a sequence of values for the output signal. Finally, if desired, the computed values are converted to analog form by a **digital-to-analog converter** (DAC) to produce the output signal $y(t)$. The generic block diagram of a DSP system is shown in Figure 6.45.

Besides filtering, many other operations, such as speech recognition, can be performed by DSP systems. DSP was used in the early days of the Space Telescope to focus blurry images resulting from an error in the telescope's design. High-definition televisions, digital cell phones, and MP3 music players are examples of products that have been made possible by DSP technology.

DSP is a large and rapidly evolving field that will continue to produce novel products. We discuss digital filters very briefly to give you a glimpse of this exciting field.

Conversion of Signals from Analog to Digital Form

Analog signals are converted to digital form by a DAC in a two-step process. First, the analog signal is sampled (i.e., measured) at periodic points in time. Then, a code word is assigned to represent the approximate value of each sample. Usually, the code words consist of binary symbols. This process is illustrated in Figure 6.46, in which each sample value is represented by a three-bit code word corresponding to the amplitude zone into which the sample falls. Thus, each sample value is converted

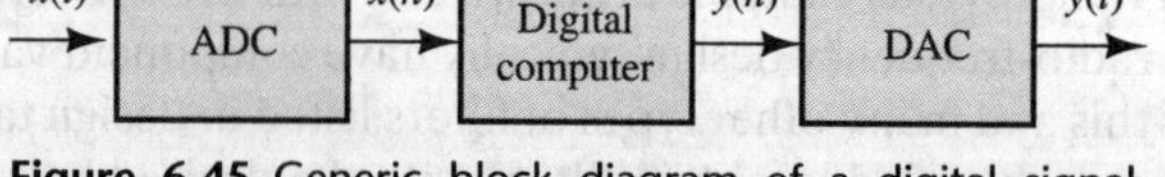

Figure 6.45 Generic block diagram of a digital signal-processing (DSP) system.

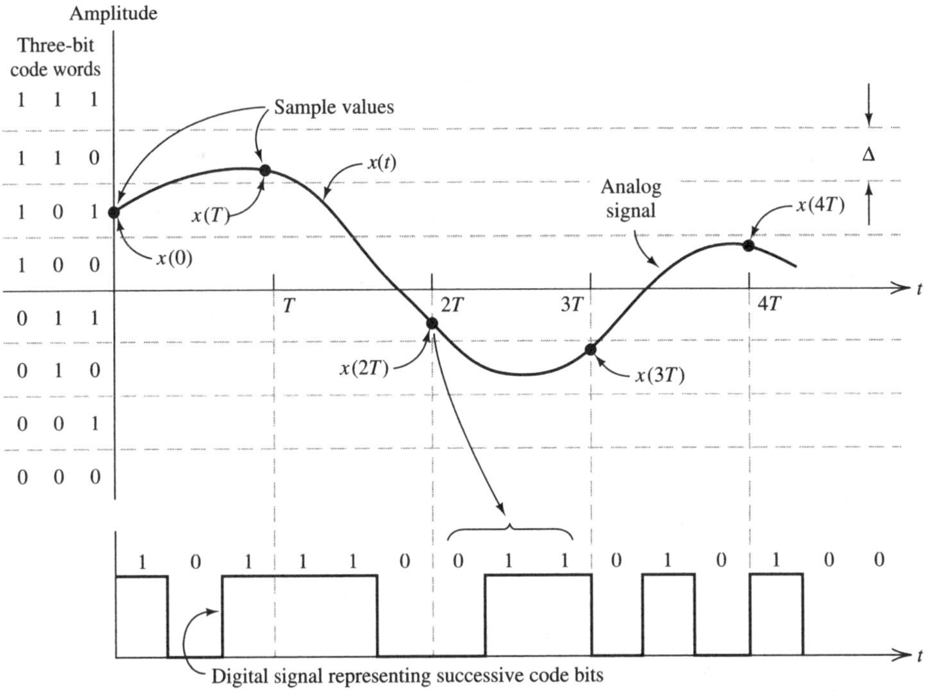

Figure 6.46 An analog signal is converted to an approximate digital equivalent by sampling. Each sample value is represented by a three-bit code word. (Practical converters use longer code words, and the width Δ of each amplitude zone is much smaller.)

into a code word, which in turn can be represented by a digital waveform as shown in the figure.

The rate f_s at which a signal must be sampled depends on the frequencies of the signal components. We have seen that all real signals can be considered to consist of sinusoidal components having various frequencies, amplitudes, and phases. If a signal contains no components with frequencies higher than f_H, the signal can (in theory) be exactly reconstructed from its samples, provided that the sampling frequency f_s is selected to be more than twice f_H:

$$f_s > 2f_H \tag{6.48}$$

If a signal contains no components with frequencies higher than f_H, the signal can be exactly reconstructed from its samples, provided that the sampling rate f_s is selected to be more than twice f_H.

For example, high-fidelity audio signals have a highest frequency of about 15 kHz. Therefore, the minimum sampling rate that should be used for audio signals is 30 kHz. Practical considerations dictate a sampling frequency somewhat higher than the theoretical minimum. For instance, audio compact-disc technology converts audio signals to digital form with a sampling rate of 44.1 kHz. Naturally, it is desirable to use the lowest practical sampling rate to minimize the amount of data (in the form of code words) that must be stored or manipulated by the DSP system.

Of course, the interval between samples T is the reciprocal of the sampling rate:

$$T = \frac{1}{f_s} \tag{6.49}$$

A second consideration important in converting analog signals to digital form is the number of amplitude zones to be used. Exact signal amplitudes cannot be

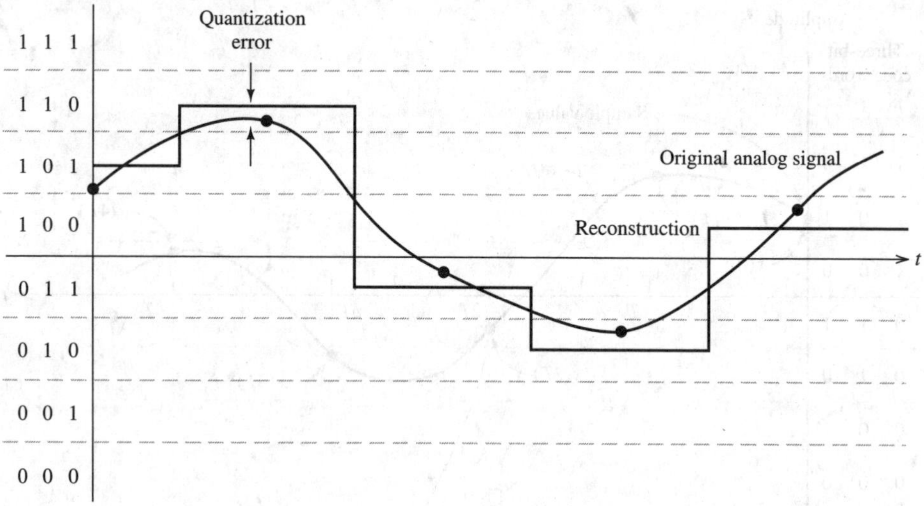

Figure 6.47 Quantization error occurs when an analog signal is reconstructed from its digital form.

represented, because all amplitudes falling into a given zone have the same code word. Thus, when a DAC converts the code words to recreate the original analog waveform, it is possible to reconstruct only an approximation to the original signal with the reconstructed voltage in the middle of each zone, which is illustrated in Figure 6.47. Thus, some **quantization error** exists between the original signal and the reconstruction. This error can be reduced by using a larger number of zones, which requires longer code words. The number N of amplitude zones is related to the number of bits k in a code word by

$$N = 2^k \tag{6.50}$$

Hence, if we are using an 8-bit ($k = 8$) ADC, there are $N = 2^8 = 256$ amplitude zones. In compact-disc technology, 16-bit words are used to represent sample values. With this number of bits, it is very difficult for a listener to detect the effects of quantization error on the reconstructed audio signal.

Often, in engineering instrumentation, we need to determine the DAC specifications needed for converting sensor signals to digital form. For example, suppose that we need to digitize a signal that ranges from −1 to +1 V with a resolution of at most $\Delta = 0.5$ mV. (Δ is illustrated in the upper right-hand corner of Figure 6.46.) Then, the minimum number of zones is the total signal range (2 V) divided by Δ, which yields $N = 4000$. However, N must be an integer power of two. Thus, we require $k = 12$. (In other words, a 12-bit ADC is needed.)

In the remainder of this section, we will ignore quantization error and assume that the exact sample values are available to the digital computer.

Digital Filters

We have seen that ADCs convert analog signals into sequences of code words that can accurately represent the amplitudes of the signals at the sampling instants. Although the computer actually manipulates code words that represent signal amplitudes, it is convenient to focus on the numbers that the code words represent. Conceptually, the

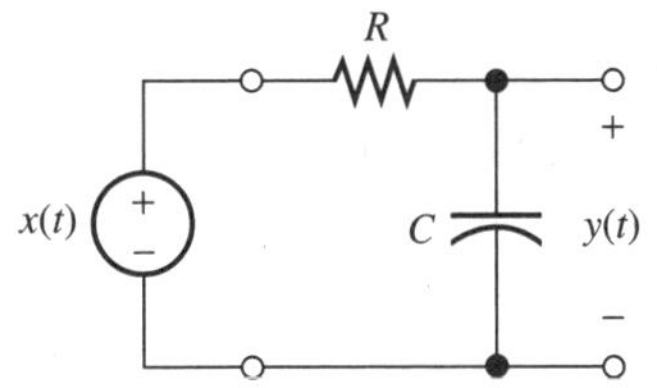

Figure 6.48 First-order RC lowpass filter.

signal $x(t)$ is converted into a list of values $x(nT)$ in which T is the interval between samples and n is a variable that takes on integer values. Often, we omit the sampling period from our notation and write the input and output samples simply as $x(n)$ and $y(n)$, respectively.

Digital Lowpass Filter

Digital filters can be designed to mimic the RLC filters that we discussed earlier in this chapter. For example, consider the first-order RC lowpass filter shown in Figure 6.48, in which we have denoted the input voltage as $x(t)$ and the output voltage as $y(t)$. Writing a Kirchhoff's current equation at the top node of the capacitance, we have

$$\frac{y(t) - x(t)}{R} + C\frac{dy(t)}{dt} = 0 \tag{6.51}$$

Multiplying each term by R and using the fact that the time constant is $\tau = RC$, we find that

$$y(t) - x(t) + \tau\frac{dy(t)}{dt} = 0 \tag{6.52}$$

We can approximate the derivative as

$$\frac{dy(t)}{dt} \cong \frac{\Delta y}{\Delta t} = \frac{y(n) - y(n-1)}{T} \tag{6.53}$$

and write the approximate equivalent to the differential equation

$$y(n) - x(n) + \tau\frac{y(n) - y(n-1)}{T} = 0 \tag{6.54}$$

This type of equation is sometimes called a **difference equation** because it involves differences between successive samples. Solving for the nth output value, we have

$$y(n) = ay(n-1) + (1-a)x(n) \tag{6.55}$$

in which wc havc defined the parameter

$$a = \frac{\tau/T}{1 + \tau/T} \tag{6.56}$$

Equation 6.55 defines the calculations that need to be carried out to perform lowpass filtering of the input $x(n)$. For each sample point, the output is a times the previous output value plus $(1 - a)$ times the present input value. Usually, we have $\tau >> T$ and a is slightly less than unity.

Example 6.10 Step Response of a First-Order Digital Lowpass Filter

Compute and plot the input and output samples for $n = 0$ to 20, given $a = 0.9$. The input is a step function defined by

$$x(n) = 0 \text{ for } n < 0$$
$$= 1 \text{ for } n \geq 0$$

Assume that $y(n) = 0$ for $n < 0$.

Solution We have

$$y(0) = ay(-1) + (1 - a)x(0) = 0.9 \times 0 + 0.1 \times 1 = 0.1$$
$$y(1) = ay(0) + (1 - a)x(1) = 0.19$$
$$y(2) = ay(1) + (1 - a)x(2) = 0.271$$
$$\ldots$$
$$y(20) = 0.8906$$

Plots of $x(n)$ and $y(n)$ are shown in Figure 6.49. Notice that the response of the digital filter to a step input is very similar to that of the RC filter shown in Figure 4.4 on page 185. ■

Exercise 6.23 **a.** Determine the value of the time constant τ, in terms of the sampling interval T corresponding to $a = 0.9$. **b.** Recall that the time constant is the time required for the step response to reach $1 - \exp(-1) = 0.632$ times its final value. Estimate the value of the time constant for the response shown in Figure 6.49.

Answer **a.** $\tau = 9T$; **b.** $\tau \cong 9T$. □

Other Digital Filters

We could develop digital bandpass, notch, or highpass filters that mimic the behavior of the RLC filters discussed earlier in this chapter. Furthermore, high-order digital

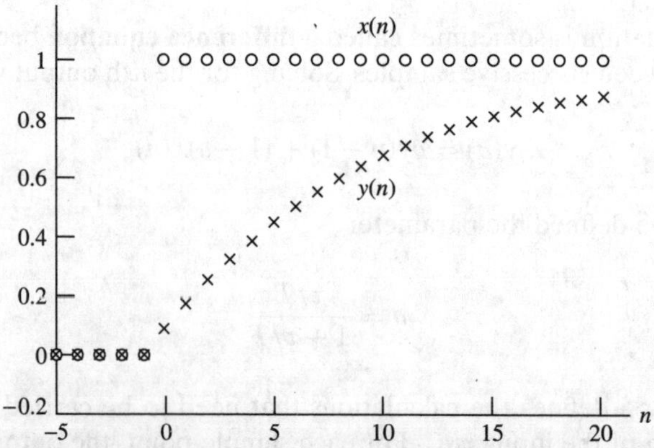

Figure 6.49 Step input and corresponding output of a first-order digital lowpass filter.

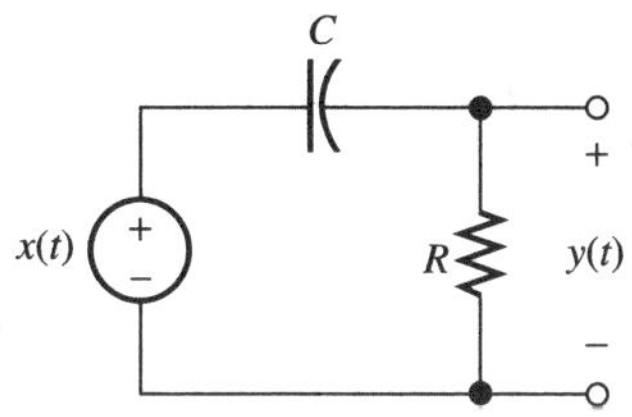

Figure 6.50 RC highpass filter. See Exercise 6.24.

filters are possible. In general, the equations defining such filters are of the form

$$y(n) = \sum_{\ell=1}^{N} a_\ell y(n-\ell) + \sum_{k=0}^{M} b_k x(n-k) \tag{6.57}$$

The type of filter and its performance depend on the values selected for the coefficients a_ℓ and b_k. For the first-order lowpass filter considered in Example 6.10, the coefficients are $a_1 = 0.9, b_0 = 0.1$, and all of the other coefficients are zero.

Exercise 6.24 Consider the RC highpass filter shown in Figure 6.50. Apply the method that we used for the lowpass filter to find an equation having the form of Equation 6.57 for the highpass filter. Give expressions for the coefficients in terms of the time constant $\tau = RC$ and the sampling interval T.
Answer $y(n) = a_1 y(n-1) + b_0 x(n) + b_1 x(n-1)$ in which

$$a_1 = b_0 = -b_1 = \frac{\tau/T}{1+\tau/T}$$

□

A Simple Notch Filter

A simple way to obtain a notch filter is to select $a_\ell = 0$ for all ℓ, $b_0 = 0.5$, $b_d = 0.5$, and to set the remaining b_k coefficients to zero. Then, the output of the digital filter is given by

$$y(n) = 0.5x(n) + 0.5x(n-d) = 0.5[x(n) + x(n-d)]$$

Thus, each input sample is delayed in time by Td and added to the current sample. Finally, the sum of the input and its delayed version is multiplied by 0.5. To see that this results in a notch filter, consider a sinewave delayed by an interval Td. We can write

$$A\cos[\omega(t - Td)] = A\cos(\omega t - \omega Td) = A\cos(\omega t - \theta)$$

Hence, a time delay of Td amounts to a phase shift of ωTd radians or $fTd \times 360°$. (Keep in mind that, in this discussion, T represents the interval between samples, *not the period of the sinewave.*) For low frequencies, the phase shift is small, so the low-frequency components of $x(n)$ add nearly in phase with those of $x(n-d)$. On the other hand, for the frequency

$$f_{\text{notch}} = \frac{1}{2Td} = \frac{f_s}{2d} \tag{6.58}$$

the phase shift is 180°. Of course, when we phase shift a sinewave by 180° and add it to the original, the sum is zero. Thus, any input component having the frequency f_{notch} does not appear in the output. The first-order lowpass filter and this simple notch filter are just two of many possible digital filters that can be realized by selection of the coefficient values in Equation 6.57.

Exercise 6.25 Suppose that the sampling frequency is $f_s = 10$ kHz, and we want to eliminate the 500-Hz component with a simple notch filter. **a.** Determine the value needed for d. **b.** What difficulty would be encountered if we wanted to eliminate the 300-Hz component?

Answer **a.** $d = 10$; **b.** Equation 6.58 yields $d = 16.67$, but d is required to be an integer value. □

Digital Filter Demonstration

Next, we will use MATLAB to demonstrate the operation of a digital filter. First, we will create samples of a virtual signal including noise and interference. The signal of interest consists of a 1-Hz sinewave and is representative of many types of real world signals such as delta waves contained in the electroencephalogram (EEG) of an individual in deep sleep, or the output of a pressure sensor submerged in the ocean with waves passing over. Part of the interference consists of a 60-Hz sinewave, which is a common real-world problem due to coupling between the ac power line and the signal sensor. The other part of the interference is random noise, which is also common in real-world data.

The MATLAB code that we use to create our simulated data is

```
t = 0:1/6000:2;
signal = cos(2*pi*t);
interference = cos(120*pi*t);
white_noise = randn(size(t));
noise = zeros(size(t));
for n = 2:12001
noise(n) = 0.25*(white_noise(n) - white_noise(n - 1));
end
x = signal + interference + noise; % This is the simulated data.
```

The first command generates a 12,001-element row vector containing the sample times for a two-second interval with a sampling frequency of $f_s = 6000$ Hz. The second and third commands set up row matrices containing samples of the signal and the 60-Hz interference. In the next line, the random-number generator feature of MATLAB generates "white noise" that contains components of equal amplitudes (on average) at all frequencies up to half of the sampling frequency. The white noise is then manipulated by the commands in the `for-end` loop, producing noise with components from dc to 3000 Hz peaking around 1500 Hz. Then, the signal, interference and noise are added to produce the simulated data $x(n)$. (Of course, in a real-world application, the data are obtained by applying the outputs of sensors, such as EEG electrodes, to analog-to-digital converters.)

Next, we use MATLAB to plot the signal, interference, noise, and the simulated data.

```
subplot(2,2,1)
plot(t, signal)
axis([0 2 -2 2])
subplot(2,2,2)
plot(t, interference)
axis([0 2 -2 2])
subplot(2,2,3)
plot(t,noise)
```

```
axis([0 2 -2 2])
subplot(2,2,4)
plot(t,x)
axis([0 2 -3 3])
```

The resulting plots are shown in Figure 6.51. The simulated data is typical of what is often obtained from sensors in real-world experiments. In a biomedical setting, for example, an electrocardiograph produces data that is the sum of the heart signal, 60-Hz power-line interference, and noise from muscle contractions, especially when the subject is moving, as in a stress test.

Actually, the plot of the 60-Hz interference appears a little uneven in Figure 6.51(b) because of finite screen resolution for the display. This is a form of distortion called aliasing that occurs when the sampling rate is too low. If you run the commands on your own computer and use the zoom tool to expand the display horizontally, you will see a smooth plot of the 60-Hz sinewave interference. An m-file named DSPdemo that contains the commands used in this demonstration of a digital filter appears in the MATLAB folder.

What we need is a digital filter that processes the data $x(n)$ of Figure 6.51(d) and produces an output closely matching the signal in Figure 6.51(a). This filter should pass the signal (1-Hz sinewave), reject the 60-Hz interference, and reject the noise, which has its largest components in the vicinity of 1500 Hz.

To achieve this, we will use a digital notch filter to remove the 60-Hz sinewave interference cascaded with a lowpass filter to remove most of the noise. The conceptual diagram of the digital filter is shown in Figure 6.52.

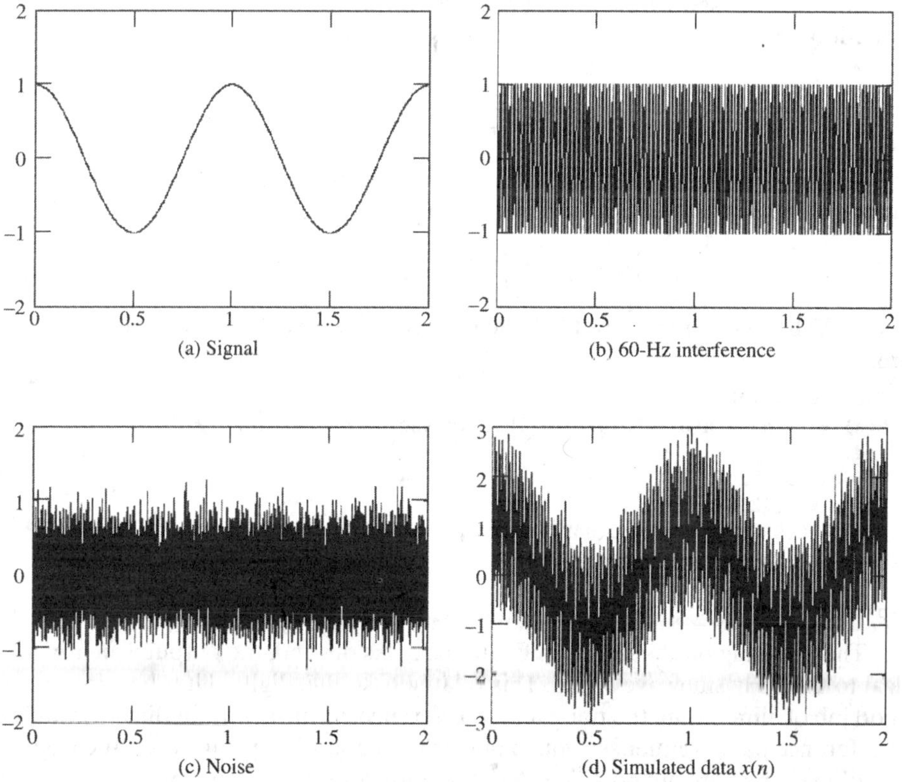

Figure 6.51 Simulated pressure-sensor output and its components.

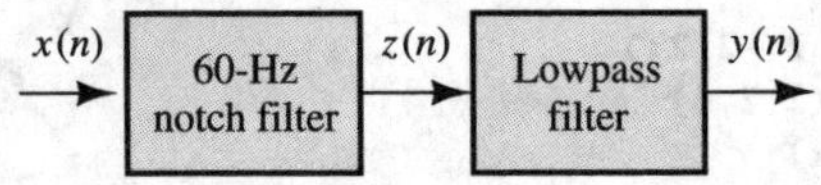

Figure 6.52 Digital filter.

Equation 6.58 reveals that by using $d = 50$ and $f_s = 6000\,\text{Hz}$, we can realize a notch filter with zero gain at precisely 60 Hz. (If 60-Hz interference is a problem, it is a good idea to pick the sampling frequency to be an even integer multiple of 60 Hz, which is one reason we picked the sampling frequency to be 6000 Hz.) The output $z(n)$ of the notch filter is given in terms of the input data $x(n)$ as

$$z(n) = \frac{1}{2}[x(n) + x(n-50)]$$

Also, we need a lowpass filter to eliminate the noise. We decide to use the first-order lowpass filter discussed earlier in this section. Because we do not want the lowpass filter to disturb the signal, we choose its break frequency to much higher than 1 Hz, say $f_B = 50\,\text{Hz}$. For an RC lowpass filter, the break frequency is

$$f_B = \frac{1}{2\pi RC}$$

Solving for the time constant and substituting values, we have

$$\tau = RC = \frac{1}{2\pi f_B} = \frac{1}{2\pi(50)} = 3.183\,\text{ms}$$

The gain constant for the (approximately) equivalent digital filter is given by Equation 6.56 in which $T = 1/f_s = 1/6000\,\text{s}$ is the sampling interval. We then have

$$a = \frac{\tau/T}{1+\tau/T} = 0.9503$$

Substituting this value into Equation 6.55 yields the equation for the present $y(n)$ output of the lowpass filter in terms of its input $z(n)$ and previous output $y(n-1)$.

$$y(n) = 0.9503y(n-1) + 0.0497z(n)$$

The MATLAB commands to filter the simulated data $x(n)$ and plot the output $y(n)$ are:

```
for n = 51:12001
z(n) = (x(n) + x(n - 50))/2; % This is the notch filter.
end
y = zeros(size(z));
for n = 2:12001
y(n) = 0.9503*y(n-1) + 0.0497*z(n); % This is the lowpass filter.
end
figure
plot(t,y)
```

The resulting plot is shown in Figure 6.53. As desired, the output is nearly identical to the 1-Hz sinewave signal. This relatively simple digital filter has done a very good job of eliminating the noise and interference because most of the noise and the interference have frequencies much higher than does the signal. When the frequencies of the signal are nearer to those of the noise and interference, we would need to resort to higher-order filters.

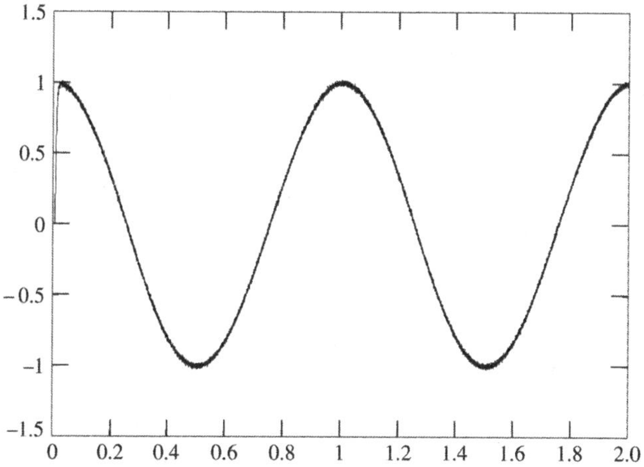

Figure 6.53 Output signal.

Comparison of Filter Technologies

We have discussed two ways to filter signals: *RLC* circuits and digital filters. There are a number of other filter types, such as **active filters** that are composed of resistances, capacitances and **operational amplifiers**, or **op amps** (which we discuss in Chapter 14). Other filters are based on mechanical resonances in piezoelectric crystals, surface acoustic waves, the propagation of electric fields in wave guides, switched capacitor networks, and transmission lines.

In all cases, the objective of a filter is to separate a desired signal from noise and interference. Radio amateurs operating in the frequency band between 28 and 29.7 MHz often need to place a band reject filter between the transmitter and antenna to eliminate second-harmonic frequency components from reaching the antenna. If they are not removed, second-harmonic components can cause some very annoying interference on their neighbor's television screens. In this application, an *RLC* filter would be the technology of choice because of the large currents and voltages involved.

The objective of a filter is to separate a desired signal from noise and interference.

On the other hand, a sleep researcher may wish to filter brain waves to separate delta waves that appear at frequencies of 4 Hz or less from higher frequency brain waves. In this case, a digital filter is appropriate.

In summary, there are many applications for filters and many technologies for implementing filters. Most of the principles we have introduced in our discussion of *RLC* circuits and digital filters apply to filters based on other technologies.

Summary

1. The fundamental concept of Fourier theory is that we can construct any signal by adding sinusoids with the proper amplitudes, frequencies, and phases.
2. In effect, a filter decomposes the input signal into its sinusoidal components, adjusts the amplitude and phase of each component, depending on its frequency, and sums the adjusted components to produce the output signal. Often, we need a filter that passes components in a given frequency range to the output, without change in amplitude

or phase, and that rejects components at other frequencies.

3. The transfer function of a filter circuit is the phasor output divided by the phasor input as a function of frequency. The transfer function is a complex quantity that shows how the amplitudes and phases of input components are affected when passing through the filter.

4. We can use circuit analysis with phasors and complex impedances to determine the transfer function of a given circuit.

5. A first-order filter is characterized by its half-power frequency f_B.

6. A transfer-function magnitude is converted to decibels by taking 20 times the common logarithm of the magnitude.

7. Two-port filters are cascaded by connecting the output of the first to the input of the second. The overall transfer function of the cascade is the product of the transfer functions of the individual filters. If the transfer functions are converted to decibels, they are added for a cascade connection.

8. On a logarithmic frequency scale, frequency is multiplied by a given factor for equal increments of length along the axis. A decade is a range of frequencies for which the ratio of the highest frequency to the lowest is 10. An octave is a two-to-one change in frequency.

9. A Bode plot shows the magnitude of a network function in decibels versus frequency, using a logarithmic scale for frequency.

10. The Bode plots for first-order filters can be closely approximated by straight-line asymptotes. In the case of a first-order lowpass filter, the transfer-function magnitude slopes downward at 20 dB/decade for frequencies that are higher than the half-power frequency. For a first-order highpass filter, the transfer-function magnitude slopes at 20 dB/decade below the break frequency.

11. At low frequencies, inductances behave as short circuits, and capacitances behave as open circuits. At high frequencies, inductances behave as open circuits, and capacitances behave as short circuits. Often, *RLC* filters can be readily analyzed at low- or high-frequencies, providing checks on computer-generated Bode plots.

12. The key parameters of series and parallel resonant circuits are the resonant frequency and quality factor. The impedance of either type of circuit is purely resistive at the resonant frequency. High-quality-factor circuits can have responses that are much larger in magnitude than the driving source.

13. Filters may be classified as lowpass, highpass, bandpass, and band-reject filters. Ideal filters have constant (nonzero) gain (transfer-function magnitude) in the passband and zero gain in the stopband.

14. The series resonant circuit can be used to form any of the four filter types.

15. A second-order filter is characterized by its resonant frequency and quality factor.

16. MATLAB is useful in deriving and plotting network functions of complex *RLC* filters.

17. In using digital signal processing (DSP) to filter a signal, the analog input signal $x(t)$ is converted to digital form (a sequence of numbers) by an analog-to-digital converter (ADC). A digital computer uses the digitized input signal to compute a sequence of values for the output signal, and, finally, (if desired) the computed values are converted to analog form by a digital-to-analog converter (DAC) to produce the output signal $y(t)$.

18. If a signal contains no components with frequencies higher than f_H, the signal can be exactly reconstructed from its samples, provided that the sampling rate f_s is selected to be more than twice f_H.

19. Approximately equivalent digital filters can be found for *RLC* filters.

***Note:* You can check many of the Bode plots in this chapter by using a computer-aided circuit-analysis program such as Multisim from National Instruments or OrCAD Capture from Cadence Inc.**

Problems

Section 6.1: Fourier Analysis, Filters, and Transfer Functions

P6.1. Briefly discuss the fundamental concept of Fourier theory.

P6.2. Define the transfer function for a filter. Describe how the transfer function of a filter can be determined using laboratory methods.

P6.3. Describe how a filter processes an input signal to produce the output signal in terms of the transfer function and the sinusoidal components.

P6.4. Fourier analysis shows that the **sawtooth waveform** of Figure P6.4 can be written as

$$v_{st}(t) = 1 - \frac{2}{\pi}\sin(2000\pi t) - \frac{2}{2\pi}\sin(4000\pi t) - \frac{2}{3\pi}\sin(6000\pi t) - \cdots - \frac{2}{n\pi}\sin(2000n\pi t) - \cdots$$

in which n assumes integer values. Use MATLAB to compute and plot the sum through $n = 3$ for $0 \le t \le 2$ ms. Repeat for the sum through $n = 50$.

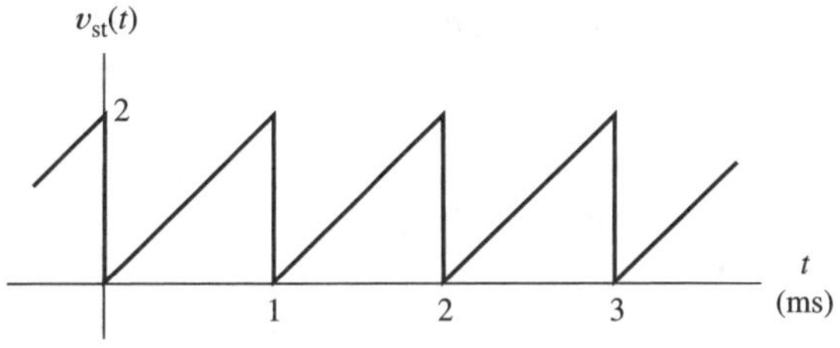

Figure P6.4

P6.5. The triangular waveform shown in Figure P6.5 can be written as the infinite sum

$$v_t(t) = 1 + \frac{8}{\pi^2}\cos(2000\pi t) + \frac{8}{(3\pi)^2}\cos(6000\pi t) + \cdots + \frac{8}{(n\pi)^2}\cos(2000n\pi t) + \cdots$$

in which n takes odd integer values only. Use MATLAB to compute and plot the sum through $n = 19$ for $0 \le t \le 2$ ms. Compare your plot with the waveform shown in Figure P6.5.

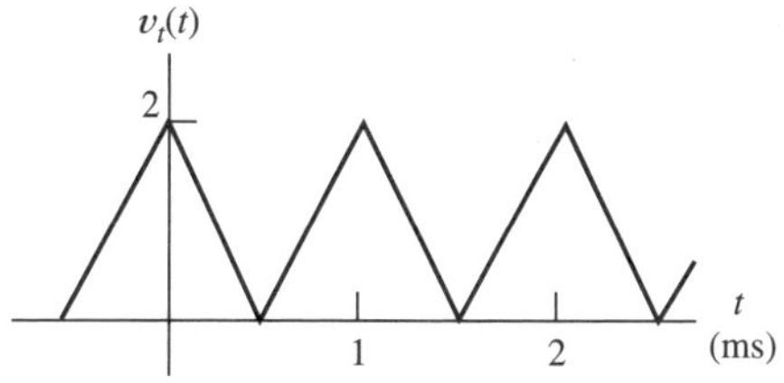

Figure P6.5

P6.6. The full-wave rectified cosine wave shown in Figure P6.6 can be written as

$$v_{fw} = \frac{2}{\pi} + \frac{4}{\pi(1)(3)}\cos(4000\pi t) - \frac{4}{\pi(3)(5)}\cos(8000\pi t) + \cdots + \frac{4(-1)^{(n/2+1)}}{\pi(n-1)(n+1)}\cos(2000n\pi t) + \cdots$$

in which n assumes even integer values. Use MATLAB to compute and plot the sum through $n = 60$ for $0 \le t \le 2$ ms. Compare your plot with the waveform shown in Figure P6.6.

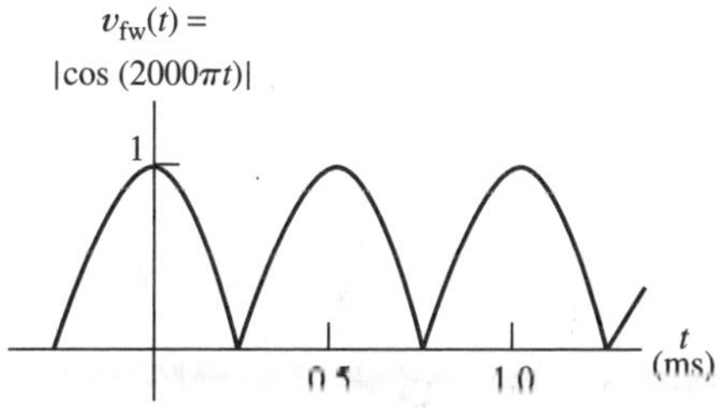

Figure P6.6

* Denotes that answers are contained in the Student Solutions files. See Appendix E for more information about accessing the Student Solutions.

P6.7. The Fourier series for the periodic waveform shown in Figure P6.7 is

$$v(t) = \sum_{n=1}^{\infty}\left[\frac{\sin(n\pi/2)}{(n\pi/2)^2} - \frac{\cos(n\pi/2)}{(n\pi/2)}\right]\sin(500n\pi t)$$

Use MATLAB to compute and plot the sum through $n = 3$ for $-4 \le t \le 4$ ms. Then, plot the sum through $n = 50$. Compare your plots with the waveform in Figure P6.7.

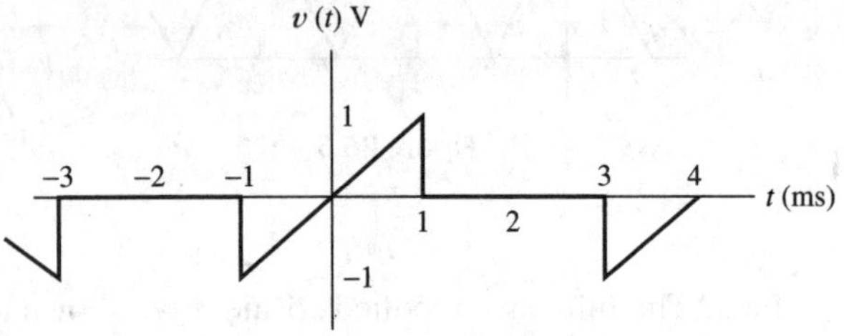

Figure P6.7

***P6.8.** The transfer function $H(f) = \mathbf{V}_{out}/\mathbf{V}_{in}$ of a filter is shown in Figure P6.8. The input signal is given by

$$v_{in}(t) = 5 + 2\cos(5000\pi t + 30°) + 2\cos(15000\pi t)$$

Find an expression (as a function of time) for the steady-state output of the filter.

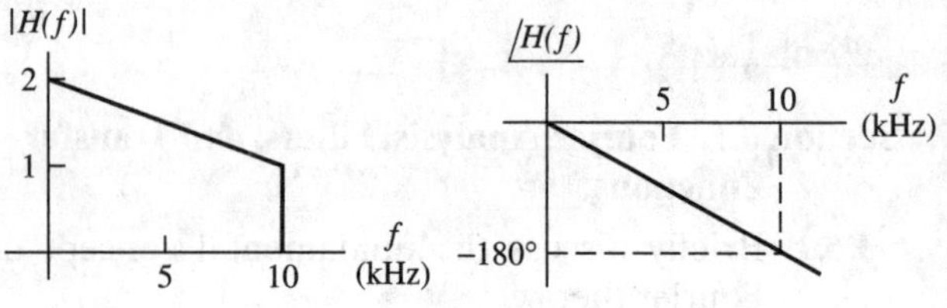

Figure P6.8

P6.9. Repeat Problem P6.8 for the input voltage given by

$$v_{in}(t) = 4 + 5\cos(10^4\pi t - 30°) + 2\sin(24000\pi t)$$

P6.10. Repeat Problem P6.8 for the input voltage given by

$$v_{in}(t) = 6 + 2\cos(6000\pi t) - 4\cos(12000\pi t)$$

***P6.11.** The input to a certain filter is given by

$$v_{in}(t) = 2\sin\left(10^4\pi t + 65°\right)$$

and the steady-state output is given by

$$v_{out}(t) = \cos\left(10^4\pi t + 20°\right).$$

Determine the (complex) value of the transfer function of the filter for $f = 5000$ Hz.

P6.12. The input and output voltages of a certain filter operating in steady state with a sinusoidal input are shown in Figure P6.12. Determine the frequency and the corresponding value of the transfer function.

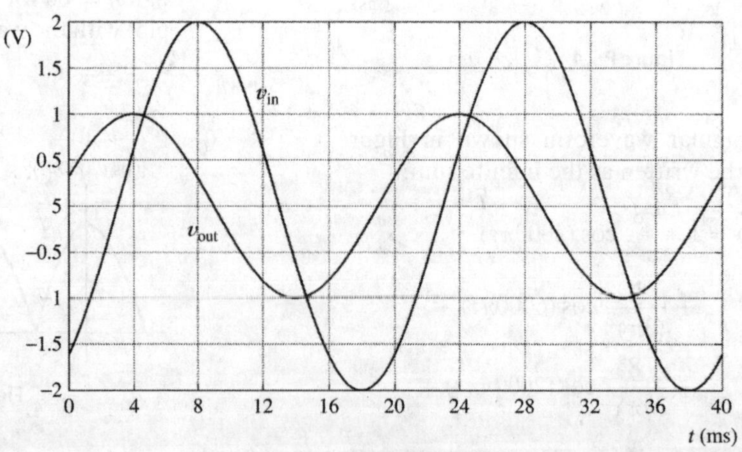

Figure P6.12

***P6.13.** The input and output voltages of a filter operating under sinusoidal steady-state conditions are observed on an oscilloscope. The peak amplitude of the input is 5 V and the output is 15 V. The period of both signals is 4 ms. The input reaches a positive peak at $t = 1$ ms, and the output reaches its positive peak at $t = 1.5$ ms. Determine the frequency and the corresponding value of the transfer function.

***P6.14.** The triangular waveform of Problem P6.5 is the input for a filter with the transfer function shown in Figure P6.14. Assume that the phase of the transfer function is zero for all frequencies. Determine the steady-state output of the filter.

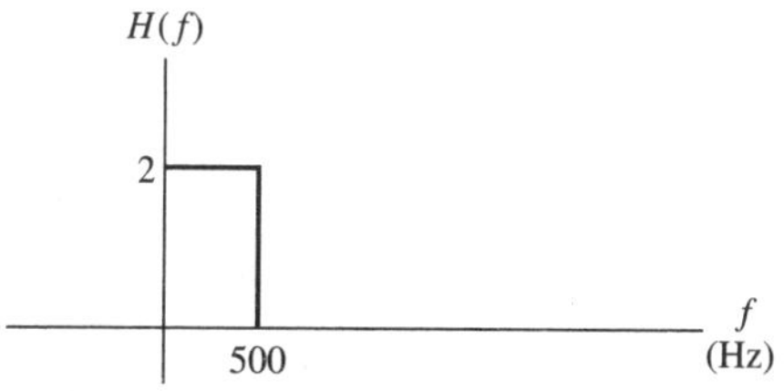

Figure P6.14

***P6.15.** Consider a circuit for which the output voltage is the running-time integral of the input voltage, as illustrated in Figure P6.15. If the input voltage is given by $v_{in}(t) = V_{max}\cos(2\pi f t)$, find an expression for the output voltage as a function of time. Then, find an expression for the transfer function of the integrator. Plot the magnitude and phase of the transfer function versus frequency.

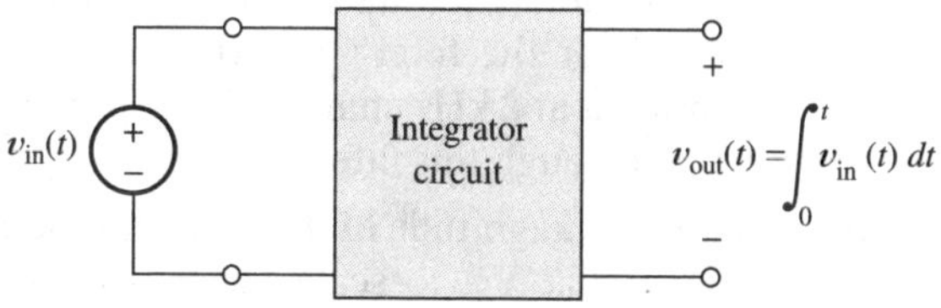

Figure P6.15

P6.16. Suppose we have a circuit for which the output voltage is the time derivative of the input voltage, as illustrated in Figure P6.16. For an input voltage given by $v_{in}(t) = V_{max}\cos(2\pi f t + 60°)$, find an expression for the output voltage as a function of time. Then, find an expression for the transfer function of the differentiator. Plot the magnitude and phase of the transfer function versus frequency.

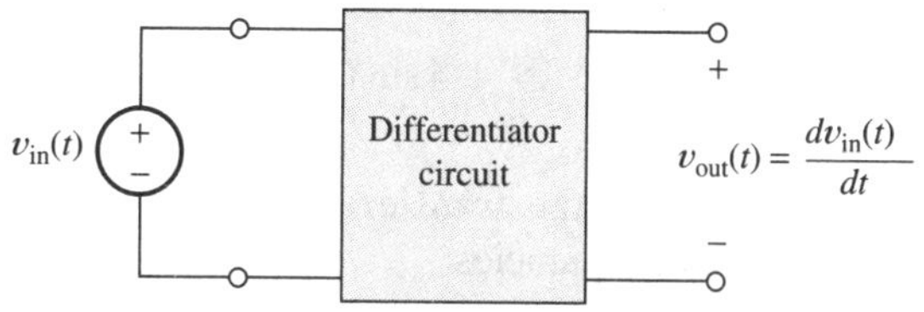

Figure P6.16

P6.17. The sawtooth waveform of Problem P6.4 is applied as the input to a filter with the transfer function shown in Figure P6.17. Assume that the phase of the transfer function is zero for all frequencies. Determine the steady-state output of the filter.

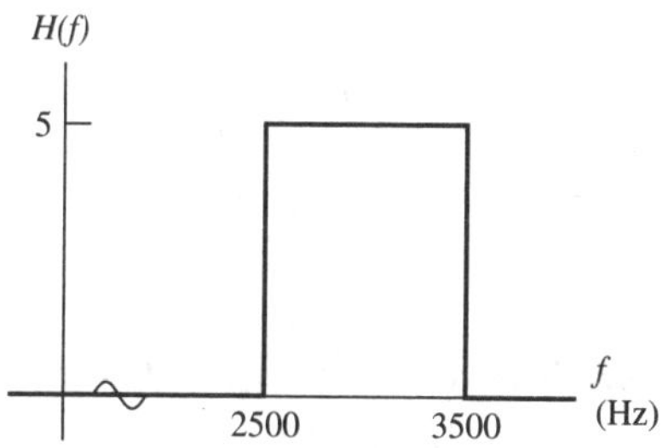

Figure P6.17

P6.18. Suppose we have a system for which the output voltage is

$$v_o(t) = 1000 \int_{t-10^{-3}}^{t} v_{in}(t)\,dt$$

Given the input voltage $v_{in}(t) = V_{max}\cos(2\pi f t)$, find an expression for the output voltage as a function of time. Then, find an expression for the transfer function of the system. Use MATLAB to plot the magnitude of the transfer function versus frequency for the range from 0 to 2000 Hz. Explain why the magnitude of the transfer function is zero for multiples of 1000 Hz.

P6.19. List the frequencies in hertz for which the transfer function of a filter can be determined given that the input to the filter is

$$v_{in}(t) = 2 + 3\cos(1000\pi t) + 3\sin(2000\pi t) + \cos(3000\pi t)\ \text{V}$$

and the output is

$$v_{out}(t) = 3 + 2\cos(1000\pi t + 30°) + 4\sin(3000\pi t)\ \text{V}$$

Compute the transfer function for each of these frequencies.

P6.20. Consider a system for which the output voltage is $v_o(t) = v_{in}(t) + v_{in}(t - 2 \times 10^{-3})$. (In other words, the output equals the input plus the input delayed by 2 ms.) Given that the input voltage is $v_{in}(t) = V_{max}\cos(2\pi ft)$, find an expression for the output voltage as a function of time. Then, find an expression for the transfer function of the system. Use MATLAB to plot the magnitude of the transfer function versus frequency for the range from 0 to 2000 Hz. Explain in terms of the phasors for $v_{in}(t)$ and for $v_{in}(t - 2 \times 10^{-3})$ why the transfer function has the magnitude it does for $f = 250$ Hz. Repeat for $f = 500$ Hz.

Section 6.2: First-Order Lowpass Filter

P6.21. Draw the circuit diagram of a first-order RC lowpass filter and give the expression for the half-power frequency in terms of the circuit components. Sketch the magnitude and phase of the transfer function versus frequency.

P6.22. Repeat Problem P6.21 for a first-order RL lowpass filter.

P6.23. In Chapter 4, we used the time constant to characterize first-order RC circuits. Find the relationship between the half-power frequency and the time constant.

P6.24. Suppose that we need a first-order RC lowpass filter with a half-power frequency of 2 kHz. Determine the value of the capacitance, given that the resistance is 15 kΩ.

***P6.25.** Consider a first-order RC lowpass filter. At what frequency (in terms of f_B) is the phase shift equal to −1°? −10°? −89°?

***P6.26.** An input signal given by

$$v_{in}(t) = 5\cos(500\pi t) + 5\cos(1000\pi t) + 5\cos(2000\pi t)$$

is applied to the lowpass RC filter shown in Figure P6.26. Find an expression for the output signal.

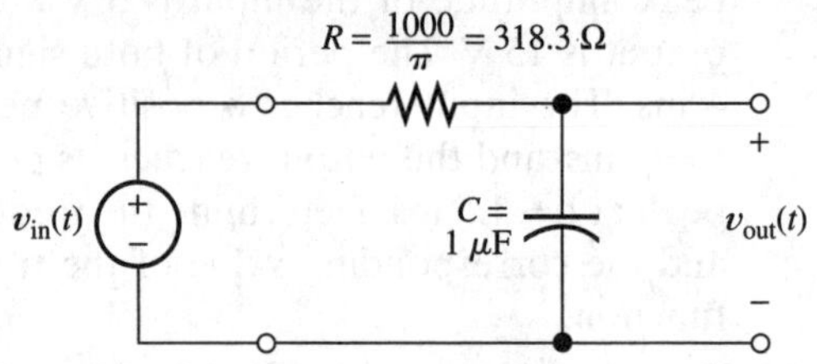

Figure P6.26

P6.27. The input signal of a first-order lowpass filter with the transfer function given by Equation 6.9 on page 308 and a half-power frequency of 400 Hz is

$$v_{in}(t) = 1 + 2\sin(800\pi t + 120°) + 3\cos\left(20 \times 10^3 \pi t\right)$$

Find an expression for the output voltage.

P6.28. Suppose we have a first-order lowpass filter that is operating in sinusoidal steady-state conditions at a frequency of 5 kHz. Using an oscilloscope, we observe that the positive-going zero crossing of the output is delayed by 20μs compared with that of the input. Determine the break frequency of the filter.

P6.29. The input signal to a filter contains components that range in frequency from 10 Hz to 20 kHz. We wish to reduce the amplitude of the 20-kHz component by a factor of 100 by passing the signal through a first-order lowpass filter. What half-power frequency is required for the filter? By what factor is a component at 2 kHz changed in amplitude in passing through this filter?

***P6.30.** Sketch the magnitude of the transfer function $H(f) = \mathbf{V}_{out}/\mathbf{V}_{in}$ to scale versus frequency for the circuit shown in Figure P6.30. What is

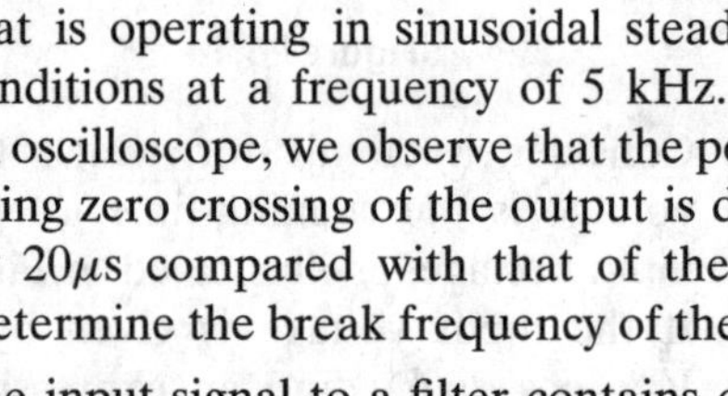

Figure P6.30

the value of the half-power frequency? (*Hint:* Start by finding the Thévenin equivalent circuit seen by the capacitance.)

P6.31. In steady-state operation, a first-order RC lowpass filter has an output signal $v_{\text{out}}(t) = 0.5\cos(20 \times 10^3 \pi t - 84.26°)$. The break frequency of the filter is 1005 Hz. Determine the input signal.

P6.32. We apply a 5-V-rms 10-kHz sinusoid to the input of a first-order RC lowpass filter, and the output voltage in steady state is 0.2 V rms. Predict the steady-state rms output voltage after the frequency of the input signal is raised to 50 kHz and the input amplitude remains constant.

P6.33. Consider the circuit shown in Figure P6.33(a). This circuit consists of a source having an internal resistance of R_s, an RC lowpass filter, and a load resistance R_L. **a.** Show that the transfer function of this circuit is given by

$$H(f) = \frac{V_{\text{out}}}{V_s} = \frac{R_L}{R_s + R + R_L} \times \frac{1}{1 + j(f/f_B)}$$

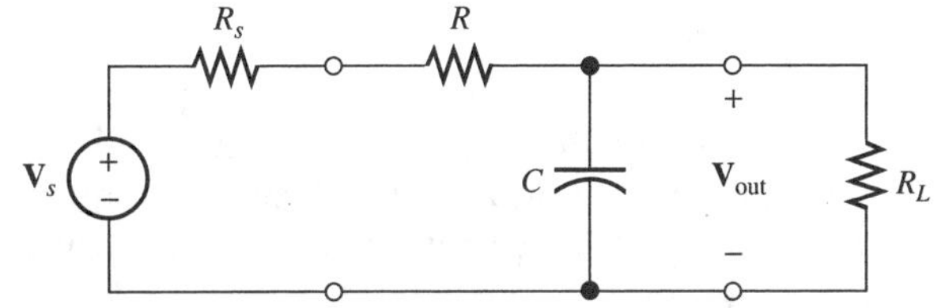

(a)

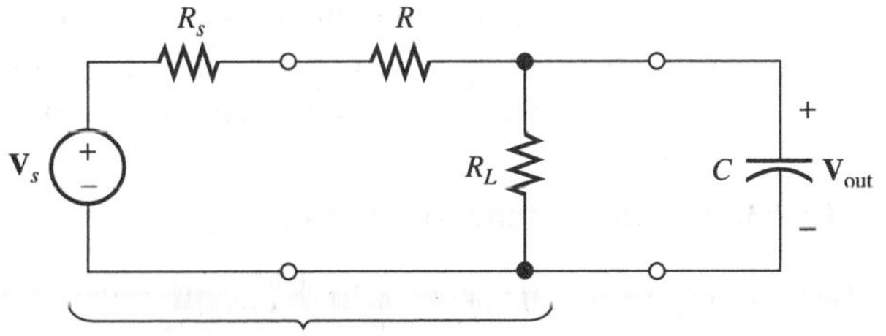

(b)

Figure P6.33

in which the half-power frequency f_B is given by

$$f_B = \frac{1}{2\pi R_t C} \quad \text{where} \quad R_t = \frac{R_L(R_s + R)}{R_L + R_s + R}$$

Notice that R_t is the parallel combination of R_L and $(R_s + R)$. [*Hint:* One way to make this problem easier is to rearrange the circuit as shown in Figure P6.33(b) and then to find the Thévenin equivalent for the source and resistances.] **b.** Given that $C = 0.2\ \mu\text{F}$, $R_s = 2\ \text{k}\Omega$, $R = 47\ \text{k}\Omega$, and $R_L = 1\ \text{k}\Omega$, sketch (or use MATLAB to plot) the magnitude of $H(f)$ to scale versus f/f_B from 0 to 3.

P6.34. **a.** Derive an expression for the transfer function $H(f) = \mathbf{V}_{\text{out}}/\mathbf{V}_{\text{in}}$ for the circuit shown in Figure P6.34. Find an expression for the half-power frequency. **b.** Given $R_1 = 50\ \Omega$, $R_2 = 50\ \Omega$, and $L = 15\ \mu\text{H}$, sketch (or use MATLAB to plot) the magnitude of the transfer function versus frequency.

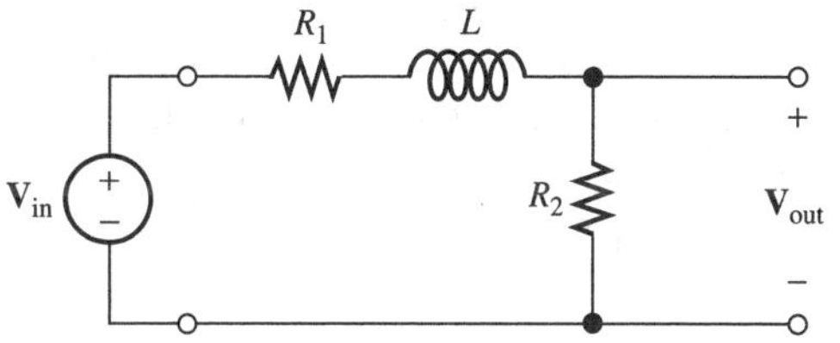

Figure P6.34

P6.35. Perhaps surprisingly, we can apply the transfer-function concept to mechanical systems. Suppose we have a mass m moving through a liquid with an applied force f and velocity v. The motion of the mass is described by the first-order differential equation

$$f = m\frac{dv}{dt} + kv$$

in which k is the coefficient of viscous friction. Find an expression for the transfer function

$$H(f) = \frac{\mathbf{V}}{\mathbf{F}}$$

Also, find the half-power frequency (defined as the frequency at which the transfer function magnitude is $1/\sqrt{2}$ times its dc value) in terms of k and m. [*Hint*: To determine the transfer function, assume a steady-state sinusoidal velocity $v = V_m \cos(2\pi ft)$, solve for the force, and take the ratio of their phasors.]

Section 6.3: Decibels, the Cascade Connection, and Logarithmic Frequency Scales

P6.36. What is the main advantage of converting transfer function magnitudes to decibels before plotting?

P6.37. What is the passband of a simple *RC* lowpass filter?

P6.38. What is a logarithmic frequency scale? A linear frequency scale?

***P6.39.** **a.** What frequency is halfway between 100 and 3000 Hz on a logarithmic frequency scale? **b.** On a linear frequency scale?

P6.40. **a.** Find the frequency that is one octave lower than 500 Hz. **b.** Two octaves higher. **c.** Two decades lower. **d.** One decade higher.

P6.41. What is a notch filter? What is one application?

***P6.42.** **a.** Given $|H(f)|_{\text{dB}} = -40$ dB, find $|H(f)|$. **b.** Repeat for $|H(f)|_{\text{dB}} = 40$ dB.

P6.43. Find the decibel equivalent for $|H(f)| = 0.5$. Repeat for $|H(f)| = 2$, $|H(f)| = 1/\sqrt{2} \cong 0.7071$, and $|H(f)| = \sqrt{2}$.

P6.44. Explain what we mean when we say that two filters are cascaded.

P6.45. We have a list of six successive frequencies, 5 Hz, f_1, f_2, f_3, f_4, and 50 Hz. Determine the values of f_1, f_2, f_3, and f_4 so that the frequencies are evenly spaced on: **a.** a linear frequency scale, and **b.** a logarithmic frequency scale.

***P6.46.** **a.** How many decades are between $f_1 = 25$ Hz and $f_2 = 10$ kHz? **b.** How many octaves?

P6.47. We have two filters with transfer functions $H_1(f)$ and $H_2(f)$ cascaded in the order 1–2. Give the expression for the overall transfer function of the cascade. Repeat if the transfer function magnitudes are expressed in decibels denoted as $|H_1(f)|_{\text{dB}}$ and $|H_2(f)|_{\text{dB}}$. What caution concerning $H_1(f)$ must be considered?

***P6.48.** Two first-order lowpass filters are in cascade as shown in Figure P6.48. The transfer functions are

$$H_1(f) = H_2(f) = \frac{1}{1 + j(f/f_B)}$$

a. Write an expression for the overall transfer function. **b.** Find an expression for the half-power frequency for the overall transfer function in terms of f_B.

(*Comment:* This filter cannot be implemented by cascading two simple *RC* lowpass filters like the one shown in Figure 6.7 on page 307 because the transfer function of the first circuit is changed when the second is connected. Instead, a buffer amplifier, such as the voltage follower discussed in Section 14.3, must be inserted between the *RC* filters.)

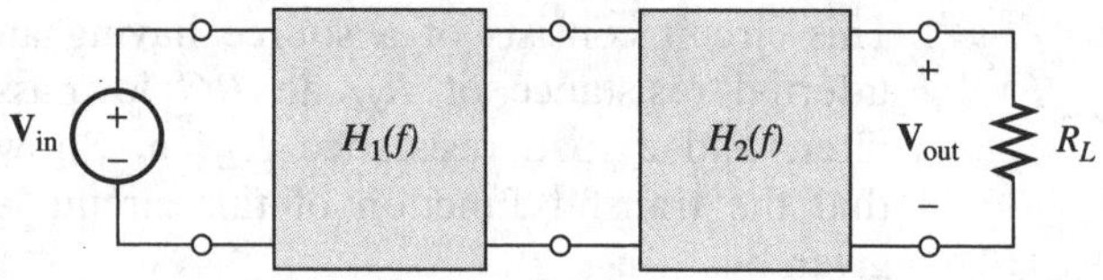

Figure P6.48

P6.49. Two filters are in cascade. At a given frequency f_1, the transfer function values are $|H_1(f_1)|_{\text{dB}} = 25$ and $|H_2(f_1)|_{\text{dB}} = +15$. Find the magnitude of the overall transfer function at $f = f_1$.

Section 6.4: Bode Plots

P6.50. What is a Bode magnitude plot? What is a Bode phase plot?

P6.51. What is the slope of the high-frequency asymptote for the Bode magnitude plot for a first-order lowpass filter? The low-frequency asymptote? At what frequency do the asymptotes meet?

P6.52. Suppose that four filters, having identical first-order lowpass transfer functions, are cascaded, what will be the rate at which the overall transfer function magnitude asymptote declines above the break frequency? Explain.

***P6.53.** A transfer function is given by

$$H(f) = \frac{100}{1 + j(f/1000)}$$

Sketch the asymptotic magnitude and phase Bode plots to scale. What is the value of the half-power frequency?

P6.54. A transfer function is given by

$$H(f) = 0.1[1 + j(f/200)]$$

Sketch the asymptotic magnitude and phase Bode plots to scale. What is the value of the break frequency?

P6.55. Sketch the asymptotic magnitude and phase Bode plots to scale for the transfer function

$$H(f) = 10\frac{1 - j(f/100)}{1 + j(f/100)}$$

P6.56. Solve for the transfer function $H(f) = \mathbf{V}_{out}/\mathbf{V}_{in}$ and sketch the asymptotic Bode magnitude and phase plots to scale for the circuit shown in Figure P6.56.

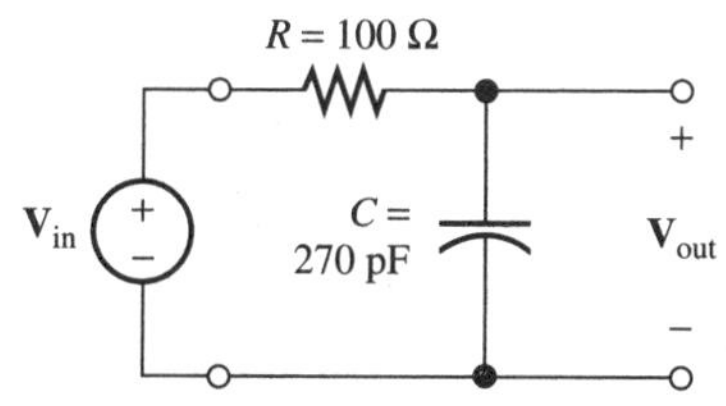

Figure P6.56

P6.57. The circuit shown in Figure 6.57 has $R_1 = R_2 = 2\ \text{k}\Omega$ and $C = (1/\pi)\mu\text{F}$. Solve for the transfer function $H(f) = \mathbf{V}_{out}/\mathbf{V}_{in}$ and draw the asymptotic Bode magnitude and phase plots.

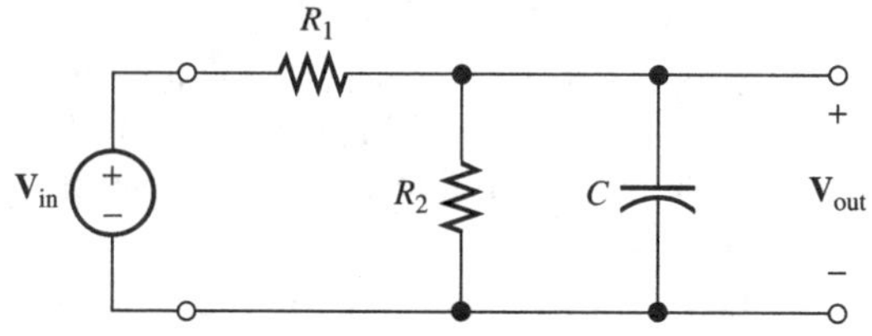

Figure P6.57

P6.58. Consider a circuit for which

$$v_{out}(t) = 10v_{in}(t) - 400\pi \int_0^t v_{out}(t)dt$$

a. Assume that $v_{out}(t) = A\cos(2\pi ft)$, and find an expression for $v_{in}(t)$. **b.** Use the results of part (a) to find an expression for the transfer function $H(f) = \mathbf{V}_{out}/\mathbf{V}_{in}$ for the system. **c.** Draw the asymptotic Bode plot for the transfer function magnitude.

P6.59. The circuit shown in Figure P6.59 has $R = 20\pi\,\Omega$ and $L = 20$ mH. Solve for the transfer function $H(f) = \mathbf{V}_{out}/\mathbf{V}_{in}$ and draw the Bode magnitude and phase plots for the circuit.

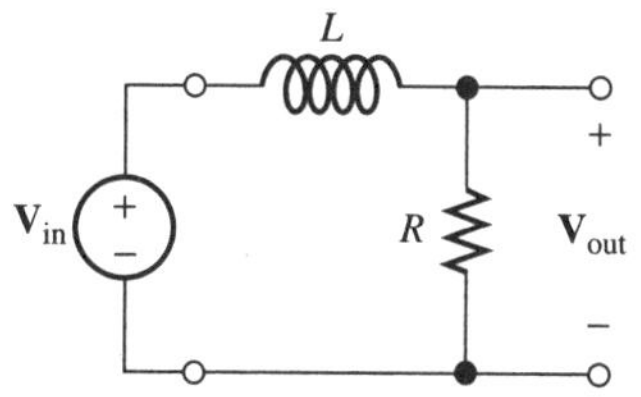

Figure P6.59

***P6.60.** In solving Problem P6.15, we find that the transfer function of an integrator circuit is given by$H(f) = 1/(j2\pi f)$. Sketch the Bode magnitude and phase plots to scale. What is the slope of the magnitude plot?

P6.61. In solving Problem P6.16, we find that the transfer function of a differentiator circuit is given by $H(f) = j4\pi f$. Sketch the Bode magnitude and phase plots to scale. What is the slope of the magnitude plot?

Section 6.5: First-Order Highpass Filter

P6.62. What is the slope of the high-frequency asymptote for the Bode magnitude plot for a first-order highpass filter? The low-frequency asymptote? At what frequency do the asymptotes meet?

P6.63. Draw the circuit diagram of a first-order RC highpass filter and give the expression for the half-power frequency in terms of the circuit components.

***P6.64.** Consider the circuit shown in Figure P6.64. Sketch the asymptotic Bode magnitude and

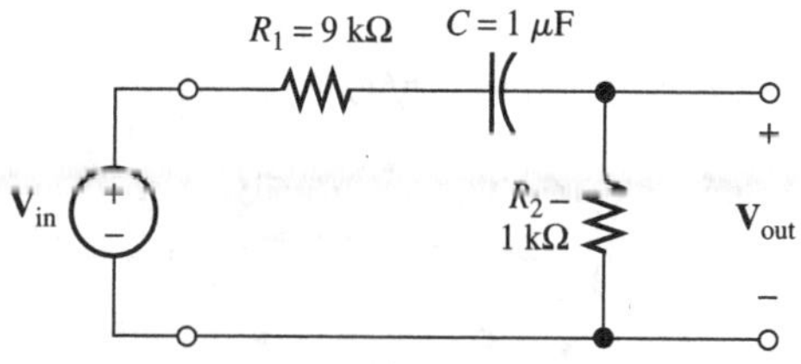

Figure P6.64

phase plots to scale for the transfer function $H(f) = \mathbf{V}_{out}/\mathbf{V}_{in}$.

***P6.65.** Consider the first-order highpass filter shown in Figure P6.65. The input signal is given by

$$v_{in}(t) = 5 + 5\cos(2000\pi t)$$

Find an expression for the output $v_{out}(t)$ in steady-state conditions.

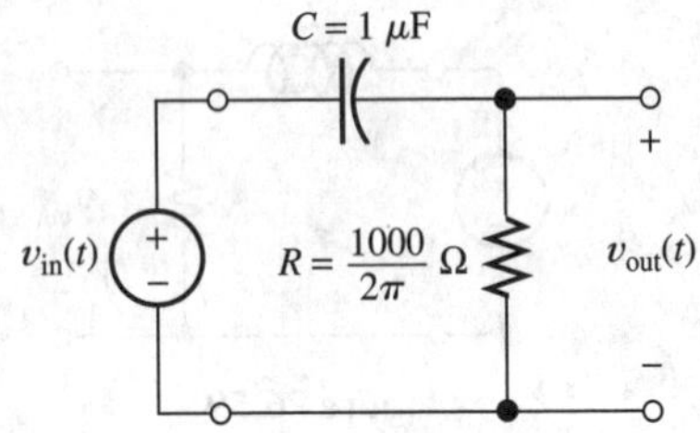

Figure P6.65

P6.66. Repeat Problem P6.65 for the input signal given by

$$v_{in}(t) = 10\cos(400\pi t + 30°) + 20\cos(4000\pi t) + 5\sin(4 \times 10^4 \pi t)$$

P6.67. The circuit shown in Figure P6.67 has $R = 200/\pi$ Ω and $C = 10$ μF. Sketch the Bode magnitude and phase plots to scale for the transfer function $H(f) = \mathbf{V}_{out}/\mathbf{V}_{in}$.

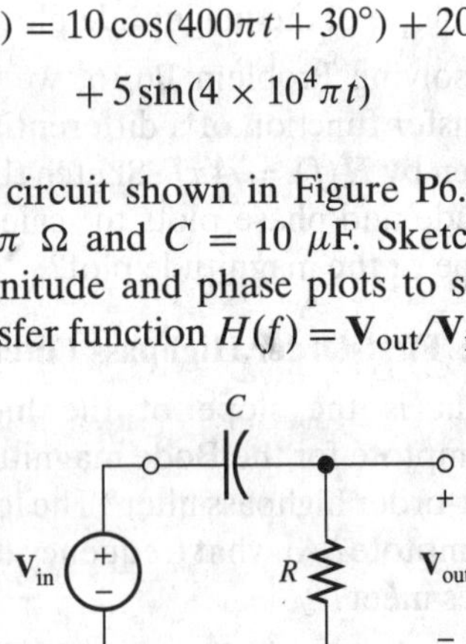

Figure P6.67

P6.68. Consider the circuit shown in Figure P6.68. Sketch the Bode magnitude and phase plots to scale for the transfer function $H(f) = \mathbf{V}_{out}/\mathbf{V}_{in}$.

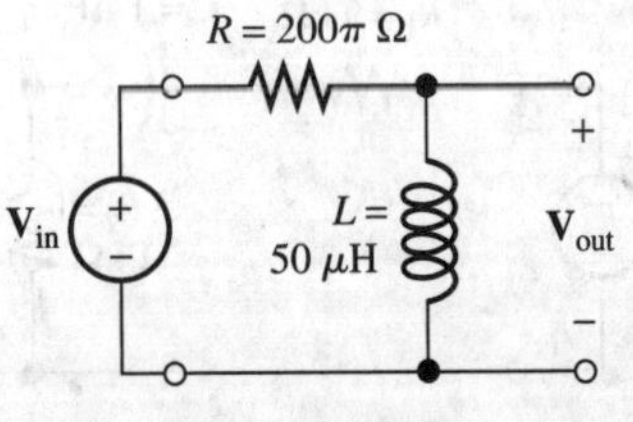

Figure P6.68

P6.69. Suppose we need a first-order highpass filter (such as Figure 6.19 on page 320) to attenuate a 60-Hz input component by 80 dB. What value is required for the break frequency of the filter? By how many dB is the 600-Hz component attenuated by this filter? If $R = 100$ Ω, what is the value of C?

Section 6.6: Series Resonance

P6.70. Define the resonant frequency and the quality factor of a series *RLC* circuit. Explain why the current in a series *RLC* circuit is maximum at the resonant frequency.

***P6.71.** Consider the series resonant circuit shown in Figure P6.71, with $L = 20$ μH, $R = 14.14$ Ω, and $C = 1000$ pF. Compute the resonant frequency, the bandwidth, and the half-power frequencies. Assuming that the frequency of the source is the same as the resonant frequency, find the phasor voltages across the elements and sketch a phasor diagram.

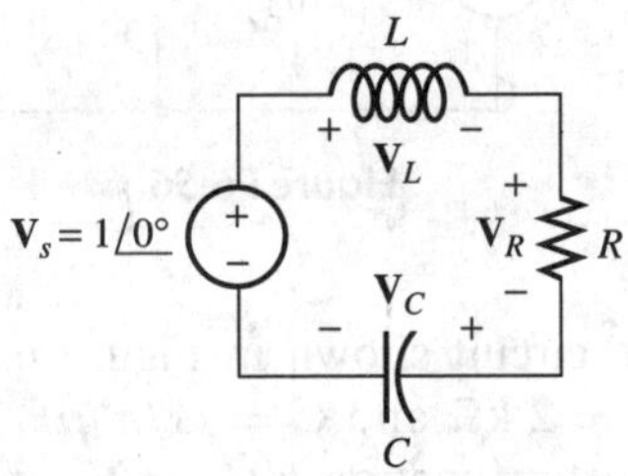

Figure P6.71

P6.72. Work Problem P6.71 for $L = 80$ μH, $R = 14.14$ Ω, and $C = 1000$ pF.

P6.73. Suppose we have a series resonant circuit for which $B = 15$ kHz, $f_0 = 300$ kHz, and $C = 663.1$ pF. Determine the values of L and R.

P6.74. Derive an expression for the resonant frequency of the circuit shown in Figure P6.74.

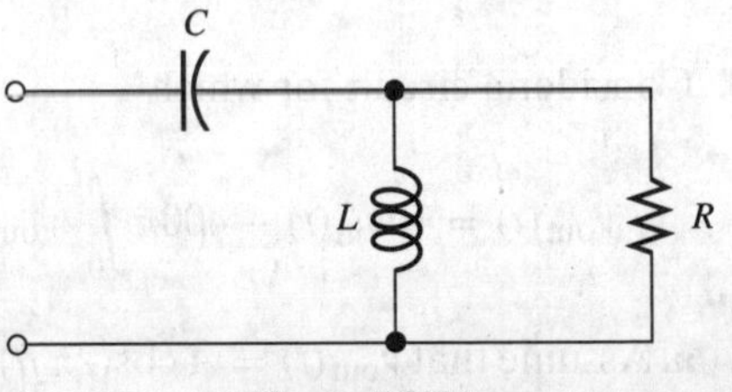

Figure P6.74

(Recall that we have defined the resonant frequency to be the frequency for which the impedance is purely resistive.)

***P6.75.** At the resonant frequency $f_0 = 1$ MHz, a series resonant circuit with $R = 50\ \Omega$ has $|\mathbf{V}_R| = 2$ V and $|\mathbf{V}_L| = 20$ V. Determine the values of L and C. What is the value of $|\mathbf{V}_C|$?

P6.76. Suppose we have a series resonant circuit for which $f_0 = 12$ MHz and $B = 400$ kHz. Furthermore, the minimum value of the impedance magnitude is 5 Ω. Determine the values of R, L, and C.

P6.77. What is a *bandpass filter*? How is its bandwidth defined?

Section 6.7: Parallel Resonance

P6.78. Define the resonant frequency of a parallel RLC circuit. Explain why the impedance of a parallel RLC circuit is maximum at the resonant frequency. How does the quality factor for a parallel resonant circuit differ from that for a series resonant circuit?

***P6.79.** A parallel resonant circuit has $R = 5$ kΩ, $L = 50\ \mu$H, and $C = 200$ pF. Determine the resonant frequency, quality factor, and bandwidth.

P6.80. A parallel resonant circuit has $f_0 = 20$ MHz and $B = 200$ kHz. The maximum value of $|Z_p|$ is 5 kΩ. Determine the values of R, L, and C.

P6.81. A parallel resonant circuit has $R = 1$ kΩ, $L = 64\ \mu$H and $C = 40$ pF. Determine the values of f_0 and B.

P6.82. Consider the parallel resonant circuit shown in Figure 6.29 on page 329. Determine the L and C values, given $R = 2$ kΩ, $f_0 = 8$ MHz, and $B = 500$ kHz. If $\mathbf{I} = 10^{-3}\angle 0°$, draw a phasor diagram showing the currents through each of the elements in the circuit at resonance.

Section 6.8: Ideal and Second-Order Filters

P6.83. Name four types of ideal filters.

***P6.84.** An ideal bandpass filter has cutoff frequencies of 9 and 11 kHz and a gain magnitude of two in the passband. Sketch the transfer-function magnitude to scale versus frequency. Repeat for an ideal band-reject filter.

P6.85. An ideal lowpass filter has a cutoff frequency of 10 kHz and a gain magnitude of two in the passband. Sketch the transfer-function magnitude to scale versus frequency. Repeat for an ideal highpass filter.

P6.86. Suppose that sinewave interference has been inadvertently added to an audio signal that has frequency components ranging from 20 Hz to 15 kHz. The frequency of the interference slowly varies in the range 950 to 1050 Hz. A filter that attenuates the interference by at least 20 dB and passes most of the audio components is desired. What type of filter is needed? Sketch the magnitude Bode plot of a suitable filter, labeling its specifications.

P6.87. In an electrocardiograph, the heart signals contain components with frequencies ranging from dc to 100 Hz. During exercise on a treadmill, the signal obtained from the electrodes also contains noise generated by muscle contractions. Most of the noise components have frequencies exceeding 100 Hz. What type of filter should be used to reduce the noise? What cutoff frequency is appropriate?

P6.88. Each AM radio signal has components ranging from 10 kHz below its carrier frequency to 10 kHz above its carrier frequency. Various radio stations in a given geographical region are assigned different carrier frequencies so that the frequency ranges of the signals do not overlap. Suppose that a certain AM radio transmitter has a carrier frequency of 980 kHz. What type of filter should be used if we want the filter to pass the components from this transmitter and reject the components of all other transmitters? What are the best values for the cutoff frequencies?

***P6.89.** Draw the circuit diagram of a second-order highpass filter. Suppose that $L = 1.6$ mH, $Q_s = 1$, and $f_0 = 100$ kHz. Determine the values of R and C.

P6.90. Draw the circuit diagram of a second-order lowpass filter. Given that $R = 50\ \Omega$, $Q_s = 0.5$, and $f_0 = 30$ MHz, determine the values of L and C.

Section 6.9: Transfer Functions and Bode Plots with MATLAB

P6.91. Consider the filter shown in Figure P6.91. **a.** Derive an expression for the transfer function $H(f) = \mathbf{V}_{out}/\mathbf{V}_{in}$. **b.** Use MATLAB to obtain a Bode plot of the transfer-function magnitude for $R_1 = 9$ kΩ, $R_2 = 1$ kΩ, and $C = 0.01\mu$F. Allow frequency to range from 10 Hz to 1 MHz. **c.** At very low frequencies, the capacitance becomes an open circuit. In this case, determine an expression for the transfer function and evaluate for the circuit parameters of part (b). Does the result agree with the value plotted in part (b)? **d.** At very high frequencies, the capacitance becomes a short circuit. In this case, determine an expression for the transfer function and evaluate for the circuit parameters of part (b). Does the result agree with the value plotted in part (b)?

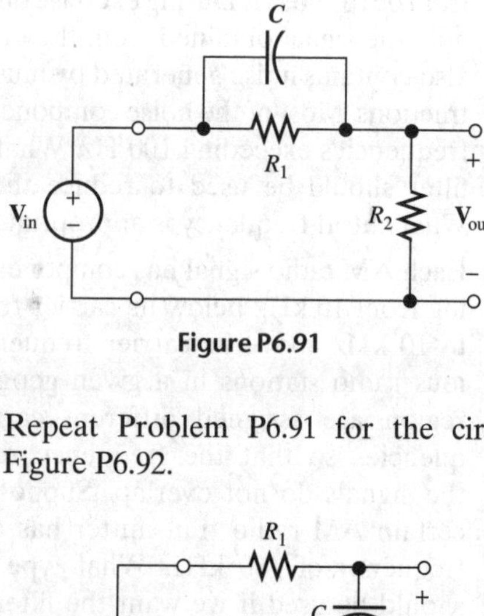

Figure P6.91

P6.92. Repeat Problem P6.91 for the circuit of Figure P6.92.

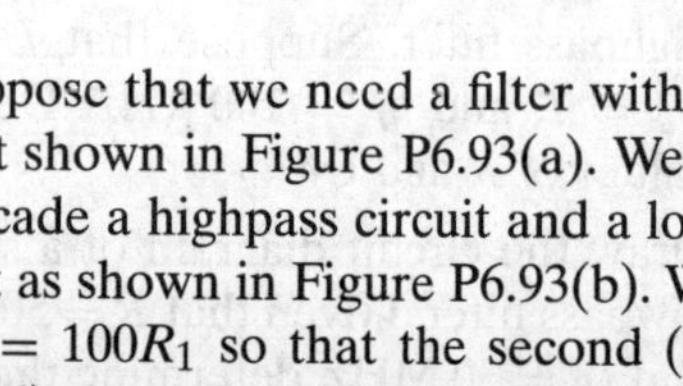

Figure P6.92

P6.93. Suppose that we need a filter with the Bode plot shown in Figure P6.93(a). We decide to cascade a highpass circuit and a lowpass circuit as shown in Figure P6.93(b). We choose $R_2 = 100R_1$ so that the second (i.e., right-hand) circuit looks like an approximate open circuit across the output of the first (i.e., left-hand) circuit. **a.** Which of the components form the lowpass filter? Which form the highpass filter? **b.** Compute the capacitances needed to achieve the desired break frequencies, making the approximation that the left-hand circuit has an open-circuit load. **c.** Write expressions that can be used to compute the exact transfer function $H(f) = \mathbf{V}_{out}/\mathbf{V}_{in}$ and use MATLAB to produce a Bode magnitude plot for f ranging from 1 Hz to 1 MHz. The result should be a close approximation to the desired plot shown in Figure P6.93(a).

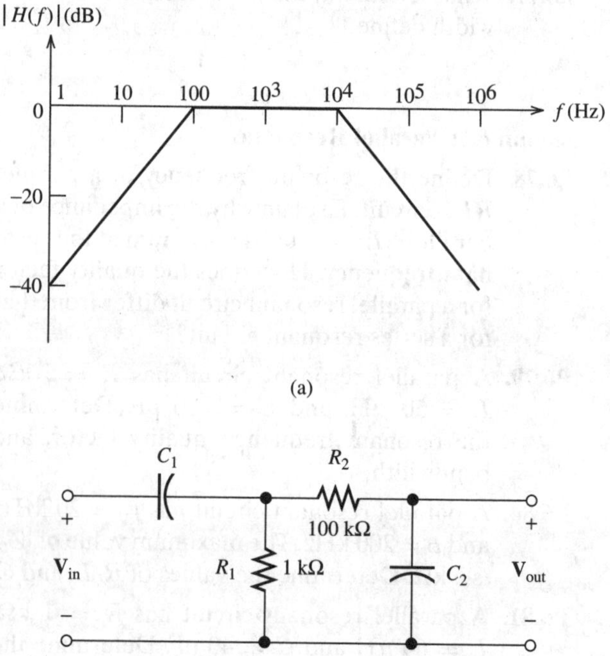

Figure P6.93

P6.94. Suppose that we need a filter with the Bode plot shown in Figure P6.93(a). We decide to cascade a highpass circuit and a lowpass circuit, as shown in Figure P6.94. We choose $C_2 = C_1/100$ so that the second (i.e., right-hand) circuit looks like an approximate open circuit across the output of the first (i.e., left-hand) circuit. **a.** Which of the components form the lowpass filter? Which form the highpass filter? **b.** Compute the resistances needed to achieve the desired break frequencies, making the approximation that the left-hand circuit has an open-circuit load. **c.** Write expressions that can be used to compute the exact transfer function $H(f) = \mathbf{V}_{out}/\mathbf{V}_{in}$ and use MATLAB to produce a Bode magnitude plot for f ranging from 1 Hz to 1 MHz. The result should be a close approximation to the desired plot shown in Figure P6.93(a).

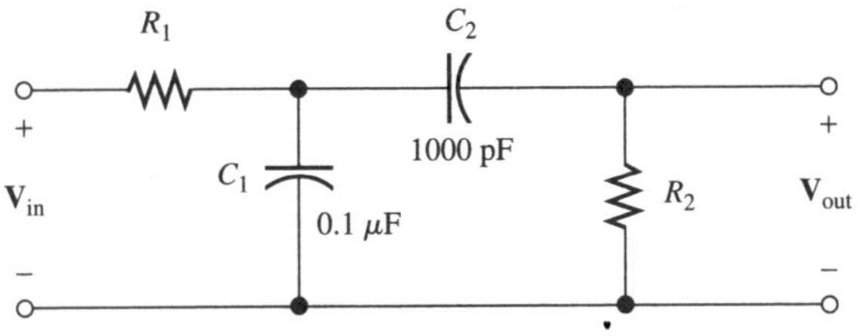

Figure P6.94

P6.95. Other combinations of R, L, and C have behaviors similar to that of the series resonant circuit. For example, consider the circuit shown in Figure P6.95. **a.** Derive an expression for the resonant frequency of this circuit. (We have defined the resonant frequency to be the frequency for which the impedance is purely resistive.) **b.** Compute the resonant frequency, given $L = 1$ mH, $R = 1000\ \Omega$, and $C = 0.25\ \mu$F. **c.** Use MATLAB to obtain a plot of the impedance magnitude of this circuit for f ranging from 95 to 105 percent of the resonant frequency. Compare the result with that of a series RLC circuit.

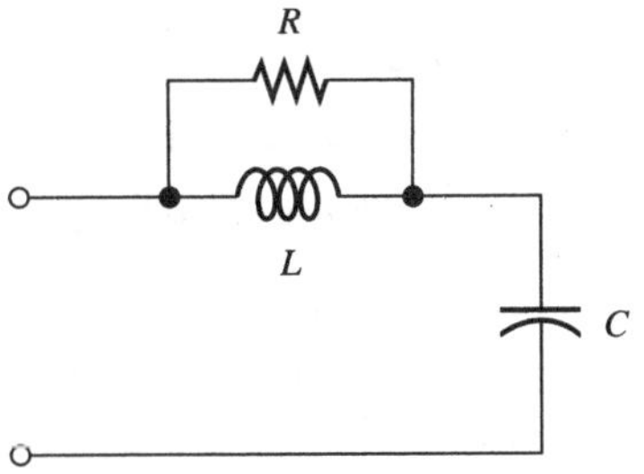

Figure P6.95

P6.96. Consider the circuit of Figure P6.74 with $R = 1$ kΩ, $L = 1$ mH, and $C = 0.25\ \mu$F. **a.** Using MATLAB, obtain a plot of the impedance magnitude of this circuit for f ranging from 9 to 11 kHz. **b.** From the plot, determine the minimum impedance, the frequency at which the impedance is minimum, and the bandwidth (i.e., the band of frequencies for which the impedance is less than $\sqrt{2}$ times the minimum value). **c.** Determine the component values for a series RLC circuit having the same parameters as those found in part (b). **d.** Plot the impedance magnitude of the series circuit on the same axes as the plot for part (a).

P6.97. Use the method of Example 6.9 to obtain a magnitude bode plot of the transfer function $H(f) = \mathbf{V}_{out}/\mathbf{V}_{in}$ for the lowpass filter of Figure P6.97 with frequency ranging from 100 kHz to 10 MHz. Manually verify the plotted values for high and low frequencies. Also, determine the half-power frequency for this filter.

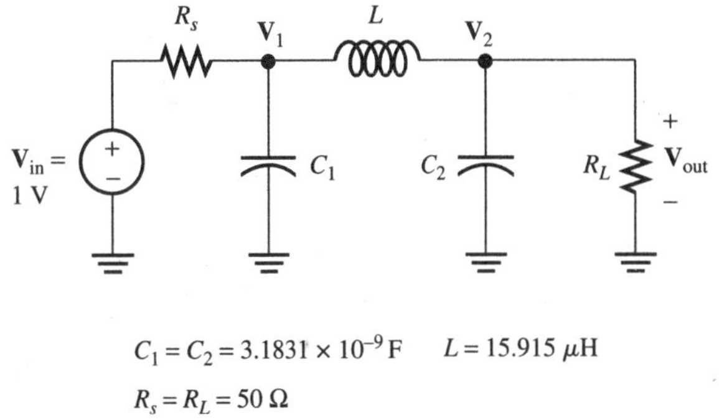

Figure P6.97

P6.98. Repeat Problem P6.97 for the highpass filter shown in Figure P6.98.

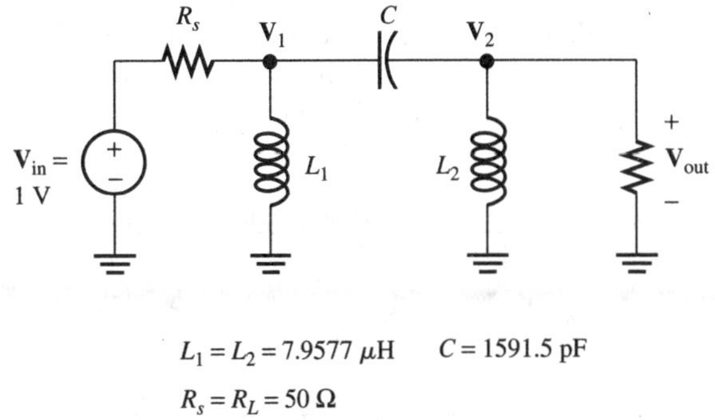

Figure P6.98

P6.99. Other combinations of R, L, and C have behaviors similar to that of the parallel circuit. For example, consider the circuit shown in Figure P6.99. **a.** Derive an expression for the resonant frequency of this circuit. (We have defined the resonant frequency to be the frequency for which the impedance is purely resistive. However, in this case you may find the algebra easier if you work with admittances.) **b.** Compute the resonant frequency, given $L = 1$ mH, $R = 1\ \Omega$, and $C = 0.25\ \mu$F. **c.** Use MATLAB to obtain a plot of the impedance magnitude of this circuit for f ranging from 95 to 105 percent of the resonant frequency. Compare the result with that of a parallel RLC circuit.

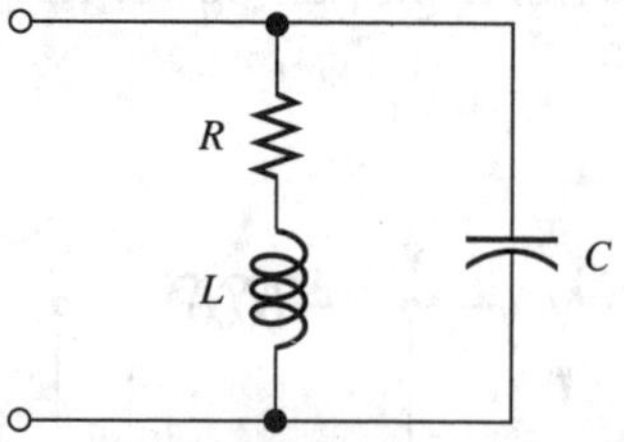

Figure P6.99

P6.100. Consider the filter shown in Figure P6.100. **a.** Derive an expression for the transfer function $H(f) = \mathbf{V}_{out}/\mathbf{V}_{in}$. **b.** Use MATLAB to obtain a Bode plot of the transfer function magnitude for $R = 10\ \Omega$, $L = 10$ mH, and $C = 0.02533\mu$F. Allow frequency to range from 1 kHz to 100 kHz. **c.** At very low frequencies, the capacitance becomes an open circuit and the inductance becomes a short circuit. In this case, determine an expression for the transfer function and evaluate for the circuit parameters of part (b). Does the result agree with the value plotted in part (b)? **d.** At very high frequencies, the capacitance becomes a short circuit and the inductance becomes an open circuit. In this case, determine an expression for the transfer function and evaluate for the circuit parameters of part (b). Does the result agree with the value plotted in part (b)?

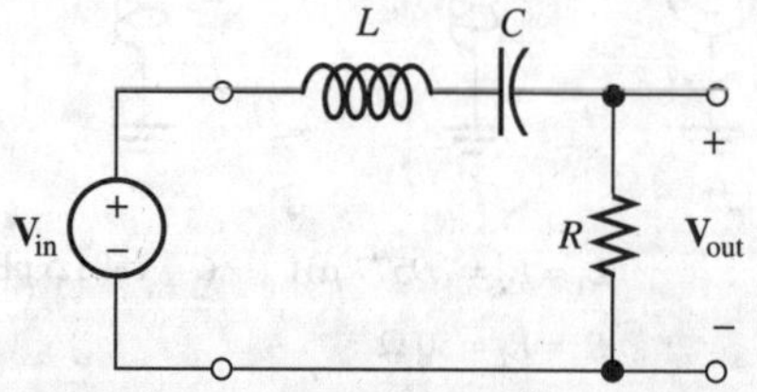

Figure P6.100

P6.101. Repeat Problem P6.100 for the circuit of Figure P6.101.

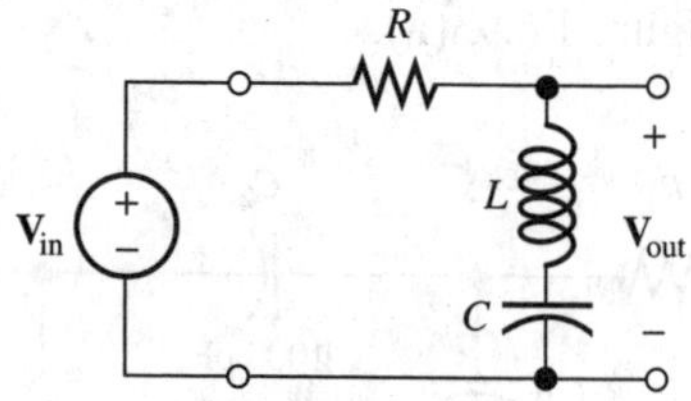

Figure P6.101

Section 6.10: Digital Signal Processing

P6.102. **a.** Develop a digital filter that mimics the action of the RL filter shown in Figure P6.102. Determine expressions for the coefficients in terms of the time constant and sampling interval T. (*Hint*: If your circuit equation contains an integral, differentiate with respect to time to obtain a pure differential equation.) **b.** Given $R = 10\ \Omega$ and $L = 200$ mH, sketch the step response of the circuit to scale. **c.** Use MATLAB to determine and plot the step response of the digital filter for several time constants. Use the time constant of part (b) and $f_s = 500$ Hz. Compare the results of parts (b) and (c).

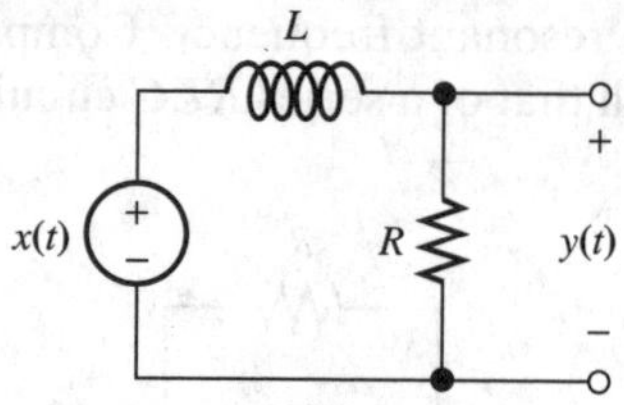

Figure P6.102

P6.103. Repeat Problem P6.102 for the filter shown in Figure P6.103.

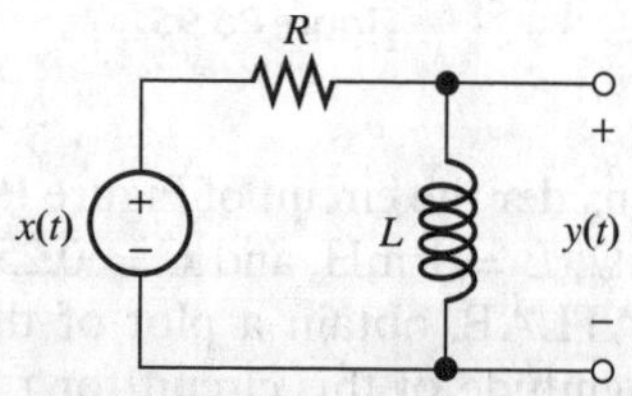

Figure P6.103

***P6.104.** Consider the second-order bandpass filter shown in Figure P6.104. **a.** Derive expressions for L and C in terms of the resonant frequency ω_0 and quality factor Q_s. **b.** Write the KVL equation for the circuit and use it to develop a digital filter that mimics the action of the RLC filter. Use the results of part (a) to write the coefficients in terms of the resonant frequency ω_0, circuit quality factor Q_s, and sampling interval T. (*Hint*: The circuit equation contains an integral, so differentiate with respect to time to obtain a pure differential equation.)

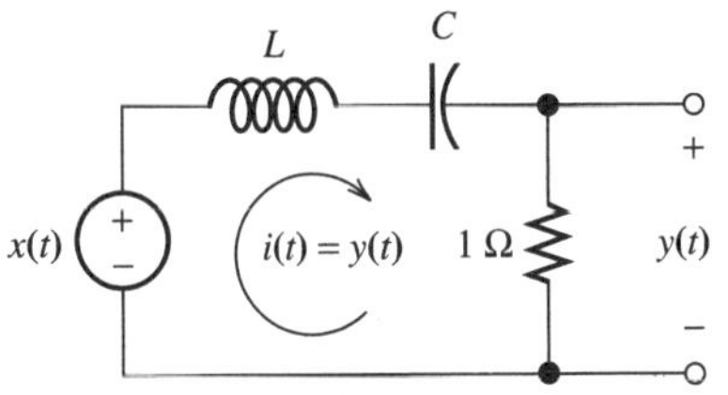

Figure P6.104

Frequently Asked Questions

1. What do you understand by Fourier analysis?
2. Explain Fourier analysis of a square wave.
3. Explain the function of filter used with a two-port network.
4. What do you understand by transfer function of a two-port filter?
5. What is a Graphic Equalizer? How it is used to adjust transfer function?
6. How can you determine the transfer function by performing an experiment?
7. What do you mean by first-order lowpass filter?
8. What is the application of transfer function?
9. What are the applications of first-order low-pass filters?
10. What do you mean by Bode-plots?
11. Analyze a first-order high pass filter.
12. Describe series resonance and resonant frequency.
13. What do you understand by band-width for a series resonant circuit?
14. What is a band pass filter?
15. Describe parallel resonance and resonant frequency.
16. What do you understand by band-width for a parallel resonant circuit?
17. What do you mean by second-order lowpass filter?
18. Make a comparison between first and second order filters.
19. What do you mean by second-order highpass filter?
20. What do you mean by second-order band-reject (Notch) filter?
21. How an analog signal is converted to digital form?
22. What is a digital filter?
23. Define and explain a digital lowpass filter.
24. Explain a simple Notch filter.
25. Make a comparsion between RLC circuit filter and digital filter.

Practice Test

Here is a practice test you can use to check your comprehension of the most important concepts in this chapter. Answers can be found in Appendix D and complete solutions are included in the Student Solutions files. See Appendix E for more information about the Student Solutions.

T6.1. What is the basic concept of Fourier theory as it relates to real-world signals? How does the transfer function of a filter relate to this concept?

T6.2. An input signal given by

$$v_{\text{in}}(t) = 3 + 4\cos(1000\pi t) + 5\cos(2000\pi t - 30°)$$

is applied to the RL filter shown in Figure T6.2. Find the expression for the output signal $v_{\text{out}}(t)$.

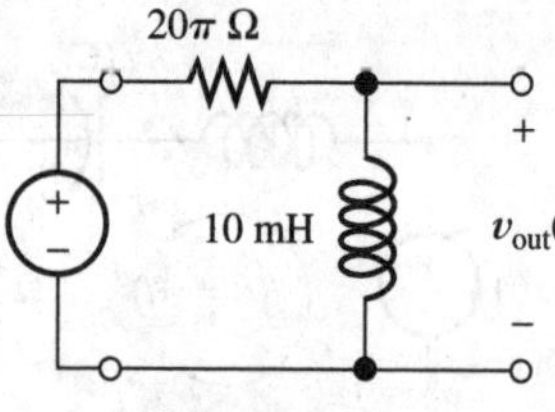

Figure T6.2

T6.3. Consider the Bode magnitude plot for the transfer function of a certain filter given by

$$H(f) = \frac{\mathbf{V}_{out}}{\mathbf{V}_{in}} = 50\frac{j(f/200)}{1 + j(f/200)}$$

a. What is the slope of the low-frequency asymptote?

b. What is the slope of the high-frequency asymptote?

c. What are the coordinates of the point at which the asymptotes meet?

d. What type of filter is this?

e. What is the value of the break frequency?

T6.4. A series resonant circuit has $R = 5\ \Omega$, $L = 20\,\text{mH}$, and $C = 1\mu\text{F}$. Determine the values of:

a. the resonant frequency in Hz.

b. Q.

c. bandwidth in Hz.

d. the impedance of the circuit at the resonant frequency.

e. the impedance of the circuit at dc.

f. the impedance of the circuit as the frequency approaches infinity.

T6.5. Repeat question T6.4 for a parallel resonant circuit with $R = 10\,\text{k}\Omega$, $L = 1\,\text{mH}$, and $C = 1000\,\text{pF}$.

T6.6. Consider the transfer function $\mathbf{V}_{out}/\mathbf{V}_{in}$ for each of the circuits shown in Figure T6.6. Classify each circuit as a first-order lowpass filter, second-order bandpass filter, etc. Justify your answers.

T6.7. Give a list of MATLAB commands to produce the magnitude Bode plot for the transfer function of question T6.3 for frequency ranging from 10 Hz to 10 kHz.

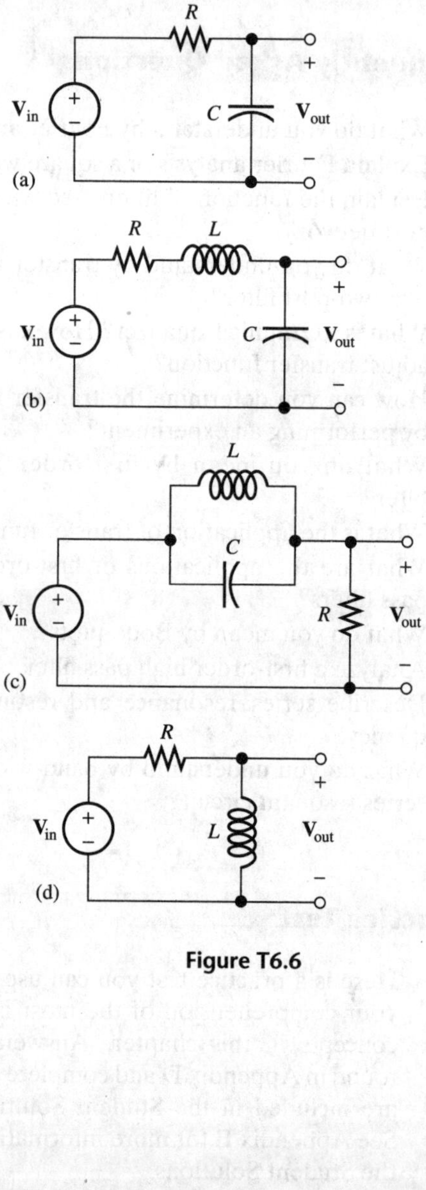

Figure T6.6

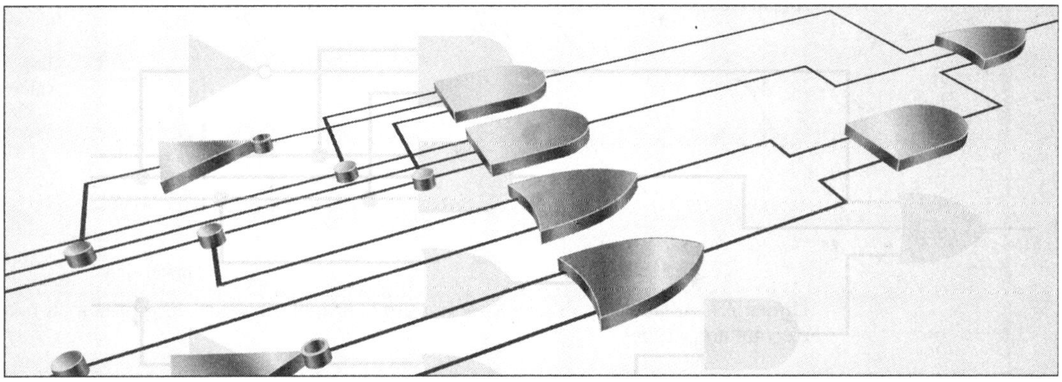

Chapter 7

Logic Circuits

Study of this chapter will enable you to:

- State the advantages of digital technology over analog technology.
- Understand the terminology of digital circuits.
- Convert numbers between decimal, binary, and other forms.
- Use the Gray code for position and angular sensors.
- Understand the binary arithmetic operations used in computers and other digital systems.
- Interconnect logic gates of various types to implement a given logic function.
- Use Karnaugh maps to minimize the number of gates needed to implement a logic function.
- Understand how gates are connected together to form flip-flops and registers.

Introduction to this chapter:

So far, we have considered circuits, such as filters, that process analog signals. For an **analog signal**, each amplitude in a continuous range has a unique significance. For example, a position sensor may produce an analog signal that is proportional to displacement. Each amplitude represents a different position. An analog signal is shown in Figure 7.1(a).

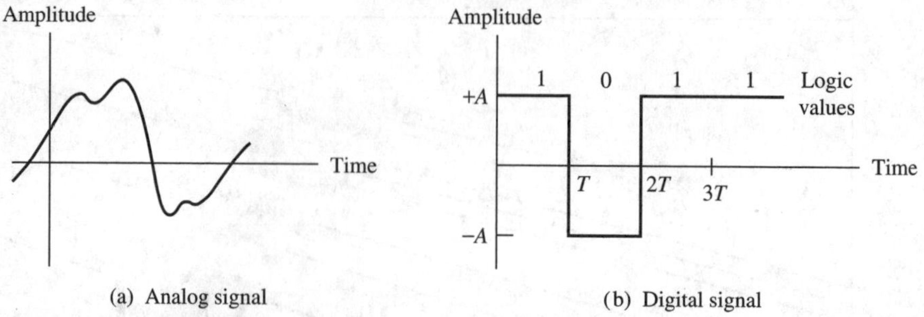

Figure 7.1 Analog signals take a continuum of amplitude values. Digital signals take a few discrete amplitudes.

In this chapter, we introduce circuits that process digital signals. For a **digital signal**, only a few restricted ranges of amplitude are allowed, and each amplitude in a given range has the same significance. Most common are **binary** signals that take on amplitudes in only two ranges, and the information associated with the ranges is represented by the **logic values** 1 or 0. An example of a digital signal is shown in Figure 7.1(b). Computers are examples of digital circuits. We will see that digital approaches have some important advantages over analog approaches.

7.1 BASIC LOGIC CIRCUIT CONCEPTS

We often encounter analog signals in instrumentation of physical systems. For example, a pressure transducer can yield a voltage that is proportional to pressure versus time in the cylinder of an internal combustion engine. In Section 6.10, we saw that analog signals can be converted to equivalent digital signals that contain virtually the same information. Then, computers or other digital circuits can be used to process this information. In many applications, we have a choice between digital and analog approaches.

Advantages of the Digital Approach

Provided that the noise amplitude is not too large, the logic values represented by a digital signal can still be determined after noise is added.

Digital signals have several important advantages over analog signals. After noise is added to an analog signal, it is usually impossible to determine the precise amplitude of the original signal. On the other hand, after noise is added to a digital signal, we can still determine the logic values—provided that the noise amplitude is not too large. This is illustrated in Figure 7.2.

For a given type of logic circuit, one range of voltages represents logic 1, and another range of voltages represents logic 0. For proper operation, a logic circuit only needs to produce a voltage somewhere in the correct range. Thus, component values in digital circuits do not need to be as precise as in analog circuits.

With modern IC technology, it is possible to manufacture exceedingly complex digital circuits economically.

It turns out that with modern integrated-circuit (IC) manufacturing technology, very complex digital logic circuits (containing millions of components) can be produced economically. Analog circuits often call for large capacitances and precise component values that are impossible to manufacture as large-scale integrated circuits. Thus, digital systems have become increasingly important in the past few decades, a trend that will continue.

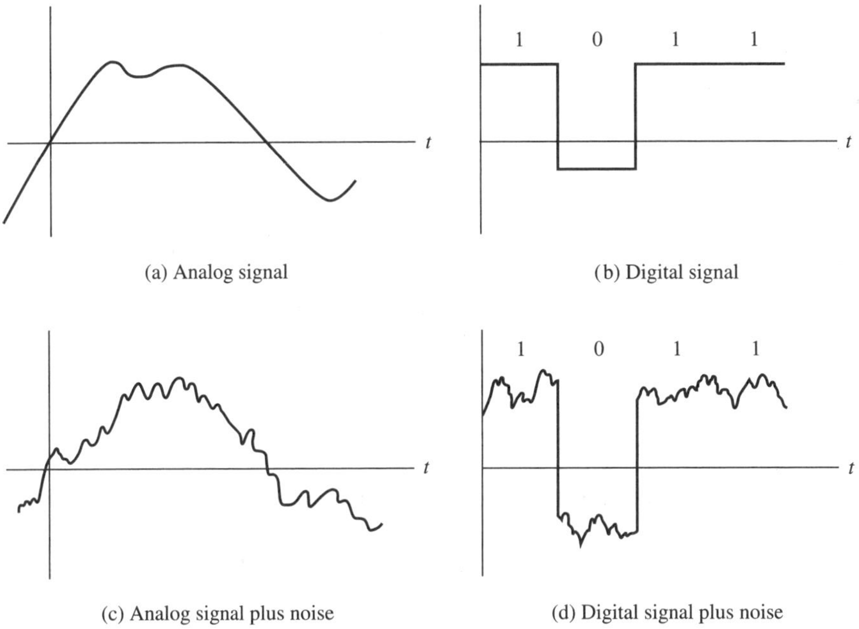

Figure 7.2 The information (logic values) represented by a digital signal can still be determined precisely after noise is added. Noise obscures the information contained in an analog signal because the original amplitude cannot be determined exactly after noise is added.

Positive versus Negative Logic

Usually, the higher amplitude in a binary system represents 1 and the lower-amplitude range represents 0. In this case, we say that we have **positive logic**. On the other hand, it is possible to represent 1 by the lower amplitude and 0 by the higher amplitude, resulting in **negative logic**. Unless stated otherwise, we assume positive logic throughout this book.

The logic value 1 is also called **high**, **true**, or **on**. Logic 0 is also called **low**, **false**, or **off**. Signals in logic systems switch between high and low as the information being represented changes. We denote these signals, or **logic variables**, by uppercase letters such as A, B, and C.

Logic Ranges and Noise Margins

Logic circuits are typically designed so that a range of input voltages is accepted as logic 1 and another nonoverlapping range of voltages is accepted as logic 0. The largest input voltage accepted as logic 0, or low, is denoted as V_{IL}, and the smallest input voltage accepted as logic 1, or high, is denoted as V_{IH}. This is illustrated in Figure 7.3. No meaning is assigned to voltages between V_{IL} and V_{IH}, which normally occur only during transitions.

Furthermore, the circuits are designed so that the output voltages fall into narrower ranges than the inputs (provided that the inputs are in the acceptable ranges). This is also illustrated in Figure 7.3. V_{OL} is the highest logic-0 output voltage, and V_{OH} is the lowest logic-1 output voltage.

Because noise can be added to a logic signal in the interconnections between outputs and inputs, it is important that the outputs have narrower ranges than the

For a logic signal, one range of amplitudes represents logic 1, a nonoverlapping range represents logic 0, and no meaning is assigned to the remaining amplitudes, which ordinarily do not occur or occur only during transitions.

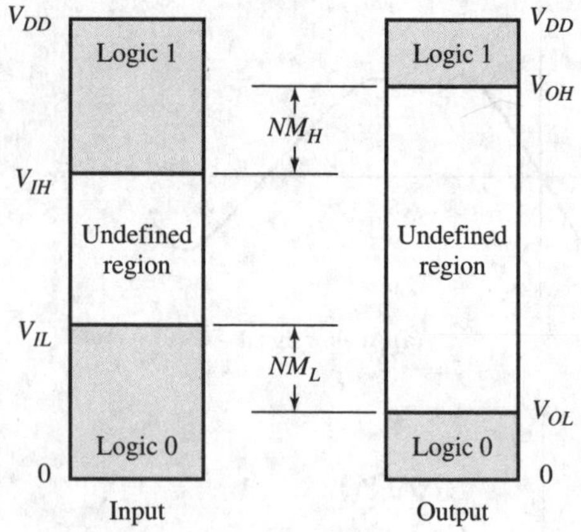

Figure 7.3 Voltage ranges for logic-circuit inputs and outputs.

acceptable inputs. The differences are called **noise margins** and are given by

$$NM_L = V_{IL} - V_{OL}$$
$$NM_H = V_{OH} - V_{IH}$$

Ideally, noise margins are as large as possible.

Digital Words

A single binary digit, called a **bit**, represents a very small amount of information. For example, a logic variable R could be used to represent whether or not it is raining in a particular location (say $R = 1$ if it is raining, and $R = 0$ if it is not raining).

To represent more information, we resort to using groups of logic variables called **digital words**. For example, the word RWS could be formed, in which R represents rain, W is 1 if the wind velocity is greater than 15 miles per hour and 0 for less wind, and S could be 1 for sunny conditions and 0 for cloudy. Then the digital word 110 would tell us that it is rainy, windy, and cloudy. A **byte** is a word consisting of eight bits, and a **nibble** is a four-bit word.

Transmission of Digital Information

In **parallel transmission**, an n-bit word is transferred on $n+1$ wires, one wire for each bit, plus a common or ground wire. On the other hand, in **serial transmission**, the successive bits of the word are transferred one after the other with a single pair of wires. At the receiving end, the bits are collected and combined into words. Parallel transmission is faster and often used for short distances, such as internal data transfer in a computer. Long-distance digital communication systems are usually serial.

Examples of Digital Information-Processing Systems

By using a prearranged 100-bit word consisting of logic values and binary numbers, we could give a rather precise report of weather conditions at a given location. Computers, such as those used by the National Weather Bureau, process words received from various weather stations to produce contour maps of temperature, wind velocity, cloud state, precipitation, and so on. These maps are useful in understanding and predicting weather patterns.

Analog signals can be reconstructed from their periodic samples (i.e., measurements of instantaneous amplitude at uniformly spaced points in time), provided that the sampling rate is high enough. Each amplitude value can be represented as a digital word. Thus, an analog signal can be represented by a sequence of digital words. In playback, the digital words are converted to the corresponding analog amplitudes. This is the principle of the compact-disc recording technique.

Thus, electronic circuits can gather, store, transmit, and process information in digital form to produce results that are useful or pleasing.

7.2 REPRESENTATION OF NUMERICAL DATA IN BINARY FORM

Binary Numbers

Digital words can represent numerical data. First, consider the decimal (base 10) number 743.2. We interpret this number as

Because digital circuits are (almost always) designed to operate with only two symbols, 0 or 1, it is necessary to represent numerical and other data as words composed of 0s and 1s.

$$7 \times 10^2 + 4 \times 10^1 + 3 \times 10^0 + 2 \times 10^{-1}$$

Similarly, the binary or base-two number 1101.1 is interpreted as

$$1 \times 2^3 + 1 \times 2^2 + 0 \times 2^1 + 1 \times 2^0 + 1 \times 2^{-1} = 13.5$$

Hence, the binary number 1101.1 is equivalent to the decimal number 13.5. Where confusion seems likely to occur, we use a subscript to distinguish binary numbers (such as 1101.1_2) from decimal numbers (such as 13.5_{10}).

With three bits, we can form 2^3 distinct words. These words can represent the decimal integers 0 through 7 as shown:

000	0
001	1
010	2
011	3
100	4
101	5
110	6
111	7

Similarly, a four-bit word has 16 combinations that represent the integers 0 through 15. (Frequently, we include leading zeros in discussing binary numbers in a digital circuit, because circuits are usually designed to operate on fixed-length words and the circuit produces the leading zeros.)

Conversion of Decimal Numbers to Binary Form

To convert a decimal integer to binary, we repeatedly divide by two until the quotient is zero. Then, the remainders read in reverse order give the binary form of the number.

		Quotient	Remainder
343/2	=	171	1
171/2	=	85	1
85/2	=	42	1
42/2	=	21	0
21/2	=	10	1
10/2	=	5	0
5/2	=	2	1
2/2	=	1	0
1/2	=	0	1

→ 101010111_2 (Read binary equivalent in reverse order)

Stop when quotient equals zero

Figure 7.4 Conversion of 343_{10} to binary form.

Example 7.1 Converting a Decimal Integer to Binary

Convert the decimal integer 343_{10} to binary.

Solution The operations are shown in Figure 7.4. The decimal number is repeatedly divided by two. When the quotient reaches zero, we stop. Then, the binary equivalent is read as the remainders in reverse order. From the figure, we see that

$$343_{10} = 101010111_2$$

■

To convert decimal fractions to binary fractions, we repeatedly multiply the fractional part by two and retain the whole parts of the results as the successive bits of the binary fraction.

Example 7.2 Converting a Decimal Fraction to Binary

Convert 0.392_{10} to its closest six-bit binary equivalent.

Solution The conversion is illustrated in Figure 7.5. The fractional part of the number is repeatedly multiplied by two. The whole part of each product is retained as a bit of the binary equivalent. We stop when the desired degree of precision has been reached. Thus, from the figure, we have

$$0.392_{10} \cong 0.011001_2$$

■

2×0.392	=	0	+	0.784
2×0.784	=	1	+	0.568
2×0.568	=	1	+	0.136
2×0.136	=	0	+	0.272
2×0.272	=	0	+	0.544
2×0.544	=	1	+	0.088

0.011001_2 (approximate binary equivalent)

Figure 7.5 Conversion of 0.392_{10} to binary.

To convert a decimal number having both whole and fractional parts, we convert each part separately and then combine the parts.

Example 7.3 Converting Decimal Values to Binary

Convert 343.392_{10} to binary.

Solution From Examples 7.1 and 7.2, we have

Convert the whole and fractional parts of the number separately and combine the results.

$$343_{10} = 101010111_2$$

and

$$0.392_{10} \cong 0.011001_2$$

Combining these results, we get

$$343.392_{10} \cong 101010111.011001_2$$

■

Exercise 7.1 Convert the following numbers to binary form, stopping after you have found six bits (if necessary) for the fractional part: **a.** 23.75; **b.** 17.25; **c.** 4.3.
Answer **a.** 10111.11; **b.** 10001.01; **c.** 100.010011. □

Exercise 7.2 Convert the following to decimal equivalents: **a.** 1101.111_2; **b.** 100.001_2.
Answer **a.** 13.875_{10}; **b.** 4.125_{10}. □

Binary Arithmetic

We add binary numbers in much the same way that we add decimal numbers, except that the rules of addition are different (and much simpler). The rules for binary addition are shown in Figure 7.6.

		Sum	Carry
0 + 0	=	0	0
0 + 1	=	1	0
1 + 1	=	0	1
1 + 1 + 1	=	1	1

Figure 7.6 Rules of binary addition.

Example 7.4 Adding Binary Numbers

Add the binary numbers 1000.111 and 1100.011.

Solution See Figure 7.7. ■

```
 0001 11  <-- Carries
 1000.111
+1100.011
---------
10101.010
```

Figure 7.7 Addition of binary numbers.

Hexadecimal and Octal Numbers

Binary numbers are inconvenient for humans because it takes many bits to write large numbers (or fractions to a high degree of precision). Hexadecimal (base 16) and octal (base 8) numbers are easily converted to and from binary numbers. Furthermore, they are much more efficient than binary numbers in representing information.

Table 7.1. Symbols for Octal and Hexadecimal Numbers and Their Binary Equivalents

Octal		Hexadecimal	
0	000	0	0000
1	001	1	0001
2	010	2	0010
3	011	3	0011
4	100	4	0100
5	101	5	0101
6	110	6	0110
7	111	7	0111
		8	1000
		9	1001
		A	1010
		B	1011
		C	1100
		D	1101
		E	1110
		F	1111

Table 7.1 shows the symbols used for hexadecimal and octal numbers and their binary equivalents. Notice that we need 16 symbols for the digits of a hexadecimal number. Customarily, the letters A through F are used to represent the digits for 10 through 15.

We can convert an octal or hexadecimal number to binary simply by substituting the binary equivalents for each digit.

Example 7.5 Converting Octal and Hexadecimal Numbers to Binary

Convert the numbers 317.2_8 and $F3A.2_{16}$ to binary.

In converting an octal or hexadecimal number to binary, use Table 7.1 to replace each digit by its binary equivalent.

Solution We simply use Table 7.1 to replace each digit by its binary equivalent. Thus, we have

$$317.2_8 = 011\ 001\ 111.\ 010_2$$
$$= 011001111.010_2$$

and

$$F3A.2_{16} = 1111\ 0011\ 1010.\ 0010$$
$$= 111100111010.0010_2$$

■

In converting binary numbers to octal, we first arrange the bits in groups of three, starting from the binary point and working outward. If necessary, we insert leading or trailing zeros to complete the groups. Then, we convert each group of three bits to its octal equivalent. Conversion to hexadecimal uses the same approach, except that the binary number is arranged in groups of four bits.

Example 7.6 Converting Binary Numbers to Octal or Hexadecimal

Convert 11110110.1_2 to octal and to hexadecimal.

Solution For conversion to octal, we first form three-bit groups, working outward from the binary point:

$$11110110.1_2 = 011\ 110\ 110.\ 100$$

Notice that we have appended leading and trailing zeros so that each group contains three bits. Next, we write the octal digit for each group. Thus, we have

$$11110110.1_2 = 011\ 110\ 110.\ 100 = 366.4_8$$

Working both directions from the binary point, group the bits into three-(octal) or four-(hexadecimal) bit words. Add leading or trailing zeros to complete the groups. Then, use Table 7.1 to replace each binary word by the corresponding symbol.

For conversion to hexadecimal, we form four-bit groups appending leading and trailing zeros as needed. Then, we convert each group to its equivalent hexadecimal integer, yielding

$$11110110.1_2 = 1111\ 0110.\ 1000 = \mathrm{F}6.8_{16}$$ ■

Exercise 7.3 Convert the following numbers to binary, octal, and hexadecimal forms: **a.** 97_{10}; **b.** 229_{10}.
Answer **a.** $97_{10} = 1100001_2 = 141_8 = 61_{16}$; **b.** $229_{10} = 11100101_2 = 345_8 = \mathrm{E}5_{16}$. □

Exercise 7.4 Convert the following numbers to binary form: **a.** 72_8; **b.** $\mathrm{FA}6_{16}$.
Answer **a.** 111010_2; **b.** 111110100110_2. □

Binary-Coded Decimal Format

Sometimes, decimal numbers are represented in binary form simply by writing the four-bit equivalents for each digit. The resulting numbers are said to be in **binary-coded decimal** (BCD) format. For example, 93.2 becomes

In converting a decimal number to BCD, each digit is replaced by its four-bit equivalent.

$$93.2 = 1001\ 0011.\ 0010_{\mathrm{BCD}}$$

Code groups 1010, 1011, 1100, 1101, 1110, and 1111 do not occur in BCD (unless an error has occurred). Calculators frequently represent numbers internally in BCD format. As each key is pressed, the BCD code group is stored. The operation

$$9 \times 3 = 27$$

would appear in BCD format as

$$1001 \times 0011 = 0010\ 0111$$

Even though binary code words are used to represent the decimal integers, the operations inside a calculator are partly decimal in nature. On the other hand, calculations are often carried out in true binary fashion in computers.

Exercise 7.5 Express 197_{10} in BCD form.
Answer $197_{10} = 000110010111_{\mathrm{BCD}}$. □

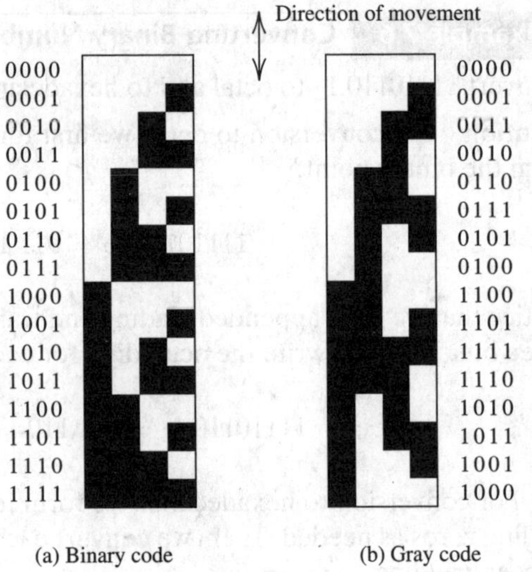

Figure 7.8 Black and white bands that can be read by a photodiode array resulting in a digital word representing the position of a robot arm.

Gray Code

Consider a transducer for encoding the position of a robot arm in which black-and-white bands are placed on the arm as illustrated in Figure 7.8. In the figure, we assume that the bands are read by light-sensitive diodes in which a black band is converted to logic 1 and a white band is converted to logic 0.

A problem occurs if the successive positions are represented by the binary code shown in Figure 7.8(a). For example, when the arm moves from the position represented by 0011 to that of 0100, three bits of the code word must change. Suppose that because the photodiode sensors are not perfectly aligned, 0011 first changes to 0001, then to 0000, and finally to 0100. During this transition, the indicated position is far from the actual position.

In a Gray code, each word differs in only one bit from each of its adjacent words.

A better scheme for coding the positions is to use the **Gray code** shown in Figure 7.8(b). *In a Gray code, each code word differs in only one bit from its neighboring code words.* Thus, erroneous position indications are avoided during transitions. Gray codes of any desired length can be constructed as shown in Figure 7.9. (Notice that successive code words differ in a single bit.)

Gray codes are also used to encode the angular position of rotating shafts. Then, the last word in the list wraps around and is adjacent to the first word. To represent angular position with a resolution better than 1 degree, we would need a nine-bit Gray code consisting of $2^9 = 512$ words. Each word would represent an angular sector of $360/512 = 0.703$ degree in width.

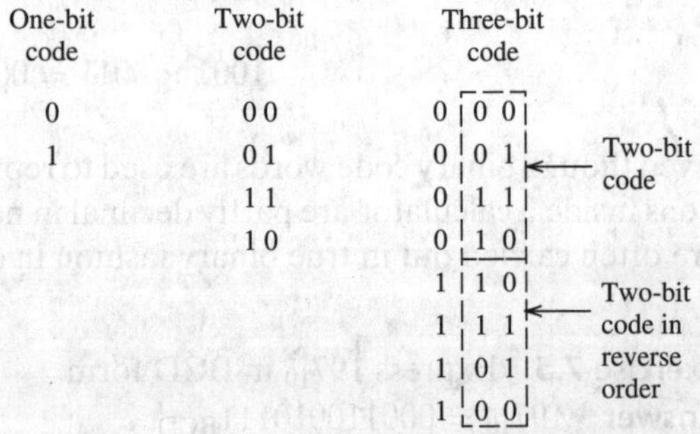

Figure 7.9 A one-bit Gray code simply consists of the two words 0 and 1. To create a Gray code of length n, repeat the code of length $n-1$ in reverse order, place a 0 on the left-hand side of each word in the first half of the list, and place a 1 on the left-hand side of each word in the second half of the list.

Exercise 7.6 We wish to represent the position of a robot arm with a resolution of 0.01 inch or better. The range of motion is 20 inches. How many bits are required for a Gray code that can represent the arm position?
Answer 11 bits. □

Complement Arithmetic

The **one's complement** of a binary number is obtained by replacing 1s by 0s, and vice versa. For example, an eight-bit binary number and its one's complement are

01001101

10110010 (one's complement)

The **two's complement** of a binary number is obtained by adding 1 to the one's complement, neglecting the carry (if any) out of the most significant bit. For example, to find the two's complement of

01001100

we first form the one's complement, which is

10110011

and then add 1. This is illustrated in Figure 7.10(a).

Another way to obtain the two's complement of a number is to copy the number—working from right to left—until after the first 1 is copied. Then, the remaining bits are inverted. An example of this process is shown in Figure 7.10(b).

10110011	One's complement
+1	
10110100	Two's complement

(a) First find the one's complement, then add 1; neglect the carry (if it occurs) out of the most significant bit

01001100	Number
Invert Copy	
10110100	Two's complement

(b) Working from right to left, copy bits until after the first 1 is copied; then invert the remaining bits

Figure 7.10 Two ways to find the two's complement of the binary number 01001100.

Complements are useful for representing negative numbers and performing subtraction in computers. Furthermore, the use of complement arithmetic simplifies the design of digital computers. Most common is the **signed two's-complement** representation, in which the first bit is taken as the sign bit. If the number is positive, the first bit is 0, whereas if the number is negative, the first bit is 1. Negative numbers are represented as the two's complement of the corresponding positive number. Figure 7.11 shows the signed two's-complement representation using eight bits. In this case, the range of numbers that can be represented runs from -128 to $+127$. Of course, if longer words are used, the range is extended.

Subtraction is performed by first finding the two's complement of the subtrahend and then adding in binary fashion and ignoring any carry out of the sign bit.

+127	01111111
	...
+2	00000010
+1	00000001
0	00000000
−1	11111111
−2	11111110
	...
−128	10000000

Sign bit (first bit)

Figure 7.11 Signed two's-complement representation using eight-bit words.

Example 7.7 Subtraction Using Two's-Complement Arithmetic

Perform the operation $29_{10} - 27_{10}$ by using eight-bit signed two's-complement arithmetic.

Solution First, we convert 29_{10} and 27_{10} to binary form. This yields

$$29_{10} = 00011101$$

and

$$27_{10} = 00011011$$

Next, we find the two's complement of the subtrahend:

$$-27_{10} = 11100101$$

Finally, we add the numbers to find the result:

$$\begin{array}{rrr} & 00011101 & 29 \\ & +\underline{11100101} & +\underline{(-27)} \\ \text{ignore carry out of sign bit} \rightarrow & 00000010 & 2 \end{array}$$

■

Of course, performing addition and subtraction in this manner is tedious for humans. However, computers excel at performing simple operations rapidly and accurately.

In performing two's-complement arithmetic, we must be aware of the possibility of **overflow** in which the result exceeds the maximum value that can be represented by the word length in use. For example, if we use eight-bit words to add

$$97_{10} = 01100001$$

and

$$63_{10} = 00111111$$

we obtain

$$\begin{array}{r} 01100001 \\ +\underline{00111111} \\ 10100000 \end{array}$$

The result is the signed two's-complement representation for −96, rather than the correct answer, which is 97 + 63 = 160. This error occurs because the signed two's-complement representation has a maximum value of +127 (assuming eight-bit words).

If the two numbers to be added have the same sign and the result has the opposite sign, overflow or underflow has occurred.

Similarly, **underflow** occurs if the result of an arithmetic operation is less than −128. Overflow and underflow are not possible if the two numbers to be added have opposite signs. If the two numbers to be added have the same sign and the result has the opposite sign, underflow or overflow has occurred.

Exercise 7.7 Find the eight-bit signed two's-complement representation of **a.** 22_{10} and **b.** -30_{10}.
Answer **a.** 00010110; **b.** 11100010. □

Exercise 7.8 Carry out $19_{10} - 4_{10}$ in eight-bit signed two's-complement form.
Answer

$$\begin{array}{rr} 19 & 00010011 \\ +\underline{(-4)} & +\underline{11111100} \\ 15 & 00001111 \end{array}$$

□

7.3 COMBINATORIAL LOGIC CIRCUITS

In this section, we consider circuits called **logic gates** that combine several logic-variable inputs to produce a logic-variable output. We focus on the external behavior of logic gates. Later, in Chapter 12, we will see how gate circuits can be implemented with field-effect transistors.

The circuits that we are about to discuss are said to be **memoryless** because their output values at a given instant depend only on the input values at that instant. Later, we consider logic circuits that are said to possess **memory**, because their present output values depend on previous, as well as present, input values.

AND Gate

An important logic function is called the AND operation. The AND operation on two logic variables, A and B, is represented as AB, read as "A and B." The AND operation is also called **logical multiplication**.

One way to specify a combinatorial logic system is to list all the possible combinations of the input variables and the corresponding output values. Such a listing is called a **truth table**. The truth table for the AND operation of two variables is shown in Figure 7.12(a). Notice that AB is 1 if and only if A and B are both 1.

For the AND operation, we can write the following relations:

$$AA = A \tag{7.1}$$

$$A1 = A \tag{7.2}$$

$$A0 = 0 \tag{7.3}$$

$$AB = BA \tag{7.4}$$

$$A(BC) = (AB)C = ABC \tag{7.5}$$

The circuit symbol for a two-input AND gate (i.e., a circuit that produces an output equal to the AND operation of the inputs) is shown in Figure 7.12(b).

It is possible to have AND gates with more than two inputs. For example, the truth table and circuit symbol for a three-input AND gate are shown in Figure 7.13.

A	B	$C = AB$
0	0	0
0	1	0
1	0	0
1	1	1

(a) Truth table

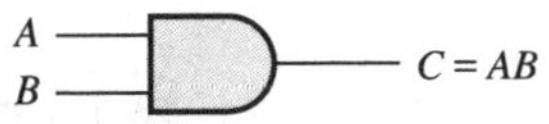

(b) Symbol for two-input AND gate

Figure 7.12 Two-input AND gate.

A	B	C	$D = ABC$
0	0	0	0
0	0	1	0
0	1	0	0
0	1	1	0
1	0	0	0
1	0	1	0
1	1	0	0
1	1	1	1

(a) Truth table

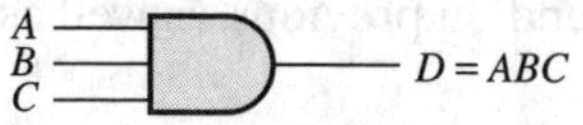

(b) Symbol for three-input AND gate

Figure 7.13 Three-input AND gate.

Logic Inverter

The NOT operation on a logic variable is represented by placing a bar over the symbol for the logic variable. The symbol $\overline{A}$ is read as "not A" or as "A inverse." If A is 0, $\overline{A}$ is 1, and vice versa.

Circuits that perform the NOT operation are called **inverters**. The truth table and circuit symbol for an inverter are shown in Figure 7.14. The *bubble* placed at the output of the inverter symbol is used to indicate inversion.

We can readily establish the following operations for the NOT operation:

A	$\overline{A}$
0	1
1	0

(a) Truth table

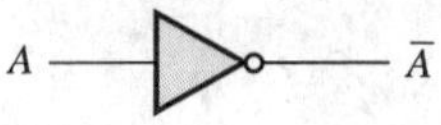

(b) Symbol for an inverter

Figure 7.14 Logical inverter.

$$A\overline{A} = 0 \tag{7.6}$$

$$\overline{\overline{A}} = A \tag{7.7}$$

OR Gate

The OR operation of logic variables is written as $A + B$, which is read as "A or B." The truth table and the circuit symbol for a two-input OR gate are shown in Figure 7.15. Notice that $A + B$ is 1 if A or B (or both) are 1. The OR operation is also called **logical addition**. The truth table and circuit symbol for a three-input OR gate are shown in Figure 7.16. For the OR operation, we can write

A	B	$C = A + B$
0	0	0
0	1	1
1	0	1
1	1	1

(a) Truth table

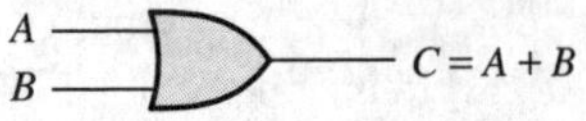

(b) Symbol for two-input OR gate

Figure 7.15 Two-input OR gate.

$$(A + B) + C = A + (B + C) = A + B + C \tag{7.8}$$

$$A(B + C) = AB + AC \tag{7.9}$$

$$A + 0 = A \tag{7.10}$$

$$A + 1 = 1 \tag{7.11}$$

A	B	C	$D = A + B + C$
0	0	0	0
0	0	1	1
0	1	0	1
0	1	1	1
1	0	0	1
1	0	1	1
1	1	0	1
1	1	1	1

(a) Truth table

(b) Circuit symbol

Figure 7.16 Three-input OR gate.

$$A + \overline{A} = 1 \tag{7.12}$$

$$A + A = A \tag{7.13}$$

Boolean Algebra

Equation 7.13 illustrates that even though we use the addition sign (+) to represent the OR operation, manipulation of logic variables by the AND, OR, and NOT operations is different from ordinary algebra. The mathematical theory of logic variables is called **Boolean algebra**, named for mathematician George Boole.

One way to prove a Boolean algebra identity is to produce a truth table that lists all possible combinations of the variables and to show that both sides of the expression yield the same results.

Example 7.8 Using a Truth Table to Prove a Boolean Expression

Prove the associative law for the OR operation (Equation 7.8), which states that

$$(A + B) + C = A + (B + C)$$

Solution The truth table listing all possible combinations of the variables and the values of both sides of Equation 7.8 is shown in Table 7.2. We can see from the truth table that $A+(B+C)$ and $(A+B)+C$ take the same logic values for all combinations of A, B, and C. Because both expressions yield the same results, the parentheses are not necessary, and we can write

$$A + (B + C) = (A + B) + C = A + B + C$$ ■

Exercise 7.9 Use truth tables to prove Equations 7.5 and 7.9.

Answer See Tables 7.3 and 7.4. □

Exercise 7.10 Prepare a truth table for the logic expression $D = AB + C$.

Answer See Table 7.5. □

Table 7.2. Truth Table Used to Prove the Associative Law for the OR Operation (Equation 7.8)

A	B	C	$(A+B)$	$(B+C)$	$A+(B+C)$	$(A+B)+C$	$A+B+C$
0	0	0	0	0	0	0	0
0	0	1	0	1	1	1	1
0	1	0	1	1	1	1	1
0	1	1	1	1	1	1	1
1	0	0	1	0	1	1	1
1	0	1	1	1	1	1	1
1	1	0	1	1	1	1	1
1	1	1	1	1	1	1	1

Table 7.3. Truth Table Used to Prove That $A(BC) = (AB)C$ (Equation 7.5)

A	B	C	(AB)	(BC)	$(AB)C$	$A(BC)$
0	0	0	0	0	0	0
0	0	1	0	0	0	0
0	1	0	0	0	0	0
0	1	1	0	1	0	0
1	0	0	0	0	0	0
1	0	1	0	0	0	0
1	1	0	1	0	0	0
1	1	1	1	1	1	1

Table 7.4. Truth Table Used to Prove That $A(B+C) = AB+AC$ (Equation 7.9)

A	B	C	$(B+C)$	AB	AC	$AB+AC$	$A(B+C)$
0	0	0	0	0	0	0	0
0	0	1	1	0	0	0	0
0	1	0	1	0	0	0	0
0	1	1	1	0	0	0	0
1	0	0	0	0	0	0	0
1	0	1	1	0	1	1	1
1	1	0	1	1	0	1	1
1	1	1	1	1	1	1	1

Implementation of Boolean Expressions

Boolean algebra expressions can be implemented by interconnection of AND gates, OR gates, and inverters.

Boolean algebra expressions can be implemented by interconnection of AND gates, OR gates, and inverters. For example, the logic expression

$$F = A\overline{B}C + ABC + (C + D)(\overline{D} + E) \tag{7.14}$$

can be implemented by the logic circuit shown in Figure 7.17.

Table 7.5. Truth Table for $D = AB + C$

A	B	C	AB	$D = AB + C$
0	0	0	0	0
0	0	1	0	1
0	1	0	0	0
0	1	1	0	1
1	0	0	0	0
1	0	1	0	1
1	1	0	1	1
1	1	1	1	1

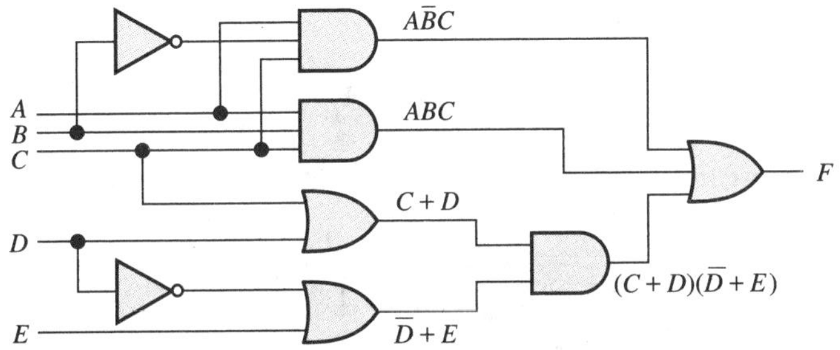

Figure 7.17 A circuit that implements the logic expression $F = A\overline{B}C + ABC + (C + D)(\overline{D} + E)$.

Sometimes, we can manipulate a logic expression to find an equivalent expression that is simpler. For example, the last term on the right-hand side of Equation 7.14 can be expanded, resulting in

$$F = A\overline{B}C + ABC + C\overline{D} + CE + D\overline{D} + DE \tag{7.15}$$

But the term $D\overline{D}$ always has the logic value 0, so it can be dropped from the expression. Factoring the first two terms on the right-hand side of Equation 7.15 results in

$$F = AC(\overline{B} + B) + C\overline{D} + CE + DE \tag{7.16}$$

However, the quantity $\overline{B} + B$ always equals 1, so we can write

$$F = AC + C\overline{D} + CE + DE \tag{7.17}$$

Factoring C from the first three terms on the right-hand side, we have

$$F = C(A + \overline{D} + E) + DE \tag{7.18}$$

This can be implemented as shown in Figure 7.18.

Thus, we can often find alternative implementations for a given logic function. Later, we consider methods for finding the implementation using the fewest gates of a given type.

We can often find alternative implementations for a given logic function.

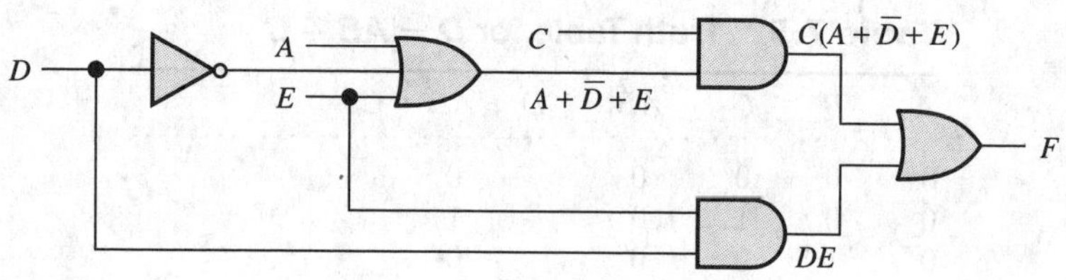

Figure 7.18 A simpler circuit equivalent to that of Figure 7.17.

De Morgan's Laws

Two important results in Boolean algebra are De Morgan's laws, which are given by

$$AB = \overline{\overline{A}+\overline{B}} \tag{7.19}$$

and

$$A+B = \overline{\overline{A}\,\overline{B}} \tag{7.20}$$

De Morgan's laws can be extended to three variables as follows:

$$ABC = \overline{\overline{A}+\overline{B}+\overline{C}} \quad \text{and} \quad A+B+C = \overline{\overline{A}\,\overline{B}\,\overline{C}}$$

Another way to state these laws is as follows: If the variables in a logic expression are replaced by their inverses, the AND operation is replaced by OR, the OR operation is replaced by AND, and the entire expression is inverted, the resulting logic expression yields the same values as before the changes.

Thus, De Morgan's laws can be used to change any (or all) of the AND operations in a logic expression to OR operations and vice versa. By changing various combinations of the operations, a variety of equivalent expressions can be found.

Example 7.9 Applying De Morgan's Laws

Apply De Morgan's laws, changing all of the ORs to ANDs and all of the ANDs to ORs, to the right-hand side of the logic expression:

$$D = AC + \overline{B}C + \overline{A}(\overline{B} + BC)$$

Solution First, we replace each variable by its inverse, resulting in the expression

$$\overline{A}\,\overline{C} + B\overline{C} + A(B + \overline{B}\,\overline{C})$$

Then, we replace the AND operation by OR, and vice versa:

$$(\overline{A}+\overline{C})(B+\overline{C})[A+B(\overline{B}+\overline{C})]$$

Finally, inverting the expression, we can write

$$D = \overline{(\overline{A}+\overline{C})(B+\overline{C})[A+B(\overline{B}+\overline{C})]}$$

Therefore, De Morgan's laws give us alternative ways to write logic expressions. ∎

Exercise 7.11 Use De Morgan's laws to find alternative expressions for

$$D = AB + \overline{B}C$$

and

$$E = \overline{\left[F(G + \overline{H}) + F\overline{G}\right]}$$

Answer

$$D = \overline{(\overline{A} + \overline{B})(B + \overline{C})}$$

$$E = (\overline{F} + \overline{G}H)(\overline{F} + G) \quad \square$$

An important implication of De Morgan's laws is that *we can implement any logic function by using AND gates and inverters*. This is true because Equation 7.20 can be employed to replace the OR operation by the AND operation (and logical inversions).

Any combinatorial logic function can be implemented solely with AND gates and inverters.

Similarly, *any logic function can be implemented with OR gates and inverters*, because Equation 7.19 can be used to replace the AND operation with the OR operation (and logical inversions). Consequently, to implement a logic function, we need inverters and either AND gates or OR gates, not both.

Any combinatorial logic function can be implemented solely with OR gates and inverters.

NAND, NOR, and XOR Gates

Some additional logic gates are shown in Figure 7.19. The NAND gate is equivalent to an AND gate followed by an inverter. Notice that the symbol is the same as for an AND gate, with a bubble at the output terminal to indicate that the output has been inverted after the AND operation. Similarly, the NOR gate is equivalent to an OR gate followed by an inverter.

The exclusive-OR (XOR) operation for two logic variables A and B is represented by $A \oplus B$ and is defined by

$$0 \oplus 0 = 0$$
$$1 \oplus 0 = 1$$
$$0 \oplus 1 = 1$$
$$1 \oplus 1 = 0$$

Notice that the XOR operation yields 1 if A is 1 or if B is 1, but yields 0 if both A and B are 1. The XOR operation is also known as **modulo-two addition**.

A **buffer** has a single input and produces an output with the same value as the input. (Buffers are commonly used to provide large currents when a logic signal must be applied to a low-impedance load.)

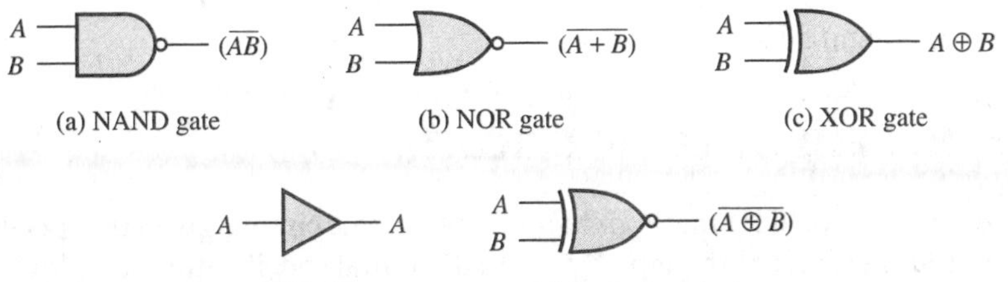

(a) NAND gate (b) NOR gate (c) XOR gate

(d) Buffer (e) Equivalence gate

Figure 7.19 Additional logic-gate symbols.

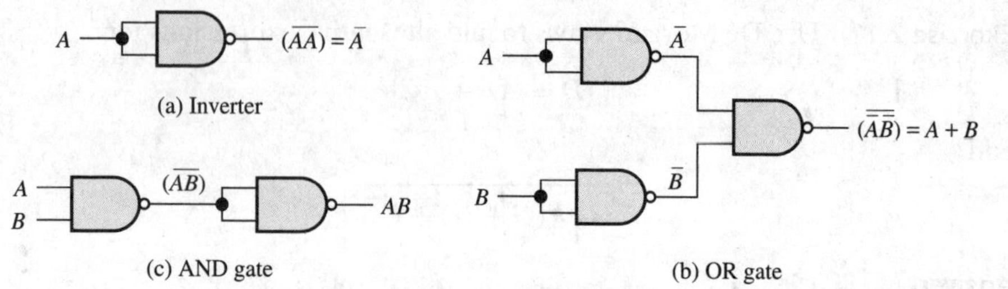

Figure 7.20 Basic Boolean operations can be implemented with NAND gates. Therefore, any Boolean function can be implemented by the use of NAND gates alone.

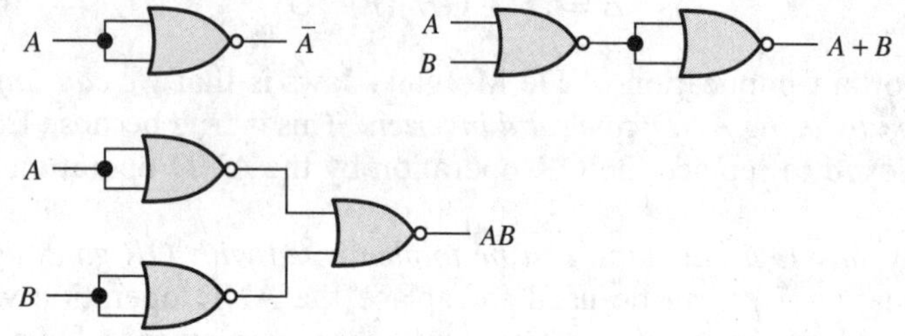

Figure 7.21 The AND, OR, and NOT operations can be implemented with NOR gates. Thus, any combinatorial logic circuit can be designed by using only NOR gates. See Exercise 7.12.

The **equivalence gate** produces a high output only if both inputs have the same value. In effect, it is an XOR followed by an inverter as the symbol of Figure 7.19(e) implies.

Logical Sufficiency of NAND Gates or of NOR Gates

As we have seen, several combinations of gates often can be found that perform the same function. For example, if the inputs to a NAND are tied together, an inverter results. This is true because

$$\overline{(AA)} = \bar{A}$$

which is illustrated in Figure 7.20(a).

Any combinatorial logic function can be implemented solely with NAND gates.

Furthermore, as shown by De Morgan's laws, the OR operation can be realized by inverting the input variables and combining the results in a NAND gate. This is shown in Figure 7.20(b), in which the inverters are formed from NAND gates. Finally, a NAND followed by an inverter results in an AND gate. *Since the basic logic functions (AND, OR, and NOT) can be realized by using only NAND gates, we conclude that NAND gates are sufficient to realize any combinatorial logic function.*

Any combinatorial logic function can be implemented solely with NOR gates.

Exercise 7.12 Show how to use only NOR gates to realize the AND, OR, and NOT functions.

Answer See Figure 7.21. □

7.4 SYNTHESIS OF LOGIC CIRCUITS

In this section, we consider methods to implement logic circuits given the specification for the output in terms of the inputs. Often, the initial specification for a logic circuit is given in natural language. This is translated into a truth table or a Boolean logic expression that can be manipulated to find a practical implementation.

Table 7.6. Truth Table Used to Illustrate SOP and POS Logical Expressions

Row	A	B	C	D
0	0	0	0	1
1	0	0	1	0
2	0	1	0	1
3	0	1	1	0
4	1	0	0	0
5	1	0	1	0
6	1	1	0	1
7	1	1	1	1

Sum-of-Products Implementation

Consider the truth table shown in Table 7.6. A, B, and C are input logic variables, and D is the desired output. Notice that we have numbered the rows of the truth table with the decimal number corresponding to the binary number formed by ABC.

Suppose that we want to find a logic circuit that produces the output variable D. One way to write a logic expression for D is to concentrate on the rows of the truth table for which D is 1. In Table 7.6, these are the rows numbered 0, 2, 6, and 7. Then, we write a logical product of the input logic variables or their inverses that equals 1 for each of these rows. Each input variable or its inverse is included in each product. In writing the product for each row, we invert the logic variables that are 0 in that row. For example, the logical product $\overline{A}\,\overline{B}\,\overline{C}$ equals logic 1 only for row 0. Similarly, $\overline{A}B\overline{C}$ equals logic 1 only for row 2, $AB\overline{C}$ equals logic 1 only for row 6, and ABC equals 1 only for row 7. Product terms that include all of the input variables (or their inverses) are called **minterms**.

Finally, we write an expression for the output variable as a logical sum of minterms. For Table 7.6, it yields

$$D = \overline{A}\,\overline{B}\,\overline{C} + \overline{A}B\overline{C} + AB\overline{C} + ABC \tag{7.21}$$

This type of expression is called a **sum of products** (SOP). Following this procedure, we can always find an SOP expression for a logic output given the truth table. A logic circuit that implements Equation 7.21 directly is shown in Figure 7.22.

In a sum-of-products expression, we form a product of all the input variables (or their inverses) for each row of the truth table for which the result is logic 1. The output is the sum of these products.

A shorthand way to write an SOP is simply to list the row numbers of the truth table for which the output is logic 1. Thus, we can write

$$D = \sum m(0, 2, 6, 7) \tag{7.22}$$

in which m indicates that we are summing the minterms corresponding to the rows enumerated.

Product-of-Sums Implementation

Another way to write a logic expression for D is to concentrate on the rows of the truth table for which D is 0. For example, in Table 7.6, these are the rows numbered 1, 3, 4, and 5. Then, we write a logical sum that equals 0 for each of these rows. Each input variable or its inverse in included in each sum. In writing the sum for each row, we invert the logic variables that are 1 in that row. For example, the logical sum

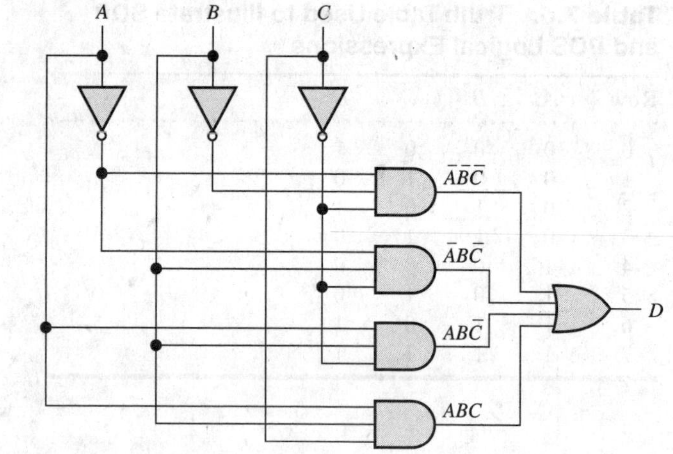

Figure 7.22 Sum-of-products logic circuit for Table 7.6.

$(A + B + \overline{C})$ equals logic 0 only for row 1. Similarly, $(A + \overline{B} + \overline{C})$ equals logic 0 only for row 3, $(\overline{A} + B + C)$ equals logic 0 only for row 4, and $(\overline{A} + B + \overline{C})$ equals 0 only for row 5. Sum terms that include all of the input variables (or their inverses) are called **maxterms**.

Finally, we write an expression for the output variable as the logical product of maxterms. For Table 7.6, it yields

$$D = (A + B + \overline{C})(A + \overline{B} + \overline{C})(\overline{A} + B + C)(\overline{A} + B + \overline{C}) \tag{7.23}$$

In a product-of-sums expression, we form a sum of all the input variables (or their inverses) for each row of the truth table for which the result is logic 0. The output is the product of these sums.

This type of expression is called a **product of sums** (POS). We can always find a POS expression for a logic output given the truth table. A circuit that implements Equation 7.23 is shown in Figure 7.23.

A shorthand way to write a POS is simply to list the row numbers of the truth table for which the output is logic 1. Thus, we write

$$D = \prod M(1, 3, 4, 5) \tag{7.24}$$

in which M indicates the maxterms corresponding to the rows enumerated.

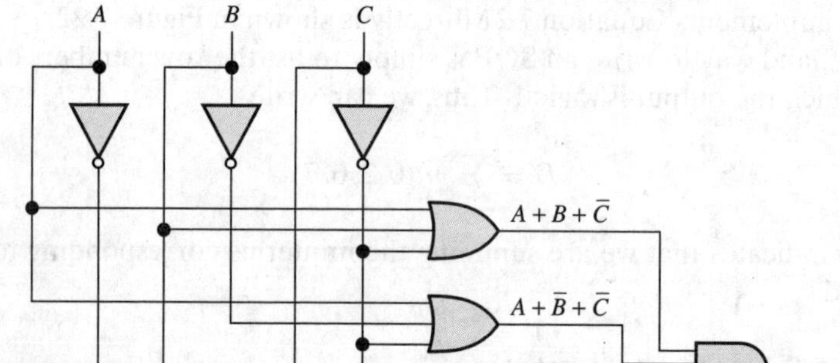

Figure 7.23 Product-of-sums logic circuit for Table 7.6.

D	H	L	F
0	0	0	1
0	0	1	0
0	1	0	×
0	1	1	0
1	0	0	1
1	0	1	1
1	1	0	×
1	1	1	0

These input combinations do not occur (rows 010 and 110); Don't cares (×)

(a) Truth table

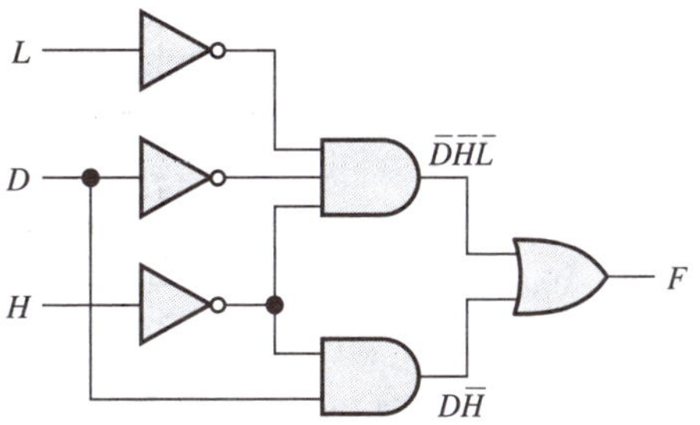

(b) Diagram of a possible implementation

Figure 7.24 See Example 7.10.

Example 7.10 Combinatorial Logic Circuit Design

The control logic for a residential heating system is to operate as follows: During the daytime, heating is required only if the temperature falls below 68°F. At night, heating is required only for temperatures below 62°F. Assume that logic signals D, L, and H are available. D is high during the daytime and low at night. H is high only if the temperature is above 68°F. L is high only if the temperature is above 62°F. Design a logic circuit that produces an output signal F that is high only when heating is required.

Some combinations of input variables may not occur; if so, the corresponding outputs are called "don't cares."

Solution First, we translate the description of the desired operation into a truth table. This is shown in Figure 7.24(a). We have listed all combinations of the inputs. However, two combinations do not occur because temperature cannot be below 62°F ($L = 0$) and also be above 68°F ($H = 1$). The output listed for these combinations is ×, which is called a **don't care** because we don't care what the output of the logic circuit is for these input combinations.

As we have seen, one way to translate the truth table into a logic expression is to write the sum of products in which there is a separate term for each high output. Applying this approach to Figure 7.24(a) yields

$$F = \overline{D}\,\overline{H}\,\overline{L} + D\overline{H}\,\overline{L} + D\overline{H}L \qquad (7.25)$$

The first term on the right-hand side $\overline{D}\,\overline{H}\,\overline{L}$ is high only for row 0 of the truth table. Also, the second term $D\overline{H}\,\overline{L}$ is high only for row 4, and the third term $D\overline{H}L$ is high only for row 5. Thus, the shorthand way to write the logic expression is

$$F = \sum m(0, 4, 5) \qquad (7.26)$$

Notice that in Equations 7.25 and 7.26, the don't cares turn out to be low.

The logic expression of Equation 7.25 can be manipulated into the form

$$F = D\overline{H} + \overline{D}\,\overline{H}\,\overline{L}$$

A logic diagram for this is shown in Figure 7.24(b).

An alternative approach is to write a product of sums with a separate sum term for each row of the truth table having an output value of 0. For the truth table of Figure 7.24(a), we have

$$F = (D + H + \overline{L})(D + \overline{H} + \overline{L})(\overline{D} + \overline{H} + \overline{L}) \qquad (7.27)$$

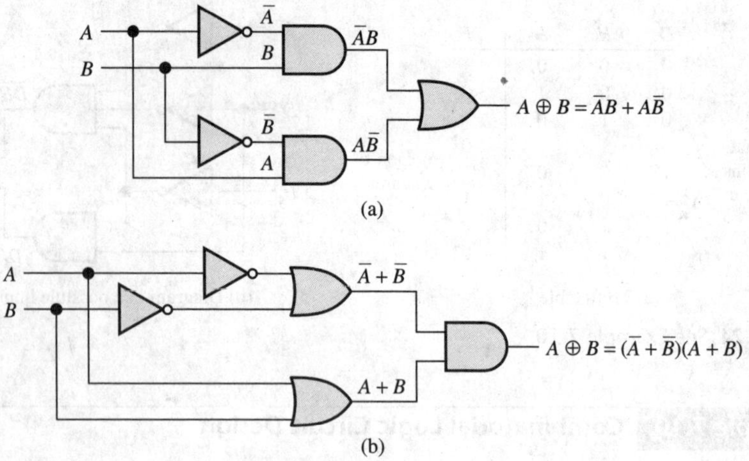

Figure 7.25 Answer for Exercise 7.13.

The first term in the product $(D + H + \overline{L})$ is low only for row 1 of the truth table. (Recall that row 1 is actually the second row because we start numbering with 0.) The second term $(D + \overline{H} + \overline{L})$ is low only for row 3, and the last term in the product is low only for the last row of the truth table.

In short form, we can write Equation 7.27 as

$$F = \prod M(1,3,7) \tag{7.28}$$

For Equations 7.27 and 7.28, the don't cares are high. Because of the different outputs for the don't cares, the expressions we have given for F in Equations 7.25 and 7.27 are not equivalent. ■

Exercise 7.13 Show two ways to realize the exclusive-OR operation by using AND, OR, and NOT gates.

Answer See Figure 7.25. □

Exercise 7.14 A traditional children's riddle concerns a farmer who is traveling with a sack of rye, a goose, and a mischievous dog. The farmer comes to a river that he must cross from east to west. A boat is available, but it only has room for the farmer and one of his possessions. If the farmer is not present, the goose will eat the rye or the dog will eat the goose.

We wish to design a circuit to emulate the conditions of this riddle. A separate switch is provided for the farmer, the rye, the goose, and the dog. Each switch has two positions depending on whether the corresponding object is on the east bank or the west bank of the river. The rules of play stipulate that no more than two switches be moved at a time and that the farmer must move (to row the boat) each time switches are moved. The switch for the farmer provides logic signal F, which is high if the farmer is on the east bank and low if he is on the west bank. Similar logic signals (G for the goose, D for the dog, and R for the rye) are high if the corresponding object is on the east bank and low if it is on the west bank.

Find a Boolean logic expression based on the sum-of-products approach for a logic signal A (alarm) that is high anytime the rye or the goose is in danger of being eaten. Repeat for the product-of-sums approach.

Table 7.7. Truth Table for Exercise 7.14

F	D	G	R	A
0	0	0	0	0
0	0	0	1	0
0	0	1	0	0
0	0	1	1	1
0	1	0	0	0
0	1	0	1	0
0	1	1	0	1
0	1	1	1	1
1	0	0	0	1
1	0	0	1	1
1	0	1	0	0
1	0	1	1	0
1	1	0	0	1
1	1	0	1	0
1	1	1	0	0
1	1	1	1	0

Answer The truth table is shown in Table 7.7. The Boolean expressions are

$$A = \sum m(3,6,7,8,9,12) = \overline{F}\,\overline{D}GR + \overline{F}DG\overline{R} + \overline{F}DGR$$
$$+ F\overline{D}\,\overline{G}\,\overline{R} + F\overline{D}\,\overline{G}R + FD\overline{G}\,\overline{R}$$

and

$$A = \prod M(0,1,2,4,5,10,11,13,14,15)$$ □

Decoders, Encoders, and Translators

Many useful combinatorial circuits known as **decoders**, **encoders**, or **translators** are available as integrated circuits. We discuss two examples. In a calculator or watch, we may represent information to be displayed in binary-coded decimal (BCD) form. Thus, 0000 is for 0, 0001 is for 1, 0010 is for 2, 0011 is for 3, and so on. Using four-bit words, 16 combinations are possible. However, only 10 combinations are used in BCD. Codes such as 1010 and 1011 do not occur in BCD.

The calculator display typically consists of liquid crystals with seven segments, as illustrated in Figure 7.26(a). The digits 0 through 9 are displayed by turning on appropriate segments as shown in Figure 7.26(b). Thus, a decoder is needed to translate the four-bit binary-coded decimal words into seven-bit words of the form $ABCDEFG$, for which A is high if segment A of the display is required to be on, B is high if segment B is required to be on, and so on. Thus, 0000 is translated to 1111110 because all segments except G are on to display the symbol for zero. Similarly, 0001 becomes 0110000, and 0010 becomes 1101101. Hence, the BCD-to-seven-segment decoder is a combinatorial circuit having four inputs and seven outputs.

Another example is the three-to-eight-line decoder that has a three-bit input and eight output lines. The three-bit input word selects one of the output lines and

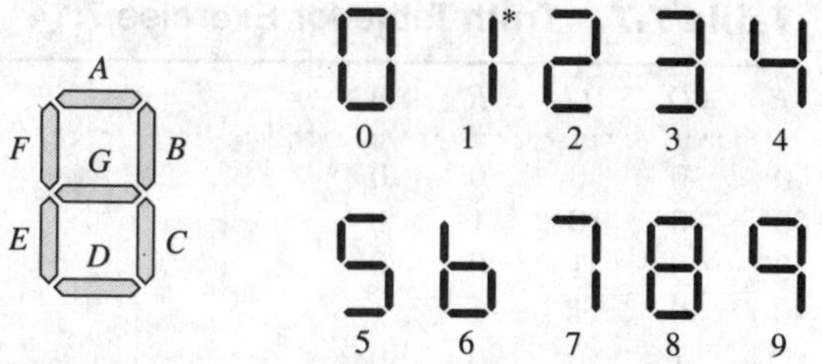

(a) Segment designations

(b) Displays for digits 0 through 9

*Segments B and C are on

Figure 7.26 Seven-segment display.

C	B	A	Y_0	Y_1	Y_2	Y_3	Y_4	Y_5	Y_6	Y_7
0	0	0	1	0	0	0	0	0	0	0
0	0	1	0	1	0	0	0	0	0	0
0	1	0	0	0	1	0	0	0	0	0
0	1	1	0	0	0	1	0	0	0	0
1	0	0	0	0	0	0	1	0	0	0
1	0	1	0	0	0	0	0	1	0	0
1	1	0	0	0	0	0	0	0	1	0
1	1	1	0	0	0	0	0	0	0	1

(a) Truth table

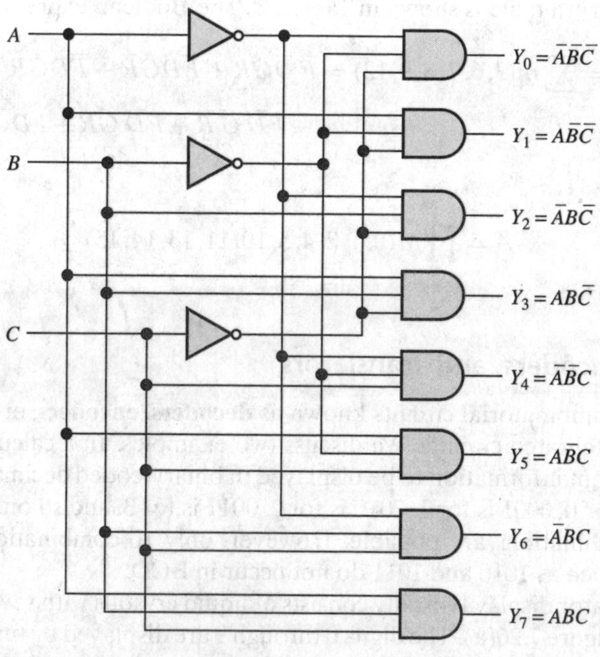

(b) Logic diagram

Figure 7.27 Three-line-to-eight-line decoder.

that output becomes high. The truth table and a circuit implementation are shown in Figure 7.27.

Decoders are available that are used to convert binary numbers to BCD, or vice versa, test whether one number is larger or smaller than another, perform arithmetic operations on binary or BCD numbers, and many similar functions.

7.5 MINIMIZATION OF LOGIC CIRCUITS

Implementations based on either a sum of minterms or a product of maxterms may not be optimum in minimizing the number of gates needed to realize a logic function.

We have seen that logic functions can be readily expressed either as a logical sum of minterms or as a logical product of maxterms. However, direct implementation of either of these expressions may not yield the best circuit in terms of minimizing the number of gates required. For example, consider the logical expression

$$F = \overline{A}\,\overline{B}D + \overline{A}BD + BCD + ABC \tag{7.29}$$

Implemented directly, this expression would require two inverters, four AND gates, and one OR gate.

Factoring the first pair of terms, we have

$$F = \overline{A}D(\overline{B} + B) + BCD + ABC$$

However, $\overline{B} + B = 1$, so we obtain

$$F = \overline{A}D + BCD + ABC$$

Of course, $BCD = 1$ only if $B = 1$, $C = 1$, and $D = 1$. In that case, either $\overline{A}D = 1$ or $ABC = 1$, because we must have either $\overline{A} = 1$ or $A = 1$. Thus, the term BCD is redundant and can be dropped from the expression. Then, we get

$$F = \overline{A}D + ABC \tag{7.30}$$

Only one inverter, two AND gates, and one OR gate are required to implement this expression.

Exercise 7.15 Create a truth table to verify that the right-hand sides of Equations 7.29 and 7.30 yield the same result.

Answer See Table 7.8. □

Table 7.8. Answer for Exercise 7.15

A	B	C	D	F
0	0	0	0	0
0	0	0	1	1
0	0	1	0	0
0	0	1	1	1
0	1	0	0	0
0	1	0	1	1
0	1	1	0	0
0	1	1	1	1
1	0	0	0	0
1	0	0	1	0
1	0	1	0	0
1	0	1	1	0
1	1	0	0	0
1	1	0	1	0
1	1	1	0	1
1	1	1	1	1

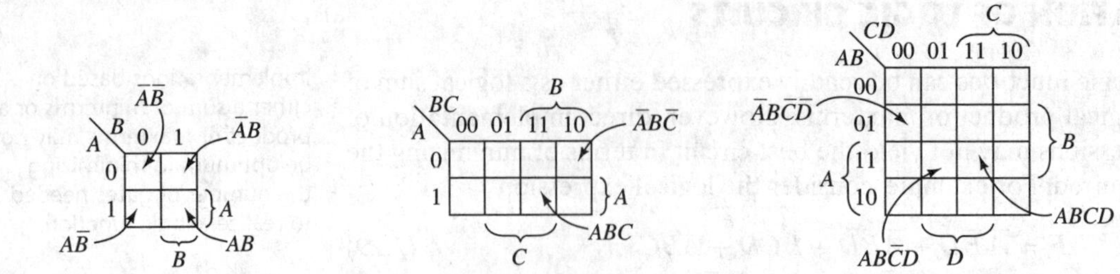

(a) Two-variable Karnaugh map (b) Three-variable Karnaugh map (c) Four-variable Karnaugh map

Figure 7.28 Karnaugh maps showing the minterms corresponding to some of the squares.

Karnaugh Maps

As we have demonstrated, logic expressions can sometimes be simplified dramatically. However, the algebraic manipulations needed to simplify a given expression are often not readily apparent. By using a graphical approach known as the **Karnaugh map**, we will find it much easier to minimize the number of terms in a logic expression.

A Karnaugh map is a rectangular array of squares, where each square represents one of the minterms of the logic variables

A Karnaugh map is an array of squares. Each square corresponds to one of the minterms of the logic variables or, equivalently, to one of the rows of the truth table. Karnaugh maps for two, three, and four variables are shown in Figure 7.28. The two-variable map consists of four squares, one corresponding to each of the minterms. Similarly, the three-variable map has eight squares, and the four-variable map has 16 squares.

The minterms corresponding to some of the squares are shown in Figure 7.28. For example, for the three-variable map, the minterm $\overline{A}B\overline{C}$ corresponds to the upper right-hand square. Also, the bit combinations corresponding to the rows of the truth table are shown down the left-hand side and across the top of the map. For example, on the four-variable map, the row of the truth table for which the four-bit word $ABCD$ is 1101 corresponds to the square in the third row (i.e., the row labeled 11) and second column (i.e., the column labeled 01). Thus, we can readily find the square corresponding to any minterm or to any row of the truth table.

In case you are wondering about the order of the bit patterns along the side or top of the four-variable Karnaugh map (i.e., 00 01 11 10), notice that this is a two-bit Gray code. Thus, the patterns for squares with a common side differ in only one bit, so that similar minterms are grouped together. For example, the minterms containing A (rather than $\overline{A}$) fall in the bottom half of each map. In the four-variable map, the minterms containing B are in the middle two rows, the minterms containing AB are in the third row, and so forth. This grouping of similar terms is the key to simplifying logic circuits.

Exercise 7.16 **a.** Write the minterm corresponding to the upper right-hand square in Figure 7.28(c). **b.** Write the minterm corresponding to the lower left-hand square.

Answer **a.** $\overline{A}\,\overline{B}C\overline{D}$; **b.** $A\overline{B}\,\overline{C}\,\overline{D}$. □

The left and right (as well as top and bottom) edges of the Karnaugh map are considered to be adjacent.

We call two squares that have a common edge a **2-cube**. Similarly, four squares with common edges are called a **4-cube**. In locating cubes, the maps should be considered to fold around from top to bottom and from left to right. Therefore, the squares on the right-hand side are considered to be adjacent to those on the left-hand side,

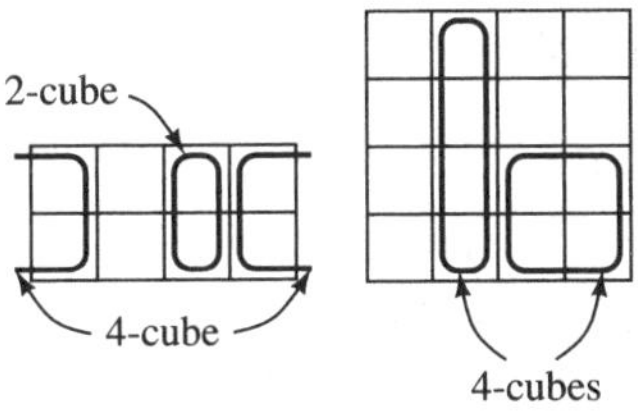

Figure 7.29 Karnaugh maps illustrating cubes.

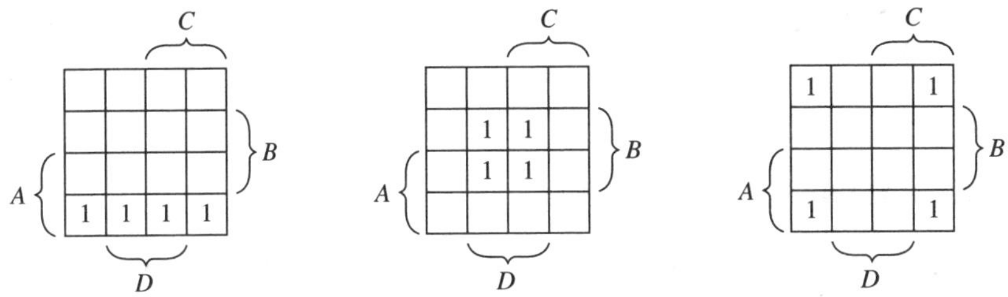

(a) Map of $A\overline{B}$ (b) Map of BD (c) Map of $\overline{B}\,\overline{D}$

Figure 7.30 Products of two variables map into 4-cubes on a 4-variable Karnaugh map.

and the top of the map is adjacent to the bottom. Consequently, the four squares in the map corners form a 4-cube. Some cubes are illustrated in Figure 7.29.

Rectangular arrays (known as cubes) of adjacent squares in a Karnaugh map represent products of logic variables or their inverses.

To map a logic function, we place 1s in the squares for which the logic function takes a value of 1. Product terms map 1s into cubes. For example, some product terms are mapped in Figure 7.30.

In a four-variable map consisting of 16 squares, a single logic variable or its inverse covers (maps into) an 8-cube. A product of two variables (such as AB or $A\overline{B}$) covers a 4-cube. A product of three variables maps into a 2-cube.

The Karnaugh map of the logic function

$$F = \overline{A}\,\overline{B}\,\overline{C}D + \overline{A}\,\overline{B}CD + \overline{A}B\overline{C}D + \overline{A}BCD + ABC\overline{D} + ABCD \qquad (7.31)$$

By finding the fewest and largest (possibly overlapping) cubes for the region in which the logic expression is one, we obtain the minimum SOP for the logic expression.

is shown in Figure 7.31. The squares containing 1s form a 4-cube corresponding to the product term $\overline{A}D$ plus a 2-cube corresponding to ABC. These are the largest cubes that cover the 1s in the map. Thus, the minimum SOP expression for F is

$$F = \overline{A}D + ABC \qquad (7.32)$$

Because it is relatively easy to spot the set of largest cubes that cover the 1s in a Karnaugh map, we can quickly minimize a logic function.

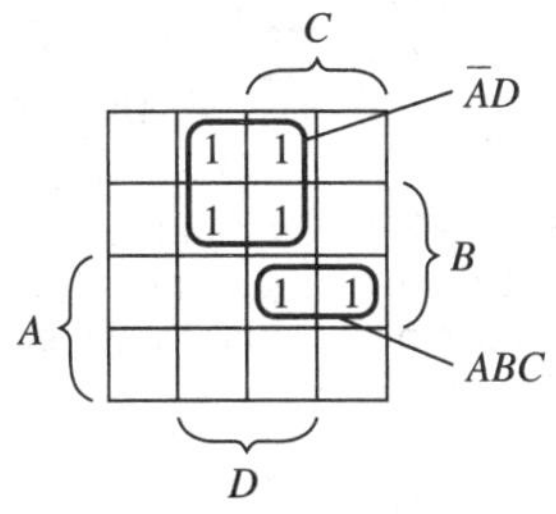

Figure 7.31 Karnaugh map for the logic function of Equation 7.31. From the map, it is evident that $F = \overline{A}D + ABC$.

Example 7.11 Finding the Minimum SOP Form for a Logic Function

A logic circuit has inputs $A, B, C,$ and D. The output of the circuit is given by

$$E = \sum m(1, 3, 4, 5, 7, 10, 12, 13)$$

Find the minimum SOP form for E.

Solution First, we construct the Karnaugh map. Because there are four input variables, the map contains 16 squares as shown in Figure 7.32. Converting the numbers

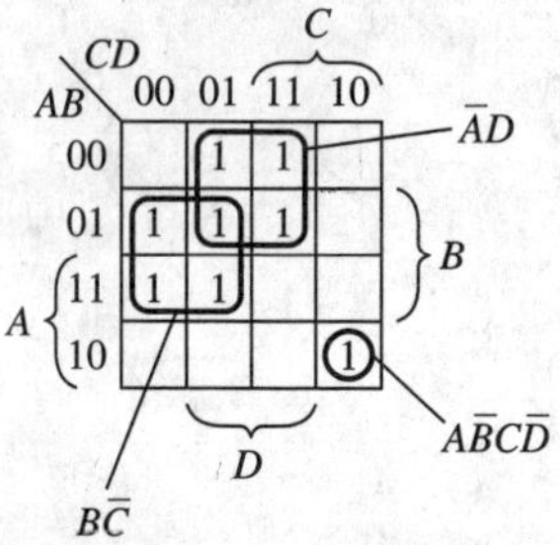

Figure 7.32 Karnaugh map for Example 7.11.

of the minterms to binary numbers, we obtain 0001, 0011, 0100, 0101, 0111, 1010, 1100, and 1101. Each of these locates a square on the map. For example, 1101 locates the square in the third row and second column, 0011 is the square in the first row and third column, and so forth. Placing a 1 in the square corresponding to each minterm results in the map shown in Figure 7.32.

Now we look for the smallest number of the largest size cubes that cover the ones in the map. To cover the ones in this map, we need two 4-cubes and a 1-cube (i.e., a single isolated square) as illustrated in the figure. Finally, the minimum SOP expression is

$$E = \overline{A}D + B\overline{C} + A\overline{B}C\overline{D}$$ ■

Minimum POS Forms

So far, we have concentrated on finding minimum SOP implementations for logic circuits. However, we can easily extend the methods to finding minimal POS circuits by following these steps:

1. Create the Karnaugh map for the desired output.
2. Invert the Karnaugh map, replacing 1s by 0s and vice versa.
3. Look for the least number of cubes of the largest sizes that cover the ones in the inverted map. Then, write the minimum SOP expression for the inverse of desired output.
4. Apply De Morgan's laws to convert the SOP expression to a POS expression. We illustrate with an example.

Example 7.12 Finding the Minimum POS Form for a Logic Function

Find the minimum POS for the logic variable E of Example 7.11.

Solution The Karnaugh map for E is shown in Figure 7.32. The map for $\overline{E}$ is obtained by replacing 1s with 0s (blank squares) and vice versa. The result is shown in Figure 7.33.

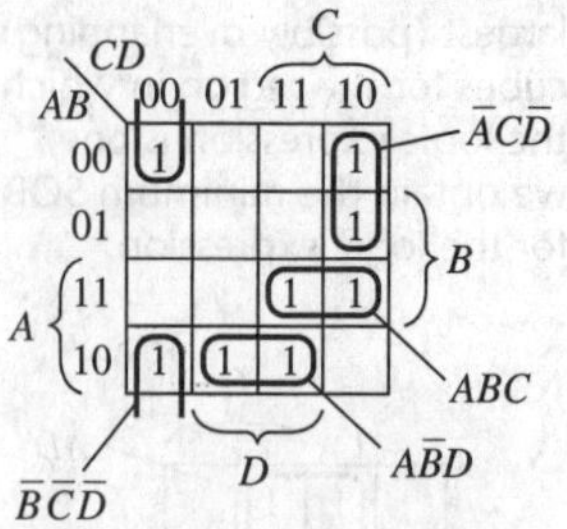

Figure 7.33 Karnaugh map of Example 7.12. (This is the inverse of the map shown in Figure 7.32.)

Now, we look for the smallest number of the largest size cubes that cover the ones in the map. Clearly, there are no 8-cubes or 4-cubes contained in Figure 7.33. A total of eight 1s appear in the map. Thus, the best we can do is to cover the map with four 2-cubes. One option is the grouping shown in the figure, which yields

$$\overline{E} = ABC + A\overline{B}D + \overline{A}C\overline{D} + \overline{B}\,\overline{C}\,\overline{D}$$

Next, we apply De Morgan's laws to obtain a minimum POS form:

$$E = (\overline{A} + \overline{B} + \overline{C})(\overline{A} + B + \overline{D})(A + \overline{C} + D)(B + C + D)$$

Choosing a different grouping in Figure 7.33 produces another equally good form, which is

$$\overline{E} = A\overline{B}\,\overline{C} + \overline{A}\,\overline{B}\,\overline{D} + ACD + BC\overline{D}$$

Then, applying De Morgan's laws gives another minimum POS form:

$$E = (\overline{A} + B + C)(A + B + D)(\overline{A} + \overline{C} + \overline{D})(\overline{B} + \overline{C} + D)$$ ■

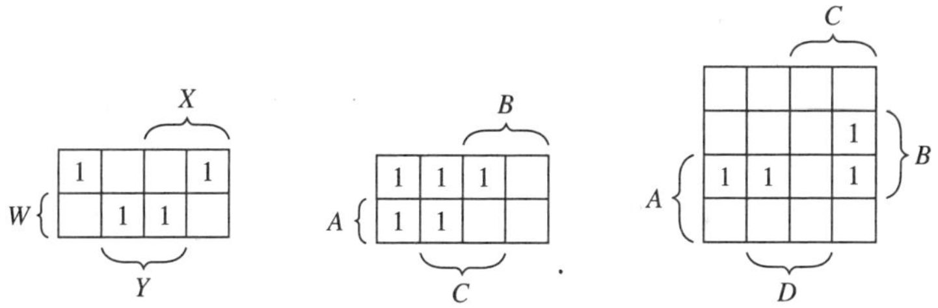

(a) $Z = \overline{W}\,\overline{Y} + WY$ (b) $D = \overline{B} + \overline{A}C$ (c) $E = AB\overline{C} + BC\overline{D}$

Figure 7.34 Answers for Exercise 7.17.

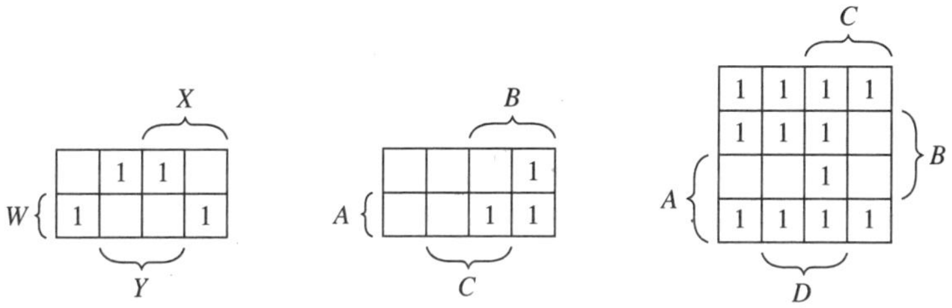

$\overline{Z} = \overline{W}Y + W\overline{Y}$ $\overline{D} = AB + B\overline{C}$ $\overline{E} = \overline{B} + \overline{A}\,\overline{C} + CD$

$Z = (W + \overline{Y})(\overline{W} + Y)$ $D = (\overline{A} + \overline{B})(\overline{B} + C)$ $E = B\,(A + C)(\overline{C} + \overline{D})$

(a) (b) (c)

Figure 7.35 Answers for Exercise 7.18.

Exercise 7.17 Construct the Karnaugh maps and find the minimum SOP expressions for each of these logic functions:

a. $Z = \overline{W}\,\overline{X}\,\overline{Y} + \overline{W}X\overline{Y} + W\overline{X}Y + WXY$

b. $D = \overline{A}\,\overline{B}\,\overline{C} + A\overline{B}\,\overline{C} + \overline{A}\,\overline{B}C + A\overline{B}C + \overline{A}BC$

c. $E = \overline{A}BC\overline{D} + AB\overline{C}\,\overline{D} + AB\overline{C}D + ABC\overline{D}$

Answer See Figure 7.34. □

Exercise 7.18 Construct the inverse maps and find the minimum POS expressions for each of the logic functions of Exercise 7.17.
Answer See Figure 7.35. □

7.6 SEQUENTIAL LOGIC CIRCUITS

So far, we have considered combinatorial logic circuits, such as gates, encoders, and decoders, for which the outputs at a given time depend only on the input values at that instant. In this section, we discuss **sequential logic circuits**, for which the outputs depend on past as well as present inputs. We say that such circuits have **memory** because they "remember" past input values.

Often, the operation of a sequential circuit is synchronized by a **clock signal** that consists of periodic logic-1 pulses, as shown in Figure 7.36. The clock signal regulates when the circuits respond to new inputs, so that operations occur in proper sequence. Sequential circuits that are regulated by a clock signal are said to be **synchronous**.

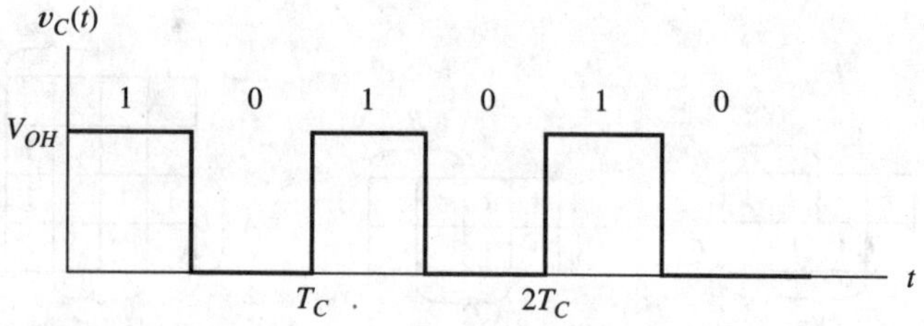

Figure 7.36 The clock signal consists of periodic logic-1 pulses.

Flip-Flops

One of the basic building blocks for sequential circuits is the **flip-flop**. A flip-flop has two stable operating states; therefore, it can store one bit of information. Many useful versions of flip-flops exist, differing in the manner that the clock signal and other input signals control the state of the flip-flop. We discuss several types shortly.

A simple flip-flop can be constructed by using two inverters, with the output of one connected to the input of the other, as shown in Figure 7.37. Two stable states are possible in the circuit. First, the output Q of the top inverter can be high and then the output of the bottom inverter is low. Thus, the output of the bottom inverter is labeled as $\overline{Q}$. Notice that Q high and $\overline{Q}$ low are consistent with the logic operation of the inverters, so the circuit can remain in that state. On the other hand, Q low and $\overline{Q}$ high are also consistent. The circuit can remain in either state indefinitely.

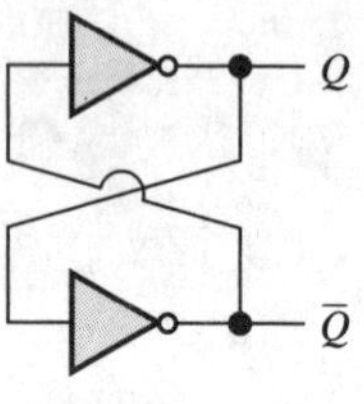

Figure 7.37 Simple flip-flop.

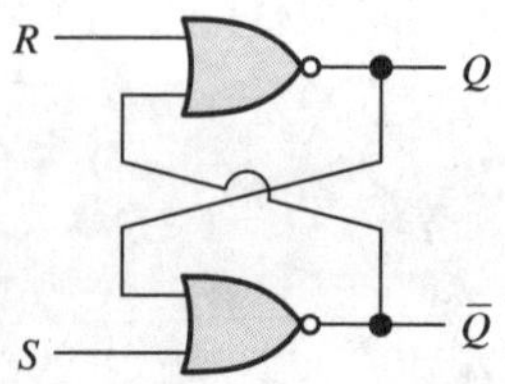

Figure 7.38 An SR flip-flop can be implemented by cross coupling two NOR gates.

***SR* Flip-Flop.** The simple two-inverter circuit of Figure 7.37 is not very useful because no provision exists for controlling its state. A more useful circuit is the **set–reset (*SR*) flip-flop**, consisting of two NOR gates, as shown in Figure 7.38. As long as the S and R inputs are low, the NOR gates act as inverters for the other input signal. Thus, with S and R both low, the SR flip-flop behaves just as the two-inverter circuit of Figure 7.37 does.

If S is high and R is low, $\overline{Q}$ is forced low and Q is high (or set). When S returns low, the flip-flop remains in the **set state** (i.e., Q stays high). On the other hand, if R becomes high and S low, Q is forced low. When R returns low, the flip-flop remains in the **reset state** (i.e., Q stays low). In normal operation, R and S are not allowed to be high at the same time. Thus, with R and S low, the SR flip-flop *remembers* which input (R or S) was high most recently.

We use subscripts on logic variables to indicate a sequence of states. For example, the flip-flop output state Q_{n-1} occurs before Q_n, which occurs before Q_{n+1}, and so on. The truth table for the SR flip-flop is shown in Figure 7.39(a). In the first row of the truth table, we see that if both R and S are logic 0, the output remains in the previous state ($Q_n = Q_{n-1}$). The symbol for the SR flip-flop is shown in Figure 7.39(b).

R	S	Q_n
0	0	Q_{n-1}
0	1	1
1	0	0
1	1	Not allowed

(a) Truth table

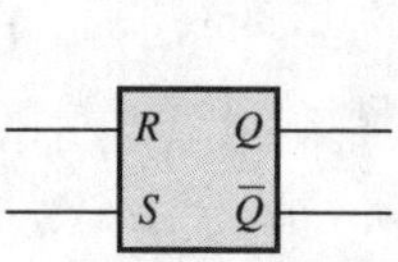

(b) Circuit symbol

Figure 7.39 The truth table and symbol for the SR flip-flop.

Using an *SR* Flip-Flop to Debounce a Switch. One application for the SR flip-flop is to *debounce* a switch. Consider the single-pole double-throw switch shown in Figure 7.40(a). When the switch is moved from position A to position B, the waveforms shown in Figure 7.40(b) typically result. At first, V_A is high because the switch is in position A. Then, the switch breaks contact, and V_A drops to zero. Next, the switch makes initial contact with B and V_B goes high. Contact bounce at B again causes V_B to drop to zero, then back high several times, until finally it ends up high. Later, when the switch is returned to A, contact bounce occurs again.

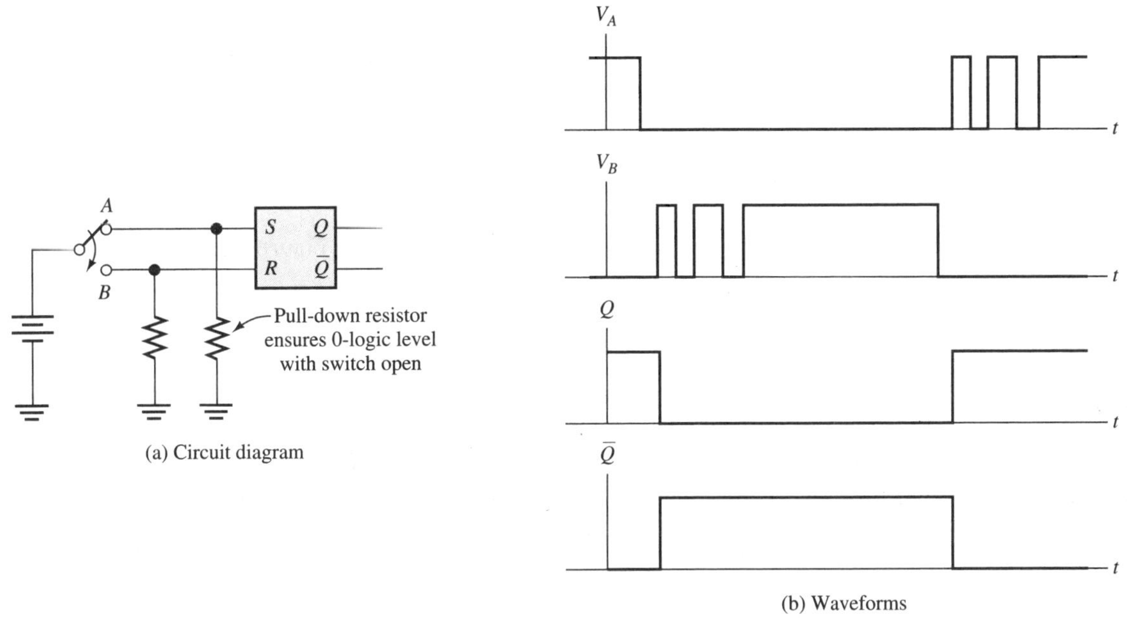

Figure 7.40 An *SR* flip-flop can be used to eliminate the effects of switch bounce.

This kind of behavior can be troublesome. For example, a computer keyboard consists of switches that are depressed to select a character. Contact bounce could cause several characters to be accepted by the computer or calculator each time a key is depressed.

An *SR* flip-flop can eliminate the effects of contact bounce. The switch voltages V_A and V_B are connected to the S and R inputs as shown in Figure 7.40(a). At first, when the switch is at position A, the flip-flop is in the set state, and Q is high. When contact is broken with A, V_A drops to zero, but the flip-flop does not change state until the first time V_B goes high. As contact bounce occurs, the flip-flop stays in the reset state with Q low. The waveforms for the flip-flop outputs Q and $\overline{Q}$ are shown in Figure 7.40(b).

Exercise 7.19 The waveforms present at the input terminals of an *SR* flip-flop are shown in Figure 7.41. Sketch the waveforms for Q versus time.
Answer See Figure 7.42. □

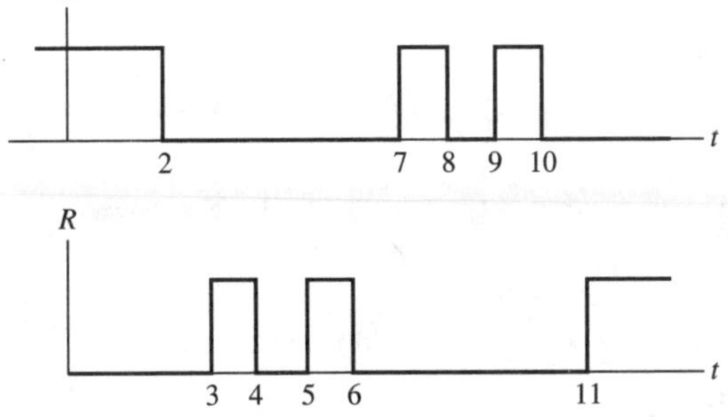

Figure 7.41 See Exercise 7.19.

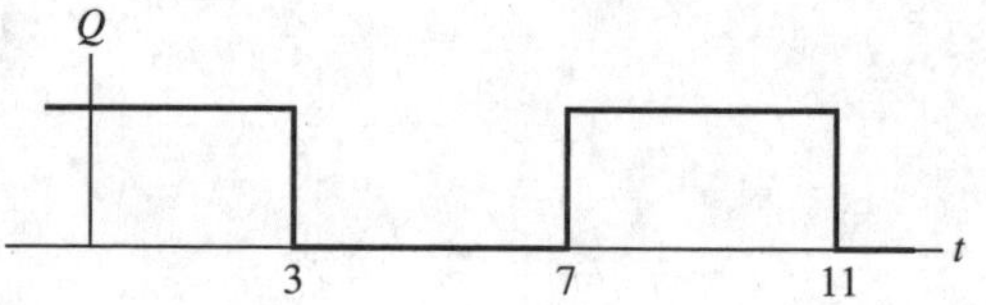

Figure 7.42 Answer for Exercise 7.19.

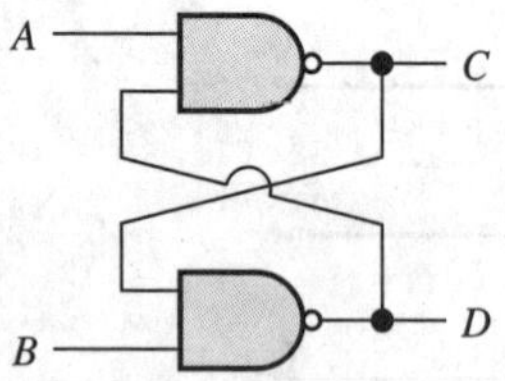

Figure 7.43 A flip-flop implemented with NAND gates. See Exercise 7.20.

Exercise 7.20 Prepare a truth table similar to that of Figure 7.39(a) for the circuit of Figure 7.43.

Answer See Table 7.9. □

Clocked *SR* Flip-Flop. Often, it is advantageous to control the point in time that a flip-flop responds to its inputs. This is accomplished with the **clocked *SR* flip-flop** shown in Figure 7.44. Two AND gates have been added at the inputs of an *SR* flip-flop. If the clock signal C is low, the inputs to the *SR* flip-flop are both low, and the state cannot change. The clock signal must be high for the R and S signals to be transmitted to the input of the *SR* flip-flop.

The truth table for the clocked *SR* flip-flop is shown in Figure 7.44(b), and the circuit symbol is shown in Figure 7.44(c). We say that a high clock level **enables** the inputs to the flip-flop. On the other hand, the low clock level **disables** the inputs.

Usually, we design digital systems so that R, S, and C are not all high at the same time. If all three signals are high and then C goes low, the state of the flip-flop settles either to $Q = 1$ or to $Q = 0$ unpredictably. Usually, systems that behave in an unpredictable manner are not useful.

Sometimes, a clocked *SR* flip-flop is needed, but it is also necessary to be able to set or clear the flip-flop state independent of the clock. A circuit having this feature is shown in Figure 7.45(a). If the **preset input** Pr is high, Q becomes high even if the clock is low. Similarly, the **clear input** Cl can force Q low. The Pr and Cl inputs are called **asynchronous inputs** because their effect is not synchronized by the clock

Table 7.9. Truth Table for Exercise 7.20

A	B	C_n	D_n
0	0	1	1
0	1	1	0
1	0	0	1
1	1	C_{n-1}	D_{n-1}

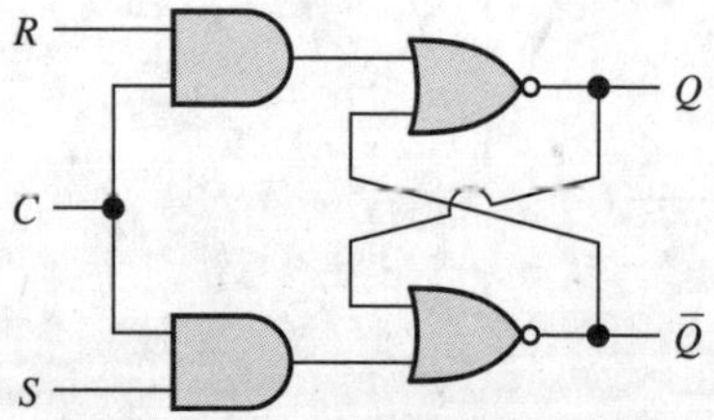

(a) Circuit diagram

R	S	C	Q_n
0	0	×	Q_{n-1}
0	1	1	1
1	0	1	0
1	1	1	Not allowed
×	×	0	Q_{n-1}

(b) Truth table

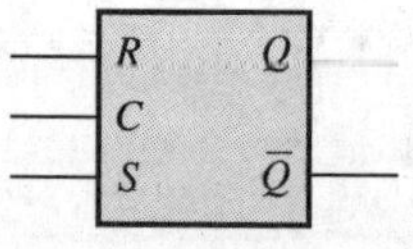

(c) Circuit symbol

Figure 7.44 A clocked *SR* flip-flop.

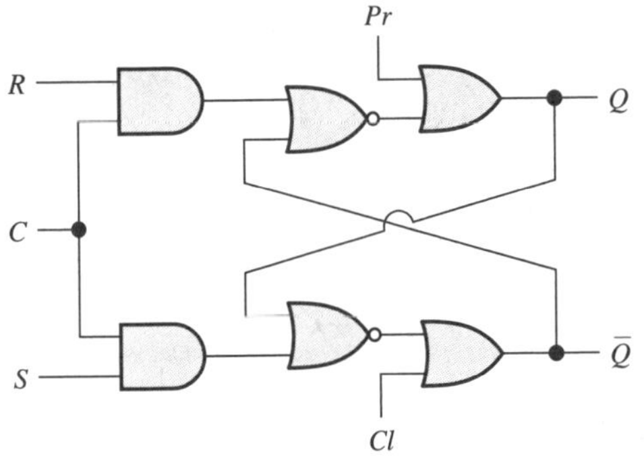

(a) Circuit diagram

Pr	Cl	R	S	C	Q_n
0	0	0	0	×	Q_{n-1}
0	0	0	1	1	1
0	0	1	0	1	0
×	×	1	1	1	Not allowed
0	1	×	×	×	0
1	0	×	×	×	1
1	1	×	×	×	Not allowed

(b) Truth table

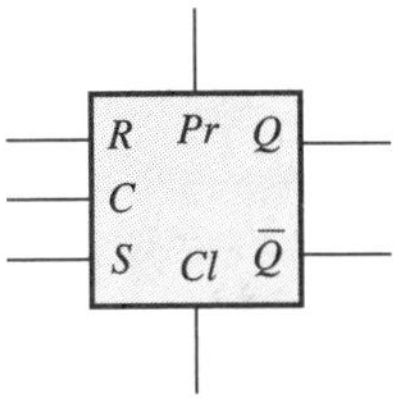

(c) Circuit symbol

Figure 7.45 A clocked *SR* flip-flop with asynchronous preset and clear inputs.

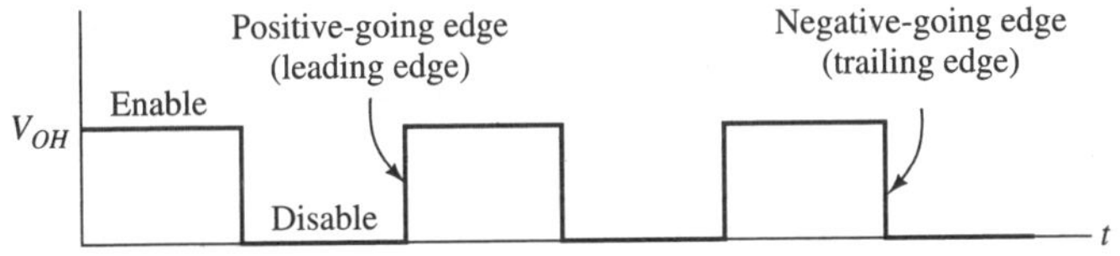

Figure 7.46 Clock signal.

signal. On the other hand, the *R* and *S* inputs are recognized only if the clock signal is high, and are therefore called **synchronous inputs.**

Edge-Triggered *D* Flip-Flop. So far, we have considered circuits for which the level of the clock signal *enables* or *disables* other input signals. On the other hand, **edge-triggered** circuits respond to their inputs only at a transition in the clock signal. If the clock signal is steady, either high or low, the inputs are disabled. At the clock transition, the flip-flop responds to the inputs present just prior to the transition. **Positive-edge-triggered** circuits respond when the clock signal switches from low to high. Conversely, **negative-edge-triggered** circuits respond on the transition from high to low. The positive-going edge of the clock is also called the **leading edge**, and the negative-going edge is called the **trailing edge**. A clock signal illustrating these points is shown in Figure 7.46. Thus, clocked flip-flops can be sensitive either to the level of the clock or to transitions.

An example of an edge-triggered circuit is the ***D* flip-flop**, which is also known as the **delay flip-flop**. Its output takes the value of the input that was present just prior to the triggering clock transition. The circuit symbol for the edge-triggered *D* flip-flop is shown in Figure 7.47(a). The "knife edge" symbol at the *C* input indicates that

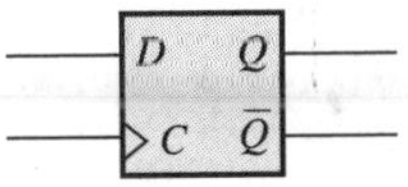

(a) Circuit symbol

C	D	Q_n
0	×	Q_{n-1}
1	×	Q_{n-1}
↑	0	0
↑	1	1

(b) Truth table
↑ indicates a transition from low to high

Figure 7.47 A positive-edge-triggered *D* flip-flop.

the flip-flop is edge triggered. The truth table for a positive-edge-triggered version is shown in Figure 7.47(b). Notice the symbols in the clock column of the truth table, indicating transitions of the clock signal from low to high.

Exercise 7.21 The input signals to a positive-edge-triggered D flip-flop are shown in Figure 7.48. Sketch the output Q to scale versus time. (Assume that Q is low prior to $t = 2$.)
Answer See Figure 7.49. □

***JK* Flip-Flop.** The circuit symbol and truth table for a negative-edge-triggered ***JK* flip-flop** are shown in Figure 7.50. Its operation is very similar to that of an SR flip-flop except that if both control inputs (J and K) are high, the state changes on the next negative-going clock edge. Thus when both J and K are high, the output of the flip-flop **toggles** on each cycle of the clock—switching from high to low on one negative-going clock transition, back to high on the next negative transition, and so on.

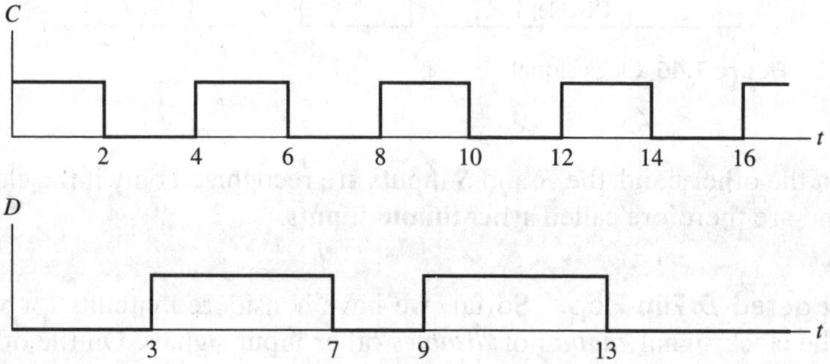

Figure 7.48 See Exercise 7.21.

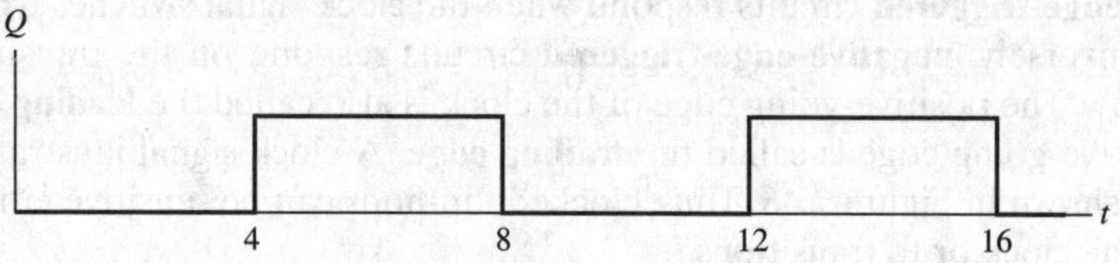

Figure 7.49 Answer for Exercise 7.21.

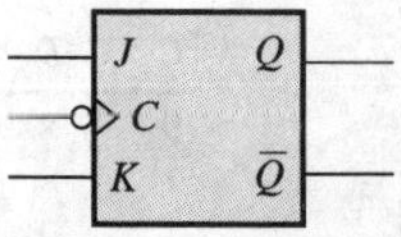

(a) Circuit symbol

C	J	K	Q_n	Comment
0	×	×	Q_{n-1}	Memory
1	×	×	Q_{n-1}	Memory
↓	0	0	Q_{n-1}	Memory
↓	0	1	0	Reset
↓	1	0	1	Set
↓	1	1	$\overline{Q}_{n-1}$	Toggle

(b) Truth table indicates a transition from low to high

Figure 7.50 Negative-edge-triggered JK flip-flop.

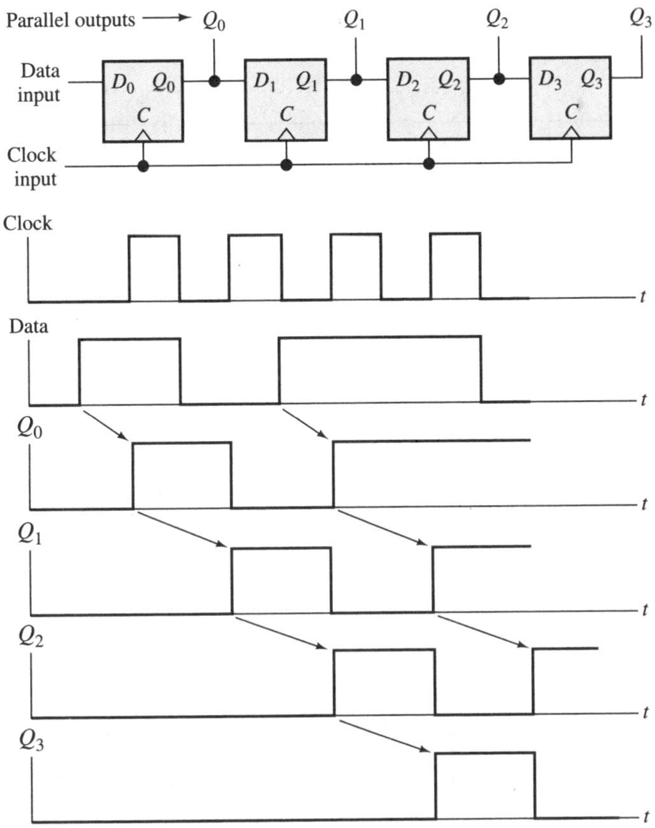

Figure 7.51 Serial-input parallel-output shift register.

Serial-In Parallel-Out Shift Register

A **register** is an array of flip-flops that is used to store or manipulate the bits of a digital word. For example, if we connect several positive-edge-triggered D flip-flops as shown in Figure 7.51, a **serial-in parallel-out shift register** results. As the name implies, the digital input word is shifted through the register moving one stage for each clock pulse.

The waveforms shown in Figure 7.51 illustrate the operation of the shift register. We assume that the flip-flops are initially ($t = 0$) all in the reset state ($Q_0 = Q_1 = Q_2 = Q_3 = 0$). The input data are applied to the input of the first stage serially (i.e., one bit after another). On the leading edge of the first clock pulse, the first data bit is transferred into the first stage. On the second clock pulse, the first bit is transferred to the second stage, and the second bit is transferred into the first stage. After four clock pulses, four bits of input data have been transferred into the shift register. Thus, serial data applied to the input are converted to parallel form available at the outputs of the stages of the shift register.

Parallel-In Serial-Out Shift Register

Sometimes, we have parallel data that we wish to transmit serially. Then, the **parallel-in serial-out shift register** shown in Figure 7.52 is useful. This register consists of four positive-edge-triggered D flip-flops with asynchronous preset and clear inputs. First, the register is cleared by applying a high pulse to the clear input. (The clear input

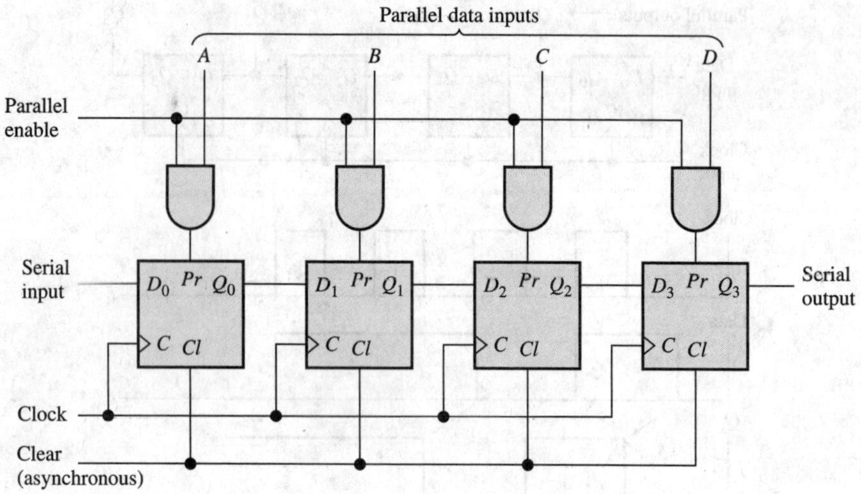

Figure 7.52 Parallel-input serial-output shift register.

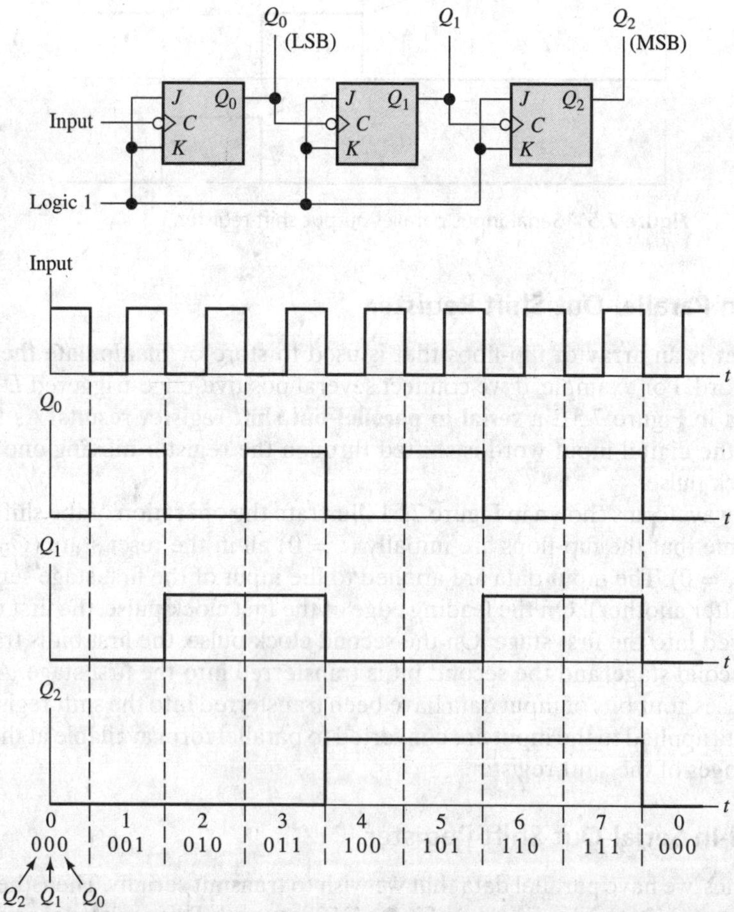

Figure 7.53 Ripple counter.

is asynchronous, so a clock pulse is not necessary to clear the register.) Parallel data are applied to the A, B, C, and D inputs. Then, a high pulse is applied to the parallel enable (PE) input. The result is to set each flip-flop for which the corresponding data line is high. Thus, four parallel bits are loaded into the stages of the register. Then, application of clock pulses produces the data in serial form at the output of the last stage.

Counters

Counters are used to count the pulses of an input signal. An example is the **ripple counter** shown in Figure 7.53. It consists of a cascade of JK flip-flops. Reference to Figure 7.50 shows that with the J and K inputs high, the Q-output of the flip-flop toggles on each falling edge of the clock input. The input pulses to be counted are connected to the clock input of the first stage, and the output of the first stage is connected to the clock input of the second stage.

Assume that the flip-flops are initially all in the reset state ($Q = 0$). When the falling edge of the first input pulse occurs, Q_0 changes to logic 1. On the falling edge of the second pulse, Q_0 toggles back to logic 0, and the resulting falling input to the second stage causes Q_1 to become high. As shown by the waveforms in Figure 7.53, after seven pulses, the shift register is in the 111 state. On the eighth pulse, the counter returns to the 000 state. Thus, we say that this is a modulo-8 or mod-8 counter.

PRACTICAL APPLICATION 7.1

Biomedical Engineering Application of Electronics: Cardiac Pacemaker

In certain types of heart disease, the biological signals that should stimulate the heart to beat are blocked from reaching the heart muscle. When this blockage occurs, the heart muscle may spontaneously beat at a very low rate, so death does not occur. However, the afflicted person is not able to function at a normal level of activity because of the low heart rate. The application of electrical pacemaker pulses to force beating at a higher rate is dramatically helpful in many of these cases.

Sometimes, the blockage of the natural pacemaking is not complete. In this case, the heart beats normally part of the time but experiences missed beats sporadically. A demand pacemaker can be useful for the patient with partial blockage. The demand pacemaker contains circuits that sense natural heartbeats and apply an electrical pulse to the heart muscle only if a beat does not occur within a predetermined interval. If natural beats are detected, no pulses are applied. This type of circuit is called a *demand* pacemaker because pulses are issued only when needed.

It turns out to be advantageous to the patient for the heart to beat naturally, provided that its natural rate is above some limit. On the other hand, if artificial pulses are required, a slightly higher rate is better. Typical values are a natural limit of 66.7 beats per minute (corresponding to 0.9 s between beats) and 75 beats per minute (corresponding to 0.8 s between beats) for forced pacing. Thus, in a typical situation, the circuit waits 0.9 s after a natural beat before applying a pacing pulse but waits only 0.8 s after an artificial pulse before applying another artificial pulse.

Another feature of the pacemaker is that it should ignore signals from the heart for a short period (about 0.4 s) after detection of a natural beat or after issuing a pacemaker pulse. This is because natural signals occur during the contraction and relaxation of the heart muscle. These signals should not cause the timing functions of the pacemaker to be reset. Thus, when the start of a contraction is sensed (or is stimulated by the circuit), the timing circuits are reset, but cannot be reset again until the contraction and relaxation is over.

The electrical signals present at the terminals of the pacemaker are shown for a typical case in Figure PA7.1. At the left side of the tracing, the

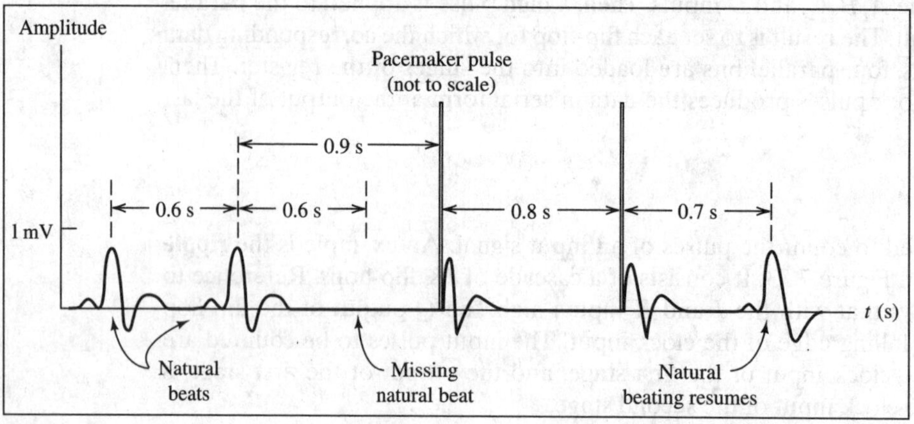

Figure PA7.1 Typical electrical signals at the terminals of a demand cardiac pacemaker.

signals occurring during natural beating are shown. Then, blockage of the natural beating occurs, and a pacemaker pulse is issued 0.9 s after the last natural beat. The amplitude of these pulses is typically 5 V and their durations are 0.7 ms. After the pacemaker pulse, natural signals occur from the contraction and relaxation of the heart. These are ignored by the circuit. After two forced cycles, the heart again begins natural beating.

The pacemaker circuitry and battery are enclosed in a metal case. This is implanted under the skin on the chest of the patient. A wire (enclosed in an insulating tube known as a catheter) leads from the pacemaker through an artery into the interior of the heart. The electrical terminals of the pacemaker are the metal case and the tip of the catheter. A pacemaker and catheter are shown in Figure PA7.2.

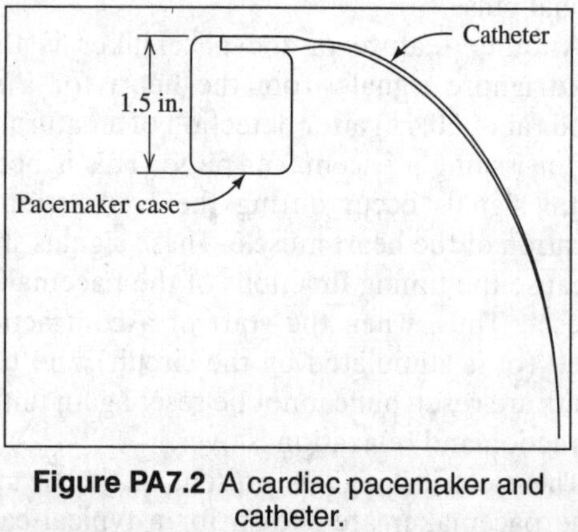

Figure PA7.2 A cardiac pacemaker and catheter.

The block diagram of a typical demand pacemaker is shown in Figure PA7.3. Notice that the electrical terminals serve both as the input to the amplifier and the output terminals for the pulse generator. The input amplifier increases the amplitude of the natural signals. Amplification is necessary because the natural heart signals have a very small amplitude (on the order of 1 mV), which must be increased before a comparator circuit can be employed to decide on the presence or absence of a natural heartbeat. Filtering to eliminate certain frequency components is employed in the amplifier to enhance the detectability of the heartbeats. Furthermore, proper filtering eliminates the possibility that radio or power-line signals will interfere with the pacemaker. Thus, the important specifications of the amplifier are its gain and frequency response.

Amplifiers and comparators are discussed at length in Part III of this book.

The output of the amplifier is applied to an analog comparator circuit that compares the amplified and filtered signal to a threshold value. If the input signal becomes higher than the threshold, the output of the comparator becomes high. Thus, the comparator output is a digital signal that indicates the detection of either a natural heartbeat or an output-pacing pulse.

This detection decision is passed through an AND gate to the counting and timer circuitry. The second input to the AND gate comes from the counter circuit and is low for 0.4 s after a detected

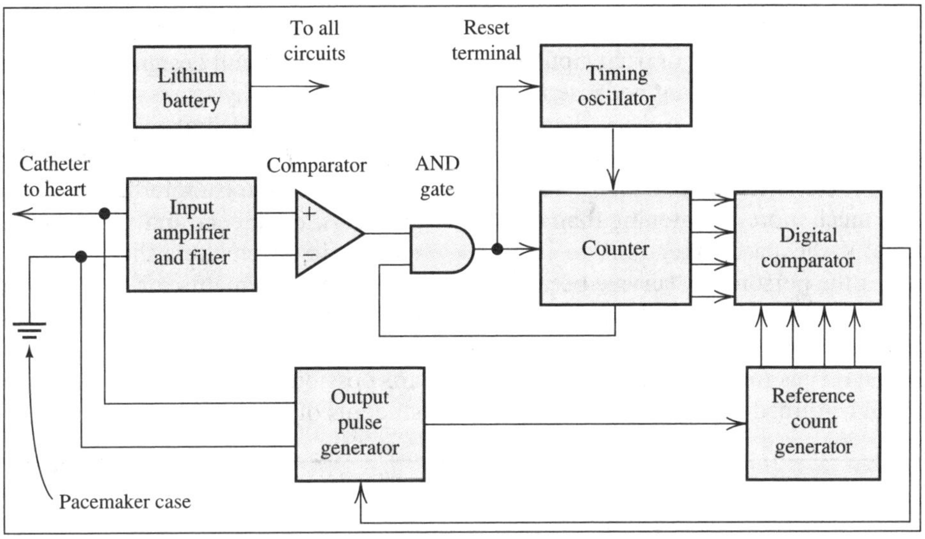

Figure PA7.3 Block diagram of a demand cardiac pacemaker.

beat. Thus, the AND gate prevents another decision from passing through for 0.4 s after the first. In this manner, the pacemaker ignores input signals for 0.4 s after the start of a natural or forced beat.

The timing functions are accomplished by counting the output cycles of a timing oscillator. The timing oscillator generates a square wave with a period of 0.1 s. When a heartbeat is detected, the timing oscillator is reset to the beginning of a cycle, so the completion of each oscillator cycle occurs exactly at an integer multiple of 0.1 s after the heartbeat. The timing oscillator must maintain a precise period because the proper operation of the circuit depends on accurate timing. Thus, the frequency stability of the timing oscillator is its primary specification.

The counter is a digital circuit that counts the output cycles of the timing oscillator. The counter is also capable of being reset to zero when a heartbeat occurs or when a pacemaker pulse is issued. The digital signals produced by the counter are applied to a digital comparator. Signals from a reference circuit are also applied to the digital comparator. The reference count is nine if the last beat was natural, but the reference count is eight if the last beat was forced. When the counter input to the digital comparator agrees with the reference count, the output of the digital comparator goes to a high level. This causes the pulse generator to issue an output pulse.

The pulse generator must produce output pulses of a specified amplitude and duration. In some designs, the output pulse amplitude is required to be higher than the battery voltage. This can be accomplished by charging capacitors in parallel with the battery and then switching them to series to generate the higher voltage.

What we have described so far is a relatively simple demand pacemaker readily implemented with registers, counters, and gates. By substituting a **microcontroller** and **software** (discussed in Chapter 8) for the digital functions of the pacemaker many additional useful features can be accommodated. For example, an accelerometer can sense the physical activity of the patient and the software can use this information to adjust heart rate. Communicating through magnetic fields linked to a coil in the pacemaker, the physician can instruct the software to alter the operating characteristics appropriately for each patient.

Extremely low power consumption is an important requirement for all pacemaker circuits. This is because the circuit must operate from a small battery for many years. After all, replacement of the battery requires a surgical procedure. When pacing pulses are not needed, a typical circuit can function with a few microamperes from a 2.5-V battery. When pacing pulses are required, the average current drain increases to a few tens of microamperes. This higher current consumption is unavoidable because of the output power required in the form of pacing pulses.

High reliability is very important because malfunction can be life-threatening. A very detailed failure-mode analysis must be performed for every component in the circuit. This is necessary because some failures are much more threatening than others. For example, if the pacemaker fails to issue pacemaking pulses, the person may survive because of the natural (low-rate) pacing of the heart muscle. On the other hand, if the timing generator fails in such a manner that it runs too fast, the unfortunate person's heart will be forced to beat much too fast. This can be quickly fatal, especially for those in a weakened condition from heart disease.

Clearly, circuit design is not the total solution to this problem. Physicians must provide the specifications for the pacemaker. Mechanical and chemical engineers must be involved in selecting the materials and form of the catheter and the case. By working in teams, engineers and physicians have designed electronic pacemakers that provide very dramatic health improvements for many people. Those who have contributed can be most proud of their achievements. Nevertheless, many further improvements are possible and may be achieved by some of the students of this book.

CONCLUSIONS

In this chapter, we have seen that complex combinatorial logic functions can be achieved simply by interconnecting NAND gates (or NOR gates). Furthermore, logic gates can be interconnected to form flip-flops. Interconnections of flip-flops form registers. A complex digital system, such as a computer, consists of many gates, flip-flops, and registers. Thus, logic gates are the basic building blocks for complex digital systems.

Summary

1. Digital signals are more immune to the effects of noise than analog signals. The logic levels of a digital signal can be determined after noise is added, provided that the noise amplitude is not too high.
2. Component values in digital circuits do not need to be as precise as in analog circuits.
3. Digital circuits are more amenable than analog circuits to implementation as large-scale integrated circuits.
4. In positive logic, the higher voltage represents logic 1.
5. Numerical data can be represented in decimal, binary, octal, hexadecimal, or binary-coded-decimal forms.
6. In the Gray code, each word differs from adjacent words in only a single bit. The Gray code is useful for representing position or angular displacement.
7. In computers, numbers are frequently represented in signed two's-complement form. (See Figure 7.11 on page 377.)
8. Logic variables take two values, logic 1 or logic 0. Logic variables may be combined by the AND, OR, and inversion operations according to the rules of Boolean algebra. A truth table lists all combinations of input variables and the corresponding output.
9. De Morgan's laws state that

$$AB = \overline{\overline{A} + \overline{B}}$$

and

$$A + B = \overline{\overline{A}\,\overline{B}}$$

10. NAND (or NOR) gates are sufficient to realize any combinatorial logic function.

11. Any combinatorial logic function can be written as a Boolean expression consisting of a logical sum of products (SOP). Each product is a minterm corresponding to a line of the truth table for which the output variable is logic 1.
12. Any combinatorial logic function can be written as a Boolean expression consisting of a logical product of sums (POS). Each sum is a maxterm corresponding to a line of the truth table for which the output variable is logic 0.
13. Many useful combinatorial circuits, known as decoders, encoders, or translators, are available as integrated circuits.
14. Karnaugh maps can be used to minimize the number of gates needed to implement a given logic function.
15. Sequential logic circuits are said to have memory because their outputs depend on past as well as present inputs. Synchronous or clocked sequential circuits are regulated by a clock signal.
16. Various types of flip-flops are the *SR* flip-flop, the clocked flip-flop, the *D* flip-flop, and the *JK* flip-flop.
17. Flip-flops can be combined to form registers that are used to store or manipulate digital words.
18. Logic gates can be interconnected to form flip-flops. Interconnections of flip-flops form registers. A complex digital system, such as a computer, consists of many gates, flip-flops, and registers. Thus, logic gates are the basic building blocks for complex digital systems.

Problems

Section 7.1: Basic Logic Circuit Concepts

***P7.1.** State three advantages of digital technology compared with analog technology.

P7.2. Explain noise margins and why they are important.

P7.3. Define these terms: *bit*, *byte*, and *nibble*.

P7.4. Discuss the difference between positive logic and negative logic.

P7.5. A certain digital system produces logic signals at location *A* that are transmitted to a second part of the system at location *B*. The logic circuits in use have $V_{IH} = 3.7$ V, $V_{IL} = 2.2$ V, $V_{OH} = 4.7$ V, and $V_{OL} = 0.5$ V. Noise voltages in the range from −0.3 V to +0.6 V are added to the logic signals as they travel from location *A* to location *B*. Is this system reliable? Why or why not? What is the most extreme range of noise voltages for which the system is reliable?

P7.6. How is the analog signal reconstructed from a digital word in a compact disk recording technique?

Section 7.2: Representation of Numerical Data in Binary Form

P7.7. How many bits per word are needed to represent the decimal integers 0 through 100? 0 through 1000? 0 through 10^6?

P7.8. Convert the following binary numbers to decimal form: **a.*** 101.101; **b.*** 10101.011; **c.** 1110.001; **d.** 1101.001; **e.** 1001.0101; **f.** 1111.11.

P7.9. Add these pairs of binary numbers: **a.*** 1101.11 and 101.111; **b.** 1011.01 and 1010.11; **c.** 10101.101 and 1101.001; **d.** 1011000.1000 and 10001001.1001.

P7.10. Express the following decimal numbers in binary form and in binary-coded decimal form: **a.*** 9.75; **b.** 6.5; **c.** 11.75; **d.** 63.03125; **e.** 67.375.

P7.11. Express each of the following octal numbers in binary, decimal, and hexadecimal forms: **a.*** 77.7_8; **b.** 36.5_8; **c.** 123.4_8; **d.** 57.5_8

P7.12. Find the result (in BCD format) of adding the BCD numbers: **a.*** 10010011.0101

* Denotes that answers are contained in the Student Solutions files. See Appendix E for more information about accessing the Student Solutions.

and 00110111.0001; **b.** 01010100.1000 and 01001001.1001; **c.** 011110010110.0011 and 00110101.1001.

P7.13. What number follows 377 when counting in **a.** decimal; **b.** octal; **c.** hexadecimal?

P7.14. Express the following decimal numbers in binary, octal, and hexadecimal forms: **a.*** 313.0625; **b.** 253.25; **c.** 349.75; **d.** 835.25; **e.** 212.5.

P7.15. Write each of the following decimal numbers as an eight-bit signed two's-complement number: **a.*** 75; **b.*** −87; **c.** 19; **d.** −19; **e.** −95; **f.** 125.

P7.16. Express each of the following octal numbers in binary, hexadecimal, and decimal forms: **a.*** 777.7_8; **b.** 123.5_8; **c.** 24.4_8; **d.** 644.2_8.

P7.17. What range of decimal integers can be represented by: **a.** four-bit binary numbers; **b.** four-digit octal numbers; **c.** four-digit hexadecimal numbers?

***P7.18.** Compare 4-bit binary code with 4-bit gray code. Enlist the application areas in which use of gray code is advantageous. Comment.

P7.19. Find the one's and two's complements of the binary numbers: **a.*** 11101000; **b.** 00000000; **c.** 01010101; **d.** 01111100; **e.** 11000000.

P7.20. Convert the following numbers to decimal form: **a.*** $EB4.5_{16}$; **b.*** 657.7_8; **c.** $DC.9_{16}$; **d.** 235.43_8; **e.** $BA.C_{16}$

P7.21. Demonstrate these operations using 8-bit signed two's compliment arithmetic: **a.*** $43_{10} - 45_{10}$; **b.** $27_{10} + 15_{10}$; **c.** $34_{10} - 45_{10}$; **d.** $25_{10} - 39_{10}$; **e.** $59_{10} - 34_{10}$.

P7.22. Describe how to test whether overflow or underflow has occurred in adding signed two's-complement numbers.

P7.23. Sometimes we cannot express a value exactly with a finite number of decimal or binary symbols. For example, we have $1/3 = 0.33333\overline{3}$ or $4/33 = 0.121212\overline{12}$ in which the overbar indicates that the pattern repeats indefinitely. **a.** Convert 0.1_{10} to binary form indicating the repeating pattern with an overbar. **b.** Repeat for 0.6_{10}. **c.** Perform the calculation $(0.1 \times 1024 - 102) \times 10 - 4$ on your calculator. This should give the exact result, namely zero, because most calculators use BCD and the arithmetic is essentially decimal in nature. **d.** Repeat the calculation of part (c) with MATLAB. The result is not exactly zero because the calculations are carried out in binary with a finite number of bits, resulting in an approximate representation with round-off error for 0.1. This error is magnified in the calculation.

Section 7.3: Combinatorial Logic Circuits

P7.24. What is a truth table?

P7.25. State De Morgan's laws in English. Also, use them to give the equivalent mathematical forms for ABC and for $A + B + C$.

P7.26. Draw the circuit symbol and list the truth table for the following: an AND gate; an OR gate; an inverter; a NAND gate; a NOR gate; and an XOR gate. Assume two inputs for each gate (except the inverter).

P7.27. Verify the following Boolean expressions with the help of truth tables:

a.* $E = A\overline{B} + ABC + C\overline{D}$

b. $D = \overline{AC} + A\overline{B}C$

c. $Z = \overline{W}Y + \overline{(\overline{X} + Y + W)}$

d. $D = \overline{AC}B + \overline{B}$

e. $D = \overline{(\overline{C} + AB)}$

P7.28. How do we use a truth table to prove the validity of a Boolean algebra identity?

***P7.29.** Prove the following identity using a truth table:

$$AB + \overline{B}C = \overline{(\overline{A} + \overline{B})(B + \overline{C})}$$

P7.30. Use truth tables to prove De Morgan's laws for three logic variables, which are

$$ABC = \overline{\overline{A} + \overline{B} + \overline{C}} \quad \text{and} \quad A + B + C = \overline{\overline{A}\,\overline{B}\,\overline{C}}$$

P7.31. Prove the following identity using a truth table:

$$\overline{(A + B)(\overline{A} + AB)} = \overline{B}$$

P7.32. Use a truth table to prove the identity

$$A + \overline{A}B = A + B$$

P7.33. Use a truth table to prove the identity

$$ABC + AB\overline{C} + A\overline{B}\,\overline{C} + A\overline{B}C = A$$

P7.34. Write a Boolean expression for the output of each of the logic circuits shown in

Figure P7.34. Also, give the truth table for each circuit.

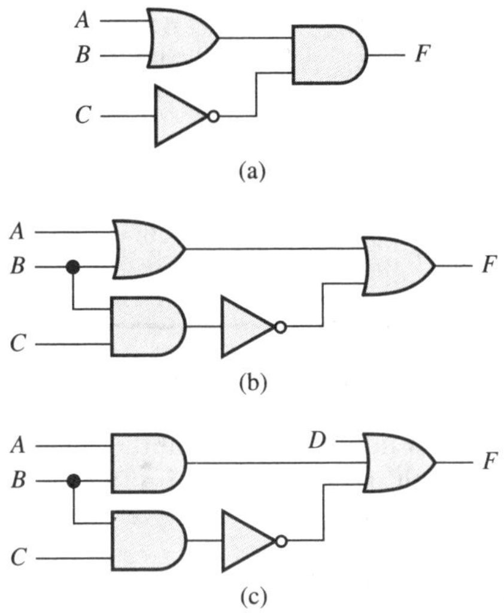

Figure P7.34

P7.35. Draw a circuit to realize each of the following expressions using AND gates, OR gates, and inverters:

a.* $F = \left(\overline{A} + \overline{B} + C\right)\left(A + B + \overline{C}\right)\left(A + \overline{B} + C\right)$

b. $F = A\overline{B}C + AB\overline{C} + \overline{A}BC$

c. $F = A + \overline{B}C$

P7.36. Replace the AND operations by ORs and vice versa by applying De Morgan's laws to each of these expressions:

a.* $F = (A + B + C)\left(A + \overline{B} + C\right)\left(\overline{A} + B + \overline{C}\right)$

b.* $F = ABC + A\overline{B}C + \overline{A}B\overline{C}$

c. $F = AB + \left(\overline{C} + A\right)\overline{D}$

d. $F = A\left(\overline{B} + C\right) + D$

e. $F = A\overline{B}C + A(B + C)$

P7.37. Sometimes "bubbles" are used to indicate inverters on the input lines to a gate, as illustrated in Figure P7.37. What are the equivalent gates for those of Figure P7.37? Justify your answers.

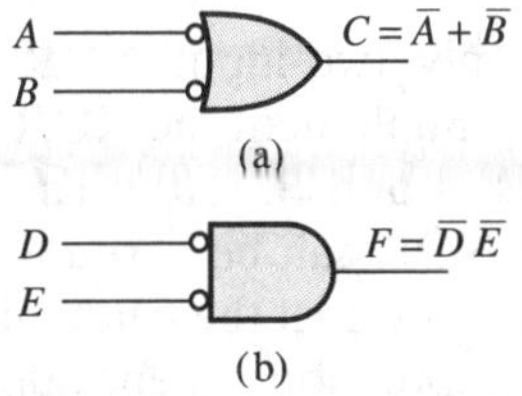

Figure P7.37

P7.38. Consider the addition of binary numbers as illustrated in Figure 7.7 on page 373. A **half adder** is a logic circuit that can be used to perform the addition of the right-most column of the addition. A symbol for the half adder is shown in Figure P7.38. The two inputs are the right-most bits, denoted as A and B, of the two numbers to be added. The two outputs are the sum bit S and the carry C to the next column. Write the truth table and show how to implement the half adder using an exclusive OR gate and an AND gate.

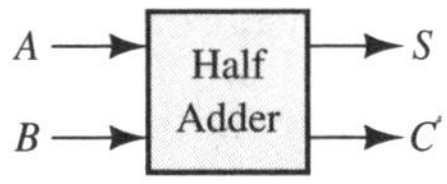

Figure P7.38 Half adder.

P7.39. Consider the addition of binary numbers as illustrated in Figure 7.7 on page 373. A **full adder** is a logic circuit that can be used to perform the operations for the columns that have carries into them. A symbol for the full adder is shown in Figure P7.39. The inputs are corresponding bits, denoted as A and B, of the two numbers to be added and the carry from the previous column C_{in}. The two outputs are the sum bit S and the carry C_{out} to the next column. Write the truth table for the full adder. Show how to implement a full adder using two half adders of Problem P7.38 and an OR gate.

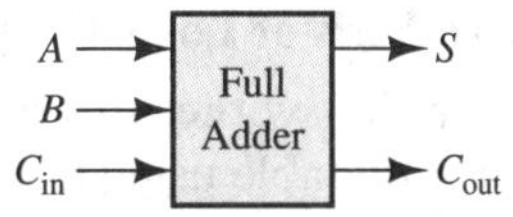

Figure P7.39 Full adder.

P7.40. Why are NAND gates said to be *sufficient* for combinatorial logic? What other type of gate is sufficient?

P7.41. Consider the circuit shown in Figure P7.41. The switches are controlled by logic variables such that, if A is high, switch A is closed, and if A is low, switch A is open. Conversely, if B is high, the switch labeled $\overline{B}$ is open, and if B is low, the switch labeled $\overline{B}$ is closed. The output variable is high if the output voltage is 5 V, and the output variable is low if the output voltage is zero. Write a logic expression for the output variable. Construct the truth table for the circuit.

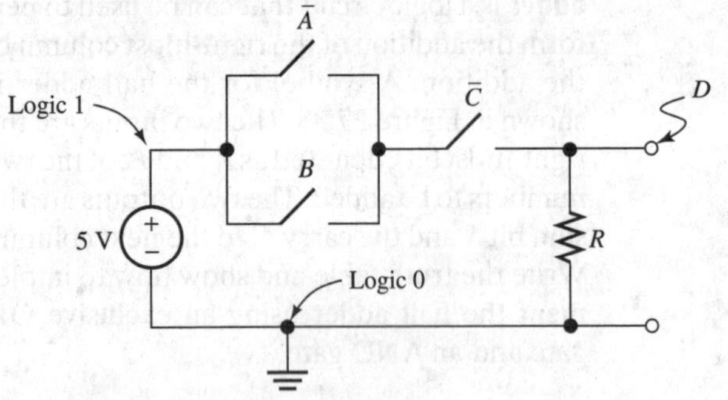

Figure P7.41

P7.42. Draw a circuit similar to that of Fig. P7.41 so the output variable is given by

$$D = \overline{A}\,\overline{B} + \overline{C}$$

P7.43. Repeat Problem 7.41 for the circuit shown in Figure P7.43.

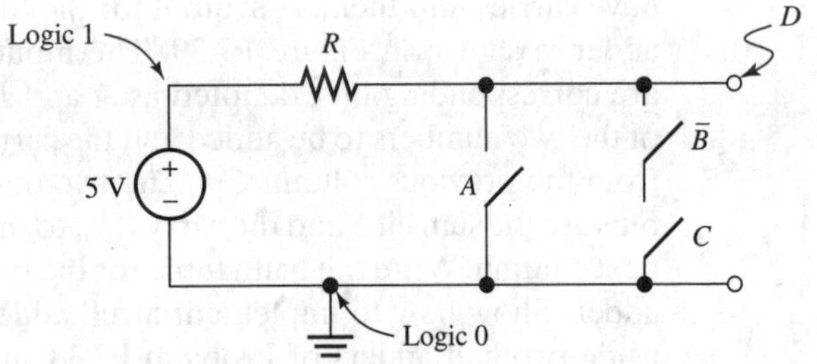

Figure P7.43

Section 7.4: Synthesis of Logic Circuits

P7.44. Describe the synthesis of a logic expression from a truth table using the sum-of-products approach. Repeat for the product-of-sums approach.

***P7.45.** Consider Table P7.45. *A*, *B*, and *C* represent logic-variable input signals; *F* through *K* are outputs. Using the sum-of-products approach, write a Boolean expression for *F* in terms of the inputs. Repeat by using the product-of-sums approach.

P7.46. Repeat Problem P7.45 for *G*.

P7.47. Repeat Problem P7.45 for *H*.

P7.48. Repeat Problem P7.45 for *I*.

P7.49. Repeat Problem P7.45 for *J*.

P7.50. Repeat Problem P7.45 for *K*.

Table P7.45

A	*B*	*C*	*F*	*G*	*H*	*I*	*J*	*K*
0	0	0	1	1	1	0	0	0
0	0	1	0	1	1	0	1	0
0	1	0	1	0	1	1	0	1
0	1	1	0	1	0	1	1	1
1	0	0	0	0	0	0	0	1
1	0	1	1	0	0	0	0	1
1	1	0	0	0	1	1	1	0
1	1	1	1	0	1	1	1	0

P7.51. Show how to implement the sum-of-products circuit shown in Figure P7.51 using only NAND gates.

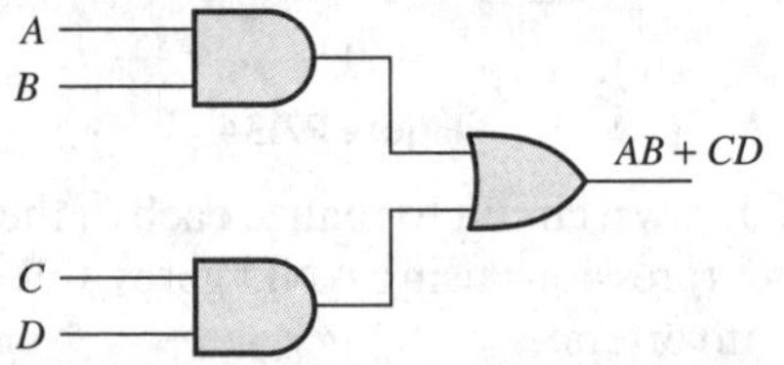

Figure P7.51

P7.52. Show how to implement the product-of-sums circuit shown in Figure P7.52 using only NOR gates.

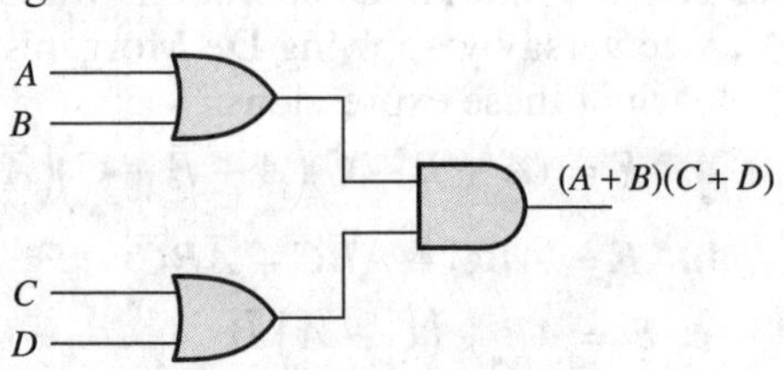

Figure P7.52

***P7.53.** Use only dual input NAND gates to find a way to implement the XOR function for two inputs: *A* and *B*. (*Hint*: The inputs of a two-input NAND can be wired together to obtain an inverter. List the truth table and write the SOP expression. Then, apply De Morgan's laws to convert the OR operation to AND.)

P7.54. Use only two-input NOR gatcs to find a way to implement the XOR function for two inputs: *A* and *B*. (*Hint:* The inputs of a two-input NOR can be wired together to obtain an inverter. List the truth table and write the POS expression. Then, apply De Morgan's laws to convert the AND operation to OR.)

P7.55. Give an example of a decoder.

***P7.56.** Suppose that two numbers in signed two's-complement form have been added. S_1 is the sign bit of the first number, S_2 is the sign bit of the second number, and S_T is the sign bit of the total. Suppose that we want a logic circuit with output E that is high if either overflow or underflow has occurred; otherwise, E is to remain low. **a.** Write the truth table. **b.** Find an SOP expression composed of minterms for E. **c.** Draw a circuit that yields E, using AND, OR, and NOT gates.

P7.57. Design a logic circuit to control electrical power to the engine ignition of a speed boat. Logic output I is to become high if ignition power is to be applied and is to remain low otherwise. Gasoline fumes in the engine compartment present a serious hazard of explosion. A sensor provides a logic input F that is high if fumes are present. Ignition power should not be applied if fumes are present. To help prevent accidents, ignition power should not be applied while the outdrive is in gear. Logic signal G is high if the outdrive is in gear and is low otherwise. A blower is provided to clear fumes from the engine compartment and is to be operated for five minutes before applying ignition power. Logic signal B becomes high after the blower has been in operation for five minutes. Finally, an emergency override signal E is provided so that the operator can choose to apply ignition power even if the blower has not operated for five minutes and if the outdrive is in gear, but not if gasoline fumes are present. **a.** Prepare a truth table listing all combinations of the input signals B, E, F, and G. Also, show the desired value of I for each row in the table. **b.** Using the sum-of-products approach, write a Boolean expression for I. **c.** Using the product-of-sums approach, write a Boolean expression for I. **d.** Try to manipulate the expressions of parts (b) and (c) to obtain a logic circuit having the least number of gates and inverters. Use AND gates, OR gates, and inverters.

P7.58. Consider the BCD-to-seven-segment decoder discussed in conjunction with Figure 7.26 on page 392. Suppose that the BCD data are represented by the logic variables B_8, B_4, B_2, and B_1. For example, the decimal number 7 is represented in BCD by the word 0111 in which the leftmost bit is $B_8 = 0$, the second bit is $B_4 = 1$, and so forth. **a.** Find a logic circuit based on the product of maxterms having output A that is high only if segment A of the display is to be on. **b.** Repeat for segment B. **c.** Repeat for segment C. **d.** Repeat for segment D.

Section 7.5: Minimization of Logic Circuits

P7.59. **a.** Construct a Karnaugh map for the logic function

$$F = \overline{A}BCD + A\overline{B}C\overline{D} + \overline{A}B\overline{C}D + AB\overline{C}\overline{D} + \overline{A}BC\overline{D} + ABC\overline{D}$$

b. Find the minimum SOP expression.
c. Find the minimum POS expression.

P7.60. A logic circuit has inputs A, B, and C. The output of the circuit is given by

$$D = \sum m(0, 2, 3)$$

a. Construct the Karnaugh map for D.
b. Find the minimum SOP expression.
c. Find two equally good minimum POS expressions.

P7.61. A logic circuit has inputs A, B, and C. The output of the circuit is given by

$$D = \sum m(0, 2, 4, 6, 7)$$

a. Construct the Karnaugh map for D.
b. Find the minimum SOP expression.
c. Find the minimum POS expressions.

P7.62. A logic circuit has inputs A, B, and C. The output of the circuit is given by

$$D = \prod M(1, 2, 3, 5)$$

a. Construct the Karnaugh map for D.
b. Find the minimum SOP expression.
c. Find the minimum POS expression.

P7.63. **a.** Construct a Karnaugh map for the logic function

$$D = ABC + \overline{A}BC + AB\overline{C} + BC$$

b. Find the minimum SOP expression and realize the function using AND, OR, and NOT gates. **c.** Find the minimum POS expression and realize the function using AND, OR, and NOT gates.

P7.64. **a.** Construct a Karnaugh map for the logic function

$$F = AB\overline{C}\,\overline{D} + ABCD + ABC\overline{D} + \overline{A}BCD$$

b. Find the minimum SOP expression. **c.** Realize the minimum SOP function using AND, OR, and NOT gates. **d.** Find the minimum POS expression.

***P7.65.** Consider Table P7.65 in which A, B, C, and D are input variables. E, F, G, H, and I are the output variables. **a.** Construct a Karnaugh map for the output variable E. **b.** Find the minimum SOP expression for this logic function. **c.** Use AND, OR, and NOT gates to realize the minimum SOP function. **d.** Find the minimum POS expression.

Table P7.65

A	B	C	D	E	F	G	H	I
0	0	0	0	0	0	0	1	0
0	0	0	1	1	0	0	1	1
0	0	1	0	0	0	0	1	0
0	0	1	1	0	0	0	0	0
0	1	0	0	0	0	0	1	0
0	1	0	1	1	0	0	1	1
0	1	1	0	0	0	1	1	0
0	1	1	1	0	1	1	0	0
1	0	0	0	0	0	0	0	1
1	0	0	1	1	0	1	0	1
1	0	1	0	0	0	0	0	1
1	0	1	1	0	0	1	0	1
1	1	0	0	1	0	0	0	0
1	1	0	1	1	0	1	0	1
1	1	1	0	1	1	1	0	0
1	1	1	1	1	1	1	0	0

P7.66. Repeat Problem P7.65 for output variable F.

P7.67. Repeat Problem P7.65 for output variable G.

P7.68. Repeat Problem P7.65 for output variable H.

P7.69. Repeat Problem P7.65 for output variable I.

P7.70. We need a logic circuit that gives an output X that is high when an error in the form of an unused code occurs in a given BCD codeword. The inputs to the logic circuit are the bits B_8, B_4, B_2, and B_1 of the BCD codeword. (The most significant bit is B_8, and the least significant bit is B_1.) Construct the Karnaugh map and write the minimized SOP and POS expressions for X.

P7.71. We need a logic circuit that gives an output X that is high only if a given hexadecimal digit is even (including 0) and less than 7. The inputs to the logic circuit are the bits B_8, B_4, B_2, and B_1 of the binary equivalent for the hexadecimal digit. (The most significant bit is B_8, and the least significant bit is B_1.) Construct a truth table and the Karnaugh map; then, write the minimized SOP expression for X.

P7.72. In an apartment we have two power supplies: one from the electric supply utility, and the second from an in-house Diesel Generator (DG) set. Design a logic circuit which turns the DG set on when the supply from the utility is cut off. The DG set should be turned off as soon as the supply from the utility is restored. At any given time, only one power supply is to be used to supply power to the residents.

P7.73. We need a logic circuit that gives a high output if a given hexadecimal digit is 4, 6, C, or E. The inputs to the logic circuit are the bits B_8, B_4, B_2, and B_1 of the binary equivalent for the hexadecimal digit. (The most significant bit is B_8, and the least significant bit is B_1.) Construct the Karnaugh map and write the minimized SOP and POS expressions for X.

P7.74. A city council has three members; A, B, and C. Each member votes on a proposition (1 for yes, 0 for no). Find a minimized SOP logic expression having inputs A, B, and C, and output X that is high when the majority vote is yes and low otherwise. Show that the minimized logic circuit checks to see if any pair of the three board members have voted yes. Repeat for a council with five members. (*Hint*: In this case, the circuit checks to see if any group of three has all voted yes.)

P7.75. A city council has four members; A, B, C, and D. Each member votes on a proposition (1 for yes, 0 for no). Find a minimized SOP logic expression having inputs A, B, C, and D, and output X that is high when the vote is tied and low otherwise.

P7.76. One way to help ensure that data are communicated correctly is to append a parity bit to each data word such that the number of 1s in the transmitted word is even. Then, if an odd number are found in the received result, we know that at least one error has occurred. **a.** Show that the circuit in Figure P7.76 produces the correct parity bit P for the nibble (four-bit data word) $ABCD$. In other words, show that the transmitted word $ABCDP$ contains an even number of 1s for all combinations of data. **b.** Determine the minimum SOP expression for P in terms of the data bits. **c.** If the received word contains a single-bit error, the number of 1s in the word will be odd. Draw a circuit using four XOR gates that outputs a 1 if the received word $ABCDP$ contains an odd number of 1s and outputs a 0 otherwise.

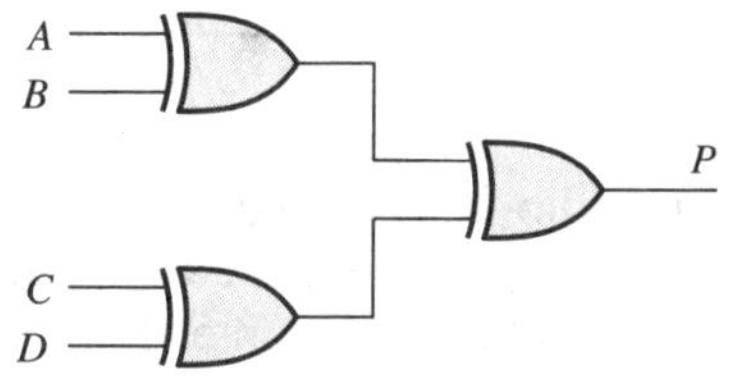

Figure P7.76

P7.77. Suppose we want circuits to convert the binary codes into the three-bit Gray codes shown in Table P7.77. Find the minimum SOP expressions for X, Y, and Z in terms of A, B, and C.

Table P7.77

Binary Code ABC	Gray Code XYZ
000	000
001	001
010	011
011	010
100	110
101	111
110	101
111	100

P7.78. Find the minimum SOP expressions for A, B, and C in terms of X, Y, and Z for the codes of Table P7.77.

***P7.79.** We have discussed BCD numbers in which the bits have weights of 8, 4, 2, and 1. Another way to represent decimal integers is the 4221 code in which the weights of the bits are 4, 2, 2, and 1. The decimal integers, the BCD equivalents, and the 4221 equivalents are shown in Table P7.79. We want to design logic circuits to convert BCD codewords to 4221 codewords. **a.** Fill in the Karnaugh map for F, placing x's (don't cares) in the squares for BCD codes that do not occur in the table. Find the minimum SOP expression allowing the various x's to be either 1s or 0s to make the expression as simple as possible. **b.** Repeat (a) for G. **c.** Repeat (a) for H. **d.** Repeat (a) for I.

Table P7.79. BCD, 4221, and excess-3 codewords for the decimal integers.

Decimal Integer	BCD Codeword $ABCD$	4221 Codeword $FGHI$	Excess-3 Codeword $WXYZ$
0	0000	0000	0011
1	0001	0001	0100
2	0010	0010	0101
3	0011	0011	0110
4	0100	1000	0111
5	0101	0111	1000
6	0110	1100	1001
7	0111	1101	1010
8	1000	1110	1011
9	1001	1111	1100

P7.80. We want to design logic circuits to convert the 4221 codewords of Problem P7.79 to BCD codewords. **a.** Fill in the Karnaugh map for A, placing x's (don't cares) in the squares for 4221 codes that do not occur in the table. Find the minimum SOP expression allowing the various x's to be either 1s or 0s to make the expression as simple as possible. **b.** Repeat (a) for B. **c.** Repeat (a) for C. **d.** Repeat (a) for D.

P7.81. Another code that is sometimes used to represent decimal digits is the **excess-3 code**. To convert a decimal digit to excess-3, we add 3 to the digit and express the sum as a four-bit binary number. For example, to convert the decimal digit 9 to excess-3 code, we have

$$9_{10} + 3_{10} = 12_{10} = 1100_2$$

Thus, 1100 is the excess-3 codeword for 9. The excess-3 codewords for the other decimal digits are shown in Table P7.79.

We want to design logic circuits to convert BCD codewords to excess-3 codewords. **a.** Fill in the Karnaugh map for W, placing x's (don't cares) in the squares for BCD codes that do not occur in the table. Find the minimum SOP expression allowing the various x's to be either 1s or 0s to make the expression as simple as possible. **b.** Repeat (a) for X. **c.** Repeat (a) for Y. **d.** Repeat (a) for Z.

P7.82. We want to design logic circuits to convert the excess-3 codewords of Problem P7.81 to BCD codewords. **a.** Fill in the Karnaugh map for A, placing x's (don't cares) in the squares for excess-3 codes that do not occur in the table. Find the minimum SOP expression allowing the various x's to be either 1s or 0s to make the expression as simple as possible. **b.** Repeat (a) for B. **c.** Repeat (a) for C. **d.** Repeat (a) for D.

Section 7.6: Sequential Logic Circuits

P7.83. Draw the circuit symbol and give the truth table for an SR flip-flop.

P7.84. Use NOR gates to draw the diagram of an SR flip-flop. Repeat using NAND gates. Label the inputs and outputs as S, R, Q, or $\overline{Q}$.

P7.85. Draw the circuit symbol and give the truth table for JK flip-flop.

P7.86. Explain the distinction between synchronous and asynchronous inputs to a flip-flop.

P7.87. What is edge triggering?

P7.88. Draw the circuit symbol and give the truth table for a positive-edge-triggered D flip-flop.

***P7.89.** Assuming that the initial state of the shift register shown in Figure P7.89 is 100 (i.e., $Q_0 = 1$, $Q_1 = 0$, and $Q_2 = 0$), find the successive states. After how many shifts does the register return to the starting state?

P7.90. Repeat Problem P7.89 if the XOR gate is replaced with: **a.** an OR gate; **b.** an AND gate.

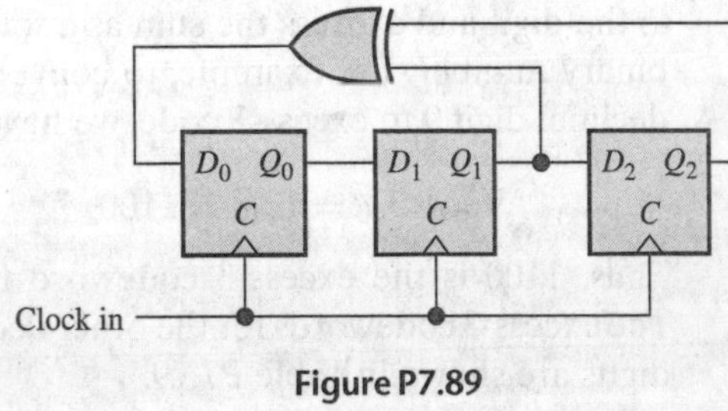

Figure P7.89

P7.91. The D flip-flops of Figure P7.91 are positive-edge triggered. Assuming that prior to $t = 0$, the states are $Q_0 = 0$ and $Q_1 = 0$, sketch the voltage waveforms at Q_0 and Q_1 versus time. Assume logic levels of 0 V and 5 V.

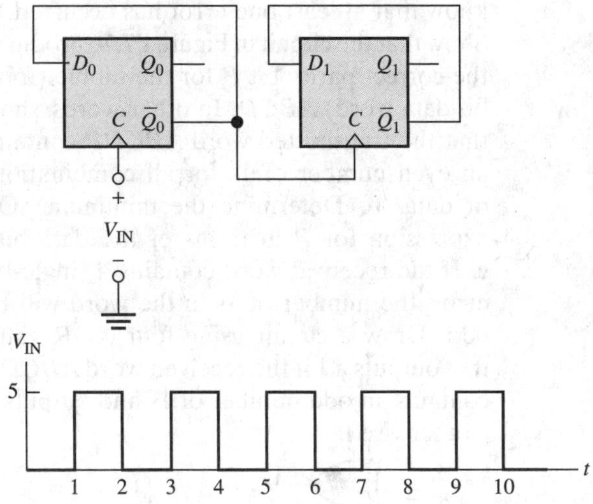

Figure P7.91

P7.92. The D flip-flops of Figure P7.92 are positive-edge triggered, and the Cl input is an asynchronous clear. Assume that the states are $Q_0 = Q_1 = Q_2 = Q_3 = 1$ at $t = 0$. The clock input V_{IN} is shown in Figure P7.91. Sketch the voltage waveforms at Q_0, Q_1, Q_2, and Q_3 versus time. Assume logic levels of 0 V and 5 V.

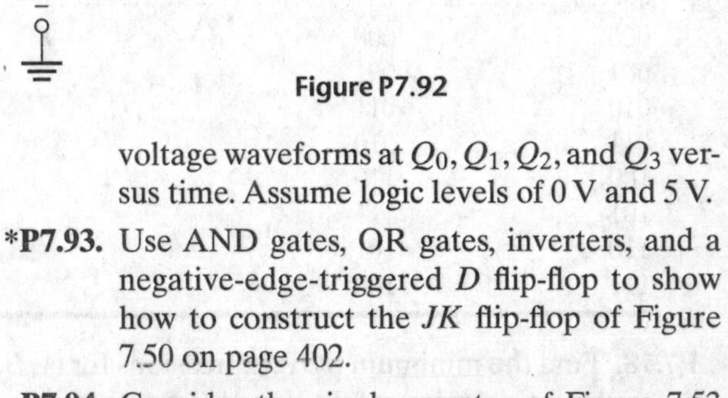

Figure P7.92

***P7.93.** Use AND gates, OR gates, inverters, and a negative-edge-triggered D flip-flop to show how to construct the JK flip-flop of Figure 7.50 on page 402.

P7.94. Consider the ripple counter of Figure 7.53 on page 404. Suppose that the flip-flops have asynchronous clear inputs. Show how to add gates so that the count resets to zero immediately when the count reaches six. This results in a modulo-six counter.

P7.95. Four light-emitting diodes (LED) are arranged at the corners of a diamond, as illustrated in Figure P7.95. When logic 1 is applied to an LED, it lights. Only one diode is to be on at a time. The on state should move from diode to diode either clockwise or counterclockwise, depending on whether S is high or low, respectively. One complete revolution should be completed in each two-second interval. **a.** What is the frequency of the clock? **b.** Draw a suitable logic circuit for the counter. **c.** Construct the truth table and use Karnaugh maps to determine the minimum SOP expressions for D_1 through D_4 in terms of S, Q_1, and Q_2.

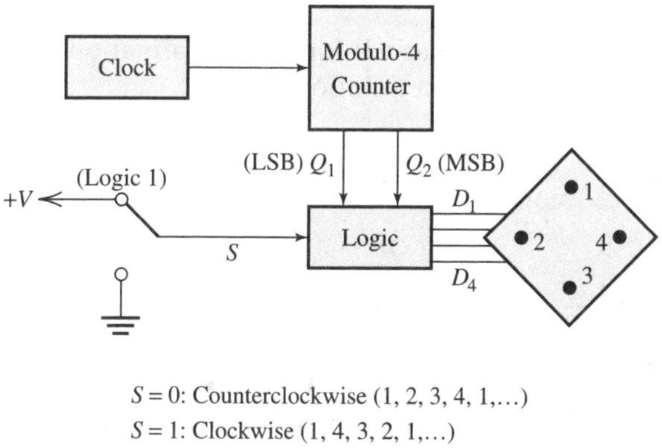

Figure P7.95

P7.96. Figure P7.96 shows the functional diagram of an electronic die that can be used in games of chance. The system contains a high-speed clock, a push-button momentary contact switch that returns to the upper (logic 1) position when released, and a counter that counts through the cycle of states: 001, 010, 011, 100, 101, and 110 (i.e., the binary equivalents of the number of spots on the various sides of the die). Q_3 is the most significant bit (MSB), and Q_1 is the least significant bit (LSB). The system has a display consisting of seven LEDs, each of which lights when logic 1 is applied to it. The encoder is a combinatorial logic circuit that translates the state of the counter into the logic signals needed by the display. Each time the switch is depressed, the counter operates, stopping in a random state when the switch is released. **a.** Use JK flip-flops having asynchronous preset and clear inputs to draw the detailed diagram of the counter. **b.** Design the encoder, using Karnaugh maps to minimize the logic elements needed to produce each of the seven output signals.

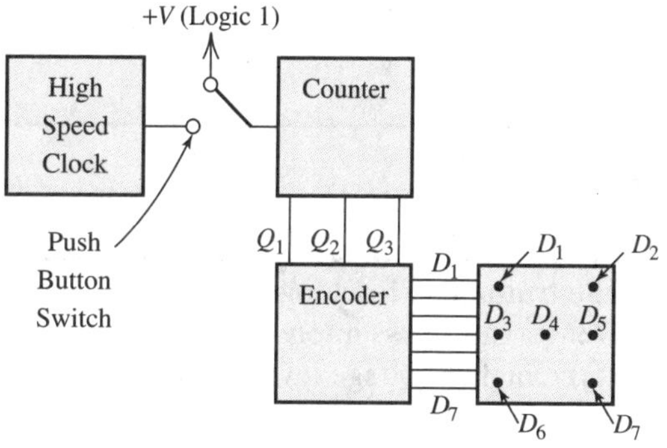

Figure P7.96 Electronic die.

Frequently Asked Questions

1. Differentiate between analog and digital signal.
2. What are the advantages of digital approach?
3. What do you mean by positive and negative logic?
4. What is binary number system?
5. What is Octal and Hexadecimal number system?
6. What do you understand by Binary Coded Decimal (BCD) system.
7. Explain Gray code number system.
8. What do you mean by logic gates?
9. Draw the logic symbol and truth-table of AND logic gate.
10. Draw the logic symbol and truth-table of OR logic gate.
11. Draw the logic symbol and truth-table of NOT logic gate.
12. Prove that A(BC) = (AB)C = ABC.
13. Using truth-table show that A + (B + C) = (A + B) + C = A + B + C.
14. State and explain De Morgan's laws.
15. Draw the logic symbol and truth table for a NAND gate.
16. Draw the logic symbol and truth table for a NOR gate.
17. Why NAND and NOR gates are called universal gates?
18. Draw the logic symbol and truth table for XOR and XNOR logic gates.

19. Write both sum-of-products and product-of-sums boolean expression for (a) two input AND gate and (b) two input EXOR gate.

20. Write both sum-of products and product-of-sums boolean expressions for (a) two input NAND gate and (b) two input NOR gate.

21. What do you mean by (a) Decoder (b) Encoder and (c) Translator

22. Explain how number of terms in a logic expression can be reduced by using Karnaugh maps.

23. What steps are taken to minimize POS circuits by using Karnaugh maps?

24. What do you understand by sequential logic circuits?

25. What is a flip-flop?

26. Draw the symbol of an SR flip-flop and explain its working with the help of a truth-table.

27. Draw the circuit and truth-table of a clocked SR flip-flop.

28. What is a register? What do you mean by serial-in parallel-out shift register?

29. How a parallel-is serial-out register works?

30. Draw the circuit for a ripple counter and explain its working.

Practice Test

Here is a practice test you can use to check your comprehension of the most important concepts in this chapter. Answers can be found in Appendix D and complete solutions are included in the Student Solutions files.

See Appendix E for more information about the Student Solutions.

T7.1. First, think of one or more correct ways to complete each statement in Table T7.1(a).

Table T7.1

Item	Best Match
(a)	

a. The truth table for a logic expression contains...
b. DeMorgan's laws imply that...
c. If the higher voltage level represents logic 1 and the lower level represents logic 0, we have...
d. For a Gray code...
e. If we invert each bit of a binary number and then add 1 to the result, we have...
f. If we add two negative signed two's complement numbers, and the left-hand bit of the result is zero, we have had...
g. Squares on the top and bottom of a Karnaugh map are considered to be...
h. If the $\overline{Q}$ output of a positive-edge-triggered D flip-flop is connected to the D input, at each positive clock edge, the state of the flip-flop...
i. Provided that it is not too large, noise can be completely eliminated from...
j. Registers are composed of...
k. An SOP logic circuit is composed of...
l. Working out from the decimal point and converting four-bit groups of a BCD number to their hexadecimal equivalents produces...

(b)

1. the decimal equivalent
2. old recordings of music
3. inverse logic
4. congruent
5. OR gates
6. AND gates and one OR gate
7. digital signals
8. inverters, AND gates, and one OR gate
9. inverses

Table T7.1 (*Cont.*)

10. flip-flops
11. overflow
12. a listing of all combinations of inputs and the corresponding outputs
13. a table of ones and zeros
14. code words appear in numerical order
15. analog signals
16. adjacent
17. each code word is rotated to form the next word
18. if AND operations are changed to OR and vice versa, and the result is inverted, the result is equivalent to the original logic expression
19. NAND gates are sufficient to implement any logic expression
20. positive logic
21. the two's complement
22. negative logic
23. adjacent words differ in a single bit
24. underflow
25. toggles

Then, select the best choice from the list given in Table T7.1(b). [Items in Table T7.1(b) may be used more than once or not at all.]

T7.2. Convert the decimal integer, 353.875_{10} to each of these forms: **a.** binary; **b.** octal; **c.** hexadecimal; **d.** binary coded decimal.

T7.3. Find the decimal equivalent for each of these eight-bit signed two's complement integers: **a.** 01100001; **b.** 10111010.

T7.4. For the logic circuit of Figure T7.4: **a.** write the logic expression for D in terms of A, B, and C directly from the logic diagram; **b.** construct the truth table and the Karnaugh map; **c.** determine the minimum SOP expression for D; **d.** determine the minimum POS expression for D.

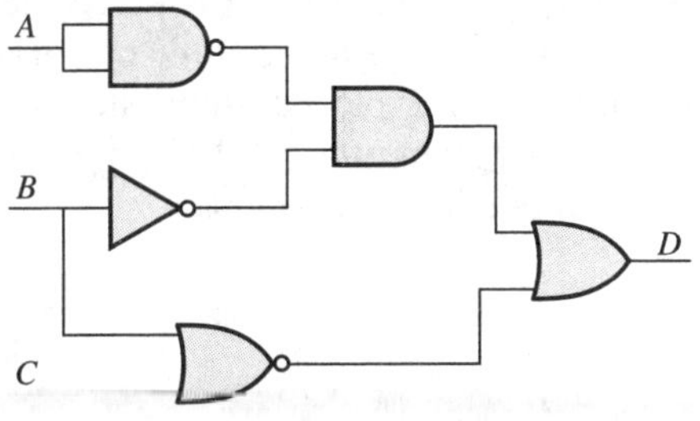

Figure T7.4

T7.5. Suppose we need a logic circuit with a logic output G that is high only if a certain hexadecimal digit is 1, 5, B, or F. The inputs to the logic circuit are the bits B_8, B_4, B_2, and B_1 of the binary equivalent for the hexadecimal digit. (The most significant bit is B_8, and the least significant bit is B_1.) **a.** Fill in the Karnaugh map shown in Figure T7.5. **b.** Determine the minimized SOP expression for G. **c.** Determine the minimum POS expression for G.

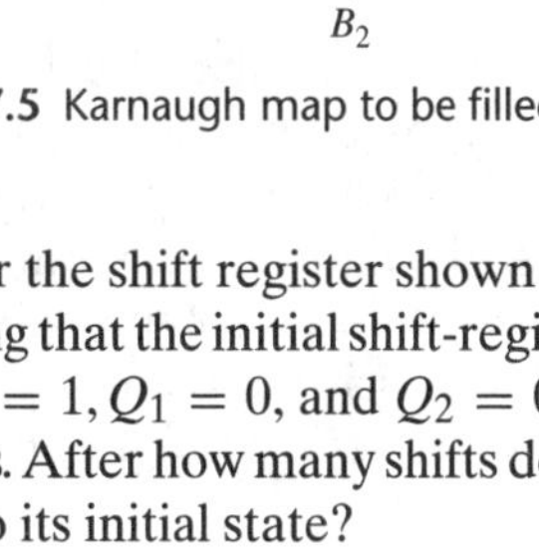

Figure T7.5 Karnaugh map to be filled in for G.

T7.6. Consider the shift register shown in Figure T7.6. Assuming that the initial shift-register state is 100 (i.e., $Q_0 = 1, Q_1 = 0$, and $Q_2 = 0$), list the next six states. After how many shifts does the register return to its initial state?

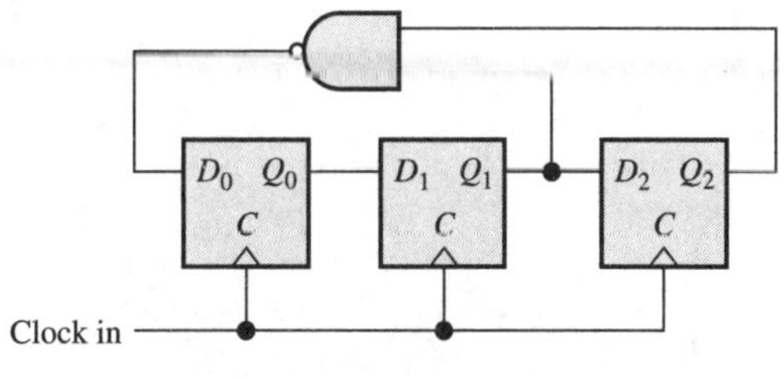

Figure T7.6

Chapter 8

Computers and Microcontrollers

Study of this chapter will enable you to:

- Identify and describe the functional blocks of a computer.
- Define the terms *microprocessor*, *microcomputer*, and *microcontroller*.
- Select the type of memory needed for a given application.
- Understand how microcontrollers can be applied in your field of specialization.
- Identify the registers in the programmer's model and their functions for the HCS12/9S12 microcontroller family from Freescale Semiconductor, Inc.
- List some of the instructions and addressing modes of HCS12/9S12 microcontrollers.
- Write simple assembly language programs, using the CPU12 instruction set.

Introduction to this chapter:

Certainly you are familiar with general-purpose electronic computers that are used for business, engineering design, word processing, and other applications. Although it is sometimes not readily apparent, special-purpose computers can be found in many products such as automobiles, appliances, cameras, fax machines, garage-door openers, and instrumentation. An **embedded computer** is part of a product that is not called a computer. Virtually any recently manufactured device that is partly electrical in nature is almost certain to contain one or more embedded computers. Typical automobiles contain over 100 embedded computers. The emphasis of this chapter is on embedded computers.

Relatively simple computers for embedded control applications can be completely implemented on a single silicon chip costing less than a dollar. This type of computer is often called a **microcontroller** (MCU) and is useful for problems such as control of a washing machine, a printer, or a toaster.

An embedded computer is part of a product, such as an automobile, printer, or bread machine, that is not called a computer.

In this chapter, we give an overview of MCU organization and instruction sets using the Freescale Semiconductor HCS12/9S12 family as an example. Hundreds of types of MCUs and their variations are in use, but most of the underlying concepts are similar from one to another. The primary intent of this chapter is to give you an understanding of these basic concepts. Space is not available in this book for the intensive coverage of a particular MCU needed to prepare you to design complex mechatronic systems.

Computer capability has advanced rapidly and costs have fallen dramatically, a trend that will continue for the foreseeable future. For the past several decades, the price for a given computer capability has been cut in half about every 18 months. You should view embedded MCUs as powerful, but inexpensive, resources that are appropriate for solving virtually any control or instrumentation problem in your field of engineering, no matter how complex or mundane the problem may be.

8.1 COMPUTER ORGANIZATION

Figure 8.1 shows the system-level diagram of a computer. The **central processing unit** (CPU) is composed of the **arithmetic/logic unit** (ALU) and the **control unit**.

The ALU carries out arithmetic and logic operations on data such as addition, subtraction, comparison, or multiplication. Basically, the ALU is a logic circuit similar to those discussed in Chapter 7 (but much more complex).

The control unit supervises the operation of the computer, such as determining the location of the next instruction to be retrieved from memory and setting up the ALU to carry out operations on data. The ALU and control unit contain various **registers** that hold operands, results, and control signals. (Recall from Section 7.6 that a register is simply an array of flip-flops that can store a word composed of binary digits.) Later in this chapter, we will discuss the functions of various CPU registers for the CPU12 which is used in the Freescale HCS12/9S12 family.

Memory

Memory can be thought of as a sequence of locations that store data and instructions. Each memory location has a unique address and typically stores one byte of data,

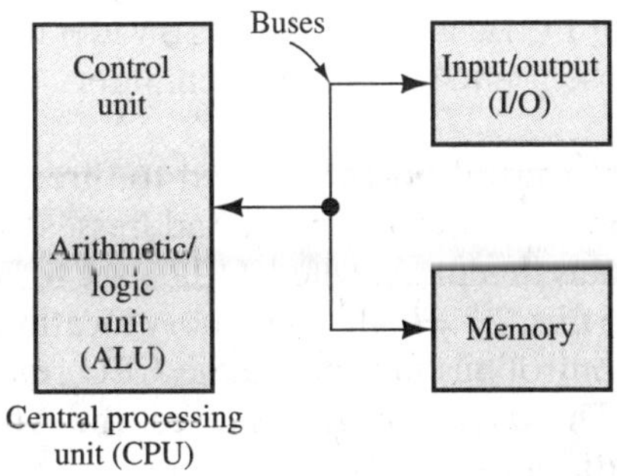

Figure 8.1 A computer consists of a central processing unit, memory, buses, and input–output devices.

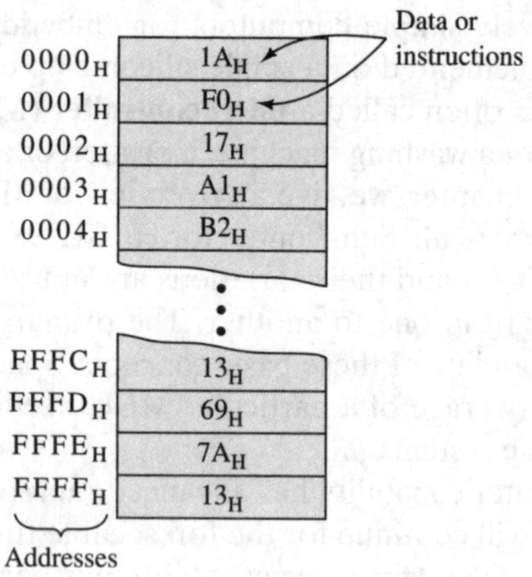

Figure 8.2 A 64-Kbyte memory that has $2^{16} = 65{,}536$ memory locations, each of which contains one byte (eight bits) of data. Each location has a 16-bit address. It is convenient to represent the addresses and data in hexadecimal form as shown. The addresses range from 0000_H to $FFFF_H$. For example, the memory location 0004_H contains the byte $B2_H = 10110010_2$.

Several notations are used for hexadecimal numbers, including the subscript 16, the subscript H, and the prefix \$. Thus, $F2_{16}$, $F2_H$, and \$F2 are alternative ways to indicate that F2 is a hexadecimal number.

which can conveniently be represented by two hexadecimal digits. (Of course, within the computer circuits, data appear in binary form.)

Usually, memory capacity is expressed in Kbytes, where $1\,K = 2^{10} = 1024$. Similarly, 1 Mbyte is $2^{20} = 1{,}048{,}576$ bytes. A 64-Kbyte memory is illustrated in Figure 8.2. Under the direction of the control unit, information can either be written to or read from each memory location. (We are assuming that we have the read/write type of memory. Later, we will see that another type, known as read-only memory, can also be very useful.)

Programs

Programs are sequences of instructions stored in memory. Typically, the controller fetches (i.e., retrieves) an instruction, determines what operation is called for by the instruction, fetches data from memory as required, causes the ALU to perform the operation, and writes results back to memory. Then, the next instruction is fetched, and the process is repeated. We will see that the CPU12 can execute a rich variety of instruction types.

Buses

The various elements of a computer are connected by **buses**, which are sets of conductors that transfer multiple bits at a time. For example, the **data bus** transfers data (and instructions) between the CPU and memory (or I/O devices). In small computers, the width of the data bus (i.e., the number of bits that can be transferred at a time) is typically eight bits. Then, one byte can be transferred between the CPU and memory (or I/O) at a time. (The bus is wider in more powerful general-purpose CPUs such as those found in personal computers, which typically have data-bus widths of 64 bits.)

Several **control buses** are used to direct the operations of the computer. For example, one control bus sends the addresses for memory locations (or I/O devices) as well as signals that direct whether data are to be read or written. With an address bus width of 16 bits, $2^{16} = 64K$ of memory locations (and I/O devices) can be addressed. Another control bus internal to the CPU transfers signals from the control unit to the ALU. These control signals direct the ALU to perform a particular operation, such as addition.

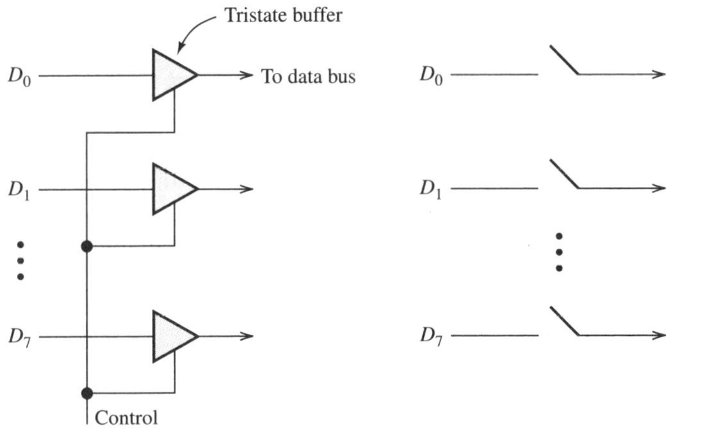

(a) Circuit symbols

(b) Functional equivalent of (a): When the control signal is low, the switches are open; when the control signal is high, the switches are closed

Figure 8.3 Data are applied to the data bus through tristate buffers, which can function either as closed or as open switches.

Buses can be bidirectional. In other words, they can transfer data in either direction. Let us consider the data bus connecting the CPU and memory. Of course, the memory and the CPU cannot apply conflicting data signals to the bus at the same time. Conflict is avoided by transferring data to the bus through **tristate buffers**, as illustrated in Figure 8.3. Depending on the control signal, the tristate buffers function either as open or as closed switches. When a byte of data is to be transferred from the CPU to memory, the tristate buffers are enabled (switches closed) on the CPU and disabled (switches open) in the memory. The data inputs of both the CPU and the memory are connected to the bus at all times, so data can be accepted from the bus as desired. Thus, the data from the CPU appear on the bus and can be stored by the memory. When data are to be transferred from memory to the CPU, the conditions of the tristate buffers are reversed.

Input–Output

Some examples of I/O devices are keyboards, display devices, and printers. An important category of input devices in control applications are **sensors**, which convert temperatures, pressures, displacements, flow rates, and other physical values to digital form that can be read by the computer. **Actuators** are output devices such as valves, motors, and switches that allow the computer to affect the system being controlled.

Some computers are said to have **memory-mapped I/O**, in which I/O devices are addressed by the same bus as memory locations. Then, the same instructions used for storing and reading data from memory can be used for I/O. Other computers have a separate address bus and instructions for I/O. We discuss primarily systems that use memory-mapped I/O.

A **microprocessor** is a CPU contained on a single integrated-circuit chip. The first microprocessor was the Intel 4004, which appeared in 1971 and cost several thousand dollars each. Subsequently, microprocessors have dramatically fallen in price and increased in performance. A **microcomputer,** such as a PC or a laptop, combines

A microcontroller is a complete computer containing the CPU, memory, and I/O on a single silicon chip.

a microprocessor with memory and I/O chips. A **microcontroller** (MCU) combines CPU, memory, buses, and I/O on a single chip and is optimized for embedded control applications. We give a more detailed overview of the HCS12/9S12 microcontroller family later in this chapter.

There are several variations of computer organization. In computers with **Harvard architecture**, there are separate memories for data and instructions. If the same memory contains both data and instructions, we have **von Neumann architecture**. The HCS12/9S12 microcontroller family uses von Neumann architecture.

Exercise 8.1 Suppose that a microprocessor has an address bus width of 20 bits. How many memory locations can it access?
Answer $2^{20} = 1{,}048{,}576 = 1024\ \text{K} = 1\ \text{M}$. □

Exercise 8.2 How many bits can be stored in a 64-Kbyte memory?
Answer 524,288. □

8.2 MEMORY TYPES

Several types of memory are used in computers: (1) Read-and-write memory (RAM), (2) Read-only memory (ROM), and (3) Mass storage. We discuss each type in turn. Then, we consider how to select the best type of memory for various applications.

RAM

Read-and-write memory (RAM) is used for storing data, instructions, and results during execution of a program. Semiconductor RAM consists of one or more silicon integrated circuits (each of which has many storage cells) and control logic so that information can be transferred into or out of the cell specified by the address.

RAM and ROM do not incur any loss of speed when the memory locations are accessed in random order. In fact, RAM originally meant random-access memory.

Usually, the information that is stored in RAM is lost when power is removed. Thus, we say that RAM is **volatile**. Originally, the acronym RAM meant random-access memory, but the term has changed its meaning over time. As the term is used now, RAM means volatile semiconductor memory. (Actually, RAM is also available with small batteries that maintain information in the absence of other power.)

The time required to access data in RAM is the same for all memory locations. The fastest RAM is capable of access times of a few nanoseconds. No time penalty is incurred by accessing locations in random order.

There are two types of RAM in common use. In **static RAM**, the storage cells are *SR* flip-flops that can store data indefinitely, provided that power is applied continuously. In **dynamic RAM**, information is stored in each cell as charge (or lack of charge) on a capacitor. Because the charge leaks off the capacitors, it is necessary to refresh the information periodically. This makes the use of dynamic RAM more complex than the use of static RAM. The advantage of dynamic RAM is that the basic storage cell is smaller, so that chips with larger capacities are available. A relatively small amount of RAM is needed in most control applications, and it is simpler to use static RAM.

An 8K-word by 8-bit static RAM chip is illustrated in Figure 8.4. The chip has 13 address lines, eight data lines, and three control lines. The "bubbles" on the control input lines indicate that they are active when low. Unless the chip select line is low, the chip neither stores data nor places data on the data bus. If both the output-enable and the chip-select inputs are low, the data stored in the location specified by the address appear on the data lines. If both the write-enable and the chip-select lines

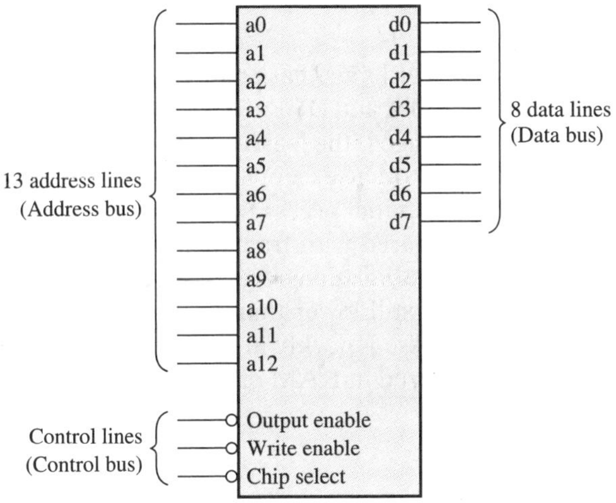

Figure 8.4 A generic 8K-word by 8-bit word RAM.

are low, the data appearing on the data bus are stored in the location specified by the address signals. In normal operation, both the output enable and write enable lines are never low at the same time.

ROM

In normal operation, **read-only memory** (ROM) can be read, but not written to. The chief advantages of ROM are that data can be read quickly in random order and that information is not lost when power is turned off. Thus, we say that ROM is **nonvolatile** (i.e., permanent). ROM is useful for storing programs such as the *boot program*, which is executed automatically when power is applied to a computer. In simple dedicated applications such as the controller for a clothes washer, all of the programs are stored in ROM.

The primary advantage of ROM is that it is nonvolatile. Information stored in RAM is lost when power is interrupted.

Several types of ROM exist. For example, in **mask-programmable ROM**, the data are written when the chip is manufactured. A substantial cost is incurred in preparing the mask that is used to write the data while manufacturing this type of ROM. However, mask-programmable ROM is the least expensive form of ROM when the mask cost is spread over a sufficiently large number of units. Mask-programmable ROM is not a good choice if frequent changes in the information stored are necessary, as in initial system development.

In **programmable read-only memory** (PROM), data can be written by special circuits that blow tiny fuses or leave them unblown, depending on whether the data bits are zeros or ones. Thus, with PROM, we write data once and can read it as many times as desired. PROM is an economical choice if a small number of units are needed.

Erasable PROM (EPROM) is another type that can be erased by exposure to ultraviolet light (through a window in the chip package) and rewritten by using special circuits. **Electrically erasable PROMs** (EEPROMs) can be erased by applying proper voltages to the chip. Although we can write data to an EEPROM, the process is much slower than for RAM.

Flash memory is a nonvolatile technology in which data can be erased and rewritten relatively quickly in blocks of locations, ranging in size from 512 bytes up to 512 Kbytes. Flash memory has a limited lifetime, typically on the order of 10 thousand to 100 thousand read/write cycles. Flash is a rapidly advancing technology and may eventually replace hard drives for mass storage in general purpose computers.

Mass Storage

Hard disks, CD-ROMs, and DVDs are examples of sequential memories in which access is faster if memory locations are accessed in order.

Mass-storage units include hard disks and flash memory, both of which are read/write memory. Another type is CD-ROM and DVD-ROM disks, which are used for storing large amounts of data. Mass storage is the least expensive type of memory per unit of capacity. With all forms of mass storage except flash, a relatively long time is required to access a particular location. Initial access times for mass storage range upward from several milliseconds, compared with fractions of a microsecond for RAM or ROM. However, if mass-storage locations are accessed sequentially, the transfer rate is considerably higher (but still lower than for RAM or ROM). Usually, data and instructions need to be accessed quickly in random order during execution of a program. Thus, programs are stored in RAM or ROM during execution.

Selection of Memory

The main considerations in choosing the type of memory to be used are:

1. The trade-off between speed and cost.
2. Whether the information is to be stored permanently or must be changed frequently.
3. Whether data are to be accessed in random order or in sequence.

In general-purpose computers, programs and data are read into RAM before execution from mass-storage devices such as hard disks. Because many different programs are used, it is not practical to store programs in semiconductor ROM, which would be too expensive for the large memory space required. Furthermore, information stored in ROM is more difficult to modify compared to data stored on a hard disk. We often find a small amount of ROM used for the startup or boot program in general-purpose computers, but most of the memory is RAM and mass storage.

On the other hand, in embedded microcontrollers, programs are usually stored in semiconductor ROM, and only a small amount of RAM is needed to store temporary results. For example, in a controller for a television receiver, the programs for operating the TV are stored in ROM, but time and channel information entered by the user is stored in RAM. In this application, power is applied to the RAM even when the TV is "turned off." However, during a power failure, the data stored in RAM are lost (unless the TV has a battery backup for its RAM). Usually, we do not find mass-storage devices used in embedded computers.

8.3 DIGITAL PROCESS CONTROL

Figure 8.5 shows the general block diagram of a control scheme for a physical process such as an internal combustion engine. Various physical inputs such as power and material flow are regulated by actuators that are in turn controlled by the microcontroller.

D/A and DAC are both acronyms for digital-to-analog converter.

Some actuators are analog and some are digital. Examples of digital actuators are switches or valves that are either on or off, depending on the logic value of their control signals. Digital actuators can be controlled directly by digital control lines. Analog actuators require an analog input. For example, the rudder of an airplane may deflect in proportion to an analog input signal. Then, **digital-to-analog** (D/A) **converters** are needed to convert the digital signals to analog form before they are applied to analog actuators.

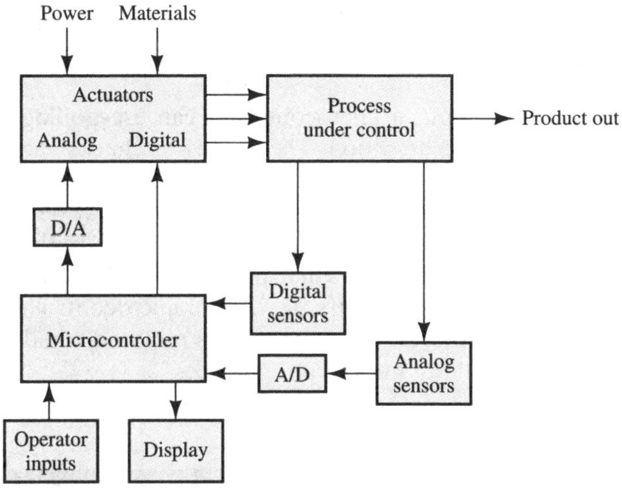

Figure 8.5 Microcontroller-based control of a physical process.

Various sensors produce electrical signals related to process parameters (such as temperature, pressure, pH, velocity, or displacement) of the process under control. Some sensors are digital and some are analog. For example, a pressure sensor may consist of a switch that closes producing a high output signal when pressure exceeds a particular value. On the other hand, an analog pressure sensor produces an output voltage that is proportional to pressure. **Analog-to-digital** (A/D) **converters** are used to convert the analog sensor signals to digital form.

Often, a display is provided so that information about the process can be accessed by the operator. A keyboard or other input device enables the operator to direct operation of the control process.

Many variations of the system shown in Figure 8.5 are possible. For example, sometimes we simply want to instrument a process and present information to the operator. This is the situation for automotive instrumentation in which sensors provide signals for speed, fuel reserve, oil pressure, engine temperature, battery voltage, and so on. These data are presented to the driver by one or more displays.

Actuators, sensors, and I/O tend to be unique to each application and do not lend themselves to integration with the MCU. A/D and D/A converters often are included within the MCU. Thus, a typical system consists of an MCU, sensors, actuators, and I/O devices. Systems may not contain all of these elements. Within a given MCU family, variations are usually available with respect to the amount and type of memory, the number of A/D channels, and so forth, that are included on the chip.

Virtually any system can be controlled or monitored by a microcontroller. Here is a short list: traffic signals, engines, chemical plants, antiskid brakes, manufacturing processes, stress measurement in structures, machine tools, aircraft instrumentation, monitoring of patients in a cardiac-care unit, nuclear reactors, and laboratory experiments.

Interrupts versus Polling

In many control applications, the microcontroller must be able to respond to certain input signals very quickly. For example, an overpressure indication in a nuclear power plant may require immediate attention. When such an event occurs, the microcontroller must **interrupt** what it is doing and start a program known as an **interrupt**

handler that determines the source of the interrupt and takes appropriate action. Many microcontrollers have hardware capability and instructions for handling these interrupts.

Instead of using interrupts, a microcontroller can use **polling** to determine if any parts of the system need attention. The processor checks each sensor in turn and takes appropriate actions as needed. However, continuous polling is wasteful of processor time. In complex applications, the processor may be required to carry out extensive, but lower-priority activities, much of the time. In this case, interrupts provide faster response to critical events than polling does. For a breadmaker, polling would be acceptable because no activity ties up the microcontroller for more than a few milliseconds. Furthermore, any of the actions required could be delayed by a few tens of milliseconds without undue consequences.

PRACTICAL APPLICATION 8.1

Fresh Bread Anyone?

Let us consider a relatively simple microcontroller application: a breadmaker. Possibly, you have had experience with this popular appliance. The chef measures and adds the ingredients (flour, water, dried milk, sugar, salt, yeast, and butter) to the bread pan, makes selections from the menu by using a keypad, and takes out a finished loaf of bread after about four hours.

The diagram of a bread machine is shown in Figure PA8.1. There are three digital actuators in the bread machine: a switch to control the heating element, a switch for the mixing and kneading motor, and a switch for the fan used to cool the loaf after baking is finished. Analog actuators are not needed in this application.

An analog sensor is used to measure temperature. The sensor output is converted to digital form by an A/D converter.

A timer circuit is part of the MCU that is initially loaded with the time needed to complete the loaf. The timer is a digital circuit that counts down, similar to the counter circuits discussed in Section 7.6.

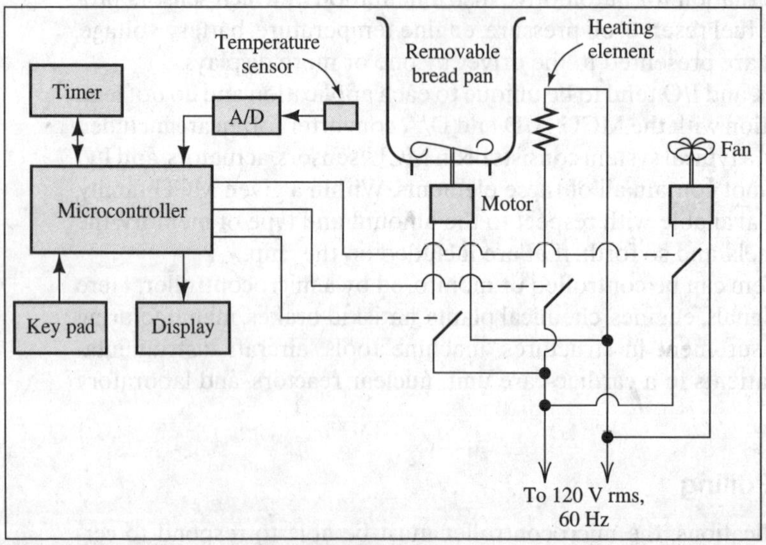

Figure PA8.1 A relatively simple application for an MCU—a breadmaking machine.

The timer indicates the number of hours and minutes remaining in the process. The MCU can read the time remaining and use it to make decisions. Time remaining to completion of the loaf is also displayed for the convenience of the chef.

The control programs are stored in ROM. The parameters entered by the chef are written into RAM (for example, whether the bread crust is to be light, medium, or dark). The MCU continually checks the time remaining and the temperature. By executing the program stored in ROM, the computer determines when the machine should mix the ingredients, turn on the heating element to warm the dough and cause it to rise, knead the dough, bake, or cool down. The duration and temperature of the various parts of the cycle depend on the initial selections made by the chef.

First, the machine mixes the ingredients for several minutes, and the heating element is turned on to warm the yeast that makes the dough rise. While the dough is rising, a warm temperature is required, say 90°F. Thus, the heating element is turned on and the MCU reads the temperature frequently. When the temperature reaches the desired value, the heating element is turned off. If the temperature falls too low, the element is again turned on.

The MCU continues to check the time remaining and the temperature. According to the programs stored in ROM and the parameters entered by the chef (which are saved in RAM), the motors and heating element are turned on and off.

In this application, about 100 bytes of RAM would be needed to store information entered by the operator and temporary data. Also, about 16 Kbytes of ROM would be needed to store the programs. Compared to the total price of the appliance, the cost of this amount of ROM is very small. Therefore, many variations of the program can be stored in ROM, and bread machines can be very versatile. In addition to finished loaves of bread, they can also bake cakes, cook rice, make jam, or prepare dough for other purposes, such as cinnamon rolls.

8.4 PROGRAMMING MODEL FOR THE HCS12/9S12 FAMILY

Earlier, we discussed a generic computer shown in Figure 8.1 on page 421. In this section, we give a more detailed internal description of the HCS12/9S12 microcontroller family from Freescale Semiconductor. Space does not allow us to discuss all of the features, instructions, and programming techniques for these microcontrollers. However, we will describe the programming model, selected instructions, and a few simple programs to give you a better understanding of how microcontrollers can be used for embedded applications that you will encounter in your field.

The HCS12/9S12 Programming Model

The ALU and the control unit contain various registers that are used to hold operands, the address of the next instruction to be executed, addresses of data, and results. For example, the programmer's model for the CPU12 is illustrated in Figure 8.6. (Actually, the microcontrollers contain many other registers—only the registers of concern to the programmer are shown in the figure; thus, Figure 8.6 is often called the programming model.)

The **accumulators** are general-purpose registers that hold one of the arguments and the result of all arithmetic and logical operations. Registers A and B each contain 8 bits with the least significant bit on the right (bit 0 in Figure 8.6) and the most significant bit on the left. Sometimes A and B are used as separate registers, and other times they are used in combination as a single 16-bit register, denoted as register D. It is important to remember that D is not separate from A and B.

It is important to remember that register D is not separate from registers A and B.

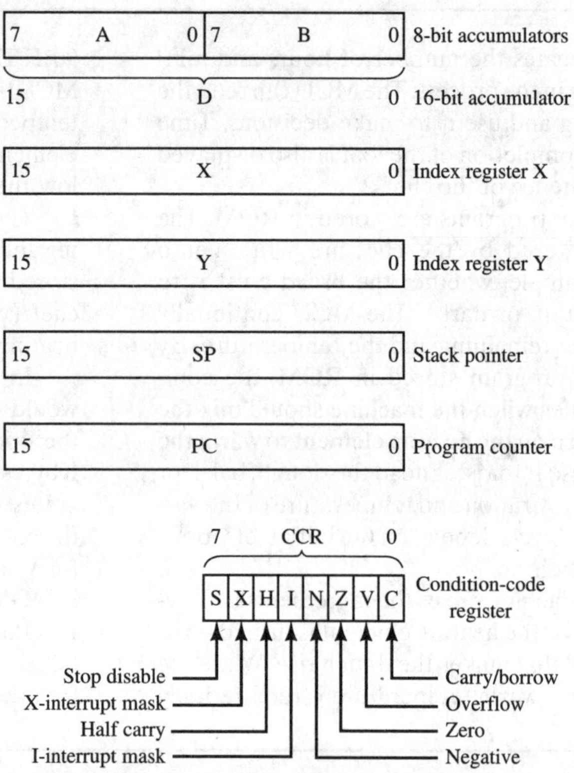

Figure 8.6 The CPU12 programmer's model.

The **program counter** (PC) is a 16-bit register that contains the address of the first byte of the next instruction to be fetched (read) from memory by the control unit. The size of the PC is the same as the size of memory addresses; thus, the memory potentially contains up to $2^{16} = 64$ K locations, each of which contains one byte of data or instructions as illustrated in Figure 8.2 on page 422.

The **index registers** X and Y are mainly used for a type of addressing (of data) known as indexed addressing, which we will discuss later.

The **condition-code register** (CCR) is an 8-bit register in which each bit depends either on a condition of the processor or on the result of the preceding logic or arithmetic operation. The details of the condition-code register are shown in Figure 8.6. For example, the carry bit C (bit 0 of the condition-code register) is set (to logic 1) if a carry (or borrow) occurred in the preceding arithmetic operation. Bit 1 (overflow or V) is set if the result of the preceding operation resulted in overflow or underflow. Bit 2 (zero or Z) is set to 1 if the result of the preceding operation was zero. Bit 3 (negative or N) is set if the result was negative. The meaning and use of the remaining bits will be discussed as the need arises.

Stacks and the Stack Pointer Register

A **stack** is a sequence of locations in memory used to store information such as the contents of the program counter and other registers when a subroutine is executed or when an interrupt occurs. (We discuss subroutines shortly.) As the name implies, information is added to (pushed onto) the top of the stack and later read out (pulled off) in the reverse order that it was written. This is similar to adding plates to the top of a stack when clearing a dinner table and then taking the plates off the top of the stack when loading a dishwasher. After data are pulled off the stack, they are considered to no longer exist in memory and are written over by later push commands. The first word pushed onto the stack is the last to be pulled off, and stacks are called **last-in first-out memories** (LIFOs).

Stacks are last-in first-out memories. Information is added to (pushed onto) the top of the stack and eventually read out (pulled off) in the reverse order that it was written.

The **stack pointer** is a CPU register that keeps track of the address of the top of the stack. Each time the content of a register is pushed onto the stack, the content of the stack pointer is decremented by one if the register contained one byte. If the register contained two bytes, the stack-pointer content is decreased by two. (Addresses are smaller in value as we progress upward in the stack.) Conversely, when data is pulled from the stack and transferred to a register, the stack-pointer content is increased by one or two (depending on the length of the register). When the content of one of the 8-bit registers (A, B, or CCR) is pushed onto the stack (by the commands PSHA, PSHB, or PSHC, respectively), these operations take place:

1. The content of the stack pointer is reduced by one.
2. The content of the 8-bit register is stored at the address corresponding to the content of the stack pointer.

When the content of one of the 16-bit registers D, X, or Y is pushed onto the stack (by the commands PSHD, PSHX, or PSHY), the following operations take place:

1. The content of the stack pointer is decremented by one and the least significant byte (bits 8 through 15) of the content of the 16-bit register is stored at the address corresponding to the content of the stack pointer.
2. The content of the stack pointer is again decremented by one, and the most significant byte of the content of the 16-bit register is stored at the address corresponding to the content of the stack pointer.

In pulling data off of the stack, the operations are reversed. For an 8-bit register (commands PULA, PULB, or PULC):

1. The data in the memory location pointed to by the stack pointer is stored in the register.
2. The content of the stack pointer is incremented by one.

For a 16-bit register (commands PULD, PULX, or PULY):

1. The data in the memory location to which the stack pointer points is stored in the high byte of the register, and the content of the stack pointer is incremented by one.
2. The data in the memory location to which the stack pointer points is stored in the low byte of the register, and the content of the stack pointer is again incremented by one.

Figure 8.7 illustrates the effects of the command sequence PSHA, PSHB, PULX. Figure 8.7(a) shows the original contents of pertinent registers and memory locations. (Memory locations always contain something; they are never blank. However, when the content of a memory location is unknown or does not matter, we have left the location blank.) Figure 8.7(b) shows the new contents after the command PSHA has been executed. Notice that the initial content of register A has been stored in location 090A and that the content of SP has been decremented by one. (Furthermore, the initial contents of A and B are unchanged.) Figure 8.7(c) shows the contents after the command PSHB has been executed. Notice that the content of register B has

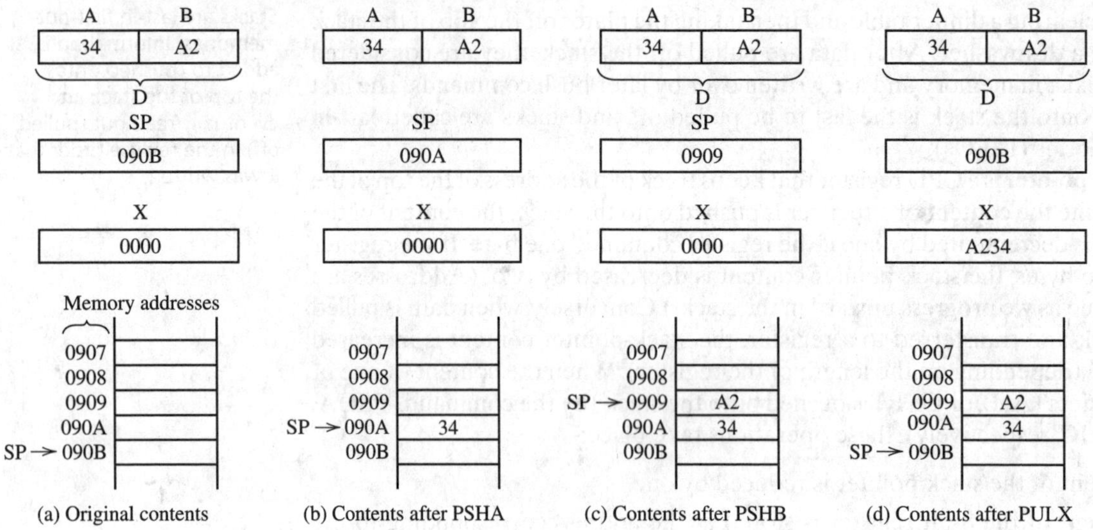

Figure 8.7 Register and memory contents for the command sequence: PSHA, PSHB, PULX.

been stored in location 0909 and that the content of SP has been decremented by one. Finally, Figure 8.7(d) shows the new contents after the command PULX has been executed. Notice that the content of memory location 0909 has been stored in the first byte of register X and the content of memory location 090A has been stored in the second byte of register X. Also, the content of SP has been increased by two.

Exercise 8.3 Starting from the initial contents shown in Figure 8.7(a), determine the content of register X after execution of the command sequence PSHB, PSHA, PULX.

Answer The content of the X register is 34A2. □

Exercise 8.4 Starting from the initial contents shown in Figure 8.7(a), determine the content of register X after the command sequence PSHX, PSHA, PULX.

Answer The content of the X register is 3400. □

Exercise 8.5 Suppose that initially the contents of all memory locations in the stack are 00, and the stack pointer register contains 0806. Then, the following operations occur in sequence:

1. The data byte $A7_H$ is pushed onto the stack.
2. 78_H is pushed onto the stack.
3. One byte is pulled from the stack.
4. FF is pushed onto the stack.

List the contents of memory locations 0800 through 0805 after each step. Also, give the content of the stack pointer (SP).

Answer After step 1, we have:

```
0800: 00    SP: 0805
0801: 00
0802: 00
0803: 00
0804: 00
0805: A7
```

After step 2, we have:

```
0800: 00    SP: 0804
0801: 00
0802: 00
0803: 00
0804: 78
0805: A7
```

After step 3, we have:

```
0800: 00    SP: 0805
0801: 00
0802: 00
0803: 00
0804: 78
0805: A7
```

After step 4, we have:

```
0800: 00    SP: 0804
0801: 00
0802: 00
0803: 00
0804: FF   (New data replaces old.)
0805: A7
```

□

8.5 THE INSTRUCTION SET AND ADDRESSING MODES FOR THE CPU12

Computers excel at executing simple instructions, such as quickly and accurately adding a number stored in a given memory location to the content of a specified register. Computers are capable of highly sophisticated and seemingly intelligent behavior by following instruction sequences called **programs** or **software**. These are prepared by a human programmer.

Unfortunately, even the smallest oversight on the part of the programmer can render a program useless until the error is corrected. A substantial part of the effort in designing an MCU-based controller is in writing software. To be effective, the programmer must be fully knowledgeable about the fine details of the instruction set for the MCU in use. Our objective in this and the next section is simply to give you a brief overview, not to make you an expert programmer.

A substantial part of the effort in designing an MCU-based controller is in writing software.

In general, instruction sets are similar between different MCU types, but details differ. Once one has mastered programming of a given machine, it is much easier to learn and make good use of the instruction set of another processor. Here again, we take the CPU12 as an example. More details about it can be readily found on the Web.

Instructions for the CPU12

A selected set of instructions for the CPU12 is listed in Table 8.1. The first column in the table gives the mnemonic for each instruction, the second column is a brief description and an equivalent Boolean expression for the instruction. For example, the ABA instruction adds the content of register B to the content of A with the result residing in A. We can indicate this operation as

(A) represents the *content* of register A.

$$(A) + (B) \rightarrow A$$

as shown in the second column of the table.

Table 8.1. Selected Instructions for the CPU12

Source Form	Operation	Addr. Mode	Machine Code	S	X	H	I	N	Z	V	C
ABA	Add Accumulators $(A) + (B) \rightarrow A$	INH	18 06	-	-	↕	-	↕	↕	↕	↕
ADDA (opr)	Add Memory to A $(A) + (M) \rightarrow A$	IMM DIR EXT IDX	8B ii 9B dd BB hh ll AB *	-	-	↕	-	↕	↕	↕	↕
ADDB (opr)	Add Memory to B $(B) + (M) \rightarrow B$	IMM DIR EXT IDX	CB ii DB dd FB hh ll EB *	-	-	↕	-	↕	↕	↕	↕
ADDD (opr)	Add Memory to D $(D) + (M{:}\,M+1) \rightarrow D$	IMM DIR EXT IDX	C3 jj kk D3 dd F3 hh ll E3 *	-	-	-	-	↕	↕	↕	↕
BCS (rel)	Branch if Carry Set (if $C = 1$)	REL	25 rr	-	-	-	-	-	-	-	-
BEQ (rel)	Branch if Equal (if $Z = 1$)	REL	27 rr	-	-	-	-	-	-	-	-
BLO (rel)	Branch if Lower^{U} (if $C = 1$)	REL	25 rr	-	-	-	-	-	-	-	-
BMI (rel)	Branch if Minus^{S} (if $N = 1$)	REL	2B rr	-	-	-	-	-	-	-	-
BNE (rel)	Branch if Not Equal (if $Z = 0$)	REL	26 rr	-	-	-	-	-	-	-	-
BPL (rel)	Branch if Plus^{S} (if $N = 0$)	REL	2A rr	-	-	-	-	-	-	-	-
BRA (rel)	Branch Always	REL	20 rr	-	-	-	-	-	-	-	-
CLRA	Clear Accumulator A $\$00 \rightarrow A$	INH	87	-	-	-	-	0	1	0	0
CLRB	Clear Accumulator B $\$00 \rightarrow B$	INH	C7	-	-	-	-	0	1	0	0
COMA	Complement Accumulator A $(\overline{A}) \rightarrow A$	INH	41	-	-	-	-	↕	↕	0	1
INCA	Increment Accumulator A $(A) + \$01 \rightarrow A$	INH	42	-	-	-	-	↕	↕	↕	-
INCB	Increment Accumulator B $(B) + \$01 \rightarrow B$	INH	52	-	-	-	-	↕	↕	↕	-
INX	Increment Index Register X $(X) + \$0001 \rightarrow X$	INH	08	-	-	-	-	-	↕	-	-
JMP (opr)	Jump Routine Address → PC	EXT IDX	06 hh ll 05 *	-	-	-	-	-	-	-	-
JSR (opr)	Jump to Subroutine (See Text)	DIR EXT IDX	17 dd 16 hh ll 15 *								
LDAA (opr)	Load Accumulator A $(M) \rightarrow A$	IMM DIR EXT IDX	86 ii 96 dd B6 hh ll A6 *	-	-	-	-	↕	↕	0	-
LDAB (opr)	Load Accumulator B $(M) \rightarrow B$	IMM DIR EXT IDX	C6 ii D6 dd F6 hh ll E6 *	-	-	-	-	↕	↕	0	-

Table 8.1. Selected Instructions for the CPU12 (*Cont.*)

Source Form	Operation	Addr. Mode	Machine Code	Condition Codes							
				S	X	H	I	N	Z	V	C
LDD (opr)	Load Accumulator D $(M):(M+1) \rightarrow D$	IMM DIR EXT IDX	CC jj kk DC dd FC hh ll EC *	-	-	-	-	↕	↕	0	-
LDX (opr)	Load Index Register X $(M):(M+1) \rightarrow X$	IMM DIR EXT IDX	CE jj kk DE dd FE hh ll EE *	-	-	-	-	↕	↕	0	-
LDY (opr)	Load Index Register Y $(M):(M+1) \rightarrow Y$	IMM DIR EXT IDX	CD jj kk DD dd FD hh ll ED *	-	-	-	-	↕	↕	0	-
MUL	Multiply A by B^U $(A) \times (B) \rightarrow D$	INH	12	-	-	-	-	-	-	-	↕
NOP	No Operation	INH	A7	-	-	-	-	-	-	-	-
PSHA	Push A onto Stack $(SP) - 1 \Rightarrow SP; (A) \Rightarrow M_{(SP)}$	INH	36	-	-	-	-	-	-	-	-
PSHB	Push B onto Stack $(SP) - 1 \Rightarrow SP; (B) \Rightarrow M_{(SP)}$	INH	37	-	-	-	-	-	-	-	-
PSHX	Push X onto Stack $(SP) - 2 \Rightarrow SP; (X_H: X_L) \Rightarrow M_{(SP)}: M_{(SP+1)}$	INH	34	-	-	-	-	-	-	-	-
PSHY	Push Y onto Stack $(SP) - 2 \Rightarrow SP; (Y_H: Y_L) \Rightarrow M_{(SP)}: M_{(SP+1)}$	INH	35	-	-	-	-	-	-	-	-
PULA	Pull A from Stack $(M_{(SP)}) \Rightarrow A;\ (SP)+1 \Rightarrow SP$	INH	32	-	-	-	-	-	-	-	-
PULB	Pull B from Stack $(M_{(SP)}) \Rightarrow B; (SP)+1 \Rightarrow SP$	INH	33	-	-	-	-	-	-	-	-
PULX	Pull X from Stack $(M_{(SP)}: M_{(SP+1)}) \Rightarrow X_H: X_L; (SP) + 2 \Rightarrow SP$	INH	30	-	-	-	-	-	-	-	-
PULY	Pull Y from Stack $(M_{(SP)}: M_{(SP+1)}) \Rightarrow Y_H: Y_L; (SP) + 2 \Rightarrow SP$	INH	31	-	-	-	-	-	-	-	-
RTS	Return from Subroutine $(M_{(SP)}: M_{(SP+1)}) \Rightarrow PC;\ (SP) + 2 \Rightarrow SP$	INH	3D	-	-	-	-	-	-	-	-
STAA (opr)	Store Accumulator A $(A) \rightarrow M$	DIR EXT IDX	5A dd 7A hh ll 6A *	-	-	-	-	↕	↕	0	-
STAB (opr)	Store Accumulator B $(B) \rightarrow M$	DIR EXT IDX	5B dd 7B hh ll 6B *	-	-	-	-	↕	↕	0	-
STD (opr)	Store Accumulator D $(A) \rightarrow M; (B) \rightarrow M+1$	DIR EXT IDX	5C dd 7C hh ll 6C *	-	-	-	-	↕	↕	0	-

Table 8.1. Selected Instructions for the CPU12 (*Cont.*)

Source Form	Operation	Addr. Mode	Machine Code	S	X	H	I	N	Z	V	C
STOP	Stop Internal Clocks. If S control bit = 1, the STOP instruction is disabled and acts like a NOP	INH	18 3E	-	-	-	-	-	-	-	-
TSTA	Test Accumulator A; (A) – 00	INH	97	-	-	-	-	↕	↕	0	0
TSTB	Test Accumulator B; (B) – 00	INH	D7	-	-	-	-	↕	↕	0	0

S indicates instructions that are intended for two's complement signed numbers
U indicates instructions that are intended for unsigned numbers
* For indexed addressing (IDX). Only the first byte of the machine coded is given. An additional one to three bytes are needed. The details are beyond the scope of our discussion.
ii 8-bit immediate data
dd low byte of a direct address
hh ll high and low bytes of an extended address
jj kk high and low bytes of 16-bit immediate data
rr signed 8-bit offset in branch instruction

Mnemonics are easy for humans to remember. However, in the microcomputer memory, the instructions are stored as machine codes (or op codes) consisting of one or more 8-bit numbers, each of which is represented in the table as a two-digit hexadecimal number. For example, in the row for the ABA instruction, we see that the op code is 1806. Thus, the ABA instruction appears in memory as the binary numbers 00011000 and 00000110.

Look at the row for the ADDA(opr) instruction in which (opr) stands for the address of a memory location. The effect of the instruction is to add the content of a memory location to the content of accumulator A with the result residing in A. This is represented by the expression

$$(A) + (M) \rightarrow A$$

in which (M) represents the content of a memory location. Several **addressing modes** can be used to select the memory location to be accessed by some instructions. For example, the ADDA instruction can use any of several addressing modes. We will discuss the CPU12 addressing modes shortly.

Table 8.1 also shows the effect of each instruction on the contents of the condition-code register. The meanings of the symbols shown for each bit of the condition code are

- \- the bit is unchanged by this instruction
- 0 the bit is always cleared by this instruction
- 1 the bit is always set by this instruction
- ↕ the bit is set or cleared depending on the result

The CPU12 has many more instructions than those listed in the table; we have just given a sample of various kinds. Next, we briefly describe each of the addressing modes used by the CPU12.

Extended (EXT) Addressing

Recall that the CPU12 uses 16 bits (usually written as four hexadecimal digits) for memory addresses. In extended addressing, the complete address of the operand is included in the instruction. Thus, the instruction

```
ADDA $CA01
```

adds the content of memory location CA01 to the content of register A. (Later, we will see that a program called an assembler is used to convert the mnemonics to op codes. The $ sign indicates to the assembler that the address is given in hexadecimal form.) The op codes appear in three successive memory locations as

In CPU12 assembly language, a prefix of $ indicates that the number is hexadecimal.

BB (op code for ADDA with extended addressing)
CA (high byte of address)
01 (low byte of address)

Notice that the high byte of the address is given first followed by the low byte.

Direct (DIR) Addressing

In **direct addressing**, only the least significant two (hexadecimal) digits of the address are given, and the most significant two digits are assumed to be zero. Therefore, the effective address falls between 0000 and 00FF. For example, the instruction

```
ADDA $A9
```

adds the content of memory location 00A9 to the content of register A. The instruction appears in two successive memory locations as

9B (the op code for ADDA with direct addressing)
A9 (the low byte of the address)

Notice that the same result could be obtained by using extended addressing, in which case the instruction would appear as

```
ADDA $00A9
```

However, the extended addressing form of the instruction would occupy three bytes of memory, rather than two with direct addressing. Furthermore, the direct addressing form is completed more quickly.

Inherent (INH) Addressing

Some instructions, such as ABA, access only the MCU registers. We say that this instruction uses **inherent addressing**. An instruction sequence that adds the numbers in locations 23A9 and 00AA, then stores the result in location 23AB is

LDAA $23A9	(extended addressing, load A from location 23A9)
LDAB $AA	(direct addressing, load B from location 00AA)
ABA	(inherent addressing, add B to A)
STAA $23AB	(extended addressing, store result in 23AB)

Immediate (IMM) Addressing

In **immediate addressing**, which is denoted by the symbol #, the address of the operand is the address immediately following the instruction. For example, the instruction ADDA #$83 adds the hexadecimal number 83 to the contents of A. It is stored in two successive memory locations as

In CPU12 assembly language, the symbol # indicates immediate addressing.

8B (op code for ADDA with immediate addressing)
83 (operand)

Because A is a single-byte register, only one byte of memory is needed to store the operand.

On the other hand, D is a double-byte (16-bit) register, and its operand is assumed to occupy two memory bytes. For example, the instruction ADDD #$A276 adds the two-byte hexadecimal number A276 to the contents of D. It is stored in three successive memory locations as

C3 (op code for ADDD with immediate addressing)
A2 (high byte of operand)
76 (low byte of operand)

Indexed (IDX) Addressing

Indexed addressing is useful when we want to access a list of items either one after another in order, or perhaps skipping forward or backward through the list by twos, threes, etc. The CPU12 has a variety of indexed addressing options. (In Table 8.1, we have grouped all of these options under the IDX label. Only the first bite of the machine code, of two to four in all, is given in the table. This has been done because of space limitations in this chapter.)

Constant-Offset Indexed Addressing. In constant-offset indexed addressing, the effective address is formed by adding a signed offset to the content of a selected CPU register (X, Y, SP, or PC). The contents of X, Y, SP, or PC are not changed in this type of addressing. Suppose that X contains $1005 and Y contains $200A. Then, some examples of source code using this type of addressing and their effects are

STAA 5,X	store the content of A in location $100A
STD −3,Y	store the content of D in locations $2007 and $2008
ADDB $A,X	add the content of location $100F to register B

Accumulator-Offset Indexed Addressing. In this form of addressing, the content of one of the accumulators (A, B, or D) is added as an unsigned number to the content of a designated register (X, Y, SP, or PC) to obtain the effective address. For example, if X contains $2000 and A contains $FF, the command

LDAB A,X

loads the content of memory location $20FF into B. The contents of A and X remain the same after the command is completed as they were before the command.

Next, we discuss four types of indexed addressing that increment or decrement the contents of the selected CPU register (X, Y, or SP) either before or after the instruction is carried out. The amount of the increment or decrement can range from 1 to 8 and the selected CPU register contains the incremented or decremented value after the instruction is completed.

Auto Pre-incremented Indexed Addressing. Suppose that X contains $1005. Then, the instruction

STAA 5,+X

pre-increments X so the content of X becomes $100A and the content of A is stored in location $100A. X contains $100A after the completion of the instruction.

Auto Pre-decremented Indexed Addressing. Again, suppose that X contains $1005. Then, the instruction

STD 5,-X

pre-decrements X so the content of X becomes $1000 and the content of D is stored in locations $1000 and $1001. X contains $1000 after the completion of the instruction.

Notice that the sign preceding X determines whether we have a pre-increment or a pre-decrement. On the other hand, if the algebraic sign comes after the register name, we have either a post-increment or post-decrement. Here again, the increment or decrement can range from 1 to 8.

Auto Post-incremented Indexed Addressing. Suppose that X contains $1005. Then, the instruction

STAA 5,X+

stores the content of A in location 1005 and then increments X so the content of X becomes $100A. As before, the increment can vary from 1 to 8.

Auto Post-decremented Indexed Addressing. Again, suppose that X contains $1005. Then, the instruction

STAA 3,X-

stores the content of A in location $1005 and then decrements X so the content of X becomes $1002.

Indirect Indexed Addressing. In this type of addressing, a 16-bit (or equivalently, four-digit hexadecimal) constant given in the command is added to the content of a selected CPU register (X, Y, SP, or PC). This results in a pointer to a location containing the address of the operand. To illustrate an example, first assume that the contents of the CPU registers and some of the memory locations are as shown in Figure 8.8. Then, if the command

LDY [$1002, X]

is executed, $1002 is added to the content of X resulting in $2003. The content of locations $2003 and $2004 contain the high byte and low byte for the starting address of the operand. Thus, the starting address of the operand is $3003. Finally, the content of locations $3003 and $3004 is written to register Y. Thus, after the execution of this command Y contains $A3F6.

Instead of specifying an offset value in the command, the content of register D can be used. Thus, in Figure 8.8, the command

LDY[D,X]

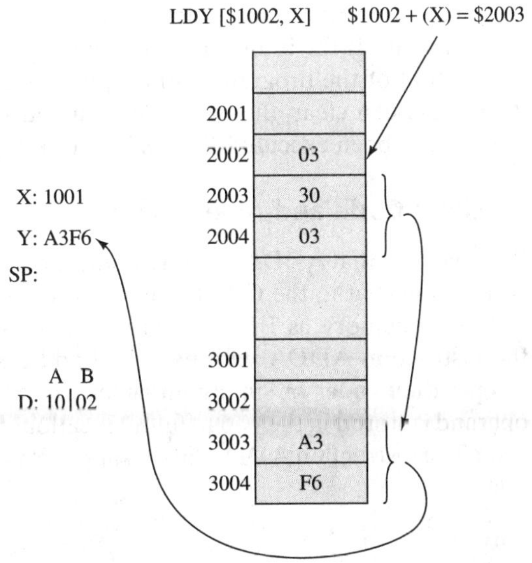

Figure 8.8 Illustration of the command LDY [$1002, X].

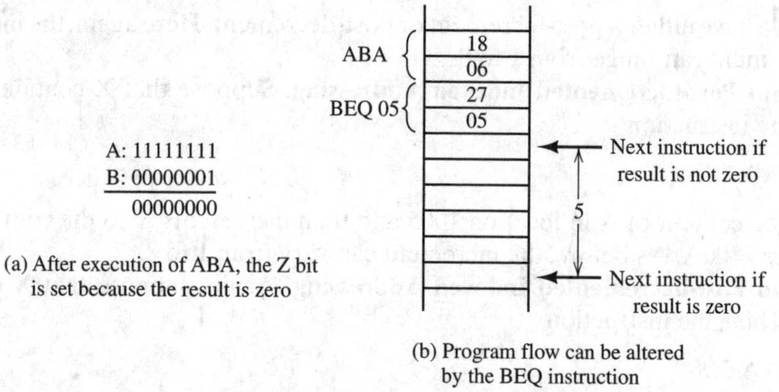

Figure 8.9 Using the BEQ instruction.

adds the content of D to the content of X again resulting in \$2003. The content of locations \$2003 and \$2004 is \$3003. This is the (initial) address of the operand. So \$A3F6 is loaded into register Y as it was for the command LDY[\$1002,X].

To recap, the steps followed by indexed indirect addressing are:

1. Add either the offset or the content of D to the content of the CPU register (X, Y, SP, or PC) specified in the command.
2. Go to the address resulting from step 1. The content of this location and the next is the address of the operand.

Relative Addressing

Branch instructions are used to alter the sequence of program flow. Recall that if we add two numbers, the Z bit of the condition code register is clear if the result was not zero and is set if the result was zero. The BEQ (branch if result equals zero) command can be used to change the program flow depending on the value of the Z bit.

In the CPU12, branch instructions use only relative addressing. Conversely, branches are the only instructions that use relative addressing.

For example, suppose that initially the A register contains FF and the B register contains 01. Then, if the instruction ABA is executed, the binary addition shown in Figure 8.9(a) is performed. Because the result is zero, the Z bit of the condition code register is set. If the branch instruction BEQ \$05 is executed, the next instruction is the content of the program counter plus the offset (which is 05 in this case). If the Z bit had been clear, the instruction immediately following the branch instruction would have been executed. This is illustrated in Figure 8.9(b).

Machine Code and Assemblers

We have seen that ADDA is a mnemonic for an instruction executed by the processor. It turns out that in the CPU12, the instruction ADDA with extended addressing is stored in memory as BB $=$ 10111011_2. We say that BB is the **machine code** for the instruction ADDA with extended addressing. Machine codes are also known as **operation codes** or simply **op codes**. In extended addressing, the address of the operand is stored in the two memory locations immediately following the instruction code. The instruction ADDA \$070A appears in three successive memory locations as

```
BB
07
0A
```

Address	Contents	Address	Contents
0200	10	0000	EF
0201	04	0001	01
0202	1A	0002	AE
0203	10	0003	F1
0204	07	0004	78
0205	FF	0005	96
0206	A3	0006	13
0207	16	0007	A4

Figure 8.10 Contents of memory for Exercise 8.6.

It would be a daunting task for a human to make conversions from instruction mnemonics to machine codes, whereas computers excel at this type of task. Furthermore, mnemonics are much easier for us to remember than machine codes. Thus, we generally start writing a program by using mnemonics. A computer program called an **assembler** is then employed to convert the mnemonics into machine code. Assemblers also help in other chores associated with programming, such as converting decimal numbers to hexadecimal and keeping track of branching addresses and operand addresses. We will have more to say about assembly language in the next section.

Exercise 8.6 Suppose that the contents of certain memory locations are as shown in Figure 8.10. Furthermore, the content of register A is zero and the X register contains 0200 prior to execution of each of these instructions:

a. LDAA $0202
b. LDAA #$43
c. LDAA $05,X
d. LDAA $06
e. LDAA $07,X-
f. LDAA $05,+X

Find the contents of registers A and X after each instruction.
Answer **a.** Register A contains 1A and X contains 0200. **b.** Register A contains 43 and X contains 0200. **c.** Register A contains FF and X contains 0200. **d.** Register A contains 13 and X contains 0200 **e.** Register A contains 10 and X contains 01F9. **f.** Register A contains FF and X contains 0205. □

Exercise 8.7 Suppose that starting in location 0200 successive memory locations contain op codes for the instructions

```
CLRA
BEQ $15
```

a. Show the memory addresses and contents (in hexadecimal form) for these instructions.
b. What is the address of the instruction executed immediately after the branch instruction?

Answer **a.** The memory addresses and contents are:

0200: 87 (op code for CLRA)
0201: 27 (op code for BEQ)
0202: 15 (offset for branch instruction)

b. The address of the next instruction is 0218. □

8.6 ASSEMBLY-LANGUAGE PROGRAMMING

A program consists of a sequence of instructions used to accomplish some task. No doubt, you have been introduced to programming that uses high-level languages such as BASIC, C, Java, MATLAB, or Pascal. When using high-level languages, a **compiler** or **interpreter** converts program statements into machine code before they are executed. It would be much too tedious to write machine-language programs for sophisticated engineering analysis. In application-oriented software, such as computer-aided design packages, even greater emphasis is placed on making the programs easy to use. However, in writing programs for embedded computers in control applications, we often need to keep the number of instructions relatively small and to minimize the time required to execute the various operations; quick response to events in the system being controlled can be highly important.

Though we may program MCUs for control applications in machine language, it is possible to relieve much of the drudgery by using an **assembler**. This provides many conveniences, such as allowing us to write instructions with mnemonics, using labels for memory addresses, and including user comments in the source program file.

In practice, we write the program as **source code** using a text editor on a general-purpose computer, called the **host computer**. The source code is then converted to **object code** (machine code) by the assembler program. Finally, the machine code is loaded into the memory of the MCU, which is called the **target system**. Sometimes, we say that the source code is written in assembly language. Assembly language code, nevertheless, is very close to the actual machine code executed by the computer.

In general, CPU12 assembly language statements take the following form:

```
LABEL    INSTRUCTION/DIRECTIVE    OPERAND    COMMENT
```

Typically each line of source code is converted into one machine instruction. Some of the source code statements, called **directives**, however, are used to give commands to the assembler. One of these is the origin directive ORG. For example,

```
         ORG    $0100
```

instructs the assembler to place the first instruction following the directive in memory location 0100 of the target system.

In CPU12 assembly language, labels must begin in the first column. The various fields are separated by spaces. Thus, when we want ORG to be treated as a directive, rather than as a label, we need to place one or more spaces ahead of it. If the first character of a line is a semicolon, the line is ignored by the assembler. Such lines are useful for comments and line spaces that make the source code more understandable to humans.

Usually, there are many ways to write a program to accomplish a given task.

In writing a program, we start by describing the algorithm for accomplishing the task. We then create a sequence of instructions to carry out the algorithm. Usually, there are many ways to write a program to accomplish a given task.

Example 8.1 An Assembly-Language Program

Suppose that we want a program starting in memory location 0400 that retrieves the number stored in location 0500, adds 5 to the number, writes the result to location 0500, and then stops. (We will use only the instructions listed in Table 8.1, even though the CPU12 has many additional instructions, which often could make our programs shorter.)

Solution The source code is:

```
;    SOURCE CODE FOR EXAMPLE 8.1
;    THIS LINE IS A COMMENT THAT IS IGNORED BY THE ASSEMBLER
;
            ORG       $0400       ;ORIGIN DIRECTIVE
BEGIN       LDAA      $0500       ;LOAD NUMBER INTO A
            ADDA      #$05        ;ADD 5, IMMEDIATE ADDRESSING
            STAA      $0500       ;STORE RESULT
            STOP
            END                   ;END DIRECTIVE
```

Comments have been included to explain the purpose of each line. BEGIN is a label that identifies the address of the LDAA instruction. (In this case, BEGIN has a value of 0400.) If we wanted to reference this location somewhere in a more complex program, the label would be useful. STOP is the mnemonic for the instruction that halts further action by the MCU. END is a directive that informs the assembler that there are no further instructions. ■

Example 8.2 Absolute Value Assembly Program

Write the source code for a program starting in location \$0300 that loads register A with the signed two's-complement number in location \$0200, computes its absolute value, returns the result to location \$0200, clears the A register, and then stops. Use the instructions listed in Table 8.1. (Assume that the initial content of location \$0200 is never $10000000 = -128_{10}$ which does not have a positive equivalent in 8-bit two's complement form.)

Solution Recall that branch instructions (also known as conditional instructions) allow different sets of instructions to be executed depending on the values of certain bits in the condition code register. For example, in Table 8.1, we see that the branch on plus instruction (BPL) causes a branch if the N bit of the condition code register is clear (i.e., logic 0).

Testing occurs automatically in many instructions. For example, in the load A instruction LDAA, the N and Z bits of the condition code register are set if the value loaded is negative or zero, respectively.

Our plan is to load the number, compute its two's complement if it is negative, store the result, clear the A register, and then stop. Recall that one way to find the two's complement is to first find the one's complement and add one. If the number is positive, no calculations are needed. The source code is:

```
;   SOURCE CODE FOR EXAMPLE 8.2
;
            ORG       $0300       ;ORIGIN DIRECTIVE
            LDAA      $0200       ;LOAD NUMBER INTO REGISTER A
            BPL       PLUS        ;BRANCH IF A IS POSITIVE
            COMA                  ;ONES'S COMPLEMENT
```

```
        INCA                ;ADD ONE TO FORM TWO'S COMPLEMENT
        STAA    $0200       ;RETURN THE RESULT TO MEMORY
PLUS    CLRA                ;CLEAR A
        STOP
        END                 ;DIRECTIVE
```

In this program, the number is first loaded into register A from memory location \$0200. If the number is negative (i.e., if the most significant bit is 1), the N bit of the condition-code register is set (logic 1); otherwise, it is not set. If the N bit is zero, the BPL PLUS instruction causes the next instruction executed to be the one starting in the location labeled PLUS. On the other hand, if the N bit is one, the next instruction is the one immediately following the branch instruction. Thus, if the content of the memory location is negative, the two's complement is computed to change its sign. Then, the result is written to the original location and A is cleared. ■

Next, to illustrate some of the chores performed by the assembler, we manually convert the source code of the previous example to machine code.

Example 8.3 Manual Conversion of Source Code to Machine Code

Manually determine the machine code for each memory location produced by the source code of Example 8.2. What is the value of the label PLUS? (*Hint:* Use Table 8.1 to determine the op codes for each instruction.)

Solution The assembler ignores the title and other comments. Because of the ORG directive, the machine code is placed in memory starting at location 0300. The memory addresses and their contents are:

```
0300: B6 Op code for LDAA with extended addressing.
0301: 02 High byte of address.
0302: 00 Low byte of address.
0303: 2A Op code for BPL which uses relative addressing.
0304: 05 Offset (On the first pass this value is unknown.)
0305: 41 Op code for COMA which computes one's complement.
0306: 42 Op code for INCA.
0307: 7A Op code for STAA with extended addressing.
0308: 02 High byte of address.
0309: 00 Low byte of address.
030A: 87 Clear A.
030B: 18 Halt processor.
030C: 3E
```

A comment has been added to explain each line; however, the assembler does not produce these comments.

Recall that branch instructions use relative addressing. The required offset for the BPL command is not known on the first pass through the source code. However, after the first pass, we see that the location corresponding to the label PLUS is 030A and that an offset of 05 is needed for the BPL command. The END directive does not produce object code. ■

Subroutines

Sometimes, certain sequences of instructions are used over and over in many different places. Memory is saved if these sequences are stored once and used wherever

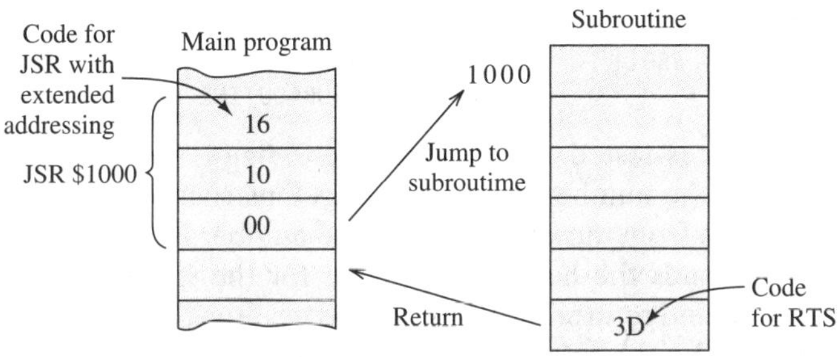

Figure 8.11 Illustration of the jump-to-subroutine command with extended addressing. (Other addressing modes are allowed with the JSR instruction.)

needed. A sequence of instructions such as this is called a **subroutine**. At any point in the main program that the subroutine needs to be executed, we place the *J*ump to *S*ub*R*outine instruction:

```
JSR address
```

in which address is a direct, extended, or indexed address of the first instruction of the subroutine. At the end of the subroutine, we place the *R*e*T*urn from *S*ubroutine command

```
RTS
```

which causes the next instruction to be taken from the location following the JSR instruction in the main program. This is illustrated in Figure 8.11.

The stack is used to keep track of where to return after the subroutine is finished. The address of the instruction following the JSR is pushed onto the system stack when the jump is executed. This address is pulled off the stack and loaded into the program counter when the return instruction is executed. After the subroutine is completed, the next instruction executed is the one following the JSR.

One of the chores that the assembler can perform is to keep track of the starting addresses of the subroutines. We simply label the first instruction of the subroutine in the source code. This is convenient because we usually don't know where subroutines will eventually be located when writing programs. After all of the source code is written, the assembler can calculate the amount of memory needed for each portion of the program and determine the subroutine starting addresses, which are then substituted for the labels.

Example 8.4 Subroutine Source Code

Assume that the content of register A is a signed two's-complement number n. Using the instructions of Table 8.1, write a subroutine called SGN that replaces the content of A with +1 (in signed two's-complement form) if n is positive, replaces it with −1 if n is negative, and does not change the content of A if n is zero.

Solution The source code for the subroutine is:

```
;  SOURCE CODE FOR SUBROUTINE OF EXAMPLE 8.4
;
SGN     TSTA            ;TEST CONTENT OF A
        BEQ     END     ;BRANCH IF A IS ZERO
        BPL     PLUS    ;BRANCH IF A IS POSITIVE
        LDAA    #$FF    ;LOAD -1, IMMEDIATE ADDRESSING
```

```
          JMP     END      ;JUMP TO END OF SUBROUTINE
PLUS      LDAA    #$01     ;LOAD +1, IMMEDIATE ADDRESSING
END       RTS              ;RETURN FROM SUBROUTINE
```

First, the number is tested. If it is zero, the Z flag is set. If it is negative, the N flag is set. Next, if the number is zero, the BEQ instruction compels a branch to END, causing a return from the subroutine. If the number is positive, the subroutine branches to PLUS, loads the hexadecimal code for the signed two's-complement representation of +1, and returns. If the number is negative, the LDA #FF instruction is executed, followed by the return. [Notice that in this subroutine, END is a label (rather than a directive) because it begins in column 1.] ■

Exercise 8.8 Write a program starting in location $0100 that adds 52_{10} to the content of location $0500, stores the result in location $0501, and then stops. Assume that all values are represented in signed two's-complement form.

Answer The source code is:

```
;  SOLUTION FOR EXERCISE 8.8
;
          ORG     $0100
          LDAA    #$34     ;LOAD HEX EQUIVALENT OF 52 BASE TEN
          ADDA    $0500    ;ADD CONTENT OF 0500
          STAA    $0501    ;STORE RESULT IN 0501
          STOP
          END              ;DIRECTIVE
```
□

Exercise 8.9 Write a subroutine named MOVE that tests the content of register A. If A is not zero, the subroutine should move the content of location 0100 to 0200. Otherwise, no move is made. The contents of A, B, X, and Y must be the same on return as before the subroutine is called.

Answer One version of the desired subroutine is:

```
;   SUBROUTINE FOR EXERCISE 8.9.
;
MOVE      TSTA             ;TEST CONTENT OF A
BEQ       END              ;BRANCH IF CONTENT OF A IS ZERO
          PSHA             ;SAVE CONTENT OF A ON STACK
          LDAA $0100       ;LOAD CONTENT OF MEMORY LOCATION 0100
          STAA $0200       ;STORE CONTENT OF A IN LOCATION 0200
          PULA             ;RETRIEVE ORIGINAL CONTENT OF A
END       RTS              ;RETURN FROM SUBROUTINE
```
□

Resources for Additional Study

In this chapter, we have given a very brief overview of microcontrollers focusing on the HCS12/9S12 family of devices from Freescale Semiconductor. Much more detail about the HCS12/9S12 can be found at www.freescale.com.

We have emphasized assembly language programming because it gives a clear picture of the inner workings of MCUs. As MCUs have gained higher speed and complexity, the trend has been away from assembly language and toward higher-level languages such as C.

If you are interested in applying MCUs to a project of your own, you should take a course devoted to MCUs exclusively or a course in mechatronics. Hands-on work is very important in learning about MCUs. Typically, one starts with a training board containing the MCU of interest, provisions for prototyping, LEDs, a small display, switches, and other components. Such boards are often equipped with an interface so programs can be downloaded to the MCU from a host computer through a USB link. An example for the HCS12 MCU is the Dragon12-Plus-USB board from EVBplus.com.

Design and construction of robots, model railroads, remotely controlled model airplanes equipped with video cameras, and so forth can become an engrossing hobby for those with an interest in combining microcontrollers, mechanical systems, and electronic elements. For numerous examples, look at *Nuts and Volts Magazine* at www.nutsvolts.com or *Make Magazine* at www.makezine.com. These may give you some good ideas for a senior project.

The Arduino microcontroller boards are very popular with artists, do-it-yourself enthusiasts, and students. You will find many articles related to these boards in *Nuts and Volts* and *Make Magazine*. Also, look at http://spectrum.ieee.org/geek-life/hands-on/the-making-of-arduino/0.

While MCUs can be useful for instrumentation applications, special purpose software combined with an interface to a general-purpose computer is often a more convenient approach for one-of-a-kind laboratory instrumentation systems. We look at this approach in the next chapter.

Summary

1. A computer is composed of a central processing unit (CPU), memory, and input–output (I/O) devices. These are connected together with bidirectional data and control buses. The CPU contains the control unit, the arithmetic/logic unit (ALU), and various registers.

2. Memory is used to store programs and data. Three types of memory are RAM, ROM, and mass storage.

3. In von Neumann computer architecture, data and instructions are stored in the same memory. In Harvard architecture, separate memories are used for data and instructions.

4. Sensors are input devices that convert physical values to electrical signals. Actuators are output devices that allow the MCU to affect the system being controlled.

5. Figure 8.5 (on page 427) shows the elements of a typical microcomputer used for process control.

6. Analog-to-digital converters (A/D) transform analog voltages into digital words. Digital-to-analog converters (D/A) transform digital words into analog voltages. Converters are needed to interface analog sensors and actuators with an MCU.

7. Figure 8.6 (on page 430) shows the register set for the CPU12.

8. In a stack memory, data is added to or read from the top of the stack. It is a last-in first-out (LIFO) memory. The stack pointer is a register that contains the address of the top of the stack.

9. Table 8.1 (on page 434) contains some of the instructions for the CPU12.

10. Six addressing modes are supported by the CPU12: extended addressing, direct addressing, inherent addressing, immediate addressing, indexed addressing, and program-relative addressing.

11. In writing programs for embedded microcontrollers, we often start by writing a source program using labels and mnemonics. An assembler converts the source program into an object program consisting of machine code that is loaded into the target system.

12. High costs are incurred in software development for microcontrollers. However, when the cost can be spread over many units, assembly language programming can be the best solution.

Problems

Section 8.1: Computer Organization

P8.1. List the functional parts of a computer.

P8.2. What are tristate buffers? What are they used for?

P8.3. Give several examples of I/O devices.

P8.4. What is I/O mapped I/O?

P8.5. What is the collection of signal lines called? How many such lines are there, function-wise?

P8.6. What is an embedded computer?

***P8.7.** Which bus width defines the memory addressing capability of a computer? Express the relationship.

P8.8. Define the terms *microprocessor*, *microcomputer*, and *microcontroller*.

P8.9. Explain the difference between Harvard computer architecture and von Neumann computer architecture.

Section 8.2: Memory Types

P8.10. What is RAM? List two types. Is it useful for storing programs in embedded computers? Explain.

***P8.11.** What is ROM? List four types. Is it useful for storing programs in embedded computers? Explain.

P8.12. List three examples of mass-storage devices.

P8.13. Which type of memory is least expensive per unit of storage? (Assume that many megabytes of capacity are required.)

P8.14. How many memory locations can be addressed if the address bus has a width of 32 bits?

***P8.15.** Which type of memory would be best in the controller for an ignition system for automobiles?

P8.16. When might we choose EEPROM rather than mask-programmed ROM?

P8.17. What types of memory are volatile? Nonvolatile?

Section 8.3: Digital Process Control

P8.18. List the elements that are potentially found in a microcomputer-based control application.

P8.19. What is a sensor? Give three examples.

P8.20. What is an actuator? Give three examples.

***P8.21.** Explain the difference between a digital sensor and an analog sensor. Give an example of each.

P8.22. List five common household products that potentially include an MCU.

P8.23. List two potential applications of MCU-based control or instrumentation in your field of specialization.

P8.24. What is an A/D? Why might one be needed in a MCU-based controller?

***P8.25.** What is a D/A? Why might one be needed in a MCU-based controller?

P8.26. What is polling? What is an interrupt? What is the main potential advantage of interrupts versus polling?

Section 8.4: Programming Model for the HCS12/9S12 Family

P8.27. What is the function of the A, B, and D registers of the CPU12?

P8.28. What is the function of the program counter register? Of the condition-code register?

***P8.29.** What are stack and SP? Compare SP and PC.

P8.30. Which memory is better—FIFO or LIFO?

P8.31. Suppose that initially the contents of the registers are

```
A:07    B:A9    SP:004F    X:34BF
```

and that memory locations 0048 through 004F initially contain all zeros. The commands PSHA, PSHB, PSHX, PULB, PULA are then executed in sequence. List the contents of the registers A, B, SP, and X, and the memory locations 0048 through 004F after each command is executed.

P8.32. Suppose that initially the contents of the registers are

```
A:A7    B:69    SP:004E    Y:B804
```

and that memory locations 0048 through 004F initially contain all zeros. The commands PSHY, PSHB, PULY, PSHA are then executed in sequence. List the contents of the

* Denotes that answers are contained in the Student Solutions files. See Appendix E for more information about accessing the Student Solutions.

registers A, B, SP, and X and the memory locations 0048 through 004F after each command is executed.

***P8.33.** Write a sequence of push and pull commands to swap the high byte and low byte of the X register. After the sequence of commands is executed, the contents of the other registers should be the same as before.

Section 8.5: The Instruction Set and Addressing Modes for the CPU12

P8.34. For each part of this problem, assume that the X register contains 2000 and the A register initially contains 01. Name the type of addressing and give the content of A after each instruction listed next. The contents of memory are shown in Figure P8.34.

a.* LDDA $2002

b. LDDA #$43

c.* LDDA $04

d. LDDA 6,X

e.* INCA

f. CLRA

g.* LDAA $2007

h. INX

0000	01	2000	37
0001	FA	2001	AF
0002	9B	2002	20
0003	61	2003	07
0004	9A	2004	20
0005	B6	2005	00
0006	73	2006	FF
0007	41	2007	F3

Figure P8.34

P8.35. Suppose that the contents of certain memory locations are as shown in Figure P8.34. Furthermore, for each part of this problem, the initial contents of the CPU registers are: (D) = $0003, (X) = $1FFF, and (Y) = $1000. Name the type of addressing and determine the contents of registers D, X, and Y after each of these instructions is executed:

a. ADDB $1002,Y

b. LDAA B,X

c. LDAB 7,+X

d. LDX [$1004,Y]

e. LDAA [D,X]

P8.36. **a.***Assume that the A register initially contains FF and that the program counter is 2000. What is the address of the instruction executed immediately after the branch command? The content of memory and the corresponding instruction mnemonics are shown in Figure P8.36(a).

b. Repeat for Figure P8.36(b).

c. Repeat for Figure P8.36(c).

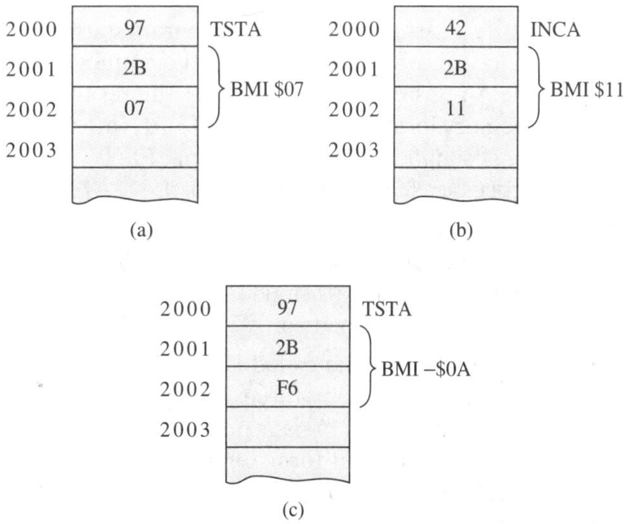

Figure P8.36

P8.37. Give the machine code for each of the following instructions:

a.* INCB

b.* ADDD $4A

c. BLO $05

d. NOP

e. ADDD #$EO

P8.38. Find the content of the A register after each instruction as the following sequence of instructions is executed:

```
LDAA #$05
ADDA #$E5
CLR A
```

Is the N bit of the condition-code register set or clear? Is the Z bit of the condition-code register set or clear?

P8.39. **a.** Suppose that the content of A is 97. Find the contents of A after the instruction COMA is executed. **b.** Suppose A contains F0. Find the contents of A after the instruction ABA is executed. (B contains 0F.)

***P8.40.** Assume that the content of A is $A7 and that the content of B is $20. Find the contents of A and B after the MUL instruction is executed. (*Hint:* The MUL instruction assumes that the contents of A and B are unsigned integer values.)

Section 8.6: Assembly-Language Programming

***P8.41.** Write an assembly-language program starting in location 200 that multiplies the content of the A register by 11_{10} and stores the result in memory locations $FF00 and $FF01, with the most significant byte in location $FF00. The processor should then be stopped.

P8.42. Write an assembly-language program starting in location 0400 that stores 00 in location 0800, 01 in 0801, 02 in 0802, 03 in 0803, and then halts the processor.

***P8.43.** Write a subroutine called DIV3 that divides the content of A by three. Assume that the initial content of A is a positive integer in two's-complement form. On return from the subroutine, the quotient should reside in B and the remainder in A.

P8.44. Change the program given in Problem P8.44 as follows: Consider the following assembly-language code for the CPU12:

```
;        PROBLEM 8.44
;
         ORG      $800;
START    LDAA     #$07;
         LDAB     #$06;
         ADDB     #$03;
         STOP;
         END;
```

Describe the effect of each line of this code and list the contents of memory location 0800 to 080A after the code has been assembled and executed.

P8.45. Write a subroutine called MUL3 that rounds the content of A to its nearest integer multiple of 3. Assume that the initial content of A is a positive integer in two's-complement form. Memory location $0A can be used for temporary storage. Include comments in your source code to explain the program and its operation to human readers. (*Hint:* Repeatedly subtract 3 until the result becomes negative. If the result is −3, the original content of A was a multiple of 3 and should not be changed. If the result is −2, the original content of A was one plus an integer multiple of 3, and we should subtract one from the original number to obtain the nearest multiple of 3. If the result is −1, the original content of A is 2 plus an integer multiple of 3, and we should add 1 to the original number to obtain the nearest multiple of 3.)

P8.46. Suppose that register B contains a two-decimal-digit BCD number *n*. Write a subroutine called BCD_HEX that replaces the content of register B by its hexadecimal equivalent. The content of the other registers (except the program counter) should be unchanged at the completion of the subroutine. Memory locations $1A, $1B, and $1C can be used for temporary storage. [*Hint:* We need to separate the upper nibble (four bits) of *n* from its lower nibble. This can be achieved by shifting *n* four bits to the left, with the result appearing in the D register. Shifting by four bits to the left is accomplished by multiplying by 2^4.]

Frequently Asked Questions

1. Draw the block diagram of a computer organization and explain the function of each block.
2. Identify and describe the functional blocks of a microcomputer.
3. What do you mean by computer memory? How memory capacity is expressed?
4. What do you mean by buses in a computer?

5. Name and explain any four input and output devices of a computer system.
6. What do you understand by RAM, ROM, PROM, EEPROM
7. What factors are considered while choosing a memory?
8. Explain microcomputer based control of a physical process with the help of a block diagram.
9. What do you mean by interrupts and polling?
10. Write a short note on direct addressing, inherent addressing, immediate addressing, indexed addressing and relative addressing.
11. What are machine codes and assemblers?
12. Write a short note on assembly language programming?

Practice Test

Here is a practice test you can use to check your comprehension of the most important concepts in this chapter. Answers can be found in Appendix D and complete solutions are included in the Student Solutions files. See Appendix E for more information about the Student Solutions.

T8.1. First, think of one or more correct ways to complete each statement in Table T8.1(a). Then, select the best choice from the list given in Table T8.1(b). [Items in Table T8.1(b) may be used more than once or not at all.]

Table T8.1

(a)
a. Tristate buffers...
b. When I/O devices are accessed by the same address and data buses as data memory locations, we have...
c. In a microcontroller, programs are usually stored in...
d. The type of memory most likely to be volatile is...
e. The type of memory most likely to be used for temporary data in a microcontroller is...
f. Arithmetic operations are carried out in the...
g. A microcontroller may be designed to respond to external events by...
h. The registers of a CPU12 that hold one of the arguments and the results of arithmetic and logical operations are...
i. The registers of a CPU12 that are used mainly for indexed addressing are...
j. The TSTA instruction may change the content of...
k. Stacks are...
l. The type of addressing used by the ABA instruction is...
m. The type of addressing used by the ADDA $0AF2 instruction is...
n. The type of addressing used by the ADDA #$0A instruction is...
o. The register that holds the address of the next instruction to be retrieved from memory is...
p. The type of addressing used by the BEQ instruction is...

Table T8.1. ***(Cont.)***

(b)
1. are composed of switches that are always closed
2. mass storage
3. B and Y
4. I/O devices have their own addresses and data buses
5. are composed of open switches
6. A and X
7. control unit
8. extended
9. A, B, and D
10. contain switches that open and close under the direction of the program counter
11. facilitate the ability to transfer data in either direction over a bus
12. the condition code register
13. ALU
14. polling
15. LIFO memories
16. inherent
17. memory-mapped I/O
18. A and D
19. interrupts
20. X and Y
21. ROM
22. direct
23. the program counter
24. dynamic RAM
25. the stack pointer
26. either interrupts or polling
27. static RAM
28. indexed
29. immediate
30. relative

T8.2. We have a CPU12 microcontroller. For each part of this problem, assume that the initial content of A is 00, the initial content of B is FF, the initial content of Y is 2004, and the initial content of selected memory locations is as shown in Figure P8.34 on page 449. Name the type of addressing used and give the content of A (in hexadecimal form) after execution of the instruction: **a.** LDAA $03; **b.** LDAA $03,Y; **c.** COMA; **d.** INCA; **e.** LDAA #$05; **f.** ADDD #A001.

T8.3. Suppose that initially the contents of the registers of a CPU12 microcontroller are

```
A:A6    B:32    SP:1039    X:1958
```

and that memory locations 1034 through 103C initially contain all zeros. List the contents of the registers A, B, SP, and X, and the contents of the memory locations 1034 through 103C after the sequence of commands, PSHX, PSHB, PULA, PSHX, has been executed.

Chapter 9

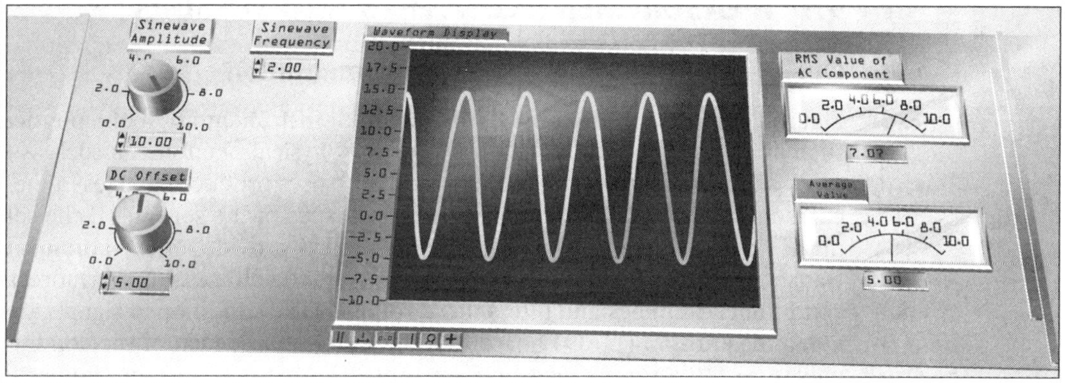

Computer-Based Instrumentation Systems

Study of this chapter will enable you to:

- Describe the operation of the elements of a computer-based instrumentation system.
- Identify the types of errors that may be encountered in instrumentation systems.
- Avoid common pitfalls such as ground loops, noise coupling, and loading when using sensors.
- Determine specifications for the elements of computer-based instrumentation systems such as data-acquisition boards.
- Know how to use LabVIEW to create virtual instruments for computer-aided test and control systems in your field of engineering.

Introduction to this chapter:

In Chapter 8, we looked at how embedded microcomputers can be programmed by using assembly language to instrument and control products such as household appliances or internal-combustion engines. High costs are associated with system development and programming for embedded microcontrollers. When these high costs are spread over a large number of units, an embedded computer and assembly-language programming can be the most cost-effective approach. On the other hand, for one-of-a-kind instrumentation and control systems, embedded computer solutions are often too time consuming and costly.

In this chapter, we will see that readily available data-acquisition (DAQ) boards, high-level software packages, and general-purpose computers can be combined to quickly create sophisticated instrumentation and process-control systems with relative ease. This approach is particularly good for one-of-a-kind systems.

9.1 MEASUREMENT CONCEPTS AND SENSORS

Overview of Computer-Based Instrumentation

Figure 9.1 shows a computer-based system for instrumentation of a physical system such as an automobile or chemical process. Physical phenomena such as temperatures, angular speeds, displacements, and pressures produce changes in the voltages, currents, resistances, capacitances, or inductances of the **sensors**. If the sensor output is not already a voltage, **signal conditioners** provide an **excitation source** that transforms the changes in electrical parameters to voltages. Furthermore, the signal conditioner amplifies and filters these voltages. The conditioned signals are input to a **data-acquisition** (DAQ) board. On the DAQ board, each of the conditioned signals is sent to a **sample-and-hold circuit** (S/H) that periodically samples the signal and holds the value steady while the **multiplexer** (MUX) connects it to the **analog-to-digital converter** (A/D or ADC) that converts the values to digital words. The words are read by the computer, which then processes the data further before storing and displaying the results. For example, the signals derived from a force sensor and a velocity sensor could be multiplied to obtain a plot of power versus time. Furthermore, the power could be integrated to show energy expended versus time. Long-term statistical analysis of a process can be carried out to facilitate quality control.

Computer-based instrumentation systems consist of four main elements: sensors, a DAQ board, software, and a general-purpose computer.

In this section, we consider sensors. Then, in the next several sections, we discuss other aspects of computer-based DAQ systems.

Sensors

We emphasize sensors or signal conditioners that produce electrical signals (usually voltages) which are analogous to the physical quantity, or **measurand**, to be measured. Often, the voltage is proportional to the measurand. Then, the sensor voltage is given by

$$V_{\text{sensor}} = Km \tag{9.1}$$

in which V_{sensor} is the voltage produced by the sensor, K is the **sensitivity constant**, and m is the measurand. For example, a **load cell** is a sensor consisting of four strain-gauge elements (see page 47) connected in a Wheatstone bridge (see page 122) and bonded to a load-bearing element. As force is applied to the load cell, a proportional voltage appears across two terminals of the bridge. **Excitation** in the form of a constant

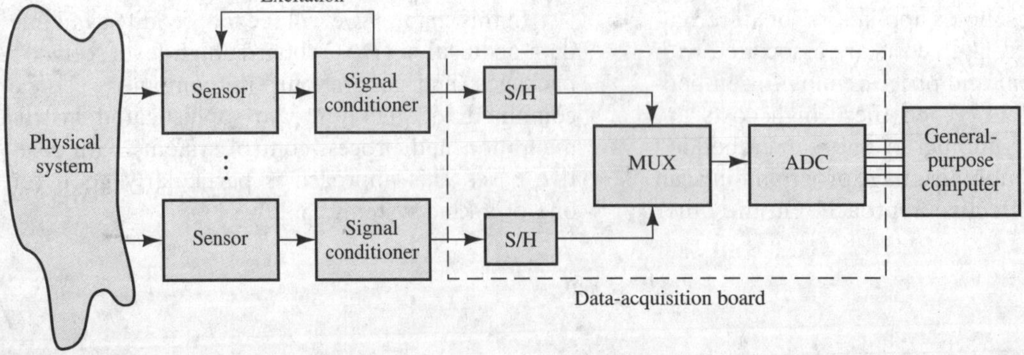

Figure 9.1 Computer-based DAQ system.

Table 9.1. Measurands and Sensor Types

Measurands	Sensor types
Acceleration	Seismic mass accelerometers Piezoelectric accelerometers
Angular displacement	Rotary potentiometers Optical shaft encoders Tachometric generators
Light	Photoconductive sensors Photovoltaic cells Photodiodes
Liquid level	Capacitance probes Electrical conductance probes Ultrasonic level sensors Pressure sensors
Linear displacement	Linear variable differential transformers (LVDTs) (see page 165) Strain gauges (see page 47) Potentiometers Piezoelectric devices Variable-area capacitance sensors
Force/torque	Load cells Strain gauges
Fluid flow	Magnetic flowmeters (see page 792) Paddle wheel sensors Constriction-effect pressure sensors Ultrasonic flow sensors
Gas flow	Hot-wire anemometers
Pressure	Bourdon tube/linear variable differential transformer combinations Capacitive pressure sensors
Proximity	Microswitches Variable-reluctance proximity sensors Hall-effect proximity sensors Optical proximity sensors Reed-switch sensors
Temperature	Diode thermometers Thermistors Thermocouples

voltage is applied to the other two terminals of the bridge. For a given excitation voltage, the sensitivity constant has units of V/N or V/lbf.

Some examples of measurands and sensor types are shown in Table 9.1. These are only a few of the many types of sensors that are available.

A good source of detailed information about computer-based instrumentation and control, including sensors, is the National Instruments website: www.ni.com.

Equivalent Circuits and Loading

An equivalent circuit that applies to many sensors is shown in Figure 9.2; the source voltage V_{sensor} is analogous to the measurand, and R_{sensor} is the Thévenin resistance. Frequently, as part of the signal conditioning, the sensor voltage must be amplified. Figure 9.2 shows the sensor connected to the input terminals of an amplifier.

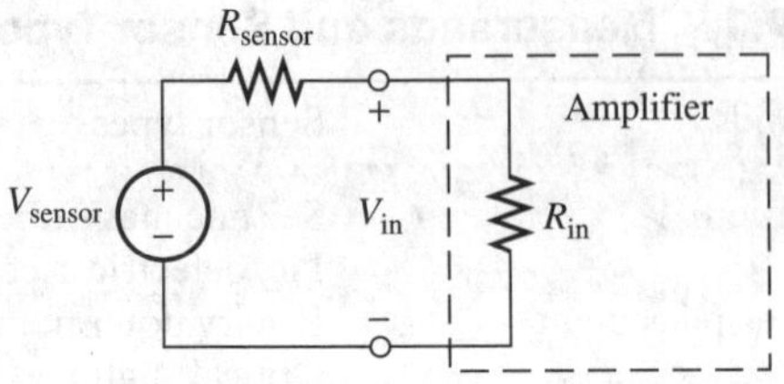

Figure 9.2 Model for a sensor connected to the input of an amplifier.

(Amplifiers are discussed in Chapter 11.) Looking into the input terminals of any amplifier, we see a finite impedance, which is represented as R_{in} in Figure 9.2. Using the voltage-division principle, we have

$$V_{in} = V_{sensor} \frac{R_{in}}{R_{in} + R_{sensor}} \tag{9.2}$$

Because of the current flowing through the circuit, the amplifier input voltage is less than the internal voltage of the sensor. This effect is known as **loading**. Loading is unpredictable and, therefore, undesirable. Provided that R_{in} is very large compared with R_{sensor}, the amplifier input voltage is nearly equal to the internal sensor voltage. Thus, when we need to measure the internal voltage of the sensor, we should specify a signal-conditioning amplifier having an input impedance that is much larger in magnitude than the Thévenin impedance of the sensor.

When we need to measure the internal voltage of the sensor, we should specify a signal-conditioning amplifier having an input impedance that is much larger in magnitude than the Thévenin impedance of the sensor.

Example 9.1 Sensor Loading

Suppose that we have a temperature sensor for which the open-circuit voltage is proportional to temperature. What is the minimum input resistance required for the amplifier so that the system sensitivity constant changes by less than 0.1 percent when the Thévenin resistance of the sensor changes from 15 kΩ to 5 kΩ?

Solution The sensitivity constant is proportional to the voltage division ratio between the input resistance and the Thévenin resistance of the sensor. We require that this ratio changes by 0.1% (or less) when the Thévenin resistance changes. Thus, with resistances in kΩ, we have

$$V_{sensor} \frac{R_{in}}{15 + R_{in}} \geq 0.999 V_{sensor} \frac{R_{in}}{5 + R_{in}}$$

Solving, we determine that R_{in} is required to be greater than 9985 kΩ. ■

Sensors with Electrical Current Output

Some types of sensors produce electrical current that is proportional to the measurand. For example, with suitable applied voltages, photodiodes produce currents that are proportional to the light intensities falling on the diodes. A photodiode is shown in Figure 9.3(a). Like the load cell, the photodiode requires a constant-voltage excitation source. Figure 9.3(b) shows diode current versus diode voltage for various light intensities. If we want the current to depend only on light intensity, the diode voltage must be held nearly constant.

Usually a **current-to-voltage converter** (also known as a transresistance amplifier) is used to produce an output voltage that is proportional to the photodiode

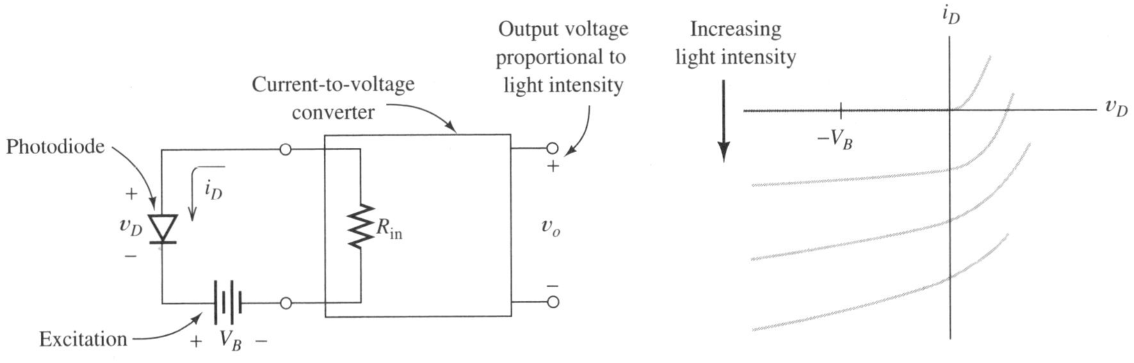

(a) Photodiode light sensor connected to a current-to-voltage converter

(b) Volt–ampere characteristic of photodiode

Figure 9.3 Photodiode light-sensing system. Because the diode voltage should be constant, R_{in} should ideally equal zero.

current. As in the case of amplifiers, we see an impedance looking into the input terminals of the current-to-voltage converter. This is shown as R_{in} in Figure 9.3(a). In order for the diode voltage to remain constant as the current varies, R_{in} must be very small (so the voltage across it is negligible). Thus, when we want to sense the current produced by a sensor, we use a current-to-voltage converter having a very small (ideally zero) input impedance magnitude.

When we want to sense the current produced by a sensor, we need a current-to-voltage converter having a very small (ideally zero) input impedance magnitude.

Variable-Resistance Sensors

Other sensors produce a changing resistance in response to changes in the measurand. For example, the resistance of a thermistor changes with temperature. Changes in resistance can be converted to changes in voltage by driving the sensor with a constant current source. To avoid loading effects, the voltage is applied to a high-input impedance amplifier as illustrated in Figure 9.4. Similar circuits that use ac excitation can convert changes in capacitance or inductance into voltage changes.

Errors in Measurement Systems

Many types of errors can occur in making measurements. We define the error of a measurement as

$$\text{Error} = x_m - x_{\text{true}} \tag{9.3}$$

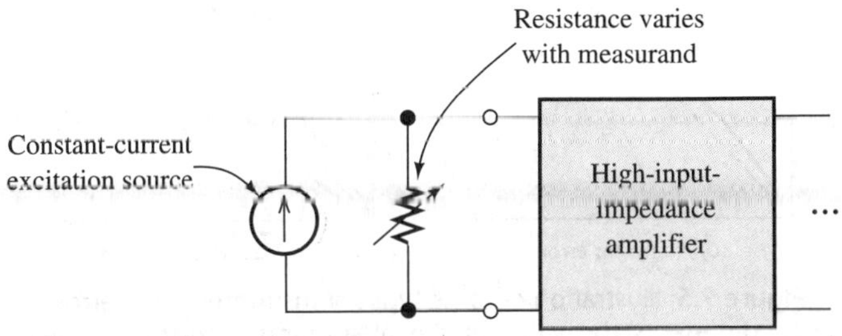

Figure 9.4 Variable-resistance sensor.

in which x_m is the measured value and x_{true} is the actual or true value of the measurand. Often, error is expressed as a percentage of the full-scale value x_{full} (i.e., the maximum value that the system is designed to measure).

$$\text{Percentage error} = \frac{x_m - x_{true}}{x_{full}} \times 100\% \tag{9.4}$$

There are many possible sources of error, some of which are specific to particular measurands and measurement systems. However, it is useful to classify the types of errors that can occur. Some are **bias errors**, also called **systematic errors**, that are the same each time a measurement is repeated under the same conditions. Sometimes, bias errors can be quantified by comparing the measurements with more accurate standards. For example, we could calibrate a weight scale by using it to measure the weights of highly accurate standards of mass. Then, the calibration data could be used to correct subsequent weight measurements.

Bias errors include **offset**, **scale error**, **nonlinearity**, and **hysteresis**, which are illustrated in Figure 9.5. Offset consists of a constant that is added to, or subtracted from, the true value. Scale error produces measurement errors that are proportional to the true value of the measurand. Nonlinearity can result from improper design or overdriving an electronic amplifier. When hysteresis error is present, the error depends on the direction and distance from which the measurand arrived at its current value. For example, hysteresis can result from static friction in measuring displacement or from materials effects in sensors that involve magnetic fields. All types of bias error are potentially subject to slow **drift** due to aging and changes in environmental factors such as temperature or humidity.

Although bias errors are the same for each measurement made under apparently identical conditions (except for drift), **random errors** are different for each instance and have zero average value. For example, in measuring a given distance, friction

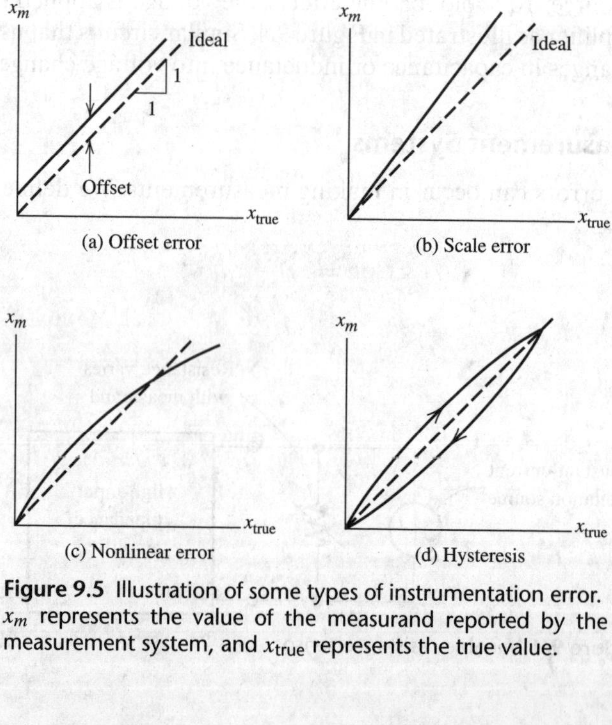

Figure 9.5 Illustration of some types of instrumentation error. x_m represents the value of the measurand reported by the measurement system, and x_{true} represents the true value.

combined with vibration may cause repeated measurements to vary. We can sometimes reduce the effect of random errors by making repeated measurements and averaging the results.

Some additional terms used in rating instrumentation performance are as follows:

1. **Accuracy:** The maximum expected difference in magnitude between measured and true values (often expressed as a percentage of the full-scale value).
2. **Precision:** The ability of the instrument to repeat the measurement of a constant measurand. More precise measurements have less random error.
3. **Resolution:** The smallest possible increment discernible between measured values. As the term is used, higher resolution means smaller increments. Thus, an instrument with a five-digit display (e.g., 0.0000 to 9.9999) is said to have higher resolution than an otherwise identical instrument with a three-digit display (e.g., 0.00 to 9.99).

Exercise 9.1 Suppose that a given magnetic flow sensor has an internal resistance (say, with variations in the electrical conductivity of the fluid) that varies from 5 kΩ to 10 kΩ. The internal (open-circuit) voltage of the sensor is proportional to the flow rate. Suppose that we want the changes in the sensitivity constant of the measurement system (including loading effects) to vary by less than 0.5 percent with changes in sensor resistance. What specification is required for the input resistance of the amplifier in this system?

Answer The input resistance of the amplifier must be greater than 990 kΩ. □

Exercise 9.2 **a.** Can a very precise instrument be very inaccurate? **b.** Can a very accurate instrument be very imprecise?

Answer **a.** Yes. Precision implies that the measurements are repeatable; however they could have large bias errors. **b.** No. If repeated measurements vary a great deal under apparently identical conditions, some of the measurements must have large errors, and therefore must be inaccurate. □

9.2 SIGNAL CONDITIONING

Some functions of signal conditioners are amplification of the sensor signals, conversion of currents to voltages, supply of (ac or dc) excitations to the sensors so that changes in resistance, inductance, or capacitance are converted to changes in voltage, and filtering to eliminate noise or other unwanted signal components. Signal conditioners are often specific to particular applications. For example, a signal conditioner for a diode thermometer may not be appropriate for use with a thermocouple.

Single-Ended versus Differential Amplifiers

Often, the signal from the sensor is very small (one millivolt or less) and an important step in signal conditioning is amplification. Thus, the sensor is often connected to the input terminals of an amplifier. In an amplifier with a **single-ended input**, one of its input terminals is grounded as shown in Figure 9.6(a), and the output voltage is a gain constant A times the input voltage.

An amplifier with a **differential input** is shown in Figure 9.6(b). Differential amplifiers have noninverting and inverting input terminals as indicated in the figure, and ideally, the output is the differential gain A_d times the difference between the input voltages.

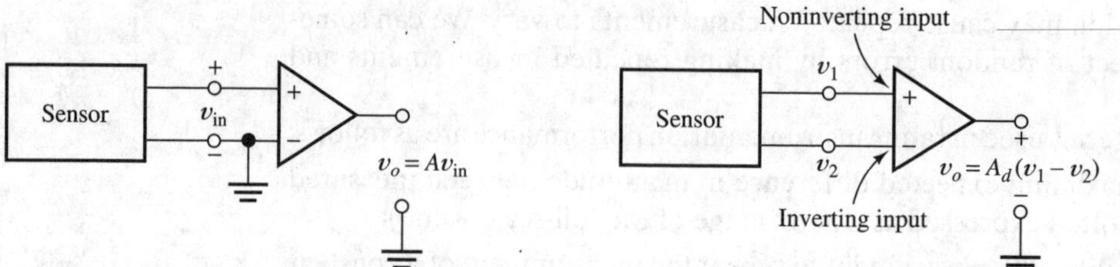

(a) Single-ended input: one amplifier input terminal is grounded

(b) Differential input: neither amplifier input is grounded and the output is gain A_d times the difference between the input voltages

Figure 9.6 Amplifiers with single-ended and differential input terminals.

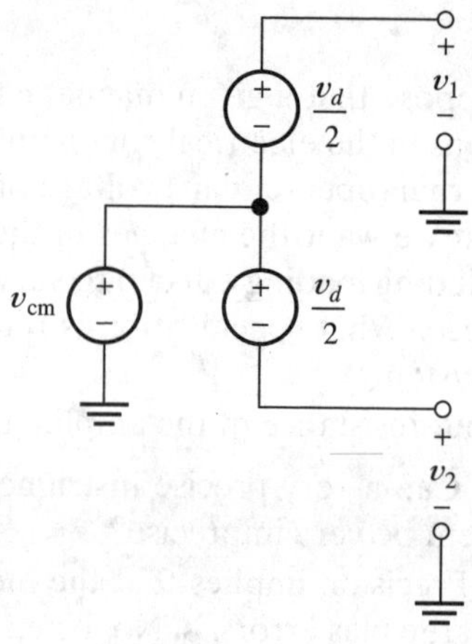

Figure 9.7 Model for a sensor with differential and common-mode components.

A model for the voltages produced by a typical sensor connected to a differential amplifier is shown in Figure 9.7. The difference between the amplifier input voltages is the **differential signal:**

$$v_d = v_1 - v_2 \tag{9.5}$$

Sometimes, a large **common-mode signal** is also present, which is given by

$$v_{cm} = \frac{1}{2}(v_1 + v_2) \tag{9.6}$$

Almost always, the differential signal is of interest, and the common-mode signal represents unwanted noise. Thus, it is often very important for the differential amplifier to respond only to the differential signal. Great care must be taken in designing amplifiers that reject large common-mode signals sufficiently. A measure of how well a differential amplifier rejects the common-mode signal is the **common-mode rejection ratio** (CMRR). When large common-mode signals are present, it is important to select a differential amplifier with a large CMRR. **Instrumentation amplifiers** are very good in this respect. Differential amplifiers, CMRR, and instrumentation amplifiers are discussed at length in Chapters 11 and 14.

When large common-mode signals are present, it is important to select a differential amplifier having a large CMRR specification.

Ground Loops

Often, the sensor and the signal-conditioning unit (such as an amplifier or current-to-voltage converter) are located some distance apart and are connected by a cable. Furthermore, the voltage (or current) produced by the sensor may be very small (less than one millivolt or one microampere). Then, several problems can occur that reduce accuracy or, in extreme cases, totally obscure the desired signal.

One of these problems is known as **ground loops**. If we have a single-ended amplifier, one of its input terminals is connected to a ground wire of the electrical distribution system. These ground wires eventually lead to the ground bus in the electrical distribution panel, which in turn is connected to a cold-water pipe or to a conducting rod driven into the earth. In instrumentation systems, we often have several pieces of equipment that are connected to ground through different wires. Ideally, the ground wires would have zero impedance, and all of the ground points would be at the same voltage. In reality, because of currents flowing through small but nonzero resistances of the various ground wires, small but significant voltages exist between various ground points.

Consider Figure 9.8, in which we have a sensor, an amplifier with a single-ended input, and a cable connecting the sensor to the amplifier. The cable wires have small resistances denoted by R_{cable}. Several ground wires are shown with their resistances R_{g1} and R_{g2}. The current source I_g represents current flowing to ground. Typically, I_g originates from the 60-Hz line voltage through power-supply circuits of the instruments. If we connect both the sensor and the amplifier input to ground, part of I_g flows through the connecting cable, and the input voltage is the sensor voltage minus the drop across R_{cable}:

$$V_{\text{in}} = V_{\text{sensor}} - I_{g1}R_{\text{cable}} \tag{9.7}$$

When the sensor voltage is very small, it can be totally obscured by the drop across R_{cable}.

On the other hand, if we break the sensor ground connection so only the amplifier is grounded, I_{g1} becomes zero and the input voltage is the sensor voltage as desired. Thus, in connecting a sensor to an amplifier with a single-ended input, we should select an ungrounded or **floating** sensor.

In connecting a sensor to an amplifier with a single-ended input, we should select a floating sensor.

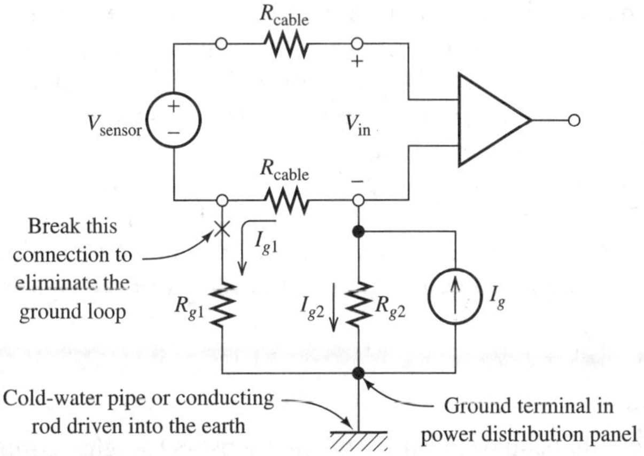

Figure 9.8 Ground loops are created when the system is grounded at several points.

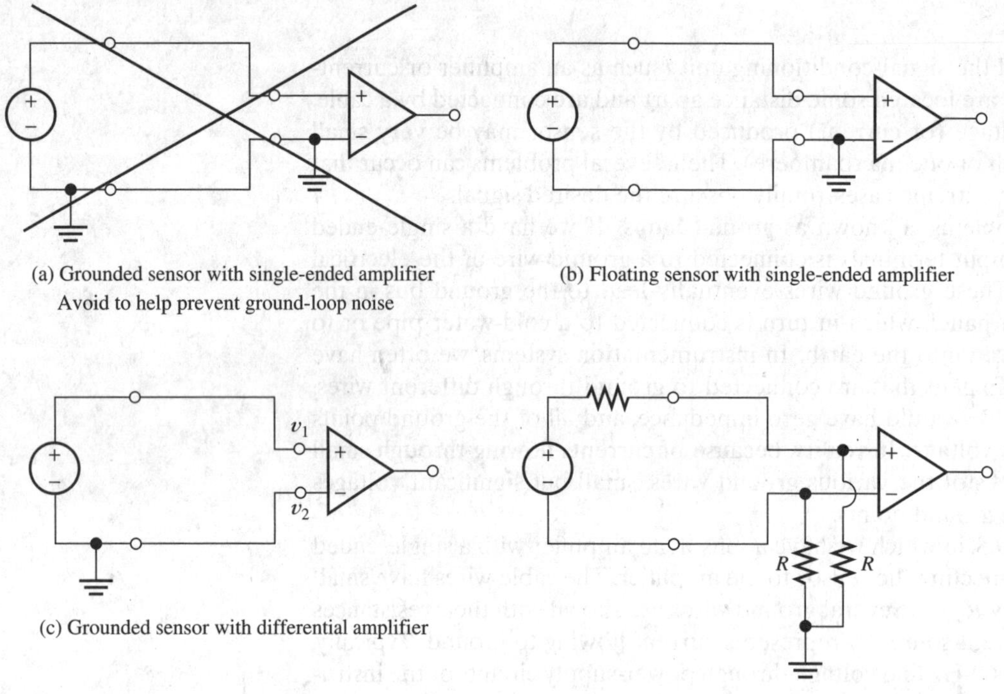

(a) Grounded sensor with single-ended amplifier
Avoid to help prevent ground-loop noise

(b) Floating sensor with single-ended amplifier

(c) Grounded sensor with differential amplifier

(d) Floating sensor with differential amplifier including resistors to provide a path for the input bias current

Figure 9.9 Four sensor–amplifier combinations.

If you have connected several audiovisual components, such as VCRs, TVs, radio tuners, CD players, stereo amplifiers, and so forth, you have probably encountered ground loops, which cause an annoying 60-Hz hum to be produced by the loudspeakers.

Alternative Connections

Figure 9.9 shows four combinations of sensors and amplifiers. As we have seen, we need to avoid the combination of grounded sensor and single-ended input shown in part (a) of the figure because of ground loops. Any of the other three connections can be used. However, for a floating sensor with a differential amplifier as shown in part (d), it is often necessary to include two high-valued (much greater than the internal impedance of the sensor to avoid loading effects) resistors to provide a path for the input bias current of the amplifier. (Input bias current is discussed further in Section 11.12.) If the resistors are not included, the common-mode voltage of the source can become so large that the amplifier does not function properly. In part (c) of the figure, the ground connection to the sensor provides a path for the bias current.

Noise

Another problem that can occur in connecting sensors to signal-conditioning units is the inadvertent addition of noise produced by the electric or magnetic fields generated by nearby circuits. (Computers are infamous for creating high-frequency

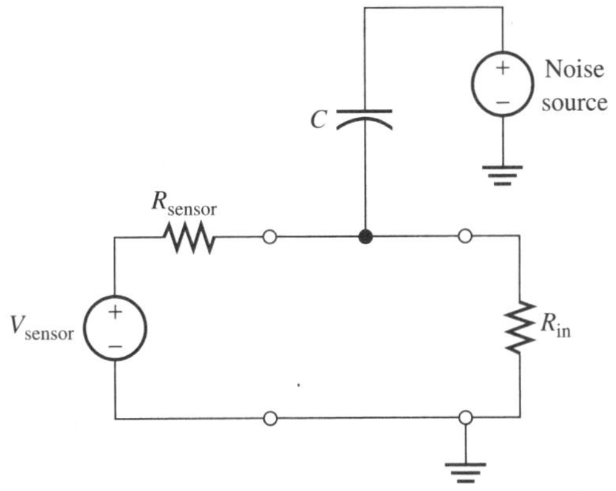

Figure 9.10 Noise can be coupled into the sensor circuit by electric fields. This effect is modeled by small capacitances between the noise source and the sensor cable.

electrical noise.) Electric field coupling can be modeled as small capacitances that are connected between nearby circuits and the cables, as shown in Figure 9.10. Currents are injected into the cable through these capacitances. This is particularly a problem with unshielded cables and when the sensor impedance is large. A shielded cable, in which an outer conductor in the form of metallic foil or braided wire encases the signal conductors, can eliminate much of the noise caused by electric fields. The shield is connected to ground, providing a low-resistance path for the capacitive currents. This is illustrated in Figure 9.11.

Electric field coupling of noise can be reduced by using shielded cables.

Noise problems can also occur due to magnetic coupling. Many circuits, particularly power-supply transformers, produce time-varying magnetic fields. When these fields pass through the region bounded by the cable conductors, voltages are induced in the cable. Magnetically coupled noise can be greatly diminished by reducing the effective area bounded by the conductors. Twisted-pair and coaxial cables (see Figure 9.12) are two good ways to accomplish this. Because the center lines of the conductors in coaxial cable are coincident, the effective bounded area is very small.

Magnetically coupled noise is reduced by using coaxial or twisted-pair cables.

Exercise 9.3 The voltages produced by a sensor are $v_1 = 5.7$ V and $v_2 = 5.5$ V. Determine the differential and common-mode components of the sensor signal.
Answer $v_d = 0.2$ V; $v_{cm} = 5.6$ V. □

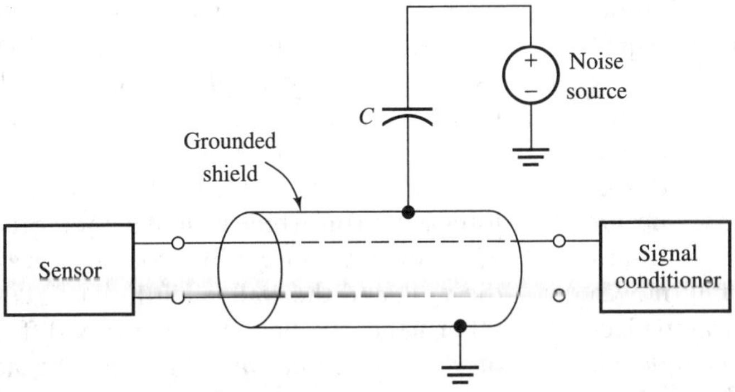

Figure 9.11 Electric field coupling can be greatly reduced by using shielded cables.

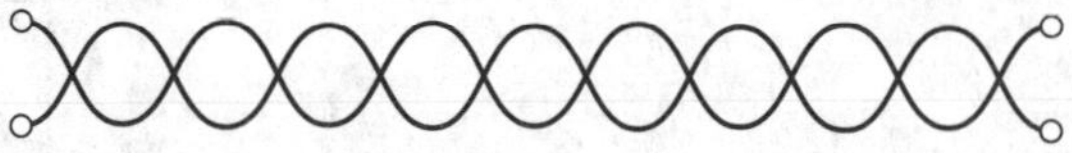

(a) Twisted-pair cable

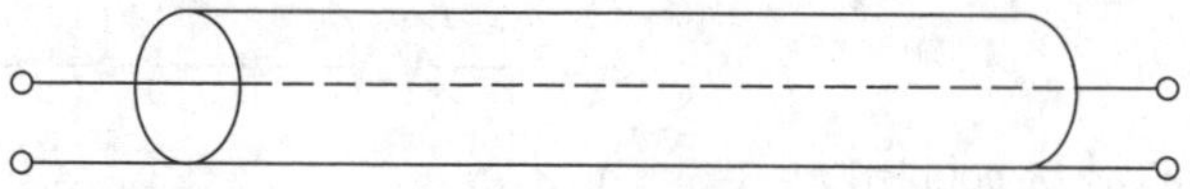

(b) Coaxial cable

Figure 9.12 Magnetic field coupling can be greatly reduced by using twisted-pair or coaxial cables.

PRACTICAL APPLICATION 9.1

The Virtual First-Down Line

In American football, the team on offense must gain 10 yards in a series of four plays to retain possession of the ball. Thus, a line marking the needed advance is of constant interest to fans viewing a game.

On September 27, 1998, in a Sunday Night Football game on ESPN, Sportvision introduced their "1st and Ten" system for electronically drawing the first-down line on television images. The system has been enthusiastically accepted and even won an Emmy award for technical innovation. In the 2003 season, 18 crews covered about 300 NCAA and NFL games.

While the concept of drawing a virtual line on a television image sounds simple, there are some formidable problems that need to be overcome to produce a result that appears to be painted on the field. Typically, three main cameras are situated above and back from the 50-yard line and both of the 25-yard lines. Each camera pans, tilts, zooms, and changes focus rapidly during the game. The virtual line needs to change its position, orientation, and width as the camera view changes. In addition, some football fields are not flat—they are crowned to ensure drainage and the yard lines are not exactly straight. If the virtual line does not closely match the curvature of the lines on the field, it will not look natural. Furthermore, the line needs to be drawn thirty times a second, once for each video frame. Of course, for realism, part of the line needs to disappear when a player, an official, or the ball moves across it. An impressive array of sophisticated electronic and computer technology has been employed by Sportvision engineers to meet these demands.

To set up the system on a given field, Sportvision starts by using laser surveying instruments to measure the elevation at a number of points along each 10-yard line. A computer uses this data to produce a virtual three-dimensional model of the field.

Sensors attached to each of the cameras measure pan, tilt, zoom, and focus. This data is fed into a computer that alters the model to match the perspective for a given camera, and a virtual map is drawn in blue lines over the image of the field seen by the camera. Finally, the virtual map is tweaked to match the real image for many combinations of pan, tilt, and zoom as illustrated in Figure PA9.1. The resulting calibration data is saved for use by the system during an actual game.

A technique known as "chroma keying" has been around for a long time and is widely used in televised weather reports. A meteorologist forecasting weather stands in front of a light blue wall. Computers substitute weather maps and graphics for all the pixels (i.e., picture elements) that are light blue. Thus, the forecaster seems to stand in front of the weather map. The same kind of technology is used to allow officials, players, and the ball to seem to move over the virtual first-down line. However, discerning which pixels are players and which are part of the field is much more difficult than separating a forecaster from a blue wall. Weather forecasters usually avoid wearing clothing that matches the color of the wall, and the wall is all the same color. On the other hand, the field can be many different shades of white (painted yard lines on the field), green (grass or artificial turf), or brown (grass or mud).

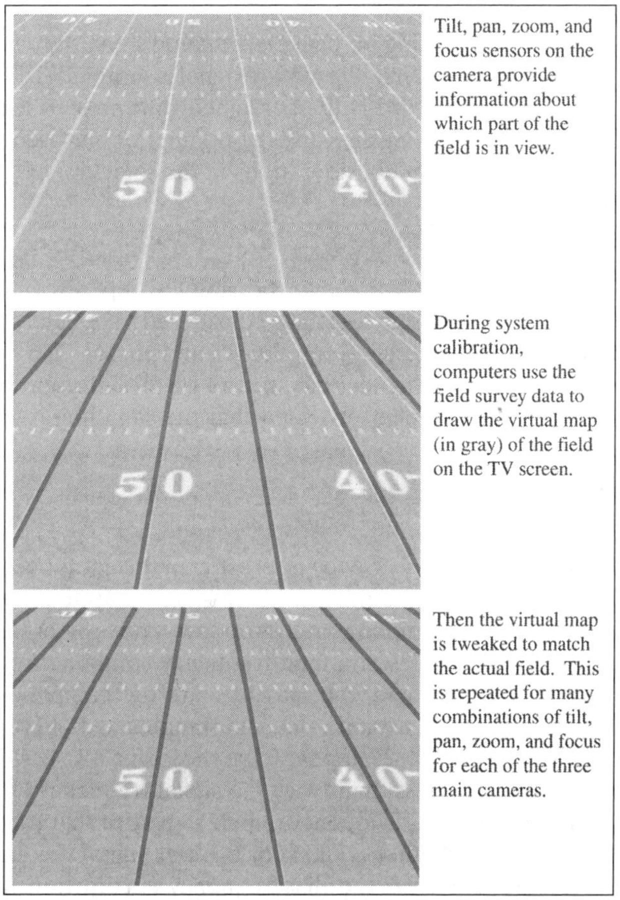

Figure PA9.1 Computers use a virtual map based on surveying the football field to draw the 1st and Ten line on the television screen in real time.

Part of the field may be sunlit while other parts are in shadow. Football teams, such as the Green Bay Packers, have uniforms that are partly green, which is especially difficult to distinguish from sunlit artificial turf. Other colors, such as brown, can also be difficult.

By constant recalibrating, the 1st and Ten system can keep track of which colors are part of the field and which are not, so the virtual down line is not drawn over players.

During a game, a team of four people operates the system, which contains five computers. The "spotter" is in the stadium and radios the location of the down line to a truck containing the equipment and two of the other team members. The "line position technician" enters the location data into the computers, monitors line position, and makes any needed adjustments. Another operator monitors changes in field colors so the chroma keying is properly accomplished. Finally, a troubleshooter looks for problems and solves them.

One of the computers receives and processes the pan, tilt, and zoom data from the cameras. Another keeps track of which camera is "on air." A third displays the on-air video and superimposes the virtual map of the field including the current down line. Another computer discerns which parts of the image are field and which are players or officials. Finally, the fifth computer places the virtual down line on the broadcast image while avoiding any superimposed graphics that the network may place on the screen.

More information about the 1st and Ten system as well as similar technology for other sports can be found at `www.sportvision.com`.

9.3 ANALOG-TO-DIGITAL CONVERSION

As discussed starting on page 342, analog signals are converted to digital form by a two-step process. First, the analog signal is sampled (i.e., measured) at periodic points in time. A code word is then assigned to represent the approximate value of each sample. The sampling rate and the number of bits used to represent each sample are two very important considerations in the selection of a DAQ system.

Sampling Rate

If a signal contains no components with frequencies higher than f_H, all of the information contained in the signal is present in its samples, provided that the sampling rate is selected to be more than twice f_H.

The rate at which a signal must be sampled depends on the frequencies of the signal's components. (All signals can be considered to be sums of sinusoidal components that have various frequencies, amplitudes, and phases.) If a signal contains no components with frequencies higher than f_H, all of the information contained in the signal is present in its samples, provided that the sampling rate is selected to be more than twice f_H.

Aliasing

Sometimes, we may only be interested in the components with frequencies up to f_H, but the signal may contain noise or other components with frequencies higher than f_H. Then, if the sampling rate is too low, a phenomenon called **aliasing** can occur. In aliasing, the samples of a high-frequency component appear to be those of a lower frequency component and may obscure the components of interest. For example, Figure 9.13 shows a 7-kHz sinusoid sampled at 10 kHz. As illustrated by the dashed line, the sample values appear to be those of a 3-kHz sinusoid. Because the sampling rate (10 kHz) is less than twice the signal frequency (7 kHz), the samples appear to be those of an alias frequency (3 kHz). (Notice that from the samples it is impossible to determine whether a 3-kHz or a 7-kHz signal was sampled.)

Figure 9.14 shows the alias frequency as a function of the signal frequency f. When the signal frequency f exceeds one-half of the sampling frequency f_s, the apparent frequency of the samples is different from the true signal frequency.

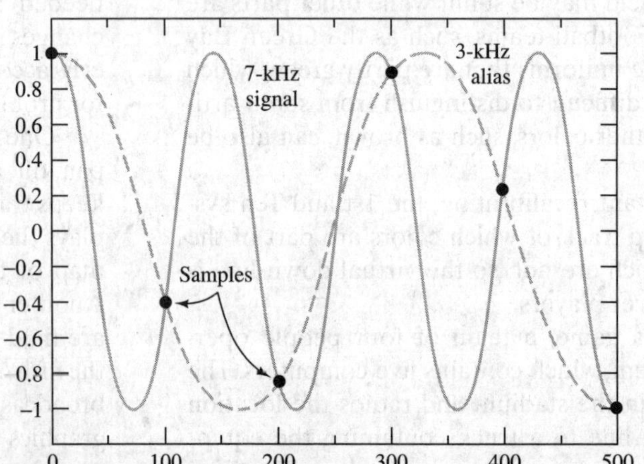

Figure 9.13 When a 7-kHz sinusoid is sampled at 10 kHz, the sample values appear to be those of a 3-kHz sinusoid.

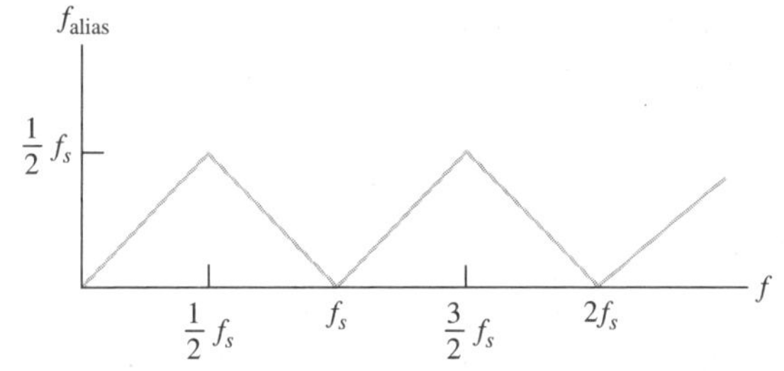

Figure 9.14 Alias or apparent frequency versus true signal frequency.

One way to avoid aliasing is to pick the sampling frequency high enough so that the alias frequencies are higher than the frequencies of interest. Then, the computer can remove the unwanted components by processing the samples with digital-filtering software. However, when high-frequency noise is present, it can result in a sampling rate that exceeds the ability of the computer to process the resulting data. Then, it is better to use an analog **antialias filter** ahead of the ADC to remove the noise above the highest frequency of interest. Typically, this is a high-order Butterworth filter implemented with operational amplifiers, such as those discussed in Section 14.10 starting on page 712. Because real filters are unable to sufficiently reject components slightly above their cutoff frequencies, it is usually necessary to select the sampling frequency at least three times the highest frequency of interest. For example, the highest audible frequency is about 15 kHz, but in CD technology a sampling rate of 44.1 kHz is used.

Quantization Noise

A second consideration important in converting analog signals to digital form is the number of amplitude zones to be used. Exact signal amplitudes cannot be represented, since all amplitudes falling into a given zone have the same codeword. Thus, when a digital-to-analog converter (DAC) converts the codewords to form the original analog waveform, it is possible to reconstruct only an approximation to the original signal—the reconstructed voltage is in the middle of each zone, which was illustrated in Figure 6.47 on page 344. Hence, some **quantization error** exists between the original signal and the reconstruction. This error can be reduced by using a larger number of zones, which requires a longer codeword for each sample. The number N of amplitude zones is related to the number of bits k in a codeword by

Analog-to-digital conversion is a two-step process. First, the signal is sampled at uniformly spaced points in time. Second, the sample values are quantized so they can be represented by words of finite length.

$$N = 2^k \tag{9.8}$$

Therefore, if we are using an 8-bit ($k = 8$) ADC, we find that there are $N = 2^8 = 256$ amplitude zones. The resolution of a computer-based measurement system is limited by the word length of the ADC. In compact-disc technology, 16-bit words are used to represent sample values. With that number of bits, it is very difficult for a listener to detect the effects of quantization error on the reconstructed audio signal. In the telephone system, 8-bit words are used and the fidelity of the reconstructed signal is relatively poor.

The effect of finite word length can be modeled as adding quantization noise to the reconstructed signal.

The effect of finite word length can be modeled as adding quantization noise to the reconstructed signal. It can be shown that the rms value of the quantization noise is approximately

$$N_{qrms} = \frac{\Delta}{2\sqrt{3}} \tag{9.9}$$

where Δ is the width of a quantization zone.

In the example that follows, we illustrate how the various factors we have discussed are used in selecting the components of a computer-based measurement system.

Example 9.2 Specifications for a Computer-Based Measurement System

Suppose that we have a single-ended (i.e., one terminal of the sensor is connected to power-system ground) piezoelectric vibration sensor that produces a signal of interest having peak values of ± 25 mV, an rms value of 3 mV, and components with frequencies up to 5 kHz. The internal impedance of the sensor is 1 kΩ. We want the system resolution to be 2 μV or better (i.e., smaller) and the accuracy to be ± 0.2 percent of the peak signal or better. (Note that the desired resolution is considerably better than the accuracy. This allows the system to discern changes in the signal that are smaller than the error.) The probe wiring is likely to be exposed to electric and magnetic field noise having components at frequencies higher than 5 kHz. An ADC having an input range from -5 V to $+5$ V is to be used. Draw the block diagram of the measurement system and give key specifications for each block.

Solution Because one end of the sensor is grounded, we need to use an instrumentation amplifier with a differential input to avoid ground-loop problems. (See Figure 9.9(c) on page 462.)

To help reduce capacitively and inductively coupled noise, we should select a shielded twisted-pair or coaxial cable to connect the sensor to the system. To avoid ground loops, the shield should be grounded at the sensor end only. Furthermore, we should use an antialias filter to reduce noise above 5 kHz. The block diagram of the system is shown in Figure 9.15.

The voltage gain of the instrumentation amplifier/antialias filter combination should be (5 V)/(25 mV) = 200; so the sensor signal is amplified to match the range of the ADC. The input impedance of the amplifier should be very large compared with the internal sensor impedance; therefore, loading effects are insignificant. If we specify a minimum input impedance of 1 MΩ, loading effects will reduce the signal

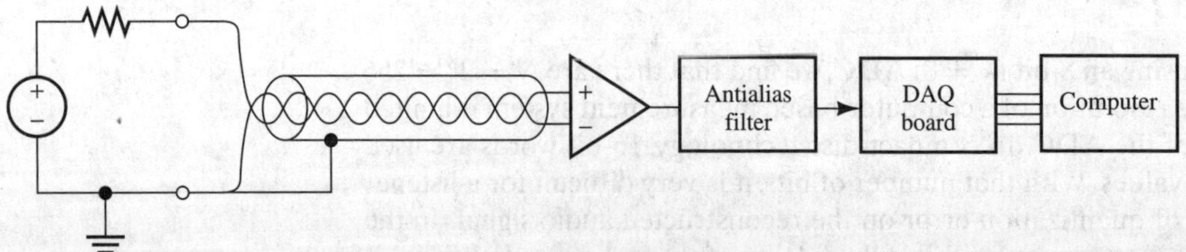

Figure 9.15 See Example 9.2.

by 0.1 percent, which is within the desired accuracy. (We need to allow for errors from other unknown sources.)

To achieve a resolution of 2 μV for a $\pm$25-mV signal, we need an ADC with at least (50 mV)/(2 μV) = 25,000 amplitude levels. This implies an ADC word length of at least $k = \log_2(25{,}000) = 14.6$. Since word length must be an integer, $k = 15$ is the smallest word length that will meet the desired specifications. It turns out that DAQ boards with 16-bit ADCs are readily available, and to provide some design margin, that is the word length we should specify.

See the National Instruments site www.ni.com to find specifications for many models of signal conditioners and DAQ boards. Using the specifications determined in our example, we could select the appropriate components.

Because the highest frequency of interest is 5 kHz, we need a sampling frequency of at least 10 kHz. However, the antialias filter will not effectively remove the components that are only slightly above 5 kHz, so a greater sampling rate (20 kHz or greater) should be chosen. ■

Exercise 9.4 A certain 8-bit ADC accepts signals ranging from -5 V to $+5$ V. Determine the width of each quantization zone and the approximate rms value of the quantization noise.

Answer $\Delta = 39.1$ mV; $N_{qrms} = 11.3$ mV. □

Exercise 9.5 A 25-kHz sinewave is sampled at 30 kHz. Determine the value of the alias frequency.

Answer $f_{\text{alias}} = 5$ kHz. □

9.4 LABVIEW

After sensor signals have been conditioned, converted to digital form, and read into a computer, we are ready to employ sophisticated programs to extract the information of interest. This information can then be displayed or used to control a physical system. Such an approach is very good for quickly developing instruments for specialized applications at relatively low cost, compared with traditional dedicated stand-alone instruments.

A Virtual Time-Frequency Vibration Analyzer

A typical example is a time-frequency vibration analyzer, which is a powerful tool for mechanical engineers. In the design of an automobile (or any other mechanical system subject to vibration), one sensor collects the vibration signal as speed is increased, while another sensor measures engine speed. Analysis software then determines the amplitude of the vibration components versus frequency and engine speed (which may be nearly proportional to time). Next, the software creates plots showing engine rpm (or time) on the horizontal scale and frequency on the vertical scale. Color is used to indicate the amplitude at each point in the rpm–frequency plane. A typical result is shown in Figure 9.16.

Typically, vibration components are generated with many frequencies, each of which increases linearly with rpm. This is evident by the streaks that slope upward from the origin of the rpm–frequency plane. In a mechanical system, the frequencies may not be harmonically related; instead, they are related by engine parameters such as gear ratios. For instance, one frequency may be 1.57 times another. Such relationships discovered from the analyzer display, coupled with knowledge of the design parameters of the system, can quickly reveal some of the sources of vibration.

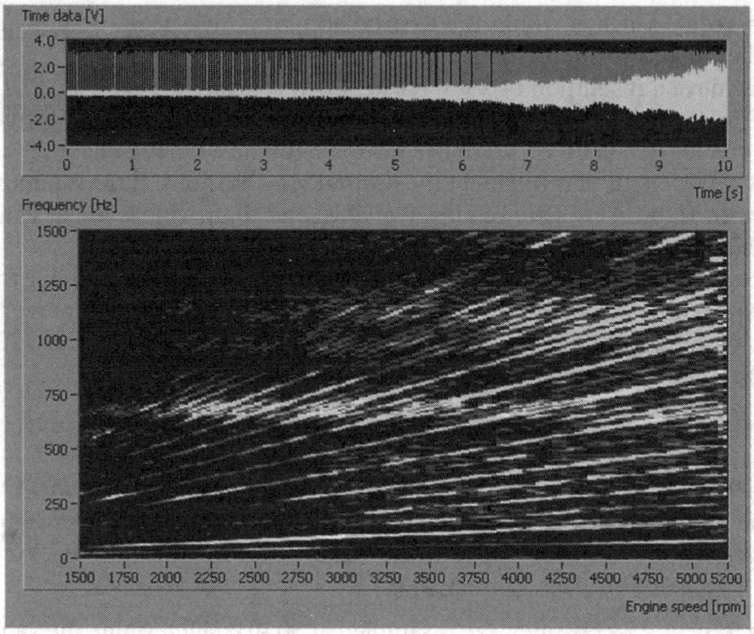

Figure 9.16 Time–frequency analyzer display of an engine-vibration signal.

Various parts of the engine or the vehicle may display mechanical resonances that accentuate vibration at specific frequencies. As the vibration frequencies pass through these resonances, the amplitudes become larger, which is evident in the horizontal streaks shown in Figure 9.16. Armed with such a powerful tool, engineers can investigate the source of annoying resonances. For example, if we press our hands against a resonant member, thereby damping the vibration, we may see one of the horizontal streaks decrease in intensity, revealing the source of the resonance. Additional tests with alternative sound-deadening features can then be used to refine the vehicle design.

LabVIEW

LabVIEW, a product of National Instruments, is an industry-standard program used by all types of engineers and scientists for developing sophisticated instrumentation systems such as the time–frequency vibration analyzer. LabVIEW is an acronym for *Lab*oratory *V*irtual *I*nstrument *E*ngineering *W*orkbench.

A great deal of useful information about sensors, signal conditioning, and applications of LabVIEW to many fields of science and engineering can be found starting from the National Instruments website `www.ni.com`.

LabVIEW uses an object-oriented graphical programming language that has a style, syntax, and data flow quite different from text-based programming languages such as C, Pascal, or Fortran. For example, the Fortran statement $X = X + 1$ means that the value of X is to be replaced by its previous value plus one. On the other hand, in LabVIEW, the new value of X is applied to a graphical shift-register icon, and the old value of X appears at the output of the shift register. Thus, the new value of X and the old value are clearly distinguished by their location in the diagram. The statement $Z = XY$ appears in LabVIEW as a graphical multiplier icon with a "wire" that carries X to one input, Y appears on another input wire, and Z appears on a wire attached to the output icon.

Instead of producing lines of code with the use of a text editor, we create LabVIEW programs, which are called **virtual instruments** (VIs), by positioning and interconnecting icons on the computer screen. LabVIEW programs appear to the programmer as a **front panel** and a **block diagram**. In general, the front panel may contain switches, push buttons, knobs, and sliders that are used by the operator to input desired analysis parameters to the virtual instrument. Furthermore, the panel contains meters, gauges, graphs, or charts that display results. For example, Figure 9.17 shows the front panel and the block diagram for a simple LabVIEW program that adds two numbers and displays the sum. The front panel contains a meter to display the result and two knobs that the operator can use to input values for the variables A and B.

The front panel is the user interface that contains the controls and displays for the virtual instrument.

The block diagram shows how the instrument analyzes sensor data and input from the front panel. LabVIEW uses **data-flow concepts**. This means that calculations of a specific block are not carried out until all of the input data are available to that block. Data flows in from sensors through various processing blocks to the display or, in the case of a controller, out through the DAQ board to actuators in the physical system.

From the programmer's point of view, virtual instruments consist of a front panel and a block diagram that are created and viewed on the computer screen.

You will get more out of this discussion by following along on your own computer. You can download an evaluation version of LabVIEW from www.ni.com and install it on your computer. Also, a folder, named **Hambley Virtual Instruments**, containing the virtual instrument files discussed in this section can be downloaded from the website for this book. See Appendix E for details on how to do this. The discussion in this section is based on LabVIEW 2012. If you are using a different version there may be some differences.

After you have installed LabVIEW on your computer and downloaded the virtual instruments folder, you can start the program and open the VI named Figure 9.17. Click on the Run Continuously button, and use the input knobs or up/down arrows to change the values of A and B. Verify that the sum is shown correctly on the meter. Then, type Control E to see the block diagram. Another simple VI named Figure 7.25a illustrates the logic circuit of Figure 7.25a on page 390.

LabVIEW contains many complicated and useful features; more than we have the space to detail in this chapter. Many books and resources are available for help when you need a complex and unique instrument in your career. In fact, you may find LabVIEW to be very useful in your senior project course.

Simulated Data Sources

LabVIEW is primarily intended to process data obtained from sensors in the real world. However, it can also be used to create an entire virtual system in which simulated data are created by the VI and then analyzed in the same way as real data. This is very useful while learning programming techniques and for debugging VIs. Often, in a new instrumentation setting, we do not know what results we will obtain from the real-world data. When the results are surprising, we may not be confident that the virtual instrument is performing its intended function. A good approach is to debug a new VI with simulated data until it seems to perform properly. After satisfactory results are obtained from a known test case, we can finally apply the VI to the real application with confidence. Furthermore, it is unrealistic to provide a test stand, sensors, signal conditioners, and DAQ boards as part of a textbook. Therefore, our examples and exercises use simulated data exclusively. However, keep in mind that LabVIEW provides a powerful way to collect and analyze data from real systems.

It is unrealistic to provide a test stand, sensors, signal conditioners, and DAQ boards as part of a textbook. Therefore, our examples and exercises use simulated data exclusively.

Using a VI to Determine Average and RMS Values

As a small part of a sophisticated instrument, we often need to analyze a signal to determine its average value (dc component) and the root-mean-square (rms) value

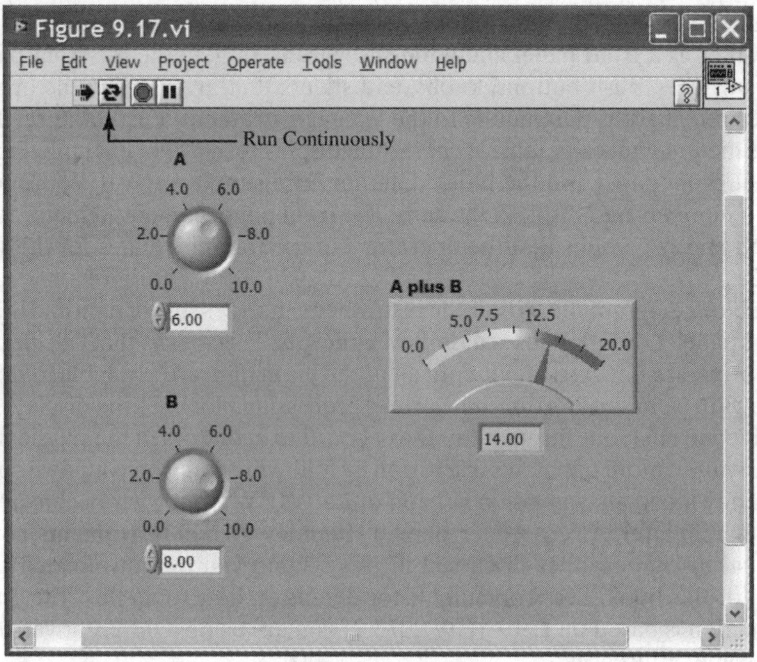

(a) Front panel

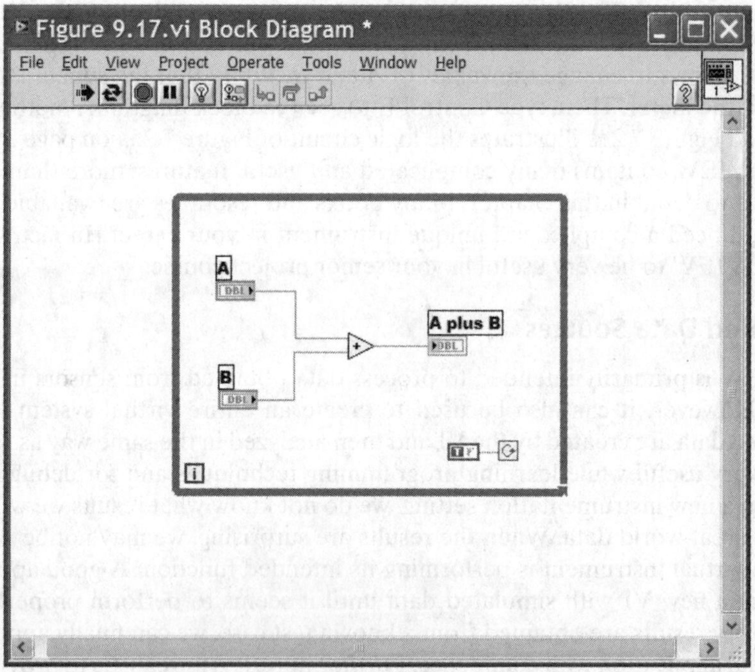

(b) Diagram

Figure 9.17 A simple LabVIEW virtual instrument.

of its ac component. These values are defined in terms of the signal $x(t)$ as:

$$X_{\text{dc}} = \frac{1}{T} \int_0^T x(t)\, dt \tag{9.10}$$

$$X_{\text{ac-rms}} = \sqrt{\frac{1}{T} \int_0^T [x(t) - X_{\text{dc}}]^2\, dt} \tag{9.11}$$

in which T is the period of the signal if it is periodic [or a long enough interval to contain a representative segment of $x(t)$ in case it is not periodic].

Next, we demonstrate how to use LabVIEW to create samples of a simulated signal and analyze the data to determine and display its average value and the rms value of its ac component. If possible, follow along on your own computer.

We have chosen a test signal of the form

$$x(t) = X_{\text{dc}} + X_{\text{peak}} \sin(2\pi f t) \tag{9.12}$$

We will place controls on the front panel so the operator can adjust the average value (also called the dc offset) X_{dc}, the peak amplitude of the ac component X_{peak}, and the frequency f. Of course, for this test signal, the rms value of the ac component is its peak value divided by the square root of 2; namely,

$$X_{\text{ac-rms}} = \frac{X_{\text{peak}}}{\sqrt{2}} \tag{9.13}$$

Constructing the Front Panel

1. Launch LabVIEW and click on **Blank VI** on the Getting Started window. This brings up two windows on your screen. The window with the gray background will eventually contain the front panel of our VI, and the window with the white background will contain the block diagram.
2. Click on the front-panel window (gray background) to make it active and, if the controls palette is not visible, click on **View>Controls Palette**. [Even if you have the same software version used here, your Controls Palette may look different from that shown in Figure 9.18(a). If so, you can make it look like the figure by using the command **Customize>View This Palette As** and check **Category (Icons and Text)**. Then, use the command **Customize>Change Visible Palettes>Select All**.] Then, click on **Express>Numeric Controls>Dial**, move the cursor to position the dial in the front-panel window, and click to drop it into place. Next, type in `Sinewave Amplitude` from your keyboard. Next, place the cursor on the dial, right-click and select **Visible Items>Digital Display**. At this point, your front-panel screen should appear as shown in Figure 9.18(a).
3. Click on the diagram window (white background) to make it active. (In case the diagram window is not open, use the **Window>Show Block Diagram** command.) Click on **View>Tools Palette**. Then, your screen should be similar to Figure 9.18(b). The block labeled Sinewave Amplitude contains a terminal that outputs the value selected by the dial on the front panel. Later, we will wire this terminal into our block diagram, so that the value selected by the operator with the dial on the front panel sets the peak amplitude of the sinusoidal test signal.

   In some earlier versions of LabVIEW, the tools palette was needed to select the appropriate tool for wiring, repositioning icons, entering text, and so forth.

Tip: The symbol > leads you through nested menu and palette items to a final selection. Thus, **View>Controls Palette** directs you to open the View menu and select Controls Palette. The window with the gray background is the front panel, and the window with the white background is the diagram.

Tip: `The Courier font is used for items that need to be entered from the keyboard.`

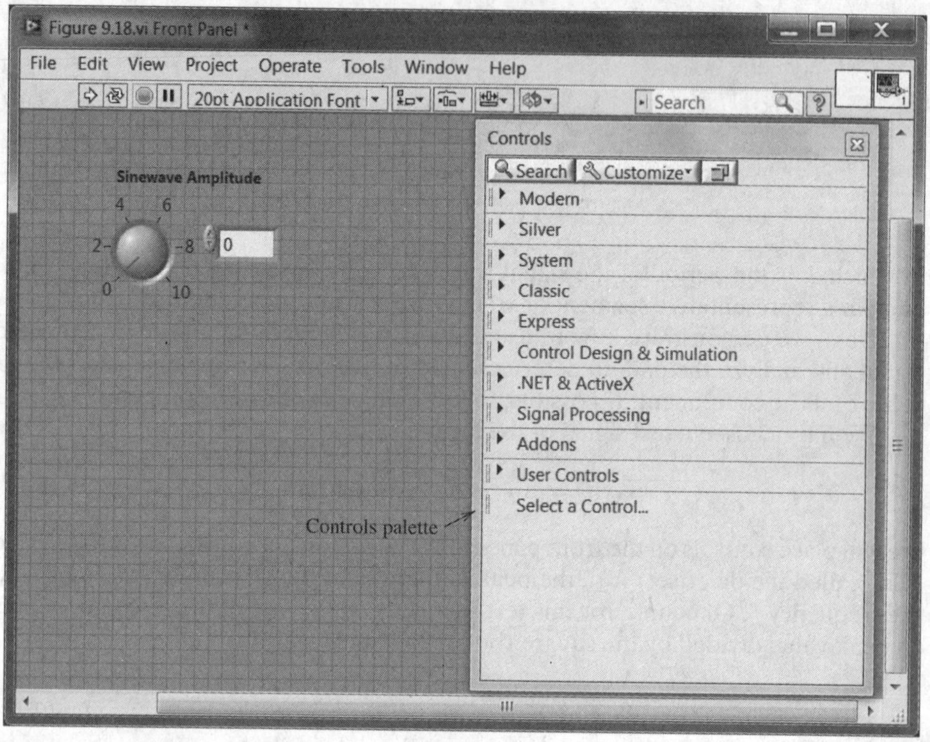

(a) Front panel

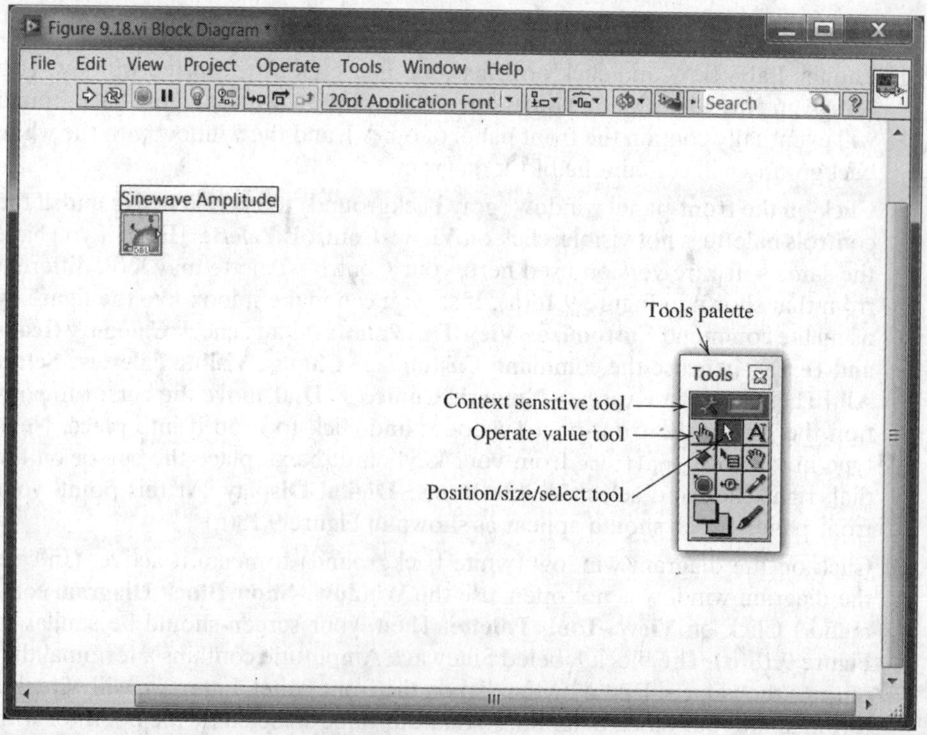

(b) Block diagram

Figure 9.18 Front panel and diagram after selecting and labeling the dial control for the sinewave amplitude.

In LabVIEW 2012, the cursor can be selected to be context driven. Then, if we place it by a node where a wire can be attached, it becomes a wiring tool. If we place the cursor in a text box it becomes a text tool, and so forth. If you find the context driven mode to be confusing, you can select the tool you need from the Palette. In the following, this manual mode is what we describe.

4. Return to the front panel (gray background) and repeat these steps to place a second dial labeled `DC Offset` and a numeric control (upper left-hand corner of the numeric controls palette) labeled `Sinewave Frequency` on the front panel.
5. Click on **Express>Graph Indicators >Waveform Chart** and place the waveform chart on the front panel. Type in the caption `Waveform Display` on your keyboard. Place the cursor on the display icon and click the right mouse button to bring up the pop-up menu. Then, use the **Visible Items>Plot Legend** and **Visible Items>X Scale** commands to hide the legend and the *x*-axis scale. Next, use the **Position/Size/Select** tool to position and resize the elements on the control panel so they appear approximately as shown in Figure 9.19. Finally, use the **Operate Value** tool to edit the lower and upper *y*-axis values, so that they range from −10 to +20. If you have trouble achieving this, you can load the file named Figure 9.19 from the folder named "Hambley Virtual Instruments." (See Appendix E for information about how to access this folder.)
6. Use the **Window>Show Block Diagram** command to switch to the block diagram window. Also, if the **Functions** menu is not visible, use the **View>Functions Palette** command to make it visible. Use the **Position/Size/Select** tool to position

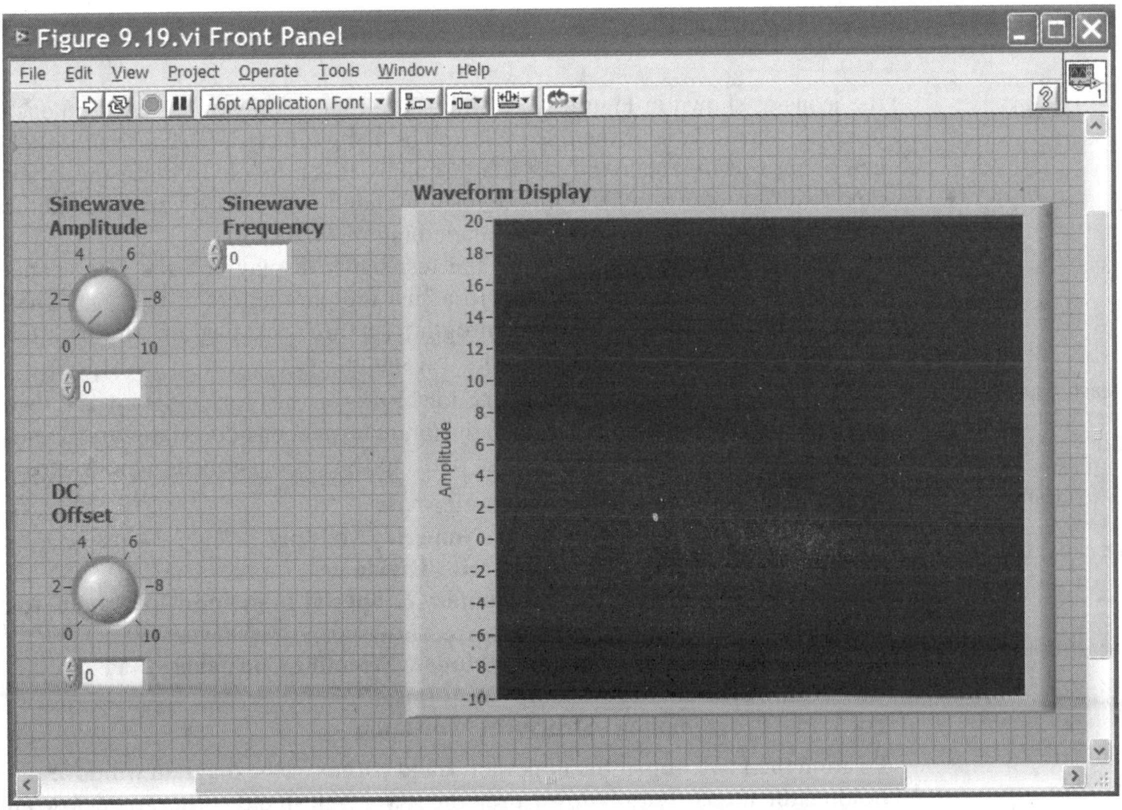

Figure 9.19 Front panel with controls and waveform display.

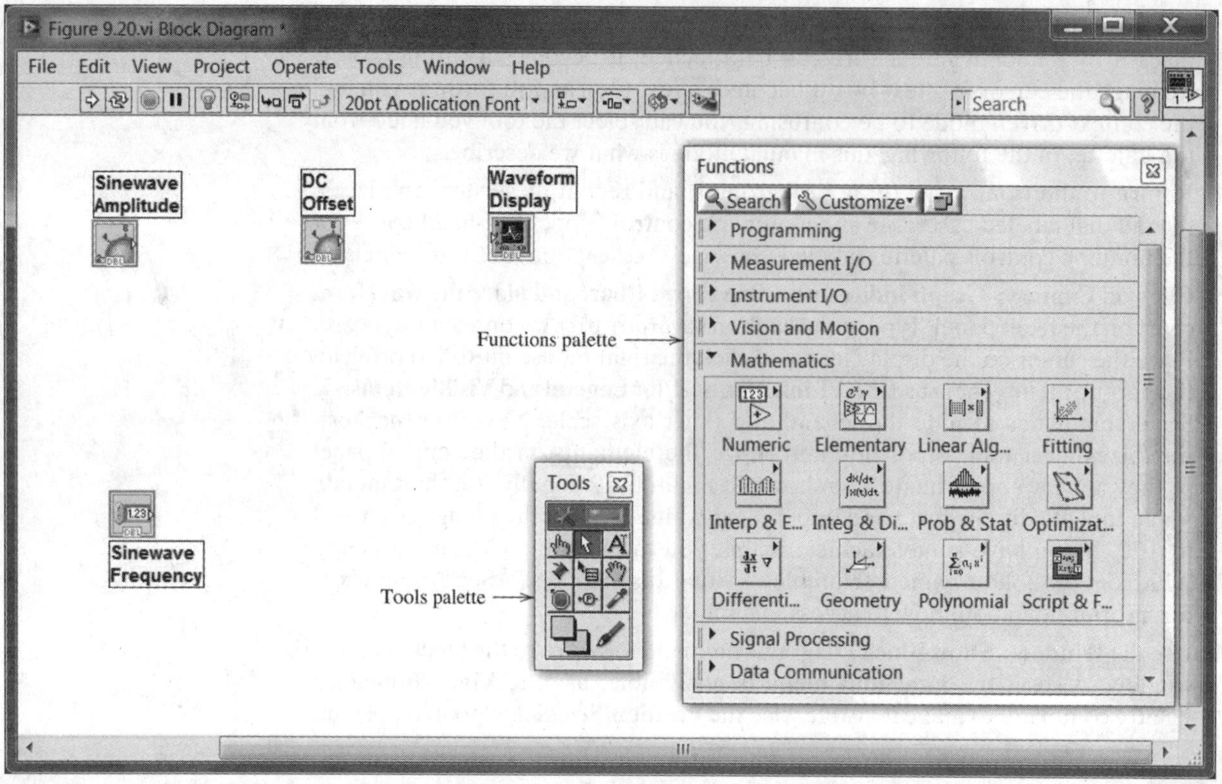

Figure 9.20 Diagram after repositioning terminals and labels.

the icons as shown in Figure 9.20. Make sure that when you left-click on each box, the nearest label is the one indicated as being selected. (Because the labels can be moved independently of the terminals, it is possible to place the labels next to different terminals from the ones intended.) As before, you can load the work up to this point from the file labeled Figure 9.20.vi.

So far, we have simply set up the test-signal display and the front-panel controls that the operator will use to adjust the parameters of the test signal. Next, we need to create a graphical program (in the block diagram window) that uses these parameters to generate the test signal.

The VI files for each figure are contained in the folder named "Hambley Virtual Instruments" that can be downloaded by following links located in www.pearsonhighered.com/hambley.

7. Start by selecting and positioning function blocks on the block diagram as shown in Figure 9.21. (Some of the labels in the figure have been added later and do not appear on the computer screen.) The paths to the various function blocks are:

Addition block: **Mathematics>Numeric>Add**
Multiplication block: **Mathematics>Numeric>Multiply**
Division block: **Mathematics>Numeric>Divide**
Sine function: **Mathematics>Elementary & Special Functions>Trigonometric Functions>Sin**
2π Block: **Mathematics>Numeric>Math & Scientific Constants>2*Pi**
Wait Icon: **Programming>Timing>Wait (ms)**
For Loop: **Programming>Structures>For Loop**

Position the **For Loop** at the upper left-hand corner of the diagram window, and holding the mouse button down, drag the lower right-hand corner to enclose all of the other icons except the Waveform Display, as shown in Figure 9.21.

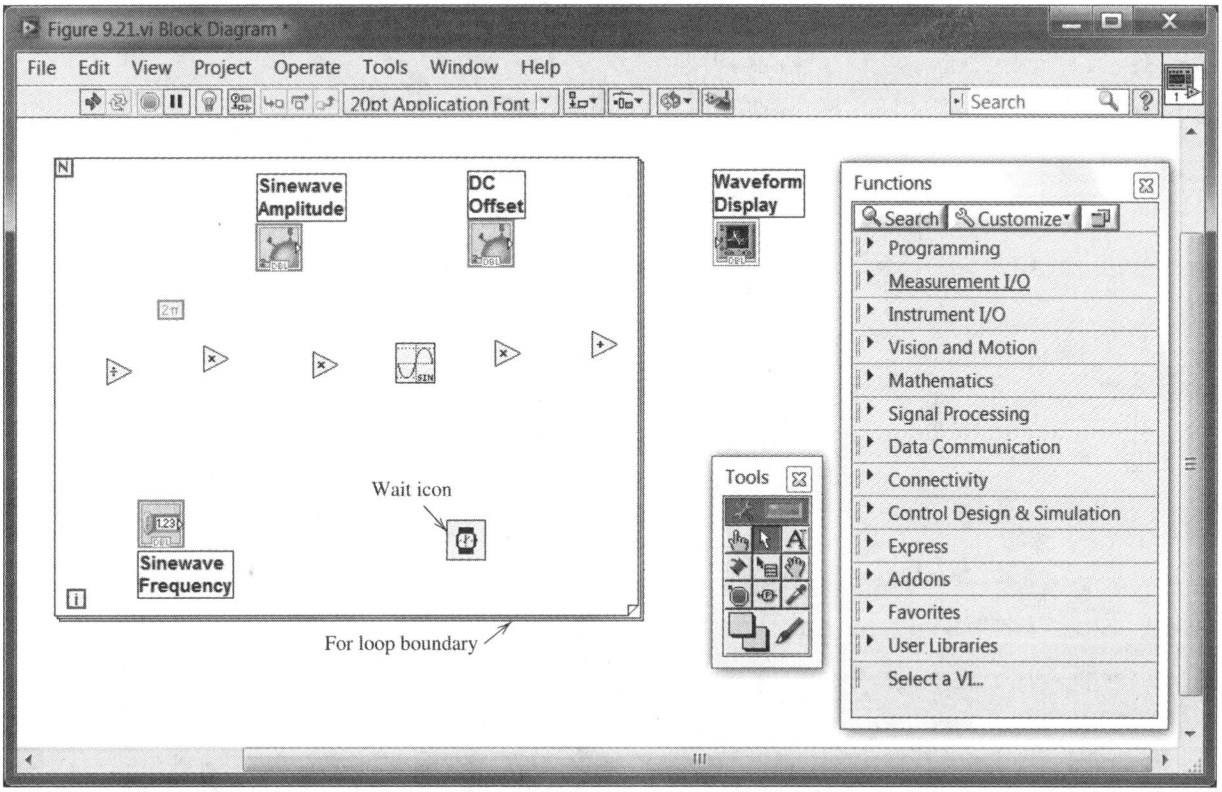

Figure 9.21 Diagram after adding function blocks.

8. Next, add constants and wire the diagram as shown in Figure 9.22. First, right-click on the upper left-hand corner of the for loop, select **Create Constant** from the pop-up menu, and type in 361. Next, select the **Connect Wire** tool from the tools palette. To make a wire connecting the for-loop index (the box labeled *i* in the lower left-hand corner of the for loop) to the divider, click on the for-loop index, move the cursor to the upper input of the divide box, and click. Next, right-click on the lower input of the divider, select **Create>Constant**, and type in 360. Use the **Connect Wire** tool to finish the wiring. As before, a VI containing the work to this point is stored in the file named Figure 9.22.
9. Now, return to the front panel and click on the run continuously button. (The location of this button is shown in Figure 9.23.) Then, adjust the controls and observe the display. You should be able to obtain a display very similar to Figure 9.23 by using the front-panel controls to set the sinewave amplitude to 10, the dc offset to 5, and the frequency to 2. (You may need to right-click on the display and use the **XScale >AutoScaleX** command before the display looks like Figure 9.23.)

   By the way, here are three different ways to adjust the amplitude and dc offset:

   A. Point to and (holding the left mouse button down) rotate the dial.

   B. Click on the up or down arrows on the left-hand side of the digital indicator under the dial.

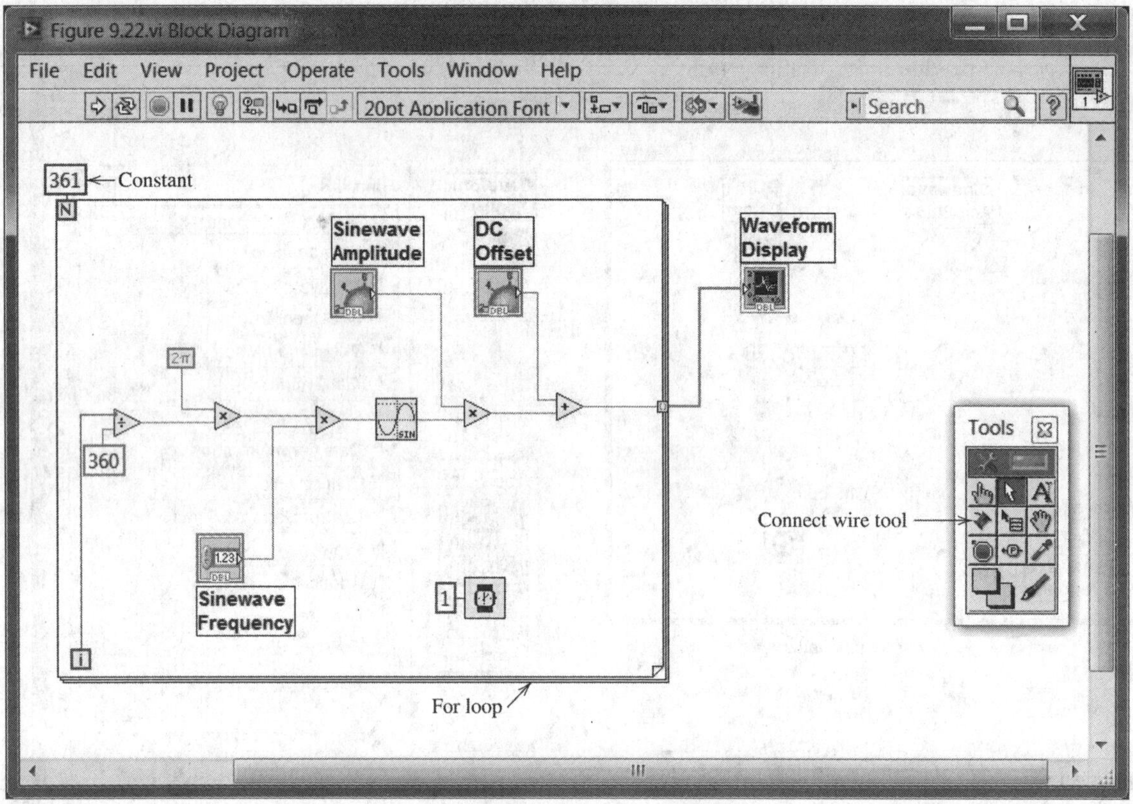

Figure 9.22 Diagram after adding constants and wires.

C. Place the cursor inside the digital indicator under the dial and type in a value, which will become effective when you click in a blank region on the front panel outside the indicator.

Program Operation

Next, let us look at the diagram (Figure 9.22) and understand the operation of this program. Each time the program is operated, all of the functions inside the for-loop structure are carried out 361 times with i starting at zero and incrementing by one after each iteration up through $i = 360$. In creating this program, we have considered the value of i to represent angles in one-degree increments. Division by 360 and multiplication by 2π converts these angles to radians. The angles are then multiplied by frequency f (which is input as an integer value from the front panel). The sine block computes the sine of each angle. Thus, each time the for loop is executed, a total of 361 points on f cycles of the sine function are calculated. Next, the amplitude input from the front panel multiplies the sinewave values, and the dc offset is added to each value.

Each time the for-loop finishes, an array of 361 data points is passed to the display. (The fact that an array is passed is indicated by the wider line used for the wire that connects the output of the for loop to the display.) Go to the front panel, click on the stop icon to stop program execution, right-click on the display to pop up the display menu, and select **Data Operations>Clear Chart**. You will see that each time you click the run button (not the run continuously button), f cycles of the signal will

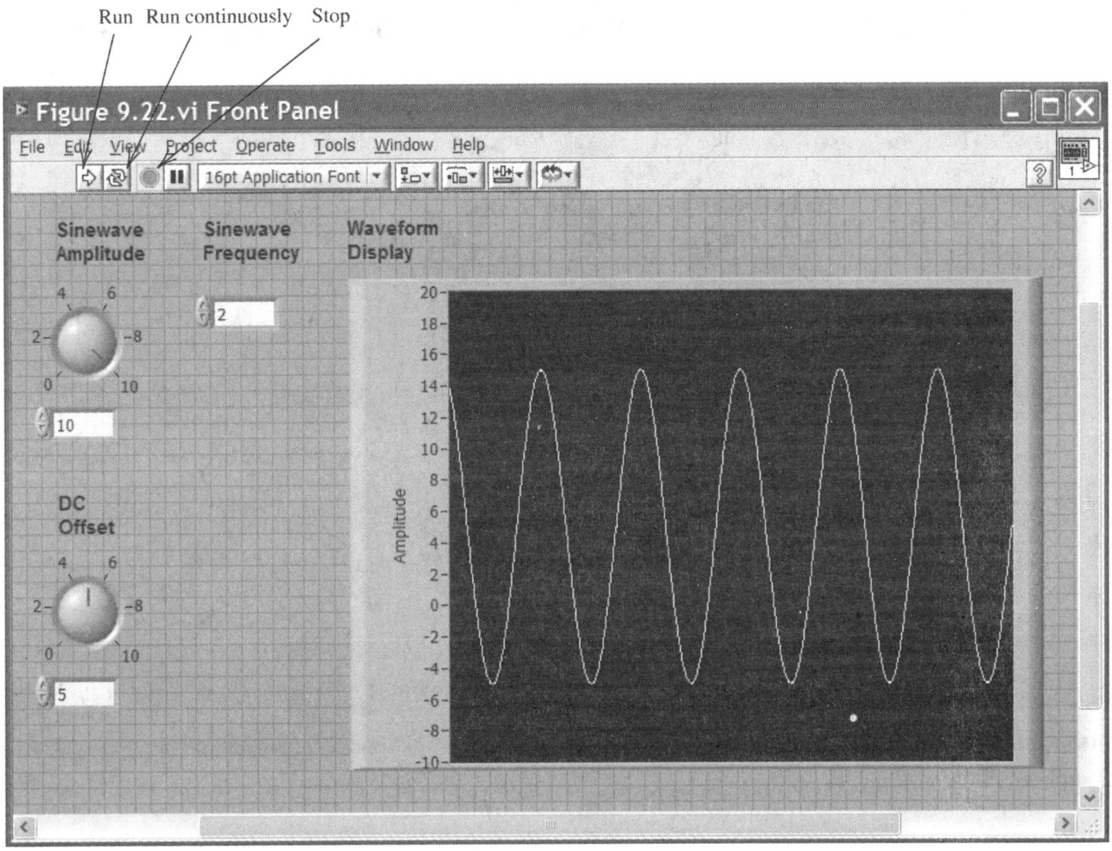

Figure 9.23 Front panel showing simulated data.

appear on the display. By default, the display shows 1024 points, so the display will be filled after clicking the run button three times. You can change the number of points displayed with the **Chart History Length . . .** item on the pop-up menu (right-click on the display). You may wish to experiment with this and other changes, such as moving the display icon inside the for-loop structure on the diagram.

Adding the DC and RMS Virtual Instruments

Now that we have some simulated data with which to work, we can place and wire additional graphical-program elements to compute the average value of the data and the rms value of the ac component. We also need to add front-panel displays for the results as shown in Figure 9.24. To obtain the front-panel meter icon, go to the **Controls Palette** and select **Express>Numeric Indicators>Meter**. Then, drop the meter icon on the front panel and label it as usual. As shown in Figure 9.24, two meters are needed. Then, we can right-click on each meter and use the **Visible Items>Digital Display** command to add the digital display shown below each meter in Figure 9.24.

Figure 9.25 shows the block diagram for obtaining the average value of each 361-point array. Equation 9.11, repeated here for convenience,

$$X_{\text{dc}} = \frac{1}{T}\int_0^T x(t)\,dt$$

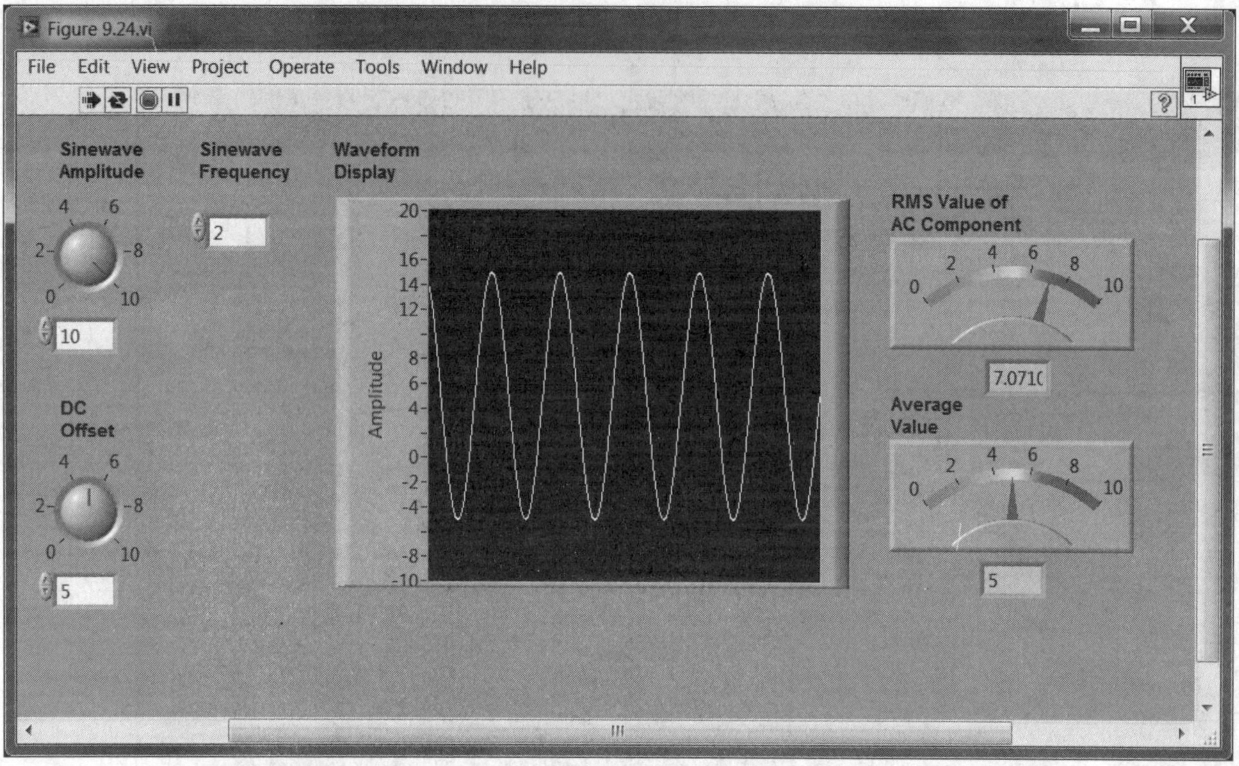

Figure 9.24 Completed front panel.

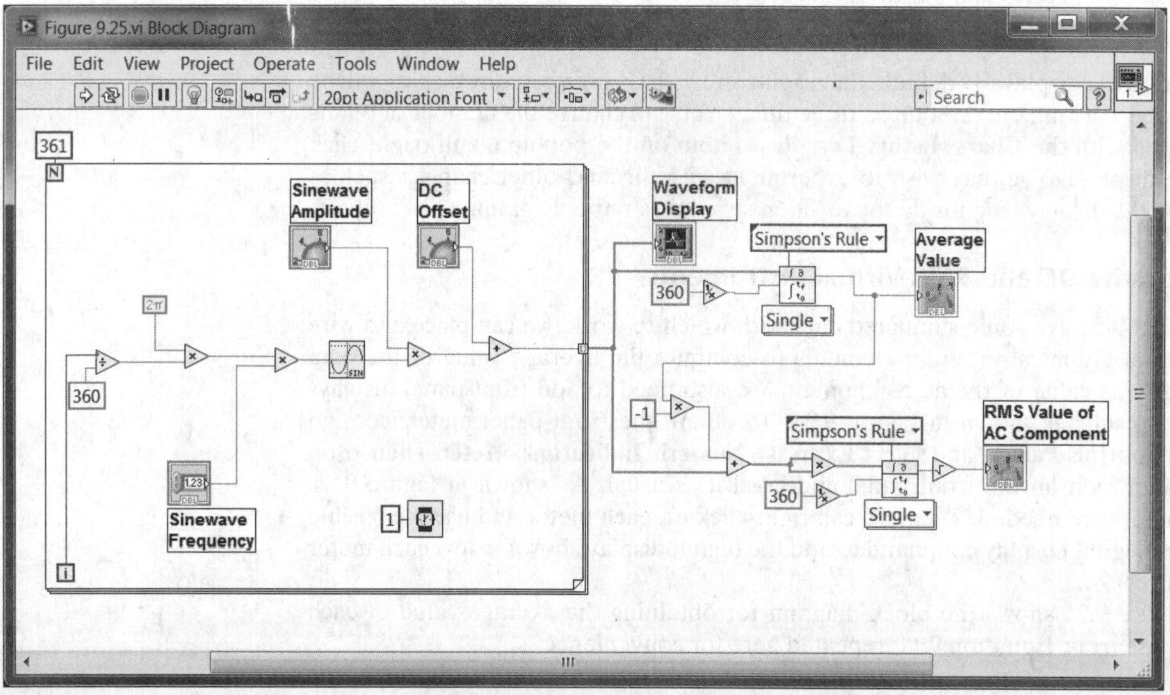

Figure 9.25 Completed diagram.

shows that the signal needs to be integrated over one cycle (from $t = 0$ to $t = T$). The result of the integration is then divided by the duration. Conceptually, we have let T equal unity, and the time increment between array values is 1/360. The integration block can be found on the functions menu by clicking on **Mathematics>Integration & Differentiation>Numeric Int....** (Actually, the integration block is a sub VI, which is the graphical-language equivalent of a subroutine. If you double-click on the integration block, you can see a little of its internal structure.) The integrator inputs are the array to be integrated, the integration algorithm (right-click on the input and use the pop-up menu to select **Create>Constant>Simpson's Rule**), and the time interval between samples.

Because we have discrete samples of $x(t)$, rather than a continuous time function, we must use numerical integration techniques such as Simpson's rule to approximate the integral.

According to Equation 9.11, the rms value of the ac component of the signal is given by

$$X_{\text{ac-rms}} = \sqrt{\frac{1}{T}\int_0^T [x(t) - X_{\text{dc}}]^2\, dt}$$

The corresponding graphical code is shown in Figure 9.25. First, the dc component (average value) is subtracted from the input array. The result is then connected to both inputs of a multiplier. When the multiplier operates on arrays, the corresponding elements of the arrays are multiplied. Thus, we obtain an array containing the squares of the values in the input array. This is integrated (again, by using the idea that $T = 1$ and $\Delta t = 1/360$). Finally, the square root is computed and displayed.

Now, if we return to the front panel, click on the run continuously button, and adjust the controls, we will see that the Average-Value meter indicates exactly the dc offset value of our simulated data. Furthermore, the rms-value meter indicates the rms value of the ac component, which is the amplitude divided by $\sqrt{2}$.

Much more sophisticated analysis can be carried out by using sensors, DAQ boards, and LabVIEW in any field of science or engineering. You can learn about many other features of LabVIEW and the graphical-programming language from sources available through `www.ni.com`. Our aim here has been mainly to acquaint you with some of the possibilities so you can go in the right direction when this approach is appropriate in your work.

Exercise 9.6 Use LabVIEW to create the VI shown in Figure 9.17.

Answer A file labeled Figure 9.17 that contains the virtual instrument can be found in the Hambley Virtual Instruments folder. □

LabVIEW MathScript

MathScript, which is an optional part of LabVIEW, contains some of the same features as those of MATLAB that we have discussed earlier in the book. To access MathScript, open LabVIEW and use the command **Tools>MathScript Window**. For example, we illustrate how to solve the node voltage equations of Equation 2.36, which are repeated here for convenience.

$$\mathbf{GV} = \mathbf{I}$$

$$\begin{bmatrix} 0.45 & -0.25 & 0 \\ -0.25 & 0.85 & -0.20 \\ 0 & -0.20 & 0.30 \end{bmatrix} \begin{bmatrix} v_1 \\ v_2 \\ v_3 \end{bmatrix} = \begin{bmatrix} -3.5 \\ 3.5 \\ 2 \end{bmatrix} \qquad (2.36)$$

To solve, we enter the commands (and optional comments following the % signs) in the MathScript Command Window as follows:

```
clear  % First we clear the work space.
% Then, we enter the coefficient matrix of Equation~2.36 with
% spaces between elements in each row and semicolons between rows.
G = [0.45 -0.25 0; -0.25 0.85 -0.2; 0 -0.2 0.30]
% Next, we enter the column vector for the right-hand side.
I = [-3.5; 3.5; 2]
% Next we compute the node voltages using the command
% V = G\I instead of using V = inv(G)*I.
V = G\I
```

These are the exact commands used to solve this system of equations with MATLAB. The result shown in the MathScript Output Window is

```
V =

   -5

    5

   10
```

which of course agrees with our earlier result using MATLAB on page 86.

Summary

1. The block diagram of a typical computer-based instrumentation system is shown in Figure 9.1 on page 454.
2. When we need to sense the internal (open-circuit) voltage of a sensor, we should specify an amplifier having an input impedance that is much larger in magnitude than the Thévenin impedance of the sensor.
3. When we want to sense the current produced by a sensor, we need a current-to-voltage converter having a very small (ideally zero) input impedance magnitude compared to the Thévenin impedance of the sensor.
4. Bias errors are the same each time a measurement is repeated under the same apparent conditions. Bias errors include offset, scale error, nonlinearity, and hysteresis.
5. Random errors are different for each measurement and have zero average value.
6. The accuracy of an instrument is the maximum expected difference in magnitude between measured and true values (often expressed as a percentage of the full-scale value).
7. Precision is the ability of the instrument to repeat the measurement of a constant measurand. More precise measurements have less random error.
8. The resolution of an instrument is the smallest increment discernible between measured values.
9. Some of the functions of signal conditioners are amplification, conversion of current signals to voltage signals, supply of excitation to the sensor, and filtering to eliminate noise or other unwanted signal components.
10. One of the input terminals of a single-ended amplifier is grounded. Neither input terminal of a differential amplifier is grounded. The output of an ideal differential amplifier is its differential gain times the difference between the input voltages.
11. If the input voltages to a differential amplifier are v_1 and v_2, the differential input signal is $v_d = v_1 - v_2$ and the common-mode signal is $v_{\text{cm}} = \frac{1}{2}(v_1 + v_2)$. Usually in instrumentation systems, the differential signal is of interest and the common-mode signal is unwanted noise.
12. When large unwanted common-mode signals are present, it is important to select a differential amplifier having a large CMRR (common-mode rejection ratio) specification.

13. Ground loops are created in an instrumentation system when several points are connected to ground. Currents flowing through the ground conductors can produce noise that makes the measurements less accurate and less precise.

14. To avoid ground-loop noise when we must connect a sensor to an amplifier with a single-ended input, we should select a floating sensor (i.e., neither terminal of the sensor should be grounded).

15. Shielded cables reduce the noise coupled by electric fields.

16. Coaxial or twisted-pair cables reduce magnetically coupled noise.

17. If a signal contains no components with frequencies higher than f_H, all of the information contained in the signal is present in its samples, provided that the sampling rate is selected to be more than twice f_H.

18. Analog-to-digital conversion is a two-step process. First, the signal is sampled at uniformly spaced points in time. Second, the sample values are quantized so they can be represented by words of finite length.

19. We model the effect of finite word length as the addition of quantization noise to the signal.

20. If a sinusoidal signal component is sampled at a rate f_s that is less than twice the signal frequency f, the samples appear to be from a component with a different frequency known as the alias frequency f_{alias}. Alias frequency is plotted versus the true signal frequency in Figure 9.14 on page 467.

21. If a sensor signal contains high-frequency components that are not of interest, we often use an analog antialias filter to remove them so a lower sampling frequency can be used without aliasing.

22. LabVIEW programs appear to the programmer as a front panel and a block diagram. The front panel is the user interface that contains the controls and displays for the virtual instrument. The block diagram, which is constructed in a graphical programming language, shows how the instrument analyzes sensor data and input from the front panel.

23. LabVIEW uses data-flow concepts. This means that calculations of a specific block are not carried out until all of the input data are available to that block.

Problems

Section 9.1: Measurement Concepts and Sensors

P9.1. Draw and explain a computer-based DAQ system.

***P9.2.** Draw the equivalent circuit of a sensor in which the open-circuit sensor voltage is proportional to the measurand. What are loading effects? How do we avoid them when we need to measure the Thévenin (i.e., open-circuit) sensor voltage?

P9.3. What signal-conditioner input impedance is best if the sensor produces short-circuit (Norton) current that is proportional to the measurand?

***P9.4.** A load cell produces an open-circuit voltage of 200 μV for a full-scale applied force of 10^4 N, and the Thévenin resistance is 1 kΩ. The sensor terminals are connected to the input terminals of an amplifier. What is the minimum input resistance of the amplifier so the overall system sensitivity is reduced by less than 1 percent by loading?

***P9.5.** A certain liquid-level sensor has a Thévenin (or Norton) resistance that varies randomly from 10 kΩ to 1 MΩ. The short-circuit current of the sensor is proportional to the measurand. What type of signal-conditioning unit is required? Suppose we allow for up to 1 percent change in overall sensitivity due to changes in the sensor resistance. Determine the specification for the input resistance of the signal conditioner.

P9.6. How are bias errors different from random errors?

* Denotes that answers are contained in the Student Solutions files. See Appendix E for more information about accessing the Student Solutions.

P9.7. List four types of bias error.

***P9.8.** An instrumentation system measures speed ranging from zero miles/hour to 200 miles/hour full scale. The system accuracy is specified as ± 0.2% of full scale. If the measured value is 140 m/hr, what is the potential range of the true value?

P9.9. Explain how precision, accuracy, and resolution of an instrument are different.

***P9.10.** Three instruments each make 10 repeated measurements of a flow rate known to be 1.500 m^3/s with the results given in Table P9.10.

a. Which instrument is most precise? Least precise? Explain.

b. Which instrument has the best accuracy? Worst accuracy? Explain.

c. Which instrument has the best resolution? Worst resolution? Explain.

Table P9.10

Trial	Instrument A	Instrument B	Instrument C
1	1.5	1.73	1.552
2	1.3	1.73	1.531
3	1.4	1.73	1.497
4	1.6	1.73	1.491
5	1.3	1.73	1.500
6	1.7	1.73	1.550
7	1.5	1.73	1.456
8	1.7	1.73	1.469
9	1.6	1.73	1.503
10	1.5	1.73	1.493

Section 9.2: Signal Conditioning

P9.11. Comment on the necessity of signal conditioners, and the functions performed by them.

P9.12. What value addition does a differential amplifier provide over a single-ended one?

P9.13. Suppose that the input voltages to an ideal differential amplifier are equal. Determine the output voltage.

***P9.14.** The input voltages to a differential amplifier are

$$v_1(t) = 0.002 + 5\cos(\omega t)$$

and

$$v_2(t) = -0.002 + 5\cos(\omega t)$$

Determine the differential input voltage and the common-mode input voltage. Assuming that the differential amplifier is ideal with a differential gain $A_d = 1000$, determine the output voltage of the amplifier.

P9.15. A sensor produces a differential signal of 6 mV dc and a 2-V-rms 60-Hz ac common-mode signal. Write expressions for the voltages between the sensor output terminals and ground.

***P9.16.** Suppose we have a sensor that has one terminal grounded. The sensor is to be connected to a DAQ board in a computer 5 meters away. What type of amplifier should we select? To mitigate noise from electric and magnetic fields, what type of cable should we use? Draw the schematic diagram of the sensor, cable, and amplifier.

P9.17. What is an ungrounded sensor called? When should we use ungrounded sensors?

***P9.18.** Suppose that the data collected from a sensor is found to contain an objectionable 60-Hz ac component. What potential causes for this interference would you look for? What are potential solutions for each cause of the interference?

Section 9.3: Analog-to-Digital Conversion

P9.19. In principle, analog-to-digital conversion involves two operations. What are they?

P9.20. What is aliasing? Under what conditions does it occur?

P9.21. What causes quantization noise?

***P9.22.** We need to use the signal from a piezoelectric vibration sensor in a computer-aided instrumentation system. The signal is known to contain components with frequencies up to 30 kHz. What is the minimum sampling rate that should be specified? Suppose that we want the resolution of the sampled values to be 0.1 percent (or better) of the full range of the ADC. What is the fewest number of bits that should be specified for the ADC?

***P9.23.** A 3-V-peak sinewave signal is converted to digital form by an 8-bit ADC that has been designed to accept signals ranging from −5 V to 5 V. (In other words, codewords are assigned for equal increments of amplitude for amplitudes between −5 V and +5 V.)

a. Determine the width Δ of each quantization zone.

b. Determine the rms value of the quantization noise and the power that the quantization noise would deliver to a resistance R.

c. Determine the power that the 3-V sinewave signal would deliver to a resistance R.

d. Divide the signal power found in part (c) by the noise power determined in part (b). This ratio is called the signal-to- noise ratio (SNR). Express the SNR in decibels, using the formula $\mathrm{SNR_{dB}} = 10\log(P_{\mathrm{signal}}/P_{\mathrm{noise}})$.

***P9.24.** We need an ADC that can accept input voltages ranging from -5 to $+5$ V. How many bits must the code word have for a resolution of 0.01 V and 0.02 V respectively.

***P9.25.** A 10-kHz sinewave is sampled. Determine the apparent frequency of the samples. Has aliasing occurred? The sampling frequency is **a.** 11 kHz; **b.** 8 kHz; **c.** 40 kHz.

P9.26. A 60-Hz sinewave $x(t) = A\cos(120\pi t + \phi)$ is sampled at a rate of 360 Hz. Thus, the sample values are $x(n) = A\cos(120\pi n T_s + \phi)$, in which n assumes integer values and $T_s = 1/360$ is the time interval between samples. A new signal is computed by the equation

$$y(n) = \frac{1}{2}[x(n) + x(n-3)]$$

a. Show that $y(n) = 0$ for all n.

b. Now suppose that $x(t) = V_{\mathrm{signal}} + A\cos(120\pi t + \phi)$, in which V_{signal} is constant with time and again find an expression for $y(n)$.

c. When we use the samples of an input $x(n)$ to compute the samples for a new signal $y(n)$, we have a **digital filter**. Describe a situation in which the filter of parts (a) and (b) could be useful.

Section 9.4: LabVIEW

P9.27. What does LabVIEW stand for? What is VI—a code or a graphical tool?

P9.28. Explain the data-flow concept used by LabVIEW.

***P9.29.** Use LabVIEW to create a virtual instrument that converts a complex number of the form $x + jy$ into polar form $A\angle\theta$. Provide front-panel controls for the operator to input the values of x and y, each of which can range from -10 to $+10$. Also, provide displays for A and θ (in degrees) on the front panel.

P9.30. Use LabVIEW to create a virtual instrument that converts a complex number in polar form $A\angle\theta$ (with θ in degrees ranging from $-360°$ to $+360°$ and A ranging from 0 to 20) into the form $x + jy$. Provide front-panel controls for the operator to input the values for A and θ. Also, provide displays for x and y on the front panel.

***P9.31.** Open the VI stored in the file named Figure 9.24.vi from the folder named **Hambley Virtual Instruments**. Click on the run continuously button. Set the sinewave amplitude to 10 and the frequency to 1. Then, gradually increase the frequency in unit steps several times and observe the changes in the waveform. Next, set the frequency to 355 and again gradually increase the frequency in unit steps to 365, observing the changes in the waveform after each step. Discuss and explain your observations.

P9.32. Create a VI to demonstrate a concept or solve a problem from your own field. Write a short description of the concept or problem and of the vi.

Frequently Asked Questions

1. Describe the operation of a computer based instrumentation system with the help at a block-diagram.

2. What do you mean by a sensor and sensitivity constant?

3. Which type of sensor is used for the following measurements
   (i) Acceleration
   (ii) Light
   (iii) Linear displacement
   (iv) Fluid flow
   (v) Pressure
   (vi) Temperature
4. What do you mean by loading of a sensor?
5. How photo-diode is used as a sensor for measurements?
6. What can be the errors in measurement system?
7. What do you mean by accuracy, precision and resolution?
8. What do you mean by signal conditioning? What is its need?
9. Differentiate between single-ended and differential amplifiers.
10. Define and explain common-mode rejection ratio (CMRR).
11. What do you understand by noise in measuring system?
12. Define and explain
    (i) Sampling rate
    (ii) Aliasing
    (iii) Quantization noise
13. What is a virtual Time-Frequency vibration analyzer?
14. How data-flow concept is used by LabVIEW?
15. How virtual instruments are created?

Practice Test

Here is a practice test you can use to check your comprehension of the most important concepts in this chapter. Answers can be found in Appendix D and complete solutions are included in the Student Solutions files. See Appendix E for more information about the Student Solutions.

T9.1. Name the four elements of a computer-based instrumentation system.

T9.2. Name four types of systematic errors in measurement systems.

T9.3. How are bias errors different from random errors?

T9.4. What causes ground loops in an instrumentation system? What are the effects of a ground loop?

T9.5. If a sensor must have one end connected to ground, what type of amplifier should we choose? Why?

T9.6. What types of cable are best for connecting a sensor to an instrumentation amplifier to avoid coupling of noise by electrical and magnetic fields?

T9.7. If we need to sense the open-circuit voltage of a sensor, what specification is important for the instrumentation amplifier?

T9.8. How do we choose the sampling rate for an ADC? Why?

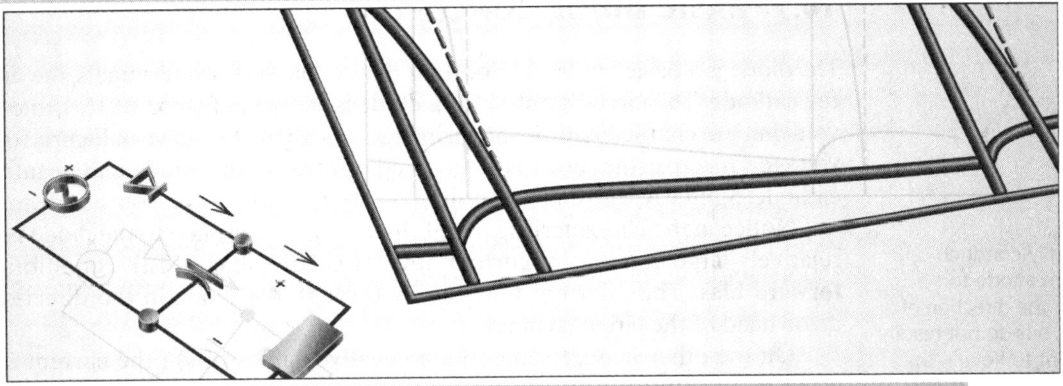

Chapter 10

Diodes

Study of this chapter will enable you to:

- Understand diode operation and select diodes for various applications.
- Use the graphical load-line technique to analyze nonlinear circuits.
- Analyze and design simple voltage-regulator circuits.
- Use the ideal-diode model and piecewise-linear models to solve circuits.
- Understand various rectifier and wave-shaping circuits.
- Understand small-signal equivalent circuits.

Introduction to this chapter:

Electronic circuits are useful for processing information and controlling energy. Some applications of electronic circuits are computers, radio, television, navigation systems, light dimmers, calculators, appliances, controls for machines, motion sensors, and surveying equipment. A basic understanding of electronic circuits will help you in working with instrumentation in any field of engineering. In the next several chapters, we introduce the most important electronic devices, their basic circuit applications, and several important analysis techniques. In this chapter, we discuss the diode.

10.1 BASIC DIODE CONCEPTS

The diode is a basic but very important device that has two terminals, the **anode** and the **cathode**. The circuit symbol for a diode is shown in Figure 10.1(a), and a typical volt–ampere characteristic is shown in Figure 10.1(b). As shown in Figure 10.1(a), the voltage v_D across the diode is referenced positive at the anode and negative at the cathode. Similarly, the diode current i_D is referenced positive from anode to cathode.

Diodes readily conduct current from anode to cathode (in the direction of the arrow), but do not readily allow current to flow in the opposite direction.

Notice in the characteristic that if the voltage v_D applied to the diode is positive, relatively large amounts of current flow for small voltages. This condition is called **forward bias**. Thus, current flows easily through the diode in the direction of the arrowhead of the circuit symbol.

On the other hand, for moderate negative values of v_D, the current i_D is very small in magnitude. This is called the **reverse-bias region**, as shown on the diode characteristic. In many applications, the ability of the diode to conduct current easily in one direction, but not in the reverse direction, is very useful. For example, in an automobile, diodes allow current from the alternator to charge the battery when the engine is running. However, when the engine stops, the diodes prevent the battery from discharging through the alternator. In these applications, the diode is analogous to a one-way valve in a fluid-flow system, as illustrated in Figure 10.1(d).

If a sufficiently large reverse-bias voltage is applied to the diode, operation enters the reverse-breakdown region of the characteristic, and currents of large magnitude flow.

If a sufficiently large reverse-bias voltage is applied to the diode, operation enters the **reverse-breakdown region** of the characteristic, and currents of large magnitude flow. Provided that the power dissipated in the diode does not raise its temperature too high, operation in reverse breakdown is not destructive to the device. In fact, we

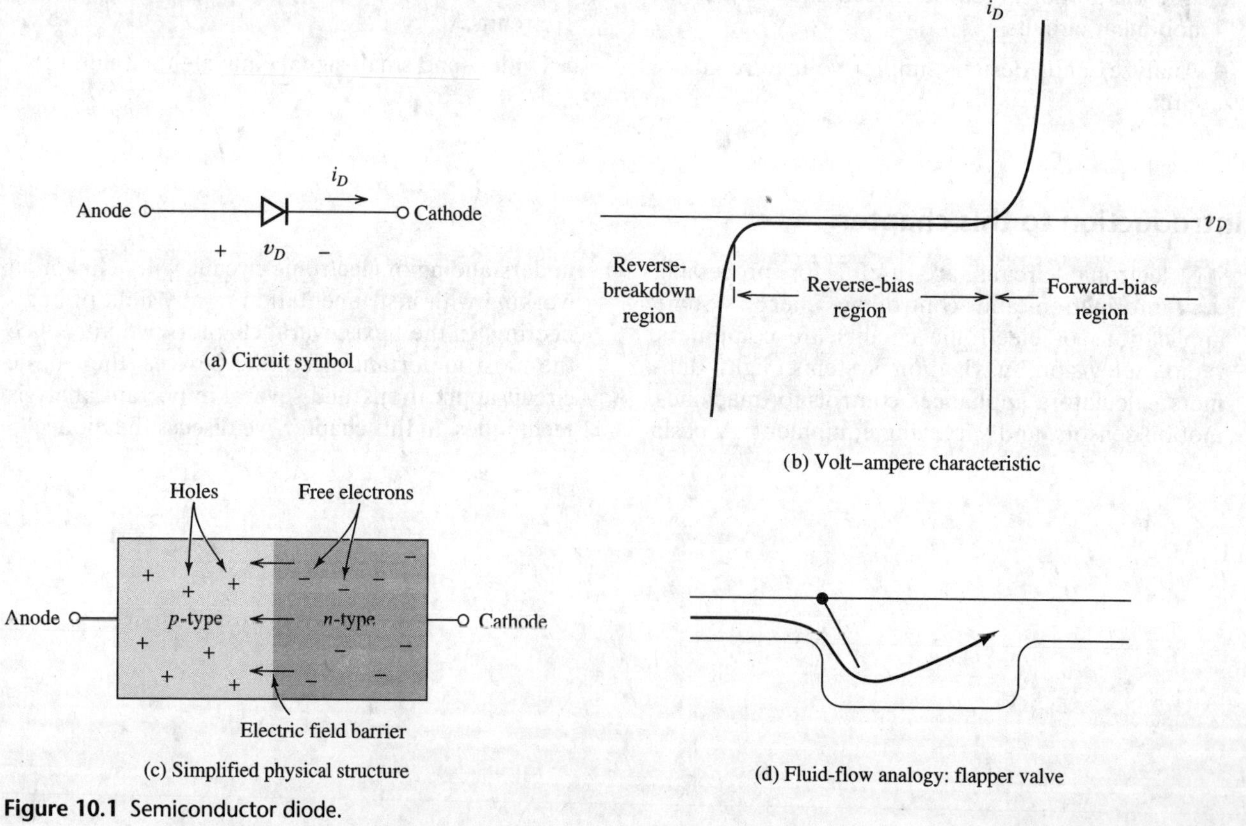

(a) Circuit symbol

(b) Volt–ampere characteristic

(c) Simplified physical structure

(d) Fluid-flow analogy: flapper valve

Figure 10.1 Semiconductor diode.

will see that diodes are sometimes deliberately operated in the reverse-breakdown region.

Brief Sketch of Diode Physics

We concentrate our discussion on the external behavior of diodes and some of their circuit applications. However, at this point, we give a thumbnail sketch of the internal physics of the diode.

The diodes that we consider consist of a junction between two types of semiconducting material (usually, silicon with carefully selected impurities). On one side of the junction, the impurities create ***n*-type material**, in which large numbers of electrons move freely. On the other side of the junction, different impurities are employed to create (in effect) positively charged particles known as **holes**. Semiconductor material in which holes predominate is called ***p*-type material**. Most diodes consist of a junction between *n*-type material and *p*-type material, as shown in Figure 10.1(c).

Even with no external applied voltage, an electric-field **barrier** appears naturally at the *pn* junction. This barrier holds the free electrons on the *n*-side and the holes on the *p*-side of the junction. If an external voltage is applied with positive polarity on the *n*-side, the barrier is enhanced and the charge carriers cannot cross the junction. Thus, virtually no current flows. On the other hand, if a voltage is applied with positive polarity on the *p*-side, the barrier is reduced and large currents cross the junction. Thus, the diode conducts very little current for one polarity and large current for the other polarity of applied voltage. The anode corresponds to the *p*-type material and the cathode is the *n*-side.

Small-Signal Diodes

Various materials and structures are used to fabricate diodes. For now, we confine our discussion to small-signal silicon diodes, which are the most common type found in low- and medium-power electronic circuits.

The characteristic curve of a typical small-signal silicon diode operated at a temperature of 300 K is shown in Figure 10.2. Notice that the voltage and current scales

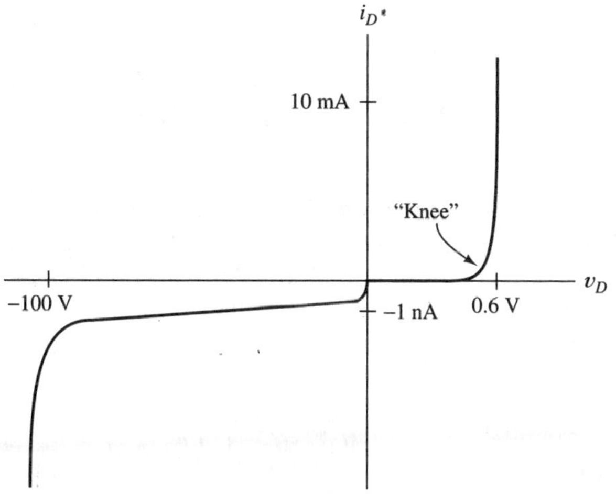

Figure 10.2 Volt–ampere characteristic for a typical small-signal silicon diode at a temperature of 300 K. Notice the change of scale for negative current and voltage.

for the forward-bias region are different than for the reverse-bias region. This is necessary for displaying details of the characteristic, because the current magnitudes are much smaller in the reverse-bias region than in the forward-bias region. Furthermore, the forward-bias voltage magnitudes are much less than typical breakdown voltages.

In the forward-bias region, small-signal silicon diodes conduct very little current (much less than 1 mA) until a forward voltage of about 0.6 V is applied (assuming that the diode is at a temperature of about 300 K). Then, current increases very rapidly as the voltage is increased. We say that the forward-bias characteristic displays a *knee* in the forward bias characteristic at about 0.6 V. (The exact value of the knee voltage depends on the device, its temperature, and the current magnitude. Typical values are 0.6 or 0.7 V.) As temperature is increased, the knee voltage decreases by about 2 mV/K. (Because of the linear change in voltage with temperature, diodes are useful as temperature sensors. The diode is operated at a fixed current, and the voltage across the diode depends on its temperature. Electronic thermometers used by physicians contain a diode sensor, amplifiers, and other electronic circuits that drive the liquid-crystal temperature display.)

In the reverse-bias region, a typical current is about 1 nA for small-signal silicon diodes at room temperature. As temperature increases, reverse current increases in magnitude. A rule of thumb is that the reverse current doubles for each 10-K increase in temperature.

When reverse breakdown is reached, current increases in magnitude very rapidly. The voltage for which this occurs is called the **breakdown voltage**. For example, the breakdown voltage of the diode characteristic shown in Figure 10.2 is approximately −100 V. Breakdown-voltage magnitudes range from several volts to several hundred volts. Some applications call for diodes that operate in the forward-bias and nonconducting reverse-bias regions without entering the breakdown region. Diodes intended for these applications have a specification for the minimum magnitude of the breakdown voltage.

Shockley Equation

Under certain simplifying assumptions, theoretical considerations result in the following relationship between current and voltage for a junction diode:

$$i_D = I_s\left[\exp\left(\frac{v_D}{nV_T}\right) - 1\right] \tag{10.1}$$

This is known as the **Shockley equation**. The **saturation current**, I_s, has a value on the order of 10^{-14} A for small-signal junction diodes at 300 K. (I_s depends on temperature, doubling for each 5-K increase in temperature for silicon devices.) The parameter n, known as the **emission coefficient**, takes values between 1 and 2, depending on details of the device structure. The voltage V_T is given by

$$V_T = \frac{kT}{q} \tag{10.2}$$

and is called the **thermal voltage**. The temperature of the junction in kelvin is represented by T. Furthermore, $k = 1.38 \times 10^{-23}$ J/K is Boltzmann's constant, and

$q = 1.60 \times 10^{-19}$ C is the magnitude of the electrical charge of an electron. At a temperature of 300 K, we have $V_T \cong 0.026$ V.

If we solve the Shockley equation for the diode voltage, we find that

$$v_D = nV_T \ln[\left(\frac{i_D}{I_s}\right) + 1] \tag{10.3}$$

For small-signal junction diodes operated at forward currents between 0.01 μA and 10 mA, the Shockley equation with n taken as unity is usually very accurate. Because the derivation of the Shockley equation ignores several phenomena, the equation is not accurate for smaller or larger currents. For example, under reverse bias, the Shockley equation predicts $i_D \cong -I_s$, but we usually find that the reverse current is much larger in magnitude than I_s (although still small). Furthermore, the Shockley equation does not account for reverse breakdown.

With forward bias of at least several tenths of a volt, the exponential in the Shockley equation is much larger than unity; with good accuracy, we have

$$i_D \cong I_s \exp\left(\frac{v_D}{nV_T}\right) \tag{10.4}$$

This approximate form of the equation is often easier to use.

Occasionally, we are able to derive useful analytical results for electronic circuits by use of the Shockley equation, but much simpler models for diodes are usually more useful.

Zener Diodes

Diodes that are intended to operate in the breakdown region are called **Zener diodes**. Zener diodes are useful in applications for which a constant voltage in breakdown is desirable. Therefore, manufacturers try to optimize Zener diodes for a nearly vertical characteristic in the breakdown region. The modified diode symbol shown in Figure 10.3 is used for Zener diodes. Zener diodes are available with breakdown voltages that are specified to a tolerance of $\pm 5\%$.

Figure 10.3 Zener-diode symbol.

Exercise 10.1 At a temperature of 300 K, a certain junction diode has $i_D = 0.1$ mA for $v_D = 0.6$ V. Assume that n is unity and use $V_T = 0.026$ V. Find the value of the saturation current I_s. Then, compute the diode current at $v_D = 0.65$ V and at 0.70 V.

Answer $I_s = 9.50 \times 10^{-15}$ A, $i_D = 0.684$ mA, $i_D = 4.68$ mA. □

Exercise 10.2 Consider a diode under forward bias so that the approximate form of the Shockley equation (Equation 10.4) applies. Assume that $V_T = 0.026$ V and $n = 1$. **a.** By what increment must v_D increase to double the current? **b.** To increase the current by a factor of 10?

Answer **a.** $\Delta v_D = 18$ mV; **b.** $\Delta v_D = 59.9$ mV. □

10.2 LOAD-LINE ANALYSIS OF DIODE CIRCUITS

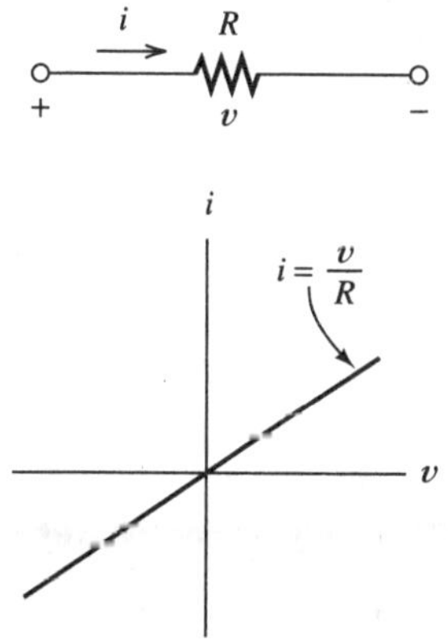

Figure 10.4 In contrast to diodes, resistors have linear volt–ampere characteristics.

In Section 10.1, we learned that the volt–ampere characteristics of diodes are nonlinear. We will see shortly that other electronic devices are also nonlinear. On the other hand, resistors have linear volt–ampere characteristics, as shown in Figure 10.4.

Because of this nonlinearity, many of the techniques that we have studied for linear circuits in Chapters 1 through 6 do not apply to circuits involving diodes. In fact, much of the study of electronics is concerned with techniques for analysis of circuits containing nonlinear elements.

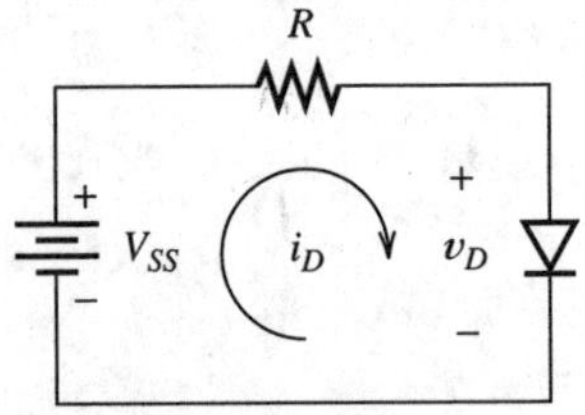

Figure 10.5 Circuit for load-line analysis.

Graphical methods provide one approach to analysis of nonlinear circuits. For example, consider the circuit shown in Figure 10.5. By application of Kirchhoff's voltage law, we can write the equation

$$V_{SS} = Ri_D + v_D \tag{10.5}$$

We assume that the values of V_{SS} and R are known and that we wish to find i_D and v_D. Thus, Equation 10.5 has two unknowns, and another equation (or its equivalent) is needed before a solution can be found. This is available in graphical form in Figure 10.6, which shows the volt–ampere characteristic of the diode.

We can obtain a solution by plotting Equation 10.5 on the same set of axes used for the diode characteristic. Since Equation 10.5 is linear, it plots as a straight line, which can be drawn if two points satisfying the equation are located. A simple method is to assume that $i_D = 0$, and then Equation 10.5 yields $v_D = V_{SS}$. This pair of values is shown as point A in Figure 10.6. A second point results if we assume that $v_D = 0$, for which the equation yields $i_D = V_{SS}/R$. The pair of values is shown as point B in Figure 10.6. Then, connecting points A and B results in a plot called the **load line**. The **operating point** is the intersection of the load line and the diode characteristic. This point represents the simultaneous solution of Equation 10.5 and the diode characteristic.

Example 10.1 Load-Line Analysis

If the circuit of Figure 10.5 has $V_{SS} = 2\,\text{V}, R = 1\,\text{k}\Omega$, and a diode with the characteristic shown in Figure 10.7, find the diode voltage and current at the operating point.

Solution First, we locate the ends of the load line. Substituting $v_D = 0$ and the values given for V_{SS} and R into Equation 10.5 yields $i_D = 2$ mA. These values plot as point B in Figure 10.7. Substitution of $i_D = 0$ and circuit values results in

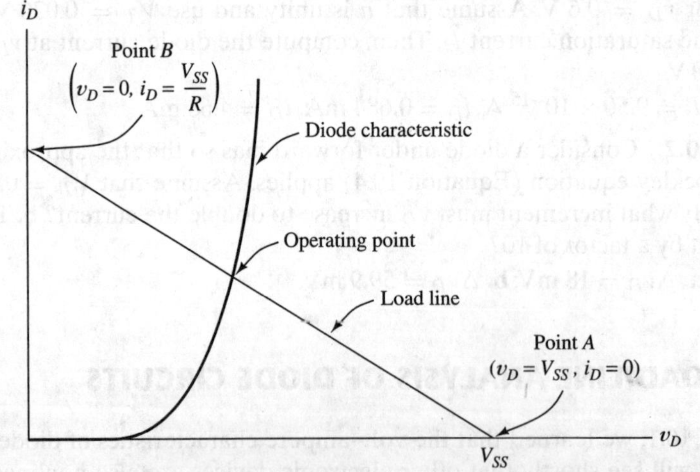

Figure 10.6 Load-line analysis of the circuit of Figure 10.5.

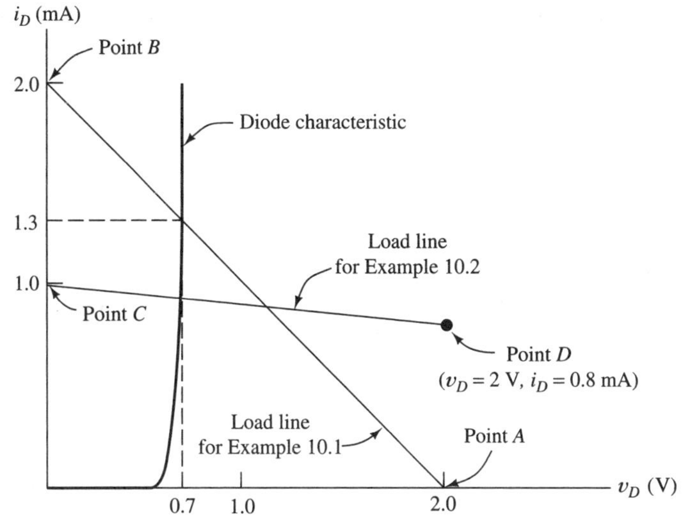

Figure 10.7 Load-line analysis for Examples 10.1 and 10.2.

$v_D = 2$ V. These values plot as point A in the figure. Constructing the load line results in an operating point of $V_{DQ} \cong 0.7$ V and $I_{DQ} \cong 1.3$ mA, as shown in the figure. ■

Example 10.2 Load-Line Analysis

Repeat Example 10.1 if $V_{SS} = 10$ V and $R = 10$ kΩ.

Solution If we let $v_D = 0$ and substitute values into Equation 10.5, we find that $i_D = 1$ mA. This is plotted as point C in Figure 10.7.

If we proceed as before by assuming that $i_D = 0$, we find that $v_D = 10$ V. This is a perfectly valid point on the load line, but it plots at a point far off the page. Of course, we can use any other point satisfying Equation 10.5 to locate the load line. Since we already have point C on the i_D axis, a good point to use would be on

When an intercept of the load line falls off the page, we select a point at the edge of the page.

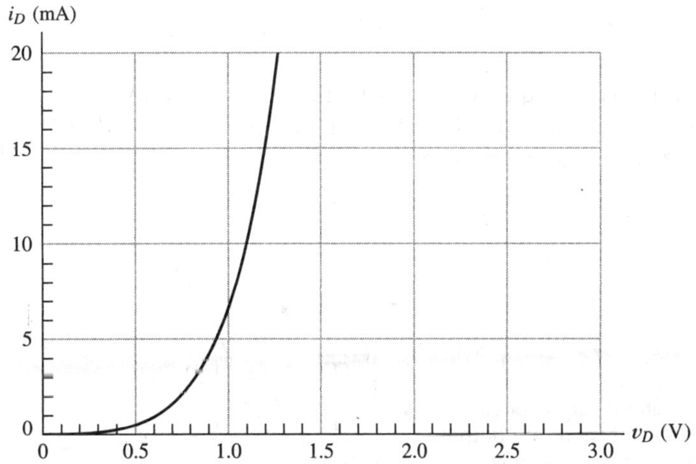

Figure 10.8 Diode characteristic for Exercise 10.3.

the right-hand edge of Figure 10.7. Thus, we assume that $v_D = 2$ V and substitute values into Equation 10.5, resulting in $i_D = 0.8$ mA. These values plot as point D. Then, we can draw the load line and find that the operating-point values are $V_{DQ} \cong 0.68$ V and $I_{DQ} \cong 0.93$ mA. ■

Exercise 10.3 Find the operating point for the circuit of Figure 10.5 if the diode characteristic is shown in Figure 10.8 and: **a.** $V_{SS} = 2$ V and $R = 100\ \Omega$; **b.** $V_{SS} = 15$ V and $R = 1\ \text{k}\Omega$; **c.** $V_{SS} = 1.0$ V and $R = 20\ \Omega$.

Answer **a.** $V_{DQ} \cong 1.1$ V, $I_{DQ} \cong 9.0$ mA; **b.** $V_{DQ} \cong 1.2$ V, $I_{DQ} \cong 13.8$ mA; **c.** $V_{DQ} \cong 0.91$ V, $I_{DQ} \cong 4.5$ mA. □

10.3 ZENER-DIODE VOLTAGE-REGULATOR CIRCUITS

A voltage regulator circuit provides a nearly constant voltage to a load from a variable source.

Sometimes, a circuit that produces constant output voltage while operating from a variable supply voltage is needed. Such circuits are called **voltage regulators**. For example, if we wanted to operate computer circuits from the battery in an automobile, a voltage regulator would be needed. Automobile battery voltage typically varies between about 10 and 14 V (depending on the state of the battery and whether or not the engine is running). Many computer circuits require a nearly constant voltage of 5 V. Thus, a regulator is needed that operates from the 10 to 14 V supply and produces a nearly constant 5-V output.

In this section, we use the load-line technique that we introduced in Section 10.2 to analyze a simple regulator circuit. The regulator circuit is shown in Figure 10.9. (For proper operation, it is necessary for the minimum value of the variable source voltage to be somewhat larger than the desired output voltage.) The Zener diode has a breakdown voltage equal to the desired output voltage. The resistor R limits the diode current to a safe value so that the Zener diode does not overheat.

Assuming that the characteristic for the diode is available, we can construct a load line to analyze the operation of the circuit. As before, we use Kirchhoff's voltage law to write an equation relating v_D and i_D. (In this circuit, the diode operates in the breakdown region with negative values for v_D and i_D.) For the circuit of Figure 10.9, we obtain

$$V_{SS} + Ri_D + v_D = 0 \tag{10.6}$$

Once again, this is the equation of a straight line, so location of any two points is sufficient to construct the load line. The intersection of the load line with the diode characteristic yields the operating point.

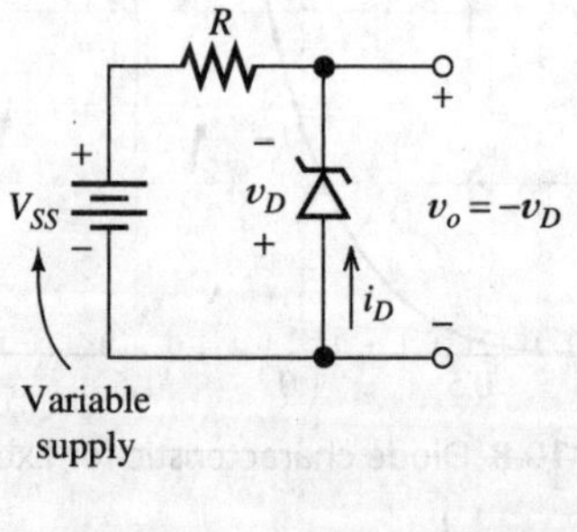

Figure 10.9 A simple regulator circuit that provides a nearly constant output voltage v_o from a variable supply voltage.

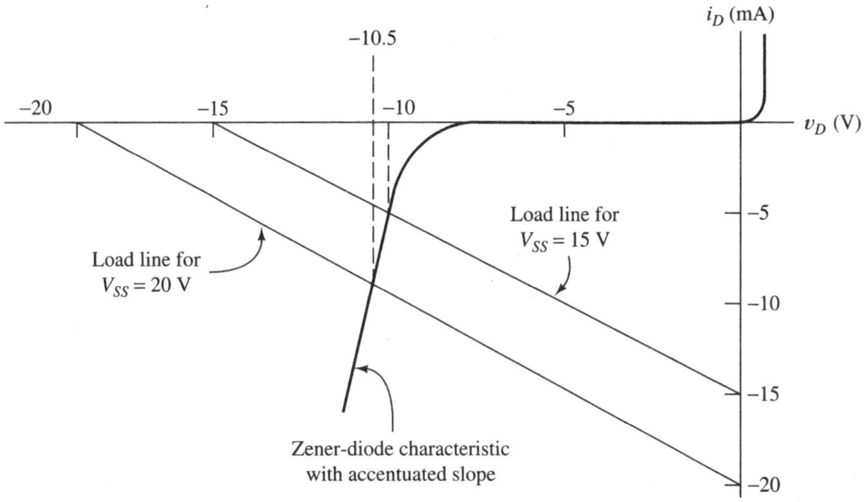

Figure 10.10 See Example 10.3.

Example 10.3 Load-Line Analysis of a Zener-Diode Voltage Regulator

The voltage-regulator circuit of Figure 10.9 has $R = 1\ \text{k}\Omega$ and uses a Zener diode having the characteristic shown in Figure 10.10. Find the output voltage for $V_{SS} =$ 15 V. Repeat for $V_{SS} = 20$ V.

Solution The load lines for both values of V_{SS} are shown in Figure 10.10. The output voltages are determined from the points where the load lines intersect the diode characteristic. The output voltages are found to be $v_o = 10.0$ V for $V_{SS} =$ 15 V and $v_o = 10.5$ V for $V_{SS} = 20$ V. Thus, a 5-V change in the supply voltage results in only a 0.5-V change in the regulated output voltage.

Actual Zener diodes are capable of much better performance than this. The slope of the characteristic has been accentuated in Figure 10.10 for clarity—actual Zener diodes have a more nearly vertical slope in breakdown. ■

Slope of the Load Line

Load lines for different source voltages (but the same resistance) are parallel.

Notice that the two load lines shown in Figure 10.10 are parallel. Inspection of Equation 10.5 or Equation 10.6 shows that the slope of the load line is $-1/R$. Thus, a change of the supply voltage changes the position, but not the slope of the load line.

Load-Line Analysis of Complex Circuits

Any circuit that contains resistors, voltage sources, current sources, and a single two-terminal nonlinear element can be analyzed by the load-line technique. First, the Thévenin equivalent is found for the linear portion of the circuit as illustrated in Figure 10.11. Then, a load line is constructed to find the operating point on the characteristic of the nonlinear device. Once the operating point of the nonlinear element is known, voltages and currents can be determined in the original circuit.

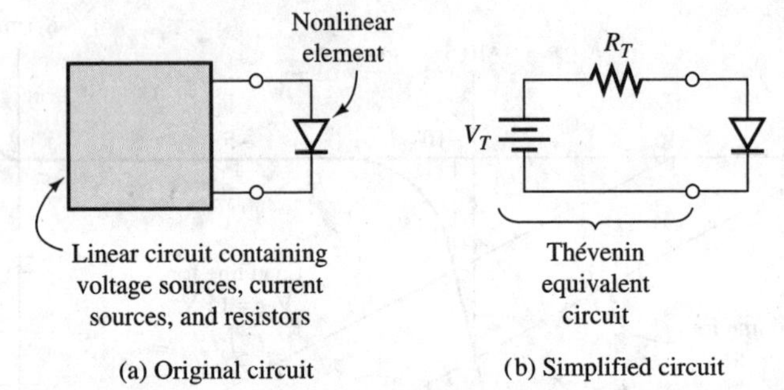

Figure 10.11 Analysis of a circuit containing a single nonlinear element can be accomplished by load-line analysis of a simplified circuit.

Example 10.4 Analysis of a Zener-Diode Regulator with a Load

Consider the Zener-diode regulator circuit shown in Figure 10.12(a). The diode characteristic is shown in Figure 10.13. Find the load voltage v_L and source current I_S if $V_{SS} = 24$ V, $R = 1.2$ kΩ, and $R_L = 6$ kΩ.

Solution First, consider the circuit as redrawn in Figure 10.12(b), in which we have grouped the linear elements together on the left-hand side of the diode. Next, we find the Thévenin equivalent for the linear portion of the circuit. The Thévenin voltage is the open-circuit voltage (i.e., the voltage across R_L with the diode replaced by an open circuit), which is given by

$$V_T = V_{SS}\frac{R_L}{R + R_L} = 20 \text{ V}$$

The Thévenin resistance can be found by zeroing the voltage source and looking back into the circuit from the diode terminals. This is accomplished by reducing V_{SS}

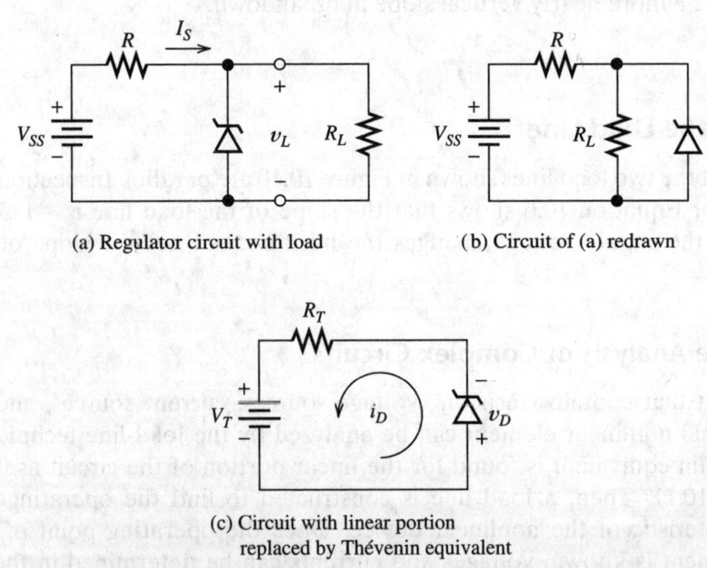

Figure 10.12 See Example 10.4.

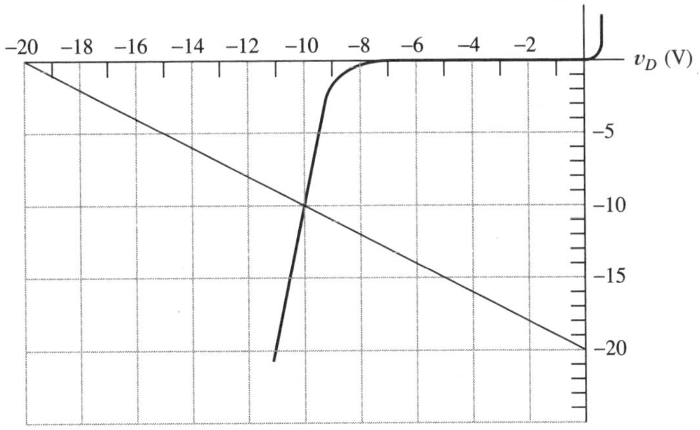

Figure 10.13 Zener-diode characteristic for Example 10.4 and Exercise 10.4.

to zero so that the voltage source becomes a short circuit. Then, we have R and R_L in parallel, so the Thévenin resistance is

$$R_T = \frac{RR_L}{R + R_L} = 1\ \text{k}\Omega$$

The resulting equivalent circuit is shown in Figure 10.12(c).

Now, we can use Kirchhoff's voltage law to write the load-line equation from the equivalent circuit as

$$V_T + R_T i_D + v_D = 0$$

Using the values found for V_T and R_T, we can construct the load line shown in Figure 10.13 and locate the operating point. This yields $v_L = -v_D = 10.0$ V.

Once v_L is known, we can find the voltages and currents in the original circuit. For example, using the output voltage value of 10.0 V in the original circuit of Figure 10.12(a), we find that $I_S = (V_{SS} - v_L)/R = 11.67$ mA. ■

Exercise 10.4 Find the voltage across the load in Example 10.4 if: **a.** $R_L = 1.2\ \text{k}\Omega$; **b.** $R_L = 400\ \Omega$.

Answer **a.** $v_L \cong 9.4$ V; **b.** $v_L \cong 6.0$ V. (Notice that this regulator is not perfect because the load voltage varies as the load current changes.) □

Exercise 10.5 Consider the circuit of Figure 10.14(a). Assume that the breakdown characteristic is vertical, as shown in Figure 10.14(b). Find the output voltage v_o for: **a.** $i_L = 0$; **b.** $i_L = 20$ mA; **c.** $i_L = 100$ mA. [*Hint:* Applying Kirchhoff's voltage law to the circuit, we have

$$15 - 100(i_L - i_D) - v_D$$

Construct a different load line for each value of i_L.]

Answer **a.** $v_o = 10.0$ V; **b.** $v_o = 10.0$ V; **c.** $v_o = 5.0$ V. (Notice that the regulator is not effective for large load currents.) □

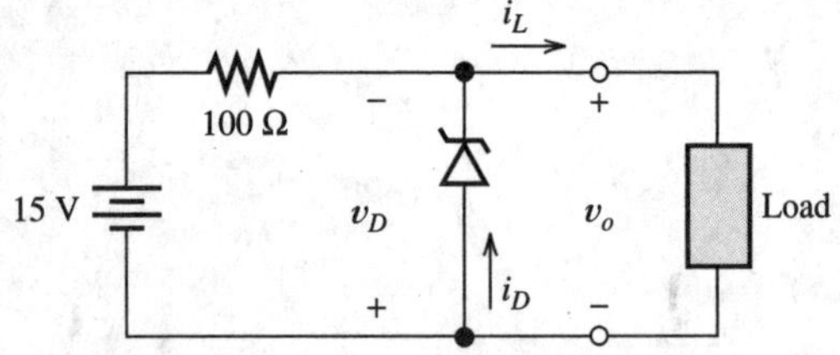

(a) Circuit diagram

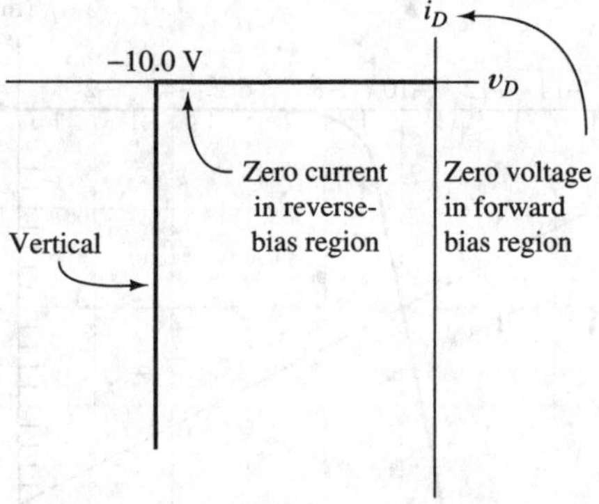

(b) Ideal Zener-diode characteristic

Figure 10.14 See Exercise 10.5.

10.4 IDEAL-DIODE MODEL

Graphical load-line analysis is useful for some circuits, such as the voltage regulator studied in Section 10.3. However, it is too cumbersome for more complex circuits. Instead, we often use simpler models to approximate diode behavior.

The ideal diode acts as a short circuit for forward currents and as an open circuit with reverse voltage applied.

One model for a diode is the **ideal diode**, which is a perfect conductor with zero voltage drop in the forward direction. In the reverse direction, the ideal diode is an open circuit. We use the ideal-diode assumption if our judgment tells us that the forward diode voltage drop and reverse current are negligible, or if we want a basic understanding of a circuit rather than an exact analysis.

The volt–ampere characteristic for the ideal diode is shown in Figure 10.15. If i_D is positive, v_D is zero, and we say that the diode is in the *on* state. On the other hand, if v_D is negative, i_D is zero, and we say that the diode is in the *off* state.

Assumed States for Analysis of Ideal-Diode Circuits

In analysis of a circuit containing ideal diodes, we may not know in advance which diodes are on and which are off. Thus, we are forced to make a considered guess. Then, we analyze the circuit to find the currents in the diodes assumed to be on and the voltages across the diodes assumed to be off. If i_D is positive for the diodes assumed to be on and if v_D is negative for the diodes assumed to be off, our assumptions are correct, and we have solved the circuit. (We are assuming that i_D is referenced positive in the forward direction and that v_D is referenced positive at the anode.)

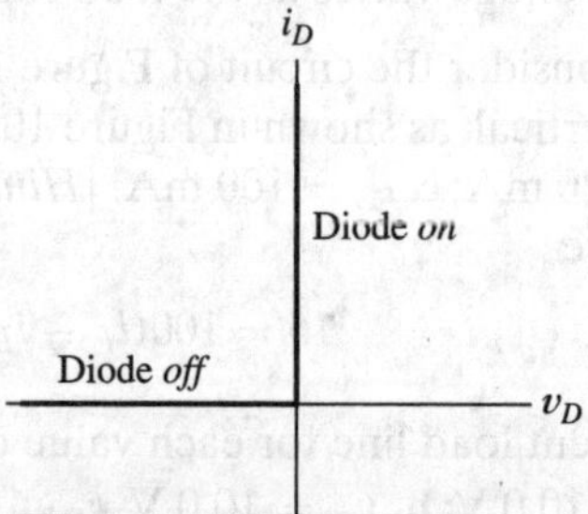

Figure 10.15 Ideal-diode volt–ampere characteristic.

Otherwise, we must make another assumption about the diodes and try again. After a little practice, our first guess is usually correct, at least for simple circuits.

A step-by-step procedure for analyzing circuits that contain ideal diodes is to:

1. Assume a state for each diode, either on (i.e., a short circuit) or off (i.e., an open circuit). For n diodes there are 2^n possible combinations of diode states.
2. Analyze the circuit to determine the current through the diodes assumed to be on and the voltage across the diodes assumed to be off.
3. Check to see if the result is consistent with the assumed state for each diode. Current must flow in the forward direction for diodes assumed to be on. Furthermore, the voltage across the diodes assumed to be off must be positive at the cathode (i.e., reverse bias).
4. If the results are consistent with the assumed states, the analysis is finished. Otherwise, return to step 1 and choose a different combination of diode states.

Example 10.5 Analysis by Assumed Diode States

Use the ideal-diode model to analyze the circuit shown in Figure 10.16(a). Start by assuming that D_1 is off and D_2 is on.

Solution With D_1 off and D_2 on, the equivalent circuit is shown in Figure 10.16(b). Solving results in $i_{D2} = 0.5$ mA. Since the current in D_2 is positive, our assumption that D_2 is on seems to be correct. However, continuing the solution of the circuit of Figure 10.16(b), we find that $v_{D1} = +7$ V. This is not consistent with the assumption that D_1 is off. Therefore, we must try another assumption.

This time, we assume that D_1 is on and D_2 is off. The equivalent circuit for these assumptions is shown in Figure 10.16(c). We can solve this circuit to find that $i_{D1} = 1$ mA and $v_{D2} = -3$ V. These values are consistent with the assumptions about the diodes (D_1 on and D_2 off) and, therefore, are correct. ■

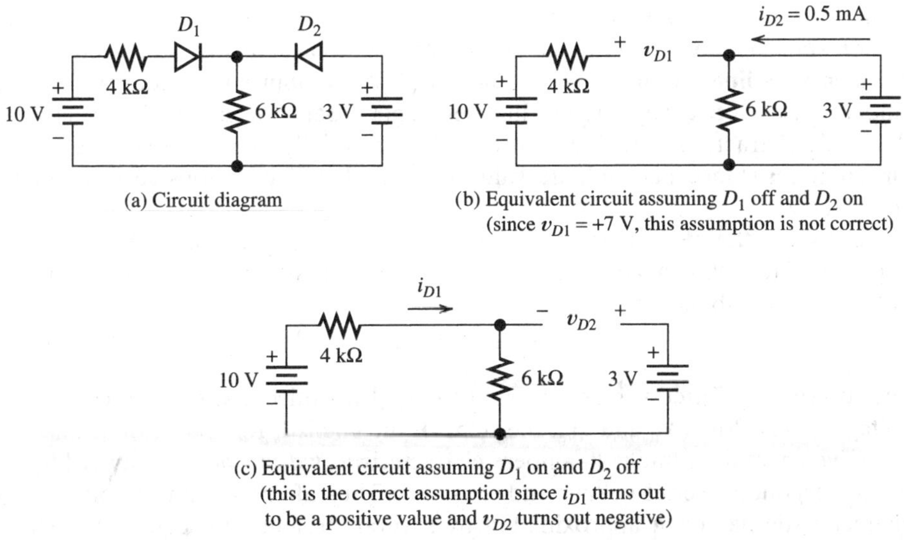

Figure 10.16 Analysis of a diode circuit, using the ideal-diode model. See Example 10.5.

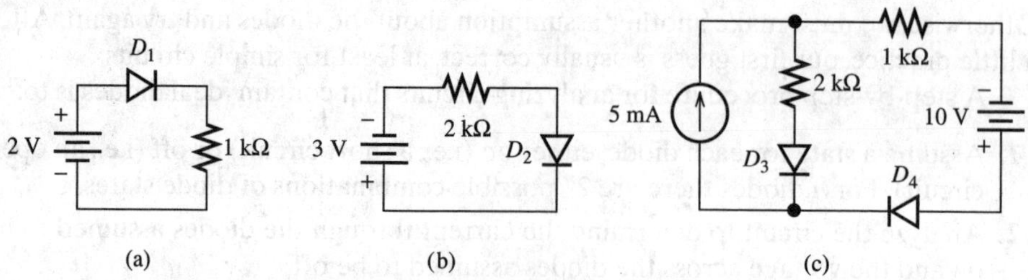

Figure 10.17 Circuits for Exercise 10.8.

In general, we cannot decide on the state of a particular diode until we have found a combination of states that works for all of the diodes in the circuit.

Notice in Example 10.5 that even though current flows in the forward direction of D_2 for our first guess about diode states (D_1 off and D_2 on), the correct solution is that D_2 is off. Thus, in general, we cannot decide on the state of a particular diode until we have found a combination of states that works for all the diodes in the circuit.

For a circuit containing n diodes, there are 2^n possible states. Thus, an exhaustive search eventually yields the solution for each circuit.

Exercise 10.6 Show that the condition D_1 off and D_2 off is not valid for the circuit of Figure 10.16(a). □

Exercise 10.7 Show that the condition D_1 on and D_2 on is not valid for the circuit of Figure 10.16(a). □

Exercise 10.8 Find the diode states for the circuits shown in Figure 10.17. Assume ideal diodes.

Answer **a.** D_1 is on; **b.** D_2 is off; **c.** D_3 is off and D_4 is on. □

10.5 PIECEWISE-LINEAR DIODE MODELS

Sometimes, we want a more accurate model than the ideal-diode assumption, but do not want to resort to nonlinear equations or graphical techniques. Then, we can use **piecewise-linear models** for the diodes. First, we approximate the actual volt–ampere characteristic by straight-line segments. Then, we model each section of the diode characteristic with a resistance in series with a constant-voltage source. Different resistance and voltage values are used in the various sections of the characteristic.

Consider the resistance R_a in series with a voltage source V_a shown in Figure 10.18(a). We can write the following equation, relating the voltage and current of the series combination:

$$v = R_a i + V_a \tag{10.7}$$

The current i is plotted versus v in Figure 10.18(b). Notice that the intercept on the voltage axis is at $v = V_a$ and that the slope of the line is $1/R_a$.

Given a straight-line volt–ampere characteristic, we can work backward to find the corresponding series voltage and resistance. Thus, after a nonlinear volt–ampere characteristic has been approximated by several straight-line segments, a circuit model consisting of a voltage source and series resistance can be found for each segment.

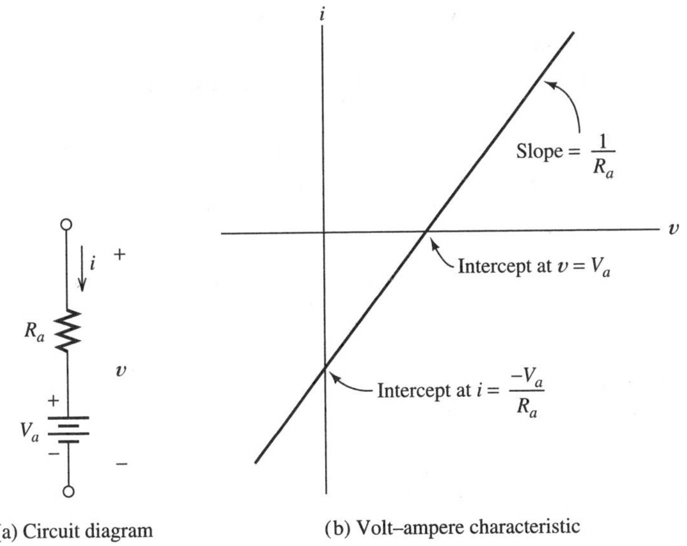

Figure 10.18 Circuit and volt–ampere characteristic for piecewise-linear models.

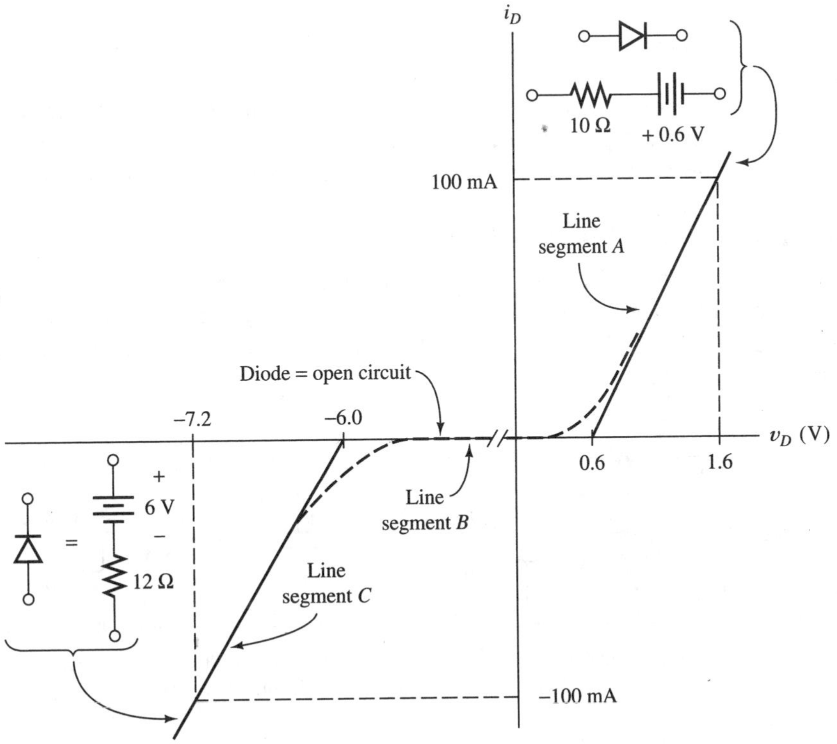

Figure 10.19 Piecewise-linear models for the diode of Example 10.6.

Example 10.6 Piecewise-Linear Model for a Zener Diode

Find circuit models for the Zener-diode volt–ampere characteristic shown in Figure 10.19. Use the straight-line segments shown.

Solution For line segment A of Figure 10.19, the intercept on the voltage axis is 0.6 V and the reciprocal of the slope is 10 Ω. Hence, the circuit model for the diode on this segment is a 10-Ω resistance in series with a 0.6-V source, as shown in the figure. Line segment B has zero current, and therefore, the equivalent circuit for segment B is an open circuit, as illustrated in the figure. Finally, line segment C has an intercept of −6 V and a reciprocal slope of 12 Ω, resulting in the equivalent circuit shown. Thus, this diode can be approximated by one of these linear circuits, depending on where the operating point is located. ■

Example 10.7 Analysis Using a Piecewise-Linear Model

Use the circuit models found in Example 10.6 to solve for the current in the circuit of Figure 10.20(a).

Solution Since the 3-V source has a polarity that results in forward bias of the diode, we assume that the operating point is on line segment A of Figure 10.19. Consequently, the equivalent circuit for the diode is the one for segment A. Using this equivalent circuit, we have the circuit of Figure 10.20(b). Solving, we find that $i_D = 80$ mA. ■

Exercise 10.9 Use the appropriate circuit model from Figure 10.19 to solve for v_o in the circuit of Figure 10.21 if: **a.** $R_L = 10$ kΩ; and **b.** $R_L = 1$ kΩ. (*Hint:* Be sure that your answers are consistent with your choice of equivalent circuit for the diode—the various equivalent circuits are valid only for specific ranges of diode voltage and current. The answer must fall into the valid range for the equivalent circuit used.)

Answer **a.** $v_o = 6.017$ V; **b.** $v_o = 3.333$ V. □

Exercise 10.10 Find a circuit model for each line segment shown in Figure 10.22(a). Draw the circuit models identifying terminals a and b for each equivalent circuit.

Answer See Figure 10.22(b). Notice the polarity of the voltage sources with respect to terminals a and b. □

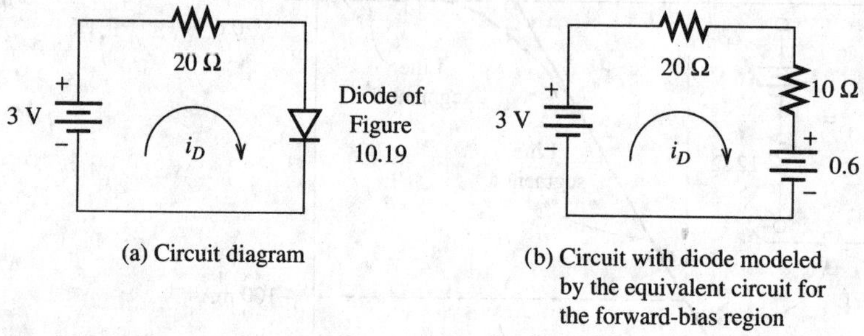

Figure 10.20 Circuit for Example 10.7.

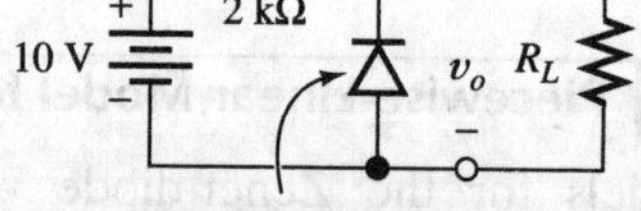

Figure 10.21 Circuit for Exercise 10.9.

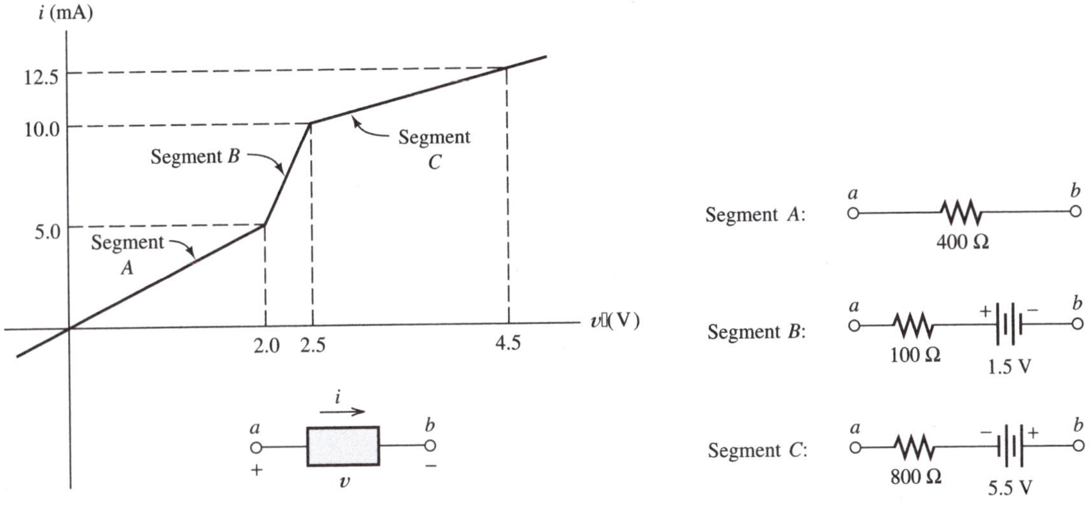

(a) Volt–ampere characteristic (b) Equivalent circuits

Figure 10.22 Hypothetical nonlinear device for Exercise 10.10.

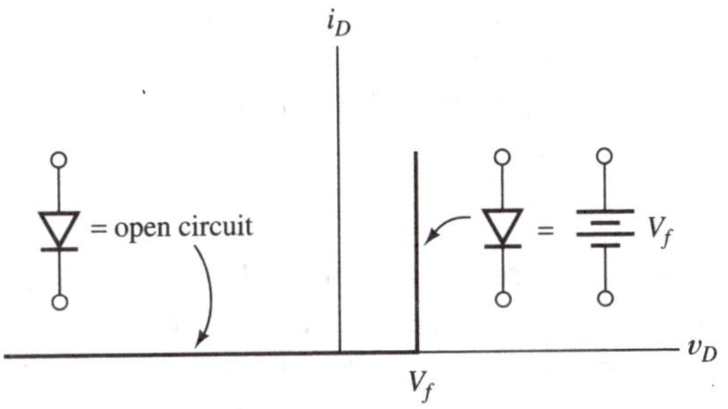

Figure 10.23 Simple piecewise-linear equivalent for the diode.

Simple Piecewise-Linear Diode Equivalent Circuit

Figure 10.23 shows a simple piecewise-linear equivalent circuit for diodes that is often sufficiently accurate. It is an open circuit in the reverse-bias region and a constant voltage drop in the forward direction. This model is equivalent to a battery in series with an ideal diode.

10.6 RECTIFIER CIRCUITS

Now that we have introduced the diode and some methods for analysis of diode circuits, we consider some additional practical circuits. First, we consider several types of **rectifiers**, which convert ac power into dc power. These rectifiers form the basis for electronic **power supplies** and battery-charging circuits. Typically, a power supply takes power from a raw source, which is often the 60-Hz ac power line, and delivers steady dc voltages to a load such as computer circuits or television circuits. Other applications for rectifiers are in signal processing, such as demodulation of a radio signal. (*Demodulation* is the process of retrieving the message, such as a voice or video signal.) Another application is precision conversion of an ac voltage to dc in an electronic voltmeter.

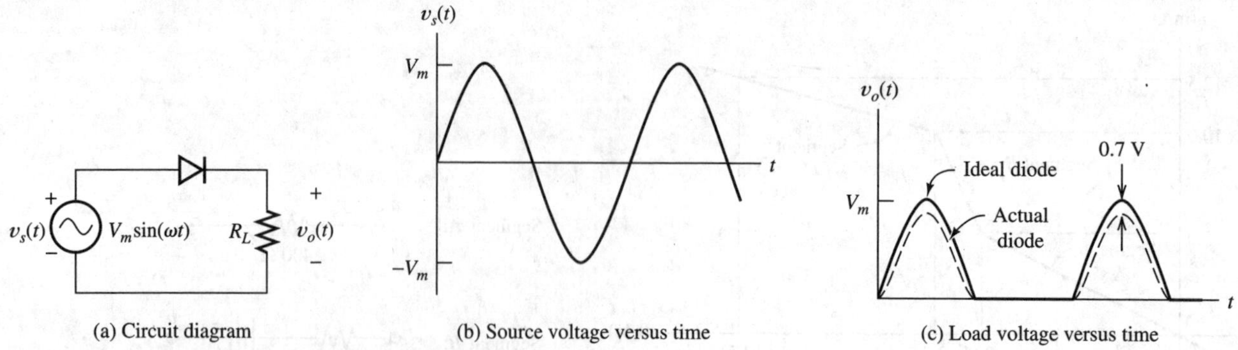

Figure 10.24 Half-wave rectifier with resistive load.

Half-Wave Rectifier Circuits

A **half-wave rectifier** with a sinusoidal source and resistive load R_L is shown in Figure 10.24. When the source voltage $v_s(t)$ is positive, the diode is in the forward-bias region. If an ideal diode is assumed, the source voltage appears across the load. For a typical real diode, the output voltage is less than the source voltage by an amount equal to the drop across the diode, which is approximately 0.7 V for silicon diodes at room temperature. When the source voltage is negative, the diode is reverse biased and no current flows through the load. Even for typical real diodes, only a very small reverse current flows. Thus, only the positive half-cycles of the source voltage appear across the load.

Battery-Charging Circuit. We can use a half-wave rectifier to charge a battery as shown in Figure 10.25. Current flows whenever the instantaneous ac source voltage is higher than the battery voltage. As shown in the figure, it is necessary to add resistance to the circuit to limit the magnitude of the current. When the ac source voltage is less than the battery voltage, the diode is reverse biased and the current is zero. Hence, the current flows only in the direction that charges the battery.

Half-Wave Rectifier with Smoothing Capacitor. Often, we want to convert an ac voltage into a nearly constant dc voltage to be used as a power supply for electronic circuits. One approach to smoothing the rectifier output voltage is to place a large

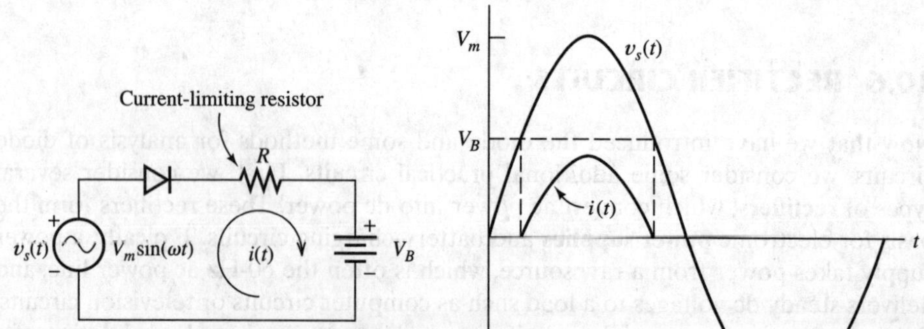

Figure 10.25 Half-wave rectifier used to charge a battery.

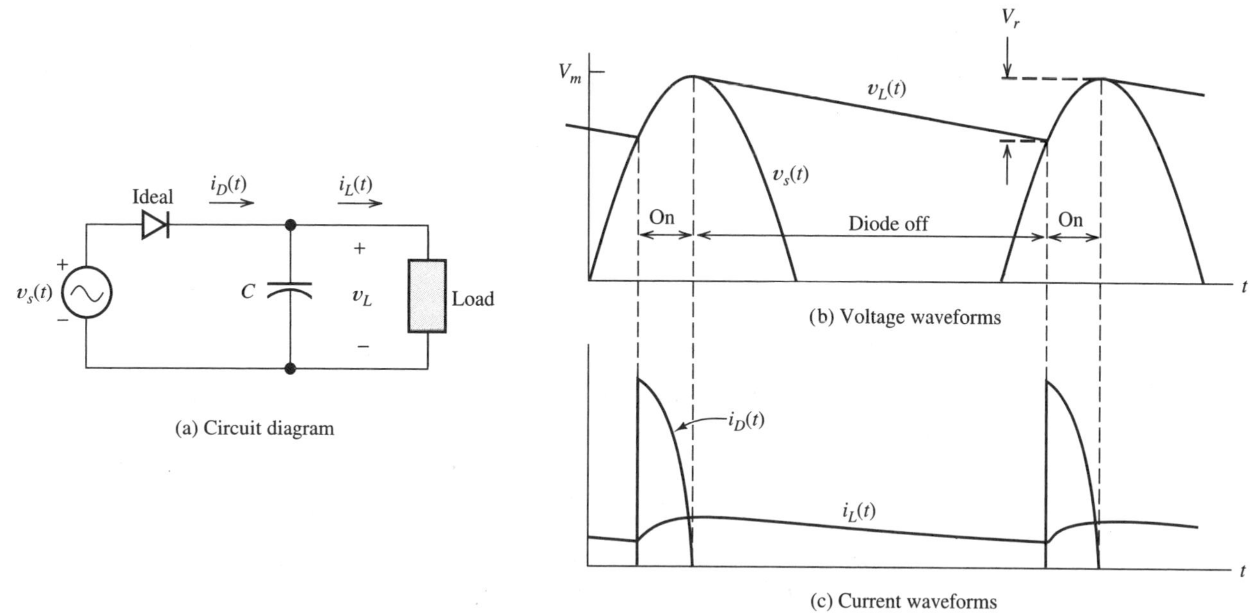

Figure 10.26 Half-wave rectifier with smoothing capacitor.

capacitance across the output terminals of the rectifier. The circuit and waveforms of current and voltage are shown in Figure 10.26. When the ac source reaches a positive peak, the capacitor is charged to the peak voltage (assuming an ideal diode). When the source voltage drops below the voltage stored on the capacitor, the diode is reverse biased and no current flows through the diode. The capacitor continues to supply current to the load, slowly discharging until the next positive peak of the ac input. As shown in the figure, current flows through the diode in pulses that recharge the capacitor.

Because of the charge and discharge cycle, the load voltage contains a small ac component called **ripple**. Usually, it is desirable to minimize the amplitude of the ripple, so we choose the largest capacitance value that is practical. In this case, the capacitor discharges for nearly the entire cycle, and the charge removed from the capacitor during one discharge cycle is

$$Q \cong I_L T \tag{10.8}$$

where I_L is the average load current and T is the period of the ac voltage. Since the charge removed from the capacitor is the product of the change in voltage and the capacitance, we can also write

$$Q = V_r C \tag{10.9}$$

where V_r is the peak-to-peak ripple voltage and C is the capacitance. Equating the right-hand sides of Equations 10.8 and 10.9 allows us to solve for C:

$$C = \frac{I_L T}{V_r} \tag{10.10}$$

In practice, Equation 10.10 is approximate because the load current varies and because the capacitor does not discharge for a complete cycle. However, it gives a good starting value for calculating the capacitance required in the design of power-supply circuits.

The average voltage supplied to the load if a smoothing capacitor is used is approximately midway between the minimum and maximum voltages. Thus, referring to Figure 10.26, the average load voltage is

$$V_L \cong V_m - \frac{V_r}{2} \tag{10.11}$$

Peak Inverse Voltage

An important aspect of rectifier circuits is the **peak inverse voltage** (PIV) across the diodes. Of course, the breakdown specification of the diodes should be greater in magnitude than the PIV. For example, in the half-wave circuit with a resistive load, shown in Figure 10.24, the PIV is V_m.

The addition of a smoothing capacitor in parallel with the load increases the PIV to (approximately) $2V_m$. Referring to Figure 10.26, for the negative peak of the ac input, we see that the reverse bias of the diode is the sum of the source voltage and the voltage stored on the capacitor.

Full-Wave Rectifier Circuits

A wire or other conductor, not shown explicitly in the diagram, connects all of the points that are connected to ground symbols.

Several **full-wave rectifier** circuits are in common use. One approach uses two ac sources and two diodes, as shown in Figure 10.27(a). One feature of this diagram is the **ground symbol**. Usually in electronic circuits, many components are connected to a common point known as *ground*. Often, the chassis containing the circuit is the electrical ground. Therefore, in Figure 10.27(a), the lower end of R_L and the point between the voltage sources are connected together.

When the upper source applies a positive voltage to the left-hand end of diode A, the lower source applies a negative voltage to the left-hand end of diode B, and vice versa. We say that the sources are **out of phase**. Thus, the circuit consists of two half-wave rectifiers with out-of-phase source voltages and a common load. The diodes conduct on alternate half-cycles.

Usually, the two out-of-phase ac voltages are provided by a **transformer**. (Transformers are discussed in Chapter 15.) Besides providing the out-of-phase ac voltages,

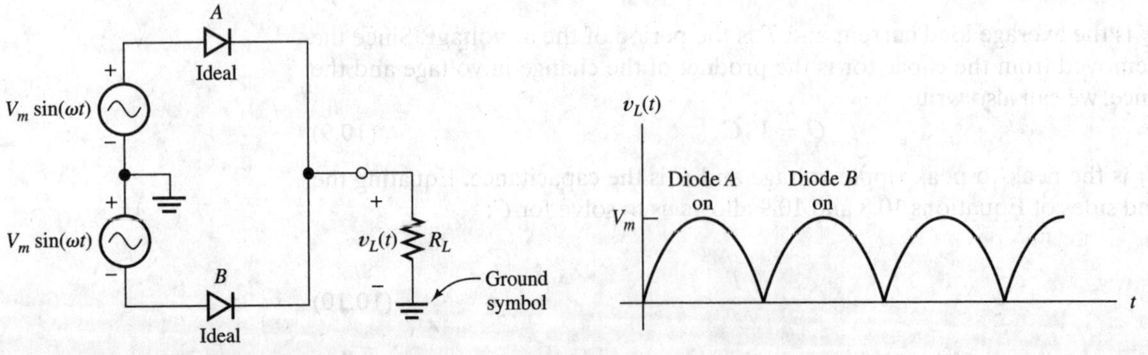

(a) Circuit diagram (b)

Figure 10.27 Full-wave rectifier.

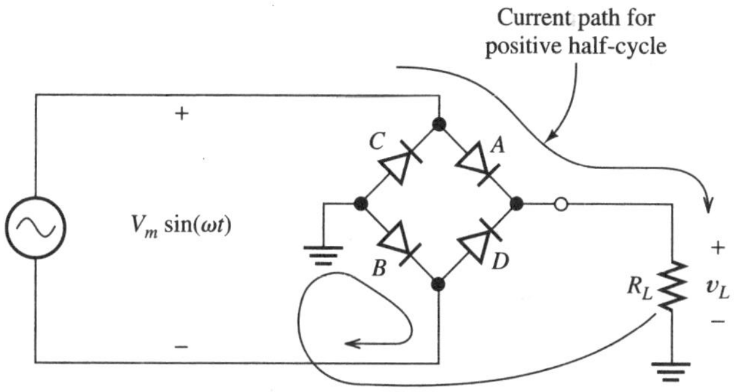

Figure 10.28 Diode-bridge full-wave rectifier.

the transformer also allows the designer to adjust V_m by selection of the turns ratio. This is important, because the ac voltage available is often not of a suitable amplitude for direct rectification—usually either a higher or lower dc voltage is required.

A second type of full-wave rectifier uses the **diode bridge** shown in Figure 10.28. When the ac voltage, $V_m \sin(\omega t)$, is positive, current flows through diode A, then through the load, and returns through diode B, as shown in the figure. For the opposite polarity, current flows through diodes C and D. Notice that in either case, current flows in the same direction through the load.

Usually, neither of the ac source terminals is connected to ground. This is necessary if one side of the load is to be connected to ground, as shown in the figure. (If both the ac source and the load have a common ground connection, part of the circuit is shorted.)

If we wish to smooth the voltage across the load, a capacitor can be placed in parallel with the load, similar to the half-wave circuit discussed earlier. In the full-wave circuits, the capacitor discharges for only a half-cycle before being recharged. Hence, the capacitance required is only half as much in the full-wave circuit as for the half-wave circuit. Therefore, we modify Equation 10.10 to obtain

$$C = \frac{I_L T}{2V_r} \tag{10.12}$$

for the full-wave rectifier with a capacitive filter.

Exercise 10.11 Consider the battery-charging circuit of Figure 10.25 with $V_m = 20$ V, $R = 10\ \Omega$, and $V_B = 14$ V. **a.** Find the peak current assuming an ideal diode. **b.** Find the percentage of each cycle for which the diode is in the on state.

Answer **a.** $I_{\text{peak}} = 600$ mA; **b.** the diode is on for 25.3 percent of each cycle. □

Exercise 10.12 A power-supply circuit is needed to deliver 0.1 A and 15 V (average) to a load. The ac source has a frequency of 60 Hz. Assume that the circuit of Figure 10.26 is to be used. The peak-to-peak ripple voltage is to be 0.4 V. Instead of assuming an ideal diode, allow 0.7 V for forward diode drop. Find the peak ac voltage V_m needed and the approximate value of the smoothing capacitor. (*Hint:* To achieve an average load voltage of 15 V with a ripple of 0.4 V, design for a peak load voltage of 15.2 V.)

Answer $V_m = 15.9$, $C = 4166\ \mu$F. □

Exercise 10.13 Repeat Exercise 10.12 using the circuit of Figure 10.28 with the smoothing capacitor in parallel with the load R_L.

Answer $V_m = 16.6, C = 2083\ \mu\text{F}$. □

10.7 WAVE-SHAPING CIRCUITS

A wide variety of **wave-shaping circuits** are used in electronic systems. These circuits are used to transform one waveform into another. Numerous examples of wave-shaping circuits can be found in transmitters and receivers for television or radar. In this section, we discuss a few examples of wave-shaping circuits that can be constructed with diodes.

Clipper Circuits

A clipper circuit "clips off" part of the input waveform to produce the output waveform.

Diodes can be used to form **clipper circuits**, in which a portion of an input signal waveform is "clipped" off. For example, the circuit of Figure 10.29 clips off any part of the input waveform above 6 V or less than −9 V. (We are assuming ideal diodes.) When the input voltage is between −9 and +6 V, both diodes are off and no current flows. Then, there is no drop across R and the output voltage v_o is equal to the input voltage v_{in}. On the other hand, when v_{in} is larger than 6 V, diode A is on and the output voltage is 6 V, because the diode connects the 6-V battery to the

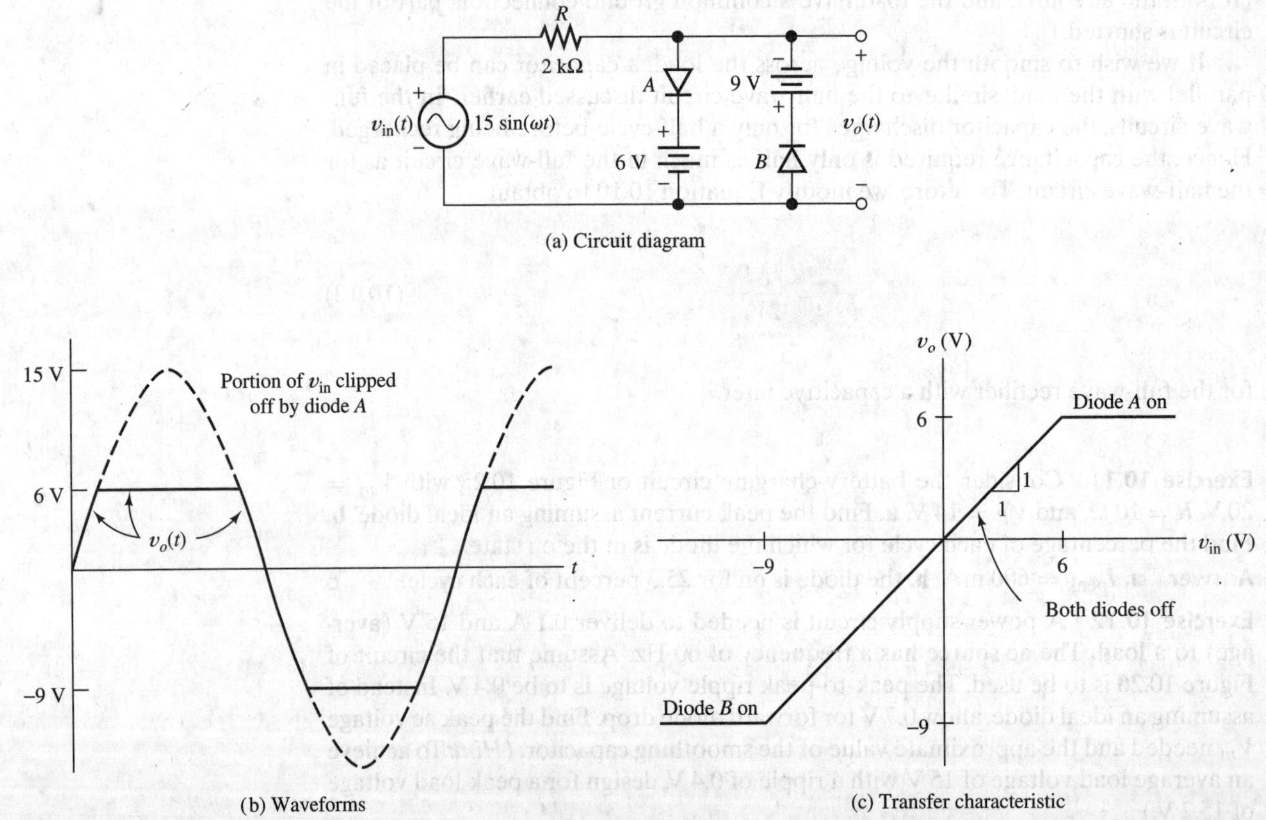

Figure 10.29 Clipper circuit.

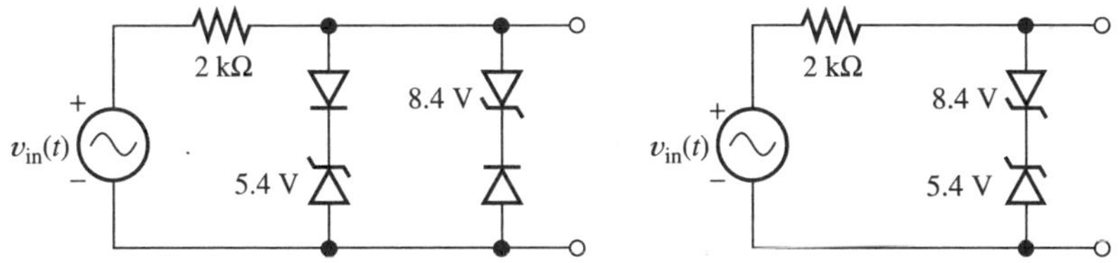

(a) Circuit of Figure 10.29 with batteries replaced by Zener diodes and allowance made for a 0.6-V forward diode drop

(b) Simpler circuit

Figure 10.30 Circuits with nearly the same performance as the circuit of Figure 10.29.

output terminals. Similarly, when v_{in} is less than -9 V, diode B is on and the output voltage is -9 V. The output waveform resulting from a 15-V-peak sinusoidal input is shown in Figure 10.29(b), and the transfer characteristic of the circuit is shown in Figure 10.29(c).

The resistance R is selected large enough so that the forward diode current is within reasonable bounds (usually, a few milliamperes), but small enough so that the reverse diode current results in a negligible voltage drop. Often, we find that a wide range of resistance values provides satisfactory performance in a given circuit.

In Figure 10.29, we have assumed ideal diodes. If small-signal silicon diodes are used, we expect a forward drop of 0.6 or 0.7 V, so we should reduce the battery voltages to compensate. Furthermore, batteries are not desirable for use in circuits if they can be avoided, because they may need periodic replacement. Thus, a better design uses Zener diodes instead of batteries. Practical circuits equivalent to Figure 10.29 are shown in Figure 10.30. The Zener diodes are labeled with their breakdown voltages.

Exercise 10.14 **a.** Sketch the transfer characteristics to scale for the circuits of Figure 10.31(a) and (b). Allow a 0.6-V forward drop for the diodes. **b.** Sketch the output waveform to scale if $v_{in}(t) = 15\sin(\omega t)$.

Answer **a.** See Figure 10.31(c); **b.** see Figure 10.31(d). □

Exercise 10.15 Design clipper circuits that have the transfer characteristics shown in **a.** Figure 10.32(a) and **b.** Figure 10.32(b). Allow for a 0.6-V drop in the forward direction for the diodes. [*Hint for part (b):* Include a resistor in series with the diode that begins to conduct at $v_{in} = 3$ V to achieve the slope required for the section between $v_{in} = 3$ V and 6 V.]

Answer **a.** See Figure 10.32(c); **b.** see Figure 10.32(d). □

Clamp Circuits

Another diode wave-shaping circuit is the **clamp circuit**, which is used to add a dc component to an ac input waveform so that the positive (or negative) peaks are forced to take a specified value. In other words, the peaks of the waveform are "clamped" to a specified voltage value. An example circuit is shown in Figure 10.33. In this circuit, the positive peaks are clamped to -5 V.

In a clamp circuit, a variable dc voltage is added to the input waveform so that one of the peaks of the output is clamped to a specified value.

The capacitance is a large value, so it discharges only very slowly and we can consider the voltage across the capacitor to be constant. Because the capacitance is large, it has a very small impedance for the ac input signal. The output voltage of the circuit is given by

$$v_o(t) = v_{in}(t) - V_C \tag{10.13}$$

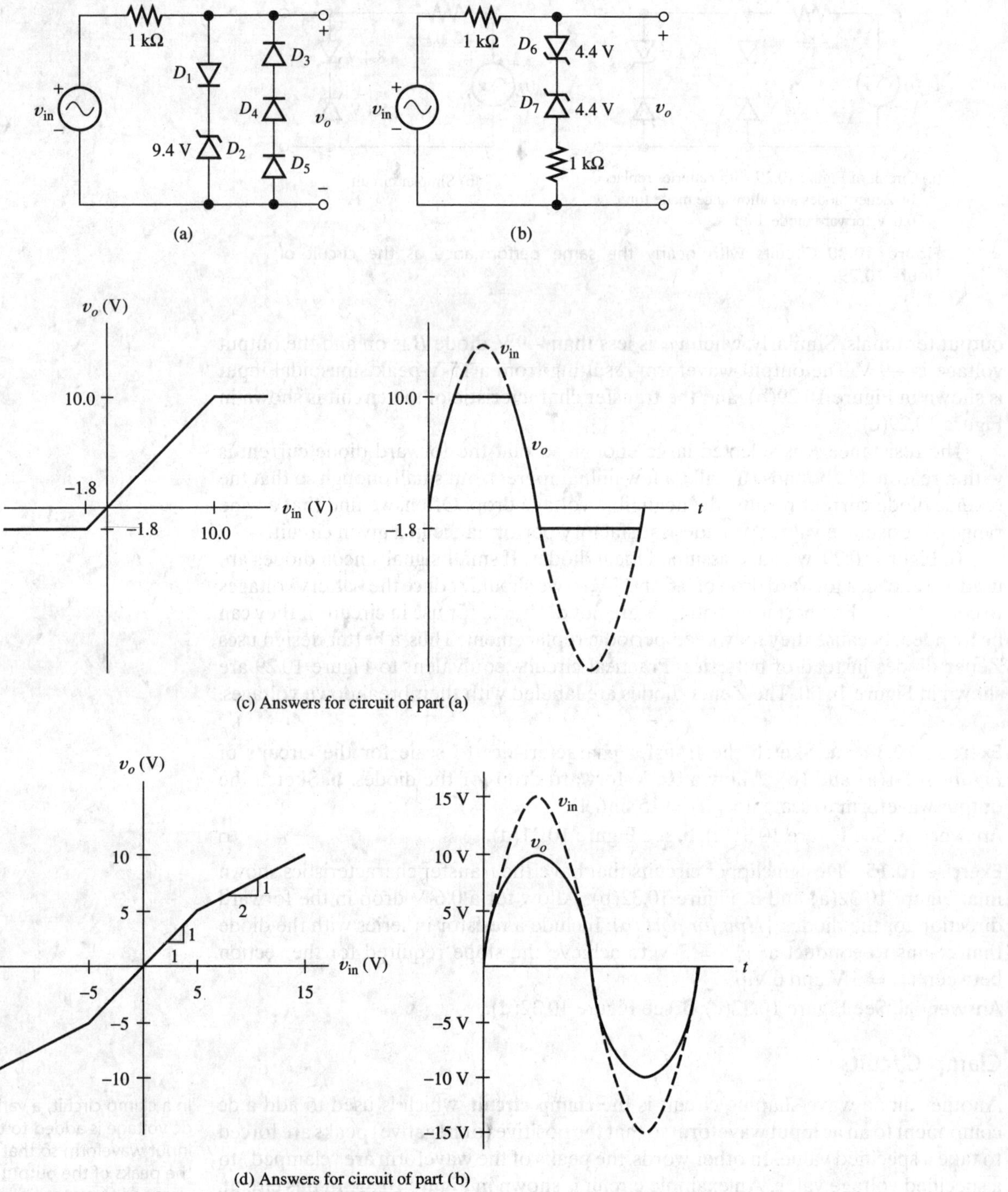

Figure 10.31 See Exercise 10.14.

If a positive swing of the input signal attempts to force the output voltage to become greater than −5 V, the diode conducts, increasing the value of V_C. Thus, the capacitor is charged to a value that adjusts the maximum value of the output voltage to −5 V. A large resistance R is provided so that the capacitor can discharge slowly. This

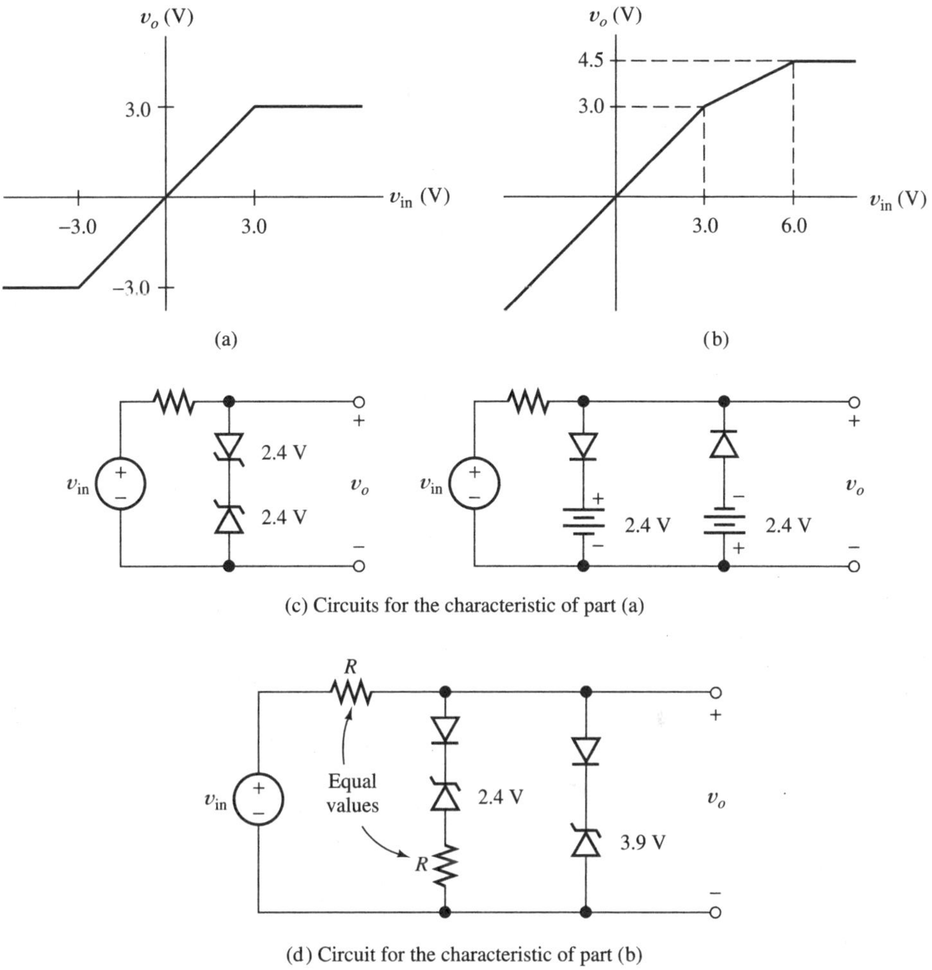

Figure 10.32 See Exercise 10.15.

is necessary so the circuit can adjust if the input waveform changes to a smaller peak amplitude.

Of course, we can change the voltage to which the circuit clamps by changing the battery voltage. Reversing the direction of the diode causes the negative peak to be clamped instead of the positive peak. If the desired clamp voltage requires the diode to be reverse biased, it is necessary to return the discharge resistor to a suitable dc supply voltage to ensure that the diode conducts and performs the clamping operation. Furthermore, it is often more convenient to use Zener diodes rather than batteries. A circuit including these features is shown in Figure 10.34.

Exercise 10.16 Consider the circuit of Figure 10.34(a). Assume that the capacitance is large enough so that the voltage across it does not discharge through R appreciably during one cycle of input. **a.** What is the steady-state output voltage if $v_{in}(t) = 0$? **b.** Sketch the steady-state output to scale versus time if $v_{in}(t) = 2\sin(\omega t)$. **c.** Suppose that the resistor is returned directly to ground instead of −15 V (i.e., replace the 15-V source by a short circuit). In this case, sketch the steady-state output versus time if $v_{in}(t) = 2\sin(\omega t)$.

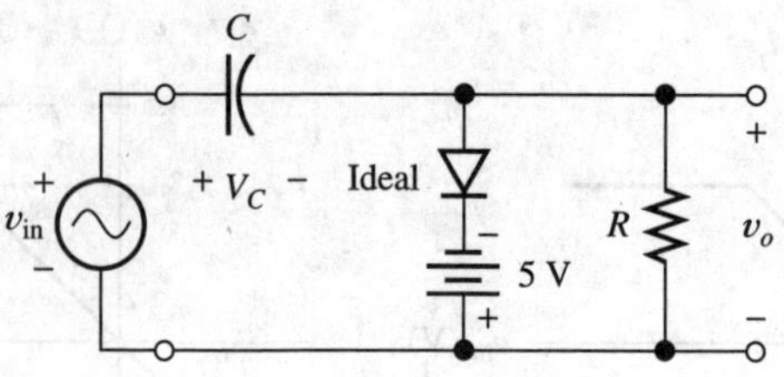

(a) Circuit diagram

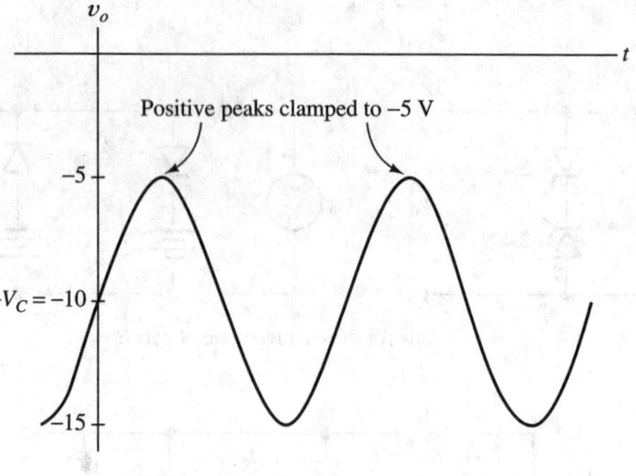

(b) Output waveform for $v_{in} = 5\sin(\omega t)$

Figure 10.33 Example clamp circuit.

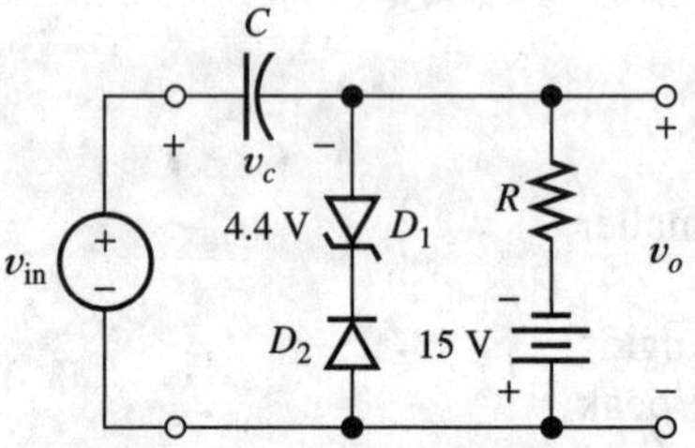

(a) A circuit that clamps the negative peaks to −5 V (an allowance of 0.6-V forward drop is made)

(b) Output for $v_{in} = 2\sin(\omega t)$

(c) Output for $v_{in} = 2\sin(\omega t)$ and R returned directly to ground (diodes do not turn on)

Figure 10.34 See Exercise 10.16.

Answer **a.** For $v_{in}(t) = 0$, we have $v_o = -5$ V; **b.** see Figure 10.34(b); **c.** see Figure 10.34(c). □

Exercise 10.17 Design a circuit that clamps the negative peaks of an ac signal to +6 V. You can use batteries, resistors, and capacitors of any value desired in addition to Zener or conventional diodes. Allow 0.6 V for the forward drop.

Answer A solution is shown in Figure 10.35. Other solutions are possible. □

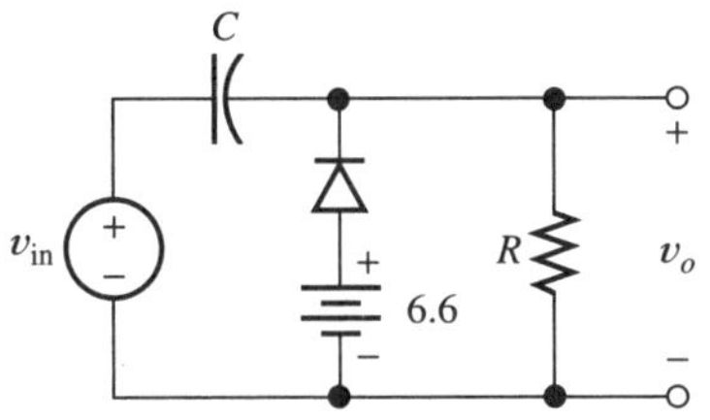

Figure 10.35 Answer for Exercise 10.17.

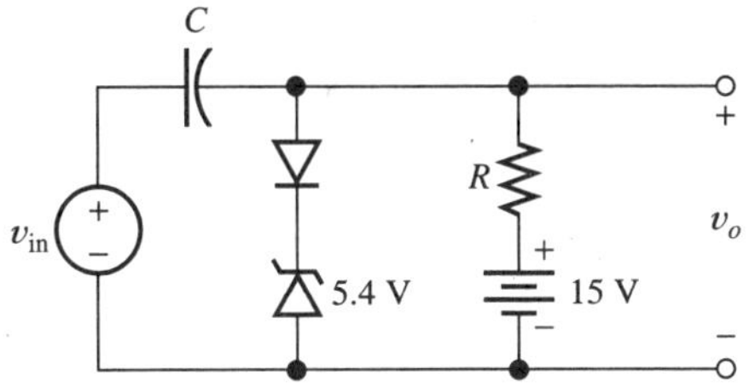

Figure 10.36 Answer for Exercise 10.18.

Exercise 10.18 Repeat Exercise 10.17 for a circuit that clamps the positive peaks to +6 V.

Answer A solution is shown in Figure 10.36. Other solutions are possible. □

10.8 LINEAR SMALL-SIGNAL EQUIVALENT CIRCUITS

We will encounter many examples of electronic circuits in which dc supply voltages are used to **bias** a nonlinear device at an operating point, and a small ac signal is injected into the circuit. We often split the analysis of such circuits into two parts. First, we analyze the dc circuit to find the operating point. In this analysis of bias conditions, we must deal with the nonlinear aspects of the device. In the second part of the analysis, we consider the small ac signal. Since virtually any nonlinear characteristic is approximately linear (straight) if we consider a sufficiently small portion, we can find a **linear small-signal equivalent circuit** for the nonlinear device to use in the ac analysis.

Often, the main concern in the design of such circuits is what happens to the ac signal. The dc supply voltages simply bias the device at a suitable operating point. For example, in a portable radio, the main interest is the signal being received, demodulated, amplified, and delivered to the speaker. The dc currents supplied by the battery are required for the devices to perform their intended function on the ac signals. However, most of our design time is spent in consideration of the small ac signals to be processed.

The small-signal linear equivalent circuit is an important analysis approach that applies to many types of electronic circuits. In this section, we demonstrate the principles with a simple diode circuit. In Chapters 12 and 13, we use similar techniques for transistor amplifier circuits.

Now, we show that in the case of a diode, the small-signal equivalent circuit consists simply of a resistance. Consider the diode characteristic shown in Figure 10.37. Assume that the dc supply voltage results in operation at the **quiescent point**, or Q **point**, indicated on the characteristic. Then, a small ac signal injected into the circuit swings the instantaneous point of operation slightly above and below the Q point.

The small-signal equivalent circuit for a diode is a resistance.

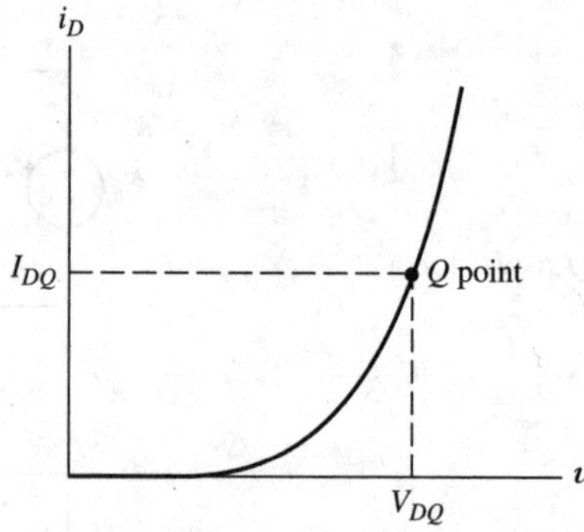

Figure 10.37 Diode characteristic, illustrating the Q point.

For a sufficiently small ac signal, the characteristic is straight. Thus, we can write

$$\Delta i_D \cong \left(\frac{di_D}{dv_D}\right)_Q \Delta v_D \tag{10.14}$$

where Δi_D is the small change in diode current from the Q-point current caused by the ac signal, Δv_D is the change in the diode voltage from the Q-point value, and $(di_D/dv_D)_Q$ is the slope of the diode characteristic evaluated at the Q point. Notice that the slope has the units of inverse resistance.

Hence, we define the **dynamic resistance** of the diode as

$$r_d = \left[\left(\frac{di_D}{dv_D}\right)_Q\right]^{-1} \tag{10.15}$$

and Equation 10.14 becomes

$$\Delta i_D \cong \frac{\Delta v_D}{r_d} \tag{10.16}$$

We find it convenient to drop the Δ notation and denote changes of current and voltage from the Q-point values as v_d and i_d. (Notice that lowercase subscripts are used for the small changes in current and voltage.) Therefore, for these small ac signals, we write

$$i_d = \frac{v_d}{r_d} \tag{10.17}$$

As shown by Equation 10.15, we can find the equivalent resistance of the diode for the small ac signal as the reciprocal of the slope of the characteristic curve. The current of a junction diode is given by the Shockley equation (Equation 10.1), repeated here for convenience:

$$i_D = I_s\left[\exp\left(\frac{v_D}{nV_T}\right) - 1\right]$$

The slope of the characteristic can be found by differentiating the Shockley equation, resulting in

$$\frac{di_D}{dv_D} = I_s\frac{1}{nV_T}\exp\left(\frac{v_D}{nV_T}\right) \tag{10.18}$$

Substituting the voltage at the Q point, we have

$$\left(\frac{di_D}{dv_D}\right)_Q = I_s \frac{1}{nV_T} \exp\left(\frac{V_{DQ}}{nV_T}\right) \tag{10.19}$$

For forward-bias conditions with V_{DQ} at least several times as large as V_T, the -1 inside the brackets of the Shockley equation is negligible. Thus, we can write

$$I_{DQ} \cong I_s \exp\left(\frac{V_{DQ}}{nV_T}\right) \tag{10.20}$$

Substituting this into Equation 10.19, we have

$$\left(\frac{di_D}{dv_D}\right)_Q = \frac{I_{DQ}}{nV_T} \tag{10.21}$$

Taking the reciprocal and substituting into Equation 10.15, we have the dynamic small-signal resistance of the diode at the Q point:

$$r_d = \frac{nV_T}{I_{DQ}} \tag{10.22}$$

To summarize, for signals that cause small changes from the Q point, we can treat the diode simply as a linear resistance. The value of the resistance is given by Equation 10.22 (provided that the diode is forward biased). As the Q-point current I_{DQ} increases, the resistance becomes smaller. Thus, an ac voltage of fixed amplitude produces an ac current that has higher amplitude as the Q point moves higher. This is illustrated in Figure 10.38.

Notation for Currents and Voltages in Electronic Circuits

Perhaps we should review the notation we have used for the diode currents and voltages, because we use similar notation throughout this book:

- v_D and i_D represent the total instantaneous diode voltage and current. At times, we may wish to emphasize the time-varying nature of these quantities, and then we use $v_D(t)$ and $i_D(t)$.
- V_{DQ} and I_{DQ} represent the dc diode current and voltage at the quiescent point.
- v_d and i_d represent the (small) ac signals. If we wish to emphasize their time-varying nature, we use $v_d(t)$ and $i_d(t)$.

This notation is illustrated for the waveform shown in Figure 10.39.

Exercise 10.19 Compute the dynamic resistance of a junction diode having $n = 1$ at a temperature of 300 K for $I_{DQ} =$ **a.** 0.1 mA; **b.** 1 mA; **c.** 10 mA.
Answer **a.** 260 Ω; **b.** 26 Ω; **c.** 2.6 Ω. □

Voltage-Controlled Attenuator

Now, we consider an example of linear-equivalent-circuit analysis for the relatively simple, but useful circuit shown in Figure 10.40. The function of this circuit is to

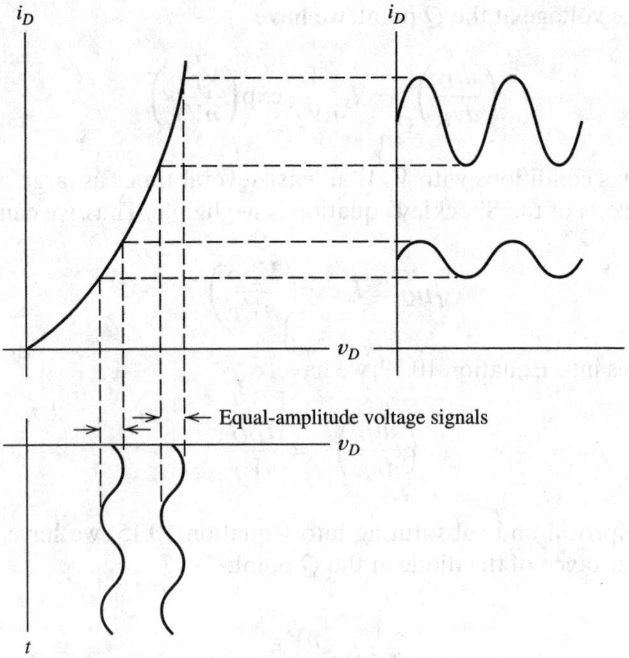

Figure 10.38 As the Q point moves higher, a fixed-amplitude ac voltage produces an ac current of larger amplitude.

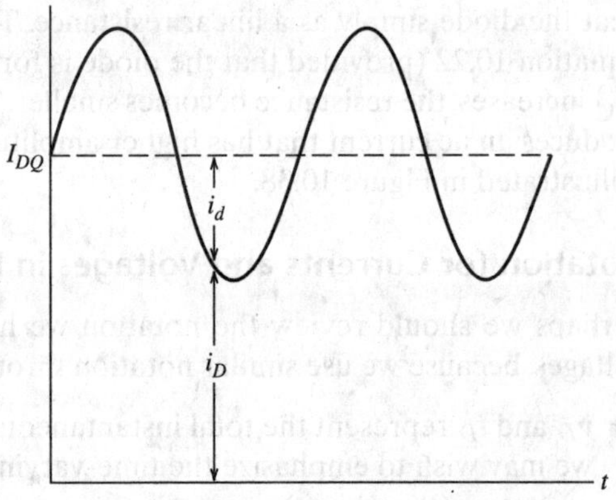

Figure 10.39 Illustration of diode currents.

produce an output signal $v_o(t)$ that is a variable fraction of the ac input signal $v_{in}(t)$. It is similar to the resistive voltage divider (see Section 2.3), except that in this case, we want the division ratio to depend on another voltage V_C called the **control signal**. We refer to the process of reduction of the amplitude of a signal as **attenuation**. Thus, the circuit to be studied is called a **voltage-controlled attenuator**. The degree of attenuation depends on the value of the dc control voltage V_C.

Notice that the ac signal to be attenuated is connected to the circuit by a **coupling capacitor**. The output voltage is connected to the load R_L by a second coupling capacitor. Recall that the impedance of a capacitance is given by

$$Z_C = \frac{1}{j\omega C}$$

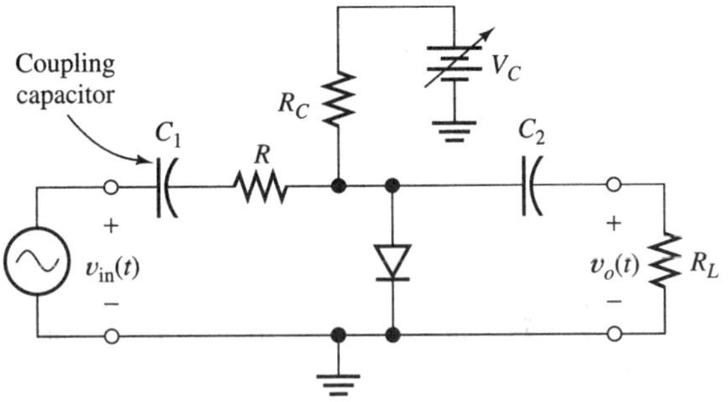

Figure 10.40 Variable attenuator using a diode as a controlled resistance.

in which ω is the angular frequency of the ac signal. We select the capacitance values large enough so that they are effectively short circuits for the ac signal. However, the coupling capacitors are open circuits for dc. Thus, the quiescent operating point (Q point) of the diode is unaffected by the signal source or the load. This can be important for a circuit that must work for various sources and loads that could affect the Q point. Furthermore, the coupling capacitors prevent (sometimes undesirable) dc currents from flowing in the source or the load.

Because of the coupling capacitors, we only need to consider V_C, R_C, and the diode to perform the bias analysis to find the Q point. Hence, the dc circuit is shown in Figure 10.41. We can use any of the techniques discussed earlier in this chapter to find the Q point. Once it is known, the Q-point value of the diode current I_{DQ} can be substituted into Equation 10.22 to determine the dynamic resistance of the diode.

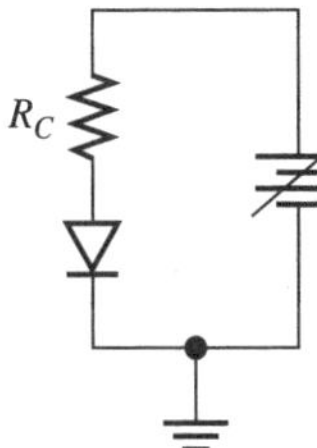

Figure 10.41 Dc circuit equivalent to Figure 10.40 for Q-point analysis.

Now, we turn our attention to the ac signal. The dc control source should be considered as a short circuit for ac signals. The signal source causes an ac current to flow through the V_C source. However, V_C is a dc voltage source, and by definition, the voltage across it is constant. *Since the dc voltage source has an ac component of current, but no ac voltage, the dc voltage source is equivalent to a short circuit for ac signals.* This is an important concept that we will use many times in drawing ac equivalent circuits.

Dc sources and coupling capacitors are replaced by short circuits in small-signal ac equivalent circuits. Diodes are replaced with their dynamic resistances.

The equivalent circuit for ac signals is shown in Figure 10.42. The control source and the capacitors have been replaced by short circuits, and the diode has been replaced by its dynamic resistance. This circuit is a voltage divider and can be analyzed by ordinary linear-circuit analysis. The parallel combination of R_C, R_L, and r_d is denoted as R_p, given by

$$R_p = \frac{1}{1/R_C + 1/R_L + 1/r_d} \tag{10.23}$$

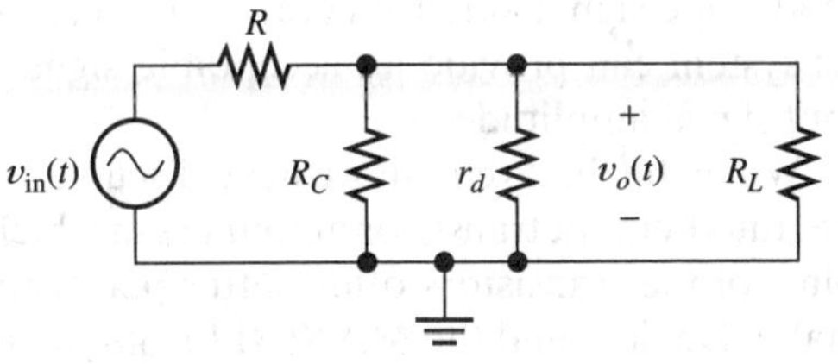

Figure 10.42 Small-signal ac equivalent circuit for Figure 10.40.

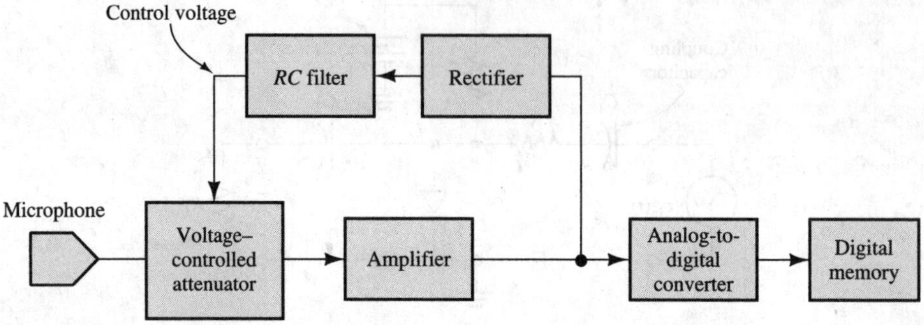

Figure 10.43 The voltage-controlled attenuator is useful in maintaining a suitable signal amplitude at the analog-to-digital converter.

Then, the **voltage gain** of the circuit is

$$A_v = \frac{v_o}{v_{\text{in}}} = \frac{R_p}{R + R_p} \tag{10.24}$$

(Of course, A_v is less than unity.)

Exercise 10.20 Suppose that the circuit of Figure 10.40 has $R = 100\ \Omega$, $R_C = 2\ \text{k}\Omega$, and $R_L = 2\ \text{k}\Omega$. The diode has $n = 1$ and is at a temperature of 300 K. For purposes of Q-point analysis, assume a constant diode voltage of 0.6 V. Find the Q-point value of the diode current and A_v for $V_C =$ **a.** 1.6 V; **b.** 10.6 V.
Answer **a.** $I_{DQ} = 0.5$ mA and $A_v = 0.331$; **b.** $I_{DQ} = 5$ mA and $A_v = 0.0492$. □

An application for voltage-controlled attenuators occurs in digital voice recorders in which the audio signal from a microphone is amplified to a suitable level, converted to digital form in an analog-to-digital converter (ADC), and stored in a digital memory. (Analog-to-digital conversion is discussed in Section 6.10, starting on page 342.) A problem frequently encountered in recording audio is that some persons speak quietly, while others speak loudly. Furthermore, some may be far from the microphone, while others are close. If an amplifier with fixed gain is used between the microphone and the ADC, either the weak signals are small compared with the quantization error or the strong signals exceed the maximum limits of the ADC so that severe distortion occurs.

A solution is to use a voltage-controlled attenuator in a system such as the one shown in Figure 10.43. The attenuator is placed between the microphone and a high-gain amplifier. When the signal being recorded is weak, the control voltage is small and very little attenuation occurs. On the other hand, when the signal is strong, the control voltage is large so that the signal is attenuated, preventing distortion. The control voltage is generated by rectifying the output of the amplifier. The rectified signal is filtered by a long-time constant *RC* filter so that the attenuation responds to the average signal amplitude rather than adjusting too rapidly. With proper design, this system can provide an acceptable signal at the converter for a wide range of input signal amplitudes.

While the diode circuit we have discussed is convenient for illustrating principles, integrated-circuit transistor amplifiers in which gain is controlled by changing the Q-points of the transistors offer better performance. Examples are the AN-934 from Analog Devices and the MAX9814 from Maxim Integrated Products.

Summary

1. A *pn*-junction diode is a two-terminal device that conducts current easily in one direction (from anode to cathode), but not in the opposite direction. The volt–ampere characteristic has three regions: forward bias, reverse bias, and reverse breakdown.
2. The Shockley equation relates current and voltage in a *pn*-junction diode.
3. Nonlinear circuits, such as those containing a diode, can be analyzed by using the load-line technique.
4. Zener diodes are intended to be operated in the reverse-breakdown region as constant-voltage references.
5. Voltage regulators are circuits that produce a nearly constant output voltage while operating from a variable source.
6. The ideal-diode model is a short circuit (on) if current flows in the forward direction and an open circuit (off) if voltage is applied in the reverse direction.
7. In the method of assumed states, we assume a state for each diode (on or off), analyze the circuit, and check to see if the assumed states are consistent with the current directions and voltage polarities. This process is repeated until a valid set of states is found.
8. In a piecewise-linear model for a nonlinear device, the volt–ampere characteristic is approximated by straight-line segments. On each segment, the device is modeled as a voltage source in series with a resistance.
9. Rectifier circuits can be used to charge batteries and to convert ac voltages into constant dc voltages. Half-wave rectifiers conduct current only for one polarity of the ac input, whereas full-wave circuits conduct for both polarities.
10. Wave-shaping circuits change the waveform of an input signal and deliver the modified waveform to the output terminals. Clipper circuits remove that portion of the input waveform above (or below) a given level. Clamp circuits add or subtract a dc voltage, so that the positive (or negative) peaks have a specified voltage.
11. The small-signal (incremental) equivalent circuit of a diode consists of a resistance. The value of the resistance depends on the operating point (Q point).
12. Dc sources and coupling capacitors are replaced by short circuits in small-signal ac equivalent circuits. Diodes are replaced with their dynamic resistances.

Problems

Section 10.1: Basic Diode Concepts

P10.1. Describe a fluid-flow analogy for a diode.

P10.2. Draw the circuit symbol for a diode, labeling the anode and cathode.

P10.3. Sketch the volt–ampere characteristic of a typical diode and label the various regions.

P10.4. Write the Shockley equation and define all of the terms.

P10.5. Determine the values of V_T for temperatures of 20°C, 30°C, and 200°C. Comment.

P10.6. For a junction diode, $i_D = 0.6$ mA for $v_D = 0.6$ V is specified. Find the diode current for $v_D = 0.7$ V and $v_D = 0.8$ V if $n = 1$ and $V_T = 25$ mV. [Use the Shockley equation]

***P10.7.** The operation of a diode in forward bias is expressed by Equation 10.4 with $V_T = 25$ mV. For $v_{D1} = 0.65$ V the current i_{D1} is 1.2 mA. For $v_{D2} = 0.70$ V the current is $i_{D2} = 12$ mA. Determine the value of I_s and n.

* Denotes that answers are contained in the Student Solutions files. See Appendix E for more information about accessing the Student Solutions.

***P10.8.** Sketch i versus v to scale for the circuits shown in Figure P10.8. The reverse-breakdown voltages of the Zener diodes are shown. Assume voltages of approximately 0.6 V for all diodes including the Zener diodes when current flows in the forward direction.

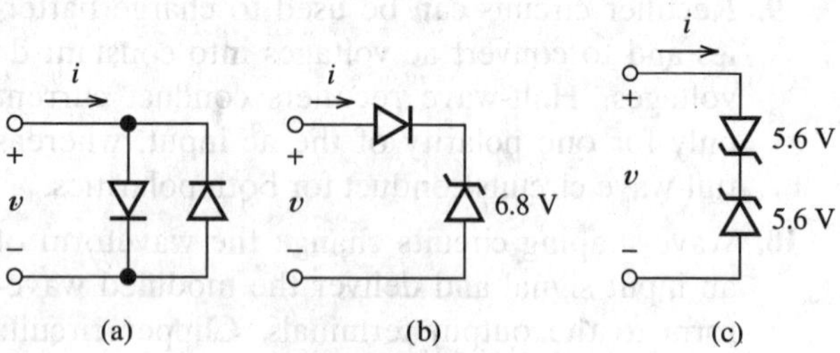

Figure P10.8

P10.9. Repeat Problem P10.8 for the circuits shown in Figure P10.9.

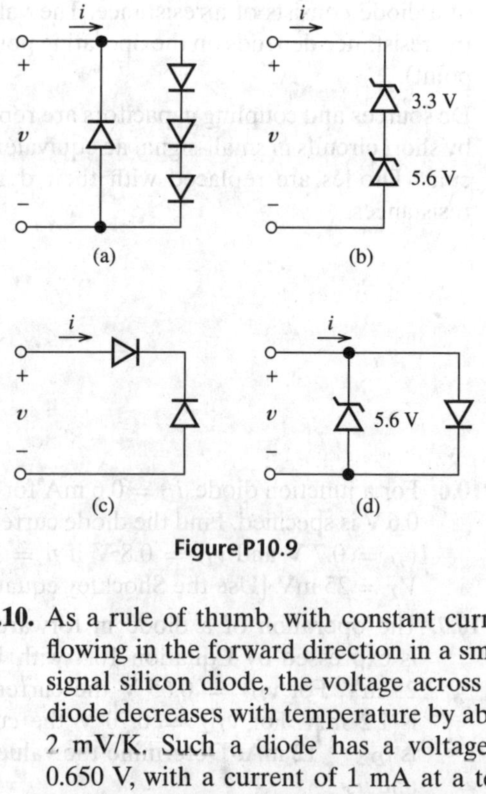

Figure P10.9

P10.10. As a rule of thumb, with constant current flowing in the forward direction in a small-signal silicon diode, the voltage across the diode decreases with temperature by about 2 mV/K. Such a diode has a voltage of 0.650 V, with a current of 1 mA at a temperature of 25°C. Find the diode voltage at 1 mA and a temperature of 150°C.

P10.11. We have a silicon diode described by the Shockley equation. The diode has $n = 2$ and operates at 100°C with a current of 1 mA and voltage of 0.5 V. Determine the current after the voltage is increased to 0.6 V.

P10.12. Consider a junction diode operating at a constant temperature of 300 K. With a forward current of 1 mA, the voltage is 600 mV. Furthermore, with a current of 10 mA, the voltage is 700 mV. Find the value of n for this diode.

P10.13. Assume a diode with $n = 2$, $I_s = 20$ nA, and $V_T = 26$ mV. **a.** Using a computer program of your choice, obtain a plot of i_D versus v_D for i_D ranging from 10 μA to 10 mA. Choose a logarithmic scale for i_D and a linear scale for v_D. What type of curve results? **b.** Place a 100-Ω resistance in series with the diode, and plot current versus voltage across the series combination on the same axes used for part (a). Compare the two curves. When is the added series resistance significant?

***P10.14.** The diodes shown in Figure P10.14 are identical and have $n = 2$. V_T is specified as 25 mV at 25°C. Before the switch is closed, the voltage v is 0.7 V. Find v after the switch is closed. Repeat for $n = 1$.

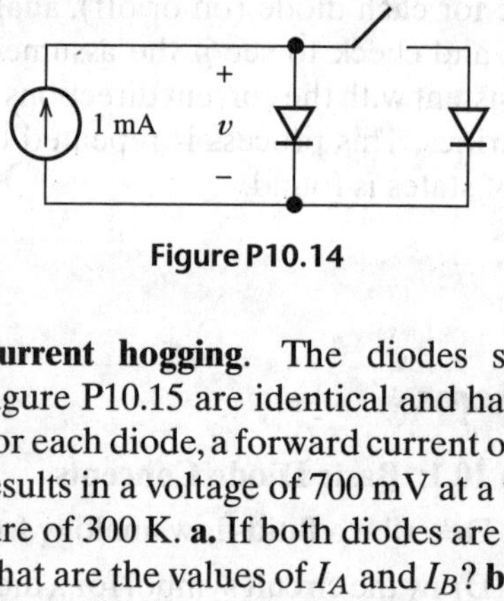

Figure P10.14

***P10.15. Current hogging.** The diodes shown in Figure P10.15 are identical and have $n = 1$. For each diode, a forward current of 100 mA results in a voltage of 700 mV at a temperature of 300 K. **a.** If both diodes are at 300 K, what are the values of I_A and I_B? **b.** If diode A is at 300 K and diode B is at 305 K, again find I_A and I_B, given that I_s doubles in value for a 5-K increase in temperature. [*Hint:* Answer part (a) by use of symmetry. For part (b), a transcendental equation for the voltage across the diodes can be found. Solve by trial and error. An important observation to be made from this problem is that, starting at the same temperature, the diodes should theoretically each conduct half of the total current. However, if one diode conducts slightly more, it becomes warmer, resulting in even more current. Eventually, one of the diodes "hogs" most of the current. This is particularly noticeable for devices that are thermally isolated from one another with large currents, for which significant heating occurs.]

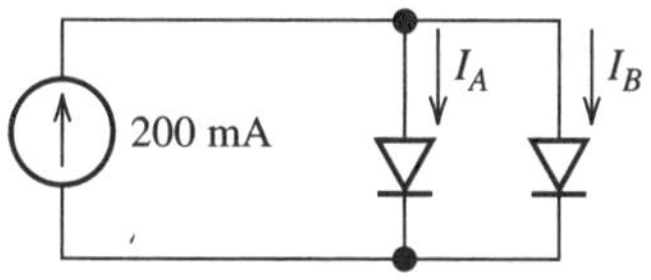

Figure P10.15

Section 10.2: Load-Line Analysis of Diode Circuits

***P10.16.** The nonlinear circuit element shown in Figure P10.16 has $i_x = [\exp(v_x) - 1]/10$, in which the units of v_x and i_x are V and A, respectively. Also, we have $V_s = 5$ V and $R_s = 2\ \Omega$. Use graphical load-line techniques to solve for i_x and v_x. (You may prefer to use a computer program to plot the characteristic and the load line.)

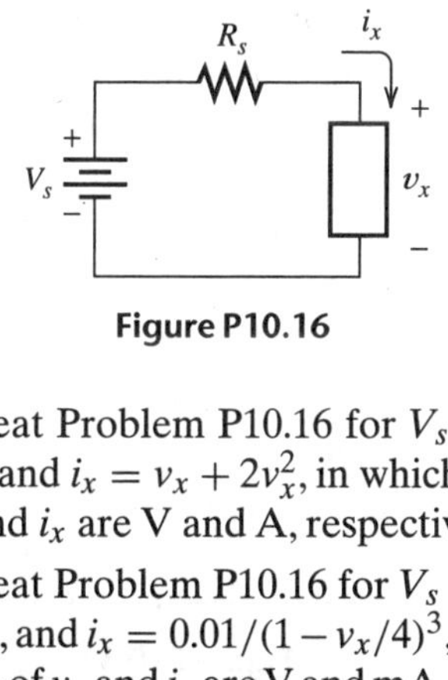

Figure P10.16

P10.17. Repeat Problem P10.16 for $V_s = 4$ V, $R_s = 1\ \Omega$, and $i_x = v_x + 2v_x^2$, in which the units of v_x and i_x are V and A, respectively.

P10.18. Repeat Problem P10.16 for $V_s = 15$ V, $R_s = 3\ \text{k}\Omega$, and $i_x = 0.01/(1 - v_x/4)^3$, in which the units of v_x and i_x are V and mA, respectively.

P10.19. Repeat Problem P10.16 for $V_s = 8$ V, $R_s = 2\ \Omega$, and $i_x = v_x^3/4$, in which the units of v_x and i_x are V and A, respectively.

P10.20. Several types of special-purpose diodes exist. One is the constant-current diode for which the current is constant over a wide range of voltage. The circuit symbol and volt–ampere characteristic for a particular constant-current diode are shown in Figure P10.20(a). Another special type is the LED for which the circuit symbol and a typical volt–ampere characteristic are shown in Figure P10.20(b). Sometimes, the series combination of these two devices is used to provide constant current to the LED from a variable voltage source. **a.** Sketch the overall volt–ampere characteristic to scale for the series combination shown in Figure P10.20(c). **b.** Sketch the overall volt–ampere characteristic to scale for the parallel combination shown in Figure P10.20(d).

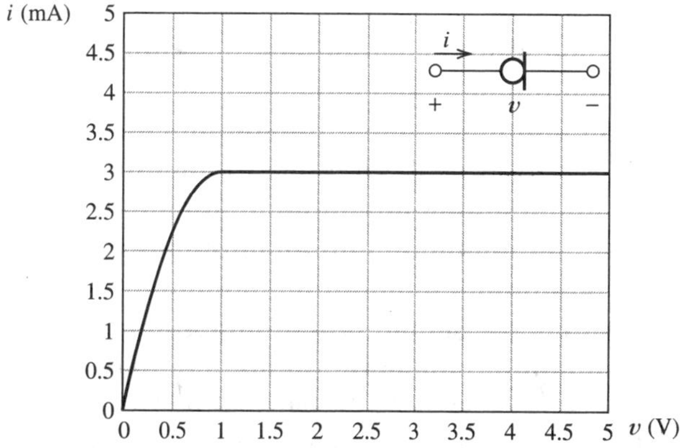

(a) Volt–ampere characteristic of a constant-current diode

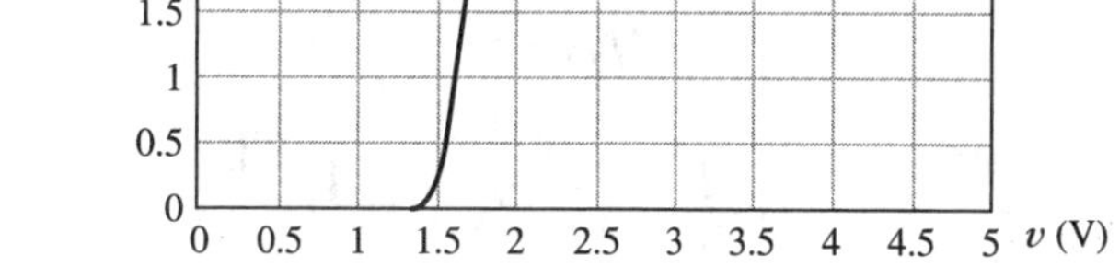

(b) Volt–ampere characteristic of a light-emitting diode (LED)

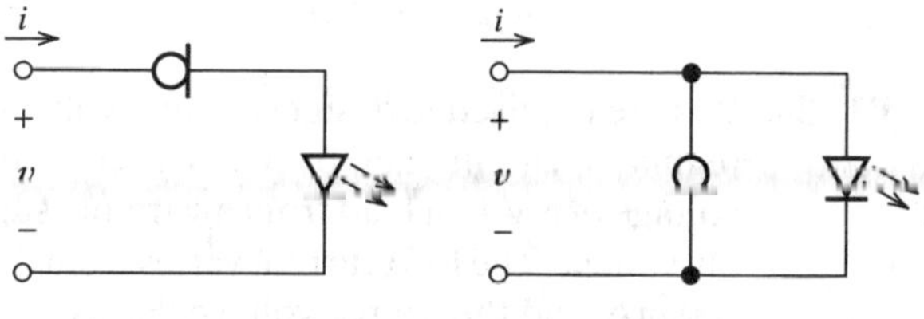

(c) (d)

Figure P10.20

P10.21. Determine the values for i and v for the circuit of Figure P10.21. The diode is the LED having the characteristic shown in Figure P10.20(b).

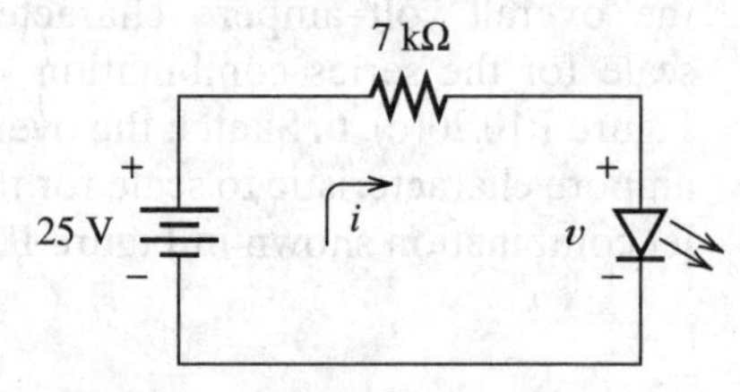

Figure P10.21

Section 10.3: Zener-Diode Voltage-Regulator Circuits

P10.22. What is a Zener diode? For what is it typically used? Draw the volt–ampere characteristic of an ideal 5.8-V Zener diode.

***P10.23.** Draw the circuit diagram of a simple voltage regulator.

P10.24. Consider the Zener-diode regulator shown in Figure 10.14 on page 498. What is the smallest load resistance for which v_o is 10 V?

P10.25. A simple voltage regulator is shown in Figure P10.25. The source voltage V_s varies from 8 to 12 V, and the load current i_L varies from 50 to 150 mA. Assume that the Zener diode is ideal. Determine the largest value allowed for the resistance R_s so that the load voltage v_L remains constant with variations in load current and source voltage. Determine the maximum power dissipation in R_s.

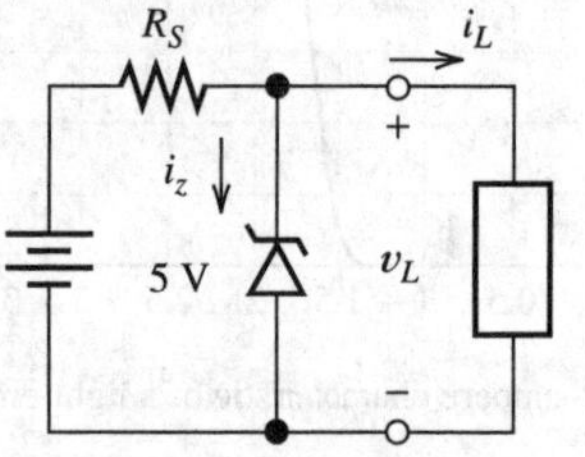

Figure P10.25

P10.26. You are required to design a simple voltage-regulator circuit that provides a constant voltage of 5 V to a load from a variable supply voltage. The load current varies from 0 to 100 mA, and the source voltage varies from 8 to 10 V. You may assume that ideal Zener diodes are available. Resistors of any value may be specified. Draw the circuit diagram of your regulator, and specify the value of each component. Also, find the worst case (maximum) power dissipated in each component in your regulator. Try to use good judgment in your design.

P10.27. Repeat Problem P10.26 given that the supply voltage ranges from 6 to 10 V.

P10.28. Repeat Problem P10.26 given that the load current varies from 0 to 1 A.

P10.29. Explain in general terms the method for solving a circuit that contains a single nonlinear element plus resistors, dc voltage sources, and dc current sources, given the volt–ampere characteristic of the nonlinear device.

***P10.30.** A certain linear two-terminal circuit has terminals a and b. Under open-circuit conditions, we have $v_{ab} = 10$ V. A short circuit is connected across the terminals, and a current of 2 A flows from a to b through the short circuit. Determine the value of v_{ab} when a nonlinear element that has $i_{ab} = \sqrt[3]{v_{ab}}$ is connected across the terminals.

P10.31. Determine the values for i_1 and i_2 for the circuit of Figure P10.31. The device is the constant-current diode having the characteristic shown in Figure P10.20(a).

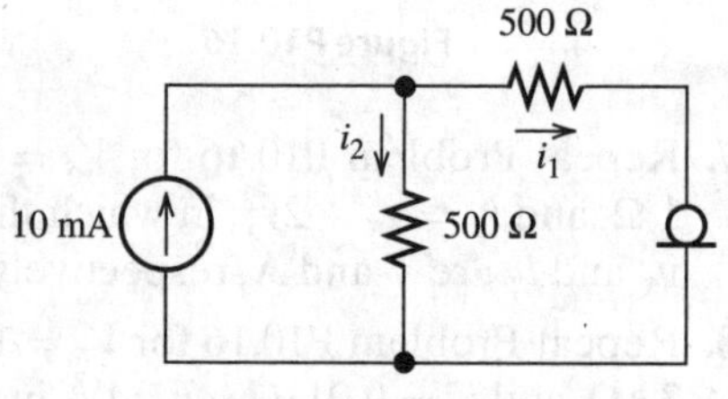

Figure P10.31

P10.32. Determine the values for i and v for the circuit of Figure P10.32. The diode is the LED having the characteristic shown in Figure P10.20(b).

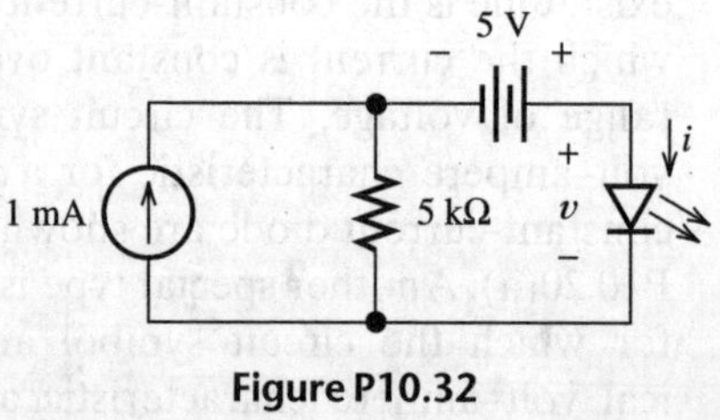

Figure P10.32

P10.33. Repeat Problem P10.32 for the circuit of Figure P10.33. The diode is the LED having the characteristic shown in Figure P10.20(b).

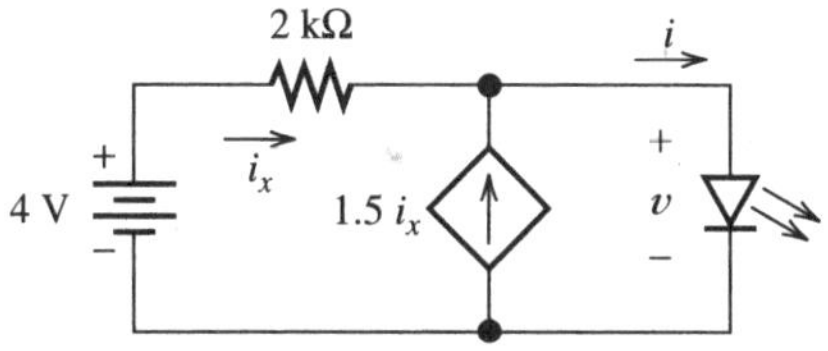

Figure P10.33

Section 10.4: Ideal-Diode Model

P10.34. What is an ideal diode? Draw its volt–ampere characteristic. After solving a circuit with ideal diodes, what check is necessary for diodes initially assumed to be on? Off?

P10.35. Consider two ideal diodes in series, pointing in opposite directions. What is the equivalent circuit for the combination? What is the equivalent circuit if the diodes are in parallel and pointing in opposite directions?

***P10.36.** Find the values of I and V for the circuits of Figure P10.36, assuming that the diodes are ideal.

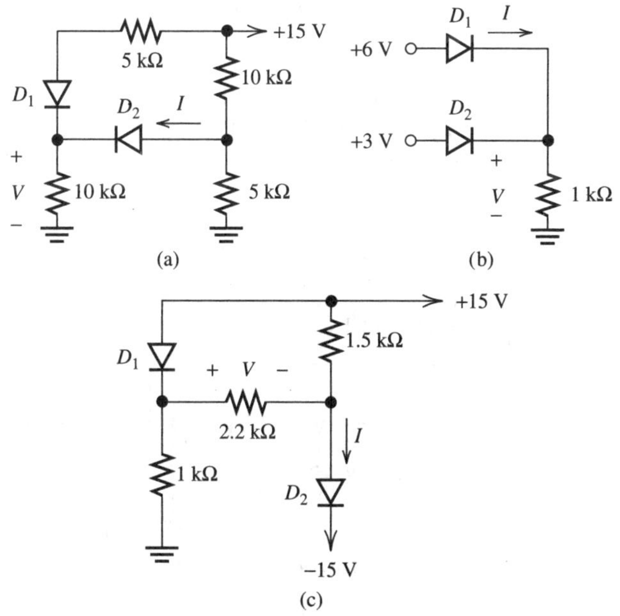

Figure P10.36

P10.37. Find the values of I and V for the circuits of Figure P10.37, assuming that the diodes are ideal.

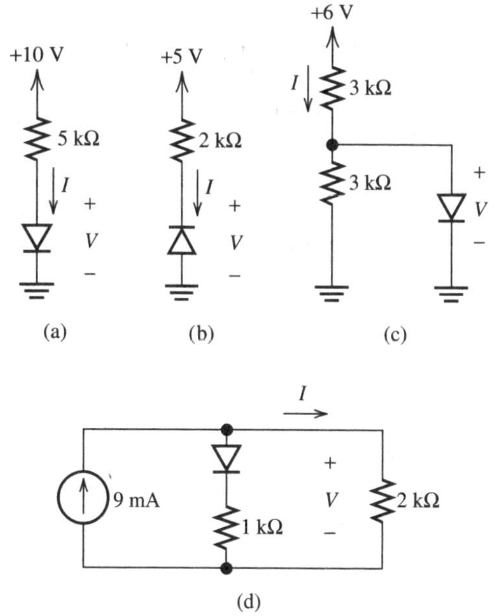

Figure P10.37

P10.38. Find the values of I and V for the circuits of Figure P10.38, assuming that the diodes are ideal. For part (b), consider $V_{in} = 0, 5, 10,$ and 20 V. Also, for part (b) of the figure, plot V versus V_{in} for V_{in} ranging from −10 V to 10 V. Find the currents in diodes D_1 and D_3.

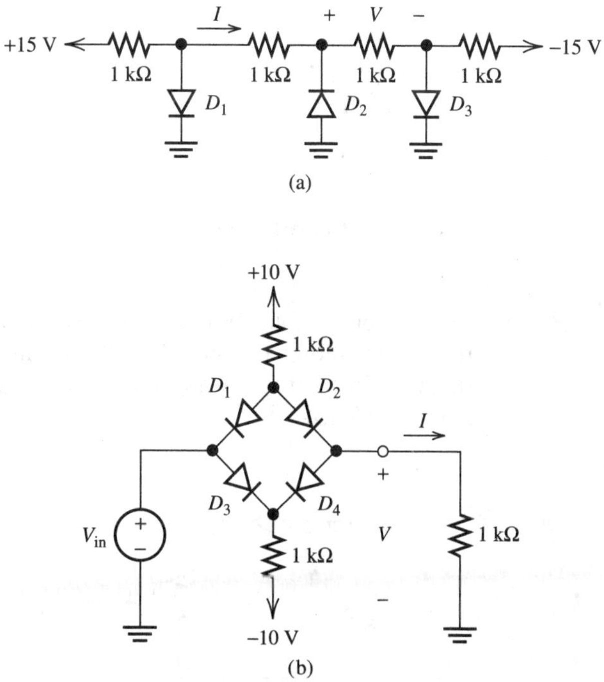

Figure P10.38

P10.39. **a.** Figure P10.39(a) shows a type of logic gate. Assume that the diodes are ideal. The voltages V_A and V_B independently have values of either 0 V (for logic 0, or low) or 5 V (for logic 1, or high). For which of the four combinations of input voltages is the output high (i.e., $V_o = 5$ V)? What type of logic gate is this? **b.** Repeat for Figure P10.39(b).

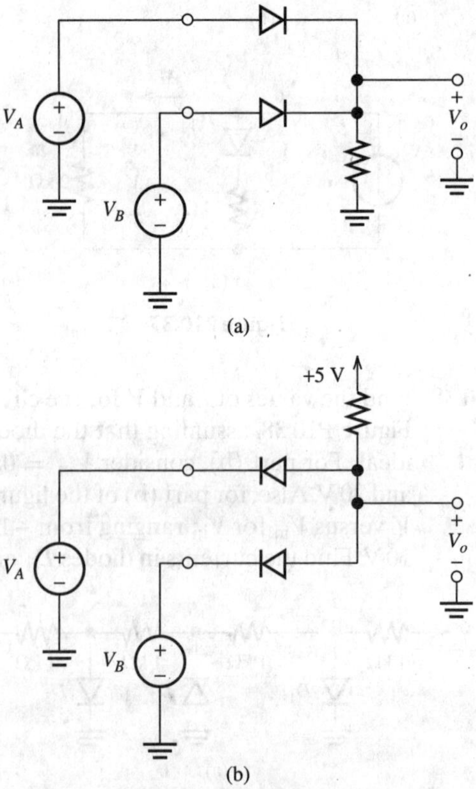

Figure P10.39

P10.40. Sketch a plot of i versus v to scale for each of the circuits shown in Figure P10.40. Assume that the diodes are ideal and allow v to range from −10 V to +10 V.

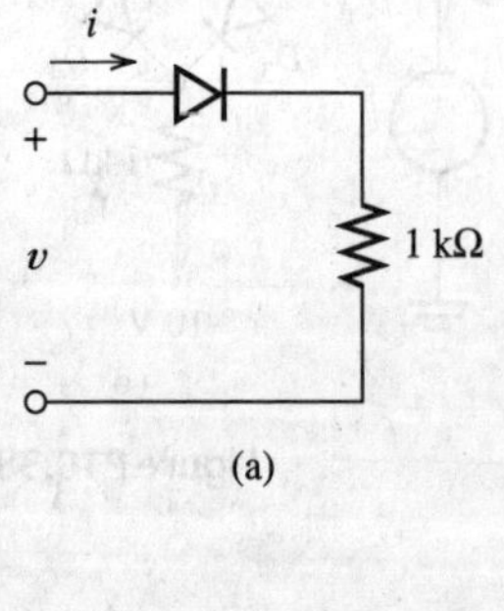

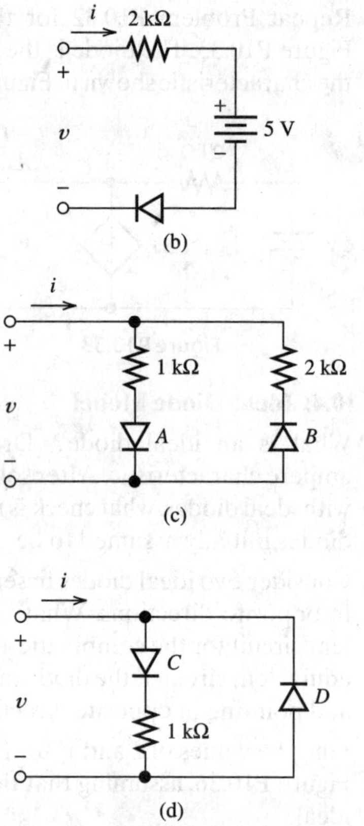

Figure P10.40

P10.41. Sketch $v_o(t)$ to scale versus time for the circuit shown in Figure P10.41. Assume that the diodes are ideal.

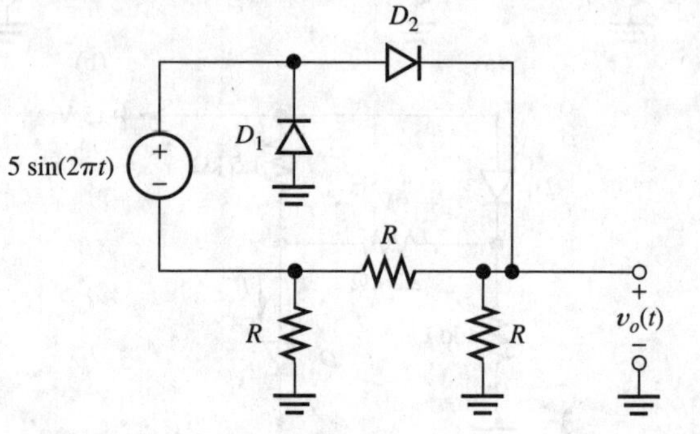

Figure P10.41

Section 10.5: Piecewise-Linear Diode Models

P10.42. If a nonlinear two-terminal device is modeled by the piecewise-linear approach, what is the equivalent circuit of the device for each linear segment?

P10.43. A resistor R_a is in series with a voltage source V_a. Draw the circuit. Label the voltage across the combination as v and the current as i. Draw and label the volt–ampere characteristic (i versus v).

P10.44. The volt–ampere characteristic of a certain two-terminal device is a straight line that passes through the points (2 V, 5 mA) and (3 V, 15 mA). The current reference points into the positive reference for the voltage. Determine the equivalent circuit for this device.

***P10.45.** Assume that we have approximated a nonlinear volt–ampere characteristic by the straight-line segments shown in Figure P10.45(c). Find the equivalent circuit for each segment. Use these equivalent circuits to find v in the circuits shown in Figure P10.45(a) and (b).

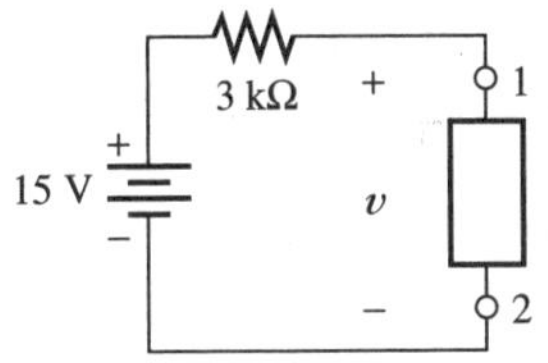

(a)

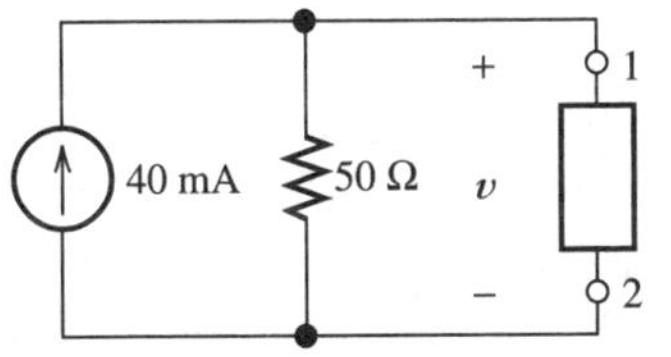

(b)

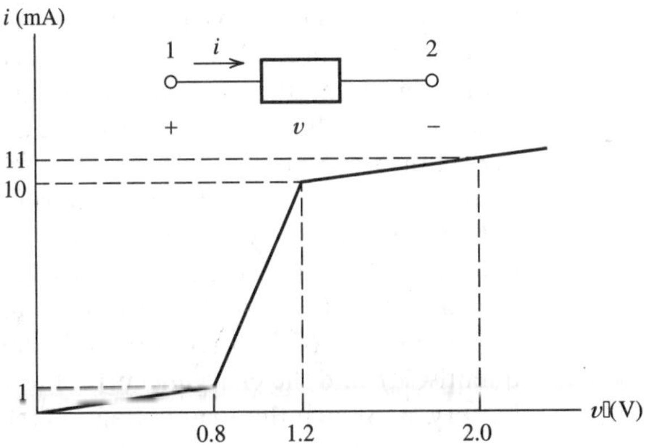

(c)

Figure P10.45

P10.46. Consider the volt–ampere characteristic of an ideal 10-V Zener diode shown in Figure 10.14 on page 498. Determine the piecewise-linear equivalent circuit for each segment of the characteristic.

P10.47. The Zener diode shown in Figure P10.47 has a piecewise-linear model shown in Figure 10.19 on page 501. Plot load voltage v_L versus load current i_L for i_L ranging from 0 to 200 mA.

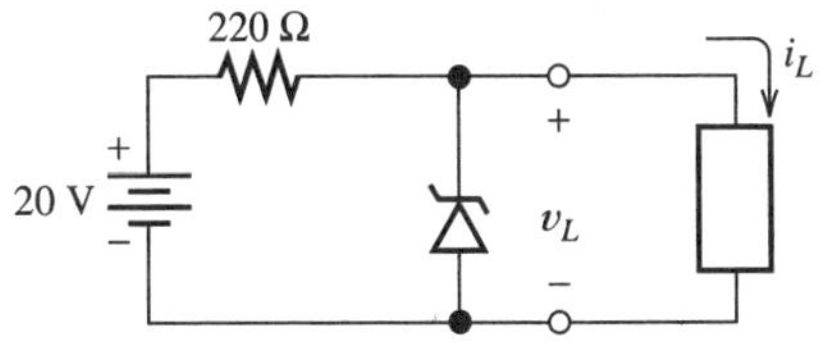

Figure P10.47

P10.48. In this problem, we will assume that the diode shown in Figure P10.48 can be represented by the model of Figure 10.23 on page 503, with $V_f = 0.7$ V. **a.** Assume that the diode operates as an open-circuit and solve for the node voltages v_1 and v_2. Are the results consistent with the model? Why or why not? **b.** Repeat part (a), assuming that the diode operates as a 0.7-V voltage source.

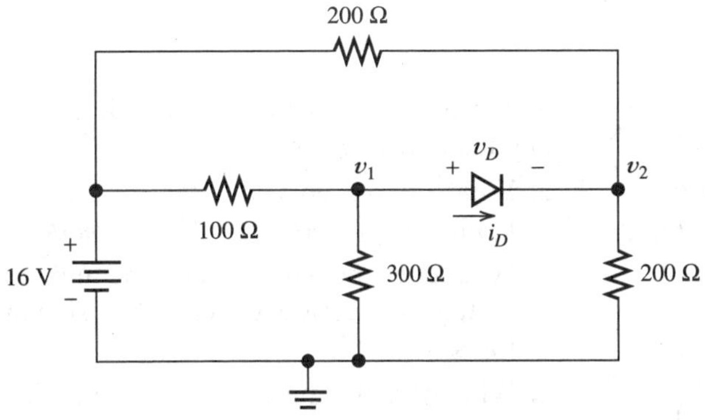

Figure P10.48

Section 10.6: Rectifier Circuits

P10.49. Draw the circuit diagram of a half-wave rectifier for producing a nearly steady dc voltage from an ac source. Draw two different full-wave circuits (one of which requires two ac sources).

P10.50. A 25-V rms 60-Hz ac source is in series with an ideal diode and a 100-Ω resistance. Determine the peak current and peak inverse voltage (PIV) for the diode.

P10.51. This problem relates to the half-wave rectifier shown in Figure 10.26 on page 505. The ac source has an rms value of 25 V and a frequency of 60 Hz. The diodes are ideal, and the capacitance is very large, so the ripple voltage V_r is very small. The load is a 1 kΩ resistance. Determine the peak inverse voltage across the diode and the charge that passes through the diode per cycle. What can you say about the peak current through the diode?

P10.52. This problem relates to the battery-charging circuit shown in Figure 10.25 on page 504. The ac source voltage is $v_s(t) = V_m \sin(\omega t) = 24 \sin(120\pi t)$. The resistance is 0.5 Ω, the diode is ideal, and $V_B = 12$ V. Determine the average current (i.e., the value of the charge that passes through the battery in 1 second). Suppose that the battery starts from a totally discharged state and has a capacity of 100 ampere hours. How long does it take to fully charge the battery?

P10.53. Dc voltmeters produce readings equal to the average values of the voltages measured. The average value of a periodic waveform is

$$V_{\text{avg}} = \frac{1}{T}\int_0^T v(t)dt$$

in which T is the period of the voltage $v(t)$ applied to the meter.

a. What does a dc voltmeter read if the applied voltage is $v(t) = V_m \sin(\omega t)$?

b. What does the meter read if the applied voltage is a half-wave rectified version of the sinewave?

c. What does the meter read if the applied voltage is a full-wave rectified version of the sinewave?

***P10.54.** A half-wave rectifier is needed to supply 15-V dc to a load that draws an average current of 250 mA. The peak-to-peak ripple is required to be 0.2 V or less. What is the minimum value allowed for the smoothing capacitance? If a full-wave rectifier is needed?

P10.55. Design a half-wave rectifier power supply to deliver an average voltage of 12 V with a peak-to-peak ripple of 1 V to a load. The average load current is 0.2 A. Assume that ideal diodes and 60-Hz ac voltage sources of any amplitudes needed are available. Draw the circuit diagram for your design. Specify the values of all components used.

P10.56. Repeat Problem P10.55 with a full-wave bridge rectifier.

P10.57. Repeat Problem P10.55 with two diodes and out-of-phase voltage sources to form a full-wave rectifier.

P10.58. Repeat Problem P10.55, assuming that the diodes have forward drops of 0.8 V.

P10.59. Consider the battery-charging circuit shown in Figure 10.25 on page 504, in which $v_s(t) = 110 \sin(200\pi t)$, $R = 1$ k Ω, $V_B = 12$ V, and the diode is ideal. **a.** Sketch the current $i(t)$ to scale versus time. **b.** Determine the average charging current for the battery. (*Hint*: The average current is the charge that flows through the battery in one cycle, divided by the period.)

P10.60. a. Consider the full-wave rectifier shown in Figure 10.27 on page 506, with a large smoothing capacitance placed in parallel with the load R_L and $V_m = 12$ V. Assuming that the diodes are ideal, what is the approximate value of the load voltage? What peak inverse voltage (PIV) appears across the diodes? **b.** Repeat for the full-wave bridge shown in Figure 10.28 on page 507.

P10.61. Figure P10.61 shows the equivalent circuit for a typical automotive battery charging system. The three-phase delta-connected source represents the stator coils of the alternator. (Three-phase ac sources are discussed in Section 5.7. Actually, the alternator stator is usually wye connected, but the terminal voltages are the same as for the equivalent delta.) Not shown in the figure is a voltage regulator that controls the current applied to the rotor coil of the alternator and, consequently, V_m and the charging current to the battery. **a.** Sketch the load voltage $v_L(t)$ to scale versus time. Assume ideal diodes and that V_m is large enough that current flows into the battery at all times. (*Hint*: Each source and four of the diodes form a full-wave bridge rectifier.) **b.** Determine the peak-to-peak ripple and the average load voltage in terms of V_m. **c.** Determine the value of V_m needed to provide an average charging current of 30 A. **d.** What additional factors would need to be considered in a realistic computation of V_m?

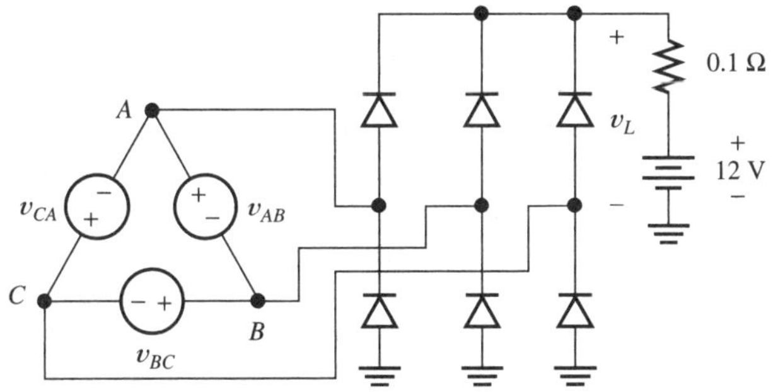

$v_{AB} = V_m \cos(\omega t)$
$v_{BC} = V_m \cos(\omega t - 120°)$
$v_{CA} = V_m \cos(\omega t + 120°)$

Figure P10.61 Idealized model of an automotive battery-charging system.

Section 10.7: Wave-Shaping Circuits

P10.62. Illustrate the working of a clipper circuit which clips the negative half-cycle of a sinusoidal waveform with a circuit diagram.

P10.63. Sketch the transfer characteristic (v_o versus v_{in}) to scale for the circuit shown in Figure P10.63. Allow v_{in} to range from −10 V to +10 V and assume that the diodes are ideal.

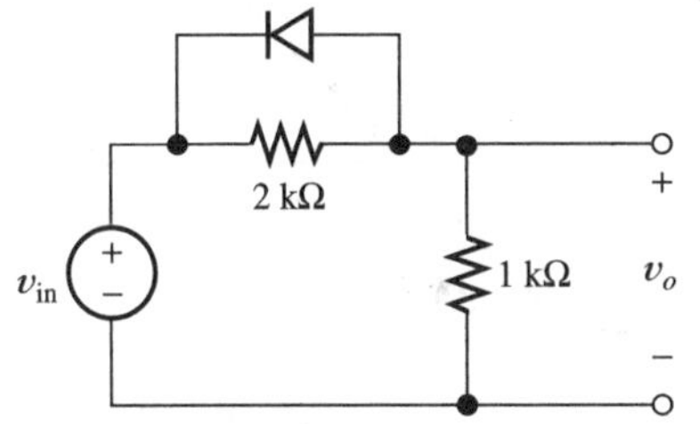

Figure P10.63

P10.64. Sketch to scale the output waveform for the circuit shown in Figure P10.64. Assume that the diodes are ideal.

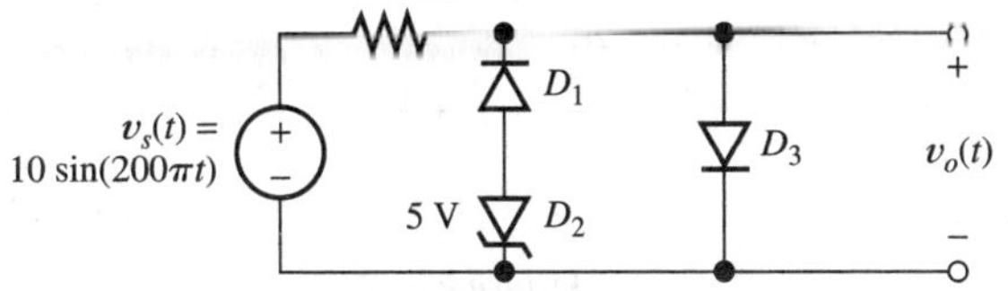

Figure P10.64

P10.65. Sketch the transfer characteristic (v_o versus v_{in}) to scale for the circuit shown in Figure P10.65. Allow v_{in} to range from −15 V to +15 V and assume the diodes have a forward drop of 0.7 V.

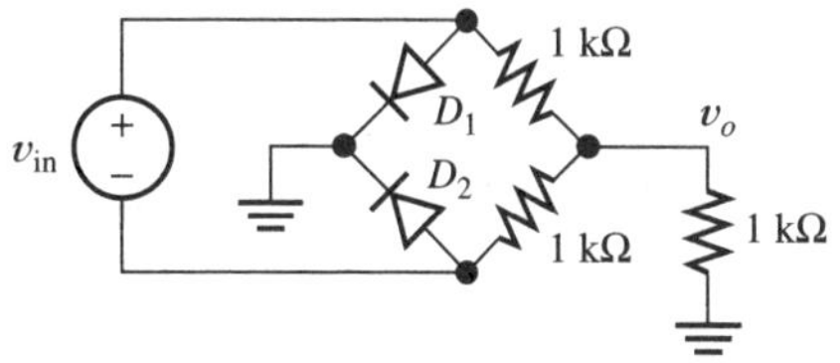

Figure P10.65

P10.66. Sketch the transfer characteristic (v_o versus v_{in}) to scale for the circuit shown in Figure P10.66. Allow v_{in} to range from −15 V to +15 V and the diodes have a forward drop of 0.7 V.

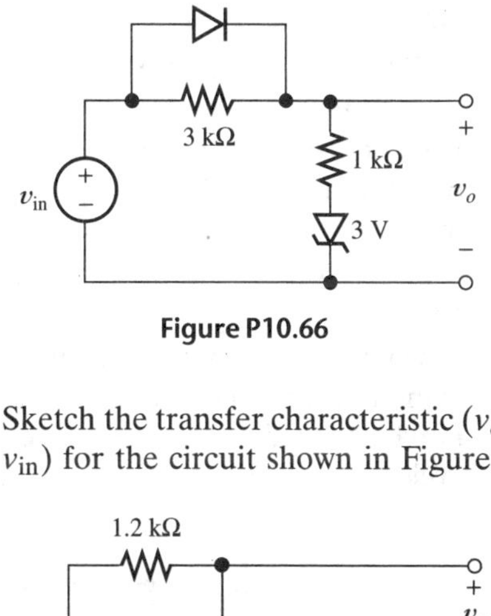

Figure P10.66

P10.67. Sketch the transfer characteristic (v_o versus v_{in}) for the circuit shown in Figure P10.67,

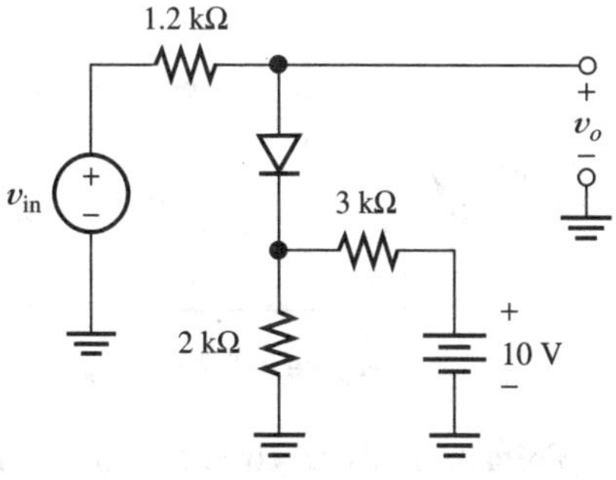

Figure P10.67

carefully labeling the breakpoint and slopes. Allow v_{in} to range from −10 V to +10 V and assume that the diodes are ideal.

P10.68. What is a clamp circuit? Draw an example circuit diagram, including component values, an input waveform, and the corresponding output waveform.

***P10.69.** Sketch to scale the steady-state output waveform for the circuit shown in Figure P10.69. Assume that RC is much larger than the period of the input voltage and that the diodes are ideal.

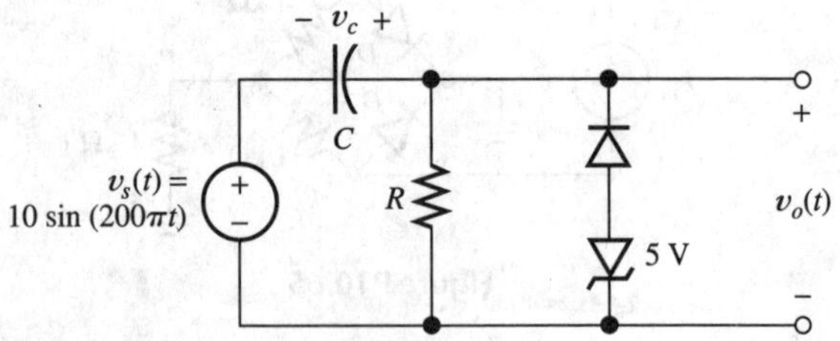

Figure P10.69

***P10.70.** Design a clipper circuit to clip off the portions of an input voltage that fall above 3 V or below −5 V. Assume that diodes having a constant forward drop of 0.7 V are available. Ideal Zener diodes of any breakdown voltage required are available. Dc voltage sources of any value needed are available.

P10.71. Repeat Problem P10.70, with clipping levels of +5 V and +10 V (i.e., every part of the input waveform below +5 or above +10 is clipped off).

P10.72. Consider the circuit shown in Figure P10.72, in which the RC time constant is very long compared with the period of the input and in which the diode is ideal. Sketch several cycles of $v_o(t)$ to scale versus time.

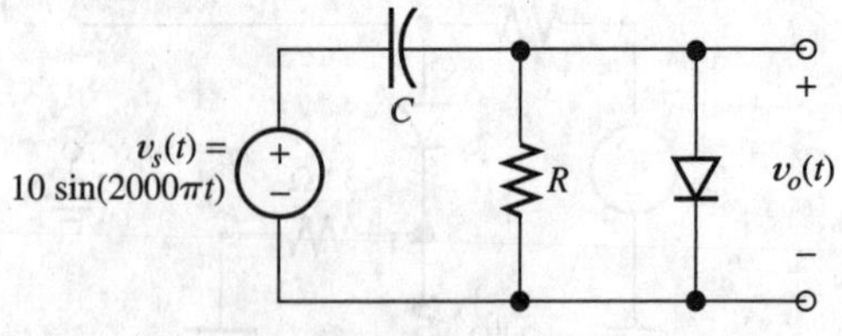

Figure P10.72

P10.73. Voltage-doubler circuit. Consider the circuit of Figure P10.73. The capacitors are very large, so they discharge only a very small amount per cycle. (Thus, no ac voltage appears across the capacitors, and the ac input plus the dc voltage of C_1 must appear at point A.) Sketch the voltage at point A versus time. Find the voltage across the load. Why is this called a voltage doubler? What is the peak inverse voltage across each diode?

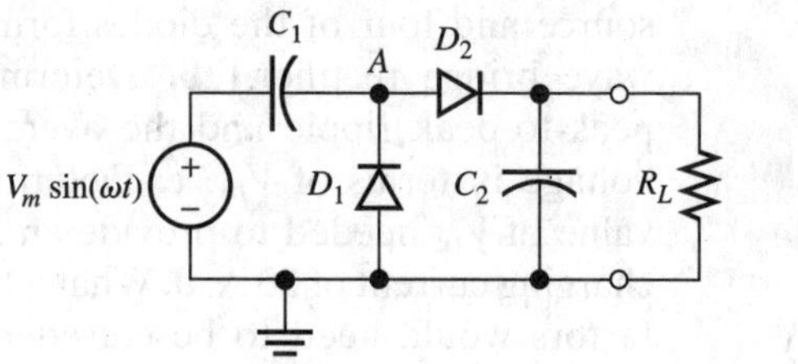

Figure P10.73 Voltage doubler

***P10.74.** Design a clamp circuit to clamp the negative extreme of a periodic input waveform to −5 V. Use diodes, Zener diodes, and resistors of any values required. Assume a 0.6-V forward drop for all diodes and that the Zener diodes have an ideal characteristic in the breakdown region. Power-supply voltages of ±15 V are available.

P10.75. Repeat Problem P10.74 for a clamp voltage of +5 V.

P10.76. Design circuits that have the transfer characteristics shown in Figure P10.76. Assume that v_{in} ranges from −10 to +10 V. Use diodes, Zener diodes, and resistors of any values needed. Assume a 0.6-V forward drop for all diodes and that the Zener diodes have an ideal characteristic in the breakdown region. Power-supply voltages of ±15 V are available.

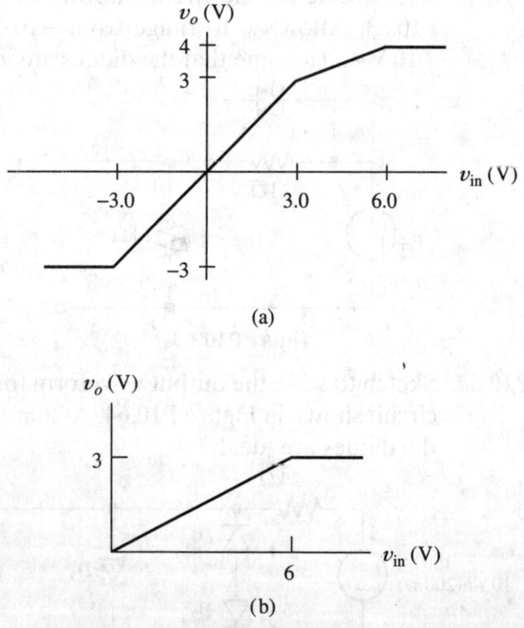

Figure P10.76

Section 10.8: Linear Small-Signal Equivalent Circuits

P10.77. A certain diode has $I_{DQ} = 6$ mA and $i_d(t) = 1.2\cos(200\pi t)$ mA. Find an expression for $i_D(t)$, and sketch several cycles to scale versus time.

P10.78. Of what does the small-signal equivalent circuit of a diode consist? How is the dynamic resistance of a nonlinear circuit element determined at a given operating point?

P10.79. With what are dc voltage sources replaced in a small-signal ac equivalent circuit? Why?

P10.80. With what should we replace a dc current source in a small-signal ac equivalent circuit? Justify your answer.

***P10.81.** A certain nonlinear device has $i_D = v_D^3/8$. Sketch i_D versus v_D to scale for v_D ranging from −2 V to +2 V. Is this device a diode? Determine the dynamic resistance of the device and sketch it versus v_D to scale for v_D ranging from −2 V to +2 V.

P10.82. A breakdown diode has

$$i_D = \frac{-10^{-6}}{(1 + v_D/7)^2} \quad \text{for } -7\text{ V} < v_D < 0$$

where i_D is in amperes. Plot i_D versus v_D for the ranges $-10\text{ mA} \le i_D \le 0$ and $-7\text{ V} \le v_D \le 0$. Find the dynamic resistance of this diode at $I_{DQ} = -0.5$ mA and at $I_{DQ} = -10$ mA.

P10.83. A diode is operating with an applied voltage given by

$$v_D(t) = 5 + 0.03\cos(\omega t)\text{ V}$$

The current is given by

$$i_D(t) = 10 + 0.2\cos(\omega t)\text{ mA.}$$

Determine the dynamic resistance and Q point of the diode under the conditions given.

P10.84. The voltage of an ideal Zener diode is constant in the breakdown region. What does this imply about the dynamic resistance in the breakdown region for an ideal Zener diode?

***P10.85.** Consider the voltage-regulator circuit shown in Figure P10.85. The ac ripple voltage is 1 V peak to peak. The dc (average) load voltage is 5 V. What is the Q-point current in the Zener diode? What is the maximum dynamic resistance allowed for the Zener diode if the output ripple is to be less than 10 mV peak to peak?

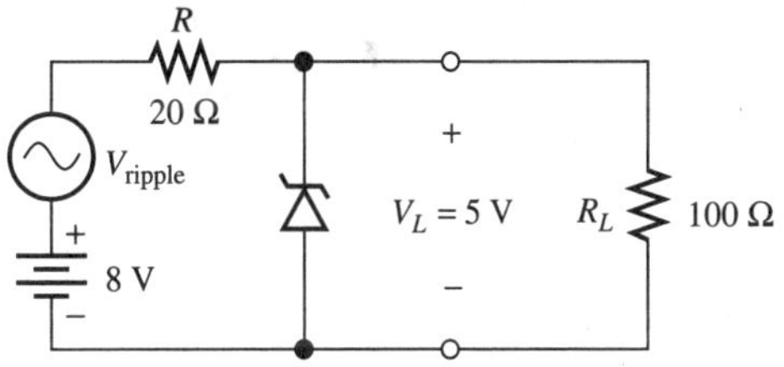

Figure P10.85

Frequently Asked Questions

1. Draw the symbol of a crystal diode and explain its significance. How can you identify the polarity of a crystal diode?
2. Draw the static characteristics of a crystal diode and label clearly the knee voltage and avalanche breakdown voltage.
3. What do you mean by reverse bias region and reverse breakdown region?
4. Draw and explain the characteristics of a small signal silicon diode.
5. Write down the Shockley equation for a junction diode. What is its significance?
6. Define and explain the following terms:

   (i) Knee voltage
   (ii) Peak inverse voltage
   (iii) Maximum forward current
   (iv) Maximum power rating
7. What is a Zener diode? Draw the equivalent circuit of an ideal and actual Zener diode.
8. Explain how a Zener diode maintains a constant voltage across the load.
9. What do you mean by ideal diode? Draw its V-I characteristics and explain its switching action.
10. Draw the circuit diagram of a half-wave rectifier and explain how it works.
11. Draw the circuit of a full-wave center-tap rectifier and explain its working.
12. Draw the circuit of a full-wave bridge rectifier and explain its working.

13. Draw the circuit diagram of a clipper using diodes and explain its working.
14. How diodes are used for clipping the ac wave?
15. Draw the circuit diagram of a clipper using Zener diodes. Explain its working with the help of wave diagrams.
16. Draw and explain the working of a damping circuit.
17. Draw and explain the working of a variable attenuator using a diode as a controlled resistance.
18. A 5 KΩ resistance is connected in series with a silicon diode (barrier potential V_0 = 0.7 V) across 12V supply. What will be the circuit current when the diode is forward biased?

Solution:

$$I = \frac{V - V_o}{R} = \frac{12 - 0.7}{5 \times 10^3} = \mathbf{2.26\,mA} \qquad \textbf{(Ans)}$$

Practice Test

Here is a practice test you can use to check your comprehension of the most important concepts in this chapter. Answers can be found in Appendix D and complete solutions are included in the Student Solutions files. See Appendix E for more information about the Student Solutions.

T10.1. Determine the value of i_D for each of the circuits shown in Figure T10.1. The characteristic for the diode is shown in Figure 10.8 on page 493.

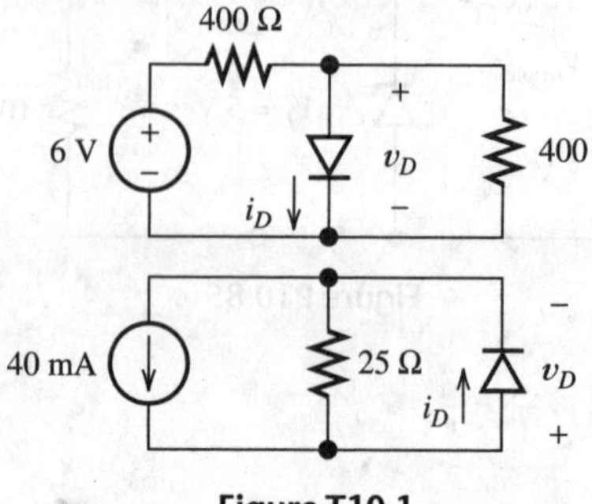

Figure T10.1

T10.2. The diode shown in Figure T10.2 is ideal. Determine the state of the diode and the values of v_x and i_x.

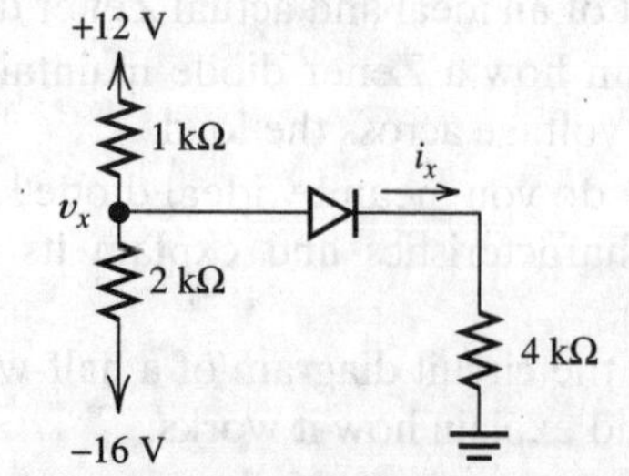

Figure T10.2

T10.3. The current versus voltage characteristic of a certain two-terminal device passes through the points (5 V, 2 mA) and (10 V, 7 mA). The reference for the current points into the positive reference for the voltage. Determine the values for the resistance and voltage source for the piecewise linear equivalent circuit for this device between the two points given.

T10.4. Draw the circuit diagram for a full-wave bridge rectifier with a resistance as the load.

T10.5. Suppose we have a 10-V-peak sinusoidal voltage source. Draw the diagram of a circuit that clips off the part of the sinusoid above 5 V and below −4 V. The circuit should be composed of ideal diodes, dc voltage sources, and other components as needed. Be sure to label the terminals across which the clipped output waveform $v_o(t)$ appears.

T10.6. Suppose we have a 10-Hz sinusoidal voltage source, $v_{\text{in}}(t)$. Draw the diagram of a circuit that clamps the positive peaks to −4 V. The circuit should be composed of ideal diodes, dc voltage sources, and other components as needed. List any constraints that should be observed in selecting component values. Be sure to label the terminals across which the clamped output waveform $v_o(t)$ appears.

T10.7. Suppose we have a silicon diode operating with a bias current of 5 mA at a temperature of 300 K. The diode current is given by the Shockley equation with $n = 2$. Draw the small-signal equivalent circuit for the diode including numerical values for the components.

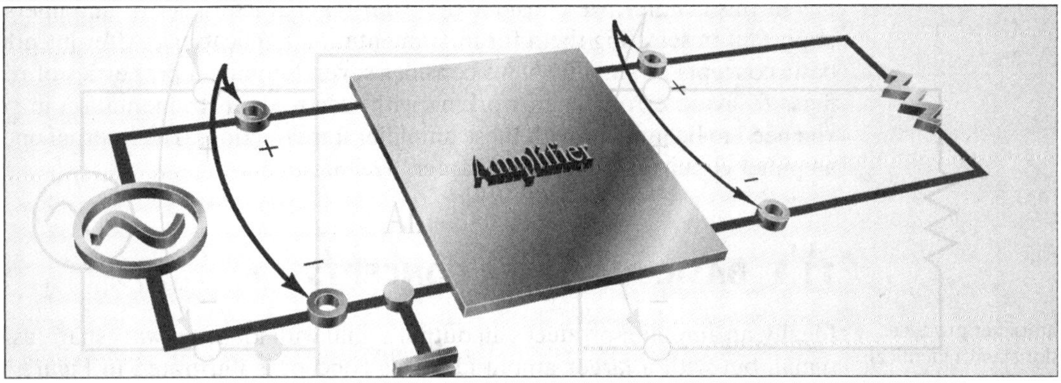

Chapter 11

Amplifiers: Specifications and External Characteristics

Study of this chapter will enable you to:

- Use various amplifier models to calculate amplifier performance for given sources and loads.
- Compute amplifier efficiency.
- Understand the importance of input and output impedances of amplifiers.
- Determine the best type of ideal amplifier for various applications.
- Specify the frequency-response requirements for various amplifier applications.
- Understand linear and nonlinear distortion in amplifiers.
- Specify the pulse-response parameters of amplifiers.
- Work with differential amplifiers and specify common-mode rejection requirements.
- Understand the various sources of dc offsets and design balancing circuits.

Introduction to this chapter:

The most important analog blocks found in electronic systems are amplifiers. Basically, amplifiers are used to increase the amplitudes of electrical signals. For example, the signals from most sensors (such as strain gauges used in mechanical engineering or flow meters used in chemical processes) are small in amplitude and need to be amplified before they can be utilized.

In this chapter, we consider the external characteristics of amplifiers that are important in selecting them for instrumentation applications. After introducing the basic concepts of amplifiers, we consider several nonideal properties of real amplifiers. To avoid errors when working with electronic instrumentation in your field, you need to be familiar with these amplifier imperfections. The internal operation of amplifier circuits is treated in Chapters 12, 13, and 14.

11.1 BASIC AMPLIFIER CONCEPTS

Ideally, an amplifier produces an output signal with identical waveshape as the input signal, but with a larger amplitude.

Ideally, an amplifier produces an output signal with identical waveshape as the input signal, but with a larger amplitude. This concept is illustrated in Figure 11.1. The signal source produces a voltage $v_i(t)$ that is applied to the input terminals of the amplifier, which generates an output voltage

$$v_o(t) = A_v v_i(t) \tag{11.1}$$

across a **load resistance** R_L connected to the output terminals. The constant A_v is called the **voltage gain** of the amplifier. Often, the voltage gain is much larger in magnitude than unity, but we will see later that useful amplification can take place even if A_v is less than unity.

An example of a signal source is a microphone that typically produces a signal of 1-mV peak as we speak into it. This small signal can be used as the input to an amplifier with a voltage gain of 10,000 to produce an output signal with a peak value of 10 V. If this larger output voltage is applied to a loudspeaker, a much louder version of the sound entering the microphone results—the principle of operation for the electronic megaphone.

Sometimes, A_v is a negative number, so the output voltage is an inverted version of the input, and the amplifier is then called an **inverting amplifier**. On the other hand, if A_v is a positive number, we have a **noninverting amplifier**. A typical input waveform and the corresponding output waveforms for a noninverting amplifier and for an inverting amplifier are shown in Figure 11.2.

Inverting amplifiers have negative voltage gain, and the output waveform is an inverted version of the input waveform. Noninverting amplifiers have positive voltage gain.

For monaural audio signals, it does not matter whether the amplifier is inverting or noninverting because the sounds produced by the loudspeaker are perceived the same either way. However, in a stereo system, it is important that the amplifiers for the left and right channels are the same (i.e., either both inverting or both noninverting), so that the signals applied to the two loudspeakers have the proper phase relationship.

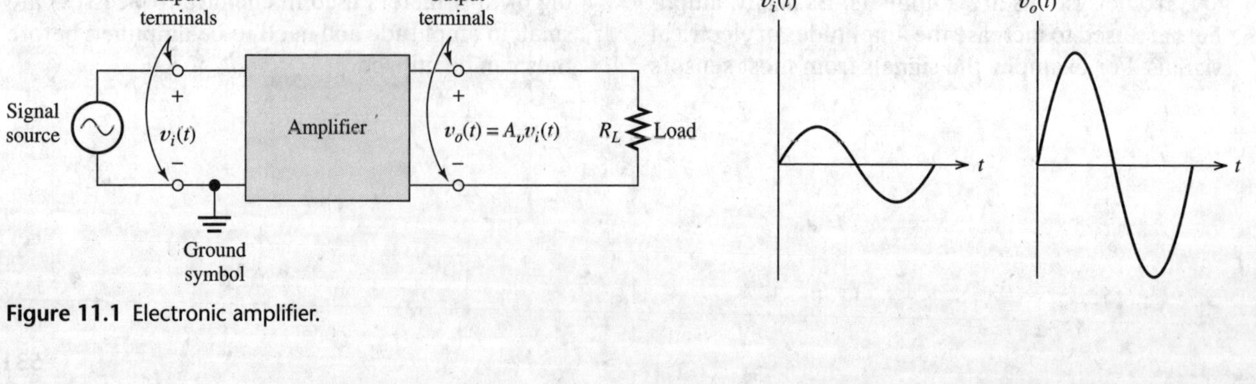

Figure 11.1 Electronic amplifier.

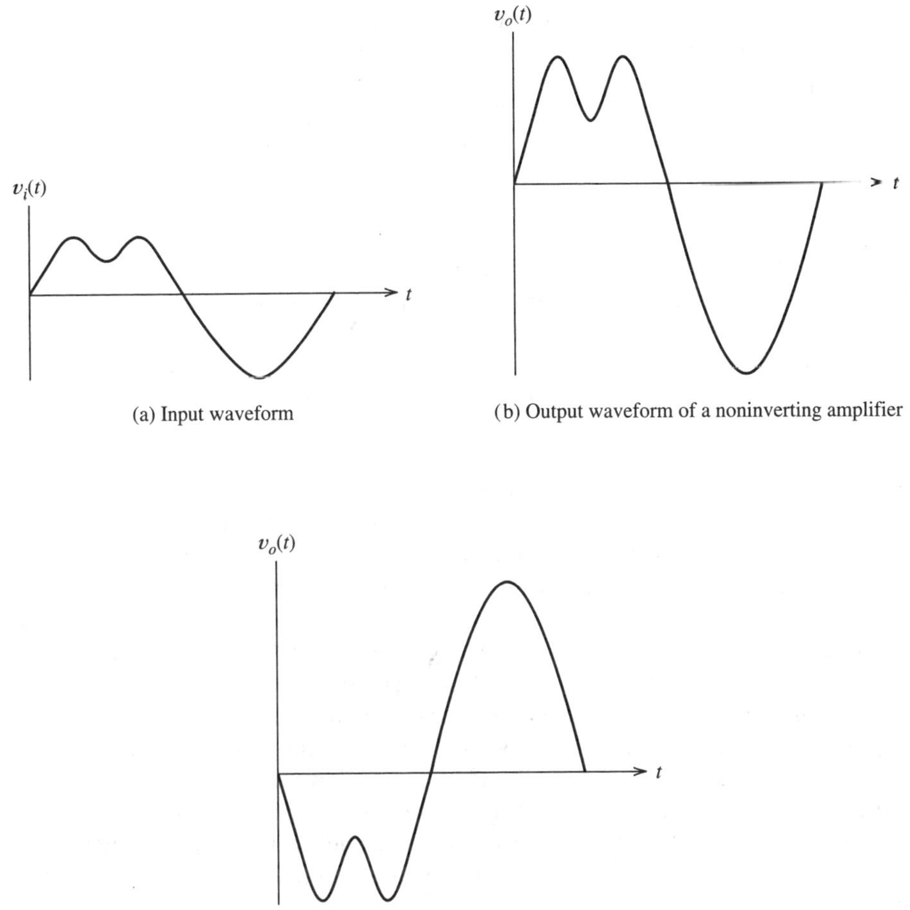

(a) Input waveform

(b) Output waveform of a noninverting amplifier

(c) Output waveform of an inverting amplifier

Figure 11.2 Input waveform and corresponding output waveforms.

If video signals are inverted, a negative image with black and white interchanges results; hence, it is important whether video amplifiers are inverting or noninverting.

Common Ground Node

Often, one of the amplifier input terminals and one of the output terminals are connected to a common **ground**. Notice the ground symbol shown in Figure 11.1. Typically, the ground terminal consists of the metal chassis that contains the circuit as well as circuit-board conductors. This common ground serves as the return path for signal currents and, as we will see later, the dc power-supply currents in electronic circuits.

You may be familiar with the concept of electrical grounds in automobile wiring. Here, the ground conductor consists of the frame, fenders, and other conductive parts of the car. For example, current is carried to the taillights by a wire, but may return through the ground conductors, consisting of the fenders and frame. Similarly, residential 60-Hz power distribution systems are grounded, often to a cold-water pipe. However, in this case, return currents are not intended to flow through the ground conductors because that could pose safety hazards.

Always be careful in working with electrical circuits.

Sometimes, *but not always*, the chassis ground is connected through the line cord to the 60-Hz power-system ground. *Always be careful in working with electrical circuits.* In some types of electronic circuits, the chassis ground can be at 120 V ac with respect to the power-system ground. Touching the chassis while in contact with the power-system ground (through a water pipe or a damp concrete floor for example) *can be fatal.*

Exercise 11.1 A certain noninverting amplifier has a voltage-gain magnitude of 50. The input voltage is $v_i(t) = 0.1\sin(2000\pi t)$. **a.** Find an expression for the output voltage $v_o(t)$. **b.** Repeat for an inverting amplifier.

Answer **a.** $5\sin(2000\pi t)$; **b.** $-5\sin(2000\pi t)$. □

Voltage-Amplifier Model

Amplifiers are characterized by their input impedance, output impedance, and a gain parameter.

Amplification can be modeled by a controlled source as illustrated in Figure 11.3. Because real amplifiers draw some current from the signal source, a realistic model of an amplifier must include a resistance R_i across the input terminals. Furthermore, a resistance R_o must be included in series with the output terminals to account for the fact that the output voltage of a real amplifier is reduced when load current flows. The complete amplifier model shown in Figure 11.3 is called the **voltage-amplifier model**. Later, we will see that other models can be used for amplifiers.

The **input resistance** R_i of the amplifier is the equivalent resistance seen when looking into the input terminals. As we will find later, the input circuitry can sometimes include capacitive or inductive effects, and we would then refer to the **input impedance**. For example, the input amplifiers of typical oscilloscopes have an input impedance consisting of a 1-MΩ resistance in parallel with a 47-pF capacitance. In this chapter, we assume that the input impedance is purely resistive, unless stated otherwise.

The resistance R_o in series with the output terminals is known as the **output resistance**. Real amplifiers are not able to deliver a fixed voltage to an arbitrary load resistance. Instead, the output voltage becomes smaller as the load resistance becomes smaller, and the output resistance accounts for this reduction. When the load draws current, a voltage drop occurs across the output resistance, resulting in a reduction of the output voltage.

The voltage-controlled voltage source models the amplification properties of the amplifier. Notice that the voltage produced by this source is simply a constant A_{voc} times the input voltage v_i. If the load is an open circuit, there is no drop across

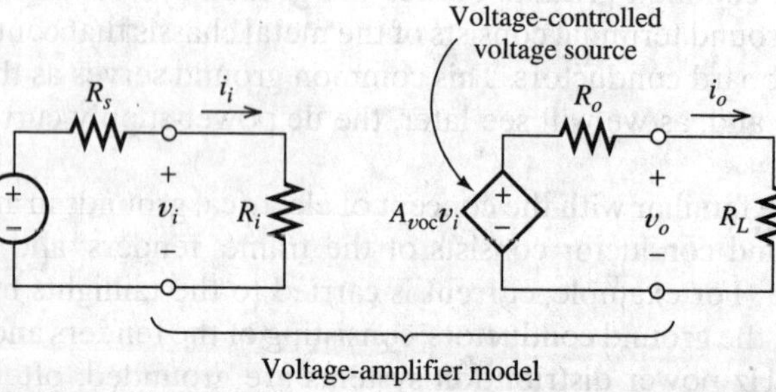

Figure 11.3 Model of an electronic amplifier, including input resistance R_i and output resistance R_o.

the output resistance, and then, $v_o = A_{voc}v_i$. For this reason, A_{voc} is called the **open-circuit voltage gain.**

In sum, the voltage-amplifier model includes the input impedance, the output impedance, and the open-circuit voltage gain in an equivalent circuit for the amplifier.

The voltage-amplifier model includes the input impedance, the output impedance, and the open-circuit voltage gain in an equivalent circuit for the amplifier.

Current Gain

As shown in Figure 11.3, the input current i_i is the current delivered to the input terminals of the amplifier, and the output current i_o is the current flowing through the load. The **current gain** A_i of an amplifier is the ratio of the output current to the input current:

$$A_i = \frac{i_o}{i_i} \tag{11.2}$$

The input current can be expressed as the input voltage divided by the input resistance, and the output current is the output voltage divided by the load resistance. Thus, we can find the current gain in terms of the voltage gain and the resistances as

$$A_i = \frac{i_o}{i_i} = \frac{v_o/R_L}{v_i/R_i} = A_v\frac{R_i}{R_L} \tag{11.3}$$

in which

$$A_v = \frac{v_o}{v_i}$$

is the voltage gain with the load resistance connected. Usually, A_v is smaller in magnitude than the open-circuit voltage gain A_{voc} because of the voltage drop across the output resistance.

A_v is the voltage gain with the load attached, whereas A_{voc} is the voltage gain with the output terminals open circuited.

Power Gain

The power delivered to the input terminals by the signal source is called the input power P_i, and the power delivered to the load by the amplifier is the output power P_o. The **power gain** G of an amplifier is the ratio of the output power to the input power:

$$G = \frac{P_o}{P_i} \tag{11.4}$$

Because we are assuming that the input impedance and load are purely resistive, the average power at either set of terminals is simply the product of the root-mean-square (rms) current and rms voltage. Thus, we can write

$$G = \frac{P_o}{P_i} = \frac{V_oI_o}{V_iI_i} = A_vA_i = (A_v)^2\frac{R_i}{R_L} \tag{11.5}$$

Notice that we have used uppercase symbols, such as V_o and I_o, for the rms values of the currents and voltages. We use lowercase symbols, such as v_o and i_o, for the instantaneous values. Of course, since we have assumed so far that the instantaneous output is a constant times the instantaneous input, the ratio of the rms voltages is the same as the ratio of the instantaneous voltages, and both ratios are equal to the voltage gain of the amplifier.

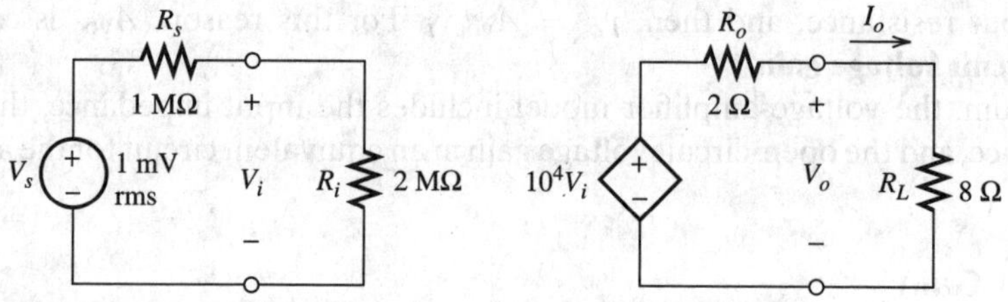

Figure 11.4 Source, amplifier, and load for Example 11.1.

Example 11.1 Calculating Amplifier Performance

A source with an internal voltage of $V_s = 1$ mV rms and an internal resistance of $R_s = 1$ MΩ is connected to the input terminals of an amplifier having an open-circuit voltage gain of $A_{voc} = 10^4$, an input resistance of $R_i = 2$ MΩ, and an output resistance of $R_o = 2$ Ω. The load resistance is $R_L = 8$ Ω. Find the voltage gains $A_{vs} = V_o/V_s$ and $A_v = V_o/V_i$. Also, find the current gain and power gain.

Solution First, we draw the circuit containing the source, amplifier, and load as shown in Figure 11.4. We can apply the voltage-divider principle to the input circuit to write

$$V_i = \frac{R_i}{R_i + R_s} V_s = 0.667 \text{ mV rms}$$

The voltage produced by the voltage-controlled source is given by

$$A_{voc} V_i = 10^4 V_i = 6.67 \text{ V rms}$$

Next, the output voltage can be found by using the voltage-divider principle, resulting in

$$V_o = A_{voc} V_i \frac{R_L}{R_L + R_o} = 5.33 \text{ V rms}$$

Now, we can find the required voltage gains:

$$A_v = \frac{V_o}{V_i} = A_{voc} \frac{R_L}{R_o + R_L} = 8000$$

and

$$A_{vs} = \frac{V_o}{V_s} = A_{voc} \frac{R_i}{R_i + R_s} \frac{R_L}{R_o + R_L} = 5333$$

Using Equations 11.3 and 11.5, we find that the current gain and power gain are

$$A_i = A_v \frac{R_i}{R_L} = 2 \times 10^9$$

$$G = A_v A_i = 16 \times 10^{12}$$

Notice that the current gain is very large, because the high input resistance allows only a small amount of input current to flow, whereas the relatively small load resistance allows the output current to be relatively large. ■

Loading Effects

Notice that not all of the internal voltage of the source appears at the input terminals of the amplifier in Example 11.1. This is because the finite input resistance of the amplifier allows current to flow into the input terminals, resulting in a voltage drop across the internal resistance R_s of the source. Similarly, the voltage produced by the controlled source does not all appear across the load. These reductions in voltage are called **loading effects**. Because of loading effects, the voltage gains (A_v or A_{vs}) realized are less than the internal gain A_{voc} of the amplifier.

Because of loading effects, the voltage gains realized (A_v or A_{vs}) are smaller in magnitude than the internal gain A_{voc} of the amplifier.

Exercise 11.2 An amplifier has an input resistance of 2000 Ω, an output resistance of 25 Ω, and an open-circuit voltage gain of 500. The source has an internal voltage of $V_s = 20$ mV rms and a resistance of $R_s = 500\ \Omega$. The load resistance is $R_L = 75\ \Omega$. Find the voltage gains $A_v = V_o/V_i$ and $A_{vs} = V_o/V_s$. Find the current gain and the power gain.

Answer $A_v = 375, A_{vs} = 300, A_i = 10^4, G = 3.75 \times 10^6$. □

Exercise 11.3 Assume that we can change the load resistance in Exercise 11.2. What value of load resistance maximizes the power gain? What is the power gain for this load resistance?

Answer $R_L = 25\ \Omega, G = 5 \times 10^6$. □

11.2 CASCADED AMPLIFIERS

Sometimes, we connect the output of one amplifier to the input of another as shown in Figure 11.5. This is called a **cascade connection** of the amplifiers. The overall voltage gain of the cascade connection is given by

When the output of one amplifier is connected to the input of another amplifier, we say that the amplifiers are cascaded.

$$A_v = \frac{v_{o2}}{v_{i1}}$$

By multiplying and dividing by v_{o1}, this becomes

$$A_v = \frac{v_{o1}}{v_{i1}} \times \frac{v_{o2}}{v_{o1}}$$

Moreover, referring to Figure 11.5, we see that $v_{i2} = v_{o1}$. Therefore, we can write

$$A_v = \frac{v_{o1}}{v_{i1}} \times \frac{v_{o2}}{v_{i2}}$$

However, $A_{v1} = v_{o1}/v_{i1}$ is the gain of the first stage, and $A_{v2} = v_{o2}/v_{i2}$ is the gain of the second stage, so we have

$$A_v = A_{v1}A_{v2} \tag{11.6}$$

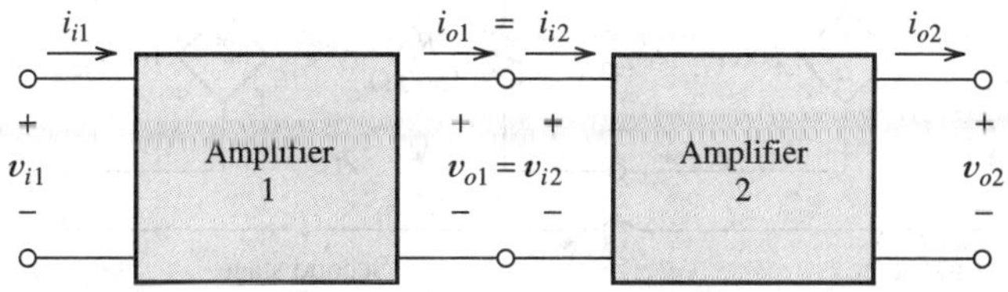

Figure 11.5 Cascade connection of two amplifiers.

It is necessary to include loading effects in computing the gain of each stage.

Thus, the overall voltage gain of cascaded amplifier stages is the product of the voltage gains of the individual stages. (Of course, it is necessary to include loading effects in computing the gain of each stage. Notice that the input resistance of the second stage loads the first stage.)

Similarly, the overall current gain of a cascade connection of amplifiers is the product of the current gains of the individual stages. Furthermore, the overall power gain is the product of the individual power gains.

Example 11.2 Calculating Performance of Cascaded Amplifiers

Consider the cascade connection of the two amplifiers shown in Figure 11.6. Find the current gain, voltage gain, and power gain of each stage and for the overall cascade connection.

Solution Considering loading by the input resistance of the second stage, the voltage gain of the first stage is

$$A_{v1} = \frac{v_{o1}}{v_{i1}} = \frac{v_{i2}}{v_{i1}} = A_{voc1}\frac{R_{i2}}{R_{i2}+R_{o1}} = 150$$

where we have used the fact that $A_{voc1} = 200$, as indicated in Figure 11.6. Similarly,

$$A_{v2} = \frac{v_{o2}}{v_{i2}} = A_{voc2}\frac{R_L}{R_L+R_{o2}} = 50$$

The overall voltage gain is

$$A_v = A_{v1}A_{v2} = 7500$$

Because R_{i2} is the load resistance for the first stage, we can find the current gain of the first stage by use of Equation 11.3:

$$A_{i1} = A_{v1}\frac{R_{i1}}{R_{i2}} = 10^5$$

Similarly, the current gain of the second stage is found as

$$A_{i2} = A_{v2}\frac{R_{i2}}{R_L} = 750$$

The overall current gain is

$$A_i = A_{i1}A_{i2} = 75 \times 10^6$$

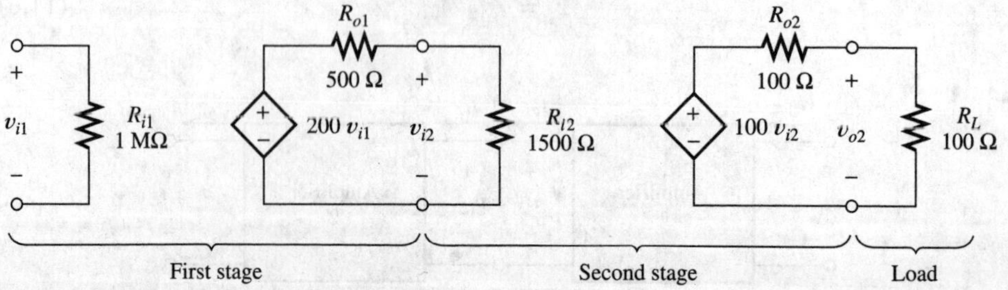

Figure 11.6 Cascaded amplifiers of Examples 11.2 and 11.3.

Now, the power gains can be found as

$$G_1 = A_{v1}A_{i1} = 1.5 \times 10^7$$
$$G_2 = A_{v2}A_{i2} = 3.75 \times 10^4$$

and

$$G = G_1G_2 = 5.625 \times 10^{11}$$ ■

Simplified Models for Cascaded Amplifier Stages

Sometimes, we will want to find a simplified model for a cascaded amplifier. The input resistance of the cascade is the input resistance of the first stage, and the output resistance of the cascade is the output resistance of the last stage. The open-circuit voltage gain of the cascade is computed with an open-circuit load on the last stage. However, loading effects of each stage on the preceding stage must be considered. Once the open-circuit voltage gain of the overall cascade connection is found, a simplified model can be drawn.

Simplified models can be found for cascaded amplifiers.

Example 11.3 Simplified Model for an Amplifier Cascade

Find the overall simplified model for the cascade connection of Figure 11.6.

Solution The voltage gain of the first stage, accounting for the loading of the second stage, is

First, determine the voltage gain of the first stage accounting for loading by the second stage.

$$A_{v1} = A_{voc1}\frac{R_{i2}}{R_{i2} + R_{o1}} = 150$$

With an open-circuit load, the gain of the second stage is

$$A_{v2} = A_{voc2} = 100$$

The overall open-circuit voltage gain is

The overall voltage gain is the product of the gains of the separate stages.

$$A_{voc} = A_{v1}A_{v2} = 15 \times 10^3$$

The input resistance of the cascade amplifier is

$$R_i = R_{i1} = 1\ \text{M}\Omega$$

and the output resistance is

The input impedance is that of the first stage, and the output impedance is that of the last stage.

$$R_o = R_{o2} = 100\ \Omega$$

The simplified model for the cascade is shown in Figure 11.7. ■

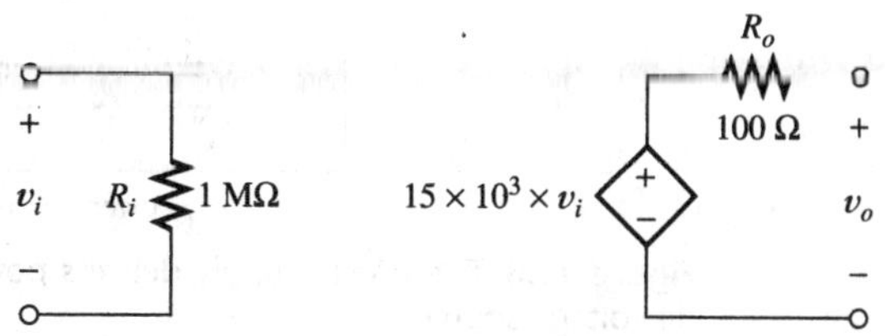

Figure 11.7 Simplified model for the cascaded amplifiers of Figure 11.6. See Example 11.3.

Exercise 11.4 Three amplifiers with the following characteristics are cascaded:

$$\text{Amplifier 1:} \quad A_{voc1} = 10, R_{i1} = 1\text{ k}\Omega, R_{o1} = 100\ \Omega$$
$$\text{Amplifier 2:} \quad A_{voc2} = 20, R_{i2} = 2\text{ k}\Omega, R_{o2} = 200\ \Omega$$
$$\text{Amplifier 3:} \quad A_{voc3} = 30, R_{i3} = 3\text{ k}\Omega, R_{o3} = 300\ \Omega$$

Find the parameters for the simplified model of the cascaded amplifier. Assume that the amplifiers are cascaded in the order 1, 2, 3.

Answer $R_i = 1\text{ k}\Omega, R_o = 300\ \Omega, A_{voc} = 5357.$ □

Exercise 11.5 Repeat Exercise 11.4 if the order of the amplifiers is 3, 2, 1.

Answer $R_i = 3\text{ k}\Omega, R_o = 100\ \Omega, A_{voc} = 4348.$ □

11.3 POWER SUPPLIES AND EFFICIENCY

Power is supplied to the internal circuitry of amplifiers from **power supplies.** The power supply typically delivers current from several dc voltages to the amplifier; an example configuration is shown in Figure 11.8. The average power supplied to the amplifier by each voltage source is the product of the average current and the voltage. The total power supplied is the sum of the powers supplied by each voltage source. For example, the total average power supplied to the amplifier of Figure 11.8 is

The number of terms in this equation depends on the number of supply voltages applied to the amplifier.

$$P_s = V_{AA}I_A + V_{BB}I_B \tag{11.7}$$

Notice that we have assumed that the current directions in the supply voltages are such that both sources deliver power to the amplifier. Rarely, a condition occurs for which some of the power taken from one supply source is returned to another source. We may only have a single supply voltage or there can be several, so the number of terms in a supply-power calculation such as Equation 11.7 is variable. It is customary

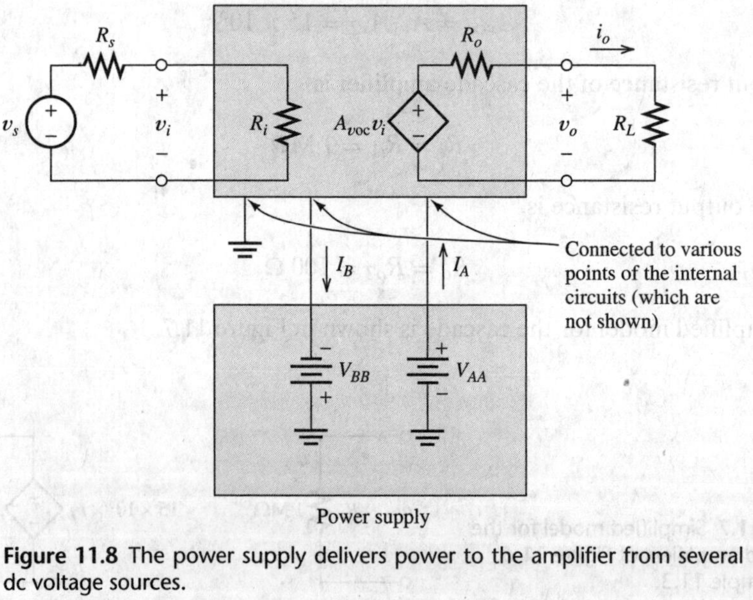

Figure 11.8 The power supply delivers power to the amplifier from several dc voltage sources.

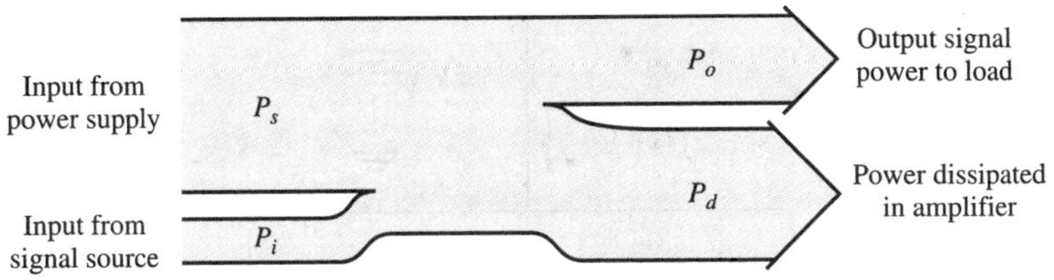

Figure 11.9 Illustration of power flow.

to use uppercase symbols with repeated uppercase subscripts, such as V_{CC}, for dc supply voltages in electronic circuits.

We have seen that the power gain of typical amplifiers can be very large. Thus, the output power delivered to the load is much greater than the power taken from the signal source. This additional power is taken from the power supply. Power taken from the power supply can also be **dissipated** as heat in the internal circuits of the amplifier. Such dissipation is an undesirable effect that we usually try to minimize when designing the internal circuitry of an amplifier.

Power flows into an amplifier from dc power supplies and from the signal source. Part of this power is delivered to the load as a useful signal, and part is dissipated as heat.

The sum of the power entering the amplifier from the signal source P_i and the power from the power supply P_s must be equal to the sum of the output power P_o and the power dissipated P_d:

$$P_i + P_s = P_o + P_d \tag{11.8}$$

This is illustrated in Figure 11.9. Often, the input power P_i from the signal source is insignificant compared with the other terms in this equation.

To summarize, we can view an amplifier as a system that takes power from the dc power supply and converts part of this power into output signal power. For example, a stereo audio system converts part of the power taken from the power supply into signal power that is finally converted to sound by the loudspeakers.

Efficiency

The **efficiency** η of an amplifier is the percentage of the power supplied that is converted into output power, or

$$\eta = \frac{P_o}{P_s} \times 100 \text{ percent} \tag{11.9}$$

Example 11.4 Amplifier Efficiency

Find the input power, output power, supply power, and power dissipated in the amplifier shown in Figure 11.10. Also, find the efficiency of the amplifier. (The values given in this example are typical of one channel of a stereo amplifier under high-output test conditions.)

Solution The average signal power delivered to the amplifier is given by

$$P_i = \frac{V_i^2}{R_i} = 10^{-11} \text{ W} = 10 \text{ pW}$$

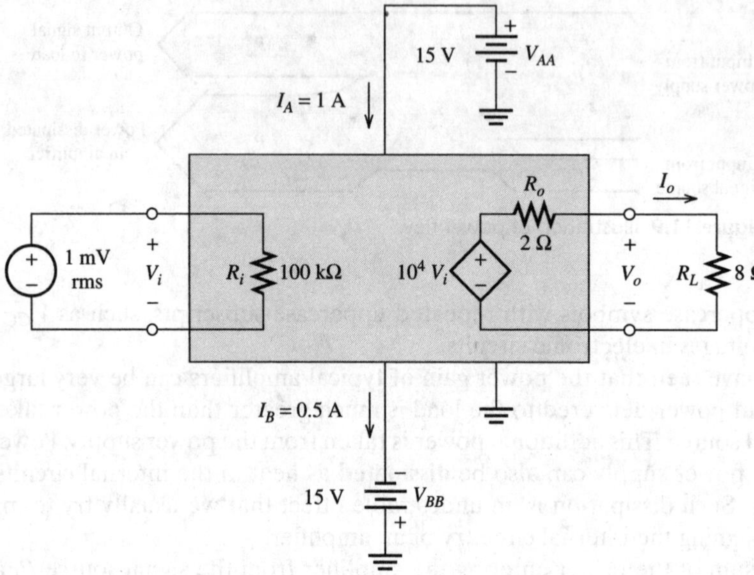

Figure 11.10 Amplifier of Example 11.4.

(Recall that 1 pW = 1 picowatt = 10^{-12} W.) The output voltage is

$$V_o = A_{voc} V_i \frac{R_L}{R_L + R_o} = 8 \text{ V rms}$$

Then, we find the average output power as

$$P_o = \frac{V_o^2}{R_L} = 8 \text{ W}$$

The supply power is given by

$$P_s = V_{AA} I_A + V_{BB} I_B = 15 + 7.5 = 22.5 \text{ W}$$

Notice that (as often happens) the power of the input signal is insignificant compared with the output and supply powers. The power dissipated as heat in the amplifier is

$$P_d = P_s + P_i - P_o = 14.5 \text{ W}$$

and the efficiency of the amplifier is

$$\eta = \frac{P_o}{P_s} \times 100 \text{ percent} = 35.6 \text{ percent}$$ ■

Exercise 11.6 A certain amplifier is supplied with 1.5 A from a 15-V supply. The output signal power is 2.5 W, and the input signal power is 0.5 W. Find the power dissipated in the amplifier and the efficiency.

Answer $P_d = 20.5$ W, $\eta = 11.1$ percent. □

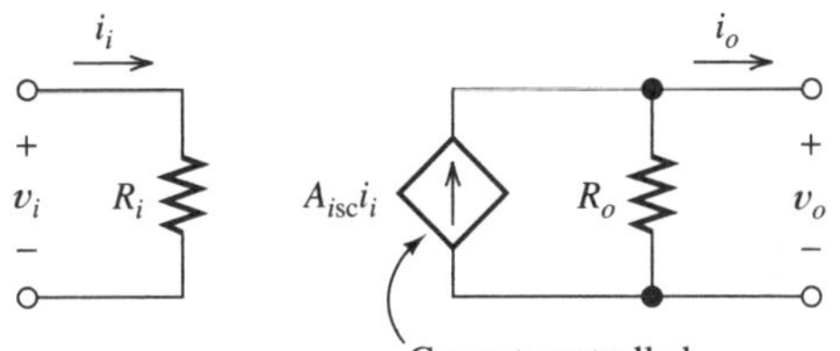

Figure 11.11 Current-amplifier model.

11.4 ADDITIONAL AMPLIFIER MODELS

Current-Amplifier Model

Until now, we have modeled amplifiers as shown in Figure 11.3, in which the gain property of the amplifier is represented by a voltage-controlled voltage source. An alternative model known as a **current amplifier** is shown in Figure 11.11. In this model, the gain property is modeled by a current-controlled current source. As before, the input resistance accounts for the current that the amplifier draws from the signal source. The output resistance is now in parallel with the controlled source and accounts for the fact that the amplifier cannot supply a fixed current to an arbitrarily high-load resistance.

If the load is a short circuit, no current flows through R_o, and the ratio of output current to input current is A_{isc}. For this reason, A_{isc} is known as the **short-circuit current gain**. An amplifier, initially modeled as a voltage amplifier, can also be modeled as a current amplifier. The input resistance and output resistance are the same for both models. The short-circuit current gain can be found from the voltage-amplifier model by connecting a short circuit to the output and computing the current gain.

A_{isc} is the current gain of the amplifier with the output short circuited.

Notice that we have converted the Thévenin circuit of the voltage-amplifier model to a Norton circuit in the current-amplifier model.

Example 11.5 Determining the Current-Amplifier Model from the Voltage-Amplifier Model

A certain amplifier is modeled by the voltage-amplifier model shown in Figure 11.12. Find the current-amplifier model.

Solution To find the short-circuit current gain, we connect a short circuit to the output terminals of the amplifier as shown in Figure 11.12. Then, we find that

Connect a short circuit across the output terminals and analyze the circuit to determine A_{isc}.

$$i_i = \frac{v_i}{R_i} \quad \text{and} \quad i_{osc} = \frac{A_{voc}v_i}{R_o}$$

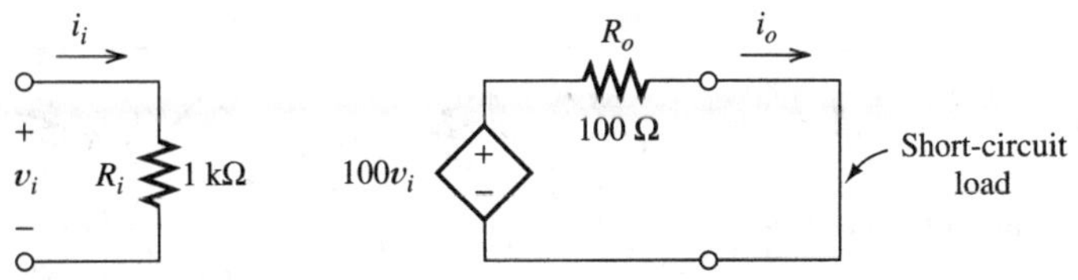

Figure 11.12 Voltage amplifier of Examples 11.5, 11.6, and 11.7.

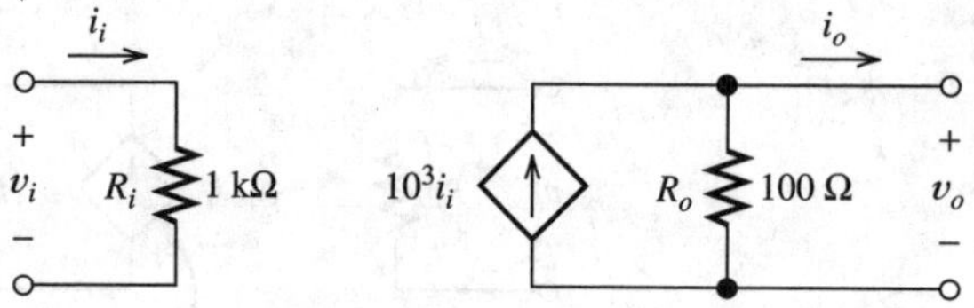

Figure 11.13 Current-amplifier model equivalent to the voltage-amplifier model of Figure 11.12.

The short-circuit current gain is

$$A_{isc} = \frac{i_{osc}}{i_i} = A_{voc}\frac{R_i}{R_o} = 10^3$$

The resulting current-amplifier model is shown in Figure 11.13. ■

Exercise 11.7 A certain amplifier modeled as a current amplifier has an input resistance of 1 kΩ, an output resistance of 20 Ω, and a short-circuit current gain of 200. Find the parameters for the voltage-amplifier model.
Answer $A_{voc} = 4, R_i = 1\ \text{k}\Omega, R_o = 20\ \Omega$. □

Transconductance-Amplifier Model

Another model for an amplifier, known as a **transconductance amplifier**, is shown in Figure 11.14. In this case, the gain is modeled by a voltage-controlled current source, and the gain parameter G_{msc} is called the **short-circuit transconductance gain**. G_{msc} is the ratio of the short-circuit output current i_{osc} to the input voltage v_i:

$$G_{msc} = \frac{i_{osc}}{v_i}$$

The units of transconductance gain are siemens. The input resistance and output resistance model the same effects as they do in the voltage-amplifier and current-amplifier models. A given amplifier can be modeled as a transconductance amplifier if the input resistance, output resistance, and short-circuit transconductance gain can be found.

The input resistance is the resistance seen looking into the input terminals. It has the same value for all models of a given amplifier. Similarly, the output resistance is the Thévenin resistance seen looking back into the output terminals and is the same for all the models.

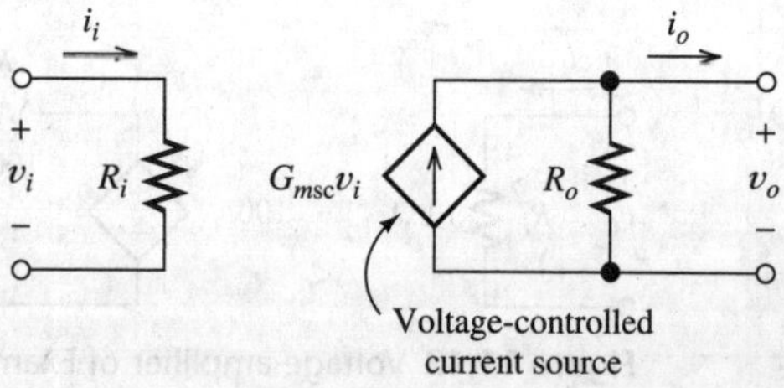

Figure 11.14 Transconductance-amplifier model.

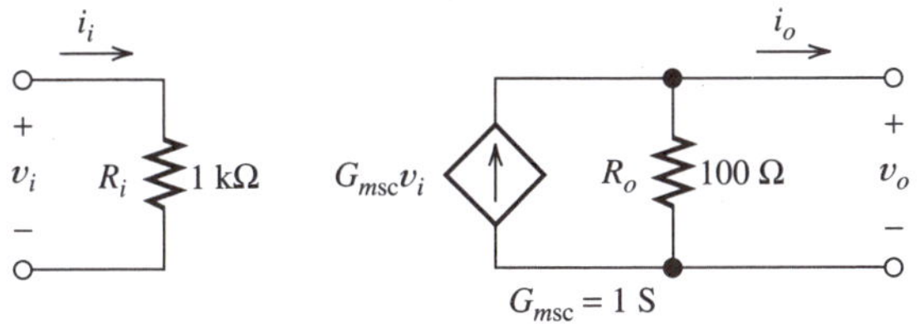

Figure 11.15 Transconductance-amplifier equivalent to the voltage amplifier of Figure 11.12. See Example 11.6.

Example 11.6 Determining the Transconductance-Amplifier Model

Find the transconductance model for the amplifier of Figure 11.12.

Solution The short-circuit transconductance gain is given by

$$G_{msc} = \frac{i_{osc}}{v_i}$$

The output current for a short-circuit load is

Connect a short circuit across the output terminals, and analyze the circuit to determine G_{msc}.

$$i_{osc} = \frac{A_{voc} v_i}{R_o}$$

Thus, we find that

$$G_{msc} = \frac{A_{voc}}{R_o} = 1.0 \text{ S}$$

The resulting amplifier model is shown in Figure 11.15. ■

Exercise 11.8 A current amplifier has an input resistance of 500 Ω, an output resistance of 50 Ω, and a short-circuit current gain of 100. Find the parameters for the transconductance-amplifier model.

Answer $G_{msc} = 0.2$ S, $R_i = 500$ Ω, and $R_o = 50$ Ω. □

Transresistance-Amplifier Model

Finally, we can model an amplifier as a **transresistance amplifier** as shown in Figure 11.16. In this case, the gain property is modeled by a current-controlled voltage source. The gain parameter R_{moc} is called the **open-circuit transresistance gain** and has units of ohms. It is the ratio of the open-circuit output voltage v_{ooc} to the input current i_i:

$$R_{moc} = \frac{v_{ooc}}{i_i}$$

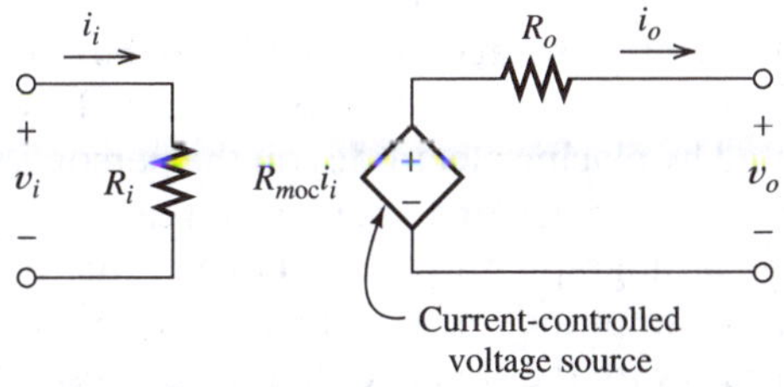

Figure 11.16 Transresistance-amplifier model.

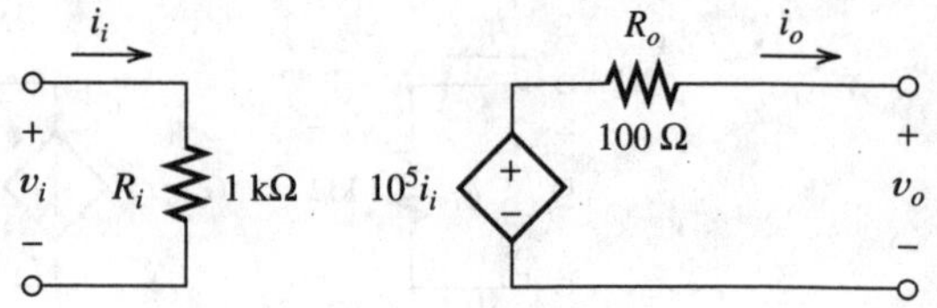

Figure 11.17 Transresistance amplifier that is equivalent to the voltage amplifier of Figure 11.12. See Example 11.7.

The values of the input resistance and output resistance are the same as in any of the other amplifier models.

Example 11.7 Determining the Transresistance-Amplifier Model

Find the transresistance-amplifier model for the amplifier shown in Figure 11.12.

Open circuit the output terminals, and analyze the circuit to determine R_{moc}.

Solution With an open-circuit load, the output voltage is

$$v_{ooc} = A_{voc} v_i$$

and the input current is

$$i_i = \frac{v_i}{R_i}$$

Thus, we find the transresistance gain as

$$R_{moc} = \frac{v_{ooc}}{i_i} = A_{voc} R_i = 100 \text{ k}\Omega$$

The resulting transresistance model of the amplifier is shown in Figure 11.17. ■

Exercise 11.9 An amplifier has an input resistance of 1 MΩ, an output resistance of 10 Ω, and $G_{msc} = 0.05$ S. Find R_{moc} for this amplifier.

Answer $R_{moc} = 500$ kΩ. □

We have seen that an amplifier can be modeled by any of the four models: voltage amplifier, current amplifier, transconductance amplifier, or transresistance amplifier. However, in cases for which either of the resistances (input or output) is zero or infinity, it is not possible to make conversions to all the models because the gain parameter is not defined for all the models. For example, if $R_i = 0$, then $v_i = 0$, and the voltage gain $A_{voc} = v_o/v_i$ is not defined.

11.5 IMPORTANCE OF AMPLIFIER IMPEDANCES IN VARIOUS APPLICATIONS

Applications Calling for High or Low Input Impedance

Some applications call for amplifiers with high input impedance, while others call for low input impedance.

Sometimes, we have an application for an amplifier that calls for the internal voltage produced by the source to be amplified. For example, an electrocardiograph amplifies and records the small voltages generated by a person's heart. These voltages are detected by placing electrodes on the person's skin. The impedance of the electrodes is variable from person to person and can be quite high. If the input impedance of the electrocardiograph is low, a variable reduction in voltage occurs because of loading. Thus, the amplitude of the signal can be affected by the contact resistance of the electrodes with the skin and therefore does not truly represent the electrical activity

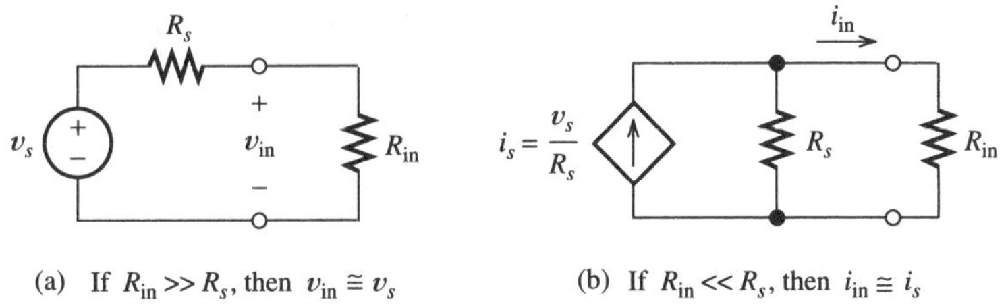

Figure 11.18 If we want to sense the open-circuit voltage of a source, the amplifier should have a high input resistance, as in (a). To sense the short-circuit current of the source, low input resistance is called for, as in (b).

of the heart. On the other hand, if the input impedance of the electrocardiograph is much higher than the source impedance, the actual voltage produced by the heart appears at the input terminals. Thus, the input impedance of an electrocardiograph amplifier should be very high.

Other applications call for the amplifier to respond to the short-circuit current of a source. Then a very low input impedance is needed. An example is an electronic ammeter inserted in series with a circuit to measure current. Usually, we do not want the ammeter to change the current that is being measured. This is accomplished by designing the ammeter to have a low enough input impedance so that it does not change the impedance of the circuit significantly.

To summarize, if the input impedance of an amplifier is much higher than the internal impedance of the source, the voltage produced across the input terminals is nearly the same as the internal source voltage. This is illustrated in Figure 11.18(a). On the other hand, if the input impedance is very low, the input current is nearly equal to the short-circuit current of the source. This is illustrated in Figure 11.18(b).

Applications Calling for High or Low Output Impedance

Diverse requirements for output impedance also occur. For example, we could have an audio amplifier that supplies background music to loudspeakers in many rooms of an office building, as shown in Figure 11.19. A switch is provided so that each loudspeaker can be turned off independent of the others (by opening its switch). Therefore, the load impedance presented to the amplifier is quite variable, depending on the number of loudspeakers turned on. If the amplifier output impedance is high compared with the load, the voltage supplied depends on the load impedance. Thus, as loudspeakers are turned off, the voltage applied to the others becomes higher, resulting in louder music. This effect could be undesirable. On the other hand, if the output impedance of the amplifier is very low compared with the load impedance,

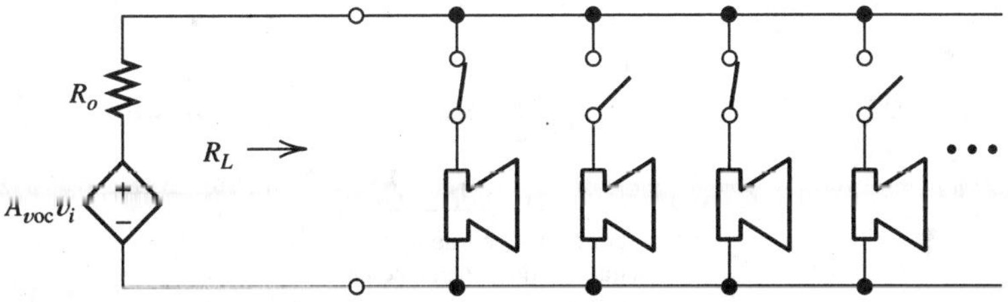

Figure 11.19 If the amplifier output resistance R_o is much less than the (lowest) load resistance, the load voltage is nearly independent of the number of switches closed.

the output voltage is nearly independent of the load. Thus, in this situation, a low output impedance is better.

Some applications call for amplifiers with high output impedance while other applications call for low output impedance.

Another example occurs in optical communication systems, where a *light-emitting diode* (LED) can be used to produce a light wave with intensity proportional to a message signal such as a voice waveform. Over a certain range of operation, the intensity of the light output of an LED is proportional to the current through it. Because LEDs have a nonlinear relationship between voltage and current, light intensity is *not* proportional to the voltage across the LED. Thus, it is desirable to force a current proportional to the message waveform to flow through the diode. This can be achieved by designing an amplifier with a very high output impedance to drive the LED. (On the other hand, if a very low output impedance were used, the voltage supplied to the diode would be proportional to the input signal to the amplifier, but because of the nonlinear relationship between current and voltage for the diode, the light output would no longer be proportional to the message.)

To summarize, we can force a desired voltage waveform to appear across a variable load by designing the amplifier to have a very low output impedance compared with the load impedance. On the other hand, we can force a given current waveform through a variable load by designing the amplifier to have a very high output impedance compared with the load impedance.

Applications Calling for a Particular Impedance

Some applications call for amplifiers that have specific input and output impedances.

Not all applications call for amplifiers with either very small or very large impedances. For example, consider an amplifier whose input is connected to a source by a **transmission line** as shown in Figure 11.20. An example of a transmission line with which you may be familiar is *coaxial cable*, commonly used to connect TV sets to cable systems or to digital television (DTV) antennas. Each type of transmission line has a **characteristic impedance**, which is typically 75 Ω for the coaxial line used in television applications. Unless the transmission line is terminated in its characteristic impedance, a signal traveling along a transmission line is partially reflected and travels back toward the source. This is illustrated in Figure 11.20. When connecting a signal source, such as a set-top box or antenna, to a television set, the reflections from the TV set can be reflected again at the source, so the signal arrives a second time at the set. These extra signals are delayed because of the round-trip travel along the transmission line and can cause degradation of picture quality. In the case of analog signals, the effect of the reflection is a faint image known as a "ghost" slightly to the right-hand side of the main image. Digital signals can become so corrupted

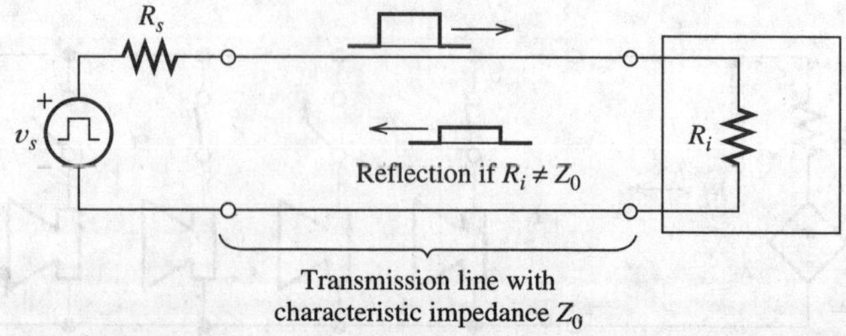

Figure 11.20 To avoid reflections, the amplifier input resistance R_i should equal the characteristic resistance Z_0 of the transmission line.

by these reflections that the set cannot decode them. Therefore, it is important for the input impedance of the television set and the output impedance of the source to be nearly equal to the characteristic impedance of the transmission line so that significant reflections do not occur.

The output impedance of an audio amplifier is another situation that sometimes calls for an intermediate value. The frequency response of a loudspeaker depends on the output impedance of the amplifier driving it. Thus, if high fidelity is a primary consideration, the amplifier should be designed so that it has the output impedance that gives the most nearly constant response versus frequency.

11.6 IDEAL AMPLIFIERS

We have seen in Section 11.5 that certain applications call for amplifiers with very high or very low input impedance (compared with the source impedance) and very high or very low output impedance (compared with the load). Such amplifiers can be classified as follows:

1. An **ideal voltage amplifier** senses the open-circuit voltage of the source and produces an amplified voltage across the load—independent of the load impedance. Thus, the ideal voltage amplifier has infinite input impedance (so the open-circuit voltage appears across the input terminals) and zero output impedance (so the output voltage is independent of the load impedance).
2. An **ideal current amplifier** senses the short-circuit current of the source and forces an amplified version of this current to flow through the load. Thus, the ideal current amplifier has zero input impedance and infinite output impedance.
3. An **ideal transconductance amplifier** senses the open-circuit voltage of the source and forces a current proportional to this voltage to flow through the load. Thus, the ideal transconductance amplifier has infinite input impedance and infinite output impedance.
4. An **ideal transresistance amplifier** senses the short-circuit current of the source and causes a voltage proportional to this current to appear across the load. Thus, the ideal transresistance amplifier has zero input impedance and zero output impedance. Table 11.1 shows the input impedance, output impedance, and gain parameter for each type of ideal amplifier.

According to their input and output impedances, ideal amplifiers can be classified into four types. These are the ideal voltage amplifier, the ideal current amplifier, the ideal transconductance amplifier, and the ideal transresistance amplifier. The best amplifier type to select depends on the application.

Classifying Real Amplifiers

In practice, amplifiers do not have either zero or infinite impedances. However, real amplifiers can often be classified as approximately ideal amplifiers. For example, if the input impedance is very large (compared with the source impedance) and the

Table 11.1. Characteristics of Ideal Amplifiers

Amplifier Type	Input Impedance	Output Impedance	Gain Parameter
Voltage	∞	0	A_{voc}
Current	0	∞	A_{isc}
Transconductance	∞	∞	G_{msc}
Transresistance	0	0	R_{moc}

output impedance is very small (compared with the load), we have an approximate ideal voltage amplifier.

Notice that a given amplifier cannot be classed as an approximate ideal amplifier unless the source and load impedances to be encountered are known in advance. For example, an amplifier with an input impedance of 1000 Ω and an output impedance of 100 Ω would be classed as an approximate ideal voltage amplifier if the source impedances to be encountered are much less than 1000 Ω and the load impedances are much greater than 100 Ω. On the other hand, if the source impedances are on the order of 1 MΩ and the load impedances are on the order of 1 Ω, the same amplifier would be properly classed as an approximate ideal current amplifier.

The proper classification of a given amplifier depends on the ranges of source and load impedances with which the amplifier is used.

In general, the "middle range" of impedances in low-power electronic circuits runs from 1 kΩ to 100 kΩ. Impedances less than 100 Ω are usually considered to be "small," and impedances greater than 1 MΩ are classed as "large." Thus, we would usually be inclined to classify an amplifier with an input impedance of 10 Ω and an output impedance of 2 MΩ as an approximate ideal current amplifier. However, we might want to change this classification, depending on the actual load and source impedances.

Exercise 11.10 A certain amplifier has an input resistance of $R_i = 1$ kΩ and an output resistance of $R_o = 1$ kΩ. R_s is the source resistance, and R_L is the load. Classify the amplifier if: **a.** R_s is less than 10 Ω and R_L is greater than 100 kΩ; **b.** R_s is greater than 100 kΩ and R_L is less than 10 Ω; **c.** R_s is less than 10 Ω and R_L is less than 10 Ω; **d.** R_s is greater than 100 kΩ and R_L is greater than 100 kΩ; **e.** R_s is approximately 1 kΩ and R_L is less than 10 Ω.

Answer **a.** Approximate ideal voltage amplifier; **b.** approximate ideal current amplifier; **c.** approximate ideal transconductance amplifier; **d.** approximate ideal transresistance amplifier; **e.** for this source resistance, the amplifier does not fit into any ideal amplifier category. □

Exercise 11.11 A particular transducer is to be used in measuring liquid level in a chemical process. The short-circuit current of the transducer is proportional to the liquid level. (However, the open-circuit voltage of the transducer is nearly independent of the level.) An amplifier is needed to deliver a voltage signal proportional to liquid level to a resistive load that may vary in value between 1 kΩ and 10 kΩ. What type of ideal amplifier is needed?

Answer A transresistance amplifier is needed, because we want the input impedance of the amplifier to be small enough so it responds to the short-circuit transducer current. Furthermore, to deliver an output voltage independent of the load, the output impedance of the amplifier must be very small compared with the load impedance. □

11.7 FREQUENCY RESPONSE

So far, we have considered the gain parameter of an amplifier to be a constant. However, if we apply a variable-frequency sinusoidal input signal to an amplifier, we will find that gain is a function of frequency. Moreover, the amplifier affects the phase as well as the amplitude of the sinusoid. Therefore, we now give a more general definition of amplifier gain. We define complex gain to be the ratio of the phasor for the output signal to the phasor for the input signal:

$$A_v = \frac{\mathbf{V}_o}{\mathbf{V}_i} \tag{11.10}$$

Complex voltage gain is the output-voltage phasor divided by the input-voltage phasor. Complex gain has magnitude and phase.

We use uppercase bold symbols to stand for the phasors of the input and output voltages. Similarly, we define complex current gain, transconductance gain, and transresistance gain as the ratio of the appropriate phasor quantities. We have used the term *complex gain* to emphasize the fact that these gains have both magnitude and phase. Shortly, we will drop the word *complex*, for simplicity. Actually, complex gain is the same concept as that of a transfer function, which we discussed in Chapter 6. Recall that to express a transfer function in decibels, we take 20 times the common logarithm of its magnitude.

Example 11.8 Determining Complex Gain

The input voltage to a certain amplifier is

$$v_i(t) = 0.1\cos(2000\pi t - 30°)$$

and the output voltage is

$$v_o(t) = 10\cos(2000\pi t + 15°)$$

Find the complex voltage gain of the amplifier and express the magnitude of the gain in decibels.

Solution Recall that the phasor for the input voltage is a complex number whose magnitude is the peak value of the sinusoidal signal and whose angle is the phase angle of the sinusoidal signal. Thus,

$$\mathbf{V}_i = 0.1\angle{-30°}$$

Similarly,

$$\mathbf{V}_o = 10\angle{15°}$$

Now, we can find the complex voltage gain as

$$A_v = \frac{\mathbf{V}_o}{\mathbf{V}_i} = \frac{10\angle{15°}}{0.1\angle{-30°}}$$
$$= 100\angle{45°}$$

The meaning of this complex voltage gain is that the output signal is 100 times larger in amplitude than the input signal. Furthermore, the output signal is phase shifted by 45° relative to the input signal.

To express gain in decibels, we first find the magnitude of the gain by dropping the angle and then compute decibel gain:

$$|A_v|_{\text{dB}} = 20\log|A_v| = 20\log(100) = 40 \text{ dB}$$

■

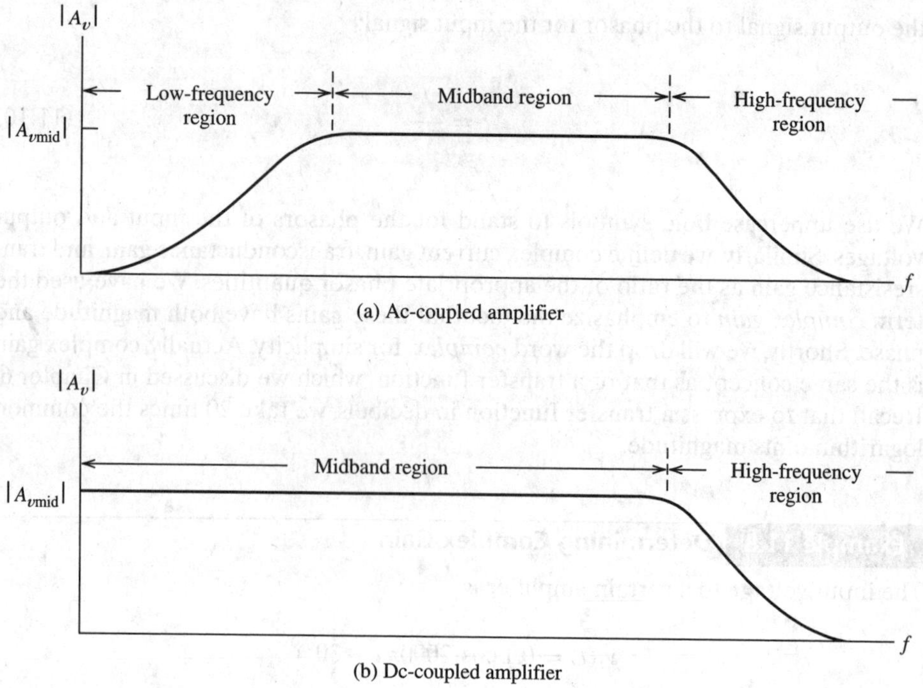

Figure 11.21 Gain magnitude versus frequency.

Gain as a Function of Frequency

Many amplifiers have a midband range in which the gain magnitude is constant.

If we plot the magnitude of the gain of a typical amplifier versus frequency, a plot such as the one shown in Figure 11.21 results. Notice that the gain magnitude is constant over a wide range of frequencies known as the **midband region**.

AC Coupling versus Direct Coupling

In some cases, such as the one shown in Figure 11.21(a), the gain drops to zero at dc (zero frequency). Such amplifiers are said to be **ac coupled** because only ac signals are amplified. These amplifiers are often constructed by cascading several amplifier circuits or stages that are connected together by **coupling capacitors** so that the dc voltages of the amplifier circuits do not affect the signal source, adjacent stages, or the load. This is illustrated in Figure 11.22. (Sometimes, transformers are used to couple individual stages together, which also leads to an ac-coupled amplifier with zero gain at dc. Transformers are discussed in Chapter 15.)

Amplifiers may be ac coupled or dc coupled.

Other amplifiers have constant gain all the way down to dc, as shown in Figure 11.21(b). They are said to be **dc coupled**, or **direct coupled**. Amplifiers that are realized as integrated circuits are often dc coupled because the capacitors or transformers needed for ac coupling cannot be fabricated in integrated form.

Audio amplifiers are almost always ac coupled because audible sounds span the frequency range from about 20 Hz to 15 kHz. Therefore, there is no need to provide gain down to dc, and furthermore, it is not desirable to apply dc voltages to the loudspeakers.

Electrocardiograph amplifiers are deliberately ac coupled because a dc voltage of nearly a volt often occurs at the input due to electrochemical potentials developed by the electrodes. The ac signal generated by the heart is on the order of 1 mV,

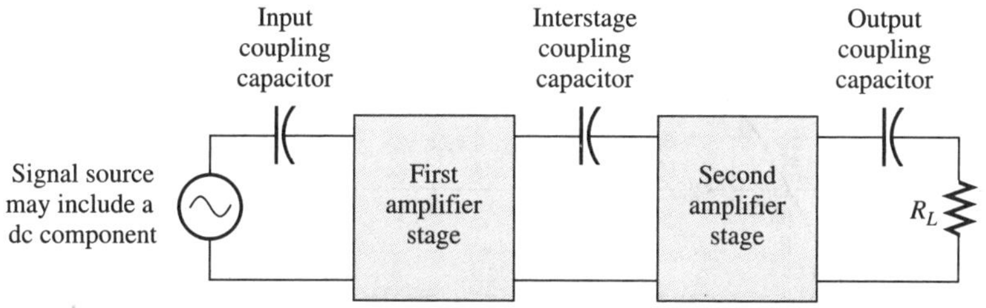

Figure 11.22 Capacitive coupling prevents a dc input component from affecting the first stage, dc voltages in the first stage from reaching the second stage, and dc voltages in the second stage from reaching the load.

and therefore, the gain of the amplifier is high—typically, 1000 or more. A 1-V dc input would cause the amplifier to try to produce an output of 1000 V. It would be difficult (and highly undesirable) to design an amplifier capable of such large outputs. Therefore, it is necessary to ac couple the input circuit of an electrocardiograph to prevent the dc component from overloading the amplifier.

Amplifiers for (analog) video signals need to be dc coupled because video signals have frequency components from dc into the MHz range. Dark pictures result in a different dc component than bright pictures. To obtain pictures with the proper brightness, it is necessary to use a dc-coupled amplifier to preserve the dc component.

High-Frequency Region

Gain magnitude declines for all amplifiers at sufficiently high frequencies.

As indicated in Figure 11.21(a) and (b), the gain of an amplifier always drops off at high frequencies. This is caused either by small amounts of capacitance in parallel with the signal path or by small inductances in series with the signal path in the amplifier circuitry, as illustrated in Figure 11.23. Recall that the impedance of a capacitor is inversely proportional to frequency, resulting in an effective short circuit at sufficiently high frequencies. The impedance of an inductor is proportional to frequency, so it becomes an open circuit at very high frequencies.

Some of these small capacitances occur because of stray wiring capacitance between signal-carrying conductors and ground. Other capacitances are integral parts of the active devices (transistors) necessary for amplification. Small inductances result from the magnetic fields surrounding the conductors in the circuit. For example, a critically placed piece of wire one-half inch long can have enough inductance to limit severely the frequency response of an amplifier intended to operate at several gigahertz.

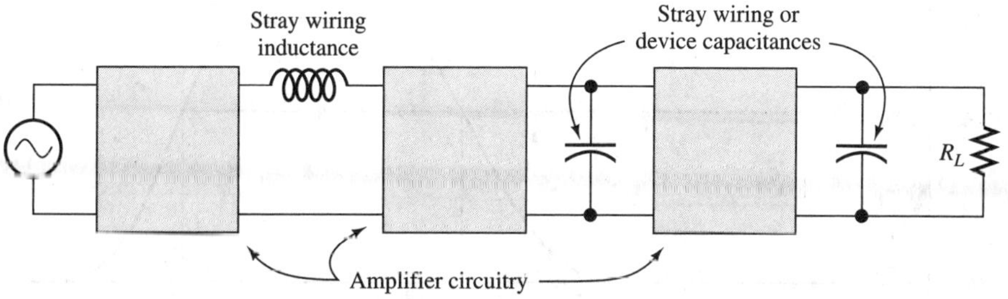

Figure 11.23 Capacitance in parallel with the signal path and inductance in series with the signal path reduce the gain in the high-frequency region.

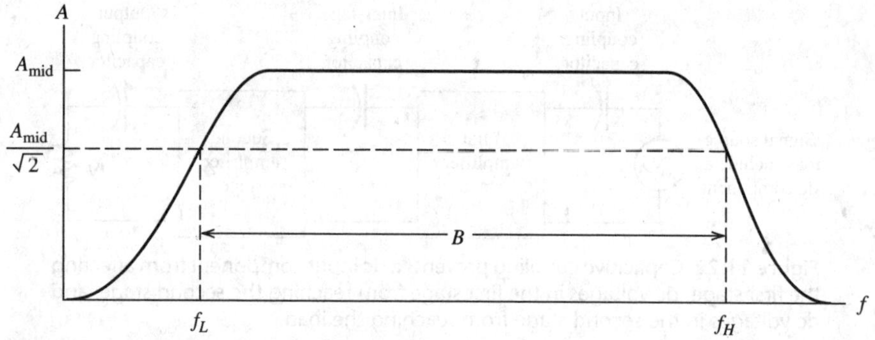

Figure 11.24 Gain versus frequency for a typical amplifier showing the upper and lower half-power (3-dB) frequencies (f_H and f_L) and the half-power bandwidth B.

Half-Power Frequencies and Bandwidth

At the half-power frequencies, voltage-gain magnitude is $1/\sqrt{2}$ times the midband value.

Usually, we specify the approximate useful frequency range of an amplifier by giving the frequencies for which the voltage (or current) gain magnitude is $1/\sqrt{2}$ times the midband gain magnitude. These are known as the **half-power frequencies** because the output power level is half the value for the midband region if a constant-amplitude variable-frequency input test signal is used. Expressing the factor $1/\sqrt{2}$ in decibels, we have $20 \log(1/\sqrt{2}) = -3.01$ dB. Thus, at the half-power frequencies, the voltage (or current) gain is approximately 3 dB lower than the midband gain. The bandwidth B of an amplifier is the distance between the half-power frequencies. These definitions are illustrated in Figure 11.24.

Wideband versus Narrowband Amplifiers

Amplifiers that are either dc coupled or have a lower half-power frequency that is a small fraction of the upper half-power frequency are called **wideband** or **baseband amplifiers**. Wideband amplifiers are used for signals that occupy a wide range of frequencies, such as audio signals (20 Hz to 15 kHz) or video signals (dc into the MHz range).

On the other hand, the frequency response of an amplifier is sometimes deliberately limited to a small bandwidth compared with the center frequency. Such an amplifier is called a **narrowband** or **bandpass amplifier**. The gain versus frequency response of a bandpass amplifier is shown in Figure 11.25. Bandpass amplifiers are

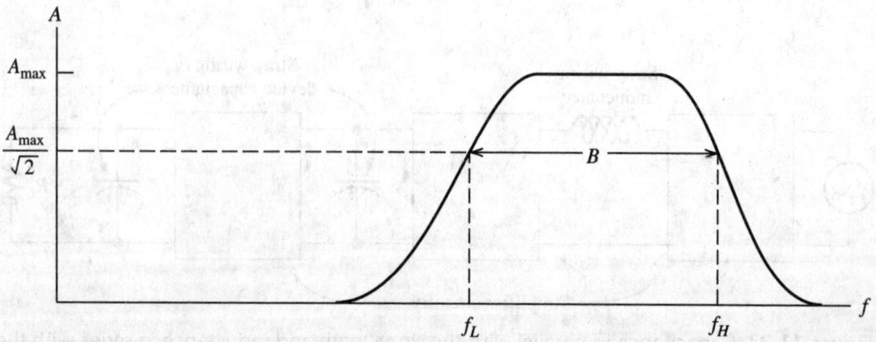

Figure 11.25 Gain magnitude versus frequency for a bandpass amplifier.

used in radio receivers, where it is desired to amplify the signal from one transmitter and reject the signals from other transmitters in adjacent frequency ranges.

11.8 LINEAR WAVEFORM DISTORTION

Amplitude Distortion

If the gain of an amplifier has a different magnitude for the various frequency components of the input signal, a form of distortion known as **amplitude distortion** occurs.

Audio systems often suffer from amplitude distortion because the amplifier, and particularly the loudspeakers, tend to reduce the amplitudes of the high-pitched and low-pitched components relative to the midband components. This is especially true for telephone systems. Hence, the music we hear while on hold is of poor quality.

Example 11.9 Amplitude Distortion

The input signal to a certain amplifier contains two frequency components and is given by

$$v_i(t) = 3\cos(2000\pi t) - 2\cos(6000\pi t)$$

The gain of the amplifier at 1000 Hz is $10\angle 0^\circ$, and the gain at 3000 Hz is $2.5\angle 0^\circ$. Plot the input waveform and output waveform to scale versus time.

Solution The first term of the input signal is at a frequency of 1000 Hz, so it experiences a gain of $10\angle 0^\circ$, whereas the second term of the input is at a frequency of 3000 Hz, so the gain for it is $2.5\angle 0^\circ$. Applying these gains and phase shifts to the terms of the input signal, we find the output:

$$v_o(t) = 30\cos(2000\pi t) - 5\cos(6000\pi t)$$

Plots of the input and output waveforms are shown in Figure 11.26. Notice that the output waveform has a different shape than the input waveform because of amplitude distortion. ∎

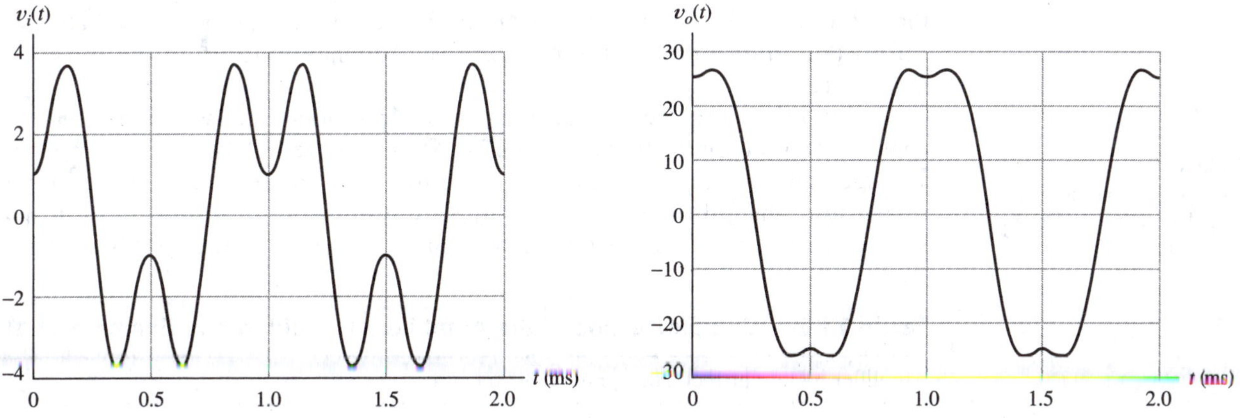

(a) Input waveform

(b) Output distorted because of unequal gain magnitude for various frequency components

Figure 11.26 Linear amplitude distortion. See Example 11.9.

Phase Distortion

If the phase shift of an amplifier is not proportional to frequency, **phase distortion** occurs. Zero phase at all frequencies results in an output waveform identical to the input. On the other hand, if the phase shift of the amplifier is proportional to frequency, the output waveform is a time-shifted version of the input. However, we do not say that distortion has occurred because the shape of the waveform is unchanged. If phase is not proportional to frequency, the waveform shape is changed in passing through the amplifier and phase distortion has occurred.

Example 11.10 Phase Distortion

Suppose that the input signal given by

$$v_i(t) = 3\cos(2000\pi t) - \cos(6000\pi t)$$

is applied to the inputs of three amplifiers having the gains shown in Table 11.2. Find and plot the output of each amplifier.

Solution Applying the gains and phase shifts to the input signal, we find the output signals for the amplifiers to be

$$v_{oA}(t) = 30\cos(2000\pi t) - 10\cos(6000\pi t)$$

$$v_{oB}(t) = 30\cos(2000\pi t - 45°) - 10\cos(6000\pi t - 135°)$$

$$v_{oC}(t) = 30\cos(2000\pi t - 45°) - 10\cos(6000\pi t - 45°)$$

Plots of the output waveforms are shown in Figure 11.27. Amplifier A produces an output waveform identical to the input, and amplifier B produces an output waveform identical to the input, except for a time delay. For amplifier A, the phase shift is zero for both frequency components, whereas the phase shift of amplifier B is proportional to frequency. (The phase shift for the 3000-Hz component is three times the phase shift for the 1000-Hz component.) Amplifier C produces a distorted output waveform because its phase response is not proportional to frequency. ∎

Amplitude and phase distortion are sometimes called **linear distortion** because they occur even though the amplifier is linear (i.e., obeys superposition). Later, we will see that another type of distortion, known as nonlinear distortion, can also occur in amplifiers.

If adding input signals always corresponds to adding the output signals, an amplifier is said to be linear.

Let us briefly consider the meaning of superposition, which we discussed in the context of resistive circuits in Section 2.7 starting on page 119. For a given amplifier, suppose that the input $v_{\text{in}A}$ results in the output v_{oA} and input $v_{\text{in}B}$ results in the output v_{oB}. Then, if the input $v_{\text{in}A} + v_{\text{in}B}$ always produces the output $v_{oA} + v_{oB}$, we say that the amplifier obeys superposition or is linear. In other words, if adding

Table 11.2. Complex Gains of the Amplifiers Considered in Example 11.10

Amplifier	Gain at 1000 Hz	Gain at 3000 Hz
A	$10\angle 0°$	$10\angle 0°$
B	$10\angle -45°$	$10\angle -135°$
C	$10\angle -45°$	$10\angle -45°$

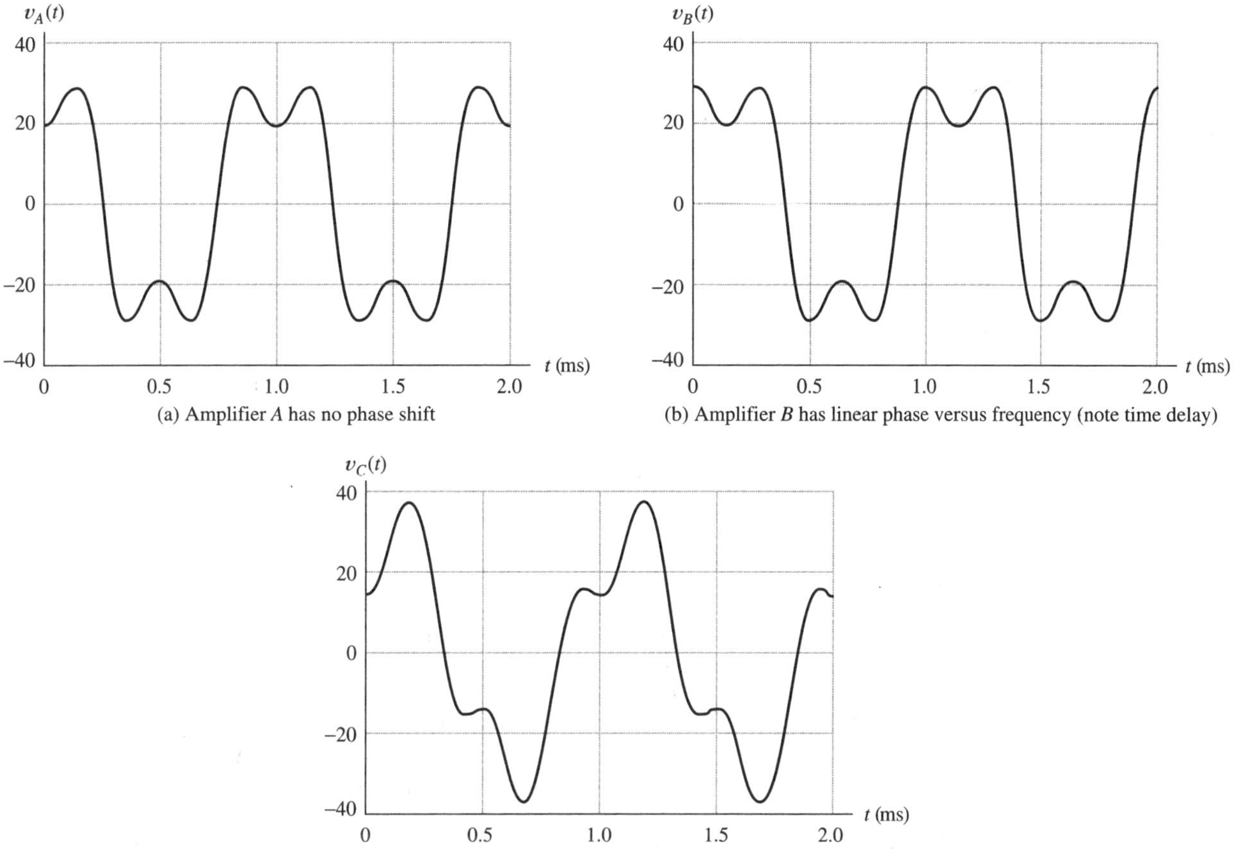

(a) Amplifier A has no phase shift

(b) Amplifier B has linear phase versus frequency (note time delay)

(c) Amplifier C has phase that is not proportional to frequency (note waveform distortion)

Figure 11.27 Effect of amplifier phase response. See Example 11.10. [*Note:* Input waveform has the same shape as $v_A(t)$.]

input signals always corresponds to adding the output signals, an amplifier is said to be linear. For example, an amplifier for which $v_o(t) = 10v_{\text{in}}(t)$ is linear because $10(v_{\text{in}A} + v_{\text{in}B}) = 10v_{\text{in}A} + 10v_{\text{in}B}$. However, an amplifier for which $v_o(t) = [v_{\text{in}}(t)]^2$ is not linear because $(v_{\text{in}A} + v_{\text{in}B})^2 \neq v_{\text{in}A}^2 + v_{\text{in}B}^2$.

Requirements for Distortionless Amplification

To avoid linear waveform distortion, an amplifier should have constant gain magnitude and a phase response that is linear versus frequency for the range of frequencies contained in the input signal. Of course, departure from these requirements outside the frequency range of the input signal components does not result in distortion. These requirements for distortionless amplification are illustrated in Figure 11.28.

To avoid linear waveform distortion, an amplifier should have constant gain magnitude and a phase response that is linear versus frequency for the range of frequencies contained in the input signal.

In the examples we have given, the input signals contained only a few components at specific frequencies. However, most signals of interest in electronic systems contain components spread over a continuous range of frequencies. For example, audio signals contain components from about 20 Hz to about 15 kHz. Thus, we require an audio amplifier to have nearly constant gain magnitude over that range. (However, since it turns out that the ear is not sensitive to phase distortion, we would not require the phase response of an audio amplifier to be proportional to frequency.)

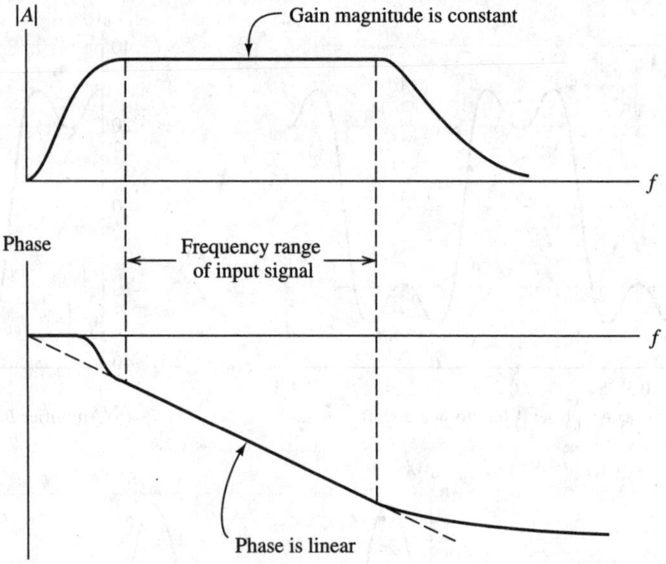

Figure 11.28 Linear distortion does not occur if the gain magnitude is constant and the phase is proportional to frequency over the frequency range of the input signal.

Analog video signals contain significant frequency components from dc to several MHz. Since the shape of the waveform ultimately determines the brightness of various points in the picture, either phase distortion or amplitude distortion would affect the image. Therefore, we require the gain magnitude to be constant and the phase response to be proportional to frequency over the stated range for a video amplifier.

Definition of Gain Revisited

As a final comment, recall that we originally defined the gain of an amplifier to be the ratio of the output signal to the input:

$$A_v = \frac{v_o(t)}{v_i(t)}$$

However, if linear waveform distortion occurs (or even a time delay), the ratio of output to input is a function of time, rather than a constant. Thus, we should not try to find the gain of an amplifier by taking the ratio of the instantaneous output and input. Instead, we recognize that gain is a function of frequency and take the ratio of the phasors for a sinusoidal input signal to find the (complex) gain for each frequency.

Exercise 11.12 Suppose that an input signal is given by

$$v_i(t) = \sin(1000\pi t) + \cos(2000\pi t) + 2\cos(3000\pi t)$$

and the gain of an amplifier at 1000 Hz is $5\angle 30^\circ$. What are the required amplifier gain and phase shift at the frequencies of the other components if both types of linear waveform distortion are to be avoided?

Answer $5\angle 15^\circ$ for the 500-Hz component and $5\angle 45^\circ$ for the 1500-Hz component. □

Exercise 11.13 The output of a certain amplifier is given by $v_o(t) = 10v_{\text{in}}(t - 0.01)$. Consider a sinusoidal input signal $v_{\text{in}}(t) = V_m \cos(\omega t)$, and find the complex gain (magnitude and phase) as a function of ω.

Answer $10\angle -0.01\omega$. □

11.9 PULSE RESPONSE

Often, we need to amplify a pulse signal such as the one shown in Figure 11.29(a). Pulses contain components spread over a wide range of frequencies; therefore, amplification of pulses calls for a wideband amplifier. A typical amplified output pulse is shown in Figure 11.29(b). The output waveform differs from the input in several important respects: The pulse displays **overshoot** and **ringing**, the leading and trailing edges are gradual rather than abrupt, and if the amplifier is ac coupled, the top of the output pulse is **tilted**.

Rise Time

The gradual rise of the leading edge of the amplifier response is often quantified by giving the **rise time** t_r, which is the time interval between the point t_{10} at which the amplifier achieves 10 percent of the eventual output amplitude and the point t_{90} at which the output is 90 percent of the final value. This is illustrated in Figure 11.30.

The rounding of the leading edge can be attributed to the roll-off of gain in the high-frequency region. A rule-of-thumb relationship between the half-power bandwidth B and the rise time t_r of a wideband amplifier is

This approximate relationship is very useful in estimating bandwidth given rise time or vice versa.

$$t_r \cong \frac{0.35}{B} \tag{11.11}$$

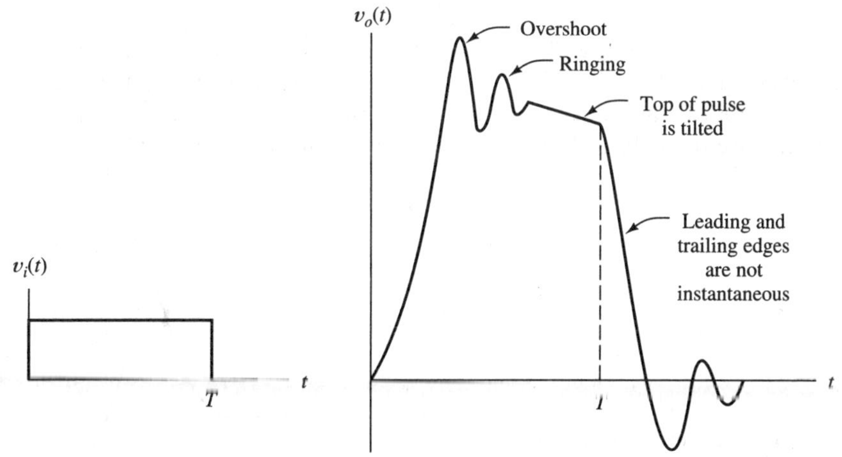

(a) Input (b) Output

Figure 11.29 Input pulse and the corresponding output of a typical ac-coupled broadband amplifier.

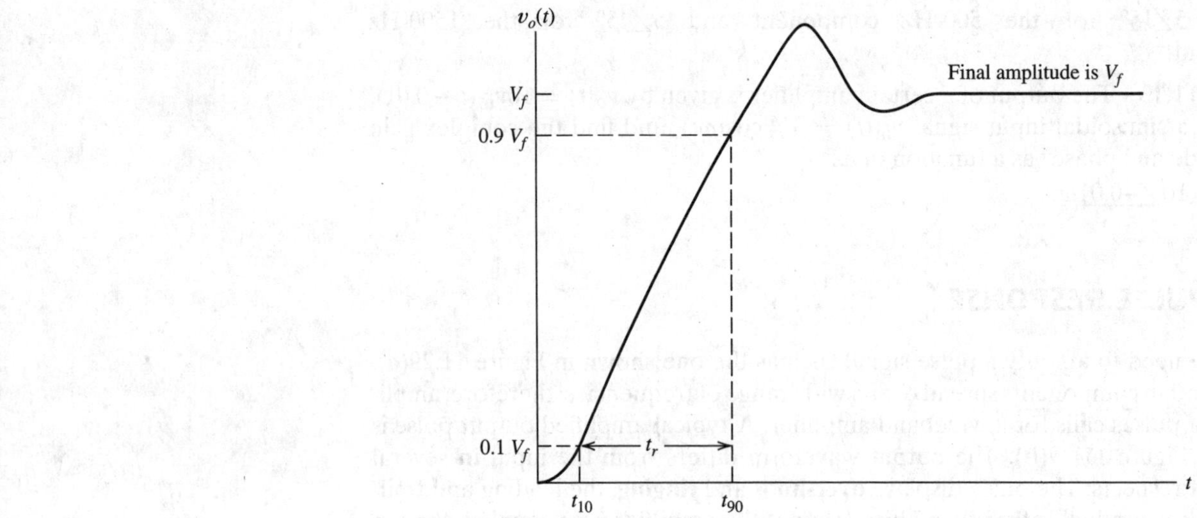

Figure 11.30 Rise time of the output pulse. (*Note:* No tilt is shown. When tilt is present, some judgment is necessary to estimate the amplitude V_f.)

This relationship is not exact for all types of wideband amplifiers, but it is a useful guide for estimating performance. (It is accurate for first-order circuits; see Problem P11.80.)

Since pulse amplifiers are broadband, the bandwidth is almost equal to the upper half-power frequency. Thus, it is mainly the high-frequency characteristics of the amplifier that restrict rise time.

Overshoot and Ringing

Another aspect of the output pulse shown in Figure 11.29 is overshoot and ringing, which are also related to the way the gain of the amplifier behaves in the high-frequency region. An amplifier that displays pronounced overshoot and ringing usually has a peak in its gain characteristic, as shown in Figure 11.31. The frequency of maximum gain approximately matches the ringing frequency.

Because both rise time and overshoot are related to the high-frequency response, there is usually some trade-off between these specifications. In a particular design, component values that reduce rise time tend to produce more overshoot and ringing. However, more than about 10 percent overshoot is usually undesirable.

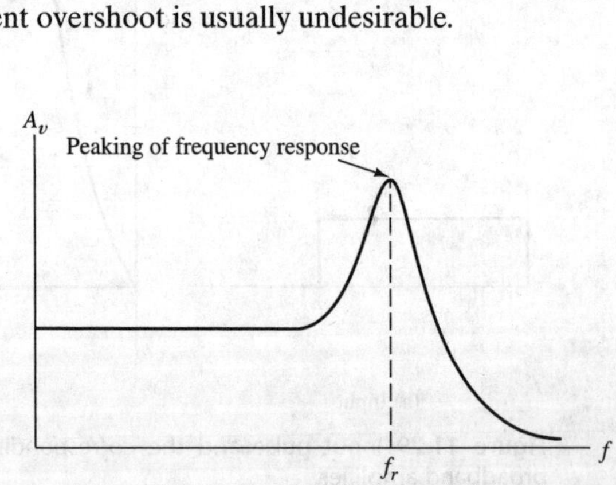

Figure 11.31 Gain versus frequency for an amplifier that displays pronounced ringing in its pulse response. The frequency of the ringing is approximately f_r.

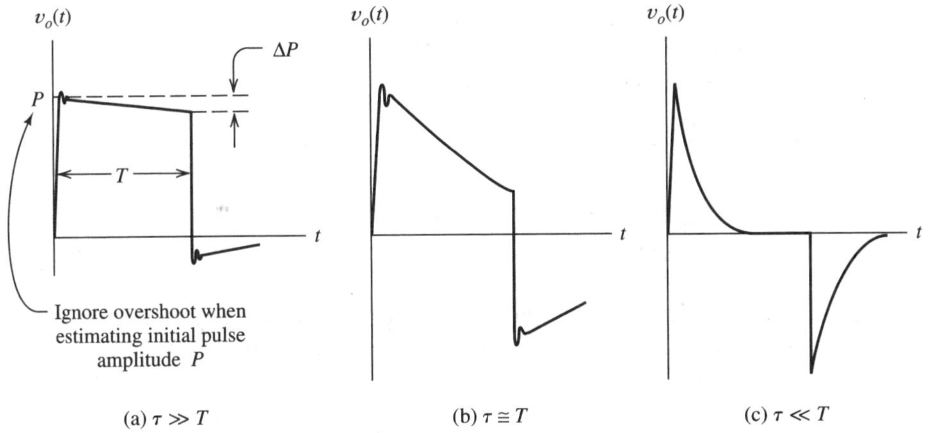

Figure 11.32 Pulse responses of ac-coupled amplifiers. T is the input pulse duration, and τ represents the shortest time constant of the coupling circuits.

Tilt

The pulse response of an amplifier may contain overshoot, ringing, and tilt. Rise time is always nonzero.

The tilt of the top of the output pulse, shown in Figure 11.32(a), occurs if the amplifier is ac coupled and arises from charging of coupling capacitors during the pulse. (After all, if the pulse lasted indefinitely, it would be the same as a new dc level at the input, and eventually the output voltage of an ac-coupled amplifier would return to zero.) Tilt is specified as a percentage of the initial pulse amplitude,

$$\text{percentage tilt} = \frac{\Delta P}{P} \times 100 \text{ percent} \tag{11.12}$$

where ΔP and P are defined in Figure 11.32(a). As the duration of the pulse is increased (or as the lower half-power frequency of the amplifier is raised by changing the coupling circuits to have shorter time constants), output waveforms such as those in Figure 11.32(b) and (c) result.

For small amounts of tilt, the percentage tilt is related to the lower half-power frequency by the approximate relation

$$\text{percentage tilt} \cong 200\pi f_L T \tag{11.13}$$

where T is the duration of the pulse and f_L is the lower half-power frequency of the amplifier. (See Problem P11.81 for a derivation of this formula for percentage tilt.)

Exercise 11.14 In a radar system, pulses of radio waves are transmitted and objects are detected by their reflected signals. After conversion of the reflected signals to baseband, they appear as pulses, and the time interval between pulses indicates the distance between objects. To distinguish objects a given distance apart, the maximum rise time allowed for the amplifiers is approximately equal to the time separation of the reflections. For example, if it is desired to distinguish objects that are 10 m apart on a line from the radar transmitter, the time separation of the echoes is 20 m (because the waves must make a round trip) divided by the speed of light. This gives a required maximum rise time of approximately 66.7 ns. Estimate the minimum bandwidth required for the amplifier.

Answer $B \cong 5.25$ MHz. □

Exercise 11.15 An amplifier is needed to amplify pulses with a duration of 100 μs with a sag (tilt) of not more than 1 percent. Estimate the highest value allowed for the lower half-power frequency of the amplifier.
Answer $f_L = 15.9$ Hz. □

11.10 TRANSFER CHARACTERISTIC AND NONLINEAR DISTORTION

The transfer characteristic is a plot of instantaneous output amplitude versus instantaneous input amplitude.

Curvature of the transfer characteristic results in nonlinear distortion.

The **transfer characteristic** of an amplifier is a plot of the instantaneous output amplitude versus the instantaneous input amplitude. For an ideal amplifier, the output is simply a larger version of the input waveform, and the transfer characteristic is a straight line whose slope is the gain. Real amplifiers have transfer characteristics that depart from straight lines, particularly at large amplitudes. This is shown in Figure 11.33. Curvature of the transfer characteristic results in an undesirable effect known as **nonlinear distortion**.

Sometimes, the departure from a straight characteristic can be very abrupt. Then the result of applying a high-amplitude input signal is **clipping** of the output waveform, as shown in Figure 11.34. However, even small departures from a straight characteristic can be very serious in some applications.

Harmonic Distortion

The input–output relationship of a nonlinear amplifier can be written as

$$v_o = A_1 v_i + A_2 (v_i)^2 + A_3 (v_i)^3 + \cdots \tag{11.14}$$

where A_1, A_2, A_3, and so on, are constants selected so that the equation matches the curvature of the nonlinear transfer characteristic.

Consider the case for which the input signal is a sinusoid given by

$$v_i(t) = V_a \cos(\omega_a t) \tag{11.15}$$

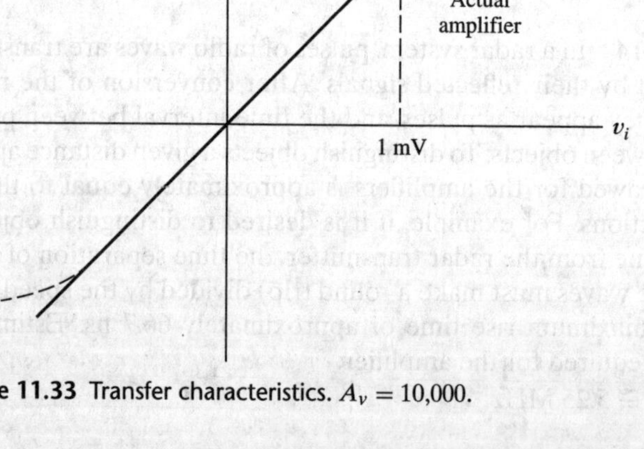

Figure 11.33 Transfer characteristics. $A_v = 10{,}000$.

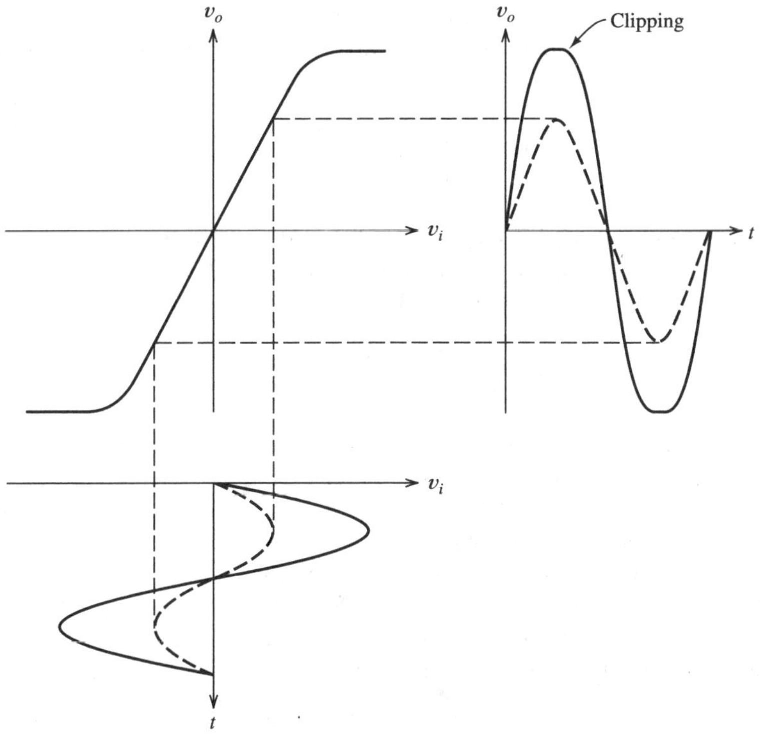

Figure 11.34 Illustration of input signal, amplifier transfer characteristic, and output signal, showing clipping for large signal amplitude.

Let us find an expression for the corresponding output signal. Substituting Equation 11.15 into 11.14, applying trigonometric identities for $[\cos(\omega_a t)]^n$, collecting terms, and defining V_0 to be equal to the sum of all of the constant terms, V_1 to be the sum of the coefficients of the terms with frequency ω_a, and so on, we find that

$$v_o(t) = V_0 + V_1 \cos(\omega_a t) + V_2 \cos(2\omega_a t) + V_3 \cos(3\omega_a t) + \cdots \quad (11.16)$$

For a sinewave input, nonlinear distortion produces output components having frequencies that are integer multiples of the input frequency.

The desired output is the $V_1 \cos(\omega_a t)$ term, which we call the **fundamental** component. The V_0 term represents a shift in the dc level (which does not appear at the load if it is ac coupled). In addition, terms at multiples of the input frequency have resulted from the second and higher power terms of the transfer characteristic. These terms are called **harmonic distortion**. The $2\omega_a$ term is called the **second harmonic**, the $3\omega_a$ term is the **third harmonic**, and so on. The higher order terms in the transfer characteristic given by Equation 11.14 produce the higher order harmonics. For example, the squared term produces the second harmonic. Similarly, the cubic term generates the third harmonic.

Harmonic distortion is objectionable in a wideband amplifier because the harmonics can fall in the frequency range of the desired signal. In an audio amplifier, harmonic distortion degrades the aesthetic qualities of the sound produced by the loudspeakers.

The **second-harmonic distortion factor** D_2 is defined as the ratio of the amplitude of the second harmonic to the amplitude of the fundamental. In equation form, we have

$$D_2 = \frac{V_2}{V_1} \quad (11.17)$$

where V_1 is the amplitude of the fundamental term of Equation 11.16 and V_2 is the amplitude of the second harmonic. Similarly, the third-harmonic distortion factor, and so on, are defined as

$$D_3 = \frac{V_3}{V_1} \quad D_4 = \frac{V_4}{V_1} \quad \cdots \tag{11.18}$$

Total harmonic distortion is a specification that indicates the degree of nonlinear distortion produced by an amplifier.

The **total harmonic distortion** (THD), denoted by D, is the ratio of the rms value of the sum of all the harmonic distortion terms to the rms value of the fundamental. The total harmonic distortion can be found from

$$D = \sqrt{D_2^2 + D_3^2 + D_4^2 + D_5^2 + \cdots} \tag{11.19}$$

We will often find THD expressed as a percentage. A well-designed audio amplifier might have a THD specification of 0.01 percent (i.e., $D = 0.0001$) at rated power output. (Some years ago, THD of 5 percent was typical for amplifiers found in inexpensive radios or phonographs.)

Notice that the THD specification of an amplifier depends on the amplitude of the output signal because the degree of nonlinearity of the transfer characteristic is amplitude dependent. Certainly, any amplifier eventually clips the output signal if the input signal becomes large enough. When severe clipping occurs, THD becomes large.

Exercise 11.16 A certain amplifier has a transfer characteristic given by

$$v_o = 100v_i + v_i^2$$

a. Find the THD rating of the amplifier for a sinusoidal input voltage $v_i(t) = \cos(\omega t)$. **b.** Repeat for $v_i(t) = 5\cos(\omega t)$. [*Hint:* Use the fact that $\cos^2 x = 1/2 + (1/2)\cos(2x)$. This amplifier produces no third or higher harmonic distortion. Thus, $D_3 = 0, D_4 = 0$, etc.]

Answer **a.** $D = 0.005$; **b.** $D = 0.025$. Notice that the THD is larger for the larger input amplitude. □

11.11 DIFFERENTIAL AMPLIFIERS

A differential amplifier has two input terminals: an inverting input and a noninverting input.

Until now, we have considered amplifiers that have only one input source. Now we consider **differential amplifiers**, which have two input sources, as shown in Figure 11.35. An ideal differential amplifier produces an output voltage proportional to the difference between the input voltages. This is demonstrated by

$$\begin{aligned} v_o(t) &= A_d[v_{i1}(t) - v_{i2}(t)] \\ &= A_d v_{i1}(t) - A_d v_{i2}(t) \end{aligned} \tag{11.20}$$

Ideally, a differential amplifier produces an output that is proportional to the difference between two input signals.

Notice that gain is positive for the voltage applied to terminal 1 and negative for the voltage applied to terminal 2. Therefore, terminal 2 is called an **inverting input**, and terminal 1 is called a **noninverting input**. Inverting input terminals are marked with a − (minus) sign, and noninverting input terminals are marked with a + (plus) sign, as indicated in Figure 11.35.

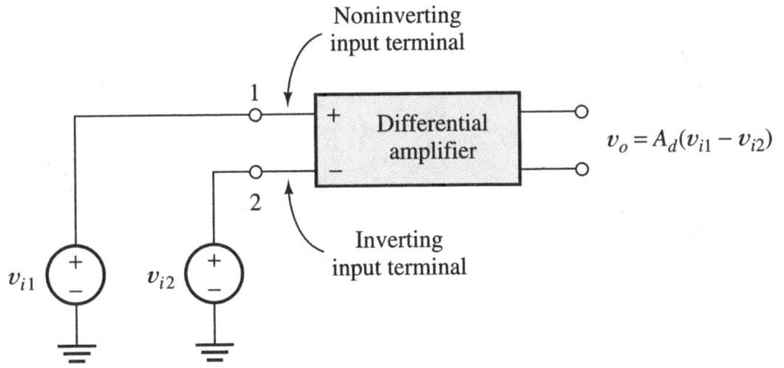

Figure 11.35 Differential amplifier with input sources.

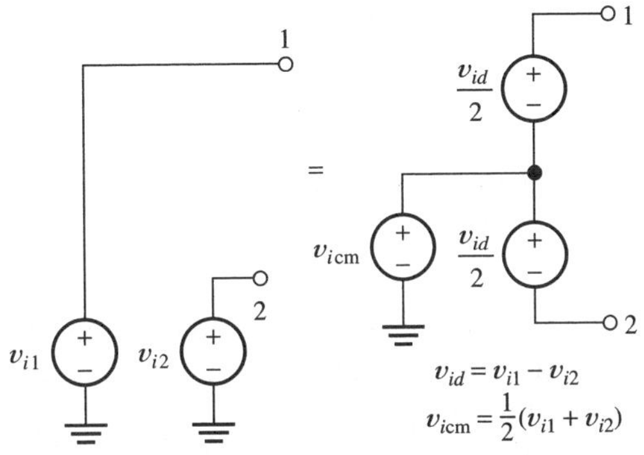

Figure 11.36 The input sources v_{i1} and v_{i2} can be replaced by the equivalent sources v_{icm} and v_{id}.

The difference between the input voltages, known as the **differential signal**, is given by

$$v_{id} = v_{i1} - v_{i2} \tag{11.21}$$

We refer to the gain A_d as the **differential gain**. Thus, we can write the output of the ideal differential amplifier as

$$v_o = A_d v_{id} \tag{11.22}$$

The **common-mode signal** v_{icm} is the average of the input voltages, given by

$$v_{icm} = \frac{1}{2}(v_{i1} + v_{i2}) \tag{11.23}$$

The original input sources v_{i1} and v_{i2} can be replaced by the equivalent system of sources shown in Figure 11.36. Thus, we can consider the inputs to the differential amplifier to be the differential signal v_{id} and the common-mode signal v_{icm}.

Sometimes, we have a small differential signal that we wish to amplify, but a large common-mode signal is also present that is of no interest. A good example of this is in recording the electrocardiogram (ECG) of a patient. Imagine a patient lying on a bed insulated from electrical ground as shown in Figure 11.37. If electrodes are placed in contact with each of the patient's arms, a differential signal generated by the patient's heart appears between the electrodes, which is the signal of interest to the cardiologist. Also, we often find a large 60-Hz common-mode signal present between each electrode and the local power-system ground. This occurs because patients are connected to the 60-Hz power line by very small incidental capacitances

Many applications exist in which a small differential signal is important and a strong interfering common-mode signal is present.

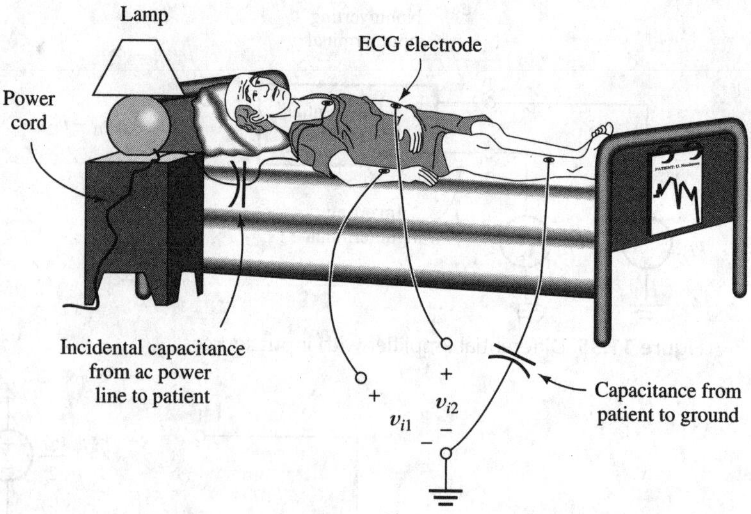

Figure 11.37 Electrocardiographs encounter large 60-Hz common-mode signals.

between their bodies and the power line. Similar small capacitances connect the patient to ground. That network of capacitances forms a voltage-divider network, so the patient's body is at a significant fraction of the power-line voltage with respect to ground. (You may have observed this 60-Hz common-mode signal in the laboratory if you have touched the input terminals of a high-input-impedance ac meter or oscilloscope.) Thus, at the input to the electrocardiograph amplifier, there exists a differential signal of about 1 mV and a 60-Hz common-mode signal of several tens of volts. Ideally, the electrocardiograph should respond only to the differential signal.

Interfacing sensors to computers is pervasive in all fields of science and engineering. The problem of large common-mode interference from the power line is also very common. Hence, it is important for you to have a good grasp of these concepts.

Common-Mode Rejection Ratio

Unfortunately, real differential amplifiers respond to both the common-mode signal and the differential signal. Recall that the gain for the differential signal is denoted as A_d. If we denote the gain for the common-mode signal as A_{cm}, the output voltage of a real differential amplifier is given by

$$v_o = A_d v_{id} + A_{\text{cm}} v_{icm} \tag{11.24}$$

For well-designed differential amplifiers, the differential gain A_d is much larger than the common-mode gain A_{cm}. A quantitative specification is the **common-mode rejection ratio** (CMRR), which is defined as the ratio of the magnitude of the differential gain to the magnitude of the common-mode gain. Often, CMRR is expressed in decibels as

Common-mode rejection ratio (CMRR) is a specification that indicates how well the common-mode signal is rejected relative to the differential signal.

$$\text{CMRR} = 20 \log \frac{|A_d|}{|A_{\text{cm}}|} \tag{11.25}$$

The CMRR of an amplifier is generally a function of frequency, becoming lower as frequency is raised. At 60 Hz, a CMRR of 120 dB is considered good.

Example 11.11 Determination of the Minimum CMRR Specification

Find the minimum CMRR for an electrocardiograph amplifier if the differential gain is 1000, the desired differential input signal has a peak amplitude of 1 mV, the common-mode signal is a 100-V-peak 60-Hz sine wave, and it is desired that the output contain a peak common-mode contribution that is 1 percent or less of the peak output caused by the differential signal.

Solution Since the peak differential input is 1 mV and the differential gain is 1000, the peak output of the desired signal is 1 V. To meet the specification required, the common-mode output signal must have a peak value of 0.01 V or less. Thus, the common-mode gain is

$$A_{\text{cm}} = \frac{0.01\text{ V}}{100\text{ V}} = 10^{-4}$$

(Therefore, the common-mode gain actually amounts to attenuation.) Now, we can find the CMRR by application of Equation 11.25:

$$\text{CMRR} = 20\log\frac{|A_d|}{|A_{\text{cm}}|} = 20\log\frac{1000}{10^{-4}} = 140\text{ dB}$$

Hence, an electrocardiograph requires a very good CMRR specification. ■

Perhaps, we should note in passing that another simpler—but dangerous—approach exists to solving the common-mode problem for the electrocardiograph: It is to short out the common-mode signal by attaching another electrode to the patient and connecting it to the power-system ground. This would reduce the 60-Hz interference to a very low level, so an amplifier with a much less stringent CMRR specification could be used. However, once the patient is in good electrical contact with the power-system ground, any contact with power-line voltages is potentially fatal. That is particularly true if the patient is too ill to protest. Even small currents imperceptible under ordinary circumstances can be fatal if conducted directly to the patient's heart. Such small currents can be conducted through other medical instrumentation or even through a surgeon's hands. *Keeping the patient isolated from ground provides some measure of protection from such problems.*

Measurement of CMRR

Measurements to find the CMRR of an amplifier are fairly straightforward. We must find both the differential and common-mode gains. The common-mode gain is found by connecting the input terminals of the amplifier together and attaching a test source, as shown in Figure 11.38. Notice that with the input terminals of the amplifier connected together, the differential signal v_{id} is zero, and any output is caused by the common-mode signal applied to both input terminals by the test source. Thus, we measure both the input voltage and output voltage, and then we compute their ratio to find the common-mode gain.

In theory, to apply a pure differential signal, we must provide two sources out of phase with each other at the amplifier input terminals, as shown in Figure 11.39(a). However, since the common-mode gain is usually much smaller than the differential gain, only a small error results if a single source is used, as shown in Figure 11.39(b). [In Figure 11.39(b), the input contains both a differential signal v_{id} and a common-mode signal $v_{icm} = v_{id}/2$.] In any case, the differential gain is found by taking the

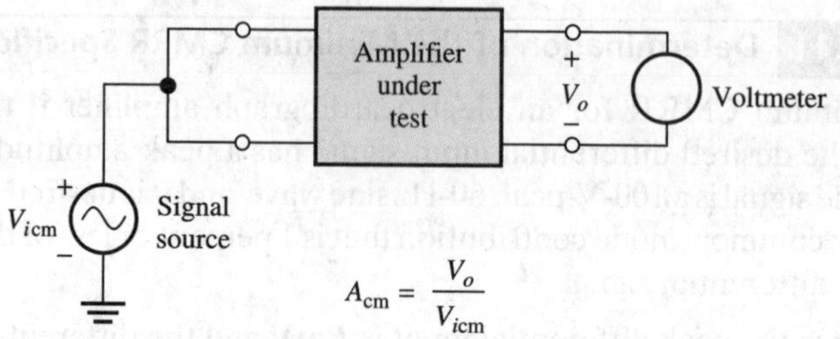

Figure 11.38 Setup for measurement of common-mode gain.

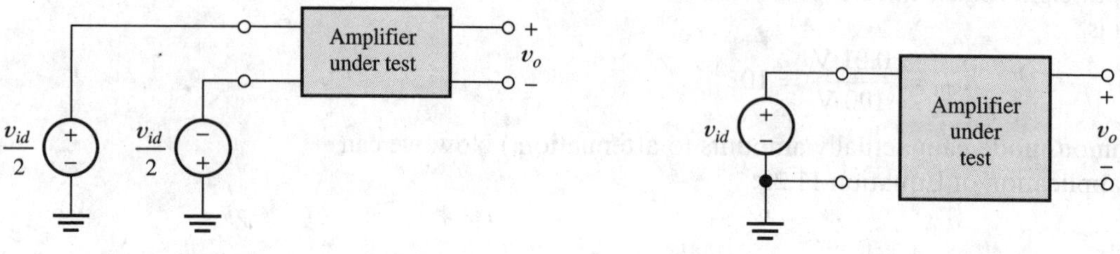

(a) Theoretically required sources to measure differential gain

(b) Practical equivalent if $A_d >> A_{cm}$

Figure 11.39 Setup for measuring differential gain. $A_d = v_o/v_{id}$.

ratio of the output voltage to the input voltage when the common-mode voltage is zero or negligible. Finally, the CMRR is found by taking the ratio of the gains.

Exercise 11.17 A certain amplifier has a differential gain $A_d = 50{,}000$. If the input terminals are connected together and a 1-V signal is applied to them, an output signal of 0.1 V results. What is the common-mode gain of the amplifier and the CMRR, both expressed in dB?

Answer $A_{cm} = -20$ dB, CMRR $= 114$ dB. □

Exercise 11.18 A certain amplifier has $v_o = A_1 v_{i1} - A_2 v_{i2}$. **a.** Assume that $v_{i1} = 1/2$ and $v_{i2} = -1/2$. Find v_{id} and v_{icm}. Find v_o and A_d in terms of A_1 and A_2. **b.** Assume that $v_{i1} = 1$ and $v_{i2} = 1$. Find v_{id} and v_{icm}. Find v_o and A_{cm} in terms of A_1 and A_2. **c.** Use the results of parts (a) and (b) to find an expression for the CMRR in terms of A_1 and A_2. Evaluate the CMRR if $A_1 = 100$ and $A_2 = 101$.

Answer **a.** $v_{id} = 1$, $v_{icm} = 0$, $v_o = A_d = (1/2)A_1 + (1/2)A_2$; **b.** $v_{id} = 0$, $v_{icm} = 1$, $v_o = A_{cm} = A_1 - A_2$; **c.** CMRR $= 20 \log |(A_1 + A_2)/2(A_1 - A_2)| = 40.0$ dB. □

11.12 OFFSET VOLTAGE, BIAS CURRENT, AND OFFSET CURRENT

Until now, we have assumed that the output of an amplifier is zero if the input sources are zero, but in real direct-coupled amplifiers that is not true. A dc output voltage is usually observed even if the input sources are zero. This is caused by undesired imbalances in the internal component values of the amplifier and because, in some types of amplifier circuits, it is necessary for the external input circuits to supply small dc currents to the amplifier input terminals. Assuming a differential amplifier, these effects can be modeled by the addition of three current sources and one voltage

PRACTICAL APPLICATION 11.1

Electronic Stud Finder

When we want to hang a heavy picture or shelf on a wall, we often need to locate a wood stud capable of bearing the weight of the picture or shelf. Usually, this can be accomplished with an electronic stud finder.

A simple electronic stud finder, illustrated in Figures PA11.1 and PA11.2, can be designed by using several of the electrical-engineering concepts discussed in this book. First, as discussed in Section 3.3, the capacitance between metal plates depends on the dielectric constants of the materials surrounding the plates. For the configuration shown in Figure PA11.1, the capacitance between plates A and B is less than the capacitance between B and C because the dielectric constant of the wood stud is higher than that of air. As the stud finder moves to the right and becomes centered on the stud, the capacitances become equal. Then, as the stud finder moves slightly past center, the capacitance between A and B becomes higher.

The second concept used in the stud finder is an ac bridge circuit similar to the Wheatstone bridge of Section 2.8. As shown in Figure PA11.2, the variable capacitances are connected in the bridge with two equal resistances and an ac source. Recall that the voltage between nodes A and C becomes zero when the bridge is balanced, which, in this case, occurs when the capacitances are equal.

The third concept is the use of a high-gain differential amplifier (such as those discussed in Section 11.11) and a beeper (which is a simple loudspeaker) to form a sensitive detector for the bridge circuit.

When the stud finder is over the stud, but not centered, the capacitances are unequal and the bridge is not balanced. Then, an ac voltage appears as the input to the differential amplifier, and a sound is emitted from the beeper. When the stud finder is centered, the bridge becomes balanced, and the sound disappears. Thus, by moving the stud finder over the surface of a wall, we can easily locate the center lines of the studs.

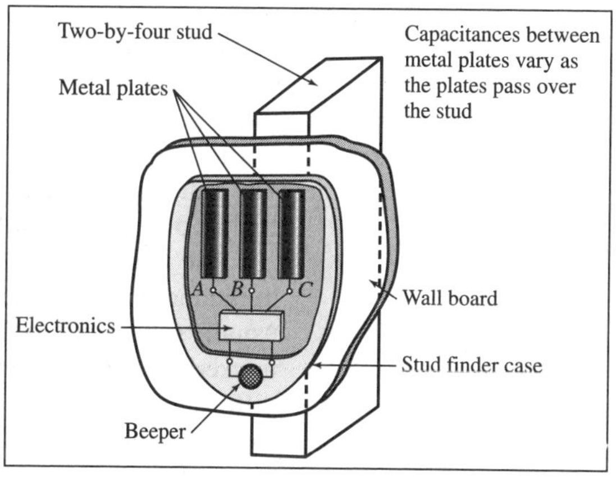

Figure PA11.1

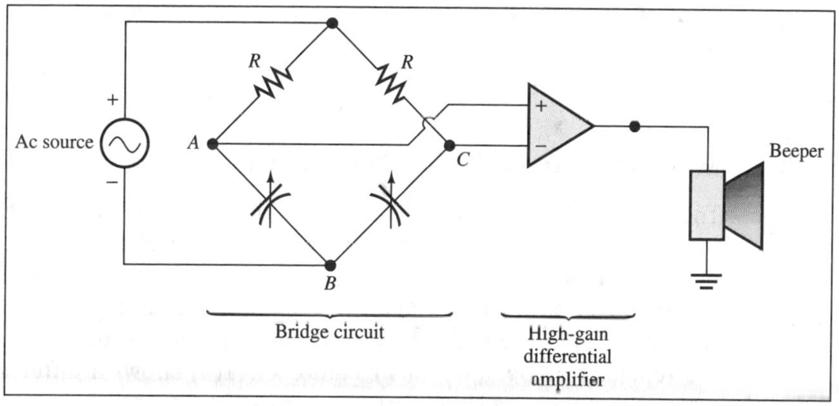

Figure PA11.2

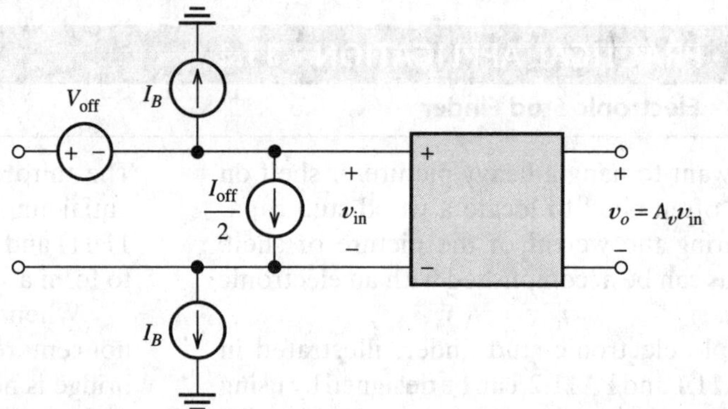

Figure 11.40 Differential amplifier, including dc sources to account for the dc output that exists even when the input signals are zero.

source to the input terminals of an otherwise ideal amplifier. These sources are shown in Figure 11.40.

The two current sources labeled I_B are known as **bias-current sources**. These sources account for the small dc currents drawn by the internal amplifier circuitry through the input terminals. The bias currents have the same value and direction (either both flow toward the amplifier input terminals or both flow toward ground). The value of the bias current I_B is a function of temperature, and it varies from unit to unit of a given amplifier type.

The current I_{off} is called **offset current**. Offset current arises from incidental imbalances in the internal components of the amplifier. The offset current value is usually somewhat smaller than the bias current. The direction of the offset current is unpredictable—it can flow toward either input terminal. The direction of flow may be different from unit to unit of a given amplifier model. Notice that the offset current source (Figure 11.40) has a value of $I_{off}/2$.

The voltage source V_{off} in series with the input terminals is called an **offset voltage**. Like the offset current, it is caused by internal circuit imbalances. The value of the offset voltage is usually a function of temperature. Furthermore, it changes in value and polarity from unit to unit. The offset voltage source can be placed in series with either input terminal.

Real differential amplifiers suffer from imperfections that can be modeled by several dc sources: two bias-current sources, an offset current source, and an offset voltage source. The effect of these sources is to add a (usually undesirable) dc term to the ideal output.

Minimizing the Effect of Bias Current

The effects of bias current can be mitigated by ensuring that the Thévenin impedances of the circuits connected to the input terminals are the same. (Recall from Section 2.6 that to find the Thévenin impedance of a network, we zero the independent sources and then compute the impedance of the network. Independent voltage sources are zeroed by replacing them with short circuits, whereas independent current sources are replaced by open circuits.) Figure 11.41(a) shows a differential amplifier with source resistances and bias-current sources. Each current source can be converted to a voltage source in series with the corresponding resistance as shown in Figure 11.41(b). If the source resistances are equal, these voltages are equal, so there is no differential signal supplied to the amplifier. Assuming that the common-mode gain is zero, the resulting output voltage is zero.

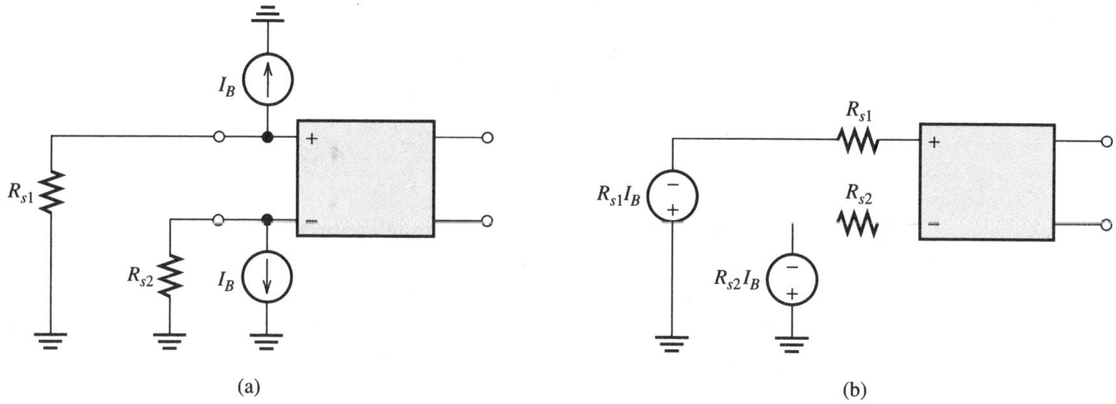

Figure 11.41 The effects of the bias-current sources cancel if $R_{s1} = R_{s2}$.

Example 11.12 Calculation of Worst-Case DC Output Voltage

A certain direct-coupled differential amplifier has a differential voltage gain of 100, an input impedance of 1 MΩ, an input bias current of 200 nA, a maximum offset current of 80 nA, and a maximum offset voltage of 5 mV. Compute the worst-case output voltage if the amplifier input terminals are connected to ground through 100-kΩ source resistances.

Solution The circuit, including the source resistances, is shown in Figure 11.42(a). Since the circuit is linear, we can use superposition, considering each source separately. Because the impedances for the two inputs are the same, the effects of the bias currents balance and, therefore, can be ignored.

Because the circuit is linear, we use superposition, thereby dividing the problem into several relatively simple problems.

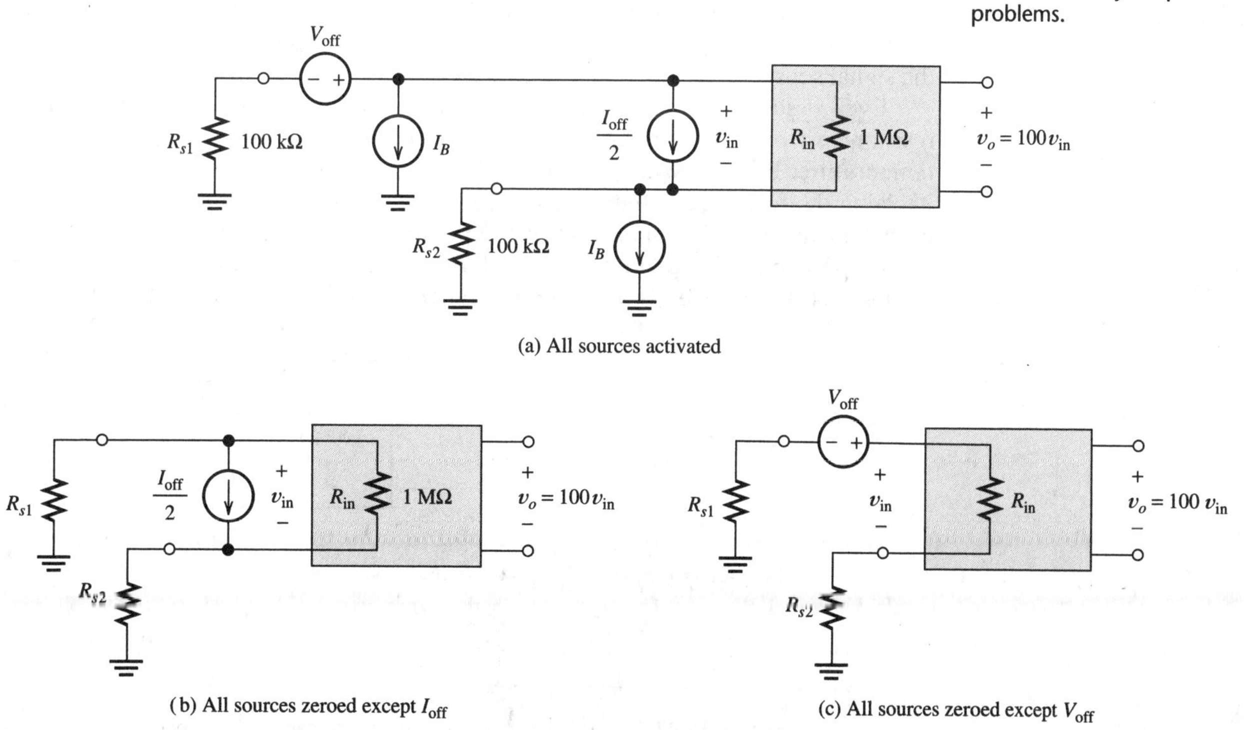

Figure 11.42 Amplifier of Example 11.13.

The offset current flows through the parallel combination of R_{in} and the sum of the source resistances as shown in Figure 11.42(b). Thus, the differential input voltage arising from the offset current has a maximum value given by

$$V_{Ioff} = \frac{I_{off}}{2} \frac{R_{in}(R_{s1} + R_{s2})}{R_{in} + R_{s1} + R_{s2}} = 6.67 \text{ mV}$$

The circuit with only the offset voltage source activated is shown in Figure 11.42(c). The differential input voltage resulting from the offset voltage is found by noting that part of the input offset source voltage appears across the input terminals and the rest appears across R_{s1} and R_{s2}. The portion across the input terminals can be computed by use of the voltage-divider principle as

$$V_{Voff} = V_{off} \frac{R_{in}}{R_{in} + R_{s1} + R_{s2}} = 4.17 \text{ mV}$$

Multiplying by the amplifier gain, we find that the maximum output voltage caused by the offset current source is 0.667 V and the maximum output voltage caused by the offset voltage source is 0.417 V. These voltages are maximum values and they can have either polarity, so the total output voltage can range between −1.084 and +1.084 V. ■

Balancing Circuits

Balancing circuits can be used to cancel the dc offset added to the output signal by amplifier imperfections.

The effects of the offset current and voltage can be canceled by the use of a balancing circuit such as that shown in Figure 11.43. The resistors R_1 and R_2 on each side of the potentiometer form voltage dividers that supply small voltages to opposite ends of the potentiometer—positive on one end and negative on the other. In use, the potentiometer is simply adjusted so that the amplifier output is zero if the input from the signal source is zero.

Even if such a balancing circuit is used, it is good practice to maintain equal resistances from both input terminals to ground, because bias current varies with temperature. Equal resistances provide balancing for bias current, independent of its value. Unfortunately, the offset current and voltage vary with temperature, so perfect balance at all temperatures is not possible with a fixed circuit.

In principle, the dividers (R_1 and R_2) could be left out of the circuit of Figure 11.43, and the ends of the potentiometer could be connected directly to the

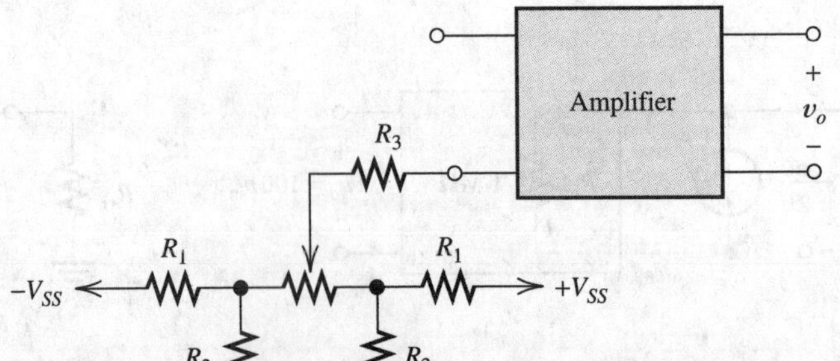

Figure 11.43 Network that can be adjusted to cancel the effects of offset and bias sources.

power-supply voltages. However, the range of adjustment would then be much larger than necessary, and the correct adjustment would be very difficult to achieve.

Some amplifiers provide separate terminals for attachment of balancing circuits so that the signal input terminals are not encumbered.

Exercise 11.19 A certain direct-coupled differential amplifier has a differential voltage gain of 500, an input impedance of 100 kΩ, an input bias current of 400 nA, a maximum offset current of 100 nA, and a maximum offset voltage of 10 mV. Compute the worst-case output voltages if the amplifier input terminals are connected to ground through 50-kΩ resistances.

Answer v_o ranges from −3.75 to +3.75 V. □

Exercise 11.20 Repeat Exercise 11.19 if the inverting input terminal is grounded directly and the noninverting input is connected to ground through a 50-kΩ resistance.

Answer v_o ranges from +2.5 V to +10.84 V. □

Summary

1. The purpose of an amplifier is to deliver a larger signal to a load than is available from the signal source.
2. Amplifiers are characterized by their input impedance, output impedance, and a gain parameter.
3. Inverting amplifiers have negative voltage gain, so the output waveform is an inverted version of the input waveform. Noninverting amplifiers have positive voltage gain.
4. Loading effects result from voltage drops across the internal source impedance and across the output impedance of the amplifier.
5. In a cascade connection, the output of each amplifier is connected to the input of the next amplifier.
6. The efficiency of an amplifier is the percentage of the supply power that is converted into output signal power.
7. Several models are useful in characterizing amplifiers. They are the voltage-amplifier model, the current-amplifier model, the transconductance-amplifier model, and the transresistance-amplifier model.
8. According to their input and output impedances, ideal amplifiers can be classified into four types: the ideal voltage amplifier, the ideal current amplifier, the ideal transconductance amplifier, and the ideal transresistance amplifier. The best amplifier type to select depends on the application.
9. Amplifiers may be direct coupled, in which case constant gain extends to dc. On the other hand, amplifiers may be ac coupled, in which case the gain falls off at low frequencies, reaching zero gain at dc. Gain magnitude falls to zero at sufficiently high frequencies for all amplifiers.
10. Linear distortion can be either amplitude distortion or phase distortion. Amplitude distortion occurs if the gain magnitude is different for various components of the input signal. Phase distortion occurs if amplifier phase shift is not proportional to frequency.
11. Amplifier pulse response is characterized by rise time, overshoot, ringing, and tilt.
12. Nonlinear distortion occurs if the transfer characteristic of an amplifier is not straight. Assuming a sinusoidal input signal, nonlinear distortion causes harmonics to appear in the output. The total harmonic distortion rating of an amplifier indicates the degree of nonlinear distortion.
13. A differential amplifier ideally responds only to the difference between its two input signals (i.e., the differential input signal).
14. The common-mode input is the average of the two inputs to a differential amplifier. Common-mode rejection ratio (CMRR) is the ratio of the differential gain to the common-mode gain.

CMRR is an important specification for many instrumentation applications.

15. Dc offset is the addition of a dc term to the signal being amplified. It is the result of bias current, offset current, and offset voltage, and it can be canceled by use of a properly designed balance circuit.

Problems

Section 11.1: Basic Amplifier Concepts

P11.1. Discuss the criteria you will use for selecting inverting and non-inverting amplifiers, citing examples.

P11.2. Draw the voltage-amplifier model and label its elements.

P11.3. In what way does the loading effect affect the voltage gain of an amplifier?

***P11.4.** A signal source with an open-circuit voltage of $V_s = 5$ mV rms and an internal resistance of 100 kΩ is connected to the input terminals of an amplifier having an open-circuit voltage gain of 150, an input resistance of 150 kΩ, and an output resistance of 10 Ω. A 8 Ω load is connected to the output terminals. Find the voltage gains $A_{vs} = V_o/V_s$ and $A_v = V_o/V_i$. Also, find the power gain and current gain.

***P11.5.** A certain amplifier operating with a 150-Ω load has a voltage gain of 75 and a power gain of 8000. Determine the current gain and input resistance of the amplifier.

P11.6. The current gain of an amplifier is 500, the load resistance is 100 Ω, and the input resistance of the amplifier is 1 MΩ. Determine the voltage gain and power gain under these conditions.

P11.7. An amplifier having $R_i = 1$ MΩ, $R_o = 1$ kΩ, and $A_{voc} = -10^4$ is operated with a 1-kΩ load. A source having a Thévenin resistance of 2 MΩ and an open-circuit voltage of $3\cos(200\pi t)$ mV is connected to the input terminals. Determine the output voltage as a function of time and the power gain.

P11.8. A certain amplifier has an open-circuit voltage gain of unity, an input resistance of 8 MΩ, and an output resistance of 150 Ω. The signal source has an internal voltage of 10 V rms and an internal resistance of 150 kΩ. The load resistance is 75 Ω. If the signal source is connected to the amplifier input terminals and the load is connected to the output terminals, find the voltage across the load and the power delivered to the load. Next, consider connecting the load directly across the signal source without the amplifier, and again find the load voltage and power. Compare the results. What do you conclude about the usefulness of a unity gain amplifier in delivering signal power to a load?

P11.9. An amplifier with $R_i = 12$ kΩ, $R_o = 1$ kΩ, and $A_{voc} = -10$ is operated with a 1-kΩ load. A source having a Thévenin resistance of 4 kΩ and a short-circuit current of $2\cos(200\pi t)$ mA is connected to the input terminals. Determine the output voltage as a function of time and the power gain.

***P11.10.** An ideal ac current source is applied to the input terminals of an amplifier, and the amplifier output voltage is 2 V rms. Then, a 2-kΩ resistance is placed in parallel with the current source and the amplifier input terminals, and the output voltage is 1.5 V rms. Determine the input resistance of the amplifier.

P11.11. An amplifier has an open-circuit voltage gain of 120. With a 15-kΩ load connected, the voltage gain is found to be only 75. Find the output resistance of the amplifier.

P11.12. Suppose we have a resistive load that varies from 5 kΩ to 10 kΩ. We connect this load to an amplifier, and we need the voltage across the load to vary by less than 1 percent with variations in the load resistance.

* Denotes that answers are contained in the Student Solutions files. See Appendix E for more information about accessing the Student Solutions.

What parameter of the amplifier is important in this situation? What range of values is allowed for the parameter?

P11.13. A certain amplifier operates with a resistive load. The current gain and the voltage gain are equal. What can you say about the input resistance and the load resistance?

P11.14. An amplifier has an open-circuit voltage gain of 1000, an input resistance of 20 kΩ, and an output resistance of 2 Ω. A signal source with an internal resistance of 10 kΩ is connected to the input terminals of the amplifier. An 8-Ω load is connected to the output terminals. Find the voltage gains $A_{vs} = V_o/V_s$ and $A_v = V_o/V_i$. Also, find the power gain and current gain.

P11.15. The output voltage v_o of the circuit of Figure P11.15 is 150 mV with the switch closed. With the switch open, the output voltage is 80 mV. Find the input resistance of the amplifier.

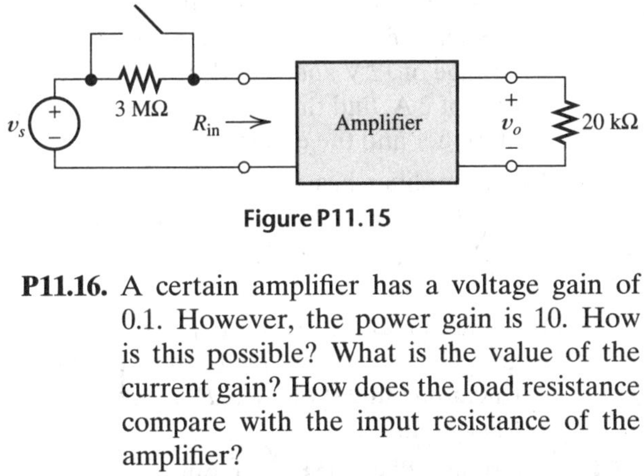

Figure P11.15

P11.16. A certain amplifier has a voltage gain of 0.1. However, the power gain is 10. How is this possible? What is the value of the current gain? How does the load resistance compare with the input resistance of the amplifier?

P11.17. Suppose we have a sensor, with a Thévenin resistance that varies from zero to 10 kΩ, connected to the input of an amplifier. We want the output voltage of the amplifier to vary by less than 2 percent with changes in the Thévenin resistance of the sensor. What parameter of the amplifier is important in this situation? What range of values is allowed for the parameter?

Section 11.2: Cascaded Amplifiers

P11.18. Draw the cascade connection of two amplifiers. Write an expression for the open-circuit voltage gain of the cascade connection in terms of the open-circuit voltage gains and impedances of the individual amplifiers.

P11.19. Given that the amplifiers having the characteristics shown in Table P11.19 are cascaded in the order $A - B$, find the input impedance, output impedance, and open-circuit voltage gain of the cascade. Repeat when the order is $B - A$.

Table P11.19. Amplifier Characteristics

Amplifier	Open-Circuit Voltage Gain	Input Resistance	Output Resistance
A	100	3 kΩ	400 Ω
B	500	1 MΩ	2 kΩ

P11.20. Three amplifiers with the following characteristics are cascaded in

$$\text{Amplifier 1: } A_{voc1} = 50, R_{i1} = 2\text{ k}\Omega, \quad R_{o1} = 1\text{ k}\Omega$$

$$\text{Amplifier 2: } A_{voc2} = 100, R_{i2} = 2\text{ k}\Omega, \quad R_{o2} = 2\text{ k}\Omega$$

$$\text{Amplifier 3: } A_{voc3} = 150, R_{i3} = 2\text{ k}\Omega, \quad R_{o3} = 3\text{ k}\Omega$$

Find the parameters for the simplified model of the cascaded amplifier.

P11.21. Repeat Problem P11.20 with the amplifiers cascaded in the order 2, 1, 3.

***P11.22.** Amplifiers having $A_{voc} = 10$, $R_i = 2$ kΩ, and $R_o = 2$ kΩ are available. How many of these amplifiers must be cascaded to attain a voltage gain of at least 1000 when operating with a 1-kΩ load?

P11.23. Three identical amplifiers having $A_{voc} = 25$, $R_i = 2$ kΩ, and $R_o = 3$ kΩ are cascaded. Determine the input resistance, the open-circuit voltage gain, and the output resistance of the cascade.

Section 11.3: Power Supplies and Efficiency

P11.24. Define efficiency of a power amplifier. Is high internal power dissipation desired?

***P11.25.** Find the net power delivered to the amplifier by the three dc supply voltages shown in Figure P11.25.

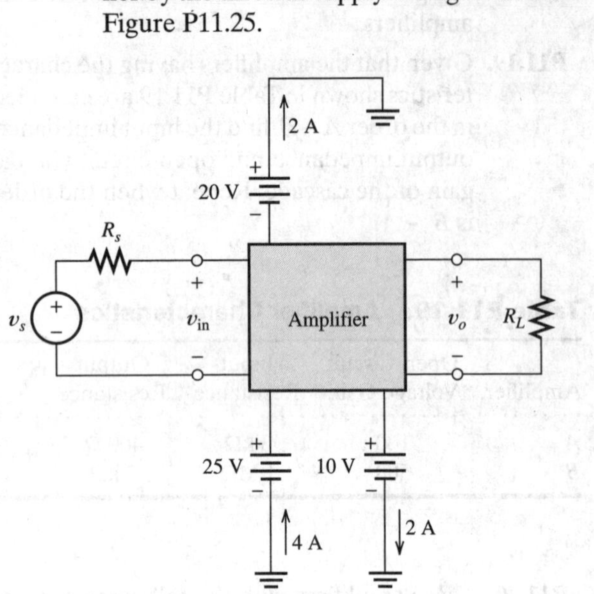

Figure P11.25

P11.26. A certain amplifier has an input voltage of 100 mV rms, an input resistance of 100 kΩ, and produces an output of 10 V rms across an 8-Ω load resistance. The power supply has a voltage of 15 V and delivers an average current of 2 A. Find the power dissipated in the amplifier and the efficiency of the amplifier.

P11.27. An amplifier operates from a 12-V power supply that supplies a current of 1.5 A. The input signal current is 1 μA rms, and the input resistance is 100 kΩ. The amplifier delivers 10 V rms to a 10-Ω load. Determine the power dissipated in the amplifier and the efficiency of the amplifier.

P11.28. Under high-signal test conditions, a certain audio amplifier supplies a 24 V rms 1-kHz sinusoidal voltage to an 8-Ω load. The power supply delivers 4 A at a voltage of 50 V to the amplifier. The signal power supplied by the input source is negligible. Determine the efficiency and the power dissipated in the amplifier.

P11.29. Two amplifiers are cascaded. The first has supply power of 2 W, an input resistance of 1 MΩ, and an input voltage of 2 mV rms. The second has a supply power of 22 W, a load resistance of 8 Ω, and output voltage of 12 V rms. Determine the overall power gain, dissipated power, and efficiency.

Section 11.4: Additional Amplifier Models

P11.30. Draw a voltage-amplifier model. Is the gain parameter measured under open-circuit or short-circuit conditions? Repeat for a current amplifier model, a transresistance-amplifier model, and a transconductance-amplifier model.

P11.31. **a.** Which amplifier model contains a current-controlled voltage source? **b.** A current-controlled current source? **c.** A voltage- controlled current source?

***P11.32.** An amplifier has an input resistance of 20 Ω, an output resistance of 10 Ω, and a short-circuit current gain of 3000. The signal source has an internal voltage of 100 mV rms and an internal impedance of 200 Ω. The load is a 5-Ω resistance. Find the current gain, voltage gain, and power gain of the amplifier. If the power supply has a voltage of 12 V and supplies an average current of 2 A, find the power dissipated in the amplifier and the efficiency.

P11.33. An amplifier has $R_i = 150\ \Omega$, $R_o = 2\ \text{k}\Omega$ and $R_{moc} = 15\ \text{k}\Omega$. Determine the values (including units) of A_{voc}, G_{msc}, and A_{isc} for this amplifier.

P11.34. An amplifier has $R_i = 10\ \text{k}\Omega$, $R_o = 100\ \Omega$, and $G_{msc} = 0.5$ S. Determine the values (including units) of A_{voc}, R_{moc}, and A_{isc} for this amplifier.

P11.35. An amplifier has an input resistance of 100 Ω, an output resistance of 10 Ω, and a short-circuit current gain of 500. Draw the voltage amplifier model for the amplifier, including numerical values for all parameters. Repeat for the transresistance and transconductance models.

P11.36. An amplifier has a short-circuit current gain of 25. When operated with a 100-Ω load, the current gain is 12. Find the output resistance of the amplifier.

P11.37. An amplifier has $R_i = 2\ \text{k}\Omega$, $R_o = 300\ \Omega$, and $A_{isc} = 200$. Determine the values (including units) of A_{voc}, R_{moc}, and G_{msc} for this amplifier.

***P11.38.** Amplifier A has an input resistance of 1 MΩ, an output resistance of 200 Ω, and an open-circuit transresistance gain of 100 MΩ. Amplifier B has an input resistance of 50 Ω, an output impedance of 500 kΩ, and a short-circuit current gain of 100. Find the voltage amplifier model for the cascade of A followed by B. Then, determine the corresponding transconductance amplifier model.

P11.39. Repeat Problem P11.38 if the order of the cascade is changed to $B - A$.

***P11.40.** An amplifier has an input resistance of 1 kΩ, an output resistance of 200 Ω, and a short-circuit transconductance gain of 0.5 S. Determine the open-circuit voltage gain, the short-circuit current gain, and the open-circuit transresistance gain.

***P11.41.** An amplifier has an input resistance of 10 kΩ, an output resistance of 2 kΩ, and an open-circuit transresistance gain of 200 kΩ. Determine the open-circuit voltage gain, the short-circuit current gain, and the short-circuit transconductance gain.

P11.42. An amplifier has an open-circuit voltage gain of 100, a short-circuit transconductance gain of 0.2 S, and a short-circuit current gain of 50. Determine the input resistance, the output resistance, and the open-circuit transresistance gain.

P11.43. An amplifier has an open-circuit transresistance gain of 200 Ω, a short-circuit transconductance gain of 0.5 S, and a short-circuit current gain of 50. Determine the input resistance, the output resistance, and the open-circuit voltage gain.

P11.44. An amplifier with $R_i = 2$ kΩ, $R_o = 500$ Ω, and $R_{moc} = -10^7$ Ω is operated with a 1-kΩ load. A source having a Thévenin resistance of 1 kΩ and an open-circuit voltage of $2\cos(200\pi t)$ mV is connected to the input terminals. Determine the output voltage as a function of time and the power gain.

Section 11.5: Importance of Amplifier Impedances in Various Applications

P11.45. Describe an application in which an amplifier with very high input impedance is needed.

P11.46. Give an application in which an amplifier with very low input impedance is needed.

P11.47. We need an amplifier to supply a constant signal to each of a variable number of loads connected in parallel. What output impedance is needed in this situation? Why? What if the loads are connected in series?

P11.48. Suppose we have a voltage source $v(t) = V_{dc} + V_m \cos(\omega t)$ connected to the input terminals of an amplifier. The load is a nonlinear device such as an LED. **a.** What output impedance is needed for the amplifier if we need the current through the load to be proportional to $v(t)$? **b.** If we need the voltage across the load to be proportional to $v(t)$?

P11.49. Give an example of a situation in which a specific input impedance is needed for an amplifier.

Section 11.6: Ideal Amplifiers

P11.50. Summarize the characteristics of ideal-voltage amplifiers. Repeat for each of the other amplifier types.

P11.51. An ideal transconductance amplifier having a short-circuit transconductance gain of 0.1 S is connected as shown in Figure P11.51. Find the resistance $R_x = v_x/i_x$ seen from the input terminals.

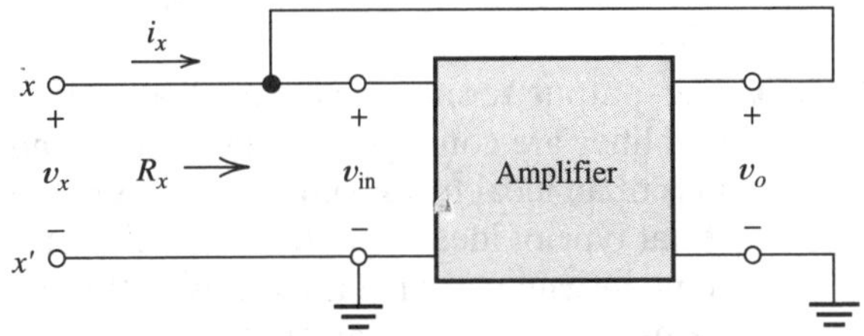

Figure P11.51

***P11.52.** Repeat Problem P11.51, assuming that the amplifier has an input resistance of 1000 Ω, an output impedance of 20 Ω, and an open-circuit transresistance gain of 10 kΩ.

P11.53. An amplifier has an input resistance of 1 Ω, an output resistance of 1 Ω, and an open-circuit voltage gain of 10. Classify this amplifier as an approximate ideal type and find the corresponding gain parameter. In deciding on an amplifier classification, assume that the source and load impedances are on the order of 1 kΩ.

P11.54. Repeat Problem P11.53 if the input impedance is 1 MΩ, the output impedance is 1 MΩ, and the open-circuit voltage gain is 100.

P11.55. A certain temperature sensor converts temperature into voltage and has a voltage range of 0 mV to 500 mV. The same needs to be fed to an ADC which has an input range of 0 to 5 V. What type of an ideal amplifier would you design? Give the details.

***P11.56.** The output terminals of an ideal transresistance amplifier are connected to the input terminals of an ideal transconductance amplifier. What type of ideal amplifier results? Determine its gain parameter in terms of the gain parameters of the separate stages.

P11.57. In recording automotive emissions, we need to sense the short-circuit current of a chemical sensor that has a variable Thévenin impedance. A voltage that is proportional to the current must be applied to the input of a data-acquisition module. What type of ideal amplifier is needed? Justify your answer.

P11.58. What type of ideal amplifier is needed if we need to sense the short-circuit current of a sensor and drive a proportional current through a variable load? Explain your answer.

P11.59. The output terminals of an ideal voltage amplifier are connected to the input terminals of an ideal transconductance amplifier. What type of ideal amplifier results? Determine its gain parameter in terms of the gain parameters of the separate stages.

***P11.60.** Suppose we have a two-stage cascaded amplifier with an ideal transconductance amplifier as the first stage and an ideal transresistance amplifier as the second stage. What type of amplifier results and what is its gain in terms of the gains of the two stages? Repeat for the amplifiers cascaded in the opposite order.

P11.61. In a certain application, an amplifier is needed to sense the open-circuit voltage of a source and force current to flow through a load. The source resistance and load resistance are variable. The current delivered to the load is to be nearly independent of both the source resistance and load resistance. What type of ideal amplifier is needed? If the source resistance increases from 1 kΩ to 2 kΩ and it causes a 1 percent decrease in load current, what is the value of the input resistance? If the load resistance increases from 100 to 300 Ω and this causes a 1 percent decrease in load current, what is the value of the output resistance?

P11.62. We need to design an amplifier for use in recording the short-circuit current of experimental electrochemical cells versus time. (For this purpose, a short circuit is any resistance less than 10 Ω.) The amplifier output is to be applied to a strip-chart recorder that deflects 1 cm ±1 percent for each volt applied. The input resistance of the recorder is unknown and likely to be variable, but it is greater than 10 kΩ. A deflection of 1 cm per milliampere of cell current with an accuracy of about ±3 percent is desired. What type of ideal amplifier is best suited for this application? Using your best judgment, find specifications for the amplifier's input impedance, output impedance, and gain parameter.

P11.63. An amplifier is needed as a part of a system for documentation of voltages in the earth created by an electrical power distribution system. Voltage waveforms occurring between probes to be placed in the earth are to be amplified before being applied to the analog-to-digital converter (ADC) inputs of laptop computers. The internal impedance of the probe can be as high as 10 kΩ in dry sand or as low as 10 Ω in muck. Because several different models of ADCs are to be used in the project, the load impedance for the amplifier varies from 10 kΩ to 1 MΩ. Nominally, the voltage applied to the ADC is required to be 1000 times the open-circuit voltage of the probe ±3 percent. What type of ideal amplifier is best suited for this application? Using your best judgment, find the specifications for the impedances and gain parameter of this amplifier.

P11.64. Repeat Problem P11.63 if, instead of an ADC, a strip-chart recorder having an unknown impedance of less than 100 Ω is used. The strip-chart recorder deflects 1 cm ±1 percent per milliampere of applied current. It is desired that the amplifier be designed so that the recorder deflects 1 cm for each 0.1 V of probe voltage.

Section 11.7: Frequency Response

P11.65. Sketch the gain magnitude of a typical dccoupled amplifier versus frequency. Repeat for an ac-coupled amplifier.

P11.66. How is a wideband amplifier different from a narrowband amplifier?

***P11.67.** The input to a certain amplifier is

$$v_{\text{in}}(t) = 0.1\cos(2000\pi t) + 0.2\cos(4000\pi t + 30°)$$

and the corresponding output voltage is

$$v_o(t) = 10\cos(2000\pi t - 20°) + 15\cos(4000\pi t + 20°)$$

Determine the values of the complex gain at $f = 1000$ Hz and at $f = 2000$ Hz.

***P11.68.** Consider the amplifier of Problem P11.62. Should this amplifier be ac coupled or dc coupled? Explain your answer.

P11.69. The output signal produced by a certain electret microphone consists of a 2-V dc term plus an ac audio signal that has an rms value of 10 mV. The frequencies of the components of the audio signal range from 20 Hz to 10 kHz. We need to amplify the audio signal to 10 V rms, which is to be applied to a loudspeaker. Should this amplifier be ac coupled or dc coupled? Explain your answer. What midband voltage gain is needed? What values are appropriate for the half-power frequencies?

***P11.70.** The gain of an amplifier is given by

$$A = \frac{5000}{[1 + j(f/3f_B)]^2}$$

Determine the lower and upper half-power frequency in terms of f_B.

P11.71. Consider Figure P11.71, in which block A is an ideal transconductance amplifier and block B is an ideal voltage amplifier. The capacitance is initially uncharged. **a.** Derive an expression for $v_o(t)$ for $t \geq 0$ in terms of the amplifier gains, $v_{\text{in}}(t)$, and the capacitance C. **b.** Derive an expression for the overall voltage gain of the system as a function of frequency. [*Hint:* Assume that $v_{\text{in}}(t) = V_m\cos(2\pi ft)$, determine the expression for $v_o(t)$, and then determine the complex voltage gain by taking the ratio of the phasors for the input and output.] **c.** Given $G_{msc} = 10^{-6}$ S, $A_{voc} = 200\pi$, and $C = 1\ \mu$F, sketch Bode plots of the voltage-gain magnitude and phase to scale for the range from 1 Hz to 1 kHz.

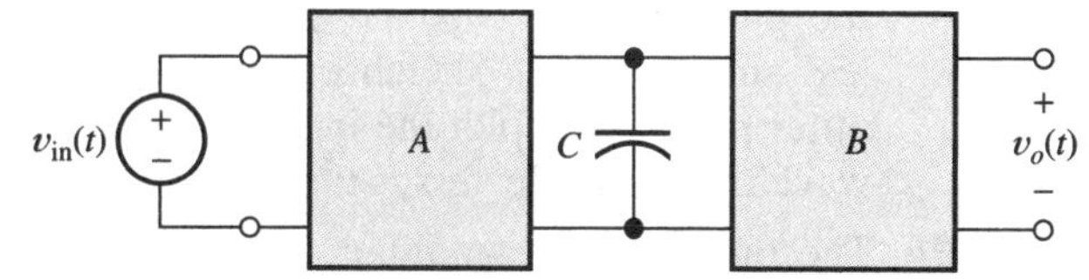

Figure P11.71

P11.72. Consider Figure P11.72, in which block A is an ideal voltage amplifier and block B is an ideal transresistance amplifier. **a.** Derive an expression for $v_o(t)$ in terms of the amplifier gains, $v_{\text{in}}(t)$, and the capacitance C. **b.** Derive an expression for the overall voltage gain of the system as a function of frequency. [*Hint:* Assume that $v_{\text{in}}(t) = V_m\cos(2\pi ft)$, determine the expression for $v_o(t)$, and then determine the complex voltage gain by taking the ratio of the phasors for the input and output.] **c.** Given $R_{moc} = 10^3\ \Omega$, $A_{voc} = 50/\pi$, and $C = 1\ \mu$F, sketch Bode plots of the voltage-gain magnitude and phase to scale for the range from 1 Hz to 1 kHz.

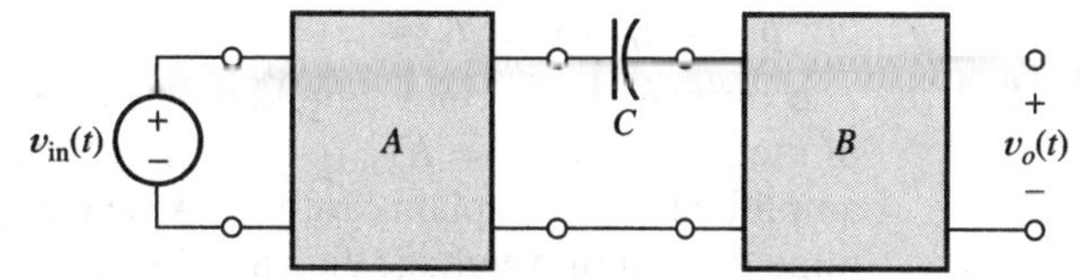

Figure P11.72

Section 11.8: Linear Waveform Distortion

P11.73. List the types of linear distortions and also discuss how their occurrence can be avoided.

P11.74. The input signal to an amplifier is $v_i(t) = 0.01\cos(2000\pi t) + 0.02\cos(4000\pi t)$. The gain of the amplifier as a function of frequency is given by

$$A = \frac{100}{1 + j(f/1000)}$$

Find an expression for the output signal of the amplifier as a function of time.

***P11.75.** The input signal to an amplifier is $v_{\text{in}}(t) = 0.01\cos(2000\pi t) + 0.02\cos(4000\pi t)$. The complex gain of the amplifier at 1000 Hz is $100\angle{-45^\circ}$. What complex value must the gain have at 2000 Hz for distortionless amplification? Sketch or write a computer program to plot the input and output waveforms to scale versus time.

P11.76. The output of an amplifier used to create special effects for audio signals is given by

$$v_o(t) = v_{\text{in}}(t) + Kv_{\text{in}}(t - t_d)$$

in which K and t_d are constants. **a.** Is this amplifier linear? Explain carefully. **b.** Determine the complex voltage gain as a function of frequency. [*Hint:* Assume that $v_{\text{in}}(t) = V_m\cos(2\pi ft)$, determine the corresponding output, and divide the phasor output by the phasor input.] **c.** Given $K = 0.5$ and $t_d = 1$ ms, use a computer to plot the gain magnitude and phase versus frequency for $0 \le f \le 10$ kHz. **d.** Does this amplifier produce amplitude distortion? Phase distortion? Explain carefully.

P11.77. Repeat Problem P11.76 for

$$v_o(t) = v_{\text{in}}(t) + K\frac{d}{dt}v_{\text{in}}(t)$$

For part (c), assume $K = 1/(2000\pi)$.

P11.78. The output of a certain amplifier in terms of the input is $v_o(t) = Kv_{\text{in}}(t - t_d)$. **a.** Is this amplifier linear? Explain carefully. **b.** Determine the complex voltage gain as a function of frequency. [*Hint:* Assume that $v_{\text{in}}(t) = V_m\cos(2\pi ft)$, determine the corresponding output, and divide the phasor output by the phasor input.] **c.** Given $K = 100$ and $t_d = 0.1$ ms, plot the gain magnitude and phase versus frequency for $0 \le f \le 10$ kHz. **d.** Does this amplifier produce amplitude distortion? Phase distortion? Explain carefully.

Section 11.9: Pulse Response

P11.79. Sketch the pulse response of an amplifier, showing the rise time, overshoot, ringing, and tilt. Give an approximate relationship between rise time and the upper half-power frequency of a broadband amplifier. Give an approximate relationship between percentage tilt and the lower half-power frequency.

P11.80. Consider the simple lowpass filter shown in Figure P11.80. **a.** Find the complex gain $A = \mathbf{V}_2/\mathbf{V}_1$ as a function of frequency. What are the magnitudes of A at dc and at very high frequencies? Find the half-power bandwidth B of the circuit in terms of R and C. **b.** Consider the case for which the capacitor is initially uncharged and $v_1(t)$ is a unit-step function. Find $v_2(t)$ and an expression for the rise time t_r of the circuit in terms of R and C. **c.** Combine the results found in parts (a) and (b) to obtain a relationship between bandwidth and rise time for this circuit. Compare your result with Equation 11.11 on page 559.

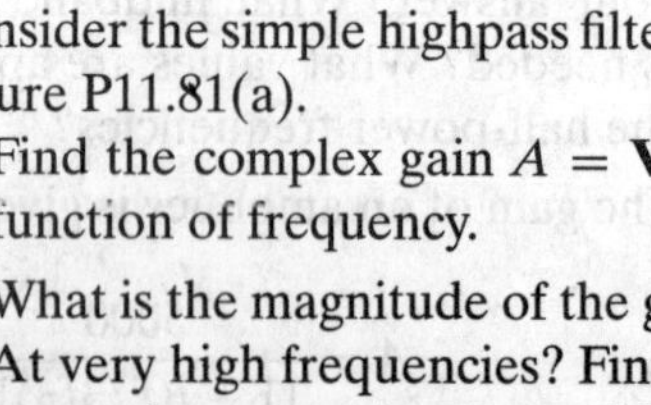

Figure P11.80

P11.81. Consider the simple highpass filter shown in Figure P11.81(a).

a. Find the complex gain $A = \mathbf{V}_2/\mathbf{V}_1$ as a function of frequency.

b. What is the magnitude of the gain at dc? At very high frequencies? Find the half-power frequency in terms of R and C.

c. Consider the input pulse shown in Figure P11.81(b). Assuming that the capacitor is initially uncharged, find an expression for the output voltage $v_2(t)$ for t between 0 and T. Assuming that RC is much greater than T, find an approximate expression for percentage tilt.

d. Combine the results of parts (b) and (c) to find a relationship between percentage tilt and the half-power frequency.

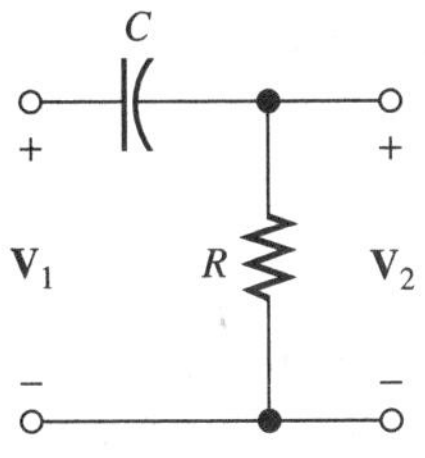

(a)

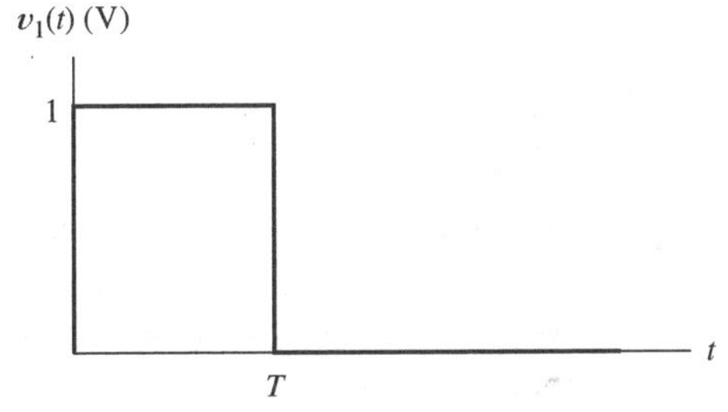

(b)

Figure P11.81

***P11.82.** An audio amplifier is specified to have half-power frequencies of 15 Hz and 15 kHz. The amplifier is to be used to amplify the pulse shown in Figure P11.81(b). Estimate the rise time and tilt of the amplifier output. The pulse width T is 2 ms.

***P11.83.** The gain magnitudes of several amplifiers are shown versus frequency in Figure P11.83. If the input to the amplifiers is the pulse shown in the figure, sketch the output of each amplifier versus time. Give quantitative estimates of as many features on each waveform sketch as you can.

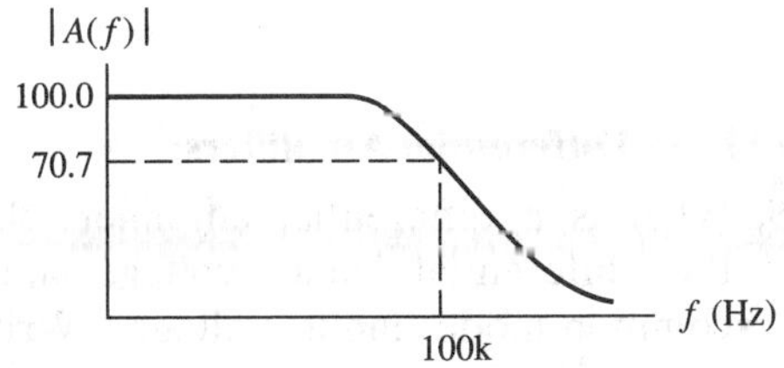

(a)

Figure P11.83

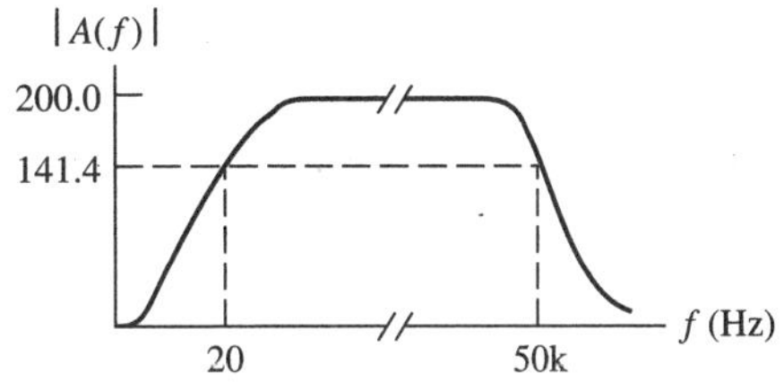

(b)

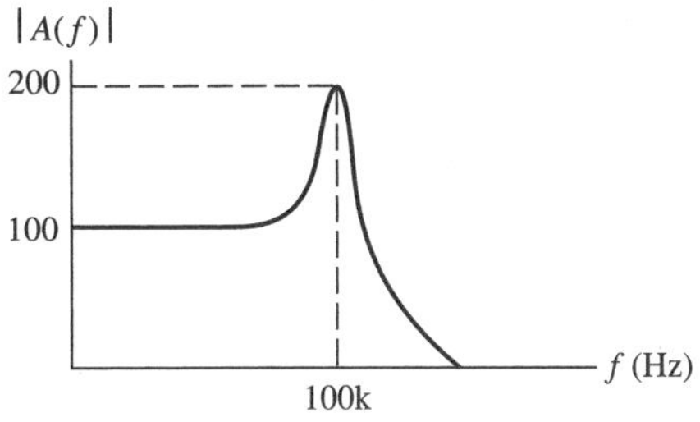

(c)

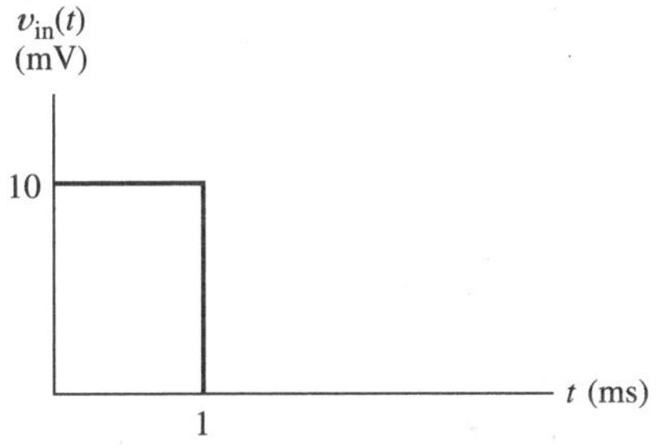

(d)

Figure P11.83 *(Cont.)*

P11.84. The input signal and corresponding output signals are shown for several amplifiers in Figure P11.84. Sketch the gain magnitude of

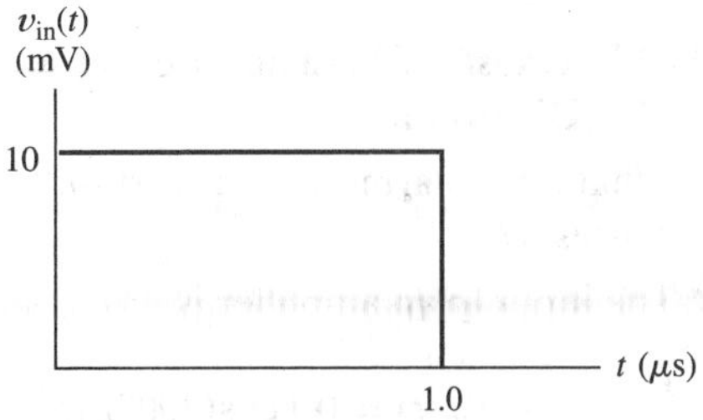

(a)

Figure P11.84

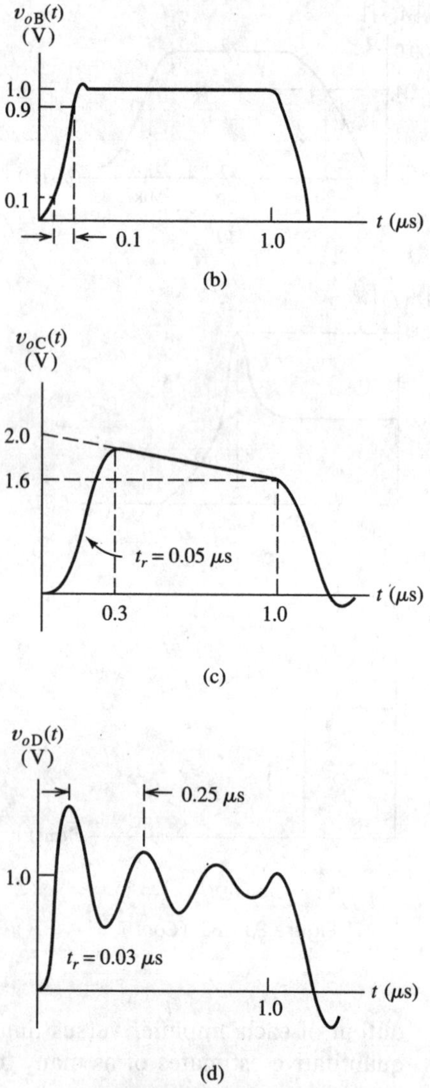

Figure P11.84 *(Cont.)*

each amplifier versus frequency. Give quantitative estimates of as many features on the gain sketches as you can.

Section 11.10: Transfer Characteristic and Nonlinear Distortion

P11.85. What is harmonic distortion? What causes it?

***P11.86.** The input to an amplifier is

$$v_{\text{in}}(t) = 0.1\cos(2000\pi t)$$

and the corresponding output is

$$v_o(t) = 10\cos(2000\pi t) + 0.2\cos(4000\pi t) + 0.1\cos(6000\pi t)$$

Determine the distortion factors D_2, D_3, and D_4. Also, determine the percentage of total harmonic distortion.

P11.87. The transfer characteristic of an amplifier is described by the equation

$$v_o(t) = 10v_{\text{in}}(t) + 0.6v_{\text{in}}^2(t) + 0.4v_{\text{in}}^3(t)$$

For the input $v_{\text{in}}(t) = 2\cos(200\pi t)$, determine the distortion factors D_2, D_3, and D_4. Also, compute the total harmonic distortion. You may find the following trigonometric identities useful:

$$\cos^2(A) = \frac{1}{2} + \frac{1}{2}\cos(2A)$$

$$\cos^3(A) = \frac{3}{4}\cos(A) + \frac{1}{4}\cos(3A)$$

P11.88. The transfer characteristic of an amplifier is described by the equation

$$v_o(t) = v_{\text{in}}(t) + 0.1v_{\text{in}}^2(t)$$

For the input $v_{\text{in}}(t) = \cos(\omega_1 t) + \cos(\omega_2 t)$, determine the frequency and amplitude of each component of the output. You may find the following trigonometric identities useful:

$$\cos^2(A) = \frac{1}{2} + \frac{1}{2}\cos(2A)$$

$$\cos(A)\cos(B) = \frac{1}{2}\cos(A - B) + \frac{1}{2}\cos(A + B)$$

Section 11.11: Differential Amplifiers

P11.89. What is a differential amplifier? Define the differential input voltage and the common-mode input voltage. Write an expression for the output in terms of the differential and common-mode input components.

P11.90. Define CMRR and discuss why high CMRR is desired in a differential amplifier.

P11.91. In your own words, describe a situation in which a small differential signal is of interest and a large common-mode signal is also present.

P11.92. The input signals v_{i1} and v_{i2} shown in Figure P11.92 are the inputs to a differential amplifier with a gain of $A_d = 10$. (Assume that the common-mode gain is zero.) Sketch the output of the amplifier to scale versus time. Sketch the common-mode input signal to scale versus time.

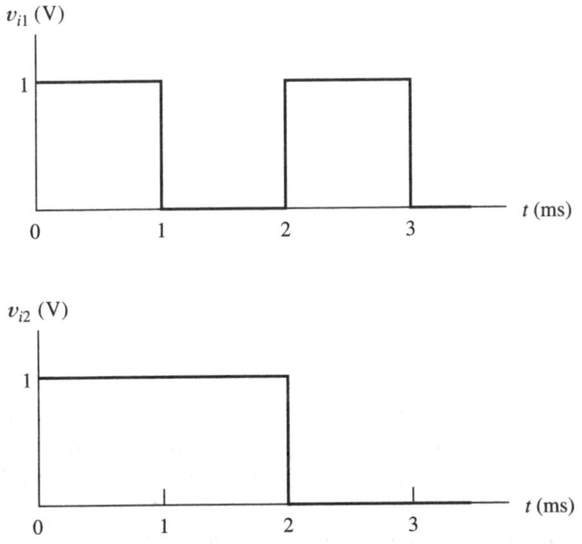

Figure P11.92

***P11.93.** A certain amplifier has a differential gain of 1000. If the two input terminals are tied together and a 5-mV-rms input signal is applied, the output signal is 30 mV rms. Find the CMRR for this amplifier.

P11.94. In a certain instrumentation amplifier, the input signal consists of a 20-mV-rms differential signal and a 5-V-rms 60-Hz interfering common-mode signal. It is desired that the common-mode contribution to the output signal be at least 60 dB lower than the contribution from the differential signal. What is the minimum CMRR allowed for the amplifier in decibels?

P11.95. The output of a certain instrumentation amplifier in terms of the inputs is $v_o(t) = 750v_{i1}(t) - 450v_{i2}(t)$. Determine the CMRR of this amplifier in decibels.

Section 11.12: Offset Voltage, Bias Current, and Offset Current

P11.96. Draw the differential amplifier symbol and show sources for the offset voltage, bias current, and offset current. What effect do these sources have on the output signal of the amplifier?

P11.97. Sketch the circuit diagram of a balancing circuit for a differential amplifier.

***P11.98.** A differential amplifier has a differential gain of 500 and negligible common-mode gain. The input terminals are tied to ground through 1-kΩ resistors having tolerances of ±5 percent. What are the extreme values of the output voltage caused by a bias current of 100 nA? What is the output voltage if the resistors are exactly equal?

***P11.99.** A differential amplifier has a bias current of 100 nA, a maximum offset current of 20 nA, a maximum offset voltage of 2 mV, an input resistance of 1 MΩ, and a differential gain of 1000. The input terminals are tied to ground through (exactly equal) 100-kΩ resistors. Find the extreme values of the output voltage if the common-mode gain is assumed to be zero.

P11.100. Repeat Problem P11.99 if the CMRR of the amplifier is 60 dB. By what percentage is the extreme output voltage increased in this case, compared with zero common-mode gain?

P11.101. A differential amplifier, including sources to model its dc imperfections, is shown under three different test conditions in Figure P11.101. The amplifier has a differential voltage gain of 100, a common-mode voltage gain of zero, and infinite input impedance. Determine the values of V_{off}, I_B, and I_{off}.

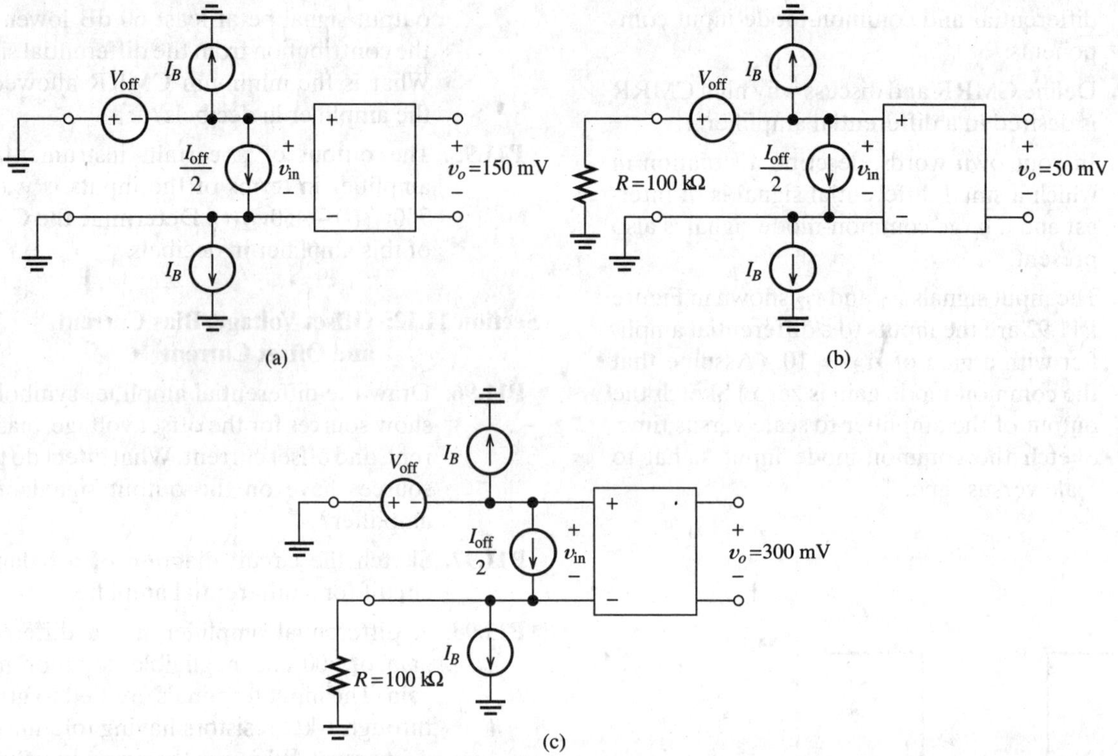

Figure P11.101

Frequently Asked Questions

1. What is an amplifier? Define voltage gain of an amplifier.
2. What do you mean by inverting and non-inverting amplifiers?
3. What do you mean by common ground node?
4. What do you understand by the following terms used in amplifiers:
   (i) Input resistance
   (ii) Input impedance
   (iii) Output resistance
   (iv) Output impedance
5. Define the following terms related to amplifiers
   (i) Open-circuit voltage gain
   (ii) Current gain
   (iii) Power gain
6. What is loading effect in an amplifier?
7. What do you mean by cascading of amplifiers? What is its use?
8. How will you obtain a simplified model for a cascaded amplifier.
9. What do you mean by a Power supply? Derive a relation for the efficiency of a power supply.
10. Draw the circuit of a current amplifier model and explain its functioning.
11. What do you mean by transconductance amplifier? How will you define the short-circuit transconductance gain of a transconductance amplifier?
12. What do you mean by an open-circuit transresistance gain?
13. What is the importance of amplifier impedances in various applications?
14. What do you understand by
    (i) An ideal voltage-amplifier
    (ii) An ideal current-amplifier
    (iii) An ideal transconductance-amplifier
    (iv) An ideal transresistance-amplifier
15. How will you classify real amplifiers?
16. What do you mean by frequency response of an amplifier?

17. Differentiate between ac coupling and direct coupling.
18. What do you understand by half-power frequencies and band-width?
19. Differentiate between wideband and narrow-band amplifiers.
20. What are the various distortions which may be introduced in the output of an amplifier?
21. What do you understand by amplitude distortion and phase distortion.
22. Why do we prefer distortionless amplification?
23. Define (i) pulse response (ii) rise time (iii) overshoot and ringing (iv) tilt
24. Define and explain harmonic distortion in amplifiers.
25. What are differentiate amplifiers? What do you mean by inverting input and non-inverting input?
26. Define the following
    (i) Differential gain
    (ii) Common-mode signal
    (iii) Common-mode rejection ratio (CMRR)
27. How will you measure CMRR?
28. What do you mean by
    (i) Offset voltage
    (ii) Bias current
    (iii) Offset current
29. How can you minimize the effect of bias current.
30. Draw and explain the balancing circuit for a differential amplifier.

Practice Test

Here is a practice test you can use to check your comprehension of the most important concepts in this chapter. Answers can be found in Appendix D and complete solutions are included in the Student Solutions files. See Appendix E for more information about the Student Solutions.

T11.1. Suppose we have two identical amplifiers each of which has $A_{voc} = 50$, $R_i = 60\ \Omega$, and $R_o = 40\ \Omega$ connected in cascade. Determine the open-circuit voltage gain, the input resistance, and the output resistance of the cascade.

T11.2. Prepare a table showing the gain parameter, the input impedance, and the output impedance for the four types of ideal amplifiers.

T11.3. Suppose we have a sensor with a variable internal impedance and a load of variable impedance. What type of ideal amplifier would be best if we need: **a.** the load current to be proportional to the Thévenin voltage of the source; **b.** the load current to be proportional to the short-circuit current of the source; **c.** the load voltage to be proportional to the open-circuit voltage of the source; **d.** the load voltage to be proportional to the short-circuit current of the source?

T11.4. Suppose we have an amplifier with $R_i = 200\ \Omega$, $R_o = 1\ \text{k}\Omega$, and $A_{isc} = 50$. Determine the values (including units) of A_{voc}, R_{moc}, and G_{msc} for this amplifier. Then draw each of the four models for the amplifier showing the value of each parameter.

T11.5. We have an amplifier that draws 2 A from a 15-V dc power supply. The input signal current is 1 mA rms, and the input resistance is 2 kΩ. The amplifier delivers 12 V rms to an 8-Ω load. Determine the power dissipated in the amplifier and the efficiency of the amplifier.

T11.6. Suppose that the input voltage to an amplifier has a peak amplitude of 100 mV and contains components with frequencies ranging from 1 to 10 kHz. We want the output voltage waveform to be nearly identical to that of the input except with larger amplitude (by a factor of 100) and possibly delayed in time. Based on this information, what specifications can you make for the amplifier?

T11.7. What is the principal effect of offset current, bias current, and offset voltage of an amplifier on the signal being amplified?

T11.8. What is harmonic distortion? What causes it?

T11.9. What is the CMRR? For what types of applications is it important?

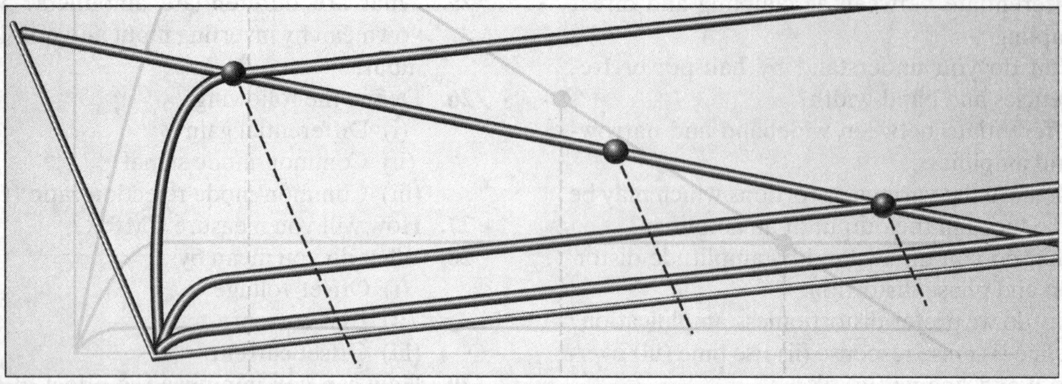

Chapter 12

Field-Effect Transistors

Study of this chapter will enable you to:

- Understand MOSFET operation.
- Use the load-line technique to analyze basic FET amplifiers.
- Analyze bias circuits.
- Use small-signal equivalent circuits to analyze FET amplifiers.
- Compute the performance parameters of several FET amplifier configurations.
- Select a FET amplifier configuration that is appropriate for a given application.
- Understand the basic operation of CMOS logic gates.

Introduction to this chapter:

Field-effect transistors (FETs) are important devices that are used in amplifiers and logic gates. In this chapter, we discuss the enhancement-mode **metal–oxide–semiconductor field-effect transistor** (MOSFET), which is the primary device that underlies the rapid advances in digital electronics over the past several decades. (Several other types of FETs exist, but to simplify our discussion, we only discuss enhancement-mode MOSFETs, which are presently the most important type.)

In the next chapter, we will discuss another device, the bipolar junction transistor, which is also used in amplifiers and logic gates. Compared to BJTs, MOSFETs can occupy less chip area and can be fabricated with fewer processing steps. Complex digital circuits such as memories and microprocessors are often implemented solely with MOSFETs. On the other hand, BJTs are capable of producing large output currents that are needed for fast switching of a capacitive load such as a circuit-board trace interconnecting digital chips. Each type of device has some applications in which it performs better than the others.

12.1 NMOS AND PMOS TRANSISTORS

Overview

The physical structure of an n-channel enhancement-mode MOSFET (also known as an NMOS transistor) is shown in Figure 12.1. It is a chip of silicon crystal with impurities added to the various regions to produce n-type and p-type material. In n-type material, conduction is due mainly to negatively charged electrons, whereas in p-type material, conduction is due mainly to positively charged holes. (Although positively charged particles called holes are a helpful concept for understanding conduction in semiconductors, they are not fundamental subatomic particles, as electrons are.)

The device terminals are the **drain** (D), **gate** (G), **source** (S), and **body** (B). (Another commonly used term for the body is substrate.) In normal operation, negligible current flows through the body terminal. Sometimes, the body is connected to the source so that we have a three-terminal device. The gate is insulated from the substrate by a thin layer of silicon dioxide, and negligible current flows through the gate terminal. When a sufficiently large (positive) voltage is applied to the gate relative to the source, electrons are attracted to the region under the gate, and a channel of n-type material is induced between the drain and the source. Then, if voltage is applied between the drain and the source, current flows into the drain through the channel and out the source. Drain current is controlled by the voltage applied to the gate.

Drain current is controlled by the voltage applied to the gate.

Although the acronym MOS stands for metal–oxide–semiconductor, the gates of modern MOSFETs are actually composed of polysilicon.

The channel length L and width W are illustrated in Figure 12.1. The trend over the past four decades has been to reduce both L and W in order to fit an ever-increasing number of transistors into a given chip area. In 1971, Intel introduced the first microprocessor, the 4004, which had a minimum feature size of 10 μm and contained 2300 transistors. By 2012, processor chips had channel lengths of 22 nm, oxide thickness on the order of 1 nm, and in excess of one billion transistors. This has led to the remarkable performance improvements that we have come to expect for

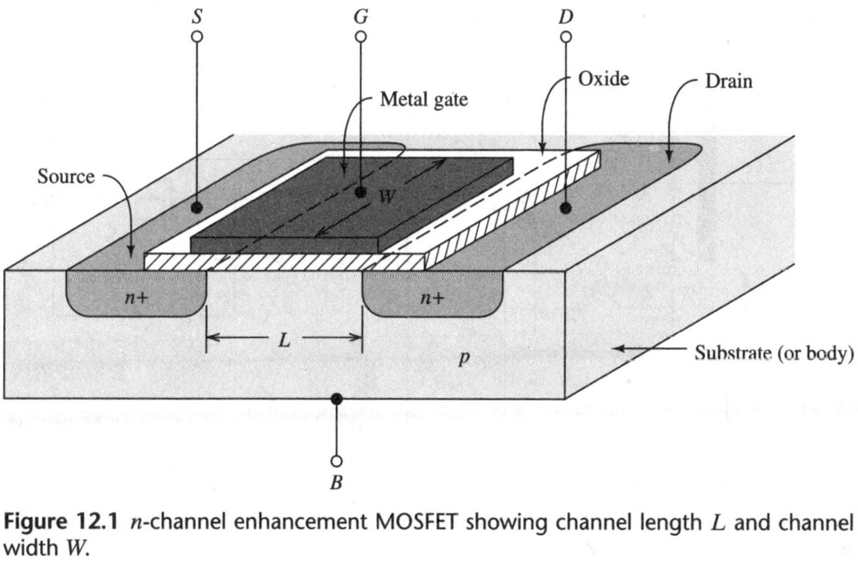

Figure 12.1 n-channel enhancement MOSFET showing channel length L and channel width W.

computers and other electronic products. Furthermore, these trends are expected to continue.

The circuit designer can adjust L and W to obtain the device best suited for a given application.

Device characteristics depend on L, W, and process parameters such as doping levels and oxide thickness. Usually, the process parameters are predetermined, but the circuit designer can adjust L and W to obtain the device best suited for a given application.

The circuit symbol for the n-channel enhancement MOSFET is shown in Figure 12.2. Next, we discuss the basic operation of this device.

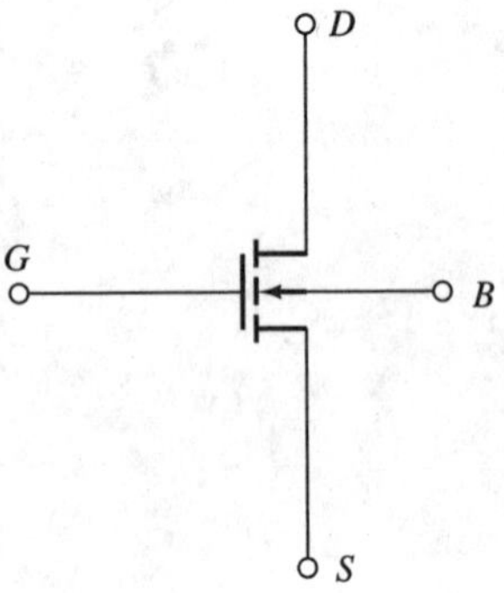

Figure 12.2 Circuit symbol for an enhancement-mode n-channel MOSFET.

Operation in the Cutoff Region

Consider the situation shown in Figure 12.3. Suppose that positive voltage v_{DS} is applied to the drain relative to the source and that we start with $v_{GS} = 0$. Notice that pn junctions (i.e., diodes) appear at the drain/body and at the source/body interfaces. Under forward bias (positive on the p-side), electrons flow easily across a pn junction, but under reverse bias (positive on the n-side), virtually no current flows. Thus, virtually no current flows into the drain terminal because the drain/body junction is reverse biased by the v_{DS} source. This is called the **cutoff region** of operation. As v_{GS} is increased, the device remains in cutoff until v_{GS} reaches a particular value called the **threshold voltage** V_{to}. Typically, the threshold voltage ranges from a fraction of a volt to one volt. Thus, in cutoff, we have

Key equation for enhancement NMOS in cutoff.

$$i_D = 0 \text{ for } v_{GS} \leq V_{to} \tag{12.1}$$

Operation in the Triode Region

The triode region is also called the linear region of operation.

For $v_{DS} < v_{GS} - V_{to}$ and $v_{GS} \geq V_{to}$, we say that the NMOS is operating in the **triode region**. Consider the situation shown in Figure 12.4, in which v_{GS} is greater than the threshold voltage. The electric field resulting from the applied gate voltage has

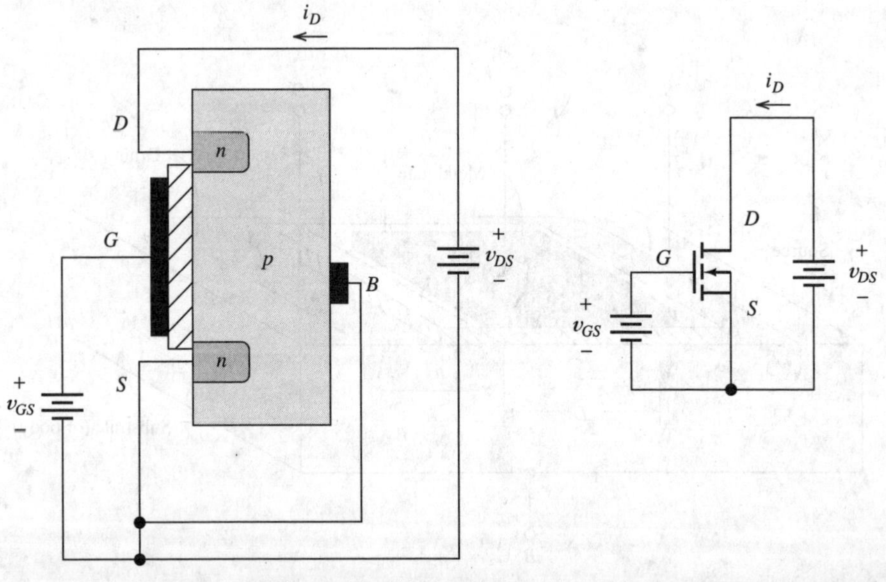

Figure 12.3 For $v_{GS} < V_{to}$, the pn junction between drain and body is reverse biased and $i_D = 0$.

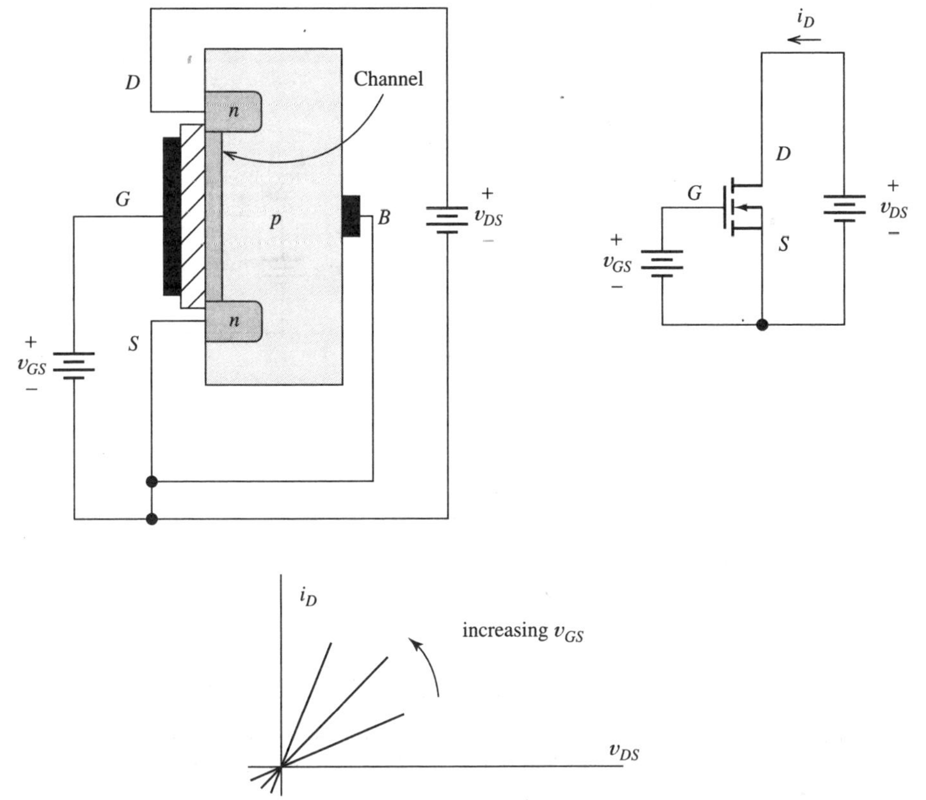

Figure 12.4 For $v_{GS} > V_{to}$, a channel of n-type material is induced in the region under the gate. As v_{GS} increases, the channel becomes thicker. For small values of v_{DS}, i_D is proportional to v_{DS}. The device behaves as a resistance whose value depends on v_{GS}.

repelled holes from the region under the gate and attracted electrons that can easily flow in the forward direction across the source/body junction. This results in an n-type channel between the drain and the source. Then, when v_{DS} is increased, current flows into the drain, through the channel, and out the source. For small values of v_{DS}, the drain current is proportional to v_{DS}. Furthermore, for a given (small) value of v_{DS}, drain current is also proportional to the **excess gate voltage** $v_{GS} - V_{to}$.

Plots of i_D versus v_{DS} are shown in Figure 12.4 for several values of gate voltage. In the triode region, the NMOS device behaves as a resistor connected between drain and source, but the resistance decreases as v_{GS} increases.

In the triode region, the NMOS device behaves as a resistor connected between drain and source, but the resistance decreases as v_{GS} increases.

Now, consider what happens if we continue to increase v_{DS}. Because of the current flow, the voltages between points along the channel and the source become greater as we move toward the drain. Thus, the voltage between gate and channel becomes smaller as we move toward the drain, resulting in tapering of the channel thickness as illustrated in Figure 12.5. Because of the tapering of the channel, its resistance becomes larger with increasing v_{DS}, resulting in a lower rate of increase of i_D.

For $v_{DS} < v_{GS} - V_{to}$ and $v_{GS} \geq V_{to}$, the device is operating in the triode region, and the drain current is given by

Key equation for enhancement NMOS in the triode region.

$$i_D = K\left[2(v_{GS} - V_{to})v_{DS} - v_{DS}^2\right] \qquad (12.2)$$

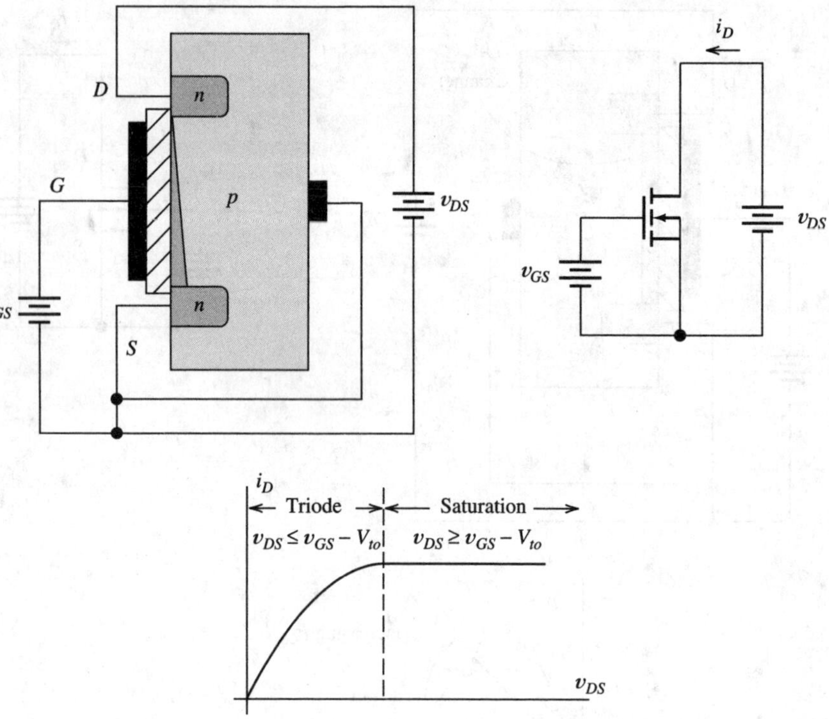

Figure 12.5 As v_{DS} increases, the channel pinches down at the drain end and i_D increases more slowly. Finally, for $v_{DS} > v_{GS} - V_{to}$, i_D becomes constant.

in which K is given by

$$K = \left(\frac{W}{L}\right)\frac{KP}{2} \tag{12.3}$$

A typical value of KP for n-channel enhancement devices is 50 $\mu A/V^2$.

As illustrated in Figure 12.1, W is the width of the channel and L is its length. The device parameter KP depends on the thickness of the oxide layer and certain properties of the channel material. A typical value of KP for n-channel enhancement devices is 50 $\mu A/V^2$.

Usually, KP is determined by the fabrication process. However, in designing a circuit, we can vary the ratio W/L to obtain transistors best suited to various parts of the circuit. The condition $v_{DS} \le v_{GS} - V_{to}$ is equivalent to $v_{GD} \ge V_{to}$. Thus, the device is in the triode region if both v_{GD} and v_{GS} are greater than the threshold voltage V_{to}.

Operation in the Saturation Region

We have seen that as v_{DS} is increased the voltage between the gate and the drain end of the channel decreases. When the gate-to-drain voltage v_{GD} equals the threshold voltage V_{to}, the channel thickness at the drain end becomes zero. For further increases in v_{DS}, i_D is constant, as illustrated in Figure 12.5. This is called the **saturation region**, in which we have $v_{GS} \ge V_{to}$ and $v_{DS} \ge v_{GS} - V_{to}$, and the current is given by

Key equation for enhancement NMOS in the saturation region.

$$i_D = K(v_{GS} - V_{to})^2 \tag{12.4}$$

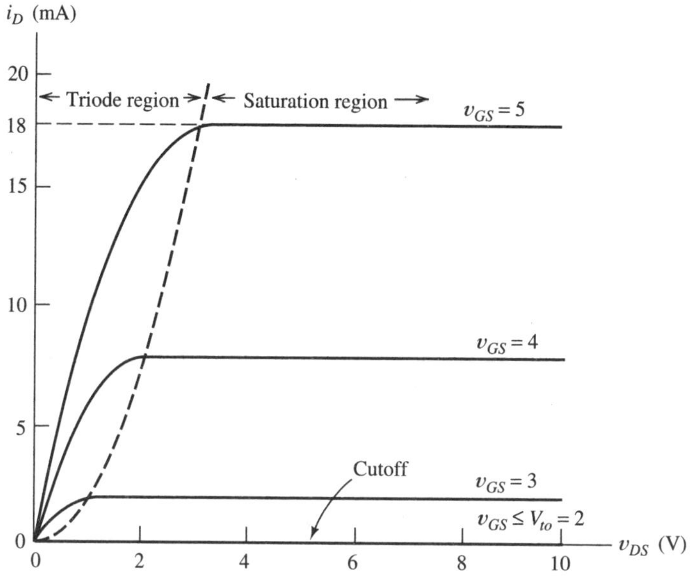

Figure 12.6 Characteristic curves for an NMOS transistor.

Keep in mind that in the saturation region, v_{GS} is greater than the threshold, but v_{GD} is less than the threshold. Figure 12.6 shows the drain characteristics of an NMOS transistor.

Boundary between the Triode and Saturation Regions

Next, we derive the equation for the boundary between the triode region and the saturation region in the i_D–v_{DS} plane. At this boundary, the channel thickness at the drain is zero, which occurs when $v_{GD} = V_{to}$. Thus, we want to find i_D in terms of v_{DS} under the condition that $v_{GD} = V_{to}$. Since $v_{GD} = v_{GS} - v_{DS}$, the condition at the boundary is given by

$$v_{GS} - v_{DS} = V_{to} \tag{12.5}$$

Solving this for v_{GS}, substituting into Equation 12.4, and reducing, we have the desired boundary equation given by

Key equation for the boundary between the triode and saturation regions.

$$i_D = Kv_{DS}^2 \tag{12.6}$$

Notice that the boundary between the triode region and the saturation region is a parabola.

Solving Equation 12.5 for v_{GS} and substituting into Equation 12.2 also produces Equation 12.6. (Equations 12.2 and 12.4 give the same values for i_D on the boundary.)

Given the values for KP, L, W, and V_{to}, we can plot the static characteristics of an NMOS transistor.

Example 12.1 Plotting the Characteristics of an NMOS Transistor

A certain enhancement-mode NMOS transistor has $W = 160\ \mu\text{m}$, $L = 2\ \mu\text{m}$, $KP = 50\ \mu\text{A/V}^2$, and $V_{to} = 2$ V. Plot the drain characteristic curves to scale for $v_{GS} = 0, 1, 2, 3, 4,$ and 5 V.

Solution First, we use Equation 12.3 to compute the device constant:

$$K = \left(\frac{W}{L}\right)\frac{KP}{2} = 2\ \text{mA/V}^2$$

Equation 12.6 gives the boundary between the triode region and the saturation region. Thus, we have

$$i_D = Kv_{DS}^2 = 2v_{DS}^2$$

where i_D is in mA and v_{DS} is in volts. The plot of this equation is the dashed line shown in Figure 12.6.

Next, we use Equation 12.4 to compute the drain current in the saturation region for each of the v_{GS} values of interest. Hence, we get

$$i_D = K(v_{GS} - V_{to})^2 = 2(v_{GS} - 2)^2$$

where again the current is in mA. Substituting values, we find that

$$i_D = 18\ \text{mA for } v_{GS} = 5\ \text{V}$$
$$i_D = 8\ \text{mA for } v_{GS} = 4\ \text{V}$$
$$i_D = 2\ \text{mA for } v_{GS} = 3\ \text{V}$$
$$i_D = 0\ \text{mA for } v_{GS} = 2\ \text{V}$$

For $v_{GS} = 0$ and 1 V, the device is in cutoff and $i_D = 0$. These values are plotted in the saturation region as shown in Figure 12.6.

Finally, Equation 12.2 is used to plot the characteristics in the triode region. For each value of v_{GS}, this equation plots as a parabola that passes through the origin ($i_D = 0$ and $v_{DS} = 0$). The apex of each parabola is on the boundary between the triode region and the saturation region. ■

Summary: In an NMOS transistor, when a sufficiently large (positive) voltage is applied to the gate relative to the source, electrons are attracted to the region under the gate, and a channel of n-type material is induced between the drain and the source. Then, if positive voltage is applied to the drain relative to the source, current flows into the drain through the channel and out the source. Drain current is controlled by the voltage applied to the gate.

Exercise 12.1 Consider an NMOS having $V_{to} = 2$ V. What is the region of operation (triode, saturation, or cutoff) if: **a.** $v_{GS} = 1$ V and $v_{DS} = 5$ V; **b.** $v_{GS} = 3$ V and $v_{DS} = 0.5$ V; **c.** $v_{GS} = 3$ V and $v_{DS} = 6$ V; **d.** $v_{GS} = 5$ V and $v_{DS} = 6$ V?

Answer **a.** cutoff; **b.** triode; **c.** saturation; **d.** saturation. □

Exercise 12.2 Suppose that we have an NMOS transistor with $KP = 50\ \mu\text{A/V}^2$, $V_{to} = 1$ V, $L = 2\ \mu$m, and $W = 80\ \mu$m. Sketch the drain characteristics for v_{DS} ranging from 0 to 10 V and $v_{GS} = 0, 1, 2, 3,$ and 4 V.

Answer The plots are shown in Figure 12.7. □

PMOS Transistors

MOSFETs can also be constructed by interchanging the n and p regions of n-channel devices, resulting in p-channel devices. The circuit symbol for the p-channel MOSFET is shown in Figure 12.8. As indicated in the figure, we usually orient the p-channel FETs with the source at the top and reference the current out of the drain. The PMOS symbol is the same as the NMOS symbol, except for orientation and the direction of the arrowhead.

The characteristics of a PMOS transistor are very similar to those for the NMOS transistor, except that voltage polarities are inverted. Because we reference the drain current into the drain for n-channel devices and out of the drain for p-channel devices,

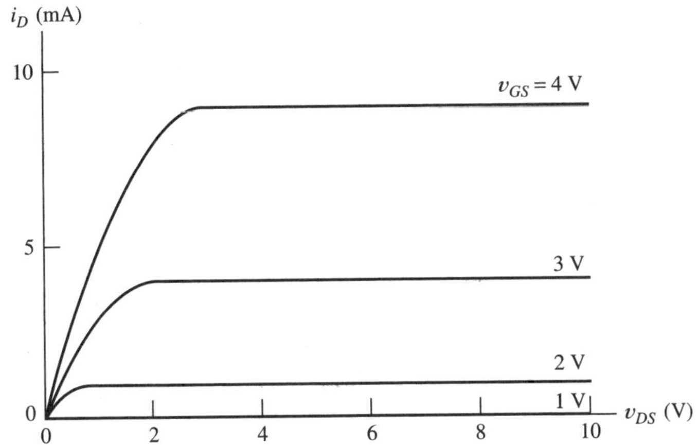

Figure 12.7 Answer for Exercise 12.2.

the drain current assumes positive values in both devices. Thus, the characteristic curves of a p-channel device are like those of an n-channel device, except that the algebraic signs of the voltages must be inverted.

Table 12.1 gives the equations of operation for both NMOS and PMOS enhancement-mode transistors. A typical value of KP for PMOS transistors is 25 $\mu A/V^2$, which is about half the value for NMOS transistors. This is due to differences in the conduction properties of electrons and holes in silicon. Notice that the threshold voltage for an enhancement-mode PMOS transistor assumes a negative value.

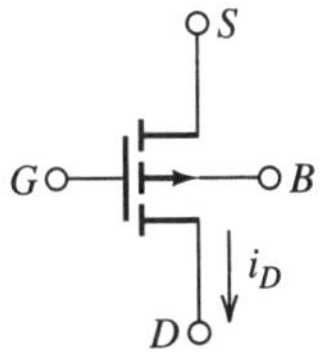

Figure 12.8 Circuit symbol for PMOS transistor.

Table 12.1. MOSFET Summary

	NMOS	PMOS
Circuit symbol	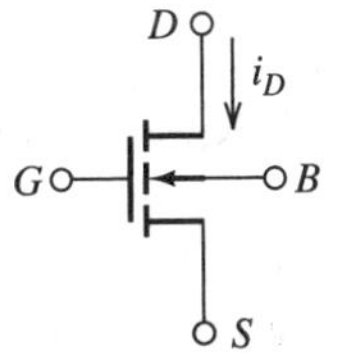	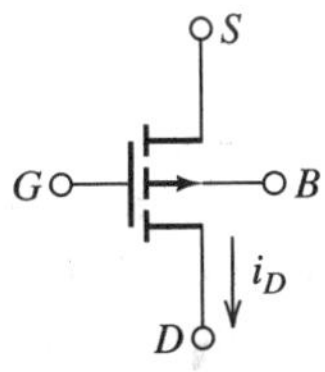
KP (typical value)	50 $\mu A/V^2$	25 $\mu A/V^2$
K	(1/2) KP (W/L)	(1/2) KP (W/L)
V_{to} (typical value)	+1 V	−1 V
Cutoff region	$v_{GS} \le V_{to}$ $i_D = 0$	$v_{GS} \ge V_{to}$ $i_D = 0$
Triode region	$v_{GS} \ge V_{to}$ and $0 \le v_{DS} \le v_{GS} - V_{to}$ $i_D = K\,[2(v_{GS} - V_{to})\,v_{DS} - v_{DS}^2]$	$v_{GS} \le V_{to}$ and $0 \ge v_{DS} \ge v_{GS} - V_{to}$ $i_D = K\,[2(v_{GS} - V_{to})\,v_{DS} - v_{DS}^2]$
Saturation region	$v_{GS} \ge V_{to}$ and $v_{DS} \ge v_{GS} - V_{to}$ $i_D = K\,(v_{GS} - V_{to})^2$	$v_{GS} \le V_{to}$ and $v_{DS} \le v_{GS} - V_{to}$ $i_D = K\,(v_{GS} - V_{to})^2$
v_{DS} and v_{GS}	Normally assume positive values	Normally assume negative values

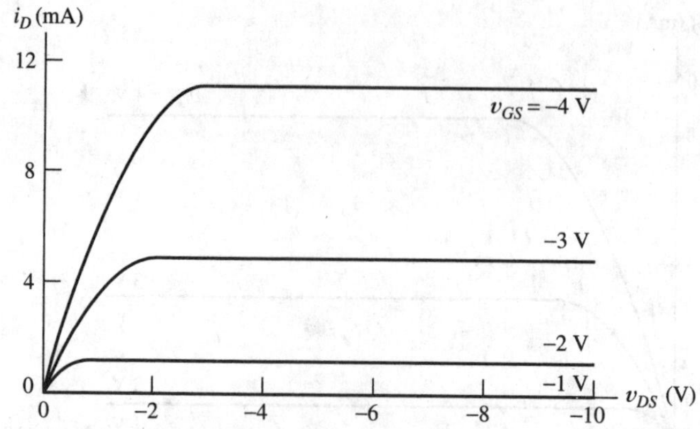

Figure 12.9 Answer for Exercise 12.3.

Exercise 12.3 Suppose that we have a PMOS transistor with $KP = 25\ \mu\text{A/V}^2$, $V_{to} = -1$ V, $L = 2\ \mu$m, and $W = 200\ \mu$m. Sketch the drain characteristics for v_{DS} ranging from 0 to −10 V and $v_{GS} = 0, -1, -2, -3$, and −4 V.
Answer The plots are shown in Figure 12.9. □

Channel-Length Modulation and Charge-Carrier-Velocity Saturation

The description of MOSFETs that we have given up to this point is reasonably accurate for devices with channel lengths in excess of about 10 μm. As the channel lengths become shorter, several effects that modify MOSFET behavior eventually come into play. One of these is channel-length modulation, which is caused because the effective length of the channel is reduced as v_{DS} increases in magnitude. Channel-length modulation causes the characteristics to slope upward in the saturation region and can be taken into account by including the factor $(1 + \lambda|v_{DS}|)$ in the expressions for i_D given in Table 12.1. The channel-length modulation factor λ is typically equal to $0.1/L$, in which L is the channel length in microns.

A second effect is caused by velocity saturation of the charge carriers. The equations we have given for i_D are based on the assumption that charge carrier velocity is proportional to electric field strength in the channel. However, at the higher field strengths encountered in devices with channels shorter than about 2 μm, carrier velocity tends to become more nearly constant as the field strength is increased. The result is that the characteristic curves enter saturation at smaller values of $|v_{DS}|$. Furthermore, i_D is more nearly linear with variations in v_{GS}, and the characteristic curves are more uniformly spaced than we have shown.

Although these and other secondary effects are important to advanced MOS designers, we can gain a basic understanding of MOSFET amplifiers and logic circuits by using the equations presented earlier in this section, except when otherwise noted.

12.2 LOAD-LINE ANALYSIS OF A SIMPLE NMOS AMPLIFIER

In this section, we analyze the NMOS amplifier circuit shown in Figure 12.10 by using a graphical load-line approach similar to the analysis we carried out for diode circuits in Section 10.2. The dc sources **bias** the MOSFET at a suitable operating point so that amplification of the input signal $v_{in}(t)$ can take place. We will see that

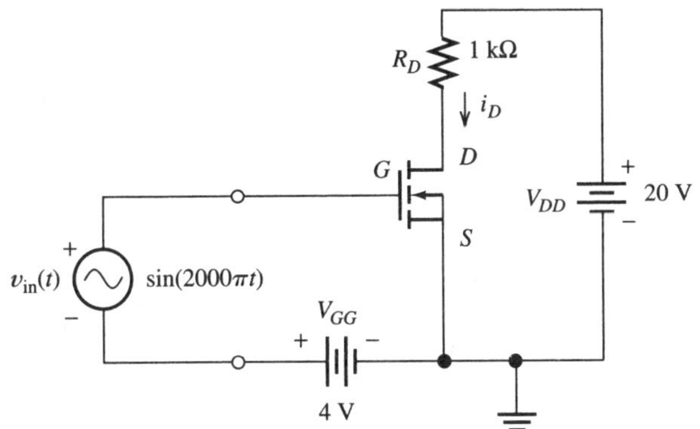

Figure 12.10 Simple NMOS amplifier circuit.

the input voltage $v_{in}(t)$ causes v_{GS} to vary with time, which in turn causes i_D to vary. The changing voltage drop across R_D causes an amplified version of the input signal to appear at the drain terminal.

The input voltage $v_{in}(t)$ causes v_{GS} to vary with time, which in turn causes i_D to vary. The changing voltage drop across R_D causes an amplified version of the input signal to appear at the drain terminal.

Applying Kirchhoff's voltage law to the input loop, we obtain the following expression:

$$v_{GS}(t) = v_{in}(t) + V_{GG} \tag{12.7}$$

For our example, we assume that the input signal is a 1-V-peak, 1-kHz sinusoid and that V_{GG} is 4 V. Then, we have

$$v_{GS}(t) = \sin(2000\pi t) + 4 \tag{12.8}$$

Writing a voltage equation around the drain circuit, we obtain

$$V_{DD} = R_D i_D(t) + v_{DS}(t) \tag{12.9}$$

Load-line equation.

For our example, we assume that $R_D = 1$ kΩ and $V_{DD} = 20$ V, so Equation 12.9 becomes

$$20 = i_D(t) + v_{DS}(t) \tag{12.10}$$

where we have assumed that $i_D(t)$ is in milliamperes. A plot of this equation on the drain characteristics of the transistor is a straight line called the **load line**.

To establish the load line, we first locate two points on it. Assuming that $i_D = 0$ in Equation 12.10, we find that $v_{DS} = 20$ V. These values plot as the lower right-hand end of the load line shown in Figure 12.11. For a second point, we assume that $v_{DS} = 0$, which yields $i_D = 20$ mA when substituted into Equation 12.10. This pair of values ($v_{DS} = 0$ and $i_D = 20$ mA) plots as the upper left-hand end of the load line.

To establish the load line, we first locate two points on it.

The operating point of an amplifier for zero input signal is called the **quiescent operating point** or ***Q* point**. For $v_{in}(t) = 0$, Equation 12.8 yields $v_{GS} = V_{GG} = 4$ V. Therefore, the intersection of the curve for $v_{GS} = 4$ V with the load line is the Q point. The quiescent values are $I_{DQ} = 9$ mA and $V_{DSQ} = 11$ V.

The term "quiescent" implies that the input signal is zero.

The maximum and minimum values of the gate-to-source voltage are $V_{GS\max} = 5$ V and $V_{GS\min} = 3$ V (see Equation 12.8). The intersections of the corresponding curves with the load line are labeled as points A and B, respectively, in Figure 12.11.

At point A, we find that $V_{DS\min} = 4$ V and $I_{D\max} = 16$ mA. At point B, we find that $V_{DS\max} = 16$ V and $I_{D\min} = 4$ mA.

Plots of $v_{GS}(t)$ and $v_{DS}(t)$ versus time are shown in Figure 12.12. Notice that the peak-to-peak swing of $v_{DS}(t)$ is 12 V, whereas the peak-to-peak swing of the input

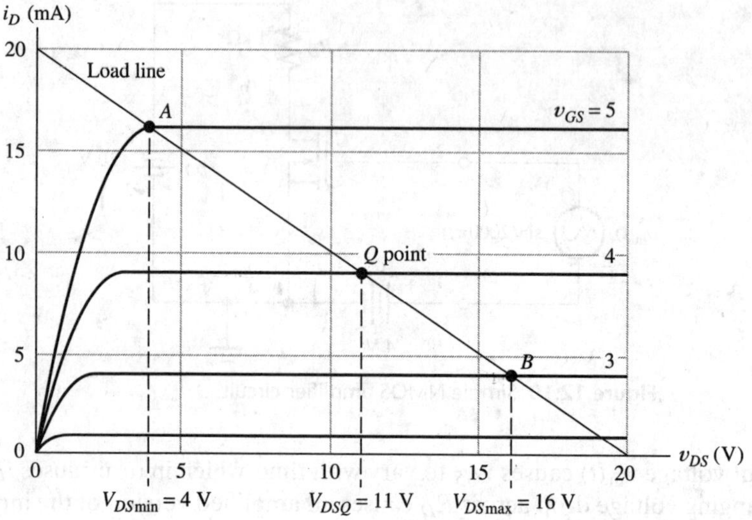

Figure 12.11 Drain characteristics and load line for the circuit of Figure 12.10.

signal is 2 V. Furthermore, the ac voltage at the drain is inverted compared to the input signal. (In other words, the positive peak of the input occurs at the same time as the minimum value of v_{DS}.) Therefore, this is an inverting amplifier. Apparently, the circuit has a voltage gain $A_v = -12\text{V}/2\text{V} = -6$, where the minus sign is due to the inversion.

Notice, however, that the output waveform shown in Figure 12.12(b) is not a symmetrical sinusoid like the input. For illustration, we see that starting from the Q

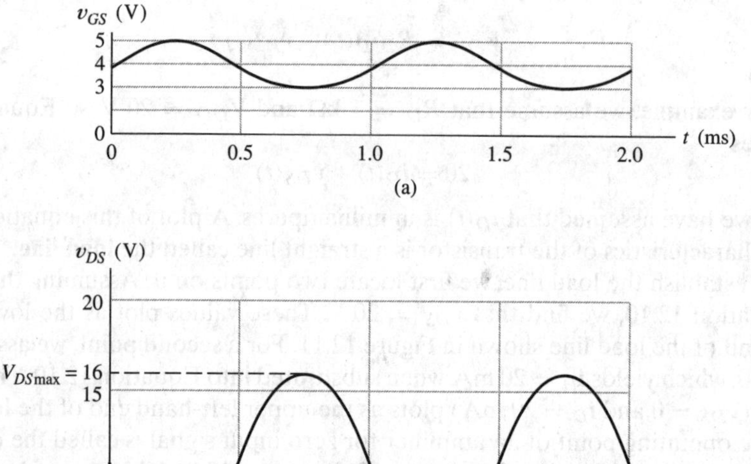

Figure 12.12 v_{GS} and v_{DS} versus time for the circuit of Figure 12.10.

point at $V_{DSQ} = 11$ V, the output voltage swings down to $V_{CE\min} = 4$ V for a change of 7 V. On the other hand, the output swings up to 16 V for a change of only 5 V from the Q point on the positive-going half cycle of the output. We cannot properly define gain for the circuit, because the ac output signal is not proportional to the ac input. Nevertheless, the output signal is larger than the input even if it is distorted. (This is an example of nonlinear distortion, which we discuss in Section 11.10.)

In this circuit, distortion is due to the fact that the characteristic curves for the FET are not uniformly spaced. If a much smaller input amplitude was applied, we would have amplification without appreciable distortion. This is true because the curves are more uniformly spaced if a very restricted region of the characteristics is considered. If we plotted the curves for 0.1-V increments in v_{GS}, this would be apparent.

Distortion is due to the fact that the characteristic curves for the FET are not uniformly spaced. If a much smaller input amplitude was applied, we would have amplification without appreciable distortion.

The amplifier circuit we have analyzed in this section is fairly simple. Practical amplifier circuits are more difficult to analyze by graphical methods. Later in the chapter, we develop a linear small-signal equivalent circuit for the FET, and then we can use mathematical circuit-analysis techniques instead of graphical analysis. Usually, the equivalent-circuit approach is more useful for practical amplifier circuits. However, graphical analysis of simple circuits provides an excellent way to understand the basic concepts of amplifiers.

Exercise 12.4 Find V_{DSQ}, $V_{DS\min}$, and $V_{DS\max}$ for the circuit of Figure 12.10 if the circuit values are changed to $V_{DD} = 15$ V, $V_{GG} = 3$ V, $R_D = 1$ kΩ, and $v_{\text{in}}(t) = \sin(2000\pi t)$. The characteristics for the MOSFET are shown in Figure 12.11.

Answer $V_{DSQ} \cong 11$ V, $V_{DS\min} \cong 6$ V, $V_{DS\max} \cong 14$ V. □

12.3 BIAS CIRCUITS

Analysis of amplifier circuits is often undertaken in two steps. First, we analyze the dc circuit to determine the Q point. In this analysis, the nonlinear device equations or the characteristic curves must be used. Then, after the bias analysis is completed, we use a linear small-signal equivalent circuit to determine the input resistance, voltage gain, and so on.

Amplifier analysis has two steps: 1. Determine the Q point. 2. Use a small-signal equivalent circuit to determine impedances and gains.

The bias circuits that we discuss are suitable for discrete-component designs in which large capacitances (for isolating the amplifier bias circuit from the source, load, and adjacent amplifier stages) and relatively tight tolerance (± 5 percent or better) resistances are practical. Close-tolerance resistances and large capacitances are not practical in integrated-circuit amplifiers; and their design is more complicated due to interaction between source, amplifier stages, and the load. Integrated-circuit design is mostly beyond the scope of this book.

The two-battery bias circuit used in the amplifier of Figure 12.10 is not practical. Usually, only one dc voltage is readily available instead of two. However, a more significant problem is that FET parameters vary considerably from device to device. In general, we want to establish a Q point near the middle of the load line so the output signal can swing in both directions without clipping. When the FET parameters vary from unit to unit, the two-battery circuit can wind up with some circuits biased near one end or the other.

The Fixed- plus Self-Bias Circuit

The **fixed- plus self-bias circuit** shown in Figure 12.13(a) is a good circuit for establishing Q points that are relatively independent of device parameters.

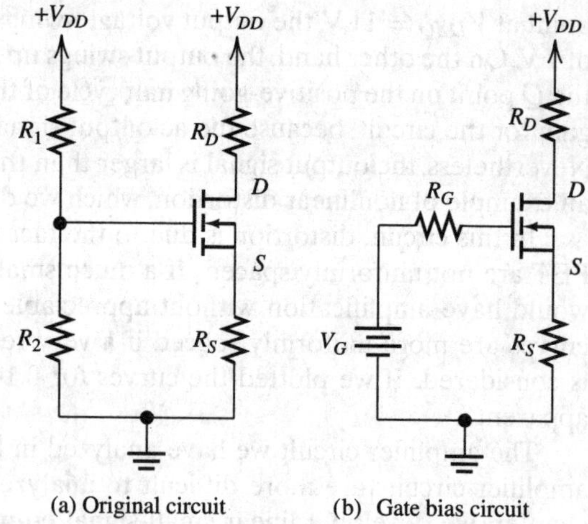

Figure 12.13 Fixed- plus self-bias circuit.

For purposes of analysis, we replace the gate circuit with its Thévenin equivalent as shown in Figure 12.13(b). The Thévenin voltage is

$$V_G = V_{DD}\frac{R_2}{R_1 + R_2} \tag{12.11}$$

and the Thévenin resistance R_G is the parallel combination of R_1 and R_2. Writing a voltage equation around the gate loop of Figure 12.13(b), we obtain

$$V_G = v_{GS} + R_S i_D \tag{12.12}$$

(We have assumed that the voltage drop across R_G is zero because the gate current of an NMOS transistor is extremely small.)

Usually, we want to bias the transistor in its saturation region, so we have

$$i_D = K(v_{GS} - V_{to})^2 \tag{12.13}$$

Simultaneous solution of Equations 12.12 and 12.13 yields the operating point (provided that it falls in the saturation region). Plots of these two equations are shown in Figure 12.14. Equation 12.12 plots as a straight line called the **bias line.** Notice that two roots will be found because Equation 12.13 plots as the dashed curve for $v_{GS} < V_{to}$. Thus, the smaller root found for v_{GS} is extraneous and should be discarded. The larger root found for v_{GS} and the smaller root for i_D are the true operating point.

Simultaneous solution of Equations 12.12 and 12.13 yields the operating point (provided that it falls in the saturation region). The larger root found for v_{GS} and the smaller root for i_D are the true operating point.

Finally, writing a voltage equation around the drain loop of Figure 12.13 gives us

$$v_{DS} = V_{DD} - (R_D + R_S)i_D \tag{12.14}$$

Example 12.2 Determination of Q Point for the Fixed- plus Self-Bias Circuit

Analyze the fixed- plus self-bias circuit shown in Figure 12.15. The transistor has $KP = 50\ \mu\text{A/V}^2$, $V_{to} = 2$ V, $L = 10\ \mu$m, and $W = 400\ \mu$m.

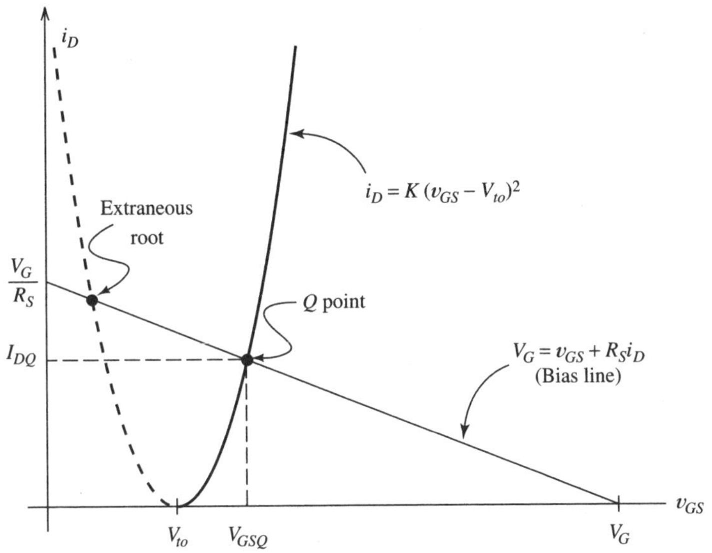

Figure 12.14 Graphical solution of Equations 12.12 and 12.13.

Solution First, we use Equation 12.3 to compute the device constant, which yields

$$K = \left(\frac{W}{L}\right)\frac{KP}{2} = 1\ \text{mA/V}^2$$

Substituting values into Equation 12.11, we have

$$V_G = V_{DD}\frac{R_2}{R_1 + R_2} = 20\frac{1}{(3+1)} = 5\ \text{V}$$

The Q-point values must satisfy Equations 12.12 and 12.13. Thus, we need to find the solution to the following pair of equations:

$$V_G = V_{GSQ} + R_S I_{DQ}$$

$$I_{DQ} = K(V_{GSQ} - V_{to})^2$$

Using the last equation to substitute for I_{DQ} in the expression for V_G, we have

$$V_G = V_{GSQ} + R_S K\left(V_{GSQ} - V_{to}\right)^2$$

Rearranging, we have

$$V_{GSQ}^2 + \left(\frac{1}{R_S K} - 2V_{to}\right)V_{GSQ} + V_{to}^2 - \frac{V_G}{R_S K} = 0$$

After values are substituted, we have

$$V_{GSQ}^2 - 3.630V_{GSQ} + 2.148 = 0$$

Solving, we find $V_{GSQ} = 2.886$ V and $V_{GSQ} = 0.744$ V. The second root is extraneous and should be discarded. Then, we have

$$I_{DQ} = K\left(V_{GSQ} - V_{to}\right)^2 = 0.784\ \text{mA}$$

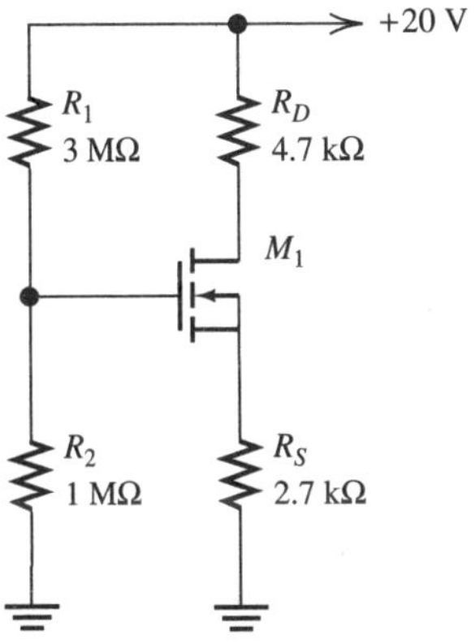

Figure 12.15 Fixed- plus self-bias circuit of Example 12.2.

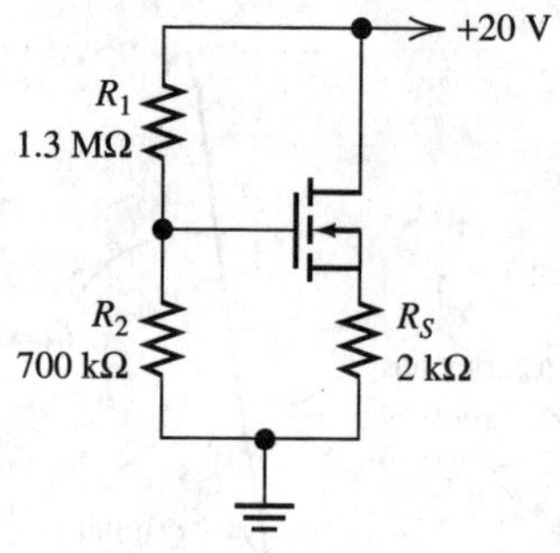

Figure 12.16 Circuit for Exercise 12.5.

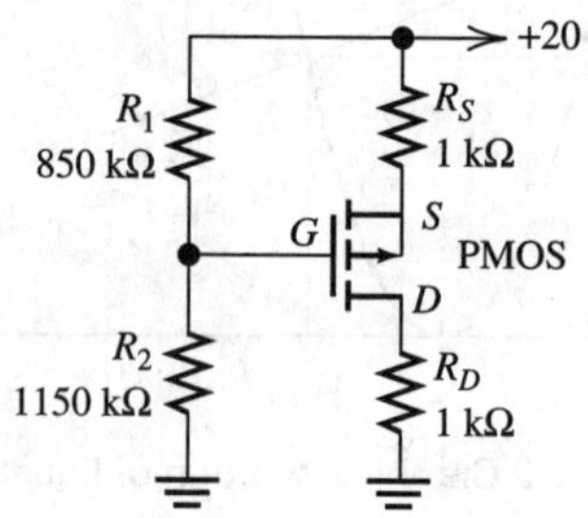

Figure 12.17 Circuit for Exercise 12.6.

Solving for the drain-to-source voltage, we get

$$V_{DSQ} = V_{DD} - (R_D + R_S) I_{DQ} = 14.2 \text{ V}$$

which is high enough to ensure that operation is in saturation as assumed in the solution. ■

Exercise 12.5 Determine I_{DQ} and V_{DSQ} for the circuit shown in Figure 12.16. The transistor has $KP = 50\ \mu\text{A/V}^2$, $V_{to} = 1$ V, $L = 10\ \mu$m, and $W = 200\ \mu$m.

Answer $I_{DQ} = 2$ mA; $V_{DSQ} = 16$ V. □

Exercise 12.6 Determine I_{DQ} and V_{DSQ} for the PMOS circuit shown in Figure 12.17. The transistor has $KP = 25\ \mu\text{A/V}^2$, $V_{to} = -1$ V, $L = 10\ \mu$m, and $W = 400\ \mu$m.

Answer $I_{DQ} = 4.5$ mA; $V_{DSQ} = -11$ V. □

12.4 SMALL-SIGNAL EQUIVALENT CIRCUITS

In the preceding section, we considered discrete-component dc bias circuits for FET amplifiers. Now, we consider the relationships between signal currents and voltages resulting in small changes from the Q point. As in Section 10.8, we denote total quantities by lowercase letters with uppercase subscripts, such as $i_D(t)$ and $v_{GS}(t)$. The dc Q point values are denoted by uppercase letters with an additional Q subscript, such as I_{DQ} and V_{GSQ}. The signals are denoted by lowercase letters with lowercase subscripts, such as $i_d(t)$ and $v_{gs}(t)$. The total current or voltage is the sum of the Q point value and the signal. Thus, we can write

$$i_D(t) = I_{DQ} + i_d(t) \tag{12.15}$$

and

$$v_{GS}(t) = V_{GSQ} + v_{gs}(t) \tag{12.16}$$

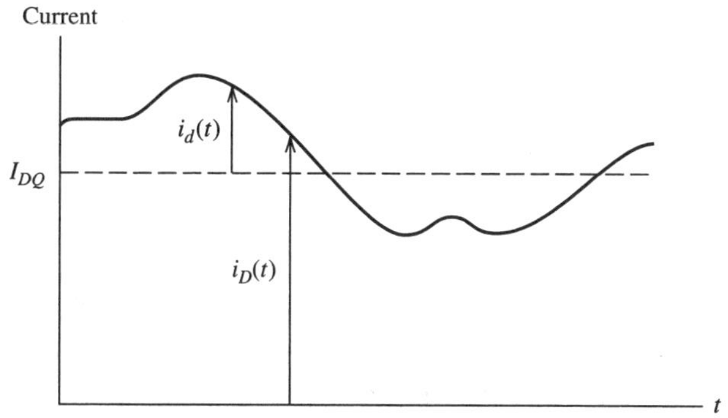

Figure 12.18 Illustration of the terms in Equation 12.15.

Figure 12.18 illustrates the terms in Equation 12.15.

In the discussion that follows, we assume that the FETs are biased in the saturation region, which is usually the case for amplifier circuits. Equation 12.4, repeated here for convenience,

$$i_D = K(v_{GS} - V_{to})^2$$

gives the total drain current in terms of the total gate-to-source voltage. Using Equations 12.15 and 12.16 to substitute into 12.4, we get

$$I_{DQ} + i_d(t) = K\left[V_{GSQ} + v_{gs}(t) - V_{to}\right]^2 \tag{12.17}$$

The right-hand side of Equation 12.17 can be expanded to obtain

$$I_{DQ} + i_d(t) = K(V_{GSQ} - V_{to})^2 + 2K(V_{GSQ} - V_{to})\, v_{gs}(t) + Kv_{gs}^2(t) \tag{12.18}$$

However, the Q point values are also related by Equation 12.4, so we have

$$I_{DQ} = K\left(V_{GSQ} - V_{to}\right)^2 \tag{12.19}$$

Therefore, the first term on each side of Equation 12.18 can be canceled. Furthermore, we are interested in small-signal conditions for which the last term on the right-hand side of Equation 12.18 is negligible and can be dropped [i.e., we assume that $|v_{gs}(t)|$ is much smaller than $|(V_{GSQ} - V_{to})|$ for all values of time]. With these changes, Equation 12.18 becomes

$$i_d(t) = 2K\left(V_{GSQ} - V_{to}\right) v_{gs}(t) \tag{12.20}$$

If we define the **transconductance** of the FET as

$$g_m = 2K\left(V_{GSQ} - V_{to}\right) \tag{12.21}$$

Equation 12.20 can be written as

$$i_d(t) = g_m v_{gs}(t) \tag{12.22}$$

The gate current for the FET is negligible, so we obtain

$$i_g(t) = 0 \tag{12.23}$$

For small signals, the FET is modeled by a voltage controlled current source connected between the drain and source terminals. The model has an open circuit between gate and source.

Equations 12.22 and 12.23 are represented by the **small-signal equivalent circuit** shown in Figure 12.19. Thus for small signals, the FET is modeled by a voltage-controlled current source connected between the drain and source terminals. The model has an open circuit between gate and source.

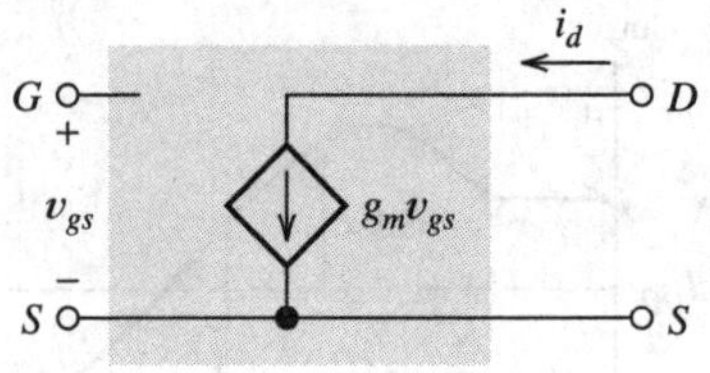

Figure 12.19 Small-signal equivalent circuit for FETs.

Dependence of Transconductance on Q Point and Device Parameters

We will see that transconductance g_m is an important parameter in the analysis of amplifier circuits. In general, better performance is obtained with higher values of g_m. Thus, it is important to know how Q point and device parameters influence transconductance.

Solving Equation 12.19 for the quantity $(V_{GSQ} - V_{to})$ and substituting into Equation 12.21, we obtain

$$g_m = 2\sqrt{KI_{DQ}} \tag{12.24}$$

We can increase g_m by choosing a higher value of I_{DQ}.

An important point to notice is that g_m is proportional to the square root of the Q point drain current. Thus, we can increase g_m by choosing a higher value of I_{DQ}.

If we use Equation 12.3 to substitute for K in Equation 12.24, we obtain

$$g_m = \sqrt{2KP}\sqrt{W/L}\sqrt{I_{DQ}} \tag{12.25}$$

We can obtain higher values of g_m for a given value of I_{DQ} by increasing the width-to-length ratio of the MOSFET.

Thus, we can obtain higher values of g_m for a given value of I_{DQ} by increasing the width-to-length ratio of the MOSFET.

More Complex Equivalent Circuits

Sometimes, additions to the equivalent circuit are needed to accurately model FETs. For example, we would need to include small capacitances between the device terminals if we considered the high-frequency response of FET amplifiers. The device equations and the equivalent circuit that we have derived from them describe only the static behavior of the device. For an accurate model with rapidly changing currents and voltages, capacitances must be considered.

Furthermore, the first-order equations we have used to obtain the equivalent circuit for the FET did not account for the effect of v_{DS} on the drain current. We have assumed that the drain characteristics are horizontal in the saturation region, but this is not exactly true—the drain characteristics of real devices slope slightly upward with increasing v_{DS}. If we wish to account for the effect of v_{DS} in the small-signal equivalent circuit, we must add a resistance r_d called the **drain resistance** between drain and source as shown in Figure 12.20. In this case, Equation 12.22 becomes

$$i_d = g_m v_{gs} + v_{ds}/r_d \tag{12.26}$$

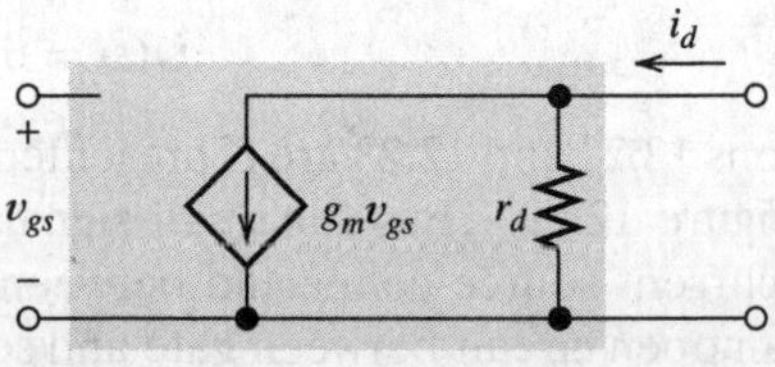

Figure 12.20 FET small-signal equivalent circuit that accounts for the dependence of i_D on v_{DS}.

Transconductance and Drain Resistance as Partial Derivatives

An alternative definition of g_m can be found by examination of Equation 12.26. Notice that if $v_{ds} = 0$, g_m is the ratio of i_d and v_{gs}. In equation form, we have

$$g_m = \left.\frac{i_d}{v_{gs}}\right|_{v_{ds}=0} \tag{12.27}$$

However, i_d, v_{gs}, and v_{ds} represent small changes from the Q point. Therefore, the condition $v_{ds} = 0$ is equivalent to requiring v_{DS} to remain constant at the Q point value, namely V_{DSQ}. Thus, we can write

$$g_m \cong \left.\frac{\Delta i_D}{\Delta v_{GS}}\right|_{v_{DS}=V_{DSQ}} \tag{12.28}$$

where Δi_D is an increment of drain current centered at the Q point. Similarly, Δv_{GS} is an increment of gate-to-source voltage centered at the Q point.

Equation 12.28 is an approximation to a partial derivative. Therefore, g_m is the partial derivative of i_D with respect to v_{GS}, evaluated at the Q point:

$$g_m = \left.\frac{\partial i_D}{\partial v_{GS}}\right|_{Q\text{ point}} \tag{12.29}$$

Similarly, the reciprocal of the drain resistance is

$$\frac{1}{r_d} \cong \left.\frac{\Delta i_D}{\Delta v_{DS}}\right|_{v_{GS}=V_{GSQ}} \tag{12.30}$$

Therefore, we can write

$$\frac{1}{r_d} = \left.\frac{\partial i_D}{\partial v_{DS}}\right|_{Q\text{ point}} \tag{12.31}$$

Given the drain characteristics, we can determine approximate values of the partial derivatives for a given Q point. Then, we can model the FET by its small-signal equivalent in analysis of an amplifier circuit and use the values found for g_m and r_d to compute amplifier gains and impedances. In the next several sections, we show examples of this process. First, we show how to determine the values of g_m and r_d starting from the characteristic curves.

Example 12.3 Determination of g_m and r_d from the Characteristic Curves

Determine the values of g_m and r_d for the MOSFET having the characteristics shown in Figure 12.21 at a Q point defined by $V_{GSQ} = 3.5$ and $V_{DSQ} = 10$ V.

Solution First, we locate the Q point as shown in the figure. Then, we use Equation 12.28 to find g_m:

$$g_m \cong \left.\frac{\Delta i_D}{\Delta v_{GS}}\right|_{v_{DS}=V_{DSQ}=10\text{ V}}$$

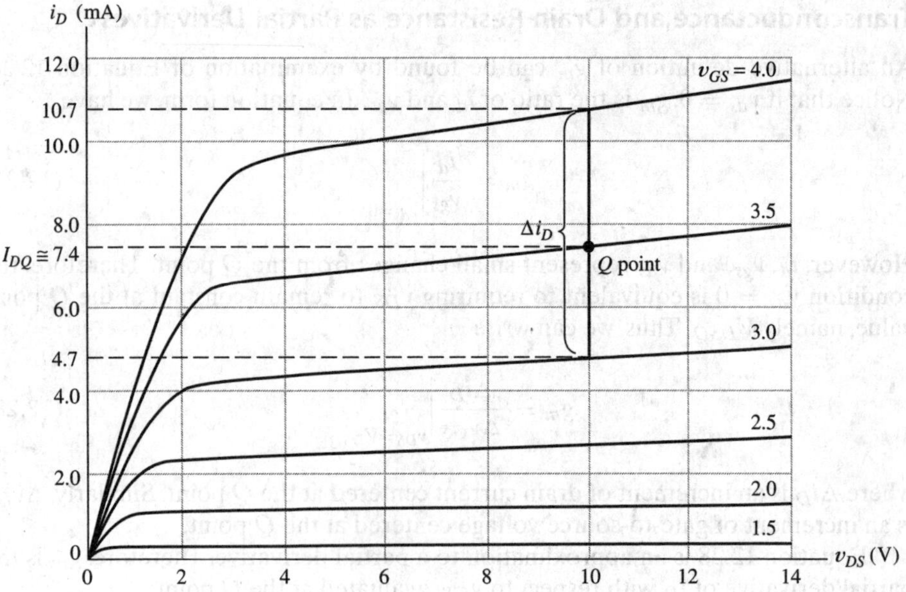

Figure 12.21 Determination of g_m and r_d. See Example 12.3.

We must make changes around the Q point while holding v_{DS} constant at 10 V. Thus, the incremental changes are made along a vertical line through the Q point. To obtain a representative value for g_m, we consider an increment centered on the Q point (rather than making the changes in one direction from the Q point). Taking the changes starting from the curve below the Q point and ending at the curve above the Q point, we have $\Delta i_D \cong 10.7 - 4.7 = 6$ mA and $\Delta v_{GS} = 1$ V. The Δi_D increment is labeled in the figure. Thus, we have

$$g_m = \frac{\Delta i_D}{\Delta v_{GS}} = \frac{6 \text{ mA}}{1 \text{ V}} = 6 \text{ mS}$$

The drain resistance is found by applying Equation 12.30:

$$\frac{1}{r_d} = \left.\frac{\Delta i_D}{\Delta v_{DS}}\right|_{v_{GS}=V_{GSQ}}$$

Because the incremental changes are to be made while holding v_{GS} constant, the changes are made along the characteristic curve through the Q point. Thus, $1/r_d$ is the slope of the curve through the Q point. For $v_{GS} = V_{GSQ} = 3.5$ V, we obtain $i_D \cong 6.7$ mA at $v_{DS} = 4$ V, and $i_D \cong 8.0$ mA at $v_{DS} = 14$ V. Thus, we get

$$\frac{1}{r_d} = \frac{\Delta i_D}{\Delta v_{DS}} \cong \frac{(8.0 - 6.7) \text{ mA}}{(14 - 4) \text{ V}} = 0.13 \times 10^{-3}$$

Taking the reciprocal, we find $r_d = 7.7$ kΩ. ■

Exercise 12.7 Find the values of g_m and r_d for the characteristics of Figure 12.21 at a Q point of $V_{GSQ} = 2.5$ V and $V_{DSQ} = 6$ V.
Answer $g_m \cong 3.3$ mS, $r_d \cong 20$ kΩ. □

Exercise 12.8 Show that Equation 12.21 results from application of Equation 12.29 to Equation 12.4. □

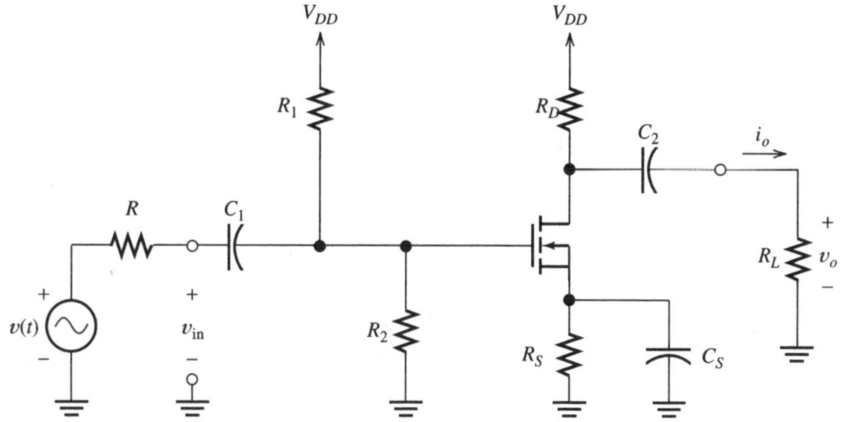

Figure 12.22 Common-source amplifier.

12.5 COMMON-SOURCE AMPLIFIERS

The circuit diagram of a **common-source amplifier** is shown in Figure 12.22. The ac signal to be amplified is $v(t)$. The **coupling capacitors** C_1 and C_2, as well as the **bypass capacitor** C_S, are intended to have very small impedances for the ac signal. In this section, we carry out a midband analysis in which we assume that these capacitors are short circuits for the signal. The resistors R_1, R_2, R_S, and R_D form the bias network, and their values are selected to obtain a suitable Q point. The amplified output signal is applied to the load R_L.

The Small-Signal Equivalent Circuit

The small-signal equivalent circuit for the amplifier is shown in Figure 12.23. The input coupling capacitor C_1 has been replaced by a short circuit. The MOSFET has been replaced by its small-signal equivalent. Because the bypass capacitor C_S is assumed to be a short circuit, the source terminal of the FET is connected directly to ground—which is why the circuit is called a *common-source amplifier.*

The dc supply voltage source acts as a short circuit for the ac signal. (Even if ac current flows through the dc source, the ac voltage across it is zero. Thus, for ac currents, the dc voltage source is a short.) Consequently, both R_1 and R_2 are connected from gate to ground in the equivalent circuit. Similarly, R_D is connected from drain to ground.

The dc supply voltage source acts as a short circuit for ac current.

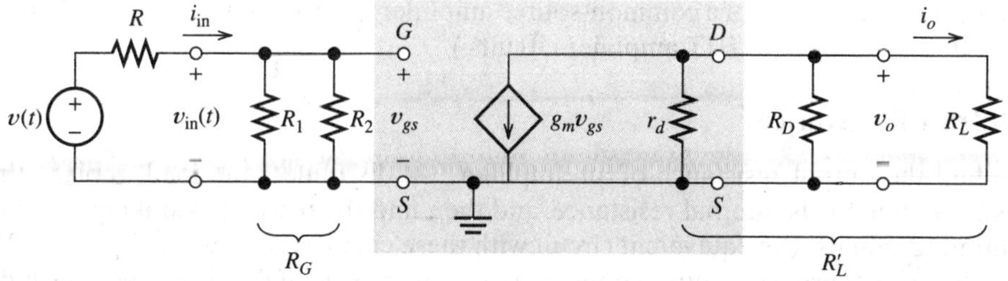

Figure 12.23 Small-signal equivalent circuit for the common-source amplifier.

Voltage Gain

Next, we derive an expression for the voltage gain of the common-source amplifier. Refer to the small-signal equivalent circuit, and notice that the resistances r_d, R_D, and R_L are in parallel. We denote the equivalent resistance by

$$R'_L = \frac{1}{1/r_d + 1/R_D + 1/R_L} \tag{12.32}$$

The output voltage is the product of the current from the controlled source and the equivalent resistance, given by

$$v_o = -(g_m v_{gs}) R'_L \tag{12.33}$$

The minus sign is needed because of the reference directions selected (i.e., the current $g_m v_{gs}$ flows out of the positive end of the voltage reference for v_o). Furthermore, the input voltage and the gate-to-source voltage are equal:

$$v_{in} = v_{gs} \tag{12.34}$$

Now if we divide the respective sides of Equation 12.33 by those of Equation 12.34, we obtain the voltage gain:

$$A_v = \frac{v_o}{v_{in}} = -g_m R'_L \tag{12.35}$$

The minus sign in the expression for the voltage gain shows that the common-source amplifier is inverting. Notice that the voltage gain is proportional to g_m.

In small-signal midband analysis of FET amplifiers, the coupling capacitors, bypass capacitors, and dc voltage sources are replaced by short circuits. The FET is replaced with its small-signal equivalent circuit. Then, we write circuit equations and derive useful expressions for gains, input impedance, and output impedance.

Input Resistance

The input resistance of the common-source amplifier is given by

$$R_{in} = \frac{v_{in}}{i_{in}} = R_G = R_1 || R_2 \tag{12.36}$$

in which $R_1 || R_2$ denotes the parallel combination of R_1 and R_2. The resistances R_1 and R_2 form part of the bias network, but their values are not critical. (See Section 12.3 for a discussion of the bias circuit.) Practical resistance values range from 0 to perhaps 10 MΩ in discrete component circuits. Thus, we have a great deal of freedom in design of the input resistance of a common-source amplifier. (We will see in the next chapter that this is not true for BJT amplifier circuits.)

We have a great deal of freedom in design of the input resistance of a common-source amplifier.

Output Resistance

To find the output resistance of an amplifier, we disconnect the load, replace the signal source by its internal resistance, and then find the resistance by looking into the output terminals.

To find the output resistance of an amplifier, we disconnect the load, replace the signal source by its internal resistance, and then find the resistance looking into the output terminals. The equivalent circuit with these changes is shown in Figure 12.24.

Because there is no source connected to the input side of the circuit, we conclude that $v_{gs} = 0$. Therefore, the controlled current source $g_m v_{gs}$ produces zero current

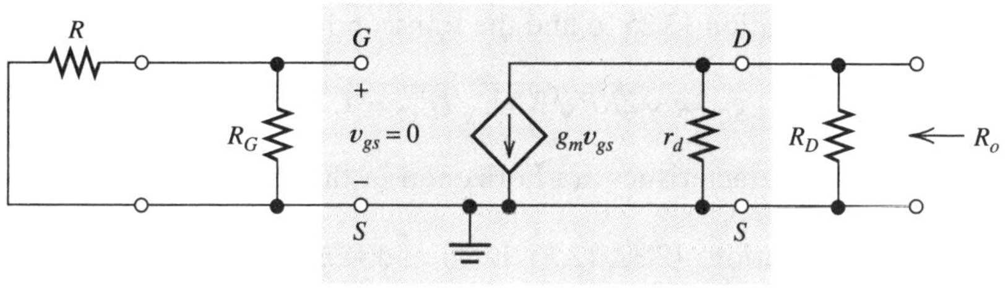

Figure 12.24 Circuit used to find R_o.

and appears as an open circuit. Consequently, the output resistance is the parallel combination of R_D and r_d:

$$R_o = \frac{1}{1/R_D + 1/r_d} \tag{12.37}$$

Example 12.4 Gain and Impedance Calculations for a Common-Source Amplifier

Consider the common-source amplifier shown in Figure 12.25. The NMOS transistor has $KP = 50\ \mu\text{A/V}^2$, $V_{to} = 2$ V, $L = 10\ \mu$m, and $W = 400\ \mu$m. Find the midband voltage gain, input resistance, and output resistance of the amplifier. Then, assuming that the input source is given by

$$v(t) = 100 \sin(2000\pi t) \text{ mV}$$

compute the output voltage. Also, assume that the frequency of the source (which is 1000 Hz) is in the midband region.

Solution First, we need to find the Q point so we can determine the value of g_m for the MOSFET. The bias circuit consists of R_1, R_2, R_D, R_S, and the MOSFET. This circuit was analyzed in Example 12.2, where we determined that $I_{DQ} = 0.784$ mA.

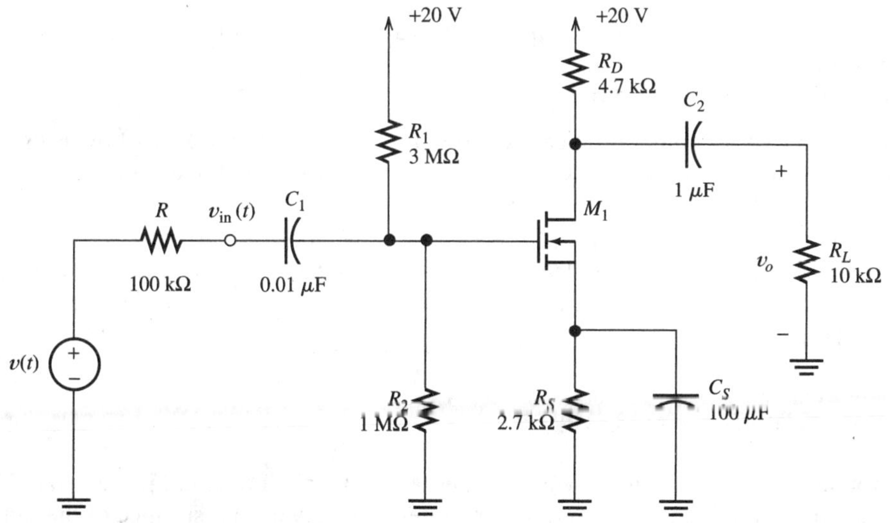

Figure 12.25 Common-source amplifier.

Next, we use Equation 12.25 to find the transconductance of the device:

$$g_m = \sqrt{2KP}\sqrt{W/L}\sqrt{I_{DQ}} = 1.77 \text{ mS}$$

Because the drain characteristics are horizontal in the saturation region, we have $r_d = \infty$.

Now, we use Equations 12.32, 12.35, 12.36, and 12.37 to find

$$R'_L = \frac{1}{1/r_d + 1/R_D + 1/R_L} = 3197\ \Omega$$

$$A_v = \frac{v_o}{v_{\text{in}}} = -g_m R'_L = -5.66$$

$$R_{\text{in}} = \frac{v_{\text{in}}}{i_{\text{in}}} = R_G = R_1 || R_2 = 750\ \text{k}\Omega$$

$$R_o = \frac{1}{1/R_D + 1/r_d} = 4.7\ \text{k}\Omega$$

The signal voltage divides between the internal source resistance and the input resistance of the amplifier. Thus, we have

$$v_{\text{in}} = v(t)\frac{R_{\text{in}}}{R + R_{\text{in}}} = 88.23 \sin(2000\pi t) \text{ mV}$$

Then, the output voltage can be found as

$$v_o(t) = A_v v_{\text{in}}(t) = -500 \sin(2000\pi t) \text{ mV}$$

Notice the phase inversion of $v_o(t)$ compared to $v_{\text{in}}(t)$. ■

Exercise 12.9 Find the voltage gain of the amplifier of Example 12.4 with R_L replaced by an open circuit.
Answer $A_{voc} = -8.32$. □

Exercise 12.10 Consider the circuit of Figure 12.22 with the bypass capacitor C_S replaced by an open circuit. Draw the small-signal equivalent circuit. Then assuming that r_d is an open circuit for simplicity, derive an expression for the voltage gain in terms of g_m and the resistances.
Answer $A_v = -g_m R'_L/(1 + g_m R_S)$. □

Exercise 12.11 Evaluate the gain expression found in Exercise 12.10 by using the values given in Example 12.4. Compare the result with the voltage gain found in the example.
Answer $A_v = -0.979$ without the bypass capacitor compared to $A_v = -5.66$ with the bypass capacitor in place. Notice that unbypassed impedance between the FET source terminal and ground strongly reduces the gain of a common-source amplifier. □

Unbypassed impedance between the FET source terminal and ground strongly reduces the gain of a common-source amplifier.

12.6 SOURCE FOLLOWERS

Another amplifier circuit known as a **source follower** is shown in Figure 12.26. The signal to be amplified is $v(t)$, and R is the internal (Thévenin) resistance of the signal source. The coupling capacitor C_1 causes the ac input signal to appear at the gate of

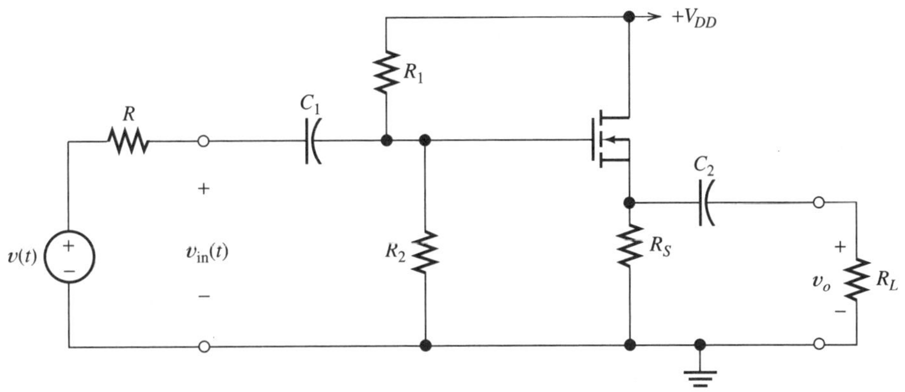

Figure 12.26 Source follower.

the FET. The capacitor C_2 connects the load to the source terminal of the MOSFET. (In the midband analysis of the amplifier, we assume that the coupling capacitors behave as short circuits.) The resistors R_S, R_1, and R_2 form the bias circuit.

The Small-Signal Equivalent Circuit

The small-signal equivalent circuit is shown in Figure 12.27. The coupling capacitors have been replaced by short circuits, and the FET has been replaced by its small-signal equivalent. Notice that the drain terminal is connected directly to ground because the dc supply becomes a short (for ac currents) in the small-signal equivalent. Here the FET equivalent circuit is drawn in a different configuration (i.e., with the drain at the bottom) from that shown earlier, but it is the same electrically.

The ability to draw the small-signal equivalent for an amplifier circuit is important. Test yourself to see if you can obtain the small-signal circuit starting from Figure 12.26.

Voltage Gain

Now, we derive an expression for the voltage gain of the source follower. Notice that r_d, R_S, and R_L are in parallel. We denote the parallel combination by

$$R'_L = \frac{1}{1/r_d + 1/R_S + 1/R_L} \tag{12.38}$$

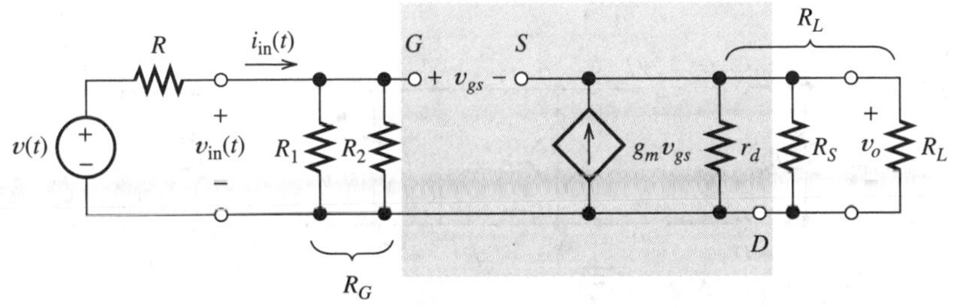

Figure 12.27 Small-signal ac equivalent circuit for the source follower.

The output voltage is given by

$$v_o = g_m v_{gs} R'_L \tag{12.39}$$

Furthermore, we can write the following voltage equation:

$$v_{\text{in}} = v_{gs} + v_o \tag{12.40}$$

Using Equation 12.39 to substitute for v_o in Equation 12.40, we have

$$v_{\text{in}} = v_{gs} + g_m v_{gs} R'_L \tag{12.41}$$

Dividing the respective sides of Equations 12.39 and 12.41, we obtain the following expression for the voltage gain:

$$A_v = \frac{v_o}{v_{\text{in}}} = \frac{g_m R'_L}{1 + g_m R'_L} \tag{12.42}$$

The source follower is a noninverting amplifier with voltage gain less than unity.

Notice that the voltage gain given in Equation 12.42 is positive and is less than unity. Thus, the source follower is a noninverting amplifier with voltage gain less than unity.

Input Resistance

The input resistance is the resistance seen looking into the input terminals of the equivalent circuit. Thus, we have

$$R_{\text{in}} = \frac{v_{\text{in}}}{i_{\text{in}}} = R_G = R_1 \| R_2 \tag{12.43}$$

in which $R_1 \| R_2$ denotes the parallel combination of R_1 and R_2.

Output Resistance

To find the output resistance, we remove the load resistance, replace the signal source with its internal resistance, and look back into the output terminals. It is helpful to attach a test source v_x to the output terminals as shown in Figure 12.28. Then, the output resistance is found as

$$R_o = \frac{v_x}{i_x} \tag{12.44}$$

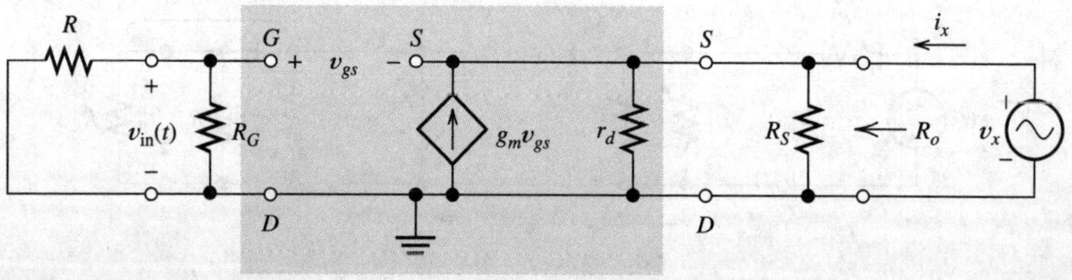

Figure 12.28 Equivalent circuit used to find the output resistance of the source follower.

where i_x is the current supplied by the test source as shown in the figure. It can be shown that the output resistance is given by

$$R_o = \frac{1}{g_m + 1/R_S + 1/r_d} \quad (12.45)$$

This can be quite low in value, and one reason for using a source follower is to obtain low output resistance.

One reason for using a source follower is to obtain low output resistance.

Example 12.5 Gain and Impedance Calculations for a Source Follower

Consider the source follower shown in Figure 12.26 given $R_L = 1\ \text{k}\Omega$ and $R_1 = R_2 = 2\ \text{M}\Omega$. The NMOS transistor has $KP = 50\ \mu\text{A/V}^2$, $L = 2\ \mu\text{m}$, $W = 160\ \mu\text{m}$, and $V_{to} = 1$ V. Find the value for R_S to achieve $I_{DQ} = 10$ mA. Then compute the voltage gain, input resistance, and output resistance.

Solution From Equations 12.3 and 12.4, we have

$$K = \left(\frac{W}{L}\right)\frac{KP}{2} = 2\ \text{mA/V}^2$$

and

$$I_{DQ} = K\left(V_{GSQ} - V_{to}\right)^2$$

Solving for V_{GSQ} and substituting values, we get

$$V_{GSQ} = \sqrt{I_{DQ}/K} + V_{to} = 3.236\ \text{V}$$

The dc voltage at the gate terminal (with respect to ground) is given by

$$V_G = V_{DD} \times \frac{R_2}{R_1 + R_2} = 7.5\ \text{V}$$

The dc voltage at the source terminal of the NMOS is

$$V_S = V_G - V_{GSQ} = 4.264\ \text{V}$$

Finally, we find the source resistance as

$$R_S = \frac{V_S}{I_{DQ}} = 426.4\ \Omega$$

(Of course, in a discrete circuit, we would choose a standard nominal value for R_S. However, we will continue this example by using the exact value computed for R_S.)

Next, we use Equation 12.25 to find the transconductance of the device:

$$g_m = \sqrt{2KP}\sqrt{W/L}\sqrt{I_{DQ}} = 8.944\ \text{mS}$$

Because the drain characteristics are horizontal in the saturation region, we have $r_d = \infty$.

Next, we substitute values into Equation 12.38 to obtain

$$R'_L = \frac{1}{1/r_d + 1/R_S + 1/R_L} = 298.9\ \Omega$$

Then, the voltage gain, given by Equation 12.42, is

$$A_v = \frac{v_o}{v_{\text{in}}} = \frac{g_m R'_L}{1 + g_m R'_L} = 0.7272$$

The input resistance is

$$R_{\text{in}} = R_1 || R_2 = 1\ \text{M}\Omega$$

The output resistance, given by Equation 12.45, is

$$R_o = \frac{1}{g_m + 1/R_S + 1/r_d} = 88.58\ \Omega$$

This is a fairly low output resistance compared with that of other single-FET amplifier configurations.

The current gain is shown by the use of Equation 11.3 to be

$$A_i = A_v \frac{R_{\text{in}}}{R_L} = 727.2$$

The source follower has voltage gain slightly less than unity, high input impedance, and low output impedance. Current gain and power gain can be larger than unity.

The power gain is given by

$$G = A_v A_i = 528.8$$

Even though the voltage gain is less than unity, the output power is much greater than the input power because of the high input resistance. ■

Exercise 12.12 Derive Equation 12.45. □

Exercise 12.13 Derive expressions for the voltage gain, input resistance, and output resistance of the **common-gate amplifier** shown in Figure 12.29, assuming that r_d is an open circuit.

Answer The small-signal equivalent circuit is shown in Figure 12.30. $A_v = g_m R'_L$; $R_{\text{in}} = 1/(g_m + 1/R_S)$; $R_o = R_D$. □

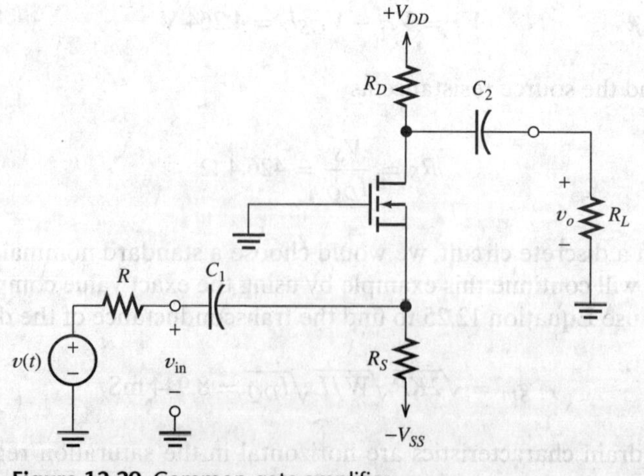

Figure 12.29 Common-gate amplifier.

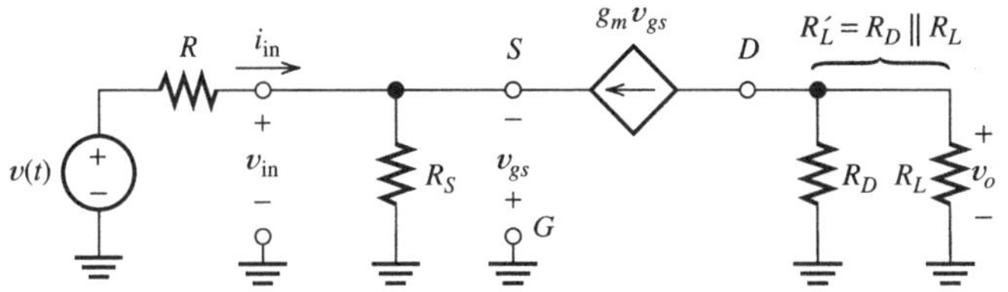

Figure 12.30 See Exercise 12.13.

12.7 CMOS LOGIC GATES

In **complementary metal–oxide–semiconductor** (CMOS) technology, both NMOS and PMOS transistors are fabricated on the same chip. Next, we describe how the basic building blocks of digital systems (NAND gates or NOR gates) are constructed using CMOS technology.

CMOS Inverter

A CMOS inverter is shown in Figure 12.31. The NMOS and PMOS FETs are constructed by adding impurities to a silicon crystal, forming regions of n-type and p-type semiconductor as shown in part (a) of the figure. Notice that the gates G are insulated from the rest of the circuit by layers of silicon dioxide (SiO_2). Thus, the input behaves as an open circuit (except for a small amount of capacitance).

PRACTICAL APPLICATION 12.1

Where *Did* Those Trout Go?

Fish biologists often need to learn about the migratory behavior of various kinds of fish to properly regulate fishing and habitat changes. One example is the effort to reestablish the coaster brook trout in various streams on the south shore of Lake Superior. Coaster brook trout migrate from their natal streams into Lake Superior, where they grow much larger than their stream-dwelling cousins. As adults, highly colored coasters return to their native streams each fall to spawn. Originally, coasters were found in nearly all of the rivers entering Lake Superior; however, over-fishing and habitat changes due mainly to logging have caused them to disappear from much of their original range. Currently, several projects are underway to attempt to reestablish these beautiful trout in south shore streams.

A powerful approach to gaining accurate information about fish migration is to implant radio frequency identification (RFID) tags in the fish. Then, antennas located in the streams can monitor movement of the fish. A wide variety of RFID systems are in use. We will describe a single representative system.

A typical RFID tag used in fisheries research consists of a coil of enameled copper wire wound on a ferrite core, a CMOS integrated circuit chip, and two capacitors, all hermetically sealed in a glass tube comparable in size to a large grain of rice. These tags are implanted in fish through hypodermic needles. The tags do not contain an internal power source, so they are sometimes called passive identification tags (PIT).

A typical streamside fish monitoring station is illustrated in Figure PA12.1. Because the important locations are often far from power lines, power for the station is provided by deep-cycle storage batteries recharged by solar panels. The antenna often consists of a loop of wire suspended above the stream on poles and weighted to the stream bottom with rocks.

A schematic diagram of a representative system is shown in Figure PA12.2. The stream antenna appears as the inductance L_1, typically 10 to 100 μH in value. A capacitance C_1, called an antenna tuner, forms a parallel resonant circuit with L_1 having a resonant frequency of 134.2 kHz. The coil in the RFID

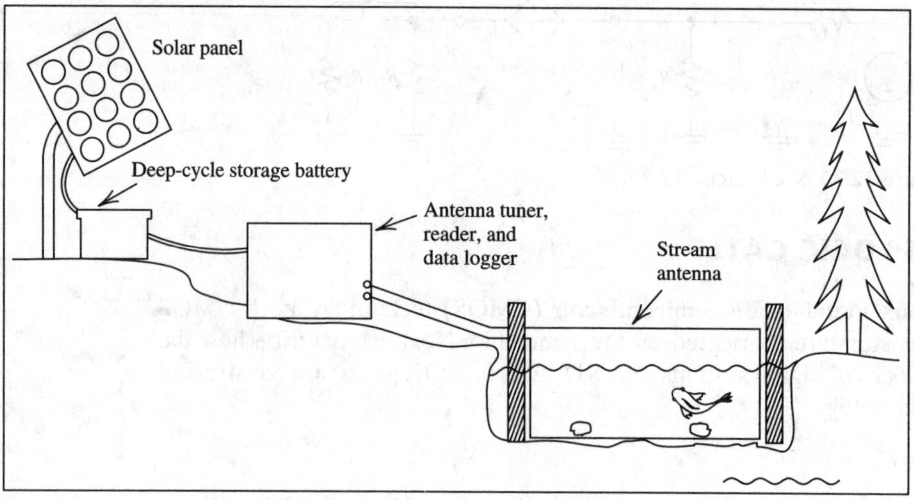

Figure PA12.1 Typical monitoring station.

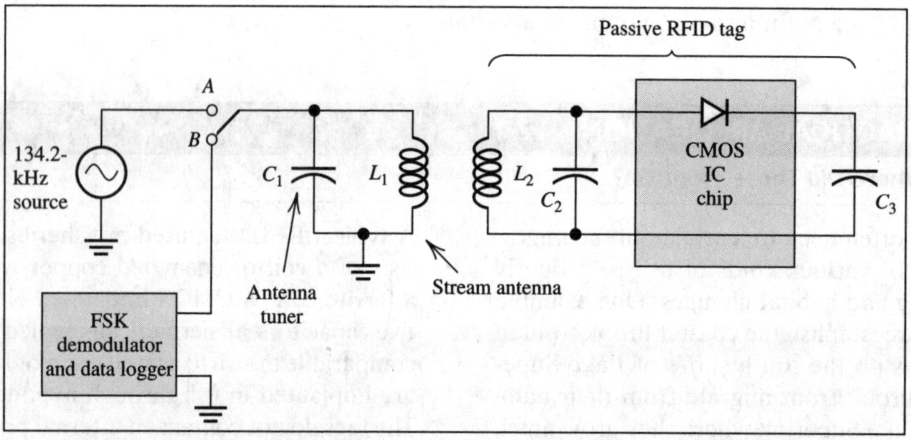

Figure PA12.2 Electrical schematic diagram of a fish monitoring system.

tag also acts as an antenna and is represented by the inductance L_2, which forms a parallel resonant circuit with C_2.

In operation, the switch periodically moves to contact A, applying a 134.2-kHz sine wave to the stream antenna, which creates an ac magnetic field in the vicinity of the antenna. When a tagged fish is present, part of the magnetic flux links L_2, resulting in a 134.2-kHz voltage at the input to the CMOS IC chip. This voltage is rectified by the diode contained in the chip charging C_3. After about 50 ms, the switch moves to contact B so that the voltage applied to the stream antenna becomes zero. Then, power is supplied to the CMOS chip by the charge stored on C_3. When the CMOS chip senses the end of the pulse, it transmits a 64-bit codeword that identifies the particular tag and, hence, the fish into which it was implanted.

Frequency shift keying (FSK) is used to encode the bits. For a 1-bit, the chip applies 16 cycles of a 123.2-kHz signal to L_2; and for a 0-bit, it applies 16 cycles of a 134.2-kHz signal to L_2. The resulting magnetic field partially links L_1, inducing a voltage that is applied to the FSK demodulator and data logger.

The FSK demodulator determines the frequency of each 16-cycle segment and the resulting bit value. The resulting code words are saved by the data logger, which is read periodically by the fish biologist. The data logger can also save additional data, such as the time of day that the fish passed through the station, the stream temperature, and thc strcam flow rate, if additional sensors are placed in the stream.

Additional information about coaster brook-trout programs and fish monitoring systems can be found at the following websites:

http://www.biomark.com/

http://www.oregonrfid.com/

http://faculty.nmu.edu/jileonar/Research.htm

http://www.bio.mtu.edu/news/Huckins_May_05.htm

http://www.ti.com/rfid/

The circuit diagram is shown in Figure 12.31(b). The dc supply voltage V_{DD} is applied at the top of the circuit. When the input voltage is high ($V_{\text{in}} = V_{DD}$), a conducting channel is induced between the drain D and the source S of the NMOS transistor. Thus, the NMOS transistor becomes a low resistance and ideally behaves as a closed switch, as shown in Figure 12.31(c). Furthermore, the PMOS transistor

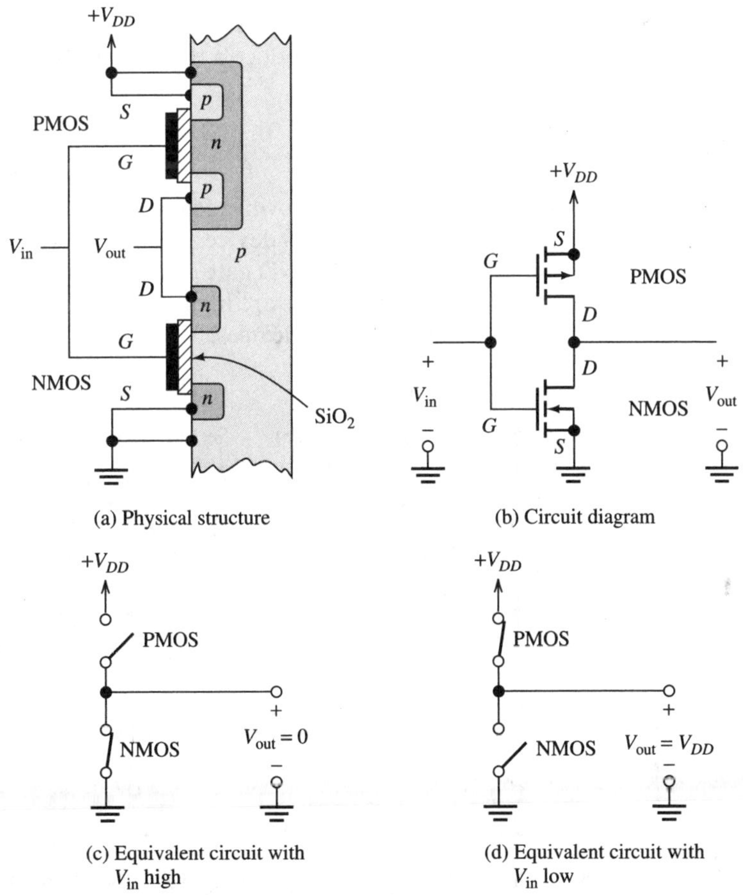

Figure 12.31 CMOS inverter.

is off, and it behaves as an open switch. Thus, with V_{in} high, the output voltage V_{out} becomes low (i.e., $V_{out} = 0$).

With the input low, the PMOS is on, the NMOS is off, and the output is high. With the input high, the situation is reversed.

On the other hand, with V_{in} low, a conducting layer is induced under the gate in the PMOS transistor, but not in the NMOS transistor. Therefore, the PMOS transistor is on and the NMOS transistor is off. This is illustrated in Figure 12.31(d). Then the output voltage V_{out} is high ($V_{out} = V_{DD}$).

Because of the switching action of the transistors, the output is low when the input is high, and vice versa. This is exactly how a logic inverter is supposed to behave.

CMOS NAND Gate

By adding transistors to the inverter circuit, we can construct a NAND gate. The circuit for a two-input NAND gate is shown in Figure 12.32(a). Notice that we have two PMOS transistors in parallel and two NMOS transistors in series.

Placing N PMOS transistors in parallel and N NMOS transistors in series produces an N-input NAND gate.

When their gates are high, the NMOS devices are on, and when their gates are low, the NMOS devices are off. The opposite is true for the PMOS transistors (i.e., the PMOS transistors are on when their gates are low, and they are off when their gates are high).

The equivalent circuit with A high and B low is shown in Figure 12.32(b). Furthermore, the equivalent circuit with both A and B high is shown in Figure 12.32(c). Notice that because of the switching action of the transistors, the output is low only if both A and B are high. This is exactly the way that a NAND gate is supposed to behave. By adding more transistors, we could produce a three-input NAND gate.

CMOS NOR Gate

Placing N PMOS transistors in series and N NMOS transistors in parallel produces an N-input NOR gate.

The circuit diagram of a two-input NOR gate is shown in Figure 12.33. In this case, we have two PMOS devices in series and two NMOS devices in parallel. The operation of the NOR gate is very similar to that of the NAND gate discussed previously. Here again, the NMOS devices are on when their gates are high and off when their gates are low. The situation is reversed for the PMOS devices.

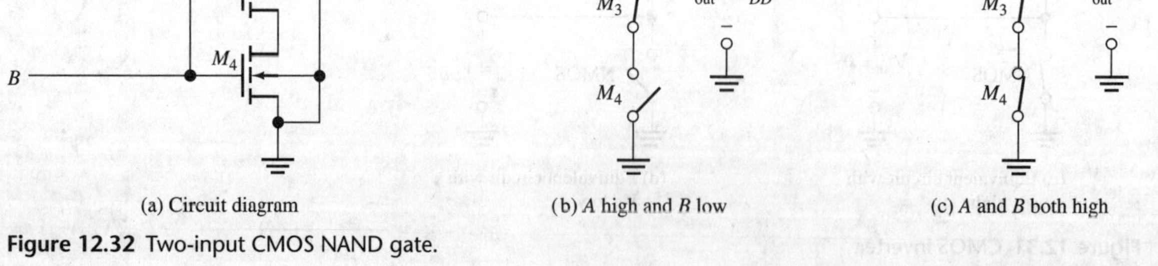

Figure 12.32 Two-input CMOS NAND gate.

Exercise 12.14 Draw the equivalent circuits [similar to Figure 12.32(b) and (c)] of the NOR gate shown in Figure 12.33 for: **a.** A high and B high; **b.** A high and B low; **c.** A low and B low. Then give a truth table for the gate.

Answer See Figure 12.34. □

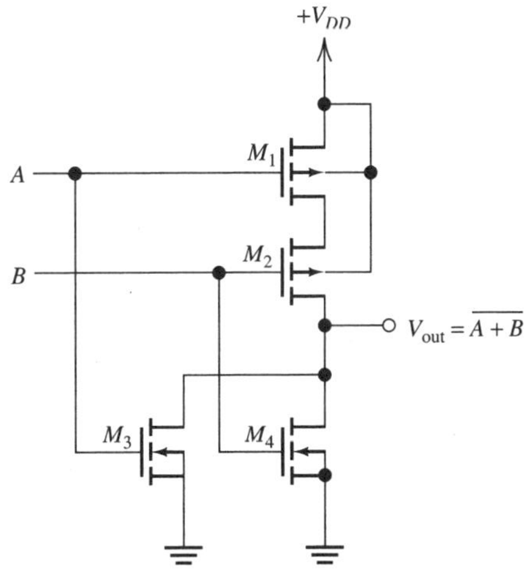

Figure 12.33 Two-input CMOS NOR gate.

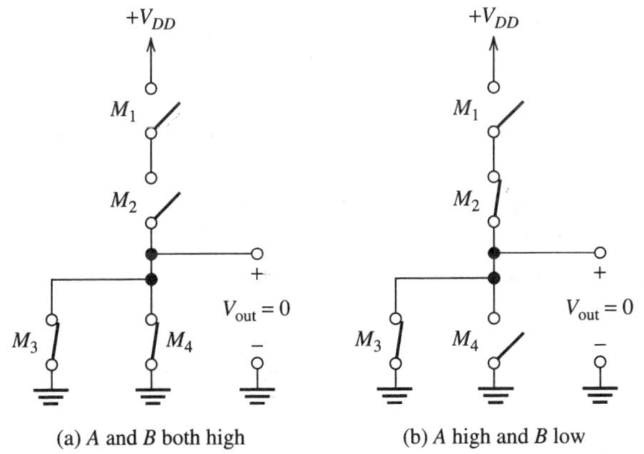

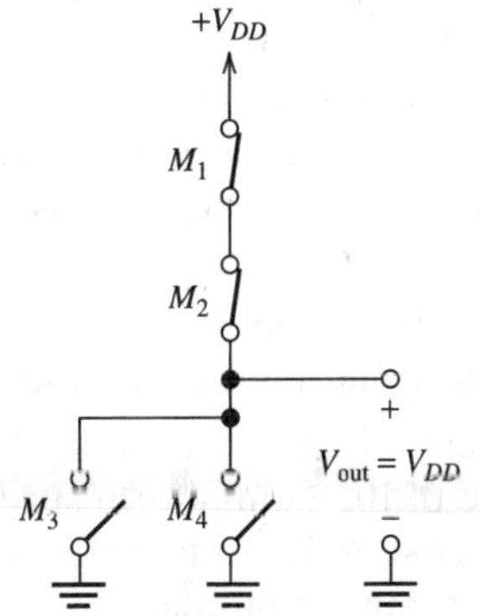

(c) A and B both low

A	B	V_{out}
Low	Low	High
Low	High	Low
High	Low	Low
High	High	Low

(d) Truth table

Figure 12.34 Answers for Exercise 12.14.

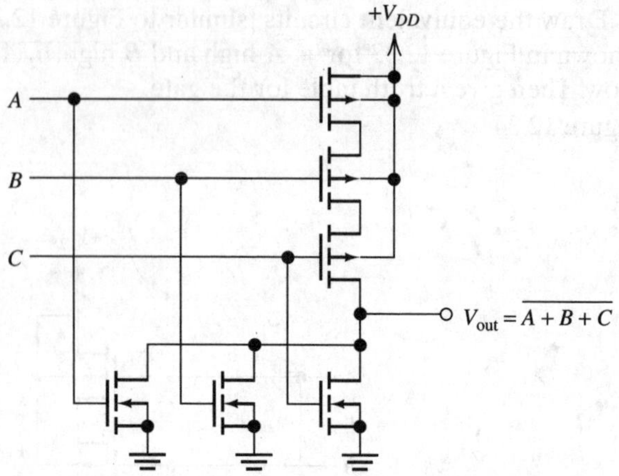

Figure 12.35 Three-input CMOS NOR gate. (Answer for Exercise 12.15.)

Exercise 12.15 Draw the circuit diagram of a three-input NOR gate.
Answer See Figure 12.35. □

Conclusions

Complex digital systems can be constructed by interconnecting billions of NMOS and PMOS transistors, all of which are fabricated on a single silicon chip by a relatively small number of processing steps.

In Chapter 7, we saw that complex combinatorial logic functions can be achieved simply by interconnecting NAND gates (or NOR gates). Furthermore, logic gates can be interconnected to form flip-flops. Interconnections of flip-flops form registers. A complex digital system, such as a computer, consists of many gates, flip-flops, and registers. Thus, logic gates are the basic building blocks for complex digital systems.

Modern technology can construct billions of CMOS gates on a silicon wafer by adding impurities, oxide layers, and metal interconnections. Relatively few (perhaps 20) steps are needed in the manufacturing process. This results in the production of powerful computers at low cost.

Summary

1. The structure of an enhancement n-channel MOS transistor is shown in Figure 12.1 on page 587.
2. The MOSFET is the primary device that underlies the rapid advances in digital electronics over the past several decades.
3. In an NMOS transistor, when a sufficiently large (positive) voltage is applied to the gate relative to the source, electrons are attracted to the region under the gate, and a channel of n-type material is induced between the drain and the source. Then if voltage is applied between the drain and the source, current flows into the drain through the channel and out the source. Drain current is controlled by the voltage applied to the gate.
4. MOSFETs operate in cutoff, in the linear region, or in saturation.
5. Usually, KP is determined by the fabrication process. However, in designing a circuit, we can vary the ratio W/L to obtain transistors best suited to various parts of the circuit.
6. Simple amplifier circuits can be analyzed by using graphical (load-line) techniques.
7. Nonlinear distortion occurs in FET amplifiers because of the nonuniform spacing of the drain characteristics. Distortion is less pronounced for smaller signal amplitudes.
8. For use as amplifiers, FETs are usually biased in the saturation region.

9. In a small-signal midband analysis, a FET can be modeled by the equivalent circuit shown in Figure 12.20 on page 602.
10. Transconductance of a FET is defined as

$$g_m = \left.\frac{\partial i_D}{\partial v_{GS}}\right|_{Q\text{ point}}$$

11. Small-signal drain resistance of a FET is defined as

$$\frac{1}{r_d} = \left.\frac{\partial i_D}{\partial v_{DS}}\right|_{Q\text{ point}}$$

12. In small-signal midband analysis of FET amplifiers, the coupling capacitors, bypass capacitors, and dc voltage sources are replaced by short circuits. The FET is replaced with its small-signal equivalent circuit. Then, we write circuit equations and derive useful expressions for gains, input impedance, and output impedance.
13. To find the output resistance of an amplifier, we disconnect the load, replace the signal source by its internal resistance, and then find the resistance looking into the output terminals.
14. The common-source amplifier is inverting and can have voltage-gain magnitude larger than unity.
15. Unbypassed impedance between the FET source terminal and ground strongly reduces the gain of a common-source amplifier.
16. The source follower has voltage gain slightly less than unity, high current gain, and relatively low output impedance. It is noninverting.
17. Complex digital systems can be constructed by interconnecting millions of NMOS and PMOS transistors, all of which are fabricated on a single chip by a relatively small number of processing steps.

Problems

Section 12.1: NMOS and PMOS Transistors

P12.1. Give the equations for the drain current and the ranges of v_{GS}, v_{DS}, and v_{GD} in terms of the threshold voltage V_{to} for each region (cutoff, saturation, and triode) of an n-channel MOSFET.

P12.2. Sketch the physical structure of an n-channel enhancement MOSFET. Label the channel length L, the width W, the terminals, and the channel region. Draw the corresponding circuit symbol.

P12.3. A certain NMOS transistor has $V_{to} = 1$ V, $KP = 50\ \mu\text{A/V}^2$, $L = 5\ \mu$m, and $W = 50\ \mu$m. For each set of voltages, state the region of operation and compute the drain current. **a.** $v_{GS} = 4$ V and $v_{DS} = 10$ V; **b.** $v_{GS} = 4$ V and $v_{DS} = 2$ V; **c.** $v_{GS} = 0$ V and $v_{DS} = 10$ V.

***P12.4.** Suppose that we have an NMOS transistor with $KP = 75\ \mu\text{A/V}^2$, $V_{to} = 1$ V, $L = 10\ \mu$m, and $W = 200\ \mu$m. Sketch the drain characteristics for v_{DS} ranging from 0 to 10 V and $v_{GS} = 0, 1, 2, 3,$ and 4 V.

P12.5. Determine the values of V_{to} and K for the NMOS transistor having the characteristics shown in Figure P12.5. If $KP = 50\ \mu\text{A/V}^2$, what is the value of W/L?

P12.6. Suppose we have an NMOS transistor that has $V_{to} = 0.8$ V. What is the region of operation (linear, saturation, or cutoff) if: **a.** $v_{GS} = 0$ V and $v_{DS} = 5$ V; **b.** $v_{GS} = 3$ and $v_{DS} = 8$ V; **c.** $v_{GS} = 5$ V and $v_{DS} = 1$ V; **d.** $v_{GS} = 3$ V and $v_{DS} = 10$ V?

P12.7. Consider an n-channel enhancement MOSFET with $V_{to} = 1$ V and $K = 0.2$ mA/V^2. Given that $v_{GS} = 3$ V, for what range of v_{DS} is the device in the saturation region? In the triode region? Plot i_D versus v_{GS} for operation in the saturation region.

P12.8. Suppose that the gate of an enhancement NMOS device is connected to the drain. What is the region of operation if a positive voltage greater than the threshold voltage is applied to the drain/gate terminal with respect to the source? If the applied voltage is less than the threshold voltage?

* Denotes that answers are contained in the Student Solutions files. See Appendix E for more information about accessing the Student Solutions.

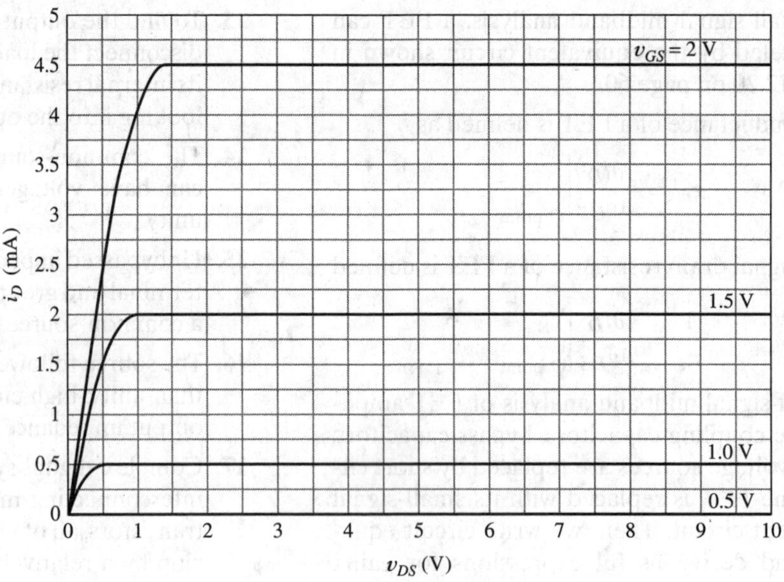

Figure P12.5

P12.9. Determine the region of operation for each of the enhancement transistors and the currents shown in Figure P12.9. All of the transistors have $K = 0.2$ mA/V^2, the NMOS transistors have $V_{to} = +1$ V, and the PMOS transistors have $V_{to} = -1$ V.

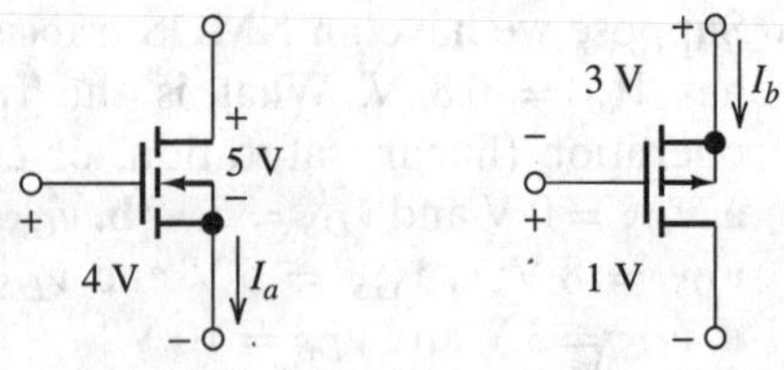

Figure P12.9

P12.10. The NMOS transistor shown in Figure P12.10 has $KP = 50\ \mu$A/V^2, $V_{to} = 1$ V, $L = 15\ \mu$m, and $W = 150\ \mu$m. Determine the drain current if: **a.** $R_1 = 4$ kΩ and $R_2 = 1$ kΩ; **b.** $R_1 = 6$ kΩ and $R_2 = 4$ kΩ.

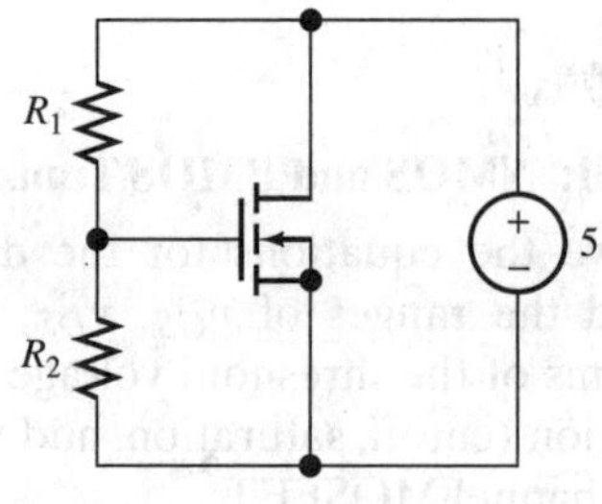

Figure P12.10

P12.11. The PMOS transistor shown in Figure P12.11 has $KP = 50\ \mu$A/V^2, $V_{to} = -0.8$ V, $L = 15\ \mu$m, and $W = 150\ \mu$m. Determine the drain current if: **a.** $R_1 = 1$ kΩ and $R_2 = 9$ kΩ; **b.** $R_1 = 3$ kΩ and $R_2 = 2$ kΩ.

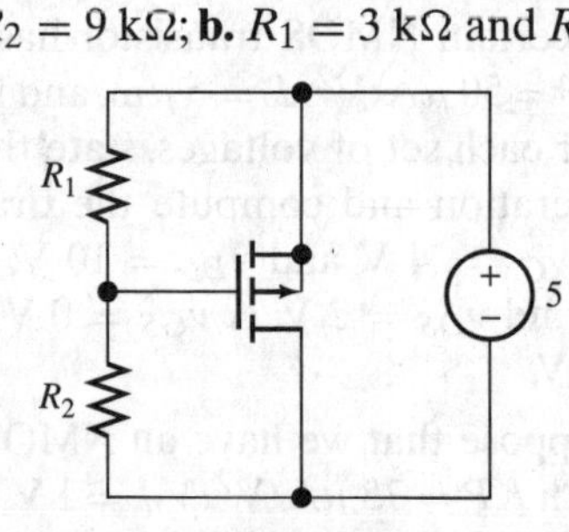

Figure P12.11

***P12.12.** Two points in the saturation region of a certain NMOS transistor are ($v_{GS} = 2$ V, $i_D = 0.2$ mA) and ($v_{GS} = 3$ V, $i_D = 1.8$ mA). Determine the values of V_{to} and K for this transistor.

*P12.13. A p-channel enhancement MOSFET has $V_{to} = -1$ V and $K = 0.2$ mA/V^2. Assuming operation in the saturation region, what value of v_{GS} is required for $i_D = 0.8$ mA?

P12.14. Suppose we need an NMOS transistor for which $i_D = 5$ mA when $v_{GS} = v_{DS} = 5$ V. Process constraints result in $KP = 50\ \mu$A/V^2 and $V_{to} = 1$ V. Determine the width-to-length ratio needed for the transistor. If $L = 2\ \mu$m, what is the value of W?

P12.15. Because of process constraints, L and W are required to be at least 0.25 μm. Furthermore, to save chip area, we do not want L or W to exceed 2 μm. How should we select L and W to obtain the least drain current for a given transistor? The greatest drain current? Assuming operation with identical voltages, what ratio of drain currents (between different transistors) can be achieved?

P12.16. Consider an NMOS transistor operating as a voltage-controlled resistance, as shown in Figure 12.4 on page 589, with $v_{DS} << v_{GS} - V_{to}$. Find an approximate expression for the resistance of the channel in terms of the device parameters and voltages. Given $V_{to} = 1$ V and $K = 0.2$ mA/V^2, compute the resistances for $v_{GS} = 1, 2, 3$, and 4 V.

P12.17. Determine the currents and the region of operation for each of the enhancement transistors shown in Figure P12.17 for $v_{in} = 0$ and for $v_{in} = 5$ V. The transistors have $|V_{to}| = 0.5$ V and $K = 0.2$ mA/V^2.

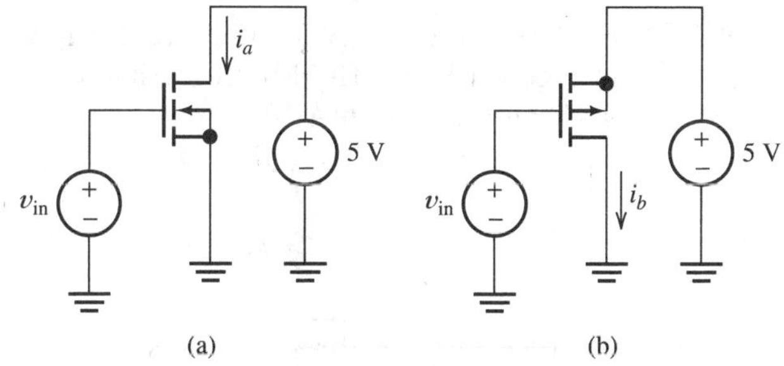

Figure P12.17

P12.18. Given that the enhancement transistor shown in Figure P12.18 has $V_{to} = 1$ V and $K = 0.5$ mA/V^2, find the value of the resistance R.

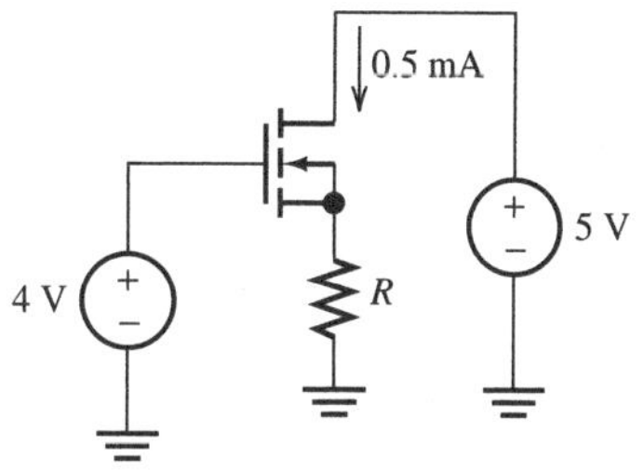

Figure P12.18

Section 12.2: Load-Line Analysis of a Simple NMOS Amplifier

P12.19. What is the principal cause of distortion in FET amplifiers?

*P12.20. Draw the load lines on the i_D–v_{DS} axes for the circuit of Figure 12.10 on page 595 for:

a. $R_D = 2$ kΩ and $V_{DD} = 10$ V

b. $R_D = 4$ kΩ and $V_{DD} = 15$ V

c. $R_D = 6$ kΩ and $V_{DD} = 20$ V

How does the position of the load line change as R_D increases in value?

P12.21. Draw the load lines on the i_D–v_{DS} axes for the circuit of Figure 12.10 on page 595 for:

a. $R_D = 3$ kΩ and $V_{DD} = 10$ V

b. $R_D = 3$ kΩ and $V_{DD} = 15$ V

c. $R_D = 3$ kΩ and $V_{DD} = 20$ V

How does the position of the load line change as V_{DD} increases in value?

*P12.22. Consider the circuit shown in Figure 12.10 on page 595. The transistor characteristics are shown in Figure 12.11. Suppose that V_{GG} is changed to 1 V. Determine the values of V_{DSQ}, $V_{DS\,\min}$, and $V_{DS\max}$. Find the gain of the amplifier. Rework with $V_{GG} = -2$ V.

P12.23. Consider the amplifier shown in Figure P12.23. The characteristic curves for the transistor are shown in Figure P12.5.

a. Find $v_{GS}(t)$, assuming that the coupling capacitor is a short circuit for the ac signal and an open circuit for dc. (*Hint*: Apply the superposition principle for the ac and dc sources.)

b. Draw the load line for the amplifier on the characteristics.

c. Find the values of V_{DSQ}, $V_{DS\min}$, and $V_{DS\max}$.

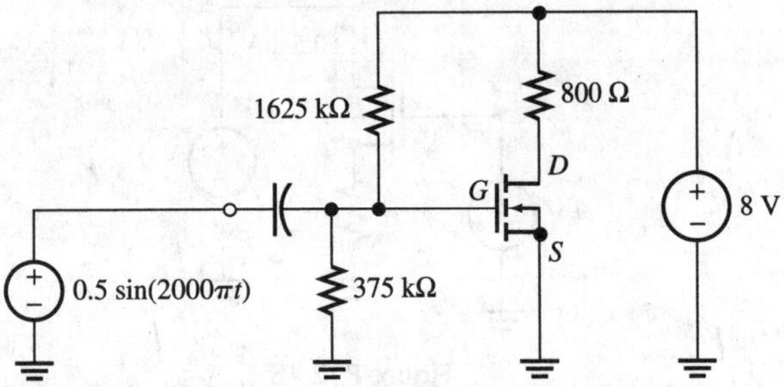

Figure P12.23

P12.24. What is the largest value of R_D allowed in the circuit of Problem P12.23 if the instantaneous operating point is required to remain in the saturation region at all times?

P12.25. Suppose that the resistance R_D in Figure 12.10 (page 595) is replaced with an unusual two-terminal nonlinear device for which $v = 0.1i_D^2$, where i_D is the current through the device in mA and v is the voltage across the device in volts (referenced positive at the end connected to V_{DD}). Carefully sketch the load line on Figure 12.11 (page 596). What shape is this load line?

P12.26. Use a load-line analysis of the circuit shown in Figure P12.26 to determine the values of V_{DSQ}, $V_{DS\min}$, and $V_{DS\max}$. The characteristics of the FET are shown in Figure 12.21 on page 604. (*Hint*: First, replace the 15-V source and the resistances by their Thévenin equivalent circuit.)

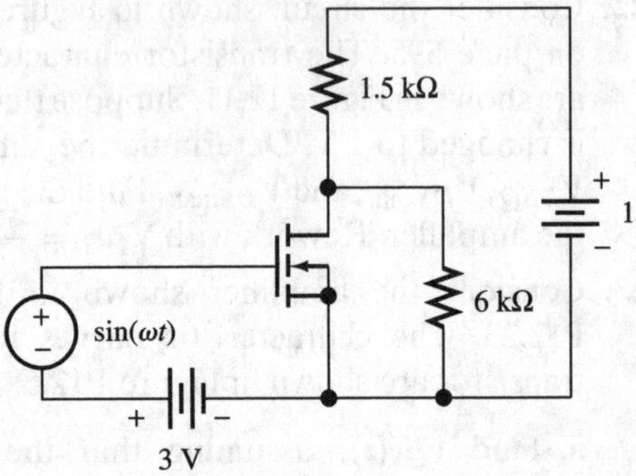

Figure P12.26

P12.27. Use a load-line analysis for the PMOS amplifier shown in Figure P12.27 to determine the maximum, minimum, and Q-point values of $v_o(t)$. The characteristics of the transistor are shown in Figure 12.9 on page 594.

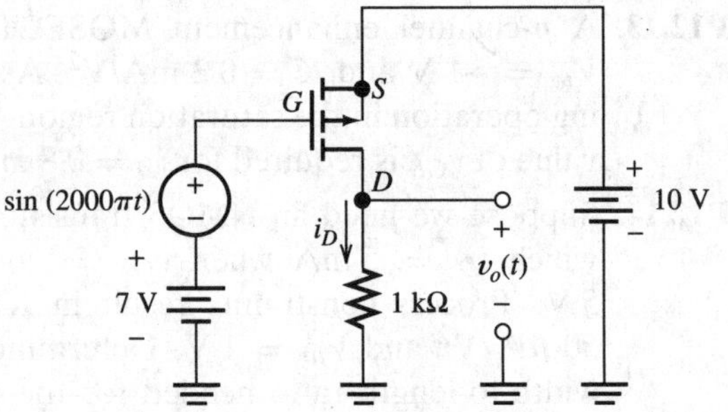

Figure P12.27

P12.28. The distorted signal shown in Figure 12.12(b) on page 596 can be written as

$$v_{DS}(t) = V_{DC} + V_{1m}\sin(2000\pi t) + V_{2m}\cos(4000\pi t)$$

The term $V_{1m}\sin(2000\pi t)$ is the desired signal. The term $V_{2m}\cos(4000\pi t)$ is distortion, which in this case has twice the frequency of the input signal, and is called second-harmonic distortion. Determine the values of V_{1m}, V_{2m}, and the percentage second-harmonic distortion, which is defined as $|V_{2m}/V_{1m}| \times 100\%$. (A high-quality audio amplifier has a distortion percentage less than 0.1 percent.)

Section 12.3: Bias Circuits

P12.29. In an amplifier circuit, why do we need to bias the MOSFET at an operating point? What would happen if the input signal peak amplitude was smaller than 1 V, the transistor had $V_{to} = 1$ V, and we biased the transistor at $V_{GSQ} = 0$?

***P12.30.** Find I_{DQ} and V_{DSQ} for the circuit shown in Figure P12.30. The MOSFET has $V_{to} = 0.8$ V and $K = 0.4$ mA/V^2.

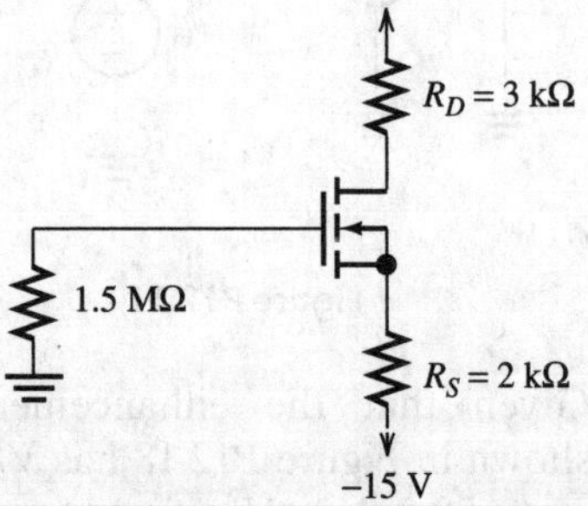

Figure P12.30

***P12.31.** We need a fixed- plus self-bias circuit for an NMOS source follower with $V_{DD} = 12$ V, $R_D = 0$, and $R_1 = 1$ MΩ. The transistor has $KP = 50\mu A/V^2$, $W = 800$ μm, $L = 10$ μm, and $V_{to} = 1$ V. The circuit is to have $V_{DSQ} = 6$ V and $I_{DQ} \cong 2$ mA. Determine the values of R_2 and R_S.

***P12.32.** The transistor of Figure P12.32 has $KP = 75$ μA/V^2, $W = 400$ μm, $L = 25$ μm, and $V_{to} = 0.9$ V. Determine the values of R_1 and R_s.

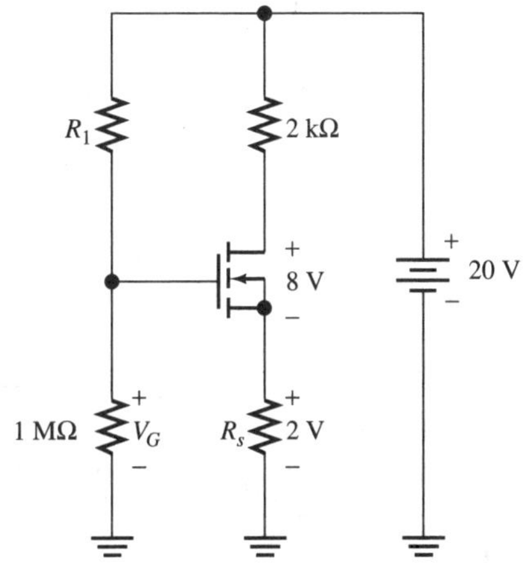

Figure P12.32

P12.33. Find I_{DQ} and V_{DSQ} for the circuit shown in Figure P12.33. The MOSFET has $V_{to} = 0.9$ V and $K = 0.5$ mA/V^2.

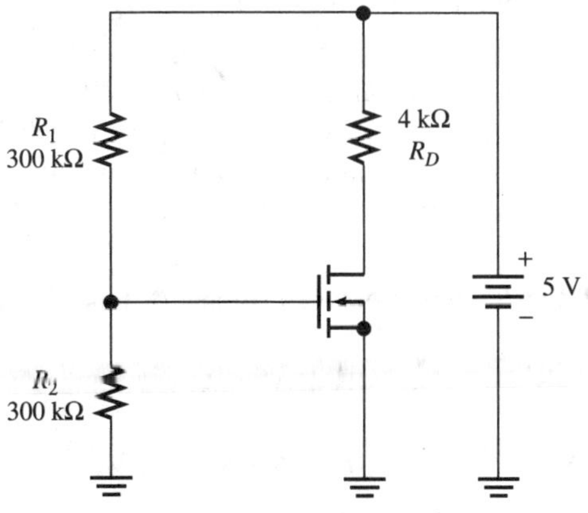

Figure P12.33

***P12.34.** Find I_{DQ} and V_{DSQ} for the circuit shown in Figure P12.34. The MOSFET has $V_{to} = 1$ V and $K = 0.25$ mA/V^2.

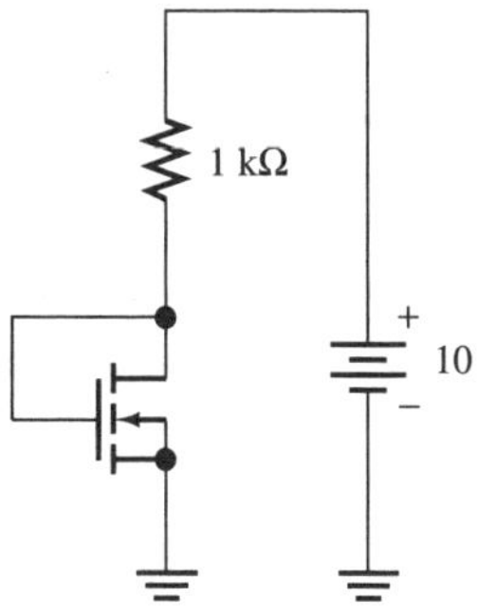

Figure P12.34

P12.35. The fixed- plus self-bias circuit of Figure 12.13 on page 598 has $V_{DD} = 12$ V, $R_1 = 2$ MΩ, $R_2 = 1$ MΩ, $R_S = 2$ k Ω, and $R_D = 2$ kΩ. The MOSFET has $V_{to} = 1$ V and $K = 0.5$ mA/V^2. Determine the Q point.

P12.36. Determine the value of I_{DQ} for the circuit shown in Figure P12.36, given that $V_{to} = 1$ V and $K = 4$ mA/V^2

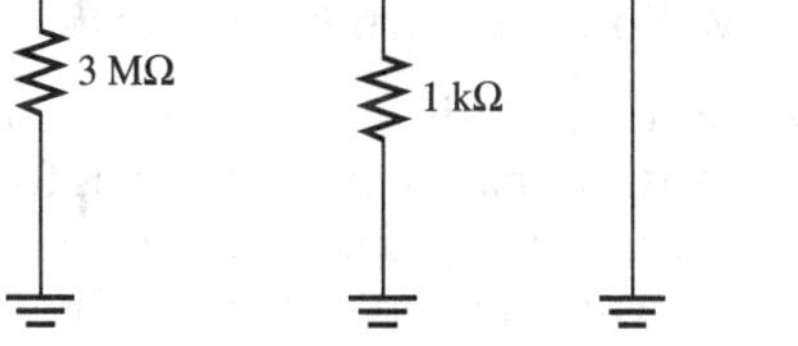

Figure P12.36

P12.37. Consider the fixed- plus self-bias circuit of Figure 12.13(a) on page 598, with $V_{DD} = 15$ V, $R_1 = 2$ MΩ, and $R_D = 3$ kΩ. Nominally, the transistor has $KP = 40$ μA/V2, $W = 125$ μm, $L = 15$ μm, and $V_{to} = 1$ V.

The circuit is to have $V_{DSQ} = 8$ V and $I_{DQ} \approx 2$ mA. Determine the values needed for R_2 and for R_S.

P12.38. Both transistors shown in Figure P12.38 have $KP = 100\ \mu A/V^2$ and $V_{to} = 1$ V. Determine the value of R needed so that $i_{D1} = 0.2$ mA. For what range of V_x is the second transistor operating in the saturation region? What is the resulting value of i_{D2}? Provided that V_x is large enough so that the second transistor operates in saturation, to what ideal circuit element is the transistor equivalent?

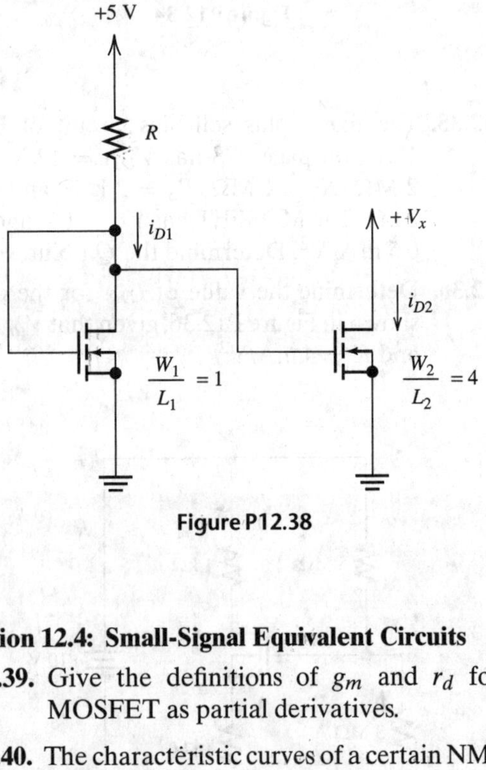

Figure P12.38

Section 12.4: Small-Signal Equivalent Circuits

P12.39. Give the definitions of g_m and r_d for a MOSFET as partial derivatives.

P12.40. The characteristic curves of a certain NMOS transistor have constant values for i_D in the saturation region. What is the value of r_d, assuming operation in the saturation region?

P12.41. Draw the small-signal equivalent circuit for a FET, including r_d.

P12.42. A certain NMOS transistor has the characteristics shown in Figure P12.42. Graphically determine the values of g_m and r_d at the operating point defined by $V_{DSQ} = 6$ V and $V_{GSQ} = 2.5$ V.

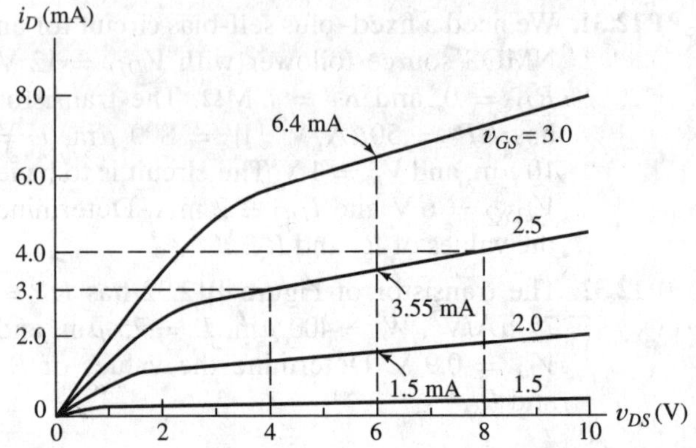

Figure P12.42

P12.43. Consider an NMOS transistor that has $g_m = 1$ mS and $r_d = 10$ kW for a Q point of $V_{GSQ} = 1$ V, $I_{DQ} = 2$ mA, and $V_{DSQ} = 5$ V. Sketch the drain characteristics to scale for a small region around the Q point, say for $v_{GS} = 0.8, 1.0,$ and 1.2 V and for $4.0 < v_{DS} < 6.0$ V.

P12.44. What is the value of g_m for $V_{DSQ} = 0$? Draw the small-signal equivalent circuit at this bias point. For what applications could the FET be used at this bias point?

***P12.45.** Derive an expression for g_m in terms of K, V_{to}, V_{GSQ}, and I_{DQ} for an NMOS transistor operating in the *triode* region.

***P12.46.** Derive an expression for r_d in terms of K, V_{to}, V_{GSQ}, and I_{DQ} for an NMOS transistor operating in the *triode* region.

P12.47. Consider an unusual type of FET for which $i_D = 0.2v_{DS} + 4v_{GS}^2$, in which i_D is in mA, v_{GS} is in volts, and v_{DS} is in volts. Determine the values of g_m and r_d for a Q point of $V_{GSQ} = 2$ V and $V_{DSQ} = 10$ V.

P12.48. Consider an unusual type of FET for which $i_D = \exp(v_{GS}) + 0.02v_{DS}^2$, in which i_D is in mA, v_{GS} is in volts, and v_{DS} is in volts. Determine the values of g_m and r_d for a Q point of $V_{GSQ} = 1$ V and $V_{DSQ} = 10$ V.

P12.49. A certain NMOS transistor has

$$v_{GS}(t) = 2 \text{ V}$$

$$v_{DS}(t) = 5 + \sin(\omega t) \text{ V},$$

$$i_D(t) = 5 + 0.08\sin(\omega t) \text{ mA}$$

Which small-signal parameter (g_m or r_d) can be determined from this information? What is its value? For what Q point (V_{GSQ}, I_{DQ}, and V_{DSQ}) does this parameter apply?

P12.50. A certain NMOS transistor has

$$v_{GS}(t) = 2 + 0.2\sin(\omega t)\ \text{V}$$

$$v_{DS}(t) = 5\ \text{V}$$

$$i_D(t) = 1 + 0.2\sin(\omega t)\ \text{mA}$$

Which small-signal parameter (g_m or r_d) can be determined from this information? What is its value? For what Q point (V_{GSQ}, I_{DQ}, and V_{DSQ}) does this parameter apply?

Section 12.5: Common-Source Amplifiers

P12.51. Draw the circuit diagram of a resistance–capacitance coupled common-source amplifier.

P12.52. Explain the function of coupling capacitors. Assuming that they are performing their intended function, how do they appear in the ac equivalent circuit? In general, what effect do coupling capacitors have on the gain of an amplifier as a function of frequency?

***P12.53.** Consider the common-source amplifier shown in Figure P12.53. The NMOS transistor has $KP = 50\ \mu\text{A/V}^2$, $L = 5\ \mu\text{m}$, $W = 500\ \mu\text{m}$, $V_{to} = 1$ V, and $r_d = \infty$.

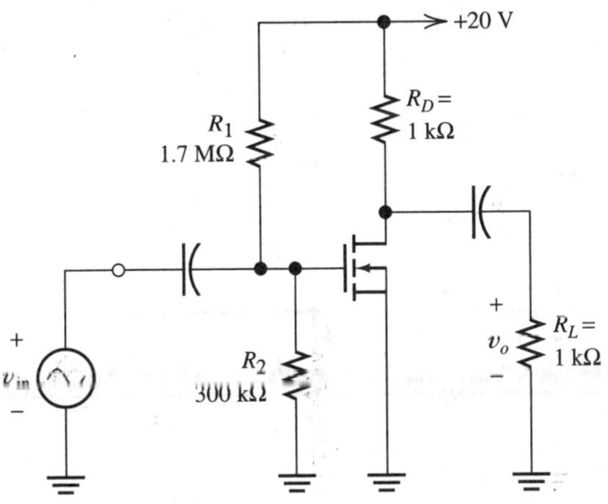

Figure P12.53

a. Determine the values of I_{DQ}, V_{DSQ}, and g_m.

b. Compute the voltage gain, input resistance, and output resistance, assuming that the coupling capacitors are short circuits for the ac signal.

P12.54. Repeat Problem P12.53 for an NMOS transistor having $KP = 50\ \mu\text{A/V}^2$, $W = 600\ \mu\text{m}$, $L = 20\ \mu\text{m}$, $V_{to} = 2$ V, and $r_d = \infty$. Compare the gain with that attained in Problem P12.53.

P12.55. Consider the amplifier shown in Figure P12.55.

a. Draw the small-signal equivalent circuit, assuming that the capacitors are short circuits for the signal.

b. Assume that $r_d = \infty$, and derive expressions for the voltage gain, input resistance, and output resistance.

c. Find I_{DQ} if $R = 100$ kΩ, $R_f = 100$ kΩ, $R_D = 3$ kΩ, $R_L = 10$ kΩ, $V_{DD} = 20$ V, $V_{to} = 5$ V, and $K = 1\ \text{mA/V}^2$. Determine the value of g_m at the Q point.

d. Evaluate the expressions found in part (b) using the values given in part (c).

e. Find $v_o(t)$ if $v(t) = 0.2\sin(2000\pi t)$.

f. Is this amplifier inverting or noninverting?

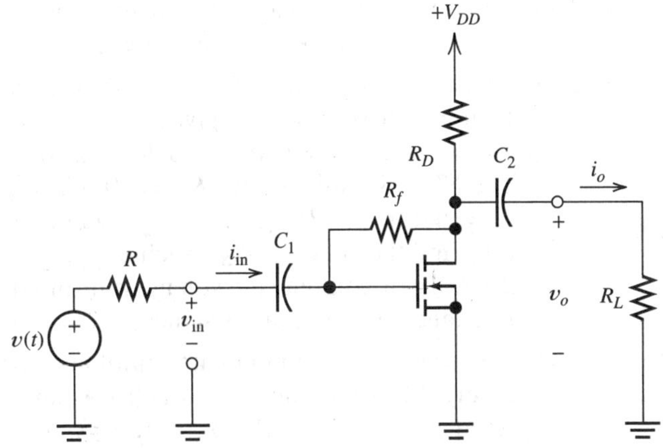

Figure P12.55

***P12.56.** Find V_{DSQ} and I_{DQ} for the FET shown in Figure P12.56, given $V_{to} = 3$ V and $K = 0.5\ \text{mA/V}^2$. Find the value of g_m at the operating point. Draw the small-signal equivalent circuit, assuming that $r_d = \infty$. Derive an expression for the resistance R_o in terms of R_D and g_m. Evaluate the expression for the values given.

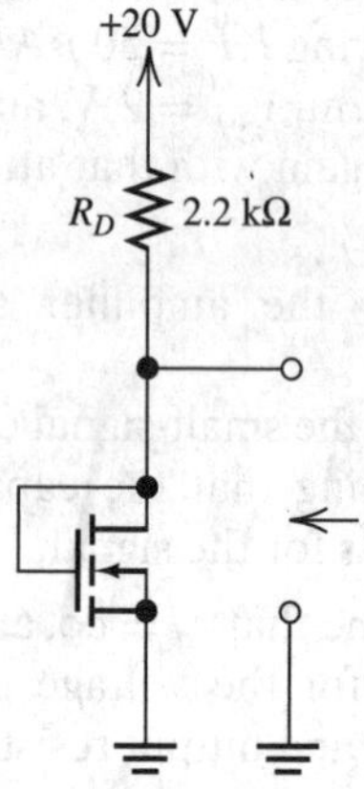

Figure P12.56

Section 12.6: Source Followers

P12.57. Consider the common-source amplifier and the source follower. Which amplifier would be used if a voltage-gain magnitude larger than unity is needed? Which would be used to obtain low output resistance?

P12.58. Draw the circuit diagram of a resistance–capacitance coupled source follower.

***P12.59.** Consider the source follower shown in Figure 12.26 on page 609, given $V_{DD} = 15$ V, $R_L = 2$ kΩ, $R_1 = 1$ MΩ, and $R_2 = 2$ MΩ. The NMOS transistor has $KP = 50\ \mu$A/V^2, $L = 10\ \mu$m, $W = 160\ \mu$m, $r_d = \infty$, and $V_{to} = 1$ V. Find the value for R_S to achieve $I_{DQ} = 2$ mA. Then, compute the voltage gain, input resistance, and output resistance.

P12.60. Consider the common-gate amplifier of Figure 12.29 on page 612, which was analyzed in Exercise 12.13 on page 612. The MOSFET has $KP = 50\ \mu$A/V2, $W = 400\ \mu$m, $L = 15\ \mu$m, $V_{to} = 1$ V, and $rd = 5$ MΩ. The supply voltages are $V_{DD} = 20$ V and $V_{SS} = 15$ V. The resistances are $R_S = 4$ kΩ, $RL = 15$ kΩ, and $RD = 2$ kΩ.

a. Determine the Q point and the value of g_m.

b. Determine the input resistance and the voltage gain.

Section 12.7: CMOS Logic Gates

P12.61. Draw the circuit diagram of a CMOS inverter. Draw its equivalent circuit (open and closed switches) if the input is high. Repeat if the input is low.

P12.62. Draw the circuit diagram of a two-input CMOS AND gate. (*Hint*: Use a two-input NAND followed by an inverter.)

P12.63. **a.** Draw the circuit diagram of a three-input CMOS NAND gate. **b.** Draw its equivalent circuit (open and closed switches) if all three inputs are high. **c.** Repeat if all three inputs are low.

P12.64. Consider the simple model for a CMOS inverter shown in Figure P12.64. Each transistor has been modeled by a switch in series with a resistance, and the input to the gates driven by the inverter is modeled as a load capacitance C_L. The output cycles from high to low and back to high at a frequency f. **a.** Find an expression for the average power taken from the V_{DD} source. Where is this power dissipated? **b.** If V_{DD} is cut in half, by how much is the power reduced? **c.** Suppose we have computer chip with $V_{DD} = 1$ V, a clock frequency of 1 GHz, and 1 billion gates, each of which cycles between states at an average rate equal to 1% of the clock frequency. Also, the average capacitive load for each gate is 2 fF. (Note: 1 fF = 1 femtofarad = 10^{-15} F.) Compute the average power consumed by this chip.

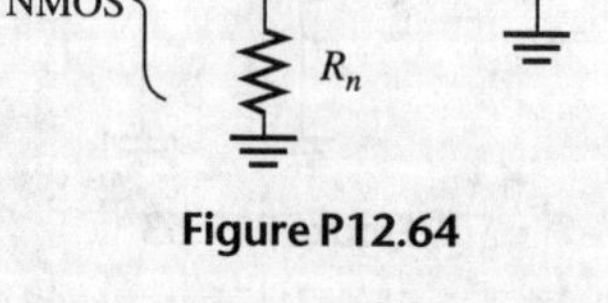

Figure P12.64

Frequently Asked Questions

1. Explain the construction details of an n-channel enhancement MOSFET.
2. How a MOSFET works? Explain.
3. Explain the operation of a MOSFET in cutoff region and saturation region.
4. Derive an equation for the boundary between the triode region and saturation region.
5. Draw the symbol of an n-channel and a n-channel MOSFET and label their terminals.
6. Draw the characteristics of PMOS and NMOS transistor.
7. Explain the load-line analysis of a simple NMOS amplifier.
8. Draw the biasing circuit for a MOSFET.
9. Draw the fixed plus self-bias circuit for a MOSFET.
10. Develop a small-signal requirement circuit of a MOSFET.
11. Define transconductance and drain resistance of a MOSFET.
12. Draw the circuit of a common source amplifier using MOSFET and explain the function of each component.
13. Derive expressions for the following:

    (i) Voltage gain
    (ii) Input resistance
    (iii) Output resistance

    for a MOSFET amplifier circuit.
14. Draw the circuit of a source follower using MOSFET and explain its working?
15. For a source follower amplifier circuit using MOSFET, define the following terms.

    (i) Voltage gain
    (ii) Input resistance
    (iii) Output resistance
16. Draw a simplified circuit for a CMOS inverter using MOSFET and explain its working.
17. Draw the circuit using CMOS for developing NAND gate.
18. Draw the circuit using CMOS for developing NOR gate.

Practice Test

Here is a practice test you can use to check your comprehension of the most important concepts in this chapter. Answers can be found in Appendix D and complete solutions are included in the Student Solutions files. See Appendix E for more information about the Student Solutions.

T12.1. An NMOS transistor has $KP = 80\,\mu A/V^2$, $V_{to} = 1\,V$, $L = 4\,\mu m$, and $W = 100\,\mu m$. Carefully sketch to scale the drain characteristics for v_{DS} ranging from 0 to 10 V and $v_{GS} = 0.5$ and 4 V.

T12.2. We have an amplifier identical to that of Figure 12.10 on page 595, except that R_D is changed to 2 kΩ and the dc sources are changed to $V_{DD} = 10\,V$ and $V_{GG} = 3\,V$. The drain characteristics for the transistor are shown in Figure 12.7 on page 593. Use load-line analysis to determine the maximum, minimum, and Q-point values of v_{DS}.

T12.3. Consider the biasing circuit shown in Figure T12.3. The transistor has $KP = 80\,\mu A/V^2$, $V_{to} = 1\,V$, $L = 4\,\mu m$, and $W = 100\,\mu m$. What value is required for R_S so the operating current is $I_{DQ} = 0.5\,mA$?

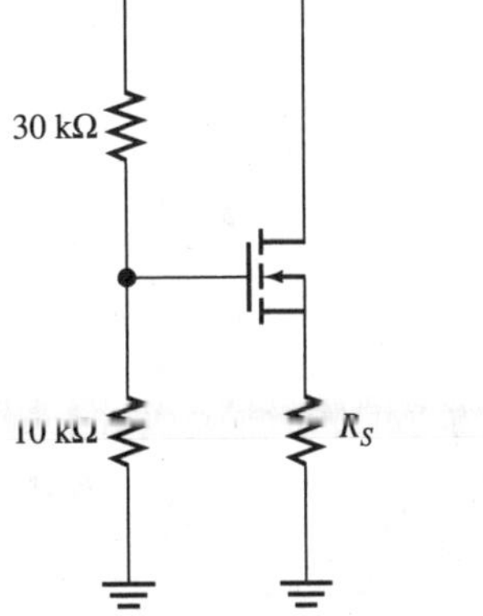

Figure T12.3

T12.4. A certain NMOS transistor has

$$v_{GS}(t) = 2 + 0.02\sin(\omega t)\text{ V}$$
$$v_{DS}(t) = 5\text{ V}$$
$$i_D(t) = 0.5 + 0.05\sin(\omega t)\text{ mA}$$

Which small-signal parameter (g_m or r_d) can be determined from this information? What is its value? For what Q point (V_{GSQ}, I_{DQ}, and V_{DSQ}) does this parameter apply?

T12.5. What replaces each of the following ele ments when we draw the mid-band small signal equivalent circuit for an amplifier **a.** a dc voltage source; **b.** a coupling capaci tor; **c.** a dc current source?

T12.6. Draw the circuit diagram of a CMOS inverter. When the input is high, which transistor is on? Which is off?

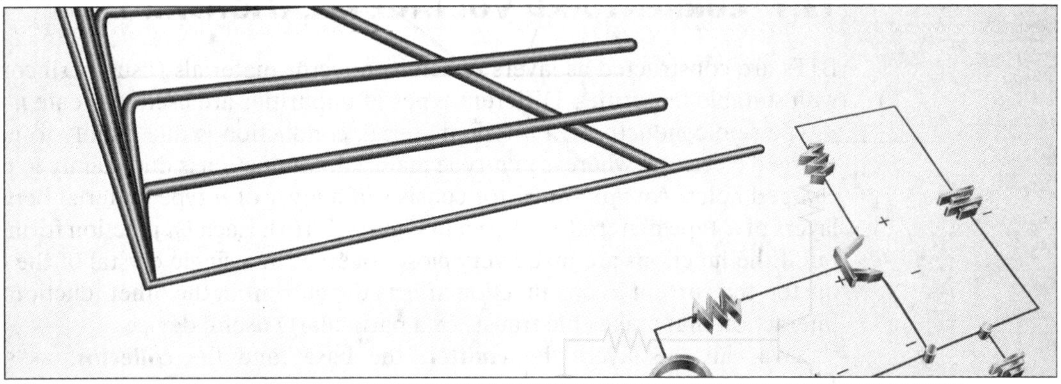

Chapter 13

Bipolar Junction Transistors

Study of this chapter will enable you to:

- Understand bipolar junction transistor operation in amplifier circuits.
- Use the load-line technique to analyze simple amplifiers and understand the causes of nonlinear distortion.
- Use large-signal equivalent circuits to analyze BJT circuits.
- Analyze bias circuits.
- Use small-signal equivalent circuits to analyze BJT amplifiers.
- Compute performance of several important amplifier configurations.
- Select an amplifier configuration appropriate for a given application.

Introduction to this chapter:

In Chapter 12, we considered the field-effect transistor, which is one of the important devices in modern electronics. Now, we turn our attention to another important device, the **bipolar junction transistor** (BJT), which is also very useful in amplifiers and digital logic circuits. These two devices are the key building blocks for modern electronics.

First, we discuss the device parameters and equations that relate the currents and voltages in the *npn* BJT. Next, we discuss the BJT common-emitter characteristics, which show the device operation graphically. Then, we use the graphical load-line technique to analyze a simple amplifier circuit. In Section 13.4, the *pnp* BJT is introduced. In the next several sections, we discuss large-signal models for the three regions of BJT operation (the active, saturation, and cutoff regions) and use the models to analyze bias circuits. Then, we develop a small-signal equivalent circuit for the BJT and use it to analyze two important amplifier configurations (common-emitter amplifiers and emitter followers).

13.1 CURRENT AND VOLTAGE RELATIONSHIPS

BJTs are constructed as layers of semiconductor materials (usually silicon) **doped** with suitable impurities. Different types of impurities are used to create *n*-type and *p*-type semiconductors. In *n*-type material, conduction is due mainly to negatively charged electrons, whereas in *p*-type material, conduction is due mainly to positively charged holes. An *npn* transistor consists of a layer of *p*-type material between two layers of *n*-type material, as shown in Figure 13.1(a). Each *pn* junction forms a diode, but if the junctions are made very close together in a single crystal of the semiconductor, the current in one junction affects the current in the other junction. It is this interaction that makes the transistor a particularly useful device.

We call the layers the **emitter**, the **base**, and the **collector**, as shown in Figure 13.1(a). The circuit symbol for an *npn* BJT is shown in Figure 13.1(b), including reference directions for the currents.

An *npn* BJT consists of a layer of *p*-type semiconductor, called the base, between two layers of *n*-type semiconductor, called the collector and the emitter. Assuming that proper voltages are applied, small amounts of current flowing into the base terminal cause much larger currents to flow from the collector to the emitter.

Fluid-Flow Analogy

We will see that the BJT is somewhat analogous to a valve in a fluid-flow system. In suitable circuits, if a small current is made to flow into the base terminal, a much larger current flows into the collector and out of the emitter terminal. We can imagine that the base current opens a valve between collector and emitter. Larger base currents open the valve wider. When a signal to be amplified is applied as a current to the base, the valve between collector and emitter opens and closes in response to changes in the signal. Thus, a current with magnified fluctuations flows between collector and emitter.

Equations of Operation

A *pn* junction is **forward biased** by applying voltage with the positive polarity on the *p*-side. On the other hand, **reverse bias** occurs if the positive polarity is applied to the *n*-side. This is illustrated in Figure 13.2.

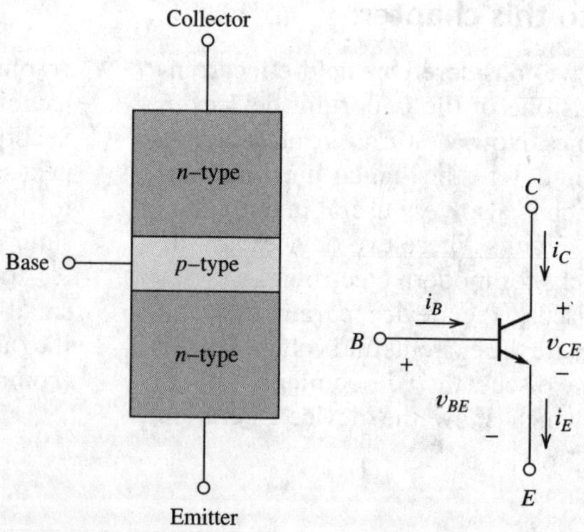

(a) Physical structure (b) Circuit symbol

Figure 13.1 The *npn* BJT.

In normal operation of a BJT as an amplifier, the base–collector junction is reverse biased and the base–emitter junction is forward biased. In the upcoming discussion, we assume that the junctions are biased in this fashion unless stated otherwise.

The Shockley equation gives the emitter current i_E in terms of the base-to-emitter voltage v_{BE}:

$$i_E = I_{ES}\left[\exp\left(\frac{v_{BE}}{V_T}\right) - 1\right] \tag{13.1}$$

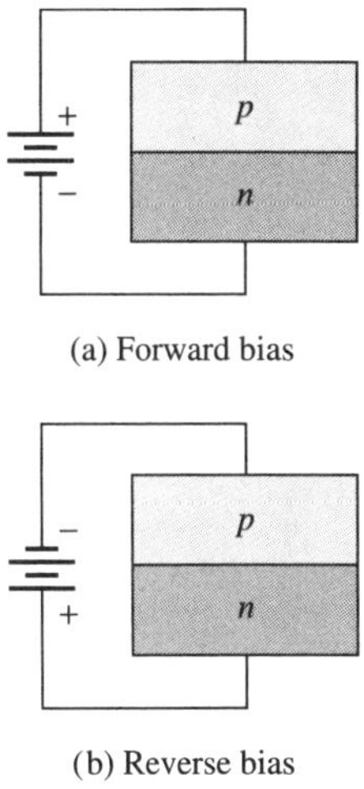

Figure 13.2 Bias conditions for *pn* junctions.

This is exactly the same equation as that for the current in a junction diode given in Equation 10.1, except for changes in notation. (We have let the emission coefficient n equal unity since that is the appropriate value for most junction transistors.) Typical values for the saturation current I_{ES} range from 10^{-12} to 10^{-16} A, depending on the physical size of the device and other factors. (Recall that at a temperature of 300 K, V_T is approximately 26 mV.)

Of course, Kirchhoff's current law requires that the current flowing out of the BJT is equal to the sum of the currents flowing into it. Thus, referring to Figure 13.1(b), we have

$$i_E = i_C + i_B \tag{13.2}$$

In normal operation as an amplifier, the base–collector junction is reverse biased and the base–emitter junction is forward biased.

(This equation is true regardless of the bias conditions of the junctions.)

We define the parameter α as the ratio of the collector current to the emitter current, displayed as

$$\alpha = \frac{i_C}{i_E} \tag{13.3}$$

Values for α range from 0.9 to 0.999, with 0.99 being very typical. Equation 13.2 indicates that the emitter current is supplied partly through the base terminal and partly through the collector terminal. However, since α is nearly unity, most of the emitter current is supplied through the collector.

Normally, α is slightly less than unity.

Substituting Equation 13.1 into 13.3 and rearranging, we have

$$i_C = \alpha I_{ES}\left[\exp\left(\frac{v_{BE}}{V_T}\right) - 1\right] \tag{13.4}$$

For v_{BE} greater than a few tenths of a volt, the exponential term inside the brackets is much larger than unity. Then, the 1 inside the brackets can be dropped. Also, we define the **scale current** as

$$I_s = \alpha I_{ES} \tag{13.5}$$

and Equation 13.4 becomes

$$i_C \cong I_s \exp\left(\frac{v_{BE}}{V_T}\right) \tag{13.6}$$

Solving Equation 13.3 for i_C, substituting into Equation 13.2, and solving for the base current, we obtain

$$i_B = (1 - \alpha) i_E \tag{13.7}$$

Since α is slightly less than unity, only a very small fraction of the emitter current is supplied through the base. Using Equation 13.1 to substitute for i_E, we obtain

$$i_B = (1-\alpha)\, I_{ES}\left[\exp\left(\frac{v_{BE}}{V_T}\right) - 1\right] \tag{13.8}$$

We define the parameter β as the ratio of the collector current to the base current. Taking the ratio of Equations 13.4 and 13.8 results in

$$\beta = \frac{i_C}{i_B} = \frac{\alpha}{1-\alpha} \tag{13.9}$$

Values for β range from about 10 to 1000, and a very common value is $\beta = 100$. We can write

$$i_C = \beta i_B \tag{13.10}$$

Because β is usually large compared to unity, the collector current is an amplified version of the base current.

Note that since β is usually large compared to unity, *the collector current is an amplified version of the base current*. Current flow in an *npn* BJT is illustrated in Figure 13.3.

Exercise 13.1 A certain transistor has $\beta = 50$, $I_{ES} = 10^{-14}$ A, $v_{CE} = 5$ V, and $i_E = 10$ mA. Assume that $V_T = 0.026$ V. Find v_{BE}, v_{BC}, i_B, i_C, and α.
Answer $v_{BE} = 0.718$ V, $v_{BC} = -4.28$ V, $i_B = 0.196$ mA, $i_C = 9.80$ mA, $\alpha = 0.980$. □

Exercise 13.2 Compute the corresponding values of β if $\alpha = 0.9, 0.99$, and 0.999.
Answer $\beta = 9, 99$, and 999, respectively. □

Exercise 13.3 A certain transistor operated with forward bias of the base–emitter junction and reverse bias of the base–collector junction has $i_C = 9.5$ mA and $i_E = 10$ mA. Find the values of i_B, α, and β.
Answer $i_B = 0.5$ mA, $\alpha = 0.95$, $\beta = 19$. □

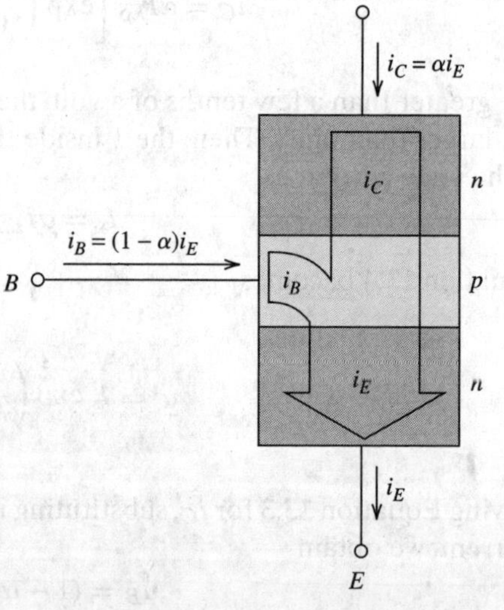

Figure 13.3 Only a small fraction of the emitter current flows into the base (provided that the collector–base junction is reverse biased and the base–emitter junction is forward biased).

13.2 COMMON-EMITTER CHARACTERISTICS

The common-emitter configuration for an *npn* BJT is shown in Figure 13.4. The voltage source connected between the base and emitter supplies a positive voltage v_{BE} that forward biases the base–emitter junction. The v_{CE} voltage source produces a positive voltage at the collector with respect to the emitter. Notice that the voltage across the base–collector junction is given by

$$v_{BC} = v_{BE} - v_{CE} \tag{13.11}$$

Thus, if v_{CE} is greater than v_{BE}, the base-to-collector voltage v_{BC} is negative (which is reverse bias).

The **common-emitter characteristics** of the transistor are plots of the currents i_B and i_C versus the voltages v_{BE} and v_{CE}. Representative characteristics for a small-signal silicon device are shown in Figure 13.5.

The **common-emitter input characteristic** shown in Figure 13.5(a) is a plot of i_B versus v_{BE}, which are related by Equation 13.8. Notice that the input characteristic takes the same form as the forward-bias characteristic of a junction diode. Thus, for appreciable current to flow at room temperature, the base-to-emitter voltage must be approximately 0.6 to 0.7 V. (The base-to-emitter voltage for a given current decreases with temperature by about 2 mV/K.)

The **common-emitter output characteristics** shown in Figure 13.5(b) are plots of i_C versus v_{CE} for constant values of i_B. The transistor illustrated has $\beta = 100$. As long as the collector–base junction is reverse biased ($v_{BC} < 0$ or equivalently, $v_{CE} > v_{BE}$),

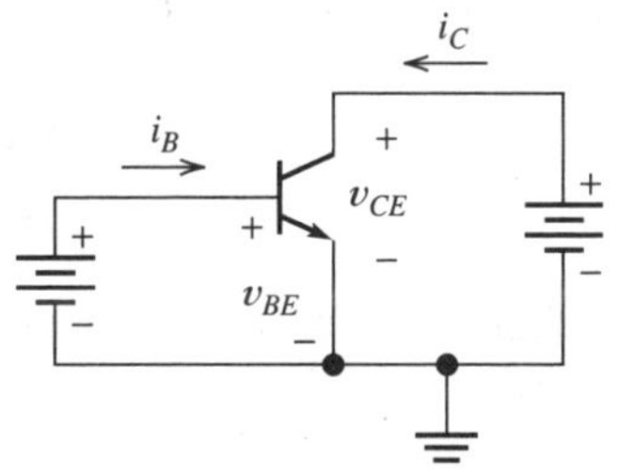

Figure 13.4 Common-emitter circuit configuration for the *npn* BJT.

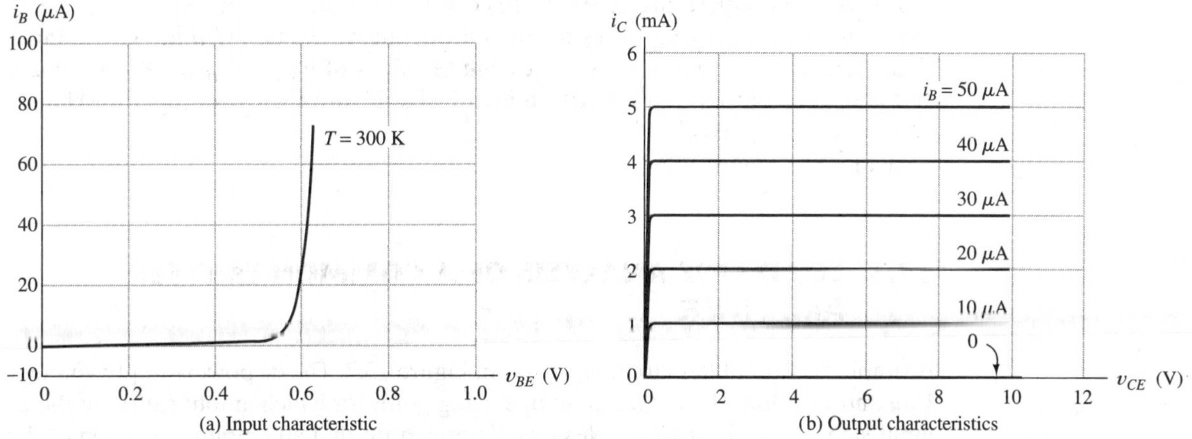

Figure 13.5 Common-emitter characteristics of a typical *npn* BJT.

we have

$$i_C = \beta i_B = 100 i_B$$

As v_{CE} becomes less than v_{BE}, the base–collector junction becomes forward biased, and eventually the collector current falls as shown at the left-hand edge of the output characteristics.

Amplification by the BJT

Refer to Figure 13.5(a), and notice that a very small change in the base-to-emitter voltage v_{BE} can result in an appreciable change in the base current i_B, particularly if the base–emitter junction is forward biased so that some current (say 40 μA) is flowing before the change in v_{BE} is made. Provided that v_{CE} is more than a few tenths of a volt, this change in base current causes a much larger change in the collector current i_C (because $i_C = \beta i_B$). In suitable circuits, the change in collector current is converted into a much larger voltage change than the initial change in v_{BE}. Thus, the BJT can amplify a signal applied to the base–emitter junction.

Example 13.1 Determining β from the Characteristic Curves

Verify that the value of β is 100 for the transistor with the characteristics shown in Figure 13.5.

Solution The value of β can be found by taking the ratio of collector current to base current provided that v_{CE} is high enough so that the collector–base junction is reverse biased. For example, at $v_{CE} = 4$ V and $i_B = 30\ \mu$A, the output characteristics yield $i_C = 3$ mA. Thus, the value of β is

$$\beta = \frac{i_C}{i_B} = \frac{3\text{ mA}}{30\ \mu\text{A}} = 100$$

(For most devices, slightly different values of β result from different points on the output characteristics.) ■

Exercise 13.4 Plot the common-emitter characteristics of an *npn* small-signal silicon transistor at a temperature of 300 K if $I_{ES} = 10^{-14}$ A and $\beta = 50$. Allow i_B to range from 0 to 50 μA in 10-μA steps for the output characteristics. [*Hints:* For the input characteristic, use Equation 13.8 to calculate values of v_{BE} for $i_B = 10\ \mu$A, 20 μA, and so on. The output characteristic is identical to Figure 13.5(b), except for a change in scale for the i_C axis.]

Answer See Figure 13.6. □

13.3 LOAD-LINE ANALYSIS OF A COMMON-EMITTER AMPLIFIER

A simple BJT amplifier circuit is shown in Figure 13.7. The dc power-supply voltages V_{BB} and V_{CC} **bias** the device at an operating point for which amplification of the ac input signal $v_{\text{in}}(t)$ is possible. Next, we demonstrate that an amplified version of the input signal voltage appears between the collector and ground.

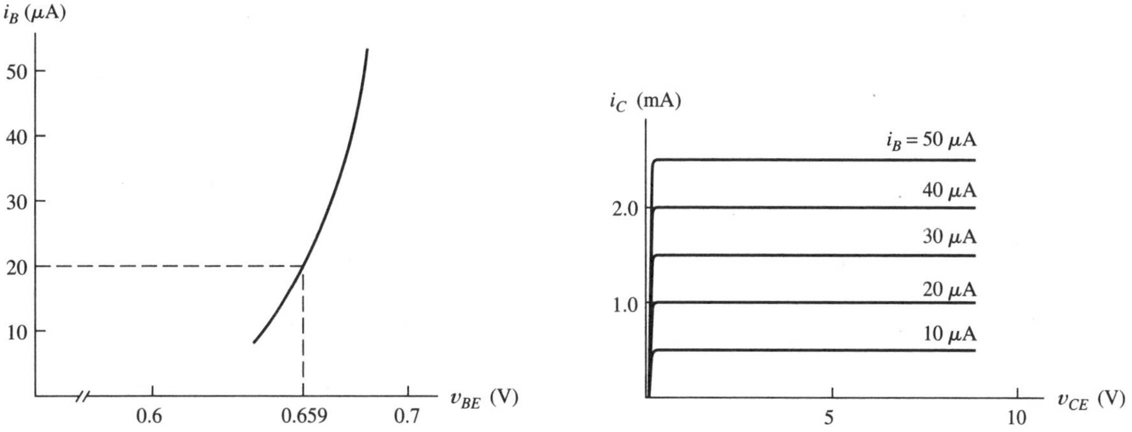

(a) Input characteristic

(b) Output characteristics

Figure 13.6 See Exercise 13.4.

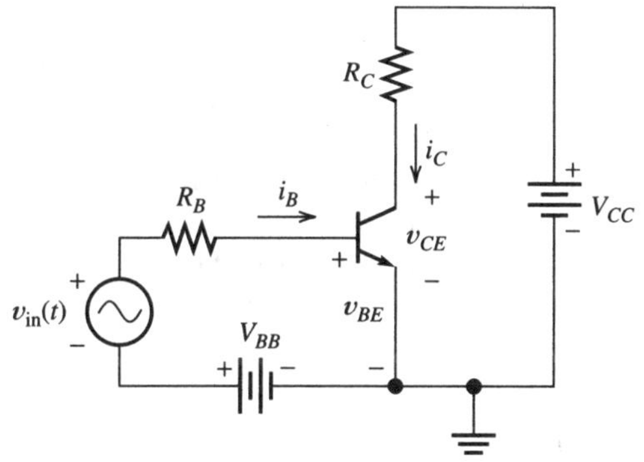

Figure 13.7 A simple common-emitter amplifier that can be analyzed by load-line techniques.

Analysis of the Input Circuit

We can analyze this circuit by use of load-line techniques similar to those that we used for diode circuits in Chapter 10. For example, if we apply Kirchhoff's voltage law to the loop consisting of V_{BB}, $v_{in}(t)$, and the base–emitter junction, we obtain

Equation for the input load line.

$$V_{BB} + v_{in}(t) = R_B i_B(t) + v_{BE}(t) \tag{13.12}$$

A plot of Equation 13.12 is shown as the **load line** on the input characteristics of the transistor in Figure 13.8(a). To establish this load line, we must locate two points. If we assume that $i_B = 0$, Equation 13.12 yields $v_{BE} = V_{BB} + v_{in}$. This establishes the point where the load line intersects the voltage axis. Similarly, assuming that $v_{BE} = 0$ results in $i_B = (V_{BB} + v_{in})/R_B$, which establishes the load-line intercept on the current axis. The load line is shown as the solid line in Figure 13.8(a).

Equation 13.12 represents the constraint placed on the values of i_B and v_{BE} by the external circuit. In addition, i_B and v_{BE} must fall on the device characteristic. The values that satisfy both constraints are the values at the intersection of the load line and the device characteristic.

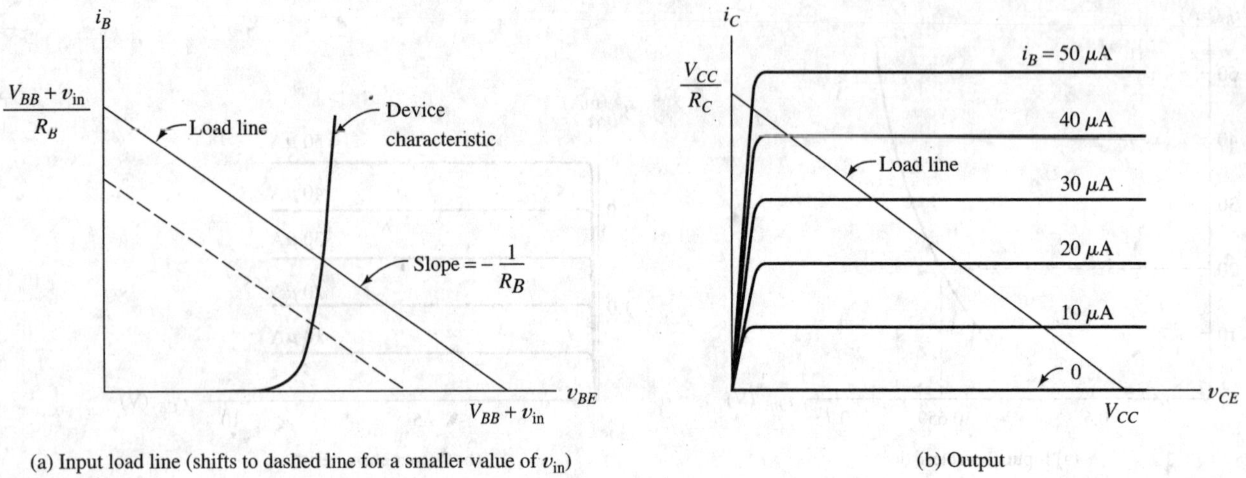

Figure 13.8 Load-line analysis of the amplifier of Figure 13.7.

The input load line shifts position but maintains constant slope as $v_{in}(t)$ changes.

The slope of the load line is $-1/R_B$. Thus, the load line shifts position but maintains a constant slope as v_{in} changes in value. For example, the dashed load line in Figure 13.8(a) is for a smaller value of v_{in} than that for the solid load line.

The **quiescent operating point**, or **Q point**, corresponds to $v_{in}(t) = 0$. Thus, as the ac input signal $v_{in}(t)$ changes in value with time, the instantaneous operating point swings above and below the Q-point value. Values of i_B can be found from the intersection of the load line with the input characteristic for each value of v_{in}.

Analysis of the Output Circuit

After the input circuit has been analyzed to find values of i_B, a load-line analysis of the output circuit is possible. Referring to Figure 13.7, we can write a voltage equation for the loop through V_{CC}, R_C, and the transistor from collector to emitter. Thus, we have

Equation for the output load line.

$$V_{CC} = R_C i_C + v_{CE} \tag{13.13}$$

This is plotted on the output characteristics of the transistor in Figure 13.8(b).

As v_{in} swings through a range of values, i_B changes, and the instantaneous operating point swings up and down the load line on the output characteristics.

Now, with the values of i_B found by prior analysis of the input circuit, we can locate the intersection of the corresponding output curve with the load line to find values for i_C and v_{CE}. Thus, as v_{in} swings through a range of values, i_B changes, and the instantaneous operating point swings up and down the load line on the output characteristics. Usually, the ac component of v_{CE} is much larger than the input voltage, and amplification has taken place.

Examination of Figure 13.8(a) shows that as $v_{in}(t)$ swings positive, the input load line moves upward and to the right, and the value of i_B increases (i.e., the intersection of the load line with the input characteristic moves upward). This in turn causes the instantaneous operating point to move upward on the output load line, and v_{CE} decreases in value. Thus, a swing in the positive direction for v_{in} results in a (much larger) swing in the negative direction for v_{CE}. Therefore, as well as being amplified, the signal is inverted. Thus, the common-emitter circuit is an **inverting amplifier**.

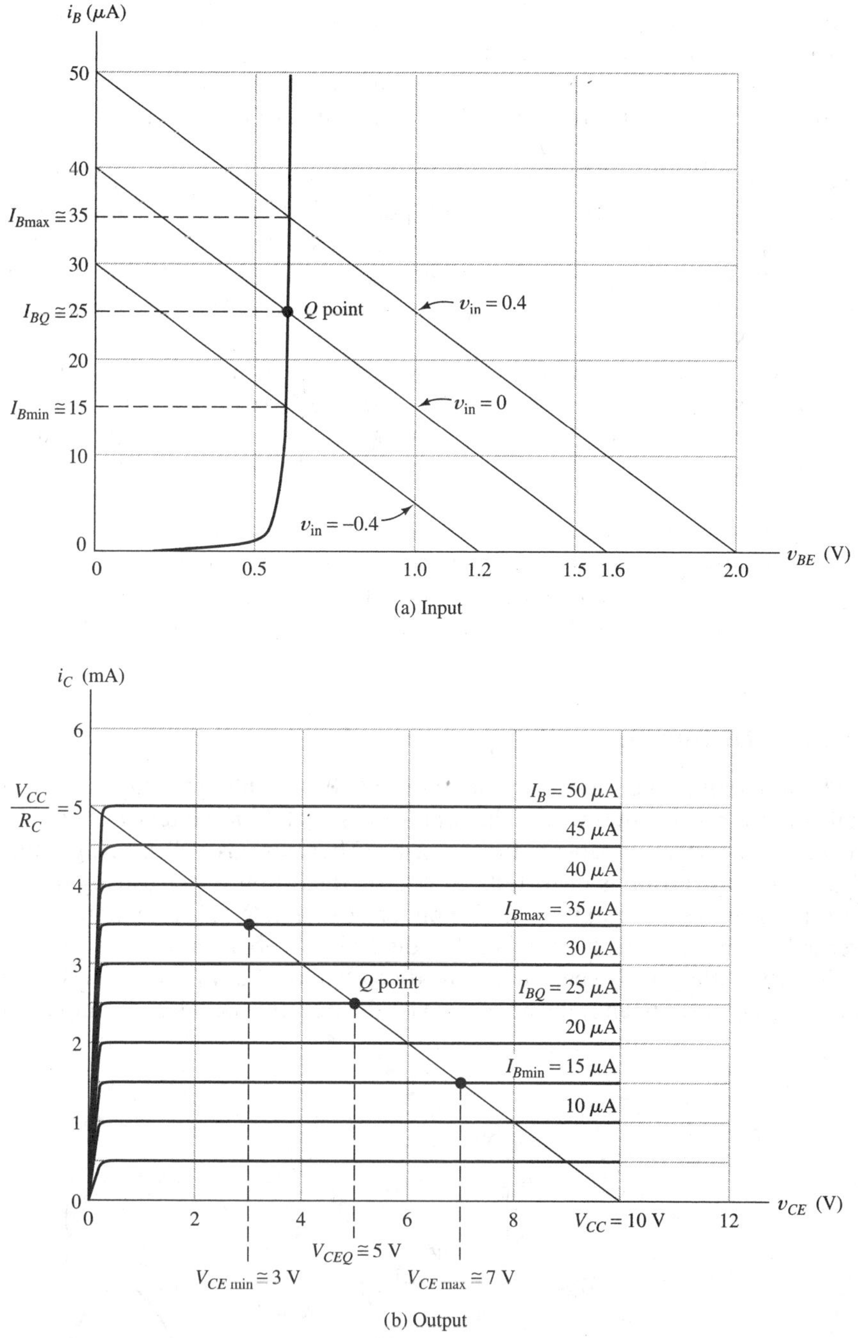

Figure 13.9 Load-line analysis for Example 13.2.

Example 13.2 Load-Line Analysis of a BJT Amplifier

Assume that the circuit of Figure 13.7 has $V_{CC} = 10$ V, $V_{BB} = 1.6$ V, $R_B = 40$ kΩ, and $R_C = 2$ kΩ. The input signal is a 0.4-V-peak 1-kHz sinusoid given by $v_{in}(t) = 0.4 \sin(2000\pi t)$. The common-emitter characteristics for the transistor are shown in Figure 13.9. Find the maximum, minimum, and Q-point values for v_{CE}.

Solution First, we must find values for i_B. The load lines for $v_{\text{in}} = 0$ (to find the Q point), $v_{\text{in}} = 0.4$ (positive extreme), and $v_{\text{in}} = -0.4$ (negative extreme) are shown in Figure 13.9(a). The values for the base current are found at the intersection of the load lines with the input characteristic. The (approximate) values are $I_{B\max} \cong 35\ \mu\text{A}$, $I_{BQ} \cong 25\ \mu\text{A}$, and $I_{B\min} \cong 15\ \mu\text{A}$.

Next, the load line is constructed on the output characteristic as shown in Figure 13.9(b). The intersection of the output load line with the characteristic for $I_{BQ} = 25\ \mu\text{A}$ establishes the Q point on the output characteristics. The values are $I_{CQ} = 2.5$ mA and $V_{CEQ} = 5$ V. Similarly, the intersection of the load line with the characteristic for $I_{B\max} = 35\ \mu\text{A}$ yields $V_{CE\min} \cong 3$ V. The opposite extreme is $I_{B\min} \cong 15\ \mu\text{A}$ resulting in $V_{CE\max} \cong 7$ V.

If more points are found as v_{in} varies with time, we can eventually plot the v_{CE} waveform versus time. The waveforms for $v_{\text{in}}(t)$ and $v_{CE}(t)$ are shown in Figure 13.10. Notice that the ac component of $v_{CE}(t)$ is inverted compared to the input signal [i.e., the minimum of $v_{CE}(t)$ occurs at the same instant as the maximum of $v_{\text{in}}(t)$, and vice versa].

The peak-to-peak value of the input voltage is 0.8 V and the peak-to-peak value of the ac component of v_{CE} is 4 V. Thus, the voltage-gain magnitude is 5 (i.e., the ac component of v_{CE} is five times larger in amplitude than v_{in}). Usually, we would state the gain as -5 to emphasize the fact that the amplifier inverts the input signal. ■

Nonlinear Distortion

It is not apparent in the waveforms of Figure 13.10, but unlike the input, the output signal is not a precise sine wave. The amplifier is slightly nonlinear because of the curvature of the characteristics of the transistor. Therefore, as well as being amplified and inverted, the signal is distorted. Of course, distortion is not usually desirable. Figure 13.11 shows the output of the amplifier of Example 13.2 if the input signal is increased in amplitude to 1.2-V peak. The distortion is obvious.

Notice that the positive peak of v_{CE} has been "clipped" at $V_{CC} = 10$ V. This occurs when i_B and i_C have been reduced to zero by the negative peaks of the input signal,

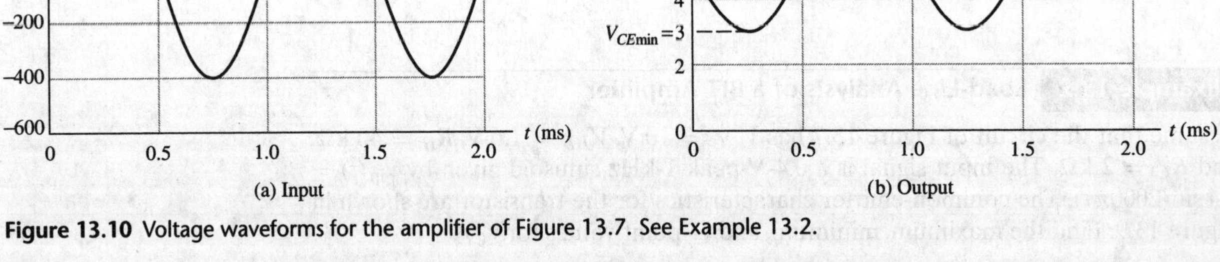

Figure 13.10 Voltage waveforms for the amplifier of Figure 13.7. See Example 13.2.

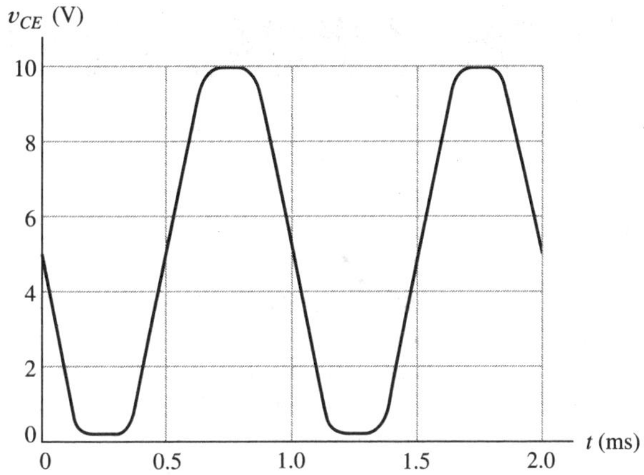

Figure 13.11 Output of the amplifier of Example 13.2 for $v_{in}(t) = 1.2\sin(2000\pi t)$ showing gross nonlinear distortion.

and the instantaneous operating point moves down to the voltage-axis intercept of the output load line. When this happens, we say that the transistor has been driven into **cutoff**.

When i_C becomes zero, we say that the transistor is cut off.

The negative-going peak of the output waveform in Figure 13.11 is clipped at $v_{CE} \cong 0.2$ V. This occurs because i_B becomes large enough so that operation is driven into the region at the upper end of the output load line, where the characteristic curves are crowded together. We call this the **saturation region**.

When $v_{CE} \cong 0.2$ V, we say that the transistor is in saturation.

Reasonably linear amplification occurs only if the signal swing remains in the **active region** between saturation and cutoff on the load line. An output load line is shown in Figure 13.12, including labels for the cutoff, saturation, and active regions.

Exercise 13.5 Repeat Example 13.2 if $v_{in}(t) = 0.8\sin(2000\pi t)$. Find the values of $V_{CE\max}$, V_{CEQ}, and $V_{CE\min}$.
Answer $V_{CE\max} \cong 9.0$ V, $V_{CEQ} \cong 5.0$ V, $V_{CE\min} \cong 1.0$ V. □

Exercise 13.6 Repeat Example 13.2 if $v_{in}(t) = 0.8\sin(2000\pi t)$ and $V_{BB} = 1.2$ V. Find the values of $V_{CE\max}$, V_{CEQ}, and $V_{CE\min}$.
Answer $V_{CE\max} \cong 9.8$ V, $V_{CEQ} \cong 7.0$ V, $V_{CE\min} \cong 3.0$ V. □

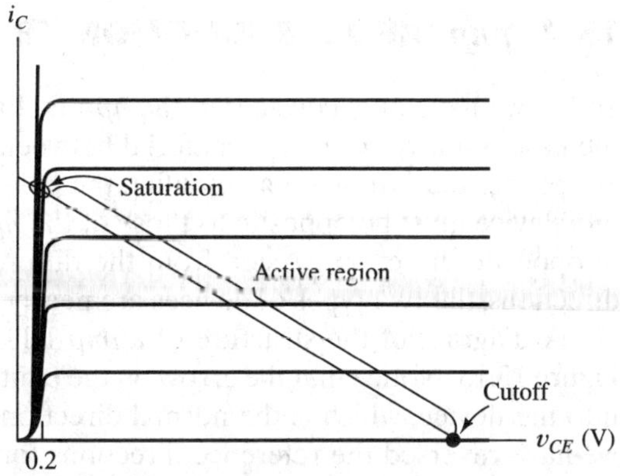

Figure 13.12 Amplification occurs in the active region. Clipping occurs when the instantaneous operating point enters saturation or cutoff. In saturation, $v_{CE} \cong 0.2$ V.

PRACTICAL APPLICATION 13.1

Soup Up Your Automobile by Changing Its Software?

Early automobiles contained no electronics and very little in the way of electrical circuits. Perhaps the most important initial application was electrical ignition, an early version of which is described briefly on page 208. The first step in modernizing the ignition system was to replace the "points" (a mechanically operated switch) with a bipolar junction transistor. The transistor cycles between saturation (in which it behaves as a closed switch) and cutoff (in which it behaves as an open switch). The ignition spark is produced as the result of rapidly switching off the current flowing through the coil.

A significant advantage of electronic switching is that the BJT does not wear out as points do. However, the use of an electronic switch in the place of mechanically operated points has paved the way for numerous additional improvements in ignition control. Optimum ignition timing relative to engine rotation varies with engine speed, throttle setting, air temperature, engine temperature, fuel quality, and load on the engine, as well as the design goals (good fuel economy, long engine life, or highest racing performance). Early ignition systems used mechanical and pneumatic systems to adjust timing, but such systems cannot achieve optimum performance under all conditions. Modern engine-control systems employ electrical sensors to determine operating conditions, various electronic circuits to process the sensor signals, and a special-purpose computer (including software) to compute the optimum firing instant for each cylinder. The computer switches the BJT from saturation to cutoff, creating the ignition spark.

In the 1950s, souping up an engine involved boring out cylinders and milling heads. Today, an equally important factor is to modify the engine-control software. Hot-rod magazines abound with advertisements offering ROMs (read-only memories, discussed in Chapter 8) loaded with engine control software optimized for high performance (as opposed to fuel economy or long engine life).

A milestone event underscores the importance of electronics in what was originally an almost purely mechanical system. To commemorate the 100th anniversary of the first mass production of the automobile, *Automotive Engineering* magazine surveyed the Society of Automotive Engineers Fellows Committee to ascertain the top 10 significant events in automotive history. What came out at the top of the list? The answer is as follows: "automotive electronics, including applications in engine controls, brakes, steering, and stability control" (*Automotive Engineering*, February 1996, p. 4). Of course, advances in electronics applications for automobiles did not stop in 1996; instead, reports of automotive electronic innovations appear at an ever increasing rate. A few more recent examples are collision avoidance systems, automatic parallel parking, night vision systems, and even cars that drive themselves. Certainly, today's mechanical engineers need to be familiar with the capabilities and limitations of the electronics, as well as with mechanical design and materials issues.

13.4 *pnp* BIPOLAR JUNCTION TRANSISTORS

So far, we have only considered the *npn* BJT, but an equally useful device results if the base is a layer of *n*-type material between *p*-type emitter and collector regions. For proper operation as an amplifier, the polarities of the dc voltages applied to the *pnp* device must be opposite to those of the *npn* device. Furthermore, currents flow in opposite directions. Aside from the differences in voltage polarity and current directions, the two types of devices are nearly identical.

A diagram of the structure of a *pnp* BJT and the circuit symbol are shown in Figure 13.13. Notice that the arrow on the emitter of the *pnp* transistor symbol points into the device, which is the normal direction of the emitter current. Furthermore, we have reversed the reference directions for the currents to agree with the actual direction of current flow for the *pnp* in the active region.

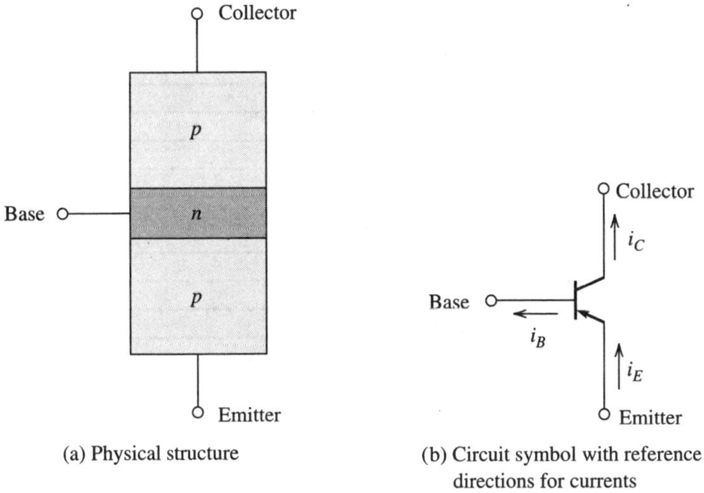

Figure 13.13 The *pnp* BJT.

For the *pnp* transistor, we can write the following equations, which are exactly the same as for the *npn* transistor:

$$i_C = \alpha i_E \tag{13.14}$$

$$i_B = (1 - \alpha) i_E \tag{13.15}$$

$$i_C = \beta i_B \tag{13.16}$$

and

$$i_E = i_C + i_B \tag{13.17}$$

Equations 13.14 through 13.16 are valid only if the base–emitter junction is forward biased (v_{BE} negative for a *pnp*) and the collector–base junction is reverse biased (v_{BC} positive for a *pnp*). As for the *npn* transistor, typical values are $\alpha \cong 0.99$ and $\beta \cong 100$.

Except for reversal of current directions and voltage polarities, the *pnp* BJT is almost identical to the *npn* BJT.

For the *pnp* transistor in the active region, we have

$$i_E = I_{ES}\left[\exp\left(\frac{-v_{BE}}{V_T}\right) - 1\right] \tag{13.18}$$

and

$$i_B = (1 - \alpha)\, I_{ES}\left[\exp\left(\frac{-v_{BE}}{V_T}\right) - 1\right] \tag{13.19}$$

These equations are identical to Equations 13.1 and 13.8 for the *npn* transistor, except that $-v_{BE}$ has been substituted for v_{BE} (because v_{BE} takes negative values for the *pnp* device). As for the *npn* device, typical values for I_{ES} range from 10^{-12} to 10^{-16} A, and at 300 K, we have $V_T \cong 0.026$ V.

The common-emitter characteristics of a *pnp* transistor are exactly the same as for the *npn* transistor, except that the values on the voltage axes are negative. A typical set of characteristics is shown in Figure 13.14.

Exercise 13.7 Find the values of α and β for the transistor having the characteristics shown in Figure 13.14.

Answer $\alpha = 0.980, \beta = 50.$ □

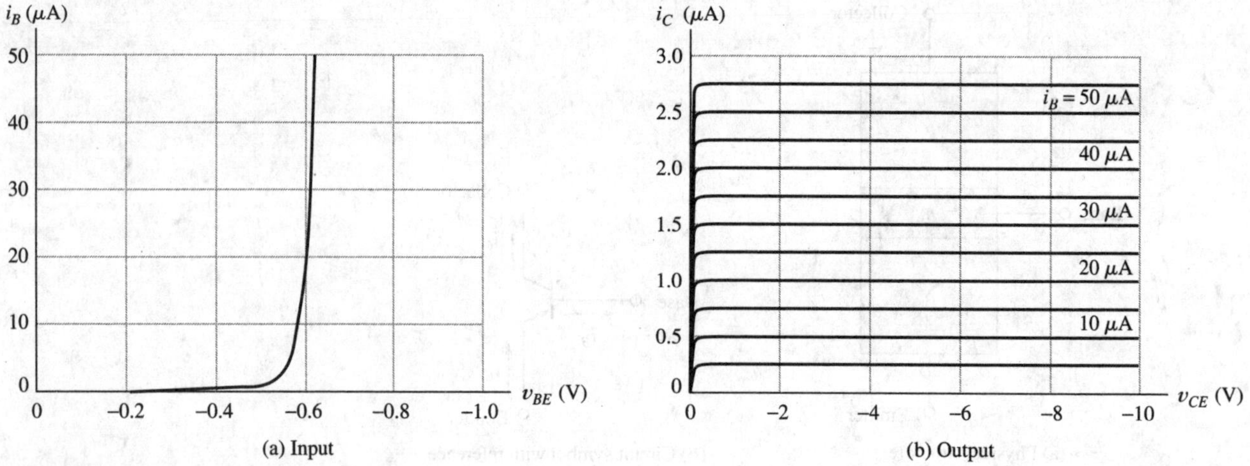

Figure 13.14 Common-emitter characteristics for a *pnp* BJT.

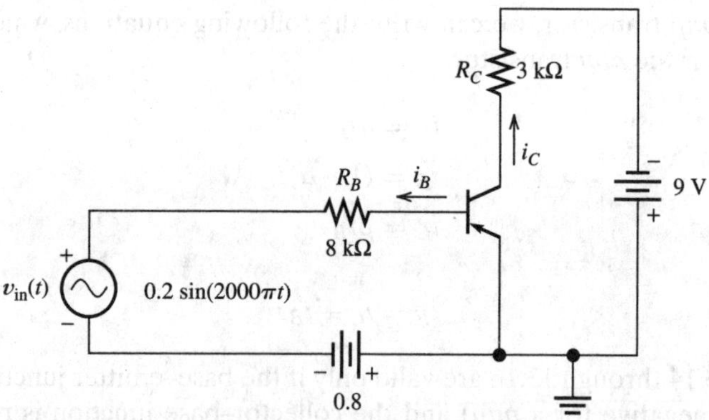

Figure 13.15 Common-emitter amplifier for Exercise 13.8.

Exercise 13.8 **a.** Use load-line analysis to find the maximum, minimum, and Q-point values of i_B and v_{CE} for the amplifier circuit shown in Figure 13.15. Use the characteristics shown in Figure 13.14. **b.** Does this *pnp* common-emitter amplifier invert the signal?

Answer **a.** $I_{B\max} \cong 48\ \mu A$, $I_{BQ} \cong 24\ \mu A$, $I_{B\min} \cong 5\ \mu A$, $V_{CE\max} \cong -1.8$ V, $V_{CEQ} \cong -5.3$ V, $V_{CE\min} \cong -8.3$ V. **b.** Yes, the output signal is inverted. [If you are in doubt about this, try sketching the $v_{in}(t)$, $i_B(t)$, and $v_{CE}(t)$ waveforms to scale versus time.] □

13.5 LARGE-SIGNAL DC CIRCUIT MODELS

In the analysis or design of BJT amplifier circuits, we often consider the dc operating point (Q point) separately from the analysis of the signals. (This was illustrated for a voltage-controlled attenuator in Section 10.8.) Usually, we consider the dc operating point first. Then, we turn our attention to the signal to be amplified. In this section, we present models for large-signal dc analysis of BJT circuits. Then, in the next section, we show how to use these models to analyze bias circuits for BJT amplifiers. Later, we consider small-signal models used to analyze circuits for the signals being amplified.

It is customary to use uppercase symbols with uppercase subscripts to represent large-signal dc currents and voltages in transistor circuits. Thus, I_C and V_{CE} represent the dc collector current and collector-to-emitter voltage, respectively. Similar notation is used for the other currents and voltages.

As we have seen, BJTs can operate in the active region, in saturation, or in cutoff. In the active region, the base–emitter junction is forward biased, and the base–collector junction is reverse biased.

Active-Region Model

Circuit models for BJTs in the active region are shown in Figure 13.16(a). A current-controlled current source models the dependence of the collector current on the base

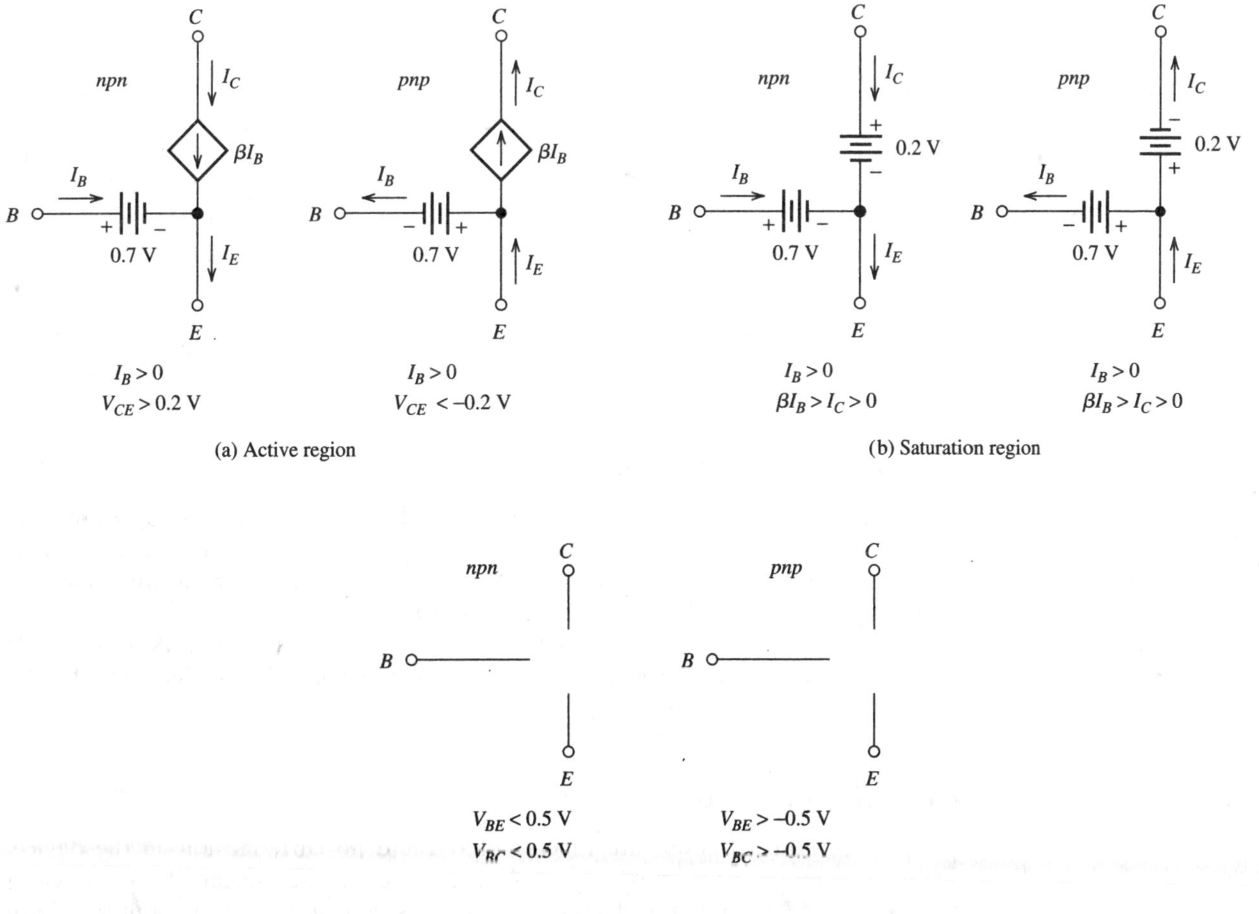

(c) Cutoff region

Figure 13.16 BJT large-signal models. (*Note:* Values shown are appropriate for typical small-signal silicon devices at a temperature of 300 K.)

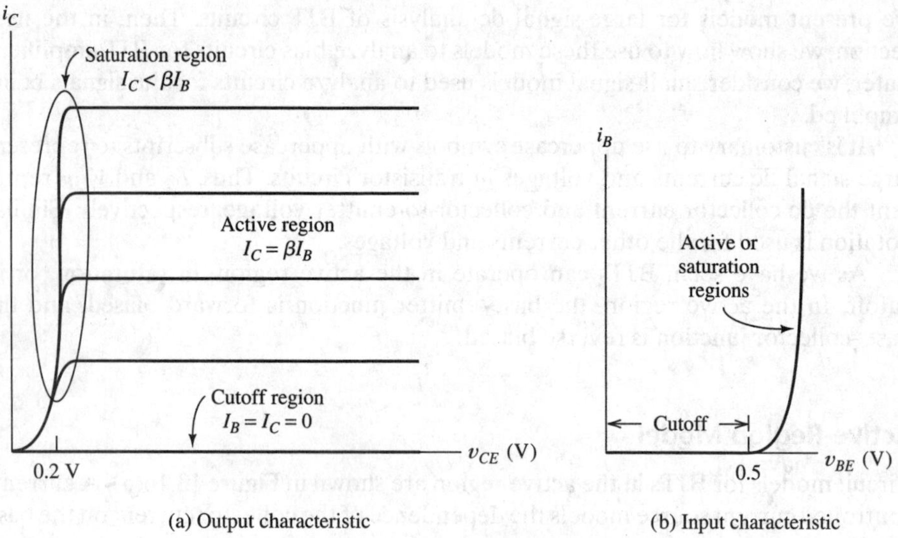

Figure 13.17 Regions of operation on the characteristics of an *npn* BJT.

current. The constraints given in the figure for I_B and V_{CE} must be satisfied to ensure validity of the active-region model.

Let us relate the active-region model to the device characteristics. Figure 13.17 shows the characteristic curves of an *npn* transistor. The base current I_B is positive and $V_{BE} \cong 0.7$ V for forward bias of the base-to-emitter junction, as shown in Figure 13.17(b). Also notice in Figure 13.17(a) that V_{CE} must be greater than about 0.2 V to ensure that operation is in the active region (i.e., above the *knees* of the characteristic curves).

Similarly, for the *pnp* BJT, we must have $I_B > 0$ and $V_{CE} < -0.2$ V for validity of the active-region model. (As usual, we assume that I_B is referenced positive out of the base for the *pnp* BJT.)

Saturation-Region Model

The BJT models for the saturation region are shown in Figure 13.16(b). In the saturation region, both junctions are forward biased. Examination of the collector characteristics in Figure 13.17(a) shows that $V_{CE} \cong 0.2$ V for an *npn* transistor in saturation. Thus, the model for the saturation region includes a 0.2-V source between collector and emitter. As in the active region, I_B is positive. Also, we see in Figure 13.17(a) that for operation below the knee of the collector characteristic, the constraint is $\beta I_B > I_C > 0$.

Cutoff-Region Model

In cutoff, both junctions are reverse biased and no currents flow in the device. Thus, the model consists of open circuits among all three terminals, as shown in Figure 13.16(c). (Actually, if small forward-bias voltages of up to about 0.5 V are applied, the currents are often negligible and we still use the cutoff-region model.) The constraints on the voltages for the BJT to be in the cutoff region are shown in the figure.

Example 13.3 Determining the Operating Region of a BJT

A given *npn* transistor has $\beta = 100$. Determine the region of operation if **a.** $I_B = 50\,\mu\text{A}$ and $I_C = 3\,\text{mA}$; **b.** $I_B = 50\,\mu\text{A}$ and $V_{CE} = 5\,\text{V}$; **c.** $V_{BE} = -2\,\text{V}$ and $V_{CE} = -1\,\text{V}$.

Solution

a. Since I_B and I_C are positive, the transistor is either in the active or saturation region. The constraint for saturation

$$\beta I_B > I_C$$

is met, so the device is in saturation.

b. Since we have $I_B > 0$ and $V_{CE} > 0.2$, the transistor is in the active region.

c. We have $V_{BE} < 0$ and $V_{BC} = V_{BE} - V_{CE} = -1 < 0$. Therefore, both junctions are reverse biased, and operation is in the cutoff region. ■

Exercise 13.9 A certain *npn* transistor has $\beta = 100$. Determine the region of operation if **a.** $V_{BE} = -0.2\,\text{V}$ and $V_{CE} = 5\,\text{V}$; **b.** $I_B = 50\,\mu\text{A}$ and $I_C = 2\,\text{mA}$; **c.** $V_{CE} = 5\,\text{V}$ and $I_B = 50\,\mu\text{A}$.

Answer **a.** cutoff; **b.** saturation; **c.** active. □

13.6 LARGE-SIGNAL DC ANALYSIS OF BJT CIRCUITS

In Section 13.5, we presented large-signal dc models for the BJT. In this section, we use those models to analyze circuits. In dc analysis of BJT circuits, we first assume that the operation of the transistor is in a particular region (i.e., active, cutoff, or saturation). Then, we use the appropriate model for the device and solve the circuit. Next, we check to see if the solution satisfies the constraints for the region assumed. If so, the analysis is complete. If not, we assume operation in a different region and repeat until a valid solution is found. This is very similar to the analysis of diode circuits using the ideal-diode model or a piecewise-linear model. The step-by-step procedure is as follows:

1. Choose one of the three BJT operating regions: saturation, cutoff, or active.
2. Analyze the circuit to determine I_C, I_B, V_{BE}, and V_{CE}, by using the transistor model for the region chosen.
3. Check to see if the constraints for the chosen region are met. If so, the analysis is completed. If not, return to Step 1, and choose a different region.

This approach is particularly useful in the analysis and design of bias circuits for BJT amplifiers. The objective of the bias circuit is to place the operating point in the active region so that signals can be amplified. Because transistors show considerable variation of parameters, such as β, from unit to unit and with temperature, it is important for the bias point to be independent of these variations.

The next several examples illustrate the technique and provide some observations that are useful in bias-circuit design.

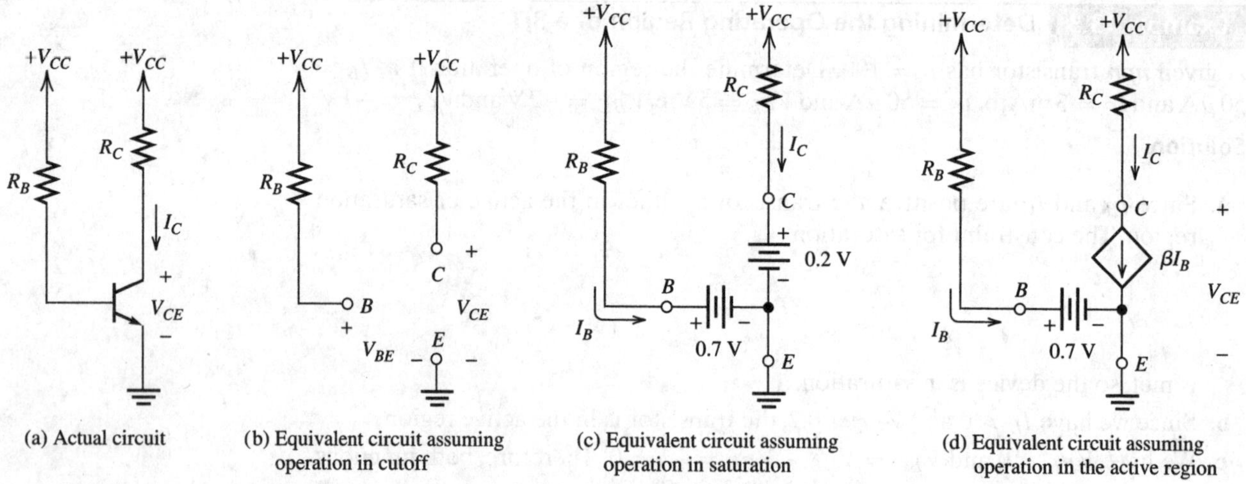

Figure 13.18 Bias circuit of Examples 13.4 and 13.5.

Example 13.4 Analysis of the Fixed Base Bias Circuit

The dc bias circuit shown in Figure 13.18(a) has $R_B = 200\text{ k}\Omega$, $R_C = 1\text{ k}\Omega$, and $V_{CC} = 15$ V. The transistor has $\beta = 100$. Solve for I_C and V_{CE}.

Solution We will eventually see that the transistor is in the active region, but we start by assuming that the transistor is cut off (to illustrate how to test the initial guess of operating region). Since we assume operation in cutoff, the model for the transistor is shown in Figure 13.16(c), and the equivalent circuit is shown in Figure 13.18(b). We reason that $I_B = 0$ and that there is no voltage drop across R_B. Hence, we conclude that $V_{BE} = 15$ V. However, in cutoff, we must have $V_{BE} < 0.5$ for an *npn* transistor. Therefore, we conclude that the cutoff assumption is invalid.

Next, let us assume that the transistor is in saturation. The transistor model is shown in Figure 13.16(b). Then, the equivalent circuit is shown in Figure 13.18(c). Solving, we find that

$$I_C = \frac{V_{CC} - 0.2}{R_C} = 14.8\text{ mA}$$

and

$$I_B = \frac{V_{CC} - 0.7}{R_B} = 71.5\ \mu\text{A}$$

Checking the conditions required for saturation, we find that $I_B > 0$ is met, but $\beta I_B > I_C$ is not met. Therefore, we conclude that the transistor is not in saturation.

Finally, if we assume that the transistor operates in the active region, we use the BJT model of Figure 13.16(a), and the equivalent circuit is shown in Figure 13.18(d). Solving, we find that

$$I_B = \frac{V_{CC} - 0.7}{R_B} = 71.5\ \mu\text{A}$$

(We have assumed a forward bias of 0.7 V for the base–emitter junction. Some authors assume 0.6 V for small-signal silicon devices at room temperature; others assume 0.7 V. In reality, the value depends on the particular device and the current level. Usually, the difference is not significant.) Now, we have

$$I_C = \beta I_B = 7.15 \text{ mA}$$

Finally,

$$V_{CE} = V_{CC} - R_C I_C = 7.85 \text{ V}$$

The requirements for the active region are $V_{CE} > 0.2$ V and $I_B > 0$, which are met. Thus, the transistor operates in the active region. ■

Example 13.5 **Analysis of the Fixed Base Bias Circuit**

Repeat Example 13.4 with $\beta = 300$.

Solution First, we assume operation in the active region. This leads to

$$I_B = \frac{V_{CC} - 0.7}{R_B} = 71.5\ \mu\text{A}$$

$$I_C = \beta I_B = 21.45 \text{ mA}$$

$$V_{CE} = V_{CC} - R_C I_C = -6.45 \text{ V}$$

The requirements for the active region are $V_{CE} > 0.2$ V and $I_B > 0$, which are not met. Thus, the transistor is not operating in the active region.

Next, we assume that the transistor is in saturation. This leads to

$$I_C = \frac{V_{CC} - 0.2}{R_C} = 14.8 \text{ mA}$$

and

$$I_B = \frac{V_{CC} - 0.7}{R_B} = 71.5\ \mu\text{A}$$

Now, we find that the conditions for saturation ($I_B > 0$ and $\beta I_B > I_C$) are met. Thus, we have solved the circuit, and $V_{CE} = 0.2$ V. ■

Implications for Bias-Circuit Design

It is instructive to consider the load-line constructions that are shown in Figure 13.19 for Examples 13.4 and 13.5. For $\beta = 100$, the operating point (Q point) is approximately in the center of the load line. On the other hand, for $\beta = 300$, the operating point has moved up into saturation.

To use this circuit as an amplifier, we would want a Q point in the active region, where changes in base current cause the instantaneous operating point to move up and down the load line. In saturation, the operating point does not move significantly for small changes in base current, and amplification is not achieved. Thus, a suitable Q point is achieved for $\beta = 100$ but not for $\beta = 300$. Since we often find unit-to-unit variations in β of this magnitude, this circuit is not suitable as an amplifier bias circuit for mass production. (We could consider adjusting R_B to compensate for unit-to-unit variations in β, but this is usually not practical.)

Sometimes this circuit [Figure 13.18(a)] is called a **fixed base bias circuit** because the base current is fixed by V_{CC} and R_B and does not adjust for changes in β. (Notice

The amplifiers that we discuss in this book need to be biased near the center of the active region.

The fixed base bias circuit shown in Figure 13.18(a) is not suitable for amplifiers because unit-to-unit variations in β cause some of the circuits to operate in saturation or close to cutoff.

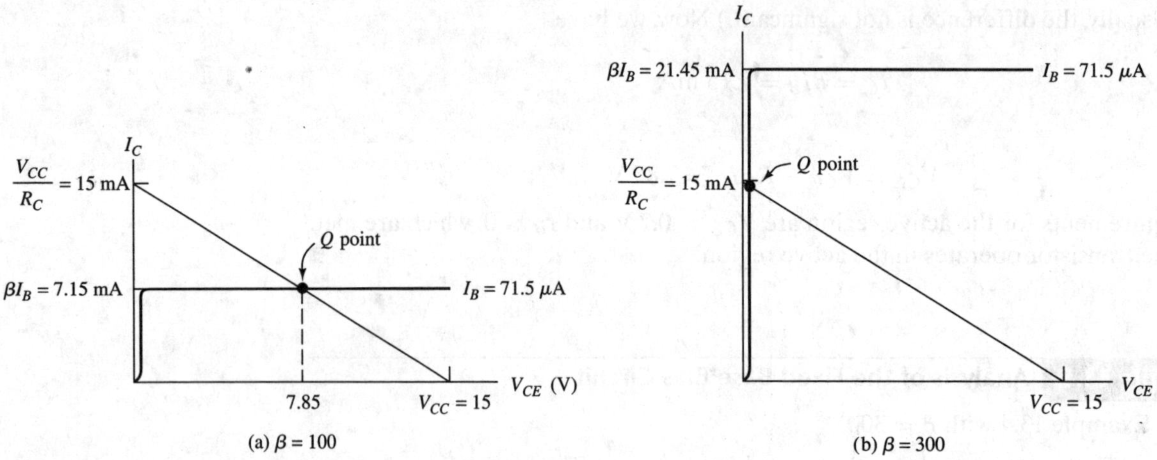

Figure 13.19 Load lines for Examples 13.4 and 13.5.

that if we want a circuit that achieves a particular operating point on the collector load line, the base current must change when β changes.)

Exercise 13.10 Repeat Example 13.4 for **a.** $\beta = 50$; **b.** $\beta = 250$.

Answer **a.** $I_C = 3.575$ mA, $V_{CE} = 11.43$ V; **b.** $I_C = 14.8$ mA, $V_{CE} = 0.2$ V. □

Exercise 13.11 Assume that $R_C = 5$ kΩ, $V_{BE} = 0.7$ V, and $V_{CC} = 20$ V in the circuit of Figure 13.18(a). Solve for the value of R_B needed to place the operating point exactly in the middle of the output load line (i.e., the Q point should fall at $V_{CE} = V_{CC}/2 = 10$ V) for: **a.** $\beta = 100$; **b.** $\beta = 300$.

Answer **a.** $R_B = 965$ kΩ; **b.** $R_B = 2.90$ MΩ. □

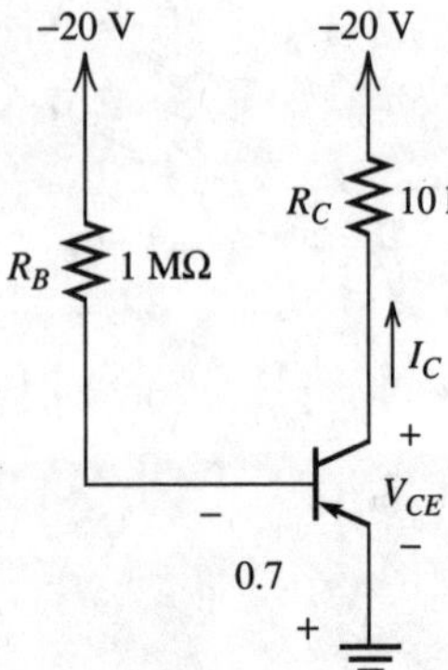

Figure 13.20 Circuit for Exercise 13.12.

Exercise 13.12 Solve the circuit shown in Figure 13.20 to find I_C and V_{CE} if: **a.** $\beta = 50$; **b.** $\beta = 150$.

Answer **a.** $I_C = 0.965$ mA, $V_{CE} = -10.35$ V; **b.** $I_C = 1.98$ mA, $V_{CE} = -0.2$ V (transistor in saturation). □

In the next example, we consider a circuit that achieves an emitter current that is relatively independent of β.

Example 13.6 Analysis of a BJT Bias Circuit

Solve for I_C and V_{CE} in the circuit of Figure 13.21(a) given that $V_{CC} = 15$ V, $V_{BB} = 5$ V, $R_C = 2$ kΩ, $R_E = 2$ kΩ, and $\beta = 100$. Repeat for $\beta = 300$.

Solution We assume that the transistor is in the active region and use the equivalent circuit shown in Figure 13.21(b). Writing a voltage equation through V_{BB}, the base–emitter junction, and R_E, we have

$$V_{BB} = 0.7 + I_E R_E$$

This can be solved for the emitter current:

$$I_E = \frac{V_{BB} - 0.7}{R_E} = 2.15 \text{ mA}$$

Notice that the emitter current does not depend on the value of β.

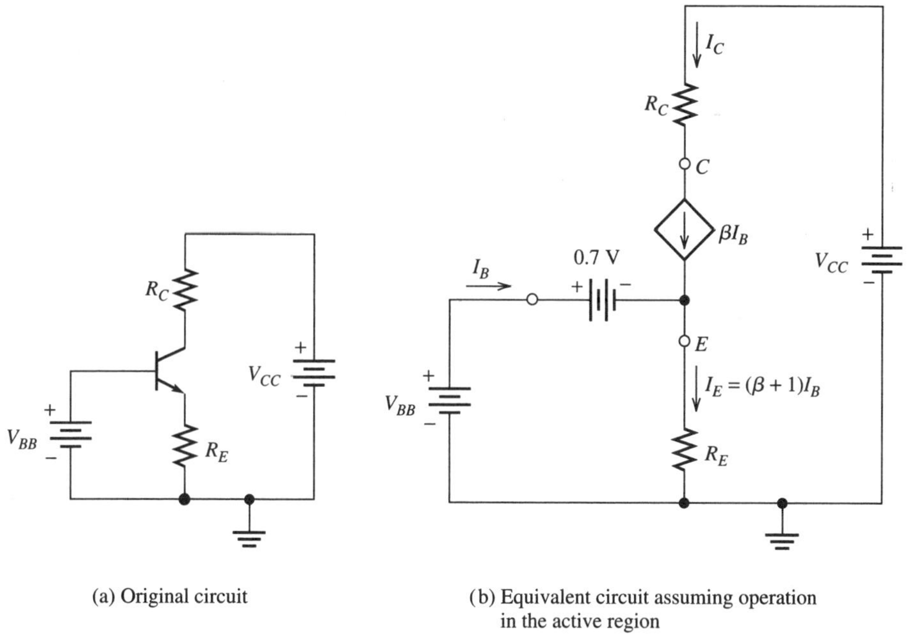

Figure 13.21 Circuit for Example 13.6.

Next, we can compute the base current and collector current using Equations 13.2 and 13.10.

$$I_C = \beta I_B$$

$$I_E = I_B + I_C$$

Using the first equation to substitute for I_C in the second equation, we have

$$I_E = I_B + \beta I_B = (\beta + 1)I_B$$

Solving for the base current, we obtain

$$I_B = \frac{I_E}{\beta + 1}$$

Substituting values, we obtain the results given in Table 13.1. Notice that I_B is lower for the higher β transistor, and I_C is nearly constant.

Now, we can write a voltage equation around the collector loop to find V_{CE}:

$$V_{CC} = R_C I_C + V_{CE} + R_E I_E$$

Substituting values found previously, we find that $V_{CE} = 6.44$ V for $\beta = 100$ and $V_{CE} = 6.42$ V for $\beta = 300$. ■

Table 13.1. Results for the Circuit of Example 13.6

β	$I_B (\mu A)$	I_C (mA)	V_{CE} (V)
100	21.3	2.13	6.44
300	7.14	2.14	6.42

The Q point for the circuit of Figure 13.21(a) is almost independent of β. However, the circuit is not practical for use in most amplifier circuits. First, it requires two voltage sources V_{CC} and V_{BB}, but often only one source is readily available. Second, we may want to inject the signal into the base (through a coupling capacitor), but the base voltage is fixed with respect to ground by the V_{BB} source. Because the V_{BB} source is constant, it acts as a short circuit to ground for ac signal currents (i.e., the V_{BB} source does not allow an ac voltage to appear at the base).

Analysis of the Four-Resistor Bias Circuit

The four-resistor BJT bias circuit is practical for amplifiers composed of discrete components. However, it is not practical in integrated circuits.

A circuit that avoids these objections is shown in Figure 13.22(a). We call this the **four-resistor BJT bias circuit**. The resistors R_1 and R_2 form a voltage divider that is intended to provide a nearly constant voltage at the base of the transistor (independent of transistor β). As we saw in Example 13.6, constant base voltage results in nearly constant values for I_C and V_{CE}. Because the base is not directly connected

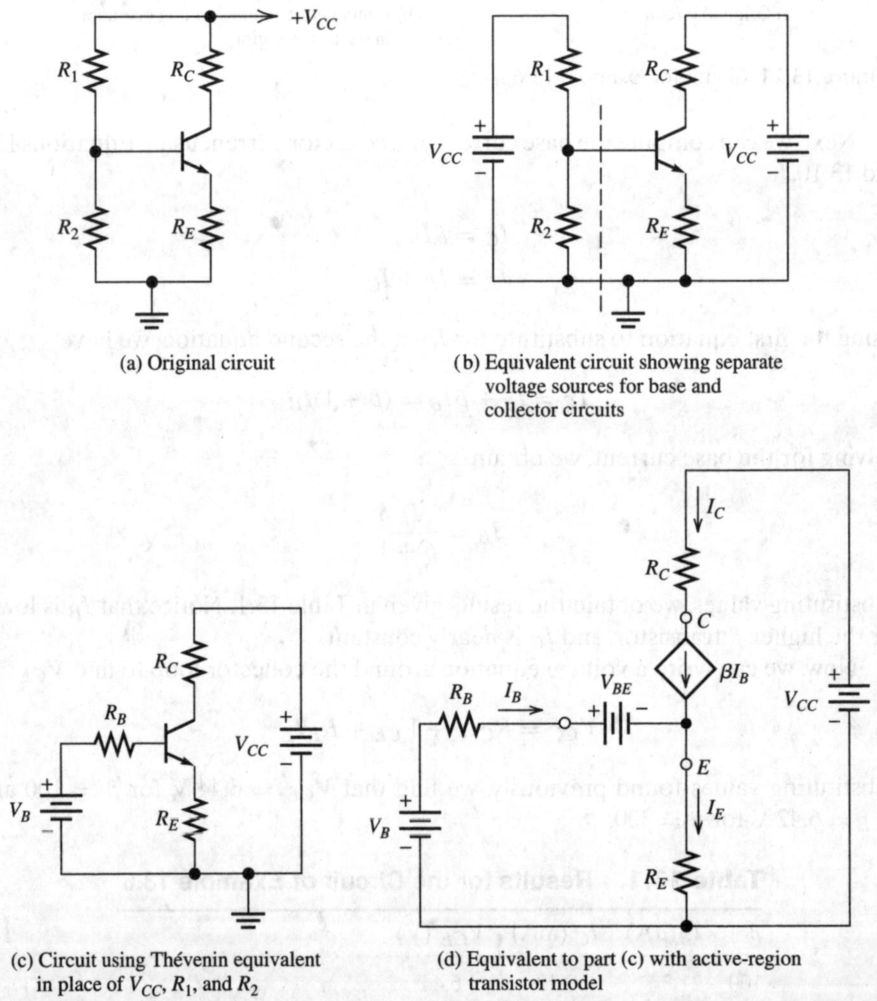

(a) Original circuit

(b) Equivalent circuit showing separate voltage sources for base and collector circuits

(c) Circuit using Thévenin equivalent in place of V_{CC}, R_1, and R_2

(d) Equivalent to part (c) with active-region transistor model

Figure 13.22 Four-resistor bias circuit.

to the supply or ground in the four-resistor bias circuit, it is possible to couple an ac signal to the base through a coupling capacitor.

The circuit can be analyzed as follows. First, the circuit is redrawn as shown in Figure 13.22(b). Two separate voltage supplies are shown as an aid in the analysis to follow, but otherwise, the circuits in parts (a) and (b) of the figure are identical. Next, we find the Thévenin equivalent for the circuit to the left of the dashed line in Figure 13.22(b). The Thévenin resistance R_B is the parallel combination of R_1 and R_2 given by

$$R_B = \frac{1}{1/R_1 + 1/R_2} = R_1 || R_2 \tag{13.20}$$

The Thévenin voltage V_B is

$$V_B = V_{CC}\frac{R_2}{R_1 + R_2} \tag{13.21}$$

The circuit with the Thévenin equivalent replacement is shown in Figure 13.22(c). Finally, the transistor is replaced by its active-region model, as shown in Figure 13.22(d).

Now, we can write a voltage equation around the base loop of Figure 13.22(d), resulting in

$$V_B = R_B I_B + V_{BE} + R_E I_E \tag{13.22}$$

Of course, for small-signal silicon transistors at room temperature, we have $V_{BE} \cong 0.7$ V. Now, we can substitute

$$I_E = (\beta + 1)\, I_B$$

and solve to find that

$$I_B = \frac{V_B - V_{BE}}{R_B + (\beta + 1)\, R_E} \tag{13.23}$$

Once I_B is known, I_C and I_E can easily be found. Then, we can write a voltage equation around the collector loop of Figure 13.22(d) and solve for V_{CE}. This yields

$$V_{CE} = V_{CC} - R_C I_C - R_E I_E \tag{13.24}$$

Example 13.7 Analysis of the Four-Resistor Bias Circuit

Find the values of I_C and V_{CE} in the circuit of Figure 13.23 for $\beta = 100$ and $\beta = 300$. Assume that $V_{BE} = 0.7$ V.

Solution Substituting into Equations 13.20 and 13.21, we find that

$$R_B = \frac{1}{1/R_1 + 1/R_2} = 3.33\ \text{k}\Omega$$

$$V_B = V_{CC}\frac{R_2}{R_1 + R_2} = 5\ \text{V}$$

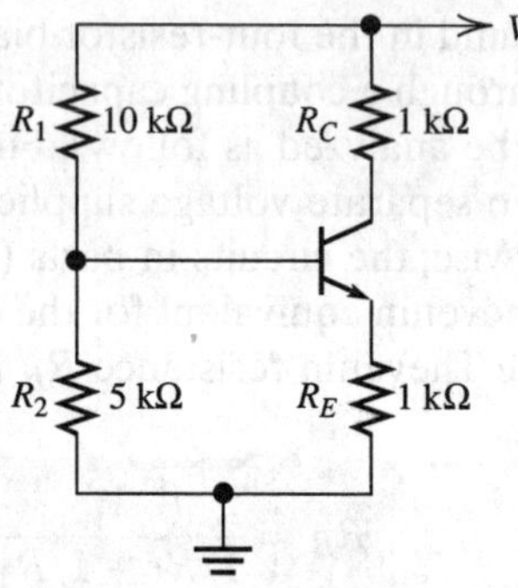

Figure 13.23 Circuit for Example 13.7.

Then, substituting into Equation 13.23 and using $\beta = 100$, we have

$$I_B = \frac{V_B - V_{BE}}{R_B + (\beta + 1)R_E} = 41.2\ \mu\text{A}$$

For $\beta = 300$, we find that $I_B = 14.1\ \mu$A. Notice that the base current is significantly smaller for the higher β.

Now, we can compute the collector current by using $I_C = \beta I_B$. For $\beta = 100$, we find that $I_C = 4.12$ mA, and for $\beta = 300$, we have $I_C = 4.24$ mA. For a 3:1 change in β, the collector current changes by less than 3 percent. The emitter current is given by $I_E = I_C + I_B$. The results are $I_E = 4.16$ mA for $\beta = 100$ and $I_E = 4.25$ mA for $\beta = 300$.

Finally, Equation 13.24 can be used to find V_{CE}. The results are $V_{CE} = 6.72$ for $\beta = 100$ and $V_{CE} = 6.51$ for $\beta = 300$. ■

Thus, we have seen that for proper resistance values, the four-resistor bias circuit can achieve a Q point that is nearly independent of β. Because of this fact, the circuit is commonly used for biasing BJT amplifiers (except in integrated circuits, for which resistors tend to be impractical).

Exercise 13.13 Repeat Example 13.7 for $R_1 = 100$ kΩ and $R_2 = 50$ kΩ. Compute the ratio of I_C for $\beta = 300$ to I_C for $\beta = 100$, and compare to the ratio of the currents found in Example 13.7. Comment.

Answer For $\beta = 100, I_C = 3.20$ mA, and $V_{CE} = 8.57$ V. For $\beta = 300, I_C = 3.86$ mA, and $V_{CE} = 7.27$ V. The ratio of the collector currents is 1.21. On the other hand, in the example, the ratio of the collector currents is only 1.03. Larger values of R_1 and R_2 lead to larger changes in I_C with changes in β. □

13.7 SMALL-SIGNAL EQUIVALENT CIRCUITS

In this section, we develop a BJT equivalent circuit that is useful in analyzing amplifier circuits.

Now, we turn our attention to the signal currents and voltages in the BJT. First, we establish the notation used in amplifier circuits. We denote the total currents and voltages by lowercase symbols with uppercase subscripts. Thus, $i_B(t)$ is the total base current as a function of time.

The dc Q-point currents and voltages are denoted by uppercase symbols with uppercase subscripts. Thus, I_{BQ} is the dc base current if the input signal is set to zero.

Finally, we denote the changes in currents and voltages from the Q point (due to the input signal being amplified) by lowercase symbols with lowercase subscripts.

Thus, $i_b(t)$ denotes the signal component of the base current. Since the total base current is the sum of the Q-point value and the signal component, we can write

$i_b(t)$ denotes the signal current flowing into the base, I_{BQ} is the dc current that flows when the signal is absent, and $i_B(t)$ is the total base current. Similar notation is used for the other currents and voltages.

$$i_B(t) = I_{BQ} + i_b(t) \tag{13.25}$$

These quantities are illustrated in Figure 13.24. Similarly, we can write

$$v_{BE}(t) = V_{BEQ} + v_{be}(t) \tag{13.26}$$

The Q point is established by the bias circuit, as discussed in Section 13.6. Now, we consider how the (small) signal components are related in the BJT. The total base current is given in terms of the total base-to-emitter voltage by Equation 13.8, repeated here for convenience:

$$i_B = (1-\alpha)I_{ES}\left[\exp\left(\frac{v_{BE}}{V_T}\right) - 1\right]$$

We are concerned with operation in the active region, for which the 1 inside the brackets is negligible and can be dropped.

We substitute Equations 13.25 and 13.26 into 13.8 to obtain

$$I_{BQ} + i_b(t) = (1-\alpha)I_{ES}\exp\left[\frac{V_{BEQ} + v_{be}(t)}{V_T}\right] \tag{13.27}$$

This can be written as

$$I_{BQ} + i_b(t) = (1-\alpha)I_{ES}\exp\left(\frac{V_{BEQ}}{V_T}\right)\exp\left[\frac{v_{be}(t)}{V_T}\right] \tag{13.28}$$

Equation 13.8 also relates the Q-point values, so we can write

$$I_{BQ} = (1-\alpha)I_{ES}\exp\left(\frac{V_{BEQ}}{V_T}\right) \tag{13.29}$$

Substituting into Equation 13.28, we have

$$I_{BQ} + i_b(t) = I_{BQ}\exp\left(\frac{v_{be}(t)}{V_T}\right) \tag{13.30}$$

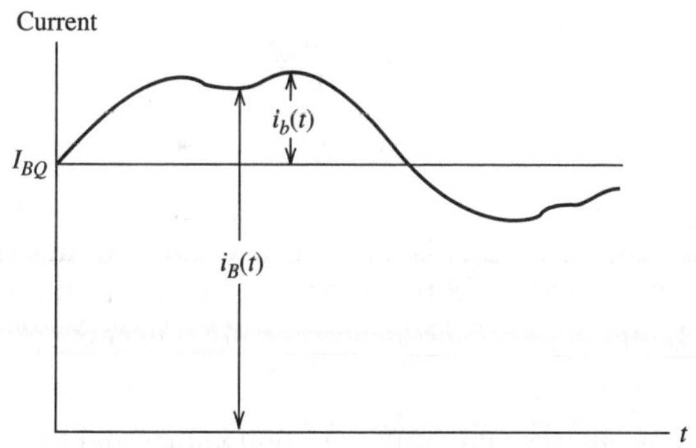

Figure 13.24 Illustration of the Q-point base current I_{BQ}, signal current $i_b(t)$, and total current $i_B(t)$.

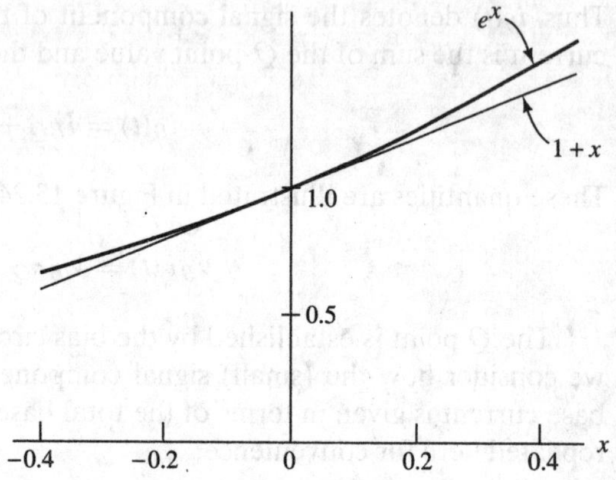

Figure 13.25 Comparison of e^x and its approximation $1+x$.

We are interested in small signals for which the magnitude of $v_{be}(t)$ is much smaller than V_T at all times. [Thus, the magnitude of $v_{be}(t)$ is confined to be a few millivolts or less.]

For $|x| << 1$, the following approximation holds:

$$\exp(x) \cong 1 + x \tag{13.31}$$

This approximation is illustrated in Figure 13.25. Hence, we can write Equation 13.30 as

$$I_{BQ} + i_b(t) \cong I_{BQ}\left[1 + \frac{v_{be}(t)}{V_T}\right] \tag{13.32}$$

If we subtract I_{BQ} from both sides and define $r_\pi = V_T/I_{BQ}$, we have

$$i_b(t) = \frac{v_{be}(t)}{r_\pi} \tag{13.33}$$

Therefore, for small-signal variations around the Q point, the base-to-emitter junction of the transistor appears to be a resistance r_π, given by

$$r_\pi = \frac{V_T}{I_{BQ}} \tag{13.34}$$

Substituting $I_{BQ} = I_{CQ}/\beta$, we have an alternative formula:

$$r_\pi = \frac{\beta V_T}{I_{CQ}} \tag{13.35}$$

At room temperature, $V_T \cong 0.026$ V. A typical value of β is 100, and a typical bias current for a small-signal amplifier is $I_{CQ} = 1$ mA. These values yield $r_\pi = 2600\ \Omega$.

The total collector current is β times the total base current:

$$i_C(t) = \beta i_B(t) \tag{13.36}$$

But, the total currents are the sum of the Q-point values and the signal components, so we have

$$I_{CQ} + i_c(t) = \beta I_{BQ} + \beta i_b(t) \tag{13.37}$$

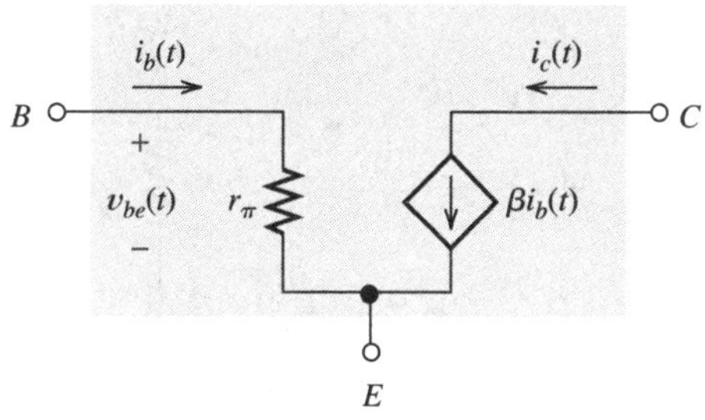

Figure 13.26 Small-signal equivalent circuit for the BJT.

Thus, the signal components are related by

$$i_c(t) = \beta i_b(t) \tag{13.38}$$

Small-Signal Equivalent Circuit for the BJT

Equations 13.33 and 13.38 relate the small-signal currents and voltages in a BJT. It is convenient to represent the BJT by the **small-signal equivalent circuit** shown in Figure 13.26. Notice that the circuit embodies the relationships of Equations 13.33 and 13.38.

The small-signal BJT equivalent circuit shown in Figure 13.26 is very useful in amplifier analysis.

It turns out that the *pnp* transistor has exactly the same small-signal equivalent circuit as the *npn*—even the reference directions for the *signal* currents are the same. The resistance r_π is given by Equation 13.35 for both types of transistors. (We assume that I_{CQ} is referenced out of the collector of the *pnp*, so it has a positive value.) In the next several sections, we will find that the small-signal equivalent circuit is very useful in analysis of BJT amplifier circuits.

13.8 COMMON-EMITTER AMPLIFIERS

In a BJT amplifier circuit, the power supply biases the transistor at an operating point in the active region for which amplification can take place. For example, we can use the four-resistor bias circuit discussed in Section 13.6. Coupling capacitors can be used to connect the load and the signal source without affecting the dc bias point.

Coupling capacitors can be used to connect the signal source and the load to a BJT amplifier without affecting the dc bias point.

We can analyze amplifier circuits to find gain, input resistance, and output resistance by use of the small-signal equivalent circuit. In this section and the next, we illustrate this procedure for two important BJT amplifier circuits.

Figure 13.27(a) shows the circuit diagram of a **common-emitter amplifier**. The resistors R_1, R_2, R_E, and R_C form the four-resistor biasing network. The capacitor C_1 couples the signal source to the base of the transistor, and C_2 couples the amplified signal at the collector to the load R_L. The capacitor C_E is called a **bypass capacitor**. It provides a low-impedance path for the ac emitter current to ground.

The coupling and bypass capacitors are chosen large enough so they have very low ac impedances at the signal frequencies. Thus, for simplicity in our small-signal ac analysis, we treat the capacitors as short circuits. However, at sufficiently low frequencies, the capacitors reduce the gain of the amplifier.

In small-signal ac analysis, we treat the coupling and bypass capacitors as short circuits.

Because the bypass capacitor grounds the emitter for ac signals, the emitter terminal is common to the input source and to the load. This is the origin of the name *common-emitter amplifier*.

The analysis we give here is valid for the **midband region** of frequency. In the **low-frequency region**, the effects of the coupling and bypass capacitors must be

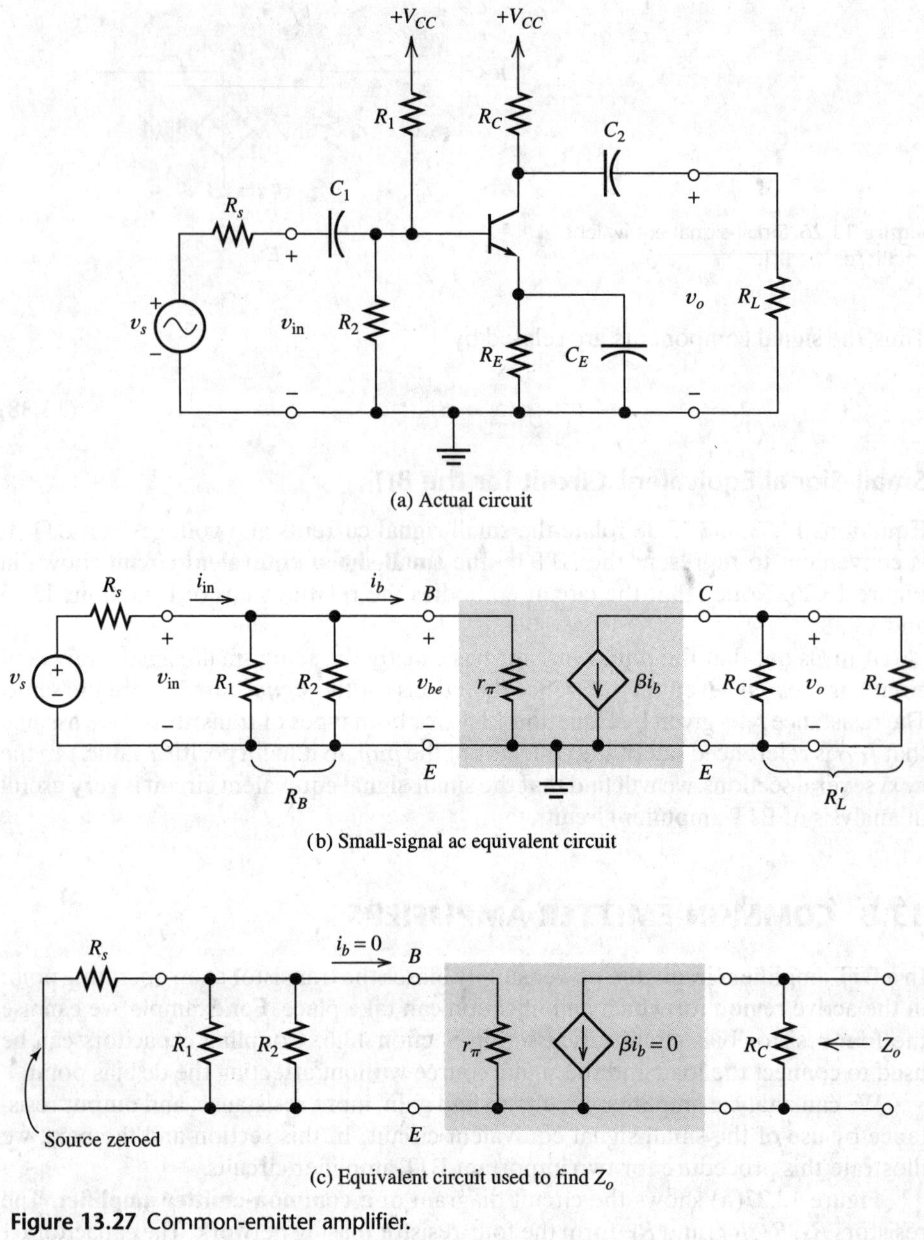

Figure 13.27 Common-emitter amplifier.

considered. In the **high-frequency region**, a more complex transistor model would be needed that includes its frequency limitations. Analysis for the low- and high-frequency regions is beyond the scope of this book.

The Small-Signal Equivalent Circuit

Before we analyze the amplifier, it is very helpful to draw its small-signal ac equivalent circuit. This is shown in Figure 13.27(b). The coupling capacitors have been replaced by short circuits, and the transistor has been replaced by its small-signal equivalent, which was discussed in the preceding section.

The dc power supply is replaced by a short circuit. This is appropriate because no ac voltage can appear across an ideal dc voltage source that is assumed to have zero internal impedance.

Dc voltage sources act as short circuits for ac signals.

Carefully compare the actual circuit of Figure 13.27(a) with the small-signal ac equivalent shown in Figure 13.27(b). Notice that the signal source is connected directly to the base terminal because C_1 has been treated as a short circuit. Similarly, the emitter is connected directly to ground, and the load is connected to the collector.

Notice also that the top end of R_1 connects to the supply in the original circuit, but R_1 is connected from base to ground in the equivalent circuit, because the power-supply voltage is treated as a short circuit to ground for ac signals.

Notice that R_1 ends up in parallel with R_2. Similarly, R_C and R_L are in parallel. We find it convenient to define R_B as the parallel combination of R_1 and R_2:

$$R_B = R_1 || R_2 = \frac{1}{1/R_1 + 1/R_2} \tag{13.39}$$

Similarly, R'_L is the parallel combination of R_C and R_L:

$$R'_L = R_L || R_C = \frac{1}{1/R_L + 1/R_C} \tag{13.40}$$

These parallel combinations are indicated in Figure 13.27(b).

Voltage Gain

Now, we analyze the equivalent circuit to find an expression for the voltage gain of the amplifier. First, the input voltage is equal to the voltage across r_π, given by

$$v_{\text{in}} = v_{be} = r_\pi i_b \tag{13.41}$$

The output voltage is produced by the collector current flowing through R'_L, given by

$$v_o = -R'_L \beta i_b \tag{13.42}$$

The minus sign is necessary because of the reference directions for the current and voltage—the current flows out of the positive voltage reference. Dividing the respective sides of Equation 13.42 by those of 13.41 gives the voltage gain:

$$A_v = \frac{v_o}{v_{\text{in}}} = -\frac{R'_L \beta}{r_\pi} \tag{13.43}$$

Notice that the gain is negative showing that the common-emitter amplifier is inverting. The gain magnitude can be quite large—several hundred is not unusual.

The expression for gain given in Equation 13.43 is the gain with the load connected. We found the open-circuit voltage gain useful to characterize amplifiers in Chapter 11. With R_L replaced by an open circuit, the voltage gain becomes

$$A_{voc} = \frac{v_o}{v_{\text{in}}} = -\frac{R_C \beta}{r_\pi} \tag{13.44}$$

Input Impedance

Another important amplifier specification is the input impedance, which in this case can be obtained by inspection of the equivalent circuit. The input impedance is the impedance *seen* looking into the input terminals. For the equivalent circuit of Figure 13.27(b), it is the parallel combination of R_B and r_π, given by

$$Z_{\text{in}} = \frac{v_{\text{in}}}{i_{\text{in}}} = \frac{1}{1/R_B + 1/r_\pi} \tag{13.45}$$

(In this case, the input impedance is a pure resistance. Therefore, we can find the input impedance by dividing the *instantaneous* voltage v_{in} by the *instantaneous* current i_{in}. Of course, if there were capacitances or inductances in the equivalent circuit, it would be necessary to obtain the impedance as the ratio of the *phasor* voltage and the *phasor* current.)

Current Gain and Power Gain

The current gain A_i can be found by use of Equation 11.3. With some changes in notation, the equation is

$$A_i = \frac{i_o}{i_{\text{in}}} = A_v \frac{Z_{\text{in}}}{R_L} \tag{13.46}$$

The power gain G of the amplifier is the product of the current gain and the voltage gain (assuming that the input and load impedances are pure resistive).

$$G = A_i A_v \tag{13.47}$$

Output Impedance

The output impedance is the impedance seen looking back from the load terminals with the source voltage v_s set to zero. This situation is shown in Figure 13.27(c). With v_s set to zero, there is no driving source for the base circuit, so i_b is zero. Therefore, the controlled source βi_b produces zero current and appears as an open circuit. Thus, the impedance seen from the output terminals is simply R_C.

$$Z_o = R_C \tag{13.48}$$

Example 13.8 Common-Emitter Amplifier

Find A_v, A_{voc}, Z_{in}, A_i, G, and Z_o for the amplifier shown in Figure 13.28. If $v_s(t) = 0.001\sin(\omega t)$, find and sketch $v_o(t)$ versus time. Assumc that the circuit operates at a temperature for which $V_T = 26$ mV.

Solution First, we need to know I_{CQ} to be able to find the value of r_π. Hence, we start by analyzing the dc conditions in the circuit. Only the dc supply, the transistor, and the resistors R_1, R_2, R_C, and R_E need to be considered in the bias-point analysis. The capacitors, the signal source, and the load resistance have no effect on the Q point (because the capacitors behave as open circuits for dc currents).

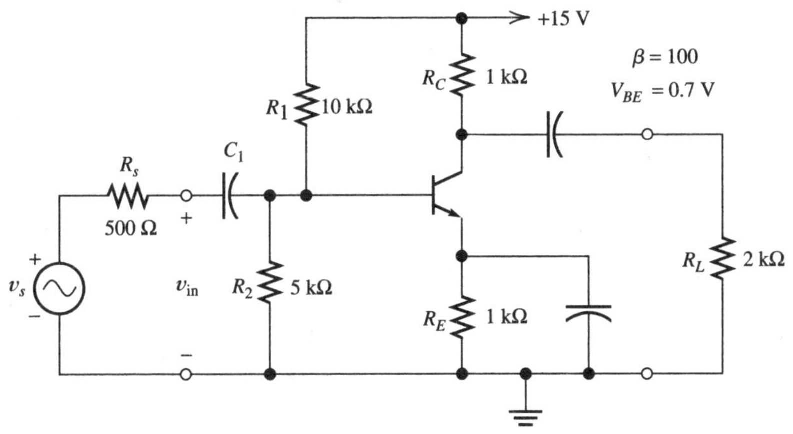

Figure 13.28 Common-emitter amplifier of Example 13.8.

The dc circuit was shown earlier in Figure 13.23 and was analyzed in Example 13.7. For $\beta = 100$, the resulting Q point was found to be $I_{CQ} = 4.12$ mA and $V_{CE} = 6.72$ V. Substituting values into Equation 13.35, we have

$$r_\pi = \frac{\beta V_T}{I_{CQ}} = 631\ \Omega$$

Using Equations 13.39 and 13.40, we find that

$$R_B = R_1 || R_2 = \frac{1}{1/R_1 + 1/R_2} = 3.33\ \text{k}\Omega$$

$$R'_L = R_L || R_C = \frac{1}{1/R_L + 1/R_C} = 667\ \Omega$$

Equations 13.43 through 13.48 yield

$$A_v = \frac{v_o}{v_{\text{in}}} = -\frac{R'_L \beta}{r_\pi} = -106$$

$$A_{voc} = \frac{v_o}{v_{\text{in}}} = -\frac{R_C \beta}{r_\pi} = -158$$

$$Z_{\text{in}} = \frac{v_{\text{in}}}{i_{\text{in}}} = \frac{1}{1/R_B + 1/r_\pi} = 531\ \Omega$$

$$A_i = \frac{i_o}{i_{\text{in}}} = A_v \frac{Z_{\text{in}}}{R_L} = -28.1$$

$$G = A_i A_v = 2980$$

$$Z_o = R_C = 1\ \text{k}\Omega$$

Notice that A_v is somewhat smaller in magnitude than A_{voc}. This is due to loading of the amplifier by R_L as discussed in Chapter 11. Power gain is quite large for the common-emitter amplifier, and primarily for this reason, it is a commonly used configuration.

The common-emitter amplifier is inverting and has large voltage gain magnitude, large current gain, and large power gain.

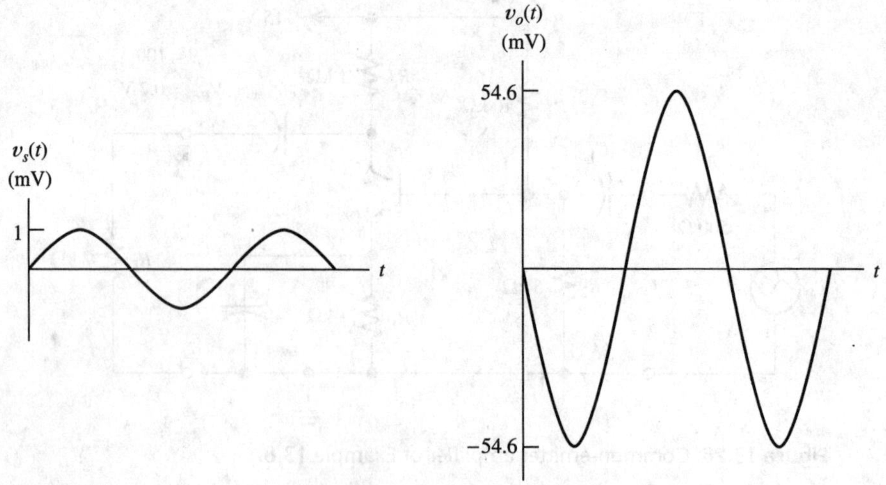

Figure 13.29 Source and output voltages for Example 13.8.

The source voltage divides between the internal source resistance and the input impedance of the amplifier. Thus, we can write

$$v_{\text{in}} = v_s \frac{Z_{\text{in}}}{Z_{\text{in}} + R_s} = 0.515 v_s$$

Now, with the load connected, we have

$$v_o = A_v v_{\text{in}} = -54.6 v_s$$

But, we are given that $v_s(t) = \sin(\omega t)$ mV, so we have

$$v_o(t) = -54.6 \sin(\omega t) \text{ mV}$$

The source voltage $v_s(t)$ and the output voltage are shown in Figure 13.29. Notice the phase inversion. ■

Exercise 13.14 Repeat Example 13.8 if $\beta = 300$. [*Hint:* Do not forget that the Q point changes (slightly) when β changes.]
Answer $A_v = -109$, $A_{voc} = -163$, $Z_{\text{in}} = 1186\ \Omega$, $A_i = -64.4$, $G = 7004$, $Z_o = 1\ \text{k}\Omega$, $v_o(t) = -76.5 \sin(\omega t)$ mV. □

13.9 EMITTER FOLLOWERS

The circuit diagram of another type of BJT amplifier called an **emitter follower** is shown in Figure 13.30(a). The resistors R_1, R_2, and R_E form the bias circuit. The collector resistor R_C (used in the common-emitter amplifier) is not needed in this circuit. Thus, we have a version of the four-resistor bias circuit with $R_C = 0$. Analysis of this bias circuit is very similar to the examples that we considered in Section 13.6.

The input signal is applied to the base through the coupling capacitor C_1. The output signal is coupled from the emitter to the load by the coupling capacitor C_2.

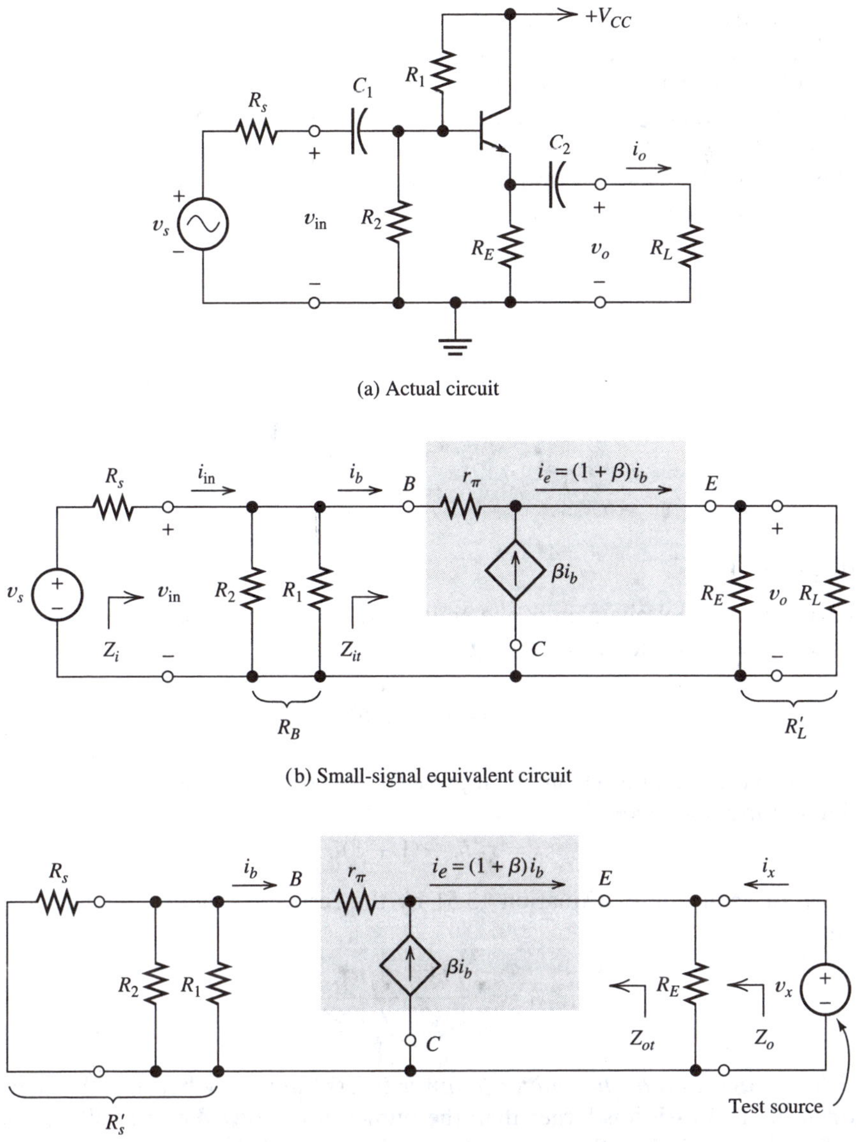

(a) Actual circuit

(b) Small-signal equivalent circuit

(c) Equivalent circuit used to find output impedance Z_o

Figure 13.30 Emitter follower.

Small-Signal Equivalent Circuit

The ac small-signal equivalent circuit is shown in Figure 13.30(b). As before, we replace the capacitors and power supply with short circuits. The transistor is replaced by its small-signal equivalent.

Notice that as a result, the collector terminal is connected directly to ground in the equivalent circuit. The transistor equivalent circuit is oriented with the collector at the bottom in Figure 13.30(b), but it is electrically the same as the transistor equivalent circuit we have used before. Because the collector is connected to ground, this circuit is sometimes called a **common-collector amplifier**.

Try to draw the small-signal equivalent on your own, referring to only Figure 13.30(a).

The ability to draw small-signal equivalent circuits is an important skill for analyzing electronic circuits. Carefully compare the small-signal equivalent in Figure 13.30(b) to the original circuit. Better still, try to draw the small-signal equivalent circuit on your own, starting from the original circuit.

Notice that R_1 and R_2 are in parallel in the equivalent circuit. We denote the combination by R_B. Also, R_E and R_L are in parallel, and we denote the combination by R'_L. In equation form, we have

$$R_B = R_1 || R_2 = \frac{1}{1/R_1 + 1/R_2} \tag{13.49}$$

and

$$R'_L = R_L || R_E = \frac{1}{1/R_L + 1/R_E} \tag{13.50}$$

Voltage Gain

Next, we find the voltage gain of the emitter follower. The current flowing through R'_L is $i_b + \beta i_b$. Thus, the output voltage is given by

$$v_o = R'_L(1+\beta)i_b \tag{13.51}$$

Writing a voltage equation from the input terminals through r_π and then through the load to ground, we have

$$v_{\text{in}} = r_\pi i_b + (1+\beta)i_b R'_L \tag{13.52}$$

Division of each side of Equation 13.51 by the respective side of Equation 13.52 results in

$$A_v = \frac{(1+\beta)R'_L}{r_\pi + (1+\beta)R'_L} \tag{13.53}$$

Emitter followers have voltage gains that are less than unity.

The voltage gain of the emitter follower is less than unity because the denominator of the expression is larger than the numerator. However, the voltage gain is usually only slightly less than unity. An amplifier with voltage gain less than unity can sometimes be useful because it can have a large current gain.

Also, notice that the voltage gain is positive. In other words, the emitter follower is noninverting. Thus, if the input voltage changes, the output at the emitter changes by almost the same amount and in the same direction as the input. Thus, the output voltage *follows* the input voltage. This is the reason for the name *emitter follower*.

Input Impedance

The input impedance of the emitter follower is relatively large compared with that of other BJT amplifiers.

The input impedance Z_i can be found as the parallel combination of R_B and the input impedance seen looking into the base of the transistor, which is indicated as Z_{it} in Figure 13.30(b). Thus, we can write

$$Z_i = \frac{1}{1/R_B + 1/Z_{it}} = R_B || Z_{it} \tag{13.54}$$

The input impedance looking into the base can be found by dividing both sides of Equation 13.52 by i_b:

$$Z_{it} = \frac{v_{\text{in}}}{i_b} = r_\pi + (1+\beta)R'_L \tag{13.55}$$

The input impedance of the emitter follower is relatively high compared to other BJT amplifier configurations. (However, in Chapter 12, we saw that field-effect transistors are capable of providing much higher input impedance than BJTs.) Once we have found the voltage gain and input impedance of the emitter follower, the current gain and power gain can be found by use of Equations 11.3 and 11.5.

Output Impedance

The output impedance of an amplifier is the Thévenin impedance seen from the output terminals. To find the output impedance of the emitter follower, we remove the load resistance, zero the signal source, and look back into the output terminals of the equivalent circuit. This is shown in Figure 13.30(c). We have attached a test source v_x, which delivers a current i_x to the impedance that we want to find. The output impedance is given by

$$Z_o = \frac{v_x}{i_x} \tag{13.56}$$

(Here again, the impedance can be found as the ratio of instantaneous time-varying quantities because the circuit is purely resistive. Otherwise, we should use phasors.)

To find this ratio, we write equations involving v_x and i_x. For example, summing currents at the top end of R_E, we have

$$i_b + \beta i_b + i_x = \frac{v_x}{R_E} \tag{13.57}$$

We must eliminate i_b from this equation before we can find the desired expression for the output impedance. *We do not want any circuit variables such as i_b in the result—only transistor parameters and resistance values.* Thus, we need to write another circuit equation.

First, we denote the parallel combination of R_s, R_1, and R_2 as

$$R'_s = \frac{1}{1/R_s + 1/R_1 + 1/R_2} \tag{13.58}$$

The additional equation needed can now be obtained by applying Kirchhoff's voltage law to the loop consisting of v_x, r_π, and R'_s:

$$v_x + r_\pi i_b + R'_s i_b = 0 \tag{13.59}$$

If we solve Equation 13.59 for i_b, substitute into Equation 13.57, and rearrange the result, we obtain the output impedance:

$$Z_o = \frac{v_x}{i_x} = \frac{1}{(1+\beta)/(R'_s + r_\pi) + 1/R_E} \tag{13.60}$$

This can be recognized as the parallel combination of R_E and the impedance

$$Z_{ot} = \frac{R'_s + r_\pi}{1 + \beta} \tag{13.61}$$

[It can be shown that Z_{ot} is the impedance seen looking into the emitter of the transistor, as indicated in Figure 13.30(c).]

The output impedance of the emitter follower is relatively small compared with that of other BJT amplifiers.

The output impedance of the emitter follower tends to be smaller than that of other BJT amplifier configurations.

Example 13.9 Emitter-Follower Performance

Compute the voltage gain, input impedance, current gain, power gain, and output impedance for the emitter-follower amplifier shown in Figure 13.31. Assume that the circuit operates at a temperature for which $V_T = 26$ mV.

First, we must determine the bias point so the value of r_π can be calculated.

Solution First, we must find the bias point so that the value of r_π can be calculated. The dc circuit is shown in Figure 13.31(b). Because the coupling capacitors act as open circuits for dc, R_s and R_L do not appear in the dc bias circuit.

Replacing the base bias circuit by its Thévenin equivalent, we obtain the equivalent circuit shown in Figure 13.31(c). Now, if we assume operation in the active

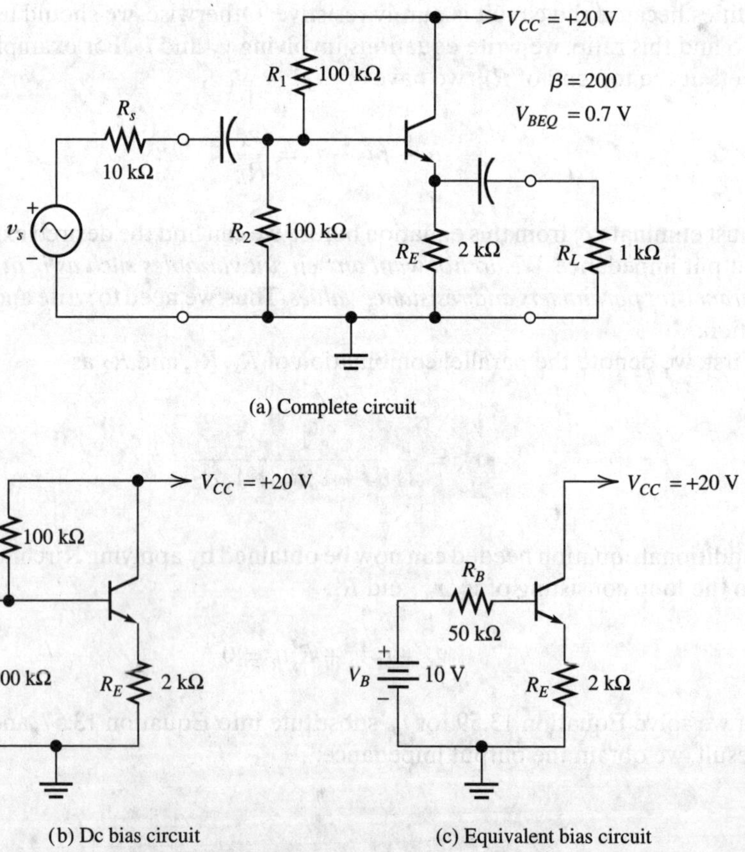

Figure 13.31 Emitter follower of Example 13.9.

region, we can write the following voltage equation around the base loop:

$$V_B = R_B I_{BQ} + V_{BEQ} + R_E(1+\beta)I_{BQ}$$

Substituting values, we find $I_{BQ} = 20.6\ \mu\text{A}$. Then, we have

$$I_{CQ} = \beta I_{BQ} = 4.12\ \text{mA}$$

$$V_{CEQ} = V_{CC} - I_{EQ}R_E = 11.7\ \text{V}$$

Since V_{CEQ} is greater than 0.2 V and I_{BQ} is positive, the transistor is indeed operating in the active region.

Equation 13.35 yields

$$r_\pi = \frac{\beta V_T}{I_{CQ}} = 1260\ \Omega$$

Now that we have established that the transistor operates in the active region and found the value of r_π, we can proceed in finding the amplifier gains and impedances. Substituting values into Equations 13.49 and 13.50, we discover that

$$R_B = R_1 || R_2 = \frac{1}{1/R_1 + 1/R_2} = 50\ \text{k}\Omega$$

$$R'_L = R_L || R_E = \frac{1}{1/R_L + 1/R_E} = 667\ \Omega$$

Equation 13.53 gives the voltage gain:

$$A_v = \frac{(1+\beta)R'_L}{r_\pi + (1+\beta)R'_L} = 0.991$$

Equations 13.54 and 13.55 give the input impedance:

$$Z_{it} = r_\pi + (1+\beta)R'_L = 135\ \text{k}\Omega$$

$$Z_i = \frac{1}{1/R_B + 1/Z_{it}} = 36.5\ \text{k}\Omega$$

Equations 13.58 and 13.60 yield

$$R'_s = \frac{1}{1/R_s + 1/R_1 + 1/R_2} = 8.33\ \text{k}\Omega$$

$$Z_o = \frac{1}{(1+\beta)/(R'_s + r_\pi) + 1/R_E} = 46.6\ \Omega$$

From Equation 11.3, we can find the current gain:

$$A_i = A_v \frac{Z_i}{R_L} = 36.2$$

Using Equation 11.5, we find that the power gain is

$$G = A_v A_i = 35.8$$

Even though the voltage gain of the emitter follower is less than unity, the current gain and power gain can be large.

Notice that even though the voltage gain is less than unity, the current gain is large (compared with unity). Thus, the output power is larger than the input power, and the circuit is effective as an amplifier. ■

In general, the output impedance of the emitter follower is much lower and the input impedance is higher than those of other single-stage BJT amplifiers. *Thus, we use an emitter follower if high input impedance or low output impedance is needed.*

If the emitter follower is cascaded with common-emitter stages, amplifiers with many useful combinations of parameters are possible. Furthermore, there are several other useful amplifier configurations using the BJT.

Exercise 13.15 Repeat Example 13.9 for $\beta = 300$. Compare the results to those of the example.

Answer $A_v = 0.991, Z_i = 40.1\ \text{k}\Omega, Z_o = 33.2\ \Omega, A_i = 39.7, G = 39.4.$ □

Summary

1. An *npn* bipolar junction transistor (BJT) consists of a layer of *p*-type material (the base) between two layers of *n*-type material (the collector and emitter, respectively).
2. In the active region, the collector current is an amplified version of the base current. In equation form, we have $i_C = \beta i_B$. A typical value of β is 100.
3. Typical common-emitter characteristics for an *npn* BJT are shown in Figure 13.5 on page 633.
4. Load-line analysis provides a basic understanding of amplifier circuits.
5. Besides being amplified, signals are often distorted by BJT amplifiers. In the active region, distortion is due to curvature of the input characteristic or unequal spacing of the output characteristics. For large signals, a more severe type of distortion, known as clipping, occurs when the device operation swings either into saturation or cutoff.
6. The *pnp* BJT is similar to the *npn* except that the current directions and voltage polarities are reversed.
7. BJTs can operate in three regions: the active region, the saturation region, and the cutoff region. Circuit models for BJTs in each of the three regions of operation are shown in Figure 13.16 on page 643.
8. In large-signal analysis of BJT circuits, we assume operation in a given region, use the corresponding circuit model to solve for currents and voltages, and then check to see if the results are consistent with the assumed operating region. The process is repeated until a self-consistent solution is found.
9. To be effective as an amplifier, the BJT must be biased in the active region. The four-resistor bias circuit [Figure 13.22(a) on page 650] is commonly used for simple amplifiers.
10. The small-signal equivalent circuit for the BJT is shown in Figure 13.26 on 655.
11. To analyze an amplifier in the midrange of frequencies, we first draw the small-signal equivalent circuit. The BJT is replaced by its equivalent circuit, the coupling and bypass capacitors are replaced by short circuits, and the dc power voltage sources are replaced by short circuits. Then, circuit equations are written, and algebra is used to find expressions for the gains and impedances of interest.
12. Two important BJT amplifier circuits are the common-emitter amplifier [Figure 13.27(a) on page 656] and the emitter follower [Figure 13.30(a) on page 661]. The common-emitter amplifier is inverting, has relatively large voltage and current gain magnitudes, and has moderate input impedance. The emitter follower is noninverting and has nearly unity voltage gain, large current gain, and moderate to high input impedance.

Problems

Section 13.1: Current and Voltage Relationships

P13.1. Which side of a *pn* junction should be connected to the positive voltage for forward bias? In normal operation, which type of bias (forward or reverse) is applied to the emitter–base junction of a BJT? To the collector–base junction?

P13.2. Write the Shockley equation for the emitter current of an *npn* transistor.

P13.3. Give the definitions of α and β for a BJT. What bias conditions for each junction are assumed in these definitions?

P13.4. Draw the circuit symbol for an *npn* BJT. Label the terminals and the currents. Choose reference directions that agree with the true current directions for operation in the active region.

P13.5. Sketch the construction of an *npn* BJT and label the three regions. In normal operation, does current flow into or out of the base terminal? The collector terminal? The emitter terminal?

***P13.6.** An *npn* transistor is operating with the base–emitter junction forward biased and the base–collector junction reverse biased. Given $i_C = 9$ mA for $i_B = 50$ mA, compute i_E, α, and β.

***P13.7.** Suppose that we have an *npn* transistor at room temperature, with $I_{ES} = 10^{-12}$ A, $I_B =$ 0.5 mA, $v_{CE} = 10$ V, and $i_E = 15$ mA. Determine $v_{BE}, v_{BC}, i_B, i_C, \alpha$ and β (assume that $V_T = 26$ mV at room temperature).

P13.8. Consider an *npn* transistor that is operating with the base–emitter junction forward biased and the base–collector junction reverse biased. Given that the transistor has $i_C = 10$ mA and $i_E = 10.5$ mA, determine the values of i_B, α, and β.

P13.9. If a transistor has $\alpha = 0.989$, what is its value for β?

P13.10. Suppose that an *npn* transistor is operating with the base–emitter junction forward biased and the base–collector junction reverse biased. Given $\beta = 200$ and $i_B =$ 10 μA, determine the values of i_C and i_E.

P13.11. Suppose that a certain *npn* transistor has $V_{BE} = 0.7$ V for $I_E = 15$ mA. Compute V_{BE} for $I_E = 1.5$ mA. Repeat for $I_E = 15$ μA. Assume that $V_T = 26$ mV.

P13.12. At an absolute temperature of $T = 300$ K, a certain transistor has $i_E = 10$ mA and $v_{BE} = 0.600$ V. Determine the value of I_{ES}. At 310 K, the transistor has $i_E = 10$ mA and $v_{BE} = 0.580$ V. Determine the new value of I_{ES}. By what factor did I_{ES} change for this 10 K increase in temperature? (Recall that Equation 10.2 on page 490 gives $V_T = kT/q$, in which $k = 1.38 \times 10^{-23}$ J/K is Boltzmann's constant and $q = 1.60 \times 10^{-19}$ C is the magnitude of the charge on an electron.)

P13.13. Determine the value of β for the transistor of Figure P13.13.

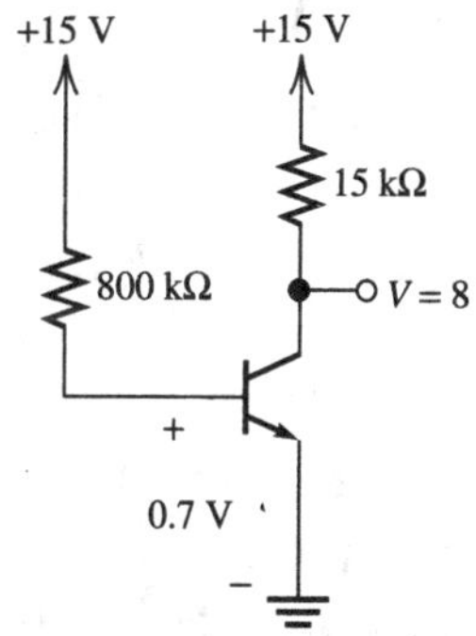

Figure P13.13

***P13.14.** Two transistors Q_1 and Q_2 connected in parallel are equivalent to a single transistor, as indicated in Figure P13.14. If the individual transistors have $I_{ES1} = I_{ES2} = 10^{-13}$ A and $\beta_1 = 150$ $\beta_2 = 50$, determine I_{ES} and β_{eq} for the equivalent transistor. Assume that all transistors have the same temperature.

* Denotes that answers are contained in the Student Solutions files. See Appendix E for more information about accessing the Student Solutions.

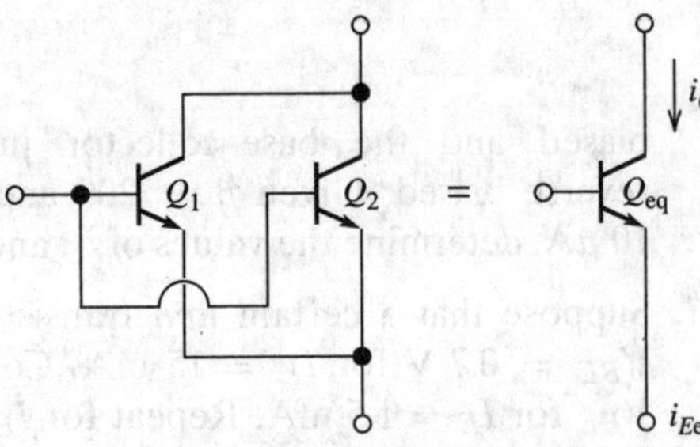

Figure P13.14

P13.15. Consider the circuit shown in Figure P13.15. The transistors Q_1 and Q_2 are identical, both having $I_{ES} = 10^{-14}$ A and $\beta = 100$. Calculate V_{BE} and I_{C2}. Assume that $V_T = 26$ mV for both transistors. (*Hint:* Both transistors are operating in the active region. Because the transistors are identical and have identical values of V_{BE}, their collector currents are equal.)

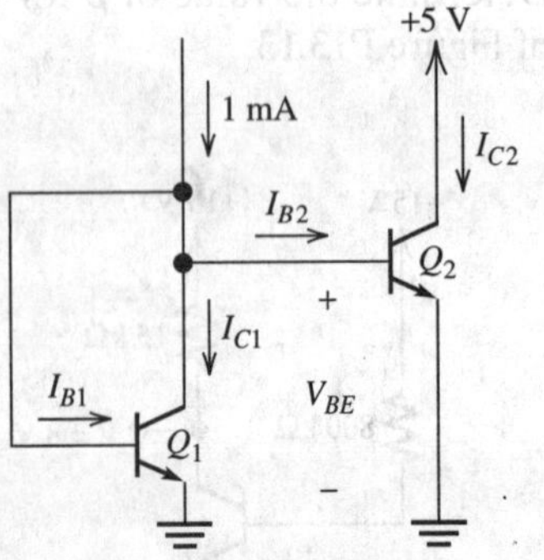

Figure P13.15

P13.16. The transistors Q_1 and Q_2 shown in Figure P13.16 are said to be **Darlington connected** and can be considered to be equivalent to a single transistor, as indicated. Find an expression for β_{eq} of the equivalent transistor in terms of β_1 and β_2.

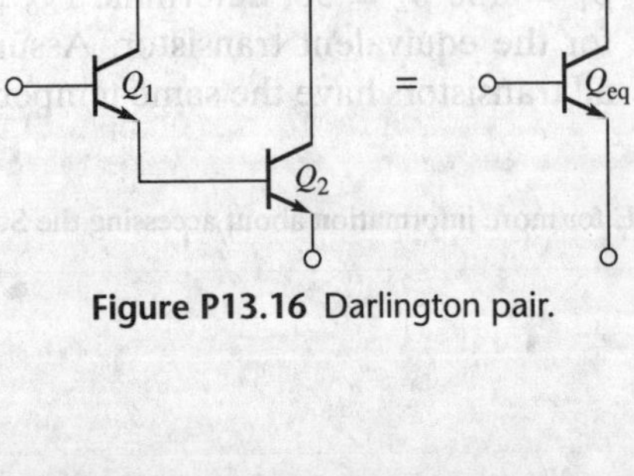

Figure P13.16 Darlington pair.

P13.17. Consider the circuit of Figure P13.17 in which Q_1 has $I_{ES1} = 10^{-14}$ A and $\beta_1 = 100$, while Q_2 has $I_{ES2} = 10^{-13}$ A and $\beta_2 = 100$. Determine the values of V_{BE} and I_{C2}. Assume that both transistors have $V_T = 26$ mV and are operating in the active region.

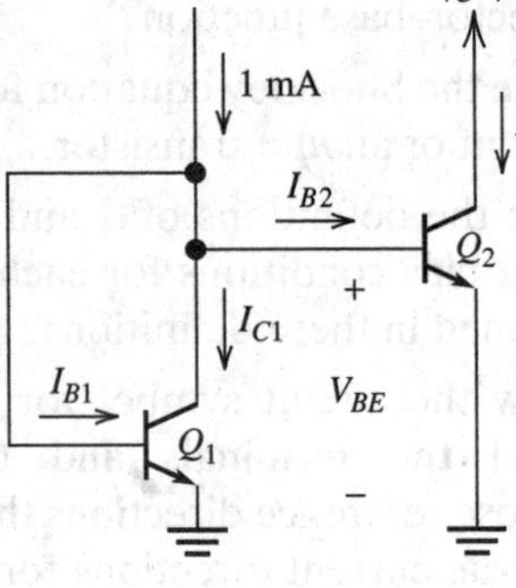

Figure P13.17

Section 13.2: Common-Emitter Characteristics

***P13.18.** Determine the values of α and β for the transistor whose characteristics are shown in Figure P13.18.

***P13.19.** A certain *npn* silicon transistor has $v_{BE} = 0.7$ V for $i_B = 0.1$ mA at a temperature of 30°C. Sketch the input characteristic to scale at 30°C. What is the approximate value of v_{BE} for $i_B = 0.1$ mA at 180°C? (Use the rule of thumb that v_{BE} is reduced in magnitude by 2 mV per degree increase in temperature.) Sketch the input characteristic to scale at 180°C.

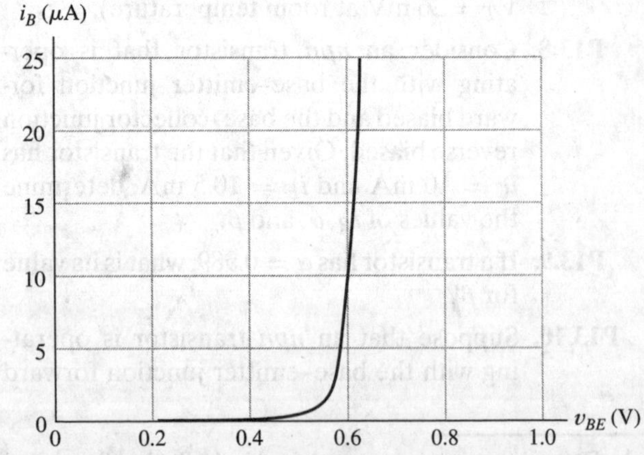

(a) Input

Figure P13.18

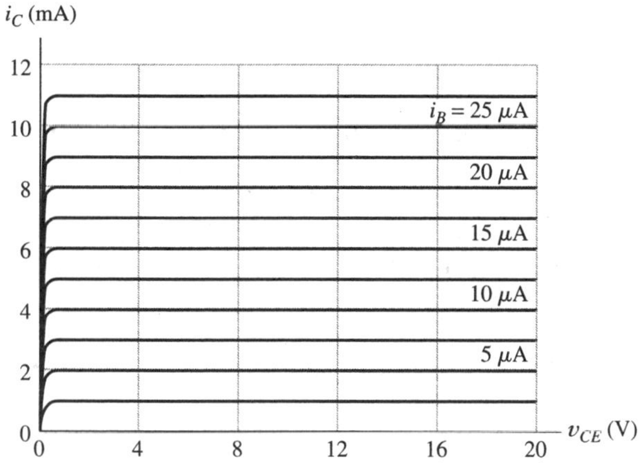

(b) Output

Figure P13.18 (*Cont.*)

P13.20. The transistor having the characteristics shown in Figure P13.18 is operating with $i_C = 8$ mA and $v_{CE} = 12$ V. Locate the point of operation on both the input and output characteristics.

P13.21. Consider an *npn* silicon transistor that has $\beta = 75$ and $i_B = 0.2$ mA. Sketch i_C versus v_{CE}, for v_{CE} ranging from 0 to 5 V. Repeat for $\beta = 200$.

Section 13.3: Load-Line Analysis of a Common-Emitter Amplifier

***P13.22.** Consider the circuit of Figure 13.7 on page 635. Assume that $V_{CC} = 20$ V, $V_{BB} = 0.8$ V, $R_B = 40$ kΩ, and $R_C = 2$ kΩ. The input signal is a 0.2-V-peak 1-kHz sinusoid given by $v_{in}(t) = 0.2 \sin(2000\pi t)$. The common-emitter characteristics for the transistor are shown in Figure P13.18. Determine the maximum, minimum, and *Q*-point values for v_{CE}. What is the approximate voltage gain for this circuit?

P13.23. List several reasons that distortion occurs in BJT amplifiers.

P13.24. Consider the circuit of Figure 13.7 on page 635. Given $V_{CC} = 5$ V and $R_C = 1$ kΩ, construct the load line on the i_C versus v_{CE} axes. Repeat for $V_{CC} = 9$ V. How does the slope of the load line change when V_{CC} changes?

P13.25. Consider the circuit of Figure 13.7 on page 635. Assume that $V_{CC} = 20$ V, $V_{BB} = 0.3$ V, $R_B = 40$ kΩ, and $R_C = 2$ kΩ. The input signal is a 0.2-V-peak 1-kHz sinusoid given by $v_{in}(t) = 0.2 \sin(2000\pi t)$. The common-emitter characteristics for the transistor are shown in Figure P13.18. Determine the maximum, minimum, and *Q*-point values for v_{CE}. What is the approximate voltage gain for this circuit? Why is the gain so small in magnitude?

P13.26. Consider the circuit of Figure 13.7 on page 635. Assume that $V_{CC} = 20$ V, $V_{BB} = 0.8$ V, $R_B = 40$ kΩ, and $R_C = 10$ kΩ. The input signal is a 0.2-V-peak 1-kHz sinusoid given by $v_{in}(t) = 0.2 \sin(2000\pi t)$. The common-emitter characteristics for the transistor are shown in Figure P13.18. Determine the maximum, minimum, and *Q*-point values for v_{CE}. What can you say about the waveform for $v_{CE}(t)$? Why isn't voltage gain an appropriate concept in this case?

Section 13.4: *pnp* Bipolar Junction Transistors

P13.27. Draw the circuit symbol for a *pnp* BJT. Label the terminals and the currents. Choose reference directions that agree with the true current direction for operation in the active region.

***P13.28.** The circuit shown in Figure P13.28 has $i_s(t) = 15 + 5\sin(2000\pi t)$ μA. The transistor has $\alpha = 0.989$.

a. Sketch the output characteristics for $i_B = 0, 5, 10, 15, 20,$ and 25 μA with v_{CE} ranging from zero to −20 V.

b. Draw the output load line on the characteristics sketched in part (a).

c. Determine the values for $I_{C\max}$, I_{CQ}, and $I_{C\min}$.

d. Sketch $v_{CE}(t)$ to scale versus time.

e. Repeat parts (c) and (d) for $i_s(t) = 25 + 5\sin(2000\pi t)$ μA.

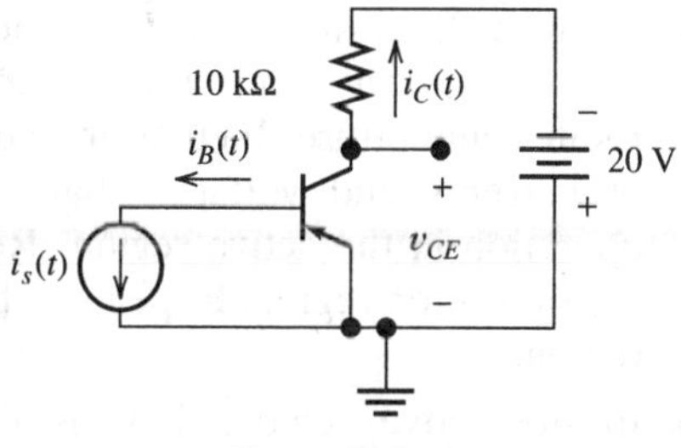

Figure P13.28

***P13.29.** A certain *pnp* silicon transistor has $\beta = 100$ and $i_B = 50\ \mu A$. Sketch i_C versus v_{CE}, for v_{CE} ranging from 0 to −5 V. Repeat for $\beta = 300$.

P13.30. Figure P13.30 shows an *npn* transistor and a *pnp* transistor connected as a **Sziklai pair**, which is equivalent to a single *npn* transistor, as indicated. Find an expression for β_{eq} of the equivalent transistor in terms of β_1 and β_2.

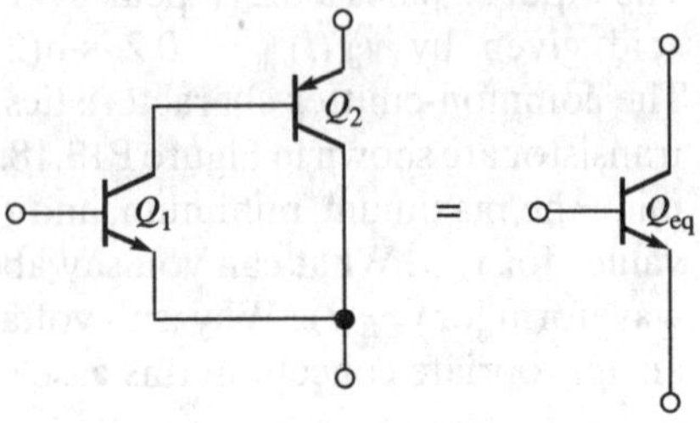

Figure P13.30 Sziklai pair.

P13.31. Suppose we have a certain *pnp* BJT that has $V_{CE} = -5$ V, $I_C = 0.98$ mA, and $\alpha = 0.95$. Determine the values for I_E and β for this transistor.

P13.32. At a temperature of 150°C, a particular *pnp* transistor has $V_{BE} = 0.5$ V for $I_E = 1.5$ mA. Estimate V_{BE} for $I_E = 1.5$ mA at a temperature of 30°C.

Section 13.5: Large-Signal DC Circuit Models

P13.33. Sketch the input characteristic curve to scale for a typical small-signal silicon *npn* BJT at room temperature, allowing i_B to range from 0 to 40 μA. Sketch the output characteristic curves to scale, with i_B ranging from 0 to 40 μA in 10 μA steps, given that $\beta = 100$. Finally, label the cutoff, active, and saturation regions.

P13.34. Draw the large-signal dc circuit model for a silicon *npn* transistor in the active region at room temperature. Include the constraints of currents and voltages that guarantee operation in the active region. Repeat for the saturation region. Repeat for the cutoff region.

***P13.35.** In the active region, how is the base–collector junction biased (forward or reverse)? How is the base–emitter junction biased? Repeat for the saturation region. Repeat for the cutoff region.

P13.36. Each of the transistors shown in Figure P13.36 has $\beta = 100$, $|V_{CE}| = 0.2$ V in saturation, and $|V_{BE}| = 0.6$ V in the active and saturation regions. For each transistor, determine the region of operation and the values of $V_{CE}, I_B, I_E,$ and I_C.

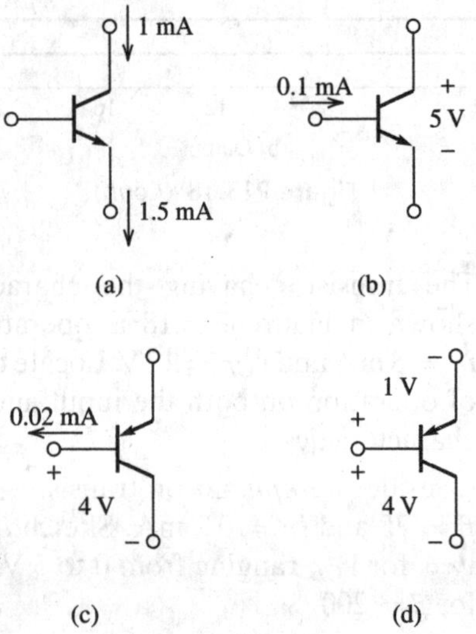

Figure P13.36

P13.37. Shown in Figure P13.16 is the Darlington pair, which is equivalent to a single transistor, as indicated in the figure. Assuming that both Q_1 and Q_2 are operating in the active region, with $|V_{BE}| = 0.7$ V, determine the value for V_{BE} for the equivalent transistor in the active region. Repeat for the Sziklai pair shown in Figure P13.30.

P13.38. Determine the region of operation for a room-temperature silicon *pnp* transistor that has $\beta = 150$ if: **a.** $V_{CE} = -5$ V and $V_{BE} = -0.6$ V; **b.** $I_C = 25$ mA and $I_B = 2$ mA; **c.** $I_B = 1$ mA and $V_{CE} = -5$ V.

P13.39. Determine the region of operation for a room-temperature silicon *npn* transistor that has $\beta = 100$ if: **a.** $V_{CE} = 10$ V and $I_B = 20\ \mu A$; **b.** $I_C = I_B = 0$; **c.** $V_{CE} = 3$ V and $V_{BE} = 0.4$ V; **d.** $I_C = 1$ mA and $I_B = 50\mu A$.

P13.40. Draw the large-signal dc circuit model for a silicon *pnp* transistor in the active region at room temperature. Include the constraints of currents and voltages that guarantee operation in the active region. Repeat for the saturation region. Repeat for the cutoff region.

Section 13.6: Large-Signal DC Analysis of BJT Circuits

***P13.41.** List the step-by-step procedure for dc analysis of a BJT circuit, using the large-signal circuit models.

***P13.42.** Use the large-signal models shown in Figure 13.16 on page 643 for the transistors to determine I_C and V_{CE} for the circuits of Figure P13.42. Assume that $\beta = 100$ and $|V_{BE}| = 0.7$ V in both the active and saturation regions. Repeat for $\beta = 300$, and compare the results for both values.

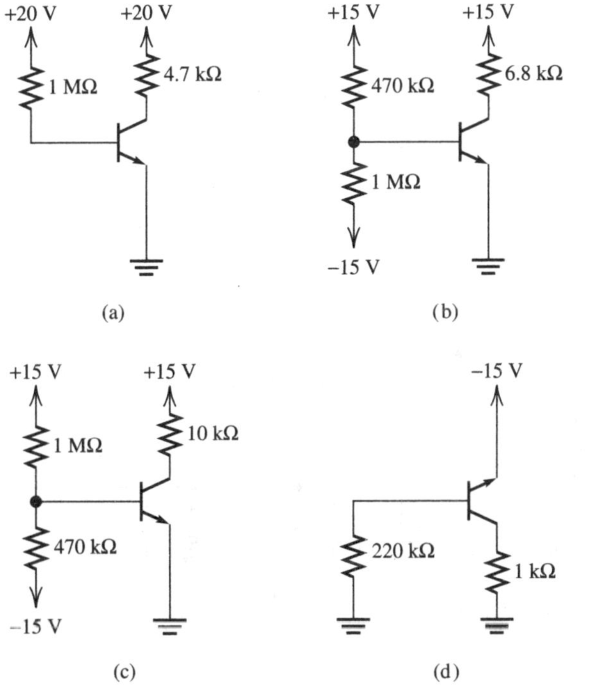

Figure P13.42

P13.43. Draw and explain the voltage-divider bias circuit for an *npn* transisitor.

P13.44. Draw the fixed base bias circuit. What is the principal reason that this circuit is unsuitable for mass production of amplifier circuits?

***P13.45.** Consider the circuit shown in Figure P13.45. A *Q*-point value for I_C between a minimum of 4 mA and a maximum of 5 mA is required. Assume that resistor values are constant and that β ranges from 100 to 300. It is desired for R_B to have the largest possible value while meeting the other constraints. Determine the values of R_B and R_E.

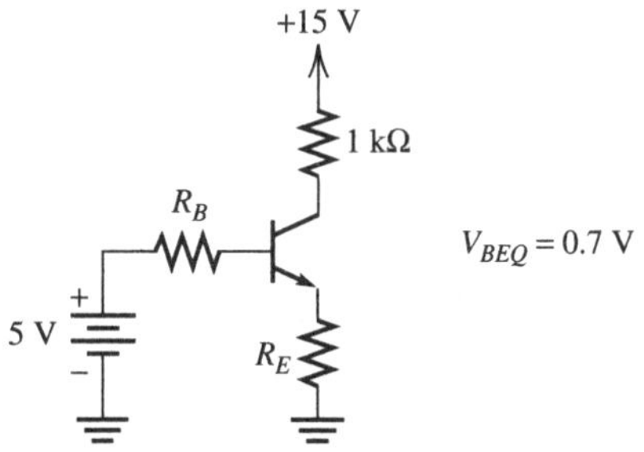

Figure P13.45

P13.46. Analyze the circuits shown in Figure P13.46 to determine I and V. For all transistors,

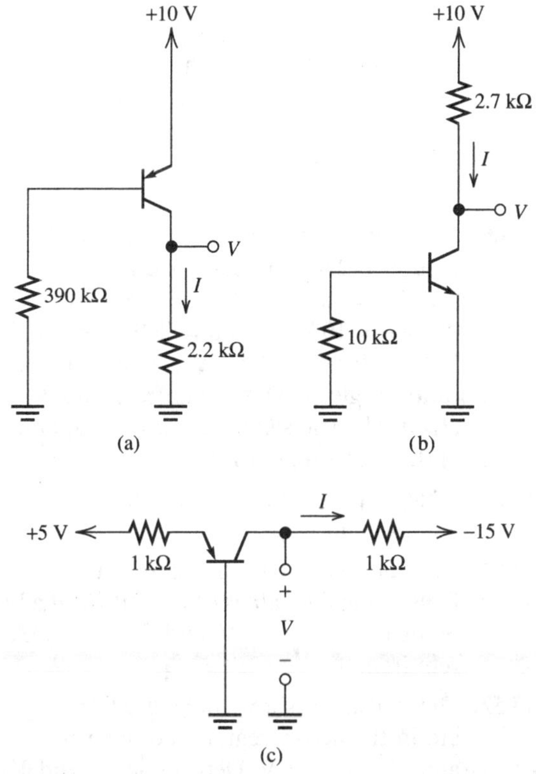

Figure P13.46

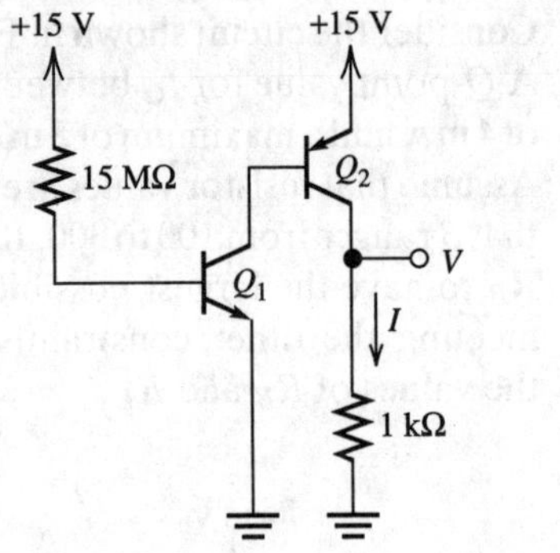

(d)

Figure P13.46 *(Cont.)*

assume that $\beta = 100$ and $|V_{BE}| = 0.7$ V in both the active and saturation regions. Repeat for $\beta = 300$.

P13.47. Consider the four-resistor bias network of Figure 13.22(a) on page 650, with $R_1 = 100$ kΩ, $R_2 = 200$ kΩ, $V_{CC} = 15$ V, $R_C = 20$ kΩ, $R_E = 20$ kΩ, and $\beta = 150$. Assume that $V_{BE} = 0.7$ V. Determine I_{CQ} and V_{CEQ}.

***P13.48.** The four-resistor bias network of Figure 13.22(a) on page 650 has $V_{CC} = 15$ V, $R_1 = 100$ kΩ, $R_2 = 47$ kΩ, $R_C = 4.7$ kΩ, and $R_E = 4.7$ kΩ. Suppose that β ranges from 50 to 200, $V_{BE} = 0.7$ V, and the resistors have tolerances of ±5 percent. Calculate the maximum and minimum values for I_C.

P13.49. The four-resistor bias network of Figure 13.22(a) on page 650 has $R_1 = 100$ kΩ, $R_2 = 200$ kΩ, $V_{CC} = 15$ V, $R_C = 10$ kΩ, $R_E = 10$ kΩ, and $\beta = 200$. Assume that $V_{BE} = 0.7$ V in both the active and saturation regions. Determine I_{CQ} and V_{CEQ}. (*Hint:* The transistor may *not* be operating in the active region.)

P13.50. Analyze the circuit of Figure P13.50 to determine I_C and V_{CE}.

P13.51. Consider the circuit shown in Figure P13.51. Determine the values of R_1 and R_C if a bias point of $V_{CE} = 7$ V and $I_C = 5$ mA is required.

P13.52. The transistors shown in Figure P13.52 operate in the active region and have $\beta = 100$ and $V_{BE} = 0.7$ V. Determine I_C and V_{CE} for each transistor.

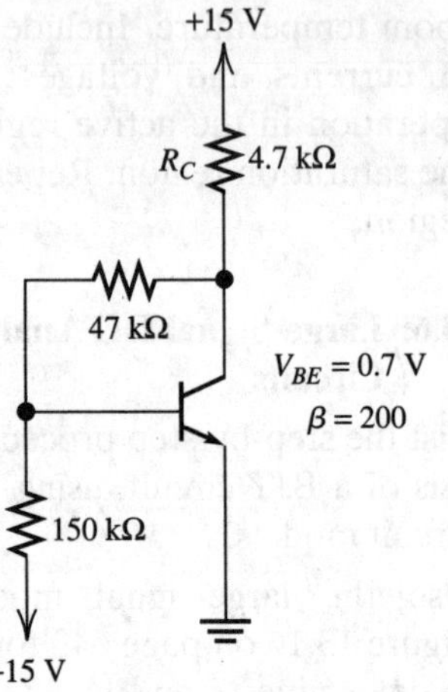

Figure P13.50

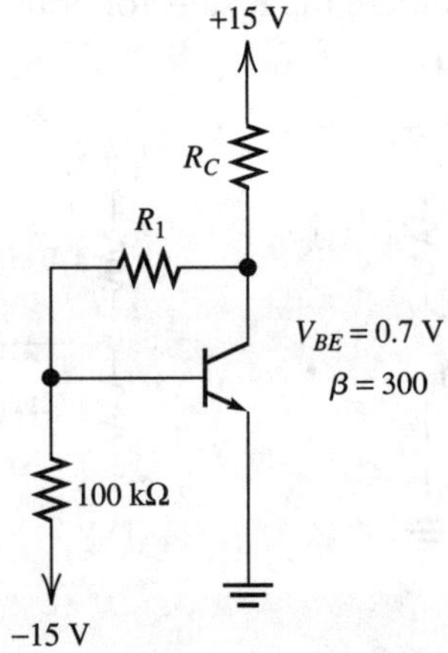

Figure P13.51

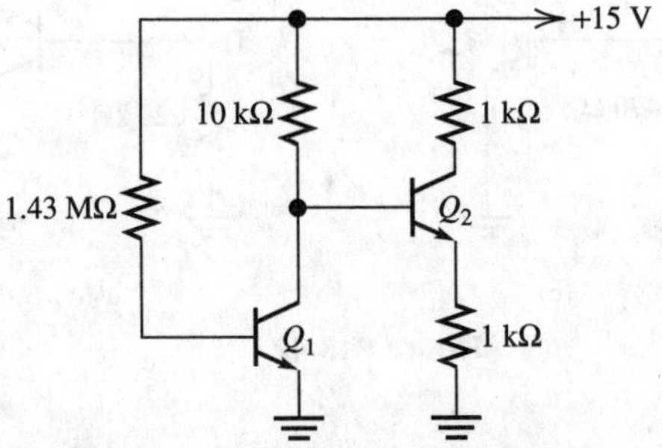

Figure P13.52

Section 13.7: Small-Signal Equivalent Circuits

P13.53. Give the formula for determination of r_π assuming that β and the Q point are known.

P13.54. Draw the small-signal equivalent circuit for a BJT.

***P13.55.** Suppose that a new type of BJT has been invented for which $i_B = 10^{-5}v_{BE}^2$, where i_B is in amperes and v_{BE} is in volts. Also, this new transistor has $i_C = 100i_B$. The small-signal equivalent circuit for the transistor is shown in Figure 13.26 on page 655. Find an equation relating r_π of this new transistor to I_{CQ}. Evaluate r_π for $I_{CQ} = 1$ mA.

P13.56. A certain *npn* silicon transistor at room temperature has $\beta = 200$. Find the corresponding values of r_π if $I_{CQ} = 3$ mA, 0.3 mA, and 3 μA. Assume operation in the active region and that $V_T = 26$ mV.

P13.57. Shown in Figure P13.30 is the Sziklai pair, which is equivalent to a single transistor, as indicated in the figure. Draw the small-signal equivalent circuits for the pair and for the equivalent transistor. Find an expression for $r_{\pi\text{eq}}$ in terms of $r_{\pi 1}, r_{\pi 2}, \beta_1$, and β_2.

P13.58. Shown in Figure P13.16 is the Darlington pair, which is equivalent to a single transistor, as indicated in the figure. Draw the small-signal equivalent circuits for the pair and for the equivalent transistor. Find an expression for $r_{\pi\text{eq}}$ in terms of $r_{\pi 1}$, $r_{\pi 2}$, β_1, and β_2.

Section 13.8: Common-Emitter Amplifiers

P13.59. Draw the circuit diagram of a common-emitter amplifier circuit that uses the four-resistor biasing network. Include a signal source and a load resistance.

P13.60. Why are coupling capacitors often used to connect the signal source and the load to amplifier circuits? Should coupling capacitors be used if it is necessary to amplify dc signals? Explain.

P13.61. Are common-emitter amplifiers inverting or noninverting? What can you say about the magnitudes of the voltage and current gains?

***P13.62.** Consider the common-emitter amplifier of Figure P13.62. Draw the dc circuit and find I_{CQ}. Find the value of r_π. Then, calculate values for $A_v, A_{voc}, Z_{\text{in}}, A_i, G$, and Z_o. Assume operation in the frequency range for which the coupling and bypass capacitors are short circuits.

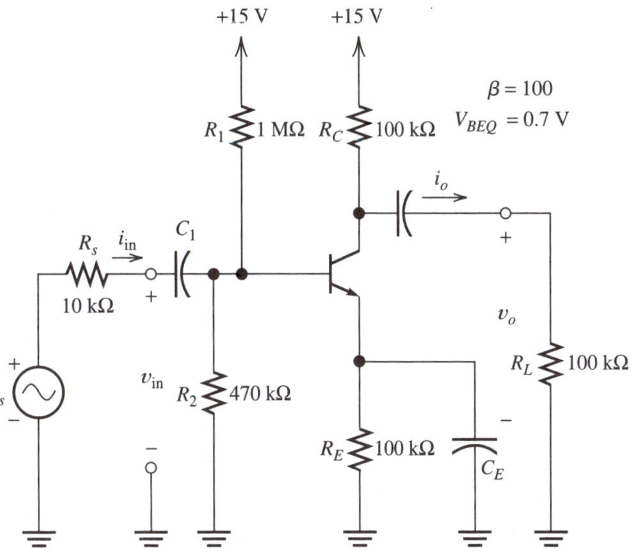

Figure P13.62

P13.63. Consider the common-emitter amplifier shown in Figure P13.63. **a.** Draw the small-signal equivalent circuit by assuming that the coupling capacitors are short circuits. **b.** Derive an expression for the voltage gain in terms of resistor values, r_π, and β. **c.** Derive an expression for the input resistance in terms of resistor values, r_π, and β. **d.** Suppose that $\beta = 100$, $V_{BEQ} = 0.7$ V, $R_C = 2$ kΩ, $R_L = 2$ kΩ, $R_E = 100$ Ω, and $V_{CC} = 20$ V. Determine the value required for R_B so that $I_{CQ} = 5$ mA. **e.** Evaluate the expressions found in parts (c) and (d) for the values given in part (d).

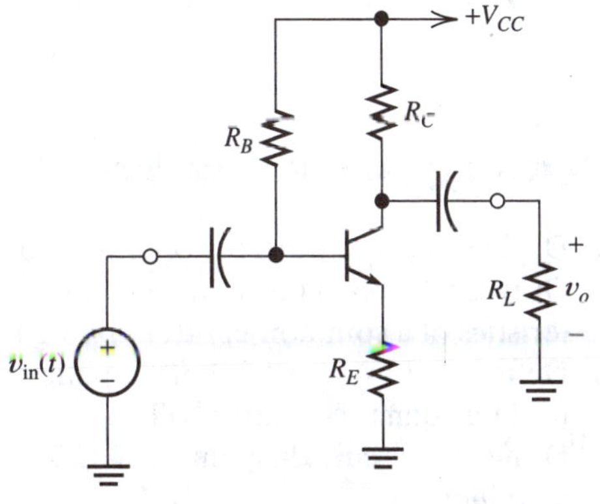

Figure P13.63

P13.64. Consider the common-emitter amplifier of Figure P13.64. Draw the dc circuit and find I_{CQ}. Find the value of r_π. Then, calculate values for $A_v, A_{voc}, Z_{in}, A_i, G,$ and Z_o. Assume operation in the frequency range for which the coupling and bypass capacitors are short circuits.

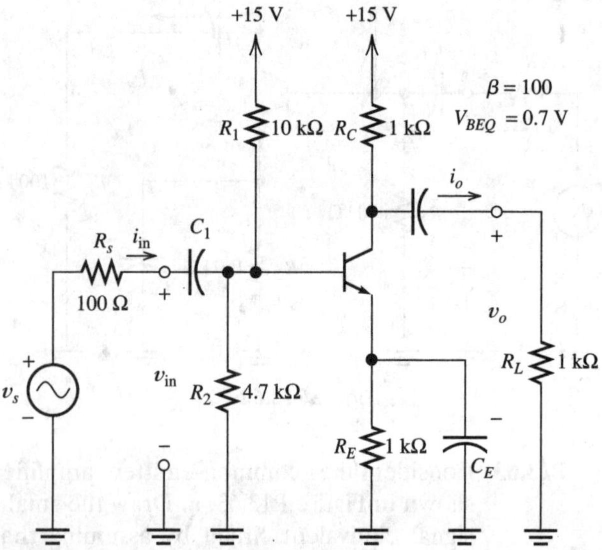

Figure P13.64

Section 13.9: Emitter Followers

P13.65. Draw the circuit diagram of an emitter follower. Include a signal source and a load resistance.

***P13.66.** Consider the emitter-follower amplifier of Figure P13.66. Draw the dc circuit and find I_{CQ}. Next, determine the value of r_π. Then, calculate midband values for A_v, A_{voc}, Z_{in}, A_i, G, and Z_o.

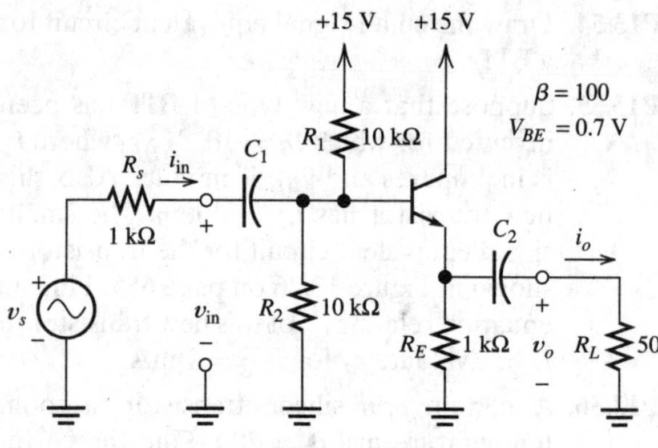

Figure P13.66

P13.67. What can you say about the voltage gain of an emitter follower? The current gain? The power gain?

P13.68. Consider the emitter-follower amplifier of Figure P13.68. Draw the dc circuit and find I_{CQ}. Next, determine the value of r_π. Then, calculate midband values for A_v, A_{voc}, Z_{in}, A_i, G, and Z_o.

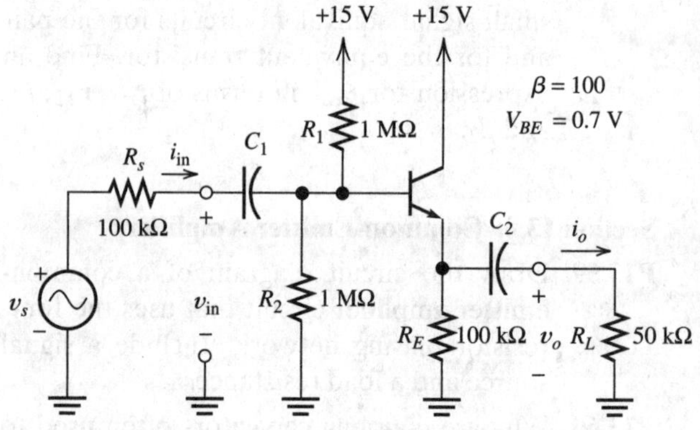

Figure P13.68

Frequentlty Asked Questions

1. Derive the equations for operation of a BJT.
2. Draw and explain the input and output characteristics of a common-emitter *npn* BJT.
3. Define and explain current amplification factor of a common-emitter BJT.
4. Draw the circuit diagram of a CE transistor amplifier.
5. Explain the working of a Common-Emitter transistor amplifier and analyse it with the help of load-line
6. What do you mean by non-linear distortion?
7. Explain cut off, saturation and active region of a bipolar function transistor (BJT).
8. Explain the working of a *pnp* bipolar junction transistor.

9. Draw and explain active-region model, saturated region model and cutoff-region model of a BJT.
10. Explain with circuit diagrams, the large signal dc analysis of BJT amplifier circuit.
11. What are the requirements for bias-circuit design?
12. How will you design a fixed basc bias-circuit for a BJT amplifier circuit.
13. How will you design a four-resistor bias circuit for a BJT amplifier circuit?
14. Draw and explain small-signal equivalent circuit of BJT amplifier.
15. Draw and explain a common-emitter, small-signal amplifier circuit and derive relations for voltage gain, current gain and power gain.
16. Draw the circuit and explain the working of an emitter follower.

Practice Test

Here is a practice test you can use to check your comprehension of the most important concepts in this chapter. Answers can be found in Appendix D and complete solutions are included in the Student Solutions files. See Appendix E for more information about the Student Solutions.

T13.1. Match each entry in Table T13.1(a) with the best choice from the list given in Table T13.1(b). [Items in Table T13.1(b) may be used more than once or not at all.]

T13.2. The simple amplifier of Figure 13.7 on page 635 has $R_B = 10\,\text{k}\Omega$, $V_{BB} = 0.8\,\text{V}$, $v_{\text{in}}(t) = 0.2\sin(2000\pi t)$ V, $R_C = 2.5\,\text{k}\Omega$, and

Table T13.1

(a)
a. For a BJT in the active region, the collector–base junction is _____.
b. For a BJT in the active region, the emitter–base junction is _____.
c. For a BJT in the active region, the _____ current is usually much smaller than the current in the other two terminals.
d. Clipping occurs when the BJT in an amplifier circuit reaches _____ or _____.
e. In a BJT amplifier, nonuniform spacing of the collector characteristics can cause _____.
f. The large signal model for a BJT consists of two voltage sources in the _____ region.
g. The large signal model for a BJT has open circuits between all three terminals in the _____ region.
h. A BJT with $\beta = 50, I_C = 1\,\text{mA}$, and $I_E = 1.5\,\text{mA}$ is operating in the _____ region.
i. The voltage gain of an emitter follower is typically _____.
j. The voltage gain of a common-emitter amplifier is typically _____.
k. In a midband small-signal equivalent circuit, a coupling capacitance appears as _____.

Table T13.1 (Cont.)

(b)
1. cutoff
2. forward biased
3. reverse biased
4. emitter
5. base
6. collector
7. saturation
8. large gain
9. small gain
10. distortion
11. active
12. large in magnitude compared to unity and negative
13. small in magnitude compared to unity and negative
14. inverting
15. slightly less than unity
16. noninverting
17. large in magnitude compared to unity and positive
18. an open circuit
19. a short circuit

$V_{CC} = 10$ V. The characteristics are shown in Figure 13.5 on page 633. Use load-line analysis to determine the values for $V_{CE\,\mathrm{min}}$, V_{CEQ}, and $V_{CE\,\mathrm{max}}$.

T13.3. An *npn* BJT is operating in the active region with $I_{CQ} = 1.0$ mA and $I_{EQ} = 1.04$ mA. Determine the values for α, β, and r_π. Draw the small-signal equivalent circuit for the BJT. Assume that $V_T = 26$ mV.

T13.4. The BJT shown in Figure T13.4 has $\beta = 50$ and $V_{BE} = 0.7$ V. **a.** Determine the values of I_C and V_{CE}; **b.** Repeat for $\beta = 250$.

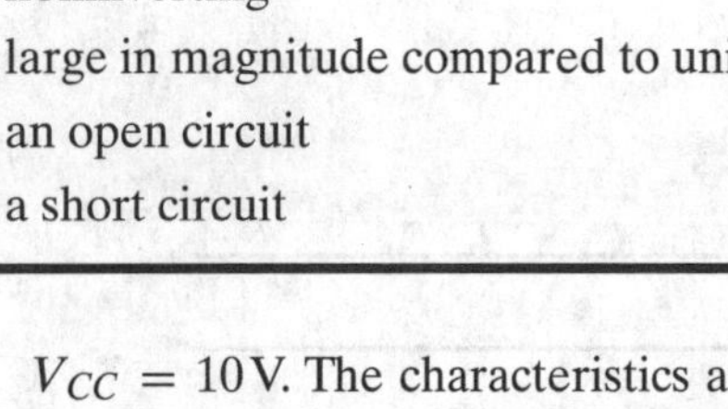

Figure T13.4

T13.5. Draw the midband small-signal equivalent circuit for the amplifier shown in Figure T13.5. Be sure to label each element in your equivalent circuit.

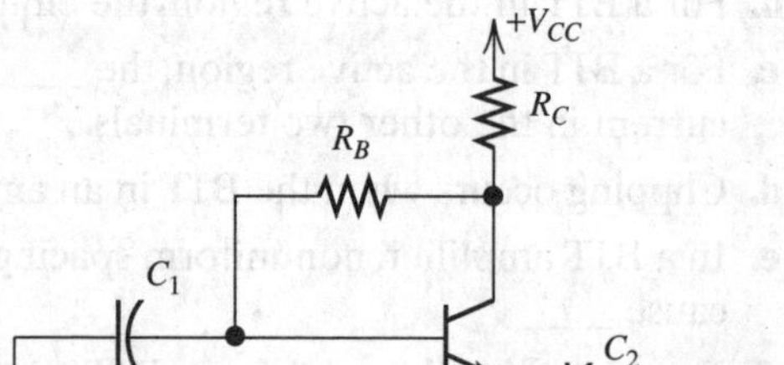

Figure T13.5

T13.6. The common-emitter amplifier shown in Figure 13.27(a) on page 656 has $R_1 = 100\,\mathrm{k}\Omega$, $R_2 = 47\,\mathrm{k}\Omega$, $R_C = 2.2\,\mathrm{k}\Omega$, $R_L = 5.6\,\mathrm{k}\Omega$, $\beta = 120$, $V_T = 26$ mV, and $I_{CQ} = 4$ mA. Determine the value of the voltage gain $A_v = v_o/v_{in}$ and the input impedance.

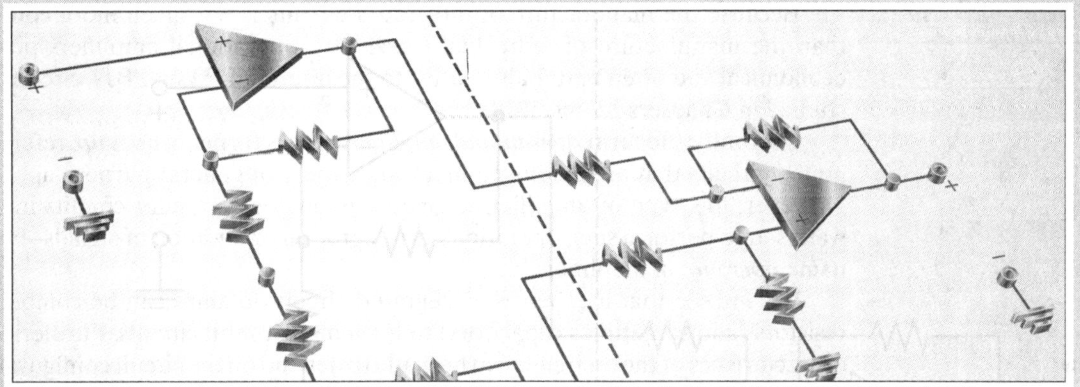

Chapter 14

Operational Amplifiers

Study of this chapter will enable you to:

- List the characteristics of ideal op amps.
- Identify negative feedback in op-amp circuits.
- Use the summing-point constraint to analyze ideal op-amp circuits that have negative feedback.
- Select op-amp circuit configurations suitable for various applications.
- Use op amps to design useful circuits.
- Identify practical op-amp limitations and recognize potential inaccuracies in instrumentation applications.
- Work with instrumentation amplifiers.
- Apply integrators, differentiators, and active filters.

Introduction to this chapter:

In Chapter 11, we discussed the external characteristics of amplifiers in general. In Chapters 12 and 13, we saw how basic amplifiers can be built by using FETs or BJTs. In this chapter, we introduce an important device known as the **operational amplifier**, which finds application in a wide range of engineering instrumentation.

An operational amplifier is a circuit composed of perhaps 30 BJTs or FETs, 10 resistors, and several capacitors. These components are manufactured concurrently on a single piece of silicon crystal (called a chip) by a sequence of processing steps. Circuits manufactured in this way are called **integrated circuits** (ICs).

Because the manufacture of integrated circuits is not much more complicated than the manufacture of individual transistors, operational amplifiers provide an economical and often better alternative to the discrete FET and BJT circuits that we studied in Chapters 12 and 13.

Currently, the term *operational amplifier*, or less formally *op amp*, refers to integrated circuits that are employed in a wide variety of general-purpose applications. However, this type of amplifier originated in analog-computer circuits in which it was used to perform such operations as integration or addition of signals—hence, the name *operational* amplifier.

We will see that inexpensive integrated-circuit op amps can be combined with resistors (and sometimes capacitors) to form many useful circuits. Furthermore, the characteristics of these circuits can be made to depend on the circuit configuration and the resistor values but only weakly on the op amp—which can have large unit-to-unit variations in some of its parameters.

14.1 IDEAL OPERATIONAL AMPLIFIERS

The circuit symbol for the operational amplifier is shown in Figure 14.1. The operational amplifier is a differential amplifier having both inverting and noninverting input terminals. (We discussed differential amplifiers in Section 11.11.) The input signals are denoted as $v_1(t)$ and $v_2(t)$. (As usual, we use lowercase letters to represent general time-varying voltages. Often, we will omit the time dependence and refer to the voltages simply as v_1, v_2, and so on.)

The input signal of a differential amplifier consists of a differential component and a common-mode component.

Recall that the average of the input voltages is called the **common-mode signal** and is given by

$$v_{icm} = \frac{1}{2}(v_1 + v_2)$$

Also, the difference between the input voltages is called the **differential signal**, given by

$$v_{id} = v_1 - v_2$$

An ideal operational amplifier has the following characteristics:

Characteristics of ideal op amps.

- Infinite input impedances
- Infinite gain for the differential input signal
- Zero gain for the common-mode input signal

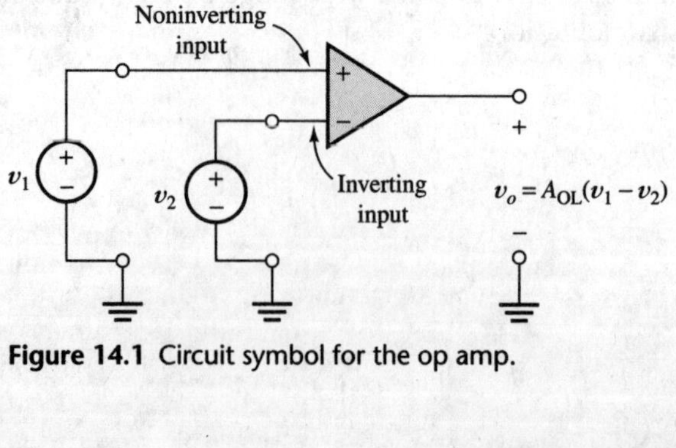

Figure 14.1 Circuit symbol for the op amp.

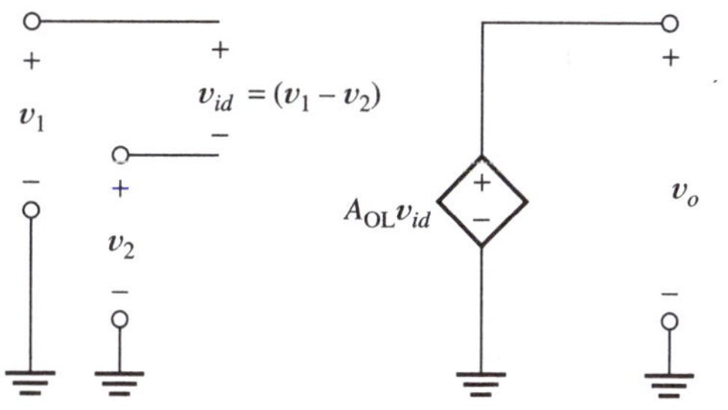

Figure 14.2 Equivalent circuit for the ideal op amp. The open-loop gain A_{OL} is very large (approaching infinity).

- Zero output impedance
- Infinite bandwidth

An equivalent circuit for the ideal operational amplifier consists simply of a controlled source as shown in Figure 14.2. The **open-loop gain** A_{OL} is very large in magnitude—ideally, infinite.

As we will shortly see, op amps are generally used with feedback networks that return part of the output signal to the input. Thus, a *loop* is created in which signals flow through the amplifier to the output and back through the feedback network to the input. A_{OL} is the gain of the op amp without a feedback network. That is why we call it the *open-loop gain.*

For now, we assume that the open-loop gain A_{OL} is constant. Thus, there is no distortion, either linear or nonlinear, and the output voltage v_o has a waveshape identical to that of the differential input $v_{id} = v_1 - v_2$. (Later, we will see that A_{OL} is actually a function of frequency. Furthermore, we will learn that real op amps suffer from nonlinear imperfections.)

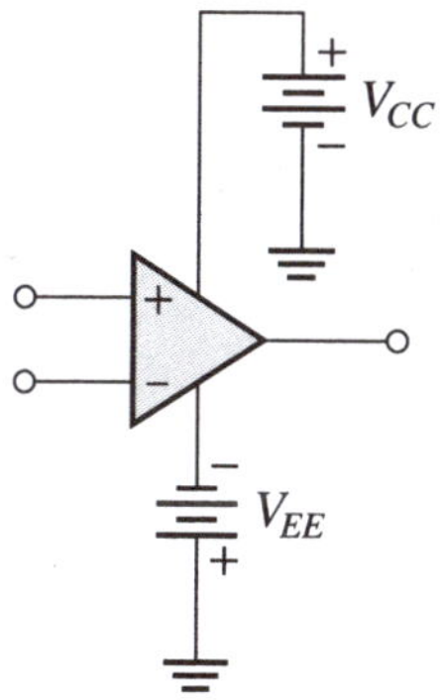

Figure 14.3 Op-amp symbol showing the dc power supplies, V_{CC} and V_{EE}.

Power-Supply Connections

For a real op amp to function properly, one or more dc supply voltages must be applied, as shown in Figure 14.3. Often, however, we do not explicitly show the power-supply connections in circuit diagrams. (As indicated in the figure, it is standard practice to use uppercase symbols with repeated uppercase subscripts, such as V_{CC} and V_{EE}, to represent dc power-supply voltages.)

14.2 INVERTING AMPLIFIERS

Operational amplifiers are almost always used with **negative feedback**, in which part of the output signal is returned to the input in opposition to the source signal. (It is also possible to have *positive* feedback, in which the signal returned to the input *aids* the original source signal. However, negative feedback turns out to be more useful in amplifier circuits.) Frequently, we analyze op-amp circuits by assuming an ideal op amp and employing a concept that we call the summing-point constraint.

Operational amplifiers are almost always used with negative feedback, in which part of the output signal is returned to the input in opposition to the source signal.

For an ideal op amp, the open-loop differential gain is assumed to approach infinity, and even a very tiny input voltage results in a very large output voltage. In a negative-feedback circuit, a fraction of the output is returned to the inverting input terminal. This forces the differential input voltage toward zero. If we assume infinite gain, the differential input voltage is driven to zero exactly. Since the differential input voltage of the op amp is zero, the input current is also zero. The fact that the differential input voltage and the input current are forced to zero is called the **summing-point constraint**.

In a negative feedback system, the ideal op-amp output voltage attains the value needed to force the differential input voltage and input current to zero. We call this fact the summing-point constraint.

Ideal op-amp circuits are analyzed by the following steps:

1. Verify that *negative* feedback is present.
2. Assume that the differential input voltage and the input current of the op amp are forced to zero. (This is the summing-point constraint.)
3. Apply standard circuit-analysis principles, such as Kirchhoff's laws and Ohm's law, to solve for the quantities of interest.

Next, we illustrate this type of analysis for some important circuits that are commonly used in engineering and scientific instrumentation.

The Basic Inverter

An op-amp circuit known as the **inverting amplifier** is shown in Figure 14.4. We will determine the voltage gain $A_v = v_o/v_{in}$ by assuming an ideal op amp and employing the summing-point constraint. However, before starting analysis of an op-amp circuit, we should always check to make sure that negative feedback is present rather than positive feedback.

In Figure 14.4, the feedback is negative, as we shall demonstrate. For example, suppose that due to the input source v_{in}, a positive voltage v_x appears at the inverting input. Then a negative output voltage of large (theoretically infinite) magnitude results at the output. Part of this output voltage is returned to the inverting input by the feedback path through R_2. Thus, the initially positive voltage at the inverting input is driven toward zero by the feedback action. A similar chain of events occurs for the appearance of a negative voltage at the inverting input terminal. Hence, the output voltage of the op amp takes precisely the value needed to oppose the source and produce (nearly) zero voltage at the op-amp input. Since we assume that the gain of the op amp is infinite, a negligible (theoretically zero) input voltage v_x is needed to produce the required output.

Figure 14.5 shows the inverting amplifier, including the conditions of the summing-point constraint at the input of the op amp. Notice that the input voltage v_{in} appears across R_1. Thus, the current through R_1 is

$$i_1 = \frac{v_{in}}{R_1} \tag{14.1}$$

Because the current flowing into the op-amp input terminals is zero, the current flowing through R_2 is

$$i_2 = i_1 \tag{14.2}$$

Thus, from Equations 14.1 and 14.2, we have

$$i_2 = \frac{v_{in}}{R_1} \tag{14.3}$$

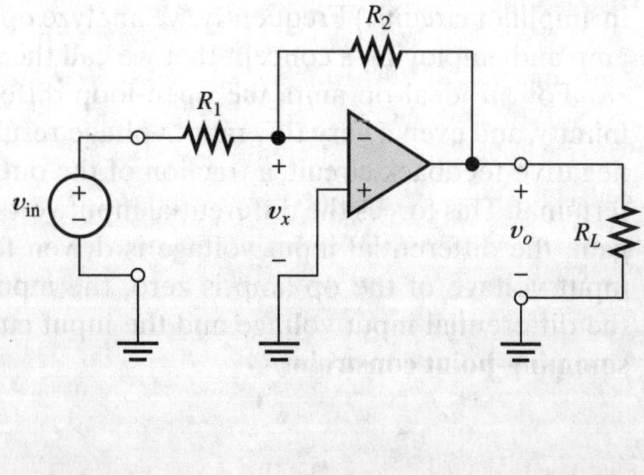

Figure 14.4 The inverting amplifier.

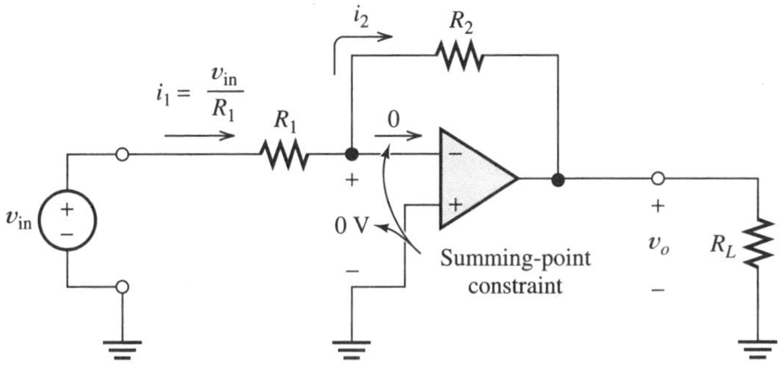

Figure 14.5 We make use of the summing-point constraint in the analysis of the inverting amplifier.

Writing a voltage equation around the loop by including the output terminals, the resistor R_2, and the op-amp input, we obtain

$$v_o + R_2 i_2 = 0 \tag{14.4}$$

Using Equation 14.3 to substitute for i_2 in Equation 14.4 and solving for the circuit voltage gain, we have

$$A_v = \frac{v_o}{v_{in}} = -\frac{R_2}{R_1} \tag{14.5}$$

We refer to A_v as the **closed-loop gain** because it is the gain of the circuit with the feedback network in place.

Under the ideal-op-amp assumption, the closed-loop voltage gain is determined solely by the ratio of the resistances. This is a very desirable situation because resistors are available with precise and stable values. Notice that the voltage gain is negative, indicating that the amplifier is inverting (i.e., the output voltage is out of phase with the input voltage).

Under the ideal op-amp assumption, the closed-loop voltage gain of the inverter is determined solely by the ratio of the resistances.

The input impedance of the inverting amplifier is

$$Z_{in} = \frac{v_{in}}{i_1} = R_1 \tag{14.6}$$

Thus, we can easily control the input impedance of the circuit by our choice of R_1.

Rearranging Equation 14.5, we have

$$v_o = -\frac{R_2}{R_1} v_{in} \tag{14.7}$$

Consequently, we see that the output voltage is independent of the load resistance R_L. We conclude that the output acts as an ideal voltage source (as far as R_L is concerned). *In other words, the output impedance of the inverting amplifier is zero.*

The inverter has a closed-loop voltage gain $A_v = -R_2/R_1$, an input impedance equal to R_1, and zero output impedance.

Later, we will see that the characteristics of the inverting amplifier are influenced by nonideal properties of the op amp. Nevertheless, in many applications, the departure of actual performance from the ideal is insignificant.

Virtual-Short-Circuit Concept

Sometimes, the condition at the op-amp input terminals of Figure 14.5 is called a **virtual short circuit**. That terminology is used because even though the differential input voltage of the op amp is forced to zero (as if by a short circuit to ground), the op-amp input current is also zero.

This terminology can be confusing unless it is realized that it is the action at the output of the op amp acting through the feedback network that enforces zero differential input voltage. (Possibly, it would be just as valid to call the condition at the op-amp input terminals a "virtual open circuit" because no current flows.)

Variations of the Inverter Circuit

Several useful versions of the inverter circuit exist. Analysis of these circuits follows the same pattern that we have used for the basic inverter: Verify that *negative* feedback is present, assume the summing-point constraint, and then apply basic circuit laws.

Example 14.1 Analysis of an Inverting Amplifier

Figure 14.6 shows a version of the inverting amplifier that can have high gain magnitude without resorting to as wide a range of resistor values as is needed in the standard inverter configuration. Derive an expression for the voltage gain under the ideal-op-amp assumption. Also, find the input impedance and output impedance. Evaluate the results for $R_1 = R_3 = 1\ \text{k}\Omega$ and $R_2 = R_4 = 10\ \text{k}\Omega$. Then, consider the standard inverter configuration of Figure 14.5 with $R_1 = 1\ \text{k}\Omega$, and find the value of R_2 required to achieve the same gain.

Generally, if a network of resistors is connected between the inverting input and the output, negative feedback exists.

Solution First, we verify that negative feedback is present. Assume a positive value for v_i, which results in a negative output voltage of very large magnitude. Part of this negative voltage is returned through the resistor network and opposes the original input voltage. Thus, we conclude that negative feedback is present.

Next, we assume the conditions of the summing-point constraint:

$$v_i = 0 \quad \text{and} \quad i_i = 0$$

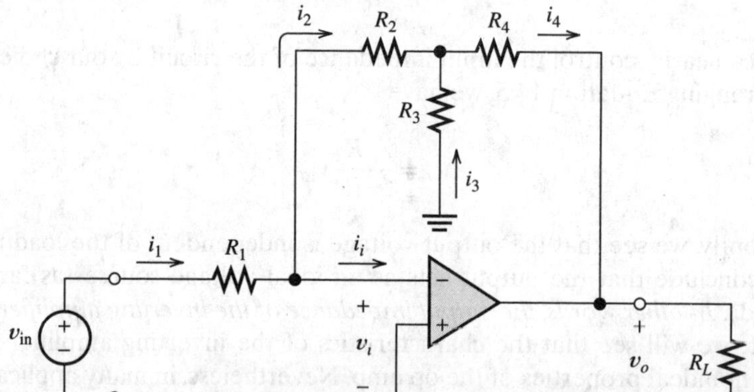

Figure 14.6 An inverting amplifier that achieves high gain magnitude with a smaller range of resistance values than required for the basic inverter. See Example 14.1.

Then, we apply Kirchhoff's current law, Kirchhoff's voltage law, and Ohm's law to analyze the circuit. To begin, we notice that v_{in} appears across R_1 (because $v_i = 0$). Hence, we can write

$$i_1 = \frac{v_{\text{in}}}{R_1} \tag{14.8}$$

Next, we apply Kirchhoff's current law to the node at the right-hand end of R_1, to obtain

$$i_2 = i_1 \tag{14.9}$$

(Here, we have used the fact that $i_i = 0$.)

Writing a voltage equation around the loop through v_i, R_2, and R_3, we obtain

$$R_2 i_2 = R_3 i_3 \tag{14.10}$$

(In writing this equation, we have used the fact that $v_i = 0$.) Applying Kirchhoff's current law at the top end of R_3 yields

$$i_4 = i_2 + i_3 \tag{14.11}$$

Writing a voltage equation for the loop containing v_o, R_4, and R_3 gives

$$v_o = -R_4 i_4 - R_3 i_3 \tag{14.12}$$

Next, we use substitution to eliminate the current variables (i_1, i_2, i_3, and i_4) and obtain an equation relating the output voltage to the input voltage. First, from Equations 14.8 and 14.9, we obtain

$$i_2 = \frac{v_{\text{in}}}{R_1} \tag{14.13}$$

Then, we use Equation 14.13 to substitute for i_2 in Equation 14.10 and rearrange terms to obtain

$$i_3 = v_{\text{in}} \frac{R_2}{R_1 R_3} \tag{14.14}$$

Using Equations 14.13 and 14.14 to substitute for i_2 and i_3 in Equation 14.11, we find that

$$i_4 = v_{\text{in}} \left(\frac{1}{R_1} + \frac{R_2}{R_1 R_3} \right) \tag{14.15}$$

Finally, using Equations 14.14 and 14.15 to substitute into 14.12, we obtain

$$v_o = -v_{\text{in}} \left(\frac{R_2}{R_1} + \frac{R_4}{R_1} + \frac{R_2 R_4}{R_1 R_3} \right) \tag{14.16}$$

Therefore, the voltage gain of the circuit is

$$A_v = \frac{v_o}{v_{\text{in}}} = -\left(\frac{R_2}{R_1} + \frac{R_4}{R_1} + \frac{R_2 R_4}{R_1 R_3} \right) \tag{14.17}$$

The input impedance is obtained from Equation 14.8:

$$R_{\text{in}} = \frac{v_{\text{in}}}{i_1} = R_1 \tag{14.18}$$

Inspection of Equation 14.16 shows that the output voltage is independent of the load resistance. Thus, the output appears as an ideal voltage source to the load. In other words, the output impedance of the amplifier is zero.

Evaluating the voltage gain for the resistor values given ($R_1 = R_3 = 1\ \text{k}\Omega$ and $R_2 = R_4 = 10\ \text{k}\Omega$) yields

$$A_v = -120$$

In the basic inverter circuit of Figure 14.5, the voltage gain is given by Equation 14.5, which states that

$$A_v = -\frac{R_2}{R_1}$$

Therefore, to achieve a voltage gain of -120 with $R_1 = 1\ \text{k}\Omega$, we need $R_2 = 120\ \text{k}\Omega$. Notice that a resistance ratio of 120 : 1 is required for the basic inverter, whereas the circuit of Figure 14.6 has a ratio of only 10 : 1. Sometimes, there are significant practical advantages in keeping the ratio of resistances in a circuit as close to unity as possible. Then, the circuit of Figure 14.6 is preferable to the basic inverter shown in Figure 14.5. ■

Now that we have demonstrated how to make use of the summing-point constraint in analysis of ideal-op-amp circuits having negative feedback, we provide some exercises for you to practice applying the technique. Each of these circuits has negative feedback, and if we assume ideal op amps, the summing-point constraint can be used in analysis.

Exercise 14.1 A circuit known as a summer is shown in Figure 14.7. **a.** Use the ideal-op-amp assumption to solve for the output voltage in terms of the input voltages and resistor values. **b.** What is the input resistance seen by v_A? **c.** By v_B? **d.** What is the output resistance seen by R_L?

Answer **a.** $v_o = -(R_f/R_A)v_A - (R_f/R_B)v_B$; **b.** the input resistance for v_A is equal to R_A; **c.** the input resistance for v_B is equal to R_B; **d.** the output resistance is zero. □

Exercise 14.2 Solve for the currents and voltages labeled in the circuits of Figure 14.8.

Answer **a.** $i_1 = 1$ mA, $i_2 = 1$ mA, $i_o = -10$ mA, $i_x = -11$ mA, $v_o = -10$ V; **b.** $i_1 = 5$ mA, $i_2 = 5$ mA, $i_3 = 5$ mA, $i_4 = 10$ mA, $v_o = -15$ V. □

Exercise 14.3 Find an expression for the output voltage of the circuit shown in Figure 14.9.

Answer $v_o = 4v_1 - 2v_2$. □

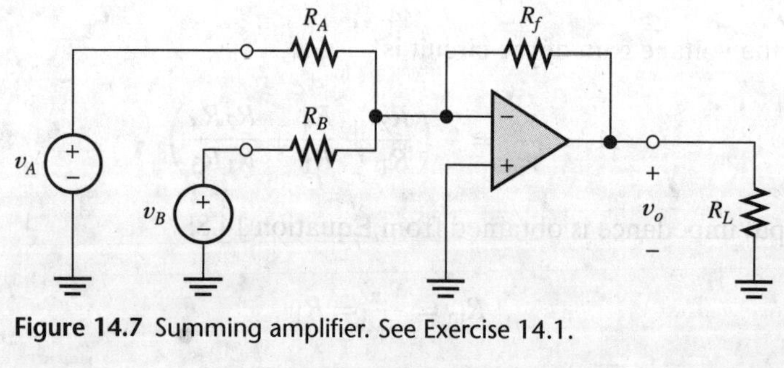

Figure 14.7 Summing amplifier. See Exercise 14.1.

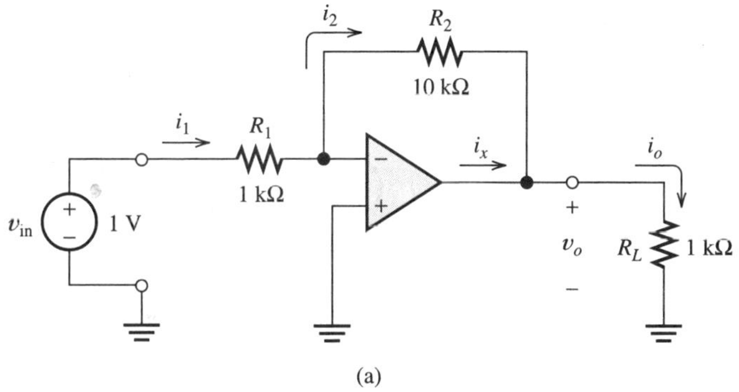

(a)

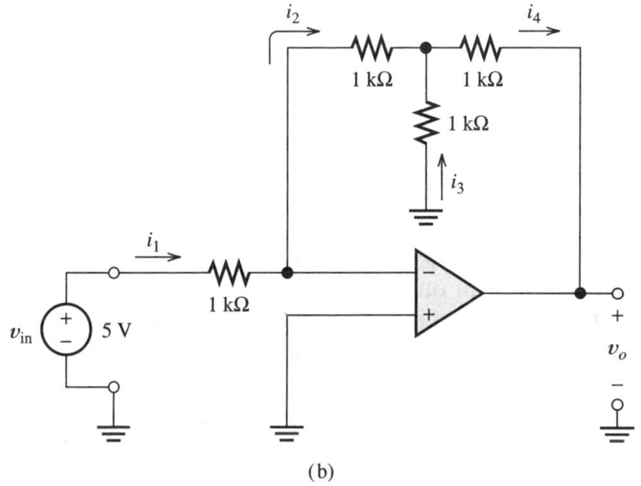

(b)

Figure 14.8 Circuits for Exercise 14.2.

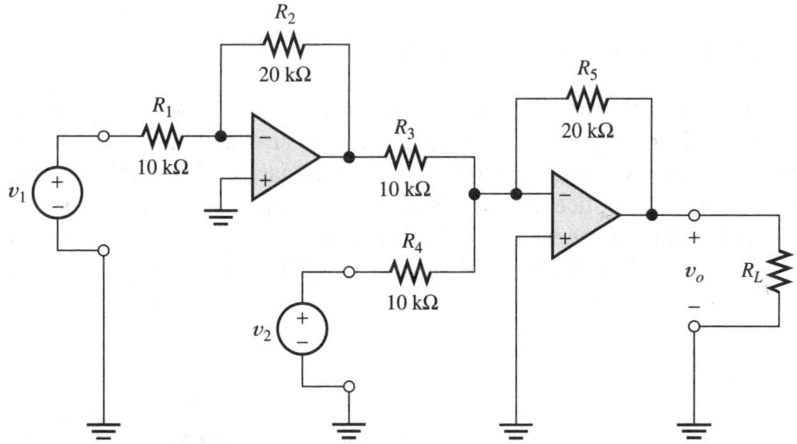

Figure 14.9 Circuit for Exercise 14.3.

Positive Feedback

It is interesting to consider the inverting amplifier configuration with the input terminals of the op amp interchanged as shown in Figure 14.10. In this case, the feedback is positive—in other words, the feedback signal *aids* the original input signal. For

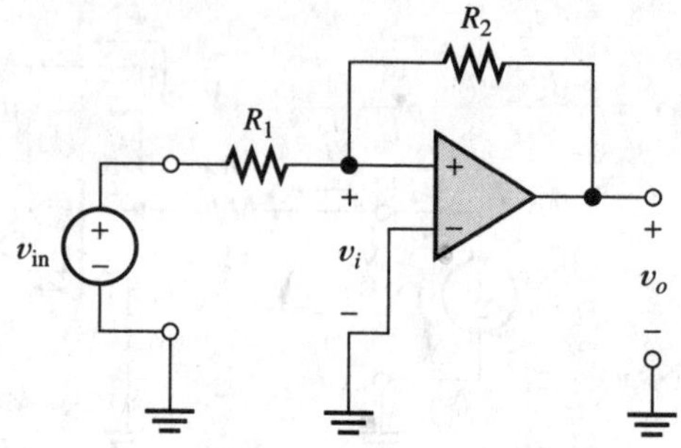

Figure 14.10 Circuit with positive feedback.

example, if the input voltage v_i is positive, a very large positive output voltage results. Part of the output voltage is returned to the op-amp input by the feedback network. Thus, the input voltage becomes larger, causing an even larger output voltage. The output quickly becomes saturated at the maximum possible voltage that the op amp can produce.

With positive feedback, the op-amp's input and output voltages increase in magnitude until the output voltage reaches one of its extremes.

If an initial negative input voltage is present, the output saturates at its negative extreme. Hence, the circuit does not function as an amplifier—the output voltage is stuck at one extreme or the other and does not respond to the input voltage v_{in}. (However, if the input voltage v_{in} becomes sufficiently large in magnitude, the output can be forced to switch from one extreme to the other. We saw in Chapter 7 that positive-feedback circuits [i.e., flip-flops] are useful for memory in digital systems.)

If we were to ignore the fact that the circuit of Figure 14.10 has positive rather than negative feedback and to apply the summing-point constraint erroneously, we would obtain $v_o = -(R_2/R_1)v_{in}$, just as we did for the circuit with negative feedback. This illustrates the importance of verifying that negative feedback is present before using the summing-point constraint.

14.3 NONINVERTING AMPLIFIERS

The circuit configuration for a noninverting amplifier is shown in Figure 14.11. We assume an ideal op amp to analyze the circuit. First, we check to see whether the feedback is negative or positive. In this case, it is negative. To see this, assume that v_i becomes positive and notice that it produces a very large positive output voltage. Part of the output voltage appears across R_1. Since $v_i = v_{in} - v_1$, the voltage v_i becomes smaller as v_o and v_1 become larger. Thus, the amplifier and feedback network act to

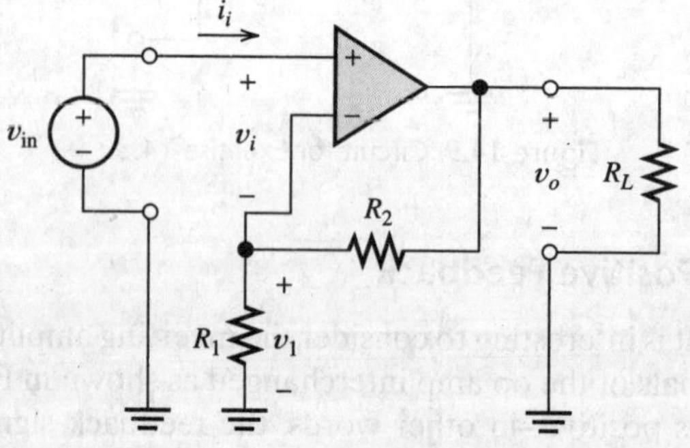

Figure 14.11 Noninverting amplifier.

drive v_i toward zero. This is negative feedback because the feedback signal opposes the original input.

Having verified that negative feedback is present, we utilize the summing-point constraint: $v_i = 0$ and $i_i = 0$. Applying Kirchhoff's voltage law and the fact that $v_i = 0$, we can write

$$v_{in} = v_1 \tag{14.19}$$

Since i_i is zero, the voltage across R_1 is given by the voltage-division principle:

$$v_1 = \frac{R_1}{R_1 + R_2} v_o \tag{14.20}$$

Using Equation 14.20 to substitute into 14.19 and rearranging, we find that the closed-loop voltage gain is:

$$A_v = \frac{v_o}{v_{in}} = 1 + \frac{R_2}{R_1} \tag{14.21}$$

Notice that the circuit is a noninverting amplifier (A_v is positive), and the gain is set by the ratio of the feedback resistors.

The input impedance of the circuit is theoretically infinite because the input current i_i is zero. Since the voltage gain is independent of the load resistance, the output voltage is independent of the load resistance. Thus, the output impedance is zero. *Therefore, under the ideal-op-amp assumption, the noninverting amplifier is an ideal voltage amplifier.* (Ideal amplifiers are discussed in Section 11.6.)

Under the ideal-op-amp assumption, the noninverting amplifier is an ideal voltage amplifier having infinite input resistance and zero output resistance.

Voltage Follower

Notice from Equation 14.21 that the minimum gain magnitude is unity, which is obtained with $R_2 = 0$. Usually, we choose R_1 to be an open circuit for unity gain. The resulting circuit, called a **voltage follower**, is shown in Figure 14.12.

Exercise 14.4 Find the voltage gain $A_v = v_o/v_{in}$ and input impedance of the circuit shown in Figure 14.13: **a.** with the switch open; **b.** with the switch closed.

Answer **a.** $A_v = +1, R_{in} = \infty$; **b.** $A_v = -1, R_{in} = R/2$. □

Exercise 14.5 Assume an ideal op amp and use the summing-point constraint to find an expression for the output current i_o in the circuit of Figure 14.14. Also find the input and output resistances of the circuit.

Answer $i_o = v_{in}/R_F, R_{in} = \infty, R_o = \infty$ (because the output current is independent of the load resistance). □

Exercise 14.6 **a.** Derive an expression for the voltage gain v_o/v_{in} of the circuit shown in Figure 14.15. **b.** Evaluate for $R_1 = 10$ kΩ and $R_2 = 100$ kΩ. **c.** Find the input resistance of this circuit. **d.** Find the output resistance.

Answer **a.** $A_v = 1 + 3(R_2/R_1) + (R_2/R_1)^2$; **b.** $A_v = 131$; **c.** $R_{in} = \infty$; **d.** $R_o = 0$. □

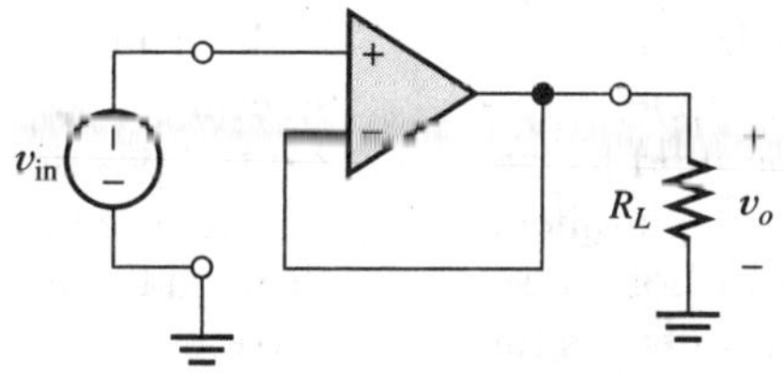

Figure 14.12 The voltage follower which has $A_v = 1$.

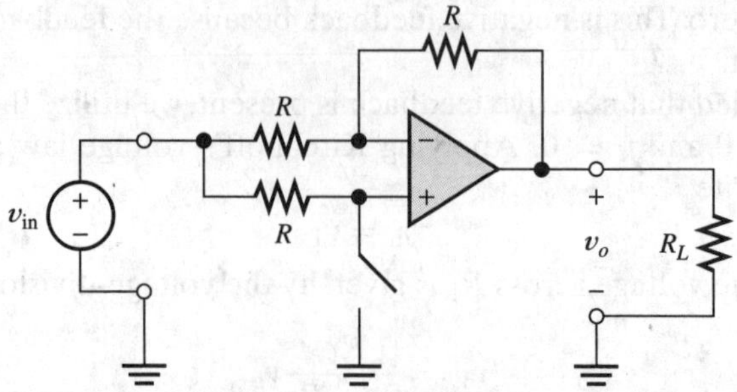

Figure 14.13 Inverting or noninverting amplifier. See Exercise 14.4.

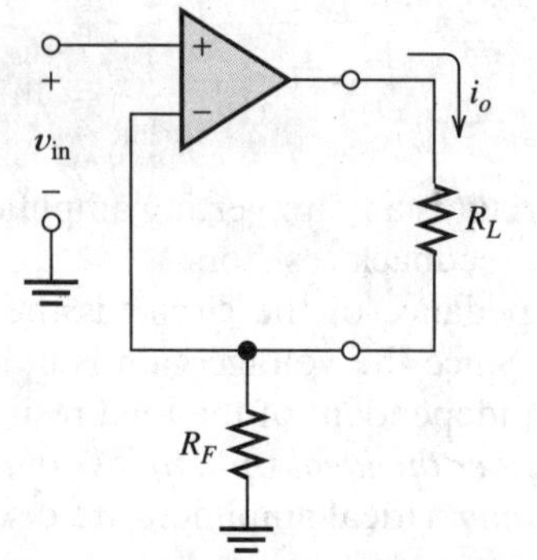

Figure 14.14 Voltage-to-current converter (also known as a transconductance amplifier). See Exercise 14.5.

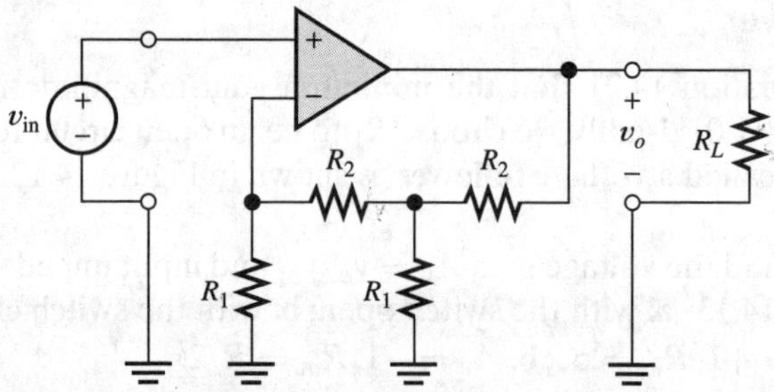

Figure 14.15 Circuit for Exercise 14.6.

PRACTICAL APPLICATION 14.1

Mechanical Application of Negative Feedback: Power Steering

Besides its use in op-amp circuits, negative feedback has many other applications in engineering—automotive power-assisted steering systems provide one example. A simplified diagram of a typical system is shown in Figure PA14.1. A hydraulic pump driven by the engine continuously supplies pressure to a control valve that routes the fluid to two sides of the booster cylinder. For straight-ahead steering, pressure is applied equally to both sides of the cylinder and no turning force results. As the steering wheel is moved by the driver, more pressure is applied to one side of the cylinder or the other to help turn the wheels in the desired direction. (The fluid path for one direction is illustrated in the figure.) A mechanical feedback arm from the steering linkage causes the valve to return to its neutral position as the wheels turn. Thus, there is a negative feedback path from the booster cylinder through the mechanical linkage back to the control valve.

Negative feedback is an important aspect of the power-steering system. Consider what would happen if the mechanical linkage between the output of the booster cylinder and the control valve is removed. Then, if the steering wheel were moved slightly off center, pressure would be applied continuously to the booster cylinder and the wheels would eventually move all the way to their extreme position. It would be very difficult for the driver to make a gradual turn.

On the other hand, with the feedback linkage in place, the wheels move only far enough to return the valve (nearly) to its neutral position. As the steering wheel is turned, the wheels move a proportional amount.

Notice that the control valve responds to the difference between the input from the steering wheel and the position of the steering linkage. This is similar to the way that the op amp responds to its differential input signal. The pump is analogous to the power supply in an op-amp circuit. Furthermore, the booster cylinder position is analogous to the op-amp output signal, and the mechanical linkage back to the control valve is analogous to a feedback circuit.

Steer-by-wire systems are a more modern alternative for mechanical/hydraulic steering systems and are under intense development. Here, the mechanical/hydraulic steering components are replaced by electrical sensors, microcontrollers, software, and electrical motors. Just as in the mechanical system, negative feedback is used in steer-by-wire systems. Actual and desired wheel positions are derived from sensors connected to the wheels and to the steering wheel. The software compares the actual wheel position to the desired position and uses the difference to control motors that turn the wheels. There is no steering column or actual mechanical path from the steering wheel to the vehicle wheels. These new electronic steering systems have the potential to significantly reduce weight and increase fuel economy of a vehicle. However, they also have some new safety issues. After all, loss of steering due to a software glitch could be very serious.

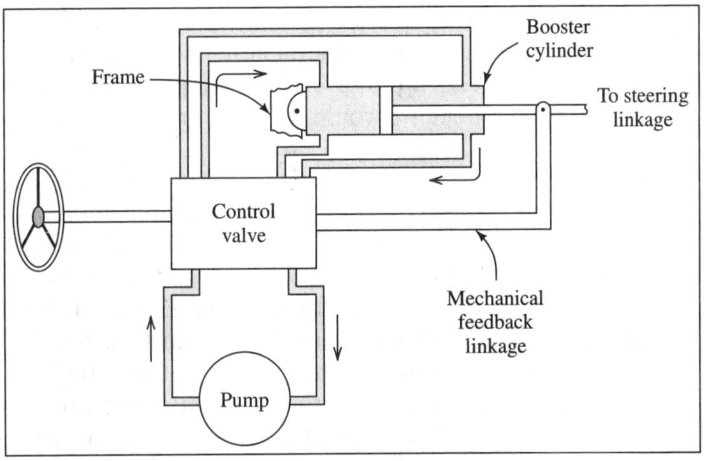

Figure PA14.1 A simplified diagram of an automotive power-assisted steering system illustrating the importance of negative feedback.

14.4 DESIGN OF SIMPLE AMPLIFIERS

Many useful amplifiers can be designed by using resistive feedback networks with op amps.

Many useful amplifiers can be designed by using resistive feedback networks with op amps. For now, we consider the op amps to be ideal. Later, we consider the effects of the nonideal properties of real op amps. Often in practice, the performance requirements of the circuits to be designed are not extreme, and design can be carried out assuming ideal op amps.

Amplifier design using op amps mainly consists of selecting a suitable circuit configuration and values for the feedback resistors.

We illustrate design by using the op-amp circuits that we have considered in previous sections (including the exercises). For these circuits, design consists primarily of selecting a suitable circuit configuration and values for the feedback resistors.

Example 14.2 Design of a Noninverting Amplifier

Design a noninverting amplifier that has a voltage gain of 10, using an ideal op amp. The input signals lie in the range −1 to +1 V. Use standard 5-percent-tolerance resistors in the design. (See Appendix B for a list of standard 5-percent-tolerance resistor values.)

Solution We use the noninverting amplifier configuration of Figure 14.11. The gain is given by Equation 14.21. Thus, we have

$$A_v = 10 = 1 + \frac{R_2}{R_1}$$

Theoretically, any resistor values would provide the proper gain, provided that $R_2 = 9R_1$. However, very small resistances are not practical because the current through the resistors must be supplied by the output of the op amp, and ultimately, by the power supply. For example, if $R_1 = 1\ \Omega$ and $R_2 = 9\ \Omega$, for an output voltage of 10 V, the op amp must supply 1 A of current. This is illustrated in Figure 14.16. Most integrated-circuit op amps are not capable of such a large output current, and even if they were, the load on the power supply would be unwarranted. In the circuit at hand, we would want to keep $R_1 + R_2$ large enough so the current that must be supplied to them is reasonable. For general design with power supplies operated from the ac power line, currents up to several milliamperes are usually acceptable. (In battery-operated equipment, we would try harder to reduce the current and avoid having to replace batteries frequently.)

If the resistances are too small, an impractical amount of current and power will be needed to operate the amplifier.

On the other hand, very large resistances, such as $R_1 = 10\ \text{M}\Omega$ and $R_2 = 90\ \text{M}\Omega$, also present problems. Such large resistances are unstable in value, particularly in a humid environment. Later, we will see that large resistances lead to problems due to an op-amp imperfection known as bias current. Furthermore, high-impedance circuits are prone to injection of unwanted signals from nearby circuits through stray capacitive coupling. This is illustrated in Figure 14.17.

Very large resistance may be unstable in value and lead to stray coupling of undesired signals.

Generally, resistance values between about 100 Ω and 1 MΩ are suitable for use in op-amp circuits. Since the problem statement calls for standard 5-percent-tolerance resistors (see Appendix B), we look for a pair of resistor values such that

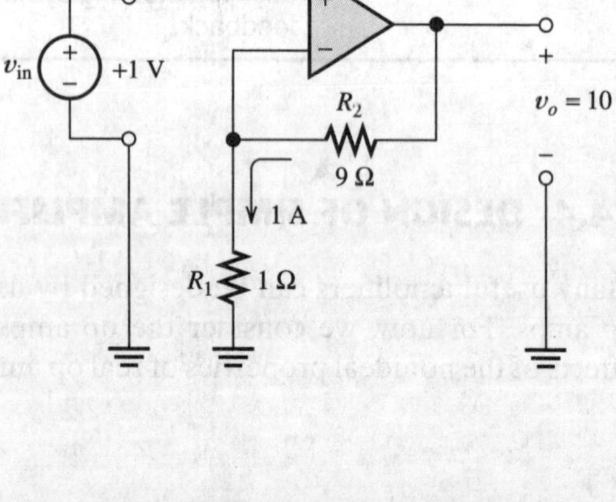

Figure 14.16 If low resistances are used, an excessively large current is required.

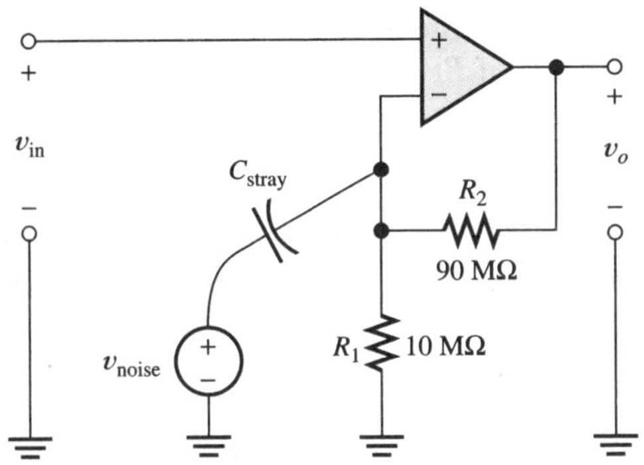

Figure 14.17 If very high resistances are used, stray capacitance can couple unwanted signals into the circuit.

the ratio R_2/R_1 is 9. One possibility is $R_2 = 180\ \text{k}\Omega$ and $R_1 = 20\ \text{k}\Omega$. However, for many applications, we would find that $R_2 = 18\ \text{k}\Omega$ and $R_1 = 2\ \text{k}\Omega$ would work just as well. Of course, if 5-percent-tolerance resistors are used, we can expect unit-to-unit variations of about ±10 percent in the ratio R_2/R_1. This is because R_2 could be 5-percent low while R_1 is 5 percent high, or vice versa. Thus, the gain of the amplifier (which is $A_v = 1 + R_2/R_1$) varies by about ±9 percent.

If more precision is needed, 1-percent-tolerance resistors can be used. Another possibility is an adjustable resistor to set the gain to the desired value. ■

Example 14.3 Amplifier Design

A transducer for instrumentation of vibrations in a forge hammer has an internal impedance that is always less than 500 Ω but is variable over time. An amplifier that produces an amplified version of the internal source voltage v_s is required. The voltage gain should be -10 ± 5 percent. Design an amplifier for this application.

Solution Since an inverting gain is specified, we choose to use the inverting amplifier of Figure 14.4. The proposed amplifier and the signal source are shown in Figure 14.18.

Using the summing-point constraint and conventional circuit analysis, we can show that

$$v_o = -\frac{R_2}{R_1 + R_s} v_s$$

Hence, we must select resistance values so that

$$\frac{R_2}{R_1 + R_s} = 10 \pm 5\%$$

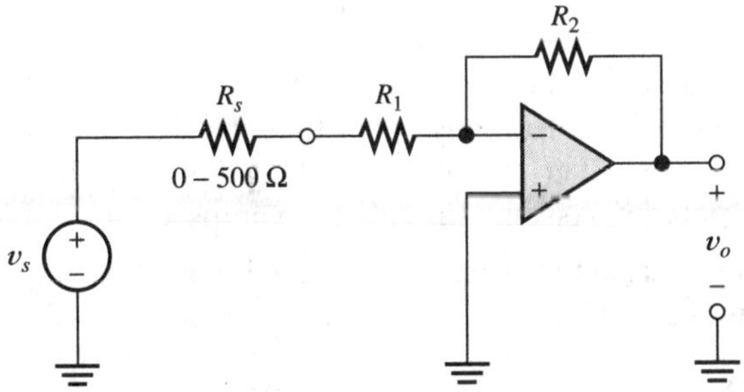

Figure 14.18 Circuit of Example 14.3.

Because the value of R_s is variable, we must choose R_1 much greater than the maximum value of R_s. Thus, we are led to choose $R_1 \cong 100R_{s\,\max} = 50$ kΩ. (Then, as R_s ranges from zero to 500 Ω, the sum $R_1 + R_s$ varies by only 1 percent.) To achieve the desired gain, we require that $R_2 \cong 500$ kΩ.

Since a gain tolerance of ±5 percent is specified, we resort to the use of 1 percent resistors. This is necessary because gain variations occur due to variations in R_s, variations in R_1, and variations in R_2. If each of these causes a ±1 percent gain variation, the gain varies by about ±3 percent, which is within the allowed range.

Consulting a table of standard values for 1 percent resistors (see Appendix B), we choose $R_1 = 49.9$ kΩ and $R_2 = 499$ kΩ. As well as ensuring that the gain does not vary outside the specified limits, these values are not so small that large currents occur or so large that undue coupling of unwanted signals into the circuit is likely to be a problem.

Another solution would be to use 5-percent-tolerance resistors, but choose $R_1 =$ 51 kΩ and R_2 as the series combination of a 430-kΩ fixed resistor and a 200-kΩ adjustable resistor. Then the gain could be set initially to the desired value. Some gain fluctuation would occur in operation due to variation of R_s and drift of the other resistance values due to aging, temperature changes, and so on. ■

Exercise 14.7 Find the maximum and minimum values of the gain $A_{vs} = v_o/v_s$ for the circuit designed in Example 14.3. The nominal resistor values are $R_1 = 49.9$ kΩ and $R_2 = 499$ kΩ. Assume that the resistors R_1 and R_2 range as far as ±1 percent from their nominal values and that R_s ranges from 0 to 500 Ω.

Answer The extreme gain values are −9.71 and −10.20. □

Close-Tolerance Designs

When designing amplifiers with tight gain tolerances (1 percent or better), it is necessary to employ adjustable resistors. We might be tempted to use lower cost 5-percent-tolerance resistors rather than 1-percent-tolerance resistors and use the adjustable resistor to offset the larger variations. However, this is not good practice because 5-percent-tolerance resistors tend to be less stable than 1-percent-tolerance resistors. Furthermore, fixed resistors tend to be more stable than adjustable resistors. The best approach from the standpoint of long-term precision is to use 1-percent-tolerance fixed resistors and design for only enough adjustment to overcome the resulting gain variations.

Often, we combine various types of op-amp circuits in the design of a desired function. These points are illustrated in the next example.

Example 14.4 Summing Amplifier Design

Two signal sources have internal voltages $v_1(t)$ and $v_2(t)$, respectively. The internal resistances of the sources are known always to be less than 1 kΩ, but the exact values are not known and are likely to change over time. Design an amplifier for which the output voltage is $v_o(t) = A_1v_1(t) + A_2v_2(t)$. The gains are to be $A_1 = 5 \pm 1$ percent and $A_2 = -2 \pm 1$ percent. Assume that ideal op amps are available.

Solution The summer circuit of Figure 14.7 can be used to form the weighted sum of the input voltages given by

$$v_o = -\frac{R_f}{R_A}v_A - \frac{R_f}{R_B}v_B$$

in which the gains for both input signals are negative. However, the problem statement calls for a positive gain for v_1 and a negative gain for v_2. Thus, we first pass v_1 through an inverting amplifier. The output of this inverter and v_2 are then applied to the summer. The proposed circuit diagram is shown in Figure 14.19.

It can be shown that the output voltage of this circuit is given by

$$v_o = \frac{R_2}{R_{s1} + R_1}\frac{R_f}{R_A}v_1 - \frac{R_f}{R_{s2} + R_B}v_2 \tag{14.22}$$

We must select values for the resistances so that the gain for the v_1 input is +5 and the gain for the v_2 input is −2. Many combinations of resistances can be used to meet these specifications. However, we should keep the input impedances seen by the sources much larger than the internal source impedances, to avoid gain variations due to loading. This implies that we should choose large values for R_1 and R_B. (However, keep in mind that extremely large values are not practical.) Since we want the gain values to remain within ±1 percent of the design values, we choose $R_1 = R_B \cong 500$ kΩ. Then, as the source impedances change, the gains change by only about 0.2 percent (because the input impedances are approximately 500 times larger than the highest value of the source impedances).

Even if we choose to use 1-percent-tolerance resistors, we must use adjustable resistors to trim the gain. For example, with 1 percent resistors, the gain

$$A_1 = \frac{R_2}{R_{s1} + R_1}\frac{R_f}{R_A}$$

varies by about ±4 percent, due to the resistance tolerances. Thus, to provide for adjustment of A_1, we use the combination of a fixed resistance in series with a variable resistance for R_1. Similarly, to provide for adjustment of A_2, we use a second variable resistance in series with a fixed resistance for R_B.

Suppose that we select R_1 as a 453-kΩ (this is a standard nominal value for 1-percent-tolerance resistors) fixed resistor in series with a 100-kΩ trimmer (i.e., an adjustable resistor having a maximum value of 100 kΩ). We use the same combination for R_B. (Recall that we plan to design for nominal values of R_1 and R_B of 500 kΩ each.) The trimmers allow for approximately a ±10 percent adjustment, which is more than adequate to allow for variations of the fixed resistors.

The gain for the v_2 input is

$$A_2 = -\frac{R_f}{R_{s2} + R_B}$$

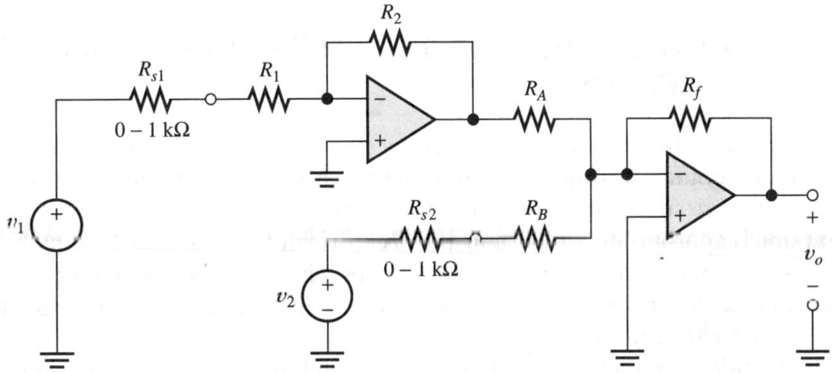

Figure 14.19 Amplifier designed in Example 14.4.

Because $R_{s2} + R_B$ has a nominal value 500 kΩ and we want to have $A_2 = -2$, R_f is selected to be a 1-MΩ 1-percent-tolerance resistor. Now, since we want to achieve

$$A_1 = \frac{R_2}{R_{s1} + R_1} \frac{R_f}{R_A} = 5$$

and the values we have already selected result in $R_f/(R_{s1} + R_1) = 2$, we must choose values of R_2 and R_A such that $R_2/R_A \cong 2.5$. Thus, we choose R_2 as a 1-MΩ resistor and R_A as a 402-kΩ resistor. This completes the design. The following values are selected:

R_1 = a 453-kΩ fixed resistor in series with a 100-kΩ trimmer (500-kΩ nominal design value).

R_B is the same as R_1.

$R_2 = 1$ MΩ.

$R_A = 402$ kΩ.

$R_f = 1$ MΩ.

Usually, design problems have many "right" answers.

These are by no means the only values that can be used to meet the specifications. Usually, design problems have many "right" answers. ■

Exercise 14.8 Derive Equation 14.22. □

Exercise 14.9 A certain source has an internal impedance of 600 Ω ± 20 percent. Design an amplifier whose output voltage is $v_o = A_{vs} v_s$, where v_s is the internal voltage of the source. Assume ideal op amps and design for $A_{vs} = 20 \pm 5$ percent.

Answer Many answers are possible. A good solution is the circuit of Figure 14.11 with $R_2 \cong 19 \times R_1$. For example, we could use 1-percent-tolerance resistors having nominal values of $R_1 = 1$ kΩ and $R_2 = 19.1$ kΩ. □

Exercise 14.10 Repeat Exercise 14.9 for $A_{vs} = -25 \pm 3$ percent.

Answer Many answers are possible. A good solution is the circuit of Figure 14.18 with $R_1 \geq 20R_s$ and with $R_2 \cong 25(R_1 + R_s)$. For example, we could use 1-percent-tolerance resistors having nominal values of $R_1 = 20$ kΩ and $R_2 = 515$ kΩ. □

Exercise 14.11 Repeat Example 14.4 if $A_1 = +1 \pm 1$ percent and $A_2 = -3 \pm 1$ percent.

Answer Many answers can be found by following the approach taken in Example 14.4. □

14.5 OP-AMP IMPERFECTIONS IN THE LINEAR RANGE OF OPERATION

In Sections 14.1 through 14.4, we introduced the op amp, learned how to use the summing-point constraint to analyze negative-feedback amplifier circuits, and learned how to design simple amplifiers. So far, we have assumed ideal op amps. This assumption is appropriate for learning the basic principles of op-amp circuits, but not for high-performance circuits using real op amps. Therefore, in this and the next few sections, we consider the imperfections of real op amps and how to allow for these imperfections in circuit design.

Real op amps have several categories of imperfections compared with ideal op amps.

The nonideal characteristics of real op amps fall into three categories: (1) nonideal properties in linear operation, (2) nonlinear characteristics, and (3) dc offsets.

We discuss the imperfections for the linear range of operation in this section. In the next several sections, we consider nonlinear operation and dc offsets.

Input and Output Impedances

Real op amps have finite input impedance and nonzero output impedance.

An ideal op amp has infinite input impedance and zero output impedance. However, a real op amp has finite input impedance and nonzero output impedance. The input impedances of IC op amps having BJT input stages are usually about 1 MΩ. Op amps having FET input stages have much higher input impedances, as much as 10^{12} Ω. Output impedance is ordinarily between 1 and 100 Ω for an IC op amp, although it can be as high as several thousand ohms for a low-power op amp.

In circuits with negative feedback, the impedances are drastically altered by the feedback action, and the input or output impedances of the op amps rarely place serious limits on closed-loop circuit performance.

Gain and Bandwidth Limitations

Real op amps have finite open-loop gain and finite bandwidth.

Ideal op amps have infinite gain magnitude and unlimited bandwidth. Real op amps have finite open-loop gain magnitude, typically between 10^4 and 10^6. (We are referring to the open-circuit voltage gain of the op amp without feedback resistors.) Furthermore, the bandwidth of real op amps is limited. The gain of a real op amp is a function of frequency, becoming smaller in magnitude at higher frequencies.

Usually, the bandwidth of an IC op amp is intentionally limited by the op-amp designer. This is called *frequency compensation* and is necessary to avoid oscillation in feedback amplifiers. An in-depth discussion of frequency compensation is beyond the scope of this book. However, it turns out that the open-loop gain of most integrated-circuit op amps is of the form

$$A_{OL}(f) = \frac{A_{0OL}}{1 + j(f/f_{BOL})} \tag{14.23}$$

in which A_{0OL} is the dc open-loop gain of the amplifier, and f_{BOL} is the open-loop break frequency.

In a Bode plot for $A_{OL}(f)$, the gain magnitude is approximately constant up to f_{BOL}. Above f_{BOL}, the gain magnitude falls at 20 dB per decade. This is illustrated in Figure 14.20. (Bode plots are discussed in Section 6.4.)

Closed-Loop Bandwidth

We will show that negative feedback reduces the dc gain of an op amp and extends its bandwidth. Consider the noninverting amplifier circuit shown in Figure 14.21. The phasor output voltage $\mathbf{V}_o$ is the open-loop gain times the phasor differential input voltage $\mathbf{V}_{id}$:

$$\mathbf{V}_o = A_{OL}(f)\mathbf{V}_{id} \tag{14.24}$$

We assume that the input impedance of the op amp is infinite, so the input current is zero. Then, the voltage across R_1 can be found by applying the voltage-division principle to the feedback network (which is composed of R_1 in series with R_2). The voltage across R_1 is shown as $\beta\mathbf{V}_o$, in which β is the voltage-division ratio of R_1 and R_2:

$$\beta = \frac{R_1}{R_1 + R_2} \tag{14.25}$$

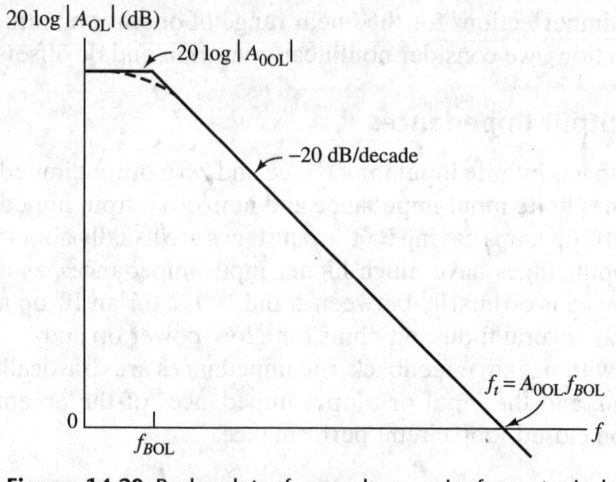

Figure 14.20 Bode plot of open-loop gain for a typical op amp.

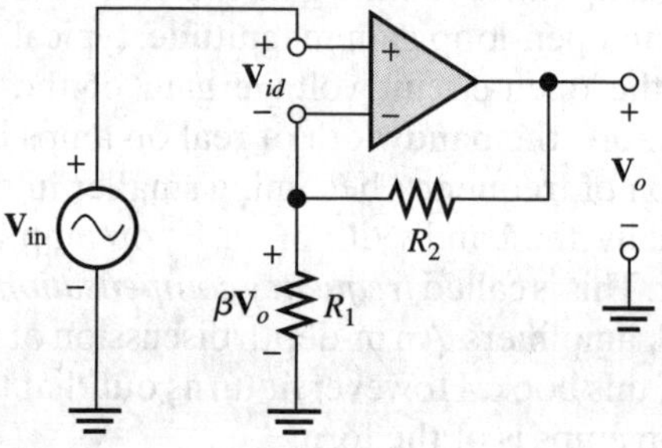

Figure 14.21 Noninverting amplifier circuit used for analysis of closed-loop bandwidth.

Applying Kirchhoff's voltage law to Figure 14.21, we have

$$\mathbf{V}_{\text{in}} = \mathbf{V}_{id} + \beta \mathbf{V}_o$$

Solving for $\mathbf{V}_{id}$ and substituting into Equation 14.24, we have

$$\mathbf{V}_o = A_{\text{OL}}(\mathbf{V}_{\text{in}} - \beta \mathbf{V}_o) \tag{14.26}$$

Now, we can solve for the gain of the circuit including the feedback resistors, which is called the **closed-loop gain**. This is given by

$$A_{\text{CL}} = \frac{\mathbf{V}_o}{\mathbf{V}_{\text{in}}} = \frac{A_{\text{OL}}}{1 + \beta A_{\text{OL}}} \tag{14.27}$$

Using Equation 14.23 to substitute for the open-loop gain, we get

$$A_{\text{CL}}(f) = \frac{A_{0\text{OL}}/[1 + j(f/f_{\text{BOL}})]}{1 + \{\beta A_{0\text{OL}}/[1 + j(f/f_{\text{BOL}})]\}} \tag{14.28}$$

This can be put into the form

$$A_{\text{CL}}(f) = \frac{A_{0\text{OL}}/(1 + \beta A_{0\text{OL}})}{1 + \{jf/[f_{\text{BOL}}(1 + \beta A_{0\text{OL}})]\}} \tag{14.29}$$

Now, we define the closed-loop dc gain as

$$A_{0CL} = \frac{A_{0OL}}{1+\beta A_{0OL}} \tag{14.30}$$

and the closed-loop bandwidth as

$$f_{BCL} = f_{BOL}(1+\beta A_{0OL}) \tag{14.31}$$

Using these definitions in Equation 14.29, we obtain

$$A_{CL}(f) = \frac{A_{0CL}}{1+j(f/f_{BCL})} \tag{14.32}$$

Comparing this with Equation 14.23, we see that the closed-loop gain takes exactly the same form as the open-loop gain. The dc open-loop gain A_{0OL} is very large, and we usually have $(1+\beta A_{0OL}) >> 1$. Thus, from Equation 14.30, we see that the closed-loop gain is much smaller than the open-loop gain. Furthermore, Equation 14.31 shows that the closed-loop bandwidth is much greater than the open-loop bandwidth. *In sum, we see that negative feedback reduces the gain magnitude and increases bandwidth.*

Negative feedback reduces gain magnitude and increases bandwidth.

Gain–Bandwidth Product

Now, consider the product of the closed-loop gain and closed-loop bandwidth. From Equations 14.30 and 14.31, we have

$$A_{0CL}f_{BCL} = \frac{A_{0OL}}{1+\beta A_{0OL}} \times f_{BOL}(1+\beta A_{0OL}) = A_{0OL}f_{BOL} \tag{14.33}$$

Hence, we see that the product of dc gain and bandwidth is independent of the feedback ratio β. We denote the **gain–bandwidth** as f_t. Thus, we have

$$f_t = A_{0CL}f_{BCL} = A_{0OL}f_{BOL} \tag{14.34}$$

As indicated in Figure 14.20, it turns out that f_t is the frequency at which the Bode plot of the open-loop gain crosses 0 dB. Recall that 0 dB corresponds to unity-gain magnitude. Consequently, f_t is also called the **unity-gain bandwidth**. General-purpose IC op amps have gain–bandwidth products of several megahertz.

The gain–bandwidth product is constant for the noninverting amplifier. As we reduce the gain (by choosing a lower value for $1R_2/R_1$), the bandwidth becomes greater.

Example 14.5 Open-Loop and Closed-Loop Bode Plots

A certain op amp has a dc open-loop gain of $A_{0OL} = 10^5$ and $f_{BOL} = 40$ Hz. Find the closed-loop bandwidth if this op amp is used with feedback resistors to form a noninverting amplifier having a closed-loop dc gain of 10. Then, construct a Bode plot of the open-loop gain and a Bode plot of the closed-loop gain.

Solution The gain–bandwidth product is

$$f_t = A_{0OL}f_{BOL} = 10^5 \times 40 \text{ Hz} = 4 \text{ MHz} = A_{0CL}f_{BCL}$$

Thus, if feedback is used to reduce the gain to $A_{0CL} = 10$, the bandwidth is $f_{BCL} = 400$ kHz.

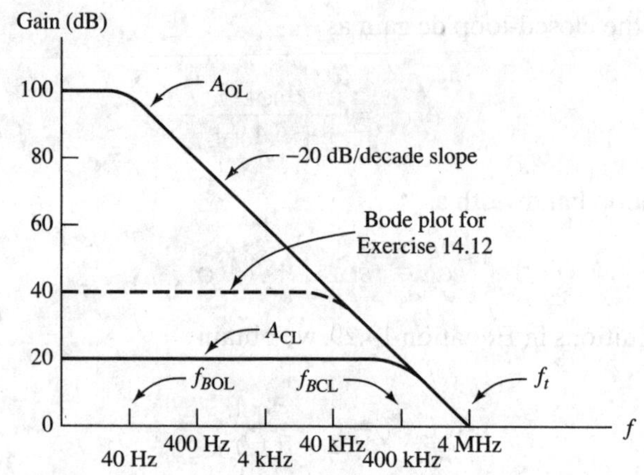

Figure 14.22 Bode plots for Example 14.5 and Exercise 14.12.

In decibels, the dc open-loop gain becomes

$$A_{0OL} = 20\log(10^5) = 100 \text{ dB}$$

The break frequency is $f_{BOL} = 40$ Hz. Recall from our discussion of Bode plots in Section 6.4 that this gain function is approximated as being constant below the break frequency and falls at 20 dB/decade above the break frequency. The Bode plot is shown in Figure 14.22.

Converting $A_{0CL} = 10$ to decibels, we have

$$A_{0CL} = 10\log(10) = 20 \text{ dB}$$

and the break frequency is $f_{BCL} = 400$ kHz. The resulting Bode plot is shown in Figure 14.22. Notice that the closed-loop gain plot is constant until it reaches the open-loop plot and then the closed-loop plot rolls off. ■

Exercise 14.12 Repeat Example 14.5 for $A_{0CL} = 100$.
Answer $f_{BCL} = 40$ kHz. The Bode plot is shown in Figure 14.22. □

14.6 NONLINEAR LIMITATIONS

Output Voltage Swing

The output voltage of a real op amp is limited to the range between certain limits that depend on the internal design of the op amp. When the output voltage tries to go beyond these limits, clipping occurs.

There are several nonlinear limitations of the outputs of real op amps. First, the output voltage range is limited. If an input signal is sufficiently large that the output voltage would be driven beyond these limits, clipping occurs.

The range of allowed output voltage depends on the type of op amp in use, on the load resistance value, and on the values of the power-supply voltages. For example, with supply voltages of +15 V and −15 V, the LM741 op amp (LM741 is the manufacturer's type number for a popular op amp) is capable of producing output voltages in the range from approximately −14 to +14 V. If smaller power-supply voltages are used, the linear range is reduced. (These are *typical* limits for load resistances greater than 10 kΩ. The *guaranteed* output range for the LM741 is only from −12 to +12 V. Smaller load resistances further restrict the range.)

Consider the noninverting amplifier with a sinusoidal input signal shown in Figure 14.23. Assuming an ideal op amp, the voltage gain is given by Equation 14.21, which is repeated here for convenience:

$$A_v = 1 + \frac{R_2}{R_1}$$

Substituting the values shown in Figure 14.23 ($R_1 = 1\ \text{k}\Omega$ and $R_2 = 3\ \text{k}\Omega$), we find $A_v = 4$. The output waveform for $R_L = 10\ \text{k}\Omega$ and $V_{im} = 1$ V is shown in Figure 14.24. The output waveform is sinusoidal because none of the nonlinear limits of the op amp have been exceeded. On the other hand, for $V_{im} = 5$ V, the output reaches the maximum output voltage limits and the output waveform is clipped. This is shown in Figure 14.25.

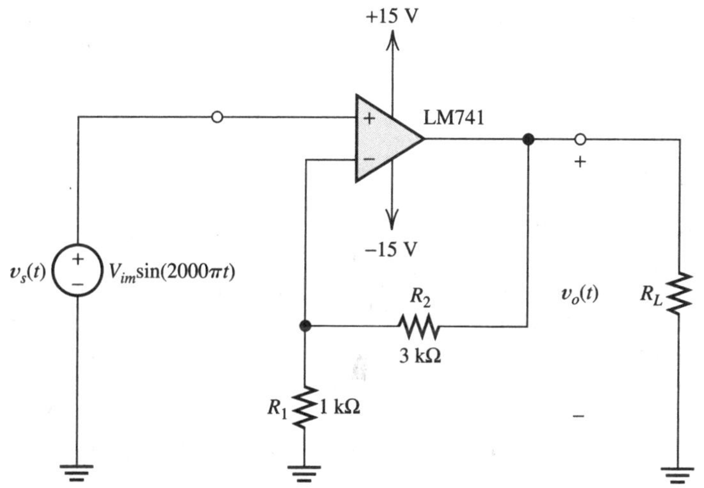

Figure 14.23 Noninverting amplifier used to demonstrate various nonlinear limitations of op amps.

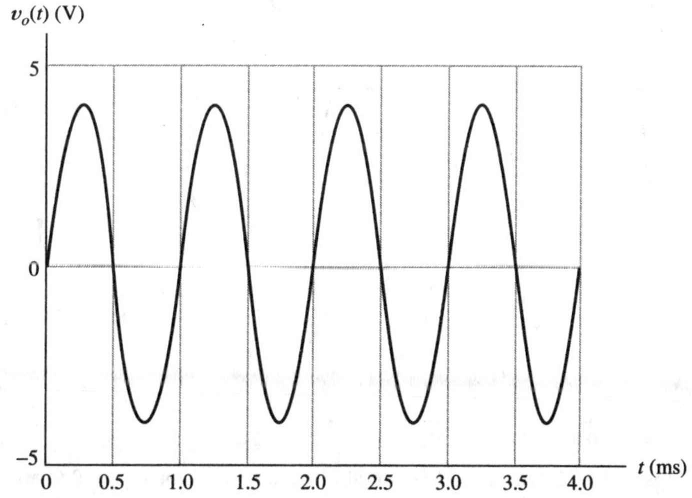

Figure 14.24 Output of the circuit of Figure 14.23 for $R_L = 10\ \text{k}\Omega$ and $V_{im} = 1$ V. None of the limitations are exceeded, and $v_o(t) = 4v_s(t)$.

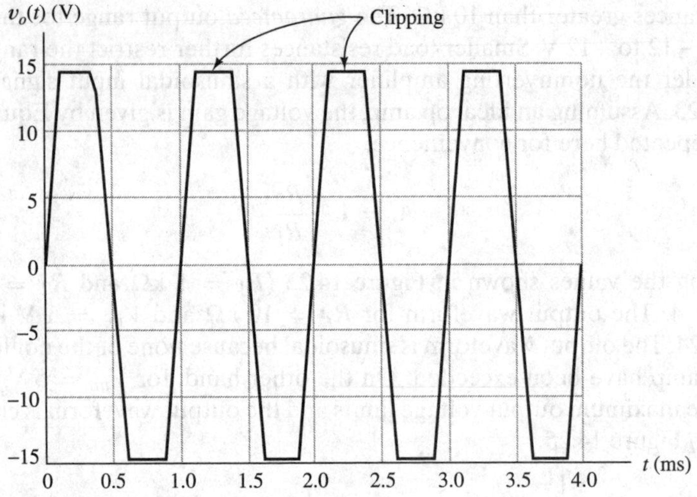

Figure 14.25 Output of the circuit of Figure 14.23 for $R_L = 10$ kΩ and $V_{im} = 5$ V. Clipping occurs because the maximum possible output voltage magnitude is reached.

Output Current Limits

The output current range of a real op amp is limited. If an input signal is sufficiently large that the output current would be driven beyond these limits, clipping occurs.

A second limitation is the maximum current that an op amp can supply to a load. For the LM741, the limits are ±40 mA. If a small-value load resistance would draw a current outside these limits, the output waveform becomes clipped.

For example, suppose that we set the peak input voltage to $V_{im} = 1$ V and adjust the load resistance to $R_L = 50$ Ω for the circuit of Figure 14.23. For an ideal op amp, we would expect a peak output voltage of $V_{om} = 4$ V and a peak load current of $V_{om}/R_L = 80$ mA. However, output current magnitude of the LM741 is limited to 40 mA. Therefore, clipping occurs due to current limiting. The output voltage

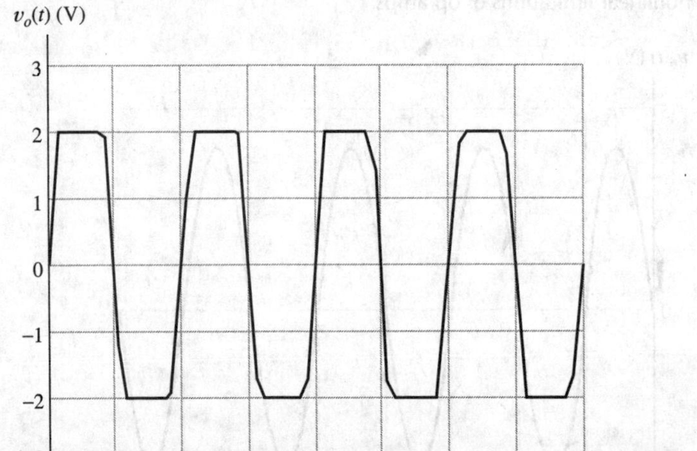

Figure 14.26 Output of the circuit of Figure 14.23 for $R_L = 50$ Ω and $V_{im} = 1$ V. Clipping occurs because the maximum output current limit is reached.

waveform of the circuit is shown in Figure 14.26. Notice that the peak output voltage is 40 mA $\times R_L = 2$ V.

Slew-Rate Limitation

Another nonlinear limitation of actual op amps is that the magnitude of the rate of change of the output voltage is limited. This is called the **slew-rate limitation**. The output voltage cannot increase (or decrease) in magnitude at a rate exceeding this limit. In equation form, the slew-rate limit is

Another nonlinear limitation of actual op amps is that the magnitude of the rate of change of the output voltage is limited.

$$\left|\frac{dv_o}{dt}\right| \leq \text{SR} \tag{14.35}$$

For various types of IC op amps, the slew-rate limit ranges from SR $= 10^5$ V/s to SR $= 10^8$ V/s. For the LM741 with $\pm$15-V supplies and $R_L > 2$ kΩ, the typical value is 5×10^5 V/s (which is often stated as 0.5 V/μs).

For example, consider the circuit of Figure 14.23, except that the input source voltage is changed to a 2.5-V-peak 50-kHz sine wave given by

$$v_s(t) = 2.5\sin(10^5\pi t)$$

starting at $t = 0$. [$v_s(t)$ is assumed to be zero prior to $t = 0$.] The output waveform is shown in Figure 14.27. Also plotted in the figure is four times the input voltage, which is the output assuming an ideal op amp. At $t = 0$, the output voltage is zero. The ideal output increases at a rate exceeding the slew-rate limit of the LM741, so the LM741 output increases at its maximum rate, which is approximately 0.5 V/μs. At point A, the actual output finally "catches up" with the ideal output, but by then, the ideal output is decreasing at a rate that exceeds the slew-rate limit. Thus, at point A, the output of the LM741 begins to decrease at its maximum possible rate. Notice that because of slew-rate limiting, the actual op-amp output is a triangular waveform rather than a sinusoid.

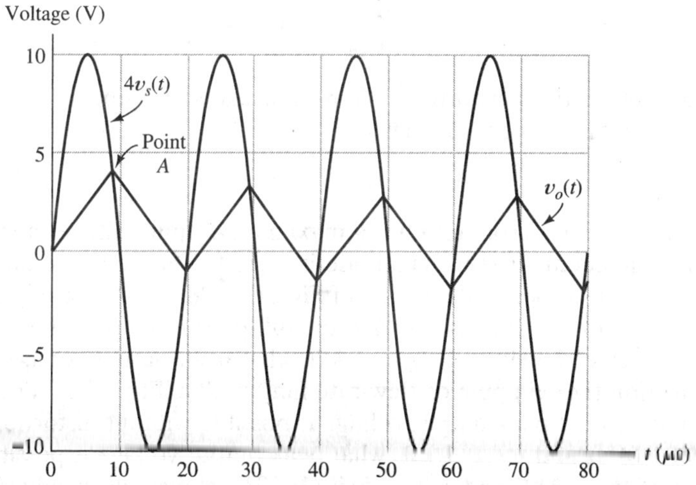

Figure 14.27 Output of the circuit of Figure 14.23 for $R_L = 10$ kΩ and $v_s(t) = 2.5\sin(10^5\pi t)$. The output waveform is a triangular waveform because the slew-rate limit is exceeded. The output for an ideal op amp, which is equal to $4v_s(t)$, is shown for comparison.

Full-Power Bandwidth

The full-power bandwidth of an op amp is the range of frequencies for which the op amp can produce an undistorted sinusoidal output with peak amplitude equal to the guaranteed maximum output voltage.

The **full-power bandwidth** of an op amp is the range of frequencies for which the op amp can produce an undistorted sinusoidal output with peak amplitude equal to the guaranteed maximum output voltage.

Next, we derive an expression for the full-power bandwidth in terms of the slew rate and peak amplitude. The output voltage is given by

$$v_O(t) = V_{om} \sin(\omega t)$$

Taking the derivative with respect to time, we have

$$\frac{dv_o(t)}{dt} = \omega V_{om} \cos(\omega t)$$

The maximum magnitude of the rate of change is $\omega V_{om} = 2\pi f V_{om}$. Setting this equal to the slew-rate limit, we get

$$2\pi f V_{om} = \text{SR}$$

Solving for frequency, we obtain

$$f_{\text{FP}} = \frac{\text{SR}}{2\pi V_{om}} \tag{14.36}$$

where we have denoted the full-power bandwidth as f_{FP}. An undistorted full-amplitude sinusoidal output waveform is possible only for frequencies less than f_{FP}.

Example 14.6 Full-Power Bandwidth

Find the full-power bandwidth of the LM741 op amp given that the slew rate is $\text{SR} = 0.5\ \text{V}/\mu\text{s}$ and the guaranteed maximum output amplitude is $V_{om} = 12$ V.

Solution We substitute the given data into Equation 14.36 to obtain

$$f_{\text{FP}} = \frac{\text{SR}}{2\pi V_{om}} \cong 6.63 \text{ kHz}$$

Thus, we can obtain an undistorted 12-V-peak sinusoidal output from the LM741 only for frequencies less than 6.63 kHz. ■

Exercise 14.13 A certain op amp has a maximum output voltage range from −4 to +4 V. The maximum current magnitude is 10 mA. The slew-rate limit is SR = $5\ \text{V}/\mu\text{s}$. This op amp is used in the circuit of Figure 14.28. Assume a sinusoidal input signal for all parts of this exercise. **a.** Find the full-power bandwidth of the op amp. **b.** For a frequency of 1 kHz and $R_L = 1\ \text{k}\Omega$, what peak output voltage is possible without distortion (i.e., clipping or slew-rate limiting)? **c.** For a frequency of 1 kHz and $R_L = 100\ \Omega$, what peak output voltage is possible without distortion? **d.** For a frequency of 1 MHz and $R_L = 1\ \text{k}\Omega$, what peak output voltage is possible without distortion? **e.** If $R_L = 1\ \text{k}\Omega$ and $v_s(t) = 5\sin(2\pi 10^6 t)$, sketch the steady-state output waveform to scale versus time.

Answer **a.** $f_{\text{FP}} = 199$ kHz; **b.** 4 V; **c.** 1 V; **d.** 0.796 V; **e.** the output waveform is a triangular wave with a peak amplitude of 1.25 V. □

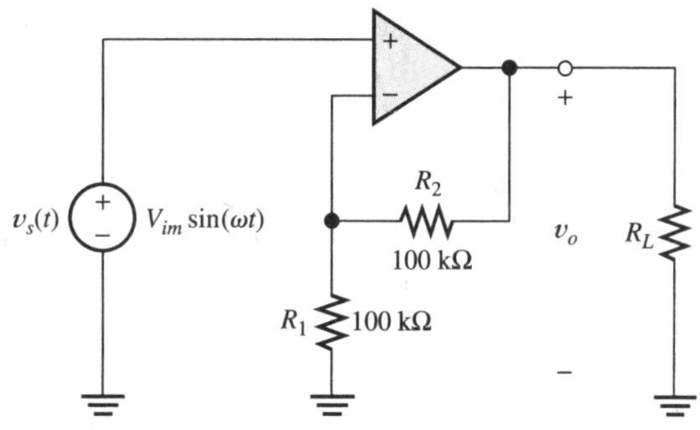

Figure 14.28 Circuit for Exercise 14.13.

14.7 DC IMPERFECTIONS

Op amps have direct-coupled input circuits. Thus, dc bias currents that flow into (or from) the input devices of the op amp must flow through the elements that are connected to the input terminals, such as the signal source or feedback resistors.

The dc current flowing into the noninverting input is denoted as I_{B+}, and the dc current flowing into the inverting input is I_{B-}. The average of the dc currents is called **bias current** and is denoted as I_B. Thus, we have

$$I_B = \frac{I_{B+} + I_{B-}}{2} \tag{14.37}$$

Nominally, the input circuit of the op amp is symmetrical, and the bias currents flowing into the inverting and noninverting inputs are equal. However, in practice, the devices are not perfectly matched, and the bias currents are not equal. The difference between the bias currents, called the **offset current**, is denoted as

$$I_{\text{off}} = I_{B+} - I_{B-} \tag{14.38}$$

Another dc imperfection of op amps is that the output voltage may not be zero for zero input voltage. The op amp behaves as if a small dc source known as the **offset voltage** is in series with one of the input terminals.

The three dc imperfections (bias current, offset current, and offset voltage) can be modeled by placing dc sources at the input of the op amp as shown in Figure 14.29. The I_B current sources model the bias current. The $I_{\text{off}}/2$ current source models the offset current, and the V_{off} voltage source models the offset voltage. (These sources

The three dc imperfections (bias current, offset current, and offset voltage) can be modeled by placing dc sources at the input of the op amp as shown in Figure 14.29.

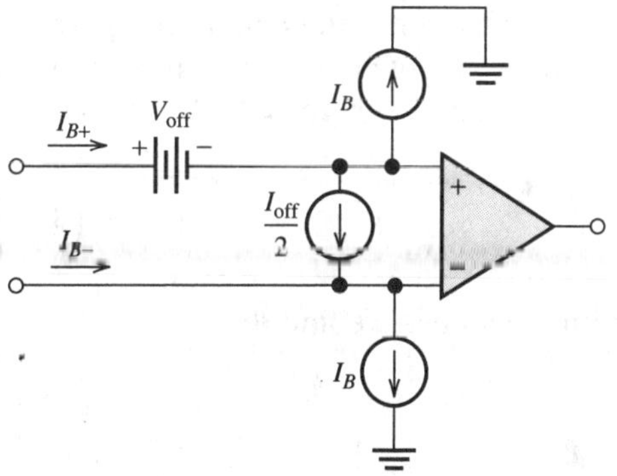

Figure 14.29 Three current sources and a voltage source model the dc imperfections of an op amp.

were discussed in Section 11.12, and the discussion given there applies to op-amp circuits as well as to amplifiers in general.)

The bias-current sources are equal in magnitude and are referenced in the same direction (which is away from the input terminals in Figure 14.29). In some op amps, the bias current can have a negative value, so the currents flow toward the input terminals. Usually, the direction of the bias current is predictable for a given type of op amp. For example, if the input terminals of an op amp are the base terminals of *npn* BJTs, the bias current I_B is positive (assuming the reference directions shown in Figure 14.29). On the other hand, *pnp* BJTs would result in a negative value for I_B.

Since the bias-current sources are matched in magnitude and direction, it is possible to design circuits in such a way that their effects cancel. On the other hand, the polarity of the offset voltage and the direction of the offset current are unpredictable—varying from unit to unit. For example, if the offset voltage of a given type of op amp is specified as a maximum of 2 mV, the value of V_{off} ranges from -2 to $+2$ mV from unit to unit. Usually, most units have offset values close to zero, and only a few have values close to the maximum specification. A typical specification for the maximum offset voltage magnitude for IC op amps is several millivolts.

Bias currents are usually on the order of 100 nA for op amps with BJT input devices. Bias currents are much lower for op amps with FET inputs—a typical specification is 100 pA at 25°C for a device with JFET input devices. Usually, offset current specifications range from 20 to 50 percent of the bias current.

The effect of bias current, offset current, and offset voltage on inverting or noninverting amplifiers is to add a (usually undesirable) dc voltage to the intended output signal.

The effect of bias current, offset current, and offset voltage on inverting or noninverting amplifiers is to add a (usually undesirable) dc voltage to the intended output signal. We can analyze these effects by including the sources shown in Figure 14.29 and assuming an otherwise ideal op amp.

Example 14.7 Determining Worst-Case DC Output

Find the worst-case dc output voltage of the inverting amplifier shown in Figure 14.30(a), assuming that $v_{\text{in}} = 0$. The maximum bias current of the op amp is 100 nA, the maximum offset current magnitude is 40 nA, and the maximum offset-voltage magnitude is 2 mV.

First, we calculate the output voltage resulting from each of the dc sources acting individually. Then, we use superposition.

Solution Our approach is to calculate the output voltage due to each of the dc sources acting individually. Then by using superposition, the worst-case output can be found by adding the outputs due to the various sources.

First, we consider the offset voltage. The circuit, including the offset voltage source, is shown in Figure 14.30(b). The offset voltage source can be placed in series with either input. We have elected to place it in series with the noninverting input. Then, the circuit takes the form of a noninverting amplifier. [Notice that although it is drawn differently, the circuit of Figure 14.30(b) is electrically equivalent to the noninverting amplifier of Figure 14.11.] Thus, the output voltage is the gain of the noninverting amplifier, given by Equation 14.21, times the offset voltage:

$$V_{o,\text{voff}} = -\left(1 + \frac{R_2}{R_1}\right) V_{\text{off}}$$

Substituting values, we find that

$$V_{o,\text{voff}} = -11 V_{\text{off}}$$

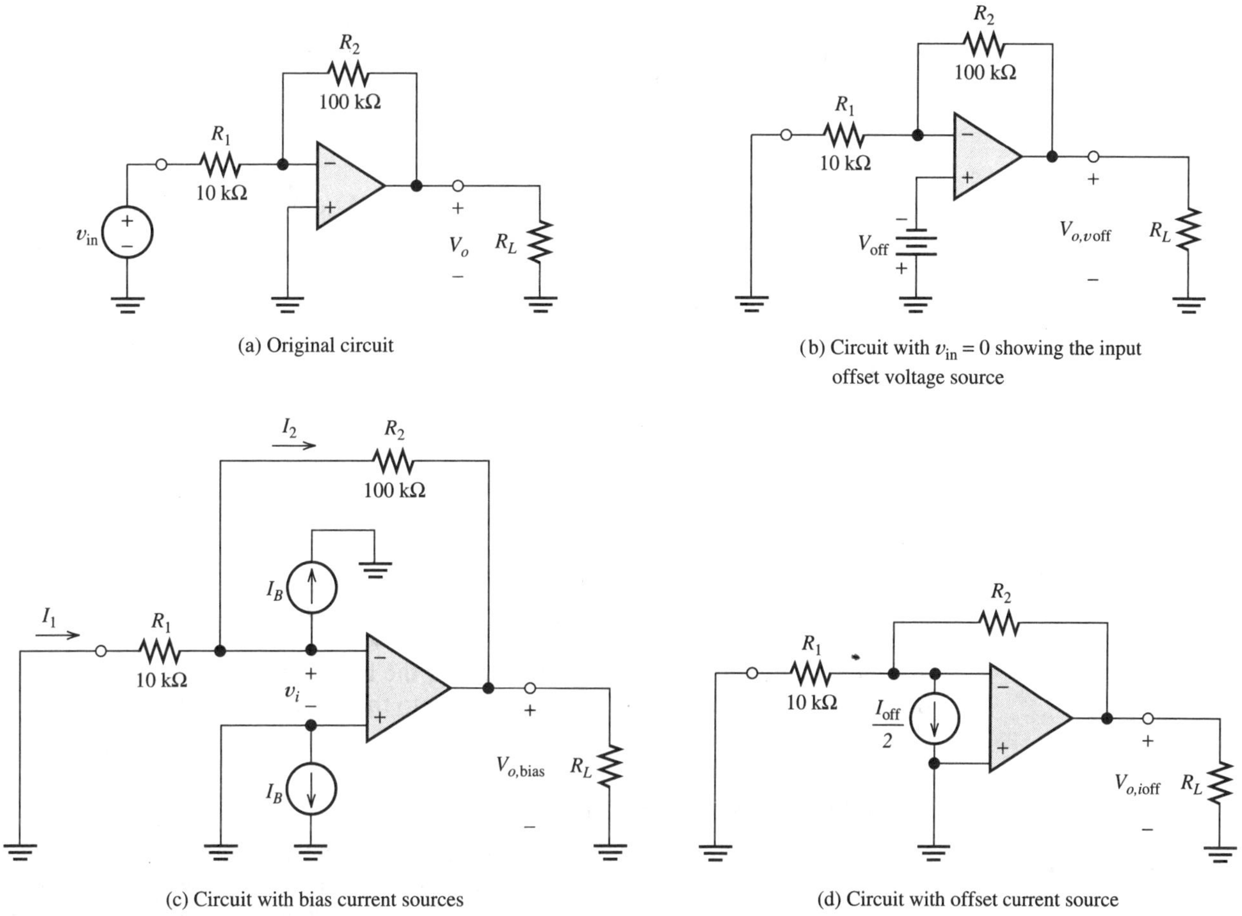

Figure 14.30 Circuits of Example 14.7.

Since the offset voltage V_{off} is specified to have a maximum value of 2 mV, the value of $V_{o,voff}$ ranges between extremes of −22 and +22 mV. However, most units would have $V_{o,voff}$ closer to zero.

Next, we consider the bias-current sources. The circuit, including the bias-current sources, is shown in Figure 14.30(c). Because the noninverting input is connected directly to ground, one of the bias-current sources is short circuited and has no effect. Since we assume an ideal op amp (aside from the dc sources), the summing-point constraint applies, and $v_i = 0$. Thus, the current I_1 is zero. Applying Kirchhoff's current law, we have $I_2 = -I_B$. Writing a voltage equation from the output through R_2 and R_1, we have

$$V_{o,\text{bias}} = -R_2 I_2 - R_1 I_1$$

Substituting $I_1 = 0$ and $I_2 = -I_B$, we obtain

$$V_{o,\text{bias}} = R_2 I_B$$

Because the maximum value of I_B is 100 nA, the maximum value of $V_{o,\text{bias}}$ is 10 mV. As is often the case, the maximum value of I_B is specified, but the minimum is not. Thus, $V_{o,\text{bias}}$ ranges from some small indeterminate voltage (perhaps a few millivolts) up to 10 mV. (We will conservatively assume that the minimum value of $V_{o,\text{bias}}$ is zero.)

Next, we consider the offset-current source. The circuit is shown in Figure 14.30(d). By an analysis similar to that for the bias current, we can show that

$$V_{o,\text{ioff}} = R_2\left(\frac{I_{\text{off}}}{2}\right)$$

The specification for the maximum magnitude of I_{off} is 40 nA. Therefore, the value of $V_{o,\text{ioff}}$ ranges between extremes of −2 and +2 mV.

By superposition, the dc output voltage is the sum of the contributions of the various sources acting individually.

By superposition, the dc output voltage is the sum of the contributions of the various sources acting individually, yielding

$$V_o = V_{o,\text{voff}} + V_{o,\text{bias}} + V_{o,\text{ioff}}$$

Hence, the extreme values of the output voltage are

$$V_o = 22 + 10 + 2 = 34 \text{ mV}$$

and

$$V_o = -22 + 0 - 2 = -24 \text{ mV}$$

Thus, the output voltage ranges from −24 to +34 mV from unit to unit. (We have assumed a minimum contribution of zero for the bias current.) Typical units would have total output voltages closer to zero than to these extreme values. ■

Cancellation of the Effects of Bias Currents

As mentioned earlier, it is possible to design circuits in which the effects of the two bias-current sources cancel. For example, consider the inverting amplifier configuration. Adding a resistor R_{bias} in series with the noninverting op-amp input, as shown in Figure 14.31, does not affect the gain of the amplifier, but results in cancellation of the effects of the I_B sources. Notice that the value of R_{bias} is equal to the parallel combination of R_1 and R_2.

Exercise 14.14 Consider the amplifier shown in Figure 14.31. **a.** Assume an ideal op amp, and derive an expression for the voltage gain v_o/v_{in}. Notice that the result is the same as Equation 14.5, which was derived for the inverting amplifier without the bias-current-compensating resistor R_{bias}. **b.** Redraw the circuit with $v_{\text{in}} = 0$, but include the bias-current sources. Show that the output voltage is zero. **c.** Assume that $R_1 = 10$ kΩ, $R_2 = 100$ kΩ, and a specification of 3 mV for the maximum magnitude

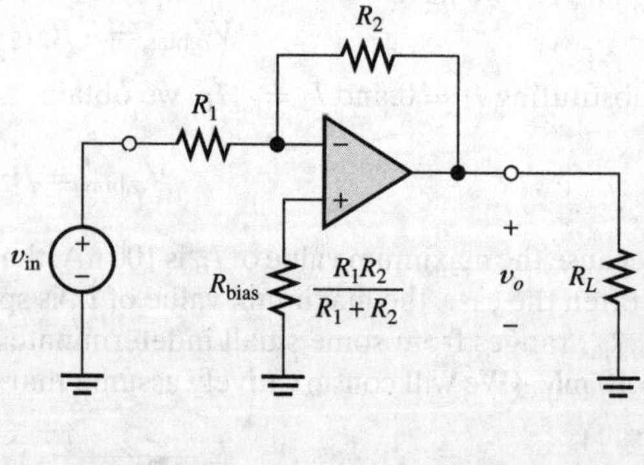

Figure 14.31 Adding the resistor R_{bias} to the inverting amplifier circuit causes the effects of bias currents to cancel.

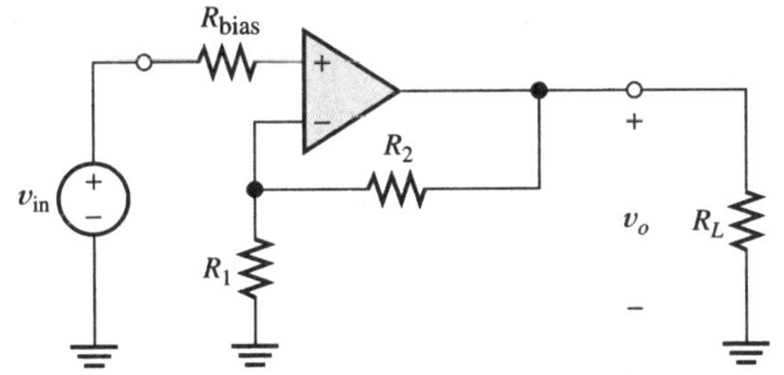

Figure 14.32 Noninverting amplifier, including resistor R_{bias} to balance the effects of the bias currents. See Exercise 14.15.

of V_{off}. Find the range of output voltages resulting from the offset voltage source V_{off}. **d.** Assume that $R_1 = 10\ k\Omega$, $R_2 = 100\ k\Omega$, and a specification of 40 nA for the maximum magnitude of I_{off}. Find the range of output voltages resulting from the offset current. **e.** Assuming the values given in parts (c) and (d), what range of output voltages results from the combined action of the bias current, offset voltage, and offset current?

Answer **a.** $v_o/v_{in} = -R_2/R_1$; **b.** ±33 mV; **c.** ±4 mV; **d.** ±37 mV. □

Exercise 14.15 Consider the noninverting amplifier shown in Figure 14.32. **a.** Derive an expression for the voltage gain v_o/v_{in}. Does the gain depend on the value of R_{bias}? Explain. **b.** Derive an expression for R_{bias} in terms of the other resistance values so that the output voltage due to the bias currents is zero.

Answer **a.** $v_o/v_{in} = 1+R_2/R_1$. The gain is independent of R_{bias} because the current through R_{bias} is zero (assuming an ideal op amp). **b.** $R_{bias} = R_1||R_2 = 1/(1/R_1 + 1/R_2)$. □

14.8 DIFFERENTIAL AND INSTRUMENTATION AMPLIFIERS

Differential amplifiers are widely used in engineering instrumentation.

Figure 14.33 shows a differential amplifier. Assuming an ideal op amp and that $R_4/R_3 = R_2/R_1$, the output voltage is a constant times the differential input signal $(v_1 - v_2)$. The gain for the common-mode signal is zero. (See Section 11.11 for a discussion of common-mode signals.) To minimize the effects of bias current, we should choose $R_2 = R_4$ and $R_1 = R_3$.

The output impedance of the circuit is zero. The input impedance for the v_1 source is $R_3 + R_4$.

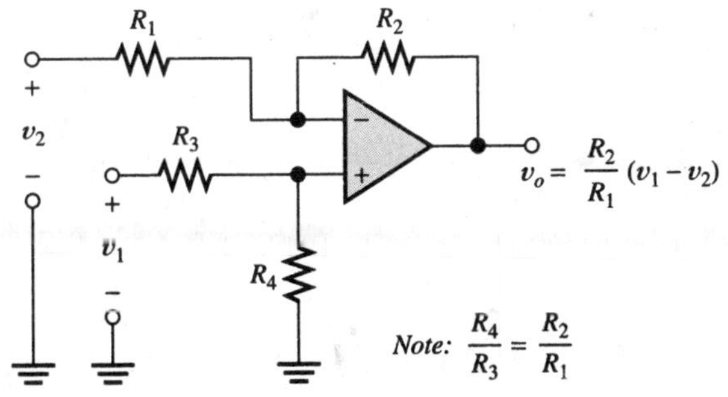

Figure 14.33 Differential amplifier.

A current that depends on v_1 flows back through the feedback network (R_1 and R_2) into the input source v_2. Thus, as seen by the v_2 source, the circuit does not appear to be passive. Hence, the concept of input impedance does not apply for the v_2 source (unless v_1 is zero).

In some applications, the signal sources contain internal impedances, and the desired signal is the difference between the internal source voltages. Then, we could design the circuit by including the internal source resistances of v_2 and v_1 as part of R_1 and R_3, respectively. However, to obtain very high common-mode rejection, it is necessary to match the ratios of the resistances closely. This can be troublesome if the source impedances are not small enough to be neglected and are not predictable.

Instrumentation-Quality Differential Amplifier

Figure 14.34 shows an improved differential amplifier circuit for which the common-mode rejection ratio is not dependent on the internal resistances of the sources. Because of the summing-point constraint at the inputs of X_1 and X_2, the currents drawn from the signal sources are zero. Hence, the input impedances seen by both sources are infinite, and the output voltage is unaffected by the internal source impedances. This is an important advantage of this circuit compared to the simpler differential amplifier of Figure 14.33. Notice that the second stage of the instrumentation amplifier is a unity-gain version of the differential amplifier.

A subtle point concerning this circuit is that the differential-mode signal experiences a higher gain in the first stage (X_1 and X_2) than the common-mode signal does. To illustrate this point, first consider a pure differential input (i.e., $v_1 = -v_2$). Then, because the circuit is symmetrical, point A remains at zero voltage. Hence, in the analysis for a purely differential input signal, point A can be considered to be grounded. In this case, the input amplifiers X_1 and X_2 are configured as noninverting

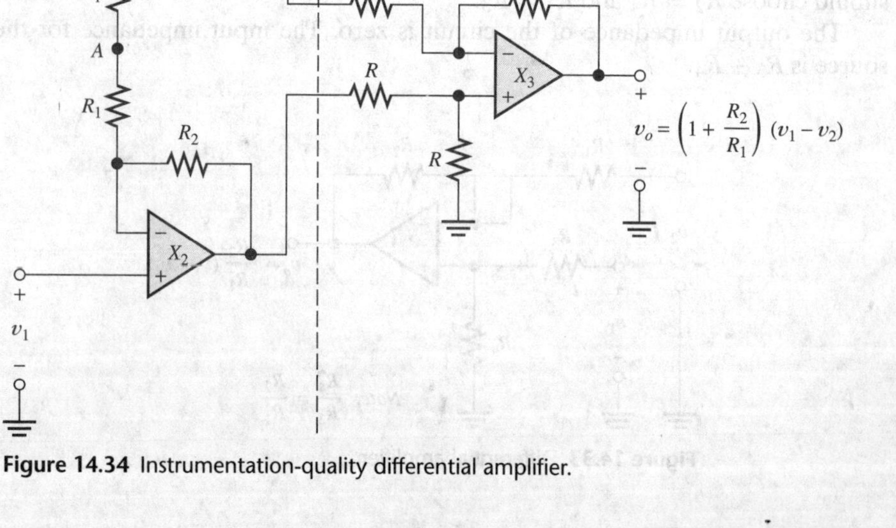

Figure 14.34 Instrumentation-quality differential amplifier.

amplifiers having gains of $(1 + R_2/R_1)$. The differential gain of the second stage is unity. Thus, the overall gain for the differential signal is $(1 + R_2/R_1)$.

Now, consider a pure common-mode signal (i.e., $v_1 = v_2 = v_{cm}$). Because of the summing-point constraint, the voltage between the input terminals of X_1 (or X_2) is zero. Thus, the voltages at the inverting input terminals of X_1 and X_2 are both equal to v_{cm}. Hence, the voltage across the series-connected R_1 resistors is zero, and no current flows through the R_1 resistors. Therefore, no current flows through the R_2 resistors. Thus, the output voltages of X_1 and X_2 are equal to v_{cm}, and we have shown that the first-stage gain is unity for the common-mode signal. On the other hand, the differential gain of the first stage is $(1 + R_2/R_1)$, which can be much larger than unity, thereby achieving a reduction of the common-mode signal amplitude relative to the differential signal. [Notice that if point A were actually grounded, the gain for the common-mode signal would be the same as for the differential signal, namely $(1 + R_2/R_1)$.]

In practice, the series combination of the two R_1 resistors is implemented by a single resistor (equal in value to $2R_1$) because it is not necessary to have access to point A. Thus, matching of component values for R_1 is not required. Furthermore, it can be shown that close matching of the R_2 resistors is not required to achieve a higher differential gain than common-mode gain in the first stage. Since the first stage reduces the relative amplitude of the common-mode signal, matching of the resistors in the second stage is not as critical.

Thus, although it is more complex, the differential amplifier of Figure 14.34 has better performance than that of Figure 14.33. Specifically, the common-mode rejection ratio is independent of the internal source resistances, the input impedance seen by both sources is infinite, and resistor matching is not as critical.

Exercise 14.16 Assume an ideal op amp, and derive the expression shown for the output voltage of the differential amplifier of Figure 14.33. Assume that $R_4/R_3 = R_2/R_1$. □

14.9 INTEGRATORS AND DIFFERENTIATORS

Figure 14.35 shows the diagram of an **integrator**, which is a circuit that produces an output voltage proportional to the running-time integral of the input voltage. (By the term *running time integral*, we mean that the upper limit of integration is t.)

Integrators produce output voltages that are proportional to the running-time integral of the input voltages. In a running time integral, the upper limit of integration is t.

The integrator circuit is often useful in instrumentation applications. For example, consider a signal from an accelerometer that is proportional to acceleration. By integrating the acceleration signal, we obtain a signal proportional to velocity. Another integration yields a signal proportional to position.

In Figure 14.35, negative feedback occurs through the capacitor. Thus, assuming an ideal op amp, the voltage at the inverting op-amp input is zero. The input current is given by

$$i_{in}(t) = \frac{v_{in}(t)}{R} \tag{14.39}$$

The current flowing into the input terminal of the (ideal) op amp is zero. Therefore, the input current i_{in} flows through the capacitor. We assume that the reset switch is opened at $t = 0$. Therefore, the capacitor voltage is zero at $t = 0$. The voltage across the capacitor is given by

$$v_C(t) = \frac{1}{C}\int_0^t i_{in}(t)\,dt \tag{14.40}$$

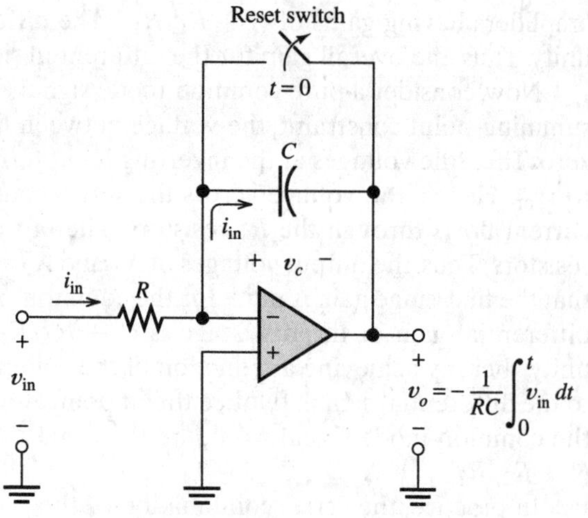

Figure 14.35 Integrator.

Writing a voltage equation from the output terminal through the capacitor and then to ground through the op-amp input terminals, we obtain

$$v_o(t) = -v_C(t) \tag{14.41}$$

Using Equation 14.39 to substitute into 14.40 and the result into 14.41, we obtain

$$v_o(t) = -\frac{1}{RC}\int_0^t v_{in}(t)\, dt \tag{14.42}$$

Thus, the output voltage is $-1/RC$ times the running integral of the input voltage. If an integrator having positive gain is desired, we can cascade the integrator with an inverting amplifier.

The magnitude of the gain can be adjusted by the choice of R and C. Of course, in selecting a capacitor, we usually want to use as small a value as possible to minimize cost, volume, and mass. However, for a given gain constant $(1/RC)$, smaller C leads to larger R and smaller values of i_{in}. Therefore, the bias current of the op amp becomes more significant as the capacitance becomes smaller. As usual, we try to design for the best compromise.

Exercise 14.17 Consider the integrator of Figure 14.35 with the square-wave input signal shown in Figure 14.36. **a.** If $R = 10$ kΩ, $C = 0.1$ μF, and the op amp is ideal, sketch the output waveform to scale. **b.** If $R = 10$ kΩ, what value of C is required for the peak-to-peak output amplitude to be 2 V?

Answer **a.** See Figure 14.37; **b.** $C = 0.5$ μF. □

Exercise 14.18 Consider the circuit of Figure 14.35 with $v_{in} = 0$, $R = 10$ kΩ, and $C = 0.01$ μF. As indicated in the figure, the reset switch opens at $t = 0$. The op amp is ideal except for a bias current of $I_B = 100$ nA. **a.** Find an expression for the output voltage of the circuit as a function of time. **b.** Repeat for $C = 1$ μF.

Answer **a.** $v_o(t) = 10t$; **b.** $v_o(t) = 0.1t$. □

Exercise 14.19 Add a resistance R in series with the noninverting input of the op amp in Figure 14.35 and repeat Exercise 14.18.

Answer **a.** $v_o(t) = -1$ mV; **b.** $v_o(t) = -1$ mV. □

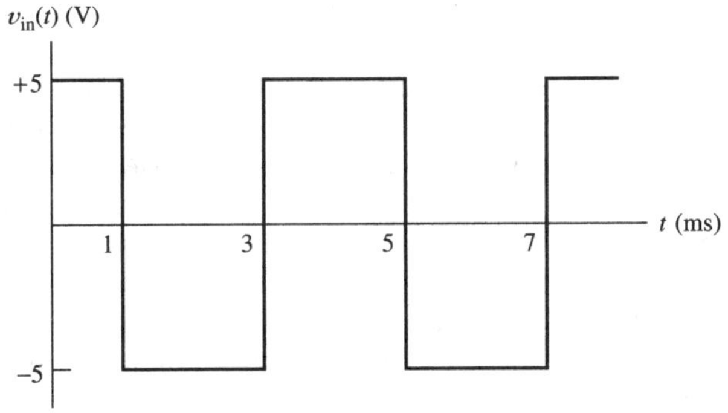

Figure 14.36 Square-wave input signal for Exercise 14.17.

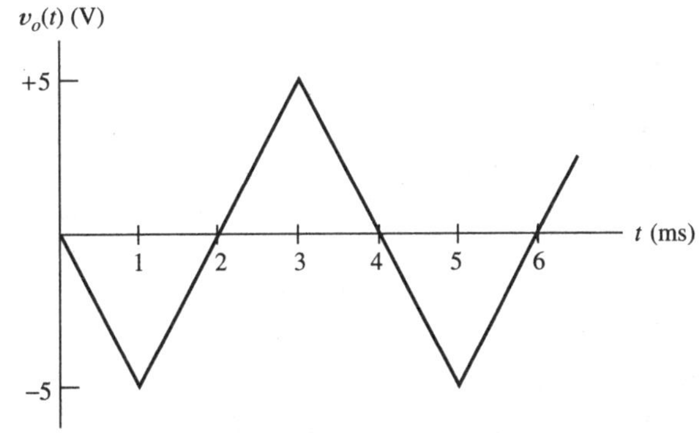

Figure 14.37 Answer for Exercise 14.17.

Differentiator Circuit

Figure 14.38 shows a **differentiator** that produces an output voltage proportional to the time derivative of the input voltage. By an analysis similar to that used for the integrator, we can show that the circuit produces an output voltage given by

$$v_o(t) = -RC\frac{dv_{in}}{dt} \tag{14.43}$$

Exercise 14.20 Derive Equation 14.43. □

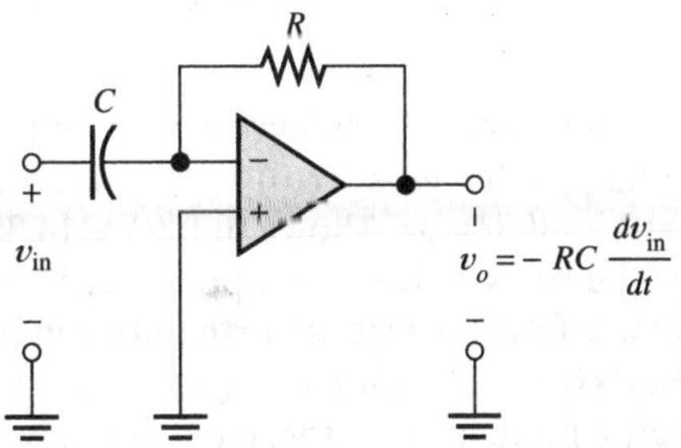

Figure 14.38 Differentiator.

14.10 ACTIVE FILTERS

Filters are circuits designed to pass input components with frequencies in one range to the output and prevent input components with frequencies in other ranges from reaching the output. For example, a lowpass filter passes low-frequency input components to the output but not high-frequency components. A common application for filters is to separate a signal of interest from other signals and noise. For example, in an electrocardiograph, we need a filter that passes the heart signals, which have frequencies below about 100 Hz, and rejects higher frequency noise that can be created by contraction of other muscles. We might use a lowpass filter to remove noise from historical phonograph recordings. In radio receivers, filters separate one station from the others. In digital instrumentation systems, a lowpass filter is often needed to remove noise and signal components that have frequencies higher than half of the sampling frequency in order to avoid a type of distortion, known as aliasing, during sampling and analog-to-digital conversion.

Filters can be very useful in separating desired signals from noise.

In Sections 6.2 and 6.8, we considered a few examples of passive-filter design. In this section, we show how to design lowpass filters composed of resistors, capacitors, and op amps. Filters composed of op amps, resistors, and capacitors are said to be **active filters**. In many respects, active filters have improved performance compared to passive circuits.

Active filters have been studied extensively and many useful circuits have been found. Ideally, an active filter circuit should:

1. Contain few components.
2. Have a transfer function that is insensitive to component tolerances.
3. Place modest demands on the op amp's gain–bandwidth product, output impedance, slew rate, and other specifications.
4. Be easily adjusted.
5. Require a small spread of component values.
6. Allow a wide range of useful transfer functions to be realized.

Various circuits have been described in the literature that meet these goals to varying degrees. Many complete books have been written that deal exclusively with active filters. In this section, we confine our attention to a particular (but practical) means for implementing lowpass filters.

Butterworth Transfer Function

The magnitude of the **Butterworth transfer function** is given by

$$|H(f)| = \frac{H_0}{\sqrt{1 + (f/f_B)^{2n}}} \tag{14.44}$$

in which the integer n is the *order* of the filter and f_B is the 3-dB cutoff frequency. Substituting $f = 0$ yields $|H(0)| = H_0$; thus, H_0 is the dc gain magnitude. Plots of this transfer function are shown in Figure 14.39. Notice that as the order of the filter increases, the transfer function approaches that of an ideal lowpass filter.

An active lowpass Butterworth filter can be implemented by cascading modified **Sallen–Key circuits**, one of which is shown in Figure 14.40. In this version of the Sallen–Key circuit, the resistors labeled R have equal values. Similarly, the capacitors

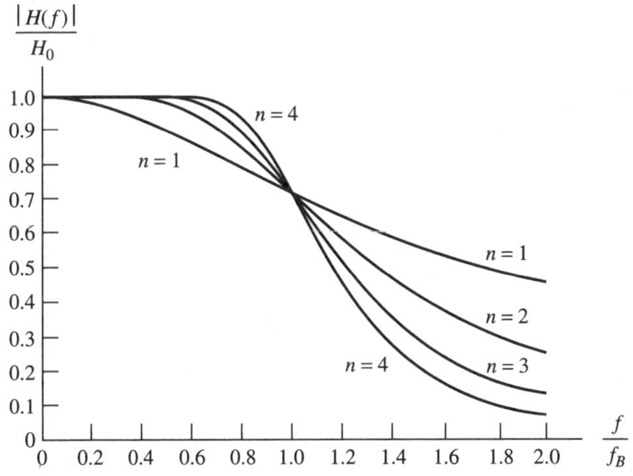

Figure 14.39 Transfer-function magnitude versus frequency for lowpass Butterworth filters.

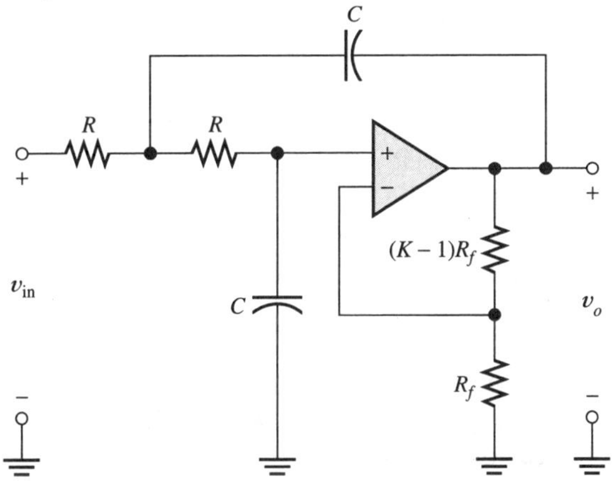

Figure 14.40 Equal-component Sallen–Key lowpass active-filter section.

labeled C have equal values. Useful circuits having unequal components are possible, but equal components are convenient.

The Sallen–Key circuit shown in Figure 14.40 is a second-order lowpass filter. To obtain an nth-order filter, $n/2$ circuits must be cascaded. (We assume that n is even.)

The 3-dB cutoff frequency of the overall filter is related to R and C by

$$f_B = \frac{1}{2\pi RC} \tag{14.45}$$

Usually, we wish to design for a given cutoff frequency. We try to select small capacitance values because this leads to small physical size and low cost. However, Equation 14.45 shows that as the capacitances become small, the resistance values become larger (for a given cutoff frequency). If the capacitance is selected too small, the resistance becomes unrealistically large. Furthermore, stray wiring capacitance can easily affect a high-impedance circuit. Thus, we select a capacitance value that is small, but not too small (say not smaller than 1000 pF).

In selecting the capacitor, we should select a value that is readily available in the tolerance required. Then, we use Equation 14.45 to compute the resistance. It is helpful to select the capacitance first and then compute the resistance, because resistors are commonly available in more finely spaced values than capacitors. Possibly, we cannot find nominal values of R and C that yield exactly the desired break frequency; however, it is a rare situation for which the break frequency must be controlled to an accuracy less than a few percent. Thus, 1-percent-tolerance resistors usually result in a break frequency sufficiently close to the value desired.

Notice in the circuit of Figure 14.40 that the op amp and the feedback resistors R_f and $(K - 1)R_f$ form a noninverting amplifier having a gain of K. At dc, the capacitors act as open circuits. Then, the resistors labeled R are in series with the input terminals of the noninverting amplifier and have no effect on gain. Thus, the dc gain of the circuit is K. As K is varied from zero to three, the transfer function displays more and more peaking (i.e., the gain magnitude increases with frequency and reaches a peak before falling off). For $K = 3$, infinite peaking occurs. It turns out that for K greater than three, the circuit is unstable—it oscillates.

The most critical issue in selection of the feedback resistors R_f and $(K - 1)R_f$ is their ratio. If desired, a precise ratio can be achieved by including a potentiometer, which is adjusted to yield the required dc gain for each section. To minimize the effects of bias current, we should select values such that the parallel combination of R_f and $(K - 1)R_f$ is equal to $2R$. However, with FET input op amps, input bias current is often so small that this is not necessary.

An nth-order Butterworth lowpass filter is obtained by cascading $n/2$ stages having proper values for K. (Here again, we assume that n is even.) Table 14.1 shows the required K values for filters of various orders. The dc gain H_0 of the overall filter is the product of the K values of the individual stages.

Table 14.1. *K* Values for Lowpass or Highpass Butterworth Filters of Various Orders

Order	K
2	1.586
4	1.152
	2.235
6	1.068
	1.586
	2.483
8	1.038
	1.337
	1.889
	2.610

Example 14.8 Lowpass Active Filter Design

Design a fourth-order lowpass Butterworth filter having a cutoff frequency of 100 Hz.

Solution We arbitrarily choose capacitor values of $C = 0.1\ \mu F$. This is a standard value and not prohibitively large. (Perhaps we could achieve an equally good design by using smaller capacitances, say $0.01\ \mu F$. However, as we have mentioned earlier, there is a practical limit to how small the capacitances can be.)

Next, we solve Equation 14.45 for R. Substituting $f_B = 100$ Hz and $C = 0.1\ \mu$F results in $R = 15.92$ kΩ. In practice, we would select a 15.8-kΩ 1-percent-tolerance resistor. This results in a nominal cutoff frequency slightly higher than the design objective.

Active lowpass filters such as this are useful as antialias filters in computer-based instrumentation systems as discussed in Section 9.3.

Consulting Table 14.1, we find that a fourth-order filter requires two sections having gains of $K = 1.152$ and 2.235. This results in an overall dc gain of $H_0 = 1.152 \times 2.235 \cong 2.575$. We arbitrarily select $R_f = 10$ kΩ for both sections. The complete circuit diagram is shown in Figure 14.41. The resistors R_3 and R_{13} consist of fixed resistors in series with small trimmers that can be adjusted to obtain the required gain for each stage. ■

A Bode plot of the overall gain magnitude for the filter designed in Example 14.8 is shown in Figure 14.42. It can be verified that the dc gain in decibels is $20 \log H_0 \cong 8.2$ dB. As desired, the 3-dB frequency is very nearly 100 Hz.

Figure 14.43 shows the gain of each section normalized by its dc gain. The figure also shows the normalized overall gain. Of course, the overall normalized gain is the product of the normalized gains of the individual stages. (Notice that the gains are plotted as ratios rather than in decibels.) The transfer function of the first stage—which is the low-gain stage—rolls off without peaking. However, considerable peaking occurs in the second stage. It is this peaking that squares up the shoulder of the overall transfer characteristic.

Exercise 14.21 Show that for frequencies much greater than f_B, the magnitude of the lowpass Butterworth transfer function given in Equation 14.44 rolls off at $20 \times n$ decibels per decade. □

Exercise 14.22 Design a sixth-order Butterworth lowpass filter having a cutoff frequency of 5 kHz.

Answer Many answers are possible. For a sixth-order filter, three stages like Figure 14.40 need to be cascaded. A good choice is to use capacitors in the range

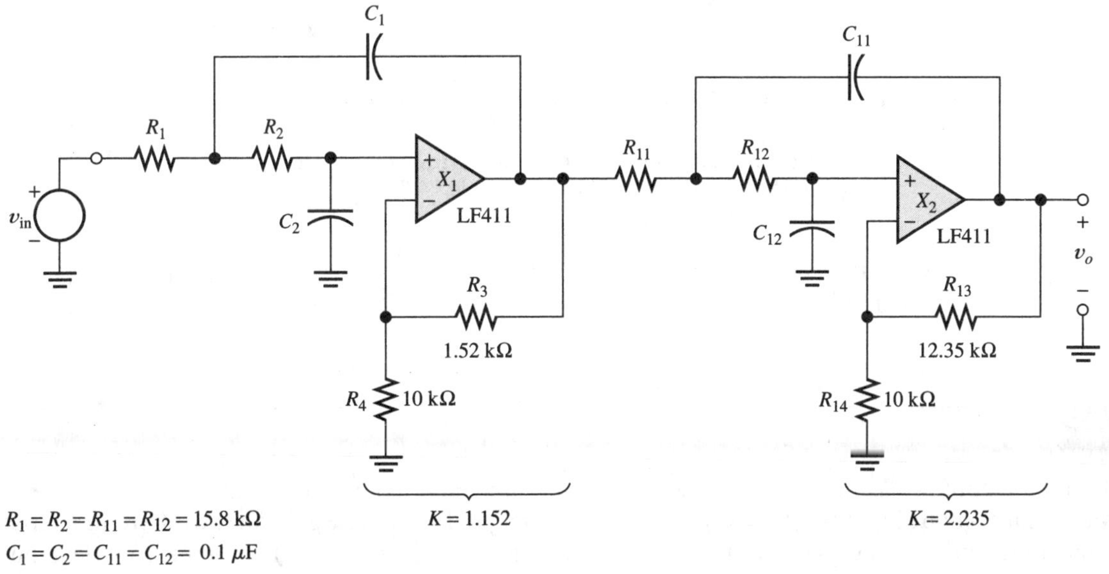

Figure 14.41 Fourth-order Butterworth lowpass filter designed in Example 14.8.

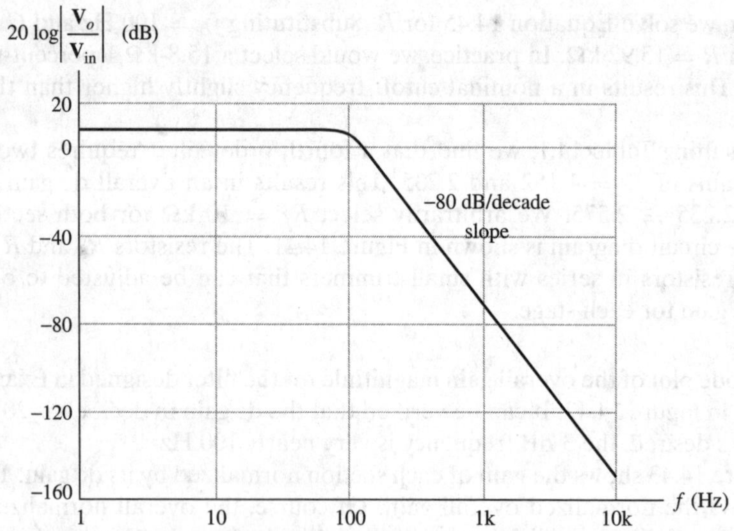

Figure 14.42 Bode magnitude plot of the gain for the fourth-order lowpass filter of Example 14.8.

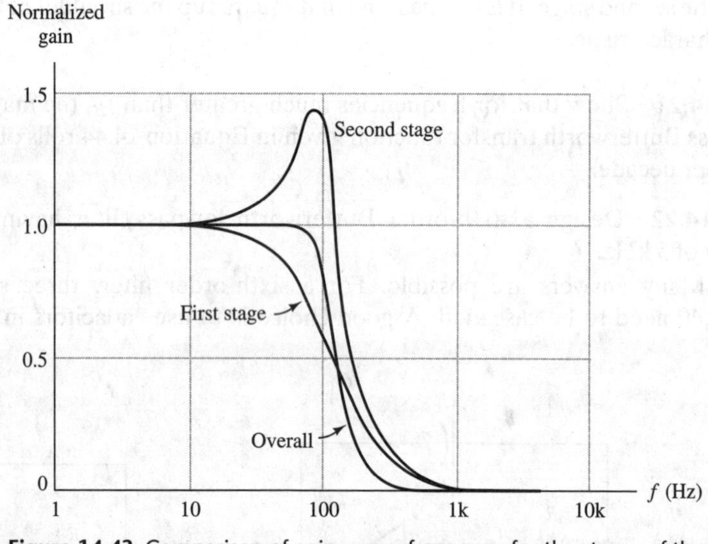

Figure 14.43 Comparison of gain versus frequency for the stages of the fourth-order lowpass filter of Example 14.8.

from 1000 pF to 0.01 μF. With $C = 0.01\ \mu$F, we need $R = 3.183$ kΩ. $R_f = 10$ kΩ is a good choice. From Table 14.1, we find the gain values to be 1.068, 1.586, and 2.483. □

Summary

1. If a differential amplifier has input voltages v_1 and v_2, the common-mode input is $v_{icm} = \frac{1}{2}(v_1 + v_2)$ and the differential input signal is $v_{id} = v_1 - v_2$.

2. An ideal operational amplifier has infinite input impedance, infinite gain for the differential input signal, zero gain for the common-mode input

signal, zero output impedance, and infinite bandwidth.

3. In an amplifier circuit with negative feedback, part of the output is returned to the input. The feedback signal opposes the input source.

4. To analyze ideal op-amp circuits with negative feedback, we assume that the differential input voltage and the input current of the op amp are driven to zero (this is the summing-point constraint), and then we use basic circuit principles to analyze the circuit.

5. The basic inverting amplifier configuration is shown in Figure 14.4 on page 680. Its closed-loop voltage gain is $A_v = -R_2/R_1$.

6. The basic noninverting amplifier configuration is shown in Figure 14.11 on page 686. Its closed-loop voltage gain is $A_v = 1 + R_2/R_1$.

7. Many useful amplifier circuits can be designed with the use of op amps. First, we select a suitable circuit configuration, and then we determine the resistor values that achieve the desired gain values.

8. In the design of op-amp circuits, very large resistances are unsuitable because their values are unstable and because high-impedance circuits are vulnerable to capacitive coupling of noise. Very low resistances are unsuitable because large currents flow in them for the voltages typically encountered in op-amp circuits.

9. In the linear range of operation, the imperfections of real op amps include finite input impedance, nonzero output impedance, and finite open-loop gain magnitude, which falls off with increasing frequency.

10. Negative feedback reduces gain magnitude and extends bandwidth. For the noninverting amplifier, the product of dc gain magnitude and bandwidth is constant for a given op-amp type.

11. The output voltage range and the output current range of any op amp are limited. If the output waveform reaches (and tries to exceed) either of these limits, clipping occurs.

12. The rate of change of the output voltage of any op amp is limited in magnitude. This is called the slew-rate limitation. The full-power bandwidth is the highest frequency for which the op amp can produce a full-amplitude sinusoidal output signal.

13. Dc imperfections of op amps are bias current, offset current, and offset voltage. These effects can be modeled by the sources shown in Figure 14.29 on page 703. The effect of dc imperfections is a (usually undesirable) dc component added to the intended output signal.

14. A single op amp can be used as a differential amplifier as shown in Figure 14.33 on page 707. However, the instrumentation amplifier shown in Figure 14.34 on page 708 has better performance.

15. The integrator circuit shown in Figure 14.35 on page 710 produces an output voltage that is proportional to the running time integral of the input voltage. A differentiator circuit is shown in Figure 14.38 on page 711.

16. Active filters often have better performance than passive filters. Active Butterworth lowpass filters can be obtained by cascading several Sallen–Key circuits having the proper gains.

Problems

Section 14.1: Ideal Operational Amplifiers

P14.1. Suppose that a real op amp has five terminals. List the probable functions of the terminals.

P14.2. List the characteristics of an ideal operational amplifier.

P14.3. Differentiate between **a.** open loop & closed loop gain; and **b.** inverting and non-inverting amplifiers.

P14.4. Give the definitions of the differential input voltage and of the common-mode input voltage for a differential amplifier having input voltages v_1 and v_2.

---

* Denotes that answers are contained in the Student Solutions files. See Appendix E for more information about accessing the Student Solutions.

P14.5. The input voltages (in units of volts) of a differential amplifier are

$$v_1(t) = 0.7\cos(2000\pi t) + 30\cos(120\pi t)$$
$$v_2(t) = -0.5\cos(2000\pi t) + 30\cos(120\pi t)$$

Find expressions for the common-mode and differential components of the input signal.

Section 14.2: Inverting Amplifiers

***P14.6.** What are the steps in analyzing an amplifier containing an ideal op amp?

P14.7. Draw the circuit diagram of the basic inverting amplifier configuration. Give an expression for the closed-loop voltage gain of the circuit in terms of the resistances, assuming an ideal op amp. Give expressions for the input impedance and output impedance of the circuit.

P14.8. Define the term *summing-point constraint.* Does it apply when negative feedback is present? When positive feedback is present?

***P14.9.** Determine the closed-loop voltage gain of the circuit shown in Figure P14.9, assuming an ideal op amp.

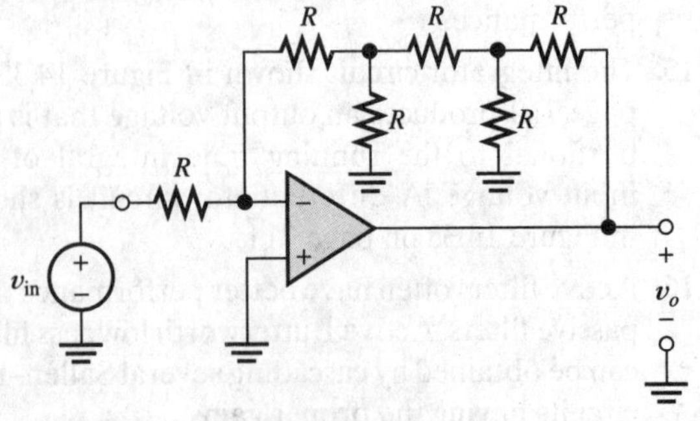

Figure P14.9

P14.10. Sketch $v_{in}(t)$ and $v_o(t)$ to scale versus time for the circuit shown in Figure P14.10. The op amp is ideal.

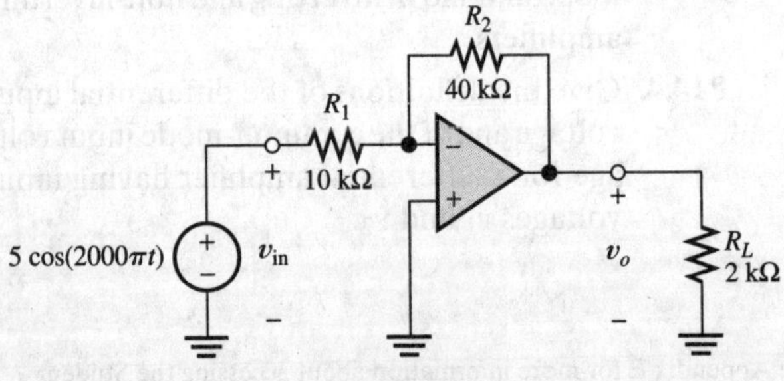

Figure P14.10

P14.11. Determine the closed-loop voltage gain of the circuit shown in Figure P14.11, assuming an ideal op amp.

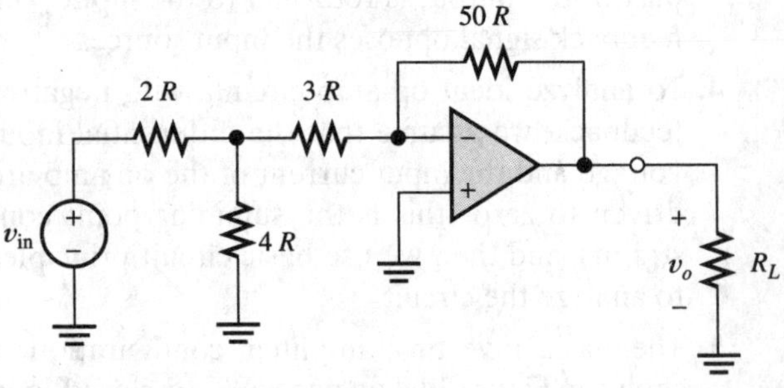

Figure P14.11

P14.12. The amplifier shown in Figure P14.12 is called an **exponential amplifier**. The circuit has negative feedback, and the summing-point constraint applies. Assume an ideal op amp, that v_{in} is positive, and that the diode current given by Equation 10.4, which states that $i_D = I_s \exp(v_D/nV_T)$. Derive an expression for v_o in terms of v_{in}, R, I_s, n, and V_T.

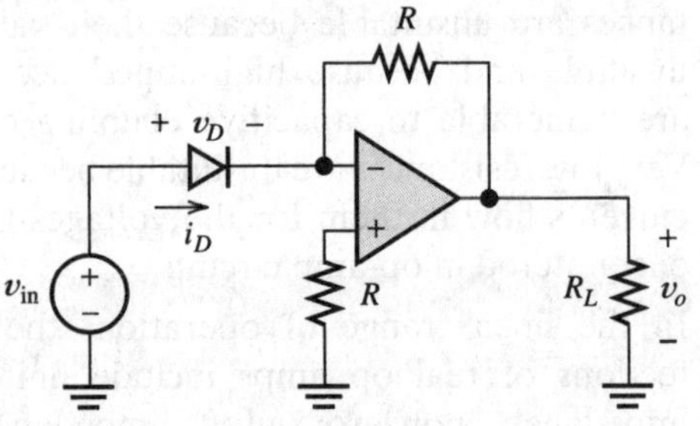

Figure P14.12

P14.13. The amplifier shown in Figure P14.13 is called a **logarithmic amplifier**. The circuit has negative feedback, and the summing-point constraint applies. Assume an ideal op amp, that v_{in} is positive, and that the

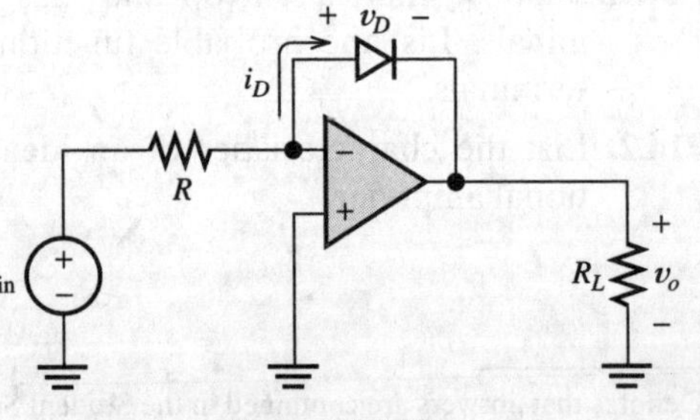

Figure P14.13

diode current given by Equation 10.4, which states that $i_D = I_s \exp(v_D/nV_T)$. Derive an expression for v_o in terms of v_{in}, R, I_s, n, and V_T.

P14.14. Consider the inverting amplifier shown in Figure P14.14. Assuming an ideal op amp, solve for the currents and voltages shown. According to Kirchhoff's current law, the sum of the currents entering a closed surface must equal the sum of the currents leaving. Explain how the law is satisfied for the closed surface shown when we use a real op amp in this circuit.

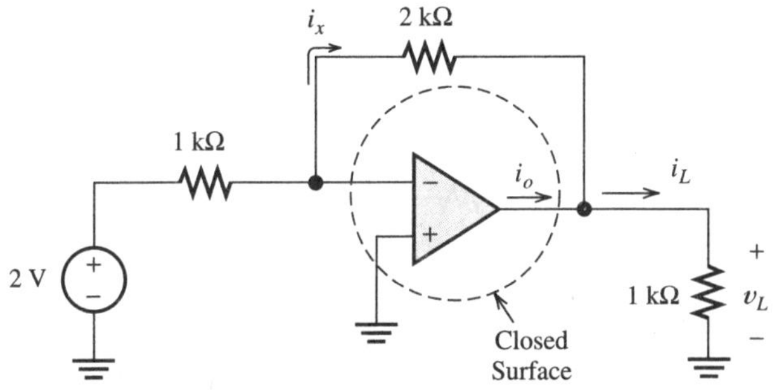

Figure P14.14

P14.15. Consider the circuit shown in Figure P14.13, with an unusual diode that has $i_D = Kv_D^3$. Assume that the summing-point constraint applies, and derive an expression for v_o in terms of v_{in}, R, and K.

P14.16. The op amp shown in Figure P14.16 is ideal, except that the extreme output voltages that it can produce are ±5 V. Determine two possible values for each of the voltages shown, assuming that the input current to the inverting input of the op amp is negligible. (*Hint*: Notice that this circuit has *positive* feedback so the summing-point constraint does not apply.)

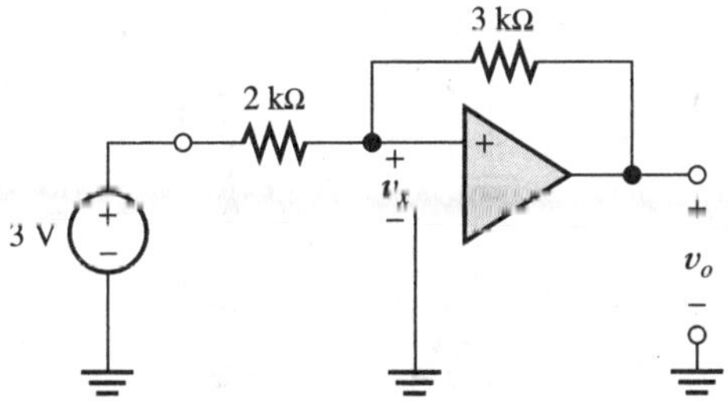

Figure P14.16

Section 14.3: Noninverting Amplifiers

***P14.17.** The voltage follower of Figure 14.12 on page 687 has unity voltage gain so that $v_o = v_{in}$. Why not simply connect the load directly to the source, thus eliminating the op amp? Give an example of a situation in which the voltage follower is particularly good compared with the direct connection.

***P14.18.** Draw the circuit diagram of an op-amp voltage follower. What value is its voltage gain? Input impedance? Output impedance?

***P14.19.** Analyze the ideal op-amp circuit shown in Figure P14.19 to find an expression for v_o in terms of v_A, v_B, and the resistance values.

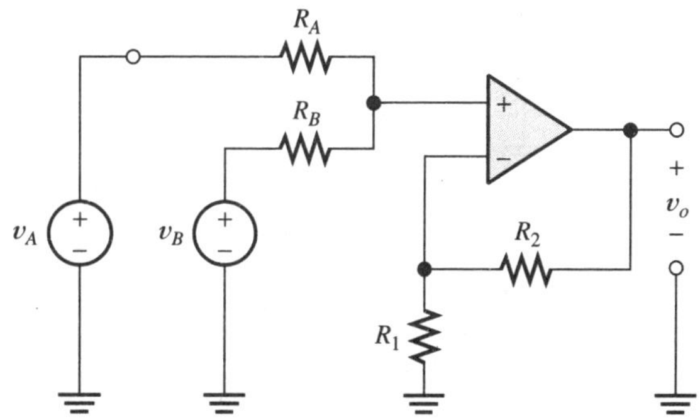

Figure P14.19

***P14.20.** Consider the circuit shown in Figure P14.20. **a.** Find an expression for the output voltage in terms of the source current and resistance values. **b.** What value is the output impedance of this circuit? **c.** What value is the input impedance of this circuit? **d.** This circuit can be classified as an ideal amplifier. What is the amplifier type? (See Section 11.6 for a discussion of various ideal-amplifier types.)

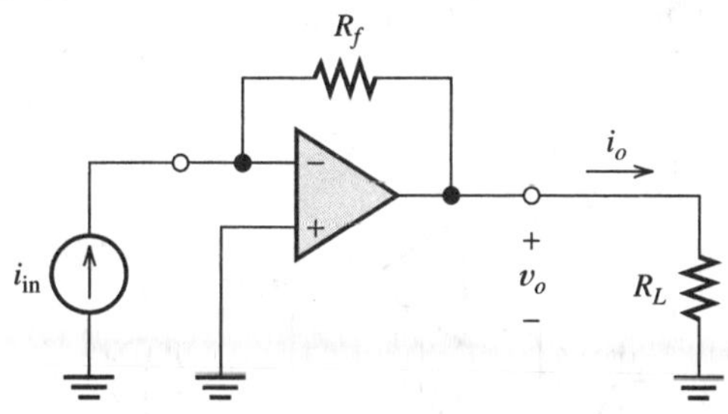

Figure P14.20

P14.21. Draw the circuit diagram of the basic noninverting amplifier configuration. Give an expression for the closed-loop voltage gain of the circuit in terms of the resistances, assuming an ideal op amp. Give expressions for the input impedance and output impedance of the circuit.

P14.22. For each of the circuits shown in Figure P14.22, assume that the op amp is ideal and find the value of v_o. Each of the circuits has negative feedback, and the summing-point constraint applies.

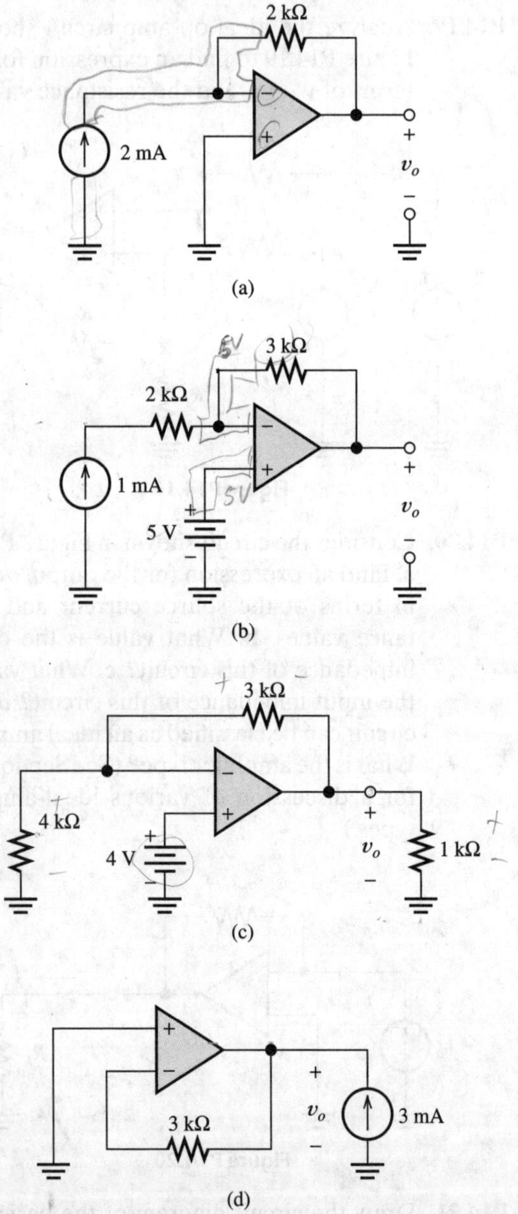

Figure P14.22

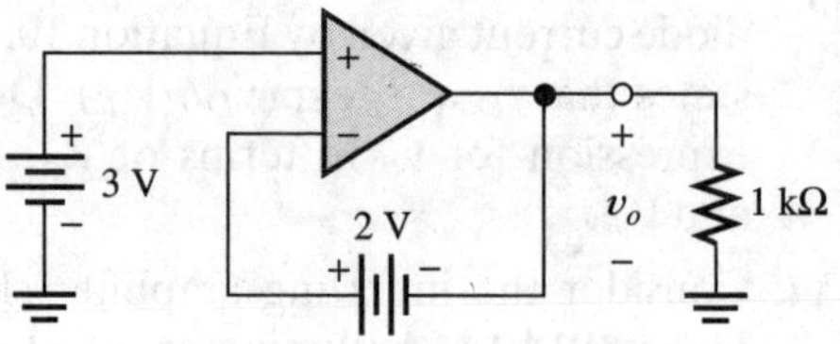

Figure P14.22 (Cont.)

P14.23. The circuit shown in Figure P14.23 has

$$v_{in}(t) = 3 + 3\cos(2000\pi t)$$

Determine the value required for R_2 so that the dc component of the output $v_o(t)$ is zero. What is the resulting output voltage?

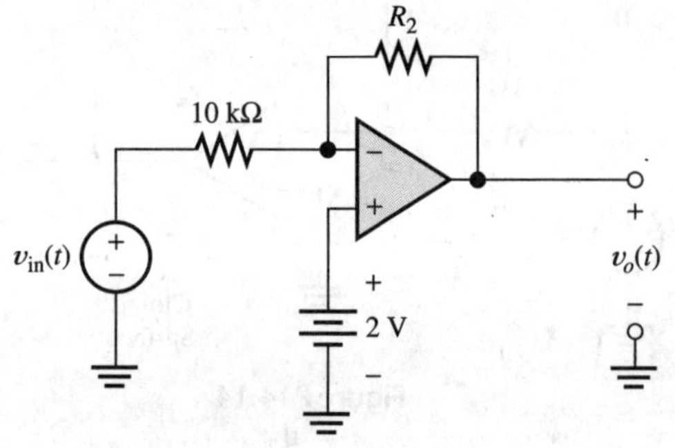

Figure P14.23

P14.24. Consider the circuit shown in Figure P14.24. **a.** Find an expression for the output current i_o in terms of the input current i_{in} and resistance values. **b.** What value is the output impedance of this circuit? **c.** What value is the input impedance of this circuit? **d.** This circuit can be classified as an ideal amplifier. What is the amplifier type? (See Section 11.6 for a discussion of various ideal-amplifier types.)

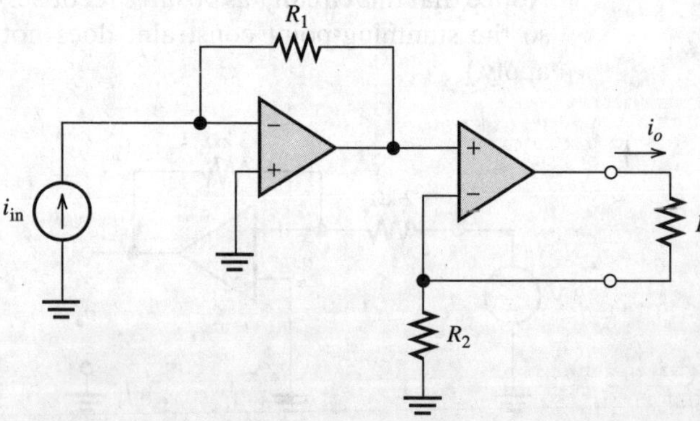

Figure P14.24

P14.25. Analyze each of the ideal-op-amp circuits shown in Figure P14.25 to find expressions for i_o. What is the value of the output impedance for each of these circuits? Why? [*Note:* The bottom end of the input voltage source is *not* grounded in part (b) of the figure. Thus, we say that this source is *floating*.]

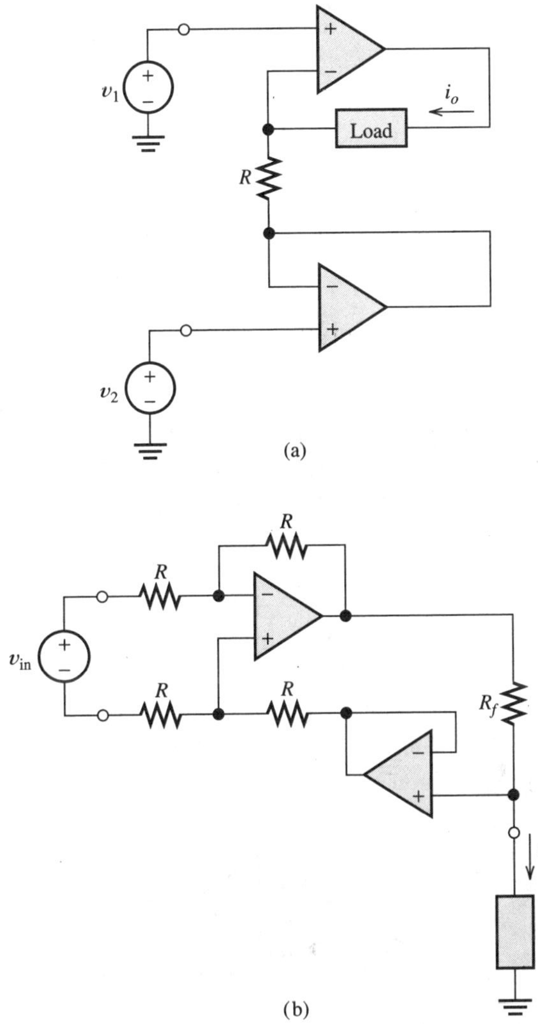

Figure P14.25

P14.26. Consider the circuit shown in Figure P14.26. **a.** Find an expression for the output voltage in terms of the source current and resistance values. **b.** What value is the output impedance of this circuit? **c.** What value is the input impedance of this circuit? **d.** This circuit can be classified as an ideal amplifier. What is the amplifier type? (See Section 11.6 for a discussion of various ideal-amplifier types.)

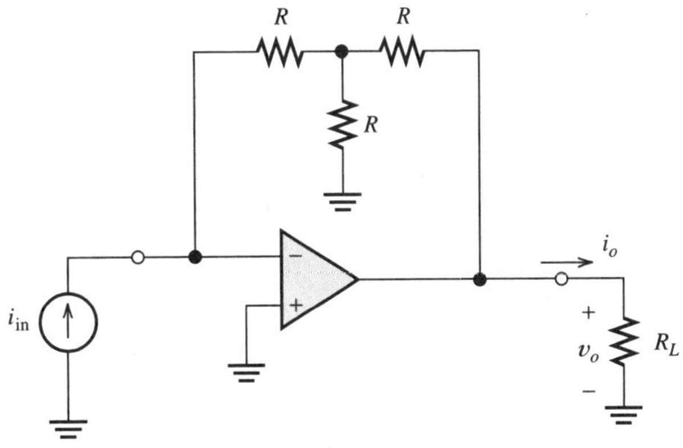

Figure P14.26

P14.27. We have an inverting amplifier using 5-percent-tolerance resistors and an ideal op amp. The nominal amplifier gain is -10. What are the minimum and maximum gains possible, assuming that the resistances are within the stated tolerance? What is the percentage tolerance of the gain?

P14.28. Consider the circuit shown in Figure P14.28. **a.** Find an expression for the output current i_o in terms of the input voltage v_{in} and resistance values. **b.** What value is the output impedance of this circuit? **c.** What value is the input impedance of this circuit? **d.** This circuit can be classified as an ideal amplifier. What is the amplifier type? (See Section 11.6 for a discussion of various ideal-amplifier types.)

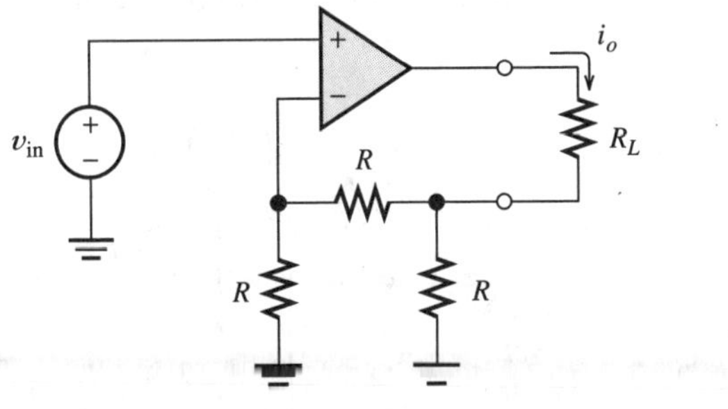

Figure P14.28

P14.29. The power gain G of an amplifier is defined to be the power delivered to the load R_L

divided by the power delivered by the signal source v_s. Find an expression for the power gain of each of the amplifiers shown in Figure P14.29. Assume ideal op amps. Which circuit has the larger power gain?

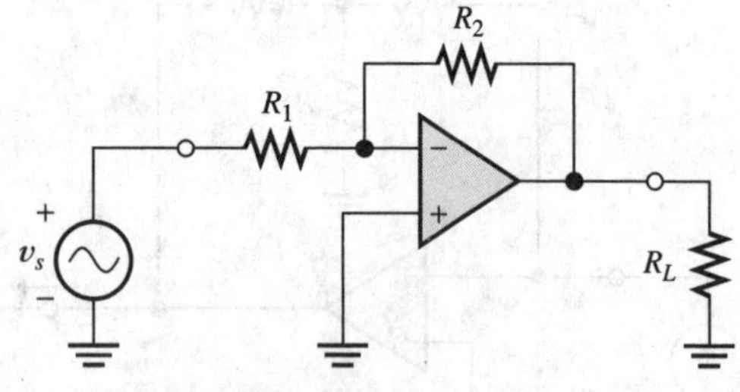

(a) Inverting amplifier

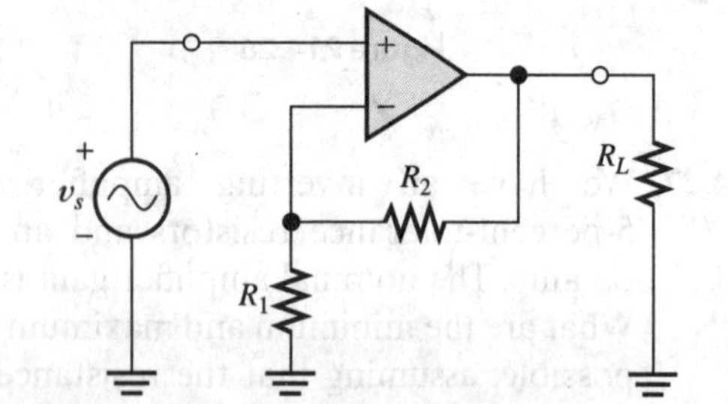

(b) Noninverting amplifier

Figure P14.29

***P14.30.** Consider the circuits shown in Figure P14.30(a) and (b). One of the circuits has negative feedback, and the other circuit has positive feedback. Assume that the op amps are ideal, except that the output voltage is limited to extremes of ±5 V. For the input voltage waveform shown in Figure P14.30(c), sketch the output voltage $v_o(t)$ to scale versus time for each circuit.

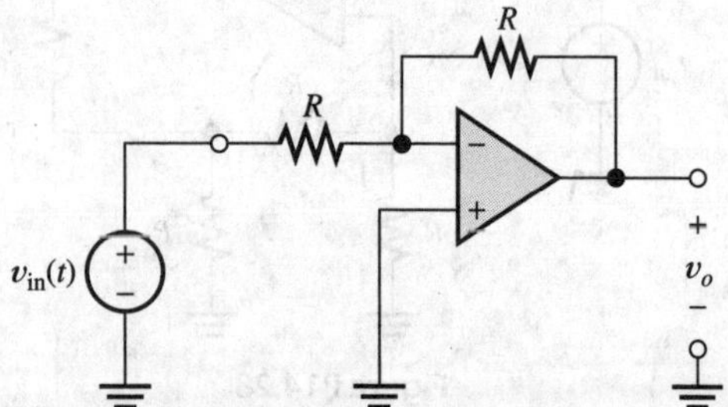

(a)

Figure P14.30

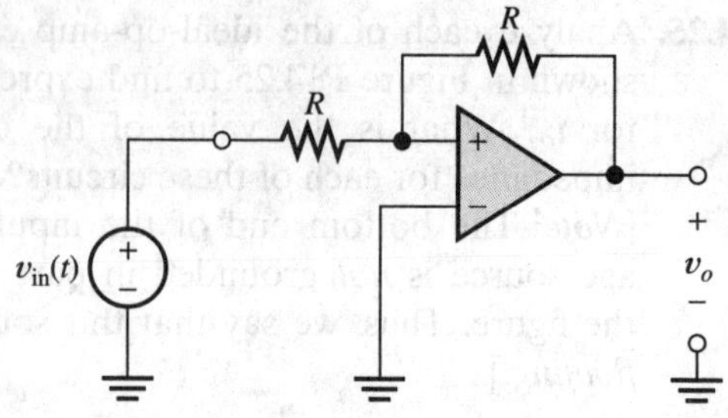

(b)

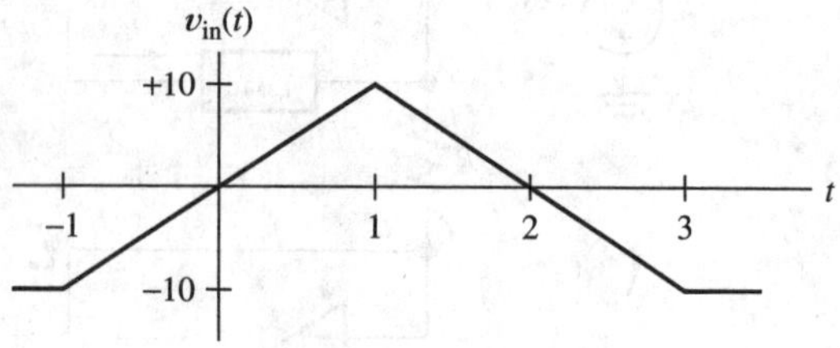

(c)

Figure P14.30 *(Cont.)*

P14.31. Repeat Problem P14.30 for the circuits of Figure P14.31(a) and (b). [The input voltage waveform is shown in Figure P14.30(c).]

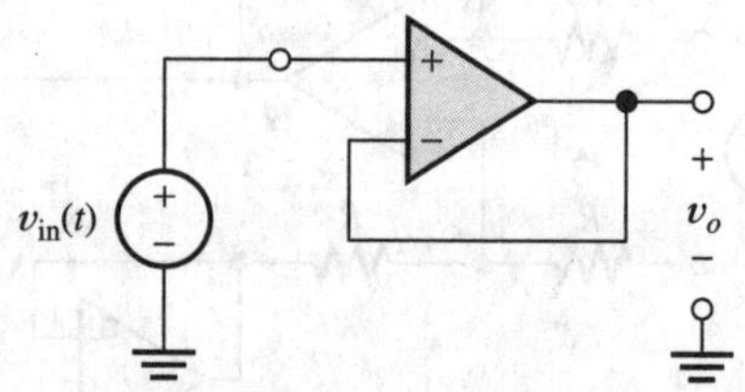

(a)

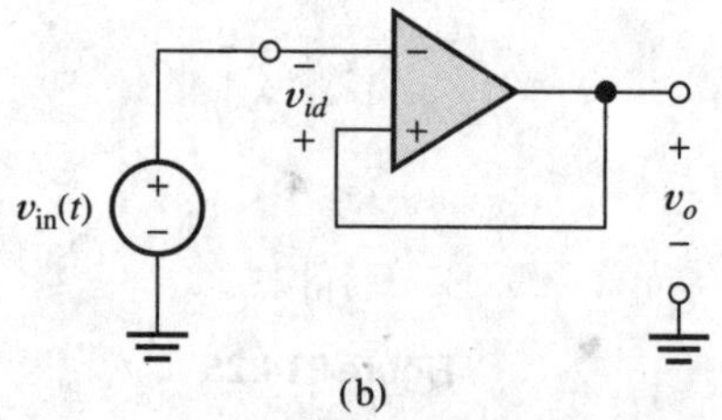

(b)

Figure P14.31

P14.32. The circuit shown in Figure P14.32 employs negative feedback. Use the summing-point constraint (for both op amps) to derive expressions for the voltage gains $A_1 = v_{o1}/v_{in}$ and $A_2 = v_{o2}/v_{in}$.

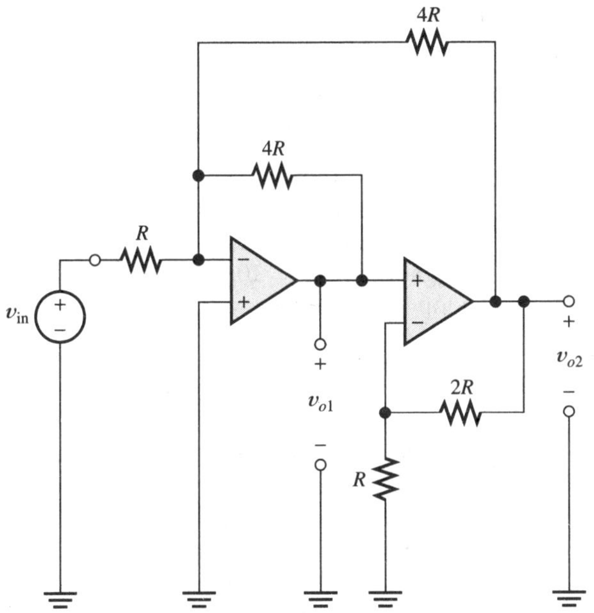

Figure P14.32

P14.33. We have a noninverting amplifier using 5-percent-tolerance resistors and an ideal op amp. The nominal amplifier gain is 10. What are the minimum and maximum gains possible, assuming that the resistances are within the stated tolerance? What is the percentage tolerance of the gain?

***P14.34.** Consider the amplifier shown in Figure P14.34. Find an expression for the output current i_o. What is the input impedance? What is the output impedance seen by R_L?

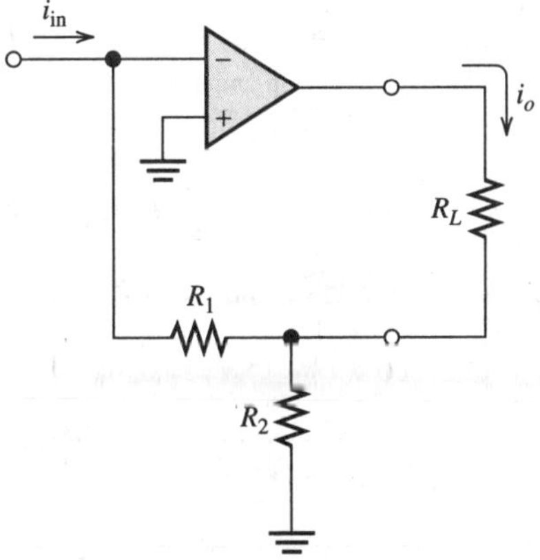

Figure P14.34

P14.35. Derive an expression for the voltage gain of the circuit shown in Figure P14.35 as a function of T, assuming an ideal op amp. (T varies from 0 to unity, depending on the position of the wiper of the potentiometer.)

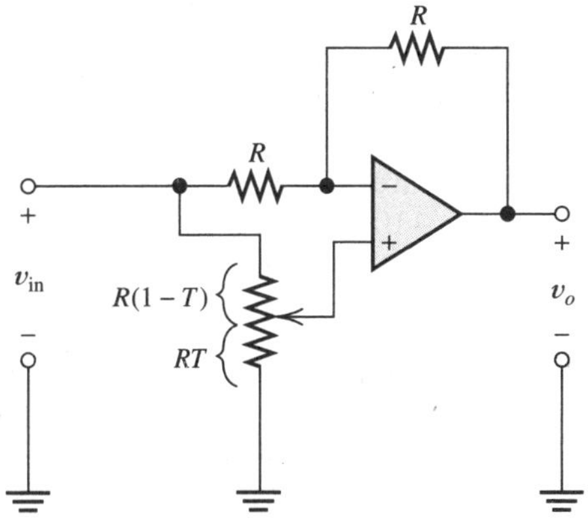

Figure P14.35

Section 14.4: Design of Simple Amplifiers

P14.36. Suppose that we are designing an amplifier, using an op amp. What problems are associated with using very small feedback resistances? With very large feedback resistances?

***P14.37.** Using the components listed in Table P14.37, design an amplifier having a voltage gain of -10 ± 20 percent. The input impedance is required to be as large as possible (ideally, an open circuit). Remember to use prac-

Table P14.37. Available Parts for Design Problems

Standard 5%-tolerance resistors. (See Appendix B.)

Standard 1%-tolerance resistors. (Don't use these if a 5%-tolerance resistor will do, because 1%-tolerance resistors are more expensive.)

Ideal op amps.

Adjustable resistors (trimmers) having maximum values ranging from 100 Ω to 1 MΩ in a 1–2–5 sequence (i.e., 100 Ω, 200 Ω, 500 Ω, 1 kΩ, etc.). Don't use trimmers if fixed resistors will suffice.

tical resistance values. (*Hint:* Cascade a noninverting stage with an inverting stage.)

***P14.38.** For Example 14.4 on page 692, it is possible to achieve a design that requires only one op amp. Find a suitable circuit configuration and resistance values. For this problem, the gain tolerances are relaxed to ±5 percent.

P14.39. Using the components listed in Table P14.37, design a circuit for which the output voltage is $v_o = A_1 v_1 + A_2 v_2$. The voltages v_1 and v_2 are input voltages. Design to achieve $A_1 = -10 \pm 5$ percent; $A_2 = 20 \pm 5$ percent. The input impedances are required to be as large as possible (ideally, open circuits).

P14.40. Using the components listed in Table P14.37, design an amplifier having an input impedance of at least 100 kΩ and a voltage gain of: **a.** 5 ± 20 percent; **b.** 5 ± 5 percent; **c.** 5 ± 0.5 percent.

P14.41. Using the components listed in Table P14.37, design a circuit for which the output voltage is $v_o = A_1 v_1 + A_2 v_2$. The voltages v_1 and v_2 are input voltages. Design to achieve $A_1 = -4 \pm 5$ percent and $A_2 = -4 \pm 5$ percent. The input impedances are required to be as large as possible (ideally, open circuits).

P14.42. Design an amplifier having a voltage gain of −5 ± 3 percent and an input impedance of 5 kΩ ± 1 percent, using the components listed in Table P14.37.

P14.43. Two signal sources have internal voltages $v_1(t)$ and $v_2(t)$, respectively. The internal (i.e., Thévenin) resistances of the sources are known always to be less than 2 kΩ, but the exact values are not known and are likely to change over time. Using the components listed in Table P14.37, design an amplifier for which the output voltage is $v_o(t) = A_1 v_1(t) + A_2 v_2(t)$. The gains are to be $A_1 = -10 \pm 1$ percent and $A_2 = 3 \pm 1$ percent.

P14.44. Suppose we have a signal source with an internal (i.e., Thévenin) impedance that is always less than 1000 Ω, but is variable over time. Using the components listed in Table P14.37, design an amplifier that produces an amplified version of the internal source voltage. The voltage gain should be −10 ± 5 percent.

Section 14.5: Op-Amp Imperfections in the Linear Range of Operation

P14.45. List the imperfections of real op amps in their linear range of operation.

***P14.46.** A certain op amp has a unity-gain bandwidth of $f_t = 15$ MHz. If this op amp is used in a noninverting amplifier having a closed-loop dc gain of $A_{0CL} = 10$, determine the closed-loop break frequency f_{BCL}. Repeat for a dc gain of 100.

***P14.47.** A certain op amp has an open-loop dc gain of $A_{0OL} = 200{,}000$ and an open-loop 3-dB bandwidth of $f_{BOL} = 5$ Hz. Sketch the Bode plot of the open-loop gain magnitude to scale. If this op amp is used in a noninverting amplifier having a closed-loop dc gain of 100, sketch the Bode plot of the closed-loop gain magnitude to scale. Repeat for a closed-loop dc gain of 10.

P14.48. A certain op amp has an open-loop dc gain of $A_{0OL} = 5 \times 10^6$ and an open-loop 3-dB bandwidth of $f_{BOL} = 25$ Hz. Find the open-loop gain magnitude at a frequency of: **a.** 100 Hz; **b.** 10 kHz; **c.** 5 MHz.

P14.49. The objective of this problem is to investigate the effects of finite open-loop gain, finite input impedance, and nonzero output impedance of the op amp on the voltage follower. The circuit, including the op-amp model, is shown in Figure P14.49. **a.** Derive an expression for the circuit voltage gain v_o/v_s. Evaluate for $A_{OL} = 10^5$, $R_{in} = 1$ MΩ,

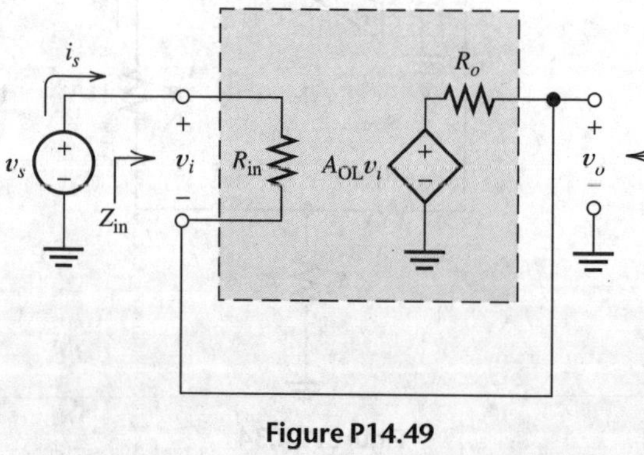

Figure P14.49

and $R_o = 25\ \Omega$. Compare this result to the gain with an ideal op amp. **b.** Derive an expression for the circuit input impedance $Z_{\text{in}} = v_s/i_s$. Evaluate for $A_{\text{OL}} = 10^5$, $R_{\text{in}} = 1\ \text{M}\Omega$, and $R_o = 25\ \Omega$. Compare this result to the input impedance with an ideal op amp. **c.** Derive an expression for the circuit output impedance Z_o. Evaluate for $A_{\text{OL}} = 10^5$, $R_{\text{in}} = 1\ \text{M}\Omega$, and $R_o = 25\ \Omega$. Compare this result to the output impedance with an ideal op amp.

P14.50. We need a noninverting amplifier that has a dc gain of 10, and the gain magnitude at 5 kHz must be not less than 9.5. Determine the minimum gain–bandwidth specification required for the op amp.

P14.51. We need a noninverting amplifier that has a dc gain of 5, and the phase shift for the 100-kHz component must not exceed 10° in magnitude. Determine the minimum gain–bandwidth specification required for the op amp.

P14.52. We have the following two alternatives for designing an amplifier having a dc gain of 25. The first alternative is to use a single noninverting stage, having a gain of 25. The second alternative is to cascade two noninverting stages, each having a gain of 5. Op amps having a gain–bandwidth product of 10^6 are to be used. Draw the circuit diagram, write an expression for the gain as a function of frequency, and find the 3-dB bandwidth for each alternative. Which alternative has the greater bandwidth?

P14.53. The objective of this problem is to investigate the effects of finite gain, finite input impedance, and nonzero output impedance of the op amp on the inverting amplifier. The circuit, including the op-amp model, is shown in Figure P14.53. **a.** Derive an expression for the circuit voltage gain v_o/v_s. Evaluate for $A_{\text{OL}} = 10^5$, $R_{\text{in}} = 1\ \text{M}\Omega$, $R_o = 25\ \Omega$, $R_1 = 1\ \text{k}\Omega$, and $R_2 = 10\ \text{k}\Omega$. Compare this result to the gain with an ideal op amp. **b.** Derive an expression for the circuit input impedance $Z_{\text{in}} = v_s/i_s$. Evaluate for $A_{\text{OL}} = 10^5$, $R_{\text{in}} = 1\ \text{M}\Omega$, $R_o = 25\ \Omega$, $R_1 = 1\ \text{k}\Omega$, and $R_2 = 10\ \text{k}\Omega$. Compare this result to the input impedance with an ideal op amp.

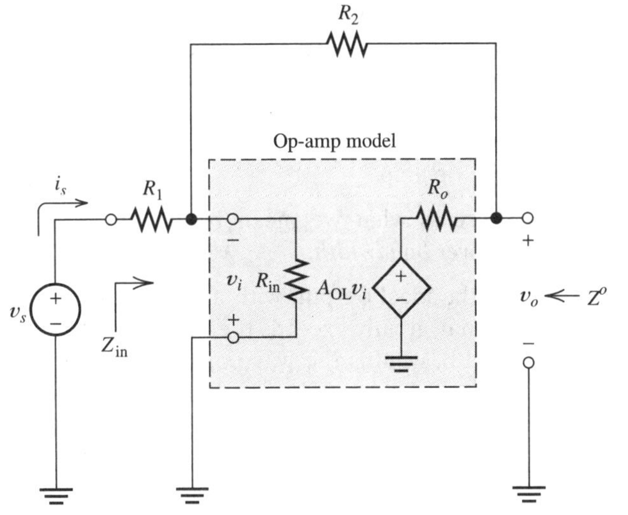

Figure P14.53

c. Derive an expression for the circuit output impedance Z_o. Evaluate for $A_{\text{OL}} = 10^5$, $R_{\text{in}} = 1\ \text{M}\Omega$, $R_o = 25\ \Omega$, $R_1 = 1\ \text{k}\Omega$, and $R_2 = 10\ \text{k}\Omega$. Compare this result to the output impedance with an ideal op amp.

Section 14.6: Nonlinear Limitations

P14.54. What are the nonlinear limitations of real op amps?

***P14.55.** Suppose that we have an op amp with a maximum output voltage range from −10 to +10 V. The maximum output current magnitude is 20 mA. The slew-rate limit is SR = 10 V/μs. This op amp is used in the circuit of Figure 14.28 on page 703. **a.** Find the full-power bandwidth of the op amp. **b.** For a frequency of 1 kHz and $R_L = 1\ \text{k}\Omega$, what peak output voltage is possible without distortion? **c.** For a frequency of 1 kHz and $R_L = 100\ \Omega$, what peak output voltage is possible without distortion? **d.** For a frequency of 1 MHz and $R_L = 1\ \text{k}\Omega$, what peak output voltage is possible without distortion? **e.** If $R_L = 1\ \text{k}\Omega$ and $v_s(t) = 5\sin(2\pi 10^6 t)$, sketch the steady-state output waveform to scale versus time.

***P14.56.** One way to measure the slew-rate limitation of an op amp is to apply a sine wave (or square wave) as the input to an amplifier and then increase the frequency until the output waveform becomes triangular. Suppose that a 1-MHz input signal produces a triangular output waveform having a peak-to-peak amplitude of 4 V. Determine the slew rate of the op amp.

P14.57. Discuss what we mean by the term *full-power bandwidth.*

P14.58. If the ideal output, with a sinusoidal input signal, greatly exceeds the full-power bandwidth, what is the waveform of the output signal? Under these conditions, if the slew rate of the op amp is 10 V/μs and the frequency of the input is 1 MHz, what is the peak-to-peak amplitude of the output signal?

P14.59. We need an amplifier that can produce a 200-kHz sine-wave output having a peak amplitude of 5 V. What is the minimum slew-rate specification allowed for the op amp?

P14.60. We need a noninverting amplifier with a dc gain of 5 to amplify an input signal, in volts, given by

$$v_{\text{in}}(t) = 0 \quad t \le 0$$
$$= 5t \quad t \ge 0$$

in which t is in ms. Determine the minimum slew-rate specification required for the op amp if distortion must be avoided.

P14.61. The circuit shown in Figure P14.61 is known as a **bridge amplifier. a.** Assuming ideal op amps, derive an expression for the voltage gain v_o/v_s. **b.** If $v_s(t) = 3\sin(\omega t)$, sketch $v_1(t)$, $v_2(t)$, and $v_o(t)$ to scale versus time. **c.** If the op amps are supplied from ± 15 V and clip at output voltages of ± 13 V, what is the peak value of $v_o(t)$ just at the threshold of clipping? (*Comment:* This circuit can be useful if a peak output voltage greater than the magnitude of the supply voltages is required.)

P14.62. A voltage follower is needed to amplify an input signal given by

$$v_{\text{in}}(t) = 0 \quad t \le 0$$
$$= t^3 \quad 0 \le t \le 2$$
$$= 8 \quad 2 \le t$$

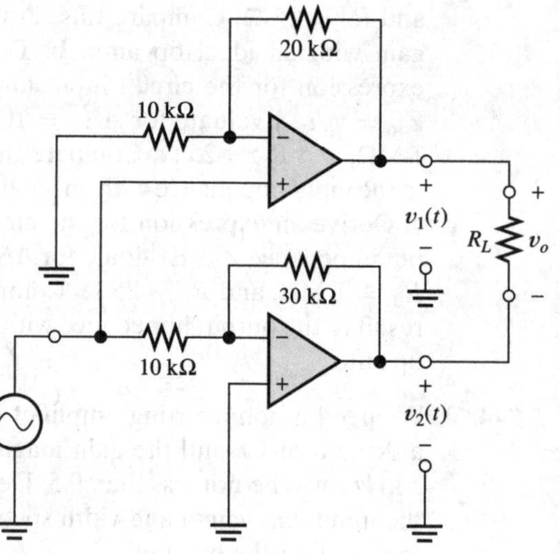

Figure P14.61

in which t is in μs. Determine the minimum slew-rate specification required for the op amp if distortion must be avoided.

P14.63. The op amp shown in Figure P14.63 has a maximum output voltage range from −5 to +5 V. The maximum output current magnitude of the op amp is 20 mA. The slew-rate limit is 5 V/μs. **a.** Find the full-power bandwidth of the op amp. **b.** For a frequency of 5 kHz and $R_L = 50\ \Omega$, what peak output voltage is possible without distortion? **c.** For a frequency of 5 kHz and $R_L = 10\ \text{k}\Omega$, what peak output voltage is possible without distortion? **d.** For a frequency of 1 MHz and $R_L = 10\ \text{k}\Omega$, what peak output voltage is possible without distortion?

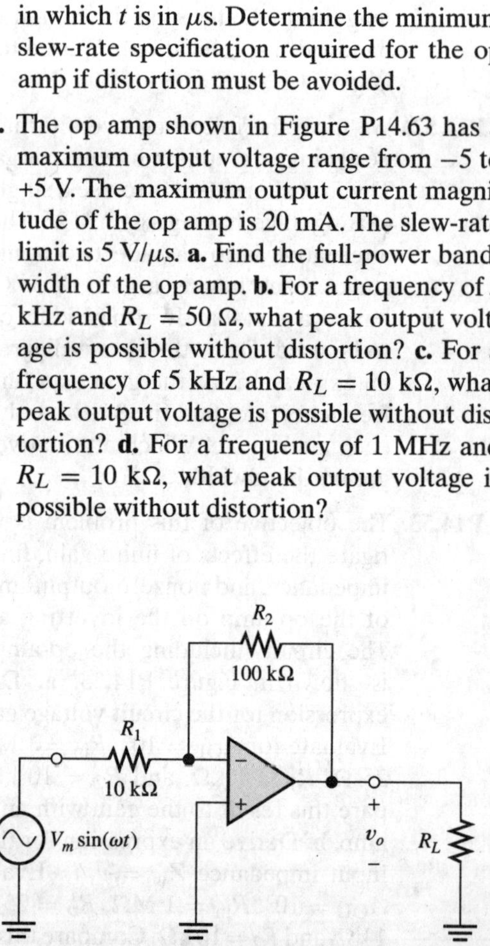

Figure P14.63

Section 14.7: DC Imperfections

***P14.64.** Draw the circuit symbol for an op amp, adding sources to account for dc imperfections.

P14.65. Name the advantage of a FET-input op amp compared with a BJT-input op amp.

P14.66. What are the dc imperfections of real op amps and the net effect of these dc imperfections?

***P14.67.** Find the worst-case dc output voltages of the inverting amplifier shown in Figure 14.30(a) on page 705 for $v_{in} = 0$. The bias current ranges from 100 to 200 nA, the maximum offset current magnitude is 50 nA, and the maximum offset voltage magnitude is 4 mV.

P14.68. Sometimes, an ac-coupled amplifier is needed. The circuit shown in Figure P14.68 is a poor way to accomplish ac coupling. Explain why. (*Hint:* Consider the effect of bias current.) Show how to add a component (including its value) so that bias current has no effect on the output voltage of this circuit.

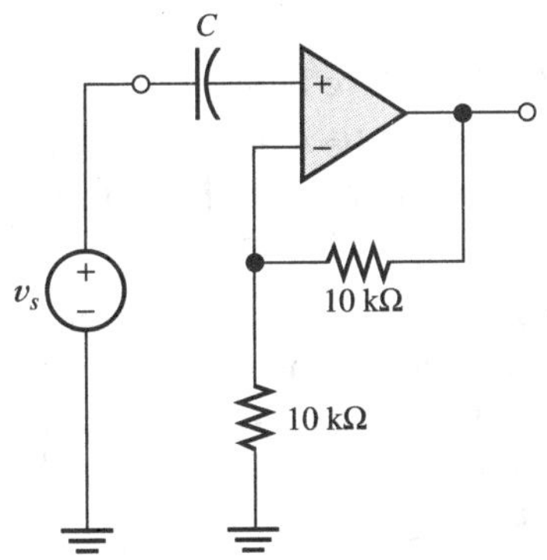

Figure P14.68

P14.69. Consider the amplifier shown in Figure P14.63. With zero dc input voltage from the signal source, it is desired that the dc output voltage be no greater than 50 mV. **a.** Ignoring other dc imperfections, what is the maximum offset voltage allowed for the op amp? **b.** Ignoring other dc imperfections, what is the maximum bias current allowed for the op amp? **c.** Show how to add a resistor to the circuit (including its value) so that the effects of the bias currents cancel. **d.** Assuming that the resistor of part (c) is in place, and ignoring offset voltage, what is the maximum offset current allowed for the op amp?

Section 14.8: Differential and Instrumentation Amplifiers

P14.70. In terms of the differential and common-mode components of a signal, what is the function of a differential amplifier?

***P14.71.** Using the parts listed in Table P14.37, design a single-op-amp differential amplifier having a nominal differential gain of 10.

P14.72. Repeat Problem P14.71, using the instrumentation-quality circuit shown in Figure 14.34 on page 708.

P14.73. We have an instrumentation-quality differential amplifier as shown in Figure 14.34 on page 708, with $R_1 = 2$ kΩ, $R_2 = 4$ kΩ, and $R = 15$ kΩ. The input signals, in volts, are given by

$$v_1(t) = 5\sin(1200\pi t) + 2\cos(120\pi t)$$

$$v_2(t) = -5\sin(2000\pi t) + 2\cos(120\pi t)$$

a. Find expressions for the differential and common-mode components of the input signal.

b. Assuming ideal op amps, find expressions for the voltages at the output terminals of X_1 and X_2.

c. Again assuming ideal op amps, find an expression for the output voltage $v_o(t)$.

Section 14.9: Integrators and Differentiators

P14.74. What do we mean by the term *running time integral*?

***P14.75.** Sketch the output voltage of the circuit shown in Figure P14.75 to scale versus time. Sometimes, an integrator circuit is used as a (approximate) pulse counter. Suppose that the output voltage is −10 V. How many input pulses have been applied (assuming that the pulses have an amplitude of 5 V and a duration of 2 ms, as shown in the figure)?

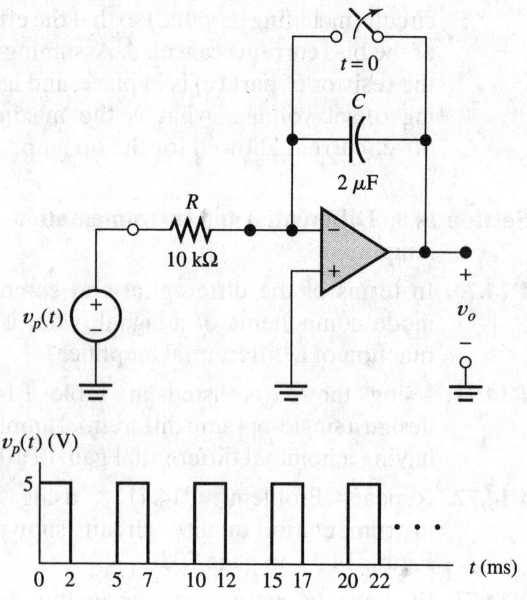

Figure P14.75

P14.76. Sketch the output voltage of the ideal op-amp circuit shown in Figure P14.76 to scale versus time.

P14.77. The displacement of a robot arm in a given direction is represented by a voltage signal $v_{in}(t)$. The voltage is proportional to displacement, and 1 V corresponds to a displacement of 10 mm from the reference position. Design a circuit that produces a voltage $v_1(t)$ that is proportional to the velocity of the robot arm such that 1 m/s corresponds to 1 V. Design an additional circuit that produces a voltage v_2 that is proportional to the acceleration of the robot arm such that 1 m/s² corresponds to 1 V. Use the components listed in Table P14.37, plus as many capacitors as needed.

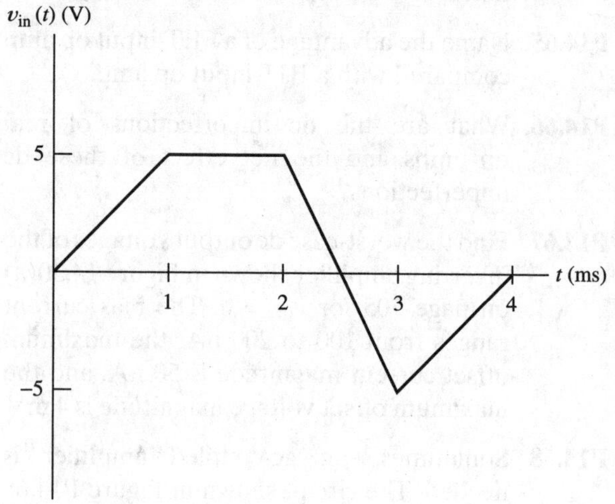

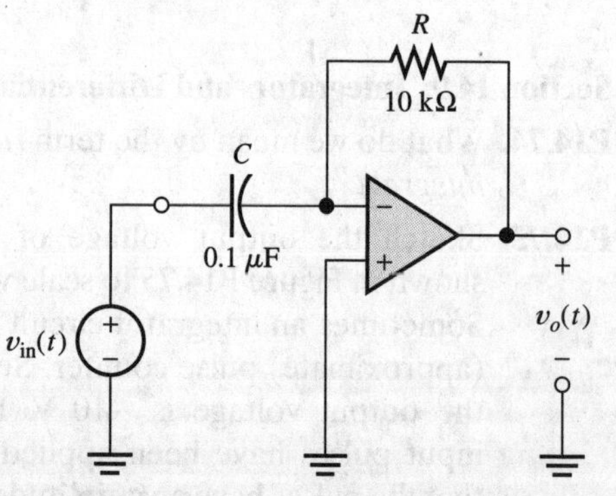

Figure P14.76

Section 14.10: Active Filters

P14.78. What are the 6 desired characteristics of an active filter? Differentiate between active and passive filters.

***P14.79.** Derive an expression for the voltage transfer ratio of each of the circuits shown in Figure P14.79. Also, sketch the magnitude Bode plots to scale. Assume that the op amps are ideal.

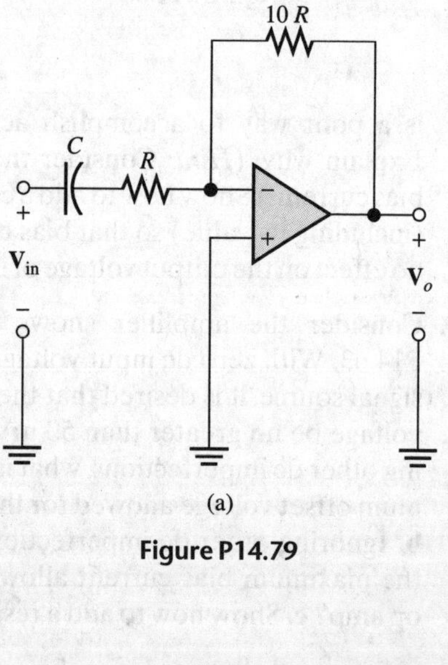

(a)

Figure P14.79

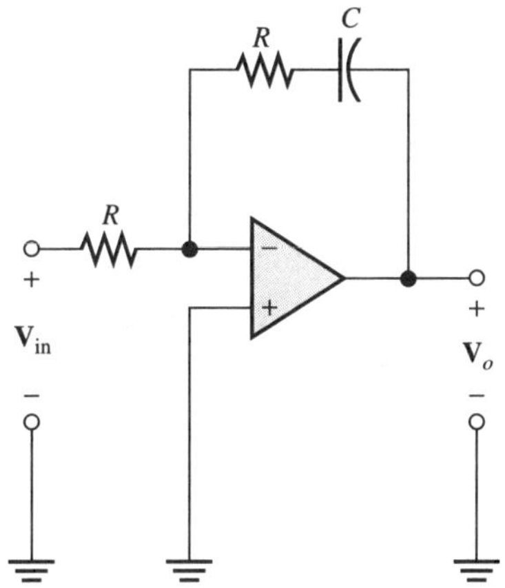

(b)

Figure P14.79 (*Cont.*)

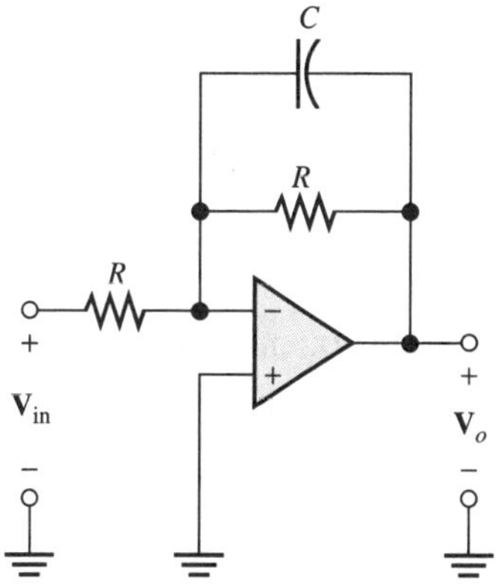

Figure P14.82

P14.80. It is illuminating to look at the integrator circuit as a filter. Derive the transfer function for the integrator of Figure P14.75, and sketch the magnitude Bode plot to scale.

P14.81. Repeat Problem P14.80 for the differentiator circuit shown in Figure P14.76.

P14.82. Derive an expression for the voltage transfer ratio for the circuit shown in Figure P14.82. Also, sketch the magnitude Bode plot to scale. Assume that the op amp is ideal.

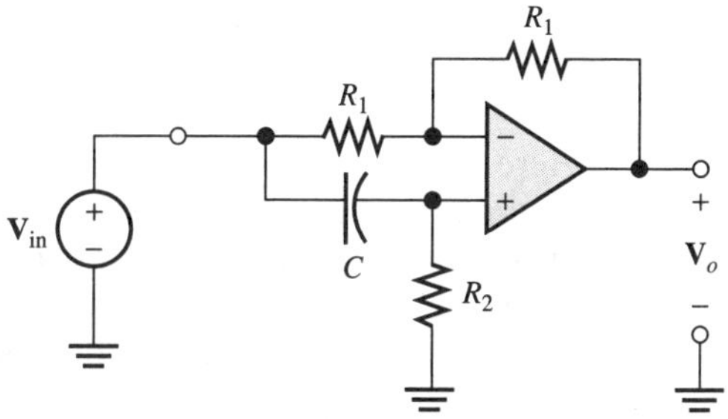

Figure P14.83

P14.83. Derive an expression for the voltage transfer ratio for the circuit shown in Figure P14.83. Then find expressions for the magnitude and phase of the transfer function. What effect does this filter have on the amplitude of a sinusoidal input? On the phase?

Frequently Asked Questions

1. What is an operational amplifier (OP amp)?
2. What do you mean by open-loop gain of an ideal op-amp?
3. How an operational amplifier is used as an inverting amplifier?
4. Derive an expression for the closed-loop gain of an inverting amplifier?
5. What do you mean by virtual-short-circuit?
6. How an op-amp is used as a non-inverting amplifier? Under what conditions it works as a voltage follower.
7. Design a simple amplifier using resistive feedback network with an op-amp.
8. Mention the imperfections when an op-amp is operated in linear range.
9. What are the dc imperfections of an op-amp?
10. For an op-amp, define
    (i) bias current
    (ii) offset current
    (iii) offset voltage

11. Differentiate between differential and instrumentation amplifiers.
12. Draw and explain the circuit of a differentiator using op-amp.
13. Draw and explain the circuit of an integrator using op-amp.
14. What do you mean by an active filter?
15. Write down the expression for Butterworth Transfer Function.
16. Draw and explain a lowpass active filter used with an operational amplifier.
17. Find the closed loop circuit gain and output voltage for an inverting op-amp having input signal V_{in} as 1 volt (peak to peak) and R_1 and R_2 as 1 kΩ and 10 kΩ respectively.

Solution: Here, $V_{in} = 1V; R_2 = 10\ k\Omega; R_1 = 1\ k\Omega$
closed loop circuit gain with negative feedback

$$A_{CL} = \frac{R_2}{R_1} = \frac{10\ k\Omega}{1\ k\Omega} = 10 \qquad \textbf{(Ans)}$$

Output voltage signal peak to peak value

$$V_{out} = A_{CL} \times V_{in} = 10 \times 1 = 10\ V \qquad \textbf{(Ans)}$$

Practice Test

Here is a practice test you can use to check your comprehension of the most important concepts in this chapter. Answers can be found in Appendix D and complete solutions are included in the Student Solutions files. See Appendix E for more information about the Student Solutions.

T14.1. Draw the circuit diagram for each of the following amplifiers. Clearly label the op amp input terminals and any resistances needed. Also, give the equation for the voltage gain of the amplifier in terms of the resistances on your diagram. **a.** The basic inverter. **b.** The noninverting amplifier. **c.** The voltage follower.

T14.2. Derive an expression for voltage gain $A_v = v_o/v_{in}$ of the circuit shown in Figure T14.2 assuming that the summing-point constraint applies. (There are both negative and positive feedback paths, but the resistances have been carefully selected so the circuit has net negative feedback and the summing-point constraint does apply.)

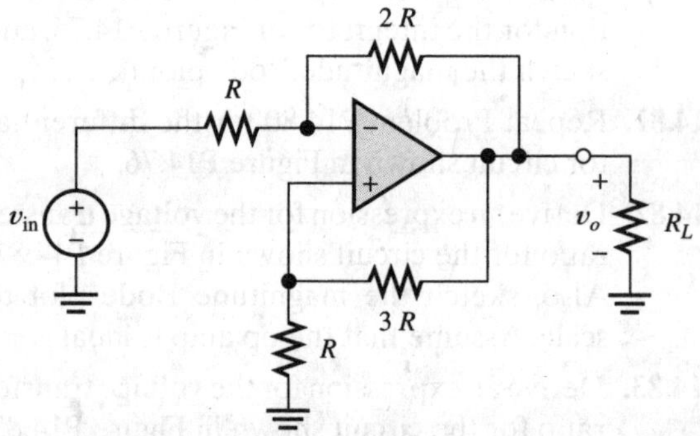

Figure T14.2

T14.3. A certain op amp has an open-loop dc gain of $A_{OOL} = 200{,}000$ and an open-loop 3-dB bandwidth of $f_{BOL} = 5$ Hz. This op amp is used in a noninverting amplifier having a closed-loop dc gain of $A_{OCL} = 100$. **a.** Determine the closed-loop break frequency f_{BCL}. **b.** Given that the input voltage to the noninverting amplifier is $v_{in}(t) = 0.05\cos(2\pi \times 10^5 t)$ V, find the expression for the output voltage.

T14.4. We have an op amp with a maximum output voltage range from −4.5 to +4.5 V. The maximum output current magnitude is 5 mA. The slew-rate limit is SR = 20 V/μs. This op amp is used in the circuit of Figure 14.28 on page 703. **a.** Find the full-power bandwidth of the op amp. **b.** For a frequency of 1 kHz and $R_L = 200\ \Omega$, what peak output voltage, V_{om}, is possible without distortion? **c.** For a frequency of 1 kHz and $R_L = 10\ k\Omega$, what peak output voltage, V_{om}, is possible without distortion? **d.** For a frequency of 5 MHz and $R_L = 10\ k\Omega$, what peak output voltage, V_{om}, is possible without distortion?

T14.5. Draw an op amp symbol including sources to account for offset voltage, offset current, and bias current. What is the principal effect of these sources in an amplifier circuit?

T14.6. Draw the circuit diagram of a differential amplifier using one op amp and resistances as needed. Give the output voltage in terms of the input voltages and resistances.

T14.7. Draw the circuit diagrams for an integrator and for a differentiator. Also give an expression for the output voltage in terms of the input voltage and component values.

T14.8. What is a filter? An active filter? Give one application for an active filter.

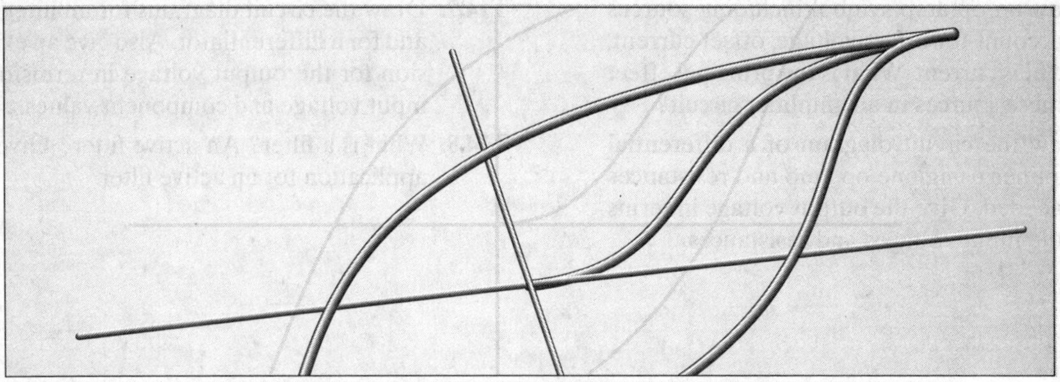

Chapter 15

Magnetic Circuits and Transformers

Study of this chapter will enable you to:

- Understand magnetic fields and their interactions with moving charges.
- Use the right-hand rule to determine the direction of the magnetic field around a current-carrying wire or coil.
- Calculate forces on moving charges and current-carrying wires due to magnetic fields.
- Calculate the voltage induced in a coil by a changing magnetic flux or in a conductor cutting through a magnetic field.
- Use Lenz's law to determine the polarities of induced voltages.
- Apply magnetic-circuit concepts to determine the magnetic fields in practical devices.
- Determine the inductance and mutual inductance of coils, given their physical parameters.
- Understand hysteresis, saturation, core loss, and eddy currents in cores composed of magnetic materials such as iron.
- Understand ideal transformers and solve circuits that include transformers.
- Use the equivalent circuits of real transformers to determine their regulations and power efficiencies.

Introduction to this chapter:

In describing interactions of matter, we often employ field concepts. For example, masses are attracted by gravitation. We envision gravitational fields produced by masses and explain the forces on other masses in terms of their interaction with these fields. Another example is stationary electrical charges. Charges of like sign repel one another, and unlike charges attract one another. Conceptually, each charge creates an electric field, and the other charge interacts with the field, resulting in a force.

In this and the next two chapters, we study some important engineering applications of **magnetic fields**, which are created by electrical charges in motion. Charges moving through magnetic fields experience forces. Furthermore, changing magnetic fields induce voltages in nearby conductors.

In this chapter, we start by reviewing basic magnetic-field concepts. Then, we consider the relationships between magnetic fields and inductance, including mutual inductance. Next, we study **transformers**, which greatly facilitate the distribution of electrical power.

Magnetic fields also form the basis of most practical devices for converting energy between electrical and mechanical forms. In subsequent chapters, we study the basic operating principles of several types of rotating energy conversion devices, collectively known as motors and generators.

15.1 MAGNETIC FIELDS

Magnetic fields exist in the space around permanent magnets and around wires that carry current. In both cases, the basic source of the magnetic field is electrical charge in motion. In an iron permanent magnet, fields are created by the spin of electrons in atoms. These fields aid one another, producing the net external field that we observe. (In most other materials, the magnetic fields of the electrons tend to cancel one another.) If a current-carrying wire is formed into a multiturn coil, the magnetic field is greatly intensified, particularly if the coil is wound around an iron core.

Magnetic flux lines form closed paths that are close together where the field is strong and farther apart where the field is weak.

We can visualize a magnetic field as **lines of magnetic flux** that form closed paths. The lines are close together where the magnetic field is strong and farther apart where the field is weaker. This is illustrated in Figure 15.1. The units of magnetic flux are webers (Wb).

The earth has a natural magnetic field that is relatively weak compared with those in typical transformers, motors, or generators. Due to interactions of the fields, magnets tend to align with the earth's field. Thus, a magnet has a north-seeking end (N) and a south-seeking end (S). Unlike ends of magnets are attracted. By convention, flux lines leave the north-seeking end (N) of a magnet and enter its south-seeking end (S). A compass can be used to investigate the direction of the lines of flux. The compass needle indicates north in the direction of the flux [i.e., the compass points toward the south-seeking (S) end of the magnet]. (Note that the earth's field lines are directed from south to north. Thus, if we were to place N and S marks on the earth as we do on a magnet, S would appear near the *north* geographic pole, because that is where the field lines enter the earth.)

Flux lines leave the north-seeking end of a magnet and enter the south-seeking end.

In equations, we represent the **magnetic flux density** as the vector quantity **B**. (Throughout our discussion, we use boldface for vector quantities. The corresponding lightface italic symbols represent the magnitudes of the vectors. Thus, B represents the magnitude of the vector **B**. We also use boldface for phasors. However, it will be clear from the context which quantities are spatial vectors and which are phasors.) Furthermore, we use the International System of Units (SI), in which the units of **B** are webers/meter^2 (Wb/m^2) or, equivalently, teslas (T). The flux density vector **B** has a direction tangent to the flux lines, as illustrated in Figure 15.1.

When placed in a magnetic field, a compass indicates north in the direction of the flux lines.

Right-Hand Rule

The direction of the magnetic field produced by a current can be determined by the right-hand rule. There are several interpretations of this rule. For example, as

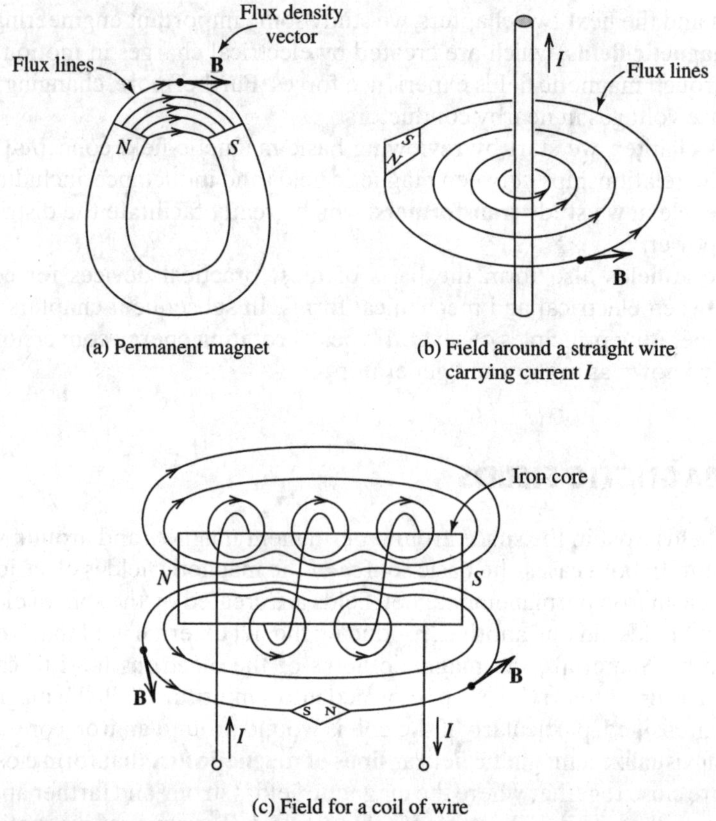

(a) Permanent magnet

(b) Field around a straight wire carrying current I

(c) Field for a coil of wire

Figure 15.1 Magnetic fields can be visualized as lines of flux that form closed paths. Using a compass, we can determine the direction of the flux lines at any point. Note that the flux density vector **B** is tangent to the lines of flux.

The right-hand rule is used to determine the directions of magnetic fields.

illustrated in Figure 15.2(a), if a wire is grasped with the thumb pointing in the direction of the current, the fingers encircle the wire, pointing in the direction of the magnetic field. Moreover, as illustrated in Figure 15.2(b), if the fingers are wrapped around a coil in the direction of current flow, the thumb points in the direction of the magnetic field that is produced inside the coil.

Exercise 15.1 A wire horizontal to the ground carries current toward the north. (Neglect the earth's field.) **a.** Directly underneath the wire, what is the direction of **B**? **b.** Directly above the wire, what is the direction of **B**?

Answer **a.** west; **b.** east. □

Exercise 15.2 A coil is wound around the periphery of a clock. If current flows clockwise, what is the direction of **B** in the center of the clock face?

Answer Into the clock face. □

Forces on Charges Moving in Magnetic Fields

An electrical charge q moving with velocity vector $\mathbf{u}$ through a magnetic field $\mathbf{B}$ experiences a force $\mathbf{f}$ as illustrated in Figure 15.3. The force vector is given by

$$\mathbf{f} = q\mathbf{u} \times \mathbf{B} \tag{15.1}$$

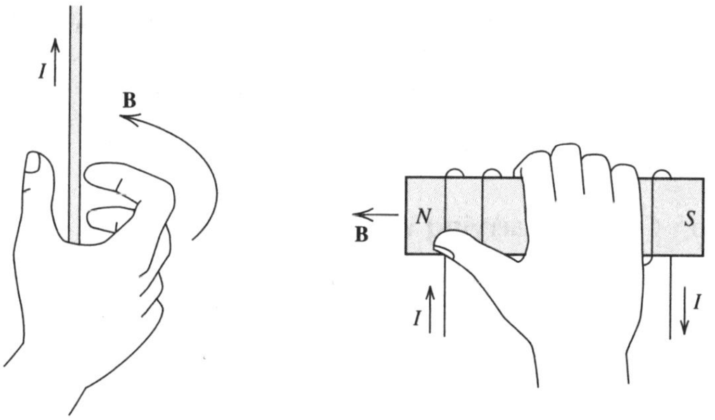

(a) If a wire is grasped with the thumb pointing in the current direction, the fingers encircle the wire in the direction of the magnetic field

(b) If a coil is grasped with the fingers pointing in the current direction, the thumb points in the direction of the magnetic field inside the coil

Figure 15.2 Illustrations of the right-hand rule.

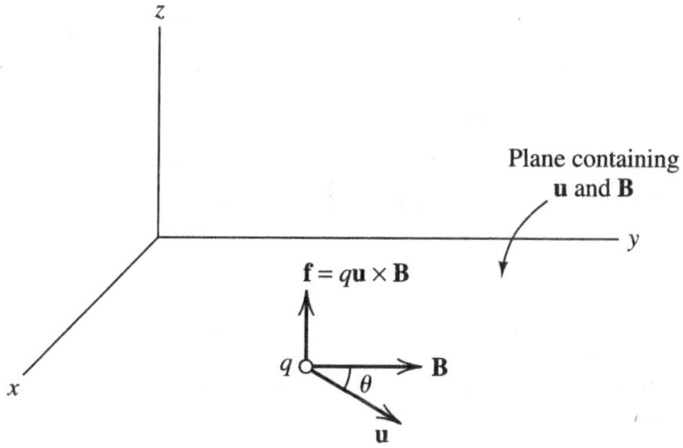

Figure 15.3 A charge moving through a magnetic field experiences a force **f** perpendicular to both the velocity **u** and flux density **B**.

in which × represents the vector cross product. Note that due to the definition of the cross product, the force is perpendicular to the plane containing the magnetic flux density **B** and the velocity **u**. Furthermore, the magnitude of the force is given by

Force is exerted on a charge as it moves through a magnetic field.

$$f = quB\sin(\theta) \tag{15.2}$$

in which θ is the angle between **u** and **B**, as illustrated in the figure.

In the SI system, the force vector **f** has units of newtons (N), the charge is in coulombs (C), and the velocity vector **u** is in meters/second (m/s). Thus, for dimensional consistency in Equations 15.1 and 15.2, the magnetic field vector **B** must have units of newton seconds per coulomb meter (Ns/Cm), which is the dimensional equivalent of the tesla (T).

Exercise 15.3 An electron ($q = -1.602 \times 10^{-19}$ C) travels at 10^5 m/s in the positive x direction. The magnetic flux density is 1 T in the positive y direction. Find the magnitude and direction of the force on the electron. (Assume a right-hand coordinate system such as that shown in Figure 15.3.)
Answer $f = 1.602 \times 10^{-14}$ N in the negative z direction. □

Forces on Current-Carrying Wires

Force is exerted on a current-carrying conductor when it is immersed in a magnetic field.

Current flowing in a conductor consists of charge (usually, electrons) in motion. Thus, forces appear on a current-carrying wire immersed in a magnetic field. The force on an incremental length of the wire is given by

$$d\mathbf{f} = i\,d\mathbf{l} \times \mathbf{B} \tag{15.3}$$

in which the direction of $d\mathbf{l}$ and the reference direction for the current are the same.

For a straight wire of length l and a constant magnetic field, we have

$$f = ilB \sin(\theta) \tag{15.4}$$

in which θ is the angle between the wire and the field. Notice that the force is maximized if the direction of the field is perpendicular to the wire.

Exercise 15.4 A wire of length $l = 1$ m carries a current of 10 A perpendicular to a field of $B = 0.5$ T. Compute the magnitude of the force on the wire.
Answer $f = 5$ N. □

Flux Linkages and Faraday's Law

The magnetic flux passing through a surface area A is given by the surface integral

$$\phi = \int_A \mathbf{B} \cdot d\mathbf{A} \tag{15.5}$$

The magnetic flux passing through a surface is determined by integrating the dot product of **B** and incremental area over the surface.

in which $d\mathbf{A}$ is an increment of area on the surface. The direction of the vector $d\mathbf{A}$ is perpendicular to the surface. If the magnetic flux density is constant and perpendicular to the surface, Equation 15.5 reduces to

$$\phi = BA \tag{15.6}$$

We say that the flux passing through the surface bounded by a coil **links** the coil. If the coil has N turns, then the total flux linkages are given by

$$\lambda = N\phi \tag{15.7}$$

Here, we have assumed that the same flux links each turn of the coil. This is a good approximation when the turns are close together on an iron form, which is often the case in transformers and electrical machines.

According to **Faraday's law of magnetic induction**, a voltage

$$e = \frac{d\lambda}{dt} \tag{15.8}$$

is induced in a coil whenever its flux linkages are changing. This can occur either because the magnetic field is changing with time or because the coil is moving relative to a magnetic field.

Lenz's law states that the polarity of the induced voltage is such that the voltage would produce a current (through an external resistance) that opposes the original change in flux linkages. (Think of the induced voltage as a voltage source.) For example, suppose that the magnetic field linking the coil shown in Figure 15.4 is pointing into the page and increasing in magnitude. (This field is the result of a coil or moving permanent magnet not shown in the figure.) Then, the voltage induced in the coil produces a counterclockwise current. According to the right-hand rule, this current produces a magnetic field directed out of the page, opposing the initial field change.

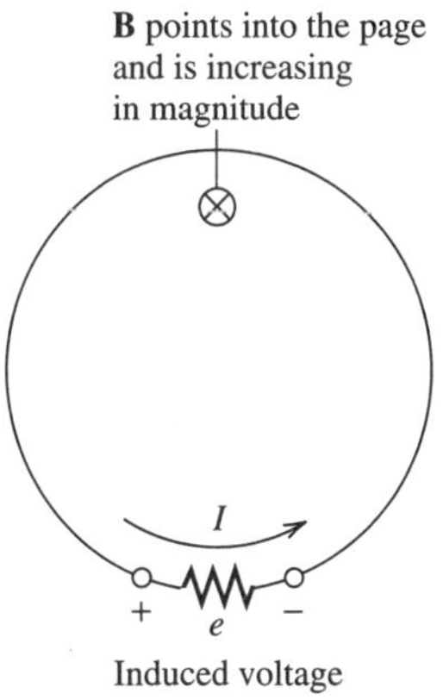

Figure 15.4 When the flux linking a coil changes, a voltage is induced in the coil. The polarity of the voltage is such that if a circuit is formed by placing a resistance across the coil terminals, the resulting current produces a field that tends to oppose the original change in the field.

Voltages Induced in Field-Cutting Conductors

Voltage is also induced in a conductor moving through a magnetic field in a direction such that the conductor cuts through magnetic lines of flux. For example, consider Figure 15.5. A uniform magnetic field is directed into the page. The sliding conductor and the stationary rails form a loop having an area of $A = lx$. The flux linkages of the coil are

$$\lambda = BA = Blx$$

According to Faraday's law, the voltage induced in the coil is given by

$$e = \frac{d\lambda}{dt} = Bl\frac{dx}{dt}$$

However, $u = dx/dt$ is the velocity of the sliding conductor, so we have

$$e = Blu \tag{15.9}$$

Voltage is induced across the terminals of a coil if the flux linkages are changing with time. Moreover, voltage is induced between the ends of a conductor moving so as to cut through flux lines.

Equation 15.9 can be used to compute the voltage induced across the ends of a straight conductor moving in uniform magnetic field, provided that the velocity, the conductor, and the magnetic-field vector are mutually perpendicular.

For example, a conductor in a typical dc generator rated for 1 kW has a length of 0.2 m, a velocity of 12 m/s, and cuts through a field of 0.5 T. This results in an

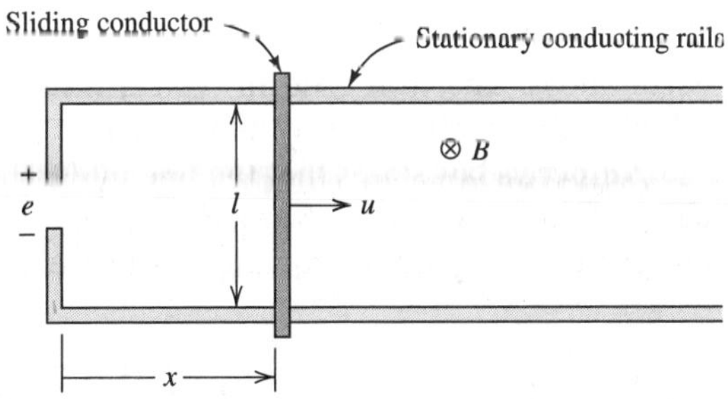

Figure 15.5 A voltage is induced in a conductor moving so as to cut through magnetic flux lines.

induced voltage of 1.2 V. (Higher output voltages are produced by connecting many such conductors in series.)

Exercise 15.5 **a.** A 10-turn circular coil has a radius of 5 cm. A flux density of 0.5 T is directed perpendicular to the plane of the coil. Evaluate the flux linking the coil and the flux linkages. **b.** Suppose that the flux is reduced to zero at a uniform rate during an interval of 1 ms. Determine the voltage induced in the coil.

Answer **a.** $\phi = 3.927$ mWb, $\lambda = 39.27$ mWb turns; **b.** $e = 39.27$ V. □

Magnetic Field Intensity and Ampère's Law

So far, we have considered the magnetic flux density **B** and its effects. To summarize, **B** produces forces on moving charges and current-carrying conductors. It also induces voltage in a coil if the flux linkages are changing with time. Furthermore, voltage is induced across a moving conductor when it cuts through flux lines.

Now, we introduce another field vector, known as the **magnetic field intensity H**, and consider how magnetic fields are established. In general, magnetic fields are set up by charges in motion. In most of the applications that we consider, the magnetic fields are established by currents flowing in coils. We will see that **H** is determined by the currents and the configuration of the coils. Furthermore, we will see that the resulting flux density **B** depends on **H**, as well as the properties of the material filling the space around the coils.

The magnetic field intensity **H** and magnetic flux density **B** are related by

$$\mathbf{B} = \mu \mathbf{H} \tag{15.10}$$

in which μ is the magnetic permeability of the material. The units of **H** are amperes/meter (A/m), and the units of μ are webers/ampere-meter (Wb/Am).

B is magnetic flux density with units of webers per square meter (Wb/m^2) or teslas (T), and **H** is magnetic field intensity with units of amperes per meter (A/m).

For free space, we have

$$\mu = \mu_0 = 4\pi \times 10^{-7} \text{ Wb/Am} \tag{15.11}$$

Some materials, most notably iron and certain rare-earth alloys, have a much higher magnetic permeability than free space. The relative permeability of a material is the ratio of its permeability to that of free space:

$$\mu_r = \frac{\mu}{\mu_0} \tag{15.12}$$

The types of iron used in typical motors and transformers have relative permeabilities of several thousand.

The value of μ_r ranges from several hundred to 1 million for various iron and rare-earth alloys. The iron used in typical transformers, motors, and generators has a relative permeability of several thousand.

Ampère's law states that the line integral of the magnetic field intensity around a closed path is equal to the algebraic sum of the currents flowing through the area enclosed by the path.

Ampère's law states that the line integral of the magnetic field intensity around a closed path is equal to the algebraic sum of the currents flowing through the area enclosed by the path. In equation form, we have

$$\oint \mathbf{H} \cdot d\mathbf{l} = \sum i \tag{15.13}$$

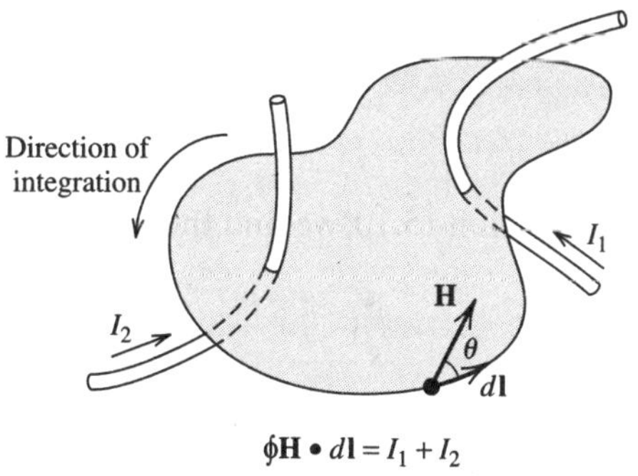

Figure 15.6 Ampère's law states that the line integral of magnetic field intensity around a closed path is equal to the sum of the currents flowing through the surface bounded by the path.

in which $d\mathbf{l}$ is a vector element of length having its direction tangent to the path of integration. Recall that the vector dot product is given by

$$\mathbf{H} \cdot d\mathbf{l} = H\,dl\cos(\theta) \tag{15.14}$$

in which θ is the angle between $\mathbf{H}$ and $d\mathbf{l}$.

Depending on its reference direction, a given current carries either a plus sign or a minus sign in the summation of Equation 15.13. If the reference direction for a current is related to the direction of integration by the right-hand rule, it carries a plus sign. (According to the right-hand rule, if you place the thumb of your right hand on the wire pointing in the reference direction, your fingers encircle the wires in the direction of integration.) Currents that are referenced in the opposite direction carry a negative sign in Equation 15.13. Ampère's law is illustrated by example in Figure 15.6, in which case the reference directions of both currents are related to the direction of integration by the right-hand rule.

If the magnetic intensity has constant magnitude and points in the same direction as the incremental length $d\mathbf{l}$ everywhere along the path, Ampère's law reduces to

$$Hl = \sum i \tag{15.15}$$

in which l is the length of the path.

In some cases, we can use Ampère's law to find formulas for the magnetic field in the space around a current-carrying wire or coil.

Example 15.1 Magnetic Field around a Long Straight Wire

Consider a long straight wire carrying current I out of the page as shown in Figure 15.7. Find expressions for the magnetic field intensity and magnetic flux density in the space around the wire. Assume that the material surrounding the wire has permeability μ.

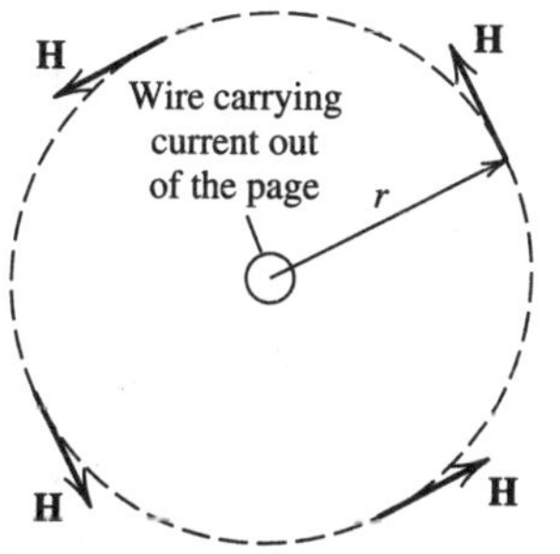

Figure 15.7 The magnetic field around a long straight wire carrying a current can be determined with Ampère's law aided by considerations of symmetry.

Solution By symmetry and the right-hand rule, we conclude that **B** and **H** fall in a plane perpendicular to the wire (i.e., in the plane of the paper) and are tangent to circles having their centers at the wire. This is illustrated in Figure 15.7. Furthermore, the magnitude of H is constant for a given radius r. Applying Ampère's law (Equation 15.15) to the circular path shown in the figure, we have

$$Hl = H2\pi r = I$$

Solving for the magnetic intensity, we obtain

$$H = \frac{I}{2\pi r}$$

Then by Equation 15.10, we find the magnetic flux density as

$$B = \mu H = \frac{\mu I}{2\pi r}$$

■

Example 15.2 Flux Density in a Toroidal Core

Consider the toroidal coil shown in Figure 15.8. Find an expression for the magnetic flux density B on the center line of the core in terms of the number of coil turns N, the current I, the permeability μ of the core, and the physical dimensions. Then, assuming that the flux density is constant throughout the core (this is approximately true if $R >> r$), find expressions for the total flux and the flux linkages.

Solution By symmetry, the field intensity is constant in magnitude along the dashed circular center line shown in the figure. (We assume that the coil is wound in a symmetrical manner all the way around the toroidal core. For clarity, only part of the coil is shown in the figure.) Applying Ampère's law to the dashed path, we obtain

$$Hl = H2\pi R = NI$$

Solving for H and using Equation 15.10 to determine B, we have

$$H = \frac{NI}{2\pi R} \tag{15.16}$$

and

$$B = \frac{\mu NI}{2\pi R} \tag{15.17}$$

Assuming that R is much greater than r, the flux density is nearly constant over the cross section of the core. Then, according to Equation 15.6, the flux is equal to the product of the flux density and the area of the cross section:

$$\phi = BA = \frac{\mu NI}{2\pi R}\pi r^2 = \frac{\mu NIr^2}{2R} \tag{15.18}$$

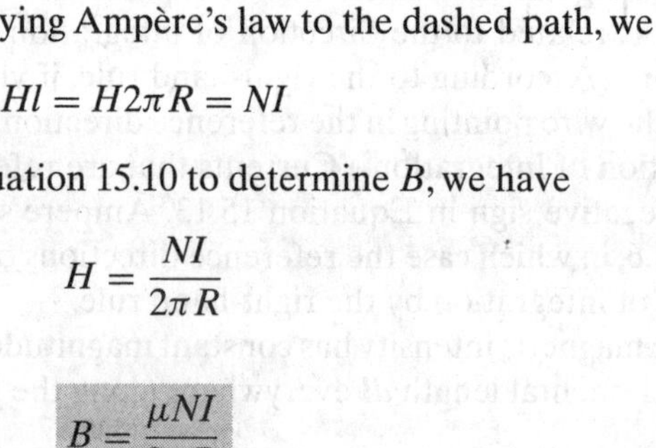

Figure 15.8 Toroidal coil analyzed in Examples 15.2, 15.3, and 15.4.

Finally, we note that all of the flux links all of the turns, and we have

$$\lambda = N\phi = \frac{\mu N^2 I r^2}{2R} \tag{15.19}$$

■

Example 15.3 Flux and Flux Linkages for a Toroidal Core

Suppose that we have a toroidal core with $\mu_r = 5000$, $R = 10$ cm, $r = 2$ cm, and $N = 100$. The current is

$$i(t) = 2\sin(200\pi t)$$

Compute the flux and the flux linkages. Then, use Faraday's law of induction to determine the voltage induced in the coil.

Solution First, the permeability of the core material is

$$\mu = \mu_r \mu_0 = 5000 \times 4\pi \times 10^{-7}$$

Using Equation 15.18, we compute the flux:

$$\phi = \frac{\mu N I r^2}{2R} = \frac{5000 \times 4\pi \times 10^{-7} \times 100 \times 2\sin(200\pi t) \times (2 \times 10^{-2})^2}{2 \times 10 \times 10^{-2}}$$
$$= (2.513 \times 10^{-3})\sin(200\pi t) \text{ Wb}$$

The flux linkages are

$$\lambda = N\phi$$
$$= 100 \times (2.513 \times 10^{-3})\sin(200\pi t)$$
$$= 0.2513\sin(200\pi t) \text{ weber turns}$$

Finally, using Faraday's law (Equation 15.8), we can find the voltage induced in the coil by the changing field:

$$e = \frac{d\lambda}{dt} = 0.2513 \times 200\pi \cos(200\pi t)$$
$$= 157.9\cos(200\pi t) \text{ V}$$

■

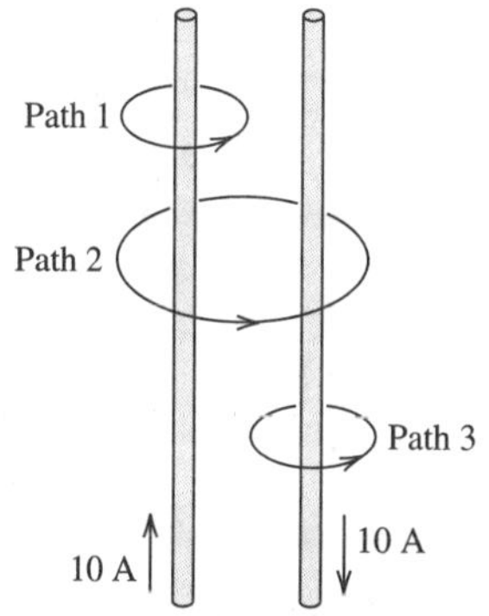

Figure 15.9 See Exercises 15.7 and 15.8.

Exercise 15.6 A long straight wire surrounded by air ($\mu_r \cong 1$) carries a current of 20 A. Compute the magnetic flux density at a point 1 cm from the wire.
Answer 4×10^{-4} T. □

Exercise 15.7 Figure 15.9 shows two wires carrying equal currents in opposite directions. Find the value of

$$\oint \mathbf{H} \cdot d\mathbf{l}$$

for each path shown in the direction indicated.
Answer Path 1, 10 A; path 2, 0; path 3, −10 A. □

Exercise 15.8 Find the force between a 1-m length of the wires shown in Figure 15.9 if the distance between the wires is 10 cm. Is this a force of attraction or of repulsion?
Answer $f = 2 \times 10^{-4}$ N; repulsion. □

15.2 MAGNETIC CIRCUITS

In many engineering applications, we need to compute the magnetic fields for structures that lack sufficient symmetry for straight-forward application of Ampère's law. Then, we use an approximate method known as magnetic-circuit analysis.

We will see that many useful devices (such as transformers, motors, and generators) contain coils wound on iron cores. In this section, we learn how to calculate the magnetic fields in these devices. A simple example, discussed in the preceding section, is the toroidal coil shown in Figure 15.8 and analyzed in Example 15.2. The toroid possesses sufficient symmetry that we readily applied Ampère's law to find an expression for the field intensity. However, in many applications, we need to analyze more complex configurations (such as cores that lack symmetry and those with multiple coils) for which the direct application of Ampère's law is not feasible. Instead, we use **magnetic circuit concepts**, which are analogous to those used to analyze electrical circuits.

The **magnetomotive force** (mmf) of an N-turn current-carrying coil is given by

$$\mathcal{F} = Ni \tag{15.20}$$

A current-carrying coil is the magnetic-circuit analog of a voltage source in an electrical circuit. Magnetomotive force is analogous to source voltage. Usually, we give the units of magnetomotive force as A·turns; however, the number of turns is actually a pure number without physical units.

The **reluctance** of a path for magnetic flux, such as the bar of iron shown in Figure 15.10, is given by

$$\mathcal{R} = \frac{l}{\mu A} \tag{15.21}$$

in which l is the length of the path (in the direction of the magnetic flux), A is the cross-sectional area, and μ is the permeability of the material. Reluctance is analogous to resistance in an electrical circuit. When the bar is not straight, the length of the path is somewhat ambiguous, and then we estimate its value as the length of the centerline. Thus, l is sometimes called the **mean length** of the path.

Magnetic flux ϕ in a magnetic circuit is analogous to current in an electrical circuit. Magnetic flux, reluctance, and magnetomotive force are related by

$$\mathcal{F} = \mathcal{R}\phi \tag{15.22}$$

which is the counterpart of Ohm's law ($V = Ri$). The units of reluctance are A·turns/Wb.

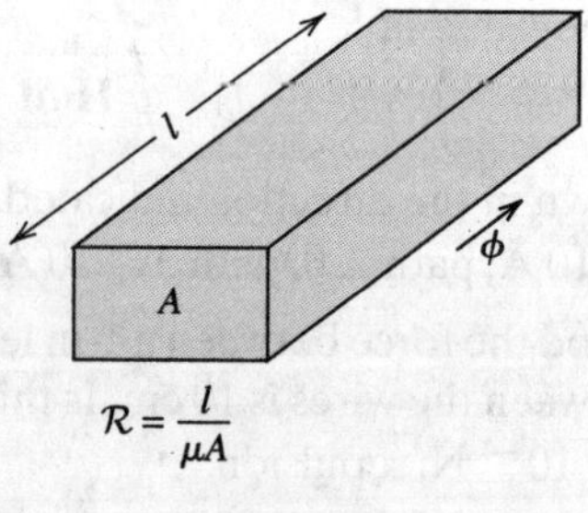

Figure 15.10 The reluctance $\mathcal{R}$ of a magnetic path depends on the mean length l, the area A, and the permeability μ of the material.

Example 15.4 The Toroidal Coil as a Magnetic Circuit

Using magnetic circuit concepts, analyze the toroidal coil shown in Figure 15.8 to find an expression for the flux.

Solution As indicated in Figure 15.11, the magnetic circuit of the toroidal coil is analogous to a simple electrical circuit with a resistance connected across a voltage source.

The mean length of the magnetic path is

$$l = 2\pi R$$

The cross section of the core is circular with radius r. Thus, the area of the cross section is

$$A = \pi r^2$$

Substituting into Equation 15.21, we find the reluctance to be

$$\mathcal{R} = \frac{l}{\mu A} = \frac{2\pi R}{\mu \pi r^2} = \frac{2R}{\mu r^2}$$

The magnetomotive force is

$$\mathcal{F} = NI$$

Solving Equation 15.22 for the flux, we have

$$\phi = \frac{\mathcal{F}}{\mathcal{R}}$$

Substituting the expressions for $\mathcal{F}$ and $\mathcal{R}$ found earlier, we get

$$\phi = \frac{\mu N r^2 I}{2R}$$

This is the same expression for the flux that we obtained in Examples 15.2 and 15.3 by applying Ampère's law. ■

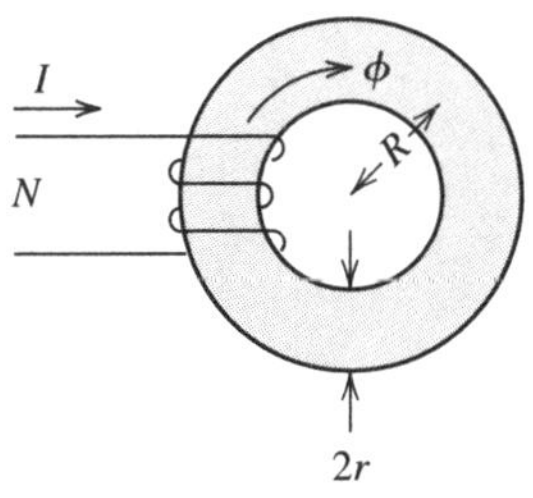

(a) Coil on a toroidal iron core

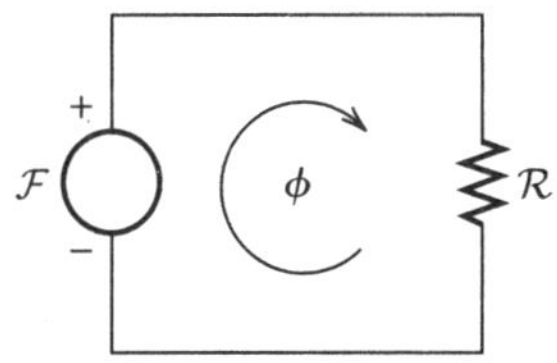

(b) Magnetic circuit

Figure 15.11 The magnetic circuit for the toroidal coil.

Advantage of the Magnetic-Circuit Approach

The advantage of the magnetic-circuit approach is that it can be applied to unsymmetrical magnetic cores with multiple coils. Coils are sources of magnetomotive forces that can be manipulated as source voltages are in an electrical circuit. Reluctances in series or parallel are combined as resistances are. Fluxes are analogous to currents. The magnetic-circuit approach is not an exact method for determining magnetic fields, but it is sufficiently accurate for many engineering applications. We illustrate these methods with a few examples.

The advantage of the magnetic-circuit approach is that it can be applied to unsymmetrical magnetic cores with multiple coils.

Example 15.5 A Magnetic Circuit with an Air Gap

Consider the magnetic core with an air gap as shown in Figure 15.12(a). The core material has a relative permeability of 6000 and a rectangular cross section 2 cm by 3 cm. The coil has 500 turns. Determine the current required to establish a flux density of $B_{gap} = 0.25$ T in the air gap.

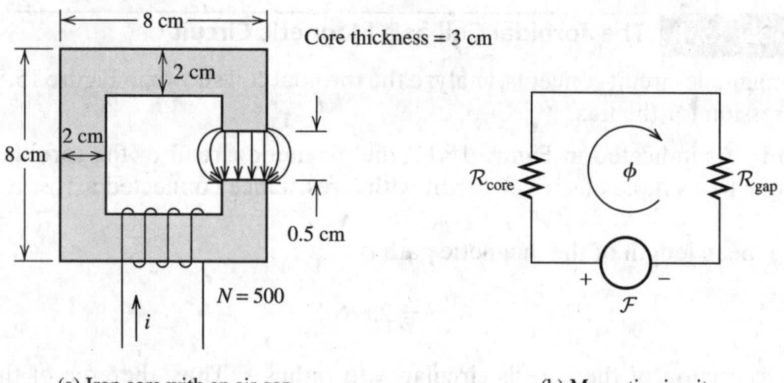

Figure 15.12 Magnetic circuit of Example 15.5.

Solution As shown in Figure 15.12(b), this magnetic circuit is analogous to an electrical circuit with one voltage source and two resistances in series. First, we compute the reluctance of the core. Notice that the centerline of the flux path is a square 6 cm by 6 cm. Thus, the mean length of the iron core is

$$l_{core} = 4 \times 6 - 0.5 = 23.5 \text{ cm}$$

The cross-sectional area of the core is

$$A_{core} = 2 \text{ cm } \times 3 \text{ cm } = 6 \times 10^{-4} \text{ m}^2$$

The permeability of the core is

$$\mu_{core} = \mu_r \mu_0 = 6000 \times 4\pi \times 10^{-7} = 7.540 \times 10^{-3}$$

Finally, the reluctance of the core is

$$\mathcal{R}_{core} = \frac{l_{core}}{\mu_{core} A_{core}} = \frac{23.5 \times 10^{-2}}{7.540 \times 10^{-3} \times 6 \times 10^{-4}}$$
$$= 5.195 \times 10^4 \text{ A·turns/Wb}$$

We approximately account for fringing by adding the length of the gap to the depth and width in computing effective gap area.

Now, we compute the reluctance of the air gap. The flux lines tend to bow out in the air gap as shown in Figure 15.12(a). This is called **fringing**. *Thus, the effective area of the air gap is larger than that of the iron core. Customarily, we take this into account by adding the length of the gap to each of the dimensions of the air-gap cross section.* Thus, the effective area of the gap is

$$A_{gap} = (2 \text{ cm } + 0.5 \text{ cm}) \times (3 \text{ cm } + 0.5 \text{ cm}) = 8.75 \times 10^{-4} \text{ m}^2$$

The permeability of air is approximately the same as that of free space:

$$\mu_{gap} \cong \mu_0 = 4\pi \times 10^{-7}$$

Thus, the reluctance of the gap is

$$\mathcal{R}_{gap} = \frac{l_{gap}}{\mu_{gap} A_{gap}} = \frac{0.5 \times 10^{-2}}{4\pi \times 10^{-7} \times 8.75 \times 10^{-4}}$$
$$= 4.547 \times 10^6 \text{ A·turns/Wb}$$

The total reluctance is the sum of the reluctance of the core and that of the gap:

$$\mathcal{R} = \mathcal{R}_{\text{gap}} + \mathcal{R}_{\text{core}} = 4.547 \times 10^6 + 5.195 \times 10^4 = 4.600 \times 10^6$$

Even though the gap is much shorter than the iron core, the reluctance of the gap is higher than that of the core because of the much higher permeability of the iron. Most of the magnetomotive force is dropped across the air gap. (This is analogous to the fact that the largest fraction of the applied voltage is dropped across the largest resistance in a series electrical circuit.)

Now, we can compute the flux:

$$\phi = B_{\text{gap}} A_{\text{gap}} = 0.25 \times 8.75 \times 10^{-4} = 2.188 \times 10^{-4} \text{ Wb}$$

The flux in the core is the same as that in the gap. However, the flux density is higher in the core, because the area is smaller. The magnetomotive force is given by

$$\mathcal{F} = \phi\mathcal{R} = 4.600 \times 10^6 \times 2.188 \times 10^{-4} = 1006 \text{ A·turns}$$

According to Equation 15.20, we have

$$\mathcal{F} = Ni$$

Solving for the current and substituting values, we get

$$i = \frac{\mathcal{F}}{N} = \frac{1006}{500} = 2.012 \text{ A}$$

■

Example 15.6 A Magnetic Circuit with Reluctances in Series and Parallel

The iron core shown in Figure 15.13(a) has a cross section of 2 cm by 2 cm and a relative permeability of 1000. The coil has 500 turns and carries a current of $i = 2$ A. Find the flux density in each air gap.

Solution The magnetic circuit is depicted in Figure 15.13(b). First, we compute the reluctances of the three paths. For the center path, we have

$$\mathcal{R}_c = \frac{l_c}{\mu_r \mu_0 A_{\text{core}}} = \frac{10 \times 10^{-2}}{1000 \times 4\pi \times 10^{-7} \times 4 \times 10^{-4}}$$
$$= 1.989 \times 10^5 \text{ A·turns/Wb}$$

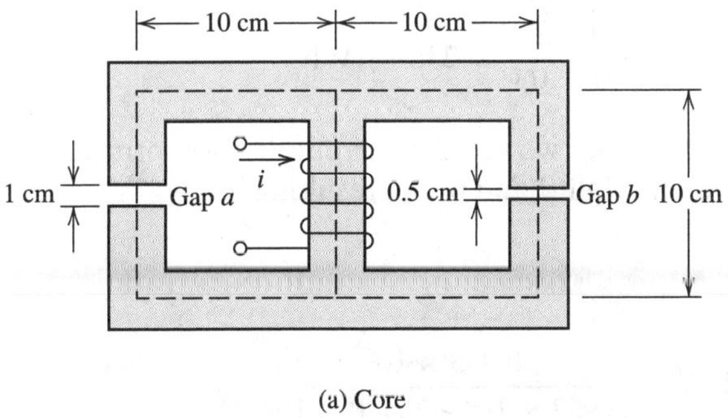

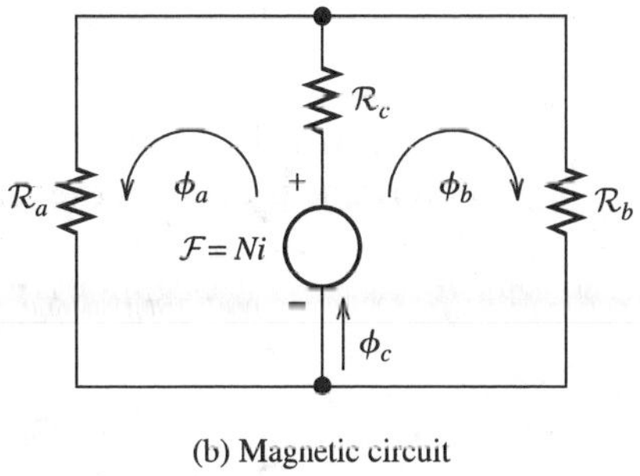

Figure 15.13 Magnetic circuit of Example 15.6.

For the left-hand path, the total reluctance is the sum of the reluctance of the iron core plus the reluctance of gap a. We take fringing into account by adding the gap length to its width and depth in computing area of the gap. Thus, the area of gap a is $A_a = 3 \text{ cm} \times 3 \text{ cm} = 9 \times 10^{-4}\ \text{m}^2$. Then, the total reluctance of the left-hand path is

$$\begin{aligned}\mathcal{R}_a &= \mathcal{R}_{\text{gap}} + \mathcal{R}_{\text{core}} \\ &= \frac{l_{\text{gap}}}{\mu_0 A_a} + \frac{l_{\text{core}}}{\mu_r \mu_0 A_{\text{core}}} \\ &= \frac{1 \times 10^{-2}}{4\pi \times 10^{-7} \times 9 \times 10^{-4}} + \frac{29 \times 10^{-2}}{1000 \times 4\pi \times 10^{-7} \times 4 \times 10^{-4}} \\ &= 8.842 \times 10^6 + 5.769 \times 10^5 \\ &= 9.420 \times 10^6 \text{ A·turns/Wb}\end{aligned}$$

Similarly, the reluctance of the right-hand path is

$$\begin{aligned}\mathcal{R}_b &= \mathcal{R}_{\text{gap}} + \mathcal{R}_{\text{core}} \\ &= \frac{l_{\text{gap}}}{\mu_0 A_b} + \frac{l_{\text{core}}}{\mu_r \mu_0 A_{\text{core}}} \\ &= \frac{0.5 \times 10^{-2}}{4\pi \times 10^{-7} \times 6.25 \times 10^{-4}} + \frac{29.5 \times 10^{-2}}{1000 \times 4\pi \times 10^{-7} \times 4 \times 10^{-4}} \\ &= 6.366 \times 10^6 + 5.869 \times 10^5 \\ &= 6.953 \times 10^6 \text{ A·turns/Wb}\end{aligned}$$

Next, we can combine the reluctances $\mathcal{R}_a$ and $\mathcal{R}_b$ in parallel. Then, the total reluctance is the sum of $\mathcal{R}_c$ and this parallel combination:

$$\begin{aligned}\mathcal{R}_{\text{total}} &= \mathcal{R}_c + \frac{1}{1/\mathcal{R}_a + 1/\mathcal{R}_b} \\ &= 1.989 \times 10^5 + \frac{1}{1/(9.420 \times 10^6) + 1/(6.953 \times 10^6)} \\ &= 4.199 \times 10^6 \text{ A·turns/Wb}\end{aligned}$$

Now, the flux in the center leg of the coil can be found by dividing the magnetomotive force by the total reluctance:

$$\phi_c = \frac{Ni}{\mathcal{R}_{\text{total}}} = \frac{500 \times 2}{4.199 \times 10^6} = 238.1\ \mu\text{Wb}$$

Fluxes are analogous to currents. Thus, we use the current-division principle to determine the flux in the left-hand and right-hand paths, resulting in

$$\begin{aligned}\phi_a &= \phi_c \frac{\mathcal{R}_b}{\mathcal{R}_a + \mathcal{R}_b} \\ &= 238.1 \times 10^{-6} \times \frac{6.953 \times 10^6}{6.953 \times 10^6 + 9.420 \times 10^6} \\ &= 101.1\ \mu\text{Wb}\end{aligned}$$

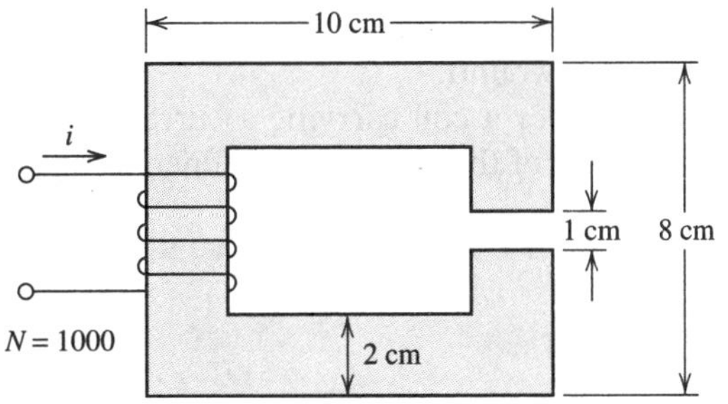

Figure 15.14 Magnetic circuit of Exercise 15.9.

Similarly, for gap b we have

$$\phi_b = \phi_c \frac{\mathcal{R}_a}{\mathcal{R}_a + \mathcal{R}_b}$$

$$= 238.1 \times 10^{-6} \frac{9.420 \times 10^6}{6.953 \times 10^6 + 9.420 \times 10^6}$$

$$= 137.0\ \mu\text{Wb}$$

As a check on these calculations, we note that $\phi_c = \phi_a + \phi_b$.

Now, we find the flux densities in the gaps by dividing the fluxes by the areas:

$$B_a = \frac{\phi_a}{A_a} = \frac{101.1\ \mu\text{Wb}}{9 \times 10^{-4}\ \text{m}^2} = 0.1123\ \text{T}$$

$$B_b = \frac{\phi_b}{A_b} = \frac{137.0\ \mu\text{Wb}}{6.25 \times 10^{-4}\ \text{m}^2} = 0.2192\ \text{T}$$

■

Typically, we find that in magnetic circuits consisting of iron cores with air gaps, the reluctance of the iron has a negligible effect on the results. Furthermore, we usually do not have a precise value of the permeability for the iron. Thus, it is often sufficiently accurate to assume zero reluctance for the iron cores. This is the counterpart of assuming zero resistance for the wires in an electrical circuit.

Exercise 15.9 Consider the magnetic circuit shown in Figure 15.14. Determine the current required to establish a flux density of 0.5 T in the air gap.
Answer $i = 4.03$ A. □

Exercise 15.10 Repeat Example 15.6, taking the reluctance of the iron paths to be zero. Determine the error as a percentage of the flux densities computed in the example.
Answer $\phi_a = 113.1\ \mu$Wb, $B_a = 0.1257$ T, 11.9 percent error; $\phi_b = 157.1\ \mu$Wb, $B_b = 0.2513$ T, 14.66 percent error. □

15.3 INDUCTANCE AND MUTUAL INDUCTANCE

We have seen that when a coil carries current, a magnetic flux is produced that links the coil. If the current changes with time, the flux also changes, inducing a voltage in the coil. This is the physical basis of inductance that we introduced in Section 3.4.

Now, we relate inductance to the physical parameters of the coil and the core upon which it is wound.

Consider a coil carrying a current i that sets up a flux ϕ linking the coil. The inductance of the coil can be defined as flux linkages divided by current:

$$L = \frac{\lambda}{i} \tag{15.23}$$

Assuming that the flux is confined to the core so that all of the flux links all of the turns, we can write $\lambda = N\phi$. Then, we have

$$L = \frac{N\phi}{i} \tag{15.24}$$

Equation 15.25 is valid only if all of the flux links all of the turns.

Substituting $\phi = Ni/\mathcal{R}$, we obtain

$$L = \frac{N^2}{\mathcal{R}} \tag{15.25}$$

Thus, we see that the inductance depends on the number of turns, the core dimensions, and the core material. Notice that inductance is proportional to the square of the number of turns.

According to Faraday's law, voltage is induced in a coil when its flux linkages change:

$$e = \frac{d\lambda}{dt} \tag{15.26}$$

Rearranging Equation 15.23, we have $\lambda = Li$. Substituting this for λ in Equation 15.26, we get

$$e = \frac{d(Li)}{dt} \tag{15.27}$$

For a coil wound on a stationary core, the inductance is constant with time, and Equation 15.27 reduces to

$$e = L\frac{di}{dt} \tag{15.28}$$

Of course, this is the equation relating voltage and current that we used to analyze circuits containing inductance in Chapters 3 through 6.

Example 15.7 Calculation of Inductance

Determine the inductance of the 500-turn coil shown in Figure 15.12 and analyzed in Example 15.5.

Solution In Example 15.5, we found that the reluctance of the magnetic path is

$$\mathcal{R} = 4.600 \times 10^6 \text{ A·turns/Wb}$$

Substituting into Equation 15.25, we obtain

$$L = \frac{N^2}{\mathcal{R}} = \frac{500^2}{4.6 \times 10^6} = 54.35 \text{ mH}$$

■

Mutual Inductance

When two coils are wound on the same core, some of the flux produced by one coil links the other coil. We denote the flux linkages of coil 2 caused by the current in coil 1 as λ_{21}. Correspondingly, the flux linkages of coil 1 produced by its own current are denoted as λ_{11}. Similarly, the current in coil 2 produces flux linkages λ_{22} in coil 2 and λ_{12} in coil 1.

The **self inductances** of the coils are defined as

$$L_1 = \frac{\lambda_{11}}{i_1} \tag{15.29}$$

and

$$L_2 = \frac{\lambda_{22}}{i_2} \tag{15.30}$$

The **mutual inductance** between the coils is

$$M = \frac{\lambda_{21}}{i_1} = \frac{\lambda_{12}}{i_2} \tag{15.31}$$

The total fluxes linking the coils are

$$\lambda_1 = \lambda_{11} \pm \lambda_{12} \tag{15.32}$$

and

$$\lambda_2 = \pm\lambda_{21} + \lambda_{22} \tag{15.33}$$

where the + sign applies if the fluxes are aiding and the − sign applies if the fluxes are opposing.

Dot Convention

It is standard practice to place a dot on one end of each coil in a circuit diagram to indicate how the fluxes interact. An example of this is shown in Figure 15.15. The dots are placed such that currents entering the dotted terminals produce aiding magnetic flux. Notice that (according to the right-hand rule) a current entering either of the dotted terminals in Figure 15.15 produces flux in a clockwise direction in the core. Thus, if both currents enter (or if both leave) the dotted terminals, the mutual flux linkages add to the self flux linkages. On the other hand, if one current enters a dotted terminal and the other leaves, the mutual flux linkages carry a minus sign.

Aiding fluxes are produced by currents entering like-marked terminals.

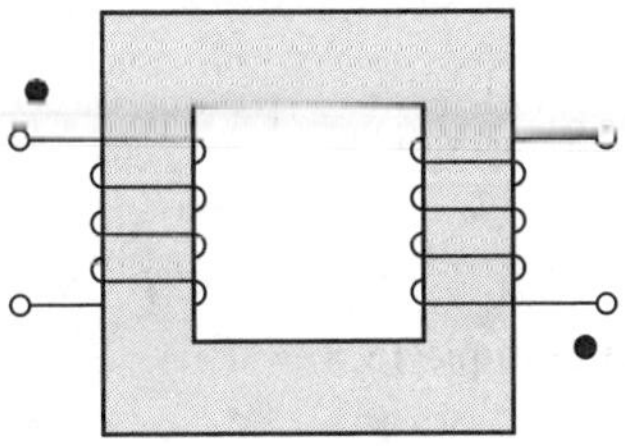

Figure 15.15 According to convention, currents entering the dotted terminals produce aiding fluxes.

Circuit Equations for Mutual Inductance

Solving Equations 15.29 through 15.31 for the flux linkages and substituting into Equations 15.32 and 15.33, we have

$$\lambda_1 = L_1 i_1 \pm M i_2 \tag{15.34}$$

and

$$\lambda_2 = \pm M i_1 + L_2 i_2 \tag{15.35}$$

Applying Faraday's law to find the voltages induced in the coils, we get

$$e_1 = \frac{d\lambda_1}{dt} = L_1 \frac{di_1}{dt} \pm M \frac{di_2}{dt} \tag{15.36}$$

and

$$e_2 = \frac{d\lambda_2}{dt} = \pm M \frac{di_1}{dt} + L_2 \frac{di_2}{dt} \tag{15.37}$$

Here again, we have assumed that the coils and core are stationary, so the inductances are constant with respect to time. These are the basic equations used to analyze circuits having mutual inductance.

Example 15.8 Calculation of Inductance and Mutual Inductance

Two coils are wound on a toroidal core as illustrated in Figure 15.16. The reluctance of the core is 10^7 (ampere-turns)/Wb. Determine the self inductances and mutual inductance of the coils. Assume that the flux is confined to the core so that all of the flux links both coils.

Solution The self inductances can be computed using Equation 15.25. For coil 1, we have

$$L_1 = \frac{N_1^2}{\mathcal{R}} = \frac{100^2}{10^7} = 1 \text{ mH}$$

Similarly, for coil 2 we get

$$L_2 = \frac{N_2^2}{\mathcal{R}} = \frac{200^2}{10^7} = 4 \text{ mH}$$

To compute the mutual inductance, we find the flux produced by i_1:

$$\phi_1 = \frac{N_1 i_1}{\mathcal{R}} = \frac{100 i_1}{10^7} = 10^{-5} i_1$$

The flux linkages of coil 2 resulting from the current in coil 1 are given by

$$\lambda_{21} = N_2 \phi_1 = 200 \times 10^{-5} i_1$$

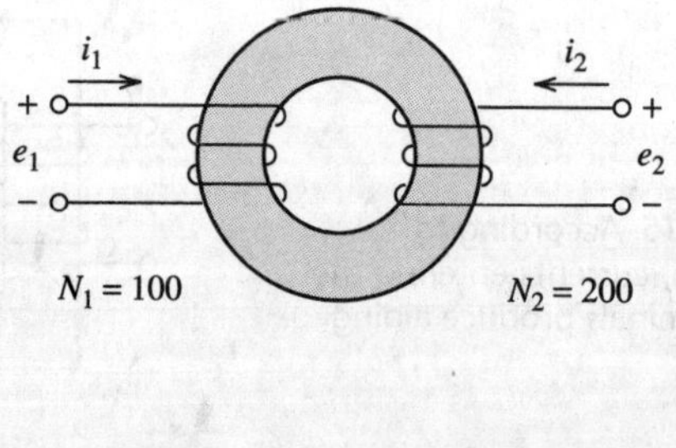

Figure 15.16 Coils of Example 15.8.

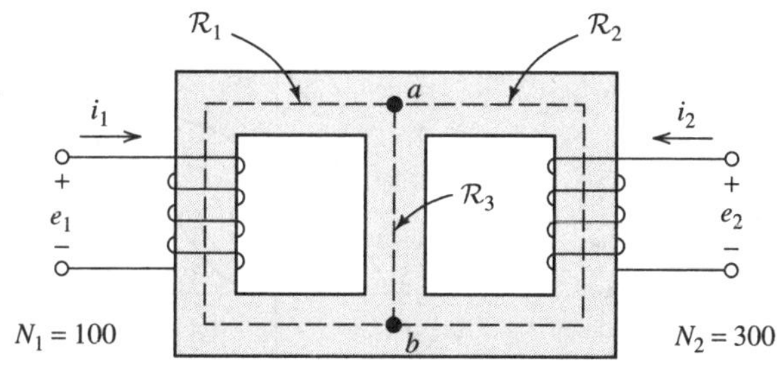

Figure 15.17 Magnetic circuit of Exercise 15.13.

Finally, the mutual inductance is

$$M = \frac{\lambda_{21}}{i_1} = 2 \text{ mH}$$

■

Exercise 15.11 Compute the mutual inductance in Example 15.8 by use of the formula $M = \lambda_{12}/i_2$.

Answer $M = 2$ mH. Notice that we get the same value from $M = \lambda_{21}/i_1$ and from $M = \lambda_{12}/i_2$. □

Exercise 15.12 Does the flux produced by i_2 aid or oppose the flux produced by i_1 for the coils shown in Figure 15.16? If a dot is placed on the top terminal of coil 1, which end of coil 2 should have a dot? Write the expressions for e_1 and e_2, taking care to select the proper sign for the mutual term.

Answer The fluxes oppose one another. The dot should be on the bottom terminal of coil 2, so the correct expressions are

$$e_1 = L_1\frac{di_1}{dt} - M\frac{di_2}{dt} \quad \text{and} \quad e_2 = -M\frac{di_1}{dt} + L_2\frac{di_2}{dt}$$

□

Exercise 15.13 For the core shown in Figure 15.17, the reluctances of all three paths between points a and b are equal.

$$\mathcal{R}_1 = \mathcal{R}_2 = \mathcal{R}_3 = 10^6 \text{ (A·turns)/Wb}$$

Assume that all of the flux is confined to the core. **a.** Do the fluxes produced by i_1 and i_2 aid or oppose one another in path 1? In path 2? In path 3? If a dot is placed on the top end of coil 1, which end of coil 2 should carry a dot? **b.** Determine the values of $L_1, L_2,$ and M. **c.** Should the mutual term for the voltages (in Equations 15.36 and 15.37) carry a plus sign or a minus sign?

Answer **a.** Aid in paths 1 and 2, oppose in path 3; the dot should be on the top end of coil 2; **b.** $L_1 = 6.667$ mH, $L_2 = 60$ mH, $M = 10$ mH; **c.** a plus sign. □

15.4 MAGNETIC MATERIALS

So far, we have assumed that the relationship between B and H is linear (i.e., $B = \mu H$). Actually, for the iron alloys used in motors, permanent magnets, and transformers, the relationship between B and H is not linear.

The relationship between B and H is not linear for the types of iron used in motors and transformers.

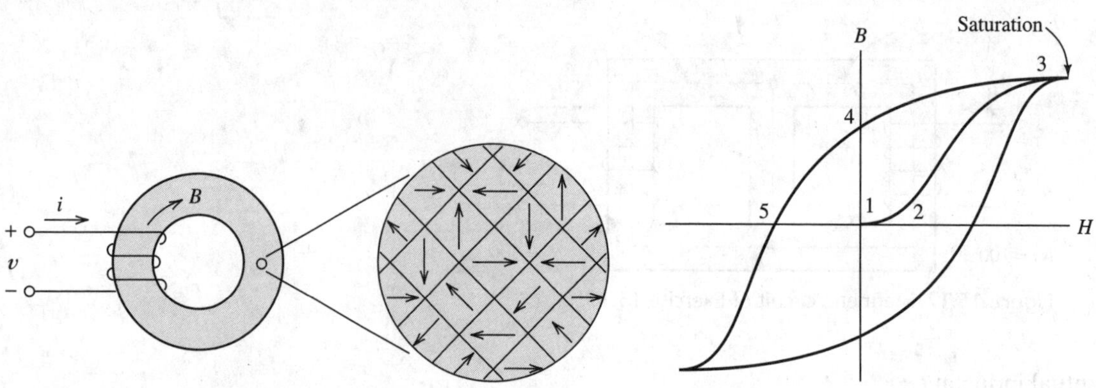

(a) Sample and coil for applying H (b) Magnetic domains (c) Hysteresis loop in the $B-H$ plane

Figure 15.18 Materials such as iron display a B–H relationship with hysteresis and saturation.

Figure 15.18(a) shows a coil used to apply magnetic field intensity H to a sample of iron. Suppose that we start with a sample that is not magnetized. If we look at the material on a microscopic scale, we see that the magnetic fields of the atoms in small **domains** are aligned. However, the field directions are random for the various domains, and the external macroscopic field is zero. This is illustrated in Figure 15.18(b).

Figure 15.18(c) shows a plot of B versus H. At point 1, both B and H are zero. As H is increased by applying current to the coil, the magnetic fields of the domains tend to align with the applied field. At first (point 1 to point 2), this is a reversible process, so that if the applied field is reduced to zero, the domains return to their original random orientations. However, for greater applied field intensities, the domains align with the applied field such that they tend to maintain their alignment even if the applied field is reduced to zero (point 2 to point 3). Eventually, for sufficiently high fields, all of the domains are aligned with the applied field and the slope of the B–H curve approaches μ_0. We say that the material is **saturated**. For typical iron core materials, saturation occurs for B in the range of 1 to 2 T.

For typical iron cores, saturation occurs for B in the range from 1 to 2 T.

If starting from point 3, the applied field H is reduced to zero, a residual flux density B remains in the core (point 4). This occurs because the magnetic domains continue to point in the direction imposed earlier by the applied field. If H is increased in the reverse direction, B is reduced to zero (point 5). Eventually, saturation occurs in the reverse direction. If an ac current is applied to the coil, a **hysteresis loop** is traced in the B–H plane.

Energy Considerations

Let us consider the energy flow to and from the coil shown in Figure 15.18(a). We assume that the coil has zero resistance. As the current is increasing, the increasing flux density induces a voltage, resulting in energy flow into the coil. The energy W delivered to the coil is the integral of power. Thus, we get

$$W = \int_0^t vi\,dt = \int_0^t N\frac{d\phi}{dt}i\,dt = \int_0^\phi Ni\,d\phi \qquad (15.38)$$

Now $Ni = Hl$ and $d\phi = AdB$, where l is the mean path length and A is the cross-sectional area. Making these substitutions in the expression on the right-hand

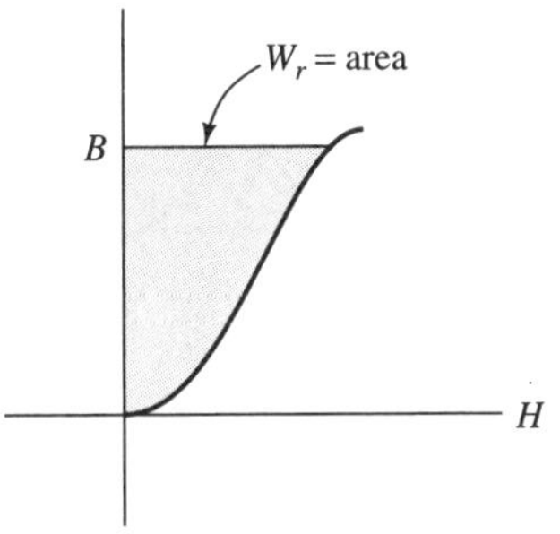

Figure 15.19 The area between the B–H curve and the B axis represents the volumetric energy supplied to the core.

side of Equation 15.38, we have

$$W = \int_0^B AlH\, dB \qquad (15.39)$$

However, the product of the cross-sectional area A and the length of the core l is the volume of the core. Dividing both sides of Equation 15.39 by the volume results in

$$W_v = \frac{W}{Al} = \int_0^B H\, dB \qquad (15.40)$$

in which W_v represents energy per unit volume of the core. As illustrated in Figure 15.19, the volumetric energy delivered to the coil is the area between the B–H curve and the B-axis. Part of this energy is returned to the circuit when H is reduced to zero, part of it remains stored in the residual field, and part of it is converted to heat in the process of magnetizing the core.

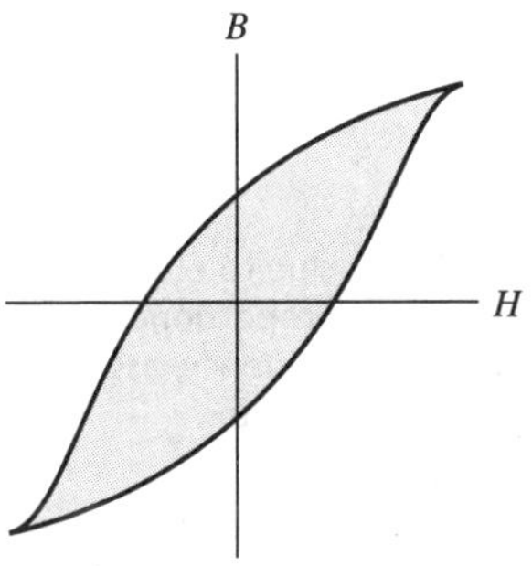

Figure 15.20 The area of the hysteresis loop is the volumetric energy converted to heat per cycle.

Core Loss

When an ac current is applied to a coil having an iron core, more energy is put into the coil on each cycle than is returned to the circuit. Part of the energy is converted to heat in reversing the directions of the magnetic domains. This is similar to the heat produced when we repeatedly bend a piece of metal. The volumetric energy converted to heat per cycle is equal to the area of the hysteresis loop as illustrated in Figure 15.20. This energy loss is called **core loss**. Since a fixed amount of energy is converted to heat for each cycle, the power loss due to hysteresis is proportional to frequency.

Power loss due to hysteresis is proportional to frequency, assuming constant peak flux.

In motors, generators, and transformers, conversion of energy into heat is undesirable. Therefore, we would choose an alloy having a thin hysteresis loop as in Figure 15.21(a). On the other hand, for a permanent magnet, we would choose a material having a large residual field, such as in Figure 15.21(b).

Eddy-Current Loss

Besides hysteresis, there is another effect that leads to core loss for ac operation. First, let us consider a solid iron core. Of course, the core itself is an electrical conductor, acting much like shorted turns. As the magnetic fields change, voltages are induced in the core, causing currents, known as **eddy currents**, to circulate in the core material. As a result, power is dissipated in the core according to $P = v^2/R$.

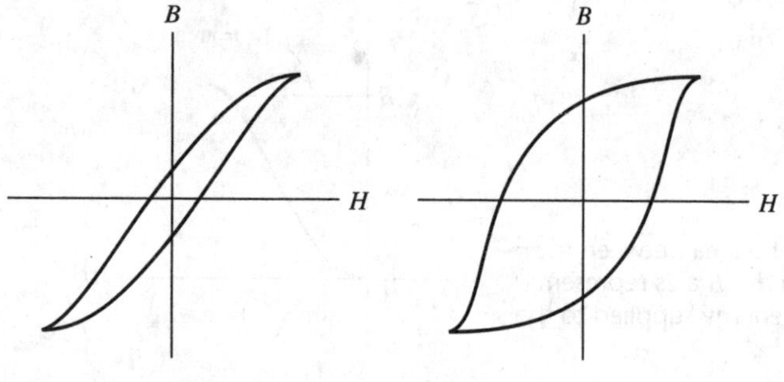

Figure 15.21 When we want to minimize core loss (as in a transformer or motor), we choose a material having a thin hysteresis loop. On the other hand, for a permanent magnet, we should choose a material with a wide loop.

A partial solution to eddy-current loss is to laminate the core with thin sheets of iron that are electrically insulated from one another. The orientation of the sheets is selected to interrupt the flow of current. Thus, the resistance is higher for eddy currents, and the loss is greatly reduced. Another approach is to make the core with powdered iron held together by an insulating binder.

Power loss due to eddy currents is proportional to the square of frequency, assuming constant peak flux.

For operation with a given peak flux density, the voltages induced in the core are proportional to frequency (because of Faraday's law). Therefore, power loss due to eddy currents increases with the square of frequency (because $P = v^2/R$).

Energy Stored in the Magnetic Field

Even though many core materials do not have a linear B–H characteristic, we often perform initial design calculations assuming that $B = \mu H$. The properties of the core material are usually not accurately known, so the calculations for motor or transformer design are approximate. The linear approximation is convenient and sufficiently accurate as long as the cores are operated below the saturation level.

Substituting $H = B/\mu$ into Equation 15.40 and integrating, we obtain

$$W_v = \int_0^B \frac{B}{\mu}\, dB = \frac{B^2}{2\mu} \tag{15.41}$$

Notice that for a given flux density, the volumetric energy stored in the field is inversely proportional to the permeability.

In a magnetic circuit having an air gap, the flux density is roughly the same in the iron core as in the air gap. (It is usually a little less in the air gap, due to fringing.) The permeability of an iron core is much greater (by a factor of several thousand or more) than that of air. Thus, the volumetric energy of the gap is much higher than that of the core. *In a magnetic circuit consisting of an iron core with a substantial air gap, nearly all of the stored energy resides in the gap.*

Exercise 15.14 Consider a coil wound on an iron core. For 60-Hz ac operation with a given applied current, the hysteresis loop of the core material has an area of 40 J/m^3. The core volume is 200 cm^3. Find the power converted to heat because of hysteresis.

Answer 0.48 W. □

Exercise 15.15 A certain iron core has an air gap with an effective area of 2 cm by 3 cm and a length of 0.5 cm. The applied magnetomotive force is 1000 ampere turns and the reluctance of the iron is negligible. Find the flux density and the energy stored in the air gap.

Answer $B = 0.2513$ T, $W = 0.0754$ J. □

15.5 IDEAL TRANSFORMERS

A transformer consists of several coils wound on a common core that usually consists of laminated iron (to reduce eddy-current loss). We will see that transformers can be used to adjust the values of ac voltages. A voltage can be *stepped up* by using a transformer. For example, 2400 V can be stepped up to 48 kV. Transformers can also be used to *step a voltage down*, such as 2400 V to 240 V.

Transformers find many applications in electric power distribution. In transporting power over long distances (from a hydroelectric power-generating station to a distant city, for example), it is desirable to use relatively large voltages, typically hundreds of kilovolts. Recall that the power delivered by an ac source is given by

$$P = V_{\text{rms}} I_{\text{rms}} \cos(\theta) \tag{15.42}$$

For a fixed power factor ($\cos\theta$), many combinations of voltage and current can be used in transferring a given amount of power. The wires that carry the current have nonzero resistances. Thus, some power is lost in the transmission lines, given by

$$P_{\text{loss}} = R_{\text{line}} I_{\text{rms}}^2 \tag{15.43}$$

in which R_{line} is the resistance of the transmission line. By designing the power distribution system with a large voltage value and a small current value, the line loss can be made to be a small fraction of the power transported. Thus, larger voltage yields higher efficiency in power distribution.

For safety and other reasons, relatively small voltages must be employed where the power is consumed. For example, in U.S. residences, the nominal voltages are either 110 or 220 V rms. Thus, transformers are useful in stepping voltage levels up or down as needed in a power distribution system.

Transformers greatly facilitate power distribution by stepping voltage up and down at various points in the distribution system.

Voltage Ratio

A transformer is illustrated in Figure 15.22. An ac voltage source is connected to the primary coil, which consists of N_1 turns of wire. Current flows into the primary side and causes an ac magnetic flux $\phi(t)$ to appear in the core. This flux induces a voltage in the N_2-turn secondary coil, which delivers power to the load. Depending on the turns ratio N_2/N_1, the rms secondary voltage can be greater or less than the rms primary voltage.

For now, we neglect the resistances of the coils and the core loss. Furthermore, we assume that the reluctance of the core is very small and that all of the flux links all of the turns of both coils.

The primary voltage is given by

$$v_1(t) = V_{1m} \cos(\omega t) \tag{15.44}$$

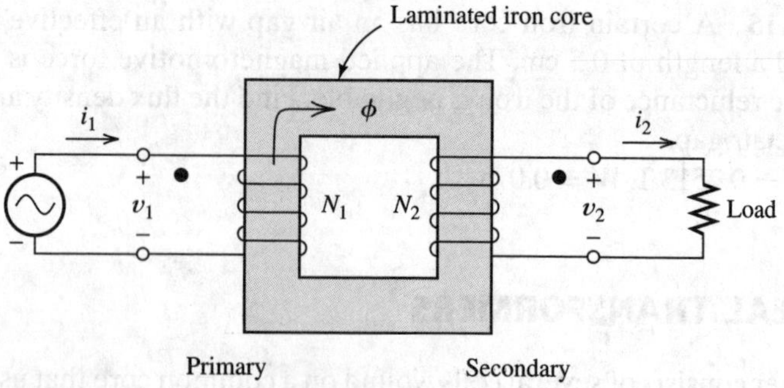

Figure 15.22 A transformer consists of several coils wound on a common core.

According to Faraday's law, we have

$$v_1(t) = V_{im}\cos(\omega t) = N_1\frac{d\phi}{dt} \tag{15.45}$$

which can be rearranged and integrated to yield

$$\phi(t) = \frac{V_{1m}}{N_1\omega}\sin(\omega t) \tag{15.46}$$

Assuming that all of the flux links all of the turns, the secondary voltage is given by

$$v_2(t) = N_2\frac{d\phi}{dt} \tag{15.47}$$

Using Equation 15.46 to substitute for $\phi(t)$, we have

$$v_2(t) = N_2\frac{V_{1m}}{N_1\omega}\frac{d}{dt}[\sin(\omega t)] \tag{15.48}$$

$$v_2(t) = \frac{N_2}{N_1}V_{1m}\cos(\omega t) \tag{15.49}$$

$$v_2(t) = \frac{N_2}{N_1}v_1(t) \tag{15.50}$$

In an ideal transformer, all of the flux links all of the turns, and the voltage across each coil is proportional to its number of turns.

Notice that the voltage across each coil is proportional to its number of turns. This is an important relationship to know when working with transformers.

Also, notice that we have included a dot on the end of each winding in Figure 15.22. As usual, the dots are placed so that if the currents entered the dotted terminals, they would produce aiding magnetic fields. Furthermore, application of Lenz's law shows that the induced voltages have positive polarity at both dotted terminals when ϕ is increasing, and negative polarity at both dotted terminals when ϕ is decreasing. *Thus in a transformer, the polarities of the voltages at the dotted terminals agree. When a voltage with positive polarity at the dotted terminal appears across coil 1, the voltage across coil 2 is also positive at the dotted terminal.*

Voltage polarities are the same at like-dotted terminals.

Hence, we have established the fact that the voltage across each winding is proportional to the number of turns. Clearly, the peak and rms values of the voltages are also related by the turns ratio:

$$V_{2\text{rms}} = \frac{N_2}{N_1}V_{1\text{rms}} \tag{15.51}$$

Example 15.9 **Determination of Required Turns Ratio**

Suppose that we have a 4700-V-rms ac source and we need to deliver 220 V rms to a load. Determine the turns ratio N_1/N_2 of the transformer needed.

Solution Rearranging Equation 15.51, we have

$$\frac{N_1}{N_2} = \frac{V_{1\text{rms}}}{V_{2\text{rms}}} = \frac{4700}{220} = 21.36$$

■

Current Ratio

Again let us consider the transformer shown in Figure 15.22. Notice that the currents i_1 and i_2 produce opposing magnetic fields (because i_1 enters a dotted terminal and i_2 leaves a dotted terminal). Thus, the total mmf applied to the core is

$$\mathcal{F} = N_1 i_1(t) - N_2 i_2(t) \quad (15.52)$$

Furthermore, the mmf is related to the flux and the reluctance of the core by

$$\mathcal{F} = \mathcal{R}\phi \quad (15.53)$$

In a well-designed transformer, the core reluctance is very small. Ideally, the reluctance is zero, and the mmf required to establish the flux in the core is zero. Then, Equation 15.52 becomes

The net mmf is zero for an ideal transformer.

$$\mathcal{F} = N_1 i_1(t) - N_2 i_2(t) = 0 \quad (15.54)$$

Rearranging this equation, we obtain

$$i_2(t) = \frac{N_1}{N_2} i_1(t) \quad (15.55)$$

This relationship also applies for the rms values of the currents:

$$I_{2\text{rms}} = \frac{N_1}{N_2} I_{1\text{rms}} \quad (15.56)$$

Compare Equation 15.51 for the voltages to Equation 15.56 for the currents. Notice that if the voltage is stepped up (i.e., $N_2/N_1 > 1$), the current is stepped down, and vice versa.

Power in an Ideal Transformer

Again consider Figure 15.22. The power delivered to the load by the secondary winding is

$$p_2(t) = v_2(t) i_2(t) \quad (15.57)$$

Using Equations 15.50 and 15.55 to substitute for $v_2(t)$ and $i_2(t)$, respectively, we have

$$p_2(t) = \frac{N_2}{N_1} v_1(t) \frac{N_1}{N_2} i_1(t) = v_1(t) i_1(t) \quad (15.58)$$

However, the power delivered to the primary winding by the source is $p_1(t) = v_1(t)i_1(t)$, and we get

$$p_2(t) = p_1(t) \tag{15.59}$$

Power is neither generated nor consumed by an ideal transformer.

Thus, we have established the fact that the power delivered to the primary winding by the source is delivered in turn to the load by the secondary winding. *Net power is neither generated nor consumed by an ideal transformer.*

Summary. Let us summarize the idealizing assumptions and their consequences for the transformer.

1. We assumed that all of the flux links all of the windings of both coils and that the resistance of the coils is zero. Thus, the voltage across each coil is proportional to the number of turns on the coil. This led to the voltage relationship

$$v_2(t) = \frac{N_2}{N_1}v_1(t)$$

2. We assumed that the reluctance of the core is negligible, so the total mmf of both coils is zero. This led to the current relationship

$$i_2(t) = \frac{N_1}{N_2}i_1(t)$$

3. A consequence of the voltage and current relationships is that all of the power delivered to an ideal transformer by the source is transferred to the load. Thus, an ideal transformer has a power efficiency of 100 percent.

The circuit symbol for the transformer is shown in Figure 15.23(a).

Mechanical Analog of the Transformer: The Lever

The lever illustrated in Figure 15.23(b) is a mechanical analog of the electrical transformer. The velocities of the ends of the lever are related by the length ratio of the lever, $v_2 = v_1(l_2/l_1)$, just as transformer voltages are related by the turns ratio. Similarly, the forces are related by $F_2 = F_1(l_1/l_2)$, which is analogous to the relationship between currents in the transformer. As in a transformer, the frictionless lever neither generates nor consumes energy. On one end of the lever, we have small force and large velocity, while on the other end the force is large and the velocity is small. This mirrors the effects of the transformer on current and voltage.

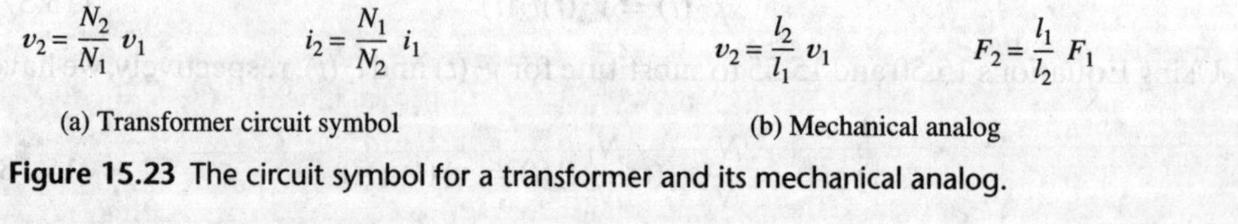

Figure 15.23 The circuit symbol for a transformer and its mechanical analog.

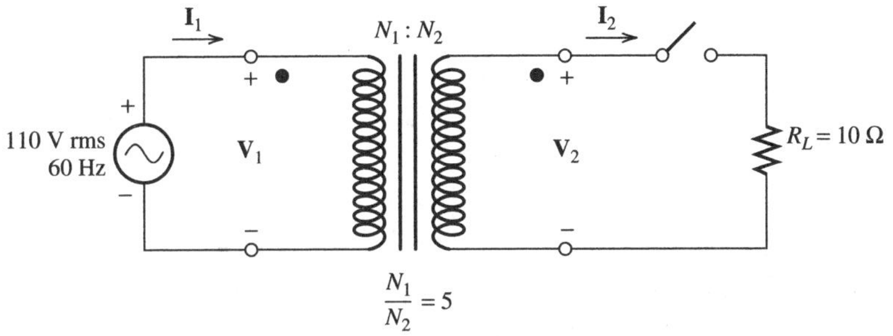

Figure 15.24 Circuit of Example 15.10.

Example 15.10 Analysis of a Circuit Containing an Ideal Transformer

Consider the source, transformer, and load shown in Figure 15.24. Determine the rms values of the currents and voltages: **a.** with the switch open; **b.** with the switch closed.

Solution Because of the applied source, the primary voltage is $V_{1\text{rms}} = 110$ V. The primary and secondary voltages are related by Equation 15.51:

$$V_{2\text{rms}} = \frac{N_2}{N_1} V_{1\text{rms}} = \frac{1}{5} \times 110 = 22 \text{ V}$$

Rearranging Equation 15.56, we have

$$I_{1\text{rms}} = \frac{N_2}{N_1} I_{2\text{rms}}$$

a. With the switch open, the secondary current is zero. Therefore, the primary current $I_{1\text{rms}}$ is also zero, and no power is taken from the source.

b. With the switch closed, the secondary current is

$$I_{2\text{rms}} = \frac{V_{2\text{rms}}}{R_L} = \frac{22}{10} = 2.2 \text{ A}$$

Then, the primary current is

$$I_{1\text{rms}} = \frac{N_2}{N_1} I_{2\text{rms}} = \frac{1}{5} \times 2.2 = 0.44 \text{ A}$$

Let us consider the sequence of events in this example. When the source voltage is applied to the primary winding, a very small primary current (ideally zero) flows, setting up the flux in the core. The flux induces the voltage in the secondary winding. Before the switch is closed, no current flows in the secondary. After the switch is closed, current flows in the secondary opposing the flux in the core. However, because of the voltage applied to the primary, the flux must be maintained in the core. (Otherwise, Kirchhoff's voltage law would not be satisfied in the primary circuit.) Thus, current must begin to flow into the primary to offset the magnetomotive force of the secondary winding. ■

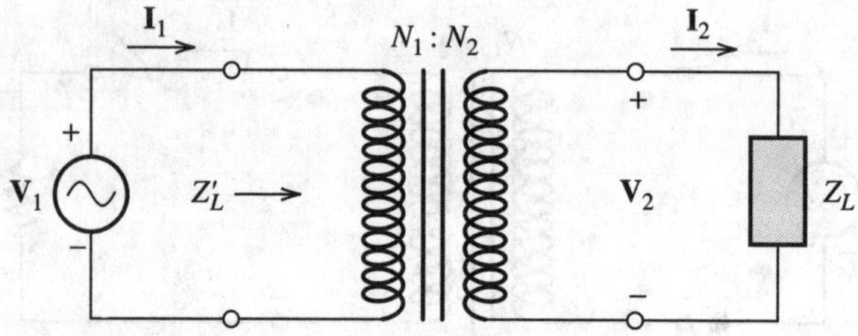

Figure 15.25 The impedance seen looking into the primary is $Z'_L = (N_1/N_2)^2 \times Z_L$.

Impedance Transformations

Consider the circuit shown in Figure 15.25. The phasor current and voltage in the secondary are related to the load impedance by

$$\frac{\mathbf{V}_2}{\mathbf{I}_2} = Z_L \tag{15.60}$$

Using Equations 15.51 and 15.56 to substitute for $\mathbf{I}_2$ and $\mathbf{V}_2$, we have

$$\frac{(N_2/N_1)\mathbf{V}_1}{(N_1/N_2)\mathbf{I}_1} = Z_L \tag{15.61}$$

Rearranging this, we get

$$Z'_L = \frac{\mathbf{V}_1}{\mathbf{I}_1} = \left(\frac{N_1}{N_2}\right)^2 Z_L \tag{15.62}$$

in which Z'_L is the impedance *seen* by the source. We say that the load impedance is *reflected* to the primary side by the square of the turns ratio.

Example 15.11 Using Impedance Transformations

Consider the circuit shown in Figure 15.26(a). Find the phasor currents and voltages. Also, find the power delivered to the load.

Solution First, we reflect the load impedance Z_L to the primary side of the transformer as shown in Figure 15.26(b). The impedance seen from the primary side is

$$Z'_L = \left(\frac{N_1}{N_2}\right)^2 Z_L = (10)^2(10 + j20) = 1000 + j2000$$

The total impedance seen by the source is

$$Z_s = R_1 + Z'_L = 1000 + 1000 + j2000 = 2000 + j2000$$

Converting to polar form, we have

$$Z_s = 2828\angle 45°$$

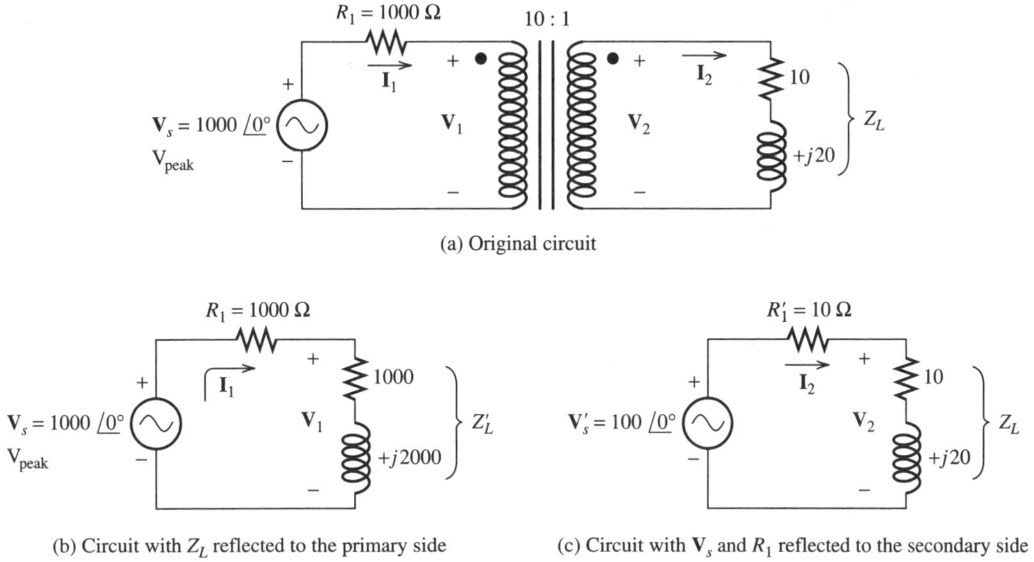

Figure 15.26 The circuit of Examples 15.11 and 15.12.

Now, we can compute the primary current and voltage:

$$\mathbf{I}_1 = \frac{\mathbf{V}_s}{Z_s} = \frac{1000\angle 0°}{2828\angle 45°} = 0.3536\angle -45° \text{ A peak}$$

$$\mathbf{V}_1 = \mathbf{I}_1 Z_L' = 0.3536\angle -45° \times (1000 + j2000)$$
$$= 0.3536\angle -45° \times (2236\angle 63.43°) = 790.6\angle 18.43° \text{ V peak}$$

Next, we can use the turns ratio to compute the secondary current and voltage:

$$\mathbf{I}_2 = \frac{N_1}{N_2}\mathbf{I}_1 = \frac{10}{1}0.3536\angle -45° = 3.536\angle -45° \text{ A peak}$$

$$\mathbf{V}_2 = \frac{N_2}{N_1}\mathbf{V}_1 = \frac{1}{10}790.6\angle 18.43° = 79.06\angle 18.43° \text{ V peak}$$

Finally, we compute the power delivered to the load:

$$P_L = I_{2\text{rms}}^2 R_L = \left(\frac{3.536}{\sqrt{2}}\right)^2 (10) = 62.51 \text{ W}$$

■

Besides transferring impedances from one side of a transformer to the other by using the square of the turns ratio, we can also reflect voltage sources or current sources by using the turns ratio.

Example 15.12 Reflecting the Source to the Secondary

Consider Figure 15.26(a). Reflect V_s and R_1 to the secondary side.

Solution The voltage is reflected by using the turns ratio. Thus, we have

$$\mathbf{V}_s' = \frac{N_2}{N_1}\mathbf{V}_s = \frac{1}{10}1000\angle 0° = 100\angle 0°$$

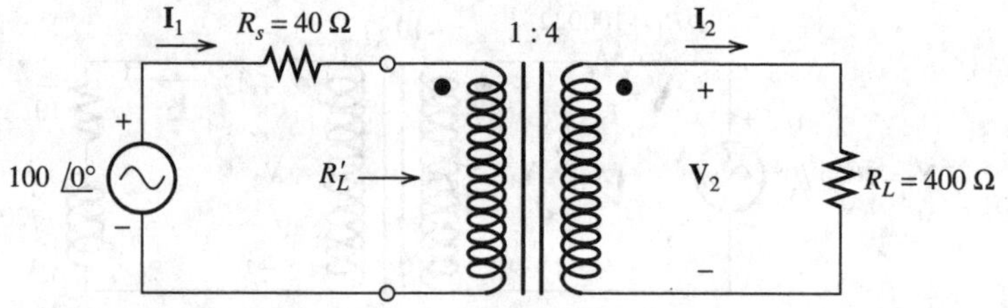

Figure 15.27 Circuit of Exercises 15.16 and 15.17.

On the other hand, the resistance is reflected using the square of the turns ratio. This yields

$$R_1' = \left(\frac{N_2}{N_1}\right)^2 R_1 = \left(\frac{1}{10}\right)^2 (1000) = 10\ \Omega$$

The circuit with $\mathbf{V}_s$ and R_1 transferred to the secondary side is shown in Figure 15.26(c). ■

Exercise 15.16 Working from the circuit of Figure 15.26(c), find the values of $\mathbf{V}_2$ and the power delivered to the load. (Of course, the answers should be the same as the values found in Example 15.11.)

Answer $\mathbf{V}_2 = 79.06\angle 18.43°$, $P_L = 62.51$ W. □

Exercise 15.17 Consider the circuit shown in Figure 15.27. Compute the values of $\mathbf{I}_1, \mathbf{I}_2, \mathbf{V}_2$, the power delivered to R_L, and R_L'.

Answer $\mathbf{I}_1 = 1.538\angle 0°$, $\mathbf{I}_2 = 0.3846\angle 0°$, $\mathbf{V}_2 = 153.8\angle 0°$, $P_L = 29.60$ W, $R_L' = 25\ \Omega$. □

Exercise 15.18 Recall that to achieve maximum power transfer from a source with an internal resistance of R_s, we want the effective load resistance R_L' to equal R_s. Find the turns ratio that would result in maximum power delivered to the load in Figure 15.27.

Answer $N_1/N_2 = 1/\sqrt{10}$. □

15.6 REAL TRANSFORMERS

Well-designed transformers approximately meet the conditions that we assumed in our discussion of the ideal transformer. Often for initial design calculations, we can assume that a transformer is ideal. However, a better model is needed for accurate calculations in the final stages of design. Moreover, a better understanding of transformers and their limitations is gained by considering a refined model.

The equivalent circuit of a real transformer is shown in Figure 15.28. The resistances R_1 and R_2 account for the resistance of the wires used to wind the coils of the transformer.

For the ideal transformer, we assumed that all of the flux links all of the turns of both coils. In fact, some of the flux produced by each coil leaves the core and does not link the other coil. We account for this **leakage flux** by adding the inductances L_1 and L_2 to the ideal transformer, as shown in Figure 15.28.

In discussing the ideal transformer, we assumed that the core reluctance was zero and ignored core loss. This meant that zero magnetomotive force was required

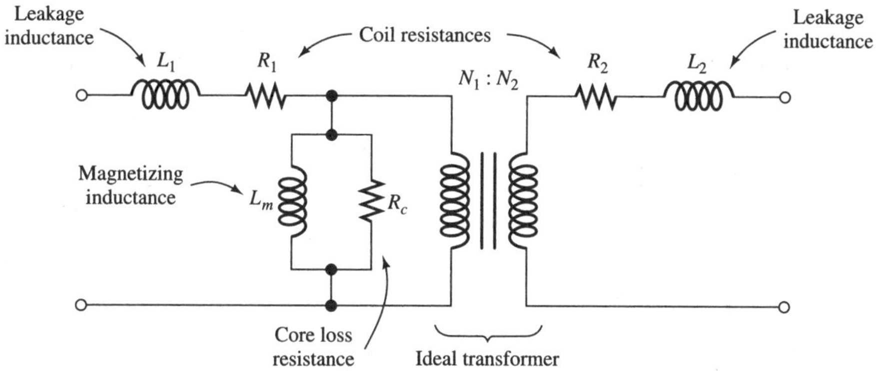

Figure 15.28 The equivalent circuit of a real transformer.

Table 15.1. Circuit Values of a 60-Hz 20-kVA 2400/240-V Transformer Compared with Those of an Ideal Transformer

Element Name	*Symbol*	*Ideal*	*Real*
Primary resistance	R_1	0	3.0 Ω
Secondary resistance	R_2	0	0.03 Ω
Primary leakage reactance	$X_1 = \omega L_1$	0	6.5 Ω
Secondary leakage reactance	$X_2 = \omega L_2$	0	0.07 Ω
Magnetizing reactance	$X_m = \omega L_m$	∞	15 kΩ
Core-loss resistance	R_c	∞	100 kΩ

to establish the flux in the core. Neither of these assumptions is exactly true. The **magnetizing inductance** L_m shown in Figure 15.28 accounts for the nonzero core reluctance. The current needed to establish the flux flows through L_m. Finally, the resistance R_c accounts for power dissipated in the core due to hysteresis and eddy currents.

Table 15.1 compares the values of the circuit elements of a real transformer with those of an ideal transformer.

Variations of the Transformer Model

Figure 15.29 shows several variations of the transformer equivalent circuit. In Figure 15.29(a), the secondary inductance and resistance have been referred to the primary side. In Figure 15.29(b), the magnetizing inductance and loss resistance have been moved to the input side of the circuit. [Actually, the circuit in Figure 15.29(b) is not precisely equivalent to that in Figure 15.29(a). However, in normal operation, the voltage drop across L_1 and R_1 is very small compared to either the input voltage or the voltage across L_m and R_m. Thus, for normal operating conditions, virtually identical results are obtained from either circuit.] Other equivalent circuits can be obtained by moving the circuit elements to the secondary side and by moving L_m and R_c to the right-hand side. Usually, we select the equivalent circuit configuration that is most convenient for the problem at hand.

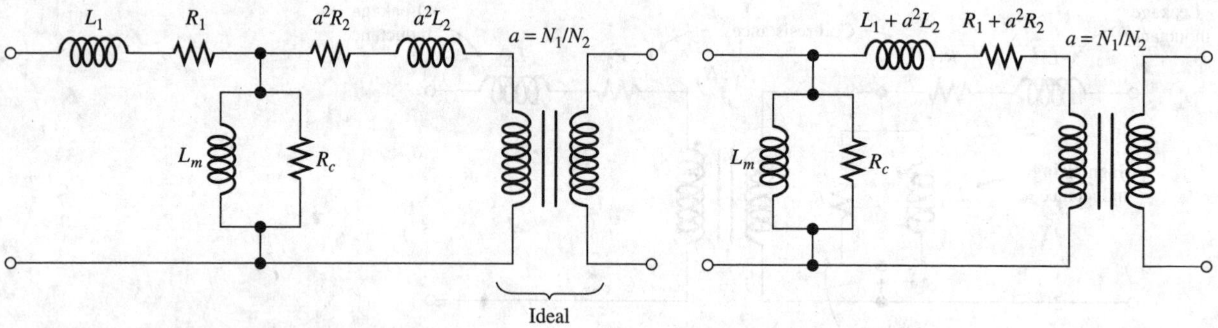

(a) All elements referred to the primary side

(b) Approximate equivalent circuit that is sometimes more convenient to use than that of part (a)

Figure 15.29 Variations of the transformer equivalent circuit. The circuit of (b) is not exactly equivalent to that of (a), but is sufficiently accurate for practical applications.

Regulation and Efficiency

Because of the elements L_1, L_2, R_1, and R_2, the voltage delivered to the load side of a transformer varies with the load current. Usually, this is an undesirable effect. The regulation of a transformer is defined as

$$\text{percent regulation} = \frac{V_{\text{no-load}} - V_{\text{load}}}{V_{\text{load}}} \times 100\%$$

in which $V_{\text{no-load}}$ is the rms voltage across the load terminals for an open-circuit load and V_{load} is the rms voltage across the actual load.

Ideally, we usually want the percentage regulation to be zero. For instance, poor regulation in a residence would mean that the lights dim when an electric clothes dryer is started. Clearly, this is not a desirable situation.

Because of the resistances in the transformer equivalent circuit, not all of the power input to the transformer is delivered to the load. We define the power efficiency as

$$\text{power efficiency} = \frac{P_{\text{load}}}{P_{\text{in}}} \times 100\% = \left(1 - \frac{P_{\text{loss}}}{P_{\text{in}}}\right) \times 100\%$$

in which P_{load} is the power delivered to the load, P_{loss} is the power dissipated in the transformer, and P_{in} is the power delivered by the source to the transformer primary terminals.

In this example, we are taking the rms values of currents and voltages (rather than peak values) as the phasor magnitudes. This is often done by power-distribution engineers. We will clearly indicate when phasors represent rms values rather than peak values.

Example 15.13 Regulation and Efficiency Calculations

Find the percentage regulation and power efficiency for the transformer of Table 15.1 for a rated load having a lagging power factor of 0.8.

Solution First, we draw the circuit as shown in Figure 15.30. Notice that we have placed the magnetizing reactance X_m and core loss resistance R_c on the left-hand side of R_1 and X_1, because this makes the calculations a bit simpler and is sufficiently accurate. We assume a zero phase reference for the load voltage. It is customary in

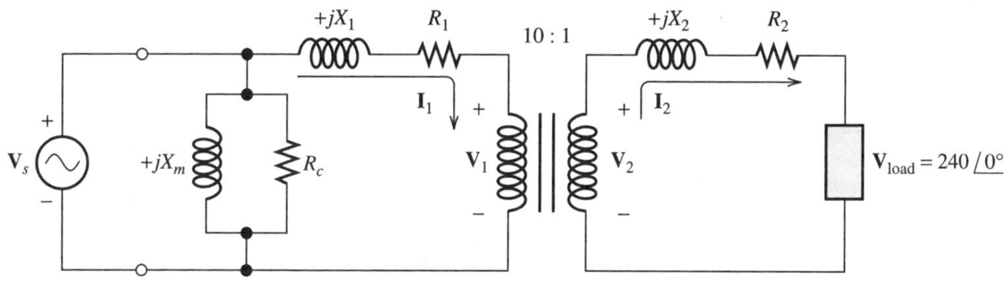

Figure 15.30 Circuit of Example 15.13.

power-system engineering to take the values of phasors as the rms values (rather than the peak values) of the currents and voltages. Thus, as a phasor, we have

$$\mathbf{V}_{\text{load}} = 240\angle 0^\circ \text{ V rms}$$

For rated load (20 kVA), the load current is

$$I_2 = \frac{20 \text{ kVA}}{240 \text{ V}} = 83.33 \text{ A rms}$$

The load power factor is

$$\text{power factor } = \cos(\theta) = 0.8$$

Solving, we find that

$$\theta = 36.87^\circ$$

Thus, the phasor load current is

$$\mathbf{I}_2 = 83.33\angle -36.87^\circ \text{ A rms}$$

where the phase angle is negative because the load was stated to have a lagging power factor.

The primary current is related to the secondary current by the turns ratio:

$$\mathbf{I}_1 = \frac{N_2}{N_1}\mathbf{I}_2 = \frac{1}{10} \times 83.33\angle -36.87^\circ = 8.333\angle -36.87^\circ \text{ A rms}$$

Next, we can compute the voltages:

$$\begin{aligned}\mathbf{V}_2 &= \mathbf{V}_{\text{load}} + (R_2 + jX_2)\mathbf{I}_2 \\ &= 240 + (0.03 + j0.07)83.33\angle -36.87^\circ \\ &= 240 + 6.346\angle 29.93^\circ \\ &= 245.50 + j3.166 \text{ V rms}\end{aligned}$$

The primary voltage is related to the secondary voltage by the turns ratio:

$$\begin{aligned}\mathbf{V}_1 &= \frac{N_1}{N_2}\mathbf{V}_2 = 10 \times (245.50 + j3.166) \\ &= 2455.0 + j31.66 \text{ V rms}\end{aligned}$$

Now, we can compute the source voltage:

$$\begin{aligned}\mathbf{V}_s &= \mathbf{V}_1 + (R_1 + jX_1)\mathbf{I}_1 \\ &= 2455.0 + j31.66 + (3 + j6.5) \times (8.333\angle -36.87^\circ) \\ &= 2508.2\angle 1.37^\circ \text{ V rms}\end{aligned}$$

Next, we compute the power loss in the transformer:

$$\begin{aligned}P_{\text{loss}} &= \frac{V_s^2}{R_c} + I_1^2 R_1 + I_2^2 R_2 \\ &= 62.91 + 208.3 + 208.3 \\ &= 479.5 \text{ W}\end{aligned}$$

The power delivered to the load is given by

$$\begin{aligned}P_{\text{load}} &= V_{\text{load}} I_2 \times \text{ power factor} \\ &= 20 \text{ kVA } \times 0.8 = 16{,}000 \text{ W}\end{aligned}$$

The input power is given by

$$\begin{aligned}P_{\text{in}} &= P_{\text{load}} + P_{\text{loss}} \\ &= 16{,}000 + 479.5 = 16{,}479.5 \text{ W}\end{aligned}$$

At this point, we can compute the power efficiency:

$$\begin{aligned}\text{efficiency} &= \left(1 - \frac{P_{\text{loss}}}{P_{\text{in}}}\right) \times 100\% \\ &= \left(1 - \frac{479.5}{16{,}479.5}\right) \times 100\% = 97.09\%\end{aligned}$$

Next, we can determine the no-load voltages. Under no-load conditions, we have

$$\begin{aligned}I_1 &= I_2 = 0 \\ V_1 &= V_s = 2508.2 \\ V_{\text{no-load}} &= V_2 = V_1 \frac{N_2}{N_1} = 250.82 \text{ V rms}\end{aligned}$$

Finally, the percentage regulation is

$$\begin{aligned}\text{percent regulation} &= \frac{V_{\text{no-load}} - V_{\text{load}}}{V_{\text{load}}} \times 100\% \\ &= \frac{250.82 - 240}{240} \times 100\% \\ &= 4.51\%\end{aligned}$$

■

Summary

1. The right-hand rule can be used to determine the direction of the magnetic field produced by a current. This is illustrated in Figure 15.2 on page 735.
2. Force is exerted on a charge moving through a magnetic field according to the equation

$$\mathbf{f} = q\mathbf{u} \times \mathbf{B}$$

Similarly, forces appear on a current-carrying wire immersed in a magnetic field. The force on an incremental length of wire is given by

$$d\mathbf{f} = id\mathbf{l} \times \mathbf{B}$$

3. According to Faraday's law of induction, voltage is induced in a coil when its magnetic flux linkages change with time. Similarly, voltages are induced in conductors that cut through magnetic flux lines. We can determine the polarity of the induced voltage by using Lenz's law.
4. Magnetic flux density **B** and the magnetic field intensity **H** are related by

$$\mathbf{B} = \mu\mathbf{H}$$

where μ is the magnetic permeability of the material. For air or vacuum, $\mu = \mu_0 = 4\pi \times 10^{-7}$.
5. According to Ampère's law, the line integral of **H** around a closed path is equal to the algebraic sum of the currents flowing through the area bounded by the path. We can use this law to find the field around a long straight wire or inside a toroidal coil.
6. Practical magnetic devices can be approximately analyzed by using circuit concepts. Magnetomotive forces are analogous to voltage sources, reluctance is analogous to resistance, and flux is analogous to current.
7. The inductance and mutual inductance of coils can be computed from knowledge of the physical parameters of the coils and the core on which they are wound.
8. The B–H relationship for iron takes the form of a hysteresis loop, which displays saturation in the neighborhood of 1 to 2 T. The area of the loop represents energy converted to heat per cycle. Eddy currents are another cause of core loss. Energy can be stored in magnetic fields. In a magnetic circuit consisting of an iron core with an air gap, most of the energy is stored in the gap.
9. In an ideal transformer, the voltage across each coil is proportional to its number of turns, the net mmf is zero, and the power efficiency is 100 percent.
10. Equivalent circuits for real transformers are shown in Figures 15.28 and 15.29 on pages 763 and 764, respectively.
11. Efficiency and regulation are important aspects of transformer operation.

Problems

Section 15.1: Magnetic Fields

P15.1. State Faraday's law of magnetic induction and Lenz's law.

P15.2. State Ampère's law, including the reference directions for the currents.

P15.3. State the right-hand rule as it applies to: **a.** a current carrying conductor; **b.** a current-carrying coil.

P15.4. What is the fundamental cause of magnetic fields?

***P15.5.** A bar magnet is inserted into a single-turn coil as illustrated in Figure P15.5. Is the voltage v_{ab} positive or negative as the bar approaches the coil?

***P15.6.** The magnetic field of the earth is approximately 3×10^{-5} T. At what distance from a

* Denotes that answers are contained in the Student Solutions files. See Appendix E for more information about accessing the Student Solutions.

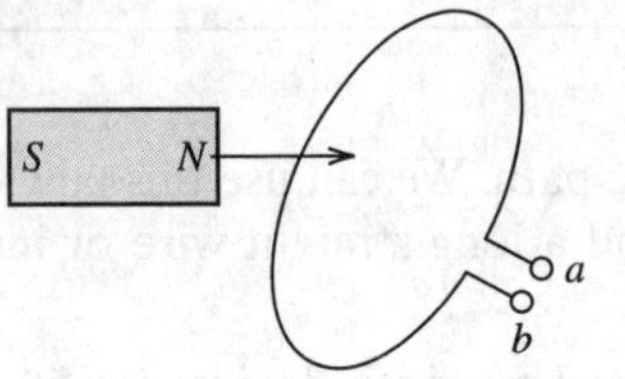

Figure P15.5

long straight wire carrying a steady current of 10 A is the field equal to 10 percent of the earth's field? Suggest at least two ways to help reduce the effect of electrical circuits on the navigation compass in a boat or airplane.

***P15.7.** A 1-m length of wire carries current perpendicular to a magnetic field. A magnetic flux density of 0.6 T results in a force of 3 N on the wire. Determine the value of the current.

P15.8. A long copper pipe carries a dc current. Is there a magnetic field inside the pipe due to the current? Outside? Justify your answers.

P15.9. An irregular loop of wire carries an electrical current as illustrated in Figure P15.9. Is there a *net* force on the loop due to the magnetic fields created? Justify your answer. (*Hint:* Consider Newton's third law of motion.)

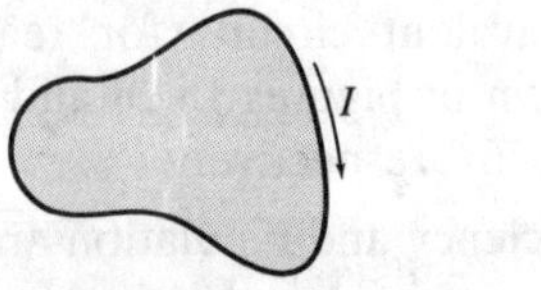

Figure P15.9

P15.10. Using equations given in Section 15.1, perform dimensional analyses to determine the units of μ, **B**, and **H** in terms of meters, kilograms, seconds, and coulombs.

P15.11. Use the right-hand rule to find the direction of the magnetic flux for each coil shown in Figure P15.11. Mark the N and S ends of each coil. Do the coils attract or repel one another?

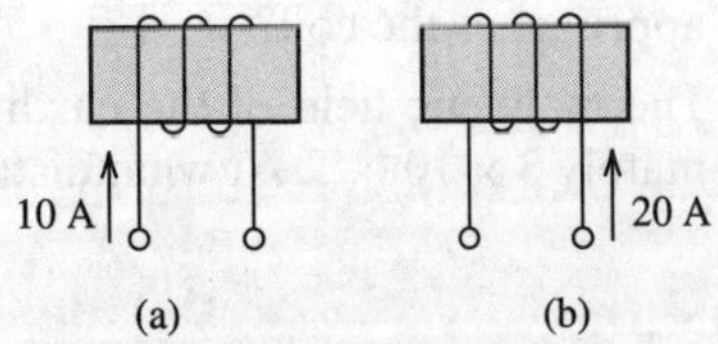

Figure P15.11

P15.12. Suppose that the flux ϕ linking the coils shown in Figure P15.12 is increasing in magnitude. Find the polarity of the voltage across each coil.

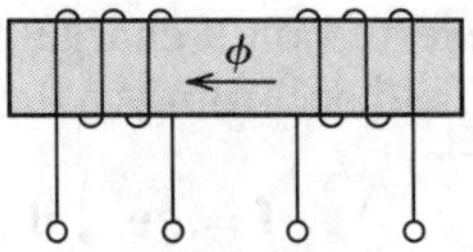

Figure P15.12

P15.13. Two very long parallel wires are 1 cm apart and carry currents of 10 A in the same direction. The material surrounding the wires has $\mu_r = 1$. Determine the force on a 0.5-m section of one of the wires. Do the wires attract or repel one another?

***P15.14.** A uniform flux density of 2 T is perpendicular to the plane of a five-turn circular coil of radius 20 cm. Find the flux linking the coil and the flux linkages. Suppose that the field is decreased to zero at a uniform rate in 10 ms. Find the magnitude of the voltage induced in the coil.

***P15.15.** Suppose that we test a material and find that $B = 0.1\ \text{Wb/m}^2$ for an applied H of 50 A/m. Compute the relative permeability of the material.

P15.16. Consider two coils which are wound on nonmagnetic forms such that part of the flux produced by each coil links the other, as shown in Figure P15.16. Assume that the

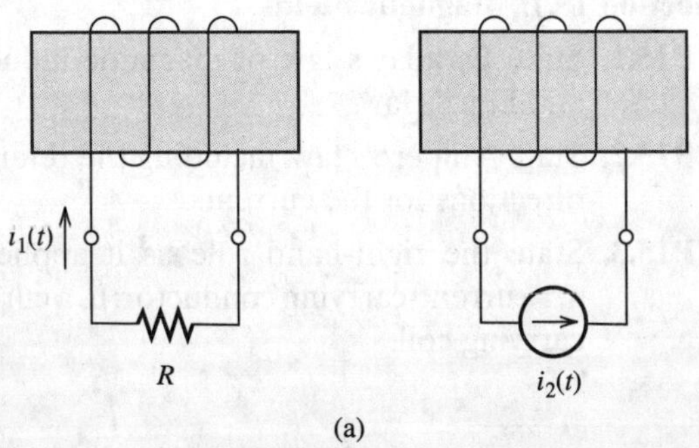

Figure P15.16

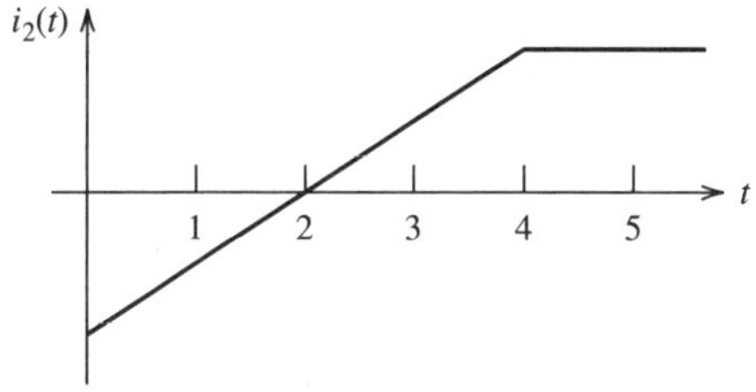

(b)

Figure P15.16 (*Cont.*)

inductance of the left-hand coil is small enough so that $i_1(t)$ is equal to the voltage induced by the magnetic field of the right-hand coil divided by the resistance. At $t =$ 1 s, is the force between the coils attraction, repulsion, or zero? Explain your reasoning. Repeat for $t = 2, 3$, and 5 s.

P15.17. A 200-turn toroidal coil (see Figure 15.8 on page 740) has $r = 1$ cm and $R = 10$ cm. When a current given by $0.05 \sin(200t)$A flows in the coil, the voltage is $0.5 \cos(200t)$A. Determine the flux ϕ as a function of time and the relative permeability of the core material.

P15.18. A uniform flux density given by $B = 0.5 \sin(377t)$ T is perpendicular to the plane of a 1000-turn circular coil of radius 10 cm. Find the flux linkages and the voltage as functions of time. Find their maximum values also.

P15.19. Suppose that, in designing an electrical generator, we need to produce a voltage of 120 V by moving a straight conductor through a uniform magnetic field of 0.5 T at a speed of 30 m/s. The conductor, its motion, and the field are mutually perpendicular. What is the required length of the conductor? It turns out that in generator design, a conductor of this length is impractical, and we must use N conductors of length 0.1 m. However, by connecting the conductors in series, we can obtain the required 120 V. What is the number N of conductors needed?

P15.20. Two infinitely long, very thin wires lie on the x and y axes and carry currents as shown in Figure P15.20. **a.** Show the direction of the forces on the wires due to the magnetic fields on the positive and negative part of each axis, assuming that I_x and I_y are both positive. **b.** Compute the torque on the wire that lies on the y axis.

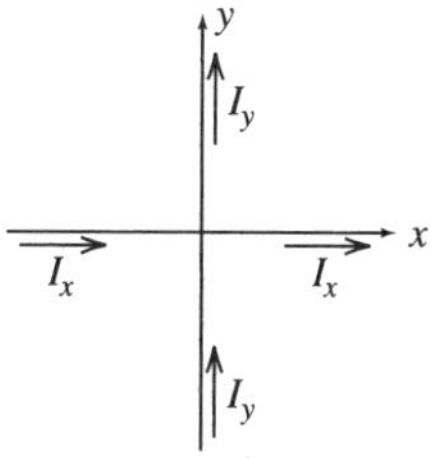

Figure P15.20

P15.21. A very long, straight wire carrying a constant current $i(t) = I_1$ and a rectangular single-turn coil lie in the same plane, as illustrated in Figure P15.22. The wire and the coil are surrounded by air. A source is applied to the coil causing a constant clockwise current I_2 to flow in the loop. **a.** Derive an expression for the net force exerted on the coil due to the magnetic field of the wire. **b.** Evaluate the force given that $I_1 = I_2 = 10$ A, $l = 10$ cm, $r_1 = 1$ cm, and $r_2 = 10$ cm. **c.** Is the loop attracted or repelled by the wire?

P15.22. A very long, straight wire carrying current $i(t)$ and a rectangular single-turn coil lie in the same plane, as illustrated in Figure P15.22. The wire and the coil are surrounded by air. **a.** Derive an expression for the flux linking the coil. **b.** Derive an expression for the voltage $v_{ab}(t)$ induced in the coil. **c.** Determine the rms value of v_{ab}, given that $i(t)$ is a 10-A rms 60-Hz sinusoid, $l = 10$ cm, $r_1 = 1$ cm, and $r_2 = 10$ cm.

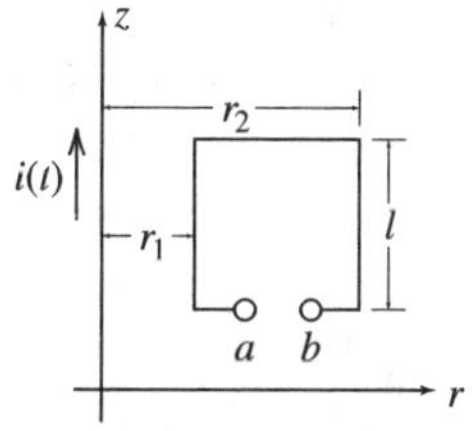

Figure P15.22

P15.23. A 120-V-rms 60-Hz sinusoidal voltage appears across a 500-turn coil. Determine the peak and rms values of the flux linking the coil.

Section 15.2: Magnetic Circuits

***P15.24.** Consider the magnetic circuit of Figure 15.13 on page 745 that was analyzed in Example 15.6. Suppose that the length of gap *a* is reduced to zero. Compute the flux in gap *b*. Why is the result less than that found in Example 15.6?

***P15.25.** An air gap and an iron core of same cross-sectional area have lengths of 1 cm and 500 cm respectively. Do the iron core and the air gap have the same reluctance? Determine the relative permeability of the iron.

***P15.26.** What quantity in a magnetic circuit is analogous to a voltage source in an electrical circuit? To resistance? To current?

P15.27. What happens to the reluctance of a magnetic path if its length is doubled? If the cross-sectional area is doubled? If the relative permeability is doubled?

P15.28. What are the physical units of reluctance in terms of kilograms, coulombs, meters, and seconds?

P15.29. Draw the electrical circuit analog for the magnetic circuit shown in Figure P15.29. Pay special attention to the polarities of the voltage sources. Determine the flux density in the core.

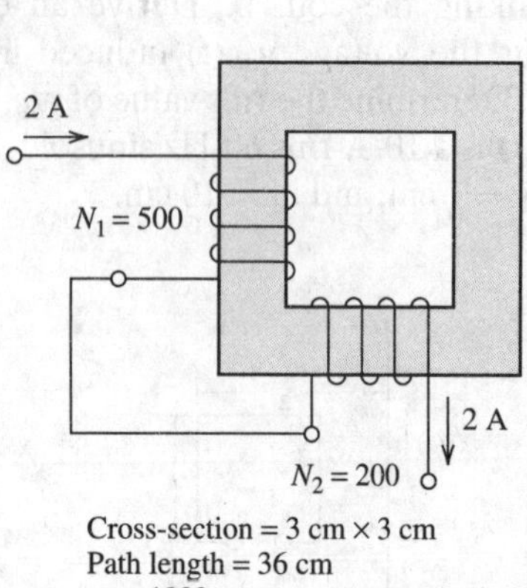

Cross-section = 3 cm × 3 cm
Path length = 36 cm
$\mu_r = 1000$

Figure P15.29

P15.30. Compute the flux in each leg of the magnetic core shown in Figure P15.30.

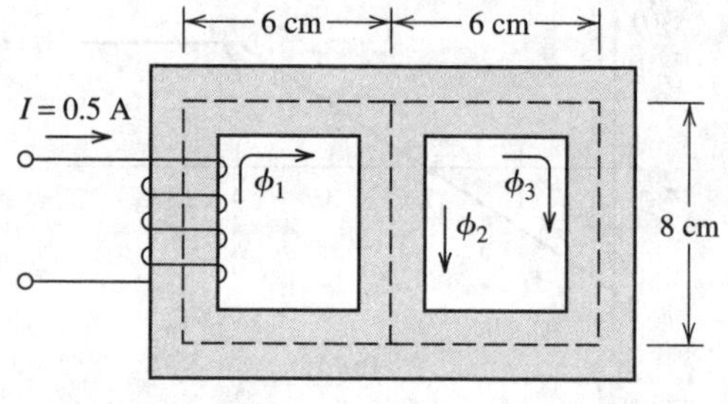

$N = 1000$ turns
Core cross-section: 2 cm × 2 cm
$\mu_r = 5000$

Figure P15.30

P15.31. Consider the magnetic circuit shown in Figure P15.31. Assume that the reluctance of the iron is small enough so it can be neglected. The lengths of the air gaps are unknown, and the effective area of each gap is 20 cm^2. The number of turns needed to produce a flux density of 1 T in the gaps is 100. Determine the length of air gaps.

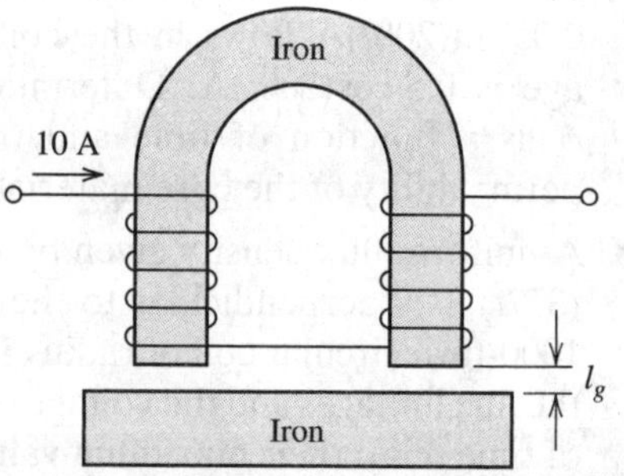

Figure P15.31

P15.32. Compute the flux in each leg of the core shown in Figure P15.32. Account for fringing by adding the gap length to each of the cross-sectional dimensions of the gap.

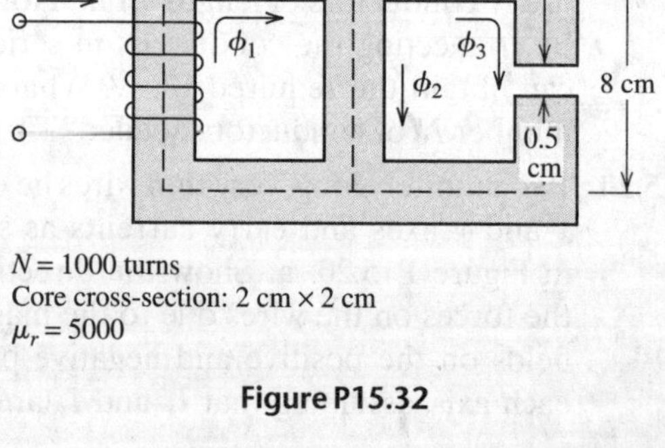

$N = 1000$ turns
Core cross-section: 2 cm × 2 cm
$\mu_r = 5000$

Figure P15.32

***P15.33.** Consider the *solenoid* shown in Figure P15.33, which is typical of those commonly used as actuators for mechanisms and for operating valves in chemical processes. Neglect fringing and the reluctance of the core. Derive an expression for the flux as a function of the physical dimensions, μ_0, the number of turns N, and the current.

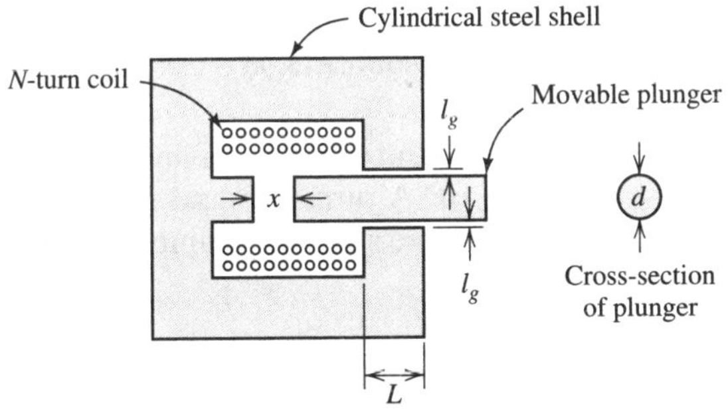

Figure P15.33

P15.34. Consider the core shown in Figure P15.34, which has two coils of N turns, each connected so that the fluxes aid in the center leg. Determine the value of N so that $I = 2$ A produces a flux density of 0.25 T in the gap. The gap and the core have square cross-sections 2 cm on each side. Account for fringing by adding the gap length to the length of each side of the gap.

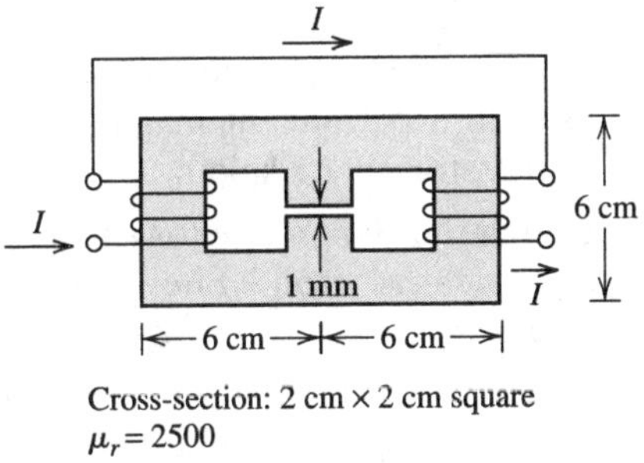

Figure P15.34

P15.35. Consider a toroidal core, as shown in Figure 15.11 on page 743, that has a relative permeability of 1000, $R = 5$ cm, and $r = 2$ cm. Two windings are wound on the core, one with 200 turns and the other with 400 turns. A voltage source described by $v(t) = 10\cos(10^5 t)$ is applied to the 200-turn coil, and the 400-turn coil is open circuited. Determine the current in the 200-turn coil and the voltage across the 400-turn coil. Assume that all of the magnetic flux is confined to the core and that the current is a pure sine wave.

Section 15.3: Inductance and Mutual Inductance

***P15.36.** A 1000-turn coil is wound on an iron core. When a 120-V-rms 60-Hz voltage is applied to the coil, the current is 2 A rms. Neglect the resistance of the coil. Determine the reluctance of the core. Given that the cross-sectional area of the core is 5 cm^2 and the length is 20 cm, determine the relative permeability of the core material.

***P15.37.** A 200-turn coil wound on a magnetic core is found to have an inductance of 100 mH. What inductance will be obtained if the number of turns is increased to 600, assuming that all of the flux links all of the turns?

P15.38. Define self-inductance and mutual inductance. Establish the relationship between reluctance, the number of turns of the coil, and the self-inductance.

***P15.39.** Consider the circuit shown in Figure P15.39. The two coils have $L_1 = 0.1$ H, $L_2 = 10$ H, and $M = 0.5$ H. Prior to $t = 0$, the currents in the coils are zero. At $t = 0$, the switch closes. Determine and sketch $i_1(t)$ and $i_2(t)$ to scale versus time.

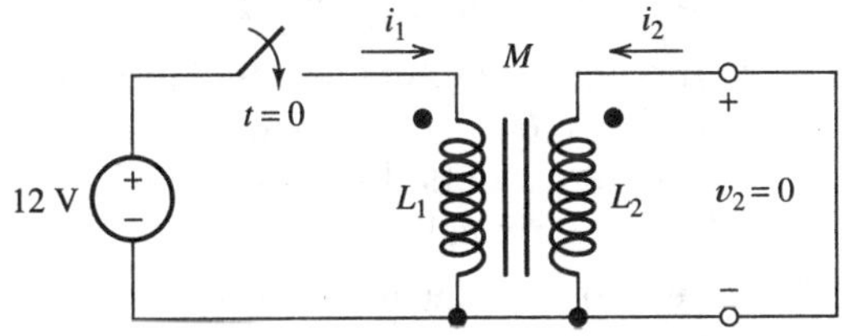

Figure P15.39

***P15.40.** Two coils wound on a common core have $L_1 = 1$ H, $L_2 = 2$ H, and $M = 0.5$ H. The currents are $i_1 = \cos(377t)$ A and $i_2 = 0.5\cos(377t)$ A. Both of the currents enter dotted terminals. Find expressions for the voltages across the coils.

P15.41. A symmetrical toroidal coil is wound on a plastic core ($\mu_r \cong 1$) and is found to have an inductance of 1 mH. What inductance will result if the core material is changed to a ferrite having $\mu_r = 200$? Assume that the entire magnetic path is composed of ferrite.

P15.42. A 100-turn coil is wound on a toroidal core and has an inductance of 100 mH. Suppose that a 200-turn coil is wound on a second toroidal core having dimensions (r and R as shown in Figure 15.11 on page 743) that are double those of the first core. Both cores are made of the same material, which has very high permeability. Determine the inductance of the second coil.

P15.43. A relay has a 100-turn coil that draws 50 mA rms when a 60-Hz voltage of 48 V rms is applied. Assume that the resistance of the coil is negligible. Determine the peak flux linking the coil, the reluctance of the core, and the inductance of the coil.

P15.44. Consider the coils shown in Figure P15.12. Suppose that a dot is placed on the leftmost terminal. Place a dot on the appropriate terminal of the right-hand coil to indicate the sense of the coupling.

P15.45. Two coils wound on a common core have $L_1 = 0.2$ H, $L_2 = 0.5$ H, and $M = 0.1$ H. The currents are $i_1 = \exp(-1000t)$ A and $i_2 = 2\exp(-1000t)$ A. Both of the currents enter dotted terminals. Find expressions for the voltages across the coils.

P15.46. Two coils wound on a common core have $L_1 = 1$ H, $L_2 = 2$ H, and $M = 0.5$ H. The currents are $i_1 = 1$ A and $i_2 = 0.5$ A. If both currents enter dotted terminals, find the flux linkages of both coils. Repeat if i_1 enters a dotted terminal and i_2 leaves a dotted terminal.

P15.47. Two coils having inductances L_1 and L_2 are wound on a common core. The fraction of the flux produced by one coil that links the other coil is called the *coefficient of coupling* and is denoted by k. Derive an expression for the mutual inductance M in terms of L_1, L_2, and k.

P15.48. Consider the circuit shown in Figure P15.48. The two coils have $L_1 = 0.1$ H, $L_2 = 10$ H, and $M = 1$ H. Prior to $t = 0$, the currents in the coils are zero. At $t = 0$, the switch closes. Determine and sketch $i_1(t)$ and $v_2(t)$ to scale versus time.

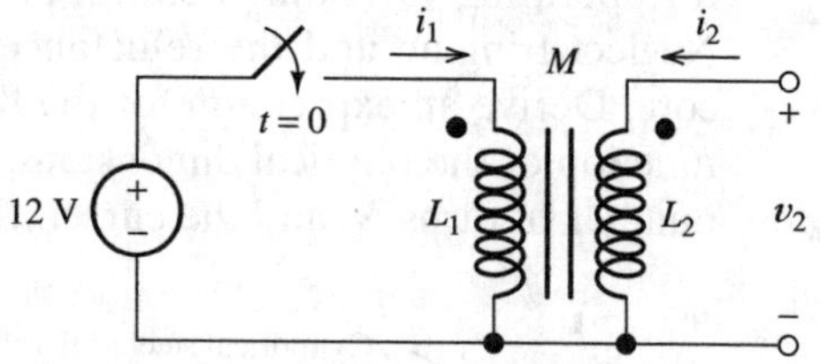

Figure P15.48

P15.49. A coil is wound on a core having a reluctance of 25 × 10^5 A·turns/Wb and inductance of 100 mH. Determine the number of turns of the coil.

Section 15.4: Magnetic Materials

***P15.50.** For operation at 60 Hz and a given peak flux density, the core loss of a given core is 1 W due to hysteresis and 0.5 W due to eddy currents. Estimate the core loss for 400-Hz operation with the same peak flux density.

***P15.51.** What are two causes of core loss for a coil with an iron core excited by an ac current? What considerations are important in minimizing loss due to each of these causes? What happens to the power loss in each case if the frequency of operation is doubled while maintaining constant peak flux density?

P15.52. What characteristic is desirable in the B–H curve for a prospective material to be used in a permanent magnet? In a motor or transformer? Explain.

P15.53. Sketch the B–H curve for a magnetic material such as iron. Show hysteresis and saturation.

P15.54. A magnetic core has a mean length of 20 cm, a cross-sectional area of 4 cm^2, and a relative permeability of 2000. A 500-turn coil wound on the core carries a dc current of 0.1 A. **a.** Determine the reluctance of the core, the flux density in the core, and the inductance. **b.** Compute the energy stored in the magnetic field as $W = (1/2)LI^2$. **c.** Use Equation 15.41 on page 754 to compute the energy density in the core. Then, compute the energy by taking the product of energy density and volume. Compare to the value found in part (b).

P15.55. At a frequency of 60 Hz, the core loss of a certain coil with an iron core is 3.6 W, and at a frequency of 120 Hz, it is 11.2 W. The peak flux density is the same for both cases. Determine the power loss due to hysteresis and that due to eddy currents for 60-Hz operation.

P15.56. A certain iron core has an air gap with an effective area of 2 cm × 3 cm and a length l_g. The applied magnetomotive force is 1000 A·turns, and the reluctance of the iron is negligible. Find the flux density and the energy stored in the air gap as a function of l_g.

P15.57. Consider a coil wound on an iron core. Suppose that for operation with a given 60-Hz ac current, the hysteresis loop of the core material takes the form of a rectangle, as shown in Figure P15.57. The core volume is 1000 cm^3. Find the power converted to heat because of hysteresis.

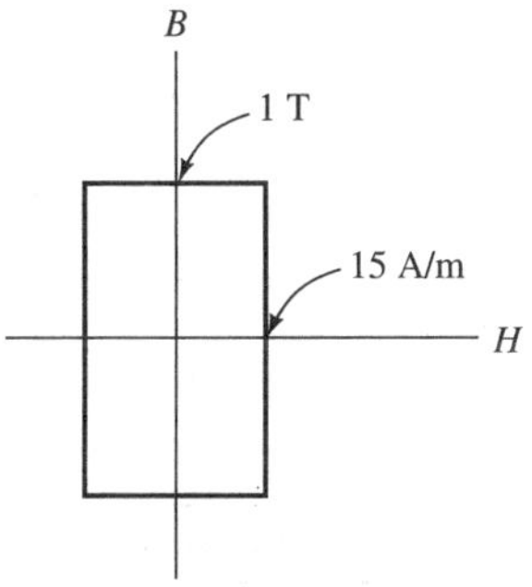

Figure P15.57

Section 15.5: Ideal Transformers

P15.58. What assumptions did we make in deriving the relationships between the voltages and currents in an ideal transformer?

***P15.59.** Suppose that we need to cause a 25-Ω load resistance to appear as a 100-Ω resistance to the source. Instead of using a transformer, we could place a 75-Ω resistance in series with the 25-Ω resistance. From the standpoint of power efficiency, which approach is better? Explain.

***P15.60.** In U.S. residences, electrical power is generally utilized at a nominal voltage of 120 V rms. What problems would become pronounced if the power distribution system and household appliances had been designed for a lower voltage (say, 12 V rms)? For a higher voltage (say, 12 kV)?

***P15.61.** Consider the transformer having three windings, as shown in Figure P15.61. **a.** Place dots on windings to indicate the sense of the coupling between coil 1 and coil 2; between coil 1 and coil 3. **b.** Assuming that all of the flux links all of the turns, determine the voltages $\mathbf{V}_2$ and $\mathbf{V}_3$. **c.** Assuming that the net mmf required to establish the core flux is zero, find an expression for $\mathbf{I}_1$ in terms of $\mathbf{I}_2$, $\mathbf{I}_3$, and the turns ratios. Then, compute the value of $\mathbf{I}_1$.

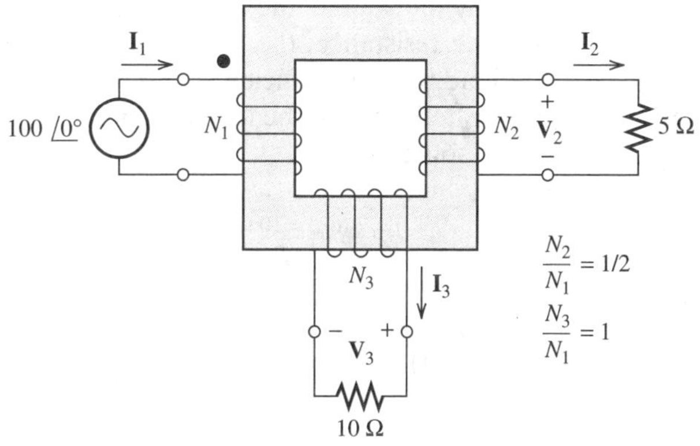

Figure P15.61

P15.62. Consider the circuit shown in Figure P15.62. Find the secondary voltage $V_{2\text{rms}}$, the secondary current $I_{2\text{rms}}$, and the power delivered to the load if the turns ratio is $N_1/N_2 = 10$. Repeat for $N_1/N_2 = 1$ and for $N_1/N_2 = 0.1$.

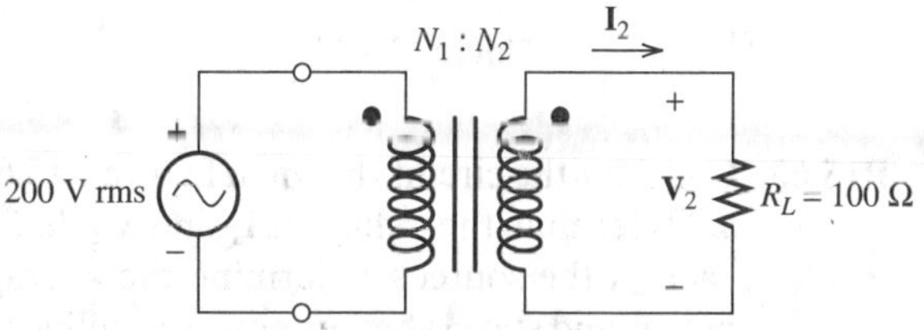

Figure P15.62

P15.63. A transformer is needed that will cause an actual load resistance of 25 Ω to appear as 100 Ω to an ac voltage source of 240 V rms. Draw the diagram of the circuit required. What turns ratio is required for the transformer? Find the current taken from the source, the current flowing through the load, and the load voltage.

P15.64. A voltage source V_s is to be connected to a resistive load $R_L = 10\ \Omega$ by a transmission line having a resistance $R_{\text{line}} = 10\ \Omega$, as shown in Figure P15.64. In part (a) of the figure, no transformers are used. In part (b) of the figure, one transformer is used to step up the source voltage at the sending end of the line, and another transformer is used to step the voltage back down at the load. For each case, determine the power delivered by the source; the power dissipated in the line resistance; the power delivered to the load; and the efficiency, defined as the power delivered to the load as a percentage of the source power.

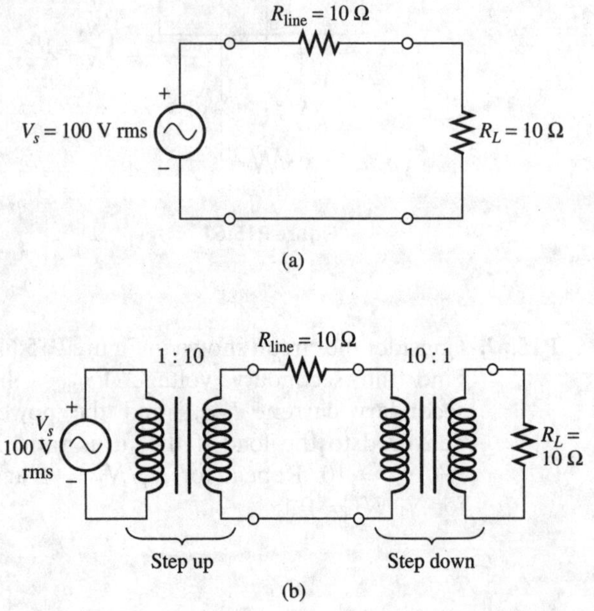

Figure P15.64

P15.65. Consider the circuit shown in Figure P15.65. **a.** Determine the values of $\mathbf{I}_1$ and $\mathbf{V}_2$. **b.** For each of the sources, determine the average power and state whether power is delivered by or absorbed by the source. **c.** Move the dot on the secondary to the bottom end of the coil and repeat parts (a) and (b).

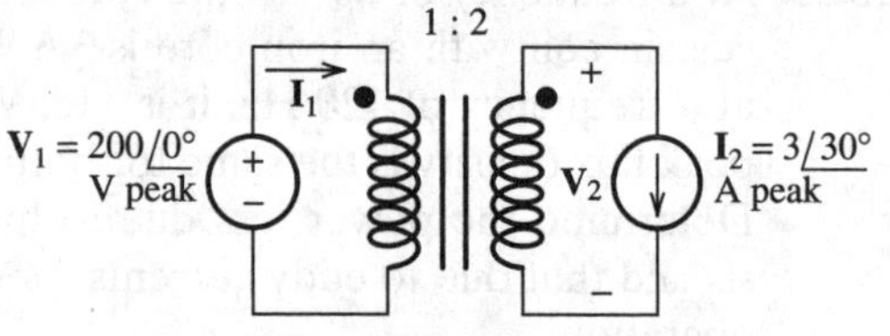

Figure P15.65

P15.66. **a.** Reflect the resistances and voltage sources to the left-hand side of the circuit shown in Figure P15.66 and solve for the current $\mathbf{I}_1$. **b.** Repeat with the dot moved to the top of the right-hand coil.

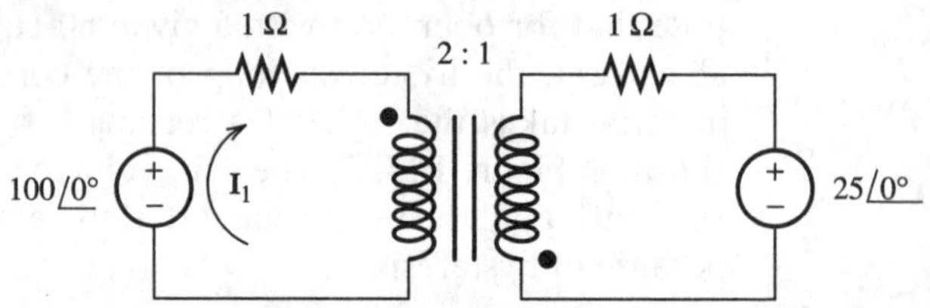

Figure P15.66

P15.67. A type of transformer known as an *autotransformer* is shown in Figure P15.67. **a.** Assuming that all of the flux links all of the turns, determine the relationship between

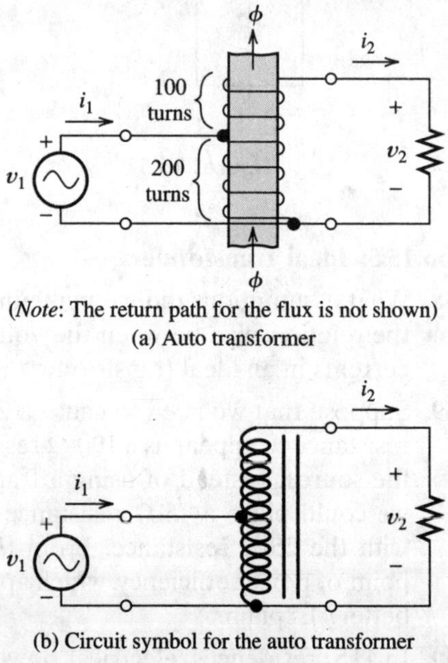

(*Note*: The return path for the flux is not shown)
(a) Auto transformer

(b) Circuit symbol for the auto transformer

Figure P15.67

v_1, v_2, and the number of turns. **b.** Assuming that the total mmf required to establish the flux is zero, find the relationship between the currents i_1 and i_2.

P15.68. Find the equivalent resistance R'_L and capacitance C'_L seen looking into the transformers in Figure P15.68. (*Hint:* Keep in mind that it is the *impedance* that is reflected by the square of the turns ratio.)

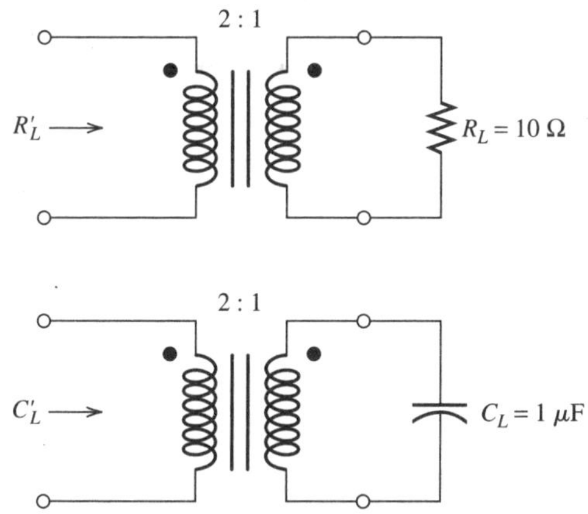

Figure P15.68

P15.69. An autotransformer is shown in Figure P15.69. Assume that all of the flux links all of the turns and that negligible mmf is needed to establish the flux. Determine the values of $\mathbf{I}_1, \mathbf{I}_2, \mathbf{I}_3$, and $\mathbf{V}_2$.

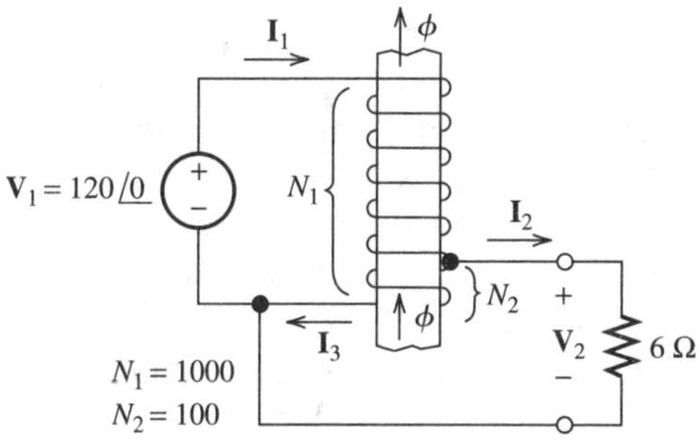

Figure P15.69

Section 15.6: Real Transformers

***P15.70.** Draw an equivalent circuit for a real transformer. Briefly discuss the reason that each element appears in the equivalent circuit.

***P15.71.** A 60-Hz 20-kVA 8000/240-V-rms transformer has the following equivalent-circuit parameters:

Primary resistance	R_1	15 Ω
Secondary resistance	R_2	0.02 Ω
Primary leakage reactance	$X_1 = \omega L_1$	120 Ω
Secondary leakage reactance	$X_2 = \omega L_2$	0.15 Ω
Magnetizing reactance	$X_m = \omega L_m$	30 kΩ
Core-loss resistance	R_c	200 kΩ

Find the percentage regulation and power efficiency for the transformer for a 2-kVA load (i.e., 10 percent of rated capacity) having a lagging power factor of 0.8.

P15.72. When operating with an open-circuit load and with rated primary voltage, a certain 60-Hz 20-kVA 8000/240-V-rms transformer has a primary current of 0.315 A rms and absorbs 360 W. Which of the elements of the equivalent circuit of Figure 15.29(b) on page 764 can be determined from this data? Find the numerical values of these elements.

***P15.73.** Usually, transformers are designed to operate with peak flux densities just below saturation of the core material. Why would we not want to design them to operate far below the saturation point? Far above the saturation point? Assume that the voltage and current ratings are to remain constant.

P15.74. Under the assumptions that we made for the ideal transformer, we determined that $v_2 = (N_2/N_1)v_1$. Theoretically, if a dc voltage is applied to the primary of an ideal transformer, a dc voltage should appear across the secondary winding. However, real transformers are ineffective for dc. Use the equivalent circuit of Figure 15.28 on page 763 to explain.

P15.75. A 60-Hz 40-kVA 8000/240-V-rms transformer has the following equivalent-circuit parameters:

Primary resistance	R_1	20 Ω
Secondary resistance	R_2	0.04 Ω
Primary leakage reactance	$X_1 = \omega L_1$	100 Ω
Secondary leakage reactance	$X_2 = \omega L_2$	0.2 Ω
Magnetizing reactance	$X_m = \omega L_m$	40 kΩ
Core-loss resistance	R_c	200 kΩ

Find the percentage regulation and power efficiency for the transformer for a rated load having a lagging power factor of 0.9.

P15.76. A certain 60-Hz, 20-kVA, 8000/240 V-rms transformer is operated with a short-circuited secondary and reduced primary voltage. It is found that, for an applied primary voltage of 500 V rms, the primary current is 2.5 A rms (i.e., this is the rated primary current) and the transformer absorbs 270 W. Consider the equivalent circuit of Figure 15.29(b) on page 764. Under the stated conditions, the current and power for L_m and R_c can be neglected. Explain why. Determine the values of the total leakage inductance referred to the primary ($L_1 + a^2L_2$) and the total resistance referred to the primary ($R_1 + a^2R_2$).

P15.77. A certain residence is supplied with electrical power by the transformer of Table 15.1 on page 763. The residence uses 400 kWh of electrical energy per month. From the standpoint of energy efficiency, which of the equivalent circuit elements listed in the table are most significant? You will need to use good judgment and make some assumptions in obtaining an answer.

P15.78. We have a transformer designed to operate at 60 Hz. The voltage ratings are 4800 V rms and 240 V rms for the primary and secondary windings, respectively. The transformer is rated for 10 kVA. Now, we want to use this transformer at 120 Hz. Discuss the factors that must be considered in setting ratings appropriate for operation at the new frequency. (Keep in mind that for best utilization of the material in the transformer, we want the peak flux density to be nearly at saturation for both frequencies.)

Frequently Asked Questions

1. Define the following terms
   (i) Magnetic field
   (ii) Magnetic flux density
2. State and explain Right-Hand thumb rule.
3. How can you determine the direction of magnetic lines of force set-up around a current carrying conductor?
4. How will you determine the direction of force acting on a current carrying conductor placed in a uniform magnetic field. What are the factors on which the magnitude of this force depends?
5. State and explain Farday's laws of electromagnetic induction.
6. State and explain magnetic field intensity and Ampere's law.
7. Derive a relation for the flux density developed in a toroidal core.
8. Define the following terms related to a magnetic circuit
   (i) Magnetomotive force
   (ii) Reluctance
9. Define and explain self and mutual inductance.
10. What do you understand by self induced emf and mutually induced emf?
11. What are hysteresis and eddy current losses?
12. What energy is stored in a magnetic field derive relation.
13. What is a transformer? Explain the working principle of a transformer.
14. What do you mean by an ideal transformer? Define voltage ratio and current ratio.
15. Explain the mechanical analogy of a transformer.
16. What are the different parameters of a real transformer? Explain the working of a real transformer.
17. What do you mean by regulation and efficiency of a transformer?

Practice Test

Here is a practice test you can use to check your comprehension of the most important concepts in this chapter. Answers can be found in Appendix D and complete solutions are included in the Student Solutions files. See Appendix E for more information about the Student Solutions.

T15.1. Consider a right-hand Cartesian coordinate system as shown in Figure 15.3 on page 735. We have a wire along the x-axis carrying 12 A in the positive x direction and a constant flux density of 0.3 T directed in the positive z direction. **a.** Determine the force and its direction on a 0.2 m length of the wire. **b.** Repeat if the field is directed in the positive x direction.

T15.2. Suppose we have a ten-turn square coil 25 cm on each side lying in the x-y plane. A magnetic flux density is directed in the positive z direction and is given by $0.7 \sin(120\pi t)$ T. The flux is constant with respect to x, y, and z. Determine the voltage induced in the coil.

T15.3. A 20-cm length of wire moves at 15 m/s in a constant flux density of 0.4 T. The wire, direction of motion, and direction of the flux are mutually perpendicular. Determine the voltage induced in the wire segment.

T15.4. Consider the magnetic circuit shown in Figure T15.4. The core has a relative permeability of 1500. **a.** Carefully estimate the flux density in the air gap. **b.** Determine the inductance of the coil.

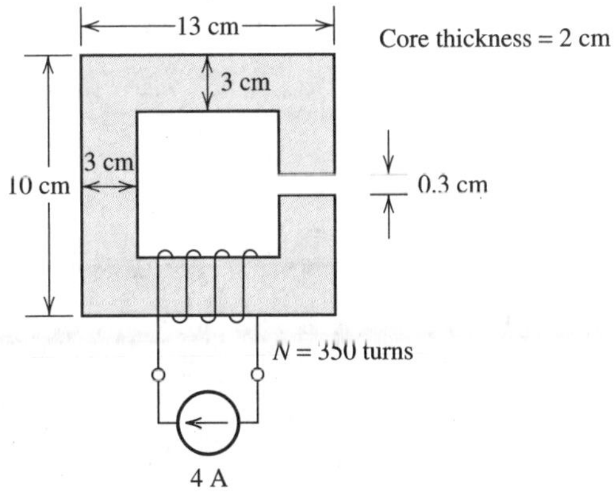

Figure T15.4

T15.5. Suppose we have an ac current flowing through a coil wound on an iron core. Name two mechanisms by which energy is converted to heat in the core material. For each, how is the core material selected to minimize the power loss? How does each power loss depend on the frequency of the ac current?

T15.6. Consider the circuit shown in Figure T15.6 which has $R_s = 0.5\,\Omega$, $R_L = 1000\,\Omega$, and $N_1/N_2 = 0.1$. **a.** Determine the rms values of the currents and voltages with the switch open. **b.** Repeat with the switch closed.

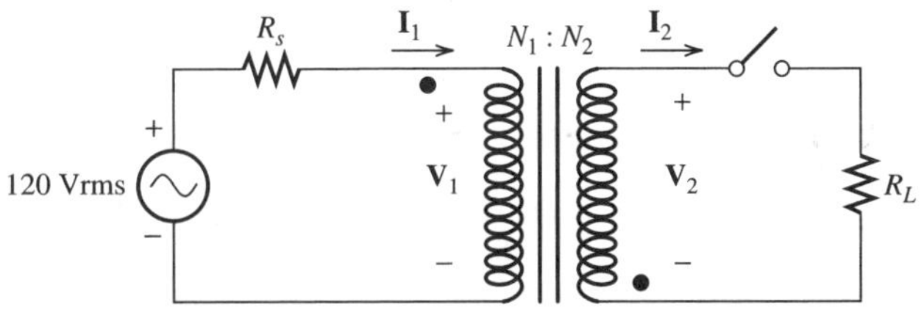

Figure T15.6

T15.7. You have been assigned to select a transformer to supply a peak power of 100 kW to a load that draws peak power only a very small percentage of the time and draws very little power the rest of the time. Two transformers, A and B, are both suitable. While both transformers have the same efficiency at peak load, most of the loss in A is due to core loss, and most of the loss in B is due to the resistances of the coils. From the standpoint of operating costs, which transformer is better? Why?

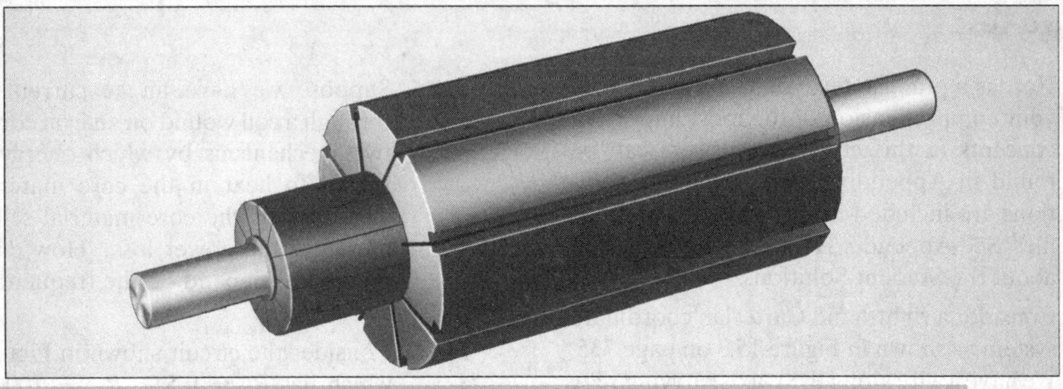

Chapter 16

DC Machines

Study of this chapter will enable you to:

- Select the proper motor type for various applications.
- State how torque varies with speed for various motors.
- Use the equivalent circuit for dc motors to compute electrical and mechanical quantities.
- Use motor nameplate data.
- Understand the operation and characteristics of shunt-connected dc motors, series-connected dc motors, and universal motors.

Introduction to this chapter:

In this chapter and the next, we consider machines that convert mechanical energy to and from electrical energy. **Motors** convert electrical energy into rotational mechanical energy. Conversely, **generators** convert mechanical energy into electrical energy. Most electrical machines can be used either as motors or as generators.

Electrical motors are used to power hundreds of the devices that we use in everyday life, such as computer disks, refrigerators, garage-door openers, washing machines, food mixers, vacuum cleaners, DVD players, ventilation fans, automotive power windows, windshield wipers, elevators, and so on. Industrial applications include bulk material handling, machining operations, pumps, rock crushers, fans, compressors, and hoists. Electrical motors are the best choice in the vast majority of stationary applications where mechanical energy is needed, whether it is a tiny fraction of a horsepower or thousands of horsepower. It is important for designers of mechanical systems that employ motors to have a good understanding of the external characteristics of various motors so they can choose the proper types to power their systems.

16.1 OVERVIEW OF MOTORS

We will see that there are many kinds of electrical motors. In this section, we give a brief overview of electrical motors, their specifications, and operating characteristics. Then, in the remainder of this chapter, we discuss dc machines in detail. In Chapter 17, we treat ac machines. We cite the three-phase ac induction motor as an example frequently in this section because it is the type in most widespread use. However, many of the concepts discussed in this section apply to other types of electrical motors as well.

You will find this section useful both as a preview of motor characteristics and as a convenient summary after you finish studying this chapter and the next.

Basic Construction

An electrical motor consists of a stationary part, or **stator**, and a **rotor**, which is the rotating part connected to a shaft that couples the machine to its mechanical load. The shaft and rotor are supported by bearings so that they can rotate freely. This is illustrated in Figure 16.1.

Depending on the type of machine, either the stator or the rotor (or both) contain current-carrying conductors configured into coils. Slots are cut into the stator and rotor to contain the windings and their insulation. Currents in the windings set up magnetic fields and interact with fields to produce torque.

Usually, the stator and the rotor are made of iron to intensify the magnetic field. As in transformers, if the magnetic field alternates in direction through the iron with time, the iron must be laminated to avoid large power losses due to eddy currents. (In certain parts of some machines, the field is steady and lamination is not necessary.)

The characteristics of several common types of machines are summarized in Table 16.1. At this point, many of the entries in the table will probably not be very meaningful to you, particularly if this is the first time that you have studied rotating

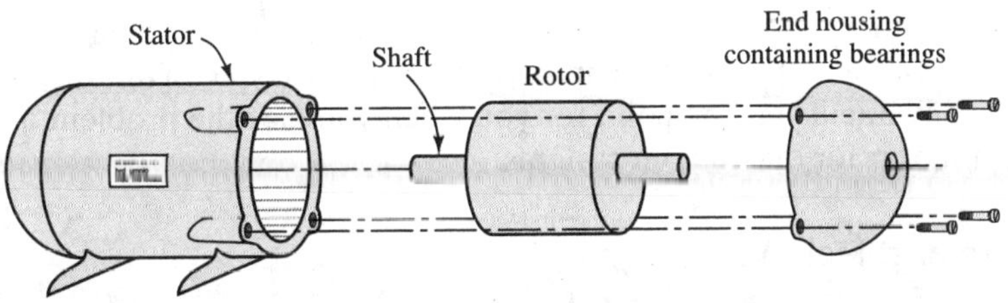

Figure 16.1 An electrical motor consists of a cylindrical rotor that spins inside a stator.

Table 16.1. Characteristics of Electrical Motors

		Type	Power Range (hp)	Rotor	Stator	Comments and Applications
Ac motors	Three phase	Induction	1–5000	Squirrel cage	Three-phase armature windings	Simple rugged construction; very common; fans, pumps
				Wound field		Adjustable speed using rotor resistance; cranes, hoists
		Synchronous	1–5	Permanent magnet		Precise speed; transport sheet materials
			1000–50,000	Dc field winding		Large constant loads; potential for power-factor correction
	Single phase	Induction	$\frac{1}{3}$–5	Squirrel cage	Main and auxiliary windings	Several types: split phase, capacitor start, capacitor run; simple and rugged; many household applications: fans, water pumps, refrigerators
		Synchronous	$\frac{1}{10}$ or less	Reluctance or hysteresis	Armature winding	Low torque, fixed speed; timing applications
Dc motors	Wound field	Shunt connected	10–200	Armature winding	Field winding	Industrial applications, grinding, machine tools, hoists
		Series connected				High torque at low speed; dangerous if not loaded; drills, automotive starting motors, (universal motor used for single-phase ac has high power/weight ratio)
		Compound connected				Can be designed to tailor torque–speed characteristic; traction motors
	Permanent-magnet field		$\frac{1}{20}$–10	Armature winding	Permanent magnets	Servo applications, machine tools, computer peripherals, automotive fans, window motors

electrical machinery. However, as we progress through this chapter and the next, the table will become a useful tool for comparing the various types of motors. Also, it will provide a convenient starting point for you when you face the problem of selecting the proper motor for one of your systems.

Armature and Field Windings

As we have mentioned, a machine may contain several sets of windings. In most types of machines, a given winding can be classed either as a **field winding** or as an **armature winding**. (We avoid classification of armature and field windings for induction motors and simply refer to stator windings and rotor conductors.) The primary purpose of a field winding is to set up the magnetic field in the machine. The current in the field winding is independent of the mechanical load imposed on the motor (except in series-connected motors). On the other hand, the armature winding carries a current that depends on the mechanical power produced. Typically, the armature current amplitude is small when the load is light and larger for heavier loads. If the machine acts as a generator, the electrical output is taken from the armature. In some machines, the field is produced by permanent magnets (PM), and a field winding is not needed.

The purpose of the field winding is to set up the magnetic field required to produce torque.

The armature windings carry currents that vary with mechanical load. When the machine is used as a generator, the output is taken from the armature windings.

Table 16.1 shows the location (stator or rotor) of the field and armature windings for some common machine types. For example, in three-phase synchronous ac machines, the field winding is on the rotor, and the armature is on the stator. In other machines, such as the wound-field dc machine, the locations are reversed. You may find it convenient to refer to Table 16.1 from time to time throughout this chapter and the next to help avoid confusion between the different types of machines.

AC Motors

Motors can be powered from either ac or dc sources. Ac power can be either single phase or three phase. (Three-phase ac sources and circuits are discussed in Section 5.7.) Ac motors include several types:

1. Induction motors, which are the most common type because they have relatively simple rugged construction and good operating characteristics.
2. Synchronous motors, which run at constant speed regardless of load torque, assuming that the frequency of the electrical source is constant, which is usually the case. Three-phase synchronous machines generate most of the electrical energy used in the world.
3. A variety of special-purpose types.

About two-thirds of the electrical energy generated in the United States is consumed by motors. Of this, well over half is used by induction motors. Thus, you are likely to encounter ac induction motors very frequently. We discuss the various types of ac motors in Chapter 17.

DC Motors

Dc motors are those that are powered from dc sources. One of the difficulties with dc motors is that nearly all electrical energy is distributed as ac. If only ac power is available and we need to use a dc motor, a rectifier or some other converter must be used to convert ac to dc. This adds to the expense of the system. Thus, ac machines are usually preferable if they meet the needs of the application.

Exceptions are automotive applications in which dc is readily available from the battery. Dc motors are employed for starting, windshield wipers, fans, and power windows.

Dc motors are common in automotive applications.

In the most common types of dc motors, the direction of the current in the armature conductors on the rotor is reversed periodically during rotation. This is accomplished with a mechanical switch composed of **brushes** mounted on the stator and a **commutator** mounted on the shaft. The commutator consists of conducting segments insulated from one another. Each commutator segment is connected to some of the armature conductors (on the rotor). The brushes are in sliding contact

A significant disadvantage of dc motors is the need for relatively frequent maintenance of brushes and commutators.

with the commutator. As the commutator rotates, switching action caused by the brushes moving from one segment to another changes the direction of current in the armature conductors. We explain this in more detail later; the important point here is that the brushes and commutator are subject to wear, and a significant disadvantage of dc motors is their relatively frequent need for maintenance.

Until recently, an important advantage of dc motors was that their speed and direction could be controlled more readily than those of ac motors. However, this advantage is rapidly disappearing because electronic systems that can vary the frequency of an ac source have become economically advantageous. These variable-frequency sources can be used with simple rugged ac induction motors to achieve speed control.

Nevertheless, dc motors are still useful in some control applications and wherever dc power is readily available, such as in vehicles. Later in this chapter, we examine the various types of dc motors in more detail.

Losses, Power Ratings, and Efficiency

Figure 16.2 depicts the flow of power from a three-phase electrical source through an induction motor to a mechanical load such as a pump. Part of the electrical power is lost (converted to heat) due to resistance of the windings, hysteresis, and eddy currents. Similarly, some of the power that is converted to mechanical form is lost to friction and windage (i.e., moving the air surrounding the rotor and shaft). Part of the power loss to windage is sometimes intentional, because fan blades to promote cooling are fabricated as an integral part of the rotor.

The electrical input power P_{in}, in watts, supplied by the three-phase source is given by

$$P_{\text{in}} = \sqrt{3}V_{\text{rms}}I_{\text{rms}}\cos(\theta) \tag{16.1}$$

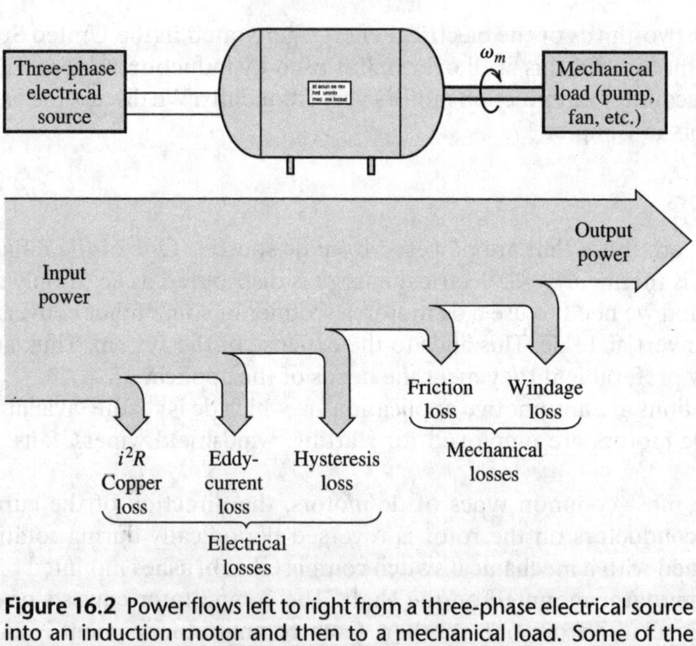

Figure 16.2 Power flows left to right from a three-phase electrical source into an induction motor and then to a mechanical load. Some of the power is lost along the way due to various causes.

where V_{rms} is the rms value of the line-to-line voltage, I_{rms} is the rms value of the line current, and $\cos(\theta)$ is the power factor.

The mechanical output power is

$$P_{out} = T_{out}\omega_m \tag{16.2}$$

in which P_{out} is the output power in watts, T_{out} is the output torque in newton-meters, and ω_m is the angular speed of the load in radians per second.

Rotational speed may be given in revolutions per minute denoted by n_m, or in radians per second denoted by ω_m. These quantities are related by

$$\omega_m = n_m \times \frac{2\pi}{60} \tag{16.3}$$

Also, torque may be given in foot-pounds instead of in newton-meters. The conversion relationship is

$$T_{\text{foot-pounds}} = T_{\text{newton-meters}} \times 0.7376 \tag{16.4}$$

In the United States, the mechanical output power for a given electric motor is frequently stated in horsepower (hp). To convert from watts to horsepower, we have

$$P_{\text{horsepower}} = \frac{P_{\text{watts}}}{746} \tag{16.5}$$

The **power rating** of a motor is the output power that the motor can safely produce on a continuous basis. For example, we can safely operate a 5-hp motor with a load that absorbs 5 hp of mechanical power. If the power required by the load is reduced, the motor draws less input power from the electrical source, and in the case of an induction motor, speeds up slightly. It is important to realize that most motors can supply output power varying from zero to several times their rated power, depending on the mechanical load. It is up to the system designer to ensure that the motor is not overloaded.

It is up to the system designer to ensure that the motor is not overloaded.

The chief output power limitation of motors is their temperature rise due to losses. Thus, a brief overload that does not cause significant rise in temperature is often acceptable.

The power **efficiency** of a motor is given by

$$\eta = \frac{P_{out}}{P_{in}} \times 100\% \tag{16.6}$$

Well-designed electrical motors operating close to their rated capacity have efficiencies in the range of 85 to 95 percent. On the other hand, if the motor is called upon to produce only a small fraction of its rated power, its efficiency is generally much lower.

Torque–Speed Characteristics

Consider a system in which a three-phase induction motor drives a load such as a pump. Figure 16.3 shows the torque produced by the motor versus speed. (In

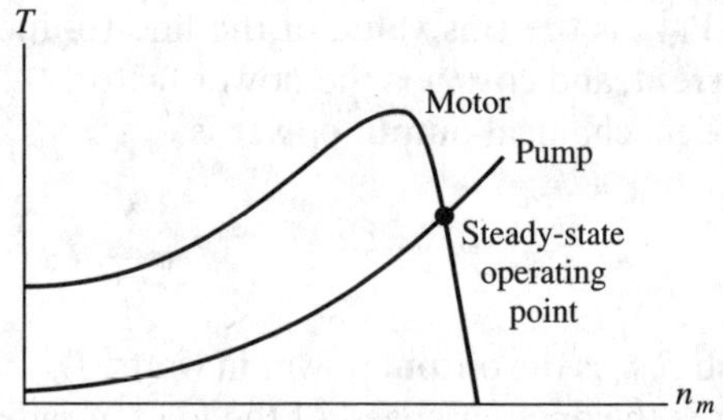

Figure 16.3 The torque–speed characteristics of an induction motor and a load consisting of a pump. In steady state, the system operates at the point for which the torque produced by the motor equals the torque required by the load.

Chapter 17, we will see why the torque–speed characteristic of the induction motor has this shape.) The torque required to drive the load is also shown. Suppose that the system is at a standstill and then a switch is closed connecting the electrical source to the motor. At low speeds, the torque produced by the motor is larger than that needed to drive the load. The excess torque causes the system to accelerate. Eventually, the speed stabilizes at the point for which the torque produced by the motor equals the torque needed to drive the load.

Now consider the torque–speed characteristics for a three-phase induction motor and a load consisting of a hoist shown in Figure 16.4. Here, the starting torque of the motor is less than that demanded by the load. Thus, if power is applied from a standing start, the system does not move. In this case, excessive currents are drawn by the motor, and unless fuses or other protection equipment disconnect the source, the motor could overheat and be destroyed.

Even though the motor cannot start the load shown in Figure 16.4, notice that the motor is capable of keeping the load moving once the speed exceeds n_1. Perhaps this could be accomplished with a mechanical clutch.

Designers must be able to choose motors having torque–speed characteristics appropriate for various loads.

The various types of motors have different torque–speed characteristics. Some examples are shown in Figure 16.5. It is important for the system designer to choose a motor suitable for the load requirements.

Speed Regulation

Depending on the torque–speed characteristics, a motor may slow down as the torque demanded by the load increases. Speed regulation is defined as the difference between the no-load speed and the full-load speed, expressed as a percentage of the full-load speed:

$$\text{speed regulation} = \frac{n_{\text{no-load}} - n_{\text{full-load}}}{n_{\text{full-load}}} \times 100\% \tag{16.7}$$

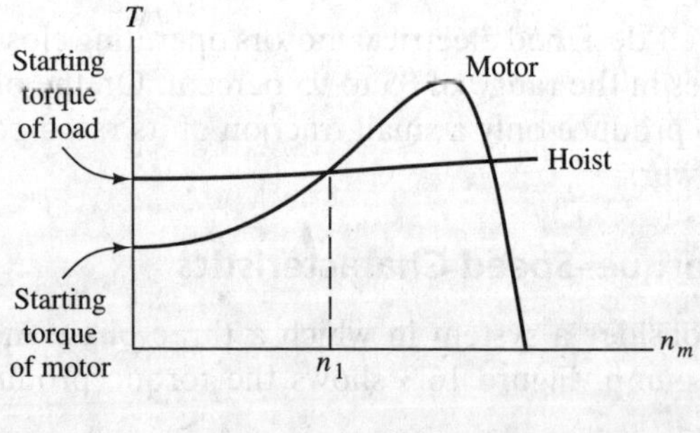

Figure 16.4 This system will not start from a standstill because the motor cannot supply the starting torque demanded by the load.

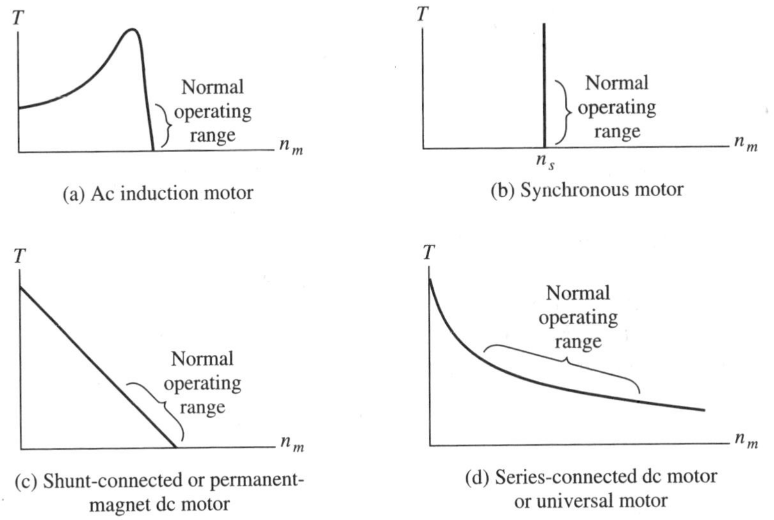

Figure 16.5 Torque versus speed characteristics for the most common types of electrical motors.

Synchronous-Motor Operating Characteristics

The torque–speed characteristic for ac synchronous motors is shown in Figure 16.5(b). The operating speed of a synchronous motor is constant and is given by

$$\omega_s = \frac{2\omega}{P} \tag{16.8}$$

in which ω is the angular frequency of the ac source and P is the number of magnetic poles possessed by the machine. (We examine the internal structure of these machines in Section 17.3.) In rpm, the synchronous speed is

$$n_s = \frac{120f}{P} \tag{16.9}$$

in which f is the frequency of the ac source in hertz.

We will see that the number of magnetic poles P is always an even integer. Substituting various values for P into Equation 16.9 and assuming 60-Hz operation, the available speeds are 3600 rpm, 1800 rpm, 1200 rpm, 900 rpm, and so on. If some other speed is required, a synchronous machine is usually not a good choice. (Electronic systems known as cycloconverters can be used to convert 60-Hz power into any desired frequency. Thus, this speed limitation can be circumvented at additional cost.)

Synchronous motors operate at $(60 \times f)$ rpm or one of its submultiples. If the frequency f is fixed and if none of the available speeds is suitable for the load, some other type of motor is needed.

As shown in Figure 16.5(b), the starting torque of a synchronous motor is zero. Therefore, special provisions must be made for starting. We will see that one approach is to operate the motor as an induction motor with reduced load until the speed approaches synchronous speed, and then to switch to synchronous operation.

The starting torque of a synchronous motor is zero.

Induction-Motor Operating Characteristics

Induction motors have starting torques that are comparable to their rated full-load torques.

The torque–speed characteristic typical of an induction motor is shown in Figure 16.5(a). The motor has good starting torque. In normal operation, the speed of an induction motor is only slightly less than synchronous speed, which is given by Equations 16.8 and 16.9. For example, at full load, a typical four-pole ($P = 4$) induction motor runs at 1750 rpm, and at no load, its speed approaches 1800 rpm. The comments given earlier about speed limitations of synchronous motors also apply to induction motors.

Induction motors operate in narrow ranges of speed that are slightly less than ($60 \times f$) rpm or one of its submultiples. If the frequency is fixed, the speeds available may not be suitable for the load.

During startup, the current drawn by an induction motor can be many times larger than its rated full-load current. To avoid excessive currents, large induction motors are usually started with reduced voltage. As you might expect, the torque produced by a motor depends on the applied voltage. At a given speed, the torque of an induction motor is proportional to the square of the magnitude of the voltage applied to the armature. When starting a motor at, say, half of its rated voltage, its torque is one-fourth of its value at rated voltage.

Shunt-Connected DC Motor Operating Characteristics

Dc motors contain field windings on the stator and armature windings located on the rotor. Depending on whether the field windings are connected in shunt (i.e., parallel) or in series with the armature windings, the torque–speed characteristics are quite different. We examine why this is true later in this chapter.

The torque–speed characteristic of the shunt-connected dc motor is shown in Figure 16.5(c). The shunt-connected motor has very high starting torque and draws very large starting currents. Usually, resistance is inserted in series with the armature during starting to limit the current to reasonable levels.

For fixed supply voltage and fixed field current, the shunt dc machine shows only a small variation in speed within its normal operating range. However, we will see that several methods can be used to shift the torque–speed characteristic of the shunt motor to achieve excellent speed control. Unlike ac induction and synchronous motors, the speeds of dc motors are not limited to specific values.

Dc motors can be designed to operate over a wide range of speeds.

Series-Connected DC Motor Operating Characteristics

The torque–speed characteristic of the series dc motor is shown in Figure 16.5(d). The series-connected dc motor has moderate starting torque and starting current. Its speed automatically adjusts over a large range as the load torque varies. Because it slows down for heavier loads, its output power is more nearly constant than for other motor types. This is advantageous because the motor can operate within its maximum power rating for a wide range of load torque. The starter motor in automobiles is a series dc motor. When the engine is cold and stiff, the starter motor operates at a lower speed. On the other hand, when the engine is warm, the starter spins faster. In either case, the current drawn from the battery remains within acceptable limits. (On the other hand, without sophisticated controls, a shunt motor would attempt to turn the load at a constant speed and would draw too much current in starting a cold engine.)

In some cases, the no-load speed of a series dc motor can be excessive—to the point of being dangerous. A control system that disconnects the motor from the electrical source is needed if the possibility of losing the mechanical load exists. We will see that a very useful type of ac motor known as a universal motor is essentially identical to the series-connected dc motor.

Example 16.1 Motor Performance Calculations

A certain 5-hp three-phase induction motor operates from a 440-V-rms (line-to-line) three-phase source and draws a line current of 6.8 A rms at a power factor of 78 percent lagging [i.e., $\cos(\theta) = 0.78$] under rated full-load conditions. The full-load speed is 1150 rpm. Under no-load conditions, the speed is 1195 rpm, and the line current is 1.2 A rms at a power factor of 30 percent lagging. Find the power loss and efficiency with full load, the input power with no load, and the speed regulation.

Solution The rated output power is 5 hp. Converting to watts, we have

$$P_{out} = 5 \times 746 = 3730 \text{ W}$$

Substituting into Equation 16.1, we find the input power under full load:

$$\begin{aligned} P_{in} &= \sqrt{3} V_{rms} I_{rms} \cos(\theta) \\ &= \sqrt{3}(440)(6.8)(0.78) = 4042 \text{ W} \end{aligned}$$

The power loss is given by

$$P_{loss} = P_{in} - P_{out} = 4042 - 3730 = 312 \text{ W}$$

The full-load efficiency is

$$\eta = \frac{P_{out}}{P_{in}} \times 100\% = \frac{3730}{4042} \times 100\% = 92.28\%$$

Under no-load conditions, we have

$$\begin{aligned} P_{in} &= \sqrt{3}(440)(1.2)(0.30) = 274.4 \text{ W} \\ P_{out} &= 0 \\ P_{loss} &= P_{in} = 274.4 \text{ W} \end{aligned}$$

and the efficiency is

$$\eta = 0\%$$

Speed regulation for the motor is given by Equation 16.7. Substituting values, we get

$$\begin{aligned} \text{speed regulation} &= \frac{n_{\text{no-load}} - n_{\text{full-load}}}{n_{\text{full-load}}} \times 100\% \\ &= \frac{1195 - 1150}{1150} \times 100\% = 3.91\% \end{aligned}$$

■

Now that we have presented an overall view of electrical motors, we will make a more detailed examination of the most common and useful types. In the remainder of this chapter, we consider dc machines, and in Chapter 17, we discuss ac machines.

Exercise 16.1 A certain 50-hp dc motor operates from a 220-V dc source with losses of 3350 W under rated full-load conditions. The full-load speed is 1150 rpm. Under no-load conditions, the speed is 1200 rpm. Find the source current, the efficiency with full load, and the speed regulation.

Answer $I_{source} = 184.8$ A, $\eta = 91.76$ percent, speed regulation = 4.35 percent. □

Exercise 16.2 Consider the torque–speed characteristics shown in Figure 16.5. **a.** Which type of motor would have the most difficulty in starting a high-inertia load from a standing start? **b.** Which type of motor would have the poorest (i.e., largest) speed regulation in its normal operating range? **c.** Which would have the best (i.e., smallest) speed regulation? **d.** Which has the best combination of high starting torque and good speed regulation? **e.** Which should not be operated without a load?

Answer **a.** The synchronous motor, because its starting torque is zero; **b.** the series-connected dc motor; **c.** the synchronous motor; **d.** the ac induction motor; **e.** the series-connected dc motor because the speed can become excessive for zero load torque. □

16.2 PRINCIPLES OF DC MACHINES

Study of the idealized linear dc machine clearly demonstrates how the principles of electromagnetism apply to dc machines in general.

In this section, we introduce the basic principles of dc machines by considering the idealized linear machine shown in Figure 16.6. Later, we will see that the operation of rotating dc machines is very similar to that of this simple linear machine. In Figure 16.6, a dc voltage source V_T is connected through a resistance R_A and a switch that closes at $t = 0$ to a pair of conducting rails. A conducting bar slides without friction on the rails. We assume that the rails and the bar have zero resistance. A magnetic field is directed into the page, perpendicular to the plane of the rails and the bar.

Suppose that the bar is stationary when the switch is closed at $t = 0$. Then, just after the switch is closed, an initial current given by $i_A(0+) = V_T/R_A$ flows clockwise around the circuit. A force given by

$$\mathbf{f} = i_A \mathbf{l} \times \mathbf{B} \tag{16.10}$$

is exerted on the bar. The direction of the current (and **l**) is toward the bottom of the page. Thus, the force is directed to the right. Because the current and the field are mutually perpendicular, the force magnitude is given by

$$f = i_A l B \tag{16.11}$$

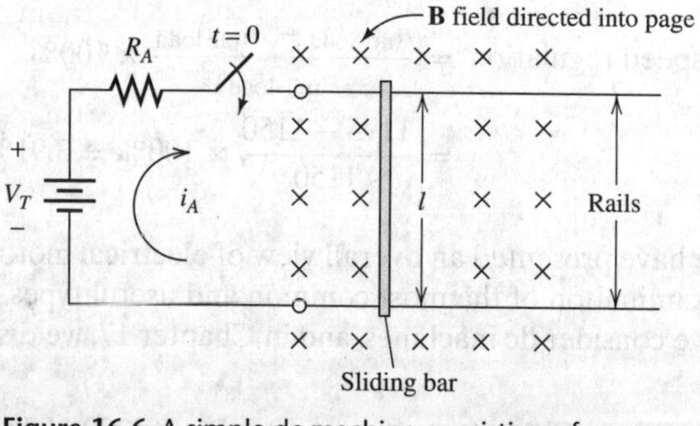

Figure 16.6 A simple dc machine consisting of a conducting bar sliding on conducting rails.

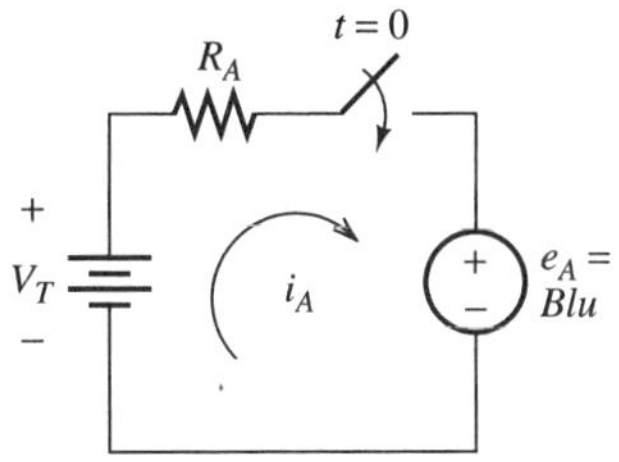

Figure 16.7 Equivalent circuit for the linear machine operating as a motor.

This force causes the bar to be accelerated toward the right. As the bar gains velocity u and cuts through the magnetic field lines, a voltage is induced across the bar. The voltage is positive at the top end of the bar and is given by Equation 15.9 (with a change in notation):

$$e_A = Blu \tag{16.12}$$

An equivalent circuit for the system is shown in Figure 16.7. Notice that the induced voltage e_A opposes the source V_T. The current is

$$i_A = \frac{V_T - e_A}{R_A} \tag{16.13}$$

As the velocity of the bar builds up, energy is absorbed by the induced voltage e_A, and this energy shows up as the kinetic energy of the bar. Eventually, the bar speed becomes high enough that $e_A = V_T$. Then, the current and the force become zero, and the bar coasts at constant velocity.

Operation as a Motor

Now, suppose that a mechanical load exerting a force to the left is connected to the moving bar. Then, the bar slows down slightly, resulting in a reduction in the induced voltage e_A. Current flows clockwise in the circuit, resulting in a magnetically induced force directed to the right. Eventually, the bar slows just enough so that the force created by the magnetic field ($f = i_A lB$) equals the load force. Then, the system moves at constant velocity.

In this situation, power delivered by the source V_T is converted partly to heat in the resistance R_A and partly to mechanical power. It is the power $p = e_A i_A$ delivered to the induced voltage that shows up as mechanical power $p = fu$.

Operation as a Generator

Again suppose that the bar is moving at constant velocity such that $e_A = V_T$ and the current is zero. Then, if a force is applied pulling the bar even faster toward the right, the bar speeds up, the induced voltage e_A exceeds the source voltage V_T, and current circulates counterclockwise as illustrated in Figure 16.8. Because the current has reversed direction, the force induced in the bar by the field also reverses and points to the left. Eventually, the bar speed stabilizes with the pulling force equal to the induced force. Then, the induced voltage delivers power $p = e_A i_A$, partly to the resistance ($p_R = R_A i_A^2$) and partly to the battery ($p_t = V_T i_A$). Thus, mechanical energy is converted into electrical energy that eventually shows up as loss (i.e., heat) in the resistance or as stored chemical energy in the battery.

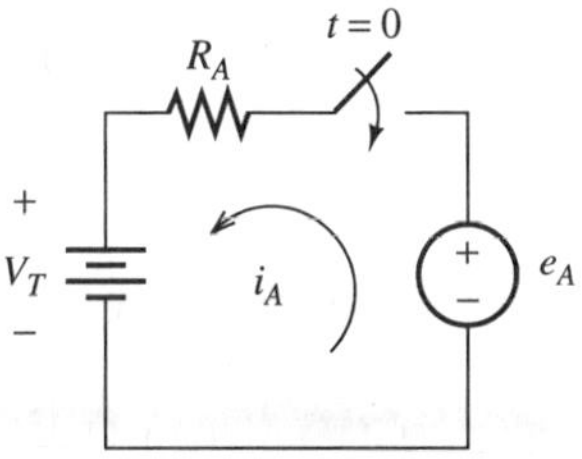

Figure 16.8 Equivalent circuit for the linear machine operating as a generator.

Example 16.2 Idealized Linear Machine

Suppose that for the linear machine shown in Figure 16.6, we have $B = 1$ T, $l = 0.3$ m, $V_T = 2$ V, and $R = 0.05\ \Omega$. **a.** Assuming that the bar is stationary at $t = 0$, compute the initial current and the initial force on the bar. Also, determine the final (i.e., steady-state) speed assuming that no mechanical load is applied to the bar. **b.** Now, suppose that a mechanical load of 4 N directed to the left is applied to the moving bar. In steady state, determine the speed, the power delivered by V_T, the power delivered to the mechanical load, the power lost to heat in the resistance R_A, and the efficiency. **c.** Now, suppose that a mechanical pulling force of 2 N directed to the right is applied to the moving bar. In steady state, determine the speed, the power taken from the mechanical source, the power delivered to the battery, the power lost to heat in the resistance R_A, and the efficiency.

Solution

a. Initially, for $u = 0$, we have $e_A = 0$, and the initial current is given by

$$i_A(0+) = \frac{V_T}{R_A} = \frac{2}{0.05} = 40 \text{ A}$$

The resulting initial force on the bar is

$$f(0+) = Bli_A(0+) = 1(0.3)40 = 12 \text{ N}$$

In steady state with no load, the induced voltage equals the battery voltage. Thus, we have

$$e_A = Blu = V_T$$

Solving for the velocity and substituting values, we get

$$u = \frac{V_T}{Bl} = \frac{2}{1(0.3)} = 6.667 \text{ m/s}$$

b. Because the mechanical force opposes the motion of the bar, we have motor action. In steady state, the net force on the bar is zero—the force created by the magnetic field equals the load force. Thus, we obtain

$$f = Bli_A = f_{\text{load}}$$

Solving for the current and substituting values, we find that

$$i_A = \frac{f_{\text{load}}}{Bl} = \frac{4}{1(0.3)} = 13.33 \text{ A}$$

From the circuit shown in Figure 16.7, we have

$$e_A = V_T - R_A i_A = 2 - 0.05(13.33) = 1.333 \text{ V}$$

Now, we can find the steady-state speed:

$$u = \frac{e_A}{Bl} = \frac{1.333}{1(0.3)} = 4.444 \text{ m/s}$$

The mechanical power delivered to the load is

$$p_m = f_{\text{load}}u = 4(4.444) = 17.77 \text{ W}$$

The power taken from the battery is

$$p_t = V_T i_A = 2(13.33) = 26.67 \text{ W}$$

The power dissipated in the resistance is

$$p_R = i_A^2 R = (13.33)^2 \times 0.05 = 8.889 \text{ W}$$

As a check, we note that $p_t = p_m + p_R$ to within rounding error. Finally, the efficiency of converting electrical power from the battery into mechanical power is

$$\eta = \frac{p_m}{p_t} \times 100\% = \frac{17.77}{26.67} \times 100\% = 66.67\%$$

c. With a pulling force applied to the bar to the right, the bar speeds up, the induced voltage exceeds V_T, and current circulates counterclockwise, as illustrated in Figure 16.8. Thus, the machine operates as a generator. In steady state, the force induced by the field is directed to the left and equals the pulling force. Thus, we have

$$f = Bli_A = f_{\text{pull}}$$

Solving for i_A and substituting values, we find that

$$i_A = \frac{f_{\text{pull}}}{Bl} = \frac{2}{1(0.3)} = 6.667 \text{ A}$$

From the circuit shown in Figure 16.8, we obtain

$$e_A = V_T + R_A i_A = 2 + 0.05(6.67) = 2.333 \text{ V}$$

Now, we can find the steady-state speed:

$$u = \frac{e_A}{Bl} = \frac{2.333}{1(0.3)} = 7.778 \text{ m/s}$$

The mechanical power delivered by the pulling force is

$$p_m = f_{\text{pull}}u = 2(7.778) = 15.56 \text{ W}$$

The power absorbed by the battery is

$$p_t = V_T i_A = 2(6.667) = 13.33 \text{ W}$$

The power dissipated in the resistance is

$$p_R = i_A^2 R = (6.667)^2 \times 0.05 = 2.222 \text{ W}$$

As a check, we note that $p_m = p_t + p_R$ to within rounding error. Finally, the efficiency of converting mechanical power into electrical power charging the battery is

$$\eta = \frac{p_t}{p_m} \times 100\% = \frac{13.33}{15.56} \times 100\% = 85.67\%$$

■

In Example 16.2, we have seen that only modest forces (12 N) were produced on a conductor carrying a fairly large current (40 A). The force could be increased by using a longer conductor, but this increases the size of the machine. Another option would be to increase the field strength. However, because of the fact that magnetic materials used in motors saturate in the neighborhood of 1 T, it is not practical to increase the forces on conductors greatly by increasing the field.

On the other hand, a cylindrical rotor containing many conductors is a practical way to obtain large forces in a compact design. Furthermore, rotary motion is more useful than translation in many applications. Thus, most (but not all) practical motors are based on rotational motion. We study rotating dc machines in the remaining sections of this chapter.

Exercise 16.3 Repeat the calculations of Example 16.2 if the field strength is doubled to 2 T.

Answer **a.** $i_A(0+) = 40$ A, $f(0+) = 24$ N, $u = 3.333$ m/s; **b.** $i_A = 6.667$, $e_A = 1.667$, $u = 2.778$ m/s, $p_m = 11.11$ W, $p_t = 13.33$ W, $p_R = 2.22$ W, $\eta = 83.33\%$; **c.** $i_A = 3.333$ A, $e_A = 2.167$ V, $u = 3.612$ m/s, $p_m = 7.222$ W, $p_t = 6.667$ W, $p_R = 0.555$ W, $\eta = 92.3\%$. □

PRACTICAL APPLICATION 16.1

Magnetic Flowmeters, Faraday, and *The Hunt for Red October*

Flowmeters measure the flow rate of liquids through pipes and are very important sensors in chemical-process-control systems. A commonly used type is the magnetic flowmeter (also called a *magflow*), which operates on the same principles as the linear machine discussed in Section 16.2.

The basic operation of a magflow is illustrated in Figure PA16.1. Coils set up a vertical magnetic field in the fluid, and electrodes are located at opposite sides of the pipe, which is lined with an electrical insulating material such as ceramic or epoxy resin. Thus, the magnetic field, the direction of flow, and the line between the electrodes are mutually perpendicular. As the conductive fluid moves through the magnetic field, a voltage proportional to velocity is induced between the electrodes. The flow rate can be determined by multiplying the cross-sectional area of the pipe times the velocity. Hence, the meter measures the induced voltage, but can be calibrated in units of volumetric flow.

Faraday realized the potential for using his law of electromagnetic induction to measure water flow, and attempted to measure the flow rate of the Thames River with a device suspended from a bridge. However, lacking the advantages of modern electronics, he was not successful.

In modern meters, an electronic amplifier is used to amplify the induced voltage. In many units, this voltage is converted to digital form by an analog-to-digital converter and processed by a microcomputer that drives a display or sends data to a central plant-control computer.

If the fluid has low electrical conductivity, the Thévenin resistance seen looking into the electrode terminals is very high. Then, it is important for the amplifier input impedance to be very high; otherwise, the observed voltage would vary with changes in electrical conductivity of the fluid, leading to gross inaccuracies in flow rate. Of course, it is important for chemical engineers employing magflows to understand this limitation. Even with their limitations, well-designed meters are available for a wide range of applications.

As a meter, the magflow acts as a generator. However, it can also act as a motor if an electrical current is passed through the fluid between the electrodes. Then force is exerted directly on the fluid by the interaction of the magnetic field with the electrical current. Of course, if we wanted to build a pump based on this approach, we would want a fluid with high conductivity, such as seawater. We would also need a strong magnetic field and high current. By making these modifications, a powerful pump can be constructed. This is the operating principle of the ultraquiet submarine propulsion system mentioned by Tom Clancy in *The Hunt for Red October*. Such a system can be very quiet because force is applied smoothly and directly to the seawater without rotating parts or valves that could cause vibration.

Sources: Ian Robertson, "Magnetic flowmeters: the whole story," *The Chemical Engineer*, February 24, 1994, pp. 17–18; The Magmeter Flowmeter Homepage, `http://www.magmeter.com`.

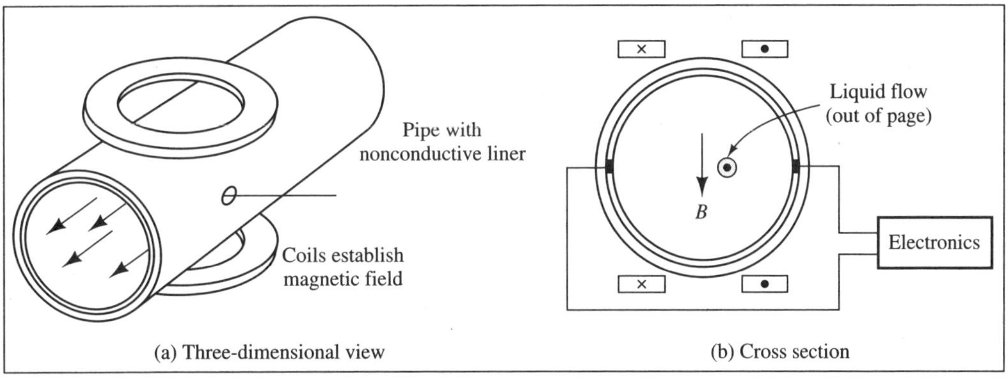

Figure PA16.1 A magnetic flowmeter.

16.3 ROTATING DC MACHINES

We have gained some familiarity with basic principles of dc machines from our analysis of the linear machine in the preceding section. In this section, we will see that the same principles apply to rotating dc machines.

The basic principles of rotating dc machines are the same as those of the linear dc machine.

Structure of the Rotor and Stator

The most common type of dc machine contains a cylindrical stator with an even number P of magnetic poles that are established by field windings or by permanent magnets. The poles alternate between north and south around the periphery of the stator.

Inside the stator is a rotor consisting of a laminated iron cylinder mounted on a shaft that is supported by bearings so that it can rotate. Slots cut lengthwise into the surface of the rotor contain the armature windings. A rotor with armature conductors (and other features to be discussed shortly) is illustrated in Figure 16.9.

The cross section of a two-pole machine showing the flux lines in the air gap is illustrated in Figure 16.10. Magnetic flux tends to take the path of least reluctance. Because the reluctance of air is much higher than that of iron, the flux takes the shortest path from the stator into the rotor. Thus, the flux in the air gap is perpendicular to the surface of the rotor and to the armature conductors. Furthermore, the flux density is nearly constant in magnitude over the surface of each pole face. Between poles, the gap flux density is small in magnitude.

In a motor, external electrical sources provide the currents in the field windings and in the armature conductors. The current directions shown in Figure 16.10 result

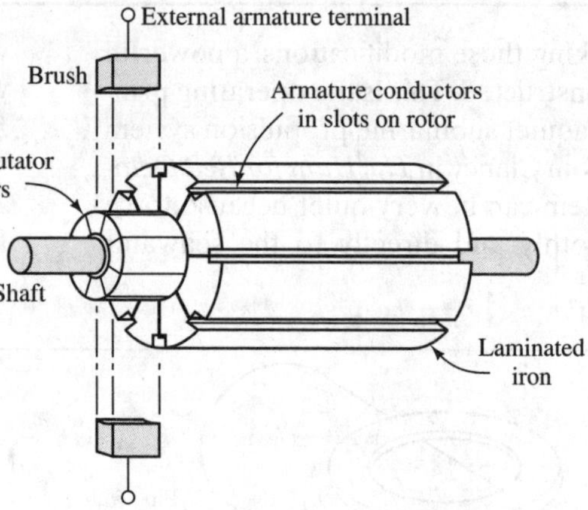

Figure 16.9 Rotor assembly of a dc machine.

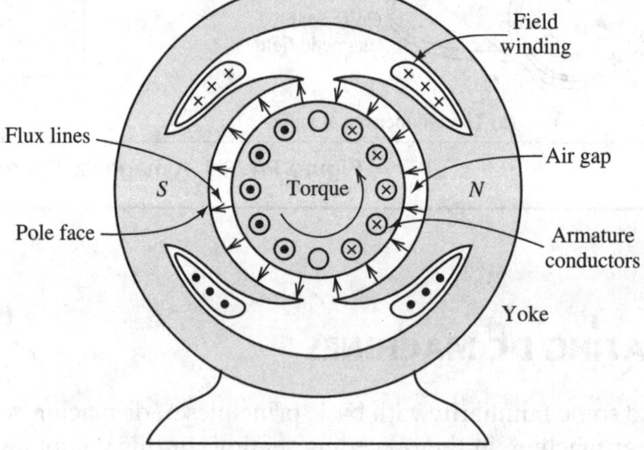

Figure 16.10 Cross section of a two-pole dc machine.

in a counterclockwise torque. This can be verified by applying the equation $\mathbf{f} = i\mathbf{l} \times \mathbf{B}$ that gives the force on a current-carrying conductor.

The cross section of a four-pole machine is shown in Figure 16.11. Notice that the directions of the currents in the armature must be reversed under south poles relative to the direction under the north poles to achieve aiding contributions to total torque.

Induced EMF and Commutation

As the rotor turns, the conductors move through the magnetic field produced by the stator. Under the pole faces, the conductors, the field, and the direction of motion are mutually perpendicular, just as in the linear machine discussed in Section 16.2. Thus, a nearly constant voltage is induced in each conductor as it moves under a pole. However, as the conductors move between poles, the field direction reverses. Therefore, the induced voltages fall to zero and build up with the opposite polarity. A mechanical switch known as a **commutator** reverses the connections to the conductors as they move between poles so that the polarity of the induced voltage seen from the external machine terminals is constant.

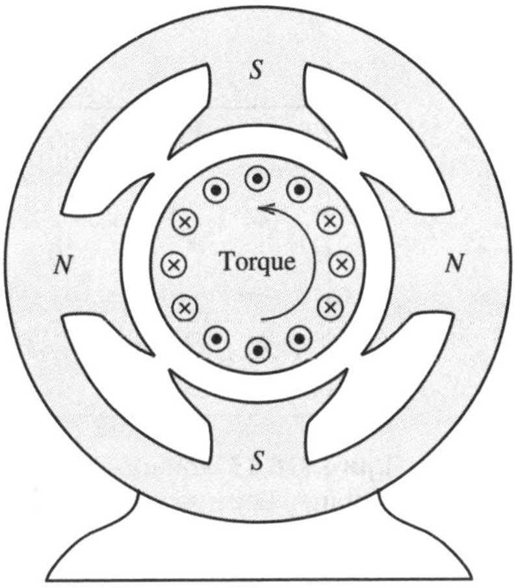

Figure 16.11 Cross section of a four-pole dc machine.

Let us illustrate these points with a two-pole machine containing one armature coil, as shown in Figure 16.12. In this case, the ends of the coil are attached to a two-segment commutator mounted on the shaft. The segments are insulated from one another and from the shaft. Brushes mounted to the stator make electrical contact with the commutator segments. (For clarity, we have shown the brushes inside the commutator, but in a real machine, they ride on the outside surface of the commutator. A more realistic version of the commutator and brushes was shown in Figure 16.9.)

The commutator and brushes form a mechanical switch that reverses the external connections to conductors as they move from pole to pole.

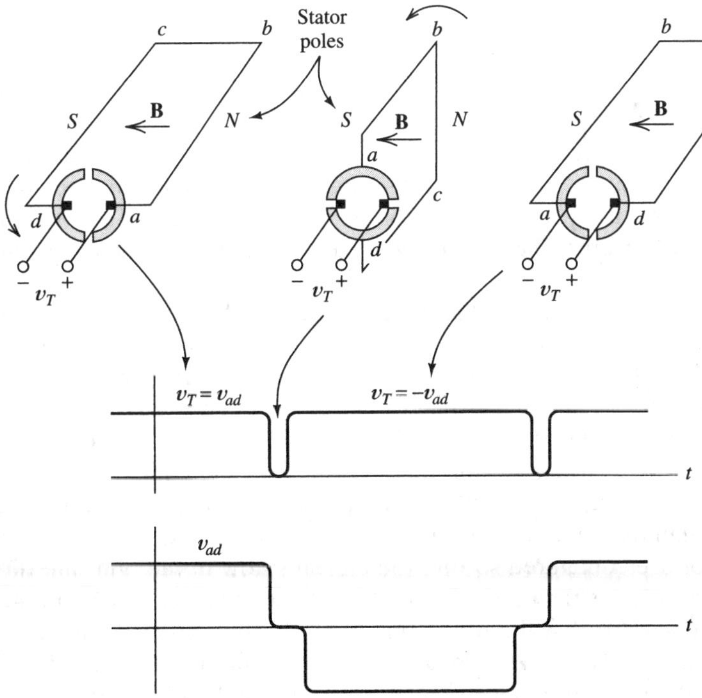

Figure 16.12 Commutation for a single armature winding.

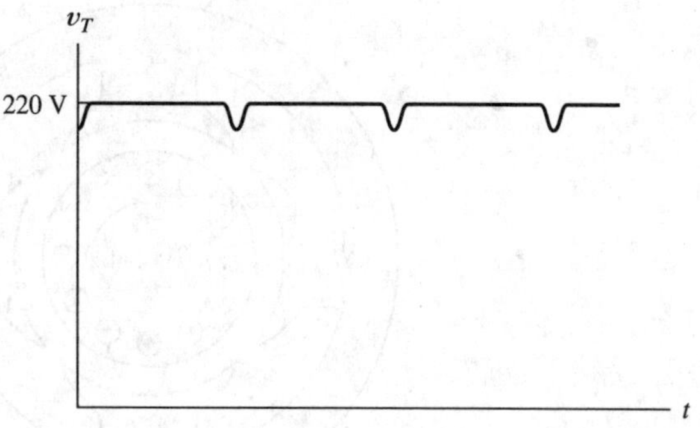

Figure 16.13 Voltage produced by a practical dc machine. Because only a few (out of many) conductors are commutated (switched) at a time, the voltage fluctuations are less pronounced than in the single-loop case illustrated in Figure 16.12.

Notice that as the rotor turns in Figure 16.12, the left-hand brush is connected to the conductor under the south stator pole, and the right-hand brush is connected to the conductor under the north stator pole.

The voltage v_{ad} induced across the terminals of the coil is an ac voltage, as shown in the figure. As mentioned earlier, this voltage passes through zero when the conductors are between poles where the flux density goes to zero. While the conductors are under the pole faces where the flux density is constant, the induced voltage has nearly constant magnitude. Because the commutator reverses the external connections to the coil as it rotates, the voltage v_T seen at the external terminals is of constant polarity.

Notice that the brushes short the armature winding briefly during the switching process. This occurs because the brushes are wider than the insulation between commutator segments. This shorting is not a problem, provided that the voltage is small when it occurs. (Actual machines have various provisions to ensure that the coil voltage is close to zero during commutation for all operating conditions.)

Commutators in typical machines contain 20 to 50 segments. Because only part of the coils are commutated at a time, the terminal voltage of a real machine shows relatively little fluctuation compared to the two-segment example that we used for the illustration of concepts. The terminal voltage of an actual dc machine is shown in Figure 16.13.

Generally, the commutator segments are copper bars insulated from one another and from the shaft. The brushes contain graphite that lubricates the sliding contact. Even so, a significant disadvantage of dc machines is the need to replace brushes and redress the commutator surface because of mechanical wear.

Actual armatures consist of a large number of conductors placed around the circumference of the rotor. To attain high terminal voltages, many conductors are placed in series, forming coils. Furthermore, there are usually several parallel current paths through the armature. The armature conductors and their connections to the commutator are configured so that the currents flow in the opposite direction under south stator poles than they do under north stator poles. As mentioned earlier, this is necessary so that the forces on the conductors produce aiding torques. The construction details needed to produce these conditions are beyond the scope of our discussion. As a user of electrical motors, you will find the external behavior of machines more helpful than the details of their internal design.

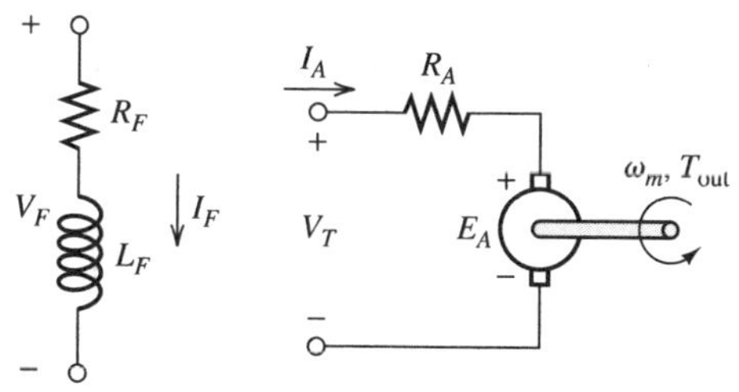

Figure 16.14 Equivalent circuit for the rotating dc machine.

Equivalent Circuit of the DC Motor

The equivalent circuit of the dc motor is shown in Figure 16.14. The field circuit is represented by a resistance R_F and an inductance L_F in series. We consider steady-state operation in which the currents are constant, and we can neglect the inductance because it behaves as a short circuit for dc currents. Thus, for dc field currents, we have

$$V_F = R_F I_F \tag{16.14}$$

The voltage E_A shown in the equivalent circuit represents the average voltage induced in the armature due to the motion of the conductors relative to the magnetic field. In a motor, E_A is sometimes called a **back emf** (electromotive force) because it opposes the applied external electrical source. The resistance R_A is the resistance of the armature windings plus the brush resistance. (Sometimes, the drop across the brushes is estimated as a constant voltage of about 2 V rather than as a resistance. However, in this book, we lump the brush drop with the armature resistance.)

The induced armature voltage is given by

$$E_A = K\phi\omega_m \tag{16.15}$$

in which K is a **machine constant** that depends on the design parameters of the machine, ϕ is the magnetic flux produced by each stator pole, and ω_m is the angular velocity of the rotor.

The torque developed in the machine is given by

$$T_{dev} = K\phi I_A \tag{16.16}$$

in which I_A is the armature current. (We will see that the *output* torque of a dc motor is less than the *developed* torque because of friction and other rotational losses.)

The **developed power** is the power converted to mechanical form, which is given by the product of developed torque and angular velocity:

$$P_{dev} = \omega_m T_{dev} \tag{16.17}$$

This is the power delivered to the induced armature voltage, and therefore, is also given by

$$P_{dev} = E_A I_A \tag{16.18}$$

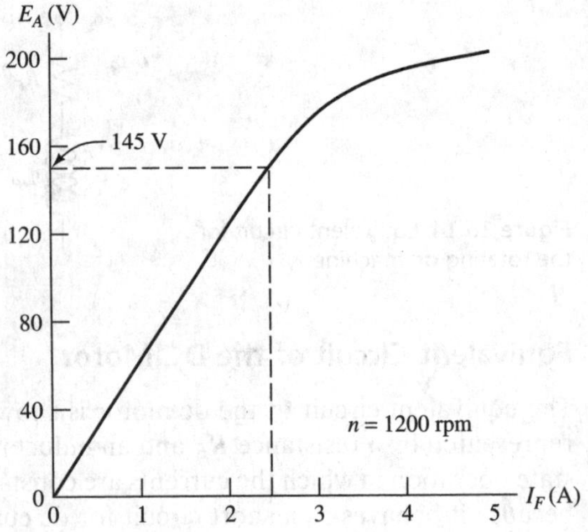

Figure 16.15 Magnetization curve for a 200-V 10-hp dc motor.

Magnetization Curve

The magnetization curve of a dc machine is a plot of E_A versus the field current I_F with the machine being driven at a constant speed.

The **magnetization curve** of a dc machine is a plot of E_A versus the field current I_F with the machine being driven at a constant speed. (E_A can be found by measuring the open-circuit voltage at the armature terminals.) A typical magnetization curve is shown in Figure 16.15.

Because E_A is proportional to the flux ϕ, the magnetization curve has the same shape as a ϕ versus I_F plot, which depends on the parameters of the magnetic circuit for the field. The magnetization curve flattens out for high field currents due to magnetic saturation of the iron. Of course, different machines usually have differently shaped magnetization curves.

The equivalent circuit, Equations 16.14 through 16.19, and the magnetization curve are the key elements needed to solve dc machine problems.

As Equation 16.15 shows, the induced armature voltage E_A is directly proportional to speed. If E_{A1} represents the voltage at speed n_1, and E_{A2} is the voltage at a second speed n_2, we have

$$\frac{E_{A1}}{E_{A2}} = \frac{n_1}{n_2} = \frac{\omega_1}{\omega_2} \tag{16.19}$$

Equations 16.14 through 16.19, in combination with the equivalent circuit shown in Figure 16.14 and the magnetization curve, provide the basis for analyzing a dc machine.

Example 16.3 DC Machine Performance Calculations

The machine having the magnetization curve shown in Figure 16.15 is operating as a motor at a speed of 800 rpm with $I_A = 30$ A and $I_F = 2.5$ A. The armature resistance is 0.3 Ω and the field resistance is $R_F = 50$ Ω. Find the voltage V_F applied to the field circuit, the voltage V_T applied to the armature, the developed torque, and the developed power.

Solution Equation 16.14 allows us to find the voltage for the field coil:

$$V_F = R_F I_F = 50 \times 2.5 = 125 \text{ V}$$

From the magnetization curve, we see that the induced voltage is $E_{A1} = 145$ V at $I_F = 2.5$ A and $n_1 = 1200$ rpm. Rearranging Equation 16.19, we can find the induced voltage E_{A2} for $n_2 = 800$ rpm:

$$E_{A2} = \frac{n_2}{n_1} \times E_{A1} = \frac{800}{1200} \times 145 = 96.67 \text{ V}$$

The machine speed in radians per second is

$$\omega_m = n_2 \times \frac{2\pi}{60} = 800 \times \frac{2\pi}{60} = 83.78 \text{ rad/s}$$

Rearranging Equation 16.15, we have

$$K\phi = \frac{E_A}{\omega_m} = \frac{96.67}{83.78} = 1.154$$

From Equation 16.16, the developed torque is

$$T_{\text{dev}} = K\phi I_A = 1.154 \times 30 = 34.62 \text{ Nm}$$

The developed power is

$$P_{\text{dev}} = \omega_m T_{\text{dev}} = 2900 \text{ W}$$

As a check, we can also compute the developed power by using Equation 16.18:

$$P_{\text{dev}} = I_A E_A = 30 \times 96.67 = 2900 \text{ W}$$

Applying Kirchhoff's voltage law to the armature circuit in Figure 16.14, we have

$$V_T = R_A I_A + E_A = 0.3(30) + 96.67 = 105.67$$ ■

Exercise 16.4 Find the voltage E_A for the machine having the magnetization curve shown in Figure 16.15 for $I_F = 2$ A and a speed of 1500 rpm.

Answer $E_A \cong 156$ V. □

Exercise 16.5 The machine having the magnetization curve shown in Figure 16.15 is operating as a motor at a speed of 1500 rpm with a developed power of 10 hp and $I_F = 2.5$ A. The armature resistance is 0.3 Ω and the field resistance is $R_F = 50$ Ω. Find the developed torque, the armature current I_A, and the voltage V_T applied to the armature circuit.

Answer $T_{\text{dev}} = 47.49$ Nm, $I_A = 41.16$ A, $V_T = 193.6$ V. □

In the next several sections, we will see that different torque–speed characteristics can result, depending on how the field windings and the armature are connected to the dc source.

16.4 SHUNT-CONNECTED AND SEPARATELY EXCITED DC MOTORS

In a shunt-connected dc machine, the field circuit is in parallel with the armature, as shown in Figure 16.16. The field circuit consists of a rheostat having a variable resistance, denoted as R_{adj}, in series with the field coil. Later, we will see that the rheostat can be used to adjust the torque–speed characteristic of the machine.

By designing the field windings to be connected either in parallel or in series with the armature, we can obtain drastically different torque–speed characteristics.

We assume that the machine is supplied by a constant voltage source V_T. The resistance of the armature circuit is R_A, and the induced voltage is E_A. We denote the mechanical shaft speed as ω_m and the developed torque as T_{dev}.

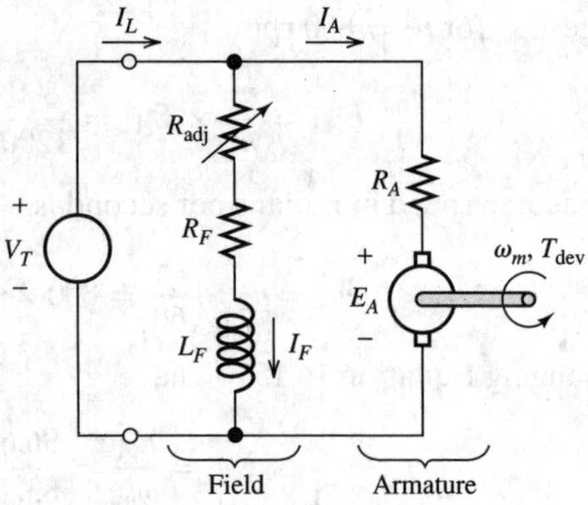

Figure 16.16 Equivalent circuit of a shunt-connected dc motor. R_{adj} is a rheostat that can be used to adjust motor speed.

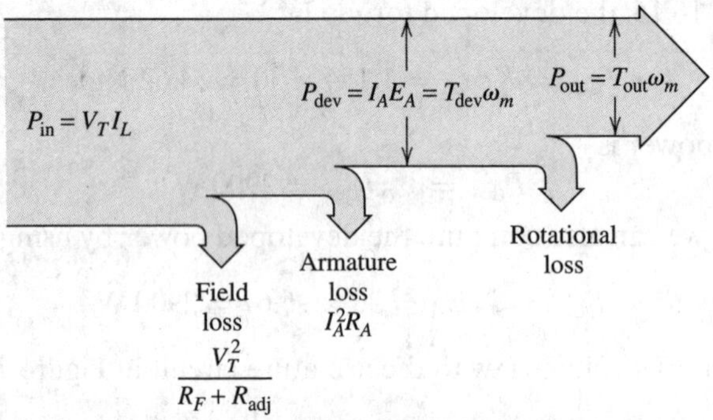

Figure 16.17 Power flow in a shunt-connected dc motor.

Power Flow

Figure 16.17 shows the flow of power in the shunt-connected dc machine. The electrical source supplies an input power given by the product of the supply voltage and the line current I_L:

$$P_{in} = V_T I_L \tag{16.20}$$

Some of this power is used to establish the field. The power absorbed by the field circuit is converted to heat. The **field loss** is given by

$$P_{\text{field-loss}} = \frac{V_T^2}{R_F + R_{adj}} = V_T I_F \tag{16.21}$$

Furthermore, **armature loss** occurs due to heating of the armature resistance:

$$P_{\text{arm-loss}} = I_A^2 R_A \tag{16.22}$$

Sometimes, the sum of the field loss and armature loss is called **copper loss.**

The power delivered to the induced armature voltage is converted to mechanical form and is called the **developed power**, given by

$$P_{dev} = I_A E_A = \omega_m T_{dev} \tag{16.23}$$

in which T_{dev} is the developed torque.

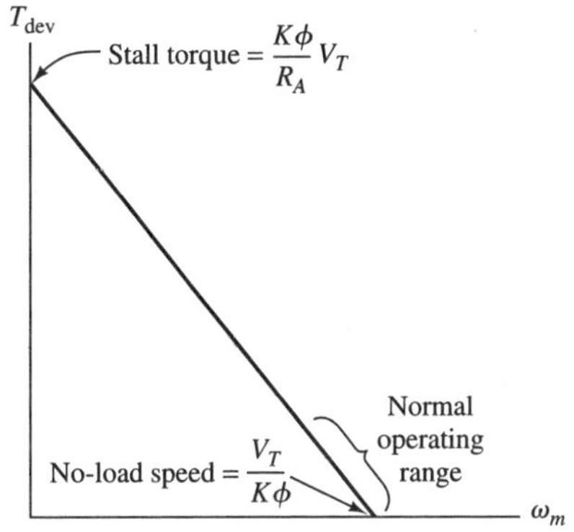

Figure 16.18 Torque–speed characteristic of the shunt dc motor.

The output power P_{out} and output torque T_{out} are less than the developed values because of **rotational losses**, which include friction, windage, eddy-current loss, and hysteresis loss. Rotational power loss is approximately proportional to speed.

Torque–Speed Characteristic

Next, we derive the torque–speed relationship for the shunt-connected machine. Applying Kirchhoff's voltage law to the equivalent circuit shown in Figure 16.16, we obtain

$$V_T = R_A I_A + E_A \tag{16.24}$$

Next, rearranging Equation 16.16 yields

$$I_A = \frac{T_{dev}}{K\phi} \tag{16.25}$$

Then, using Equation 16.15 to substitute for E_A and Equation 16.25 to substitute for I_A in Equation 16.24, we obtain

$$V_T = \frac{R_A T_{dev}}{K\phi} + K\phi\omega_m \tag{16.26}$$

Finally, solving for the developed torque, we get

$$T_{dev} = \frac{K\phi}{R_A}(V_T - K\phi\omega_m) \tag{16.27}$$

which is the torque–speed relationship that we desire. Notice that this torque–speed relationship plots as a straight line, as illustrated in Figure 16.18. The speed for no load (i.e., $T_{dev} = 0$) and the stall torque are labeled in the figure. The normal operating range for most motors is on the lower portion of the torque–speed characteristic, as illustrated in the figure.

The starting or stall torque of a shunt-connected machine is usually many times higher than the rated full-load torque.

Example 16.4 Shunt-Connected DC Motor

A 50-hp shunt-connected dc motor has the magnetization curve shown in Figure 16.19. The dc supply voltage is $V_T = 240$ V, the armature resistance is $R_A = 0.065\ \Omega$, the field resistance is $R_F = 10\ \Omega$, and the adjustable resistance is

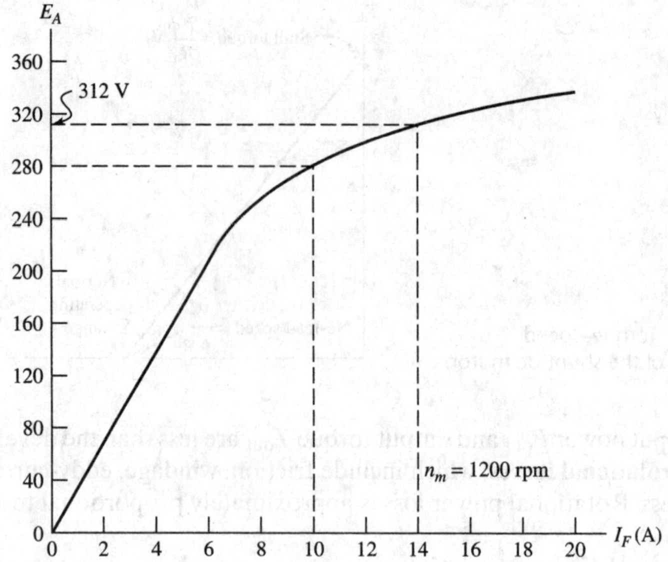

Figure 16.19 Magnetization curve for the motor of Example 16.4.

$R_{adj} = 14\ \Omega$. At a speed of 1200 rpm, the rotational loss is $P_{rot} = 1450$ W. If this motor drives a hoist that demands a torque of $T_{out} = 250$ Nm independent of speed, determine the motor speed and efficiency.

Solution The equivalent circuit is shown in Figure 16.20. The field current is given by

$$I_F = \frac{V_T}{R_F + R_{adj}} = \frac{240}{10 + 14} = 10 \text{ A}$$

Next, we use the magnetization curve to find the machine constant $K\phi$ for this value of field current. From the curve shown in Figure 16.19, we see that the induced armature voltage is $E_A = 280$ V at $I_F = 10$ A and $n_m = 1200$ rpm. Thus, rearranging Equation 16.15 and substituting values, we find that the machine constant is

$$K\phi = \frac{E_A}{\omega_m} = \frac{280}{1200(2\pi/60)} = 2.228$$

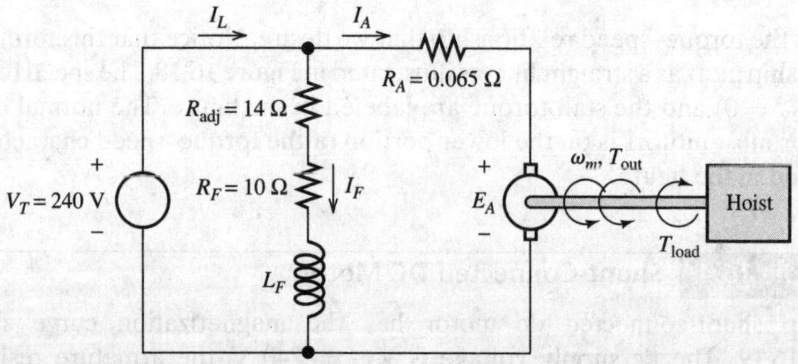

Figure 16.20 Equivalent circuit for Example 16.4.

We assume that the rotational power loss is proportional to speed. This is equivalent to assuming constant torque for the rotational loss. The rotational-loss torque is

$$T_{rot} = \frac{P_{rot}}{\omega_m} = \frac{1450}{1200(2\pi/60)} = 11.54 \text{ Nm}$$

Thus, the developed torque is

$$T_{dev} = T_{out} + T_{rot} = 250 + 11.54 = 261.5 \text{ Nm}$$

Now, we use Equation 16.16 to find the armature current:

$$I_A = \frac{T_{dev}}{K\phi} = \frac{261.5}{2.228} = 117.4 \text{ A}$$

Then, applying Kirchhoff's voltage law to the armature circuit, we have

$$E_A = V_T - R_A I_A = 240 - 0.065(117.4) = 232.4 \text{ V}$$

Solving Equation 16.15 for speed and substituting values, we get

$$\omega_m = \frac{E_A}{K\phi} = \frac{232.4}{2.228} = 104.3 \text{ rad/s}$$

or

$$n_m = \omega_m \left(\frac{60}{2\pi}\right) = 996.0 \text{ rpm}$$

To find efficiency, we first compute the output power and the input power, given by

$$P_{out} = T_{out}\omega_m = 250(104.3) = 26.08 \text{ kW}$$

$$P_{in} = V_T I_L = V_T(I_F + I_A) = 240(10 + 117.4) = 30.58 \text{ kW}$$

$$\eta = \frac{P_{out}}{P_{in}} \times 100\% = \frac{26.08}{30.58} \times 100\% = 85.3\%$$

■

Exercise 16.6 Repeat Example 16.4 with the supply voltage increased to 300 V while holding the field current constant by increasing R_{adj}. What is the new value of R_{adj}? What happens to the speed?
Answer $R_{adj} = 20\ \Omega$; the speed increases to $\omega_m = 131.2$ rad/s or $n_m = 1253$ rpm. □

Exercise 16.7 Repeat Example 16.4 if the adjustable resistance R_{adj} is increased in value to 30 Ω while holding V_T constant at 240 V. What happens to the speed?
Answer $I_F = 6$ A, $E_A = 229.3$ V, $I_A = 164.3$ A, $\omega_m = 144.0$, $\eta = 88.08$ percent, $n_m = 1376$ rpm; thus, speed increases as R_{adj} is increased. □

Separately Excited DC Motors

A separately excited dc motor is similar to a shunt-connected motor except that different sources are used for the armature and field circuits. The equivalent circuit for a separately excited dc machine is shown in Figure 16.21. Analysis of a separately excited machine is very similar to that of a shunt-connected machine. The chief reason for using two separate sources for the armature and field is to be able to control speed by varying one of the two sources.

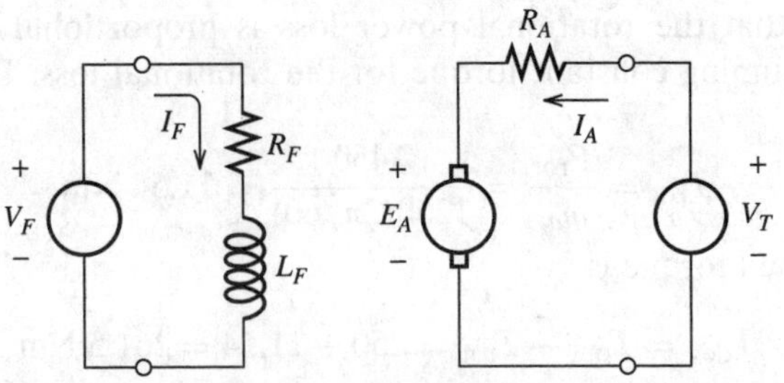

Figure 16.21 Equivalent circuit for a separately excited dc motor. Speed can be controlled by varying either source voltage (V_F or V_T).

Permanent-Magnet Motors

Separately excited and permanent-magnet motors have similar characteristics to those of shunt-connected motors.

In a permanent-magnet (PM) dc motor, the field is supplied by magnets mounted on the stator rather than by field coils. Its characteristics are similar to those of the separately excited machine except that the field cannot be adjusted. PM motors have several advantages compared to motors with field windings. First, no power is required to establish the field—leading to better efficiency. Second, PM motors can be smaller than equivalent machines with field windings. PM motors are common in applications calling for fractional- or subfractional-horsepower sizes. Typical applications include fan and power-window motors in automobiles.

PM motors also have some disadvantages. The magnets can become demagnetized by overheating or because of excessive armature currents. Also, the flux density magnitude is smaller in PM motors than in wound-field machines. Consequently, the torque produced per ampere of armature current is smaller in PM motors than in wound-field motors with equal power ratings. PM motors are confined to operation at lower torque and higher speed than wound-field motors with the same power rating.

16.5 SERIES-CONNECTED DC MOTORS

The equivalent circuit of a **series-connected dc motor** is shown in Figure 16.22. Notice that the field winding is in series with the armature. In this section, we will see that the series connection leads to a torque–speed characteristic that is useful in many applications.

Field windings are designed differently for series-connected machines than they are for shunt-connected machines.

In series dc motors, the field windings are made of larger diameter wire and the field resistances are much smaller than those of shunt machines of comparable size. This is necessary to avoid dropping too much of the source voltage across the field winding.

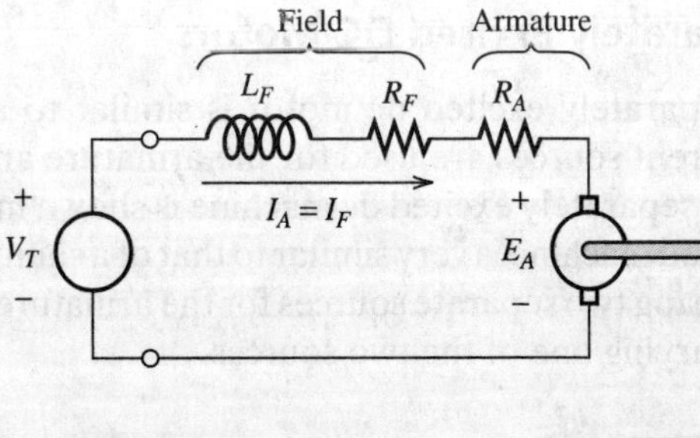

Figure 16.22 Equivalent circuit of the series-connected dc motor.

Next, we derive the relationship between torque and speed for the series motor. We use a linear equation to approximate the relationship between magnetic flux and field current. In equation form, we have

$$\phi = K_F I_F \tag{16.28}$$

in which K_F is a constant that depends on the number of field windings, the geometry of the magnetic circuit, and the B–H characteristics of the iron. Of course, the actual relationship between ϕ and I_F is nonlinear, due to magnetic saturation of the iron. (A plot of ϕ versus I_F has exactly the same shape as the magnetization curve of the machine.) However, Equation 16.28 will give us insight into the behavior of the series dc motor. Later, we consider saturation effects.

Because $I_A = I_F$ in the series machine, we have

$$\phi = K_F I_A \tag{16.29}$$

Using Equation 16.29 to substitute for ϕ in Equations 16.15 and 16.16, we obtain

$$E_A = KK_F \omega_m I_A \tag{16.30}$$

and

$$T_{\text{dev}} = KK_F I_A^2 \tag{16.31}$$

If we apply Kirchhoff's voltage law to the equivalent circuit shown in Figure 16.22, we get

$$V_T = R_F I_A + R_A I_A + E_A \tag{16.32}$$

As usual, we are assuming steady-state conditions so that the voltage across the inductance is zero.

Then using Equation 16.30 to substitute for E_A in Equation 16.32 and solving for I_A, we have

$$I_A = \frac{V_T}{R_A + R_F + KK_F \omega_m} \tag{16.33}$$

Finally, using Equation 16.33 to substitute for I_A in Equation 16.31, we obtain the desired relationship between torque and speed:

$$T_{\text{dev}} = \frac{KK_F V_T^2}{(R_A + R_F + KK_F \omega_m)^2} \tag{16.34}$$

Torque–speed relationship for series-connected dc machines.

A plot of torque versus speed for the series dc motor is shown in Figure 16.23. The figure shows a plot of Equation 16.34 as well as an actual curve of torque versus speed, illustrating the effects of rotational loss and magnetic saturation. Equation 16.34 predicts infinite no-load speed. (In other words, for $T_{\text{dev}} = 0$, the speed must be infinite.) Yet, at high speeds, rotational losses due to windage and eddy currents become large, and the motor speed is limited.

However, in some cases, *the no-load speed can become large enough to be dangerous. It is important to have protection devices that remove electrical power to a series machine when the load becomes disconnected.*

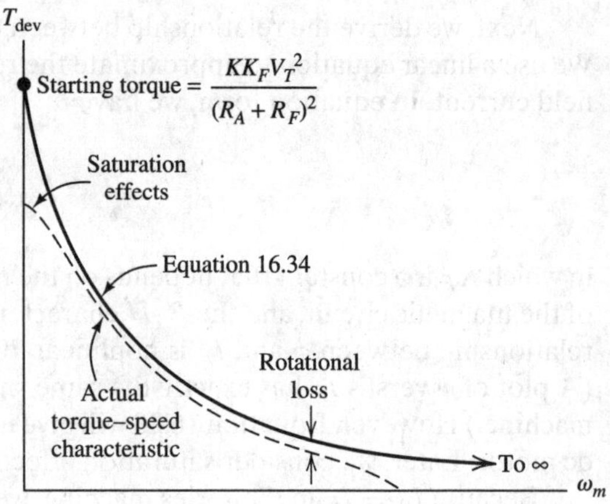

Figure 16.23 Torque–speed characteristic of the series-connected dc motor.

At very low speeds, Equation 16.33 shows that the current $I_F = I_A$ becomes large. Then, magnetic saturation occurs. Therefore, the starting torque is not as large as predicted by Equation 16.34.

Example 16.5 Series-Connected DC Motor

A series-connected dc motor runs at $n_{m1} = 1200$ rpm while driving a load that demands a torque of 12 Nm. Neglect the resistances, rotational loss, and saturation effects. Find the power output. Then, find the new speed and output power if the load torque increases to 24 Nm.

Solution Since we are neglecting losses, the output torque and power are equal to the developed torque and power, respectively. First, the angular speed is

$$\omega_{m1} = n_{m1} \times \frac{2\pi}{60} = 125.7 \text{ rad/s}$$

and the output power is

$$P_{\text{dev1}} = P_{\text{out1}} = \omega_{m1} T_{\text{out1}} = 1508 \text{ W}$$

Setting $R_A = R_F = 0$ in Equation 16.34 gives

$$T_{\text{dev}} = \frac{KK_F V_T^2}{(R_A + R_F + KK_F\omega_m)^2} = \frac{V_T^2}{KK_F\omega_m^2}$$

Thus, for a fixed supply voltage V_T, torque is inversely proportional to speed squared, and we can write

$$\frac{T_{\text{dev1}}}{T_{\text{dev2}}} = \frac{\omega_{m2}^2}{\omega_{m1}^2}$$

Solving for ω_{m2} and substituting values, we have

$$\omega_{m2} = \omega_{m1}\sqrt{\frac{T_{\text{dev1}}}{T_{\text{dev2}}}} = 125.7\sqrt{\frac{12}{24}} = 88.88 \text{ rad/s}$$

which corresponds to

$$n_{m2} = 848.5 \text{ rpm}$$

Finally, the output power with the heavier load is

$$P_{\text{out2}} = T_{\text{dev2}}\omega_{m2} = 2133 \text{ W}$$ ■

Exercise 16.8 Find the speed and output power in Example 16.5 for a load torque of $T_{\text{dev3}} = 6$ Nm.

Answer $P_{\text{out3}} = 1066$ W, $\omega_{m3} = 177.8$ rad/s, $n_{m3} = 1697$ rpm. □

Exercise 16.9 Repeat Example 16.5 for a shunt-connected motor. (In an ideal shunt-connected motor, the field resistance would be very large rather than zero.)

Answer With $R_A = 0$ and fixed V_T, the shunt motor runs at constant speed, independent of load. Thus, $n_{m1} = n_{m2} = 1200$ rpm, $P_{\text{out1}} = 1508$ W, and $P_{\text{out2}} = 3016$ W. □

Notice that by comparing the results of Exercise 16.9 with those of Example 16.5 we find that the output power variation is larger for the shunt-connected motor than for the series-connected motor.

Universal Motors

Equation 16.34 shows that the torque produced by the series dc motor is proportional to the square of the source voltage. Thus, the direction of the torque is independent of the polarity of the applied voltage. The series-connected machine can be operated from a single-phase ac source, provided that the stator is laminated to avoid excessive losses due to eddy currents. Since the field and armature inductances have nonzero impedances for ac currents, the current is not as large for an ac source as it would be for a dc source of the same average magnitude.

The universal motor is an important type of ac motor.

Series motors that are intended for use with ac sources are called **universal motors** because in principle they can operate from either ac or dc. Any time that you examine an ac motor and find brushes and a commutator, you have a universal motor. Compared with other types of single-phase ac motors, the universal motor has several advantages:

1. For a given weight, universal motors produce more power than other types. This is a large advantage for hand-held tools and small appliances, such as drills, saws, mixers, and blenders.
2. The universal motor produces large starting torque without excessive current.
3. When load torque increases, the universal motor slows down. Hence, the power produced is relatively constant, and the current magnitude remains within reasonable bounds. (In contrast, shunt dc motors or ac induction motors tend to run at constant speed and are more prone to drawing excessive currents for high-torque loads.) Thus, the universal motor is more suitable for loads that demand a wide range of torque, such as drills and food mixers. (For the same reason, series dc motors are used as starter motors in automobiles.)
4. Universal motors can be designed to operate at very high speeds, whereas we will see that other types of ac motors are limited to 3600 rpm, assuming a 60-Hz source.

One disadvantage of universal motors (as well as dc machines in general) is that the brushes and commutators wear out relatively quickly. Thus, the service life is much less than for ac induction motors. Induction motors are better choices than universal motors in applications that need to run often over a long life, such as refrigerator compressors, water pumps, or furnace fans.

16.6 SPEED CONTROL OF DC MOTORS

Speed control is an important consideration in the design of many systems.

Several methods can be used to control the speed of dc motors:

1. Vary the voltage supplied to the armature circuit while holding the field constant.
2. Vary the field current while holding the armature supply voltage constant.
3. Insert resistance in series with the armature circuit.

In this section, we discuss briefly each of these approaches to speed control.

Variation of the Supply Voltage

This method is applicable to separately excited motors and PM motors. For the shunt motor, varying the supply voltage is not an appropriate method of speed control, because the field current and flux vary with V_T. The effects of increasing both armature supply voltage and the field current tend to offset one another, resulting in little change in speed.

In normal operation, the drop across the armature resistance is small compared to E_A, and we have

$$E_A \cong V_T$$

Since we also have

$$E_A = K\phi\omega_m$$

we can write

$$\omega_m \cong \frac{V_T}{K\phi} \tag{16.35}$$

Thus, the speed of a separately excited motor with constant field current or of a PM motor is approximately proportional to the source voltage.

Variation of the supply voltage is also appropriate for control of a series-connected dc motor; however, the flux does not remain constant in this case. Equation 16.34 shows that the torque of a series machine is proportional to the square of the source voltage at any given speed. Thus, depending on the torque–speed characteristic of the load, the speed varies with applied voltage. Generally, higher voltage produces higher speed.

Variable DC Voltage Sources

Historically, variable dc voltages were obtained from dc generators. For example, one popular approach was the **Ward Leonard system**, in which a three-phase induction motor drives a dc generator that in turn supplies a variable dc voltage to the motor to be controlled. The magnitude and polarity of the dc supply voltage are controlled by using a rheostat or switches to vary the field current of the dc generator. A disadvantage of this scheme is that three machines are needed to drive one load.

Since the advent of high-power electronics, a more economical approach is to use a rectifier to convert three-phase ac into dc, as illustrated in Figure 16.24. The resulting dc voltage v_L has some ripple, but a smoother voltage can be obtained with a full-wave version of the rectifier using six diodes. In any case, it is not necessary for the dc source supplying motors to be absolutely free of ripple, because the inductances and inertia tend to smooth the response.

Once a constant dc source has been created, an electronic switching circuit can be used to control the average voltage delivered to a load, as illustrated in Figure 16.25. (In Chapters 12 and 13, we showed how electronic devices such as BJTs and FETs

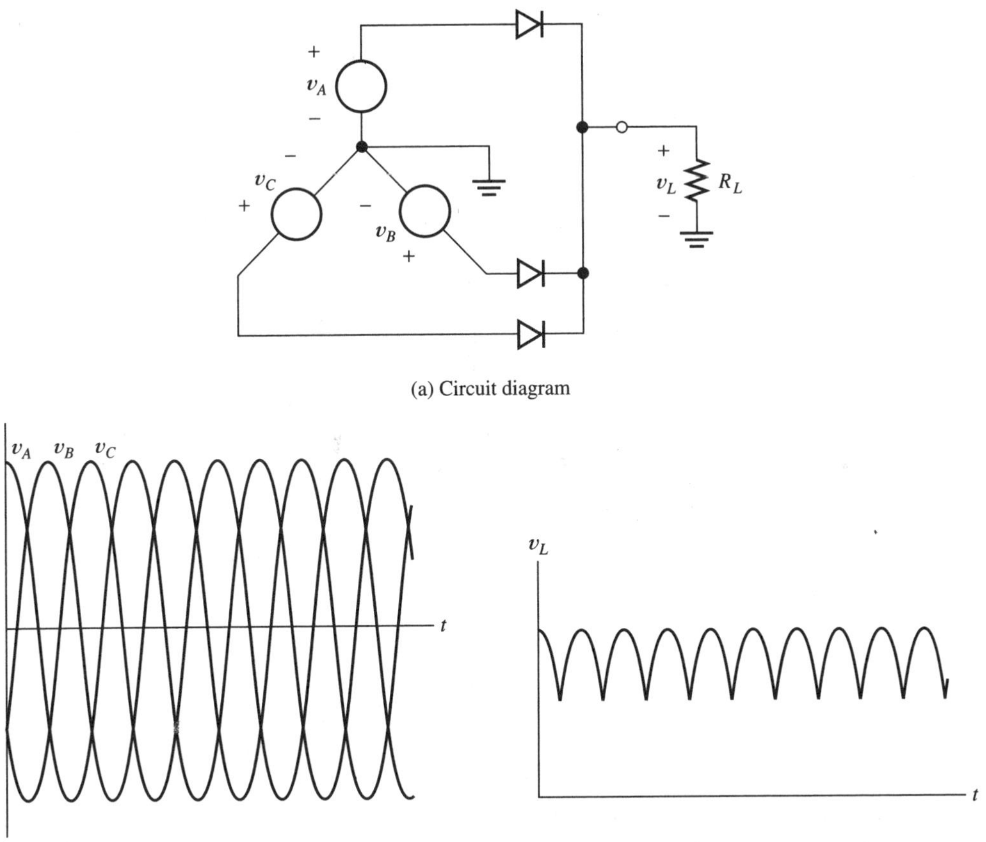

(a) Circuit diagram

(b) Three-phase ac voltages

(c) Rectified output voltage

Figure 16.24 Three-phase half-wave rectifier circuit used to convert ac power to dc.

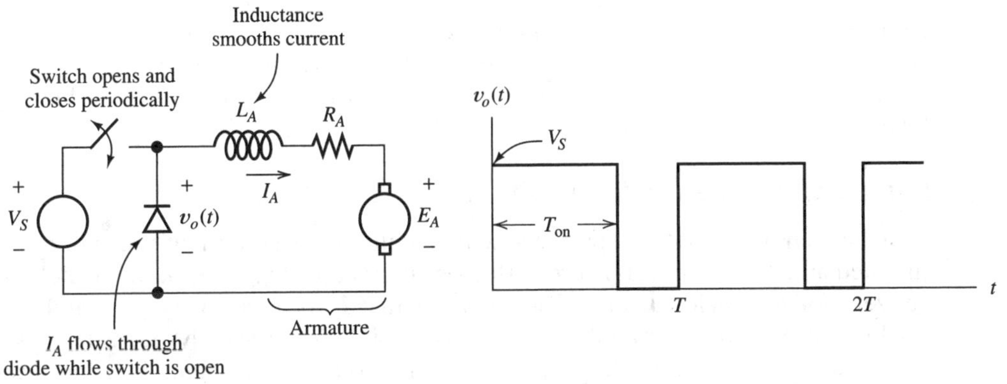

Figure 16.25 An electronic switch that opens and closes periodically can efficiently supply a variable dc voltage to a motor from a fixed dc supply voltage.

can be used as switches. In high-power electronics, similar devices such as silicon-controlled rectifiers perform the switching function.)

The switch periodically opens and closes with period T, spending T_{on} in the closed state and the remainder of the period open. The inductance L_A tends to cause the armature current to continue flowing when the switch is open. Thus, the armature current I_A is nearly constant even though the voltage $v_o(t)$ switches rapidly between zero and V_s. The diode provides a path for the armature current while the switch is open. The average value of the voltage applied to the motor is given by

$$V_T = V_s \frac{T_{\text{on}}}{T} \tag{16.36}$$

Thus, the average voltage, and therefore the speed of the motor, can be controlled by varying the fraction of the period for which the switch is closed.

Speed Control by Varying the Field Current

The speed of either a shunt-connected or a separately excited motor can be controlled by varying the field current. The circuit for the shunt-connected machine was shown in Figure 16.16, in which the rheostat R_{adj} provides the means to control field current.

On the other hand, PM motors have constant flux. In series-connected motors, the field current is the same as the armature current and cannot be independently controlled. Thus, using field current to control speed is not appropriate for either of these types of motors.

To understand the effect of field current on motor torque and speed, let us review the following equations for the shunt-connected or separately excited motor:

$$E_A = K\phi\omega_m$$

$$I_A = \frac{V_T - E_A}{R_A}$$

$$T_{\text{dev}} = K\phi I_A$$

Now, consider what happens when I_F is reduced (by increasing R_{adj}). Reducing I_F reduces the flux ϕ. Immediately, the induced voltage E_A is reduced. This in turn causes I_A to increase. In fact, the percentage increase in I_A is much greater than the percentage reduction in ϕ, because V_T and E_A are nearly equal. Thus, $I_A = (V_T - E_A)/R_A$ increases rapidly when E_A is reduced. Two of the terms in the equation for torque $T_{\text{dev}} = K\phi I_A$ change in opposite directions; specifically, ϕ falls and I_A rises. However, the change in I_A is much greater and the torque rises rapidly when I_F falls. (You can verify this by comparing your solution to Exercise 16.7 to the values in Example 16.4.)

Danger of an Open Field Circuit

What happens in a shunt or separately excited motor if the field circuit becomes open circuited and ϕ falls to nearly zero? (Because of residual magnetization, the field is not zero for zero field current.) The answer is that I_A becomes very large and the machine speeds up very rapidly. In fact, it is possible for excessive speeds to cause the armature to fly apart. Then, in a matter of seconds, the machine is reduced to a pile of useless scrap consisting of loose windings and commutator bars. Thus, it is important to operate shunt machines with well-designed protection circuits that open the armature circuit automatically when the field current vanishes.

It is important to operate shunt machines with well-designed protection circuits that automatically open the armature circuit when the field current vanishes.

Speed Control by Inserting Resistance in Series with the Armature

Another method for controlling the speed of a dc motor is to insert additional resistance in series with the armature circuit. This approach can be applied to all types of dc motors: shunt, separately excited, series, or PM. For example, a shunt-connected motor with added armature resistance is illustrated in Figure 16.26(a). We denote the

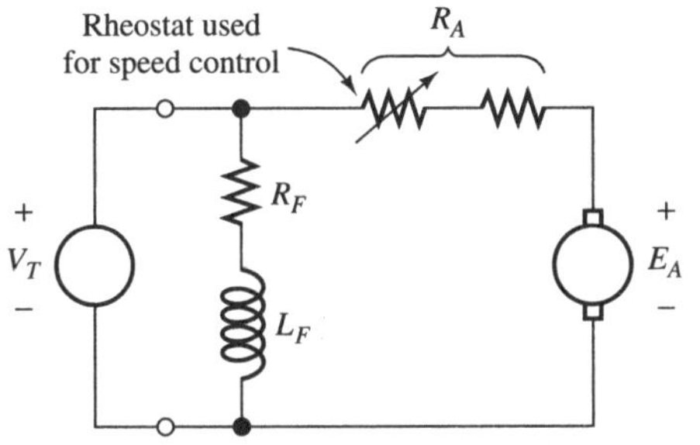

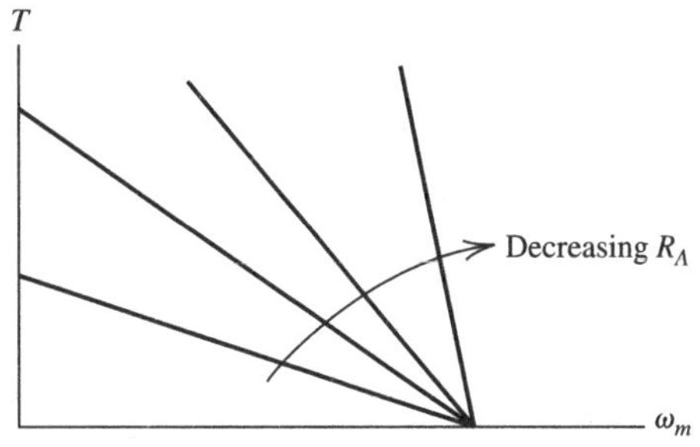

Figure 16.26 Speed can be adjusted by varying a rheostat that is in series with the armature.

total resistance as R_A, which consists of the control resistance plus the resistance of the armature winding. The torque–speed relationship for a shunt-connected motor is given by Equation 16.27, repeated here for convenience:

$$T_{\text{dev}} = \frac{K\phi}{R_A}(V_T - K\phi\omega_m)$$

Plots of the torque–speed characteristic for various resistances are shown in Figure 16.26(b). Similar results apply to separately excited and PM motors.

Starting controls for shunt or separately excited dc motors usually place resistance in series with the armature to limit armature current to reasonable values while the machine comes up to speed.

Look at Figures 16.26, 16.27, and 16.28 to see how the torque–speed characteristics of shunt-connected and separately excited dc motors can be changed by varying armature resistance, armature supply voltage, or field current to achieve speed control.

A disadvantage of inserting resistance in series with the armature to control speed is that it is wasteful of energy. When running at low speeds, much of the energy taken from the source is converted directly into heat in the series resistance.

Equation 16.34 gives the torque for series-connected machines. For convenience, the equation is repeated here:

$$T_{\text{dev}} = \frac{KK_F V_T^2}{(R_A + R_F + KK_F\omega_m)^2}$$

Notice that if R_A is made larger by adding resistance in series with the armature, the torque is reduced for any given speed.

Exercise 16.10 Why is variation of the supply voltage V_T in the shunt machine (see Figure 16.16) an ineffective way to control speed?

Answer Decreasing V_T decreases the field current, and therefore the flux. In the linear portion of the magnetization curve, the flux is proportional to the field current. Thus, reduction of V_T leads to reduction of ϕ, and according to Equation 16.35, the speed remains constant. (Actually, some variation of speed may occur due to saturation effects.) □

Exercise 16.11 Figure 16.26(b) shows a family of torque–speed curves for various values of R_A. Sketch a similar family of torque–speed characteristics for a separately excited machine (see Figure 16.21) with various values of V_T and constant field current.

Answer The torque–speed relationship is given by Equation 16.27. With constant field current, ϕ is constant. The family of characteristics is shown in Figure 16.27. □

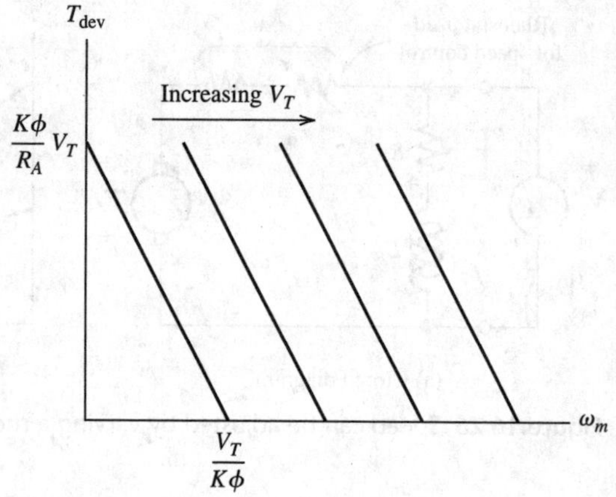

Figure 16.27 Torque versus speed for the separately excited dc motor for various values of armature supply voltage V_T.

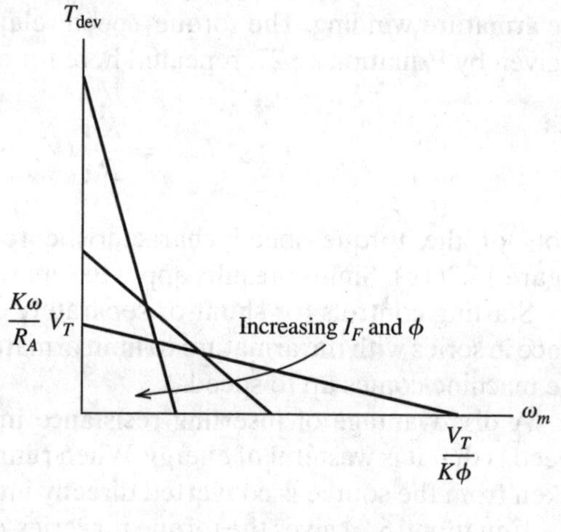

Figure 16.28 Effect of varying I_F on the torque–speed characteristics of the shunt-connected or separately excited dc motor.

Exercise 16.12 Sketch a family of torque–speed characteristics for a shunt-connected or separately excited machine with constant V_T and variable I_F.

Answer The torque–speed relationship is given in Equation 16.27. As field current is increased, the flux ϕ increases. The family of characteristics is shown in Figure 16.28. □

16.7 DC GENERATORS

Generators convert kinetic energy from a **prime mover**, such as a steam turbine or a diesel engine, into electrical energy. When dc power is needed, we can use a dc generator or an ac source combined with a rectifier. The trend is toward ac sources and rectifiers; however, many dc generators are in use, and for some applications they are still a good choice.

Several connections, illustrated in Figure 16.29, are useful for dc generators. We will discuss each type of connection briefly and conclude with an example illustrating performance calculations for the separately excited generator.

Separately Excited DC Generators

The equivalent circuit for a separately excited dc generator is shown in Figure 16.29(a). A prime mover drives the armature shaft at an angular speed ω_m, and

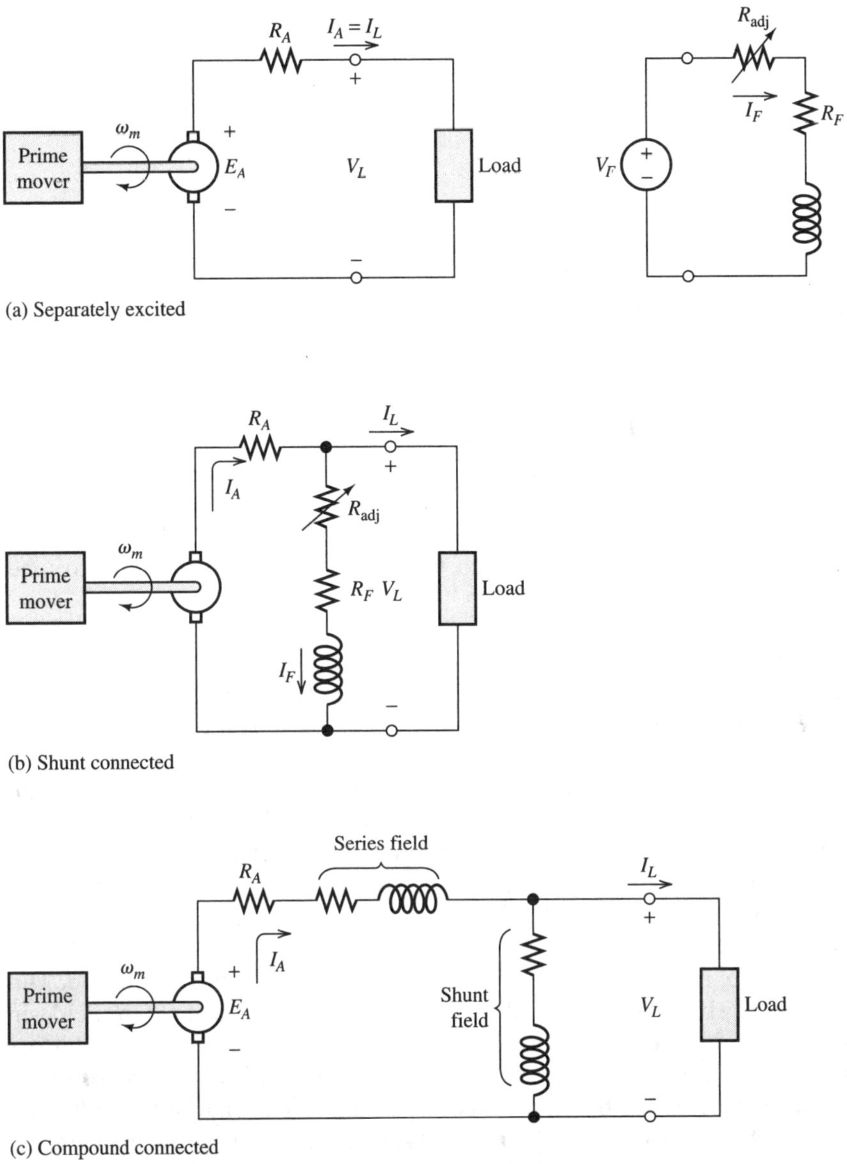

Figure 16.29 DC generator equivalent circuits.

the external dc source V_F supplies current I_F to the field coils. The induced armature voltage causes current to flow through the load. Because of the drop across the armature resistance, the load voltage V_L decreases as the load current I_L increases, assuming constant speed and field current. This is illustrated in Figure 16.30(a).

For some applications, it is desirable for the load voltage to be nearly independent of the load current. A measure of the amount of decrease in load voltage with current is the percentage load voltage regulation given by

$$\text{voltage regulation} = \frac{V_{NL} - V_{FL}}{V_{FL}} \times 100\% \tag{16.37}$$

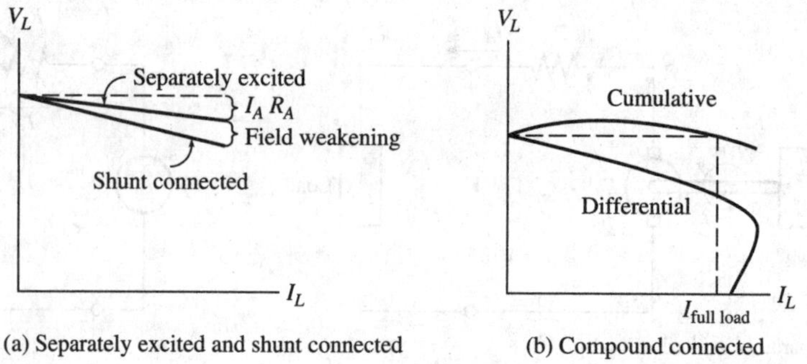

Figure 16.30 Load voltage versus load current for various dc generators.

in which V_{NL} is the no-load voltage (i.e., $I_L = 0$) and V_{FL} is the full-load voltage (i.e., with full-rated load current).

One of the advantages of the separately excited dc generator is that the load voltage can be adjusted over a wide range by varying the field current either by changing V_F or by changing R_{adj}. Also, the load voltage is proportional to speed.

Shunt-Connected DC Generators

One of the disadvantages of the separately excited dc generator is the need for a separate dc source to supply the field windings. This disadvantage is overcome in the shunt-connected machine, for which the field circuit is in parallel with the armature and the load, as shown in Figure 16.29(b). The output voltage can be adjusted by changing the resistance R_{adj} that is in series with the field winding.

Initial buildup of voltage in the shunt-connected machine usually occurs because of the residual magnetic field of the iron. (Adjusting R_{adj} to its minimum value and reversing the connections to the field winding may be needed to ensure voltage buildup.) However, if the machine becomes demagnetized, the induced armature voltage is zero, resulting in zero field current, and no output is produced. This can be remedied by briefly applying a dc source of the correct polarity to the field winding to create a residual field in the machine. Also, depending on the machine history, it is possible for the polarity of the output voltage to build up with the opposite polarity to that desired. This can be corrected by applying an external source of the correct polarity to the field winding or by reversing the connections to the machine.

The load regulation of the shunt-connected generator is poorer (i.e., larger) than that of the separately excited machine because the field current falls as the load current increases due to the drop across the armature resistance. This increased drop due to field weakening is illustrated in Figure 16.30(a).

Compound-Connected DC Generators

It is possible to design a dc generator with both series and shunt windings. When both are connected, we have a compound-connected machinc. Several variations of the connections are possible. Figure 16.29(c) illustrates a **long-shunt compound connection**. Another possibility is the **short-shunt compound connection**, in which the shunt field is directly in parallel with the armature and the series field is in series with the load. Furthermore, in either the short-shunt or the long-shunt, the field of the series field coil can either aid or oppose the field of the shunt field coil. If the fields aid, we have a **cumulative shunt connection**. On the other hand, if the fields oppose, we have a **differential shunt connection**. Thus, we have four types of connections in all.

It is possible to design a **fully compensated** cumulative-connected machine for which the full-load voltage is equal to the no-load voltage, as illustrated in Figure 16.30(b). Curvature of the voltage versus current characteristic is due to saturation effects. If the full-load voltage is less than the no-load voltage, the machine is said to be **undercompensated**. In an **overcompensated** machine, the full-load voltage is greater than the no-load voltage.

In a differential shunt connection, the output voltage falls rapidly with load current because the field of the series winding opposes that of the shunt winding. Considerable load current may flow even after the load voltage drops to zero. This is illustrated in Figure 16.30(b).

Performance Calculations

Next, we illustrate performance calculations for the separately excited generator. Analysis of the other connections is left for the problems.

As for dc motors, the following equations apply to dc generators:

$$E_A = K\phi\omega_m \tag{16.38}$$

$$T_{\text{dev}} = K\phi\omega_m \tag{16.39}$$

Referring to Figure 16.29(a), we can write:

$$E_A = R_A I_A + V_L \tag{16.40}$$

$$V_F = (R_F + R_{\text{adj}})I_F \tag{16.41}$$

Figure 16.31 illustrates the power flow of a dc generator. The efficiency is given by

$$\text{efficiency} = \frac{P_{\text{out}}}{P_{\text{in}}} \times 100\% \tag{16.42}$$

Equations 16.37 through 16.42, the magnetization curve of the machine, and Figure 16.31 are the tools for analysis of the separately excited dc generator. (Recall that the magnetization curve is a plot of E_A versus I_F for a given speed.)

Example 16.6 Separately Excited DC Generator

A separately excited dc generator has $V_F = 140$, $R_F = 10\ \Omega$, $R_{\text{adj}} = 4\ \Omega$, $R_A = 0.065\ \Omega$, the prime mover rotates the armature at a speed of 1000 rpm, and the magnetization curve is shown in Figure 16.19 on page 802. Determine the field current, the no-load voltage, the full-load voltage, and the percentage voltage regulation for a full-load current of 200 A. Assuming that the overall efficiency (not including the power supplied to the field circuit) of the machine is 85 percent, determine the input torque, the developed torque, and the losses associated with friction, windage, eddy currents, and hysteresis.

Solution The field current is

$$I_F = \frac{V_F}{R_{\text{adj}} + R_F} = \frac{140}{4 + 10} = 10\ \text{A}$$

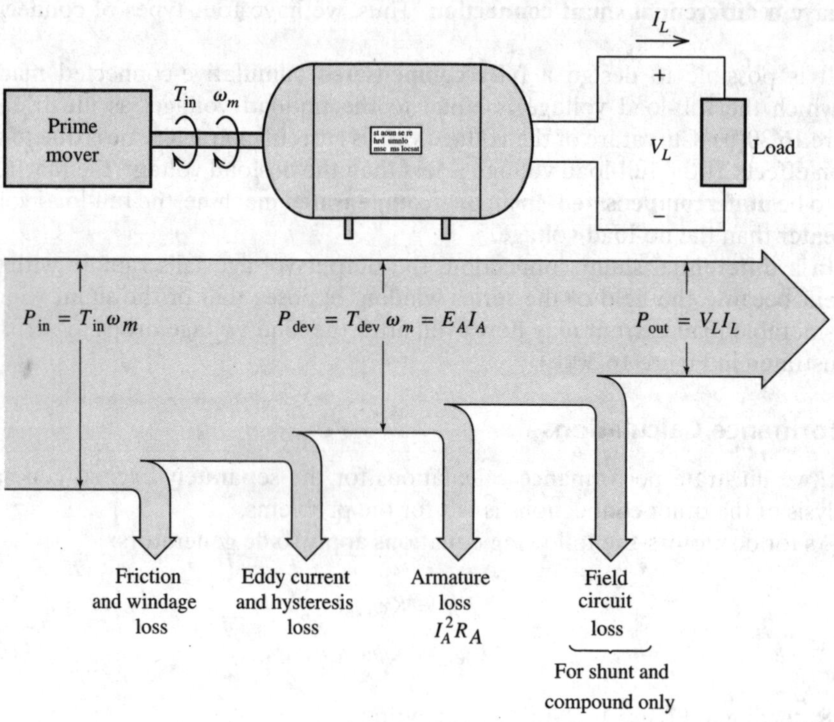

Figure 16.31 Power flow in dc generators.

Then, referring to the magnetization curve on page 802, we find that $E_A = 280$ V for a speed of 1200 rpm. Equation 16.38 shows that E_A is proportional to speed. So, for a speed of 1000 rpm, we have

$$E_A = 280\frac{1000}{1200} = 233.3 \text{ V}$$

which is also the no-load voltage of the machine. For a load current of 200 A, we get

$$V_{FL} = E_A - R_A I_A = 233.3 - 200 \times 0.065 = 220.3 \text{ V}$$

Finally, we obtain

$$\text{voltage regulation} = \frac{V_{NL} - V_{FL}}{V_{FL}} \times 100\% = \frac{233.3 - 220.3}{220.3} \times 100\% = 5.900\%$$

The output power is

$$P_{out} = I_L V_{FL} = 200 \times 220.3 = 44.06 \text{ kW}$$

The developed power is the sum of the output power and the armature loss:

$$P_{dev} = P_{out} + R_A I_A^2 = 44060 + 0.065(200)^2 = 46.66 \text{ kW}$$

The angular speed is

$$\omega_m = n_m \frac{2\pi}{60} = 104.7 \text{ rad/s}$$

The input power is

$$P_{\text{in}} = \frac{P_{\text{out}}}{0.85} = \frac{44.06}{0.85} = 51.84 \text{ kW}$$

The power losses associated with friction, windage, eddy currents, and hysteresis are

$$P_{\text{losses}} = P_{\text{in}} - P_{\text{dev}} = 51.84 - 46.66 = 5.18 \text{ kW}$$

The input and developed torques are

$$T_{\text{in}} = \frac{P_{\text{in}}}{\omega_m} = \frac{51{,}840}{104.7} = 495.1 \text{ N·m}$$

$$T_{\text{dev}} = \frac{P_{\text{dev}}}{\omega_m} = \frac{46{,}660}{104.7} = 445.7 \text{ N·m}$$

■

Exercise 16.13 Repeat Example 16.6 for $R_{\text{adj}} = 0$.

Answer $I_F = 14$ A, $V_{NL} = 260$, $V_{FL} = 247$, voltage regulation = 5.263%, $P_{\text{losses}} =$ 6.1 kW, $T_{\text{in}} = 555$ N·m, $T_{\text{dev}} = 497$ N·m. □

Summary

1. An electrical motor consists of a rotor and a stator, both of which may contain windings. A given winding may be classified as either an armature winding or as a field winding.
2. The electrical source for a motor can be dc, single-phase ac, or three-phase ac.
3. Factors to consider in selecting an electrical motor for a given application include the electrical sources available, the output power required, load torque versus speed, the service-life requirements, efficiency, speed regulation, starting current, the desired operating speed, ambient temperature, and the acceptable frequency of maintenance.
4. Dc motors in common use contain brushes and commutators that reverse the connections to the armature conductors with rotation. The need for frequent maintenance of these parts is a significant disadvantage of dc motors.
5. An advantage of dc motors is the ease with which their speed can be controlled. However, ac motors used with modern electronic systems that can change the frequencies of ac sources are taking over in many speed-control applications.
6. The speed of a motor–load combination self-adjusts to the point at which the output torque of the motor equals the torque demanded by the load. Thus, if the torque–speed characteristics of the motor and of the load are plotted on the same axes, the steady-state operating point is at the intersection of the characteristics.
7. The linear dc machine illustrates the basic principles of dc machines in an uncomplicated manner. However, it is not practical for the vast majority of applications.
8. Commonly used dc machines have field windings on the stator that establish an even number of magnetic poles. The armature windings

on the rotor carry currents, resulting in forces induced by the field. To achieve aiding contributions to torque (and induced armature voltage), a commutator and a set of brushes reverse the connections to the armature conductors as they move between poles.

9. The magnetization curve of a dc machine is a plot of the induced armature voltage versus field current with the machine being driven at a constant speed. The equivalent circuit shown in Figure 16.14 on page 797, the magnetization curve, and the following equations provide the basis for analyzing dc machines:

$$E_A = K\phi\omega_m$$

$$T_{\text{dev}} = K\phi I_A$$

$$P_{\text{dev}} = E_A I_A = \omega_m T_{\text{dev}}$$

10. The equivalent circuit of a shunt-connected dc motor is shown in Figure 16.16 on page 800. The torque–speed characteristic is shown in Figure 16.18 on page 801. Speed can be controlled either by using R_{adj} to vary the field current or by inserting a variable resistance in series with the armature. Variation of V_T is not an effective means of speed control.

11. The equivalent circuit of a separately excited dc motor is shown in Figure 16.21 on page 804. The torque–speed characteristic is identical to that of the shunt-connected machine shown in Figure 16.18 on page 801. Speed can be controlled by varying the field current, by varying the armature source voltage V_T, or by inserting additional resistance in series with the armature.

12. The characteristics of a permanent-magnet dc motor are similar to those of a separately excited motor, except that no means is available to vary the field. Speed control can be achieved by varying the armature supply voltage or by inserting additional resistance in series with the armature.

13. The equivalent circuit of the series-connected dc motor is shown in Figure 16.22 on page 804. In the normal range of operation, its torque is almost inversely proportional to the square of its speed. The series-connected dc motor is suitable for starting heavy loads. It can reach dangerous speeds if the load is totally removed.

14. The universal motor is basically a series dc motor designed for operation from an ac source. It is to be found in applications where a large power-to-weight ratio is needed. However, due to commutator wear, its service life is limited. Its speed can be controlled by varying the applied voltage or by inserting series resistance.

15. Historically, variable dc voltages for speed control of dc motors were obtained by employing dc generators. Presently, the approach of choice is to use rectifiers that first convert ac into dc. Then, electronic chopper circuits with adjustable duty factors provide variable average voltages.

16. Figures 16.26, 16.27, and 16.28 on pages 811, 812 and 812 show how the torque–speed characteristics of shunt-connected and separately excited dc motors can be changed by varying armature resistance, armature supply voltage, or field current to achieve speed control.

17. Dc generators can be separately excited, shunt connected, or compound connected. Analysis of dc generators is similar to that for dc motors.

Problems

Section 16.1: Overview of Motors

P16.1. What two types of windings are used in electrical machines? Which type is not used in permanent-magnet machines? Why not?

P16.2. In what application would dc motors be more advantageous than ac motors?

P16.3. What types of motors contain brushes and a commutator? What is the function of these parts?

P16.4. Name the principal parts of a rotating electrical machine.

P16.5. List the two principal types of three-phase ac motors. Which is in more common use?

***P16.6.** List two practical disadvantages of dc motors compared to single-phase ac induction motors for supplying power to a ventilation fan in a home or small business.

***P16.7.** A three-phase induction motor is rated at 5 hp, 1760 rpm, with a line-to-line voltage of 220 V rms. The percentage speed regulation is 5%. Determine the no-load speed of the motor.

P16.8. A three-phase induction motor is rated at 10 hp, 1760 rpm, with a line-to-line voltage of 220 V rms. The motor has a power factor of 90 percent lagging and an efficiency of 75 percent under full-load conditions. Find the electrical input power absorbed by the motor under full-load conditions. Also, find the rms line current.

P16.9. A three-phase induction motor is rated at 5 hp, 1760 rpm, with a line-to-line voltage of 220 V rms. Find the output torque and angular velocity of the motor under full-load conditions.

***P16.10.** A certain 25-hp three-phase induction motor operates from a 440-V-rms (line-to-line) three-phase source. The full-load speed is 1750 rpm. The motor has a starting torque equal to 200 percent of its full-load torque when started at rated voltage. For an engineering estimate, assume that the starting torque of an induction motor is proportional to the square of the applied voltage. To reduce the starting current of the motor, we decide to start it with a line-to-line voltage of 220 V. Estimate the starting torque with this reduced line voltage.

P16.11. We want a four-pole synchronous motor that operates at a speed of 1000 rpm. Determine the frequency of the ac source. List several other speeds that can be achieved by using synchronous motors operating from this ac source. What is the highest speed achievable?

P16.12. Consider a system consisting of a motor driving a load. The motor has the torque–speed characteristic shown in Figure P16.12. The load is a fan requiring a torque given by

$$T_{\text{load}} = K\omega_m^2$$

At a speed of $n = 1000$ rpm, the power absorbed by the load is 0.75 hp. Determine the speed of the system in radians per second and the power delivered to the fan in watts. Convert your answers to rpm and horsepower.

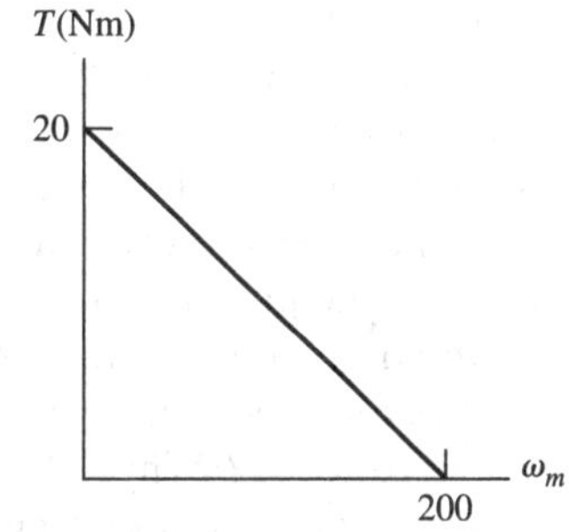

Figure P16.12

P16.13. A certain 25-hp three-phase induction motor operates from a 440-V-rms (line-to-line) three-phase source and draws a line current of 35 A rms at a power factor of 83 percent lagging under rated full-load conditions. The full-load speed is 1750 rpm. Under no-load conditions, the speed is 1797 rpm, and the line current is 6.5 A rms at a power factor of 30 percent lagging. Find the power loss and efficiency with full load, the input power with no load, and the speed regulation.

P16.14. Operating from a line-to-line voltage of 440 V rms with a line current of 14 A rms and a power factor of 80 percent, a three-phase induction motor produces an output power of 6 hp. Determine the losses in watts and the efficiency of the motor.

---

* Denotes that answers are contained in the Student Solutions files. See Appendix E for more information about accessing the Student Solutions.

***P16.15.** A 440-V-rms (line-to-line voltage) three-phase induction motor runs at 1150 rpm driving a load requiring 15 Nm of torque. The line current is 3.4 A rms at a power factor of 80 percent lagging. Find the output power, the power loss, and the efficiency.

P16.16. A motor drives a load for which the torque required is given by

$$T_{\text{load}} = \frac{800}{20 + \omega_m}\ \text{Nm}$$

The torque–speed characteristic of the motor is shown in Figure P16.12. **a.** Will this system start from zero speed? Why or why not? **b.** Suppose that the system is brought up to speed by an auxiliary driver, which is then disengaged. In principle, at what two constant speeds can the system rotate? **c.** If the system is operating at the lower of the two speeds and power to the motor is briefly interrupted so the system slows by a few radians per second, what will happen after power is restored? Why? **d.** Repeat (c) if the system is initially operating at the higher speed.

P16.17. A motor has output torque given by

$$T_{\text{out}} = 10^{-2}(60\pi - \omega_m)\omega_m$$

where ω_m is angular velocity in rad/s and T_{out} is the output torque in newton meters. **a.** Find the no-load speed of the motor. **b.** At what speed between zero and the no-load speed is the output torque maximum? What is the maximum output torque? **c.** At what speed between zero and the no-load speed is the output power maximum? What is the maximum output power? **d.** Find the starting torque of the motor. How could this motor be started?

P16.18. A 220-V-rms (line-to-line), 60-Hz, three-phase induction motor operates at 3500 rpm while delivering its rated full-load output power. Estimate the no-load speed and speed regulation for the motor.

P16.19. A 220-V-rms (line-to-line), 60-Hz, three phase induction motor operates at 3000 rpm while delivering its rated output power of 3 hp. The line current is 10 A rms and the input power are 2500 W. Find the power factor, losses and efficiency.

Section 16.2: Principles of DC Machines

***P16.20.** Consider the linear dc machine shown in Figure 16.6 on page 788 with no load force applied. What happens to the steady-state velocity of the bar if: **a.** the source voltage V_T is doubled in magnitude; **b.** the resistance R_A is doubled; **c.** the magnetic flux density B is doubled in magnitude?

***P16.21.** Consider the linear dc machine of Figure P16.21. When the switch closes, is the force on the bar toward the top of the page or toward the bottom? Determine the magnitude of the initial (starting) force. Also, determine the final velocity of the bar neglecting friction.

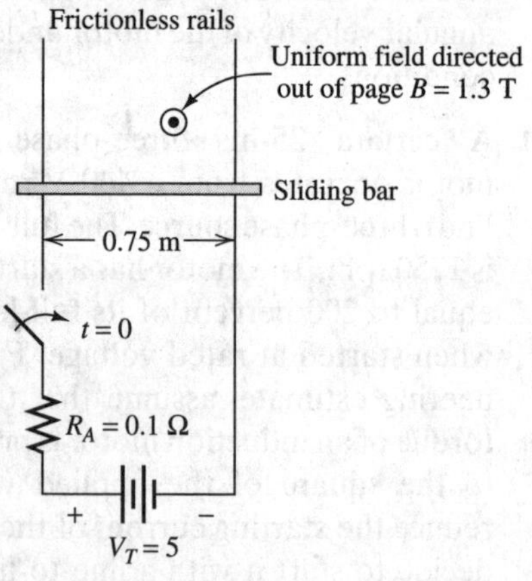

Figure P16.21

P16.22. Suppose that we wish to design a linear motor based on Figure 16.6 on page 788 that can deliver 2 hp at a steady bar velocity of 20 m/s. The flux density is limited to 1 T by the magnetic properties of the materials to be used. The length of the bar is to be 1 m, and the resistance is 0.5 Ω. Find the current i_A, the source voltage V_T, and the efficiency of the machine. Assume that the only loss is due to the resistance R_A.

P16.23. Consider the linear dc machine shown in Figure 16.6 on page 788. What happens to the initial force (i.e., starting force) induced in the bar if: **a.** the source voltage V_T is doubled in magnitude; **b.** the resistance R_A is doubled; **c.** the magnetic flux density B is doubled in magnitude?

P16.24. Suppose that an external force of 10 N directed toward the top of the page is applied to the bar as shown in Figure P16.24. In steady state, is the machine acting as a motor or as a generator? Find the power supplied by or absorbed by: **a.** the electrical voltage source V_T; **b.** R_A; **c.** the external force.

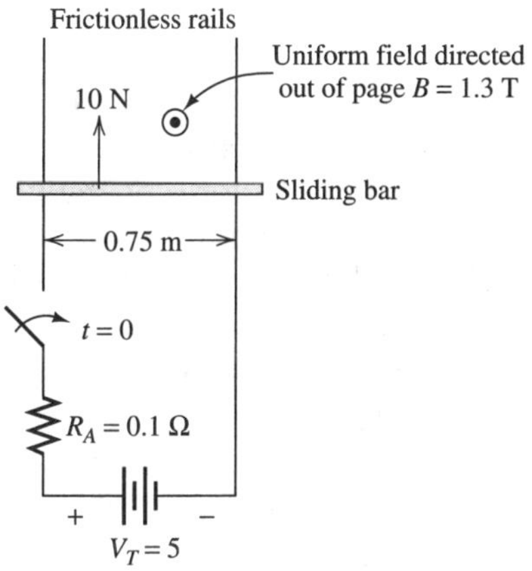

Figure P16.24

P16.25. We have presented the linear dc machine with an applied magnetic field, such as we might have in a dc motor. However, forces are exerted on the slider by the magnetic field produced by the currents in the rails. This is the principle of the **electromagnetic rail gun** for which you can find many references on the web including many practical construction tips. A version of such a rail gun is shown in Figure P16.25. **a.** When the switch closes, in which direction is force exerted on the projectile? Use physical principles discussed in this book to explain how you arrived at your answer. **b.** Suppose that the projectile mass is 3 g (which is about that of a penny). Assuming that all of the energy stored in the capacitor is transferred to kinetic energy in the slider, determine its final velocity. (*Note:* The highest velocity achievable by ordinary rifle bullets is about 1200 m/s.) **c.** List as many effects as you can that cause the velocity to be lower than the value calculated in part (b).

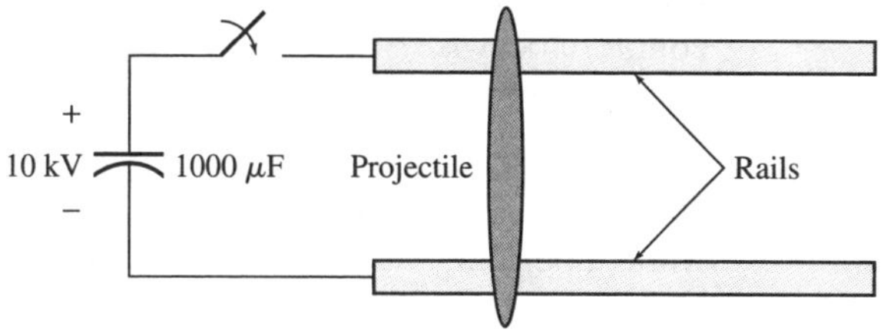

Figure P16.25

P16.26. Suppose that an external force of 10 N directed toward the bottom of the page is applied to the bar as shown in Figure P16.26. In steady state, is the machine acting as a motor or as a generator? Find the power supplied by or absorbed by: **a.** the electrical voltage source V_T; **b.** R_A; **c.** the external force.

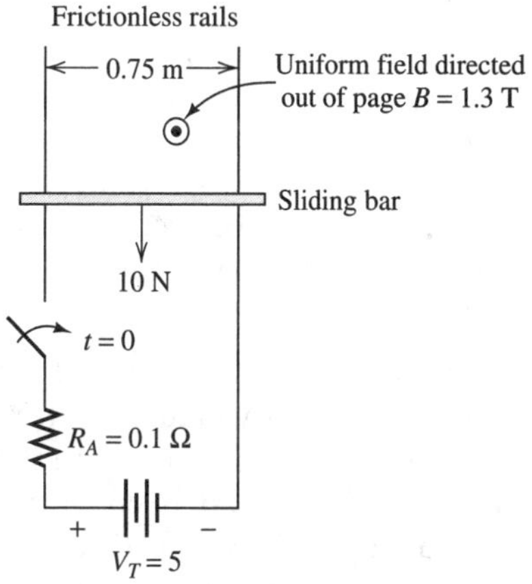

Figure P16.26

Section 16.3: Rotating DC Machines

***P16.27.** An alternative way to determine the direction of the torque produced by a dc machine is to consider the interaction of the magnetic poles produced by the armature current with the stator poles. Consider the cross section of a two-pole dc motor shown in Figure 16.10 on page 794. The magnetic poles established on the stator by the field windings are shown. For the armature current directions shown in the figure, find the locations and label the rotor poles. Of course, the rotor poles try to align with unlike stator poles. Which direction is the resulting torque, clockwise or counterclockwise? Repeat for the four-pole machine shown in Figure 16.11 on page 795.

***P16.28.** A certain dc motor has $R_A = 1.5\ \Omega$, $I_A = 20$ A, and produces a back emf $E_A = 240$ V, while operating at a speed of 1500 rpm. Determine the voltage applied to the armature, the developed torque, and the developed power.

***P16.29.** Suppose that we are designing a 1200-rpm dc motor to run from a 240-V source. We have determined that the flux density will be 1 T because smaller fluxes make inefficient use of the iron and larger fluxes result in saturation. The radius of the rotor (and thus, the torque arm for the armature conductors) is 0.1 m. The lengths of the armature conductors are 0.3 m. Approximately how many armature conductors must be placed in series in this machine?

***P16.30.** A dc motor operates with a load that demands constant developed torque. With $V_T = 200$ V, the motor operates at 1800 rpm and has $I_A = 10$ A. The armature resistance is 5 Ω and the field current remains constant. Determine the speed if V_T is increased to 300 V.

P16.31. A permanent-magnet dc motor has $R_A = 7\ \Omega$, $V_T = 240$ V, and operates under no-load conditions at a speed of 1500 rpm with $I_A = 1$ A. A load is connected and the speed drops to 1300 rpm. Determine the efficiency of the motor under loaded conditions. Assume that the losses consist solely of heating of R_A and frictional loss torque that is independent of speed.

P16.32. Consider a motor having the model shown in Figure 16.14 on page 797. The motor runs at a speed of 1200 rpm and has $R_F = 150\ \Omega$, $V_F = V_T = 180$ V, $I_A = 10$ A, and $R_A = 1.2\ \Omega$. Find E_A, T_{dev}, P_{dev}, and the power converted to heat in the resistances.

P16.33. Consider the two-pole motor shown in Figure 16.10 on page 794. The gap between the rotor and stator is 1.5 mm. Each of the two field coils has 250 turns and carries a current of 3 A. Assume that the permeability of the iron is infinite. **a.** Determine the flux density in the air gap. **b.** Each armature conductor carries a current of 30 A and has a length of 0.5 m. Find the force induced in each conductor.

P16.34. Under no-load conditions, a certain motor operates at 1200 rpm with an armature current of 0.5 A and a terminal voltage of 480 V. The armature resistance is 2 Ω. Determine the speed and speed regulation if a load demanding a torque of 50 Nm is connected to the motor. Assume that the losses consist solely of heating of R_A and frictional loss torque that is independent of speed.

P16.35. A certain motor has an induced armature voltage of 240 V at $n_{m1} = 1500$ rpm. Suppose that this motor is operating at a speed of $n_{m2} = 1800$ rpm with a developed power of 10 hp. Find the armature current and the developed torque.

P16.36. A certain dc motor produces a back emf of $E_A = 240$ V at a speed of 1200 rpm. Assume that the field current remains constant. Find the back emf for a speed of 600 rpm and for a speed of 1500 rpm.

P16.37. Sometimes the stator, particularly the yoke (shown in Figure 16.10 on page 794), of a dc machine is not laminated. However, it is always necessary to laminate the rotor. Explain.

Section 16.4: Shunt-Connected and Separately Excited DC Motors

***P16.38.** A shunt dc motor has $R_A = 0.1\ \Omega$ and $V_T = 440$ V. For an output power of 50 hp, we have $n_m = 1500$ rpm and $I_A = 103$ A. The field current remains constant for all parts of this problem. **a.** Find the developed power, power lost in R_A, and the rotational losses. **b.** Assuming that the rotational power loss is proportional to speed, find the no-load speed of the motor.

***P16.39.** A shunt-connected motor has the magnetization curve shown in Figure P16.39. Ignore rotational losses in this problem. The motor is supplied from a source of $V_T = 240$ V and has $R_A = 1.5\ \Omega$. The total field resistance is $R_F + R_{adj} = 240\ \Omega$. **a.** Find the no-load speed. **b.** A load is connected and the speed drops by 6 percent. Find the load torque, output power, armature current, field loss, and armature loss.

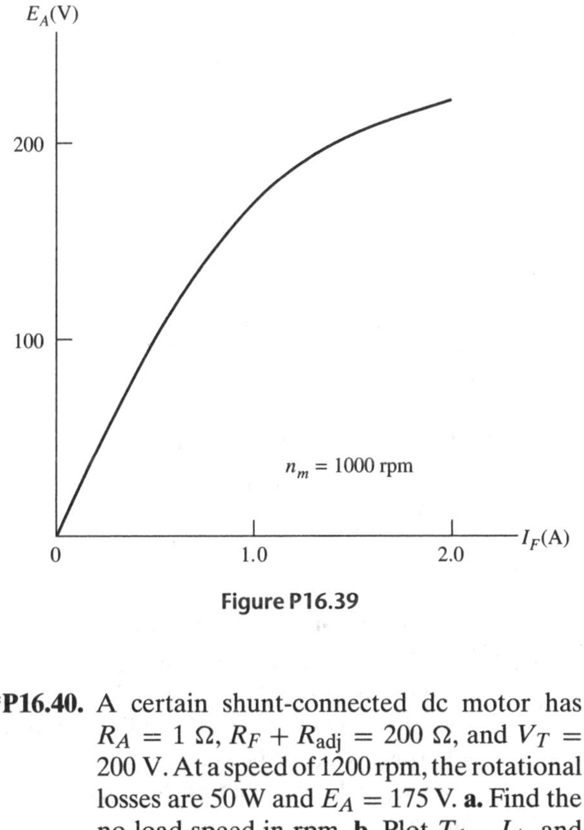

Figure P16.39

***P16.40.** A certain shunt-connected dc motor has $R_A = 1\ \Omega$, $R_F + R_{adj} = 200\ \Omega$, and $V_T = 200$ V. At a speed of 1200 rpm, the rotational losses are 50 W and $E_A = 175$ V. **a.** Find the no-load speed in rpm. **b.** Plot T_{dev}, I_A, and P_{dev} versus speed for speed ranging from zero to the no-load speed.

***P16.41.** A shunt-connected dc motor has $K\phi = 1$ V/(rad/s), $R_A = 1.2\ \Omega$, and $V_T = 200$ V. Find the two speeds for which the developed power is 5 hp. Neglect field loss and rotational loss. Find the value of I_A and efficiency for each speed. Which answer is most likely to be in the normal operating range of the machine?

***P16.42.** Is a magnetization curve needed for a permanent-magnet dc motor? Explain.

***P16.43.** A permanent-magnet dc motor has $R_A = 0.5\ \Omega$. With no load, it operates at 1000 rpm and draws 1 A from a 12.6-V source. Assume that rotational power loss is proportional to speed. Find the output power and efficiency for a load that drops the speed to 900 rpm.

P16.44. A permanent-magnet automotive fan motor draws 20 A from a 12-V source when the rotor is locked (i.e., held motionless). The motor has a speed of 800 rpm and draws 3.5 A when operating the fan with a terminal voltage of $V_T = 12$ V. Assume that the load (including the rotational losses) requires a developed torque that is proportional to the square of the speed. Find the speed for operation at 10 V. Repeat for 14 V.

P16.45. A shunt-connected 5-hp dc motor is rated for operation at $V_T = 200$ V, $I_L = 23.3$ A, and $n_m = 1500$ rpm. Furthermore, $I_F = 1.5$ A and $R_A = 0.4\ \Omega$. Under rated conditions find: **a.** the input power; **b.** the power supplied to the field circuit; **c.** the power lost in the armature resistance; **d.** the rotational loss; **e.** the efficiency.

P16.46. A separately excited dc motor (see the equivalent circuit shown in Figure 16.21 on page 804) has $R_A = 1.3\ \Omega$ and $V_T = 220$ V. For an output power of 3 hp, $n_m = 950$ rpm and $I_A = 12.2$ A. The field current remains constant for all parts of this problem. **a.** Find the developed power, developed torque, power lost in R_A, and the rotational losses. **b.** Assuming that the rotational power loss is proportional to speed, find the no-load speed of the motor.

P16.47. A permanent-magnet automotive fan motor draws 20 A from a 12-V source when the rotor is locked (i.e., held motionless). **a.** Find the armature resistance. **b.** Find the maximum developed power that this motor can produce when operated from a 12-V source. **c.** Repeat part (b) for operation at 10 V (this represents a nearly dead battery) and at 14 V (this is the terminal voltage during charging of the battery after the engine is started).

P16.48. A shunt-connected motor delivers an output power of 48 hp at 1500 rpm while operating from a source voltage of 440 V and drawing $I_L = 50$ A. The resistances are $R_A = 0.05\ \Omega$ and $R_F + R_{adj} = 100\ \Omega$. Find the developed torque and efficiency of the motor.

P16.49. A shunt-connected dc motor has zero rotational losses and $R_A = 0$. Assume that $R_F + R_{adj}$ is constant [except in part (d)] and that ϕ is directly proportional to field current. For $V_T = 200$ V and $P_{out} = 2$ hp, the speed is 1200 rpm. What is the effect on I_A and speed if: **a.** the load torque doubles; **b.** the load power doubles; **c.** V_T is changed to 100 V and P_{out} remains constant; **d.** $R_F + R_{adj}$ is doubled in value and P_{out} remains constant?

P16.50. Consider a shunt-connected dc motor that has the magnetization curve shown in Figure 16.19 on page 802. The dc supply voltage is $V_T = 200$ V, the armature resistance is $R_A = 0.085\ \Omega$, the field resistance is $R_F = 10\ \Omega$, and the adjustable resistance is $R_{adj} = 2.5\ \Omega$. At a speed of 1200 rpm, the rotational loss is $P_{rot} = 1000$ W. If this motor drives a load that demands a torque of $T_{out} = 200$ Nm independent of speed, determine the motor speed and efficiency.

P16.51. A shunt-connected dc motor has $R_A = 0.5\ \Omega$ and $R_F + R_{adj} = 400\ \Omega$. For $V_T = 200$ V and no load, the motor runs at 1150 rpm and the line current is $I_L = 1.2$ A. Find the rotational loss at this speed.

P16.52. A certain shunt-connected dc motor has $R_A = 4\ \Omega$ and $V_T = 240$ V. At a speed of 1000 rpm, the induced armature voltage is 120 V. Plot the torque–speed characteristic (T_{dev} versus n_m) to scale.

P16.53. Suppose that a dc machine is designed such that the voltage applied to the field V_F is equal to the voltage V_T applied to the armature. (Refer to Figure 16.14 on page 797.) In a well-designed machine, which current, I_F or I_A, is larger under full-load operating conditions? Why? What do you estimate as an acceptable value for the ratio I_A/I_F under full load?

P16.54. A shunt-connected motor has the magnetization curve shown in Figure P16.54. Ignore rotational losses in this problem. The motor is supplied from a source of $V_T = 240$ V and has $R_A = 2\ \Omega$. The total field resistance is $R_F + R_{adj} = 160\ \Omega$. **a.** Find the no-load speed. **b.** A load is connected and the speed drops by 3 percent. Find the load torque, output power, armature current, field loss, and armature loss.

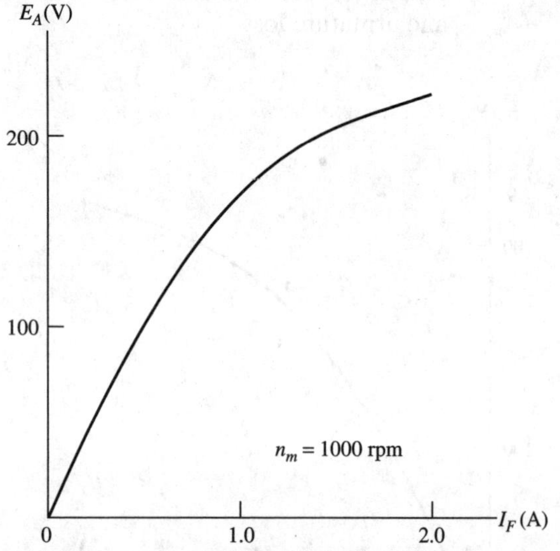

Figure P16.54

Section 16.5: Series-Connected DC Motors

***P16.55.** A series-connected dc motor has $R_F + R_A = 0.6\ \Omega$ and draws $I_A = 40$ A from the dc source voltage $V_T = 220$ V, while running at 900 rpm. What is the speed for $I_A = 20$ A? Assume a linear relationship between I_A and ϕ.

P16.56. Would a universal motor be a good choice for a clock? For a furnace fan motor? For a home coffee grinder? Give reasons for your answers.

P16.57. In examining a single-phase ac motor, what features could we look for to identify it as a universal motor?

P16.58. List four potential advantages of universal motors compared to other types of ac motors.

P16.59. Running at 1200 rpm from a 280-V source, a series-connected dc motor draws an armature current of 25 A. The field resistance is 0.2 Ω, and the armature resistance is 0.3 Ω. Rotational losses are 350 W and can be assumed to be proportional to speed. Determine the output power and developed torque. Determine the new armature current and speed if the load torque is increased by a factor of two.

P16.60. Running at 1200 rpm from a 280-V source, a series-connected dc motor draws an armature current of 25 A. The field resistance is 0.2 Ω, and the armature resistance is 0.3 Ω. Assuming that the flux is proportional to the field current, determine the speed at which the armature current is 10 A.

P16.61. A series-connected dc motor has $R_F + R_A = 0.6\ \Omega$ and draws $I_A = 40$ A from the dc source voltage $V_T = 220$ V, while running at 900 rpm. The rotational losses are 400 W. Find the output power and developed torque. Suppose that the load torque is reduced by a factor of two and that the rotational power loss is proportional to speed. Find the new values of I_A and speed.

P16.62. A series-connected dc motor is designed to operate with a variable load. The resistances R_A and R_F are negligible. To attain high efficiency, the motor has been designed to have very small rotational losses. For a load torque of 100 Nm, the machine runs at its maximum rated speed of 1200 rpm. **a.** Find the speed for a load torque of 300 Nm. **b.** What is the no-load speed? What are the potential consequences of having the load become disconnected from the motor without disconnecting the dc source?

P16.63. A series-connected dc motor has $R_A = 0.5\ \Omega$ and $R_F = 1.5\ \Omega$. In driving a certain load at 1500 rpm, the current is $I_A = 20$ A from a source voltage of $V_T = 220$ V. The efficiency is 90%. Find the output power and rotational loss.

Section 16.6: Speed Control of DC Motors

***P16.64.** Sketch the family of torque–speed characteristics for a separately excited dc motor obtained by:

a. varying the field current;

b. varying the voltage applied to the armature;

c. varying the resistance in series with the armature.

***P16.65.** A series-connected motor drives a constant torque load from a 50-V source at a speed of 1500 rpm. The resistances $R_F = R_A = 0$. Neglect rotational losses. What average source voltage is needed to achieve a speed of 1000 rpm? If this is achieved by chopping the 50-V source as illustrated in Figure 16.25 on page 809, find the duty ratio T_{on}/T.

***P16.66.** A series-connected dc motor operates at 1400 rpm from a source voltage of $V_T = 75$ V. The developed torque (load torque plus loss torque) is constant at 25 Nm. The resistance is $R_A + R_F = 0.1\ \Omega$. Determine the value of resistance that must be placed in series with the motor to reduce the speed to 1000 rpm.

P16.67. Consider a shunt-connected dc motor that has the magnetization curve shown in Figure 16.19 on page 802. The dc supply voltage is $V_T = 200$ V, the armature resistance is $R_A = 0.085\ \Omega$, the field resistance is $R_F = 10\ \Omega$, and the adjustable resistance is $R_{adj} = 2.5\ \Omega$. At a speed of 1200 rpm, the rotational loss is $P_{rot} = 1000$ W. Assume that the rotational power loss is proportional to speed.

a. With a load that demands a torque of $T_{load} = 200$ Nm independent of speed, determine the steady-state armature current.

b. Suppose that in starting this machine, the field circuit has reached steady state and the motor is not moving when power is applied to the armature circuit. What is the initial value of I_A? Determine the starting value of the developed torque. Compare these values to the steady-state values from part (a).

c. What additional resistance must be inserted in series with the armature to limit the starting current to 200 A? Find the starting torque with this resistance in place.

P16.68. Suppose that a shunt-connected machine operates at 1000 rpm on the linear portion of its magnetization characteristic. The motor drives a load that requires constant torque. Assume that $R_A = 0$. The resistances in the field circuit are $R_F = 75\ \Omega$ and $R_{adj} = 25\ \Omega$. Find a new value for R_{adj} so that the speed becomes 1500 rpm. What is the slowest speed that can be achieved by varying R_{adj}?

P16.69. Consider a PM motor that operates from a 12-V source with a no-load speed of 1800 rpm. Neglect rotational losses. What average source voltage is needed to achieve a no-load speed of 1200 rpm? If this is achieved by chopping the 12-V source as illustrated in Figure 16.25 on page 809, find the duty ratio T_{on}/T.

P16.70. List three methods that can be used to control the speed of dc motors. Which of these apply to shunt-connected motors? To separately excited motors? To permanent-magnet motors? To series-connected motors?

P16.71. A series-connected motor drives a load from a 50-V source at a speed of 1500 rpm. The load torque is proportional to speed. The resistances $R_F = R_A = 0$. Neglect rotational losses. What average source voltage is needed to achieve a speed of 1000 rpm? If this is achieved by chopping the 50-V source as illustrated in Figure 16.25 on page 809, find the duty ratio T_{on}/T.

Section 16.7: DC Generators

***P16.72.** A separately excited dc generator is rated for a load voltage of 160 V for a full load current of 10 A at 1500 rpm. With the load disconnected, the output voltage is 170 V. **a.** Determine the voltage regulation, the load resistance, the armature resistance, and the developed torque at full load. **b.** The speed of the generator is decreased to 1200 rpm, and the load resistance is unchanged. Determine the load current, the load voltage, and the developed power.

P16.73. What is the value of the voltage regulation for a fully compensated compound dc generator?

P16.74. What methods can be used to increase the load voltage of: **a.** a separately excited dc generator? **b.** a shunt-connected dc generator?

P16.75. Name the four types of compound connections for dc generators.

P16.76. Using Figure 16.30 on page 814 as a guide, list the types of connections for dc generators considered in Section 16.7, in order of percentage voltage regulation from highest to lowest.

P16.77. A separately excited dc generator has the magnetization curve shown in Figure P16.77, $V_F = 150$, $R_F = 40\ \Omega$, $R_{adj} = 60\ \Omega$, and $R_A = 1.5\ \Omega$. The prime mover rotates the armature at a speed of 1200 rpm. Determine the field current, the no-load voltage, the full-load voltage, and the percentage voltage regulation for a full-load current of 10 A. Assuming that the overall efficiency (not including the power supplied to the field circuit) of the machine is 60 percent, determine the input torque, the developed torque, and the losses associated with friction, windage, eddy currents, and hysteresis.

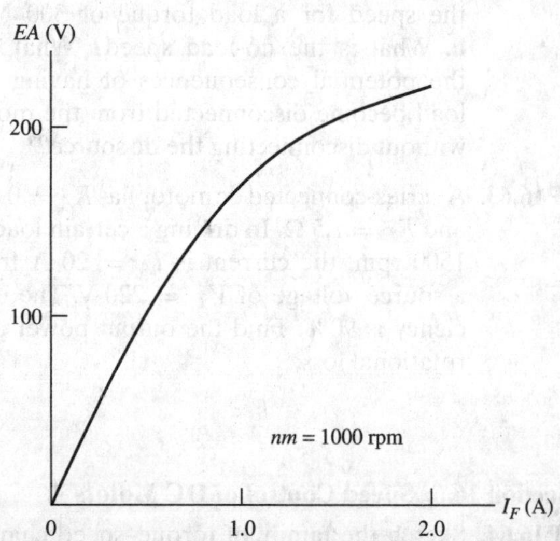

Figure P16.77

Frequently Asked Questions

1. What do you mean by an electric motor? How will you classify electric motors on the basis of supply?
2. Classify ac and dc motors and mention their important characteristics and applications
3. What do you mean by loses, power rating and efficiency of electric motors?
4. Draw and explain the speed-torque characteristics of a 3-phase induction motion.
5. What do you mean by speed regulation? State and explain speed regulation with graphs.
6. Explain the operating characteristics of synchronous motor.
7. Explain the operating characteristics of a dc shunt motor.
8. Explain the operating characteristics of a dc series motor.
9. Explain the working principle of a dc machine.
10. Explain when a machine operates as a motor or as a generator.
11. Explain the constructional features of a dc machine.
12. Explain how commutator works.
13. Draw and explain power flow diagram of a shunt connected dc motor.
14. Draw and explain the torque-speed characteristic of a dc shunt motor.
15. Write a short note on (i) Separately excited dc motors (ii) Permanent magnet dc motors
16. Draw the circuit of a dc series motor and explain its torque-speed characteristics with graph.
17. Write a note on universal motor.
18. How can you control the speed of dc motors? Explain.
19. Draw electric circuit for a separately excited dc generator and explain its characteristics.
20. Draw electric circuit for a dc shunt generator and explain its characteristics.
21. Draw electric circuit for a compound wound dc generator and explain its characteristics.
22. Draw the power flow diagram for a dc generator.

Practice Test

Here is a practice test you can use to check your comprehension of the most important concepts in this chapter. Answers can be found in Appendix D and complete solutions are included in the Student Solutions files. See Appendix E for more information about the Student Solutions.

T16.1. Consider a shunt-connected dc motor. What are the names of the two windings? Which is on the stator? On the rotor? For which does the current vary with mechanical load?

T16.2. Sketch the torque versus speed characteristic for a shunt-connected dc motor. What happens to the speed if the machine is lightly loaded and the field winding becomes disconnected?

T16.3. Sketch the torque versus speed characteristic for a series-connected dc motor.

T16.4. Give the definition of percentage speed regulation.

T16.5. Explain how the magnetization curve is measured for a dc machine.

T16.6. Name and briefly discuss the types of power loss in a shunt-connected dc motor.

T16.7. What is a universal motor? What are its advantages and disadvantages compared to other types of ac motors?

T16.8. List three methods for controlling the speed of a dc motor.

T16.9. Suppose we have a dc motor that produces a back emf of $E_A = 240$ V at a speed of 1500 rpm. **a.** If the field current remains constant, what is the back emf for a speed of 500 rpm? **b.** For a speed of 2000 rpm?

T16.10. We have a dc motor that has an induced armature voltage of 120 V at $n_{m1} =$

1200 rpm. If the field remains constant and this motor is operating at a speed of $n_{m2} =$ 900 rpm with a developed power of 4 hp, what are the values for the armature current and the developed torque?

T16.11. Consider a separately excited dc motor (see the equivalent circuit shown in Figure 16.21 on page 804) that has $R_A = 0.5\ \Omega$ and $V_T =$ 240 V. For a full-load output power of 6 hp, we have $n_m = 1200$ rpm and $I_A = 20$ A. The field current remains constant for all parts of this problem. **a.** Find the developed power, developed torque, power lost in R_A, and the rotational losses. **b.** Assuming that the rotational power loss is proportional to speed, find the speed regulation of the motor.

T16.12. We have a series-connected dc motor that draws an armature current of 20 A while running at 1000 rpm from a 240-V source. The field resistance is 0.3 Ω, and the armature resistance is 0.4 Ω. Assuming that the flux is proportional to the field current, determine the speed at which the armature current is 10 A.

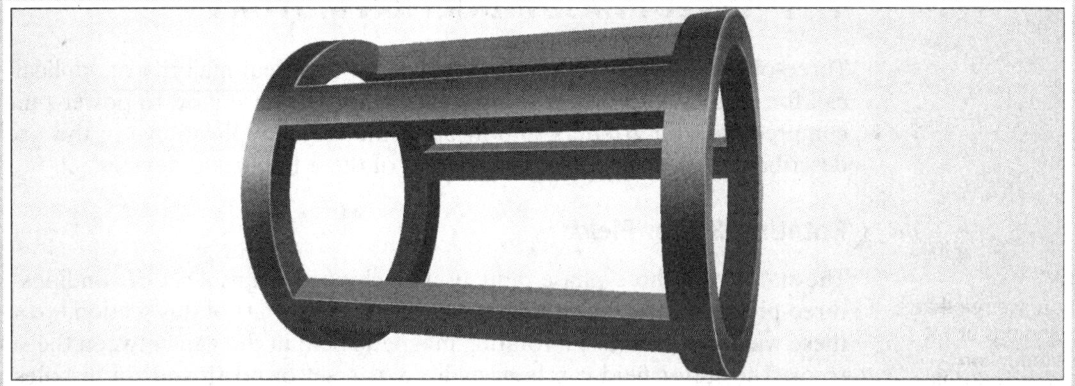

Chapter 17

AC Machines

Study of this chapter will enable you to:

- Select the proper ac motor type for various applications.
- State how torque varies with speed for various ac motors.
- Compute electrical and mechanical quantities for ac motors.
- Use motor nameplate data.
- Understand the operation and characteristics of three-phase induction motors, three-phase synchronous machines, various types of single-phase ac motors, stepper motors, and brushless dc motors.

Introduction to this chapter:

In this chapter, we continue our discussion of electrical machines. It is not necessary to have studied all of Chapter 16 before reading this chapter. However, you should read Section 16.1 to become familiar with the general concepts related to electrical machines before starting this chapter.

17.1 THREE-PHASE INDUCTION MOTORS

Three-phase induction machines account for the great majority of applications that call for motors with power ratings over 5 hp. They are used to power pumps, fans, compressors, and grinders, and in other industrial applications. In this section, we describe the construction and principles of these important devices.

Rotating Stator Field

In this section, we see that the stator windings of three-phase induction machines set up magnetic poles that rotate around the circumference of the stator.

The stator of a three-phase induction machine contains a set of windings to which three-phase electrical power is applied. In the first part of this section, we show that these windings establish a rotating magnetic field in the gap between the stator and rotor. The stator field can be visualized as a set of north and south poles rotating around the circumference of the stator. (North stator poles are where magnetic flux lines leave the stator, and south stator poles are where magnetic flux lines enter the stator.) Because north and south poles occur in pairs, the total number of poles P is always even. The field is illustrated for two-pole and four-pole machines in Figure 17.1. Similarly, it is possible for a three-phase induction motor to have six, eight, or more poles.

Next, we examine the stator windings and how the rotating field is established in a two-pole machine. The stator of the two-pole machine contains three windings embedded in slots cut lengthwise on the inside of the stator. One of the three stator windings is illustrated in Figure 17.2.

For simplicity, we represent each winding by only two conductors on opposite sides of the stator. However, each winding actually consists of a large number of conductors distributed in various slots in such a manner that the resulting air-gap flux varies approximately sinusoidally with the angle θ (which is defined in Figure 17.2). Thus, the field in the air gap due to the current $i_a(t)$ in winding a is given by

$$B_a = Ki_a(t)\cos(\theta) \tag{17.1}$$

where K is a constant that depends on the geometry and materials of the stator and rotor as well as the number of turns in winding a. B_a is taken as positive when directed from the stator toward the rotor and negative when directed in the opposite direction.

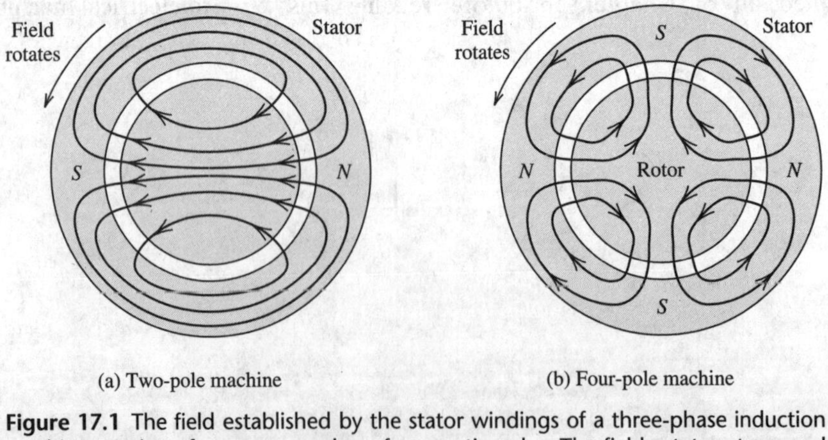

Figure 17.1 The field established by the stator windings of a three-phase induction machine consists of an even number of magnetic poles. The field rotates at a speed known as synchronous speed.

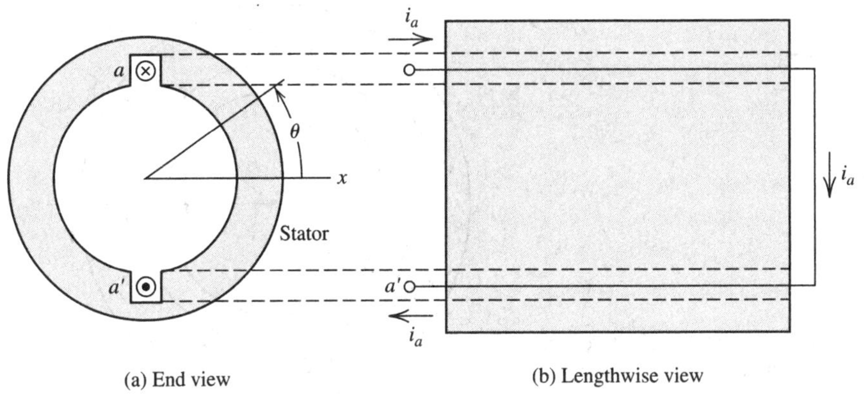

Figure 17.2 Two views of a two-pole stator showing one of the three windings. For simplicity, we represent the winding with a single turn, but in a real machine, each winding has many turns distributed around the circumference of the stator such that the air-gap flux varies sinusoidally with θ.

The field in the gap due to winding a is shown in Figure 17.3. Notice that the field is strongest at $\theta = 0$ and at $\theta = 180°$. Although it fluctuates in strength and polarity as the current changes with time, the field produced by winding a alone does not rotate. However, we are about to demonstrate that the combined field produced by all three windings does rotate.

The other two windings (b and c) are identical to winding a, except that they are rotated in space by 120° and 240°, respectively. This is illustrated in Figure 17.4. Thus, the fields in the air gap due to windings b and c are given by

Each winding sets up a field that varies sinusoidally around the circumference of the gap and varies sinusoidally with time. These fields are displaced from one another by 120° in both time and space.

$$B_b = Ki_b(t)\cos(\theta - 120°) \tag{17.2}$$

$$B_c = Ki_c(t)\cos(\theta - 240°) \tag{17.3}$$

The total field in the gap is the sum of the individual fields produced by the three coils. Thus, the total field is

$$B_{\text{gap}} = B_a + B_b + B_c \tag{17.4}$$

Using Equations 17.1 through 17.3 to substitute into Equation 17.4, we have

$$B_{\text{gap}} = Ki_a(t)\cos(\theta) + Ki_b(t)\cos(\theta - 120°) + Ki_c(t)\cos(\theta - 240°) \tag{17.5}$$

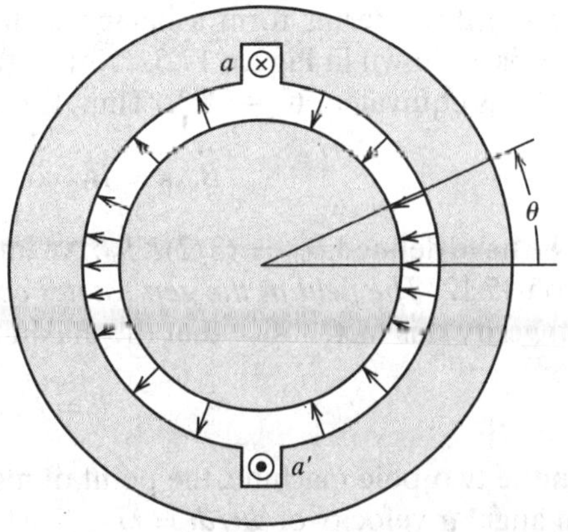

Figure 17.3 The field produced by the current in winding a varies sinusoidally in space around the circumference of the gap. The field is shown here for the positive maximum of the current $i_a(t)$. As illustrated, the field is strongest in magnitude at $\theta = 0$ and at $\theta = 180°$. Furthermore, the current and the field vary sinusoidally with time. Over time, the field dies to zero and then builds up in the opposite direction.

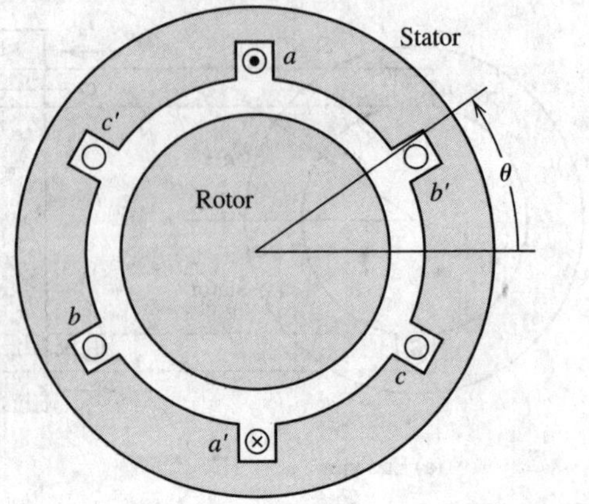

Figure 17.4 The stator of a two-pole machine contains three identical windings spaced 120° apart.

Application of a balanced three-phase source to the windings results in currents given by

$$i_a(t) = I_m \cos(\omega t) \tag{17.6}$$

$$i_b(t) = I_m \cos(\omega t - 120°) \tag{17.7}$$

$$i_c(t) = I_m \cos(\omega t - 240°) \tag{17.8}$$

Now, using Equations 17.6, 17.7, and 17.8 to substitute for the currents in Equation 17.5 yields

$$B_{\text{gap}} = KI_m \cos(\omega t)\cos(\theta) + KI_m \cos(\omega t - 120°)\cos(\theta - 120°)$$
$$+ KI_m \cos(\omega t - 240°)\cos(\theta - 240°) \tag{17.9}$$

The trigonometric identity $\cos(x)\cos(y) = (1/2)[\cos(x - y) + \cos(x + y)]$ can be used to write Equation 17.9 as

$$B_{\text{gap}} = \frac{3}{2}KI_m \cos(\omega t - \theta) + \frac{1}{2}KI_m[\cos(\omega t + \theta)$$
$$+ \cos(\omega t + \theta - 240°) + \cos(\omega t + \theta - 480°)] \tag{17.10}$$

Furthermore, we can write

$$[\cos(\omega t + \theta) + \cos(\omega t + \theta - 240°) + \cos(\omega t + \theta - 480°)] = 0 \tag{17.11}$$

because the three terms form a balanced three-phase set. A phasor diagram for these terms is shown in Figure 17.5. (Notice that −240° is equivalent to +120°, and that −480° is equivalent to −120°.) Thus, Equation 17.10 reduces to

$$B_{\text{gap}} = B_m \cos(\omega t - \theta) \tag{17.12}$$

The field in the gap rotates counterclockwise with an angular speed ω.

where we have defined $B_m = (3/2)KI_m$. An important conclusion can be drawn from Equation 17.12: *The field in the gap rotates counterclockwise with an angular speed of* ω. To verify this fact, notice that the maximum flux density occurs for

$$\theta = \omega t$$

Thus, in the two-pole machine, the point of maximum flux rotates counterclockwise with an angular velocity of $d\theta/dt = \omega$.

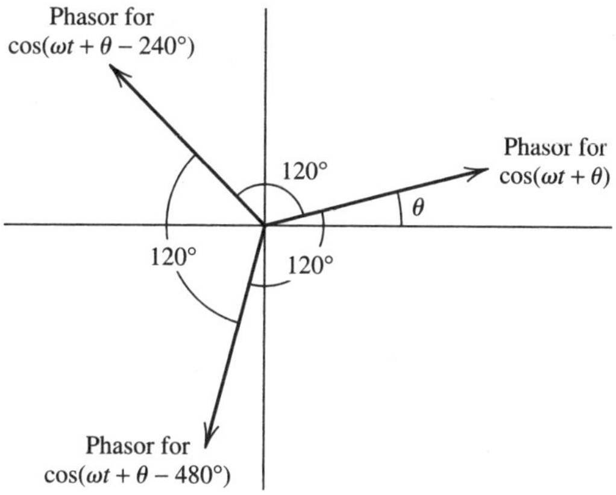

Figure 17.5 Phasor diagram for the three terms on the left-hand side of Equation 17.11. Regardless of the value of θ, the phasors add to zero.

Synchronous Speed

By a similar analysis, it can be shown that for a P-pole machine, the field rotates at an angular velocity of

In a P-pole machine, the field rotates at the synchronous speed ω_s.

$$\omega_s = \frac{\omega}{P/2} \tag{17.13}$$

which is called the **synchronous angular velocity**. In rpm, the synchronous speed is

$$n_s = \frac{120f}{P} \tag{17.14}$$

Table 17.1 gives synchronous speeds versus the number of poles assuming a frequency of 60 Hz.

In summary, we have shown that the field set up by the stator windings consists of a system of P magnetic poles that rotate at synchronous speed. These fields were illustrated in Figure 17.1 for two- and four-pole machines.

Exercise 17.1 If the connections of b and c to the three-phase source are interchanged, the currents become

The direction of rotation of a three-phase induction motor can be reversed by interchanging any two of the three line connections to the three-phase source.

$$i_a(t) = I_m \cos(\omega t)$$
$$i_b(t) = I_m \cos(\omega t - 240°)$$
$$i_c(t) = I_m \cos(\omega t - 120°)$$

Show that, in this case, the field rotates clockwise rather than counterclockwise. □

Table 17.1. Synchronous Speed versus Number of Poles for $f = 60$ Hz

P	n_s
2	3600
4	1800
6	1200
8	900
10	720
12	600

The result of Exercise 17.1 shows that the direction of rotation of the field in a three-phase induction machine can be reversed by interchanging any two of the line connections to the electrical source. We will see that this reverses the direction of mechanical rotation. You may find the fact that interchanging two of the electrical connections to the source reverses the direction of rotation to be useful in working with three-phase motors.

Squirrel-Cage Induction Machines

The rotor windings of a three-phase induction machine can take two forms. The simplest, least expensive, and most rugged is known as a **squirrel-cage rotor**. It consists simply of bars of aluminum with shorting rings at the ends, as illustrated in Figure 17.6. The squirrel cage is embedded in the laminated iron rotor by casting molten aluminum into slots cut into the rotor. In the squirrel-cage induction machine, there are no external electrical connections to the rotor. The other type of rotor construction, which we discuss later, is known as a **wound rotor**.

Next, we consider how torque is produced in squirrel-cage induction machines. We have seen earlier in this section that the stator sets up a system of P magnetic poles that rotate at synchronous speed. As this magnetic field moves past, voltages are induced in the squirrel-cage conductors. Since the field, the direction of relative motion, and the length of the conductors are mutually perpendicular, the induced voltage v_c is given by Equation 15.9, which is repeated here for convenience:

$$v_c = Blu \tag{17.15}$$

in which B is the flux density, l is the length of the conductor, and u is the relative velocity between the conductor and the field.

This voltage causes currents to flow in the conductors as illustrated in Figure 17.7. Of course, the largest voltages are induced in the conductors that are directly under the stator poles because that is where the flux density B is largest in magnitude. Furthermore, for conductors under south poles, the voltage polarity and current direction are opposite to those for conductors under north poles. Currents flow through the bars under the north pole, around the shorting ring, and back in the opposite direction through the bars under the south pole.

The rotating stator field induces voltages in the rotor conductors resulting in currents that produce magnetic poles on the rotor. The interaction of the rotor poles and stator poles produces torque.

The rotor currents establish magnetic poles on the rotor. It is the interaction of the rotor poles with the stator poles that produces torque. The north rotor pole N_r attempts to align itself with the south stator pole S_s.

If the impedances of the rotor conductors were purely resistive, the largest currents would occur directly under the stator poles S_s and N_s, as shown in Figure 17.7. Consequently, the rotor poles would be displaced by $\delta_{rs} = 90°$ with respect to the stator poles, as illustrated for a two-pole machine in Figure 17.7. This is exactly the

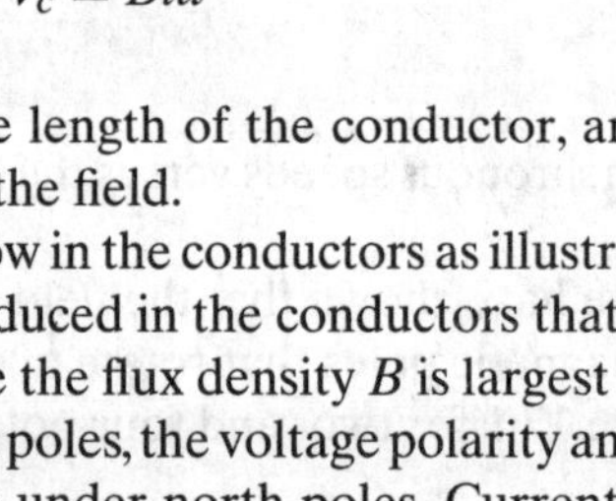

Figure 17.6 The rotor conductors of a squirrel-cage induction machine are aluminum bars connected to rings that short the ends together. These conductors are formed by casting molten aluminum into slots in the laminated iron rotor.

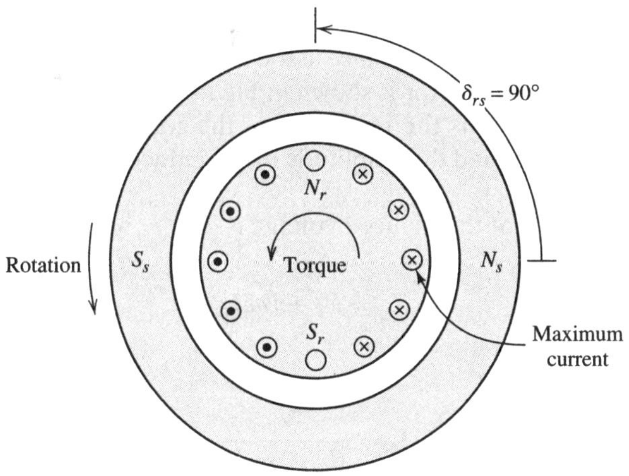

Figure 17.7 Cross section of a squirrel-cage induction motor. The rotating stator field induces currents in the conducting bars which in turn set up magnetic poles on the rotor. Torque is produced because the rotor poles are attracted to the stator poles.

angular displacement between the sets of magnetic poles that produces maximum torque.

Slip and Slip Frequency

The frequency of the voltages induced in the rotor conductors depends on the rotational speed of the stator field relative to the rotor and on the number of poles. We have seen that the stator field rotates at synchronous speed denoted as either ω_s or n_s. We denote the mechanical speed of the rotor as ω_m (or n_m). In an induction motor, the mechanical speed ω_m varies from zero to almost synchronous speed. Thus, the speed of the stator field relative to the rotor is $\omega_s - \omega_m$ (or $n_s - n_m$).

The **slip** s is defined to be the relative speed as a fraction of synchronous speed:

$$s = \frac{\omega_s - \omega_m}{\omega_s} = \frac{n_s - n_m}{n_s} \tag{17.16}$$

Slip s varies from 1 when the rotor is stationary to 0 when the rotor turns at synchronous speed.

The angular frequency of the voltages induced in the squirrel cage, called the **slip frequency**, is given by

$$\omega_{\text{slip}} = s\omega \tag{17.17}$$

The frequency of the rotor currents is called the slip frequency.

Notice that when the mechanical speed approaches the speed of the stator field (which is the synchronous speed), the frequency of the induced voltages approaches zero.

Effect of Rotor Inductance on Torque

In Figure 17.7, we saw how torque is produced in an induction motor assuming purely resistive impedances for the rotor conductors. However, the impedances of

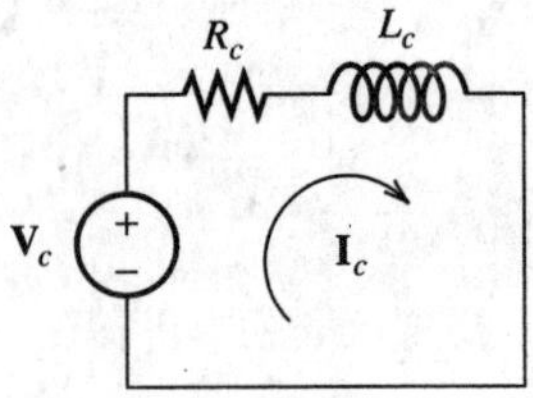

Figure 17.8 Equivalent circuit for a rotor conductor. $\mathbf{V}_c$ is the phasor for the induced voltage, R_c is the resistance of the conductor, and L_c is the inductance.

the conductors are not purely resistive. Because the conductors are embedded in iron, there is significant series inductance associated with each conductor. The equivalent circuit for a given conductor is shown in Figure 17.8, in which $\mathbf{V}_c$ is the phasor for the induced voltage, R_c is the resistance of the conductor, and L_c is its inductance. Both the frequency and the amplitude of the induced voltage are proportional to slip.

Since the frequency of the induced voltage is $\omega_{\text{slip}} = s\omega$, the impedance is

$$Z_c = R_c + js\omega L_c \tag{17.18}$$

The current is

$$\mathbf{I}_c = \frac{\mathbf{V}_c}{R_c + js\omega L_c} \tag{17.19}$$

Because of the inductances of the rotor conductors, the rotor currents lag the induced voltages.

Because of the inductance, the current lags the induced voltage. As the slip s increases, the amount of phase lag approaches 90°. Consequently, the peak current in a given rotor conductor occurs somewhat after the stator pole passes by. Furthermore, the rotor poles are displaced from the stator poles by less than 90°. This is illustrated in Figure 17.9. Because of this, the torque is reduced. (If the stator and rotor poles were aligned, no torque would be produced.)

Torque–Speed Characteristic

Now we are in a position to explain qualitatively the torque–speed characteristic shown in Figure 17.10 for the squirrel-cage induction motor. First, assume that the rotor speed n_m equals the synchronous speed n_s (i.e., the slip s equals zero). In this case, the relative velocity between the conductors and the field is zero (i.e., $u = 0$). Then according to Equation 17.15, the induced voltage v_c is zero. Consequently, the rotor currents are zero and the torque is zero.

As the rotor slows down from synchronous speed, the stator field moves past the rotor conductors. The magnitudes of the voltages induced in the rotor conductors increase linearly with slip. For small slips, the inductive reactances of the conductors, given by $s\omega L_c$, are negligible, and maximum rotor current is aligned with maximum stator field, which is the optimum situation for producing torque. Because the induced

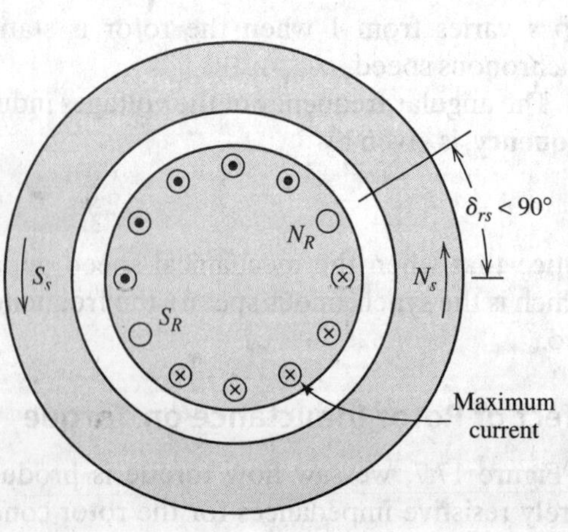

Figure 17.9 As slip s increases, the conductor currents lag the induced voltages. Consequently, the angular displacement δ_{rs} between the rotor poles and the stator poles approaches 0°.

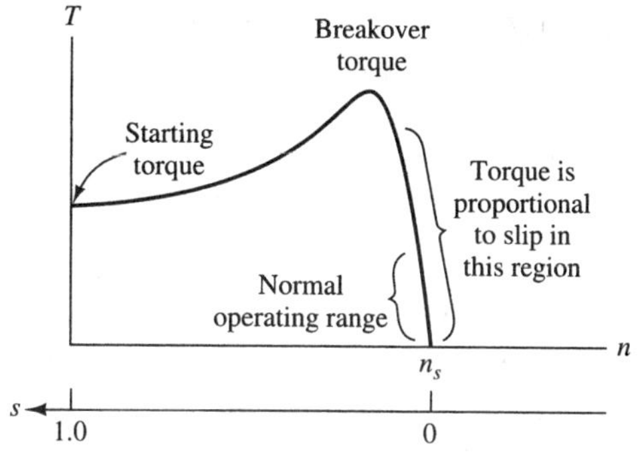

Figure 17.10 Torque-versus-speed characteristic for a typical three-phase induction motor.

voltage is proportional to slip and the impedance is independent of slip, the currents are proportional to slip. Torque is proportional to the product of the field and the current. Hence, we conclude that *torque is proportional to slip, assuming small slip.* This fact is illustrated in Figure 17.10.

For small values of slip, developed torque is proportional to slip.

As the motor slows further, the inductive reactance eventually dominates the denominator of Equation 17.19. Then, the magnitude of the current is nearly independent of slip. Thus, the torque tends to level out as the motor slows. Because the poles on the rotor tend to become aligned with the stator poles, the torque decreases as the motor slows to a stop. The torque for zero speed is called either the **starting torque** or the **stall torque**. The maximum torque is called either the **pull-out torque** or the **breakover torque**.

Our discussion has revealed the general characteristics of the three-phase induction motor. The motor designer can modify the shape of the torque–speed characteristic by variations in the dimensions and geometry of the motor and by materials selection. Some examples of torque–speed characteristics available in induction motors are shown in Figure 17.11. However, the details of motor design are beyond the scope of our discussion.

Motor designers can modify the shape of the torque–speed curve by changing various aspects of the machine design, such as the cross section and depth of the rotor conductors.

Exercise 17.2 A 5-hp four-pole 60-Hz three-phase induction motor runs at 1750 rpm under full-load conditions. Determine the slip and the frequency of the rotor currents at full load. Also, estimate the speed if the load torque drops in half.

Answer $s = 50/1800 = 0.02778, f_{\text{slip}} = 1.667$ Hz, $n = 1775$ rpm. □

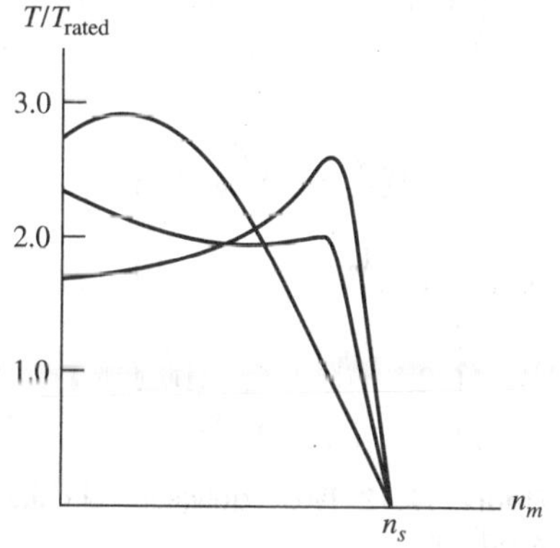

Figure 17.11 Depending on various design features, the torque–speed characteristic of the three-phase induction motor can be modified to better suit particular applications.

17.2 EQUIVALENT-CIRCUIT AND PERFORMANCE CALCULATIONS FOR INDUCTION MOTORS

In Section 17.1, we described the induction motor and its operation in qualitative terms. In this section, we develop an equivalent circuit and show how to calculate the performance of induction motors.

The equivalent circuit for each phase of an induction motor is similar to that of a transformer with the secondary winding shorted.

Consider the induction motor with the rotor locked so that it cannot turn. Then, the magnetic field of the stator links the rotor windings and causes current to flow in them. Basically, the locked-rotor induction motor is the same as a three-phase transformer with the stator windings acting as the primary. The rotor conductors act as short-circuited secondary windings. Thus, we can expect the equivalent circuit for each phase of the motor to be very similar to the transformer equivalent circuit shown in Figure 15.28 on page 763. Of course, modifications to the transformer equivalent circuit are necessary before it can be applied to the induction motor because of rotation and the conversion of electrical energy into mechanical form.

Rotor Equivalent Circuit

An equivalent circuit for one phase of the rotor windings is shown in Figure 17.12(a). (Equivalent circuits for the other two phases are identical except for the phase angles of the current and voltage.) $\mathbf{E}_r$ represents the induced voltage in phase *a* of the rotor under locked conditions. As discussed in the preceding section, the voltage induced in the rotor is proportional to slip s. Thus, the induced voltage is represented by the voltage source $s\mathbf{E}_r$. (Recall that for a stationary rotor $s = 1$.)

We have seen that the frequency of the rotor currents is $s\omega$. The rotor inductance (per phase) is denoted by L_r and has a reactance of $js\omega L_r = jsX_r$, where $X_r = \omega L_r$ is the reactance under locked-rotor conditions. The resistance per phase is denoted by R_r, and the current in one phase of the rotor is $\mathbf{I}_r$, which is given by

$$\mathbf{I}_r = \frac{s\mathbf{E}_r}{R_r + jsX_r} \tag{17.20}$$

Dividing the numerator and denominator of the right-hand side of Equation 17.20 by s, we have

$$\mathbf{I}_r = \frac{\mathbf{E}_r}{R_r/s + jX_r} \tag{17.21}$$

This can be represented by the circuit shown in Figure 17.12(b).

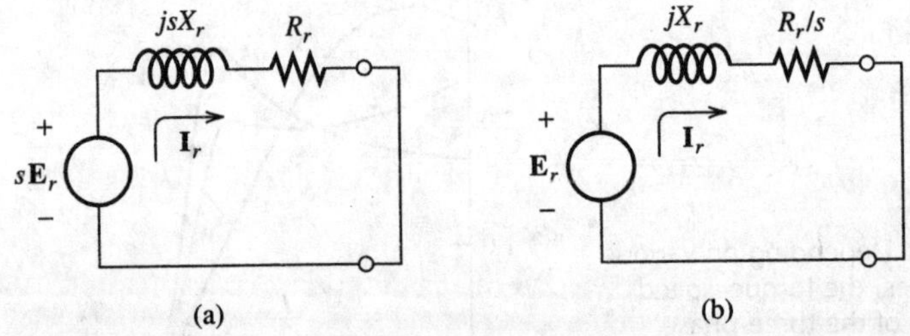

Figure 17.12 Two equivalent circuits for one phase of the rotor windings.

Figure 17.13 provides a convenient reference for the information needed to analyze induction machines.

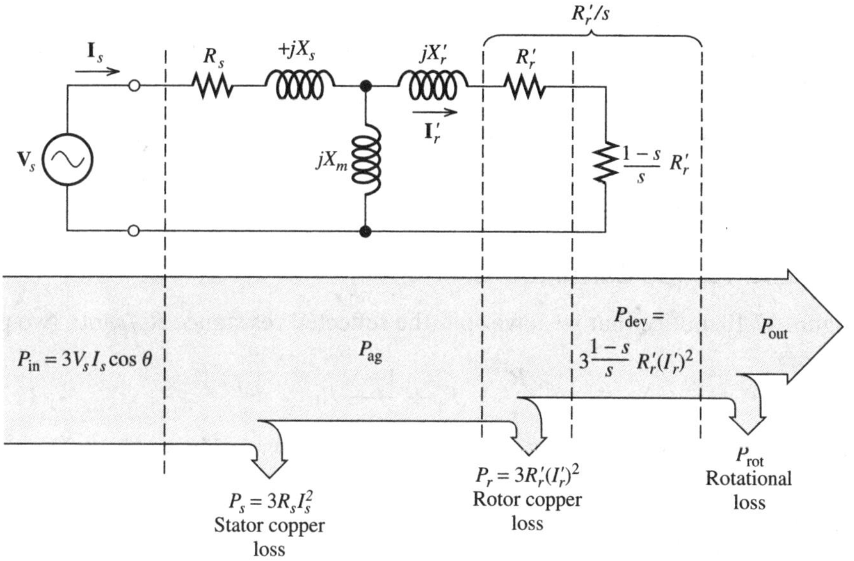

Figure 17.13 Equivalent circuit for one phase of an induction motor and the associated power-flow diagram. V_s is the rms phase voltage and I_s is the rms phase current.

Complete Induction-Motor Equivalent Circuit

As in a transformer, the induced rotor voltage E_r under locked conditions is related to the stator voltage by the turns ratio. Thus, we can reflect the rotor impedances of Figure 17.12(b) to the primary (stator) side of the equivalent circuit. We denote the reflected values of X_r and R_r/s by X_r' and R_r'/s, respectively.

The completed (per phase) induction-motor equivalent circuit is shown in Figure 17.13. The resistance of the stator winding is R_s, and the stator leakage reactance is X_s. The magnetizing reactance X_m accounts for the current needed to set up the rotating stator field. Except for changes in notation, these parts of the equivalent circuit are the same as the transformer equivalent circuit.

Phase versus Line Quantities

The voltage V_s across each winding and current I_s through each winding shown in Figure 17.13 are called the **phase voltage** and **phase current**, respectively.

The windings of an induction motor may be connected in either a delta or a wye. In the case of a delta connection, the phase voltage V_s is the same as the line-to-line voltage V_{line}. The line current I_{line} is $\sqrt{3}$ times the phase current I_s. (This is explained in Section 5.7.) In equation form for the delta connection, we have

Relationships between line and phase quantities for a delta-connected motor.

$$V_s = V_{\text{line}}$$

$$I_{\text{line}} = I_s\sqrt{3}$$

On the other hand, for a wye connection, we get

Relationships between line and phase quantities for a wye-connected motor.

$$V_s = \frac{V_{\text{line}}}{\sqrt{3}}$$

$$I_{\text{line}} = I_s$$

The voltage rating stated for a machine is invariably the line-to-line voltage.

The voltage rating stated for a machine is invariably the line-to-line voltage. For a given three-phase source, the voltage across the windings is smaller by a factor of $\sqrt{3}$ for the wye connection.

We will see that the starting current for induction motors can be very large compared to the full-load running current. Sometimes, motors are started in the wye configuration and then switched to delta as the motor approaches its running speed to lessen starting currents.

Power and Torque Calculations

In Figure 17.13, notice that we have split the reflected resistance R'_r/s into two parts as follows:

$$\frac{R'_r}{s} = R'_r + \frac{1-s}{s}R'_r \tag{17.22}$$

A power-flow diagram for induction motors is also shown in Figure 17.13. The power delivered to the resistance $[(1-s)/s]R'_r$ is the part that is converted to mechanical form. This portion of the power, called the **developed power**, is denoted by P_{dev}. The equivalent circuit shown in Figure 17.13 represents one of three phases, so the total developed power is

$$P_{\text{dev}} = 3 \times \frac{1-s}{s}R'_r(I'_r)^2 \tag{17.23}$$

On the other hand, the power delivered to the rotor resistance R'_r is converted to heat. Generally, we refer to I^2R losses as **copper losses** (even though the conductors are sometimes aluminum). The total copper loss in the rotor is

$$P_r = 3R'_r(I'_r)^2 \tag{17.24}$$

and the stator copper loss is

$$P_s = 3R_sI_s^2 \tag{17.25}$$

The input power from the three-phase source is

$$P_{\text{in}} = 3I_sV_s\cos(\theta) \tag{17.26}$$

in which $\cos(\theta)$ is the power factor.

Part of the developed power is lost to friction and windage. Another loss is core loss due to hysteresis and eddy currents. Sometimes, a resistance is included in parallel with the magnetization reactance jX_m to account for core loss. However, we will include the core loss with the rotational losses. Unless stated otherwise, we assume that the rotational power loss is proportional to speed. The output power is the developed power minus the rotational loss:

$$P_{\text{out}} = P_{\text{dev}} - P_{\text{rot}} \tag{17.27}$$

As usual, the efficiency of the machine is given by

$$\eta = \frac{P_{\text{out}}}{P_{\text{in}}} \times 100\%$$

The developed torque is

$$T_{dev} = \frac{P_{dev}}{\omega_m} \quad (17.28)$$

The power P_{ag} that crosses the air gap into the rotor is delivered to the rotor resistances. Thus, we can find the air-gap power by adding the respective sides of Equations 17.23 and 17.24:

$$P_{ag} = P_r + P_{dev} \quad (17.29)$$

$$P_{ag} = 3R'_r(I'_r)^2 + 3 \times \frac{1-s}{s} R'_r(I'_r)^2 \quad (17.30)$$

$$P_{ag} = 3 \times \frac{1}{s} R'_r(I'_r)^2 \quad (17.31)$$

Comparing Equations 17.23 and 17.31, we have

$$P_{dev} = (1-s)P_{ag} \quad (17.32)$$

Using Equation 17.32 to substitute for P_{dev} in Equation 17.28, we get

$$T_{dev} = \frac{(1-s)P_{ag}}{\omega_m} \quad (17.33)$$

However, we also have $\omega_m = (1-s)\omega_s$. Using this to substitute into Equation 17.33, we obtain

$$T_{dev} = \frac{P_{ag}}{\omega_s} \quad (17.34)$$

For speed to increase from a standing start, the initial torque or starting torque produced by the motor must be larger than the torque required by the load. We can find starting torque as follows. Under starting conditions (i.e., $\omega_m = 0$), we have $s = 1$ and $P_{ag} = 3R'_r(I'_r)^2$. Then, the starting torque can be computed by using Equation 17.34.

Equation 17.34 can be used to compute starting torque.

Example 17.1 **Induction-Motor Performance**

A certain 30-hp four-pole 440 V rms 60-Hz three-phase delta-connected induction motor has

$$R_s = 1.2\ \Omega \qquad R'_r = 0.6\ \Omega$$
$$X_s = 2.0\ \Omega \qquad X'_r = 0.8\ \Omega$$
$$X_m = 50\ \Omega$$

Under load, the machine operates at 1746 rpm and has rotational losses of 900 W. Find the power factor, the line current, the output power, copper losses, output torque, and efficiency.

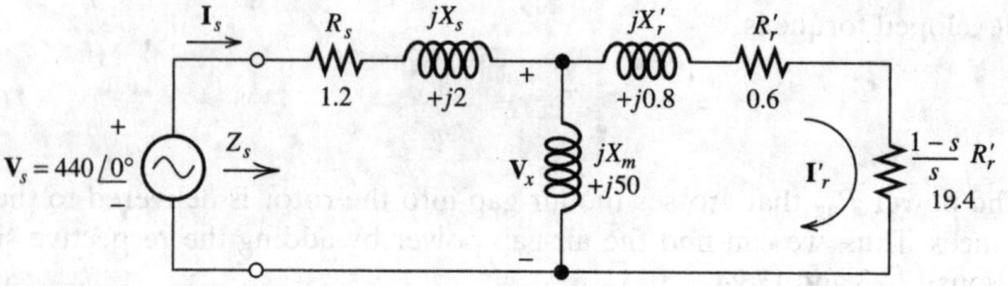

Figure 17.14 Equivalent circuit for one phase of the motor of Example 17.1.

Solution From Table 17.1, we find that synchronous speed for a four-pole motor is $n_s = 1800$ rpm. Then, we utilize Equation 17.16 to compute the slip:

$$s = \frac{n_s - n_m}{n_s} = \frac{1800 - 1746}{1800} = 0.03$$

We can use the data given to draw the equivalent circuit shown in Figure 17.14 for one phase of the motor. The impedance seen by the source is

$$Z_s = 1.2 + j2 + \frac{j50(0.6 + 19.4 + j0.8)}{j50 + 0.6 + 19.4 + j0.8}$$

$$= 1.2 + j2 + 16.77 + j7.392$$

$$= 17.97 + j9.392$$

$$= 20.28\angle 27.59° \ \Omega$$

The power factor is the cosine of the impedance angle. Because the impedance is inductive, we know that the power factor is lagging:

$$\text{power factor} = \cos(27.59°) = 88.63\% \text{ lagging}$$

For a delta-connected machine, the phase voltage is equal to the line voltage, which is specified to be 440 V rms. The phase current is

$$\mathbf{I}_s = \frac{\mathbf{V}_s}{Z_s} = \frac{440\angle 0°}{20.28\angle 27.59°} = 21.70\angle -27.59° \text{ A rms}$$

Thus, the magnitude of the line current is

$$I_{\text{line}} = I_s\sqrt{3} = 21.70\sqrt{3} = 37.59 \text{ A rms}$$

In ac machine calculations, we take the rms values of currents and voltages for the phasor magnitudes (instead of peak values as we have done previously).

The input power is

$$P_{\text{in}} = 3I_sV_s\cos\theta$$

$$= 3(21.70)440\cos(27.59°)$$

$$= 25.38 \text{ kW}$$

Next, we compute $\mathbf{V}_x$ and $\mathbf{I}'_r$:

$$\begin{aligned}\mathbf{V}_x &= \mathbf{I}_s \frac{j50(0.6 + 19.4 + j0.8)}{j50 + 0.6 + 19.4 + j0.8} \\ &= 21.70\underline{/-27.59^\circ} \times 18.33\underline{/23.78^\circ} \\ &= 397.8\underline{/-3.807^\circ} \text{ V rms} \\ \mathbf{I}'_r &= \frac{\mathbf{V}_x}{j0.8 + 0.6 + 19.4} \\ &= \frac{397.8\underline{/-3.807^\circ}}{20.01\underline{/1.718^\circ}} \\ &= 19.88\underline{/-5.52^\circ} \text{ A rms}\end{aligned}$$

The copper losses in the stator and rotor are

$$\begin{aligned}P_s &= 3R_s I_s^2 \\ &= 3(1.2)(21.70)^2 \\ &= 1695 \text{ W}\end{aligned}$$

and

$$\begin{aligned}P_r &= 3R'_r (I'_r)^2 \\ &= 3(0.6)(19.88)^2 \\ &= 711.4 \text{ W}\end{aligned}$$

Finally, the developed power is

$$\begin{aligned}P_{\text{dev}} &= 3 \times \frac{1-s}{s} R'_r (I'_r)^2 \\ &= 3(19.4)(19.88)^2 \\ &= 23.00 \text{ kW}\end{aligned}$$

As a check, we note that

$$P_{\text{in}} = P_{\text{dev}} + P_s + P_r$$

to within rounding error.

The output power is the developed power minus the rotational loss, given by

$$\begin{aligned}P_{\text{out}} &= P_{\text{dev}} - P_{\text{rot}} \\ &= 23.00 - 0.900 \\ &= 22.1 \text{ kW}\end{aligned}$$

This corresponds to 29.62 hp, so the motor is operating at nearly its rated load. The output torque is

$$\begin{aligned}T_{\text{out}} &= \frac{P_{\text{out}}}{\omega_m} \\ &= \frac{22{,}100}{1746(2\pi/60)} \\ &= 120.9 \text{ Nm}\end{aligned}$$

The efficiency is

$$\eta = \frac{P_{\text{out}}}{P_{\text{in}}} \times 100\%$$
$$= \frac{22{,}100}{25{,}380} \times 100\%$$
$$= 87.0\%$$

■

Example 17.2 Starting Current and Torque

Calculate the starting line current and torque for the motor of Example 17.1.

Solution For starting from a standstill, we have $s = 1$. The equivalent circuit is shown in Figure 17.15(a). Combining the impedances to the right of the dashed line, we have

$$Z_{\text{eq}} = R_{\text{eq}} + jX_{\text{eq}} = \frac{j50(0.6 + j0.8)}{j50 + 0.6 + j0.8} = 0.5812 + j0.7943\ \Omega$$

The circuit with the combined impedances is shown in Figure 17.15(b).

The impedance seen by the source is

$$Z_s = 1.2 + j2 + Z_{\text{eq}}$$
$$= 1.2 + j2 + 0.5812 + j0.7943$$
$$= 1.7812 + j2.7943$$
$$= 3.314\angle 57.48^\circ\ \Omega$$

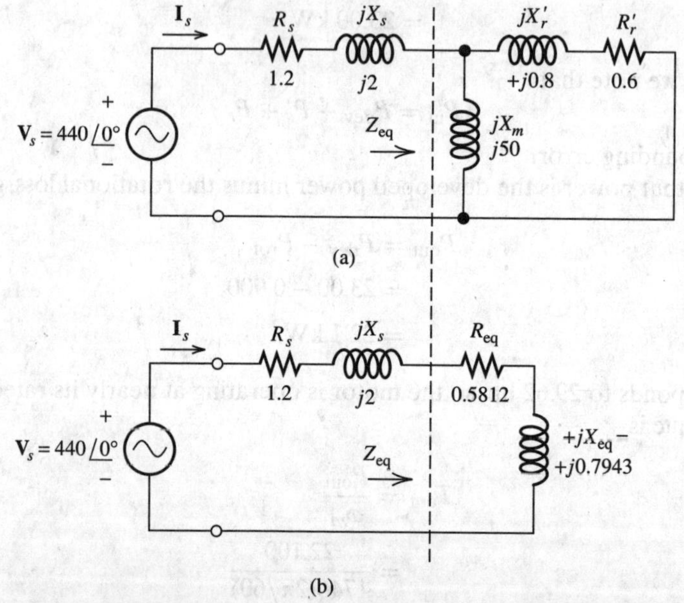

Figure 17.15 Equivalent circuit for Example 17.2.

Thus, the starting phase current is

$$\mathbf{I}_{s,\text{ starting}} = \frac{\mathbf{V}_s}{Z_s} = \frac{440\angle 0^\circ}{3.314\angle 57.48^\circ}$$
$$= 132.8\angle -57.48^\circ \text{ A rms}$$

and, because the motor is delta connected, the starting-line-current magnitude is

$$I_{\text{line, starting}} = \sqrt{3} I_{s,\text{ starting}} = 230.0 \text{ A rms}$$

In Example 17.1, with the motor running under nearly a full load, the line current is $I_{\text{line}} = 37.59$ A. Thus, the starting current is approximately six times larger than the full-load running current. This is typical of induction motors.

The power crossing the air gap is three times the power delivered to the right of the dashed line in Figure 17.15, given by

$$P_{\text{ag}} = 3R_{\text{eq}}(I_{s,\text{ starting}})^2$$
$$= 30.75 \text{ kW}$$

Finally, Equation 17.34 gives us the starting torque:

$$T_{\text{dev, starting}} = \frac{P_{\text{ag}}}{\omega_s}$$
$$= \frac{30{,}750}{2\pi(60)/2}$$
$$= 163.1 \text{ Nm}$$

Notice that the starting torque is larger than the torque while running under full-load conditions. This is also typical of induction motors. ■

Example 17.3 Induction-Motor Performance

A 220-V-rms 60-Hz three-phase wye-connected induction motor draws 31.87 A at a power factor of 75 percent lagging. For all three phases, the total stator copper losses are 400 W, and the total rotor copper losses are 150 W. The rotational losses are 500 W. Find the power crossing the air gap P_{ag}, the developed power P_{dev}, the output power P_{out}, and the efficiency.

Solution The phase voltage is $V_s = V_{\text{line}}/\sqrt{3} = 127.0$ V rms. Next, we find the input power:

$$P_{\text{in}} = 3V_s I_s \cos(\theta)$$
$$= 3(127)(31.87)(0.75)$$
$$= 9107 \text{ W}$$

The air-gap power is the input power minus the stator copper loss:

$$P_{\text{ag}} = P_{\text{in}} - P_s$$
$$= 9107 - 400$$
$$= 8707 \text{ W}$$

The developed power is the input power minus the copper losses:

$$P_{\text{dev}} = 9107 - 400 - 150 = 8557 \text{ W}$$

Next, by subtracting the rotational losses we find that the output power is

$$\begin{aligned} P_{\text{out}} &= P_{\text{dev}} - P_{\text{rot}} \\ &= 8557 - 500 \\ &= 8057 \text{ W} \end{aligned}$$

Finally, the efficiency is

$$\begin{aligned} \eta &= \frac{P_{\text{out}}}{P_{\text{in}}} \times 100\% \\ &= 94.0\% \end{aligned}$$

■

Exercise 17.3 Repeat Example 17.1 for a running speed of 1764 rpm.
Answer $s = 0.02$; power factor $= 82.62\%$; $P_{\text{in}} = 17.43$ kW; $P_{\text{out}} = 15.27$ kW; $P_s = 919$ W; $P_r = 330$ W; $T_{\text{out}} = 82.66$ Nm; $\eta = 87.61\%$. □

Exercise 17.4 Repeat Example 17.2 if the rotor resistance is increased to 1.2 Ω. Compare the starting torque with the value found in the example.
Answer $\mathbf{I}_{s,\text{ starting}} = 119.7\angle{-50°}$; $T_{\text{dev, starting}} = 265.0$ Nm. □

Wound-Rotor Induction Machine

A variation of the induction motor is the **wound-rotor machine.** The stator is identical to that of a squirrel-cage motor. Instead of a cast aluminum cage, the rotor contains a set of three-phase coils that are placed in slots. The windings are configured to produce the same number of poles on the rotor as on the stator. The windings are usually wye connected and the three terminals are brought out to external terminals through slip rings.

The results of Exercise 17.4 show that the starting torque of an induction motor can be increased by increasing the rotor resistance. By using a set of variable resistances connected to the rotor terminals, the torque–speed characteristic of the machine can be modified as illustrated in Figure 17.16. A degree of speed control can be achieved by varying the resistances. However, efficiency becomes poorer as the resistance is increased.

A disadvantage of the wound-rotor machine is that it is more expensive and less rugged than the cage machine.

Selection of Induction Motors

Some of the most important considerations in selecting an induction motor are

1. Efficiency
2. Starting torque
3. Pull-out torque
4. Power factor
5. Starting current

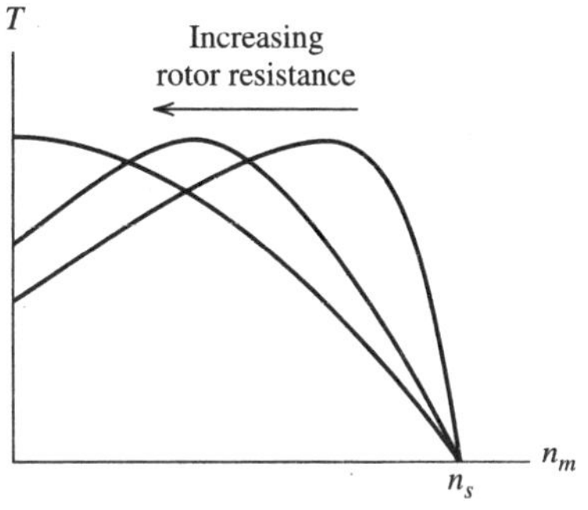

Figure 17.16 Variation of resistance in series with the rotor windings changes the torque–speed characteristic of the wound-rotor machine.

High values for the first four factors and low starting current are generally most desirable. Unfortunately, it is not possible to design a motor having the most desirable values for all of these criteria. It turns out that in the design of a motor, various tradeoffs between these criteria must be made. For example, higher rotor resistance leads to lower efficiency and higher starting torque. Larger leakage reactance X_s leads to lower starting current but poorer power factor. The design engineer must consider the various motors available and select the one that best meets the needs of the application at hand.

17.3 SYNCHRONOUS MACHINES

In this section, we discuss synchronous ac machines. These machines are used for nearly all electrical-energy generation by utility companies. As motors, they tend to be used in higher-power, lower-speed applications than those for which induction motors are used. Unlike other types of ac and dc motors that we have studied to this point, the speed of a synchronous motor does not vary with mechanical load (assuming a constant-frequency ac source). Instead, we will see that they run at synchronous speed ω_s, which is given by Equation 17.13, repeated here for convenience:

Generation of electrical energy by utility companies is done almost exclusively with synchronous machines.

Assuming a constant frequency source, the speed of a synchronous motor does not vary with load.

$$\omega_s = \frac{\omega}{P/2}$$

(Recall that ω is the angular frequency of the ac source and P is the number of magnetic poles of the stator or rotor.) Unless stated otherwise, we assume that the rotor is turning at synchronous speed throughout our discussion of synchronous machines.

The stator of a synchronous machine has the same construction as the stator of a three-phase induction motor, which was described in Section 17.1. In review, the stator contains a set of three-phase windings that establish the stator field. This field consists of P magnetic poles, alternating between north and south around the circumference of the stator and rotating at synchronous speed. In a synchronous machine, the set of stator windings is called the **armature**.

The stator windings of a synchronous machine are basically the same as those of an induction machine.

The rotor of a synchronous machine is usually a P-pole electromagnet with **field windings** that carry dc currents. (In smaller machines, the rotor can be a permanent magnet, but we will concentrate on machines with field windings.) The field current can be supplied from an external dc source through stationary brushes to **slip rings** mounted on the shaft. The slip rings are insulated from one another and from the shaft. Another method is to place a small ac generator, known as an **exciter**, on the

The rotor of a synchronous machine is a P-pole electromagnet or (in low-power machines) a permanent magnet.

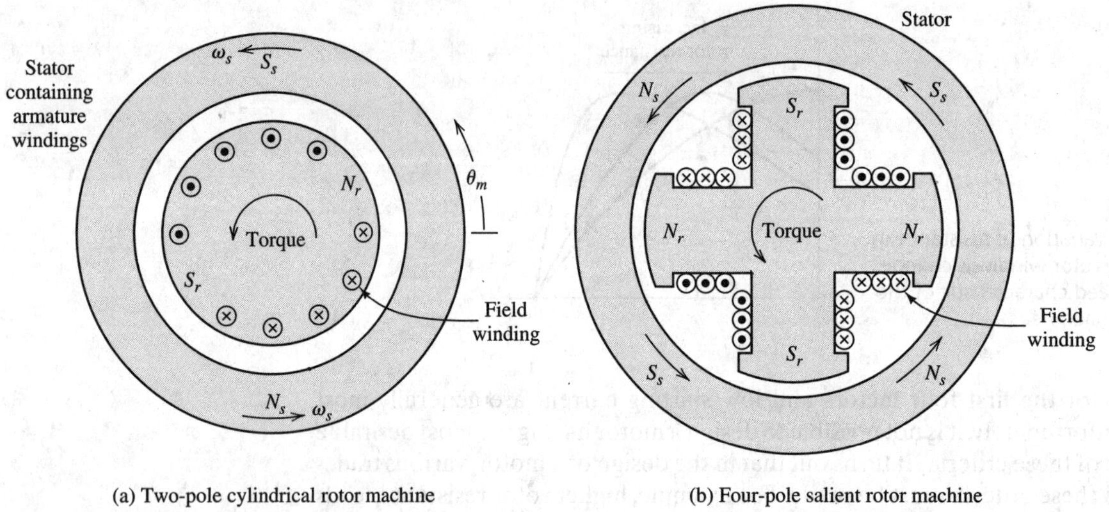

Figure 17.17 Cross-sections of two synchronous machines. The relative positions of the stator and rotor poles are shown for motor action. Torque is developed in the direction of rotation because the rotor poles try to align themselves with the opposite stator poles.

same shaft and use diodes mounted on the shaft to rectify the ac. This avoids the maintenance associated with brushes and slip rings.

Two- and four-pole synchronous machines are illustrated in Figure 17.17. The rotor can either be **cylindrical**, as shown for the two-pole machine, or it can have **salient poles** as illustrated for the four-pole machine. Generally, salient-pole construction is less costly but is limited to low-speed machines having many poles. High-speed machines usually have cylindrical rotors. Salient-pole machines are common in hydroelectric power generation, whereas cylindrical-rotor machines are common in thermal (coal, nuclear, etc.) power plants.

Automobile Alternator

The alternators found in most automobiles are basically synchronous machines, except that the armature is not connected to an independent ac source. Therefore, the speed of the alternator is not fixed. As the rotor spins, the rotating field cuts the armature conductors, inducing a set of ac voltages. The ac armature voltages are rectified, and the resulting dc is used to power the headlights, charge the battery, and so on. The frequency and amplitude of the ac voltages increase with speed. The amplitude of the induced ac armature voltages is proportional to the flux density, which in turn depends on the field current. An electronic control circuit (or regulator) varies the field current to maintain approximately 14 V dc at the output of the rectifier.

Motor Action

In using the machine as a motor, the armature is connected to a three-phase ac source. We have seen that the resulting three-phase currents in the armature windings set up a rotating stator field. The rotor turns at synchronous speed with the rotor poles lagging behind the stator poles. Torque is produced because the rotor poles attempt to align with the stator poles. This is illustrated in Figure 17.17.

Electrical Angles

We denote the angular displacement around the air gap as θ_m, which is illustrated in Figure 17.17(a). Sometimes, it is convenient to measure angular displacements in **electrical degrees**, for which 180° corresponds to the angular distance from a north pole to the adjacent south pole. Thus, a four-pole machine has 720 electrical degrees around the circumference of its air gap, a two-pole machine has 360 electrical degrees, and a six-pole machine has 3 × 360 degrees. We denote displacement in electrical degrees as θ_e. Electrical and mechanical angular displacements are related by

Angles can be measured in electrical degrees for which the angle between adjacent north and south poles is 180°.

$$\theta_e = \theta_m \frac{P}{2} \tag{17.35}$$

Field Components

The total rotating field in the air gap is due partly to dc current in the field windings on the rotor and partly to ac currents flowing in the armature windings. The air-gap flux components are functions of both time and angular displacement. The field lines cross perpendicular to the gap, because that is the path of least reluctance. Thus, at any given point, the field is directed perpendicular to the armature conductors, which lie in slots cut lengthwise into the inside face of the stator.

The total field rotating in the air gap is partly due to the dc currents in the rotor windings and partly due to the ac currents in the stator (armature) windings.

Most synchronous machines are designed so that the flux density varies sinusoidally with θ_m. Because the field rotates at a uniform rate, the flux density at any point in the gap varies sinusoidally with time. Thus, we can represent the field components at $\theta_m = 0$ by phasors denoted by $\mathbf{B}_s$, $\mathbf{B}_r$, and $\mathbf{B}_{\text{total}}$, which correspond to the stator flux component, rotor flux component, and total flux, respectively. Then, we can write

Synchronous machines are designed so the flux varies sinusoidally around the air gap.

$$\mathbf{B}_{\text{total}} = \mathbf{B}_s + \mathbf{B}_r \tag{17.36}$$

The torque developed in the rotor is given by

$$T_{\text{dev}} = KB_rB_{\text{total}} \sin(\delta) \tag{17.37}$$

in which K is a constant that depends on the dimensions and other features of the machine. B_{total} and B_r are the magnitudes of the phasors $\mathbf{B}_{\text{total}}$ and $\mathbf{B}_r$, respectively. δ is the electrical angle, called the **torque angle**, by which the rotor field lags the total field.

Equivalent Circuit

The rotating field components induce corresponding voltage components in the armature windings. We concentrate on the a phase of the armature winding. The voltages and currents in the other two armature windings are identical except for phase shifts of ±120°.

The voltage component induced by the rotor flux can be represented as a phasor that is given by

$$\mathbf{E}_r = k\mathbf{B}_r \tag{17.38}$$

in which k is a constant that depends on the machine construction features.

A second voltage component is induced in each winding by the rotating stator field. This voltage component is given by

$$\mathbf{E}_s = k\mathbf{B}_s \tag{17.39}$$

As we have seen, the stator field is established by the armature currents. The stator is a mutually coupled three-phase inductor, and the voltage due to the stator field can be written as

$$\mathbf{E}_s = jX_s\mathbf{I}_a \tag{17.40}$$

where X_s is an inductive reactance known as the **synchronous reactance**, and $\mathbf{I}_a$ is the phasor for the armature current. [Actually, the stator windings also have resistance, and more precisely, we have $\mathbf{E}_s = (R_a + jX_s)\mathbf{I}_a$. However, the resistance R_a is usually very small compared to the synchronous reactance, so Equation 17.40 is sufficiently accurate.]

$\mathbf{V}_a$ and $\mathbf{I}_a$ represent the rms phase voltage and phase current, respectively. The relationship to line voltage and line current depends on whether the machine is wye or delta connected.

The voltage observed at the terminals of the armature winding is the sum of these two components. Thus, we can write

$$\mathbf{V}_a = \mathbf{E}_r + \mathbf{E}_s \tag{17.41}$$

where $\mathbf{V}_a$ is the phasor for the terminal voltage for the a-phase winding. Using Equation 17.40 to substitute for $\mathbf{E}_s$, we have

$$\mathbf{V}_a = \mathbf{E}_r + jX_s\mathbf{I}_a \tag{17.42}$$

Throughout our discussion, we assume that the phase angle of $\mathbf{V}_a$ is 0.

Also, we can write

$$\mathbf{V}_a = k\mathbf{B}_{\text{total}} \tag{17.43}$$

because the total voltage is proportional to the total flux.

The equivalent circuit of the synchronous motor is shown in Figure 17.18. Only the a phase of the armature is shown. The three-phase source $\mathbf{V}_a$ supplies current $\mathbf{I}_a$ to the armature. The ac voltage induced in the armature by the rotor field is represented by the voltage source $\mathbf{E}_r$. The dc voltage source V_f supplies the field current I_f to the rotor. An adjustable resistance R_{adj} is included in the field circuit so that the field current can be varied. This in turn adjusts the magnitudes of the rotor field $\mathbf{B}_r$ and the resulting induced voltage $\mathbf{E}_r$.

The armature windings can be connected either in a wye or in a delta configuration. In our discussion, we do not specify the way in which the windings are connected. For either connection, $\mathbf{V}_a$ represents the voltage across the a winding. In a wye connection, $\mathbf{V}_a$ corresponds to the line-to-neutral voltage, whereas in a delta connection, $\mathbf{V}_a$ corresponds to the line-to-line voltage. Similarly, $\mathbf{I}_a$ is the current through the a winding, which corresponds to the line current in a wye connection but not in a delta connection. The important things to remember are that $\mathbf{V}_a$ is the voltage across the a winding and that $\mathbf{I}_a$ is the current through the a winding, regardless of the manner in which the machine is connected.

The phasor diagram for the current and voltages is shown in Figure 17.19(a). The corresponding phasor diagram for the fields is shown in Figure 17.19(b). Because the

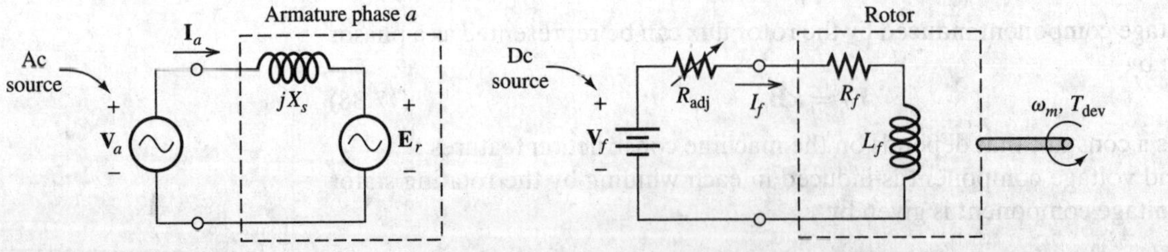

Figure 17.18 Equivalent circuit for the synchronous motor. The armature circuit is based on Equation 17.42.

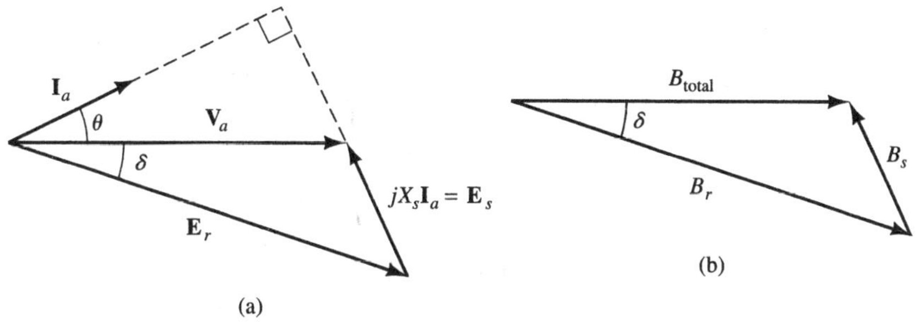

Figure 17.19 Phasor diagrams for the synchronous motor. Notice that the stator component of voltage $\mathbf{E}_s = jX_s\mathbf{I}_a$ is at right angles to the current $\mathbf{I}_a$. The developed torque is given by $T_{dev} = KB_rB_{total}\sin\delta$, and the power factor is $\cos(\theta)$.

rotor field is lagging the total field, positive torque (given by Equation 17.37) and output power are being developed. In other words, the machine is acting as a motor.

The input power taken from the three-phase ac source is given by

$$P_{dev} = P_{in} = 3V_aI_a\cos(\theta) \quad (17.44)$$

in which the factor of three accounts for the fact that there are three sets of windings. Since the equivalent circuit does not include any losses, the input power and the developed mechanical power are equal.

Potential for Power-Factor Correction

The total reactive power absorbed by the three windings is given by

$$Q = 3V_aI_a\sin(\theta) \quad (17.45)$$

in which θ is defined to be the angle by which the phase current $\mathbf{I}_a$ *lags* the phase voltage $\mathbf{V}_a$.

Notice in Figure 17.19(a) that θ takes a negative value because the phase current $\mathbf{I}_a$ *leads* the phase voltage $\mathbf{V}_a$. Therefore, the reactive power for the machine is negative, indicating that the synchronous motor can supply reactive power. This is a significant advantage because most industrial plants have an overall lagging power factor (due largely to the widespread employment of induction motors). Poor power factor leads to larger currents in the transmission lines and transformers supplying the plant. Thus, utility companies invariably charge their industrial customers more for energy supplied while the power factor is low. By using some synchronous motors in an industrial plant, part of the reactive power taken by inductive loads can be supplied locally, thereby lowering energy costs. Should you someday be employed as a plant engineer, you will need to have a good understanding of these issues.

The synchronous motor can act as a source of reactive power.

Proper use of synchronous motors can lower energy costs of an industrial plant by increasing the power factor.

Unloaded synchronous machines have sometimes been installed solely for the purpose of power-factor correction. With zero load (and neglecting losses), the rotor field and the total field align so that the torque angle δ is zero, and according to Equation 17.37, the developed torque is zero. Phasor diagrams for unloaded synchronous machines are shown in Figure 17.20.

If we have

$$V_a > E_r\cos(\delta) \quad (17.46)$$

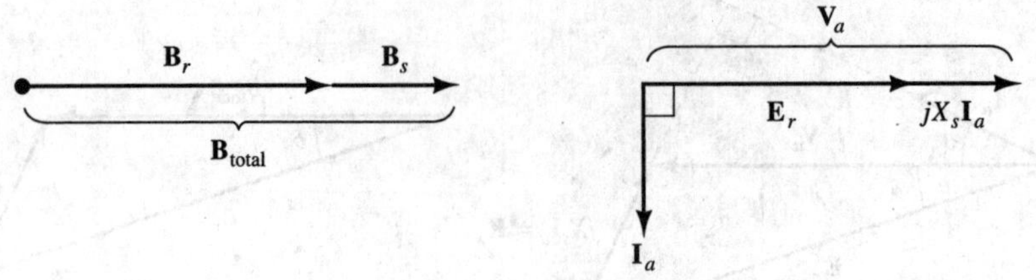

(a) Underexcited, $E_r < V_a$

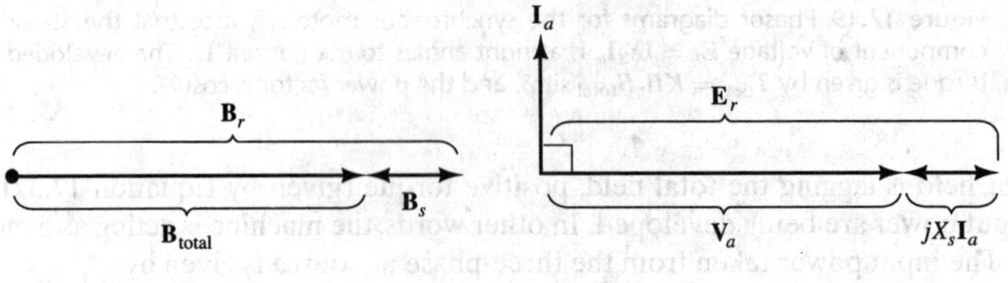

(b) Overexcited, $E_r > V_a$

Figure 17.20 Phasor diagrams for unloaded synchronous machines. When a machine has $E_r > V_a$, the current $\mathbf{I}_a$ leads the applied voltage $\mathbf{V}_a$ by 90°, and each phase of the machine is electrically equivalent to a capacitor. Thus, the machine supplies reactive power.

we say that the machine is **underexcited**. For an unloaded machine with $\delta = 0$, the machine is underexcited if the magnitude of $\mathbf{E}_r$ is less than the magnitude of the applied phase voltage $\mathbf{V}_a$. Then, the current $\mathbf{I}_a$ lags $\mathbf{V}_a$ by $\theta = 90°$. Consequently, the real power supplied (which is given by Equation 17.44) is zero, as we expect for an unloaded machine (neglecting losses). The underexcited machine absorbs reactive power. This is the opposite of the desired result for most applications.

However, if the field current is increased such that

$$V_a < E_r \cos(\delta) \tag{17.47}$$

we say that the machine is **overexcited**. The phasor diagram is shown in Figure 17.20(b) for an unloaded overexcited machine. In this case, the current leads the voltage by 90°, and the machine supplies reactive power. In the overexcited state, an unloaded synchronous machine appears as a pure capacitive reactance to the ac source. Machines used in this manner are called **synchronous capacitors**.

Operation with Variable Load and Constant Field Current

Motors are usually operated from ac voltage sources of constant magnitude and phase. This fact in combination with Equation 17.43 shows that the total flux phasor $\mathbf{B}_{\text{total}}$ is constant in magnitude and phase. Because speed is constant in a synchronous machine, power is proportional to torque, which in turn is proportional to $B_r \sin(\delta)$, as shown by Equation 17.37. Thus, we can write

$$P_{\text{dev}} \propto B_r \sin(\delta) \tag{17.48}$$

This fact is illustrated in Figure 17.21(a).

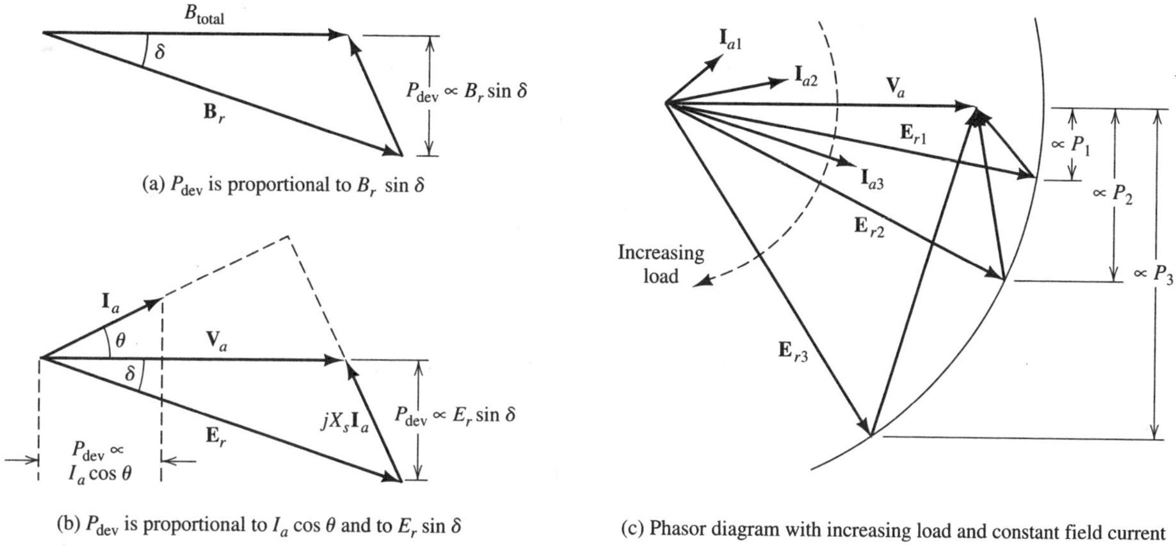

Figure 17.21 Phasor diagrams for a synchronous motor.

Furthermore, E_r is proportional to B_r. Hence, we have established that

$$P_{dev} \propto E_r \sin(\delta) \quad (17.49)$$

Since $P_{dev} = P_{in} = 3V_aI_a\cos(\theta)$ (neglecting stator copper loss) and since V_a is constant, we also have

$$P_{dev} \propto I_a \cos(\theta) \quad (17.50)$$

Equations 17.49 and 17.50 are illustrated in the phasor diagram shown in Figure 17.21(b).

Now suppose that we have a synchronous motor operating with a variable load and constant field current. Because the field current is constant, $\mathbf{E}_r$ is constant in magnitude. As the load changes, $\mathbf{E}_r$ can change in phase, but not in magnitude. Therefore, the locus formed by $\mathbf{E}_r$ is a circle. The phasor diagram for a machine with a variable load is shown in Figure 17.21(c). Notice that the power factor tends to become lagging as the load is increased.

As the load varies with constant field current, the locus of $\mathbf{E}_r$ is a circle. Notice that if $jX_s\mathbf{I}_a$ and $\mathbf{I}_a$ are extended, they meet at right angles. Careful study of Figure 7.21 shows that the power factor tends to become lagging as load is increased.

Example 17.4 Synchronous-Motor Performance

A 480-V-rms 200-hp 60-Hz eight-pole delta-connected synchronous motor operates with a developed power (including losses) of 50 hp and a power factor of 90 percent leading. The synchronous reactance is $X_s = 1.4\ \Omega$. **a.** Find the speed and developed torque. **b.** Determine the values of $\mathbf{I}_a$, $\mathbf{E}_r$, and the torque angle. **c.** Suppose that the excitation remains constant and the load torque increases until the developed power is 100 hp. Determine the new values of $\mathbf{I}_a$, $\mathbf{E}_r$, the torque angle, and the power factor.

Solution

a. The speed of the machine is given by Equation 17.14:

$$n_s = \frac{120f}{P} = \frac{120(60)}{8} = 900 \text{ rpm}$$

$$\omega_s = n_s\frac{2\pi}{60} = 30\pi = 94.25 \text{ rad/s}$$

For the first operating condition, the developed power is

$$P_{\text{dev1}} = 50 \times 746 = 37.3 \text{ kW}$$

and the developed torque is

$$T_{\text{dev1}} = \frac{P_{\text{dev1}}}{\omega_s} = \frac{37{,}300}{94.25} = 396 \text{ Nm}$$

b. The voltage rating refers to the rms line-to-line voltage. Because the windings are delta connected, we have $V_a = V_{\text{line}} = 480$ V rms. Solving Equation 17.44 for I_a and substituting values, we have

$$I_{a1} = \frac{P_{\text{dev1}}}{3V_a\cos(\theta_1)} = \frac{37{,}300}{3(480)(0.9)} = 28.78 \text{ A rms}$$

Next, the power factor is $\cos(\theta_1) = 0.9$, which yields

$$\theta_1 = 25.84°$$

Because the power factor was given as leading, we know that the phase of $\mathbf{I}_{a1}$ is positive. Thus, we have

$$\mathbf{I}_{a1} = 28.78\angle 25.84° \text{ A rms}$$

Then from Equation 17.42, we have

$$\begin{aligned}\mathbf{E}_{r1} &= \mathbf{V}_{a1} - jX_s\mathbf{I}_a = 480 - j1.4(28.78\angle 25.84°)\\ &= 497.6 - j36.3\\ &= 498.9\angle -4.168° \text{ V rms}\end{aligned}$$

Consequently, the torque angle is $\delta_1 = 4.168°$.

c. When the load torque is increased while holding excitation constant (i.e., the values of I_f, B_r, and E_r are constant), the torque angle must increase. In Figure 17.21(b), we see that the developed power is proportional to $\sin(\delta)$. Hence, we can write

$$\frac{\sin(\delta_2)}{\sin(\delta_1)} = \frac{P_2}{P_1}$$

Solving for $\sin(\delta_2)$ and substituting values, we find that

$$\sin(\delta_2) = \frac{P_2}{P_1}\sin(\delta_1) = \frac{100 \text{ hp}}{50 \text{ hp}}\sin(4.168°)$$

$$\delta_2 = 8.360°$$

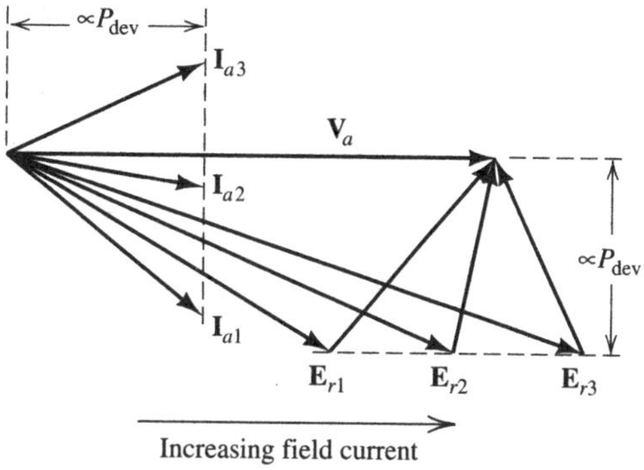

Figure 17.22 Phasor diagram for constant developed power and increasing field current.

Because E_r is constant in magnitude, we get

$$\mathbf{E}_{r2} = 498.9\angle{-8.360^\circ} \text{ V rms}$$

(We know that $\mathbf{E}_{r2}$ lags $\mathbf{V}_a = 480\angle{0^\circ}$ because the machine is acting as a motor.) Next, we can find the new current:

$$\mathbf{I}_{a2} = \frac{\mathbf{V}_a - \mathbf{E}_{r2}}{jX_s} = 52.70\angle{10.61^\circ} \text{ A rms}$$

Finally, the new power factor is

$$\cos(\theta_2) = \cos(10.61^\circ) = 98.3\% \text{ leading}$$ ■

Exercise 17.5 For the motor of Example 17.4, suppose that the excitation remains constant and the load torque increases until the developed power is $P_{dev3} = 200$ hp. Determine the new values of $\mathbf{I}_a$, $\mathbf{E}_r$, the torque angle, and the power factor.

Answer $\mathbf{I}_{a3} = 103.6\angle{-1.05^\circ}$; $\mathbf{E}_{r3} = 498.9\angle{-16.90^\circ}$; $\delta_3 = 16.90^\circ$; power factor = 99.98% lagging. □

Operation with Constant Load and Variable Field Current

When operating with constant developed power P_{dev}, Figure 17.21(b) shows that the values of $I_a\cos(\theta)$ and $E_r\sin(\delta)$ are constant. Then, if the field current increases, the magnitude of E_r increases. The resulting phasor diagram for several values of field current is shown in Figure 17.22. Notice that as the field current increases, the armature current decreases in magnitude, reaching a minimum for $\theta = 0^\circ$ (or unity power factor) and then increases with a leading power factor. The current magnitude reaches a minimum when $\mathbf{I}_a$ is in phase with $\mathbf{V}_a$ (i.e., when $\theta = 0$ and the power factor is unity). Plots of I_a versus field current are shown in Figure 17.23. These plots are called **V curves** because of their shape.

Power factor tends to become leading as field current, and consequently, $\mathbf{E}_r$ increase in magnitude.

Example 17.5 Power-Factor Control

A 480-V-rms 200-hp 60-Hz eight-pole delta-connected synchronous motor operates with a developed power (including losses) of 200 hp and a power factor of 85 percent lagging. The synchronous reactance is $X_s = 1.4\ \Omega$. The field current is $I_f = 10$ A. What must the new field current be to produce 100 percent power factor? Assume that magnetic saturation does not occur, so that B_r is proportional to I_f.

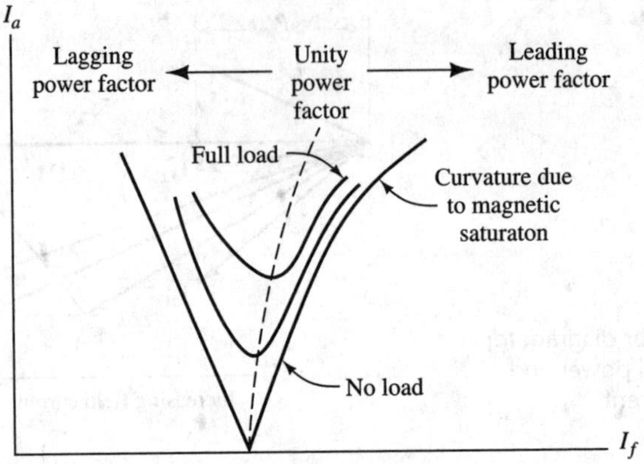

Figure 17.23 V curves for a synchronous motor with variable excitation.

Solution First, we determine the initial value of E_r. Because the initial power factor is $\cos(\theta_1) = 0.85$, we can determine that

$$\theta_1 = 31.79°$$

Then, the phase current is

$$I_{a1} = \frac{P_{\text{dev}}}{3V_a \cos(\theta_1)} = \frac{200(746)}{3(480)0.85} = 121.9 \text{ A rms}$$

Thus, the phasor current is

$$\mathbf{I}_{a1} = 121.9\angle{-31.79°} \text{ A rms}$$

The induced voltage is

$$\mathbf{E}_{r1} = \mathbf{V}_{a1} - jX_s\mathbf{I}_{a1} = 480 - j1.4(121.9\angle{-31.79°})$$
$$= 390.1 - j145.0$$
$$= 416.2\angle{-20.39°} \text{ V rms}$$

The phasor diagram for the initial excitation is shown in Figure 17.24(a).

To achieve 100 percent power factor, we need to increase the field current and the magnitude of $\mathbf{E}_r$ until $\mathbf{I}_a$ is in phase with $\mathbf{V}_a$, as shown in Figure 17.24(b). The new value of the phase current is

$$I_{a2} = \frac{P_{\text{dev}}}{3V_a \cos(\theta_2)} = \frac{200(746)}{3(480)} = 103.6 \text{ A rms}$$

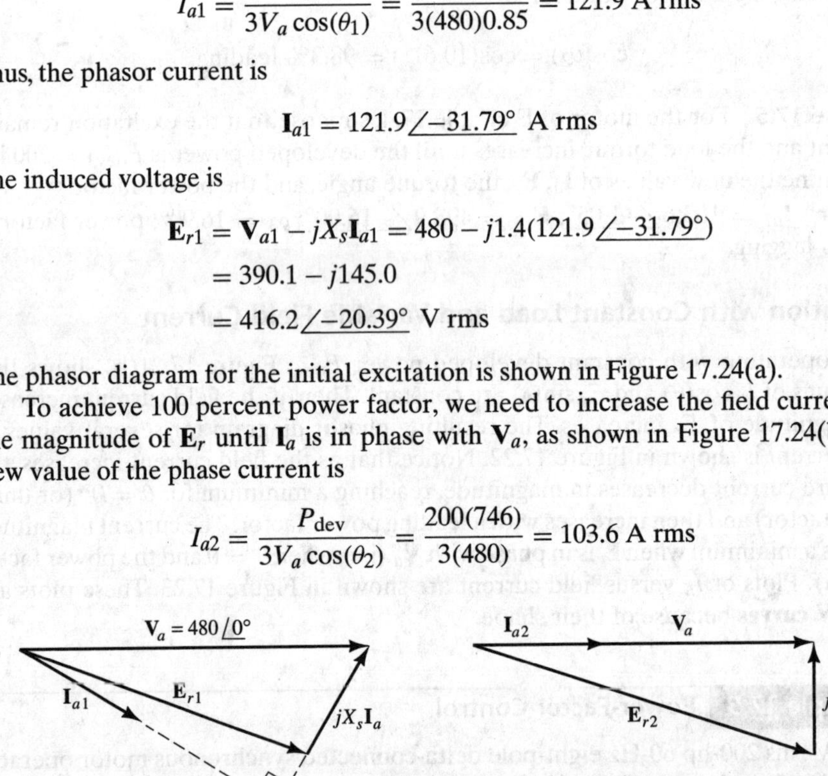

Figure 17.24 Phasor diagrams for Example 17.5.

Then, we have

$$\mathbf{E}_{r2} = \mathbf{V}_{a2} - jX_s\mathbf{I}_{a2} = 480 - j1.4(103.6)$$
$$= 480 - j145.0$$
$$= 501.4\angle{-16.81^\circ} \text{ V rms}$$

Now the magnitude of $\mathbf{E}_r$ is proportional to the field current, so we can write

$$I_{f2} = I_{f1}\frac{E_{r2}}{E_{r1}} = 10\frac{501.4}{416.2} = 12.05 \text{ A dc}$$ ■

Exercise 17.6 Find the field current needed to achieve a power factor of 90 percent leading for the motor of Example 17.5.

Answer $I_f = 13.67$ A. □

Pull-Out Torque

The developed torque of a synchronous motor is given by Equation 17.37, which states that

$$T_{\text{dev}} = KB_rB_{\text{total}}\sin(\delta)$$

This is plotted in Figure 17.25. The maximum or **pull-out torque** $T_{\max}$ occurs for a torque angle of $\delta = 90°$:

$$T_{\max} = KB_rB_{\text{total}} \quad (17.51)$$

The pull-out torque is the maximum torque that the synchronous motor can produce, which occurs for a torque angle of 90°.

Typically, the rated torque is about 30 percent of the maximum torque.

Suppose that a synchronous motor is initially unloaded. Then, it runs at synchronous speed with $\delta = 0$. As the load increases, the motor slows momentarily and δ increases just enough so that the developed torque meets the demands of the load plus losses. Then, the machine again runs at synchronous speed.

However, if the load on a synchronous machine was to exceed the pull-out torque, it would no longer be possible for the machine to drive the load at synchronous speed and δ would keep on increasing. Then, the machine would produce enormous surges in torque back and forth, resulting in great vibration. Once the rotor pulls out of synchronism with the rotating armature field, the average torque falls to zero, and the system slows to a stop.

The torque–speed characteristic of a synchronous motor is shown in Figure 17.26. Generally, it is desirable to operate synchronous motors in an overexcited state (i.e., large values of I_f, B_r, and E_r), for several reasons. First, the machine produces reactive power. Second, as shown by Equation 17.51, the pull-out torque is higher with higher B_r.

Generally, it is desirable to operate synchronous motors in the overexcited state to obtain large pull-out torque and to generate reactive power.

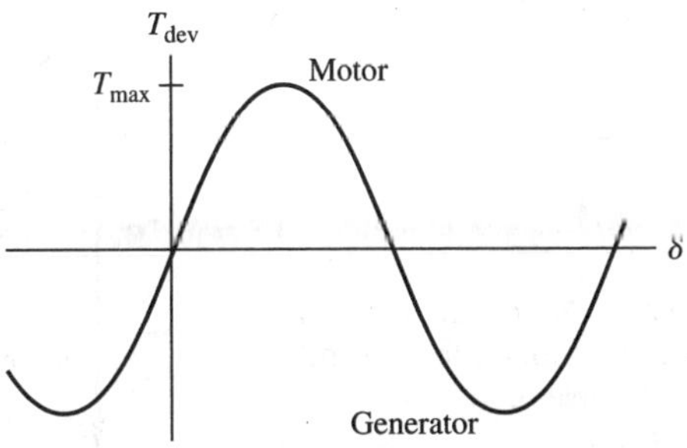

Figure 17.25 Torque versus torque angle. $T_{\max}$ is the maximum or pull-out torque.

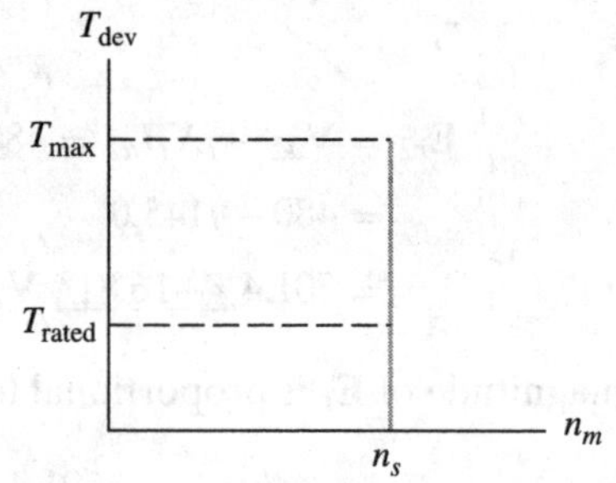

Figure 17.26 Torque–speed characteristic of synchronous motors.

Starting Methods

Because the starting torque of the synchronous motor is zero, special starting provisions are needed.

Because a synchronous motor develops zero starting torque, we need to make special provisions for starting. Several methods can be used:

1. Vary the frequency of the ac source starting very low (a fraction of a hertz) and gradually increasing to the operating speed desired. This can be accomplished with power electronic circuits known as *cycloconverters* that can convert 60-Hz ac power into three-phase power of any desired frequency. Such a system can also be used for very accurate speed control.
2. Use a prime mover to bring the synchronous motor up to speed. Then, the motor is connected to the ac source and the load is connected. Before the ac source is connected, it is important to wait until the phases of the voltages induced in the armature closely match those of the line voltages. In other words, we want the torque angle δ to be close to zero before closing the switches to the ac source. Otherwise, excessive currents and torques occur as the rotor tries to rapidly align itself with the stator field.
3. The rotors of many synchronous motors contain **amortisseur** or **damper conductors**, which are similar in structure to the squirrel-cage conductors used in induction motors. Then, the motor can be started as an induction motor with the field windings shorted and without load. After the motor has approached synchronous speed, the dc source is connected to the field and the motor pulls into synchronism. Then, the load is connected.

Damper conductors have another purpose besides use in starting. It is possible for the speed of a synchronous motor to oscillate above and below synchronous speed so that the torque angle δ swings back and forth. This action is similar to that of a pendulum. By including the damper bars, the oscillation is damped out. When running at synchronous speed, no voltage is induced in the damper bars and they have no effect.

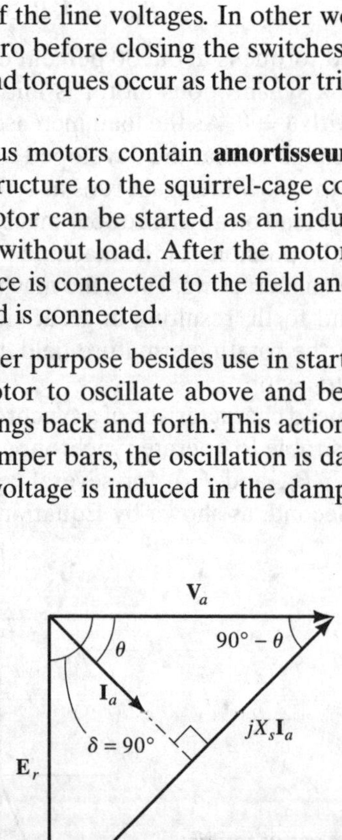

Figure 17.27 Phasor diagram for maximum developed torque and maximum power conditions. See Exercise 17.7.

Exercise 17.7 A synchronous motor produces maximum torque and maximum power for $\delta = 90°$. Draw the phasor diagram for this case and show that $P_{\max} = 3(V_a E_r / X_s)$ and $T_{\max} = 3(V_a E_r / \omega_m X_s)$.
Answer The phasor diagram is shown in Figure 17.27. □

17.4 SINGLE-PHASE MOTORS

In Chapter 16, we examined the universal motor, which can be operated from single-phase ac. In this section, we discuss briefly several additional types of single-phase ac motors. Single-phase motors are important because three-phase power is not available for homes, most offices, and many small businesses.

Compared with induction motors, universal motors have a higher power/weight ratio, but they do not have as long a service life, due to wear of the brushes. Assuming constant source frequency, induction motors are essentially constant-speed devices. On the other hand, the speed of a universal motor can be varied by changing the amplitude of the applied voltage.

Universal motors have relatively large power-to-weight ratios but short service lives.

Basic Single-Phase Induction Motor

Let us begin by considering the basic single-phase induction motor shown in Figure 17.28. The stator of this motor has a **main winding** that is connected to an ac source. (Later, we will see that an auxiliary winding is needed for starting.) It has a squirrel-cage rotor that is identical to the rotor of the three-phase induction motor shown in Figure 17.6.

Ideally, the air-gap flux varies sinusoidally in space around the circumference of the gap. Thus, the flux is given by

$$B = Ki(t)\cos(\theta) \quad (17.52)$$

which is the same, except for changes in notation, as Equation 17.1 for the flux due to winding a of a three-phase induction motor. The stator current is given by

$$i(t) = I_m \cos(\omega t) \quad (17.53)$$

Substituting this expression for the current into Equation 17.52, we have

$$B = KI_m \cos(\omega t)\cos(\theta) \quad (17.54)$$

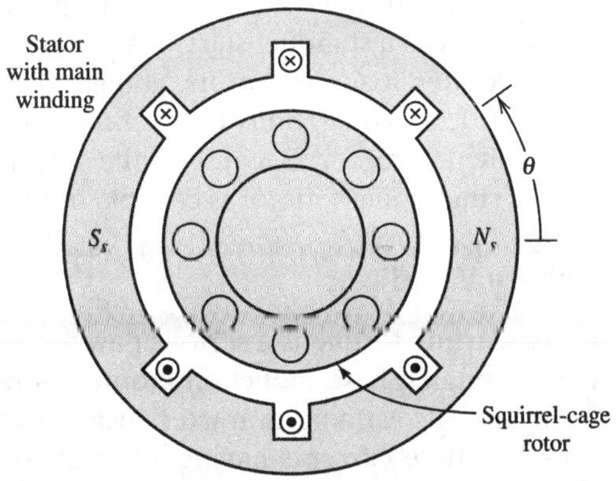

Figure 17.28 Cross-section of the basic single-phase induction motor.

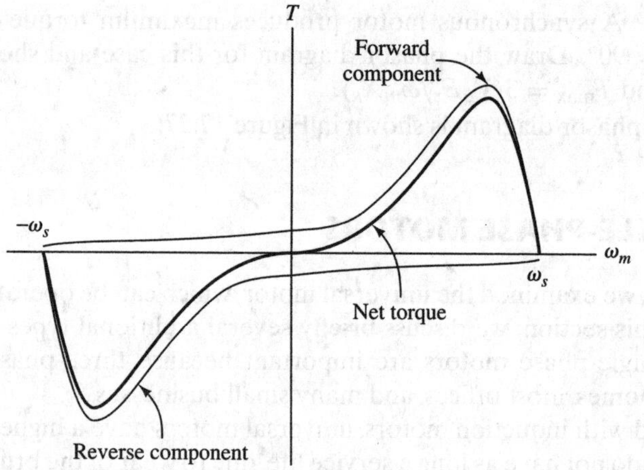

Figure 17.29 The main winding produces two counter-rotating flux components each of which induces torque in the rotor. The main winding alone induces no net starting torque.

Instead of rotating, this flux pulsates, switching direction twice per cycle.

However, by applying the trigonometric identity for the product of cosines, we can write Equation 17.54 as

$$B = \frac{1}{2}KI_m \cos(\omega t - \theta) + \frac{1}{2}KI_m \cos(\omega t + \theta) \tag{17.55}$$

The pulsating flux produced by the main winding can be resolved into two counter rotating components.

The first term on the right-hand side of Equation 17.55 represents a flux that rotates counterclockwise (i.e., in the positive θ direction), while the second term rotates clockwise. Thus, the pulsating flux in the basic single-phase induction motor can be resolved into two counter rotating components. On the other hand, the three-phase motor has flux rotating in one direction only.

We assume that the rotor spins counterclockwise with speed ω_m. The field component that rotates in the same direction as the rotor is called the **forward component**. The other component is called the **reverse component**. Each of these components produces torque, but in opposite directions. The torque versus speed characteristic for each component is similar to that of a three-phase induction motor. The torques produced by the forward component, the reverse component, and the total torque are shown in Figure 17.29.

Notice that the net starting torque is zero, and therefore the main winding will not start a load from a standing start. Once started, however, the motor develops torque and accelerates loads within its ratings to nearly synchronous speed. Its running characteristics (in the vicinity of synchronous speed) are similar to those of the three-phase induction motor. Because of the symmetry of its torque–speed characteristic, the basic single-phase motor is capable of running equally well in either direction.

Auxiliary Windings

Two windings that are 90° apart physically and carry currents 90° apart in phase produce a rotating magnetic field.

Lack of starting torque is a serious flaw for a motor in most applications. However, the basic single-phase induction motor can be modified to provide starting torque and improve its running characteristics. It can be shown (see Problem P17.11) that equal-amplitude currents having a 90° phase relationship and flowing in windings

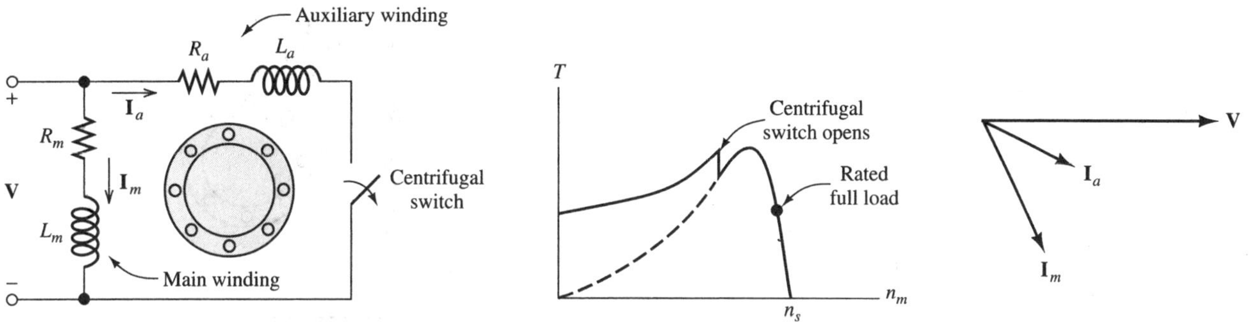

Figure 17.30 The split-phase induction motor.

that are at right angles produce only a forward rotating component of flux. (This is similar to the rotating flux produced by balanced three-phase currents flowing in windings displaced by 120° from one another.) If the two currents differ in phase by less than 90° (but by more than 0°), the forward flux component is larger than the reverse flux component and net starting torque results. Thus, nearly all single-phase induction motors have an **auxiliary winding** rotated in space by 90 electrical degrees from the main winding. Various provisions can be made to achieve the requisite phase shift between the current in the main winding and the current in the auxiliary winding.

Single-phase induction motors contain an auxiliary winding displaced by 90 electrical degrees from the main winding.

One possibility is to wind the auxiliary winding with smaller wire that has a higher ratio of resistance to inductive reactance than the main winding. Then, the current in the auxiliary winding has a different phase angle than that of the main current. Motors using this approach are called **split-phase motors** (Figure 17.30). Usually, the auxiliary winding is designed to be used only briefly during starting, and a centrifugal switch disconnects it when the motor approaches rated speed. (A common failure in this type of motor is for the switch to fail to open, and then the auxiliary winding overheats and burns out.)

When running on the main winding, the torque of a single-phase motor pulsates at twice the frequency of the ac source, because no torque is produced when the stator current passes through zero. On the other hand, torque is constant in a three-phase motor because the current is nonzero in at least two of the three windings at all instants. Thus, single-phase induction motors display more noise and vibration than do three-phase motors. Furthermore, single-phase induction motors are larger and heavier than three-phase motors of the same ratings.

Single-phase induction motors produce more noise and vibration and are larger than three-phase motors with equal power ratings.

In a **capacitor-start motor**, a capacitor is placed in series with the auxiliary winding, resulting in much higher starting torque than that of the split-phase motor, because the phase relationship between $\mathbf{I}_a$ and $\mathbf{I}_m$ is closer to 90°. In a **capacitor-run motor**, the auxiliary winding is a permanent part of the circuit, resulting in smoother torque and less vibration. Another variation is the **capacitor-start, capacitor-run motor** shown in Figure 17.31.

Shaded-Pole Motors

The least expensive approach to providing self-starting for single-phase induction motors is the **shaded-pole motor**, shown in Figure 17.32. A shorted copper band is placed around part of each pole face. As the field builds up, current is induced in this *shading ring*. The current retards changes in the field for that part of the pole face encircled by the ring. As the current in the ring decays, the center of the magnetic

Shaded-pole motors are used for inexpensive low-power applications.

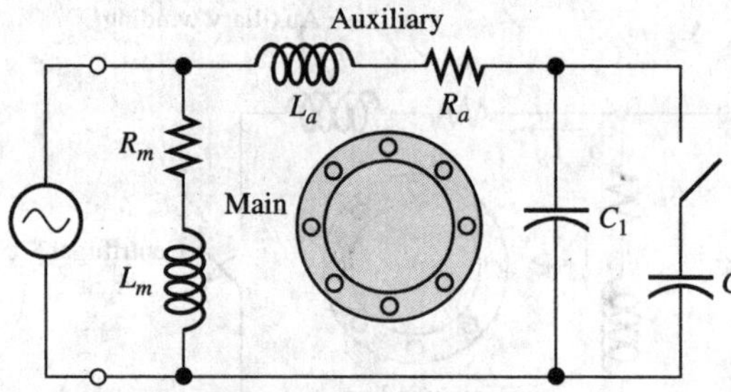

Figure 17.31 The capacitor-start, capacitor-run motor.

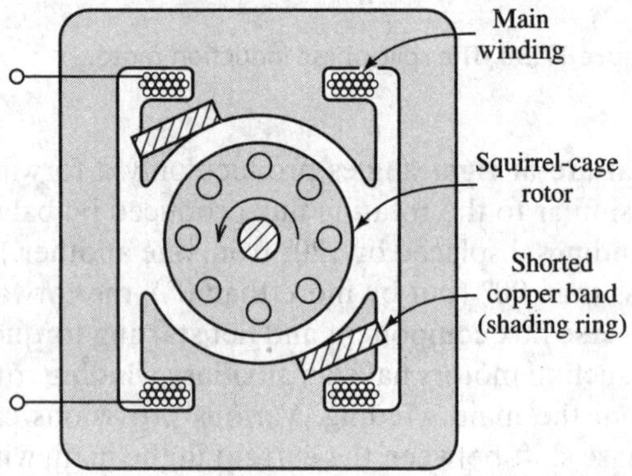

Figure 17.32 The shaded-pole motor.

pole moves in the direction of the ring. This favors rotation in one direction over the other, resulting in starting torque. This approach is used only for very small motors (1/20 hp or less).

17.5 STEPPER MOTORS AND BRUSHLESS DC MOTORS

Anytime you need accurate repeatable positioning, consider using a stepper motor.

Stepper motors are used for accurate, repeatable positioning such as machine tool applications or for moving the head in an ink-jet printer. By using an electronic controller that applies electrical pulses to the motor windings, the motor shaft can be rotated in either direction in multiples of the step angle, which can range from 0.72° (500 steps per revolution) to 15° (24 steps per revolution). Stepper motors are available with rotational accuracies on the order of 3 percent of a step, which is noncumulative as the motor is stepped back and forth. By controlling the rate at which pulses are applied to the windings of the stepper motor, speed can be varied continuously from a standing stop to a maximum that depends on the motor and load.

There are several types of stepper motors. Figure 17.33(a) shows the cross-section of the simplest, which is known as a **variable-reluctance stepper motor**. Notice that the stator has eight salient poles that are 45° apart. On the other hand, the rotor has six salient poles 60° apart. Thus, when 1 is aligned with *A* as shown, 2 is 15° counterclockwise from *B*, and 3 is 15° clockwise from *D*.

The stator contains four windings (which are not shown in the cross-section). A controller applies power to one of the coils at a time as shown in Figure 17.33(b). Coil *A* is wound partly around pole *A* and partly around *A*′, such that, when current is applied, *A* becomes a north magnetic pole and *A*′ becomes a south pole. Then, the rotor moves to shorten the air gaps between *A* (and *A*′) and the rotor. As long

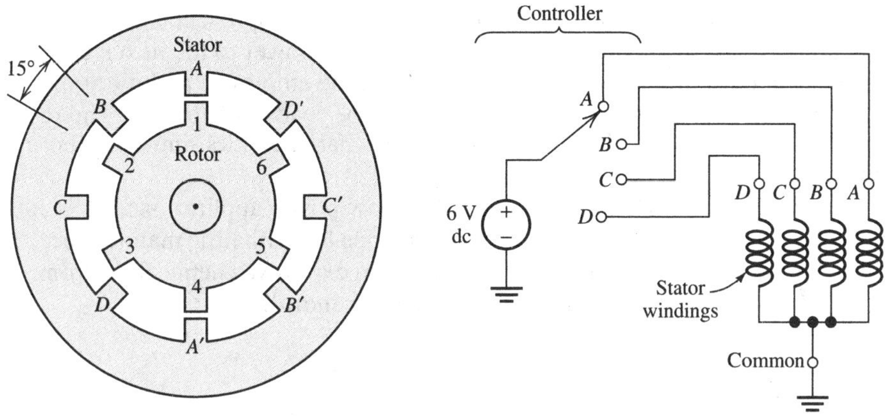

(a) Motor cross section (For simplicity, the windings on the stator poles are not shown)

(b) Circuit diagram

Figure 17.33 Variable-reluctance stepper motor.

as power is applied to coil A, the rotor is held in the position shown in the figure. However, if power is removed from A and applied to B by moving the controller switch, the rotor moves 15° clockwise so that 2 is aligned with B. Then if power is removed from B and applied to C, the shaft rotates another 15° clockwise. Thus, applying power to the coils in the sequence $ABCDABC\ldots$ results in clockwise rotation of the shaft in 15° increments. By changing the switching rate, the motor speed can be varied upward from a standing stop. Furthermore, by reversing the switching sequence to $ADCBADCB\ldots$, the direction of rotation can be reversed.

Another type is the **permanent-magnet stepper motor**, which has a cylindrical rotor (called a "tin-can rotor") that is permanently magnetized with north and south poles alternating around its circumference. The stator of the permanent magnet motor is similar to that of the reluctance motor. As in the reluctance type, the rotor position is stepped by applying a sequence of pulses to the stator windings. **Hybrid stepper motors** that combine variable reluctance with permanent magnets are also available. Of course, detailed specifications for stepper motors can be found from manufacturers' websites.

Brushless DC Motors

Conventional dc motors are particularly useful in applications that require high speeds and in those for which dc power is available, such as those in aircraft and automobiles. However, because they contain commutators and brushes, conventional dc motors have several disadvantages. These include relatively short service lives due to brush and commutator wear, particularly at very high speeds. Also, arcing as the brushes move between commutator segments can pose a hazard in explosive environments and can create severe radio interference. A relatively new development, the **brushless dc motor**, provides an excellent alternative when the disadvantages of conventional dc motors are prohibitive.

Brushless dc motors are essentially permanent-magnet stepping motors equipped with position sensors (either Hall effect or optical) and enhanced control units. As in the stepper motor, power is applied to one stator winding at a time.

When the position sensor indicates that the rotor has approached alignment with the stator field, the controller electronically switches power to the next stator winding so that smooth motion continues. By varying the amplitude and duration of the pulses applied to the stator windings, speed can be readily controlled. The result is a motor that can operate from a dc source with characteristics similar to those of a conventional shunt dc motor.

Brushless dc motors are used primarily in low-power applications. Their advantages include relatively high efficiency, long service life with little maintenance, freedom from radio interference, ability to operate in explosive chemical environments, and capability for very high speeds (50,000 rpm or more).

Summary

1. Application of a three-phase source to the stator windings of an induction motor produces a magnetic field in the air gap that rotates at synchronous speed. Interchanging any two of the connections to the three-phase source reverses the direction of rotation.
2. A squirrel-cage rotor contains aluminum conductors embedded in the rotor. As the stator field moves past, currents are induced in the rotor resulting in torque. The torque–speed characteristic takes the form shown in Figure 17.10 on page 837. In normal steady-state operation, typical motors operate with 0 to 5 percent slip, and the output power and torque are approximately proportional to slip.
3. The per-phase equivalent circuit shown in Figure 17.13 on page 839 is useful in performance calculations for induction motors.
4. Some of the most important considerations in selecting an induction motor are efficiency, starting torque, pull-out torque, power factor, and starting current.
5. Typically, the starting torque of an induction motor is 150 percent or more of the rated full-load running torque. Thus, induction motors can start all constant-torque loads that are within their full-load ratings. Starting current with rated voltage is typically five to six times the full-load running current.
6. A three-phase synchronous machine has stator windings that produce a magnetic field consisting of P poles that rotate at synchronous speed. The rotor is an electromagnet. The machine runs at synchronous speed, and the torque–speed characteristic is shown in Figure 17.26 on page 858.
7. When operated in the overexcited state, synchronous machines produce reactive power and can help to correct the power factor of industrial plants, saving on energy costs.
8. The equivalent circuit for three-phase synchronous machines shown in Figure 17.18 on page 850 can be used in performance calculations.
9. Synchronous motors have zero starting torque, and special provisions must be made for starting.
10. Single-phase induction motors have a main winding and an auxiliary winding displaced by 90 electrical degrees. With power applied to only the main winding, the motor can run but has zero starting torque. Because of differences in the resistance/reactance ratio or because a capacitor is in the circuit, the currents in the two windings have different phases and this produces starting torque. Frequently, the starting winding is disconnected from the ac source when the motor approaches rated speed.
11. Single-phase induction motors are heavier and produce more vibration than do three-phase motors of the same power rating.

12. Stepper motors are useful in applications that require accurate repeatable positioning.

13. Brushless dc motors are a good alternative to conventional dc motors for low-power applications that require long life with little maintenance, operation in explosive environments, freedom from radio interference, or very high speed.

Problems

Section 17.1: Three-Phase Induction Motors

*P17.1. A given load is to be run at 1250 rpm. A 60-Hz 6-pole and a 60 Hz 4-pole motor are available. Which one would you choose and at what slip rate should it operate? Give reasons.

*P17.2. A four-pole induction motor drives a load at 2500 rpm. This is to be accomplished by using an electronic converter to convert a 400-V dc source into a set of three-phase ac voltages. Find the frequency required for the ac voltages assuming that the slip is 4 percent. The load requires 2 hp. If the dc-to-ac converter has a power efficiency of 88 percent and the motor has a power efficiency of 80 percent, estimate the current taken from the dc source.

*P17.3. It is necessary to reduce the voltage applied to an induction motor as the frequency is reduced from the rated value. Explain why this is so.

*P17.4. The air-gap flux density of a two-pole induction motor is given by

$$B = B_m \cos(\omega t - \theta)$$

where B_m is the peak flux density, θ is the angular displacement around the air gap, and we have assumed clockwise rotation. Give the corresponding expression for the flux density of a four-pole induction motor; of a six-pole induction motor.

P17.5. Draw and explain the torque-slip characteristics of a 3-phase Induction Motor.

P17.6. Expand the table of p versus η_s given in Table 17.1 for frequencies of 50,100 and 400 Hz. Using web find the countries using 50 Hz frequency.

P17.7. The magnetic field produced in the air gap of an induction motor by the stator windings is given by $B = B_m \cos(\omega t - 2\theta)$, in which θ is angular displacement in the counterclockwise direction as illustrated in Figure 17.4 on page 832. How many poles does this machine have? Given that the frequency of the source is 50 Hz, determine the speed of rotation of the field. Does the field rotate clockwise or counterclockwise? Repeat for a field given by $B = B_m \cos(\omega t + 3\theta)$.

P17.8. Consider the induction motor shown in Figure 17.7 on page 835. Redraw the figure showing the current directions in the rotor conductors, the magnetic rotor poles, and the direction of the developed torque if a prime mover drives the rotor at a speed higher than synchronous speed. In this case, does the machine operate as a motor or as a generator?

P17.9. A 5-hp four-pole 50-Hz three-phase induction motor runs at 1440 rpm under full-load conditions. Determine the slip and the frequency of the rotor currents at full load. Also estimate the speed if the load torque drops to ¾, ½, and ¼ of the full load.

P17.10. In a proposed design for an electric automobile, the shaft of a four-pole three-phase induction motor is connected directly to the drive axle; in other words, there is

* Denotes that answers are contained in the Student Solutions files. See Appendix E for more information about accessing the Student Solutions.

no gear train. The outside diameter of the tires is 20 inches. Instead of a transmission, an electronic converter produces variable-frequency three-phase ac from a 48-V battery. Assuming negligible slip, find the range of frequencies needed for speeds ranging from 5 to 70 mph. The vehicle, including batteries and occupants, has a mass of 1000 kg. The power efficiency of the dc-to-ac converter is 85 percent, and the power efficiency of the motor is 89 percent. **a.** Find the current taken from the battery as a function of time while accelerating from 0 to 40 mph uniformly (i.e., acceleration is constant) in 10 seconds. Neglect wind load and road friction. **b.** Repeat assuming that the vehicle is accelerated with constant power.

P17.11. Consider the two-pole two-phase induction motor having two windings displaced 90° in space shown in Figure P17.11. The fields produced by the windings are given by $B_a = Ki_a(t)\cos(\theta)$ and $B_b = Ki_b(t)\cos(\theta - 90°)$. The two-phase source produces currents given by $i_a(t) = I_m \cos(\omega t)$ and $i_b(t) = I_m \cos(\omega t - 90°)$. Show that the total field rotates. Determine the speed and direction of rotation. Also find the maximum flux density of the rotating field in terms of K and I_m.

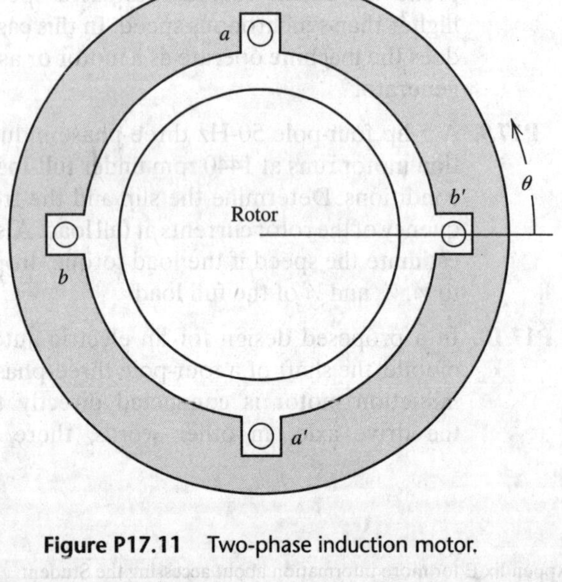

Figure P17.11 Two-phase induction motor.

P17.12. Suppose that we could use superconducting material for the rotor conductors (i.e., rotor conductors with inductance and zero resistance) of an induction motor. Would this improve the performance of the motor? Explain by considering the torque–speed characteristic.

P17.13. Consider the two-pole two-phase induction motor having two windings displaced 90° in space shown in Figure P17.11. The fields produced by the windings are given by $B_a = Ki_a(t)\cos(\theta)$ and $B_b = Ki_b(t)\cos(\theta - 90°)$. The two-phase source produces currents given by $i_a(t) = I_m \cos(2\omega t)$ and $i_b(t) = I_m \cos(2\omega t + 90°)$. Show that the total field rotates. Determine the speed and direction of rotation. Also find the maximum flux density of the rotating field in terms of K and I_m.

Section 17.2: Equivalent-Circuit and Performance Calculations for Induction Motors

***P17.14.** Sketch the torque–speed characteristic of a delta-connected 220-V-rms 5-hp four-pole 60-Hz three-phase induction motor. Estimate values and label key features for things such as the full-load running speed, the full-load torque, the pull-out torque, and the starting torque. Estimate the full-load line current and the starting line current.

***P17.15.** Sometimes, to reduce starting current to reasonable values, induction motors are started with reduced source voltage. When the motor approaches its operating speed, the voltage is increased to full rated value. Compute the starting line current and torque for the motor of Example 17.2 if it is started with a source voltage of 220 V. Compare results with the values found in the example, and comment.

***P17.16.** A certain four-pole 240-V-rms 60-Hz delta-connected three-phase induction motor has

$$R_s = 1.2\ \Omega \qquad R'_r = 0.75\ \Omega$$

$$X_s = 2\ \Omega \qquad X'_r = 0.9\ \Omega$$

$$X_m = 25\ \Omega$$

Under load, the machine operates at 1746 rpm and has rotational losses of 300 W. Neglecting the rotational losses, find the no-load speed, line current, and power factor for the motor.

***P17.17.** A certain six-pole 440-V-rms 60-Hz three-phase delta-connected induction motor has

$$R_s = 0.48\ \Omega \qquad R'_r = 0.04\ \Omega$$

$$X_s = 0.35\ \Omega \qquad X'_r = 0.25\ \Omega$$

$$X_m = 10\ \Omega$$

Neglecting the rotational losses, find the no-load speed, line current, and power factor for the motor.

***P17.18.** A 440-V-rms (line-to-line) 60-Hz three-phase wye-connected induction motor draws 16.8 A at a power factor of 80 percent lagging. The stator copper losses are 350 W, the rotor copper loss is 120 W, and the rotational losses are 400 W. Find the power crossing the air gap P_{ag}, the developed power P_{dev}, the output power P_{out}, and the efficiency.

***P17.19.** A two-pole 60-Hz induction motor produces an output power of 5 hp at a speed of 3500 rpm. With no load, the speed is 3598 rpm. Assume that the rotational torque loss is independent of speed. Find the rotational power loss at 3500 rpm.

P17.20. Compare SQIM and WRIM based on construction, speed control, cost, and maintenance. Which one is called the workhorse of the industry?

P17.21. List five important specifications to be considered (besides cost) in selecting induction motors. Indicate the optimum value or whether a high or low value is desirable for each specification.

P17.22. A 60-Hz wound-rotor induction motor operates at 25 percent slip with added rotor resistance to achieve speed control. The stator resistance is half of the rotor resistance. Neglect rotational losses. Find the efficiency of this motor.

P17.23. A four-pole 60-Hz ac induction motor produces an output power of 5 hp, and has rotational losses of 200 W at a speed of 1710 rpm. Determine the slip, the frequency of the stator currents, the frequency of the rotor currents, and the rotor copper loss.

P17.24. A certain six-pole 440-V-rms 60-Hz three-phase delta-connected induction motor has

$$R_s = 0.08\ \Omega \qquad R'_r = 0.06\ \Omega$$

$$X_s = 0.20\ \Omega \qquad X'_r = 0.15\ \Omega$$

$$X_m = 7.5\ \Omega$$

Neglecting losses, find the starting torque and starting line current for the motor.

P17.25. Another method that is used to limit starting current is to place additional resistance in series with the stator windings during starting. The resistance is switched out of the circuit when the motor approaches full speed. Compute the resistance that must be placed in series with each phase of the motor of Examples 17.1 and 17.2 to limit the starting line current to $50\sqrt{3}$ A rms. Determine the starting torque with this resistance in place. Compare the starting torque with the value found in Example 17.2 and comment.

P17.26. A certain four-pole 240-V-rms 60-Hz delta-connected three-phase induction motor has

$$R_s = 1\ \Omega \qquad R'_r = 0.5\ \Omega$$

$$X_s = 1.5\ \Omega \qquad X'_r = 0.8\ \Omega$$

$$X_m = 40\ \Omega$$

Under load, the machine operates at 1728 rpm and has rotational losses of 200 W.

Find the power factor, output power, copper losses, output torque, and efficiency.

P17.27. A certain four-pole 240-V-rms 60-Hz delta-connected three-phase induction motor has

$$R_s = 1\ \Omega \qquad R'_r = 0.5\ \Omega$$
$$X_s = 1.5\ \Omega \qquad X'_r = 0.8\ \Omega$$
$$X_m = 40\ \Omega$$

Neglecting losses, find the starting torque and starting line current for the motor.

P17.28. A certain six-pole 440-V-rms 60-Hz three-phase delta-connected induction motor has

$$R_s = 0.08\ \Omega \qquad R'_r = 0.06\ \Omega$$
$$X_s = 0.20\ \Omega \qquad X'_r = 0.15\ \Omega$$
$$X_m = 7.5\ \Omega$$

Under load, the machine operates with a slip of 4 percent and has rotational losses of 2 kW. Determine the power factor, output power, copper losses, output torque, and efficiency.

P17.29. A 5-hp four-pole 50-Hz delta-connected three-phase induction motor is rated for 1440 rpm, 220 V rms, and 6 A rms (line current) at an 90 percent lagging power factor. Find the full-load efficiency.

P17.30. The torque–speed characteristics of a 60-Hz induction motor and a load are shown in Figure P17.30. How many poles does the motor have? In steady-state operation, find the speed, the slip, the output power, and the rotor copper loss. Neglect rotational losses.

P17.31. The torque–speed characteristics of a 60-Hz induction motor and a load are shown in Figure P17.30. The rotational inertia of the motor and load is 5 kgm^2. Estimate the time required for the motor to accelerate the load from a standing start to 1000 rpm. (*Hint:* The difference between the motor output torque and the load torque is approximately 25 Nm in the range of speeds under consideration.)

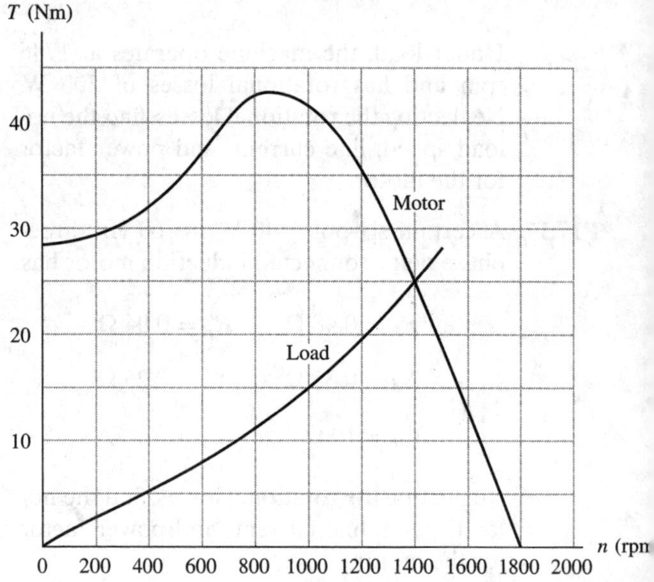

Figure P17.30

P17.32. A four-pole 60-Hz 240-V-rms induction motor operates at 1750 rpm and produces 2 hp of output power. The load is a hoist that requires constant torque versus speed. Assume that the motor is operating in the range for which torque is proportional to slip and write an equation for motor torque in terms of slip when operating at rated voltage. Then modify the equation for operation from a 220-V-rms source. Estimate the speed when operating the hoist from a 220-V-rms source.

Section 17.3: Synchronous Machines

***P17.33.** Is the synchronous motor self starting? If not, how is it started?

***P17.34.** A synchronous motor is running at 80 percent of rated load with 0.9 power factor lagging. If the load increases to the rated output power, how do the following quantities change? **a.** field current; **b.** mechanical speed; **c.** output torque; **d.** armature current; **e.** power factor; **f.** torque angle.

***P17.35.** A 10-pole 60-Hz synchronous motor operates with a developed power of 100 hp, which is its rated full load. The torque angle is 20°. Plot the torque–speed characteristic to scale, showing the values for rated torque and for the pull-out torque.

***P17.36.** A certain 440-V-rms delta-connected synchronous motor operates with zero developed power and draws a phase current of 18 A rms, which lags the voltage. The synchronous reactance is 8 Ω. The field current is 7.5 A. Assuming that the rotor field magnitude is proportional to field current, what field current is needed to reduce the armature current to zero?

***P17.37.** **a.** A 12-pole 60-Hz synchronous motor drives a 10-pole synchronous machine that acts as a generator. What is the frequency of the voltages induced in the armature windings of the generator? **b.** Suppose that we need to drive a load at exactly 1000 rpm. The power available is 60-Hz three-phase. Diagram a system of synchronous machines to drive the load, specifying the number of poles and frequency of operation for each. (Multiple correct answers exist.)

P17.38. Draw and discuss the V-curve of a synchronous motor for various power factors. What happens to the curves when magnetic saturation is reached? Label the curves properly for each power factor type.

P17.39. What is a synchronous capacitor? What is the practical benefit of using one?

P17.40. Give two situations for which a synchronous motor would be a better choice than an induction motor in an industrial application.

P17.41. An eight-pole 240-V-rms 60-Hz delta-connected synchronous motor operates with a constant developed power of 50 hp, unity power factor, and a torque angle of 15°. Then, the field current is increased such that B_r increases in magnitude by 20 percent. Find the new torque angle and power factor. Is the new power factor leading or lagging?

P17.42. A six-pole 60-Hz synchronous motor is operating with a developed power of 5 hp and a torque angle of 5°. Find the speed and developed torque. Suppose that the load increases such that the developed torque doubles. Find the new torque angle. Find the pull-out torque and maximum developed power for this machine.

P17.43. A synchronous motor is running at 100 percent of rated load with unity power factor. If the field current is increased, how do the following quantities change? **a.** output power; **b.** mechanical speed; **c.** output torque; **d.** armature current; **e.** power factor; **f.** torque angle.

P17.44. A 240-V-rms 100-hp 60-Hz six-pole delta-connected synchronous motor operates with a developed power (including losses) of 100 hp and a power factor of 85 percent lagging. The synchronous reactance is $X_s = 0.5\ \Omega$. The field current is $I_f = 10$ A. What must the new field current be to produce 100 percent power factor? Assume that magnetic saturation does not occur so that B_r is proportional to I_f.

P17.45. A six-pole 240-V-rms 60-Hz delta-connected synchronous motor operates with a developed power of 50 hp, unity power factor, and a torque angle of 15°. Find the phase current. Suppose that the load is removed so that the developed power is zero. Find the new values of the current, power factor, and torque angle.

P17.46. A 240-V-rms delta-connected 100-hp 60-Hz six-pole synchronous motor operates with a developed power (including losses) of 50 hp and a power factor of 90 percent leading. The synchronous reactance is $X_s = 0.5\ \Omega$. **a.** Find the speed and developed torque. **b.** Determine the values of $\mathbf{I}_a$, $\mathbf{E}_r$, and the torque angle. **c.** Suppose that the excitation remains constant and the load torque increases until the developed power is 100 hp. Determine the new values of $\mathbf{I}_a$, $\mathbf{E}_r$, the torque angle, and the power factor.

P17.47. Suppose that a synchronous motor is instrumented to measure its armature current, armature voltage, and field current. The field circuit contains a rheostat so that the field current can be adjusted. Discuss how to adjust the field current to obtain unity power factor.

P17.48. A 60-Hz 480-V-rms 200-hp delta-connected synchronous motor runs under no-load conditions. The field current is adjusted for minimum line current, which turns out to be 16.45 A rms. The per-phase armature impedance is $R_s + jX_s = 0.05 + j1.4$. (Until now in this chapter, we have neglected R_s. However, it is significant in efficiency calculations.) Estimate the efficiency of the machine under full-load conditions operating with 90 percent leading power factor.

Section 17.4: Single-Phase Motors

***P17.49.** A 1-hp 120-V-rms 1740-rpm 60-Hz capacitor-start induction-run motor draws a current of 10.2 A rms at full load and has an efficiency of 80 percent. Find the values of **a.** the power factor and **b.** the impedance of the motor at full load. **c.** Determine the number of poles that the motor has.

***P17.50.** A farm house is located at the end of a country road in northern Michigan. The Thévenin impedance seen looking back into the power line from the electrical distribution panel is $0.2 + j0.2\ \Omega$. The Thévenin voltage is 240 V rms 60 Hz ac. A 2-hp 240-V-rms capacitor-start motor is used for pumping water. We want to estimate the voltage drop observed in the house when the motor starts. Typically, such a motor has a power factor of 75 percent and an efficiency of 80 percent at full load. Also, the starting current can be estimated as six times the full-load current. Estimate the worst-case percentage voltage drop observed when the motor starts.

P17.51. The winding impedances under starting conditions for a 50-Hz 0.5-hp motor are shown in Figure P17.51. Determine the capacitance C needed so that the phase angle between the currents $\mathbf{I}_a$ and $\mathbf{I}_m$ is 90°.

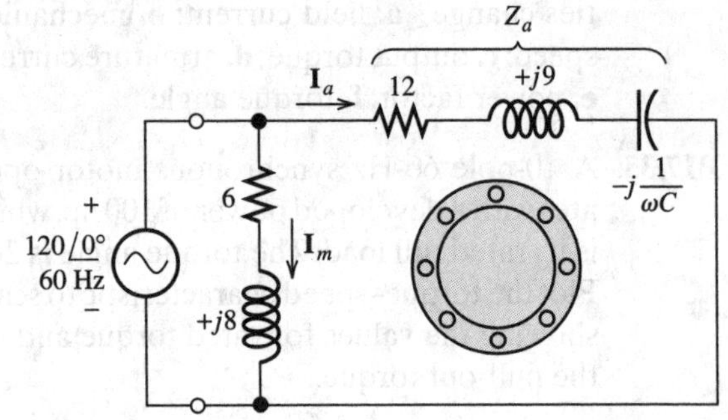

Figure P17.51

P17.52. Assuming small slip, the output power of a single-phase induction motor can be written as $P_{out} = K_1 s - K_2$, where K_1 and K_2 are constants and s is the slip. A 0.5-hp motor has a full-load speed of 3500 rpm and a no-load speed of 3595 rpm. Determine the speed for 0.2-hp output.

P17.53. How could the direction of rotation of a single-phase capacitor-start induction motor be reversed?

P17.54. Which would be more suitable for use in a portable vacuum cleaner, an induction motor or a universal motor? For the fan in a home heating system? For the compressor motor in a refrigerator? For a variable-speed hand-held drill? Give the reasons for your answer in each case.

Section 17.5: Stepper Motors and Brushless DC Motors

P17.55. Sketch the cross-section of a reluctance stepper motor that has six stator poles and eight rotor poles. Your sketch should be similar to Figure 17.33(a) on page 863. Label the windings and specify the sequence of activation for clockwise rotation. What is the rotation angle per step?

P17.56. List two universal manufacturers of stepper motors and brushless DC Motors.

P17.57. Enlist the different types of stepper motors. Compare PMSM and BLDC motors. Where is BLDC specifically used?

Frequently Asked Questions

1. Explain that a revolving magnetic field in set-up in the magnetic core of a 3-phase wound electric machine when 3-phase supply in given to it.
2. What do you mean by a squirrel cage and wound rotor induction motor?
3. Define and explain slip and slip-speed in a 3-phase induction motor.
4. Draw a torque-speed characteristics of a 3-phase induction motor. Also mark the normal operating range, breakover-torque and starting torque on the curve.
5. Draw equivalent circuit of the rotor of an induction motor.
6. Draw the power flow diagram of an induction motor.
7. Draw and explain complete equivalent circuit of an induction motor.
8. What are the considerations while selecting an induction motor for a particular job.
9. How the torque-speed curve of a wound rotor motor is affected by adding and varying the resistance connected in its rotor circuit?
10. What is a synchronous motor? How it works?
11. How a synchronous motor works as a condenser.
12. Explain the operation of a synchronous motor with constant load and variable field current.
13. Explain the operation of a synchronous motor with constant field current and variable load.
14. What do you mean by pull-out torque of a synchronous motor?
15. Explain that a synchronous motor is not self-starting motor.
16. What are the different methods of starting of synchronous motor?
17. Explain the construction and working of a single phase induction motor.
18. What is an auxiliary winding? How it helps to provide starting torque in a single-phase induction motor?
19. Draw the electrical circuit of a capacitor start 1-phase induction motor. Also draw its speed torque characteristics and applications.
20. What is a shaded pole motor, how it works?
21. Write a note on stepper motor.
22. Write a note on brushless dc motor.

Practice Test

Here is a practice test you can use to check your comprehension of the most important concepts in this chapter. Answers can be found in Appendix D and complete solutions are included in the Student Solutions files. See Appendix E for more information about the Student Solutions.

T17.1. **a.** Qualitatively describe the magnetic field set up in the air gap of a four-pole three-phase induction motor by the currents in the stator windings. **b.** Give an equation for the field intensity as a function of angular position around the gap and time, defining the terms in your equation.

T17.2. Besides low cost, what are five desirable characteristics for an induction motor?

T17.3. A certain eight-pole, 240-V-rms, 60-Hz, wye-connected, three-phase induction motor has

$$R_s = 0.5\,\Omega \qquad R'_r = 0.5\,\Omega$$
$$X_s = 2\,\Omega \qquad X'_r = 0.8\,\Omega$$
$$X_m = 40\,\Omega$$

Under load, the machine operates at 864 rpm and has rotational losses of 150 W. Find the power factor, output power, line current, copper losses, output torque, and efficiency.

T17.4. We have a 20-hp, eight-pole, 60-Hz, three-phase induction motor that runs at 850 rpm under full-load conditions. What are the values of the slip and the frequency of the rotor currents at full load? Also, estimate the speed if the load torque drops by 20 percent.

T17.5. In one or two paragraphs, describe the construction and principles of operation for a six-pole 60-Hz three-phase synchronous motor.

T17.6. We have a six-pole, 440-V-rms, 60-Hz delta-connected synchronous motor operating with a constant developed power of 20 hp, unity power factor, and a torque angle of 10°. Then, the field current is reduced such that B_r is reduced in magnitude by 25 percent. Find the new torque angle and power factor. Is the new power factor leading or lagging?

APPENDIX

A

Complex Numbers

In Chapter 5, we learnt that sinusoidal steady-state analysis is greatly facilitated if the currents and voltages are represented as complex numbers known as **phasors**. In this appendix, we review complex numbers.

Basic Complex-Number Concepts

Complex numbers involve the imaginary number $j = \sqrt{-1}$. (Electrical engineers use j to represent the square root of -1 rather than i, because i is often used for currents.) Several examples of complex numbers are

$$3 + j4 \quad \text{and} \quad -2 + j5$$

We say that a complex number $Z = x + jy$ has a **real part** x and an **imaginary part** y. We can represent complex numbers by points in the **complex plane**, in which the real part is the horizontal coordinate and the imaginary part is the vertical coordinate. We often show the complex number by an arrow directed from the origin of the complex plane to the point defined by the real and imaginary components. This is illustrated in Figure A.1.

A **pure imaginary number**, $j6$ for example, has a real part of zero. On the other hand, a **pure real number**, such as 5, has an imaginary part of zero.

We say that complex numbers of the form $x + jy$ are in **rectangular form**. The **complex conjugate** of a number in rectangular form is obtained by changing the sign of the imaginary part. For example, if

$$Z_2 = 3 - j4$$

then the complex conjugate of Z_2 is

$$Z_2^* = 3 + j4$$

(Notice that we denote the complex conjugate by the symbol *.)

We add, subtract, multiply, and divide complex numbers that are in rectangular form in much the same way as we do algebraic expressions, making the substitution $j^2 = -1$.

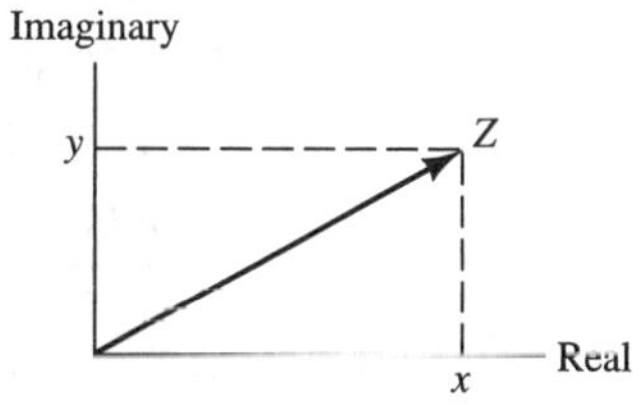

Figure A.1 Complex plane.

Example A.1 Complex Arithmetic in Rectangular Form

Given that $Z_1 = 5 + j5$ and $Z_2 = 3 - j4$, reduce $Z_1 + Z_2$, $Z_1 - Z_2$, $Z_1 Z_2$, and Z_1/Z_2 to rectangular form.

Solution For the sum, we have

$$Z_1 + Z_2 = (5 + j5) + (3 - j4) = 8 + j1$$

Notice that we add (algebraically) real part to real part and imaginary part to imaginary part.

The difference is

$$Z_1 - Z_2 = (5 + j5) - (3 - j4) = 2 + j9$$

In this case, we subtract each part of Z_2 from the corresponding part of Z_1.

For the product, we get

$$\begin{aligned} Z_1 Z_2 &= (5 + j5)(3 - j4) \\ &= 15 - j20 + j15 - j^2 20 \\ &= 15 - j20 + j15 + 20 \\ &= 35 - j5 \end{aligned}$$

Notice that we expanded the product in the usual way for binomial expressions. Then, we used the fact that $j^2 = -1$.

To divide the numbers, we obtain

$$\frac{Z_1}{Z_2} = \frac{5 + j5}{3 - j4}$$

We can reduce this expression to rectangular form by multiplying the numerator and denominator by the complex conjugate of the denominator. This causes the denominator of the fraction to become pure real. Then, we divide each part of the numerator by the denominator. Thus, we find that

$$\begin{aligned} \frac{Z_1}{Z_2} &= \frac{5 + j5}{3 - j4} \times \frac{Z_2^*}{Z_2^*} \\ &= \frac{5 + j5}{3 - j4} \times \frac{3 + j4}{3 + j4} \\ &= \frac{15 + j20 + j15 + j^2 20}{9 + j12 - j12 - j^2 16} \\ &= \frac{15 + j20 + j15 - 20}{9 + j12 - j12 + 16} \\ &= \frac{-5 + j35}{25} \\ &= -0.2 + j1.4 \end{aligned}$$

■

Exercise A.1 Given that $Z_1 = 2 - j3$ and $Z_2 = 8 + j6$, reduce $Z_1 + Z_2$, $Z_1 - Z_2$, $Z_1 Z_2$, and Z_1/Z_2 to rectangular form.

Answer $Z_1 + Z_2 = 10 + j3$, $Z_1 - Z_2 = -6 - j9$, $Z_1 Z_2 = 34 - j12$, $Z_1/Z_2 = -0.02 - j0.36$.

□

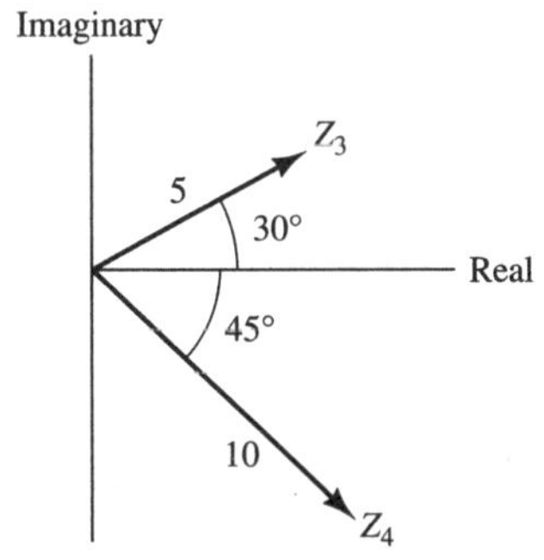

Figure A.2 Complex numbers in polar form.

Complex Numbers in Polar Form

Complex numbers can be expressed in **polar form** by giving the length of the arrow that represents the number and the angle between the arrow and the positive real axis. Examples of complex numbers in polar form are

$$Z_3 = 5\angle 30^\circ \quad \text{and} \quad Z_4 = 10\angle -45^\circ$$

These numbers are shown in Figure A.2. The length of the arrow that represents a complex number Z is denoted as $|Z|$ and is called the **magnitude** of the complex number.

Complex numbers can be converted from polar to rectangular form, or vice versa, by using the fact that the magnitude $|Z|$, the real part x, and the imaginary part y form a right triangle. This is illustrated in Figure A.3. Using trigonometry, we can write the following relationships:

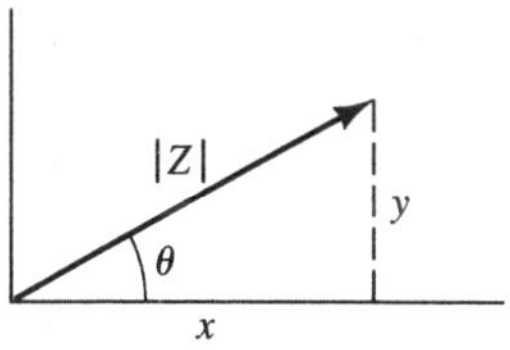

Figure A.3 Complex number representation.

$$|Z|^2 = x^2 + y^2 \tag{A.1}$$

$$\tan(\theta) = \frac{y}{x} \tag{A.2}$$

$$x = |Z|\cos(\theta) \tag{A.3}$$

$$y = |Z|\sin(\theta) \tag{A.4}$$

These equations can be used to convert numbers from polar to rectangular form, or vice versa.

Example A.2 Polar-to-Rectangular Conversion

Convert $Z_3 = 5\angle 30^\circ$ to rectangular form.

Solution Using Equations A.3 and A.4, we have

$$x = |Z|\cos(\theta) = 5\cos(30^\circ) = 4.33$$

and

$$y = |Z|\sin(\theta) = 5\sin(30^\circ) = 2.5$$

Thus, we can write

$$Z_3 = 5\angle 30^\circ = x + jy = 4.33 + j2.5$$

■

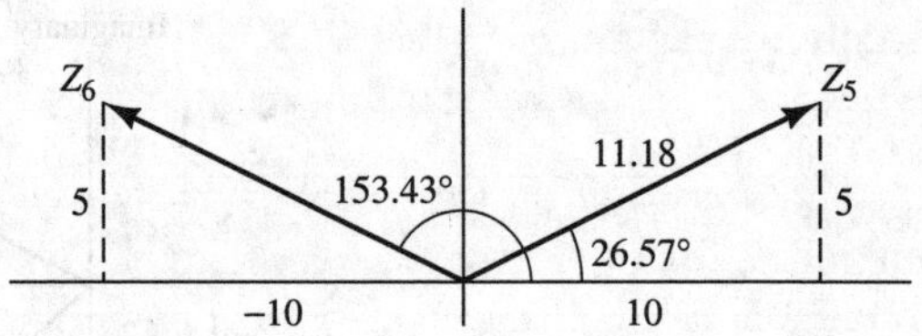

Figure A.4 Complex numbers of Example A.3.

Example A.3 Rectangular-to-Polar Conversion

Convert $Z_5 = 10 + j5$ and $Z_6 = -10 + j5$ to polar form.

Solution The complex numbers are illustrated in Figure A.4. First, we use Equation A.1 to find the magnitudes of each of the numbers. Thus,

$$|Z_5| = \sqrt{x_5^2 + y_5^2} = \sqrt{10^2 + 5^2} = 11.18$$

and

$$|Z_6| = \sqrt{x_6^2 + y_6^2} = \sqrt{(-10)^2 + 5^2} = 11.18$$

To find the angles, we use Equation A.2.

$$\tan(\theta_5) = \frac{y_5}{x_5} = \frac{5}{10} = 0.5$$

Taking the arctangent of both sides, we have

$$\theta_5 = \arctan(0.5) = 26.57°$$

Thus, we can write

$$Z_5 = 10 + j5 = 11.18\angle 26.57°$$

This is illustrated in Figure A.4.

Evaluating Equation A.2 for Z_6, we have

$$\tan(\theta_6) = \frac{y_6}{x_6} = \frac{5}{-10} = -0.5$$

Now if we take the arctan of both sides, we obtain

$$\theta_6 = -26.57°$$

However, $Z_6 = -10+j5$ is shown in Figure A.4. Clearly, the value that we have found for θ_6 is incorrect. The reason for this is that the arctangent function is multivalued. The value actually given by most calculators or computer programs is the principal value. *If the number falls to the left of the imaginary axis (i.e., if the real part is negative), we must add (or subtract)* 180° *to* $\arctan(y/x)$ *to obtain the correct angle.* Thus, the true angle for Z_6 is

$$\theta_6 = 180 + \arctan\left(\frac{y_6}{x_6}\right) = 180 - 26.57 = 153.43°$$

Finally, we can write

$$Z_6 = -10 + j5 = 11.18\underline{/153.43^\circ}$$ ■

The procedures that we have illustrated in Examples A.2 and A.3 can be carried out with a relatively simple calculator. However, if we find the angle by taking the arctangent of y/x, we must consider the fact that the principal value of the arctangent is the true angle only if the real part x is positive. If x is negative, we have

$$\theta = \arctan\left(\frac{y}{x}\right) \pm 180^\circ \tag{A.5}$$

Many scientific calculators are capable of converting complex numbers from polar to rectangular, and vice versa, in a single operation. Practice with your calculator to become proficient using this feature. *It is always a good idea to make a sketch of the number in the complex plane as a check on the conversion process.*

Exercise A.2 Convert the numbers $Z_1 = 15\underline{/45^\circ}$, $Z_2 = 10\underline{/-150^\circ}$, and $Z_3 = 5\underline{/90^\circ}$ to rectangular form.

Answer $Z_1 = 10.6 + j10.6$, $Z_2 = -8.66 - j5$, $Z_3 = j5$. □

Exercise A.3 Convert the numbers $Z_1 = 3 + j4$, $Z_2 = -j10$, and $Z_3 = -5 - j5$ to polar form.

Answer $Z_1 = 5\underline{/53.13^\circ}$, $Z_2 = 10\underline{/-90^\circ}$, $Z_3 = 7.07\underline{/-135^\circ}$. □

Euler's Identities

You may have been wondering what complex numbers have to do with sinusoids. The connection is through Euler's identities, which state that

Equations A.6 through A.9 are the bridge between sinusoidal currents or voltages and complex numbers.

$$\cos(\theta) = \frac{e^{j\theta} + e^{-j\theta}}{2} \tag{A.6}$$

and

$$\sin(\theta) = \frac{e^{j\theta} - e^{-j\theta}}{2j} \tag{A.7}$$

Another form of these identities is

$$e^{j\theta} = \cos(\theta) + j\sin(\theta) \tag{A.8}$$

and

$$e^{-j\theta} = \cos(\theta) - j\sin(\theta) \tag{A.9}$$

Thus, $e^{j\theta}$ is a complex number having a real part of $\cos(\theta)$ and an imaginary part of $\sin(\theta)$. This is illustrated in Figure A.5. The magnitude is

$$|e^{j\theta}| = \sqrt{\cos^2(\theta) + \sin^2(\theta)}$$

By the well-known identity $\cos^2(\theta) + \sin^2(\theta) = 1$, this becomes

$$|e^{j\theta}| = 1 \tag{A.10}$$

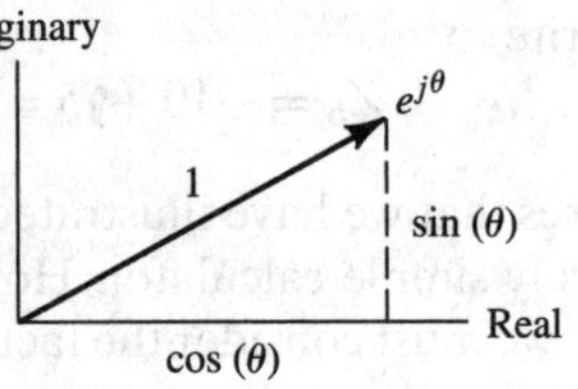

Figure A.5 Euler's identity.

Furthermore, the angle of $e^{j\theta}$ is θ. Thus, we can write

$$e^{j\theta} = 1\angle\theta = \cos(\theta) + j\sin(\theta) \tag{A.11}$$

Similarly, we have

$$e^{-j\theta} = 1\angle-\theta = \cos(\theta) - j\sin(\theta) \tag{A.12}$$

Notice that $e^{-j\theta}$ is the complex conjugate of $e^{j\theta}$.

A complex number such as $A\angle\theta$ can be written as

$$A\angle\theta = A \times (1\angle\theta) = Ae^{j\theta} \tag{A.13}$$

We call $Ae^{j\theta}$ the **exponential form** of a complex number. Hence, a given complex number can be written in three forms: the rectangular form, the polar form, and the exponential form. Using Equation A.11 to substitute for $e^{j\theta}$ on the right-hand side of Equation A.13, we obtain the three forms of a complex number:

$$A\angle\theta = Ae^{j\theta} = A\cos(\theta) + jA\sin(\theta) \tag{A.14}$$

Example A.4 Exponential Form of a Complex Number

Express the complex number $Z = 10\angle 60°$ in exponential and rectangular forms. Sketch the number in the complex plane.

Solution Conversion from polar to exponential forms is based on Equation A.13. Thus, we have

$$Z = 10\angle 60° = 10e^{j60°}$$

The rectangular form can be found by using Equation A.8:

$$Z = 10 \times (e^{j60°})$$
$$= 10 \times [\cos(60°) + j\sin(60°)]$$
$$= 5 + j8.66$$

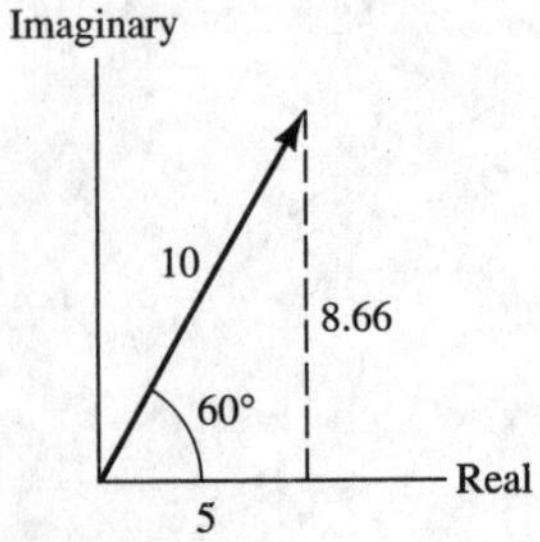

Figure A.6 See Example A.4.

The graphical representation of Z is shown in Figure A.6. ■

Exercise A.4 Express $Z_1 = 10 + j10$ and $Z_2 = -10 + j10$ in polar and exponential forms.

Answer $Z_1 = 14.14\angle 45° = 14.14e^{j45°}$, $Z_2 = 14.14\angle 135° = 14.14e^{j135°}$. □

Arithmetic Operations in Polar and Exponential Form

To add (or subtract) complex numbers, we must first convert them to rectangular form. Then, we add (or subtract) real part to real part and imaginary to imaginary.

Consider two complex numbers in exponential form given by

$$Z_1 = |Z_1|e^{j\theta_1} \quad \text{and} \quad Z_2 = |Z_2|e^{j\theta_2}$$

The polar forms of these numbers are

$$Z_1 = |Z_1|\underline{/\theta_1} \quad \text{and} \quad Z_2 = |Z_2|\underline{/\theta_2}$$

For multiplication of numbers in exponential form, we have

$$Z_1 \times Z_2 = |Z_1|e^{j\theta_1} \times |Z_2|e^{j\theta_2} = |Z_1||Z_2|e^{j(\theta_1+\theta_2)}$$

As usual, in multiplying exponentials, we add the exponents. In polar form, this is

$$Z_1 \times Z_2 = |Z_1|\underline{/\theta_1} \times |Z_2|\underline{/\theta_2} = |Z_1||Z_2|\underline{/\theta_1 + \theta_2}$$

Thus, to multiply numbers in polar form, we multiply the magnitudes and add the angles.

Now consider division:

$$\frac{Z_1}{Z_2} = \frac{|Z_1|e^{j\theta_1}}{|Z_2|e^{j\theta_2}} = \frac{|Z_1|}{|Z_2|}e^{j(\theta_1-\theta_2)}$$

As usual, in dividing exponentials, we subtract the exponents. In polar form, this is

$$\frac{Z_1}{Z_2} = \frac{|Z_1|\underline{/\theta_1}}{|Z_2|\underline{/\theta_2}} = \frac{|Z_1|}{|Z_2|}\underline{/\theta_1 - \theta_2}$$

Thus, to divide numbers in polar form, we divide the magnitudes and subtract the angle of the divisor from the angle of the dividend.

Example A.5 Complex Arithmetic in Polar Form

Given $Z_1 = 10\underline{/60^\circ}$ and $Z_2 = 5\underline{/45^\circ}$, find Z_1Z_2, Z_1/Z_2, and $Z_1 + Z_2$ in polar form.

Solution For the product, we have

$$Z_1 \times Z_2 = 10\underline{/60^\circ} \times 5\underline{/45^\circ} = 50\underline{/105^\circ}$$

Dividing the numbers, we have

$$\frac{Z_1}{Z_2} = \frac{10\underline{/60^\circ}}{5\underline{/45^\circ}} = 2\underline{/15^\circ}$$

Before we can add (or subtract) the numbers, we must convert them to rectangular form. Using Equation A.14 to convert the polar numbers to rectangular, we get

$$\begin{aligned} Z_1 &= 10\underline{/60^\circ} = 10\cos(60^\circ) + j10\sin(60^\circ) \\ &= 5 + j8.66 \\ Z_2 &= 5\underline{/45^\circ} = 5\cos(45^\circ) + j5\sin(45^\circ) \\ &= 3.54 + j3.54 \end{aligned}$$

Now, we can add the numbers. We denote the sum as Z_s:

$$Z_s = Z_1 + Z_2 = 5 + j8.66 + 3.54 + j3.54$$
$$= 8.54 + j12.2$$

Next, we convert the sum to polar form:

$$|Z_s| = \sqrt{(8.54)^2 + (12.2)^2} = 14.9$$

$$\tan \theta_s = \frac{12.2}{8.54} = 1.43$$

Taking the arctangent of both sides, we have

$$\theta_s = \arctan(1.43) = 55°$$

Because the real part of Z_s is positive, the correct angle is the principal value of the arctangent (i.e., 55° is the correct angle). Thus, we obtain

$$Z_s = Z_1 + Z_2 = 14.9\angle 55° \quad \blacksquare$$

Exercise A.5 Given $Z_1 = 10\angle 30°$ and $Z_2 = 20\angle 135°$, find Z_1Z_2, Z_1/Z_2, $Z_1 - Z_2$, and $Z_1 + Z_2$ in polar form.

Answer $Z_1Z_2 = 200\angle 165°$, $Z_1/Z_2 = 0.5\angle -105°$, $Z_1 - Z_2 = 24.6\angle -21.8°$, $Z_1 + Z_2 = 19.9\angle 106°$. □

Summary

1. Complex numbers can be expressed in rectangular, polar, or exponential forms. Addition, subtraction, multiplication, and division of complex numbers are necessary operations in solving steady-state ac circuits by the phasor method.
2. Sinusoids and complex numbers are related through Euler's identities.

Problems*

PA.1. Given that $Z_1 = 2 + j3$ and $Z_2 = 4 - j3$, reduce $Z_1 + Z_2$, $Z_1 - Z_2$, Z_1Z_2, and Z_1/Z_2 to rectangular form.

PA.2. Given that $Z_1 = 1 - j2$ and $Z_2 = 2 + j3$, reduce $Z_1 + Z_2$, $Z_1 - Z_2$, Z_1Z_2, and Z_1/Z_2 to rectangular form.

PA.3. Given that $Z_1 = 10 + j5$ and $Z_2 = 20 - j20$, reduce $Z_1 + Z_2$, $Z_1 - Z_2$, Z_1Z_2, and Z_1/Z_2 to rectangular form.

PA.4. Express each of these complex numbers in polar form and in exponential form: **a.** $Z_a = 5 - j5$; **b.** $Z_b = -10 + j5$; **c.** $Z_c = -3 - j4$; **d.** $Z_d = -j12$.

* Solutions for these problems are contained in the Student Solutions files. See Appendix F for more information about accessing the Student Solutions.

PA.5. Express each of these complex numbers in rectangular form and in exponential form: **a.** $Z_a = 5\angle 45^\circ$; **b.** $Z_b = 10\angle 120^\circ$; **c.** $Z_c = -15\angle -90^\circ$; **d.** $Z_d = -10\angle 60^\circ$.

PA.6. Express each of these complex numbers in rectangular form and in polar form: **a.** $Z_a = 5e^{j30^\circ}$; **b.** $Z_b = 10e^{-j45^\circ}$; **c.** $Z_c = 100e^{j135^\circ}$; **d.** $Z_d = 6e^{j90^\circ}$.

PA.7. Reduce each of the following to rectangular form:

a. $Z_a = 5 + j5 + 10\angle 30^\circ$

b. $Z_b = 5\angle 45^\circ - j10$

c. $Z_c = \dfrac{10\angle 45^\circ}{3 + j4}$

d. $Z_d = \dfrac{15}{5\angle 90^\circ}$

APPENDIX

B Nominal Values and the Color Code for Resistors

Several types of resistors are available for use in electronic circuits. Carbon-film and carbon-composition resistors with tolerances of 5 percent, 10 percent, or 20 percent are available with various power ratings (such as 1/8, 1/4, and 1/2 W). These resistors are used in noncritical applications such as biasing.

Metal-film 1-percent-tolerance resistors are used where greater precision is required. For example, we often choose metal-film resistors in applications such as the feedback resistors of an op amp.

Wire-wound resistors are available with high power-dissipation ratings. Wire-wound resistors often have significant series inductance because they consist of resistance wire that is wound on a form, such as ceramic. Thus, they are often not suitable for use as a resistance at high frequencies.

The value and tolerance are marked on 5-percent, 10-percent, and 20-percent-tolerance resistors by color bands as shown in Figure B.1. The first band is closest to one end of the resistor. The first and second bands give the significant digits of the resistance value. The third band gives the exponent of the multiplier. The fourth band indicates the tolerance. The fifth band is optional and indicates whether the resistors meet certain military reliability specifications.

Table B.1 shows the combinations of significant figures available as nominal values for 5-percent, 10-percent, and 20-percent-tolerance resistors. Table B.2 shows the standard nominal significant digits for 1-percent-tolerance resistors.

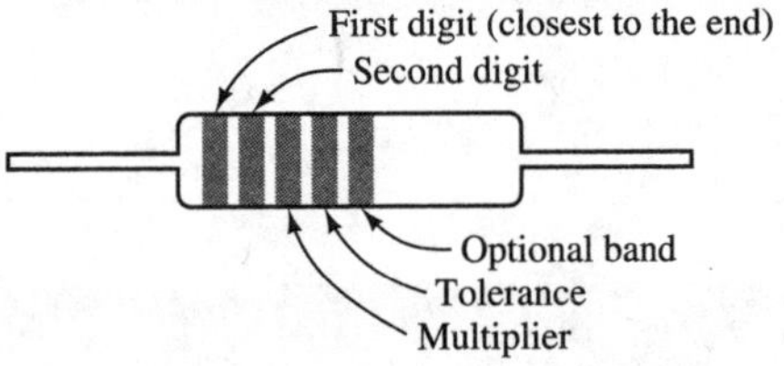

Digit	Color	Tolerance	Color
0	Black	2%	Red
1	Brown	5%	Gold
2	Red	10%	Silver
3	Orange	20%	No fourth band
4	Yellow		
5	Green		
6	Blue		
7	Violet		
8	Gray		
9	White		

Examples

Yellow	Violet	Black	$= 47 \times 10^0 = 47\ \Omega$
Yellow	Violet	Red	$= 47 \times 10^2 = 4700\ \Omega$
Brown	Black	Yellow	$= 10 \times 10^4 = 100\ \text{k}\Omega$

Figure B.1 Resistor color code.

Table B.1. Standard Nominal Values for 5-percent-Tolerance Resistors[a]

10	16	**27**	43	**68**
11	**18**	30	**47**	75
12	20	**33**	51	**82**
13	**22**	36	**56**	91
15	24	**39**	62	

[a] Resistors having tolerances of 10 percent and 20 percent are available only for the values given in boldface.

Table B.2. Standard Values for 1-percent-Tolerance Metal-Film Resistors

100	140	196	274	383	536	750
102	143	200	280	392	549	768
105	147	205	287	402	562	787
107	150	210	294	412	576	806
110	154	215	301	422	590	825
113	158	221	309	432	604	845
115	162	226	316	442	619	866
118	165	232	324	453	634	887
121	169	237	332	464	649	909
124	174	243	340	475	665	931
127	178	249	348	487	681	953
130	182	255	357	499	698	976
133	187	261	365	511	715	
137	191	267	374	523	732	

APPENDIX C The Fundamentals of Engineering Examination

Becoming licensed as a Professional Engineer (PE) is a very important step toward success in your engineering career. In the United States, a PE license is required by all 50 states for engineers whose work may affect life, health, or property, or who offer their services to the public. Thus, a license is an absolute requirement for many types of work. Furthermore, licensed engineers have more opportunities and earn higher salaries (from 15 percent to 25 percent) than other engineers.

Licensure requirements are set by the various states, but are similar from state to state. Furthermore, by reciprocal agreements, many states recognize licenses granted in other states. Generally, a degree from an ABET-accredited engineering program, four years of relevant work experience, and successful completion of two state examinations are required. A master's degree or equivalent may become a requirement for professional registration in 2020. The National Council of Examiners for Engineering and Surveying (NCEES) prepares and scores the examinations. The examinations are the Fundamentals of Engineering (FE) Examination, which can be taken any time, and the Principles and Practice of Engineering (PE) Examination, which must be taken after at least four years of experience.

Up-to-date information about the PE license can be found at the following URL: `http://www.ncees.org/`

You should plan on taking the FE Examination before or shortly after you graduate, rather than later, because it contains questions on a wide variety of subjects. To pass the FE Examination, many engineers struggle to relearn topics that they were familiar with at graduation. Most likely, if you are a mechanical, civil, or chemical engineer, you will not routinely work with electrical circuits. After 10 or more years, you may not be able to answer questions regarding electrical circuits that you could have easily answered when you finished the courses for which you used this book. Thus, taking the examination in your senior year is best. Keep this book to refresh your knowledge of electrical circuits just before taking the FE Examination.

Starting in January 2014, NCEES will make many changes to the FE exam, some of which have not been finalized at the time of this writing. Among other changes, the exam will become computer based instead of pencil and paper, there will be completely separate exams for the various engineering fields rather than a common morning session, and the exam will be administered on many dates throughout the year at Pearson VUE test centers. **If you are taking the exam in 2014 or later, you should consult the NCEES website for up-to-date information. The following description applies only for exams to be given in 2013.**

In 2013, the examination will be offered in April and October. The 2013 FE Examinations will consist of a morning session (120 questions) that is common to all disciplines and an afternoon session (60 questions) that is separate for each discipline. The morning sessions will contain 120 questions on mathematics, engineering probability and statistics, chemistry, computers, ethics and business practices, engineering economics, engineering mechanics (statics and dynamics), strength of materials, material properties, fluid mechanics, electricity and magnetism, and thermodynamics. The questions on electricity and magnetism include the following topics:

- Charge, energy, current, voltage, power
- Work done in moving a charge in an electric field (relationship between voltage and work)
- Force between charges

- Current and voltage laws (Kirchhoff, Ohm)
- Equivalent circuits (series, parallel)
- Capacitance and inductance
- Reactance and impedance, susceptance and admittance
- AC circuits
- Basic complex algebra

Except for electric fields and forces between charges, all of these topics are treated in this book.

Next, we present electrical-circuit questions typical of those found in the morning session of the examination. The correct answers are listed at the end of this Appendix and the solutions are included in the Student Solutions files.

PC1. A dc current of 3 A flows through an initially uncharged capacitor. After two microseconds, the magnitude of the net electric charge on one plate of the capacitor is most nearly:

a. 3 μC
b. 6 μC
c. 0 μC
d. 0.667 μC

PC2. For the circuit of Figure PC.2, the power dissipated in the 10-Ω resistor is most nearly:

a. 22.5 W
b. 8.27 W
c. 1.84 W
d. 7.35 W

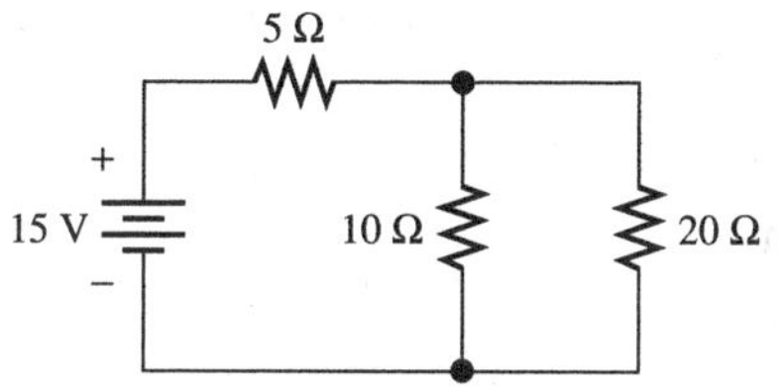

Figure PC.2

PC3. Two initially uncharged capacitors have values of 6 μF and 12 μF. The capacitors are connected in series, and a 200 V dc source is applied to the combination. The charge taken from the source is most nearly:

a. 800 μC
b. 3600 μC
c. 600 μC
d. 1200 μC

PC4. A 2-hp 220-V-rms single-phase induction motor operates at full load with 80 percent efficiency and 0.75 lagging power factor. The magnitude of the rms motor current is most nearly:

a. 4.07 A

b. 11.3 A

c. 6.35 A

d. 8.78 A

PC5. A 30-Ω resistor, a pure capacitance having a reactance magnitude of 80 Ω, and a pure inductance having a reactance magnitude of 40 Ω are in series. The impedance magnitude of the series combination is most nearly:

a. 150 Ω

b. 50 Ω

c. 14.1 Ω

d. 17.1 Ω

PC6. The apparent power supplied to a load in an ac circuit is 2000 volt-amperes with a power factor of 0.6 lagging. The reactive power is most nearly:

a. 1200 VAR

b. 3333 VAR

c. 1600 VAR

d. 2500 VAR

PC7. A 150-microfarad capacitor has been charged to a potential of 100 V. A 50-Ω resistor is placed across the capacitor. After 20 time constants, the total energy delivered to the resistor is most nearly:

a. 1.5 J

b. 0 J

c. 0.75 J

d. 15×10^{-3} J

PC8. The current through the 50-Ω resistor for the circuit shown in Figure PC.8 is most nearly:

a. 1.56 A rms

b. 1.10 A rms

c. 2.20 A rms

d. 0.52 A rms

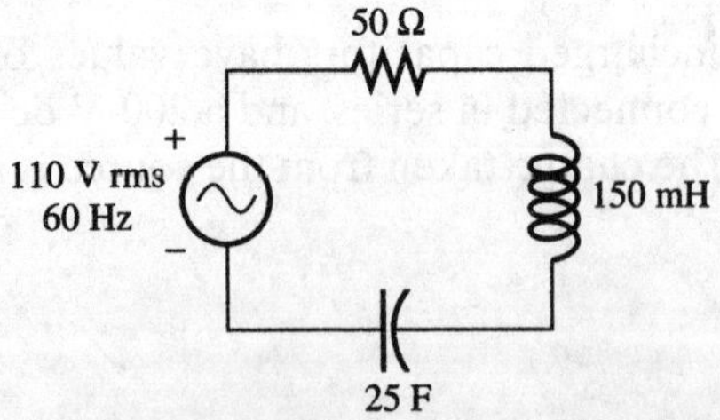

Figure PC.8

PC9. For time t greater than zero, the mathematical expression of the current through the 25-Ω resistance of Figure PC.9 is:

a. $1 - \exp(-2t)$ A

b. $1 - \exp(-t/2)$ A

c. $\exp(-2t)$ A

d. $\exp(-t/2)$ A

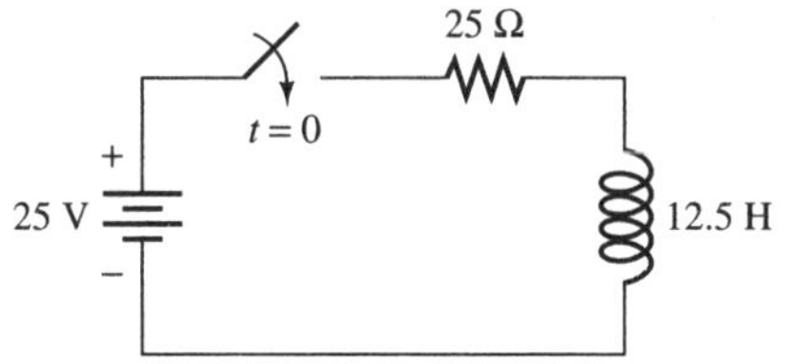

Figure PC.9

PC10. We have three points $A, B,$ and C in an electrical circuit and voltages $V_{AC} =$ 200 V and $V_{CB} = 50$ V. The energy needed to move a charge of 0.2 C from point B to point A is most nearly:

a. 30 J

b. 40 J

c. 10 J

d. 50 J

Answers:

PC1. b	PC6. c
PC2. d	PC7. c
PC3. a	PC8. a
PC4. b	PC9. a
PC5. b	PC10. d

APPENDIX

D Answers for the Practice Tests

Complete solutions for the practice tests are included in the Student Solutions files. See Appendix E for information on how to access these files.

T1.1. **a.** 4; **b.** 7; **c.** 16; **d.** 18; **e.** 1; **f.** 2; **g.** 8; **h.** 3; **i.** 5; **j.** 15; **k.** 6; **l.** 11; **m.** 13; **n.** 9; **o.** 14.

T1.2. **a.** $v_R = -6\text{ V}$. **b.** The voltage source is delivering 30 W. **c.** There are 3 nodes. **d.** The current source is absorbing 12 W.

T1.3. **a.** $v_{ab} = -8\text{ V}$. **b.** Source I_1 is supplying 24 W. Source I_2 is absorbing 8 W. **c.** $P_{R1} = 5.33\text{ W}$ and $P_{R2} = 10.67\text{ W}$.

T1.4. **a.** $v_1 = 8\text{ V}$; **b.** $i = 2\text{ A}$; **c.** $R_2 = 2\ \Omega$.

T1.5. $i_{sc} = -3\text{ A}$.

T2.1. **a.** 6; **b.** 10; **c.** 2; **d.** 7; **e.** 10 or 13; **f.** 1 or 4; **g.** 11; **h.** 3; **i.** 8; **j.** 15; **k.** 17; **l.** 14.

T2.2. $i_s = 6\text{ A}$; $i_4 = 1\text{ A}$.

T2.3.

```
G = [0.95 -0.20 -0.50; -0.20 0.30 0; -0.50 0 1.50]
I = [0; 2; -2]
V = G\ I % As an alternative, we could use V = inv(G)*I
```

T2.4. A proper set of equations consists of any two of the following three

1. KVL mesh 1:

$$R_1 i_1 - V_s + R_3(i_1 - i_3) + R_2(i_1 - i_2) = 0$$

2. KVL for the supermesh obtained by combining meshes 2 and 3:

$$R_4 i_2 + R_2(i_2 - i_1) + R_3(i_3 - i_1) + R_5 i_3 = 0$$

3. KVL around the periphery of the circuit:

$$R_1 i_1 - V_s + R_4 i_2 + R_5 i_3 = 0$$

in combination with this equation for the current source:

$$i_2 - i_3 = I_s$$

T2.5. $V_t = 24\text{ V}$, $R_t = 24\ \Omega$, and $I_n = 1\text{ A}$. The reference direction for I_n should point toward terminal b. The positive reference for V_t should be on the side of the b terminal.

T2.6. By superposition, 25 percent of the current through the 5-Ω resistance is due to the 5-V source. Superposition does not apply for power, but we can see from analysis of the complete circuit that all of the power is supplied by the 15-V source. Thus, 0 percent of the power in the 5-Ω resistance is due to the 5-V source.

T3.1. $v_{ab}(t) = 15 - 15\exp(-2000t)\text{ V}$; $w_C(\infty) = 1.125\text{ mJ}$.

T3.2. $C_{eq} = 5\ \mu\text{F}$.

T3.3. $C = 4248\text{ pF}$.

T3.4. $v_{ab}(t) = 1.2\cos(2000t)\text{ V}$; $w_{peak} = 90\ \mu\text{J}$.

T3.5. $L_{eq} = 3.208\,\text{H}$.

T3.6. $v_1(t) = -40\ \sin(500t) - 16\ \exp(-400t)\,\text{V}$;

$v_2(t) = 20\ \sin(500t) - 24\ \exp(-400t)\,\text{V}$.

T3.7. One set of commands and the result for $v_{ab}(t)$ are:

```
syms vab iab t
iab = 3*(10^5)*(t^2)*exp(-2000*t);
vab = (1/20e-6)*int(iab,t,0,t)
subplot(2,1,1)
ezplot(iab, [0 5e-3]), title('\iti_a_b\rm (A) versus \itt\rm (s)')
subplot(2,1,2)
ezplot(vab, [0 5e-3]), title('\itv_a_b\rm (V) versus \itt\rm (s)')
```

$$v_{ab} = \frac{15}{4} - \frac{15}{4}\ \exp(-2000t) - 7500t\ \exp(-2000t) - 7.5\times10^6 t^2\ \exp(-2000t)$$

You can test your commands using MATLAB to see if they produce this result for $v_{ab}(t)$ and plots like those in the Student Solutions.

T4.1. **a.** $i_1(0-) = 10\,\text{mA}$, $i_2(0-) = 5\,\text{mA}$, $i_3(0-) = 0$, $i_L(0-) = 15\,\text{mA}$, $v_C(0-) = 10\,\text{V}$;

b. $i_1(0+) = 15\,\text{mA}$, $i_2(0+) = 2\,\text{mA}$, and $i_3(0+) = -2\,\text{mA}$, $i_L(0+) = 15\,\text{mA}$, $v_C(0+) = 10\,\text{V}$;

c. $i_L(t) = 10 + 5\ \exp(-5\times 10^5 t)\,\text{mA}$;

d. $v_C(t) = 10\ \exp(-200t)\,\text{V}$.

T4.2. **a.** $2\frac{di(t)}{dt} + i(t) = 5\ \exp(-3t)$;

b. $\tau = L/R = 2\,\text{s}, i_c(t) = A\exp(-0.5t)\ \text{A}$;

c. $i_p(t) = -\exp(-3t)\,\text{A}$;

d. $i(t) = \exp(-0.5t) - \exp(-3t)\,\text{A}$.

T4.3. **a.** $\frac{d^2v_C(t)}{dt^2} + 2000\frac{dv_C(t)}{dt} + 25\times 10^6 v_C(t) = 375\times 10^6$;

b. $v_{Cp}(t) = 15\,\text{V}$;

c. Underdamped; $v_{Cc}(t) = K_1\exp(-1000t)\cos(4899t) + K_2\exp(-1000t)\sin(4899t)$;

d. $v_C(t) = 15 - 15\ \exp(-1000t)\cos(4899t) - (3.062)\exp(-1000t)\sin(4899t)\,\text{V}$.

T4.4. The commands are

```
syms vC t
S = dsolve('D2vC + 2000*DvC + (25e6)*vC = 375e6', 'vC(0) = 0, DvC(0) = 0');
simple(vpa(S,4))
```

The commands are stored in the m-file named T_4_4 that can be found in the Hambley MATLAB folder. See Appendix E for information about accessing this folder.

T5.1. $I_{rms} = \sqrt{8} = 2.828\,\text{A}$; $P = 400\,\text{W}$.

T5.2. $v(t) = 9.914\ \cos(\omega t - 37.50°)$

T5.3. **a.** $V_{1rms} = 10.61\,\text{V}$; **b.** $f = 200\,\text{Hz}$; **c.** $\omega = 400\pi$ radians/s; **d.** $T = 5\,\text{ms}$; **e.** $\mathbf{V}_1$ lags $\mathbf{V}_2$ by 15° or $\mathbf{V}_2$ leads $\mathbf{V}_1$ by 15°.

T5.4. $\mathbf{V}_R = 7.071\angle{-45°}\ \text{V}$; $\mathbf{V}_L = 10.606\angle{45°}\ \text{V}$; $\mathbf{V}_C = 5.303\angle{-135°}\ \text{V}$.

T5.5. $\mathbf{S} = 5500\angle{40°} = 4213 + j3535\,\text{VA}$;

$P = 4213\,\text{W}$; $Q = 3535\,\text{VAR}$; apparent power $= 5500\,\text{VA}$;

Power factor = 76.6 percent lagging.

T5.6. $\mathbf{I}_{aA} = 54.26 \angle -23.13^\circ$ A.

T5.7. The commands are:

```
Z = [(15+i*10) -15; -15 (15-i*5)]
V = [pin(10,45); -15]
I = inv(Z)*V
pout(I(1))
pout(I(2))
```

T6.1. All real-world signals (which are usually time-varying currents or voltages) are sums of sinewaves of various frequencies, amplitudes, and phases. The transfer function of a filter is a function of frequency that shows how the amplitudes and phases of the input components are altered to produce the output components.

T6.2. $v_{out}(t) = 1.789 \cos(1000\pi t - 63.43^\circ) + 3.535 \cos(2000\pi t + 15^\circ)$.

T6.3. **a.** The slope of the low-frequency asymptote is +20 dB/decade. **b.** The slope of the high-frequency asymptote is zero. **c.** The coordinates at which the asymptotes meet are 20 log(50) = 34 dB and 200 Hz. **d.** This is a first-order highpass filter. **e.** The break frequency is 200 Hz.

T6.4. **a.** 1125 Hz; **b.** 28.28; **c.** 39.79 Hz; **d.** 5 Ω; **e.** infinite impedance; **f.** infinite impedance.

T6.5. **a.** 159.2 kHz; **b.** 10.0; **c.** 15.92 kHz; **d.** 10 kΩ; **e.** zero impedance; **f.** zero impedance.

T6.6. **a.** First-order lowpass filter; **b.** second-order lowpass filter; **c.** second-order band-reject (or notch) filter; **d.** first-order highpass filter.

T6.7. One set of commands is:

```
f = logspace(1,4,400);
H = 50*i*(f/200)./(1 + i*f/200);
semilogx(f,20*log10(abs(H)))
```

Other sets of commands will also work. Check to see if your commands produce a plot equivalent to the one produced by the set given above.

T7.1. **a.** 12; **b.** 19 (18 is incorrect because it omits the first step, inverting the variables); **c.** 20; **d.** 23; **e.** 21; **f.** 24; **g.** 16; **h.** 25; **i.** 7; **j.** 10; **k.** 8; **l.** 1 (the binary codes for hexadecimal symbols A through F do not occur in BCD).

T7.2. **a.** 101100001.111_2; **b.** 541.7_8; **c.** $161.\mathrm{E}_{16}$; **d.** $001101010011.100001110101_{\mathrm{BCD}}$.

T7.3. **a.** $+97_{10}$; **b.** -70_{10}.

T7.4. **a.** $D = \overline{A}\,\overline{B} + \overline{(B+C)}$. **b.** Ones should appear only in the truth-table rows and map cells corresponding to $ABC = 000, 001$, and 100. **c.** $D = \overline{A}\,\overline{B} + \overline{B}\,\overline{C}$; **d.** $D = \overline{B}(\overline{A} + \overline{C})$.

T7.5. **a.** Ones should appear only in the cells corresponding to $B_8B_4B_2B_1 = 0001$, 0101, 1011, and 1111. **b.** $G = B_1\overline{B_2}\,\overline{B_8} + B_1B_2B_8$. **c.** $G = B_1(\overline{B_2} + B_8)(B_2 + \overline{B_8})$.

T7.6. The successive states are $Q_0Q_1Q_2 = 100$ (the initial state), 110, 111, 011, 001, 100, 111. The state of the register returns to its initial state after 5 shifts.

T8.1. **a.** 11; **b.** 17; **c.** 21; **d.** 24; **e.** 27; **f.** 13; **g.** 26; **h.** 9; **i.** 20; **j.** 12; **k.** 15; **l.** 16; **m.** 8; **n.** 29; **o.** 23; **p.** 30.

T8.2. **a.** direct, 61; **b.** indexed, F3; **c.** inherent, FF; **d.** inherent, 01; **e.** immediate, 05; **f.** immediate, A1.

T8.3. After the four commands have been executed, the contents of the registers and memory locations are:

A: 32	1034: 00
B: 32	1035: 19
SP: 1035	1036: 58
X: 1958	1037: 19
	1038: 58
	1039: 00
	103A: 00
	103B: 00
	103C: 00

T9.1. The four main elements are sensors, a DAQ board, software, and a general-purpose computer.

T9.2. The four types of systematic (bias) errors are offset, scale error, nonlinearity, and hysteresis.

T9.3. Bias errors are the same for measurements repeated under identical conditions, while random errors are different for each measurement.

T9.4. Ground loops occur when the sensor and the input of the amplifier are connected to ground by separate connections. The effect is to add noise (often with frequencies equal to that of the power line and its harmonics) to the desired signal.

T9.5. If we are using a sensor that has one end grounded, we should choose an amplifier with a differential input to avoid a ground loop.

T9.6. Coaxial cable or shielded twisted pair cable.

T9.7. If we need to sense the open-circuit voltage, the input impedance of the amplifier should be very large compared to the internal impedance of the sensor.

T9.8. The sampling rate should be more than twice the highest frequency of the components in the signal. Otherwise, higher frequency components can appear as lower frequency components known as aliases.

T10.1. **a.** $i_D \cong 9.6\,\text{mA}$; **b.** $i_D \cong 4.2\,\text{mA}$.

T10.2. The diode is on, $v_x = 2.286\,\text{V}$ and $i_x = 0.571\,\text{mA}$.

T10.3. The resistance is $1\,\text{k}\Omega$, and the voltage is 3 V.

T10.4. Your diagram should be equivalent to Figure 10.28. It may be correct even if it is laid out differently. Check to see that you have four diodes and that current flows from the source through a diode in the forward direction, then through the load and finally through a second diode in the forward direction back to the opposite end of the source. On the opposite half cycle, the path should be through one of the other two diodes, through the load in the same direction as before, and back through the fourth diode to the opposite end of the source.

T10.5. Your diagram should be equivalent to Figure 10.29(a) with the 6-V source changed to 5V, the 9-V source changed to 4V, and the peak ac voltage changed to 10V. However, your diagram may be somewhat different in appearance. For example, the 4-V source and diode B can be interchanged as long as the source polarity and direction of the diode don't change; similarly for the 5-V source and diode A. (In other words, the order of the elements doesn't matter in a series connection.) The parallel branches can be interchanged in position. The problem does not give enough information to properly select the value of the resistance, but any value from about 1 kΩ to 1 MΩ is acceptable.

T10.6. Your diagram should be equivalent to Figure 10.33(a) with a 4-V source in place of the 5-V source. The time constant RC should be much longer than the period of the source voltage. Thus, we should select component values so that $RC >> 0.1$ s.

T10.7. The small-signal equivalent circuit for the diode is simply a 10.4-Ω resistance.

T11.1. $A_{voc} = 1500$; $R_i = 60\ \Omega$; $R_o = 40\ \Omega$.

T11.2. Your answer should be similar to Table 11.1 on page 549.

T11.3. **a.** Transconductance amplifier; **b.** Current amplifier; **c.** Voltage amplifier; **d.** Transresistance amplifier.

T11.4. $A_{voc} = 250$ (no units or V/V); $R_{moc} = 50\,\text{k}\Omega$; $G_{msc} = 0.25$ S.

T11.5. $P_d = 12$ W; $\eta = 60$ percent.

T11.6. To avoid linear waveform distortion, the gain magnitude should be constant and the phase response should be a linear function of frequency over the frequency range from 1 to 10 kHz. Because the gain is 100 and the peak input amplitude is 100 mV, the peak output amplitude should be 10 V. The amplifier must not display clipping or unacceptable nonlinear distortion for output amplitudes of this value.

T11.7. The principal effect of offset current, bias current, and offset voltage of an amplifier is to add a dc component to the signal being amplified.

T11.8. Harmonic distortion can occur when a pure sinewave test signal is applied to the input of an amplifier. The distortion appears in the output as components whose frequencies are integer multiples of the input frequency. Harmonic distortion is caused by a nonlinear relationship between the input voltage and output voltage.

T11.9. Common mode rejection ratio (CMRR) is the ratio of the differential gain to the common mode gain of a differential amplifier. Ideally, the common mode gain is zero, and the amplifier produces an output only for the differential signal. CMRR is important when we have a differential signal of interest in the presence of a large common-mode signal not of interest. For example, in recording an electrocardiogram, two electrodes are connected to the patient; the differential signal is the heart signal of interest to the cardiologist; and the common mode signal is due to the 60-Hz power line.

T12.1. For $v_{GS} = 0.5$ V, the transistor is in cutoff, and the drain current is zero, because v_{GS} is less than the threshold voltage V_{to}. Thus, the drain characteristic for $v_{GS} = 0.5$ V lies on the horizontal axis. The drain characteristic for $v_{GS} = 4$ V is identical to that of Figure 12.11 on page 596 in the book.

T12.2. The results of the load-line analysis are $V_{DS\,\text{min}} \cong 1.0$ V, $V_{DSQ} \cong 2.05$ V, and $V_{DS\,\text{max}} \cong 8.2$ V.

T12.3. $R_S = 2.586$ kΩ.

T12.4. We can determine that $g_m = 2.5$ mS. The Q-point values are $V_{DSQ} = 5$ V, $V_{GSQ} = 2$ V, and $I_{DQ} = 0.5$ mA.

T12.5. **a.** a short circuit; **b.** a short circuit; **c.** an open circuit.

T12.6. See Figure 12.31(b) and (c) on page 615. The NMOS is on and the PMOS is off.

T13.1. **a.** 3; **b.** 2; **c.** 5; **d.** 7 and 1 (either order); **e.** 10; **f.** 7; **g.** 1; **h.** 7; **i.** 15; **j.** 12; **k.** 19.

T13.2. $V_{CE\min} \cong 0.2\,\text{V}$, $V_{CEQ} \cong 5.0\,\text{V}$, and $V_{CE\max} \cong 9.2\,\text{V}$.

T13.3. $\alpha = 0.9615$, $\beta = 25$, and $r_\pi = 650\ \Omega$. The small-signal equivalent circuit is shown in Figure 13.26 on page 655.

T13.4. **a.** $I_C = 0.8830\,\text{mA}$ and $V_{CE} = 4.850\,\text{V}$; **b.** $I_C = 1.872\,\text{mA}$ and $V_{CE} = 0.2\,\text{V}$.

T13.5. We need to replace V_{CC} by a short circuit to ground, the coupling capacitances with short circuits, and the BJT with its equivalent circuit. The result is shown in Figure T13.5.

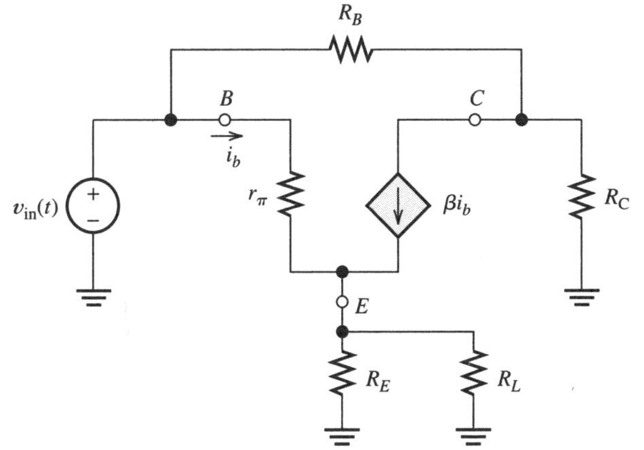

Figure T13.5

T13.6. $A_v = -243.0$, $Z_{\text{in}} = 761.4\ \Omega$.

T14.1. **a.** The circuit diagram is shown in Figure 14.4 on page 680, and the voltage gain is $A_v = -R_2/R_1$. (Of course, you could use different resistance labels, such as R_A and R_B, so long as your equation for the gain is modified accordingly.) **b.** The circuit diagram is shown in Figure 14.11 on page 686 and the voltage gain is $A_v = 1 + R_2/R_1$. **c.** The circuit diagram is shown in Figure 14.12 on page 687 and the voltage gain is $A_v = 1$.

T14.2. $A_v = -8$.

T14.3. **a.** $f_{BCL} = 10\,\text{kHz}$; **b.** $v_o(t) = 0.4975\cos(2\pi \times 10^5 t - 84.29°)$.

T14.4. **a.** $f_{FP} = 707.4\,\text{kHz}$; **b.** $V_{om} = 1\,\text{V}$. **c.** $V_{om} = 4.5\,\text{V}$; **d.** $V_{om} = 0.637\,\text{V}$.

T14.5. See Figure 14.29 on page 703 for the circuit diagram. The principal effect of bias current, offset current, and offset voltage in amplifier circuits is to add a (usually undesirable) dc voltage to the intended output signal.

T14.6. See Figure 14.33 on page 707. Usually, we would have $R_1 = R_3$ and $R_2 = R_4$.

T14.7. See Figures 14.35 and 14.38 on pages 710 and 711, respectively.

T14.8. Filters are circuits designed to pass input components with frequencies in one range to the output and prevent input components with frequencies in other ranges from reaching the output. An active filter is a filter composed of op amps, resistors, and capacitors. Several filter applications are mentioned in the first paragraph of Section 14.10 starting on page 712.

T15.1. **a.** The force is 0.72 N pointing in the negative y direction. **b.** The force is zero.

T15.2. $v = 164.9\cos(120\pi t)$ V.

T15.3. 1.2 V.

T15.4. **a.** $B_{gap} = 0.5357$ T; **b.** $L = 35.58$ mH.

T15.5. The two mechanisms by which power is converted to heat in an iron core are hysteresis and eddy currents. To minimize loss due to hysteresis, we choose a material for which the plot of B versus H displays a thin hysteresis loop. To minimize loss due to eddy currents, we make the core from laminated sheets or from powdered iron held together by an insulating binder. Hysteresis loss is proportional to frequency and eddy-current loss is proportional to the square of frequency.

T15.6. **a.** $I_{1rms} = 0$, $I_{2rms} = 0$, $V_{1rms} = 120$ V, $V_{2rms} = 1200$ V. **b.** $I_{1rms} = 11.43$ A, $I_{2rms} = 1.143$ A, $V_{1rms} = 114.3$ V, $V_{2rms} = 1143$ V.

T15.7. Transformer B is better from the standpoint of total energy loss and operating costs.

T16.1. The windings are the field winding, which is on the stator, and the armature winding, which is on the rotor. The armature current varies with mechanical load.

T16.2. See Figure 16.5(c) on page 785. If the field becomes disconnected, the speed becomes very high, and the machine can be destroyed.

T16.3. See Figure 16.5(d) on page 785.

T16.4. Speed regulation $= [(n_{\text{no-load}} - n_{\text{full-load}})/n_{\text{full-load}}] \times 100$ percent.

T16.5. To obtain the magnetization curve, we drive the machine at constant speed and plot the open-circuit armature voltage E_A versus field current I_F.

T16.6. Power losses in a shunt-connected dc motor are: 1. Field loss, which is the power consumed in the resistances of the field circuit. 2. Armature loss, which is the power converted to heat in the armature resistance. 3. Rotational losses, which include friction, windage, eddy-current loss, and hysteresis loss.

T16.7. A universal motor is an ac motor that is similar in construction to a series-connected dc motor. In principle, it can be operated from either ac or dc sources. The stator of a universal motor is usually laminated to reduce eddy-current loss. Compared to other single-phase ac motors, the universal motor has a higher power-to-weight ratio, produces a larger starting torque without excessive current, slows down under heavy loads so the power is more nearly constant, and can be designed to operate at higher speeds. A disadvantage of the universal motor is that it contains brushes and a commutator resulting in shorter service life.

T16.8. 1. Vary the voltage supplied to the armature circuit while holding the field constant. 2. Vary the field current while holding the armature supply voltage constant. 3. Insert resistance in series with the armature circuit.

T16.9. **a.** 80 V; **b.** 320 V.

T16.10. $I_A = 33.16$ A; $T_{dev} = 31.66$ Nm.

T16.11. **a.** $P_{dev} = 4600$ W; $T_{dev} = 36.60$ Nm; $P_{RA} = 200$ W; $P_{rot} = 124$ W. **b.** speed regulation = 4.25 percent.

T16.12. 2062 rpm.

T17.1. **a.** The magnetic field set up in the air gap of a four-pole three-phase induction motor consists of four magnetic poles spaced 90° from one another in alternating order (i.e., north-south-north-south). The field points from the stator toward the rotor under the north poles and in the opposite direction under the south poles. The poles rotate with time at synchronous speed around the axis of the motor. **b.** $B_{gap} = B_m \cos(\omega t - 2\theta)$ in which B_m is the peak field intensity, ω is the angular frequency of the three-phase source, and θ denotes angular position around the gap.

T17.2. Five of the most important characteristics for an induction motor are: **1.** nearly unity power factor; **2.** high starting torque; **3.** close to 100 percent efficiency; **4.** low starting current; **5.** high pull-out torque.

T17.3. power factor = 88.16 percent lagging; $P_{out} = 3.429$ kW; $I_{line} = 10.64$ A rms; $P_s = 169.7$ W; $P_r = 149.1$ W; $T_{out} = 37.90$ Nm; $\eta = 87.97$ percent.

T17.4. $s = 5.556$ percent; $f_{slip} = 3.333$ Hz; 860 rpm.

T17.5. The stator of a six-pole synchronous motor contains a set of windings (collectively known as the armature) that are energized by a three-phase ac source. These windings produce six magnetic poles spaced 60° from one another in alternating order (i.e., north-south-north-south-north-south). The field points from the stator toward the rotor under the north stator poles and in the opposite direction under the south stator poles. The poles rotate with time at synchronous speed (1200 rpm) around the axis of the motor.

The rotor contains windings that carry dc currents and set up six north and south magnetic poles evenly spaced around the rotor. When driving a load, the rotor spins at synchronous speed with the north poles of the rotor lagging slightly behind and attracted by the south poles of the stator. (In some cases, the rotor may be composed of permanent magnets.)

T17.6. $\delta_2 = 13.39°$; power factor = 56.25 percent lagging.

Index

E

F

N

O

P

W

X

Z